PUMPING MANUAL
International

Handbook of Pumps and Pumping

Edited by

Brian Nesbitt

Elsevier
Linacre House, Jordan Hill, Oxford OX2 8DP, UK
30 Corporate Drive, Suite 400, Burlington, MA 01803, USA

First edition 2006

ISBN–13: 978-1-85617-476-3
ISBN–10: 1-85617-476-X

Printed and bound in Great Britain

Published by Elsevier in association with Roles & Associates Ltd

ELSEVIER

ROLES & associates

Amsterdam Boston Heidelberg London New York Oxford
Paris San Diego San Francisco Singapore Tokyo

Foreword

Today, pumps and their associated systems play a central role in all process operations in the energy sector worldwide. These include oil, gas, petrochemical, chemical, power generation and water, together with industries such as food processing, pharmaceutical, steel, paper, agriculture and many others.

U.S. Department of Energy studies indicate that pumping systems account for nearly 20% of the world's electrical energy demand and can range from 25% to 50% of the energy usage in certain industrial plant operations. Rising costs, unease over the security of energy supplies and environmental concerns add to the pressures to optimise pumping systems, which can represent a real opportunity for companies and utilities to save money and energy, while reducing maintenance costs and increasing productivity.

One consulting group has estimated the current global market for liquid pumps to be $22 billion, growing over the next ten years by around 3% per annum. Western European and North American manufacturers can expect to see the breakneck growth of China's economy, with its lower manufacturing costs, influencing market share during the same period.

One tangible effect of this shift will be to accelerate outsourcing strategies in the West, either by seeking more low-cost parts or by own-branding of complementary pumps from other manufacturers. Global price pressures will continue intensifying.

Pumping Manual International is therefore a welcome aid to all pump users. By combining, in one convenient volume, chapters on pump types, basic flow theory, and pump applications, as well as ancillary equipment and services and a buyers' guide, users can find sensible and practical information to help them in making informed purchasing decisions. This is not simply another textbook on pumps.

Ian Leitch

Commercial Director

Energy Industries Council
(London, Aberdeen, Gateshead, Houston, Singapore, Dubai, Rio de Janeiro and Macae)

The editor

Brian Nesbitt is a highly respected consultant specialising in pump and valve technology. He works with pump manufacturers, equipment manufacturers, pump users and others who require specialist assistance. Brian regularly publishes articles, presents papers, conducts seminars and workshops, and undertakes work for industry ranging from theoretical research projects and equipment design studies, pump duty evaluations, to site visits to investigate problem pumps, valves, or problem systems. As a pump designer, with experience from 10 to 2048 barg he is admirably suitable to discuss all aspects of pumps and pumping.

He is the current chairman of the British Standard MCE/6/6 and European CEN/TC197/SC5 subcommittees which have formulated the draft European Standards for rotary and reciprocating pd pumps and is the UK Principal Technical Expert to ISO/TC 67/SC6 Joint Working Group 2 for 'Oil & Gas' reciprocating pumps. Brian was one of the UK delegates on the API task-force looking at modifications for the 3rd edition of API 674 and its conversion to ISO 13710.

Brian Nesbitt's working career began as a technical apprentice in the turbo-generator division of CA Parsons, gaining hands-on experience of building machines up to 500 MW. From the shop floor, his promotion led to periods in drawing and design offices working on rotary and reciprocating compressors and gas turbines. Large rotary compressors included gas circulators for nuclear reactors. Many machines included specially designed ancillary systems and complicated lube and seal oil systems. Mechanical seals were designed for specific compressor applications. Machine installation and interconnection was an important feature on some contracts. Process piping was considered during machine design.

A brief spell in industrial refrigeration, with reciprocating and screw compressors and considerable system design and site/installation exposure, was followed by an introduction to pumps at Ingersoll-Rand where he worked until 1985. Brian was recruited to provide engineering and application support for reciprocating pumps manufactured within Europe. Conversion from reciprocating compressors to reciprocating pumps was accomplished by an extended visit to the 'parent' factory in the USA. During this time, Brian assisted with ongoing design work on current contracts, including critical components on a batch of 2.0 MW pumps.

Once settled in the UK, Brian provided guidance to Sales for pump selection and choice of accessories and visited potential customers to discuss applications. Special pump designs were implemented for duties unsuitable for standard pumps. Emphasis was directed towards easier maintenance, low NPIPr, high pressure, high viscosity and solids handling applications. New valve designs were produced for arduous applications. North Sea pump applications and associated quality requirements proved to be a great source of development.

Many pumps required special ancillary systems for the crankcase and the stuffing boxes. On-skid process pipework featured on a considerable number of pumps where multiple pumps were assembled to create a single unit. Close liaison with the test department was essential to ensure test rigs were capable of providing the required accuracy for a wide range of operating conditions. Brian was very fortunate in that the test department had an enviable range of equipment and facilities available, together with very experienced staff. Extensive tests were conducted, in parallel with contractual tests, to explore the capabilities of the pump ranges. Novel rig designs were developed to cope with unusual operating conditions.

Brian also provided support for the after-sales departments with site visits to advise on pump and system operational problems. Exposure to complicated system problems, such as acoustic resonance, provided opportunities to work with consultants developing leading-edge technologies and investigation methodologies.

While at Ingersoll-Rand, he was surrounded by some of the world's most eminent rotodynamic pump specialists and met lots of others who visited the plant regularly. Although not knowing everything about pumps, Brian does know who will know the answer to those insoluble problems!

Using this book

Written specifically as a practical reference book for pump users, *Pumping Manual International* is intended to provide useful information about the outline design, selection and installation of pumps and how these affect performance. *Pumping Manual International* is not intended to be just 'another' textbook on pumps and pumping; rather it seeks to address the problems that exist at the interface between pump manufacturers and users. It has been compiled with the help of and benefit from the practical experience of pump users; it is aimed at everyone who has technical problems as well as these wanting to know who supplies what, and from where.

Pumping Manual International can be used in a variety of ways depending on the information required. For specific problems it is probably best used as a reference book. The detailed Contents section at the front of the book, combined with the Index at the end, will simplify finding the appropriate topic. The 'Useful references' at the end of most Chapters also provide helpful guidance, useful information and suggestions for further reading.

As a textbook though, *Pumping Manual International* may be read from cover to cover to obtain a comprehensive understanding of the subject. Of course, individual Chapters may be studied separately.

Chapter 1 is devoted to the main pump types, grouped into Rotodynamic, Positive displacement and Other types. If the user is unfamiliar with the concepts of 'rotodynamic' and 'positive displacement', then Chapter should be reviewed before Chapter 1 is examined. Which concept is better for a particular application should be known before reading Chapter 1.

The properties of liquids and liquid flow are discussed in the early Chapters. The book then follows a logical pattern with Chapters 5 to 10 covering topics such as: pumps and piping systems, flow regulation, materials, seals and sealing, shaft couplings and electric motors. Ancillary products and services are also discussed as well. Testing and quality assurance is dealt with in Chapter 12. Chapters 13 to 15 are devoted to installation, commissioning and maintenance as well as pump efficiency and economics and selection.

Chapter 16 provides details of a number of interesting pump applications that illustrate some of the problems encountered in the practical use of pumps and how these have been solved. This Chapter embraces a number of pump designs and uses, showing some of the diverse uses for pumps outside the more traditional areas. Solutions to some of the more commonly-encountered pump usage problems are however also given.

Chapter 17 gives useful guidance and information on many pump standards and important units and conversions used in pump technology.

The Buyers' Guide summarises the various pump types, divided into the three groups: Rotodynamic, Positive displacement and Other types. The Guide has been categorised in such a way to impose boundary limits on pump types and the operating conditions available, with the aim of simplifying the choice of supplier from the user's point of view.

The Buyers' Guide covers all pump types, followed by ancillary products and services. Trade names are comprehensively listed too. It is preceded by the names and addresses and contact details of all companies appearing in the Guide. They are listed alphabetically, by country.

It is strongly recommended that direct contact with the relevant companies is made to ensure that their details are clarified wherever necessary.

Contents

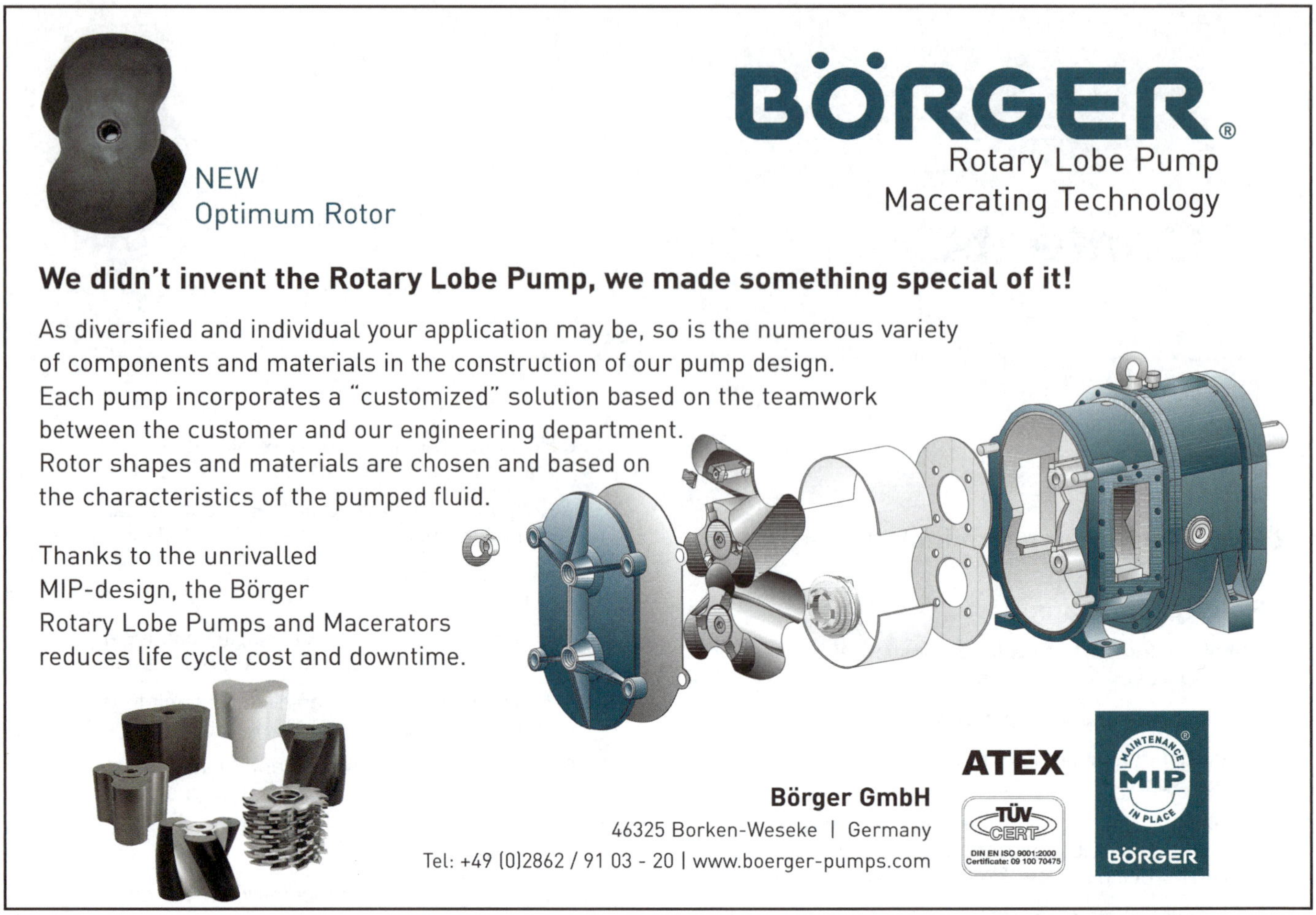
NEW
Optimum Rotor
BÖRGER®
Rotary Lobe Pump
Macerating Technology
We didn't invent the Rotary Lobe Pump, we made something special of it!
As diversified and individual your application may be, so is the numerous variety of components and materials in the construction of our pump design.
Each pump incorporates a "customized" solution based on the teamwork between the customer and our engineering department.
Rotor shapes and materials are chosen and based on the characteristics of the pumped fluid.
Thanks to the unrivalled MIP-design, the Börger Rotary Lobe Pumps and Macerators reduces life cycle cost and downtime.
ATEX
TÜV CERT
DIN EN ISO 9001:2000
Certificate: 09 100 70475
MAINTENANCE
MIP
IN PLACE
BÖRGER
Börger GmbH
46325 Borken-Weseke | Germany
Tel: +49 (0)2862 / 91 03 - 20 | www.boerger-pumps.com

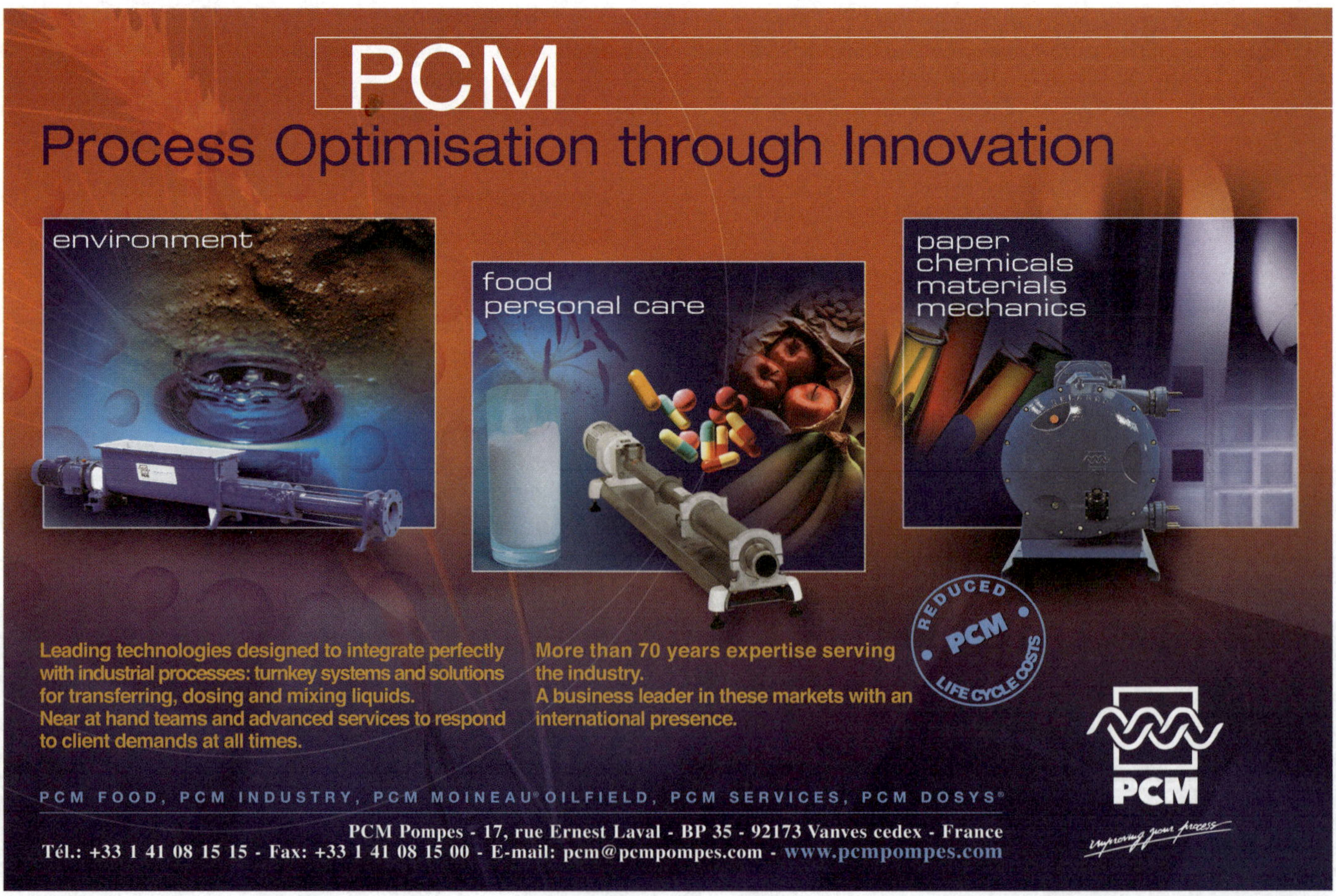
PCM
Process Optimisation through Innovation
environment
food
personal care
paper
chemicals
materials
mechanics
REDUCED
PCM
LIFE CYCLE COSTS
Leading technologies designed to integrate perfectly with industrial processes: turnkey systems and solutions for transferring, dosing and mixing liquids.
Near at hand teams and advanced services to respond to client demands at all times.
More than 70 years expertise serving the industry.
A business leader in these markets with an international presence.
PCM
PCM FOOD, PCM INDUSTRY, PCM MOINEAU® OILFIELD, PCM SERVICES, PCM DOSYS®
PCM Pompes - 17, rue Ernest Laval - BP 35 - 92173 Vanves cedex - France
Tél.: +33 1 41 08 15 15 - Fax: +33 1 41 08 15 00 - E-mail: pcm@pcmpompes.com - www.pcmpompes.com

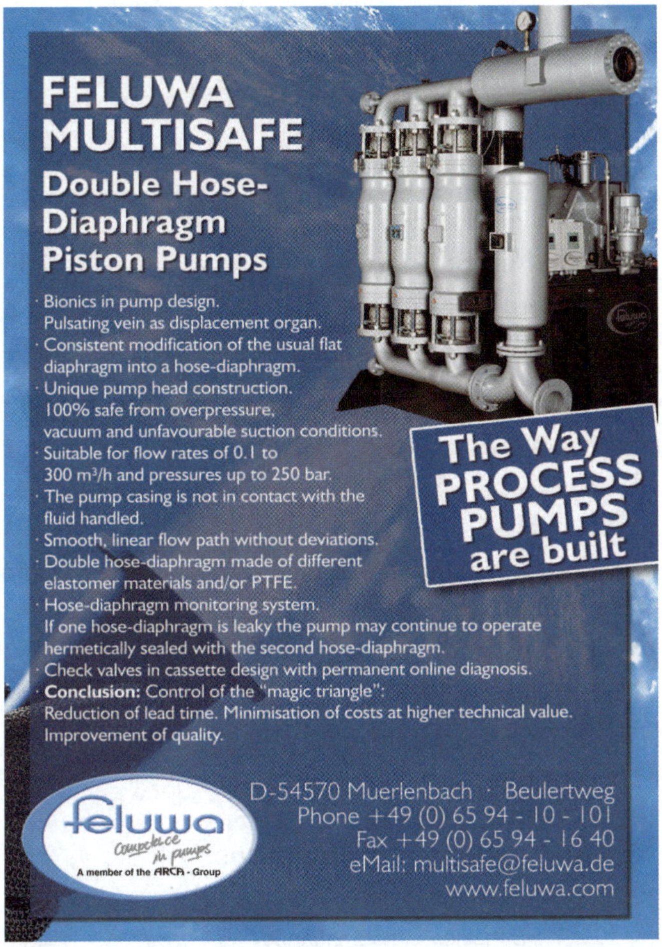
FELUWA
MULTISAFE
Double Hose-Diaphragm Piston Pumps
· Bionics in pump design.
Pulsating vein as displacement organ.
· Consistent modification of the usual flat diaphragm into a hose-diaphragm.
· Unique pump head construction.
100% safe from overpressure, vacuum and unfavourable suction conditions.
· Suitable for flow rates of 0.1 to 300 m³/h and pressures up to 250 bar.
· The pump casing is not in contact with the fluid handled.
· Smooth, linear flow path without deviations.
· Double hose-diaphragm made of different elastomer materials and/or PTFE.
· Hose-diaphragm monitoring system.
If one hose-diaphragm is leaky the pump may continue to operate hermetically sealed with the second hose-diaphragm.
· Check valves in cassette design with permanent online diagnosis.
Conclusion: Control of the "magic triangle":
Reduction of lead time. Minimisation of costs at higher technical value.
Improvement of quality.
The Way PROCESS PUMPS are built
feluwa
competence in pumps
A member of the ARCA - Group
D-54570 Muerlenbach · Beulertweg
Phone +49 (0) 65 94 - 10 - 101
Fax +49 (0) 65 94 - 16 40
eMail: multisafe@feluwa.de
www.feluwa.com

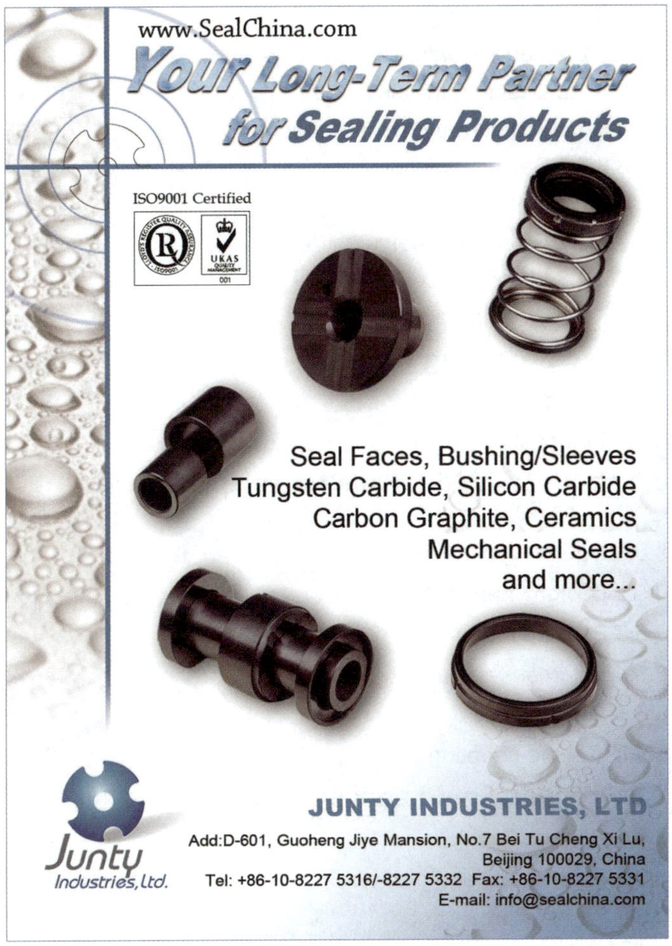
www.SealChina.com
Your Long-Term Partner for Sealing Products
ISO9001 Certified
UKAS
Seal Faces, Bushing/Sleeves
Tungsten Carbide, Silicon Carbide
Carbon Graphite, Ceramics
Mechanical Seals
and more...
Junty Industries, Ltd.
JUNTY INDUSTRIES, LTD
Add:D-601, Guoheng Jiye Mansion, No.7 Bei Tu Cheng Xi Lu, Beijing 100029, China
Tel: +86-10-8227 5316/-8227 5332 Fax: +86-10-8227 5331
E-mail: info@sealchina.com

Keep an eye on your budget !
TrueDos® helps reduce your running costs
Save chemicals – up to 25% !
✓ through more accurate dosing using stepper motor technology
✓ through continuous calibration during the process
✓ through rapid access to chemical consumption data
Save time
✓ by entering the dosing rate quickly & easily in l/h or ml/h
Save money
✓ dosing pump, flow monitoring & set point control in one single unit
Precise dosing
Can measure down to 5 ml/h
ALLDOS Ltd
Tel. +44 (0) 121 328 3336
Fax +44 (0) 121 328 4332
Email alldos.uk@alldos.com
www.alldos.com
ALLDOS
Vision for water technology

NETZSCH
NETZSCH The heart of your process
For over five decades, NETZSCH has developed, manufactured and marketed NEMO® Progressing Cavity Pumps worldwide. These accumulated capabilities are supplemented by partnerships with our national and international customers from all types of industries. In cooperation with them we optimise existing models and develop new products.
Our products for your processes
■ NEMO® Progressing Cavity Pumps
■ TORNADO® Rotary Lobe Pumps
■ NETZSCH Grinding Systems
■ NETZSCH Accessories and Optional Equipment
■ NETZSCH Genuine Spare Parts and Service
Our strength:
Innovative quality products and service to the benefit of our customers
NETZSCH
The heart of your process
NETZSCH Mohnopumpen GmbH
Geretsrieder Straße 1
D-84478 Waldkraiburg
Phone: +49 8638 63-0
Fax: +49 8638 67981
E-mail: info@nmp.netzsch.com
www.netzsch-pumps.com

Power pumps...
...for top performances.
URACA high pressure plunger pump units are the heart of all high-tech systems in the field of
■ chemical process industry
■ petrochemical industry
■ metal producing industry
■ high pressure cleaning industry
and wherever competence in high pressure technique is required.
URACA GmbH + Co. KG
P.O. Box 1260 · D-72563 Bad Urach
Phone ++49 (71 25) 133-0
Fax ++49 (71 25) 133-202
www.uraca.com
The new range
3 plunger
5 plunger
7 plunger
URACA
PIP d 3/99 dB
POWERFUL · LONG-LASTING · ECONOMIC

CLEAR SUPPORT
FOR HELSINKI.

Pump types

1

1.5.19 Metering pumps (flow within ± 0.1% to ± 3% with substantial dp changes)

1.5.20 Direct-acting reciprocating pumps (includes pneumatic, hydraulic and steam actuation)

1.5.21 Non-metallic positive displacement pumps

1.5.22 Sealless positive displacement pumps

1.5.23 Hydraulic motors

1.6 Other pump types

1.6.1 Ejectors

1.6.2 Barrel-emptying pumps (powered and hand pumps)

1.6.3 Hydraulic-ram pumps

1.6.4 Air-lift pumps

1.6.5 Contraction pumps

1.6.6 Macerators

1.1 Introduction

There are many different pump types. Pumps operate via two basic principles, rotodynamic and positive displacement, but there are a few pump designs outside this categorisation.

NOTE: If the reader is unfamiliar with the concepts of 'rotodynamic' and 'positive displacement' then Chapter 4 should be reviewed before the pump types in this chapter are examined.

Rotodynamic machines try to be 'constant energy' devices but don't quite make it work because of the variable inefficiencies and the limitations of manufacturing. Positive displacement machines try to be 'constant flow' devices but don't quite manage to make it work because of the liquid properties and manufacturing requirements. But ! there is a special case when constant flow is achieved. The reader should know which concept is better for a particular application before reading **Chapter 1**.

All pumps are not equal ! There are a wide range of pump types because some designs are better; more efficient or more consistent or last longer; at handling specific operating conditions or specific liquid properties. Each application should be assessed individually, on merit, when considering:

- Liquid properties
- Allowable leakage
- Driver type
- Installation arrangement
- Operating efficiency required
- Duty cycle
- Allowable noise level
- Operational safety
- Site facilities and local staff capabilities

Remember, sometimes the liquid properties must be augmented with the properties of entrained solids, gases or vapour. Mixtures of liquids need all components defined in detail; very small concentrations can create corrosion problems. There are plenty of benign liquids which may be allowed to drip on the floor. Some liquids are so hazardous that the vapour must not be allowed to escape. Simple seals can be very small; complicated seals may require a lot of space plus an external system piped in. Not all pumps can accommodate complicated seals. Not all pump designs can accommodate complicated sealing system piping. It is not always possible to upgrade from a relaxed approach to leakage to strict leakage control; a change of pump type may be necessary if initial leakage requirements were wrong ! Remember, the entrained/dissolved gas may be much more of a hazard than the liquid !

Most pumps are driven by standard AC squirrel-cage motors. However, it may be advantageous to consider a steam or gas turbine for fast pumps. New motor designs, like switched reluctance, are very good for high speed but require a 'black-box' of electronics. Alternatively, an engine may be better for slow pumps. Steam or compressed air power supplies can be beneficial in a hazardous area. Don't forget about the installation arrangement until the last minute. A submerged pump or a vertical pump may provide the smallest installation footprint. Also remember, that poor piping design can have an adverse affect on performance, reliability and labour. Piping which is not self-venting and not self-draining may require physical attention during each start and stop. Consider the piping arrangement while thinking about the pump configuration ! Much time is wasted troubleshooting 'pumps' when it is the piping causing all the problems !

Efficiency can be very important for large pumps; large electricity bills ! Efficiency may be important when trying to find a pump to operate from a 13A socket or on a 12V or 110V supply. Some pump types are much more efficient than others. The duty cycle must be defined as it can significantly affect pump selection. Some pumps can operate continuously for ten years; other pumps need tender-loving-care after twelve hours. The pump duty cycle limitations may necessitate the installation of standby units.

Some installations require very quiet pumps. The noise of central heating pumps can sometimes be heard all over a building. Laboratories usually require very quiet pumps. Pumps, which are in close proximity to staff or the public, may have to be inherently safe. The site services obviously affect pump and driver selection. A steam turbine may be the preferred driver for the pump; not if there's no steam available. Larger pumps may need 3.3 kV or 6.6 kV power for the best motor sizing; there may be additional cost expenditure required at the substation. If the application warrants a complicated pump unit, can the local site personnel operate it safely and maintain it ? The supplier will be very happy to maintain it, at a cost. If the local site personnel don't understand the pump, or the process, the cost of maintenance may be much higher than expected !

When considering the information to be used to define the pump application it is very easy to spot 'maximums' and forget that 'minimums' might be just as important. It is worthwhile to take the time and consider that 'normal' might be much more important than the occasional 'maximum' or 'minimum'. It might be very worthwhile to consider the duration of 'maximum' or 'minimum' in the context of long-term 'normals'. Unless specifically defined to the contrary, the pump supplier will assume all process changes occur slowly; any rapid changes must be described in detail!

When the pump application is initially considered, any applicable specifications, standards or regulations must be listed for discussion. Important factors, based on either the liquid or the installation, can have a significant impact on pump selection, pump design and other equipment selection. Remember in Europe, that ATEX regulations apply to mechanical equipment such as pumps and gearboxes, as well as motors. Size, that is much bigger size, may be required for compliance. Many specifications, standards and regulations cannot be applied retrospectively because of the effect on equipment selection.

1.2 Checklist of pump terminology and definitions

Modern communication can be very difficult. English is the de facto standard language for technical subjects BUT there are different versions of English! American English is different to UK English. Many versions of English have been developed by engineers who use English as a second language.

In the pump industry, users and manufacturers often refer to or call a pump type by different names! This is inevitable given the range of pumps and the immeasurable number of process applications in which they are used. The same applies to pump terminology. This terminology or "jargon" can be very confusing and indeed often misleading.

This Section attempts to highlight a few things to watch out for and provides an explanation of and gives guidance on some of the more important key pump terms and descriptions that occur in daily pump usage many of which are referred to within this book.

Barrel pump - radially-split centrifugal pump.

Bearing bracket - the structural component which houses the pump bearings.

Bearing pedestal - bearing bracket.

Boiler feed pump - usually a multi-stage centrifugal pump feeding treated water.

Booster pump - two different common uses. In hydraulic fluid power, a booster pump takes a high pressure and increases it significantly. In process pumps, a booster pump is used where there is insufficient pressure to supply adequate NPSH or NPIP. The booster pump provides a small increase in pressure so the main pump can operate reliably.

Dead head - closed valve head (do not assume a rotodynamic pump will run successfully at this condition!).

Duty cycle - how the pump will be operated with reference to time.

Eccentric screw pump - progressive cavity pump.

Ejector - jet pump.

Fire pump - a pump used to pressurise water for fire-fighting. (Considered an intermittent application).

Flooded suction - an imprecise term to be avoided; specify the suction head/pressure and NPSHa/NPIPa.

Fluid - not necessarily liquid; use very carefully; this term is not interchangeable with liquid!

Head - a measure of energy in a rotodynamic pump system.

Heat pump - a system with a compressor.

Helical gear pump - progressive cavity pump.

Helical rotor pump - progressive cavity pump or a screw pump.

Helical screw pump - progressive cavity pump.

Hydraulic motor - a positive displacement machine which extracts energy from liquid.

Hydraulic turbine - a rotodynamic machine which extracts energy from liquid.

Inline pump - a pump casing design where the suction and discharge pipes are on the same centre-line; concentric, but on opposite sides.

Injector - jet pump.

Liquid ring pump - a compressor.

Modular multi-stage pump - segmental pump.

Monoblock pump - a pump which utilises the motor bearings for radial support and axial alignment.

Motor stool - the structural space which supports a vertical electric motor.

Mud motor - a progressive cavity pump extracting energy from drilling mud.

NPIPa - see Section 4.4 of Chapter 4.

NPIPr - see Section 4.4 of Chapter 4.

NPSHa - see Section 4.4 of Chapter 4.

NPSHr - see Section 4.4 of Chapter 4.

Pinion pump - gear pump.

Pipeline pump - a pump used to move liquid over considerable distance, perhaps 1000 km.

Pressure - a measure of energy in a positive displacement pump system.

Priming - filling a pump with liquid prior to starting.

Pump package - sometimes used when more than one pump is built on a baseplate. Also used when multiple pumps are driven from one gearbox.

Pump unit - the pump plus any power transmission equipment and the driver; everything mounted on the baseplate.

Reciprocating pump - a type of positive displacement pump which cannot run in reverse.

Ring section pump - segmental pump.

Rotary pump - a type of positive displacement pump utilising rotating elements.

Rotodynamic pump - a type of pump which adds rotating kinetic energy to the liquid then converts to static head.

Self-priming - a pump which can remove gas from the suction piping and not run dry.

Side channel pump - peripheral pump.

Snore - the ability to run dry then self-prime when liquid is present.

Suction stage - first stage.

Siphoning - liquid flowing through the pump when the pump is stationary.

Turbining - a pump running backwards and acting as a turbine.

Vacuum pump - a compressor.

Venting - removing air and/or gas from a pump prior to starting.

Vertical turbine pump - a vertical multi-stage pump probably with mixed-flow impellers.

Waste water pump - vague! Could be surface water, dirty water or foul water (sewage).

1.3 Rotodynamic pumps

1.3.1 Small centrifugal pumps for domestic water/ heating/sanitation applications

There are specially developed pumps, so-called heating, water and sanitation pumps for the different pumping requirements in buildings. One of their general characteristics is a low noise and vibration level, the requirements varying in accordance with the size of the building from a noise level of approximately 25dB(A) for a private house, to approximately 65 dB(A) for pumps placed in a well insulated machine room in a larger building. These different pump requirements are covered by:

- Heating circulation pumps for circulating water in a central heating system. Smaller sizes up to a power requirement of normally 100W are of the wet type, i.e. wet rotor motors, Figure 1.1. All the rotating parts, including the motor rotor are sealed inside a stainless steel can. The stator windings are placed around the outside of the can and the rotating

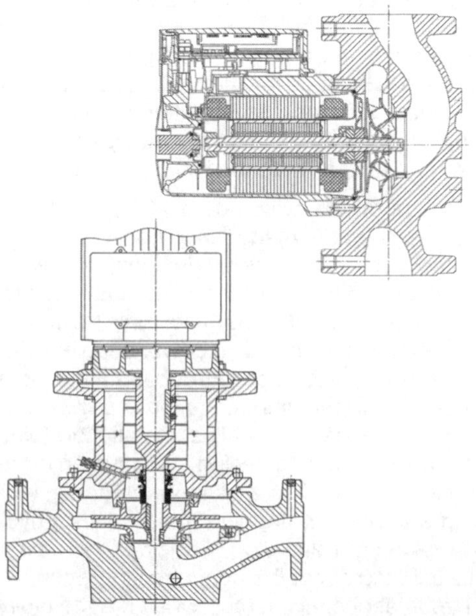

Figure 1.1 Central heating circulator pumps with inline connections. The upper illustration is the "wet" type. The lower illustration is the conventional dry motor
Courtesy of Grundfos

Figure 1.2 Heating circulation pump of the twin design
Courtesy of Grundfos

magnetic field passes through the can wall. No seals are necessary; the bearings are usually ceramic lubricated by the hot water. Pump casings are usually cast iron or aluminium alloy with bronze or stainless steel impellers. The pumps are mass produced with fixed performance, but in order to match the circulator to the system and avoid noise in the piping system, they are supplied with a two or three speed motor.

- So-called twin pump packages, i.e. a pump casing with two pumps built as a unit ready for installation in one suction and discharge pipe. The pumps can be operated together for parallel operation, see Figure 1.2, or series operation. The unit includes a non-return valve system, controlled by the flow of liquid so that either pump can be operated independently or together with the other pump.
- Hot water pumps for circulating domestic hot water in larger buildings so that the hot water is almost immediately available when the tap is turned on. In contrast to central heating pumps the parts in contact with the fluid are made of bronze or stainless steel instead of cast iron. Wet motors are available, as well as dry motors using special designs for motor and shaft seals in order to avoid blockages due to furring or deposition of the natural salts.
- Other pumps in buildings are pressure boosting pumps, (see multi-stage pumps, ground-water pumps and to some extent also standard water pumps for fire systems).
- Complete pump packages are supplied in buildings with water central heating systems, to maintain the static pressure in the heating circuit and make up any leaks.

Most of these pumps are of the inline design. The pump and motor form an integral package and the suction and discharge connections are inline so that the package can be fitted into a straight pipe run. Most pumps do not require external support but rely entirely on the rigidity of the pipework. Pumps can be mounted with the motors vertically or horizontally. In most designs the pump cannot be mounted vertically above the motor.

BS 1394 should be reviewed for relevant requirements. EN 1151 specifies the European requirements for circulating pumps. Electrical safety requirements are given in Chapter 17, Section 17.1.4.

1.3.2 Domestic water supply pump packages

Automatic water packages are used for supplying drinking water to households and properties which are not connected to the national water distribution network. The water is normally taken from a well with a suction lift for the pump of 3 to 7 m. The pump's flow and head are suitable for 1 to 3 outlets connected to the pump by relatively short pipes or hoses.

The package consists a self-priming pump, (centrifugal or liquid ring type), an electric motor, an accumulator (pressurised holding tank) and usually a pressure switch. The pressure switch ensures that the pump starts when the pressure falls due to water being drawn off and stops when the accumulator has filled up due to the corresponding increase in pressure.

There are many different makes and models available as illustrated in Figure 1.3. When purchasing, account should be taken

Figure 1.3 Automatic water supply packages - a selection of models (Top right hand illustration - *Courtesy of AquaTech Ltd*)

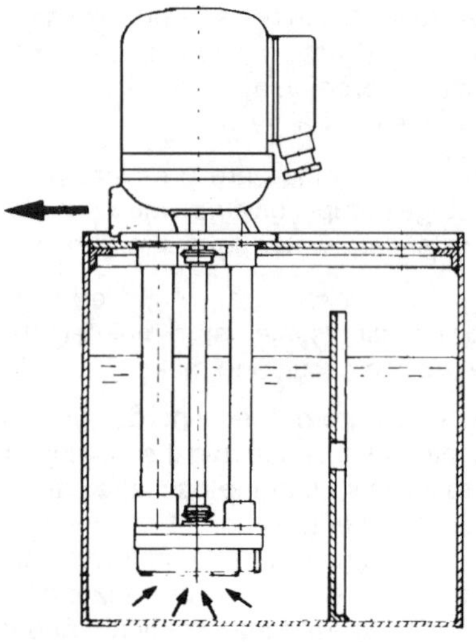

Figure 1.4 Coolant pump for machine tools

of the noise level as well as the fact that the water package should fulfil current electrical, hygienic, and legal requirements.

1.3.3 Small centrifugal pumps for machine tool cutting oil applications

Special pumps are used for pumping coolant/cutting oil, or White Water, for machine tools, lathes, grinding machines, etc. The liquid can be a specially blended oil or an emulsion of it in water, the oil content being from 2% to 15%. The flow usually varies between 0.3 and 20 m³/h with heads varying between 2 and 20m. This type of pump is standardised in Germany in accordance with DIN 5440. This standard indicates both performance and those dimensions which have an influence on interchangeability. The pump unit is constructed so that the pump casing is submerged in the liquid with the motor placed outside the tank, Figure 1.4. The immersed depth is up to 500 mm. Due to this construction, no shaft seal is required.

BS 3766:1990 specifies ten sizes in the form of dimensions and minimum performance characteristics for vertical top and side mounted units. DIN 5440 specifies six pump sizes in submerged and external forms. The French Standard, NF E44-301 incorporates DIN 5440 requirements. Normally coolant pumps are manufactured in cast iron or aluminium alloy but there are special types available in plastic. EN 12157, includes six sizes of pumps, from 1.5m³/h to 15 m³/h, for depths from 90 mm to 550 mm.

1.3.4 Horizontal single-stage end-suction overhung impeller centrifugal pumps (for general purposes includes fixed irrigation pumps, EN 733 pumps)

Standard water pumps

These are pumps designed to operate with clean water, at temperatures up to 80/120 °C. They are used for clean liquids compatible with cast iron and in some special cases bronze. Impellers may be cast iron, bronze or plastic. Small pumps will generally not have wear rings. A wide range of styles and designs is available, nevertheless, a classification into main categories can be made in accordance with Figure 1.5.

Compact pumps with the impeller mounted directly on the motor shaft end, close-coupled, are suitable for flows up to a maximum of approximately 300 m³/h and for differential heads of up to 100m. Casing pressure ratings can be up to 10 barg. Motor output can be over 100 kW however, motors exceeding 20 kW will require lifting facilities. Close coupled pumps rely on the motor bearings to absorb axial thrust and radial loads. Suction pressure may be restricted. Most of the hydraulics specified in Section 1.3.4 for EN 733 pumps are also available in close coupled units which do not comply with the dimensional part of the Standard.

Smaller vertical compact pumps can often be mounted directly in pipework without supports, this of course assumes that the pipework has sufficient rigidity. Larger vertical compact pumps have a foot to rest on a support. Normally the pump can be mounted at any angle, with the exception that the motor must not be located under the pump because of the risk of motor damage in the event of seal leakage. Mounting, other than vertical, can cause maintenance problems. Small pumps may have female screwed connections as an alternative to flanges. Space around the stuffing box is restricted and the choice of packing and seal arrangements will be limited.

Pumps with bearing brackets, housing their own radial and axial bearings, are available to cover most of the operational range of close coupled pumps. Pumps with bearing brackets will not have the same restrictions on suction pressure and sealing arrangements. Somewhat larger pumps are produced in this configuration. A coupling is required between pump and motor as is a baseplate for horizontal units. These units are slightly longer than the equivalent compact close coupled pump but more versatile. Some pumps are designed to be back-pull-out; the complete rotating assembly plus the bearing bracket can be removed leaving the casing connected to the pipework. If a spacer coupling to the motor is used, the pump can be maintained without disturbing the pipework or the motor. These pumps are capable of 2000 m³/h at differential heads of 160 m and pressure ratings up to 16 barg.

EN ISO 9908 could be useful for both horizontal and vertical pump requirements. BS 4082 could be useful for vertical pump requirements. Stuffing box and seal cavities should comply with ISO 3069.

Standard pumps to EN 733

In Germany and a few other European countries for many years, there were standardised pumps for non-hazardous liquids. The German Standard DIN 24255 was probably the most popular. The German Standard has been replaced by a very similar European Standard, EN 733. The Standard relates to foot mounted horizontal pumps with a bearing bracket, see Fig-

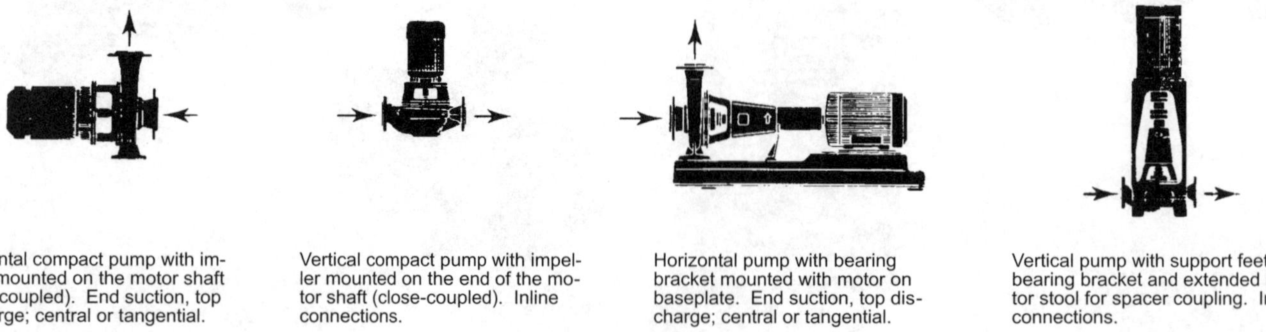

Horizontal compact pump with impeller mounted on the motor shaft (close coupled). End suction, top discharge; central or tangential.

Vertical compact pump with impeller mounted on the end of the motor shaft (close-coupled). Inline connections.

Horizontal pump with bearing bracket mounted with motor on baseplate. End suction, top discharge; central or tangential.

Vertical pump with support feet, bearing bracket and extended motor stool for spacer coupling. Inline connections.

Figure 1.5 Various designs of standard pumps

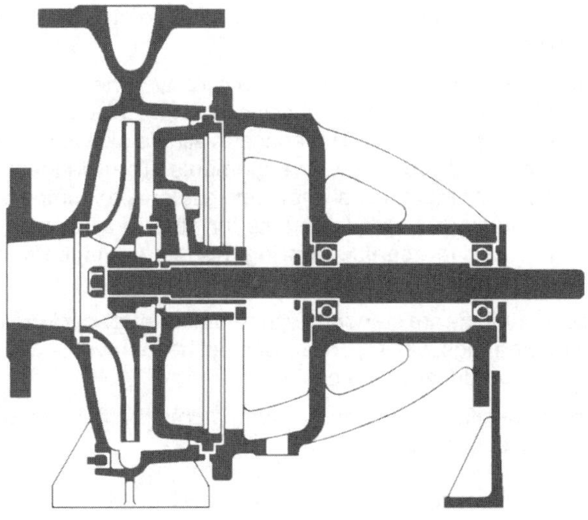

Figure 1.6 A pump in accordance with EN 733

Figure 1.7 A small pressed/fabricated stainless steel pump

ure 1.6, and a back-pull-out facility when driven via a spacer coupling.

The Standard does not define materials of construction, only performance and physical dimensions. The Standard comprises a total of 29 sizes with discharge connections from 32 up to 150 mm. The nominal performance at BEP specifies flow of 6 to 315 m^3/h and the differential head of 5 to 80 m. The pump casing must have a pressure rating of 10 barg. The pump's design, Figure 1.6, consists of a system of units based on three basic shaft sizes.

The pumps are generally built with cast iron casings, cast iron or bronze impellers and stainless steel shafts. Some manufacturers have bronze and steel casings available, as well as plastic impellers. The space, allocated for soft packing and mechanical seals, is small, limiting the use of the pumps to non-hazardous liquids. The intent of the Standard was to allow interchangeability between manufacturers, equipment and provide the user with greater choice. EN 735 lists the tolerances on pump dimensions to ensure interchangeability.

Figure 1.7 shows a small pressed/fabricated stainless steel pump. This style of construction allows very thin walls to be used and hence minimising material costs. Exotic materials can be used providing suitable rolled sheet is available. Of course, initial production costs are high because press tools are very expensive. Many pumps must be produced to recover the tooling costs. Hydraulics from popular standards, such as EN 733, are used to ensure sales into existing markets.

Fixed irrigation pumps

Irrigation and other agricultural applications make use of different pumps depending upon the location and what power sources are available. Permanent surface installations normally have single-stage pumps in accordance with Section 1.3.4 or multi-stage pumps in accordance with Sections 1.3.12 and 1.3.21. When wells have been sunk, submersible deep-well pumps in accordance with Sections 1.3.13, 1.3.18 and 1.6.3 are used. The drive is generally an electric motor although diesel or LPG operation are alternatives. Some large installations in very remote locations use crude oil burning engines or natural gas. The required pump performance depends upon the type and size of the installation. Normal duties lie within the range 10 to 100 m^3/h with heads of the order of 50 to 120m.

1.3.5 Horizontal single-stage end-suction overhung impeller centrifugal pumps (for chemical applications includes ISO 2858, ISO 3069, ISO 3661, ASME/ANSI B73.1 designs)

ANSI, the American National Standards Institute, has a philosophy more similar to DIN than ISO in that it specifies materials and mechanical seal options. B73.1 is a standard for single stage centrifugal pumps, with back pull-out, intended for chemical and corrosive duties. Construction is similar to Section 1.3.21. The Standard specifies bareshaft pump dimensions, also installation dimensions for complete packages with motor and baseplate. The Standard does not specify hydraulic duties. Mechanical seal arrangements, with various piping plans, are detailed to cover most applications.

Depending upon materials and cooling arrangements pumps must be suitable for at least 260 °C. The Standard specifies material columns with casings in:

- Cast iron
- Ductile iron (spheroidal graphite iron)
- Carbon steel
- Alloy steel
- 18-8 stainless steel
- 18-10-2 stainless steel

Many manufacturers can supply pumps in more exotic materials such as; Carpenter 20, Hastelloy B & C, duplex stainless steels; as well as non-metallic materials like glass reinforced polymer. The Standard requires a corrosion allowance of 0.125", 3 mm.

Pump sizes vary from 11/2 to 10 suction connections with flanges to ANSI B16.1, B16.5 or B16.42. Impeller diameters range from 6" to 15". At 3600 rpm, pumps are capable of 1680m^3/h at heads up to 300m. At 3000 rpm the performance falls off slightly to 1400 m^3/h at 208 m. ANSI does specify pressure ratings for the pumps; 125LB/150LB flange rating, 12 barg and 20 barg. An option for 250LB/300LB flanges is given.

The Standard specifies a minimum bearing life of 17500 hours at maximum load conditions. On test, with a speed of ± 10% of operating speed, the peak velocity of vibration shall not exceed 0.25 in/s, 6.35 mm/s.

End-suction pumps can suffer high axial thrust due to high suction pressure. Shortened bearing life may result. Manufacturers should confirm the bearing life if it is less than the standard requirement.

ISO has prepared a standard recommendation, ISO 2858 for end-suction centrifugal pumps, pressure rating PN 16 (maximum pressure 1.6 MPa = 16 barg). This is included in British Standard EN 22858 and equivalent to the German Standard DIN 24256. The Standard specifies the principal dimensions

Figure 1.8 Typical standard ISO 2858 pumps

and nominal duty point. Figure 1.8 shows typical standard pumps.

General note on ASME/ANSI Standards

The ANSI B73 Standards were produced in response to requests by users to standardise manufacturers products for the chemical industries. The end-suction pump Standard was introduced in 1974 and the vertical inline in 1975. The standards were based on recommendations by users but compiled by a committee of mostly manufacturers. There are many American chemical installations worldwide and manufacturers and users should be aware of these standards.

General note on ISO 13709 and API 610

API is the American Petroleum Institute which is predominantly a trade association of pump users concerned with oil and gas production and refining. Pump manufacturers are represented on the API committees but are outnumbered by the users.

For example, API 610 specifications have a complete section for Vendor's Data; they do not have a section for Purchaser's Data. API is a long established organisation and has a large catalogue of specifications, publications and recommended practices. Since American technology is used worldwide in oil and gas related sectors, it is logical to use proven methodology where appropriate.

API 610 is a mature specification currently up to the eighth edition published in August 1995. API is unlike the majority of DIN, and EN pump standards in that it is more a philosophy to achieve results rather than instructions. It is not a dimensional standard which would allow interchangeability between manufacturers. It does contain requirements but also many by agreements which are frowned upon by European standards authorities because of the lack of control and uniformity. However API 610 does specify materials of construction. API and ISO have collaborated on a new ISO Standard, ISO 13709.

The first two requirements are a 20 year service life and the capability of three years uninterrupted operation. API does follow these requirements with a statement that these are design criteria. From a European viewpoint both are not verifiable and are application specific. Bearing this in mind, API 610 pumps are heavier and more robust than most other pump designs, no pressure or temperature limitations are specified, and are suitable for the most arduous duties. Drain rim baseplates are standard and most pumps are centre-line mounted to reduce alignment problems caused by thermal growth. Allowable forces and moments on process nozzles are large necessitating substantial baseplates. Pump selection is guided by a requirement to have BEP between the normal operating point and the rated duty point.

API 610 specifies the size of seal cavities and the distance to the closest obstruction and specifies mechanical seals, single balanced, as a minimum. The space available allows for complex double or tandem seal arrangements. The Standard requirement is for mechanical seals and systems to comply with ISO 21049/API 682. Arrangements for other auxiliary piping, cooling water for seals, bearing housings, jackets and pedestals, are also shown in API 610.

Spacer couplings are standard with a minimum spacer length of 127mm. Bearings, seal(s) and rotor must be removable without disturbing the driver or the casing.

API 610 specifies 11 material columns covering the following range of casing materials:

- Cast iron
- Carbon steel
- 12% chrome steel
- 18-8 stainless steel
- 18-10-2 stainless steel

Although cast iron is specified as an acceptable material the lowest specification applied in practice is carbon steel. It is very unlikely that cast iron casings could support the nozzle load requirements.

1.3.6 Horizontal single-stage end-suction overhung impeller centrifugal pumps (for heavy-duty applications includes ISO 13709, API 610 designs)

These are the most popular style of process pumps. Basic pump design is similar to EN 733, Section 1.3.4 and ISO 2858, Section 1.3.21, with a back pull-out feature. In practice, the minimum material readily available is carbon steel. Some manufacturers will make specials in bronze when commercially attractive. Standard flanges are inch to ANSI standards, B16.1, B16.5 or B16.42. Both suction and discharge flanges must be the same pressure rating. Rolling bearings, typically ring oiled, are supported in a cast housing bolted to the stuffing box or casing. Bearing housings are steel for hazardous liquids.

Pumps are available with flows over 2000 m^3/h at differential heads of 300 m. Operating temperatures from -30 to 425 °C are common, down to -180 °C is possible. Popular pumps range from suction branch sizes of 2" to 12". Pressure ratings of smaller pumps are up to 100 barg reducing to 40 barg for the larger sizes. Pressure/temperature rating variations with materials are important, so check flange specifications.

1.3.7 Horizontal single-stage double-suction axially-split pumps (for general purposes)

Thirty-five years ago the double suction pump was the predominant water pump for applications where the pipe size exceeded 200 mm. The classic design with a large impeller between two

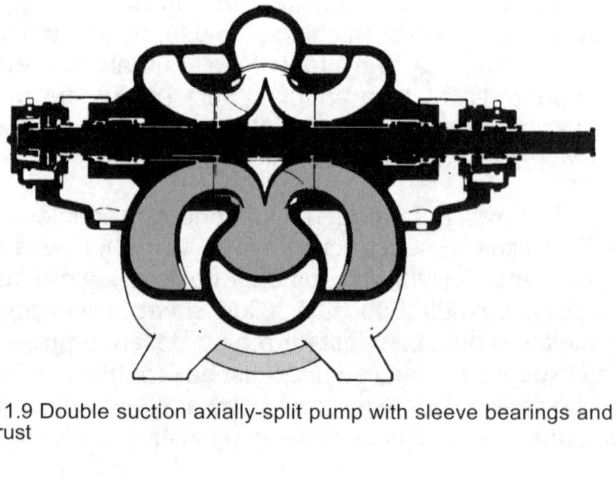

Figure 1.9 Double suction axially-split pump with sleeve bearings and tilting pad thrust

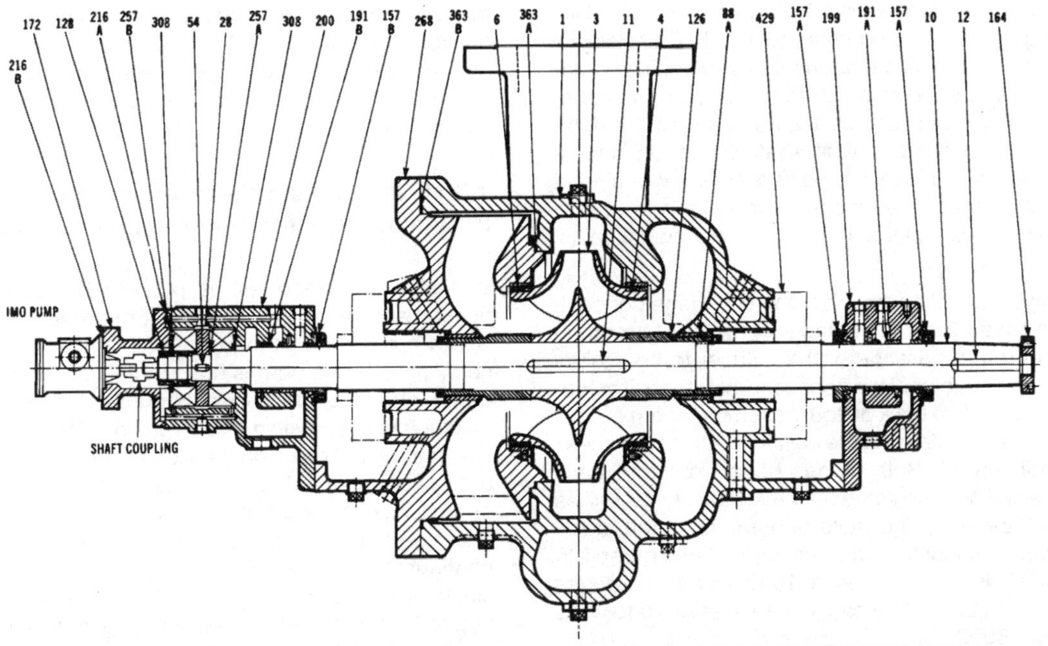

PART NO	NAME OF PART	PART NO	NAME OF PART	PART NO	NAME OF PART
172	SUPPORT HEAD	12	COUPLING KEY	308	SEALING RINGS-THRUST BAG
164	COUPLING NUT	11	IMPELLER KEY	268	SUCTION END BRACKET
157B	FLINGER-THRUST BRG.	10	SHAFT	257B	SHIMS-THRUST BRG. TO SUPPORT HEAD
157A	FLINGER-PLAIN BRG.	6	CASING RING	257A	SHIMS-THRUST COLLAR
128	LOCKNUT	4	IMPELLER RING	216B	GASKET-IMO PUMP TO SUPPORT HEAD
126	IMPELLER LOCKNUT	3	IMPELLER	216A	GASKET-SUPPORT HEAD TO THRUST BRG.
88A	STUFFING BOX BUSHING	1	CASING	200	BEARING BODY & CAP-THRUST
54	THRUST COLLAR KEY	429	MECHANICAL SEAL-COMPLETE	199	BEARING BODY & CAP-PLAIN
28	THRUST COLLAR	363B	GASKET-SUCTION END BRACKET (OUTER)	191B	BEARING LINING-THRUST
		363A	GASKET-SUCTION END BRACKET (INNER)	191A	BEARING LINING-PLAN

Figure 1.10 Double-suction process pump to API 610/ISO 13709

bearings, two stuffing boxes and axially-split pump casing with the pipe connections in the lower half, Figure 1.9, is still used for large flows at low heads. The design is normally used for large and very large water services pumps, for example, clean water pumps for large water works and distribution pumps for industrial water.

The specific speed per impeller, half for double suction impeller type pumps, is less than that of similar single suction pumps and this results in a corresponding flattening of the H-Q curve. In borderline cases the lower NPSHr value means that higher speeds can be used. In special instances where it is desirable to maintain as near pulsation-free flow as possible, as for example, fan pumps for paper machines, double suction pump impellers with staggered blades are used. Sometimes the location of the pump connection flanges can make the use of a double-suction pump advantageous.

Water temperature can be up to 120 °C with standard cast iron casings. Special bronze casings are sometimes supplied for seawater applications. Impellers can be cast iron or bronze. Pumps with 250 mm connections can handle 1200 m^3/h at 100m. Larger pumps are available for up to 40000 m^3/h at 40m. Small pumps have pressure ratings of up to 17 barg reducing to 5 barg for the largest sizes. Power requirements are up to 5.5 MW. Packed stuffing boxes or mechanical seals can be fitted as options.

This style of pump has been available in the past in two slightly different two-stage versions. A two-stage pump with single entry second stage was used as a condensate extraction pump for steam turbo-generators. The pump grew to unmanageable proportions in the 1960s as generators became much larger. The design concept changed and the condensate extraction pump became a vertical multi-stage pump. A two-stage pump with two double entry impellers has also been available, particularly in the United States. This style of pump was designed when axial thrust was not fully understood and resulted in a hydraulically balanced rotor. Neither of the two-stage designs described here are popular currently in pump applications.

1.3.8 Horizontal single-stage double-suction radially-split pumps (for heavy-duty applications includes ISO 13709, API 610 designs)

This pump, see Figure 1.10, is not as popular as the end-suction pump. It is similar in some respects to Section 1.3.7 pumps but has centre-line mounted radially split casings. Generally with top-top process connections, but side-side and other variations are available. The double entry impeller mounted between two bearings. Rolling bearings are standard, sliding bearings and various lube oil systems can be made to order. The mechanical seals can be serviced by removing the coupling and bearing housings. A spacer coupling is required. The rotor assembly is withdrawn from the casing, away from the driver.

These pumps tend to be slightly larger than the end-suction pumps. Flows over 5000 m^3/h are possible at heads up to 500 m. Suction sizes from 8" to 16" are popular with pressure ratings up to 60 barg. Materials and flanges are similar to Section 1.3.6 types.

1.3.9 Horizontal two-stage end-suction overhung impeller centrifugal pumps

These pumps are a logical extension to the single-stage pumps described earlier. The design problem to overcome is shaft deflection. The weight of both impellers is cantilevered from the front bearing. Also remember that some space must be left for shaft sealing, further increasing the overhang. Pumps for benign liquids, in days past, could be sealed by a simple soft-packed arrangement. This simple, but very effective arrangement also provided a useful degree of support and damping for the shaft. These effects helped to minimise vibration.

The primary disadvantage to soft-packing is the skill required to pack a box, run the packing-in, and then nurture the packing throughout its useful life with tender loving care. These skills are disappearing and there is no time, now, to bed packing-in properly followed by almost continuous adjustment. The modern mechanical seal is seen as a gift from heaven for accountants, but provides no shaft support and no damping. Shaft diameters must be increased to limit shaft deflection and to increase critical speeds. There is a point reached, when shaft diameter increases, that a between-bearing pump is almost cost comparable ...... except that it has two mechanical seals, **not one**!

Figure 1.11 shows a novel approach to the two-stage concept. The first stage is mixed-flow rather than truly radial. This combination allows the return passage of the diffuser to be slightly smaller than the tip diameter of the second radial stage. The design produces an effective back pullout assembly which greatly enhances maintainability, leaving the casing in situ and process connections undisturbed. Both stages have front and back wear rings; these will provide some dynamic stability as well as reducing internal leakage. This style of pump is aimed at the process industries, particularly paper making. These pumps, in super-duper 25Cr 5Ni 2Mo, can handle 1080 m^3/h and produce differential heads of 220 m. The casings are designed to cope with 25 barg and 180°C.

The two-stage overhung pump is vigorously disliked by API and many in the Oil & Gas industry. This is probably due to past failures and unreliability. The Oil & Gas industry was renowned though for getting duty conditions wrong in the past, and this pump design may have attracted an unwarranted reputation. This style of design will be susceptible to vibration, rubbing and seizure when operated well away from BEP, if gas slugs pass through and if started without being primed properly. All of these happen regularly in the Oil & Gas environment. These pumps would tend to fair much better when not used for water and selected for a minimum viscosity 4 cSt. The wear rings would perform as much better bearings.

Future developments will probably centre on enhanced bearing materials for the wear rings; the first stage wear rings at least.

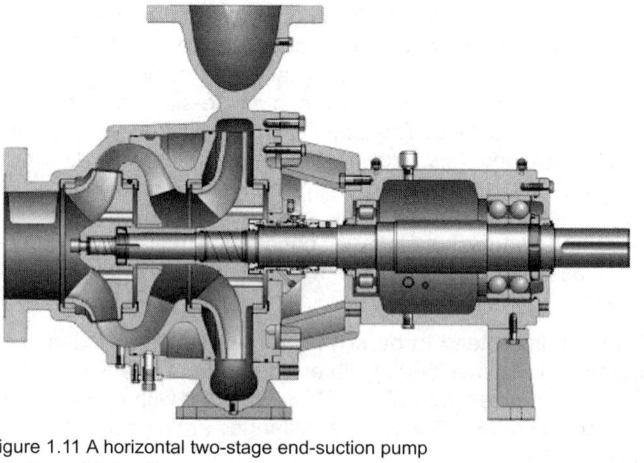

Figure 1.11 A horizontal two-stage end-suction pump
Courtesy of Sulzer Pumps

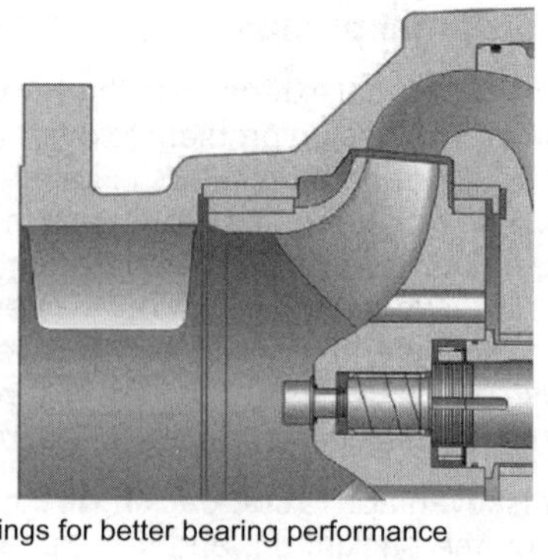

Figure 1.12 Wear rings for better bearing performance

Improved bearing performance will allow larger pumps to run successfully. Figure 1.12 shows a style of wear ring which may become common. Alternatively, a ceramic rolling-contact bearing may be a more reliable option for very clean liquid applications.

1.3.10 Vertical single-stage overhung impeller centrifugal pumps for general applications

Vertical single-stage pumps, pumps which have a suction connection and are not submerged, fall into two basic design types. Some pumps use standard electric motors and have flexible couplings. These pumps have radial and thrust bearings in the pump. The other pump design requires a special electric motor; an electric motor with a thrust bearing. The pump includes a radial bearing but relies on the electric motor for axial setting. Also available, is an almost off-the-shelf horizontal end-suction pump which is mounted vertically. A short, bent pipe section is attached to the usual suction connection to obtain the 'inline' configuration. Vertical single-stage pumps in this category are similar to central heating circulators ...... but ! the pump must be supported! The pump does not hang in the pipework; they are too heavy for that.

Pumps with a bearing bracket follow the design rules of horizontal end-suction pumps and use a number of the same parts. A motor stool is used to support the flanged motor from the pump. These pumps can save floor space but the piping arrangement must be appropriate. Small pumps can handle 400 m3/h while the largest pump can pass 3500 m^3/h. Differential head can be from 230 m to 900 m. Some pumps can cope with quite cold liquid, down to 185°C, and up to about 400°C. Pressure ratings can be light, 24 barg, to heavy-duty, 100 barg. These pumps can utilise the flexible coupling of purchaser's choice. A wide range of material combinations is possible to suit most applications. Mechanical seal chambers may comply with ISO 21049/API 682 requirements. There are mag-drive versions available when leaks are intolerable.

The pumps without a bearing bracket, like their horizontal equivalents, produce a very small, neat package. Thrust bearings in the motor absorb the pump axial thrust and locate the impeller in the casing. Motor bearings will usually be grease-lubricated and can be regreased while running. The motor can have a special extended shaft for mounting the impeller directly, or be connected via a solid muff coupling. The coupling can incorporate the pump axial setting adjustment. Space for a mechanical seal is likely to be restricted; double seals with "bells and whistles" might not be possible. Benign liquids, with just traces of hazardous/toxic compounds, might be out-of-bounds. However, it is possible to squeeze a lot out of these little packages; variable speed motors are becoming more popular. Matching the pump to the process conditions, to save energy, is worthwhile economically. Speeds up to 8000 rpm are possible.

Small pumps are limited to 25 m^3/h at differential heads up to 600 m. Larger pumps can handle 500 m3/h with slightly less head rise, 275 m. Casing design pressures of 40 bar(g) are typical and designs cope from -40 to 400°C.

One of the most prevalent problems associated with these pumps, is poor piping layout. It is not unusual to see these pumps in a dogleg from a long, straight, pipe run. The straight length of pipe on the suction connection is usually too short; 5D is the rule-of-thumb for a minimum. Increased turbulence in the impeller produces vibration leading to wear and short bearing life. In a number of cases, the vertical pump was the wrong style for the equipment layout; an end-suction pump would have better!

1.3.11 Vertical single-stage overhung impeller centrifugal pumps (for chemical applications includes ASME/ANSI B73.2 designs)

There is some confusion regarding the style of inline pumps. Inline construction requires the suction and discharge connections to be concentric; that is the longitudinal centre-line of the suction pipework is also the longitudinal centre-line of the discharge pipework. Imagine a straight length of pipe; if a section of pipe is removed an inline pump can replace the pipe section. Some standards, like BS 4082 U style, have both pump connections on one side of the pump; this cannot be an inline pump. The pump flanges may be specified as identical sizes and rating. There may be disadvantages to using the same schedule pipework on the suction as the discharge.

The basic philosophy of vertical inline pumps is identical to pumps in Section 1.3.1. The pump is mounted in the pipework and is not bolted down to foundations. ANSI B73.2 pumps are larger than these, all pumps must be mounted with the motor vertically above the pump and the pump can be optionally supported by a stool. The weight of the pump package is supported by the stool but the pump is free to move to accommodate pipe movement. The Standard is dimensional, to allow interchangeability between manufacturers, but does not specify hydraulic performance. Pump sizes with suction/discharge from 2"/1½" to 6"/4" are covered; length across flange faces is from 15" to 30".

Three pump designs are included within a common envelope. A close coupled design, VM, using a NEMA JM or JP flange mounted motor, with motor shaft and mounting dimensions and tolerances to NEMA MG 1-18.614. Obviously the motor must be moved to inspect the seal or remove the rotating assembly.

The second design, VC, is a short pump shaft rigidly coupled and spaced to a NEMA P-base flange mounted motor. Motor flanges, dimensions and tolerances are to NEMA MG 1-18.620. Removal of the pump rotating assembly or seal does not need the motor to be moved.

The third design option, VB, is for pumps with bearing housings driven through flexible spacer couplings, see Figure 1.5. These motors shall be NEMA C-face to NEMA MG 1-11.35 with solid shafts. Again removal of the pump rotating assembly or seal does not need the motor to be moved. Designs VC and VM have specified bearing L10 lives of 26000 hours at rated duty or 17500 hours at maximum load. Pumps to VB shall have bearings suitable for 17500 hours at maximum load. On test, running at 100/105% speed and ± 5% rated flow, the unfiltered vibration on the top pump or motor bearing housing shall not exceed 0.25 in/sec or 0.0025 in peak-to-peak displacement.

The Standard specifies pressure ratings of 125 LB for cast iron pumps and 150 LB for steel and stainless steel; 12 and 20 barg. An option is given for 250 LB and 300 LB flanges. Pressure/temperature ratings are in accordance with ANSI B16.1 and B16.5. Material requirements are similar to Section 1.3.5 types with a 0.125" corrosion allowance. Allowable nozzle loads must be specified by the manufacturer for the material offered at the specified operating temperature.

The Standard specifies a seal cavity which must be capable of accepting double seals with a pumping ring for forced circulation of buffer liquid. A packed stuffing box is optional.

At 60Hz pumps are capable of 320 m³/h at heads of 120 m; 50Hz performance is reduced to 265 m³/h at heads of 83 m. The Standard specifies the maximum pump absorbed power to be 172.5 hp, 128 kW.

Some manufacturers build a heavier version of the vertical inline pump to the VC design which complies with the requirements of API 610. Removal of the coupling allows seal inspection and replacement without disturbing the motor. In some designs the rotating assembly cannot be removed without detaching the motor stool. Capacities up to 1150 m³/h at heads up to 230 m are possible at 60 Hz. Suction/discharge sizes go up to 14"/8" and pressure ratings of 100 barg are available on some sizes.

Vertical inline pumps are really end-suction pumps with an extension on the suction nozzle. High suction pressure creates axial thrust which will reduce bearing life. The manufacturer should be asked to confirm the bearing life if any doubt exists about the magnitude of the suction pressure.

1.3.12 Vertical multi-stage centrifugal pumps for general applications (including segmental pumps, dH < 300 m, deep well ejector applications, wash water)

The differential head produced by a centrifugal pump is dependant upon the impeller diameter and the speed. In very simple theory, it would be possible to increase the diameter or the speed to produce any desired head. In practice life is more complicated. If a pump for a small flow and high differential was designed with a single impeller running at a 2-pole motor speed, a large diameter and narrow tip width would result. At some stage in the range progression the impeller would become impossible to cast or machine. Also the pump would be large in diameter requiring too much material. Friction losses between the sides of the impeller and pump casing increase proportionally to the fifth power of the impeller diameter resulting in low efficiency. The pump speed cannot be increased indefinitely because impellers can disintegrate under the influence of centrifugal forces which increase with the square of the speed. The solution to the problem is to divide the total differential head across several impellers working in series. The impellers, usually identical, each do a proportion of the work. The segmental pump, or ring section pump, is a popular mass-produced design to fulfil these requirements.

Very small vertical segmental pumps are produced for low flow high head applications. These pumps are similar, in some ways, to the vertical inline pumps shown in Figure 1.13. A base, with inline suction and discharge connections and a mounting foot, is used to support the stack of segments and the electric motor. Each segment consist of a diffuser, an impeller and a spacer. The diffuser forms the outer pressure casing and collects the flow from the tip of one impeller and redirects it into the eye of the next impeller. The spacer is used to separate the impellers. On small pumps the seal between diffusers is metal-to-metal; good surface finish and accurate machining being essential. Larger pumps use O rings. The bearings and

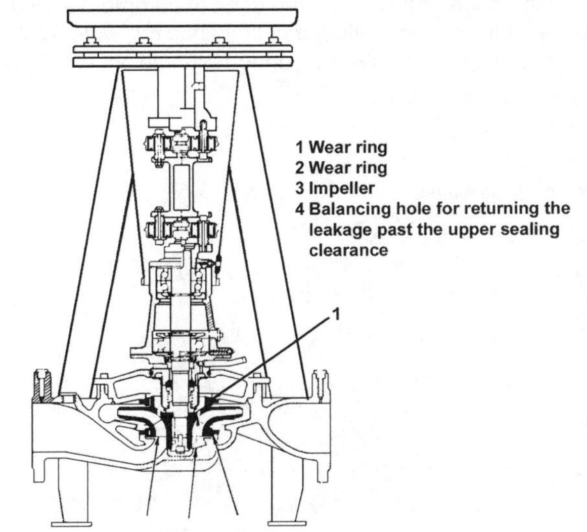

Figure 1.13 Vertical inline pump with support feet, bearing bracket and extended motor stool for spacer coupling.

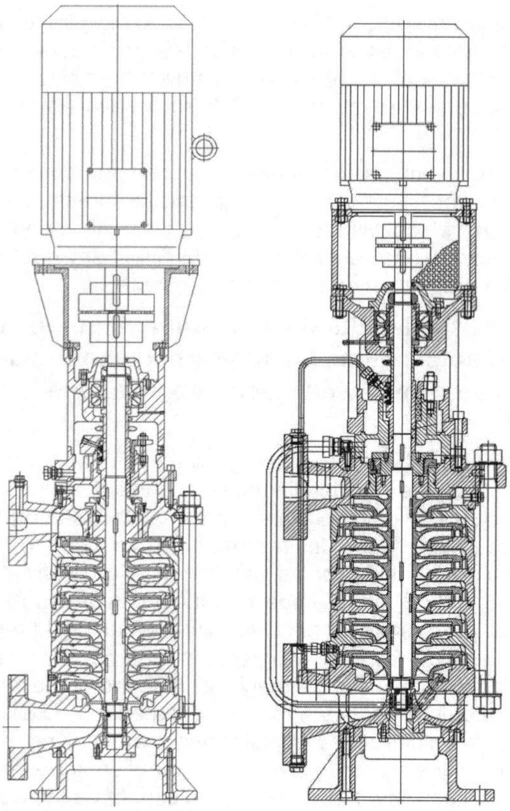

Figure 1.14 Two typical larger vertical segmental pumps
Courtesy of Sterling SIHI GmbH

seals are housed in the base and a motor mounting stool at the top. The complete pump assembly being held together by long tie-rods around the outside of the diffusers. Very small units of this style are limited to 5 to 10 kW. Pumps in cast iron and bronze with pressures up to 25 barg are standard.

Slightly larger versions of segmental pumps have the suction and discharge connections at opposite ends of the pump; horizontal and vertical variations are available in these sizes from one to twenty stages, see Figure 1.14. Power requirements would be up to 300 kW. Pumps are available in cast iron, steel, bronze and duplex stainless steel with pressure ratings up to 40 barg.

Various differential heads are achieved by varying the number of stages. In order to maintain good efficiency it is necessary, for a given head, to increase the number of stages as the volume flow decreases, Figure 1.15.

Multi-stage segmental pumps are used in commercial and domestic installations in multi-storey buildings for; water distribution, central heating, steam condensate, fixed fire fighting pumps.

One disadvantage of the segmental pump design is the number of seals and potential leak paths in the pressure casing due to the assembly of individual diffusers. For this reason maximum operating temperature is limited to around 170 °C and applications on hazardous liquids extremely restricted.

Wash water pump packages

For washing purposes, car washing for example, there are off-the-shelf packages available. Typical operating data: flow 40 litres/min with discharge pressure ≈1.5 MPa or 15 barg, equivalent to a differential head of 150 m. The pump normally supplied is an electrically driven multi-stage centrifugal pump. To prevent zero flow operation while the liquid flow is shut-off, which would cause overheating of the pump, a minimum flow sensor is included in the package. This monitors the liquid flow through the system and prevents the pump from operating in the event of low water flow. A typical arrangement of a washing package is shown in Figure 1.16. The numbers in the illustration refer to:

1. Shut-off valve. For servicing of pump and equipment.
2. Inline suction filter. Prevents contaminants from entering the minimum flow sensor and the pump.
3. Minimum flow sensor. Starts and stops the pump automatically. Connects with single-phase cable to the motor safety switch. The pump must operate with a positive supply pressure where the sensor is fitted.
4. Non-return valve. Prevents water from flowing back into the supply pipe. Essential when connected to municipal drinking water system.
5. Vacuum valve. Activated in the event of the water supply pipe failing to supply the quantity of water required by the pump.
6. Motor safety switch. Switches off electric supply to the motor and triggers alarm signal in the event of a fault arising. Must always be used when a minimum flow sensor is fitted.
7. Pressure gauge. Fitted to pump's delivery flange.
8. Solenoid valve. Water-proof. Operated by coin slotmeter.
9. Cock. To isolate hose.
10. High pressure hose. If the delivery pipeline is long, a larger diameter should be fitted in order to reduce the pressure drop.
11. Wash-gun. Completely rubber covered with shut-off and regulating trigger. Can be adjusted for high velocity jet and fine spray.
12. Coin slot-meter. For self-service vending. Can be set to 10, 15, 20, 25, 30 and 35 minutes. Coin setting as required.

Self-contained wash packages are also available using reciprocating pumps, see Sections 1.5.14 and 1.5.15..

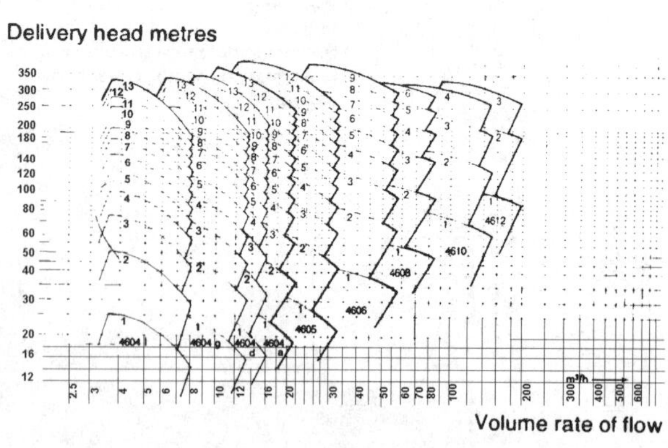

Figure 1.15 Range chart at 2900 rpm for segmental pumps

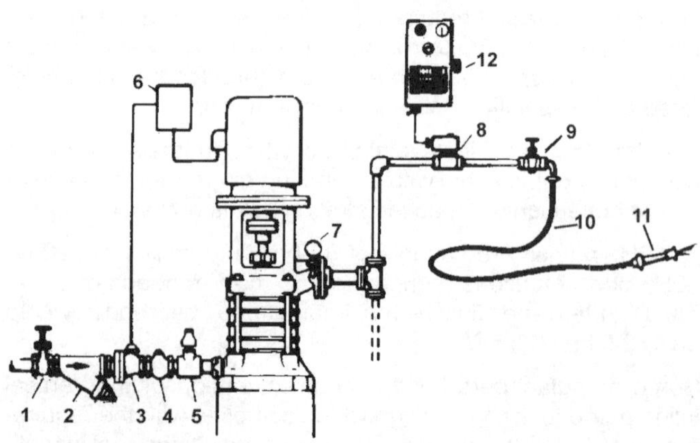

Figure 1.16 Wash water pump package

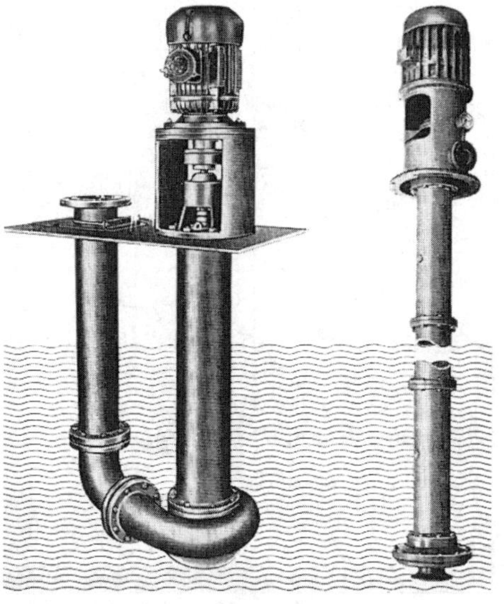

Figure 1.17 Single-stage vertical wet-pit pump

1.3.13 Vertical wet-pit pumps

This pump type comprises all those pumps designed to be suspended into the liquid with the drive motor on top. Figure 1.17 shows a typical single-stage design where the pump discharge is piped separately through the mounting plate. The pump seal is mounted in the pump casing below liquid level. The thrust bearing is included in the mounting plate assembly.

The style of construction shown in Figure 1.17 is limited to fairly low power pumps. The column surrounding the shaft is relatively small in diameter and the discharge pipe helps to increase rigidity of the assembly. As the shaft length increases intermediate bearings must be added to reduce shaft deflection and vibration. Special rubber bearings are available which can operate with water as the sole lubricant.

Larger pumps — centrifugal, mixed-flow and axial propeller, are constructed with a column which encloses the shaft and also acts as the discharge pipe. See Figure 1.18. The column diameter is increased to accommodate a reasonable liquid velocity. The column is extended through the mounting plate to allow a branch to be fitted for the discharge connection. A seal, rated for discharge pressure, must be fitted above the discharge connection. Again, intermediate bearings must be fitted as the column length extends. Pumps are available for flows in excess of 6000 m^3/h at heads up to 150m.

Vertical pumps, especially in the larger sizes, can be prone to vibration problems. The problems are caused by pump or motor shaft unbalance occurring at a frequency which coincides with a natural frequency of the column. Because of this it is essential that large pumps are tested in the manufacturer's works under conditions which duplicate, as closely as possible, the

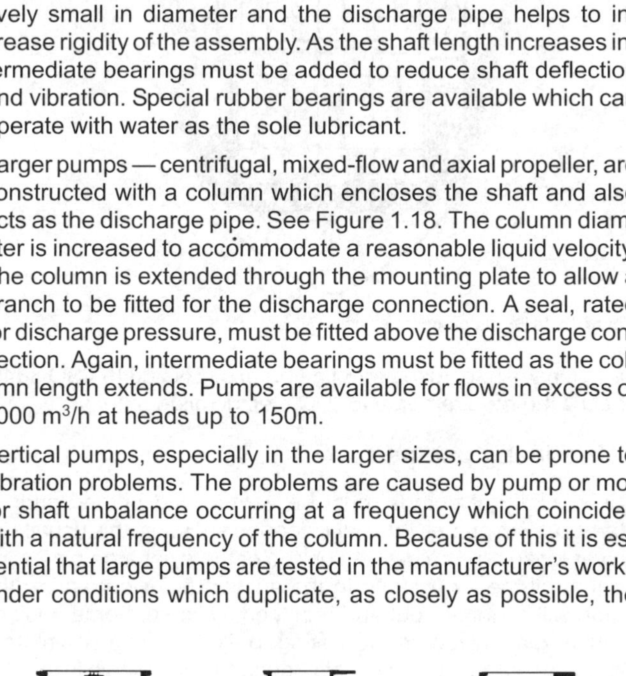

Figure 1.18 Single-stage vertical wet-pit pumps in centrifugal, mixed-flow and axial propeller designs

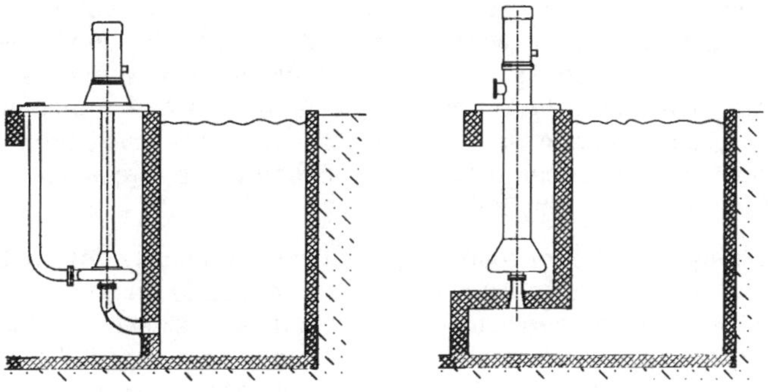

Figure 1.19 Vertical dry-pit pumps

site operating conditions. Extensive testing and modification is much easier in the works than on site.

A variation of the vertical wet-pit pump is the vertical canned pump. A vertical wet-pit pump is built inside a can or pressure vessel which is also suspended from the mounting plate. The can has a flanged pipe connection, pump suction, which may be attached above or below the mounting plate. This type of design is used for volatile liquids with low NPSHa. The liquid is taken from the suction line elevation down the can to the first stage impeller eye. The NPSHa is increased by the difference in levels between the suction line and the impeller eye. This variation is used extensively for liquified gas applications.

1.3.14 Vertical dry-pit pumps

Vertical dry-pit pumps are quite similar to wet-pit pumps but the pump casing in not immersed in the liquid. The liquid is carried from the pit to the pump suction by concrete ducts or metal pipework. Figure 1.19 shows two examples of dry-pit pump installations.

The left hand installation shows a pump with metal suction pipe and a separate discharge pipe. The right hand installation shows a concrete suction duct and the discharge within the column. Pump sizes are similar to wet-pit pumps.

1.3.15 Marine pumps

Marine pumps are included as important components in, not only ships and oil rigs, but also underwater equipment. Pumps for ships can be classified as follows:

- Systems for propulsion, e.g. pumping fuel, lubricating oil, cooling water, feed water and condensate
- Systems dependent on the cargo
- Safety systems. e.g. for bilge pumping, ballast pumping, for fire fighting
- Domestic housekeeping type systems for making and pumping drinking water, hot and cold fresh water, sewage etc

The types of pumps used for fuel is dependant upon the type of fuel. Many ships use heavy fuel which is about 3500 Redwood No 1. Rotodynamic pumps would be inappropriate. Double acting piston pumps act as transfer and low pressure feed pumps. Plunger pumps, usually built by the engine manufacturer, inject the fuel into the cylinder. Special lubricators, driven by eccentrics or geared to an engine shaft, inject lubricating oil into the cylinders. Screw or gear pumps provide pressurised lubricating oil. Fresh water for engine cooling, and seawater for cooling are provided by centrifugal pumps. Packages are available using double ended motors which supply fresh and seawater simultaneously from two pumps on a common baseplate. Vertical units have been built with fresh and seawater pumps driven by a normal motor. Steam turbine ships use multi-stage centrifugal pumps for boiler feed and condensate extraction duties.

Again, the type of cargo pump is dependant upon the cargo. Crude is a common load but is extremely variable, with low to very high viscosity. Steam driven, direct acting, vertical piston pumps are available which can operate on petrol through to the most viscous crude oil. Electrically driven piston pumps are also available but are not so versatile as the steam versions. Liquified petroleum gas and anhydrous ammonia have become popular as a sea cargo. Special centrifugal pumps are available for loading and unloading.

Safety systems for marine applications are usually designed around seawater because it is readily available. Bilge pumps largely cater for seawater. Some modern pumps have monitoring systems fitted which shut down the bilge pumps if the oil content is higher than 15 parts per million (ppm). The amount of liquid able to be pumped from the bilges is dependant upon the age of the ship. Modern ships tend to have mechanical seals on the stern tube so that very little leakage enters via that route. Soft packed stern tubes require some leakage for lubrication. In some vessels the level in the bilges must be watched very closely as it has a great effect on stability. Depending on the age of the ship the bilges may be pumped every two hours, at the end of each watch or automatically by float controls. The pumps may be reciprocating or centrifugal or ejectors powered by seawater. Reciprocating and centrifugal pumps would have a priming system to evacuate the suction line. Ballast and fire fighting pumps operate on seawater and would be centrifugal pumps.

Turbine and engine driven vessels can make fresh water from seawater by evaporation or reverse osmosis. Evaporation is a low pressure process requiring single-stage centrifugal pumps. Reverse osmosis for seawater operates at 70 barg and requires a reciprocal pump or a multi-stage segmental pump. Other domestic water circuits would have single-stage centrifugal pumps similar to those used onshore.

Ships at sea should be able to function safely and entirely independently, therefore special safety and reliability requirements are necessary. These are determined by marine insurance companies (classification societies) and each country's respective maritime authority. In certain cases additional requirements are imposed by, for example, port and canal authorities. Various types of pumps are used for the pumping systems mentioned earlier in the same manner as those land based systems.

Rotodynamic pumps on board are often of the vertical type in order to reduce space. The mechanical design is adapted, generally reinforced, to take account of the ship's movement at sea and automatic, separate or built-in air ejection systems are present to avoid operational problems in the case of long suction lines or if the suction intake is momentarily above the surface of the liquid. As a general rule, bronze or gunmetal is used when pumping seawater and the same material as in the case of land based systems is chosen for other liquids.

Figure 1.20 shows a typical vertical pump for general use on board ship. This is a double suction pump with the pump casing divided axially (in a vertical plane) in order to facilitate servicing. Single suction impellers with axially or radially split casings are used for smaller flows. In the case of a radially split casing, the pump's rotor should be able to be disassembled in a simple manner, like for example, Figure 1.5. Chapter 16, Section 16.3 shows the installation of cargo oil pumps as a practical example.

For marine use, there are many interesting pump applications, such as the steering/propulsion jet operation of non-anchored oil rigs with automatic positioning. A pump system as shown in Figure 1.21 and comprising axial pumps designed to operate with any flow direction, is used to limit heeling when loading for example; or for the intentional generation of heel angle in the case of icebreakers.

Small pleasure craft, cruisers and yachts, use flexible vane pumps for many applications. Hand-operated diaphragm pumps have been used for bilge pumps. EN 28849, ISO 8849, specify the requirements for DC electric bilge pumps.

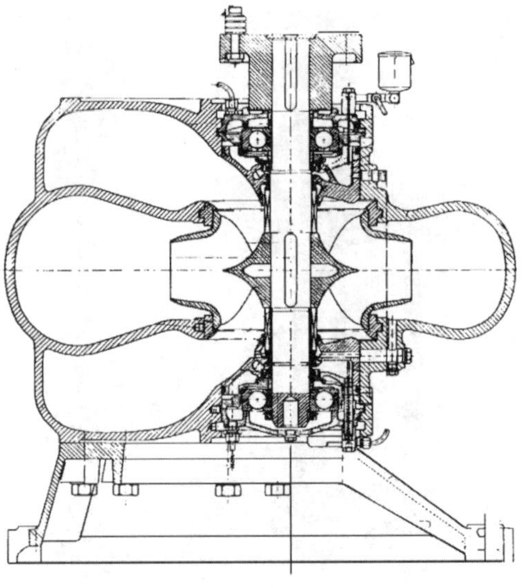

Figure 1.20 Vertical ship's pump with soluble suction impeller

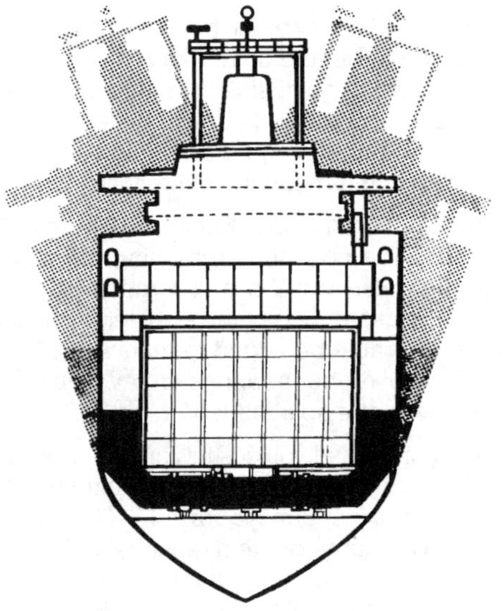

Figure 1.21 Heeling pump for ships

Self-priming pumps

The majority of pumps must be vented and primed prior to start-up. With pumps that have flooded suctions, this usually involves opening small valves on top of the discharge pipework, so that all the air trapped in the suction pipework and in the pump will be forced out and replaced by liquid. Some designs require vent valves on the casing to fill the pump completely. When the pump suction is not flooded the vent valve in the discharge pipework is coupled to an air ejector. The air is withdrawn by a partial vacuum and the liquid fills the void. Submersible pumps do not suffer these inconveniences as the pump is always full of liquid.

Self-priming pumps are designed to overcome these problems. The casing of a self-priming pump is constructed to retain enough liquid so that the impeller is always flooded. The casing needs priming only once provided the liquid does not evaporate or leak away through poor seals. The pump casing is shaped so that any gas or air bubbles become completely surrounded by liquid and are pumped from the suction to the discharge. Since there is only a limited amount of liquid during evacuation the gas and liquid must be separated, whereupon the liquid is returned to the suction side to recirculate. The internal design and shape of the pump casing to achieve this effect varies considerably depending on the manufacturer. When selecting a pump it is important to take into account the wearing effect which for example sand, may have upon the pump's evacuating abilities.

When electric power is not available, on building sites for example, other sources of power must be used, such as engines, compressed air from mobile compressors or hydraulically from a separate power pack or one which is built into a vehicle. This type of pump is also suitable for intermittent use in construction work such as road making, rock blasting and excavating or as a fire precaution when carrying out stubble burning or site clearing.

Priming of engine-driven units can be assisted by using the engine inlet manifold vacuum or a separate engine-driven exhauster.

1.3.16 Submersible pumps with electric motor

Submersible pumps placed in ponds, ditches and sumps are used to drain, mostly water from locations which are deeper than the sewage system. Three types are available:

- Simple, small, self contained for clean water
- Heavy duty for contaminated water
- Special purpose for sludge

Small submersible pumps tend to be made of plastic where ever possible. This style of construction removes many of the insulation problems associated with electrical equipment. The pumps are close-coupled and completely self-contained. The motor is totally enclosed and cooled by the water surrounding the pump. Level switches, either internally or mounted externally on the casing, detect the water level and switch the pump on and off. Low voltage versions, 110V and 240V, single phase as well as 380V three phase are available. EN 60335-1 and EN60335-2 should be consulted for relevant electrical safety requirements. The smallest units will pump 10m^3/h against a head of 7m. Pipe connections are usually screwed and a non-return valve is required on the smallest units.

Heavy duty electrically driven submersible pumps are close coupled pumps and comprise, together with external level controls, a complete unit. Their weight is the lowest possible due to the use of light metal alloys although their resistance to corrosion is low. The parts which are exposed to wear from the liquid are, on the other hand, of high quality. Typical features are a hard metal impeller, rubberised wear parts around the impeller

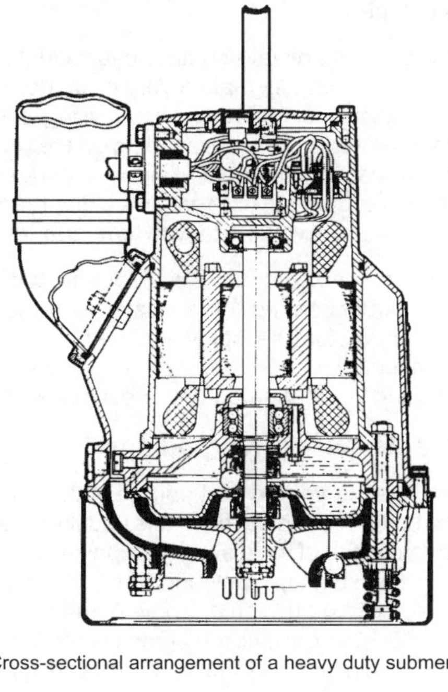

Figure 1.22 Cross-sectional arrangement of a heavy duty submersible pump

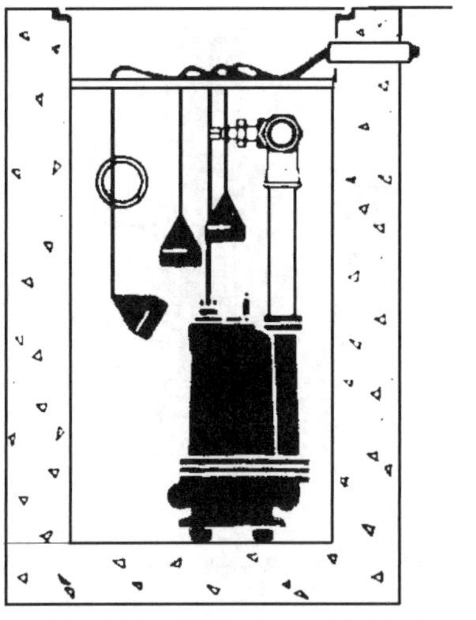

Figure 1.23 Installation of submersible pump in sump

and hard-faced seals. A non-return valve is built into the discharge connection.

The electric motor is completely encapsulated and cooled by the pumped liquid through a double jacket. In order to protect the motor in the case of blockage or operation without liquid, the motor is provided with a specially built-in motor protection unit. Normally the motor windings have thermistors which stop the motor when the temperature is too high by means of an inbuilt contactor. The shaft seal is nearly always of a double type with an intermediate oil chamber, see Figure 1.22. Normally construction pumps have suction strainers which limit the particle size to 5-10 mm, but there are special models with a through flow of up to approximately 100 mm for pumping larger solids. Figure 1.23 shows a heavy duty pump located in a sump with external level switches. Heavy duty pumps can handle up to 7000 m^3/h at heads up to 100 m.

Heavy duty submersible pumps are used for draining all types of building sites from the smallest ditch or hole to the large projects such as tunnelling in mountains or harbour building. This type of pump is practical in the event of flooding and for all other temporary pumping requirements. Of particular convenience is the fact that pump companies in this field offer both small and large pumps for hire for periods ranging from one day to several months.

A special adaptation of the heavy duty submersible pump has been developed for sludge and heavy muds. The impeller and casing are made from hard wear resistant iron alloys, not rubber coated, with an agitator mounted on the shaft end where an inducer would be fitted. The pump is intended for use on liquid solid mixtures with very high solids content where the pump can sit, initially, on the surface. The agitator beats the mixture and induces moisture to travel towards the impeller. The extra moisture locally helps the mixture to flow and the agitator action allows the pump to settle in the mixture. The novel design of the agitator allows these pumps to handle liquid solid mixtures which would otherwise be moved by shovels.

1.3.17 Non-electric submersible pumps

Electric submersible pumps are potentially very dangerous. Many small submersible pumps are used with water. Water and electricity is of course a dangerous combination. The motor manufacturer takes extreme precautions to prevent the water and electricity mixing. Large submersible pumps are used for handling hazardous liquids, such as crude oil. It is flammable but is not a pure liquid. Crude oil is a mixture of liquids, and possibly gases such as carbon dioxide and hydrogen sulphide. Hydrogen sulphide is potentially poisonous as well as flammable. Removing electricity from the pump installation therefore removes a very serious hazard!

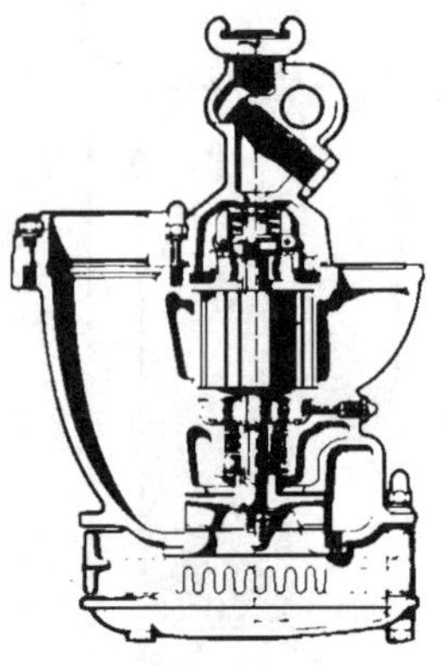

Figure 1.24 Compressed-air driven submersible pump

Sometimes electric power is not available, on building sites for example; and so other sources of power must be used, such as engines, compressed air from mobile compressors or hydraulically from a separate power pack, or one which is built into a vehicle. This type of pump is also suitable for intermittent use in construction work such as road making, rock blasting and excavating or as a fire precaution when carrying out stubble burning or site clearing.

Compressed air-driven rotodynamic submersible pumps are of similar construction to those which are electrically driven. They are equipped both with or without special heavy-duty components for pumping large solids. The high speed compressed air motor, lubricated by oil mist, can compete on performance with electrically driven units, see Figure 1.24. Also remember, the air-operated double-diaphragm pump can be treated as a submersible pump. The air exhaust should be piped to above the maximum liquid level to prevent 'flooding' and potential corrosion of the air valve.

Jet pumps can be submerged, these pumps can handle large solids providing large quantities of liquid are moved, **and** large quantities of driving fluid are provided.

1.3.18 Submersible pumps for deep well applications

Submersible deep well pumps in smaller sizes have the same range of applications as deep well pumps with ejectors. The difference here being that the pump is placed directly in the borehole connected by means of a delivery hose and electric cable.

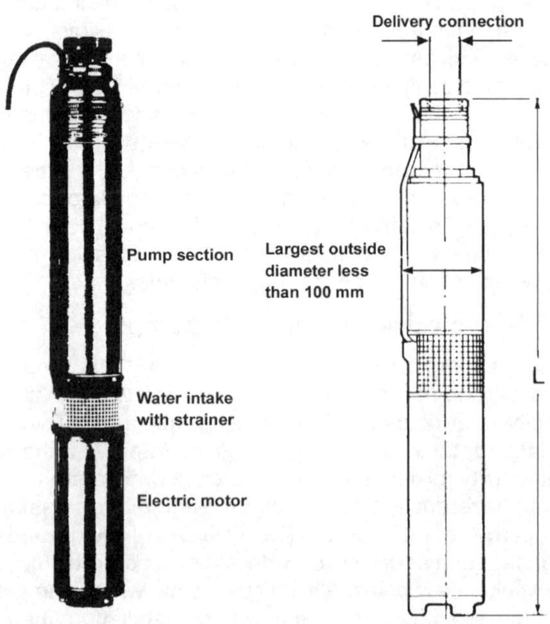

Figure 1.25 Small submersible deep well pump

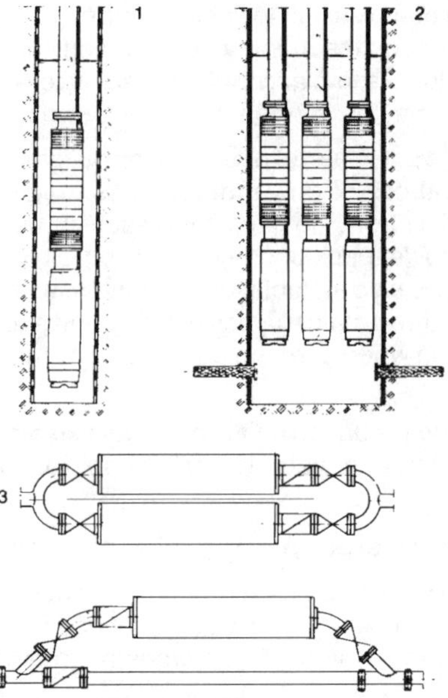

Figure 1.26 Deep well pumps placed in wells - illustrations 1 and 2, and placed in pipe section for pressure boosting - illustrations 3 and 4

Since the smallest borehole diameter is 4", the diameter of the pump must be somewhat less, see Figure 1.25. Impellers are between centrifugal and mixed-flow designs with as many as 100 stages. The pump is a multi-stage centrifugal pump driven by a special type of long thin electric motor located beneath the pump and cooled by the water in the borehole. For small pumps, suitable for domestic applications, the pump length L varies according to the depth of the borehole between 500 and 2500 mm.

Submersible deep well pumps of larger sizes have outside diameters of up to approximately 500 mm and lengths of up to 5 metres. Their hydraulic data varies considerably, the maximum values being: flow 5000 m^3/h and differential head 1000 m. Motor outputs of several megawatts are available when using high voltage motors.

These pumps are used for municipal drinking water supply, reduction of ground water levels and mine drainage. Deep well pumps find favour in oil field applications, on and offshore. Units are used for secondary oil recovery prior to water injection. Typical applications are to be found in restricted wells or shafts, Figure 1.26. They can also be used for pressure boosting in drinking water supply networks, where the pump may be integrated in a section of pipe.

The electric motors are always squirrel-cage liquid-filled induction motors, which naturally makes very great demands upon the motor's electrical insulation. The liquid filling is either clean water, water-oil emulsion or oil depending on the requirements of each specific case. The liquid filling is separated from the pumped liquid by a mechanical seal and a diaphragm to compensate for variations in volume, due to temperature changes.

The pump and motor rotors run in liquid-lubricated plain bearings. The axial thrust bearing is the most heavily loaded, which is why some manufacturers employ opposed impeller construction to reduce unbalanced axial forces. Special models are available for sand contaminated or corrosive water.

1.3.19 Portable self-priming pumps

Most pumps must be vented and primed prior to start-up. For pumps which have flooded suctions this usually involves opening small valves on top of the discharge pipework, so that all the air trapped in the suction pipework and in the pump will be forced out and replaced by liquid. Some pump designs require vent valves on the casing to fill the pump completely. When the pump suction is not flooded the vent valve in the discharge

Figure 1.27 Engine driven self-priming pump
Courtesy of Hale Products Europe

pipework is coupled to an air ejector. The air is withdrawn by a partial vacuum and the liquid fills the void.

Self-priming pumps are also designed to overcome these problems. The casing of a self-priming pump is constructed to retain sufficient liquid so that the impeller is always flooded. The casing needs priming only once provided the liquid does not evaporate or leak away through poor seals. The pump casing is shaped so that any gas or air bubbles become completely surrounded by liquid and are pumped from the suction to the discharge. Since there is only a limited amount of liquid in the pump casing during evacuation, the gas and liquid must be separated, whereupon the liquid is returned to the suction side to recirculate. The internal design and shape of the pump casing to achieve this effect varies considerably depending on the manufacturer. When choosing a pump it is important to take into account the wearing effect which sand, for example, may have upon the pump's evacuating capabilities. When considering the suction system design, the volume of gas in the suction piping should be minimised.

Engine-driven self-priming pumps are usually single-stage centrifugal pumps as shown in Figure 1.27. In order to be self-priming the pump must be filled with liquid prior to starting. The evacuation times quoted usually refer to the time taken to obtain the full pressure at the pump discharge for a specific suction lift, using a hose having the same diameter as the pump's suction connection and whose length exceeds the suction lift by 2 metres. The pump performance can be adapted to actual requirements by varying the engine speed. Figure 1.28 shows a typical small portable self-priming pump.

Engine-driven pumps can handle solid particles of up to 10 to 15 mm. The suction hose is reinforced because of the vacuum caused by pumping. The normal suction strainer can usually be replaced with a much shallower design in order to facilitate pumping down to within a few millimetres off a floor. The pump should be equipped with a protective safety frame and runners, skids or a wheeled trailer to enable it to stand on soft ground. Priming of engine driven units can be assisted by using the engine inlet manifold vacuum or a separate air pump.

Engine-driven single cylinder reciprocating pumps are also used as self-priming pumps especially for construction sites and road work. Engine-driven pumps are available with suction sizes up to 150 mm which can handle flows of 400 m^3/h at heads up to 30m. Engines sizes up to 30 kW are not unusual. Single cylinder mechanical diaphragm pumps are also available, see Figure 1.100. If compressed air is available, then the air-operated double-diaphragm pump can be considered. These pumps will not handle such large solids but are useful for 'clean' applications. Care should be taken to supervise these

Figure 1.28 A typical small portable self-priming pump
Courtesy of Varisco Srl

pumps. If the liquid supply 'dries-up' the pump will race and consume a large quantity of air while doing zero work.

Engine-driven units can be very noisy. Current designs pay special attention to noise reduction, levels of below 80 dBA at 1m are desirable. To achieve this level of noise it is usually necessary to fit silencers to both the inlet and exhaust sides. The engines are normally splash-lubricated which means that the pump must be placed on a relatively flat surface the inclination may seldom exceed 20°. Engine-driven compressors can also be extremely noisy.

1.3.20 Horizontal two-stage axially-split centrifugal pumps (includes ISO 13709, API 610 designs)

These pumps are similar in many respects to the double-suction and multi-stage pumps described later. The use of two-stages provides extra differential head. The advantage of the split casing is usually found in maintenance. Suction and discharge connections can be located in the bottom half, the top half being easily removed without disturbing piping. Process connections can be located in the top half with the obvious extra maintenance work. A radially-split design would remove this complication; see later. A radially-split design is recommended by ISO 13709 and API 610 for 'hazardous liquid' applications.

Impellers are usually mounted back-to-back. Two-stage pumps of this design are usually larger than the end-suction two-stage pumps and require the impellers to be placed between bearings. Impellers usually have a front wear ring. A common back wear ring, a centre bush, reduces leakage between the stages. The centre bush also acts as a bearing and provides some shaft stiffness. Engineering plastics can really improve the bearing effectiveness. A double-suction first stage can be used when NPSHa is low but this increases the cost significantly. The increased cost should be less than that of providing a separate booster pump and driver plus associated electrics and controls. A reorganisation of the plant layout may improve NPSHa and obviate a double-suction. Casings may be diffuser or volute. Diffuser pumps can be more efficient if the duty point is calculated accurately.

Depending upon the size of the pump, bearings may be rolling contact or sliding. Back-to-back angular contact ball bearings have become popular as a combined radial and thrust bearing. A very compact bearing housing is possible. Oil-ring lubrication is simple and effective. Larger pumps, with sliding bearings, usually have a tilting pad thrust bearing and a lube oil system. Some users prefer oil mist for rolling contact bearings and build a central supply system.

Shaft sealing is accomplished by a wide range of mechanical seals, although soft packing may be possible. Cast iron casings probably have restricted sealing space as cast iron is not recommended for handling many hazardous liquids. Connecting small bore vent and drain piping is only possible via threaded holes in many casings.

These two-stage pumps are manufactured in a wide range of materials, see Table 1.1. Pumps in super stainless steels may be available to order.

Casing	Impeller
Cast iron	Cast iron
Cast iron	Bronze
Carbon steel	Cast iron
Carbon steel	Carbon steel
Carbon steel	11-13% Cr
Carbon steel	Stainless steel
Stainless steel	Stainless steel

Table 1.1 Two-stage axially-split pumps — materials of manufacture

Two-stage pumps can usually handle flows up to about 2000 m^3/h and differential heads of up to 600 m.

1.3.21 Horizontal two-stage radially-split centrifugal pumps (includes ISO 13709, API 610 designs)

These pumps are similar in many respects to the double-suction pumps described in Section 1.3.8 and multi-stage pumps described later in Section 1.3.25. A radially-split design is recommended by ISO 13709 and API 610 for 'hazardous liquid' applications. The use of two-stages provides extra differential head. The advantage of the radially-split casing is found in application and maintenance. These pumps can be centreline-mounted to remove thermal growth problems experienced in high temperature applications.

The use of circular gaskets, for the main pressure containment closure(s), removes problems associated with large flat gaskets and bolting. Elastomer 'O'-rings can be used when temperatures permit. Hybrid fibre and metal gaskets can be used for higher temperatures. Metal 'O'-rings can be used for the most arduous applications. Suction and discharge connections can be placed at any orientation, virtually, to suit the plant layout. Some connection layouts are more popular than others; top-top is the manufacturers default arrangement; this is the shortest delivery time because there are no pattern modifications. If the plant layout is considered, practically, for piping venting and priming, then good arrangements can be accommodated in the appropriate time scale. Maintenance of the pump does not require the casing connections to be broken.

Two-stage pumps of this design are usually larger than the end-suction two-stage pumps and require the impellers to be placed between bearings. Impellers are usually mounted back-to-back to minimise axial thrust and have front and back wear rings. A double-suction first stage can be used when NPSHa is low but this increases the cost significantly. The increased cost should be less than that of providing a separate booster pump and driver plus associated electrics and controls. A reorganisation of the plant layout may improve NPSHa and obviate a double-suction. Suction pipe venting should be checked. Casings may be diffuser or volute; larger pumps may have double volutes. Diffuser pumps can be more efficient if the duty point is calculated accurately.

Depending upon the size of the pump, bearings may be rolling contact or sliding. Back-to-back angular contact ball bearings have become popular as a combined radial and thrust bearing. A very compact bearing housing is possible. Oil-ring lubrication is simple and effective. Larger pumps, with sliding bearings, usually have a tilting pad thrust bearing and a lube oil system. Some users prefer oil mist for rolling contact bearings and build a central supply system.

Shaft sealing is accomplished by a wide range of mechanical seals, usually to EN ISO 21049/API 682. These two-stage pumps are manufactured in a wide range of materials. Lowest cost is carbon steel with cast iron stationary internal parts and cast iron wear rings. Materials progress through all carbon steel, 11-13Cr, austenitic stainless steel and super stainless steels. Some manufacturers produce pumps in 19Cr 13Ni 3Mo low carbon as standard. Pumps can be built with any range of compatible materials, to order, if sufficient time is allowed in the contract timetable.

Two-stage pumps can usually handle flows up to about 2000 m^3/h and differential heads of up to 800 m with pressure ratings of 100 bar(g). Operating temperatures up to 425°C are typical.

1.3.22 Multi-stage segmental centrifugal pumps dH > 300 m

These pumps are a larger version of those in Section 1.3.12; horizontal pumps only with flows up to 3000 m^3/h at heads up to 4500 m are available requiring drivers up to 10 MW. Special designs for boiler feed pumps allow operation over 200 °C at 500 barg. Larger pumps are available in a wider range of materials; cast iron, steel, bronze, 18-10-2 austenitic stainless steel and

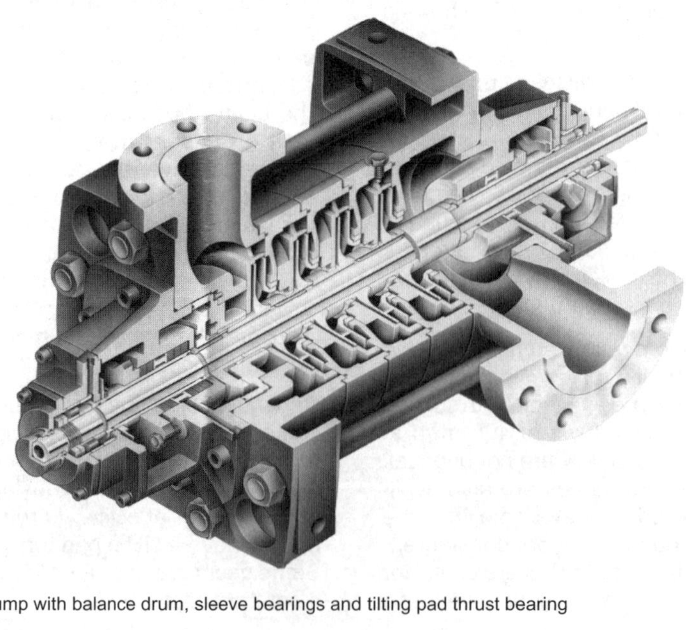

Figure 1.29 Large 4-stage segmental pump with balance drum, sleeve bearings and tilting pad thrust bearing

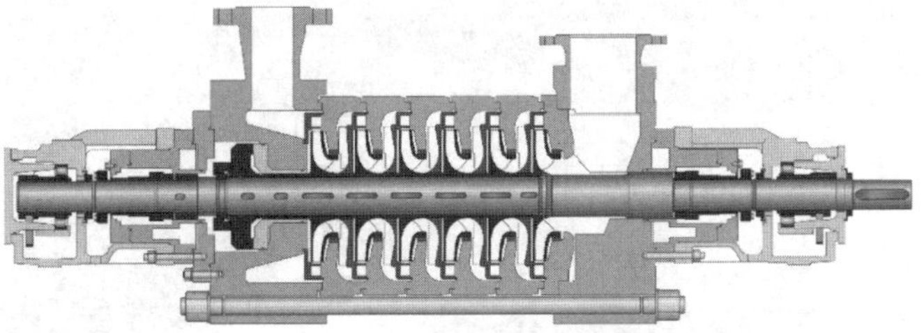

Figure 1.30 A high pressure segmental pump with balance disc and roller bearings
Courtesy of KSB AG

Figure 1.31 A typical installation of a hot high pressure segmental pump
Courtesy of KSB AG

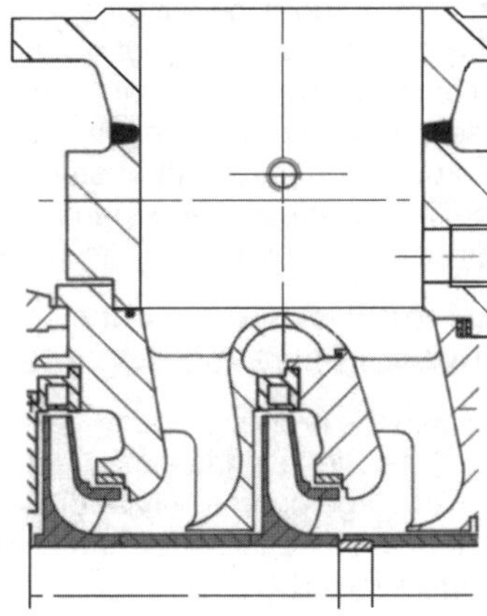

Figure 1.32 A double-suction arrangement for a high pressure segmental pump
Courtesy of KSB AG

duplex stainless steel. Some manufacturers have opted for fabricated suction and discharge heads, using thick boiler plate machined to the required profiles. Interstage connections are possible. Hydraulic axial thrust reduction can be implemented by disc, drum or double-drum (a balance drum with two diameters). Pumps can be fitted with rolling contact or sliding bearings. Tilting pad thrust bearings are available. The previous comments regarding hazards liquids still apply. Higher head segmental pumps are popular as boiler feed pumps and seawater pumps for desalination and reverse osmosis. Also useful in mining, irrigation and gas scrubbing processes. Figure 1.29 shows a large 4-stage segmental pump. Examples of high pressure pumps are shown in Figures 1.30, 1.31 and 1.32.

1.3.23 Horizontal multi-stage axially-split centrifugal pumps (includes ISO 13709, API 610 designs)

Large multi-stage axially-split pumps have a very wide range of applications, notably mine drainage, boiler feed and process industries. A sectional view of this pump type, during routine overhaul and general view, are shown in Figures 1.33 to 1.35 respectively. They are also used extensively in the refining, petrochemical industries and in oil field production. The design and construction being determined by the specific functions and components.

- Diffuser type
- First stage impeller
- Thrust balancing
- Bearing types

The theory of rotodynamic pumps shows that the kinetic energy of the liquid produced in the impeller must be converted to pressure energy by efficiently reducing the velocity of the liquid. Two methods are used; vaned diffusers and vaneless diffusers called volutes. The two main groupings of axially-split pumps are diffuser pumps, meaning vaned diffusers, and volute pumps. Modern trends are leaning towards volute pumps as these are cheaper, in theory, to produce.

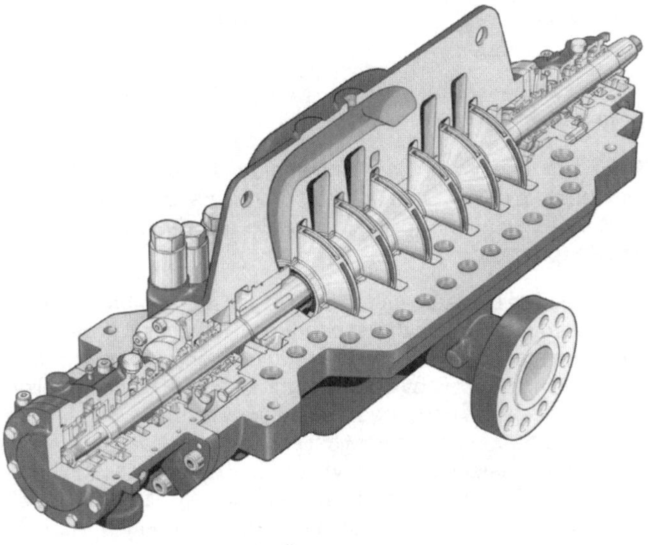

Figure 1.33 Sectional view of a multi-stage axially-split case pump, API 610 code BB3
Courtesy of Weir Pumps Ltd

Figure 1.34 Multi-stage axially-split case pump during routine overhaul
Courtesy of Weir Pumps Ltd

Figure 1.35 Multi-stage axially-split case pump, API 610 code BB3
Courtesy of Weir Pumps Ltd

There are also two choices of first stage impeller; single entry and double entry. Double entry impellers are used in low NPSHa applications to try to obviate the need for separate booster pumps.

For moderate pressure increases it is possible to absorb the axial thrust directly in the thrust bearings. However, it is generally necessary to balance out the thrust. The most common method of balancing thrust is by opposing the direction of the impellers and balance discs or by means of balance drums. Opposing impellers leads to some complications with the casing castings due to the transfer passages and can lead to poor casting integrity and weld repairs. Experts argue endlessly about the merits of balance disc and drums; personal experience with the specific application may be the best guide.

Depending upon the pump size, ball bearings are probably the standard with angular contacts, back-to-back, for the thrust. Sliding bearings with tilting pad thrust bearings may be an option or become the standard as pump speed increases. At higher speeds the life of rolling contact bearings is reduced whereas sliding bearings can have an almost unlimited life. Lube system options will cover all requirements.

Small multi-stage pumps, like borehole pumps, can have many stages, 50 or 100. Larger multi-stage pumps are limited by shaft deflection and critical speeds and the number of stages rarely exceeds 15. It is often required to run them at higher speeds than those produced directly by electric motors, this means using steam or gas turbines or geared-up electric motors. The speed is generally 5000 to 8000 rpm. Some recent development trends are towards high speed single stage pumps but this is not universal.

Pumps are available in many material combinations with casings in carbon steel, 13% Cr steel, 18-10-2 stainless steel and duplex stainless steel. Pumps at 2900 rpm can handle 2400 m^3/h at heads over 1400 m. Running at higher speeds will produce more flow at higher heads. Standard casing designs are available for pressure ratings up to 280 barg. The maximum rating to date is thought to be 430 barg. Connections are usually in the bottom half so that all maintenance can be accomplished without disturbing the process pipework. Other connection positions are sometimes optional. Stuffing boxes and seal cavities tend to be big enough to cope with most requirements. Some manufacturers can incorporate intermediate connections on the pump so that liquid can be added or extracted at an intermediate stage.

Figure 1.36 An example of capnuts on a horizontal joint
Courtesy of Weir Pumps Ltd

Pumps are available to API 610 requirements. API excludes the use of axially-split pumps when:

- Liquid temperature of 205°C or higher
- Flammable or toxic liquid with SG 0.7
- Flammable or toxic liquid over 69 barg

The main weakness in the axially-split design is the joint between the casing halves. Some designs employ gaskets and others metal-to-metal joints. As the pump pressure increases the joint bolting becomes more heavily loaded and tends to stretch reducing the interface pressure at the joint faces. The number and size of bolts or studs which can be fitted around the joint is finite. Capnuts can be used to eliminate the space required for spanners but the highly loaded stud/bolt cross section still sustains the tensile stress, see Figure 1.36. This weakness in design is cured by the pump design in the next Section.

1.3.24 Horizontal multi-stage radially-split centrifugal pumps (includes ISO 13709, API 610 designs)

As developments within the process industries have increased so the range of hazardous liquids has increased. The limitation of the axially-split multi-stage pumps was addressed to remove fears of potential leaks of hazardous liquids and the radially-split pump design evolved.

Consider the multi-stage segmental pump shown in Figure 1.29. It has already been stated that this pump design has many potential leak paths. However if the stack of segments is housed with a pressure vessel any leakage would be contained; this is the basic design principle of radially-split pumps is illustrated in Figure 1.37. Figure 1.38 shows a horizontally-split casing assembly encapsulated in a radially-split pressure containment. The process pipe connections can be flanged or prepared for direct pipe welding as shown.

The small clearance around the outside of the segments is usually at discharge pressure so that the segments do not have to be designed to cope with high tensile stresses caused by internal pressure. Unlike axially-split pumps, radially-split pumps all have vaned diffusers and usually balance compensation by disc or drum. Some very special pumps have opposed impellers. The segmental stack, together with seals and bearing assemblies, is built as a complete assembly, called the cartridge, which is inserted into the pressure vessel from the non-drive end. This allows all work on the pump to be carried out without disturbing the motor or the pressure vessel and process pipework.

Three types of construction of radially-split pumps are built as standard, the variation being in the pressure vessel which is normally called a barrel. Hence the popular name for this type of pump, the barrel pump; not to be confused with very small barrel emptying pumps described in Section 1.6.2. Small and

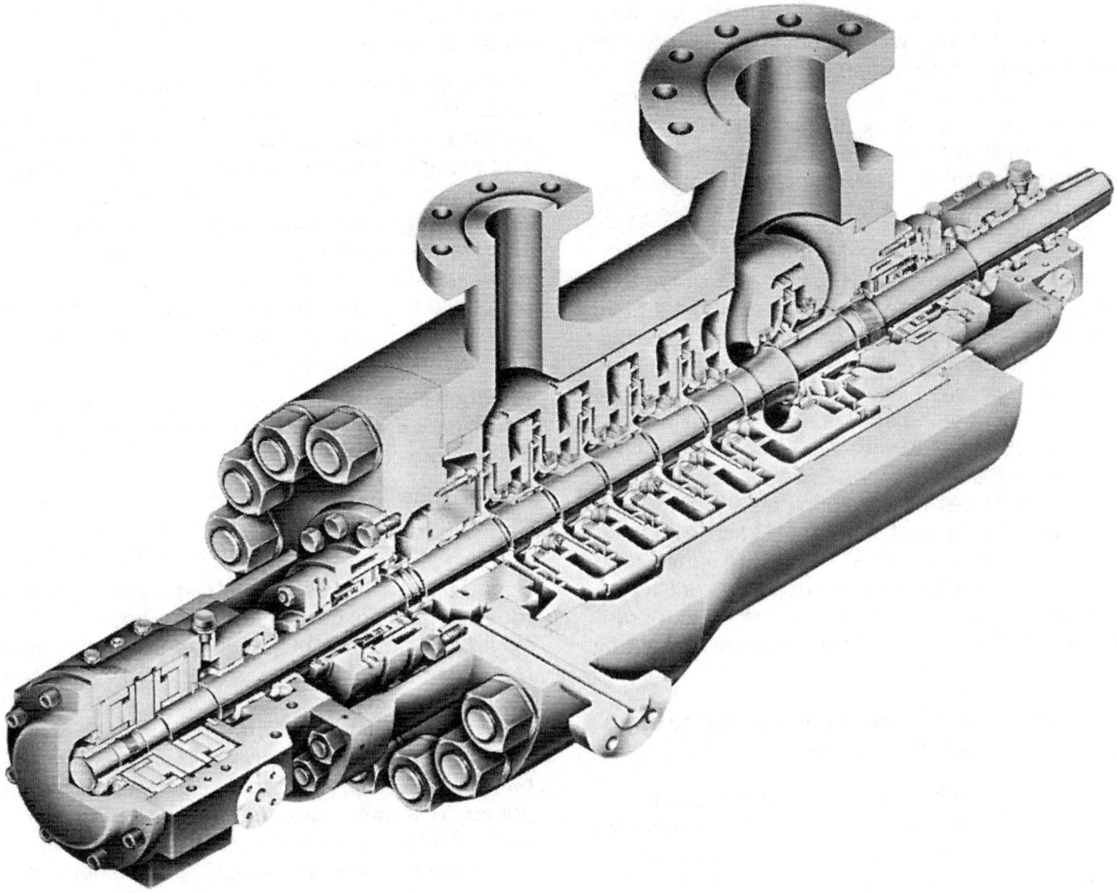

Figure 1.37 Cross-section of a multi-stage radially-split pump

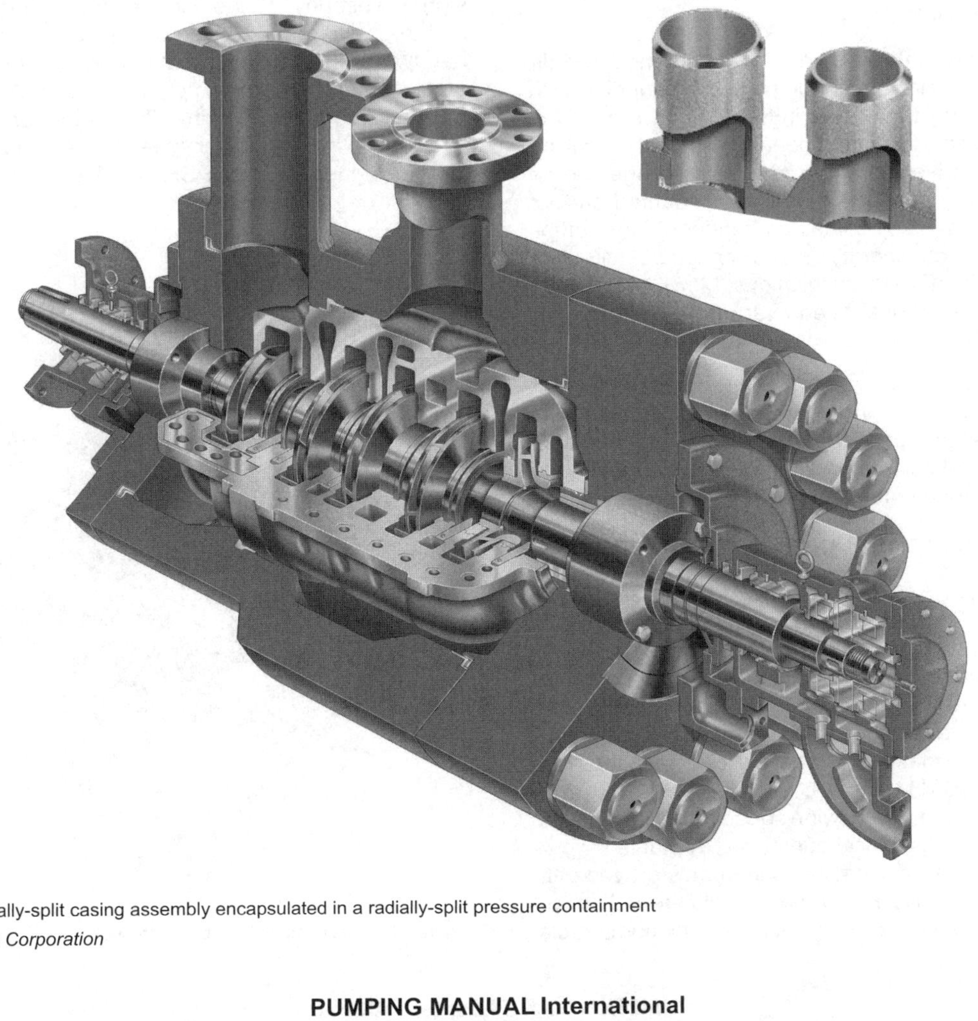

Figure 1.38 A horizontally-split casing assembly encapsulated in a radially-split pressure containment
Courtesy of Flowserve Corporation

low pressure pumps can have the barrel fabricated from standard pipe or rolled plate; this type of construction has been called segmental-pump-in-a-pipe. Larger pumps have cast or forged barrels. Forged barrels may have the process connections welded to the barrel requiring additional quality control. The barrel may be foot or centre-line mounted depending upon the application and inter-stage connections may be possible. Double suction first stage impellers are an option as is compliance with API 610.

Radially-split multi-stage pumps are used for the most arduous high pressure applications. Temperatures over 400 °C have been accommodated. Many material combinations are available to cope with temperature, pressure and corrosion. Flows over 400 m^3/h at 3000 rpm are possible; operating speeds up to 6500 rpm are not uncommon. Special pumps are available for direct drive by gas turbines. Total head rise of over 3000 m is not unusual. Pressure ratings up ANSI 2500LB, 430 barg, are available. One of the largest barrel pumps built was a boiler feed pump in a power station, rated at 55 MW. Process connections can be to any standard or special. Boiler feed pumps tend not to be flanged but butt welded directly to the pipework; as the casing is not disturbed during maintenance this is not a problem. Stuffing boxes and seal cavities can be made to suit any sealing requirements.

1.3.25 Large vertical multi-stage centrifugal pumps (includes ISO 13709, API 610 designs)

Vertical multi-stage pumps tend to be wet-pit pumps. Some versions are available where the pump is suspended within its own tank and effectively becomes a dry-pit pump. These vertical pumps are used for high flows in applications such as cooling water where one or two pumps feed a ring main which supplies a whole site. Construction details vary considerably including centrifugal and mixed-flow impeller types. Pumps are produced individually to order based on a range of standard designs. Countless material combinations are available at every conceivable level of quality.

Pumps can be built with flanged vertical motors mounted directly on top of the pump. Small pumps can be coupled with rigid couplings utilising thrust bearings built into the motor. Larger pumps have self-contained bearing assemblies, rolling or sliding, and are driven through flexible couplings. Some pumps are built with a right-angle gear box mounted on top; these pumps must have self-contained bearings. The pumps are driven via flexible couplings by motors or engines. This style of pump is popular in offshore installations. Flows over 70000 m^3/h are possible, differential heads vary from 50m to over 400m.

Vibration can be a problem on vertical pumps. It would be reasonable to purchase large pumps from manufacturers who have facilities to test pumps in the fully assembled condition. Balancing requirements on motors and rotating elements may need tightening to reduce problems.

1.3.26 Single-stage centrifugal pumps with integral gearing

The process industries have shown considerable interest in high speed pumps. Lower speed multi-stage pumps can be replaced with smaller, more compact packages. Pumps consist of a specially designed radial vane centrifugal impeller and casing which incorporates one or two step-up gears driven by a standard 2-pole motor. The maximum, practical, gear ratio for a pair of gears is 6:1 so that 60 Hz pumps could run up to 120000 rpm and 50 Hz to 105000 rpm. A very wide speed range is available at both frequencies. Only vertical pumps are available for process applications; horizontal single-stage pumps proved difficult to operate successfully. Horizontal models are only available for very small duties such as domestic or commercial water boosting. When seal flushing is fitted pumps can run dry for short periods.

Some operational problems have occurred with some installations. The high pump speed can cause mechanical seal, soft packing, and bearing problems. Due to the speed, rolling contact bearings may be replaced by sliding bearings with attendant lube systems. High speed pumps can be relatively intolerant of rapid process transients and particularly distressed by NPSHa inadequacies and solids. Liquid viscosity can be a problem. The variation between manufacturers in the pump's resilience to cope with operating problems has lead users to consolidate with a specific manufacturer for specific applications.

Although the possible speed range is great, most pumps operate in the range from 15000 to 30000 rpm. Differential heads of 1500 m are common and powers up to 500 kW are possible.

Radial vane impellers can have a smaller stable operating range than the standard backward swept vane impellers. A variable flow by-pass line may be necessary for some applications.

Speed-up gear boxes tend to give more problems than speed reducing gear boxes. Contemporary developments in electronics and electric motors may make this mechanical technology redundant. Variable frequency inverters are used routinely to vary the speed of standard squirrel cage induction motors. Cheaper electronic inverters can be used to power switched reluctance motors. The rotor of a switched reluctance motor can be made in one piece from steel; no windings or rotor bars required. This motor can run at very high speeds without problems of integrity caused by centrifugal forces on assembled components. A single-stage pump could operate at the highest speeds and be limited by impeller construction and materials.

1.3.27 Multi-stage centrifugal pumps with integral gearing

High speed multi-pump packages are an extension of the design philosophy of the high speed single-stage pumps. A speed-up gearbox is driven by an electric motor. The gearbox has two or three output shafts, each of which drives a single stage centrifugal pump. The pumps can be connected in series or parallel to produce higher flow or higher head. An option is available to have a slower speed first stage pump which is specially designed for low NPSHr. Inducers are available for some

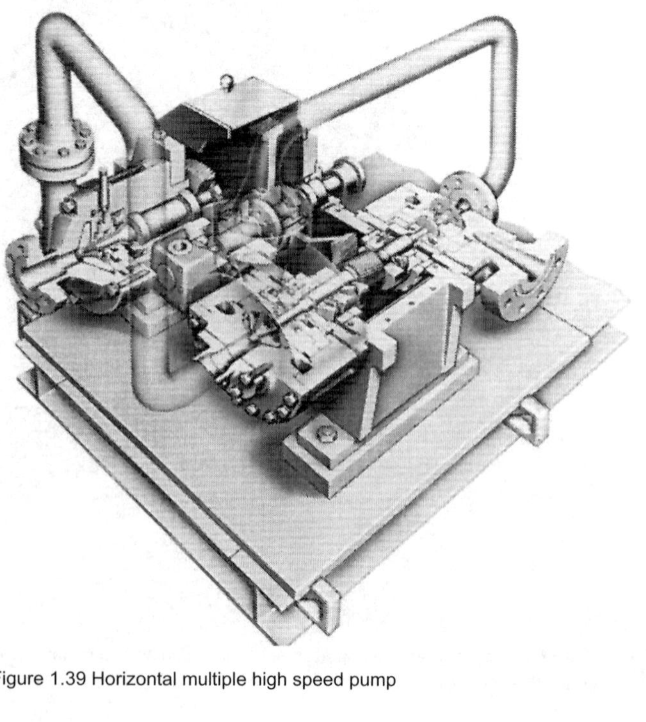

Figure 1.39 Horizontal multiple high speed pump

impellers. The multi-pump packages are restricted to horizontal pumps and an example is shown in Figure 1.39.

This type of packaging results in a compact powerful unit. Mechanical seals must be used as there is no access to the stuffing box for adjustment. Seal cavities are large enough for double and tandem seal configurations with flush and buffer liquids. The individual pump casings must be removed in order to inspect or maintain the seals; the process pipework and the inter-stage pipework must be removed to accomplish this. A special liquid injection feature allows the use of standard mechanical seals.

The individual pumps run at speeds from 12000 to 25000 rpm. Bearings may be a combination of rolling and sliding or all sliding. Lube systems can be tailored to customers requirements including API 614 compliance. Flows up to 225 m^3/h are possible at differentials up to 4500 m. High suction pressure capabilities are available up to 150 barg with casing pressure ratings up to 310 barg. Absorbed power can be to 1150 kW.

Most pump users tend to standardise on regularly used wear parts such as mechanical seals and coupling diaphragms. The choice of manufacturers for high speed seals may be very restricted. The comments at the end of Section 1.3.26 regarding by-passes and gearboxes also apply here.

High speed pump packages have replaced multi-stage centrifugal pumps and reciprocating pumps in some applications. High speed multi-pump packages are smaller but not as efficient. High speed multi-pump packages cannot reproduce the reciprocating pump characteristic but only cover the duty point.

1.3.28 Centrifugal pumps for pulp

For handling paper pulp, depending on the concentration of fibre in suspension, it is necessary to use specially designed pumps as follows:

- Up to 0.5% — Standard pumps
- 0.5 to approximately 2% — Pulp pumps or standard pumps fitted with special semi-open impellers
- Approximately 2 to 5% — Pulp pumps
- Approximately 5 to 6-7% — Pulp pumps with specially shaped inlet blading
- Above 6-7% — Positive displacement dense pulp pumps

For unscreened pulp of concentrations above 3% the pump flow passage should be at least 40 mm. The impellers for pulp pumps are semi-open with back vanes for simultaneous balancing of axial thrust and cleaning behind the impeller. Figure 1.40 shows such a pump, with a replaceable wear plate in the pump casing and axial adjustment of the rotor. The choice of material is normally stainless steel and for raw product sometimes grey cast iron. Due to the presence of entrained air in pulp suspensions it is necessary to design the blades of the impeller intake so as to prevent air locks. The concentration of air also causes rough running with increased levels of vibration and mechanical stress.

Compared with a water pump, the mechanical construction of a pulp pump requires an extra safety factor of approximately 2 in order to achieve satisfactory reliability. For pumping twigs and waste from the macerater, non-clogging free-flow pumps are best. Horizontal end-suction pumps with mixed-flow impellers are capable of 6000 m^3/h at 35 m up to 110 °C with pressure ratings of 7.5 barg.

1.3.29 Centrifugal pumps for handling solids < 10 mm

For very abrasive mixtures with particle sizes of up to approx 10 mm, the components in contact with the pumped fluid are rubberised or rubber coated. Typical solids handled would be some sands, iron and copper ore and mine tailings. Pumps which are rubber coated can have the coating repaired or replaced if the damage is not severe. Rubber lined pumps are built so that the lining of the casing can be removed and completely replaced.

Smaller pumps, 2" to 4" suctions, may limit solid sizes to 6 mm or less. Stuffing box requirements and drive arrangements are similar to those shown in Section 1.3.30. Small pumps are similar in hydraulic performance to all metal pumps. Larger pumps can handle more flow; a 16" suction will pass 5000 m^3/h at up to 48 m.

1.3.30 Centrifugal pumps for handling solids > 10 mm

Solids handling pumps are used for pumping suspensions and liquid/solid mixtures of various solid particle sizes up 70% solids by weight. Two types of construction for rotodynamic pumps are available; all metal and rubber coated or rubber lined. The particle sizes and the abrasiveness of the solids determine the choice of construction. Abrasiveness can be measured by the Miller Number according to ASTM G75-82, Test method for slurry abrasivity by Miller Number.

All metallic materials are used in the case of larger particles and higher liquid temperatures, see also Chapter 7, Section 7.5. Some manufacturers restrict the size of small particles which can be mixed with the large. With particles over 50 mm it is unlikely that 70% concentrations can be achieved. Pumps are available in various materials from plain cast iron to chrome irons with 2.5% to 27% Cr. Despite the use of extremely wear-resistant materials the operational life is relatively short - of the order of a few months and depending greatly upon the speed of flow, rpm, particle size and nature. Figures 1.41 to 1.44 show examples of typical heavy duty hard metal solids

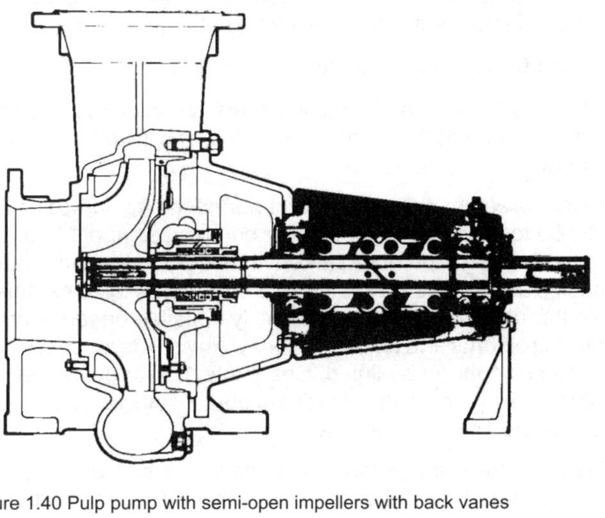

Figure 1.40 Pulp pump with semi-open impellers with back vanes

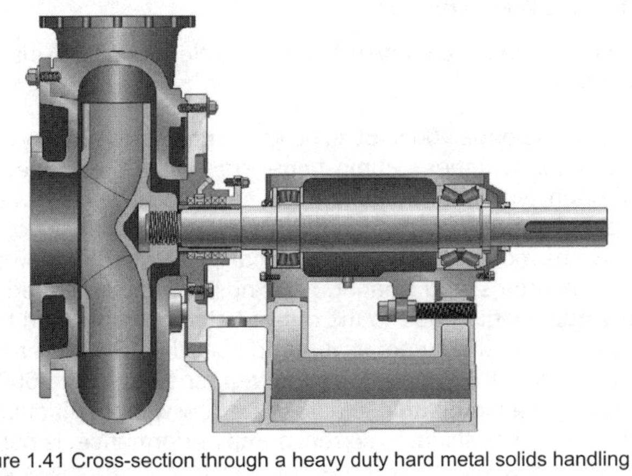

Figure 1.41 Cross-section through a heavy duty hard metal solids handling pump
Courtesy of GIW Industries Inc

Figure 1.42 Typical heavy duty hard metal solids handling pump
Courtesy of GIW Industries Inc

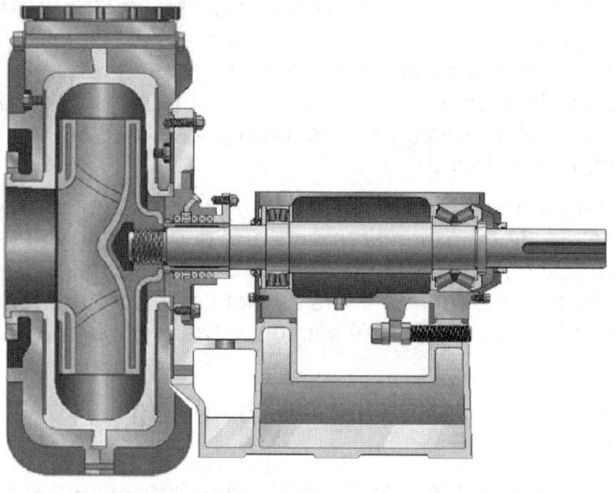

Figure 1.43 Solids handling pump with replaceable rubber lining

Figure 1.44 Typical solids handling pump with a replaceable elastomer lining
Courtesy of GIW Industries Inc

handling pumps and those with replaceable rubber and elastomer linings.

For delivery heads above 50 to 60 m it is normal to employ several pumps in series. Pump impellers are of the closed or semi-open type with suitably thick shrouds and vanes. The number of vanes are relatively few to reduce the chances of impact. Some models have adjustable casings to correct for wear. Shafts, bearings and bearing housings are dimensioned for heavy-duty loading due to the out-of-balance forces and high SG. Some manufacturers limit pump speed as SG increases. Solids pumps are usually V-belt driven, speeds from 2600 to 400 rpm, so as to be able to adjust the flow without machining the impeller. Throttling, to correct pump performance, is particularly difficult and prone to constant adjustment; variable speed is preferred.

Due to the abrasive nature of the liquid mixture particular attention has to be given to the design of the shaft seals. Design solutions such as stuffing box flush with clean liquid; hydrodynamic seals as in Chapter 8, Section 8.3; centrifugal seals with back vanes and as glandless pumps with overhung impellers for vertical models, as in Figure 1.41, are available. This pump type is particularly suitable for frothing fluids, where air can rise and not block the impeller intake. A certain degree of self-regulation is obtained by the air mixing at low level.

The maximum solid size handled continuously is a function of the pump size and impeller design; open or closed. Table 1.2 can be used as a guide for sizing pipework.

		Maximum particle size mm	
Suction nozzle	Discharge nozzle	Closed impeller	Open impeller
2"	1½"	11	24
3"	2½"	16	32
4"	3"	20	41
6"	5"	38	62
10"	8"	50	82
14"	10"	70	90
16"	12"	80	98
20"	16"	100	120

Table 1.2 Solid sizes for various pump sizes

A 2" suction pump can handle up to 30 m^3/h at heads of 60 m; a 20" suction pump can handle 2250 m^3/h at heads up to 38 m. Pumps are available which can handle solids larger than 120mm. Obviously to pump large solids in quantity, rather than an occasional piece as with a contractor's self-priming site pump, a great deal of liquid is required and a large pump. Pumps up to 36" or 900 mm suction are available. Efficiency tends to be low, because the kinetic energy given to the solids cannot be converted into increased pressure, although one pump was advertised at 90% on water between 500 and 900 rpm. It is essential to consult manufacturers very early in the system design and feasibility studies and be prepared to supply plenty of NPSHa, up to 10 m.

1.3.31 Non-clogging pumps

Non-clogging pumps are mainly used as sewage pumps for pumping untreated sewage in sewage treatment plants. This Section deals with normal models for dry installation and with only the pump section immersed in the liquid. Section 1.3.32 covers submersible models where both pump and drive motor can be operated below the surface of the liquid.

Sewage handling pumps are completely dominated by rotodynamic pump designs. The following types are used:

- Centrifugal pumps with through-flow impellers, special enclosed impellers with large almost straight passages
- Free-flow pumps, recessed open impeller
- Pumps having semi-axial impellers and axial pumps (propeller pumps) for larger flow rates and lower differential heads

The free area, the through-flow for non-clogging pumps, is normally 60 to 100mm for the smaller pump sizes and 125 to 150 mm for the larger pumps. The obviously desirable feature that particles as large as the connection diameter should be able to pass through the pump is not usually possible because of hydraulic problems and pump efficiency. Any solids should, in theory, be surrounded by liquid. See Table 1.2 for guidance. For small flow requirements, in order to obtain sufficient liquid flow, oversize pumps must be used.

All sewage handling pumps comprise single suction impellers. This is to avoid the necessity of locating the pump shaft in the intake. Pump materials chosen are usually grey cast iron or SG iron for municipal sewage and grey cast iron or stainless steel for industrial effluent.

Pumps with channel impellers have always been the most well developed and constitute a special class of their own on account of the number of different types and the numbers of pumps in use. In practice most pumps have impellers with single or double channels, although larger pumps sometimes have three channels, Figure 1.45. As implied by the description 'through- flow' impeller, the pumped liquid passes through the pump impeller and out through the delivery connection after leaving the pump intake.

Pumps with through-flow impellers are characterised by their high degree of efficiency over a large part of the H-Q curve. The overall efficiency at the nominal duty point for medium size pumps is about 60%. The wear resistance of the impeller is moderately good when used in combined systems and good in separate systems.

The shape and design of free-flow recessed impeller pumps differs from through-flow pumps in that the location of the impeller and the utilisation of the pump casing is different, Figure 1.46.

The pump impeller is symmetrical with open vanes. The profile is low and its pulled back location, recessed in the stuffing box wall, leaves the pump casing completely or partially free. The free-flow pump can be described as a centrifugal pump having very large sealing clearance. The liquid and contaminants flow freely under the pump impeller and out through the delivery connection. The absence of sealing clearance results in a reduction in efficiency compared with through flow pumps. The overall efficiency at the nominal duty point for small and medium size pumps is 40 to 42%.

A special version of the semi-axial design pump has an impeller like a corkscrew. This style of pump has been called a screw-centrifugal pump. The flow passages are large and the leading and trailing edges are sharp to cut fibrous material. These are low head machines but water efficiency can be quite high, up to 84%.

Gentle handling !!!

Pumps, similar in design to sewage handling pumps, are used for transporting larger delicate solids, such as whole fish, fruit

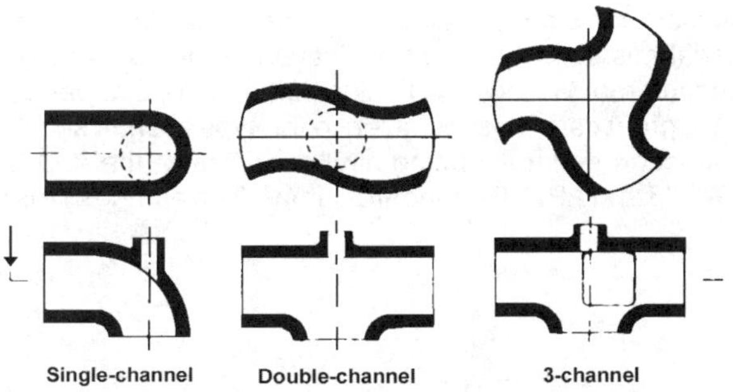

Figure 1.45 Various types of channel impeller

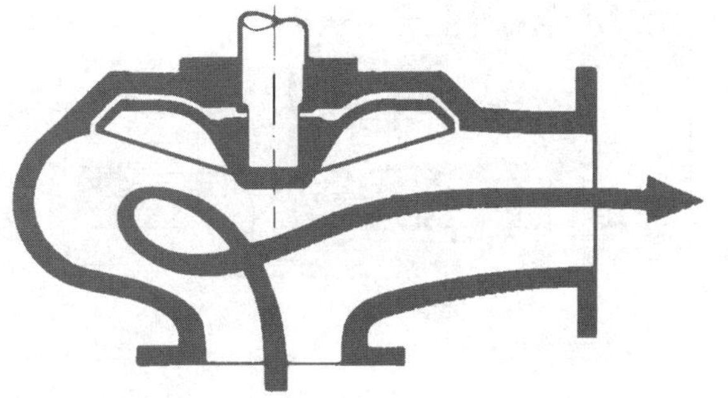

Figure 1.46 Free-flow recessed impeller pump

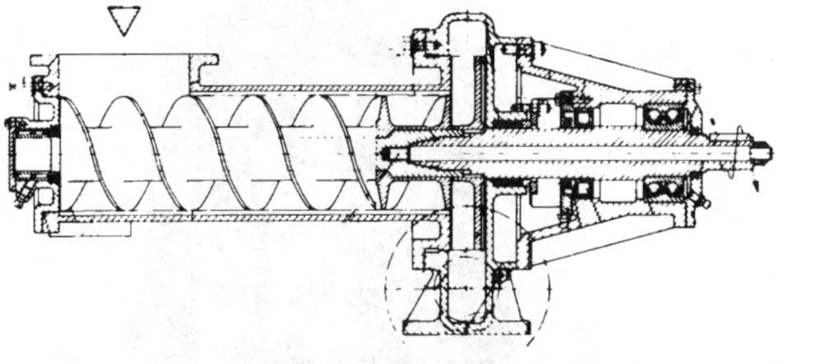

Figure 1.47 Waste pump with "compression screw" for feeding the centrifugal pump unit

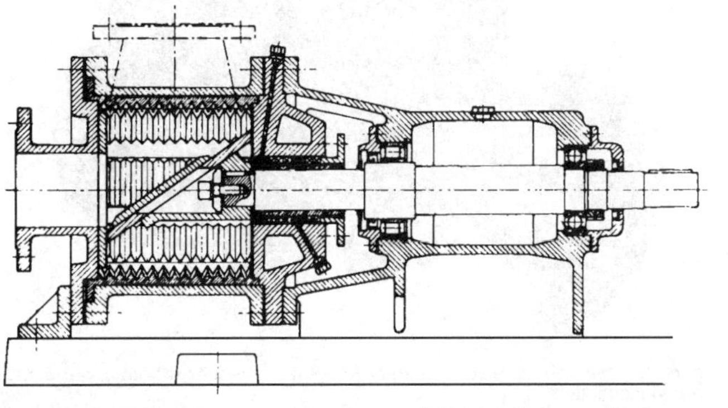

Figure 1.48 Oblique disc pump (Gorator pump) for simultaneous disintegration and pumping

and root vegetables by means of water. The channel areas and shapes are designed to cause the least amount of damage to the product being pumped. See Chapter 16, Section 16.10 for an example of the pumping of live fish and food.

In direct contrast there are other types of non-clogging or chokeless pumps where deliberate attempts are made to finely disintegrate the material being transported. Apart from sewage handling pumps with a integral fine disintegration device or macerater there are pumps specially developed for the handling of sludge and waste within the food processing industry, Figure 1.47 and Figure 1.48.

1.3.32 Submersible non-clogging pumps

Nowadays the submersible pump is predominantly more important then the conventional dry pump type. Submersible pumps are usually capable of functioning in dry locations and so can be used in low locations where there is risk of flooding, for example in the case of an electricity power cut.

Submersible pump construction is recognised by the fact that the impeller is located directly onto the combined motor and pump shaft, Figure 1.49. The drive motor, a squirrel-cage induction motor, is completely sealed by axial and radial O-rings and the pumped liquid is sealed from the drive motor by double mechanical seals which run in an oil-bath. The type of bearings, size of shaft and method of sealing are deciding factors in reliability and operational safety of the whole system. Deflection of the shaft due to radial loads should not exceed 0.05 mm at the shaft seal. The shaft seal components should be assembled in a seal cartridge, which enables pre-assembly and pressure testing of the shaft seal prior to fitting a replacement, Figure 1.50.

From the point of view of pump service and repair, it is advantageous to be able to fit separate motor units of varying sizes to a number of different pump casings. Such a range system makes it possible to maintain a complete community with a small stock of motor parts. Some pump systems also require facilities for converting on site from through-flow to free-flow impeller.

1.3.33 Mixed-flow pumps

Mixed-flow pumps are used for large flows and low differential heads. The differential head can be increased by using multiple

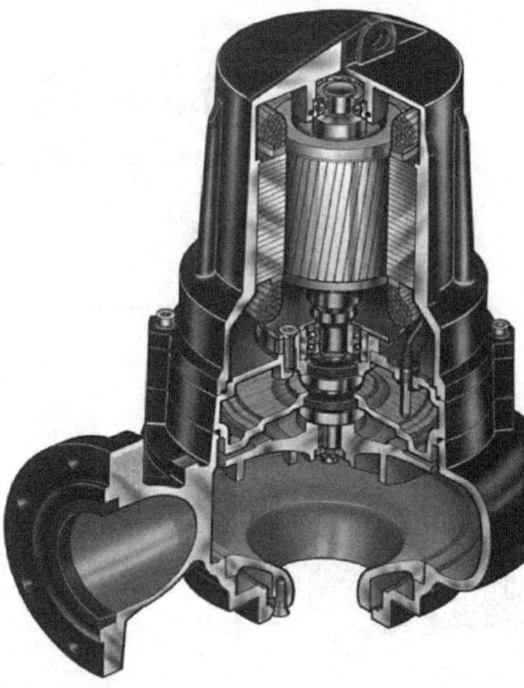

Figure 1.49 Submersible sewage handling pump suitable for both dry and submersed installation

Figure 1.50 Single cartridge mechanical seal for submersible pumps

impellers - a multi-stage mixed-flow pump. These pumps are nearly always installed with the impeller immersed in the liquid, flooded, and the motor located above, so called extended shaft pumps for immersed installation. For relatively clean fluids the central support tube also functions as a delivery pipe while for contaminated liquids the delivery pipe is separate, similar to Figure 1.17 in Section 1.3.13.

Typical of immersed installation pumps, at least for the larger sizes, is that the delivery connections and mounting plates are located according to each particular installation. If the total height of the pump exceeds the available height required for assembling and disassembling then the support and the delivery pipe as well as the shaft may be divided into segments. For long shafts it is necessary to fit intermediate bearings to prevent vibration and to increase critical speeds. There is a clear trend towards replacing extended shaft pumps with submersible pumps in the smaller sizes.

The hydraulic design and number of stages is varied according to the duty requirements. Typical applications include wells and boreholes, condenser circulating water, condenser extraction and drainage pumps. Pump sizes start at 20m^3/h and approximately 30 m differential per stage. Pumps are available in all popular pumping materials.

1.3.34 Axial flow pumps

Many of the comments in Section 1.3.33 can be applied to axial flow pumps. Most pumps are vertical, the vast majority are single stage. Some horizontal versions are available which look like pipe elbows. Axial flow pumps can be equipped with adjustable blades which can be set, either when the pump is at rest, pre-set, or during operation, variable, by means of mechanical or hydraulic actuators. The differential head produced by axial flow pumps is much less than mixed-flow or centrifugal pumps; typically 15 m per stage. Because of the simplicity of the construction they are easy to build in large sizes. Axial flow pumps are generally used for flows greater than 400 /h; flows of 30000 m^3/h are not uncommon.

A special version of multi-stage axial propeller pump has become popular in the last few years. A heavy duty version of the pump has found some applications in the multi-phase flow field. The cost of oil well exploitation can be significantly reduced if the well products can be piped or pumped to a central gathering station or pre-treatment plant. The cost of individual wellhead installations can be prohibitive if gases and liquids are separated and then pumped/compressed.

A pump which can handle crude oil, light hydrocarbon liquid fractions, sour water and natural gases is extremely useful. The pump looks like an extremely rugged version of an axial compressor. The rotor drum is of relatively large diameter and the blade height is unusually short. The rotor stages have few blades but each blade extends for 60 or 70 degrees around the circumference. The casing stages have more nozzles and resemble the high pressure nozzles from a steam turbine. Horizontally-split casings are used for ten or more stages. Good experience has been accumulated with flows of 500 m^3/h and differential pressures up to 14 bar. Gas volume fractions are generally in the region of 80 to 90% at suction but pumps have continued to operate with 100% gas for 15 minutes. These multi-stage axial flow pumps operate at speeds up to 4000 rpm.

If higher differential pressures are required, or the mixture contains solids, then twin-geared screw pumps, in Section 1.5.4, should be considered.

1.3.35 Non-metallic rotodynamic pumps

Non-metallic pumps are most suitable when used for pumping acids, alkalis and corrosive salt solutions. A general review of physical properties and corrosive resistance of various non-metallic materials is given in Chapter 7, Sections 7.2.12 and 7.3.

Larger sizes of non-metallic pumps must, because of the physical properties of the material, be equipped with an outer reinforcement casing to carry the nozzle loads and to absorb forces due to the fluid pressure, Figure 1.51. This outer casing is not necessary for smaller pumps or if the particular non-metallic material has suitably good physical properties or is fibre reinforced internally. Since the liquids handled are of a dangerous nature, great care must be given to the type of shaft seals. The construction generally being similar to that of the ISO 2858, ANSI B73 and API 610 standard pumps. The pump shaft is usu-

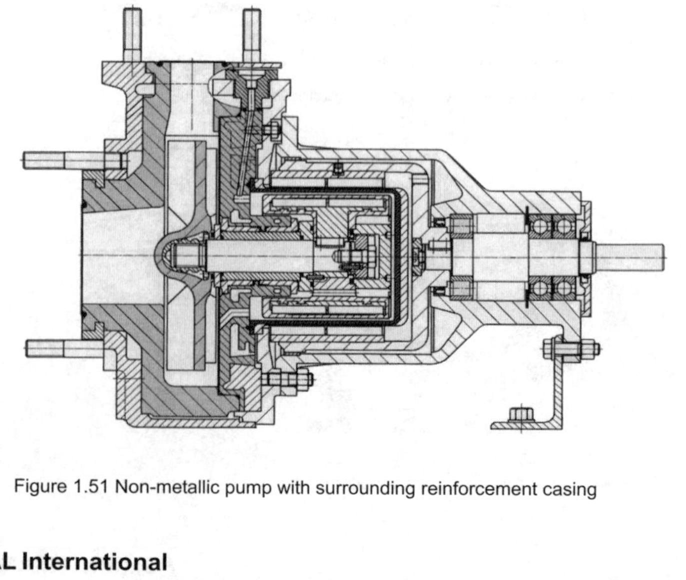

Figure 1.51 Non-metallic pump with surrounding reinforcement casing

Figure 1.52 Hygienic-quality centrifugal pump

ally 18-10-2 stainless steel minimum and can be completely isolated from the liquid by a non-metallic shaft sleeve under the mechanical seal.

Pumps are available for 400 m^3/h at differential heads up to 80m; pressure ratings of 7 bar and temperatures up to 100 °C are possible.

1.3.36 Hygienic-quality rotodynamic pumps

Within the food processing, chemical and pharmaceutical industries, special centrifugal pumps are used, Figure 1.52. Their shape and construction is distinctive and determined by the hygienic and sterile requirements.

The materials used are usually stainless steels with O-ring gaskets etc. of approved hygienic quality. Both exterior and interior surfaces must be smooth and polished to approved hygienic standards, it should also be possible to disassemble the pump quickly for cleaning and washing. When cleaning without disassembling, Cleaning In Place or CIP, is specified, it must be possible to carry out this process quickly and effectively. This means that all components and clearances must be designed for this purpose, especially with respect to temperature. To simplify external cleaning, the electric motor is often encased in a polished stainless steel casing.

For reasons of hygiene many foodstuffs pumps may not have oil lubricated shaft bearings. Food processing centrifugal pumps are normally available up to approximately 30 kW. Centrifugal pumps cannot be used for viscous and shear sensitive products such as yoghurt, cheese, liver paste etc.. In such cases lobe rotor pumps are generally used, see Section 1.5.8.

CEN Standard EN 1672 covers the requirements for hygiene for machinery used in the preparation of food for human or animal consumption. The Standard describes methods of construction and the types of acceptable fittings, couplings and flanged joints.

Single-stage end-suction pumps can handle over 400 m^3/h at heads up to 100m. Close coupled end-suction pumps can be constructed like a segmental pump, shown in Sections 1.3.12 and 1.3.21, with up to three stages. These pumps can handle 50 m^3/h at heads up to 180m. ISO is producing new standards for hygienic machinery requirements. ISO 14159 deals with machinery design. The American Standard T-02-09, which defines the requirements for 3-A Sanitary Standards originally for the dairy industry, is popular in many countries.

Traditional hygienic pumps have been used in food manufacturing processes. European standards use the term Agrifoodstuffs to include food products for animals as well as food for human consumption. The biotechnology industry has become an important user of hygienic pumps. EN 12690 deals with the risks to personnel and the environment due to the emission of micro-organisms from assembled working pumps or inadequately cleaned pumps which are disassembled. A biohazard is defined as danger or harm associated with a biological agent able to cause infection, allergy or toxicity in animals, humans, plants or the environment.

1.3.37 Magnetic drive rotodynamic pumps

Magnetic drive pumps are a type of glandless pump. Glandless pumps are characterised by the fact that their design and construction physically prevents leakage to the surrounding environment. Packing or a mechanical seal type of shaft seal cannot in this context in any way be considered as leakage free, even if it is barrier liquid or very dilute process liquid/vapour in minute quantities which escapes. The description glandless pumps is therefore associated with the alternative description hermetically sealed pumps or canned pumps.

Glandless pumps must be used for the most dangerous liquids from the point of view of toxicity, environmental hazards and radioactivity. In some cases for high or very low temperatures. Leakage may occur however when handling certain types of highly penetrative liquids due to penetration through minute faults in static seals. Magnetically driven pumps can avoid this problem by having the containment can welded to the casing. When ordering pumps for these types of liquids the level of acceptable leakage should be specified.

Two types of magnetic drive are available:

- Synchronous magnetic couplings
- Magnetic induction couplings

In both cases the pump is driven by a standard motor. However the motor does not drive the pump shaft carrying the impeller(s) directly. The motor drives a rotating magnet assembly. In synchronous magnetic couplings, Figure 1.53, the rotating magnetic field drives another set of magnets coupled to the impeller(s). In magnetic induction couplings the rotating magnets create a magnetic field in a rotor of copper bars, very similar to a standard squirrel cage motor rotor. The magnetic properties of materials decay as temperature increases so that synchronous magnetic couplings suffer from a loss in torque. Magnetic induction couplings are better at elevated temperatures because the magnets are not in contact with the liquid.

The pump shaft proper is totally immersed in the process liquid and runs in bearings which are process lubricated. Silicon carbide ceramics and some carbon are popular as bearing materials. Solids in the liquid can cause high bearing wear and in some designs can accumulate in the can and cause obstructions. This problem can be solved by using an external feed to the bearings. Process liquid can be diverted from the pump discharge, filtered and cooled if necessary, and re-injected into a lubrication circuit. This extra complexity will be worthwhile when considering the enhanced reliability and extended bearing life. Bearing wear monitoring is a useful facility on any magnetically driven pump.

A variety of magnetic materials are used of which aluminium, nickel and cobalt are probably the most widely and successfully

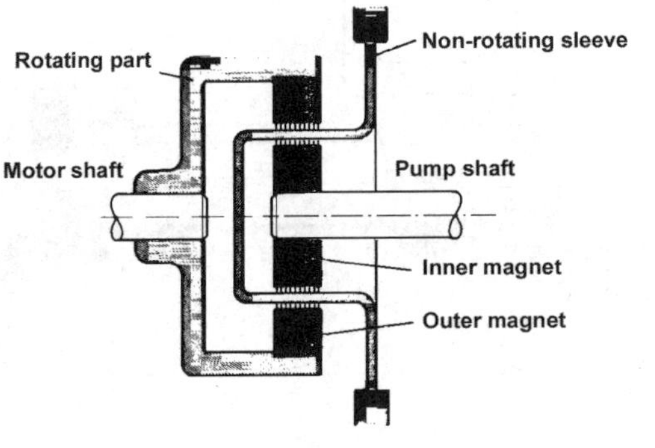

Figure 1.53 Synchronous magnetic coupling for "sealless" torque transfer

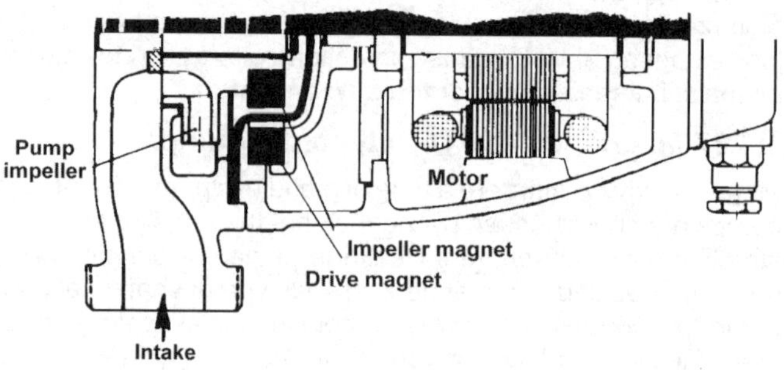

Figure 1.54 Close coupled magnetically driven pump

employed. Some designs use rare earth elements, such as samarium cobalt and neodymium iron boron for the upper end of the torque range, the higher costs involved being more than offset by improved performance.

Magnetic drive pumps cannot run dry. The pump and can must be thoroughly vented and primed before start up. The process liquid provides lubrication and cooling for the pump internal bearings. The process liquid also removes heat generated by the magnetic coupling. Failure to provide the appropriate liquid filling will cause significant wear of bearings and may cause the pump to seize before venting and priming is accomplished. Dry running monitoring could save major maintenance expenses.

Close coupled magnetically driven pumps, Figure 1.54, have become increasingly popular for critical applications where unscheduled process stoppages create additional plant problems as well as loss of revenue. Mechanical seals are seen as a maintenance intensive item, difficult to install and set-up, with costly spares and attendant inventory costs. Removing the mechanical seal and operating pumps of a modified design can save overall costs. Cases have been cited where a magnetic drive pump costs 30% of a conventional pump over a ten year operating life.

Magnetically driven pumps are available from several manufacturers, including non-metallic versions, pumps up to 350 kW are currently built. Hydraulic capacities of 550 m^3/h at 500m are possible, temperatures of -80 to 450 °C are promoted. API 610 and ISO 2858/ANSI B73.1 conditions can be met.

NOTE: Rotary positive displacement pumps can be driven by sealed magnetic couplings. Gear and triple screw pumps are typical examples. If a centrifugal pump is not the best selection for a particular liquid there may be a suitable positive displacement pump.

1.3.38 Canned motor rotodynamic pumps

Canned motor, "wet motor", driven pumps generally have the stator windings sealed from the pumped liquid by means of an annular tube - hence the name "canned", Figure 1.55. There are many types available with power outputs of up to several thousand kilowatts. Smaller sizes, up to about 500 W are generally used as domestic hot water pumps and central heating circulating pumps, see Section 1.3.1.

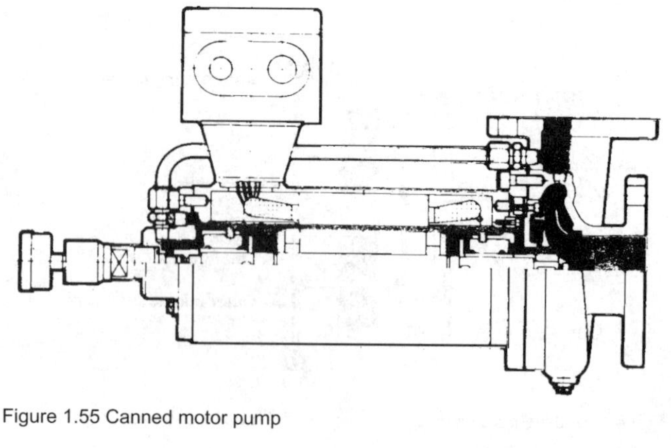

Figure 1.55 Canned motor pump

Canned motor pumps are produced as both single and multi-stage types with various types of drive motor. Liquids with temperatures ranging from -200 °C to +500 °C and pressures up to 1000 bar can also be handled. Heating jackets are available for handling liquids with high melting points, whilst built-in filters are primarily used to cope with contaminated liquids; although trickle feeding to the rotor chamber may also be necessary for liquids containing high concentrations of solid particles.

Canned motor pumps are selected when the liquid handled is hazardous and/or the user requires the highest possible availability. The canned motor pump is a sealed unit with static pressure seals; no mechanical seal. There is no coupling between motor and pump so there are no alignment problems. When selected properly the canned motor pump can provide an availability of 0.9997 which evaluates to a loss of running time of 2.6 hours per year. Very high availabilities are achieved in systems which operate continuously for months or years. Starting and stopping, which can create hydraulic, mechanical and thermal shock, tends to reduce reliability if not considered thoroughly.

Larger canned motor pumps were primarily developed for boiler circulation pumps and liquid metal circulating pumps for nuclear reactors. Circulating pumps are required on modern large multi-tube boilers to ensure a reasonable water velocity inside the tubes to improve heat transfer and eliminate hot spots and problems with localised vaporisation. The suction pressure of the pumps is boiler pressure, 150 barg or more, and the differential only 20 or 30 metres. The high suction pressure produces a large axial thrust requiring enormous, (by comparison to the standard), thrust bearings and also causes mechanical seal problems. Liquid metal pumps for nuclear reactors must be positively leak-free because of the radioactive nature of the liquid. The canned motor pump solves these problems.

The canned motor pump is also available as a chemical and process pump, very similar to magnetic drive pumps and in special variants for sewage, effluent and paper stock. Flows of 300 m^3/h at 110m with pressure ratings of 25 bar are fairly common. Some canned motor pumps are available to DIN 24256 standard pump performance and dimensions. See Figure 1.56.

Current designs include those which have no sliding bearings. The motor/pump shaft is raised magnetically and held centrally

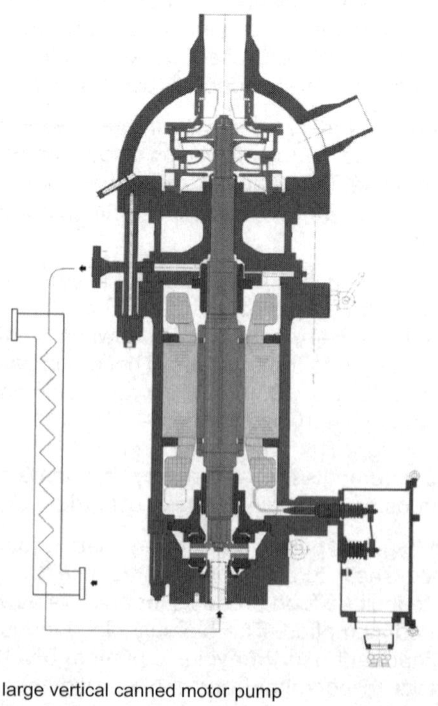

Figure 1.56 A large vertical canned motor pump
Courtesy of KSB AG

within the stator/casing. Axial location is also controlled magnetically. The mechanical efficiency is improved because mechanical bearing losses are eliminated. This style of construction allows very high pump speeds to be used without worrying about bearing performance or life.

NOTE: Magnetic drive and canned motor pumps. The American Standards body, the Hydraulic Institute, has issued a Standard covering sealless pumps, ANSI HI 5.1-5.6. This Standard is designed to eliminate misunderstandings between purchasers and manufacturers and to help the purchaser specify a pump correctly. The Standard covers; pump types, definitions, applications, installation, commissioning, operation, maintenance and testing. Details of monitoring devices unique to sealless pumps are given; testing specific to sealless pumps; can integrity and winding Integrity for canned motor pumps. The application section deals with gaskets and joints, venting and draining and internal process liquid lubricated bearings.

The Standard includes descriptions of piping arrangements for process liquid circulation and recommendations on how to prevent problems caused by solids, magnetic material and gas/vapour evolution. B73.3 covers sealless versions of the ASME/ANSI B73.1 and B73.2 style pumps.

1.3.39 Rotodynamic pumps without drivers (for specific applications including mobile agricultural applications and mobile fire pumps)

Pumps specially adapted to be driven from the power take-off of a tractor are used for temporary installations, during the preparation period for a permanent installation and for mobile applications. The speed of the tractor power take-off is too slow, approximately 500 rpm, for a small centrifugal pump. The specially adapted pumps have built-in gears to increase the speed to 3000 to 4000 rpm, see Figure 1.57.

1.4 Special rotodynamic pumps

It is possible to construct rotary machines which use the kinetic energy of all liquids but which do not obey the conventional mechanics of the centrifugal pump. Impellers do not look like conventional pump impellers. Kinetic energy can be transferred to the liquid in increments, by the use of many vanes, or all the energy can be achieved by a single stage. These pumps occupy a position between rotodynamic pumps and rotary positive displacement pumps; their theory of operation having some of the characteristics of both types.

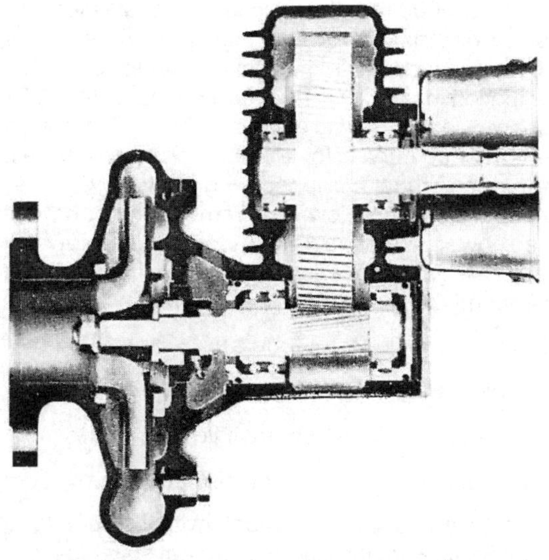

Figure 1.57 Typical tractor driven centrifugal pump with built-in gears

The positive displacement feature makes it possible to pump air or gas provided that a certain quantity of liquid is available to act as a seal. These pumps have therefore good self-priming characteristics. High heads are possible with a single-stage but efficiency is lower than other pump styles.

1.4.1 Peripheral pumps

Peripheral pumps are also called regenerative pumps. In principle, liquid within the casing is separated into elements by means of rotating vanes but the radial clearance permits a large percentage of recirculation. Energy is successively imparted to the liquid by contact with each vane in a manner which is reminiscent of rotodynamic pumps. When pumping, energy is transferred to each enclosed element of liquid by the action of the passing vanes. The increase in head for a given outer diameter and speed is 5 to 10 times that of an equivalent rotodynamic pump. There are usually many vanes mounted on the periphery of the rotating disc.

In practice the vanes are formed by making cut-outs in the periphery of the disc and supply energy during almost a complete revolution. The H-Q characteristic of peripheral pumps is characteristically steep, Figure 1.58, which is ideally suitable for automatic operation on conjunction with an accumulator. Also the NPSHr characteristic is steep. This type of pump is not usually permitted to operate at very low flow rates. Because of the pumps steep H-Q characteristic the power demand increases with reduced flow rate and increased differential head. This can cause overloading of electric drive motors designed to meet the demand at the design operating point.

Reduction of flow rates to low values should therefore be carried out by means of by-pass control, i.e. by returning liquid from the delivery side to the suction side using a control valve

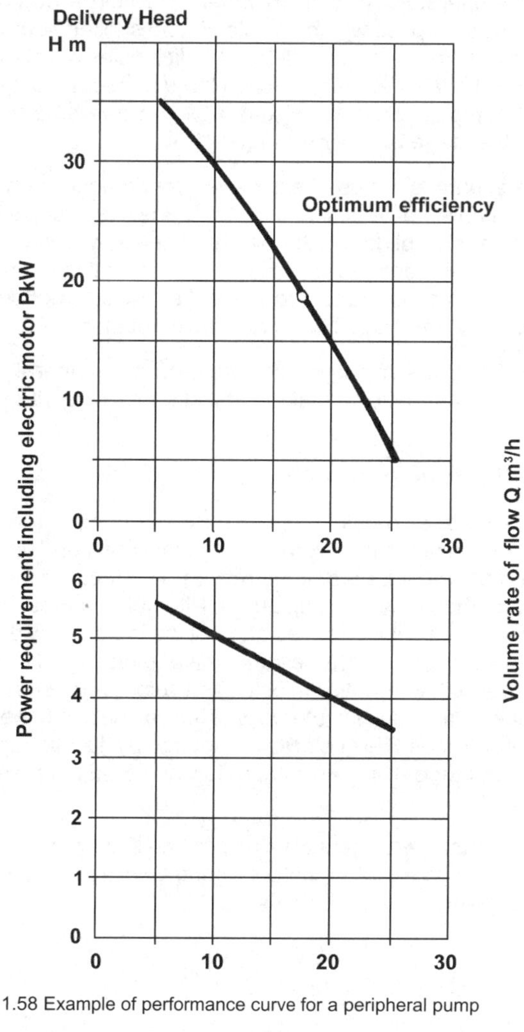

Figure 1.58 Example of performance curve for a peripheral pump

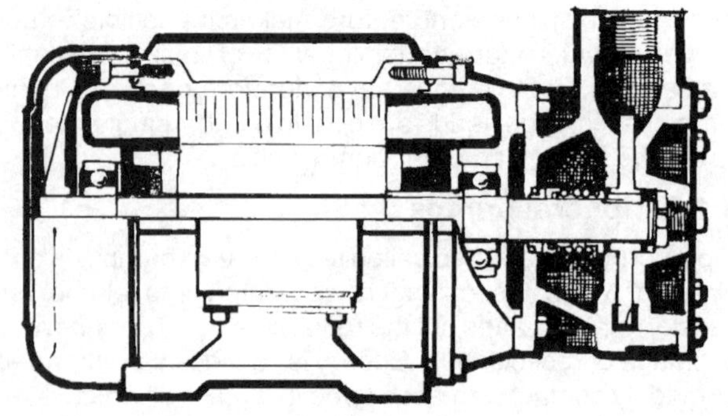

Figure 1.59 Section through a single-stage close coupled peripheral pump

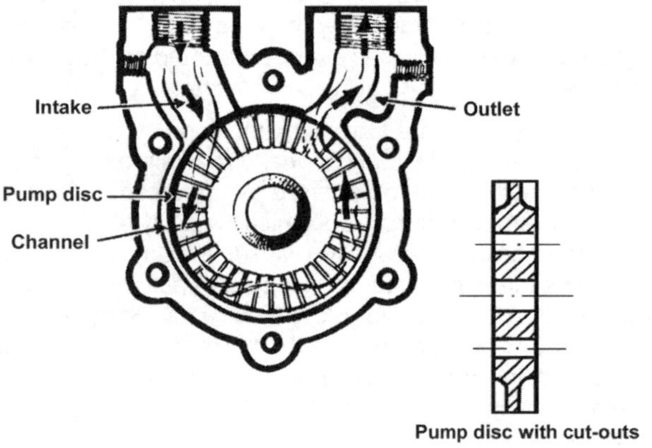

Figure 1.60 Peripheral pump stage

connected in the return pipe. Special designs with integral by-pass valves are available. However, operation under integral by-pass control should be carefully monitored to avoid excessive temperature rises. If prolonged operation at low flows is necessary, a by-pass which recycles to the suction source not the pump suction, is preferred. Pump efficiencies will generally be below 40%. A longitudinal section through a single-stage peripheral pump is shown in Figure 1.59. A transverse section through the stage is shown in Figure 1.60.

Peripheral pumps are used extensively for domestic and commercial clean water applications. Many pumps are used for marine applications: bilge, wash-down, engine seawater coupling, fuel oil transfer. Versions for direct engine mounting are readily available. Pumps in brass, bronze and stainless steel provide flows up to 70 m^3/h and heads to 70m in popular sizes.

Figure 1.61 shows an end-suction multi-stage peripheral pump. This pump us unusual in that the first stage is a centrifugal impeller.

1.4.2 Pitot tube pumps

Pitot tubes are also called, erroneously, jet pumps. If a small close-coupled pump is driven by an electric motor and does not require motive fluid input, compressed air, steam, high pressure water then, it is not a jet pump. Pitot tube pumps are unusual in that the casing rotates and not the impeller. Within the rotating casing of a Pitot tube pump the pumped liquid is accelerated to a high tangential velocity. The velocity depends upon the shape of the blades in the walls of the casing and the effects of frictional forces. This velocity is captured by a stationary tangential Pitot tube which delivers a flow at increased pressure, Figure 1.62.

Assuming that the tangential velocity of the liquid at the level of the Pitot tube is equal to the peripheral velocity of the casing then the theoretical pressure rise is:

$$\Delta p_{theor} = \rho\left(\left(\frac{\pi D n}{120}\right)^2 - \frac{V^2_{in}}{2} - \frac{V^2_{out}}{2}\right) \qquad \text{Equ 1.1}$$

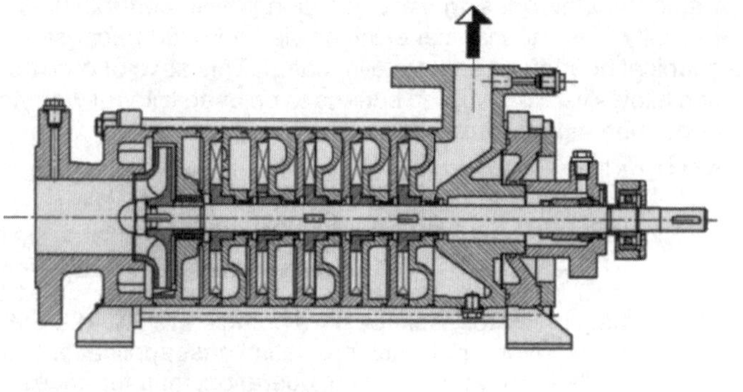

Figure 1.61 A hybrid peripheral pump
Courtesy of Sero PumpSystems GmbH

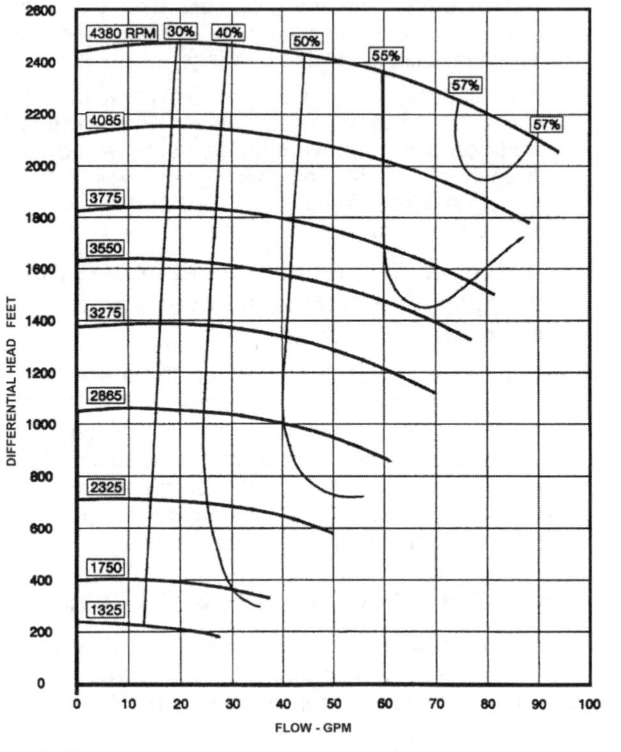

Figure 1.62 Performance curves for a Pitot tube pump

Normally pressure is reduced as the result of flow losses and due to rotational slip between the liquid and casing. The design of the Pitot tube and blades is of critical importance with regard to losses. The stationary Pitot tube is sealed inside the rotating casing by a mechanical seal. As the casing may be rotated at speeds up to 10000 rpm the mechanical seal selection is very important. For small flows and high pressures, low specific speed centrifugal pumps exhibit low efficiency due to the large impeller friction losses. The Pitot tube pump is more efficient in this respect. Maximum efficiencies of 50 to 60 % are possible. The performance curves for a Pitot tube pump are similar to those for a centrifugal pump, see Figure 1.62. Curves can be unstable and minimum flow requirements must be observed. Absorbed power increases as flow increases. Safety valves, as a protection against overloading or overpressure, are not necessary and the pump can be regulated by throttling.

The advantages of the Pitot tube are:

- Large differential head in a single stage
- Reasonable efficiency for small flows
- Non-lubricating liquids can be pumped

The Pitot tube pump cannot compete with the centrifugal pump for normal flows. Pitot tube pumps are not suitable for contaminated, abrasive or viscous liquids. Some manufacturers see their pumps as being in direct competition with piston and plunger pumps. The Pitot tube pump is not a direct competitor because it still has a rotodynamic style characteristic while being able to achieve quite high pressures in a single stage. The velocities involved are much greater that the 2 m/s from a reciprocating pump. Pitot tube pumps can achieve about 820 m or 80 bar based on water at flows up to 105 m^3/h. Small close-coupled pumps, see Figure 1.3 are used for domestic water boosting applications for private sources originating from wells.

1.4.3 Disc pumps

The disc pump is very similar in construction to the end-suction centrifugal pump however the impeller does not have vanes like a centrifugal pump impeller. Two rotating parallel discs are used to create viscous drag which generates the forces to move the liquid. The pump characteristics are very similar to centrifugal pumps. The disc pump can handle small and large solids, stringy solids and is not so susceptible to viscosity degradation as a centrifugal pump of the same size. The pump is classified as non-clogging. Impeller wear in solids handling applications is greatly reduced because solids do not impinge on rotating surfaces. The disc pump can also handle quite large volumes of entrained gas without losing prime.

Disc pumps have proved useful in the paper industry. Dense pulp, up to 18% dry solids, has been handled in existing installations. The disc pump can also be built to 3A-Sanitary requirements. These pumps can handle flows to over 2000 m^3/h at heads up to 300 m.

1.4.4 Pumps as power-recovery turbines

Rotodynamic pumps can run in reverse as turbines. This fact is long-established and is one of the reasons why non-return valves are fitted in pump discharge lines. Pumps and their drivers have suffered damage as a result of turbining due to process upsets or non-return valve failures. The power available from pumps running in reverse can be captured and harnessed for useful work.

Serious interest in power recovery turbines stems from the oil crisis of the 70s and the escalating cost of energy. Some widely used processes are ideal for considering power recovery. In oil refining, the hydrocracker process operates at over 100 barg and gas scrubbing at various pressures over 50 barg. In drinking water production by reverse osmosis from seawater, the process operates at 70 barg. In these cases the process only degrades the pressure slightly leaving a large amount of pressure energy to be throttled or recovered.

When considering power recovery from high pressure process streams there are three basic choices for installation:

- Use the power as the sole power supply to drive a pump or other machine
- Use the power as the sole power supply to drive a generator
- Use the power to assist in driving a pump or other machine

Standard mixed-flow and centrifugal pumps are popular for these applications; both types are in effect fixed geometry Francis turbines. The Francis turbine is an inward flow radial turbine and, depending upon the head to be converted, can be mixed-flow or truly radial. Power generation turbine manufacturers, in general, do not build small machines thus making standard pumps an ideal option. The slight loss of efficiency, 2 or 3%, is more than offset by the economic benefits of purchasing a standard piece of equipment.

If the power recovered is used as the sole power supply it must be remembered that the turbine will not produce power until the flow reaches 40% of design. Also, speed control may be necessary. If the recovered power is used to assist in driving then the speed control of the prime driver should suffice. Assisted driving is probably the most common application because of its simplicity but it is not the most efficient utilisation of either energy stream.

Consider a pump driven by a squirrel-cage motor which has two drive shafts, a double extended motor. The extra drive shaft is coupled, via an automatic clutch, to the turbine. The pump is started in the normal manner. If there is no flow through the turbine it remains stationary. The pump process conditions can be adjusted and allowed to stabilise. If during this period flow, and some pressure, appear at the turbine inlet the turbine will start to spin. As the turbine design conditions are approached the turbine speed will increase until it tries to run faster than the motor. As the turbine tries to overspeed, the clutch locks and power is transmitted to the motor. As the turbine transmits power into the motor, the motor speed increases slightly, unloads by the amount of power supplied by the turbine and the motor power consumption is reduced. If the turbine liquid supply fails the turbine will reduce in speed and when the motor attempts to drive the turbine the clutch will disengage leaving the motor to supply all the power.

In typical applications, such as hydrocracking and reverse osmosis, the pump and turbine are the same pump type. Hydrocrackers use radially-split, barrel machines. Depending upon the size, reverse osmosis systems use multi-stage segmental or axially-split machines. It is possible to purchase multi-stage segmental machines with the pump and turbine in the same casing; precautions must be taken to prevent the turbine running dry.

1.5 Positive displacement pumps

Positive displacement (pd) pumps operate on completely different principles to rotodynamic pumps. The most important difference is speed. Positive displacement pumps do not rely on speed to develop pressure. And now we talk of pressure, not head, because the small velocity head is not converted to static head. Velocities tend to be very low compared to rotodynamic pumps. Also differential head is replaced by discharge pressure. Positive displacement pumps do not try to increase the suction head by a fixed differential head. Positive displacement pumps develop sufficient discharge pressure to force the liquid into the discharge system.

In broad terms, positive displacement pumps compress the liquid from suction conditions to discharge conditions and can achieve this at any speed. As already stated, velocities tend to be low so there are no problems in ignoring the kinetic energies in the suction and discharge pipes. The pressures developed tend to be very large in comparison to the physical dimensions of the pumps; so there are no problems in ignoring potential energy due to elevation in the suction and discharge systems.

Positive displacement pumps can produce very high pressures, over 10000 barg is possible but uncommon. Standard pumps used for high pressure cleaning and descaling operate at 1000 barg to 2000 barg. Pumps used for water jet cutting and cold pasteurisation operate up to 4000 barg. Liquids are compressible. The liquid volume at discharge conditions may be considerably smaller than at suction conditions; less than 85%. If the discharge flow rate is important, rather than the mass flow, compressibility cannot be ignored and must be included in pump and system design calculations. Power is required to compress liquids. The standard power approximation may under estimate the power required to drive a pump when the liquid is very compressible.

NOTE: A review of the first paragraph of Section 1.1, before reading the following Sections is recommended.

1.5.1 External gear pumps

Gear pumps generally have two gears. Some designs are available with three, one gear being driven by the motor, the others by meshing with the driven gear. The driven gear usually runs in

Figure 1.63 Gear pump with external gears

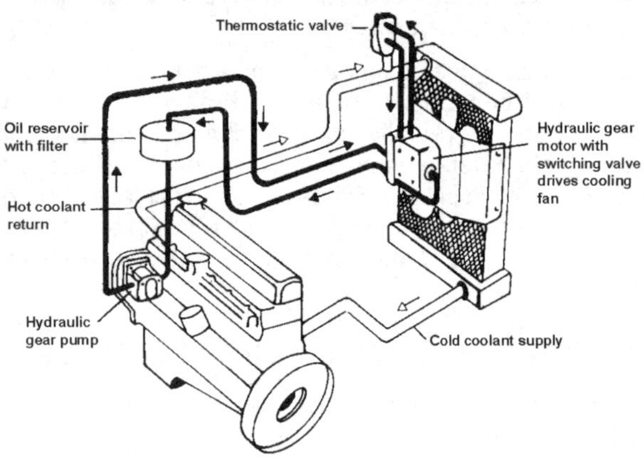

Figure 1.64 External gear pump - hydrostatic fan cooling drive

plain bearings. The bearings and shaft journals are located in the pump casing and surrounded by the pumped liquid. These bearings are thus dependent upon the lubricating qualities of the pumped liquid. Gear pumps are not therefore suitable for handling non-lubricating liquids. For reasonably acceptable pump life, gear pumps should not be used for such dry liquids as water or petrol. Paraffin and diesel oil are examples of liquids whose lubricating properties, whilst not being good, are perfectly acceptable for gear pumps.

The simplest gear pump has two external gears, Figure 1.63. This type is used for relatively free-flowing liquids and can produce quite high pressures. Pumps for process applications start at 7 bar, hydraulic power applications have a standard rating of 170 bar but pumps are available for 300 barg. Gear pumps are self-priming with a suction capacity of 4 to 8 metres. Pumps are built in cast iron, steel, stainless steel, titanium, Hastelloy and non-metallic materials.

Simple gear pumps are used for hydraulic power applications, where the motor can be a gear pump running in reverse. Figure 1.64 shows a low power application for driving a radiator fan on an engine. The thermostatic control valve provides speed control for energy saving and faster engine warm up. This type of drive is useful in hazardous areas where electrical equipment is bulky and costly.

Figure 1.65 shows an external gear pump with four stuffing boxes, external bearings and external timing gears. This style of pump is used for viscous liquids with solids, such as asphalt. External gear pumps can be equipped with heating or cooling jackets and complex sealing systems.

1.5.2 Internal gear pumps

A far greater range of application is possible with gear pumps having a small rotor mounted eccentrically within a larger external gear. A crescent shaped partition separates the two gears,

Figure 1.65 External gear pump with four stuffing boxes
Courtesy of Albany Engineering Company Ltd

Figure 1.66. When the rotor is driven, the gear also rotates. The difference between the diameters of the rotor and gear and the eccentric location means that the gear teeth engage only at one point. The tooth of the gear successively rotates out of the gear pocket in the rotor during the first half of rotation. This induces a vacuum and the gear pocket is filled with liquid from the suction line. The liquid is forced out into the discharge line by the gear teeth during the second half of rotation.

The number of teeth is kept to a minimum in order to make maximum use of material to obtain deep pockets between the teeth and achieve large displacement. By using sophisticated tooth profiles the number of teeth for external gear pumps can be as low as 10 to 15. Internal gear pumps can utilise the advantages of more favourable gear tooth engagement and therefore have even less teeth, see Figure 1.67.

The pressure of fluid causes side forces on the gear which are taken up by the bearings. The axial force is usually relatively small, although the variations in side clearance and resulting pressure variations can cause wear. Wear which causes increased gear clearance does not affect the internal leakage as much as wear at the teeth tips and sides.

It is important to know the temperature range within which a gear pump is going to operate. Working components must take up greater clearances than normal if the pump is to operate with liquids at high temperatures. Some manufacturers supply pumps suitable for operating at temperatures of up to 300 °C.

Gear pumps should be used with care for liquids containing solid particles and for abrasive liquids. Pumps are available for hard solids up to 15 mm and soft solids up to 100 mm. Slow speeds are recommended. Wear can be reduced, when handling abrasive liquids, by selecting a pump having a somewhat larger capacity enabling it to operate at lower speeds. Gear pumps are not suitable for use within the food industry where compliance with hygiene regulations is mandatory. They should not be used for handling shear-sensitive liquids.

The respective uses and applications for external and internal gear pumps varies somewhat. Greater pressure increases and increased flow is obtained with external gears, whereas internal gears are more suitable for high viscosity liquids and have better suction capabilities. To obtain a good volumetric efficiency it is always necessary to adjust pump speed or suction pressure to suit the viscosity of the liquid. A high viscosity requires a lower speed or higher suction pressure.

Shaft seals for pumps handling lube oil can be as simple as a U ring. Stuffing boxes or mechanical seals are available to cope with many liquids. To increase pump life when handling non-lubricating liquids, special precautions such as shaft journals, self-lubricating plain bearings and special gear tooth surface treatment may be necessary.

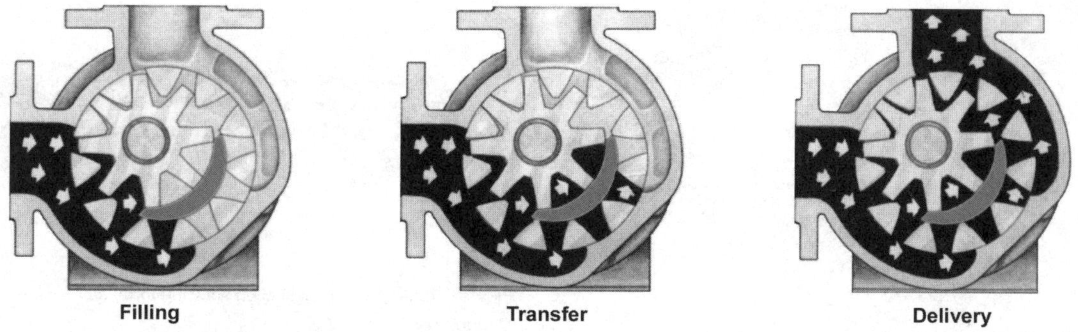

Figure 1.66 Internal gear pump showing the 3 working phases

Figure 1.67 An internal gear pump with heating jacket and double mechanical seal
Courtesy of J P Pumps Ltd

There are many models suitable for small flows designed for laboratory use, whilst larger industrial pumps have maximum flows over 3000 m^3/h.

External gear pumps offer an excellent middle-of-the-road choice between performance and cost. Gear pumps are extremely compact for the power they develop. Although unable to match the high pressure capability of piston pumps, gear pumps develop higher pressures than vane or lobe types. By the use of precision-cut gears and close tolerance assembly, particularly between the tips of the teeth and the casing, volumetric efficiencies in excess of 90% are normal.

Gear pumps must be one of the most popular pump types, since every automobile engine has at least one for lube oil. Some gearboxes and differentials have separate pumps for their lube systems. A major area of application for external gear pumps is on mobile plant and machinery. They provide a power source for a variety of lifting services, as well as acting as power-assisted steering pumps. When assembled as a tandem unit the pump becomes a compact, economic solution serving a number of circuits. In the agricultural industry, tractors often depend upon engine driven external gear pumps for powering on-board services. These pumps are also popular for use with PTO shafts, with or without assistance from a gearbox.

A number of magnetically driven sealed gear pump designs are available. These are especially useful for hazardous viscous liquids or liquids which must not come into contact with air. A disadvantage with some designs is that bearing wear can be quickly followed by damage to the magnetic coupling with the result that the complete unit is unfit for further service. With other models, should wear reach the point at which the gears come into contact with the pump casing, the magnetic coupling, by far the most expensive component, will immediately spin free without damage.

1.5.3 Archimedes screw pumps

The Archimedean screw is the oldest known type of rotating pump. The rotor is in the form of a thin helical screw. The thread form being made by plate wound around the shaft. The number of threads varying from one to three. The rotor usually rotates in an open channel having a circular cross-section, enclosing about three-quarters of the rotor periphery.

The rigid screw shaft is usually inclined at an angle of 30° to the horizontal, the length of the screw being approximately twice the lift obtained. The lift heads are rarely in excess of 10 m. Increased nominal flow is obtained by increasing the rotor diameter and reducing speed. Speeds range between 20 and 80 rpm.

Delivery heads are virtually independent of flow up to a maximum flow, maximum capacity loss factor or filling coefficient. The efficiency curve is flat as shown in Figure 1.68. The efficiency is only reduced by about 5% from maximum flow to 50% flow.

There are special types of rigid screw pumps, like that shown in Figure 1.69, where the channel is totally enclosed and rotates with the screw. Provided that the drive and shaft bearings are

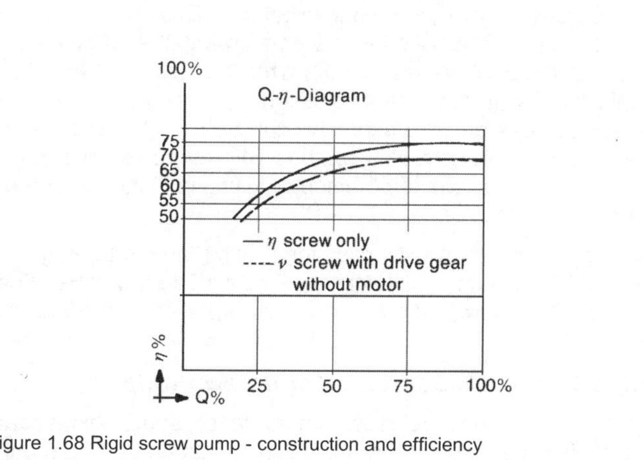

Figure 1.68 Rigid screw pump - construction and efficiency

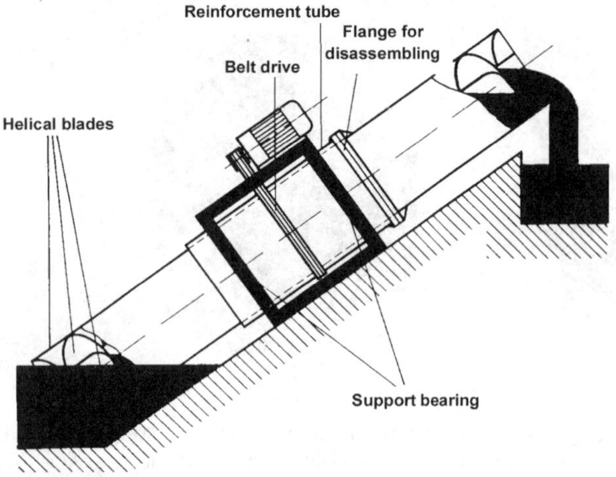

Figure 1.69 Rigid screw pump with rotating channel

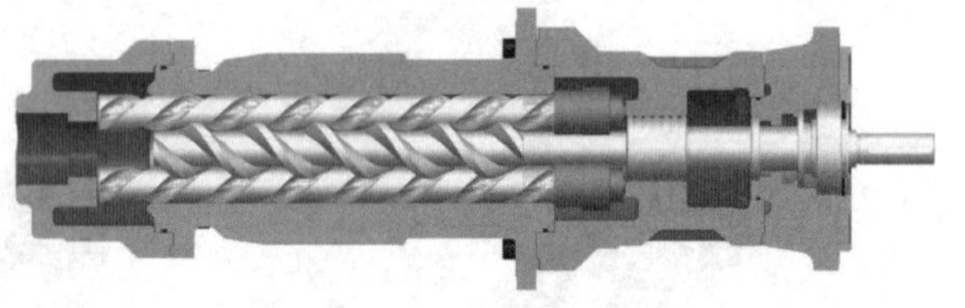

Figure 1.70 A longitudinal section showing three screws meshing
Courtesy of Allweiler AG

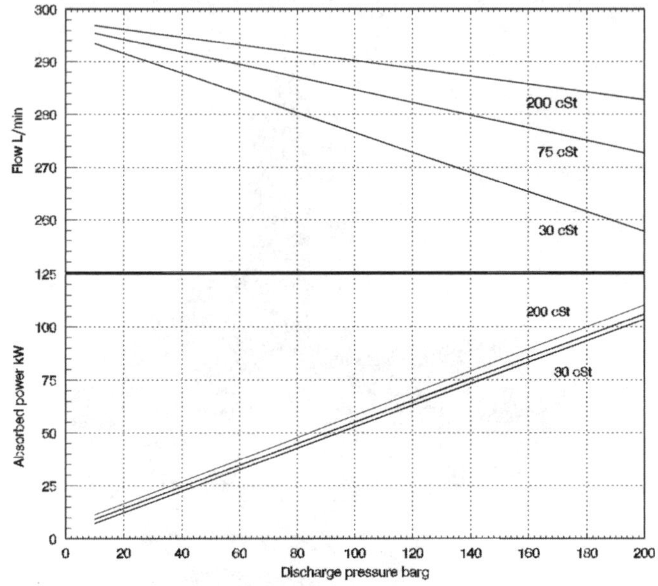

Figure 1.71 Viscous effects on triple screw pumps

completely isolated from the liquid it is possible to transport a wide variety of liquids and mixtures.

Rigid screw pumps are used for transporting liquids containing large solid contaminants, untreated sewage for example. They are also ideal for low delivery heads and large flows, up to 21000 m^3/h. As a sewage pump the rigid screw pump is self-regulating according to the supply flow. When the level in the pump sinks then the capacity loss factor, or filling coefficient, automatically reduces causing the pump flow to decrease.

The advantages of the rigid screw pump are:

- High efficiency
- Simple robust construction
- No seals
- Can run dry
- Flat efficiency curve
- Self-regulating for varying output
- Not sensitive to contaminants

1.5.4 Twin-rotor screw pumps

All screw pumps are grouped together in this section even though the operating principles of two screw pumps is different to three and five screw pumps.

Two screw pumps consist of two identical profiles meshing together and sometimes with external timing gears. Without external timing, the idler screw is driven by liquid pressure between the profiles. The two screws do not touch and a minute flow path from discharge to suction always exists.

The most common type of screw pump has three rotating screws of two different profiles, Figure 1.70. The central screw is the driver whilst the basic function of the two idler screws is to provide sealing by hydrodynamic bearing films. The screw helixes form a number of sealing elements with the pump casing which transports the liquid axially when the screw rotates. The liquid flow is smooth without disturbing flow variations. The screw pump is quiet running and can be run at speeds up to 4500 rpm. By constructing pumps with longer screws more sealed elements can be obtained resulting in higher discharge pressures.

Screw pumps rely on viscosity for sealing, and in triple screw pumps for driving the idle screws. Figure 1.71 shows the effect of viscosity on flow and power at constant speed and differential pressure.

1.5.5 Triple-rotor and 5-rotor screw pumps

Triple screw pumps are used primarily for oil, and other lubricating liquids like ethylene glycol, which should be free from contaminants. They are self-priming but must not be run dry as this can cause damage to the screws and casing. Screws and casings coated with ceramic can handle water but differential pressure is limited to about 100 bar.

Triple screw pumps are available for flows ranging from 300 l/h to over 800 l/h. Low pressure ranges cover 7 to 10 bar discharge pressure. High pressure pumps are available for 250 bar. Viscosity should not be too low and should be in excess of 1.5 cSt. The maximum permissible viscosity is about 5000 cSt; factory approval should be sort for viscous applications over 400 cSt on higher pressure applications. Screw pumps can be used for liquid temperatures ranging from -20 °C to +155 °C. Mechanical efficiency can be over 85%. Figures 1.72 and 1.73 show vertical and horizontal versions.

1.5.6 Twin-rotor geared-screw pumps

Screw pumps with two screws and an external synchronising gear can handle liquids containing small quantities of solid particles, 2 mm may be possible. Pump users have been looking for pumps which can handle mixtures of liquids and gases, two phase flow. Crude oil production, from some wells, varies considerably with water and CO_2/H_2S content. Centrifugal pumps can only accept approximately 15% gas percentage by volume before losing prime. Twin screw pumps have been developed to handle continuously variable mixtures from 100% liquid to 97% gas.

Screw pumps are made from grey cast iron, nodular cast iron and carbon steel. Special surface treatment of components can improve wear resistance. Shaft sealing is by means of stuffing boxes or mechanical seals.

Axial thrust is taken up partly by hydraulic balancing and partly by thrust bearings. Large pumps are constructed with two opposed screws on each shaft. Liquid enters at the centre of the

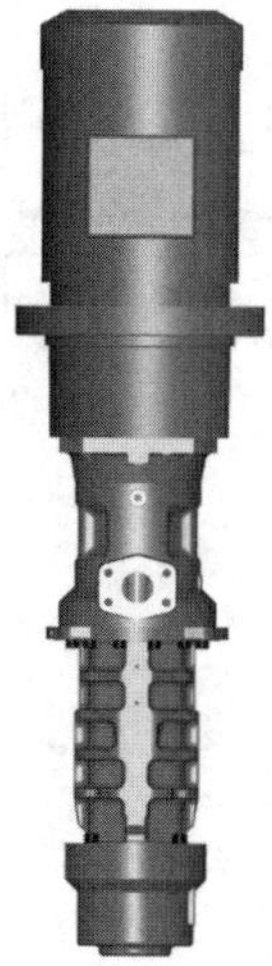

Figure 1.72 A typical vertical pump with end-suction connection
Courtesy of Allweiler AG

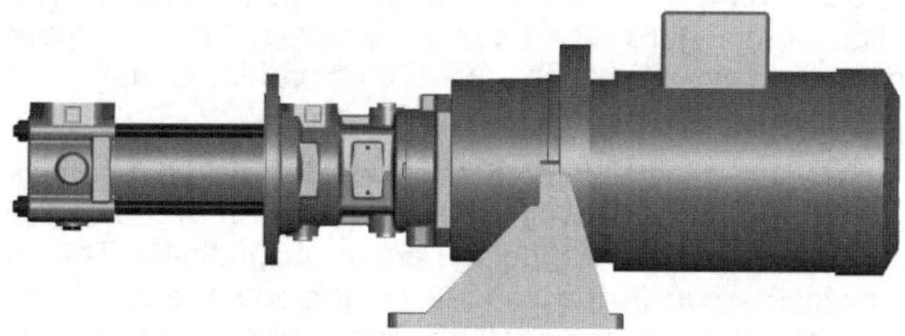

Figure 1.73 A typical horizontal monoblock "tie-bolt" pump
Courtesy of Allweiler AG

shaft and flows outward to the discharge. Equal and opposite flows creating a high degree of hydraulic balance. With regard to the construction of thrust bearings, screw pumps are normally designed for one direction of rotation and flow.

Twin synchronised screw pumps are available up to approximately 2200 m^3/h at pressures of 240 bar. These pumps can handle water and there is no maximum limit on viscosity. Gas entrained in viscous liquid tends to reduce the effective viscosity. Slugs of 100% gas and entrained abrasive solids, such as sand, can be handled.

1.5.7 Progressive cavity pumps

Progressive cavity pumps may seem to be a new pump type. This is the modern name for an old pump. In the past, these pumps have been called, eccentric screw pumps, eccentric gear pumps and helical gear pumps. The new name is better as it describes the action of the pump. The principle of operation was first established in the 1930s by Rene Moineau.

Progressive cavity pumps have only one rotor working within a rubber stator. The rotor, which looks something like an elongated corkscrew, rotates in a flexible stator having double internal helixes. The pitch of the stator helix is twice that of the rotor's. The difference in pitch forms sealed cavities between the rotor and the stator which, with rotation of the rotor, are caused to travel axially along the stator resulting in a smooth axial flow. Since the end of the rotor describes a circle when rotating, drive is usually by means of a cardan shaft with standard universal joints (Hooke's joints) at both ends which lie in the liquid flow, see Figure 1.74.

Alternative methods of flexible power transmission have been developed by different manufacturers to cope with the wide range of operating conditions, see Figure 1.75. (a) shows a simple single pin coupling. This style of coupling is suitable for small pumps when the torque requirement is low. The O ring is used to provide a flexible seal; it keeps the lubricant inside the coupling and the product outside. (b) is an open version of (a). (b) is used for small hygienic pumps when it is important to be able to clean the pump internals thoroughly. (c) and (d) show two different approaches for Hooke's joints. A crossed pin provides four bearing surfaces; two for each side of the coupling. These are the most popular types of coupling used and are suitable for any pump size and most applications. The pin bearings can be plain or the needle roller type. Lubricant is sealed in the coupling by the flexible outer cover. All of the pin joints suffer a common drawback; the possibility of cyclic angular velocity changes. If a single Hooke's joint is used to transmit rotary power between two shafts which are not parallel, the driven shaft will not rotate at a constant angular velocity. The shaft speed continuously changes as the shaft rotates. The driven shaft can transmit the effects of the cyclic velocity changes to the drive shaft as torsional oscillations. The cyclic angular velocity effects on the pumping element can be largely eliminated by using two Hooke's joints in series, one at each end of the drive shaft.

However this approach is only completely effective when the two joint angles are identical. The rotor is housed within a flexible stator and does have a degree of freedom regarding positioning. The elimination of the cyclic angular velocity effects on the pumping process, can not be guaranteed. They can be

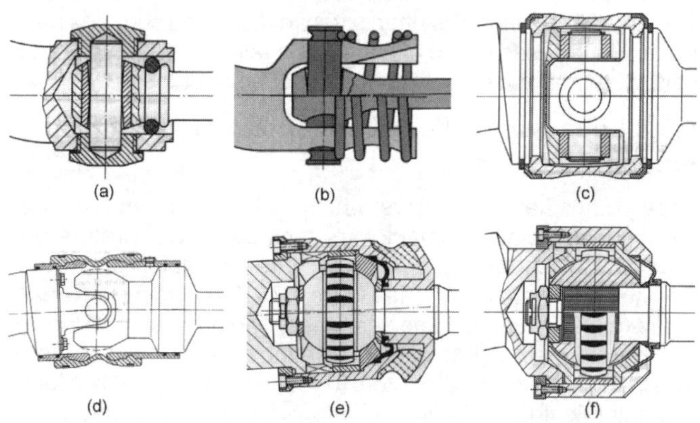

Figure 1.75 Variations of flexible couplings for progressive cavity pumps

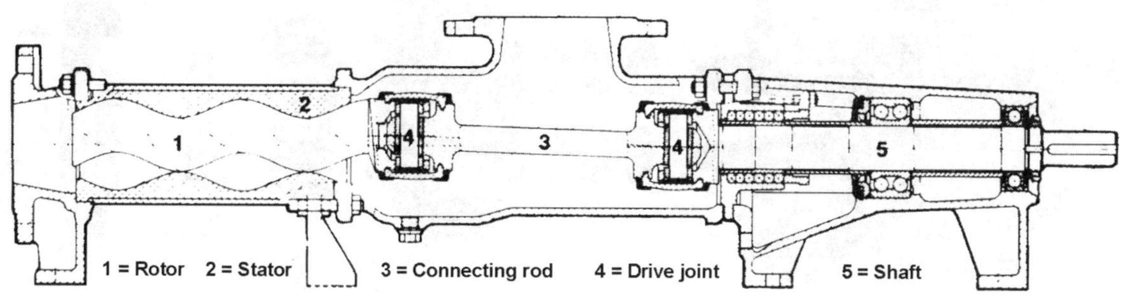

Figure 1.74 A progressive cavity pump

completely eliminated by using a different style of coupling, the constant velocity coupling, see (e) and (f). Both couplings use a crowned gear to mesh with a toothed annulus. The load is distributed over many tooth flanks and the angular motion is guided by a spherical bearing. (e) is a double-sealed version for higher operating pressures.

All coupling problems such as bearing wear, lubrication and cyclic angular velocity, can be managed by using a relatively new driving shaft style, the flexible shaft. The flexible shaft is just that. A drive shaft with proportions which are adequate to support the torsional and axial loads but thin enough to be able to bend to follow the end of the rotor. The flexible shaft is bolted or clamped between the bearing housing shaft and the rotor. In some cases the useful life of the flexible shaft may be limited by fatigue and routine replacement may be necessary. Given the full operating condition the pump manufacturer can select the most appropriate drive shaft option.

Increasing the length of the rotor and stator makes possible greater differential pressures. Pressure differential for a low pressure pump is 6 bar. Differential pressures up to 40 bar can be obtained by increasing the length. Pumps are available for flows over 300 m^3/h.

Progressive cavity pumps are used for practically all types of liquid from very fluid to very viscous. They can handle liquids containing abrasive contaminants and are relatively insensitive to solid particles. Larger pumps can allow the passage of random hard particles of 30 mm and soft particles over 100 mm. Special adaptations can be made to feed very viscous product in to the pump suction. Progressive cavity pumps have good suction capacity up to 8 m but their extreme sensitivity to dry running requires venting and priming prior to starting for long stator life.

Many different materials are used in the manufacture of the pump casings and rotors ranging from cast iron to titanium. The stator can be made from a wide variety of elastomers, for example natural rubber, nitrile rubber, Neoprene and Vitonâ. In order to select the correct elastomer it is necessary to know the chemical properties of the liquid to be pumped, the nature of any solids and the temperature at which the pump is to operate. The choice of material is often a very difficult and complex matter due to the wide range of materials available. When in doubt, consult a manufacturer who has experience of the duty.

1.5.8 Lobe pumps (including circumferential piston pumps)

Lobe pumps have two rotors, which, unlike gear pumps, operate without metallic contact with each other. Both rotors are driven by synchronising gears which are completely separated from the pump chamber, see Figure 1.76. The pump shaft bearings are also situated in the gear case. The pumped liquid does not therefore come into contact with the bearings. Synchronising drive gears cause the rotors to rotate in opposite directions. The inlet liquid flow is divided into two halves, trapped in the space formed between the rotor and the pump casing and transferred, without change in volume, towards the outlet where the rotors meet, thereby reducing the cavity and forcing the liquid out, Figure 1.77. Lobe pumps are not usually applied to pressures over 30 bar so there are no problems with volume changes due to compressibility.

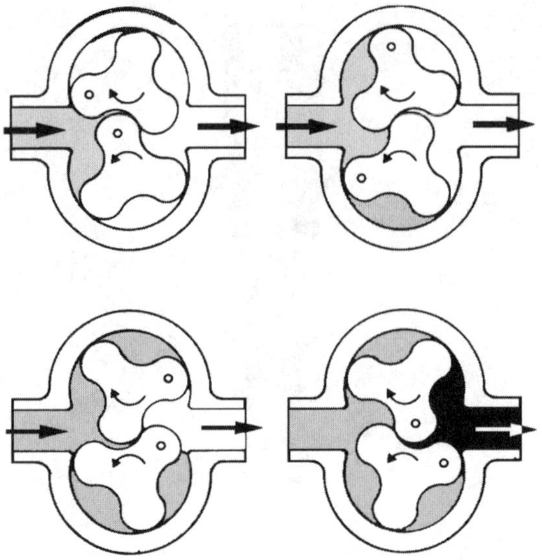

Figure 1.77 Lobe pump working principles

The absence of metallic contact between the surfaces of the rotors themselves or between the pump casing and the rotors means that wear of the rotating parts is insignificant. The only wear which occurs is due to erosion of the pumped liquid or entrained solids. Since the shaft bearings are normally mounted outside the pump casing, the shaft ends are relatively long and unsupported which imposes limits in terms of working pressure. For pressures in excess of 12 bar some manufacturers fit plain bearings inside the pump casing. Others mount an extra bearing bracket on the outboard side of the pump casing. The latter case giving rise to four shaft seals.

The profile of the rotors varies from manufacturer to manufacturer. The most usual is shown in Figure 1.78. The profile does not alter the principle of operation. It can be said however that rotors with one or two lobes give rise to greater pressure pulsations than rotors having three lobes. For gentle handling of liquids, rotors having one or two lobes should be chosen.

Rotary piston pumps are lobe pumps with greatly modified lobe profiles. The rotors are designed so that much more tip surface area is presented to the bore of the casing. This modification changes the pump viscosity characteristic and also the pumping action. The liquid is no longer squeezed out of the pockets between the lobes but is pushed directly out of the discharge. Figure 1.79 shows two typical rotor profiles.

Lobe and rotary piston pumps are suitable for both fluid and viscous products. Special feeding arrangements are necessary for extremely viscous liquids which cannot otherwise flow. Liquid temperatures of up to 200 °C can be handled if the clear-

Figure 1.76 Exploded view of a lobe pump

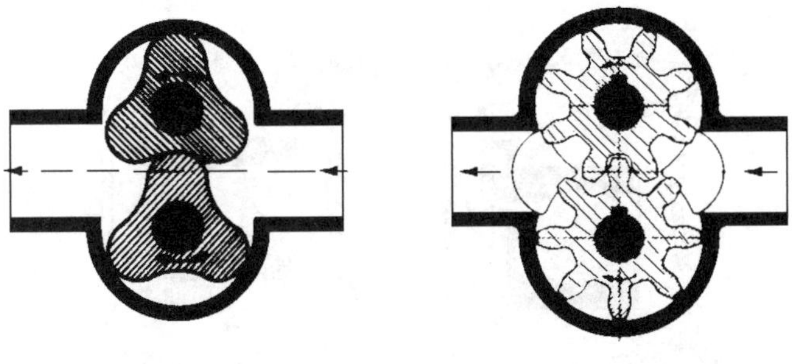

Figure 1.78 Lobe pumps - popular rotor profiles

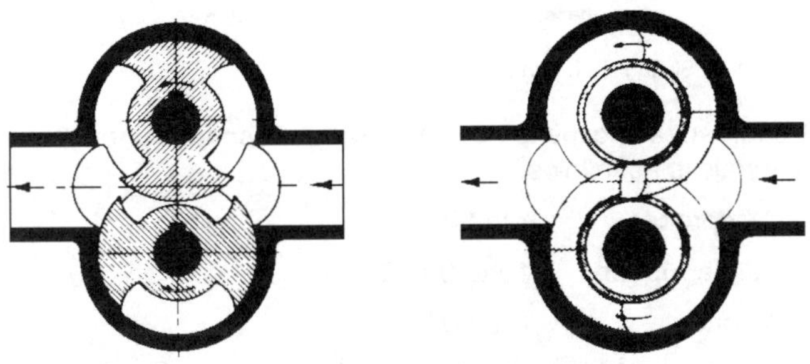

Figure 1.79 Rotary piston pumps rotor profiles

ances between the rotors and the covers are increased. By increasing the rotor side clearances it is possible to handle liquid at temperatures of down to - 40 °C. To maintain constant temperature, pumps may also be fitted with heating or cooling jackets.

Lobe pumps and rotary piston pumps handle the pumped liquid very gently. Examples of this phenomenon can be found within the foodstuffs industry where pumps are used for pumping cooked pea soup, preserves containing whole berries and other similar applications including handling whole fish. In these cases the pumps are specially constructed to fulfil the hygiene requirements; being easily dismantled and suitable for washing by hand or in accordance with the CIP method.

Pumps can also be equipped for completely aseptic pumping for use within the pharmaceutical industry. They are also extensively employed in the chemical industry for handling both corrosive and non-corrosive products. Special dense-pulp pumps have been developed for the cellulose industry, these being suitable for pulp concentrations ranging from 6 to 30%, Figure 1.80. Special derivatives are manufactured for sewage handling. These pumps can have adjustable rubber lobes.

Pumps usually operate at relatively low speeds and they are often used for high viscosity liquids. The pump is quiet running and delivers a largely pulsation-free flow. For most pumps the compon ents in contact with the liquid are manufactured in stainless steel. Pumps for non-corrosive applications can have a cast iron casing with rotors and shafts made of steel. Sometimes the rotor material can be varied in order to increase the pump's suction capacity, for example, by the use of nitrite rubber rotors. Shaft seals consist of various forms of mechanical seals and stuffing boxes. Since the pump has two shafts which pass through the pump casing, two sets of seals are needed for each pump.

The rotary piston pump is a successful variant of the lobe design. Also called the circumferential piston pump, it uses arc shaped pistons or rotor wings, as shown in Figure 1.79. The rotor wings are manufactured in a corrosion resistant, non-galling copper free nickel alloy and are designed to operate with very close clearances. This feature, combined with the geometry of the rotor wings, produces a long sealing path between inlet and outlet resulting in minimal slip when pumping low viscosity liquids. On viscous liquids and when handling products with solids in suspension, the large liquid cavities in the rotor wings combined with low operating speeds and carefully profiled anti-cavitation ports produces a smooth, pulsation free, low shear pumping action. Twin wing rotors are fitted for the majority of duties; single wing rotors are preferred for easily damaged products.

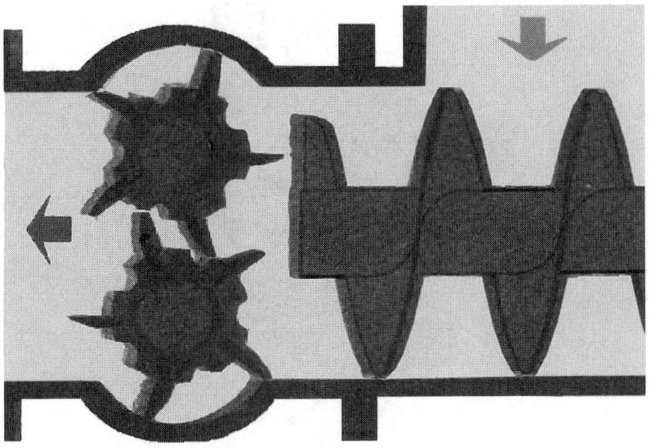

Figure 1.80 Dense-pulp pump with feed screw

Lobe pumps and rotary piston pumps are available in a wide variety of sizes from 0.1 to over 300 m^3/h. A sewage version can handle 900 m^3/h at 6 bar. Discharge pressures are normally up to 30 bar. Suction capacities vary between 1 and 5 m, depending upon internal clearances, pump size and speed. Solids handling capabilities are good with solids up to 100 mm on the largest pumps.

1.5.9 Vane pumps

In a rigid vane pump the rotor is eccentrically located in the pump casing and the vanes slide in and out to maintain contact with the casing wall. In a flexible vane pump the rotor is mounted eccentrically in the pump casing and the vanes flex to maintain contact with the casing wall. The pumping action is created by the variation in volume between the vanes, Figures 1.81 and 1.82.

In both types of pump the liquid is drawn into the pump by an increasing volume between the vanes on the suction side, transported to the discharge side whilst trapped between the vanes and a proportion forced out by a decreasing volume.

The vanes in sliding vane pumps may be controlled by springs, hydraulic pressure or rely entirely upon centrifugal forces induced by rotation. Some types use a rotating cam to guide the movement of the vanes.

There are many types of vane pump where the vanes are located in the pump casing (stator), Figure 1.83.

Another variant of vane pump is the sliding shoe pump where the vanes are U-shaped and operate against two separate sur-

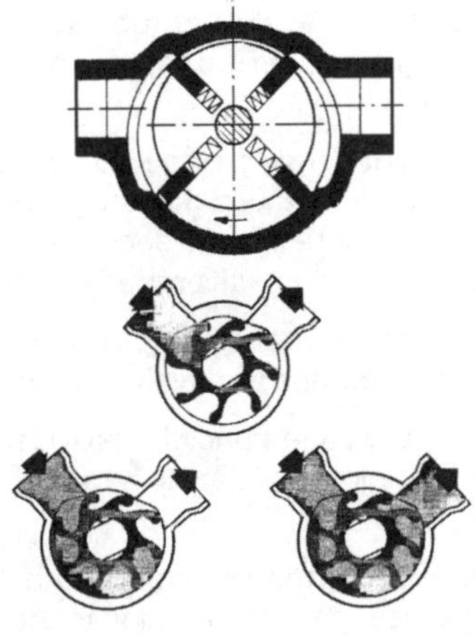

Figure 1.81 Operating principles for vane pumps

Figure 1.82 Vane pump with sliding vanes in the rotor

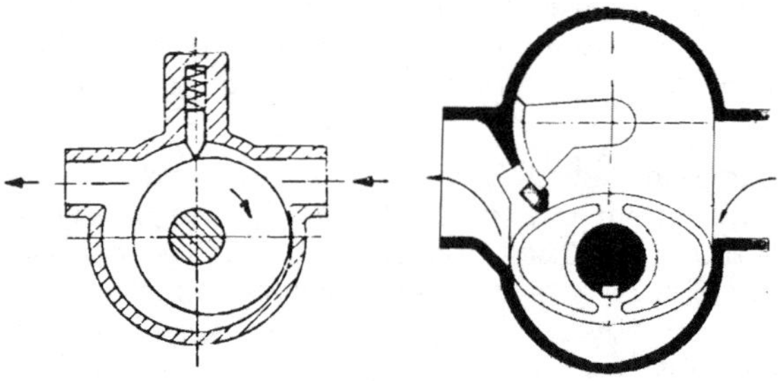

Figure 1.83 Vane pump with vanes in the stator

faces, one against the rotating cam, whilst another surface slides against a valve plate, Figure 1.84.

The pump casing and rotors for vane pumps with sliding vanes are usually made of cast iron, gunmetal or stainless steel, whilst bronze or glass fibre reinforced PTFE is used for the vanes.

Vane pumps with sliding vanes are suitable for most clean liquids and especially suited for those with entrained gases or those having a low latent heat; petrol or ammonia for example. Maintenance costs are relatively low due to the ease of replacement of the vanes, even when being used to pump liquids which are somewhat contaminated, waste oil for example. Vane pumps can operate within a large range of viscosities, although it is necessary to compensate for the speed, i.e. the higher the viscosity the lower the speed. This phenomenon is common to most displacement pumps. Vane pumps are capable of suction lifts of 2 to 5 m. Flexible vane pumps are suitable for discharge pressures up to 5 bar, sliding vane process pumps of over 10 bar. Sliding vane pumps for hydraulic power applications can run at over 3000 rpm at pressures to 200 bar.

In a flexible vane pump, the rotor (flexible impeller), made of a synthetic elastomer, creates a seal between suction and discharge side. The pumps are therefore dry self-priming with a suction lift of 4 to 5 m, i.e. do not require to be vented and primed. With liquid in the pump it is capable of a vacuum of up to 500 to 600 mm Hg.

As the rotor material depends on the liquid pumped for lubrication, it is generally recommended that the pump should not operate without liquid for more than 30 seconds. This is more than sufficient time for self priming with properly designed pipework.

Flow capacities for flexible vane pumps range from 0.1 to over 30 m^3/h, sliding vane pumps are available for over 300 m^3/h.

The single rotor, multi-vane principle, provides largely pulsation-free flow and enables the pump to handle products susceptible to damage.

Chemical resistance information, compiled by manufacturers of flexible vane pumps, provides a guide to the compatibility of various elastomers with liquids commonly used in industry.

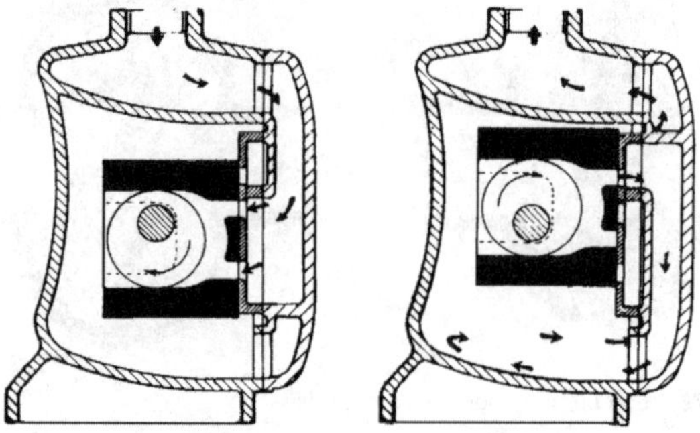

Figure 1.84 Sliding shoe pump

Broadly speaking the temperature range of some of these rotor materials is as follows:

- Neoprene 4 to 80 °C
- Nitrile rubber 10 to 90 °C
- Viton® 4 to 90 °C

However, as with any type of pump when pumping liquids which are not cold, great care must be taken to observe suction limitations and provide sufficient NPIP.

(Viton is a registered trade mark of DuPont Performance Elastomers LLC.)

1.5.10 Peristaltic pumps (including rotary peristaltic pumps)

Peristaltic hose pumps

A peristaltic hose pump transports liquid by mechanically squeezing a space enclosed by a flexible element. The most usual peristaltic pump, the flexible hose pump, operates by means of rollers or cams acting directly upon the hose containing the liquid, Figure 1.85. The number of rollers or cams vary depending upon the manufacturer; industrial pumps have 2 or 3 but small laboratory scale pumps can have as many as 7. Figure 1.86 shows an industrial size pump with two spring-loaded rollers. A pump may be fitted with one or more hoses, sometimes as many as thirty.

At one time the use of peristaltic pumps was confined largely to laboratory and similar specialised applications but developments in rubber technology have resulted in a new generation of heavy duty pumps and peristaltic units are now to be found on a wide range of industrial duties involving viscous and/or abrasive sludges and slurries as well as products with high solids content.

An important factor in favour of the peristaltic concept is the pump's ability to handle solids up to the full bore of the hose tube. With most pump designs, solids handling capacity is restricted to approximately 15% of the pump branch sizes.

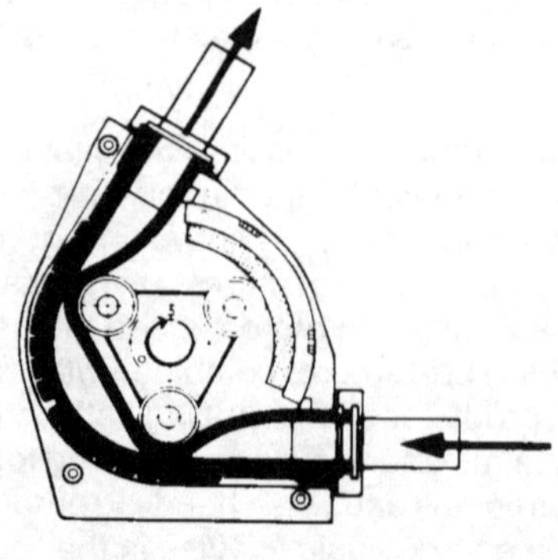

Figure 1.85 Peristaltic hose pump

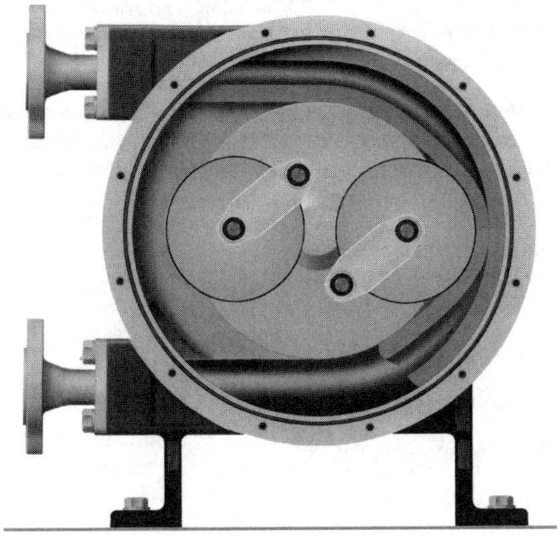

Figure 1.86 Industrial size peristaltic pump

In one modern heavy duty design the roll/squeeze peristaltic movement takes place in a sealed housing partially filled with a mixture of glycerine and glycol. This not only lubricates the cams or sliding shoes from which the roll/squeeze motion is derived; it also acts as both coolant and lubricant to the exterior of the pump hose tube. This prolongs hose life and reduces the incidence of mechanical failure. Hose life in excess of 3,000 hours is quite common.

The same design uses a hose tube with an inner core of soft natural rubber or NBR rubber with an outer covering of hard natural rubber reinforced with braided nylon. During the roll/squeeze phase of the peristaltic cycle, solids present in the pumped liquid are cushioned in the soft inner core of the tube with minimal damage or erosive effect and are released gently back into the liquid stream when the squeeze cycle is concluded. This enables delicate, shear sensitive solids to be pumped without damage and abrasive material to be handled with little or no effect on the hose interior.

The peristaltic design is inherently sealless, with the pumped liquid completely enclosed with no possibility of leakage except in the case of a hose failure. Failure from a cause other than mechanical stress is rare and, under stable operating conditions, can often be predicted with considerable accuracy. Regular hose inspection is of course advisable.

Peristaltic pumps can be used for all types of liquids and to some extent even gases. The suction capacity varies greatly for different constructions. Discharge pressures are usually below 10 bar but pumps have been used up to 40 bar; tube life is extremely short! Large pumps, with 125 mm tubes, are rated at 2.5 bar. The capacities vary from very small metering quantities of 0.001 l/h up to flows of 50 m^3/h. Special pumps are available for over 3000 m^3/h.

Rotary peristaltic pumps

Another type of peristaltic pump, sometimes called an orbital lobe pump, works by means of an eccentric rotor operating within a flexible elastomeric element with the liquid trapped between the elastomeric element and the pump casing. The rotation of the eccentric rotor transfers the liquid from the inlet to the outlet, Figure 1.87.

This pump is leak-free, conditional on the flexible elastomeric element not failing as a result of the mechanical stresses. The elastomeric element should therefore be replaced regularly to avoid unexpected failures. In order to select the correct elastomer and the correct pump casing material it is important to know the chemical properties and temperature of the liquid to be handled. Capacities range from 0.06 to over 50 m^3/h. Suction capacity 1 to 3 m. The pump creates a strongly pulsating

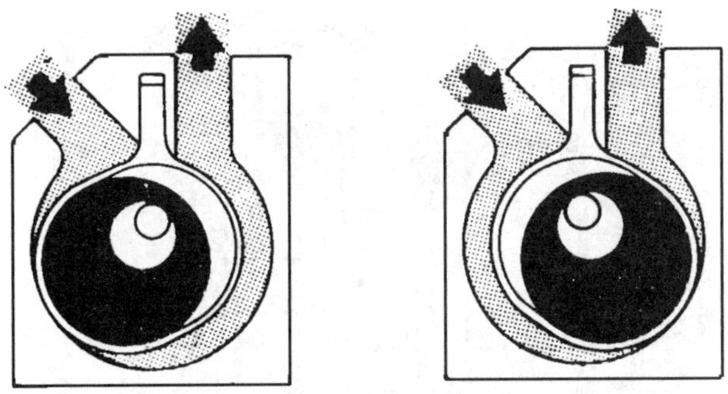

Figure 1.87 Rotary peristaltic pump

flow making it advisable to use flexible hoses in the suction and delivery lines. If fixed pipes are used some type of pulsation damper should be fitted between the pump and the pipe. Rotary peristaltic pumps are suitable for pressures over 10 bar.

1.5.11 Rotary eccentric piston pumps

The rotary eccentric piston pump can be considered as a type of single wing circumferential piston pump or a peristaltic pump without a tube. The casing forms a circular cylinder. The piston is circular but of a smaller diameter than the cylinder. The piston does not rotate but is forced to precess eccentrically around the cylinder with a minimum clearance. A viscous seal forms in the minimum clearance to reduce internal slip. Incoming liquid is swept around the circumference of the cylinder to the discharge port. The pumping action produces very little liquid shear and is extremely gentle.

There are two slightly different versions of the pump in common use. One version uses a bearing bracket very similar to an end-suction centrifugal pump. Piston eccentricity is maintained by spring loading. These pumps have a U-ring seal, mechanical seal or stuffing box to match the liquid characteristics. These pumps cannot run dry without special sealing arrangements. The other version uses a special non-rotating eccentric drive arrangement which does not require a rotating seal. Sealing is accomplished by rubber or metallic bellows which allows the pump to run dry. Both pump styles are self-priming and reversible. Figure 1.88 shows an exploded view and Figure 1.89 a longitudinal cross-section of an eccentric drive pump.

The rotary eccentric piston pump is manufactured in cast iron, steel, bronze and stainless steel and can be built to hygienic and aseptic requirements. Test have shown that CIP with 120°C water is completely effective. The pump is used extensively in the dairy industry for handling milk and milk products. Other pumped products include; alcoholic drinks, cake mixes, chocolate, ink, jam and marmalade, mineral oil, paint, resin, soap, toothpaste and varnish.

Pumps can handle flows up to 85 m^3/h at pressures up to 12 bar.

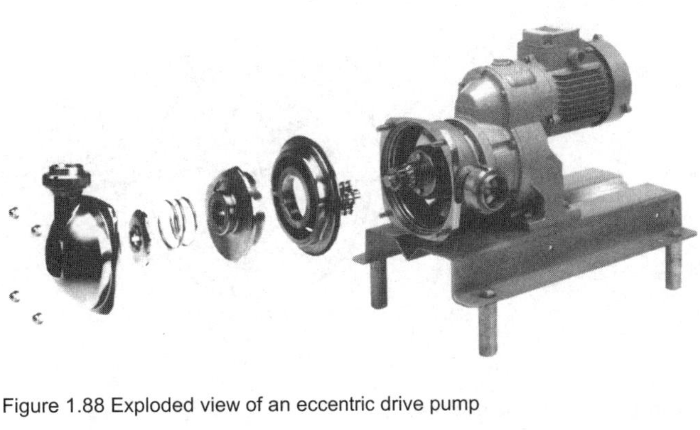

Figure 1.88 Exploded view of an eccentric drive pump

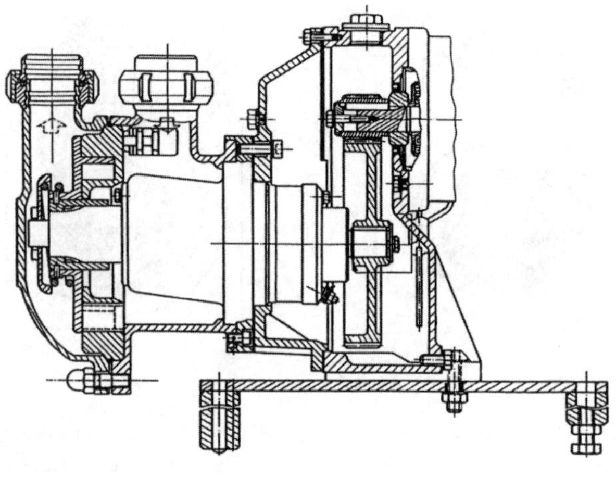

Figure 1.89 Longitudinal cross-section of an eccentric drive pump

1.5.12 Axial and radial piston pumps

These are piston pumps for hydraulic power applications **and** water pumps as well !! They are characterised by their lack of external lubrication and not having any seals on the piston. All moving parts are lubricated by the product. Early systems all used mineral oil as the motive liquid and this may have been a standard industrial lubricating oil. Installations in hazardous areas, where the problems caused by any fire were great, looked for replacement liquids to reduce fire risks. Mixtures such as water/glycerine and water/glycol were tried but equipment needed derating to cope with the poor lubrication properties. Soluble oils were developed which led eventually to 95/5 and 97/3 water/oil emulsions.

Modern oils are biodegradable and can be disposed of in normal sewers. However, they are also costly and so leaks are kept to a minimum. Piston leakage is controlled by the clearance between the piston and the cylinder bore. Any leakage returns to suction. Piston pumps for hydraulic applications fall into two broad categories; axial piston and radial piston. Inline piston pumps, the forerunner of modern axial and radial pumps, are still used for a few applications.

Axial piston pumps consist of a number of cylinder bores in a rotating block. The pistons can be anchored in various ways to produce the stroke as the cylinder block rotates. Most pumps have pistons which are driven sinusoidally. The stroke can be variable. Variable flow can be produced from fixed speed pumps. Some pumps can adjust the stroke so that the pump runs in reverse, i.e. the suction becomes the discharge. Suction and discharge valve functions are performed by ports in the cylinder block support plate which connects cylinders to the suction or discharge at the correct part of the stroke.

Radial piston pumps have multiple cylinders spaced radially around a central cam shaft. Most pumps are fixed stroke. Valves can be poppet, like engine valves, or ball; both spring loaded and automatic.

Inline pumps are similar in construction to inline car engines. Several cylinders are mounted above a cam shaft. Pumps tend to be of fixed capacity.

Pumps are available for pressures up to 670 bar for flows up to 8 l/s and operate at speeds up to 5000 rpm. Applications for these pumps are the hydraulic systems found in all parts of industry. The list below is not exhaustive:

- Agricultural machinery
- Coal mining
- Crude oil production
- Direct-acting process pumps
- Earth moving equipment
- Machine tools
- Power transmission
- Raw steel products
- Steelmaking

Pumps are now available with axial piston designs in ceramics for use on water and seawater. Low cost offshore developments are in seabed mounted wellheads with pipelines linking to centralised facilities. Wellhead hydraulic requirements serviced by seawater would have no leakage or liquid shortage problems. Optimising proportions and material combinations to reduce wear and extend operational life is a continuing priority. Current designs are limited to 160 bar.

1.5.13 Inline piston pumps

Piston pumps are similar to plunger pumps but have a moving seal on the piston. Piston pumps can be single acting or double acting. Double acting pumps have a stuffing box similar to a plunger pump stuffing box. The piston is driven by means of a rotating crank or eccentric, connecting rod and crosshead, see Figure 1.90. This Figure shows the use of Mission valves, the most popular type, and has an internal single reduction gear set built into the crankcase. (See Figure 1.96 for other valves for plunger pumps.)

Applications for piston pumps are wide ranging but duties on drilling rigs handling drill mud and crude oil production far exceed all others. Piston pumps are ideal for viscous liquids and handling solids which are not abrasive. Special pumps have been developed for pumping coal. Coal can be transported as a suspension in water or can be blended with oil to form a fuel for direct injection into boilers. Most applications are below 140 bar.

The comments regarding self-priming and cavitation in Section 1.5.15 may be helpful and apply equally to piston pumps. Piston pumps are manufactured mainly in cast iron, cast steel and abrasion resistant chrome iron.

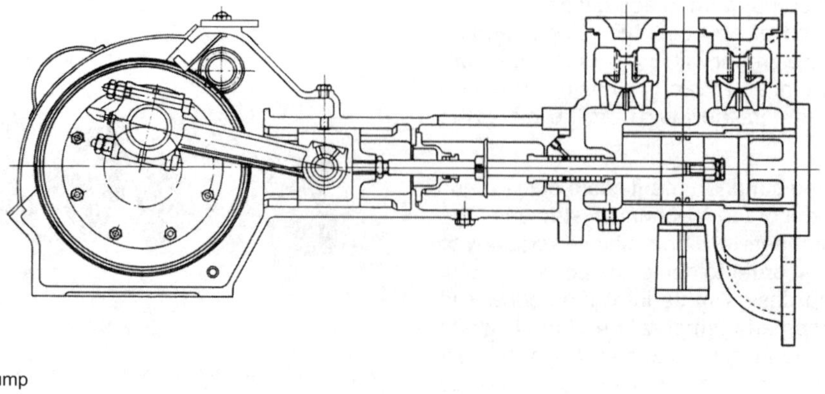

Figure 1.90 Crank drive piston pump

1.5.14 Descaling pumps

These include portable plunger pump packages for high pressure descaling/cleaning

High pressure water is used for descaling and cleaning. The high pressure produced by the plunger pump is converted to kinetic energy in a nozzle. The high velocity water jet impacts on any surface converting the kinetic energy to a destructive force. Pump packages are produced in several configurations:

- Electric motor driven
- Engine driven
- Skid-mounted
- Two or four wheel site trailer
- Two or four wheel road-going trailer with brakes and lights
- Vehicle-mounted

Packages may be complete with a water tank or have suction hoses for connection to a water supply. Facilities may be incorporated for hose reels and lance storage. Units may be capable of supplying more than one lance simultaneously.

These packages are produced to standard designs by pump manufacturers and by specialist firms who purchase all the components, including the pump. Horizontal and vertical pumps are used for these applications. If a standard package is not available to fulfil a particular requirement it is possible to buy a purpose-built unit.

The following equipment should be fitted, as a minimum, to all systems:

- Suction strainer with Dp indicator
- Pressure control valve for each operator
- Relief valve
- Low suction pressure cut-out and indicator
- Driver high temperature cut-out and indicator
- Pump high temperature cut-out and indicator

Pumps for high pressure descaling and cleaning usually run at 690 bar to 1050 bar, but manufacturers have offered pumps for 1900 bar. Portable electric motor driven packages are available from 7.5 kW. Skid-mounted engine units are available to over 1000 kW.

These packages are used for many applications:

- Concrete cleaning
- Drain cleaning
- Graffiti removal
- Heat exchanger tube cleaning
- Insulation removal
- Paint stripping
- Ship hull cleaning
- Vessel cleaning

1.5.15 Plunger pumps (includes horizontal and vertical)

Plunger pumps are available from single cylinder hand pumps, capable of pressures over 300 bar, to nine-cylinder vertical pumps absorbing over 3000 kW. Pumps are built with 1, 3, 4, 5, 6, 7 or 9 cylinders. Horizontal pumps have 1, 3, 5 or 7 cylinders. Vertical pumps have 3 or more cylinders. Increasing the number of cylinders increases the pump capacity if speed and stroke remain constant. Also increasing the number of cylinders tends to smooth the flow variations, reduces pressure pulsations and torque variations at the crankshaft. Pulsation

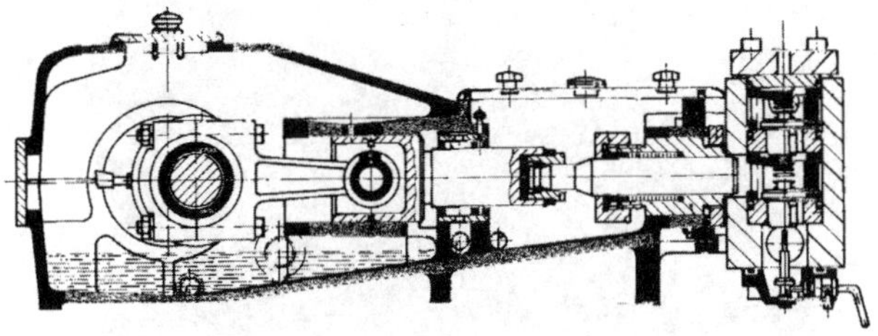

Figure 1.91 Horizontal plunger pump with monoblock liquid end and wing guided valves

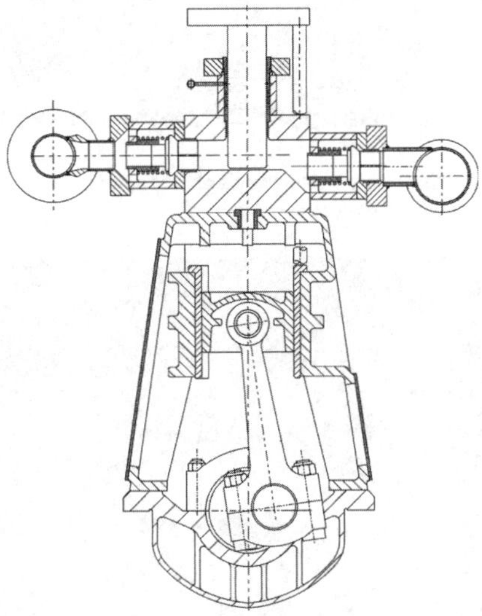

Figure 1.92 Vertical plunger pump

dampers can be fitted to both suction and discharge pipework to further attenuate pressure pulsations.

It must be stressed that flow variations and pressure pulsations are not just a function of the number of cylinders in the pump. Pump speed and the pipework system design have significant effects which cannot be ignored. Changes in pipework can eliminate the most serious problems. Acceleration head losses in the suction pipework are always considered when assessing NPIP available.

The horizontal pump shown in Figure 1.91 has suction valve unloaders fitted and would be driven through an external gear box or V-belt drive. Liquid ends can be built in different configurations for various applications. The monoblock liquid end shown is typical for standard pumps on non-corrosive applications. Other designs are available with separate manifolds and externally clamped valves.

Vertical pumps, Figure 1.92, are used for the most arduous applications. More space is available for special liquid end designs, bigger valves and longer stuffing boxes. The vertical pump shown is a modern pump design. The original vertical pump designs had the liquid end at ground level with the crankshaft above, see Figure 1.93.

Figure 1.94 shows a good horizontal pump installation. Notice the close proximity of the pulsation dampers to the pump connections; remote dampers are less effective. Also notice the suction pipework is larger than the pump connection.

Figure 1.95 shows a vertical plunger pump under construction. At the time it was believed to be the largest reciprocating pump operating in the North Sea offshore industry.

Plunger pumps can be fitted with a variety of valves to suit the application, see Figure 1.96.

- **Plate valves** General purpose valves for clean liquids with excellent dynamics, can have low NPIPr, suitable for speeds over 300 rpm. Mass produced by proprietary manu-

Figure 1.93 Low pressure vertical sewage plunger pump
Courtesy of Dawson Downie Lamont Ltd

Figure 1.94 Horizontal pump installation

facturers as well as individual pump manufacturers. Generally limited to 400 barg. Standard horizontal pumps, with monoblock liquid ends usually have plate valves.

- **Plug valves** Heavy duty valves for clean liquids, medium NPIPr characteristics, usually operated below 350 rpm. Specially made by pump manufacturers for applications up to 550 barg.

- **Wing-guided valves** Can be designed as light or heavy duty, has good dynamics, low NPIPr and suitable for all speeds. Available as a proprietary design and also as specials by pump manufacturers. Suitable for pressures up to 550 barg, some designs can have an elastomeric seal for solids handling, 5% by volume approximately up to 500 mm.

- **Plug-guided ball valves** Suitable for clean liquids at the highest pressures. Dynamics are poor and should operate below 300 rpm.

- **Mission valves** Proprietary valves manufactured by TRW Mission specifically for oil well mud pumps. Poor dynamics, high NPIPr and should operate below 200 rpm. Can handle solids, up to 1 mm in the larger valves, at concentrations up to 65% by weight. Designed for low abrasion solid mixtures, less than Miller Number 50.

- **Ball valves for solids** Special valves made to order. Almost the ideal solids handling valve. Suitable for the most abrasive solids at concentrations of 65% by weight. Solids up to 6 mm in the larger valves. Dynamics can be poor, can be very low, largest valves suitable for speeds below 150 rpm. Long parts life, but requires lots of space and is costly. Also very good for high viscosity liquids.

The choice of material for the valve and the valve seat is critical. The valves are opened by a liquid differential pressure and held open by drag forces. Closure is spring-assisted. Allowances must be made in the valve design for shock loading due to rapid closure at higher speeds and pressure pulsation loading.

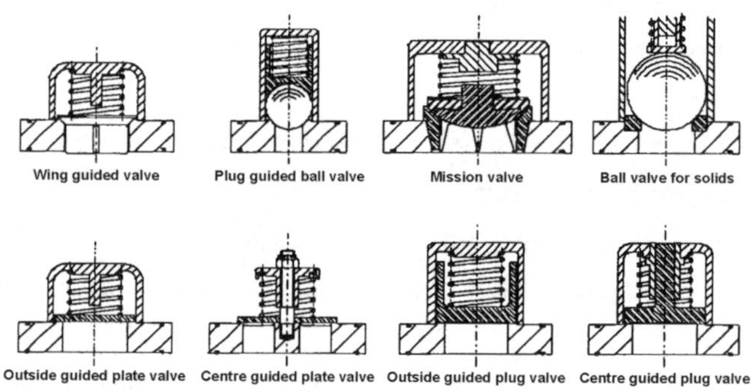

Figure 1.96 Valves for plunger pumps

Figure 1.95 A vertical plunger pump under construction *Courtesy of Flowserve Corporation*

Sometimes valves can be actuated hydraulically for special applications.

Modern pumps are designed to operate at mean plunger speeds of 1.5 to 2 m/s. Pumps may operate a lower speeds to achieve better NPIPr or to suit operating conditions. The pressure developed by any plunger is related to the plunger load; the force applied by the crankshaft/connecting rod/crosshead to the plunger. Current designs include plunger loads over 400kN.

The liquid is prevented from escaping from the liquid end by the stuffing box. Various seal designs, from simple to complex, can be fitted to suit the liquid and the environment. Pumps handling abrasive solids in suspension can have the front of the stuffing box flushed with clean liquid to protect the packing. Pumps with very low suction pressures can have a quench across the back of the packing to prevent air being induced. A quench can also be used to prevent hot liquid vapourising. Liquid can be circulated between seals to cool, lubricate and/or remove vapours. Single shot lubrication, where the lube oil mixes with the product and is lost, is the most popular stuffing box system.

Pumps are selected by adjusting all the variables; speed, stroke, plunger diameter; to suit the application. Knowledge of the liquid compressibility is essential for proper selection when the differential pressure exceeds 40 bar.

The plunger diameter is calculated from the plunger load. This diameter is rounded down to a standard seal diameter. The volumetric efficiency is calculated based on the pump proportions, the differential pressure and the liquid compressibility. The flow per revolution is then calculated based on the plunger diameter, stroke, volumetric efficiency and number of plungers. The pump speed can then be calculated for the rated flow. If the pump speed is too fast, more cylinders can be considered or a larger pump used with a higher plunger load and consequently larger plunger diameter. For pumps to run at full speed adequate NPIP must be available.

Plunger pumps can operate with high suction pressure; over 135 bar in some applications. High suction pressure is defined as a suction pressure which produces a plunger load higher than a prescribed limit. Some pump designs use 5% as the benchmark, others 20%. High suction pressure must be considered in the selection process because it affects volumetric and mechanical efficiency. Mechanical efficiency is reduce because of increased friction losses during the suction sroke.

Plunger pumps should not be used in self-priming applications unless the manufacturer has fully approved the operating conditions. Pumping air or vapour corresponds to 100% cavitation and will do no damage. The damage will occur during the transition from 100% cavitation to zero cavitation. Very high internal pressures may be created, far higher than the normal discharge pressure. This will lead to increased wear and component failure due to fatigue.

The most common failures are:

- High bearing wear and failure of bearing shells or ball/rollers
- Crankshaft failure
- Crosshead failure
- Very short packing life
- Liquid end stud/bolt fracture
- Suction valve failure, rapid seat wear
- Cylinder cracking

Plunger pumps are capable of working at the highest pressures, with the most difficult liquids and with abrasive solids if necessary. Pressures to 500 bar are common in process applications. Discharge pressures of 1000 bar are common in descaling and high pressure water jetting. Pumps for 3500 bar are available to order. Flows up to 500 m³/h are possible at pressures up to 100 bar.

Figure 1.97 A cam-operated vertical plunger pump
Courtesy of National Oilwell Varco

In addition, plunger pumps are used for low pressure, difficult applications where good suction capabilities are required and low speed is essential to reduce wear. Tank bottom residues, sediment, industrial and municipal wastes and sewage are typical mixtures handled by the original vertical pump design, shown in Figure 1.92.

The crankcase assemblies are standard using mostly cast iron. Liquid ends, stuffing boxes and valves can be made in almost any material. Standard materials include cast iron, bronze, steel, stainless steel, duplex stainless steel. High pressure pumps require materials with high fatigue endurance stress levels, such as 1% CrMo, 15-5PH and 22Cr 13Ni 5Mo.

Figure 1.97 shows one of the latest developments in plunger pump technology. Most plunger pumps are 'slider-crank', this is cam-operated. Even more unusually is that this is an axial pump. The pump shown is effectively a much larger version of the axial piston pumps used for hydraulic fluid power. The cutaway view in the Figure shows the construction. The pump is a 'vertical plunger' almost in the style of the original crankshaft pumps. Cam-driven pumps have been tried many times; not always with much success.

Currently this is the largest cam pump in production. Having six cylinders is very unusual and defies the simple (useless) theory used by most engineers. The stuffing boxes and valve assemblies are built as 'cartridges' for easy maintenance. The two drive motors, are integrated in the pump and drive through a single reduction gear. The pump is capable of about 1800 hydraulic kilowatts. Of course, there are flow and pressure limitations, but a useful pump for clean liquid applications.

Plunger pumps applications are diverse. The following list highlights some of the important areas:

- Boiler feed pumps
- Carbon dioxide injection for crude oil recover
- Chemical processing
- Crude oil/water emulsion recovery
- Domestic high pressure cleaners
- Ethylene glycol injection at wellheads
- Garage car washes
- Gas drying
- Gas sweetening

- High pressure water jetting
- Industrial waste
- LNG re-injection
- Methanol injection at wellheads
- Municipal waste
- Oil/water emulsion hydraulics
- Reverse osmosis
- Sewage
- Soap powder manufacture
- Tank residue disposal
- Viscous crude oil pipelines
- Water injection for crude oil recovery

1.5.16 Syringe pumps

Syringe pumps are different to most of the other pump types considered. Syringe pumps cannot operate continuously, only intermittently. Syringe pumps can only be used for batch processes. One of the biggest problems encountered with pump applications is ensuring an adequate supply of liquid to the pump. This problem increases as the liquid handled becomes more viscous. Syringe pumps overcome this problem by removing it from the normal pump operation completely; syringes must be filled before the pumping operation commences. As filling is divorced from discharging filling can take as long as necessary.

Syringe pumps consist of one to ten syringes and a mechanical/electric drive unit, see Figure 1.98. Normal glass and plastic medical syringes can be used, up to 21 bar, and stainless steel syringes can be used for high pressure applications, 100 bar.

1.5.17 Diaphragm pumps (includes mechanical and hydraulic actuation)

Plunger pumps and piston pumps have stuffing boxes which are potential areas of leakage. Seals, such as mechanical seals, are not yet available for reciprocating pumps. The problem of normal leakage can be completely eliminated by using a diaphragm pump, see Figure 1.99. Reciprocating motion is produced in the conventional manner (the Figure shows an eccentric shaft rather than a crankshaft). The motion from the crosshead is transmitted directly to a diaphragm.

The pump shown has ball valves. Liquid would enter through the bottom valve and be forced out of the top valve; the liquid end would be self-venting. Pumps of this horizontal style are

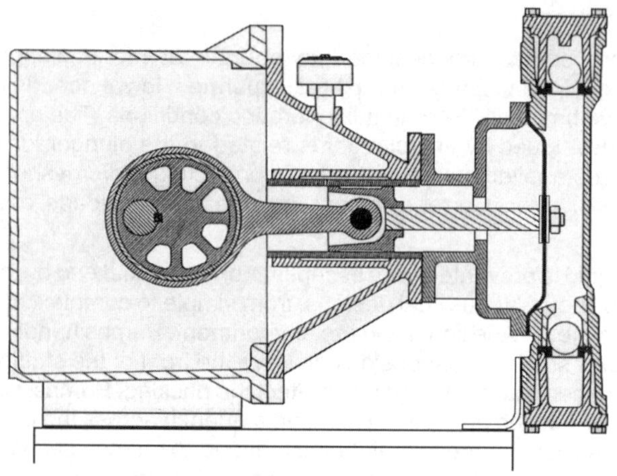

Figure 1.99 Mechanical diaphragm pump

built with one, two or three cylinders although there is no practical limit on the number of possible cylinders. Horizontal pumps are also built with a Scotch Yoke mechanism to produce the reciprocating motion. No crosshead is required if two cylinders are mounted back to back, the diaphragm motion rods can support the yoke. Scotch Yoke mechanisms tend to have more friction than crankshafts or eccentric shafts. Mechanical diaphragm pumps are also constructed as vertical pumps; these are nearly always single cylinder pumps but twin cylinder pumps are built. Single cylinder vertical style pumps are also called Single Disc pumps. Most versions of the vertical pump do not have a crosshead; the connecting rod is coupled directly to the diaphragm. This style of construction imposes extra loads on the diaphragm.

Some versions of the vertical pump have a spring-loaded coupling incorporated in the connecting rod, see Figure 1.100 which shows a portable engine driven pump. The spring loading allows the diaphragm to short stroke if large solids restrict the diaphragm motion. The vertical style is popular as a site pump and can be engine driven. Pumps can be fitted with flap, ball or duck-billed valves as appropriate for the operating conditions.

The mechanical diaphragm pump is usually a low pressure pump, discharge pressures up to 6 bar are possible. Site pumps may be only capable of 1.5 bar. As a low pressure reciprocating pump it has some useful advantages over both piston and plunger pumps. Pumps are dry self-priming and when the

Figure 1.98 Syringe pump system
Courtesy of K D Scientific Inc

Figure 1.100 Portable engine driven pump
Courtesy of Gilbert Gilkes & Gordon Ltd

valves are maintained in good condition suction lift capabilities of 6 m are achievable. Continuous dry running is acceptable.

The mechanical diaphragm pump can snore. (Snoring describes a pumps ability to run dry when the liquid supply disappears and then to self-prime and pump when the liquid supply returns.) Depending upon the type of valves fitted the pump can handle large solids and high viscosity liquids.

Mechanical diaphragm pumps can be constructed from a variety of materials, to suit many applications, see Table 1.3. Metal diaphragms can be used for high temperature applications.

Liquid ends	Diaphragms	Valves
Aluminium	Buna 'N'	Buna 'N'
Cast iron	Neoprene	Neoprene
SG iron	Viton	Polyurethane
Bronze	EPDM	PTFE
Stainless steel	PTFE	Borosilicate glass
Polypropylene		
PVDF		
PTFE		

Table 1.3 Popular mechanical diaphragm pump materials

Mechanical diaphragm pumps can handle flows up to about 80 m^3/h.

Flow variations and associated pressure pulsations may be a problem with single and twin cylinder pumps. Some pumps can be supplied with integral discharge air chambers. The manufacturers advice should be sort if cyclic flow variations could create operating problems. A section through a diaphragm pump with a pneumatic unloader valve option is shown in Figure 1.101.

A special derivative of the mechanical diaphragm pump has been developed specifically for viscous liquids and stringy solids handling. The Double Disc pump utilises two rubber diaphragms which act as pumping element and valves. The first diaphragm opens and allows product into the pump. On closing the diaphragm forces the product into the second chamber where the diaphragm is open. When the second diaphragm closes the product is pumped into the discharge system. These pumps are only capable of generating 3 barg but have found favour in the sewage and waste water treatment industries. Flow rates up to $25m^3/h$ are possible.

1.5.18 Air-operated double-diaphragm pumps

Air-driven diaphragm pumps are a style of direct-acting pump which utilise compressed air to deflect the diaphragm. This type of actuation imposes much lower stresses on the diaphragm

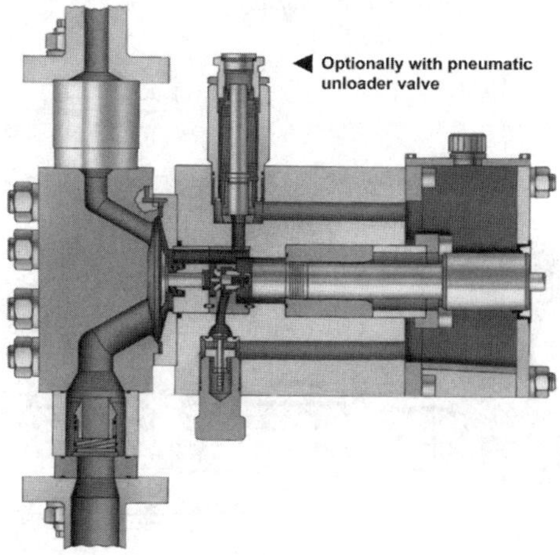

Figure 1.101 Diaphragm pump
Courtesy of Uraca Pumpenfabrik GmbH & Co KG

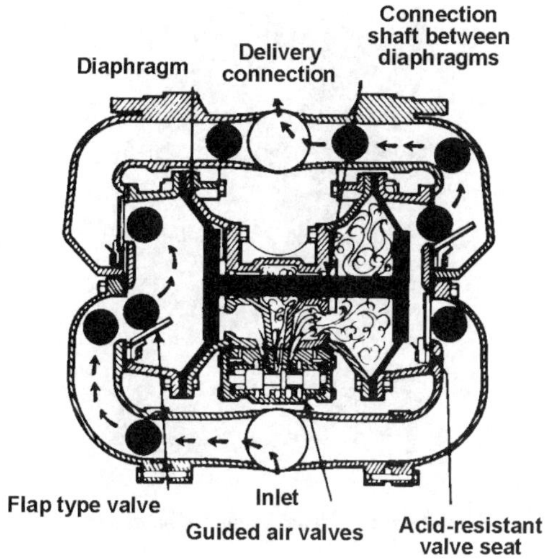

Figure 1.102 Compressed air-driven diaphragm pump

because of the high degree of pressure balancing. The basic pump construction is very similar to the pumps in Section 1.5.17, suction valves at the bottom and discharge at the top, but two diaphragms are mounted back-to-back, see Figure 1.102.

Compressed air is applied to the inside face of the diaphragms, alternatively, to create the stroke cycling. Three styles of valves are popular; flap, plate or cone, or ball. The choice of valve is dictated by the type of product to be handled and the size of solids. Flap valves can handle the largest solids.

Diaphragm pumps are leak-free since there are no plungers or piston rods which pass through the process pressure boundary. Leakage can occur however if the diaphragm fails. It is therefore necessary to replace the diaphragms regularly, as part of a preventative maintenance programme. Some models are available with a double-diaphragm incorporating a rupture detection system. The material used for the diaphragm is usually an elastomer such as nitrile rubber, Neoprene or fluorocarbon. Pumps suitable for hygienic applications are available. Figure 1.103 shows a non-metallic pump. This particular design is manufactured in polypropylene, Acetal or PVDF and suitable for pressures up to 8.3 barg.

The air-driven diaphragm pump is not a fixed speed pump like an electric motor driven pump. Because the power supply is compressed air the pump is not fixed to an electricity supply frequency. Diaphragm pump speed is controlled by the differential pressure between the compressed air and the liquid. Increasing the air pressure increases pump speed. Also, reducing the liquid pressure increases the pump speed. Compressed air flow control, as well as pressure control, may be required. Using compressed air in place of electricity means the pump is inherently safe and can be installed in any hazardous area. Also the pump can be submerged without creating a hazard.

Air-driven diaphragm pumps have good self-priming and suction lift capabilities. Suction lift can be up to 8m depending upon the pump size, speed and type and condition of the valves. Flap valves can allow the passage of solids which are 90% of the suction connection size. Industrial pumps are manufactured in a wide range of metallic and non-metallic materials; aluminium, cast iron, stainless steel, polypropylene, PVC, PVDF, PTFE. Industrial pumps can handle 60 m^3/h at pressures up to 7 barg when powered by factory air at 8.3 barg. Hygienic quality pumps can handle flows up to 28 m^3/h. Air-driven diaphragm pumps always produce a pulsating flow because the pump design is twin cylinder. Problems caused by pulsating flow can be alleviated by fitting suction and discharge pulsation dampers.

Figure 1.103 A non-metallic air-operated double-diaphragm pump

1.5.19 Metering pumps (flow within ± 0.1% to ± 3% with substantial dp changes)

Metering pumps are also called dosing pumps and proportioning pumps. Metering pumps are used to add small quantities of liquid to other liquid streams or vessels. Table 1.4 shows examples of processes which use metering pumps.

Pressure bar	Flow L/h	Temp °C	Liquid	Process	Problem
5	300	room	hydrochloric acid	water preparation	corrosion
5	1	room	mercury	chemical	high density
5	500	500	lead	metallurgical	high temperature
10	10	120	sodium	chemical	explosive
10	100	room	plutonium salt	nuclear fuel recovery	radioactive
10	20000	room	sugar solution	soft drink manufacture	hygienic
100	200	room	tungsten carbide suspensions	spray drying	abrasion
200	1500	-30	chlorine	chlorination	toxic
300	500	-200	oxygen	chemical	low temperature explosion cleanliness
500	1000	room	vinyl acetate		toxic high pressure
3000	30	room	catalyst	polythene manufacture	high pressure

Table 1.4 Processes in which metering pumps are used

Other typical applications include the injection of biocide into water and seawater pipelines to prevent the growth of organisms. Injection of methanol or ethylene glycol into wellheads to prevent the formation of hydrates. The injection of chemicals into seawater prior to reverse osmosis treatment.

The basic requirements of a metering pump are to deliver measured volumes of liquids accurately without significant changes in volume as discharge pressure varies. Typical accuracies would be ±0.25% to ± 3% of the rated flow. Positive displacement pumps are ideally qualified and specific variants of reciprocating pumps have evolved to fulfil the functions. Metering pumps are usually small plunger pumps, diaphragm pumps or peristaltic pumps for low pressure applications.

Plunger pumps used as metering pumps are modified versions of the pumps described in Section 1.5.15. The main modification involves providing a variable stroke mechanism so that the flow can be adjusted while the pump runs at constant speed. In theory, pumps can operate from zero to maximum flow. In practice, accuracy can only be maintained from 10% to 100% flow. Figure 1.104 shows a plunger pump type metering pump fitted with ball valves on suction and discharge.

Raising or lowering the Z-shaped crankshaft changes the length of the stroke. Metering pumps are single cylinder pumps. However, pumps can be ganged together on one drive shaft to form multiple units. This type of construction is useful when more than one liquid is metered and when several metering rates must be adjusted, in the same proportions, simultaneously. Stroke lengths and plunger diameters of each cylinder can be different. The stroke of each cylinder is adjusted individually to deliver the correct volume; the speed of all cylinders is varied simultaneously for collective flow adjustment.

Diaphragm metering pumps are of two different designs. Small pumps can have the diaphragm driven directly by an electric solenoid. Stroke length is adjusted mechanically and stroke speed electrically. Larger diaphragm metering pumps have the diaphragm actuated hydraulically, the hydraulic power being provided by a plunger pump. Figure 1.105 illustrates the modifications necessary to add a diaphragm.

The movement of the diaphragm is limited by supporting walls with slots on each side of the diaphragm or by profiled cylinder walls. Hydraulically actuated diaphragm pumps can become complicated. Provision must be made to maintain the oil volume behind the diaphragm. Leakage through the stuffing box must be replenished to preserve accuracy. Also overpressure of the hydraulic oil may present problems.

Ball valves are the most popular choice of suction and discharge valve for metering pumps. Ball valves have the longest possible life while preserving good sealing capabilities. Metering pumps always run slowly and ball dynamics are not a problem. Some designs incorporate soft seals within the valve seat to improve accuracy. Pumps are often fitted with double spring-loaded ball valves in both the suction and discharge in order to reduce the risk of backflow through the valve after each respective suction and delivery stroke.

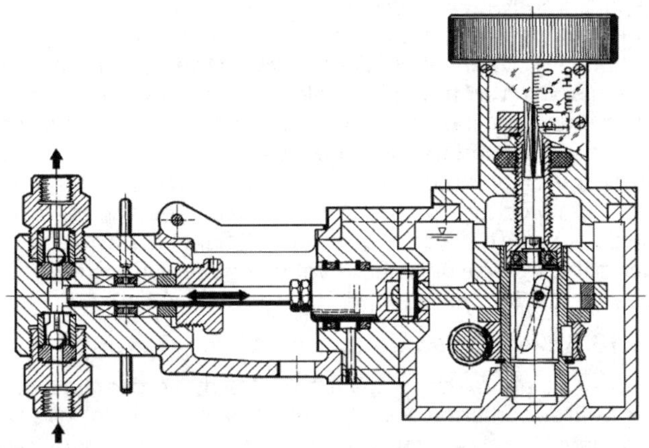

Figure 1.104 Plunger type metering pump

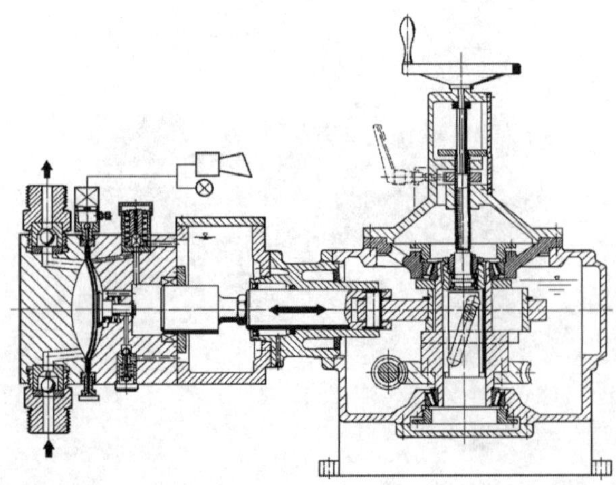

Figure 1.105 Hydraulic diaphragm type metering pump

Plunger style pumps are used for all pressure ranges when the liquid is not toxic and there is no risk of corrosion of the pump external components. Diaphragm pumps are used when any leakage from the pump is unacceptable. Because metering pumps run slowly there are no real problems with viscous liquids. The ability to handle solids in suspension is dependent upon the pump size and should be confirmed with the manufacturer.

Metering pumps can cause pressure pulsations. Pulsation dampers may be required on the suction to preserve the NPIPa. Dampers may be required on the discharge to eliminate pipework vibration and smooth out the dosing rate.

Metering pumps are available in a wide range of materials including complete construction in PTFE. Flow rates can be as low as 0.1 l/h and over 10 m^3/h per head.

1.5.20 Direct-acting reciprocating pumps (includes pneumatic, hydraulic and steam actuation)

Direct-acting, d-a pumps are one of the oldest styles of positive displacement pumps. Direct-acting pumps are fluid powered reciprocating pumps which have no rotary mechanisms; a reciprocating motor provides the power to drive a reciprocating pump. The air-driven diaphragm pump, in Section 1.5.18, is technically a direct-acting pump but all the pumps within Section 1.5.20 have piston motors. Pumps within this category can be divided into three subgroups which are listed in order of popularity:

- Air-driven pumps
- Hydraulically-driven pumps
- Steam-driven pumps

The steam-driven pump is the oldest style of d-a pump and became popular as a boiler feed pump around 1841 when Henry Rossiter Worthington of New York patented the direct-acting simplex steam piston pump. The duplex pump followed in 1859. The motion of d-a pumps differs from slider-crank pumps because it does not tend to be sinusoidal. The motion of d-a pumps tends to be trapezoidal. The acceleration and deceleration of the pumping element(s) can be controlled independently of the (fairly) constant velocity which is maintained for most of the stroke. The effects of acceleration and deceleration can be alleviated by multiplexing or by pulsation dampers.

The air-driven pump is currently the most popular style of d-a pump because of its primary area of application which is pressure testing components and assemblies. All the d-a pumps have the ability to stall and maintain discharge pressure. This means the pumps can produce zero flow but maintain a predetermined pressure. A double-acting air cylinder drives one or two single-acting plunger pumps.

Figure 1.106 shows an air-driven pump with two plunger pumps, one at either end of the double-acting air cylinder and has the air filter/regulator fitted. This is an old-fashioned design; about 35 years old; but is very useful to show the basic construction of the pump. Modern pumps are unitised; it can be difficult to distinguish the air and liquid ends. When air is applied to one side of the piston the liquid plunger is withdrawn from the cylinder; suction stroke. At the end of the stroke the air valve diverts the air supply to the other side of the piston and exhausts the pressurised side. The piston moves in the other direction and the liquid plunger forces liquid into the discharge system. The pump continues to operate until the air and liquid forces are balanced. The air-hydro produces a discharge pressure which is a function of the air supply pressure and the pump ratio. The pump ratio is the ratio of the air piston area to the liquid piston/plunger area. Most pumps are designed to work with factory air at 7 barg. Air-hydros can produce pressures from 2.7 to 6900 barg.

Figure 1.106 Typical high pressure air-driven pump

Some modern designs incorporate a two-stage effect. The pump can operate with either a large or small diameter plunger. Under low liquid pressure conditions the large diameter plunger is used for fast filling. At a predetermined pressure the large diameter plunger stalls and the small diameter plunger continues pressurisation. The d-a air-driven pump is capable of water flows of 100 l/min at low pressure and 0.35l/min at 2000 barg. There are many pump sizes available and flow is infinitely adjustable from zero to maximum. The high pressure d-a air-driven pump can be used for other applications such as clamping or water cutting. Special versions are available to work with refrigerants.

NOTE: Investigate the availability of high pressure pipework and fittings before spending a lot of time designing a system for over 420 barg. Pulsation dampers, or the size of pulsation dampers, may be an additional problem.

Air-driven pumps are also popular for barrel and container emptying applications. It is possible to construct a long, slender piston pump with a cylinder outer diameter small enough to pass through the standard connection on drums and intermediate bulk containers (IBCs). Pumps are manufactured in standard diameters and lengths as shown in Table 1.5.

Pump diameter mm	Standard lengths mm	Flow range l/min
27	490	1
34	210, 460, 1000, 1090	2, 7.5
42	230, 1020	4, 6.7
51	225, 1170	6, 10.5
55	278, 335, 1000, 1015	7.4, 12.5
76	330, 1110	14, 24
84	327, 1100	14, 24
96	396, 1130	20, 33
120	465, 900	31, 52

Table 1.5 Standard sizes of container emptying pumps

Pumps normally have ball valves so viscous liquids can be handled, up to 10000 Poise. Spring-loaded chevron packing is an option to reduce routine maintenance. Pistons have PTFE rings. Most pump parts are 304 or 316 stainless steel. Pumps have ratios between 1 and 33 to allow pressures up to 150 barg to be developed. Pressure pulsations may be a problem when pumps operate at high speed and high pressure. Some pump manufacturers produce their own pulsation dampers.

Hydraulically-driven pumps are used for two vastly different types of application; high viscosity/large solid mixtures and high pressure liquids. Hydraulically-driven reciprocating pumps are ideal for very arduous applications because the pumps can operate at very low speed to reduce wear and pressure pulsation problems.

One of the most popular applications for hydraulically-driven pumps is handling cement on building sites. For viscous liquids with small solids a standard double-acting horizontal piston liquid end can be used driven by a double-acting hydraulic cylinder. Hydraulic power is provided by a power pack. An axial piston swashplate pump is normally used to provide a variable hydraulic flow from a constant speed pump. Control is via a small programmable logic controller (PLC). Process liquid valves can be flexible elastomer for low pressure pumps, plate, tapered plug or ball. Cylinder liners and piston rods are normally protected by a substantial layer of chrome plate. Other coatings can be applied for improved wear/abrasion resistance. A single-acting piston liquid end can be used for extremely viscous liquids. A plunger pump liquid end, with actuated valves, can be used for the most abrasive products or very large solids. This style of horizontal pump is very versatile because the stroke length is easily adjustable to produce larger or smaller pumps.

Hydraulically-driven pumps are produced in low pressure, 2barg, and higher pressures, up to 160 barg, versions. There are no practical limits for the design pressures of special pumps. Standard pumps can maintain flows of 480 m^3/h. NPIPr can be as low as 0.1 bar. In applications where the process medium does not flow easily a screw conveyor can be attached to the pump suction to provide a positive supply. Special provision can be made for self-priming and dry running. A vertical plunger pump version is very popular in the sewage treatment industry; these pumps are built as simplexes and duplexes. Standard pumps can handle 216 m^3/h at pressures up to 20 barg.

A special version of the hydraulically-driven pump uses the process liquid as the motive power to produce a higher pressure. This type of pump is usually called an intensifier. Intensifiers are used to produce the highest possible pressures, 10300 barg. Intensifiers are used in water cutting applications and in oil well servicing applications.

Many pump users think the steam pump is obsolete and has been replaced by modern technology; this is not completely correct. New steam pumps are still installed in refineries and on board ships. Steam pumps are very good with high viscosity liquids and for tank emptying duties. Steam pumps, like other d-a pumps, can operate extremely slowly and can have very low NPIPr. Also, at low speed the d-a pump can handle liquid/gas or liquid/vapour mixtures if the differential pressure is low. Steam pumps are ideal for tank stripping. Large tanks would be emptied quickly by a centrifugal pump. As the tank level reduced the NPSHa for the pump would reduce. At some point the centrifugal pump would cavitate and have to stop. A steam pump is then used to strip the dregs out of the tank. Steam pumps are usually double-acting piston pumps and are built as simplexes or duplexes; horizontal and vertical styles are available. Horizontal pumps are preferred for land installations and some marine users prefer vertical pumps.

Figure 1.107 shows horizontal duplex piston pump with discharge pulsation damper. The suction connection, at the bottom, and the valve access covers are easily seen. Pumps are capable of flows up to 545 m^3/h and rated pressures are from 10 to 55 barg. Steam conditions do not normally exceed 17barg and 380 °C. Figure 1.108 shows a special quadruplex plunger steam pump for a high pressure application. Pumps of this style are capable of 500 barg.

1.5.21 Non-metallic positive displacement pumps

Positive displacement pumps are often described as 'low-flow, high-pressure' pumps; this description completely ignores the pumps which are designed for 2.5 barg discharge pressure. In simplistic terms, high-pressure pumps cannot be non-metallic because non-metallic materials are not strong enough to support the stresses generated. However, non-metallic materials do possess exceptional corrosion resistance. Providing the strength limitations are strictly observed, low strength materials can make very useful pumps. The low strength of materials can be partially compensated by external reinforcing. This technique is used on centrifugal pumps, see Figures 1.52 and 7.9 in Chapter 7. External reinforcing substantially increases the allowable forces and moments which can be applied by piping. Metallic pumps can have wetted surfaces coated with non-metallic material. These pumps should be considered as a special case of metallic pumps.

Figure 1.107 A typical horizontal duplex steam pump
Courtesy of Dawson Downie Lamont Ltd

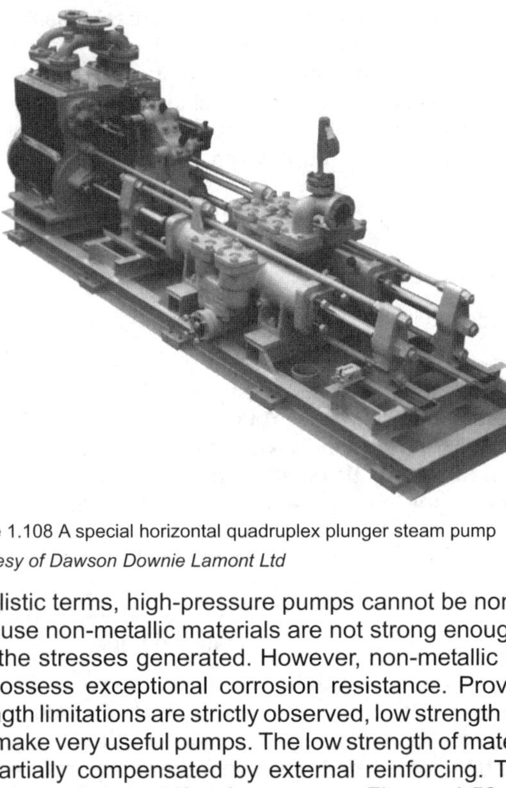

Figure 1.108 A special horizontal quadruplex plunger steam pump
Courtesy of Dawson Downie Lamont Ltd

The most popular non-metallic positive displacement pump is the air-operated double-diaphragm pump; this is one of the best selling pump types of all categories!! Typical casing materials include Acetal, nylon, polypropylene (PP), polyvinylidene (PVDF) and polytetrafluoroethylene (PTFE). Diaphragms are slightly more complicated. Because of the flexure, fatigue due to stressing becomes a problem. Popular materials are chloroprene, neoprene, nitrile rubber, polyurethane, ethylene propylene diene (EPDM), Santoprene® and fluorocarbon alloys. The diaphragm can be reinforced with fibres, such as nylon mesh, to increase strength. Diaphragms can be coated, or moulded in layers, so that a good flexible material could have its corrosion resistance improved. Neoprene can be coated with a Viton ®/PTFE alloy or nitrile rubber faced with PTFE.

Direct-acting pumps, driven by compressed air, are fairly easy to manufacture without metal components. Pressure ratings would be much reduced compared to their metallic relatives.

Some rotary pd pumps are easy to manufacture from 'plastics' which can be extruded. Pd pumps don't necessarily require complicated castings. Figure 1.63 shows an external gear pump with a casing machined from a solid block; this manufacturing route can be applied to plastic bar of the appropriate size. Other pump types which are amenable to this treatment include; internal gear pumps, triple-rotor and 5-rotor screw pumps, lobe pumps, circumferential piston pumps, vane pumps, piston pumps, plunger pumps, syringe pumps and metering pumps. The peristaltic pump is always a non-metallic pump. The only component in contact with the liquid is an elastomer tube. The pump structure is usually metallic, but not normally 'in-contact'.

When considering a non-metallic pump for a specific application, it is important think about the piping at the same time. Flexible pipes, hoses, do not transmit significant forces and moments to the pump connections. But, flexible pipes do try to straighten out when pressurised. Also they can 'kick' a lot when the liquid flow pulsates. Non-metallic pulsation dampers are available to alleviate these problems.

1.5.22 Sealless positive displacement pumps

The term 'sealless' is being used irrationally by some pump manufacturers. The original intent of 'sealless' was to describe a pump which did not have a dynamic leak path requiring a dynamic seal of some description. Sealless meant a static seal; a stationary boundary. Most pump developments are driven by centrifugal pump requirements and new ideas are tried on them first. There are two solutions to the 'sealless' problem, canned drive motors and magnetic drives. Canned motors cannot be used on all applications because the process liquid may damage the motor. Magnetic drives can be used on a wide range of applications.

Manufacturers of peristaltic pumps are claiming them to be 'sealless'. This claim is true to the extent that the pump does not have a rotating shaft seal or a reciprocating rod seal. But! the pump does have a dynamic seal, the hose! The hose may fail due to wear or fatigue. Static seals don't usually wear and do not suffer from fatigue problems. The peristaltic pump is a very useful pump but it isn't sealless! (One manufacturer claims his peristaltic pumps are contactless! Be very wary of pump manufacturers' literature!)

Reciprocating pumps cannot be sealless. A diaphragm pump can be sealless to the same extent as the peristaltic pump; the seal is dynamic.

Rotary positive displacement pumps can be sealless in the same ways as rotodynamic pumps. Canned motor pumps are very rare. Small sliding vane pumps powered by 12 V DC are used as petrol pumps on some vehicles; this is the only canned motor pd pump seen to date. Some rotary pd pumps are available as magnetic drive versions: internal gear pumps (see Figure 1.109), external gear pumps, triple screw pumps and progressive cavity pumps. Pump materials are not affected. Just one last thought ..... the jet pump is sealless. Of course it is a mixer as well. So why not multi-task and save time and money? It can be a heater as well ..... if you use steam !

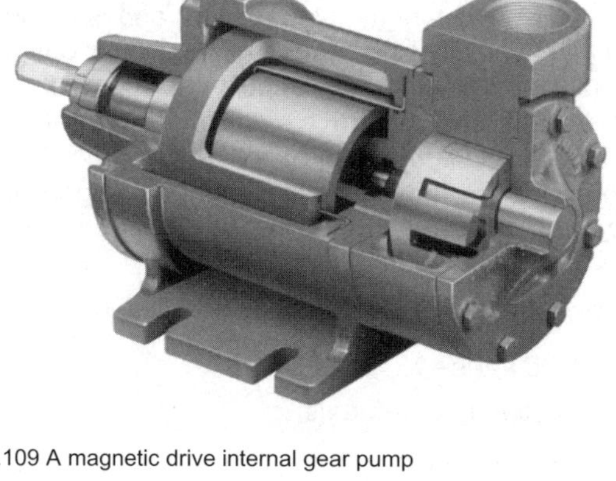

Figure 1.109 A magnetic drive internal gear pump

Figure 1.109 raises a very interesting question. Why would an exotic shaft seal be necessary if threaded process connections were adequate?

1.5.23 Hydraulic motors

Most positive displacement pumps can run in reverse and act as a motor. Reciprocating pumps with valves; diaphragm, piston, plunger; can't run backwards unless there is a major leakage problem with the valves. Both suction and discharge valves must be worn to allow this to happen. The pump may not rotate; this depends upon the speed ratio to the driver. Reciprocating pumps don't usually require a non-return valve in the discharge piping because the internal valves prevent reverse flow. Rotary pumps can usually run in both directions; not necessarily with the same effectiveness. Whether a pump can operate as a hydraulic motor, intentionally or unintentionally, is dependant upon the speed ratio to the driver and the nature of the power transmission and driver. Many rotary positive displacement pumps are installed without non-return valves and the possibility of 'turbining' exists. The level of static pressure in the discharge system is a critical factor.

Hydraulic motors are used extensively in hydraulic fluid power applications. The most popular styles are vane, gear and piston. This book is concerned with the process applications of pumps and hydraulic fluid power is mentioned only in passing. Process applications for hydraulic motors are relatively limited but this is likely to change as the move towards greater operating efficiency, to reduce energy consumption and energy costs, concentrates attention on process detail design. Hydraulic motors are used in applications where the inclusion produces obvious operational advantages.

One area of pump use has adopted hydraulic motors to improve operations, well drilling, most notably oil well drilling. Drilling rigs, for oil wells, use 'mud' to lubricate the drill bit and to remove cuttings/debris from the hole bottom. The mud is pressurised by pumps on the rig. The potential energy of the mud can be used to drive a motor. Progressing cavity pumps have been adapted to provide rotary power for the drill bit and other tools, reamers, rereamers and underreamers. These motors are usually called 'mud motors'.

Mud motors driving drill bits are used for three basic reasons: speed, reach and accuracy. Typically, a mud motor drill head will penetrate up to four times faster than a drill-string assembly. Tests have been conducted with high pressure mud, 690 barg, so that the rotary drill action can be augmented by high pressure jets. Tests showed a speed increase up to eight times for drilling sandstone and limestone. Tests were very short-term due wear problems experienced by the high pressure nozzles. Remember, drilling mud is abrasive; it's not like clean water. Faster drilling speeds reduce production costs and allow a rig to drill more wells. Mud motors have been used to drill much longer angled and horizontal wells. This allows more reservoir to be coupled to a central base; cheaper installation. Mud motor assemblies have been controlled to within ± 1.5 metres; this might have been at the end of a 8000 metre well. Instrument packages can analyse the mud, close to the drill bit, and confirm the pressures and formation composition.

Mud motors must be designed to fit inside standard well casing bores. Typical sizes range from $1\frac{7}{16}$" (36.5 mm) to $11\frac{1}{4}$" (286 mm). Motor assemblies range from about 2.2 m long to 7 m. Stator designs follow the pump pattern, normal variable thickness elastomer and constant thickness elastomer. Power transmission from the rotor is accomplished by flexi-shaft, sometimes in titanium, or a shaft with two sealed flexible coupling. Bearings, radial and thrust, are sealed and lubricated. Tests have been conducted with diamond-faced thrust bearings to cope with higher loads and produce smaller bearings. Typically, motors run with a 15 to 35 bar differential pressure and rotate, when loaded, between 50 and 150 rpm. A small motor will require about 10 m^3/h.

1.6 Other pump types

1.6.1 Ejectors

Jet pumps or injectors operate by converting the potential energy of a motive fluid to kinetic energy of the driving jet thus reducing the static pressure and moving the pumped liquid directly without the assistance of moving mechanical components. The motive medium can be gas, air for example, steam or a liquid, e.g. water. There are different combinations of motive-pumped media. The most usual are water/steam - air, water/steam - water, air - water, water - water. Jet pumps working with gases as the pumped medium are called ejectors.

By applying the momentum equation, equation 3.7, in Chapter 3, to a control volume, which surrounds the medium in the mixing tube:

$$p_4 \cdot A - p_6 \cdot A - \pi \cdot D \cdot l \cdot \tau_0 = \left(\dot{m}_p + \dot{m}_d\right) \cdot c_6 - \dot{m}_d \cdot c_4 - \dot{m}_p \cdot c_5 \qquad \text{Equ 1.2}$$

or

$$\left(p_6 - p_4\right) \cdot A = \dot{m}_d\left(c_4 - c_6\right) - \dot{m}_p\left(c_6 - c_5\right) - \pi \cdot D \cdot l \cdot \tau_0 \qquad \text{Equ 1.3}$$

Certain characteristics of jet pumps can be learned by studying equation 1.3. In order that the pressure p_6, exiting the mixing tube, should be greater than p_4 the velocity leaving the jet nozzle c_4 must be considerably higher than c_6 and c_5. Considerable mixing losses are therefore also unavoidable. Pressure increase is greatest when p = 0. The shear stresses at the wall of the mixing tube have a tendency to reduce the pressure increase. A diffuser is mounted after the mixing tube where velocity is reduced and static pressure increases.

The relationship between the pumped mass flow p and the motive fluid mass flow p is called the flow coefficient and is designated:

$$q = \frac{\dot{m}_p}{\dot{m}_d} = \frac{\rho_p \cdot Q_p}{\rho_d \cdot Q_d} \qquad \text{Equ 1.4}$$

The pressure relationship for a jet pump can be defined further as:

$$z = \frac{\text{total pressure increase of the pumped media}}{\text{total pressure increase of the motive media}} =$$

$$= \frac{p_{03} - p_{02}}{p_{01} - p_{03}} = \frac{H_p}{H_d} \qquad \text{Equ 1.5}$$

Using these designations, jet pump efficiency can be expressed as:

$$\eta = q \cdot z \qquad \text{Equ 1.6}$$

Jet pump losses consist of flow losses in the nozzle, intake chamber, mixing tube and diffuser. The mixing losses constituting the greatest loss. Mixing losses are primarily dependent upon the area relationship:

$$a = \frac{\text{motive fluid nozzle outlet area}}{\text{mixing tube crosss-sectional area}} = \left(\frac{d}{D}\right)^2 \qquad \text{Equ 1.7}$$

For each combination of pressure relationship z and flow relationship q there is an optimum area relationship a, see Figure 1.110.

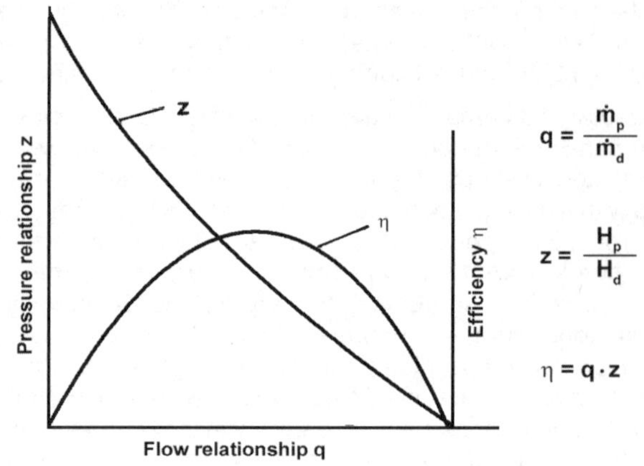

Figure 1.110 Performance curve for a jet pump

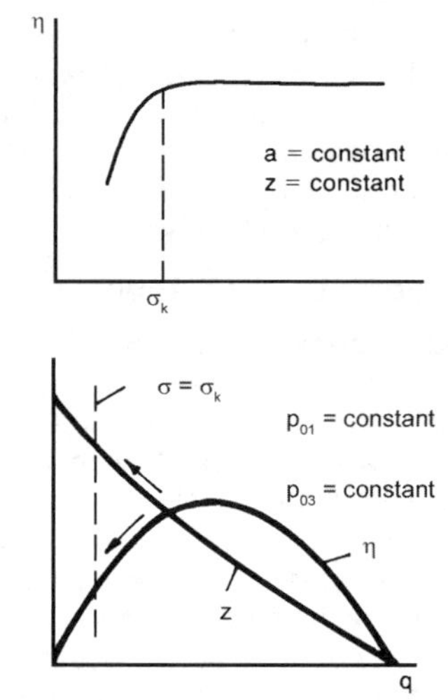

Figure 1.111 Cavitation in liquid jet pumps

Other important design parameters are the distance between the mouth of the nozzle and the start of the mixing tube; the length of the mixing tube and the angle of the diffuser. The lowest pressure in a liquid jet pump occurs in the upstream section of the mixing tube. If the lowest pressure reaches the liquid's vapour pressure then cavitation will occur. The cavitation number is defined as:

$$\sigma = \frac{p_{02} - p_v}{p_{01} - p_{02}} \qquad p_0 = \text{total pressure}, \quad p_v = \text{vapour pressure} \qquad \text{Equ 1.8}$$

If, for a given jet pump, p_{02} is reduced at the same time as p_{01} and p_{03} is adjusted so that the pressure relationship z is maintained constant, the cavitation number s will reduce without initially changing the value of q and h. Further reduction of p_{02} causes successively more cavitation in the mixing tube and a rapid reduction in efficiency. This value of cavitation number is designated σk, see Figure 1.111.

Another way of illustrating the onset of cavitation in a liquid jet pump is that at constant motive pressure and back-pressure, p_{01} and p_{03} respectively, p_{02} reduces. The pump's optimum operating point will then follow down the performance curve, z increases, q and s reduce. For $\sigma > \sigma k$ the pump operates without any effects of cavitation. At $\sigma = \sigma k$ the performance is reduced in relation to the cavitation free operation. Prolonged operation where cavitation is present can cause damage to the mixing tube and the diffuser.

Jet pumps have certain fundamental advantages:

- No moving parts
- No lubrication requirements
- No sealing problems

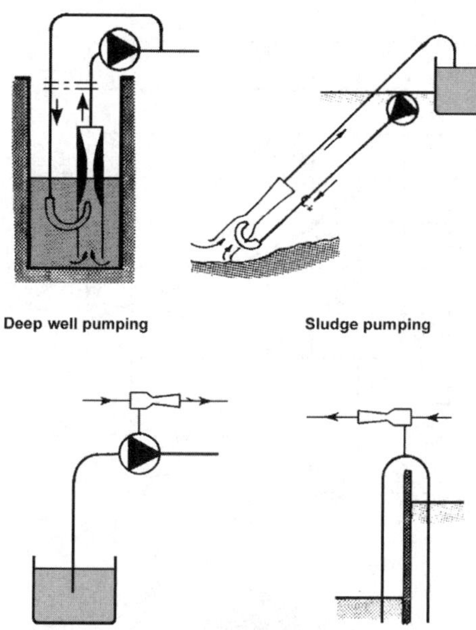

Figure 1.112 Liquid jet pumps - practical examples

- Self-priming, can evacuate the suction line
- Non-electrical, no temperature or sparking problems

The most obvious disadvantage is the low efficiency. Maximum efficiency being 25 to 30%. But with small flows the power absorbed will be small therefore low efficiency may be of no consequence.

Steam jet pumps are normally used to achieve low pressure at the inlet side. For inlet pressures of down to 103 Pa absolute, water steam is generally used as the motive medium. For even lower pressures, down to hundredths Pa, oil steam is used as the motive medium.

Practical examples include de-airing condensers, evacuating flammable gases and liquid transportation coupled with simultaneous heating requirements. Compressed air is often available and is the most usual motive medium for gas jet pumps. Some common practical examples of liquid jet pumps, usually using water as the motive medium, are illustrated in Figure 1.112.

Jet pumps are especially useful if the various combinations of motive and pumped fluid can be mixed simultaneously. Examples of this are steam jet pumps used in ventilation systems where simultaneous humidifying is required and liquid jet pumps for transporting liquids which require simultaneous dilution.

An interesting application is where the use of normal refrigerants is prohibited or a green policy is imposed. Here, water is the refrigerant and water is the motive fluid. Water temperatures down to 5 °C could be produced without difficulty. Leakage would be easily observed, without any hazards, and topping up would be simple.

Deep well pumps with ejector

Deep well pumps with ejectors complement the previously described automatic water supply packages, in Section 1.3.2, when the level of the water in a well or primarily in a borehole, lies more than 5 to 7 metres below the pump. An ejector, or jet pump is placed below the surface of the water in the borehole, which is supplied with motive water from a pump located at ground level. By means of the ejector, the water from the borehole together with the motive water is transported to the pump. It follows that it is necessary to have two hoses or pipes, as shown in Figure 1.113, between the pump and the ejector. Discharge water is taken from a separate outlet on the pump. For

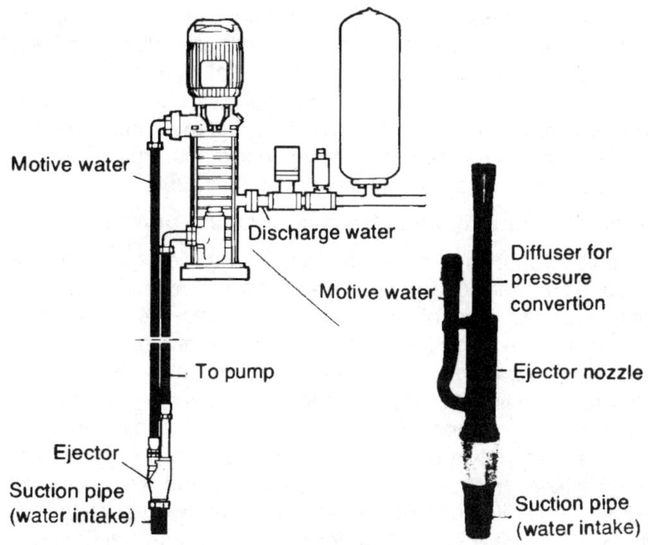

Figure 1.113 Deep well pump with ejector

multi-stage pumps this outlet is located approximately in the middle of the pump at a suitable pressure.

By using an ejector the level of the water can be more than 100 metres below the pump. Inasmuch as the water supply to the borehole is small, dry running of the pump system can be avoided if the ejector is equipped with a suction pipe which is approximately 10 m long. A self-regulating effect is thus created due to cavitation in the ejector. The pump system is primed prior to initial start by filling a small reservoir on the pump. During normal operation, start and stop is performed automatically by the pressure in an accumulator.

1.6.2 Barrel-emptying pumps (powered and hand pumps)

Barrel-emptying pumps are designed for emptying small containers and tanks. They consist either of a motor section or immersion tube and a pump section. See Figure 1.114.

The pump section is located in the lower end of the immersion tube and is driven by the motor via an extended shaft. The shaft is protected by a sealed column. The pumped liquid flows between the column and the extended shaft to the pump's outlet at the motor end. For obvious reasons, a barrel-emptying pump should be lightweight and easy to transport to the next container to be emptied.

These pumps are manufactured with a variety of immersion tube lengths, between 210 and 1200 mm. They can be made

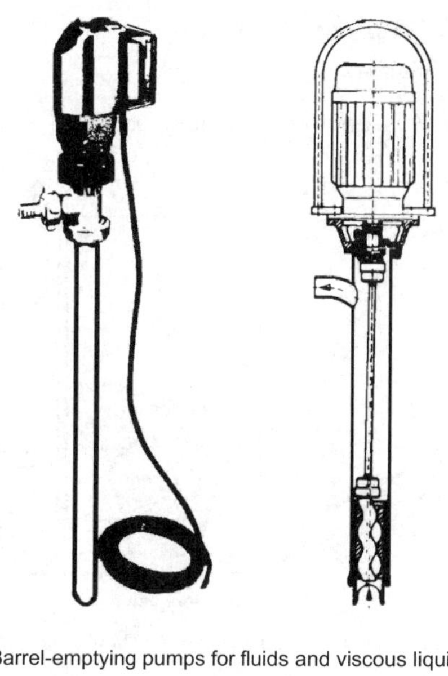

Figure 1.114 Barrel-emptying pumps for fluids and viscous liquids

from many materials having good resistance to chemical corrosion. Pumps can be fitted with low voltage, totally enclosed electric motors, motors for use in potentially hazardous atmospheres and air motors. The pump design for fluid liquids is of the centrifugal or mixed-flow type whilst screw or progressing cavity pump types are used for viscous liquids. Pumps are capable of flows up to 14.5 m^3/h and heads of 25m.

Semi-rotary hand pumps and rotary hand pumps

When power of any type is not available muscle-power must be used. Semi-rotary hand pumps are similar to vane pumps but with valves. One set of valves are in the vanes, the others in the casing. On the filling stroke, the valve in the vane opens to allow liquid to change sides. When the vane is moved in the opposite direction, the liquid is squeezed in the casing opening the casing valve. Pumps normally have two chambers providing a suction and discharge stroke for each action. Semi-rotary hand pumps have good suction lift capabilities and work well with fairly viscous liquids. As well as barrel-emptying applications, these pumps are often used for priming lube and fuel oil systems on large engines, pumps, compressors and other equipment.

1.6.3 Hydraulic-ram pumps

The principle of a hydraulic-ram is illustrated in Figure 1.115. The supply line must be of a certain length, usually 5 to 20 m. The air vessel is located 0.5 to 3 m below the water surface and pumps water up to a higher level, usually a tank for fresh water supply.

If the waste valve suddenly closes, the mass of water in the supply pipe is subjected to rapid deceleration which causes a rise in pressure at the waste valve and delivery valve. The delivery valve opens and water is forced up into the air vessel. When the water in the supply line momentarily changes direction, a partial vacuum is created at the waste valve causing it to open, whereupon water begins to flow in the supply line again. The action of water flowing through the waste valve causes it to close again and the cycle is repeated. The following applies for the hydraulic ram:

$$Q_p \cdot H_p = \eta \cdot Q_d \cdot H_d \qquad \text{Equ 1.9}$$

or

$$Q_d = \frac{Q_p}{\eta} \cdot \frac{H_p}{H_d} \qquad \text{Equ 1.10}$$

where:

Q_d = Motive water velocity in supply line (m/s)

Q_p = Delivered velocity in delivery line (m/s)

H_d = Supply head(m)

H_p = Delivery head(m)

η = Efficiency (non-dimensional)

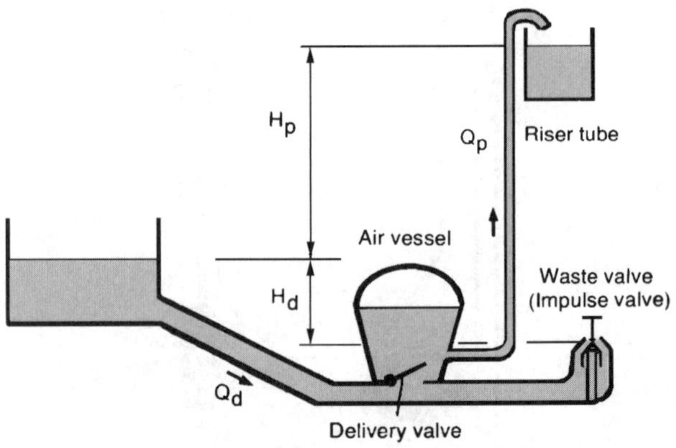

Figure 1.115 Diagrammatic arrangement of the hydraulic-ram pump

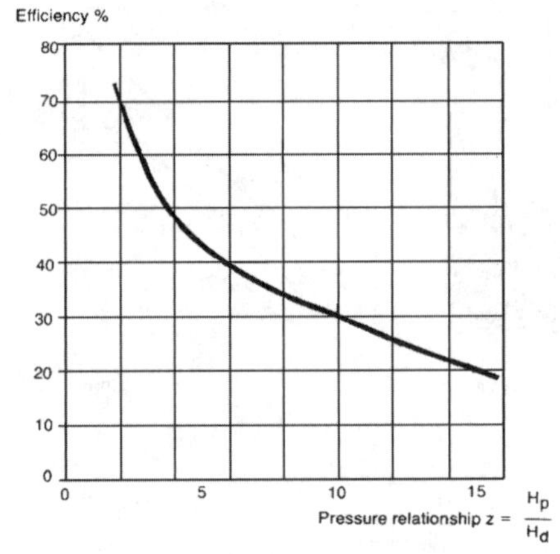

Figure 1.116 Efficiency of hydraulic-ram pump

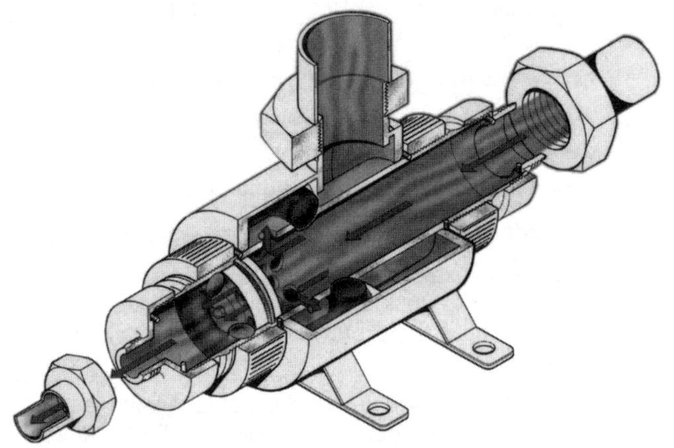

Figure 1.117 Hydraulic-ram pump

Courtesy of Papa Pump Ltd

The efficiency depends upon the quotient z, H_p/H_d according to Figure 1.116. By means of comparison, it may be noted that for large scale water delivery via a water turbine, electric grid system and water pump, efficiencies of up to about 50% can be obtained, whereas for domestic use via electric grid and water pump the efficiency is usually less than 10%.

The hydraulic ram has long been used for transporting fresh water for domestic and agricultural purposes. There are installations which have been in use for more than 50 years. Hydraulic rams are sold in many sizes for supply water flows Qd from 60 l/h to 9 m^3/h with total delivery heads of up to about 40 m. Figure 1.117 shows a section through a typical pump.

1.6.4 Air-lift pumps

The air-lift pump consists of a riser tube immersed in the pumped liquid, a compressed-air line and a pressure chamber where air forces the liquid, via small holes, into the riser tube, as shown in Figure 1.118. The air/liquid mixture is lighter than the surrounding liquid and therefore rises up the tube. The flow of liquid up the riser tube increases as the flow of air is increased up to a certain maximum value, after which it begins to decrease.

The air supply must be compressed to a pressure which is equivalent to the immersed depth plus the losses in the piping and inlet holes. The product of the work done during compression and the air mass flow, neglecting the efficiency of the compressor, is equal to the power input to the pump P_{input}.

The pump efficiency

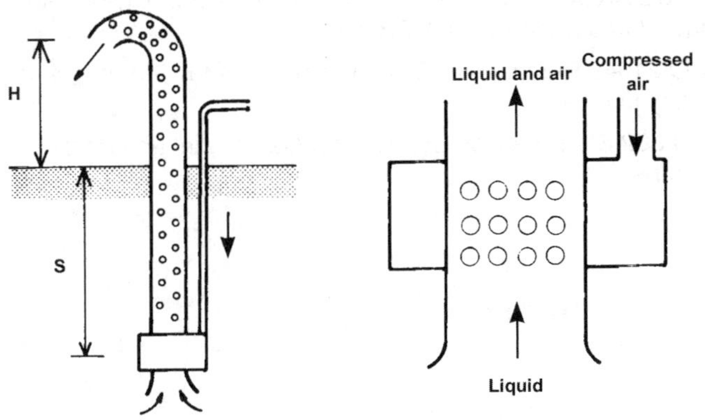

Figure 1.118 Air-lift pump

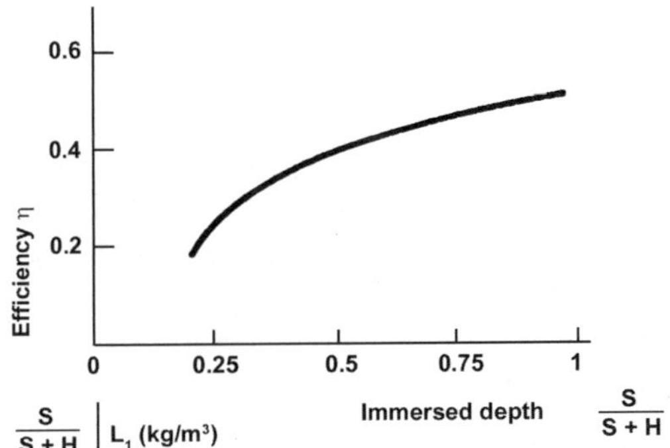

$\frac{S}{S+H}$	L_1 (kg/m³)
0.3	10
0.5	4
0.7	1.5

Figure 1.119 Examples of maximum efficiency and air requirement for air-lift pump - Pump efficiency neglects compressor efficiency

$$\eta = \frac{\dot{m}_{Liquid} \cdot H}{P_{input}} \qquad \text{Equ 1.11}$$

is greatly dependent upon the immersed depth in relation to the delivery head, Figure 1.119. The quantity of air required in kg per cubic metre of pumped liquid, designated L, is greatly affected by the immersed depth.

The advantages of the air-lift pump are:

- Simple construction, no moving parts
- No sealing problems
- Small risk of blockage
- Not sensitive to temperature

The disadvantages are:

- Low efficiency
- Immersed depth requirements, S/H > l
- Compressed air can be expensive

Air-lift pumps are used for pumping sludge, contaminated liquid, large particles, sugar beets, and also hot or corrosive fluids.

1.6.5 Contraction pumps

A contraction pump consists of a special rubber hose which is reinforced in such a way that when it is stretched by means of applying an axial force, the diametral contraction is such that the volume is reduced. Pumping effect is achieved by fitting non-return valves to each end of the hose, suction and discharge valves. Suction is achieved as a result of the hose returning to its original length and volume as a result of releasing the axial force.

Efficiency is very good since the only losses are valve losses and hysteresis losses in the rubber, which can be reduced to a minimum within the pump's most suitable operating range. Typical values are 95% to 98%.

Applications include hand operated or motor driven deep well pumps for well depths of 30 to 100 m and borehole diameters of 50 to 150 mm. Pump performance depends upon the type of hose. Normally hose stretch is 10% of its length and the transported volume is 5 to 20% of the internal hose volume. For hand operated pumps the flow volume is within the region of 5 to 20 l/min depending upon the depth of the well. A particular advantage is the simplicity of installation and removal and the single tube, which functions both as a pull-rod and a delivery pipe, resulting in considerable weight and cost savings.

1.6.6 Macerators

Some solids handling pumps include a macerator in the pump inlet. A macerator is a type of solids grinding or cutting machine which ensures solids entering the pump are below a certain size. Macerators cannot normally be used for hard, strong solids such as gravel, copper ore or tailings. They are most often used for long stringy solids as found in pulp, sewage and rubbish handling. The macerator ensures that solids entering the pump will pass right through the pump without fouling. Some macerators may be capable of handling soft coal. Extra power must be available to cover the macerator's requirements.

1.7 Useful references

Fire & Explosion ATEX Regulations Part 1. The Regulations implement a European Directive called, the ATEX Directive No.1992/92, concerned with the risks from fire and explosion arising from flammable substances stored or used in the workplace. The Regulations apply from 1st July 2003 to new workplaces or workplaces which undergo modifications, extensions or restructuring after July 2003. Existing workplaces must comply by 30th June 2006, but it should be noted that the requirements of this legislation are replicated in other legislation without such a time provision and reference should be made to the Safety, Health and Welfare at Work (General Application) Regulations the Safety, Health and Welfare at Work (Chemical Agents) Regulations, 2001.

ATEX Manufacturers Directive 94/9/EC. This directive defines each of the following product groups for use in potentially explosive atmospheres: Electrical and non-electrical equipment, Electrical and non-electrical protective systems, Electrical and non-electrical components, Electrical and non-electrical safety devices, It places responsibilities on the manufacturer of these products.

ATEX User Directive 99/92/EC. This directive is concerned with the health and safety of workers with relation to potentially explosive atmospheres. It places responsibilities on an employer.

Dangerous Substances and Explosive Atmospheres Regulations (DSEAR). From July 2006 employers must have completed the risk assessment, classification and documentation of their workplace and personnel.

ISO 2858 End-suction centrifugal pumps (rating 16 bar) - Designation, nominal duty point and dimensions.

ISO 3069 End-suction centrifugal pumps - Dimensions of cavities for mechanical seals and for soft packing.

ISO 2858End-suction centrifugal pumps (rating 16 bar) — Designation, nominal duty point and dimensions.

ASME/ANSI B73.1 Specification for Horizontal End Suction Centrifugal Pumps for Chemical Process.

ISO 13709 Centrifugal pumps for petroleum, petrochemical and natural gas industries.

ANSI/API STD 610 Centrifugal Pumps for Petroleum, Petrochemical and Natural Gas Industries.

ISO 13709 Centrifugal pumps for petroleum, petrochemical and natural gas industries.

ASME/ANSI B73.2 Specifications for Vertical In-Line Centrifugal Pumps for Chemical Process.

ISO 21049 Pumps - Shaft sealing systems for centrifugal and rotary pumps.

API 682 Shaft Sealing Systems for Centrifugal and Rotary Pumps.

Properties of liquids 2

2.1 Explanation of terms

2.1.1 Introduction

Liquids, together with solids and gases, are the forms in which substances occur in nature: the solid state, the liquid state and the gaseous state. The three physical states are sometimes called phases. Liquids and gases can be combined in the general group called fluids. Fluids differ from solids in that they will readily change shape to suit the container.

A solid body subjected to a small shear force undergoes a small elastic deformation and returns to its original shape when the force is removed. When subjected to larger shear force the shape may be permanently changed due to plastic deformation.

A fluid when subjected to an arbitrarily small shear force undergoes a continuous deformation. This happens regardless of the inertia of the fluid. For a fluid the magnitude of the shear force and the speed of deformation are directly related. In a solid body it is the deformation itself which is related to the shear force.

A fluid may be either a liquid or a gas. A gas differs from a liquid in that it will expand to completely fill the container. A gas at conditions very close to boiling point or in contact with the liquid state is usually called a vapour. Fluids are compressible; gases being much more compressible than liquids.

A substance can exist in all three states. A typical example of this is ice, water and steam. When ice is heated at constant pressure, the ice converts to water at the melting point and to steam at the boiling point. If the steam pressure is increased at constant temperature, the steam converts to water at the saturation (vapour) pressure.

Solid particles can be suspended or mixed in a liquid. Such a combination, liquid plus particles, is called a suspension. When the particles distribute themselves evenly through the liquid, we speak of a homogeneous mixture. When concentration gradients occur, we speak of a heterogeneous mixture.

The word solution refers to an otherwise pure liquid in which a solid, another liquid or gas has been dissolved.

Two liquids which are not soluble in each other, can be mixed by mechanical action. Such a mixture is called an emulsion. Emulsions can be very difficult to separate.

The properties of liquids vary greatly.. For the purposes of pumping the following characteristics of liquids should generally be known:-

- changes of state
- viscosity
- density
- compressibility
- pH value
- hazards

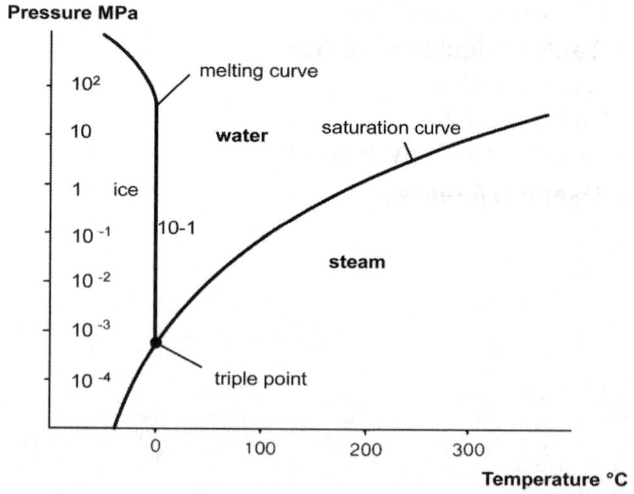

Figure 2.1 Phase diagram for water

2.1.2 Changes of state

Melting point

SI unit °C

The melting point is the temperature at which a substance changes from a solid to a liquid state and also solidifies from a liquid to a solid. The melting point in most substances is pressure dependent only to a very limited degree. In those cases where the pressure dependence has to be taken into account, the boundary between solid and liquid state is shown by the melting curve in a pressure-temperature diagram.

Boiling point

SI unit °C

The boiling point is the temperature at which a liquid converts to vapour or gas at a particular local pressure. The boiling point is usually stated at a standardised atmospheric pressure, 101.325 kPa (760 mm Hg). The boiling point of water at this pressure is 100°C. The boiling point of all liquids is heavily dependent upon pressure.

Vapour pressure

SI unit Pa, kPa, MPa

Preferred unit bar

All liquids have a tendency to evaporate. Vapour or gas accumulates above the free surface of a liquid by reason of interchange of molecules. The partial pressure of the vapour rises to a point at which as many molecules are being returned to the liquid as are leaving it. At this equilibrium state the partial pressure of the liquid is called the vapour or saturation pressure.

Vapour pressure depends on temperature alone and increases with the temperature of the liquid. At a particular temperature, the equilibrium pressure above the surface of a liquid can never be less than the vapour pressure. Any attempt to lower the pressure of the vapour (by means of a vacuum pump for example) immediately results in increased evaporation, i.e. the liquid boils. If the pressure in a liquid decreases locally to the vapour pressure at the actual temperature, vapour bubbles are generated in the liquid. In a pump installation, the formation of vapour bubbles (cavitation) can cause serious mechanical damage and can seriously diminish the performance of the installation.

Various liquids display widely differing values of vapour pressure as a function of temperature. Some substances in a solid state can also change directly to the gaseous state without passing through the liquid state. As an example, a complete pressure-temperature diagram, showing the phases for water is shown in Figure 2.1. At the triple point, all three states may exist simultaneously. In practice the only substance which exhibits the solid-gas transformation is carbon dioxide.

NOTE: The SI unit of pressure is very small. In most practical applications it may be necessary to use Pa or kPa for suction pressures and MPa for discharge pressures. The pump industry is concerned that the change in prefix could confuse operators and lead to potentially dangerous situations. The MPa is a very large unit; small numbers representing large pressures. The bar is a much more suitable unit and does not require prefixes.

Whenever a pressure value is quoted it must be qualified as absolute or gauge.

2.1.2.1 Comments

Melting point (freezing or solidifying point) must be known when assessing the pumpability of a liquid and the risk of blockage, freezing or solidifying.

Information regarding the vapour pressure is required for calculating the permitted suction lift and $NPSH_a/NPIP_a$. Vapour pressure is a critical factor in the choice of pump type, speed and shaft sealing requirements.

2.1.3 Viscosity

SI units

Absolute viscosity	Ns/m^2
Kinematic viscosity	m^2/s

Other units

Absolute viscosity	cP
Kinematic viscosity	cSt

Viscosity (the ability to flow) is a property of liquids treated under the heading of rheology. The word rheology derives from the Greek "rheos" meaning flow.

Between two layers of liquid flowing at different speeds, a tangential resistance, a shear stress, is developed because of molecular effects. We say that the shear stress is caused by the internal friction of the liquid, or conversely that the liquid transmits shear forces by reason of its internal friction.

A liquid in motion is continuously deformed by the effects of the shear forces. The magnitude of the stress depends on the rate of shear deformation and the sluggishness of the liquid, i.e. the viscosity.

2.1.3.1 Newton's Law of Viscosity

Viscosity is defined for flow in layers, laminar flow, by Newton's Law of Viscosity, see Figure 2.2.

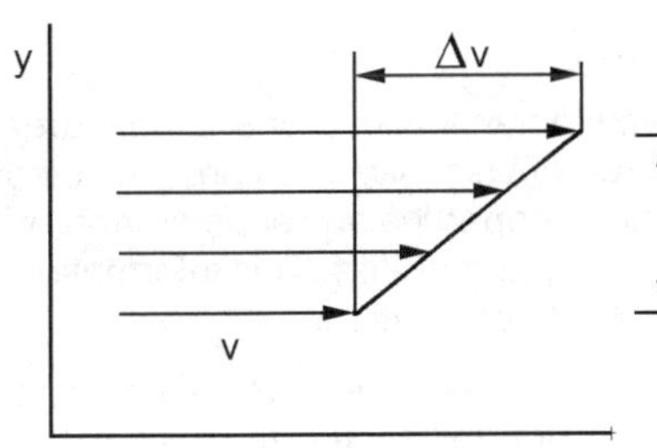

Figure 2.2 Definition of viscosity

$$\tau = \mu \frac{\Delta v}{\Delta y} \qquad \text{Equ 2.1}$$

where:

τ = shear stress (N/m^2)

μ = dynamic viscosity (kg/ms)

Δv = change in viscosity (m/s)

Δy = distance between layers (m)

2.1.3.2 Absolute viscosity

In the SI system the unit is:

1 kg/ms = 1 Ns/m^2

Other units are:

1 Poise = 1P = 0.1 kg/ms

or

1 centipoise = 1 cP = 0.01 P = 0.001 kg/ms

2.1.3.3 Kinematic viscosity

In viscous flow equations the dynamic viscosity divided by the density of the liquid is given the symbol ν. This parameter is called kinematic viscosity.

$$\nu = \frac{\mu}{\rho} \qquad \text{Equ 2.2}$$

where

ν = kinematic viscosity (m^2/s)

μ = dynamic viscosity (kg/ms)

ρ = density (kg/m^3)

The SI unit for kinematic viscosity is 1 m^2/s.

Sometimes, for convenience, the following units are used:

Stoke = 1 St = 0.0001 m^2/s

or more usually

1 cSt = 0.01 St = 0.000001 m^2/s = 1 mm^2/s

Water at 20°C and 0.1 MPa has a kinematic viscosity of 1 cSt.

2.1.3.4 Newtonian liquids

A liquid which follows Newton's Law of Viscosity in laminar flow and has constant viscosity, regardless of shear rate and time, is known as a Newtonian liquid, see Figure 2.3.

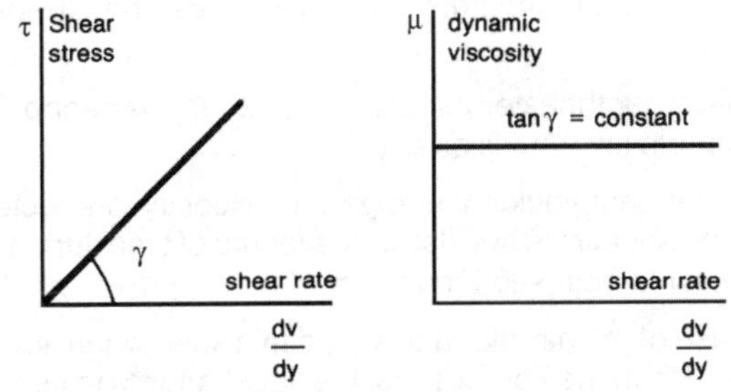

Figure 2.3 Newtonian liquid

Examples: water, aqueous solutions, low molecular liquids, oils and oil distillates. Black liquor, liquid resin and resinous sebacic acid also behave like Newtonian liquids.

2.1.3.5 Non-Newtonian liquids

Liquids which do not fulfil the requirements for Newtonian liquids are called non-Newtonian liquids. Most high molecular liquids, suspensions and emulsions display non-Newtonian properties. Non-Newtonian liquids usually fall into three main groups:

I) Non time-dependent

pseudo-plastic

dilatant

plastic

ii) Time-dependent

thixotropic

rheopectic

irreversible

iii) Visco-elastic

Non time-dependent liquids are not affected by the length of time of the flow process, see Figure 2.4. The shear stress in the case of laminar flow at a given temperature is determined entirely by the shear rate. In analogy with Newtonian liquids, it may be stated:

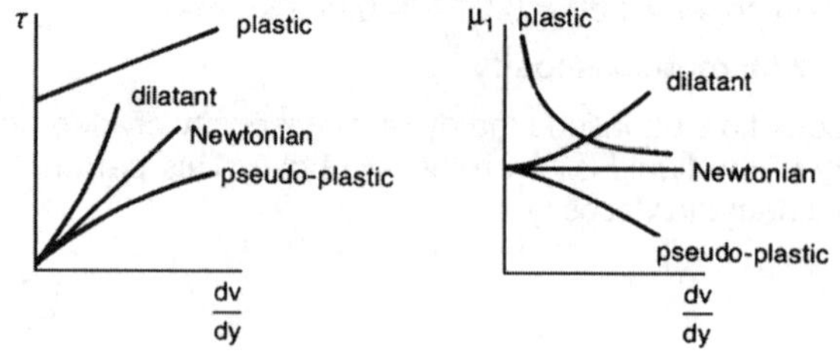

Figure 2.4 Non time-dependent non-Newtonian liquids

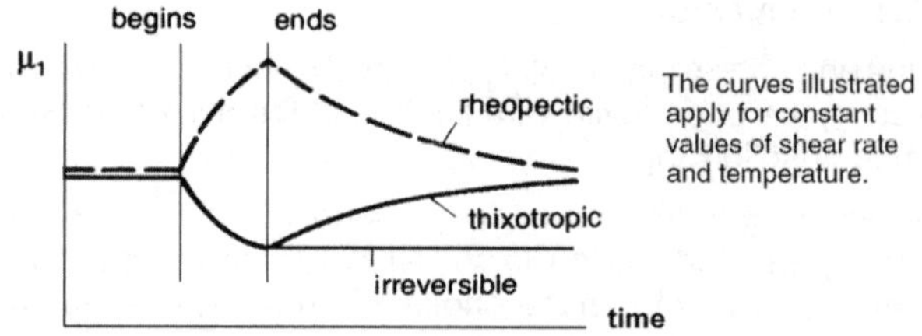

Figure 2.5 Illustrations of time-dependent non-Newtonian liquids

$$\tau = \mu_1 \frac{dv}{dy} \qquad \text{Equ 2.3}$$

where:

μ_1 = apparent dynamic viscosity

In the case of pseudo-plastic liquids, the apparent viscosity decreases with increasing shear rate.

Examples: high molecular solutions, rubber, latex, certain molten materials, mayonnaise.

In the case of dilatant liquids the apparent viscosity increases with increasing shear rate.

Examples: base in oil paints, suspensions with high concentrations of small particles: cement, lime, sand, starch.

Plastic liquids require a certain minimum shear stress - yield stress - in order to initiate flow. The apparent viscosity decreases from an infinitely high value as the shear rate increases.

Examples: toothpaste, ointments, grease, margarine, paper pulp, printers ink, emulsions.

Time-dependent liquids,the apparent viscosity is affected not only by the shear rate but also by the length of time during which the flow continues, see Figure 2.5.

In the case of thixotropic liquids, μ_1 diminishes when flow commences. When the flow ceases, the liquid returns to its original viscosity after a certain period of time.

Examples: paint, gelatinous foodstuffs.

Rheopectic liquids display increasing viscosity under mechanical influence and resume their original viscosity when the influence ceases.

Examples: certain suspensions of gypsum.

Irreversible liquids do not resume their original viscosity at all or perhaps only after a very long time after the removal of the influence. These liquids must be pumped carefully.

Examples: cheese coagulates, yoghurt, marmalade.

visco-elastic liquids, contain liquids exhibiting both elastic and viscous properties. Visco-elastic liquids undergo both elastic and viscous deformations. When the flow ceases there is a certain reversal of elastic deformation.

Examples: asphalt, liquid nylon, rubber, polymer solutions.

2.1.3.6 Comments

- Dynamic viscosity is temperature dependent. Increasing temperature results in diminishing viscosity. Certain liquids can be pumped easier when heated.
- Viscosity is also an indication of the lubricating qualities of the liquid. At low values, less than 1 cSt, additional lubrication may be necessary.
- Temperature dependence on density has also to be considered in kinematic viscosity. At pressures of 10 MPa a certain pressure dependence can be observed in the viscosity.
- Loss coefficients in pipe flows are dependent upon Reynold's Number which is in turn dependent upon dynamic viscosity.
- The performance of rotodynamic pumps depends upon Reynold's Number. Standard pump data is always for water and this has to be corrected when pumping other liquids. Rotary positive displacement pumps are better for high viscosities but there is no specific value to aid selection. The viscosity effects are dependent upon the pump size. For rotodynamic pumps, increased viscosity reduces efficiency and increases NPSHr. The effects of viscosity on positive displacement pumps are dependent upon the pump type as well as size.
- Viscosity is defined as the ratio of shear stress and shear rate in laminar flow. In the case of turbulent flow, this ratio is affected by exchange of momentum between the layers caused by the random motion of the liquid particles.
- In the past viscosity has been specified by the results of proprietary test methods such as Engler, Saybolt and Redwood. These tests measured the time taken for an oil sample to flow through an orifice. In Chapter 17, Units and Conversion factors, the necessary basis is given for conversion to SI units.

2.1.3.7 Penetration

For some non-Newtonian liquids, for example grease and asphalt, determination of viscosity does not provide sufficient information about flow properties and supplementary information regarding consistency is required. This information is obtained from determination of penetration.

For greases, penetration is a measure of consistency. This is the depth, in tenths of a millimetre, to which a cone forces its way down into a test receptacle containing a grease sample heated to 25°C. The penetration depends on whether the consistency is changed by stirring, shaking and so on. Hard greases have low penetration figures whilst soft greases have high penetration. Manufacturers of greases state the penetration figure for each quality, e.g. 240 - 325. Two lubricant greases, furthermore, may have the same penetration figure but may still have different flow capabilities depending upon the viscosity of their respective component oils. The oil grades and thickeners should also be specified.

Determination of penetration for asphalt is carried out in principle at 25°C in the same way, as with lubricant grease, but using a loaded pin instead of a cone. The values measured are used to classify asphalt, typical penetration figures being 10 - 50. See BS 2000- 50 : 1993, Methods of tests for petroleum and its products, cone penetration of lubricating grease - which is equivalent to ISO 2137.

2.1.4 Density and relative density

2.1.4.1 Density

Unit, kg/m^3, symbol ρ

The density of a liquid is the relationship between the mass of the liquid and its volume. The density of liquids changes slightly with temperature and very slightly with pressure unless the liquid is very compressible. The density of a liquid is to a very slight extent affected by the quantity of dissolved gases. Since the ability to dissolve gases is temperature and pressure dependent, there is an indirect dependence on these parameters. The affect of dissolved gases on density can generally be neglected however.

2.1.4.2 Relative density

Unit, non-dimensional, symbol d

Relative density, formerly called specific gravity, is the ratio of a liquid's density to water at standard conditions, atmospheric pressure, 101.325 kPa and 4°C.

NOTE: Information about the density of a liquid, or its relative density, is required when converting pressure to head. Rotodynamic pumps calculate power using differential head and specific gravity. Positive displacement pumps calculate power using differential pressure therefore density or specific gravity has no influence.

2.1.5 Compressibility

All liquids are compressible to some extent. This compressibility can generally be neglected when considering rotodynamic pumps. However when considering reciprocating positive displacement pumps its value must be quantified. It may be noted for example that water is about 100 times more elastic than steel and about 0.012 times as elastic as air. Compressibility is very temperature dependent and slightly pressure dependent. Any values used must relate to the operating conditions. Classically, compressibility is expressed in terms of the bulk modulus defined by the relationship:

$$\text{Compressibility} = \frac{1}{K} \qquad \text{Equ 2.4}$$

$$K = \rho \cdot \frac{\Delta p}{\Delta \rho} = -V\frac{\Delta p}{\Delta V} \qquad \text{Equ 2.5}$$

where:

K = bulk modulus of the liquid (N/m^2)

p = liquid pressure (Pa)

ρ = density of the liquid (kg/m^3)

V = volume of liquid (m^3)

Δ = change of magnitude

The change in volume due to a change in pressure can be calculated directly from the definition

$$\Delta V = -\frac{V\Delta p}{K} \qquad \text{Equ 2.6}$$

where the minus sign indicates that the volume decreases with increasing pressure.

Densities and specific volumes of water under a wide range of pressures and temperatures are readily available in good steam tables. However compressibility data for other liquids can be difficult to obtain. Some data is available in the form of "contraction per unit volume per unit pressure increase" which is designated by χ. In the cases where definitive data is not available the compressibility can be approximated as a multiple of water compressibility.

2.1.5.1 Acoustic velocity

In the case of acoustics however, compressibility is of critical importance. The acoustic velocity, or wave speed, is directly related to the bulk modulus and compressibility. If acoustic resonance occurs in the pipework the acoustic velocity must be known to affect a successful cure. Acoustic resonance can be a very serious problem creating destructive piping vibrations and large pressure pulsations. In suction pipework the resonance can create cavitation conditions leading to pump damage and reduced performance.

The acoustic velocity calculated from the bulk modulus applies to pure clean liquid. If the liquid contains gas bubbles or solid particles the acoustic velocity will be greatly reduced from the theoretical value. Testing may be the only approach to find the true value. See Table 2.1 for indicative values of bulk modulus and acoustic velocity.

2.1.6 pH value

The concentration of hydrogen ions in an aqueous solution is a measure of the acidity of the solution and is expressed as the pH value.

The definition of the pH value is

$$pH = \log_{10}\frac{1}{H^+} \qquad \text{Equ 2.7}$$

where:

H^+ = concentration of hydrogen ions (mol/l).

If, for example, $H^+ = 10^{-4}$ mol/l, then the pH value = 4.

The hydrogen exponent is another name for the pH value.

The numerical value of pH in Figure 2.6 varies between the limits 0 and 14. Acid solutions have pH values between 0 and 6.5, neutral solutions pH = 6.5 to 7.5 and the alkaline or base solutions pH = 7.5 to 14. Acid solutions turn blue litmus red and alkaline solutions turn red litmus blue.

All aqueous solutions reacting as acids contain a surfeit of hydrogen ions H^+. All aqueous solutions reacting as alkalis have a surfeit of hydroxide ions OH^-. The product of the hydrogen ion

Liquid	Bulk modulus N/m^2	Wave propagation speed m/s	Estimated wave propagation speed in pipe	
			Steel	PVC
Acetone	$0.8 \cdot 10^9$	1000	850	280
Petrol	$1.0 \cdot 10^9$	1170	1000	320
Ethylene glycol	$2.7 \cdot 10^9$	1560	1150	300
Glycerol	$4.5 \cdot 10^9$	1890	1400	380
Chloroform	$1.0 \cdot 10^9$	820	700	230
Mercury	$24 \cdot 10^9$	1330	430	-
Methanol	$0.8 \cdot 10^9$	1000	850	280
Oil	$1.5 \cdot 10^9$	1300	1000	300
Turpentine	$1.4 \cdot 10^9$	1270	1000	300
Water, distilled	$2.2 \cdot 10^9$	1480	1100	300
Water, sea	$2.4 \cdot 10^9$	1530	1100	300

Table 2.1 Elastic properties of some liquids at 25°C and 0.1 MPa (1bar)

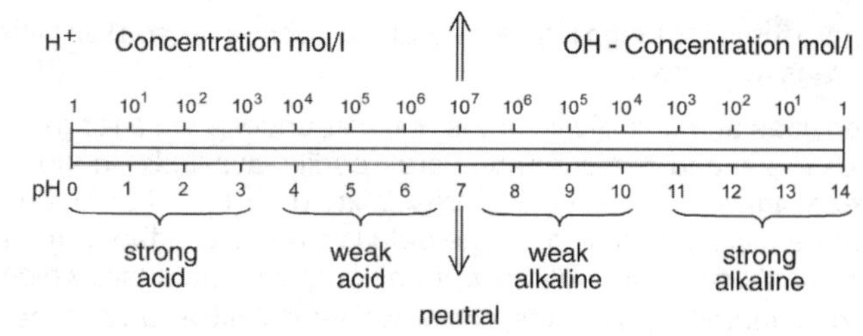

Figure 2.6 Ion concentration, pH value, degree of acidity and alkalinity

concentration and the hydroxide ion concentration is always a constant at any fixed temperature.

At 22°C for example the ion product is:

$$H^+ \times OH^- = 10^{-14} \text{ mol/l}$$

When the hydrogen ion concentration and the hydroxide ion concentration are equal, i.e.

$H^+ = OH^- = 10^{-7}$ then

pH = 7 and the solution reacts neutrally.

Pure water is a very poor electrical conductor. Water only acquires good conductivity in solutions of so-called electrolytes such as salts, acids or bases. When dissolved in water, these substances split into two components with opposing electrical charges, by means of which an electric current may be carried. This act of splitting is called electrolytic dissociation. The charged particles so formed are called ions. Distinction is made between the positively charged cation particles and the negatively charged anion particles. The positively charged particles are symbolised by $^+$ and the negatively charged particles by $^-$. Metals and hydrogen in acids produce cations by electrolytic dissociation. The acid remnants and the alkaline hydroxyl groups on the other hand generate negatively- charged anions.

For liquids with pH values of less than 4, the hydrogen ion concentration is a powerful factor affecting the speed of the dissolving process for most metals. The positively-charged hydrogen ions relinquish their charge at the surface of the metal to the metal atoms and break down their solid connections, whilst they themselves, robbed of their ion construction, return to the atomic state and unite into molecules leaving the metal surface in the form of gas bubbles. The dissolving process is faster, therefore, when the concentration of hydrogen ions is greater, i.e. the lower the pH value of the liquid.

Within the pH range of about 4 to 9, the oxygen present in the liquid, from air for example, has an influence on the corrosion process, say, for iron. The discharged hydrogen does not leave in the form of gas bubbles, as is the case with pH values in the region of 0 to 4, but combines immediately with the oxygen to produce water. At the same time the iron is oxidised to rust. The affect of pH on non-metallic materials must be considered on a case by case basis, see Figure 2.7. See also Chapter 7, Pump materials.

Although pH is an important factor in the attack of metals, this is by no means the only factor which needs to be considered. The concentration (as distinct from pH) of the chemical, particularly of acids and alkalis, has an important bearing on the degree of corrosion or attack. Dissolved oxygen plays an important part in corrosion of metal components. There is now a whole range of organic chemical compounds and biological compounds which will give rise to chemical attack, and many of these are completely unrelated to pH. Even very small concentrations of some modern chemicals can create serious hazards both from the point of view of chemical attack, leakage and toxic hazards.

2.1.7 Hazards

The word "hazard" is in common use in the English language. Its use in *Pumping Manual International* is defined as follows:

Hazard: — a physical situation with a potential for human injury, damage to property, damage to the environment or some combination of these.

It can be seen from the definition that three distinct types of affect are considered but in some cases one hazard may lead to others. Fire, for example, can be a serious health hazard.

A hazardous substance: — a substance which, by virtue of its chemical properties, constitutes a hazard.

2.1.7.1 Health hazards

The pump user must consider the affects of the liquid, and its vapour, on the health of the operators and employees. Most countries have legislation limiting the exposure of employees to substances judged to be hazardous. If the liquid to be pumped is listed in local regulations the pump manufacturer must be informed. The type of health hazard must be specified.

Another health hazard, sometimes not recognised as such, is noise. Some countries have regulations stipulating the acceptable noise levels and exposure times. Some pump types are inherently noisy. Large equipment, in general, is noisy. Noise levels can be attenuated by fitting acoustic enclosures however these tend to drastically diminish the maintainability of the equipment by hindering access. In some instances, costly acoustic enclosures have been removed at site and scrapped in order to achieve acceptable access. One easy solution to this hazard is to declare certain areas as "Ear Protection Zones".

2.1.7.2 Physical hazards

Physical hazards include fire and explosions as well as corrosion and temperature. The degree of risk attached to the hazard is dependent upon the properties of the liquid and the vapour; if the liquid evaporates quickly at ambient temperature and whether the vapour is lighter or heavier than air. A light vapour can disperse easily in an open installation whereas a heavy vapour can become trapped in sumps, depressions and pipe trenches. The upper and lower limits of flammability are important when considering allowable concentrations.

The pumping equipment itself may pose a physical hazard. Within the EU, the Machinery Directive, 89/392/EEC, (amended 98/37/EEC), places the responsibility for safety on the machine designer. The machine must be designed to be safe in all aspects; installation, commissioning, operation and maintenance. If the designer is unable to devise a completely safe machine the areas of concern must be documented and recommended precautions communicated to the user. Because this is a legal requirement in all EU countries the machine designer may not be relieved of the obligations by a third party.

2.1.7.3 Environmental hazards

The Earth's resources and waste-disposal capabilities are finite. Stricter limitations are being imposed gradually on the amount of pollutant which can be released while the list of pollutants is becoming longer. The pump user must be aware of the full consequences of leakage of liquid/vapour from the pump and installation. The environment can be considered in two separate identities:

- local
- global

If the site is surrounded or close to a town what risk is likely to the population, structures or habitat in the event of a failure? In the global sense, what are the likely cumulative affects of product leakage?

2.1.7.4 Installation hazard assessment

The user and system designer are in full possession of all the relevant available facts regarding the liquid and the installation. Any assumptions made should be passed to the pump manu-

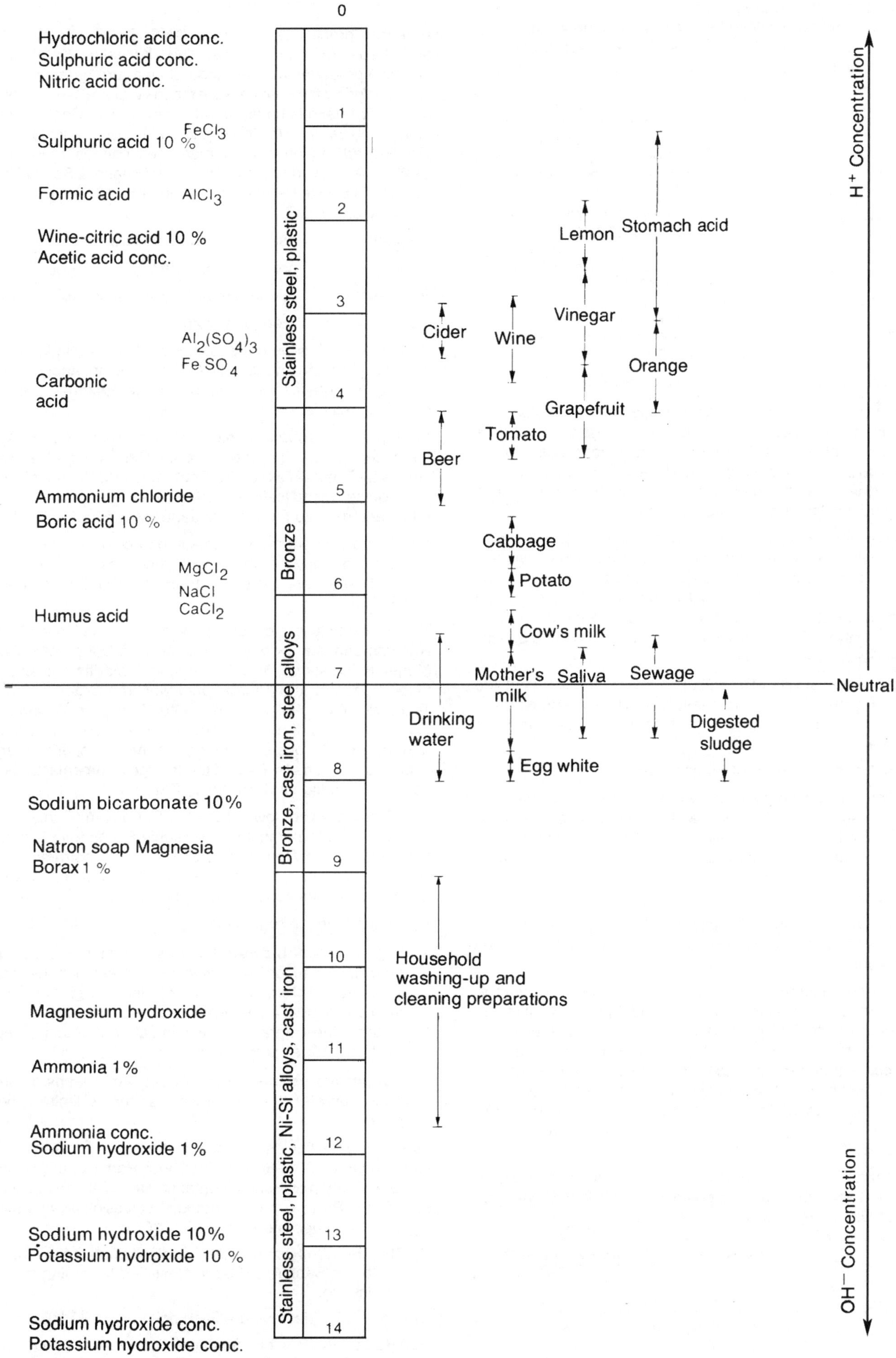

Figure 2.7 Effect of pH values on non-metallic materials. See also Chapter 7

facturer and identified as such. The user must assess the risks attached to all the possible hazards and decide what, if any, leakage is acceptable. Liquid properties reviewed during the assessment, see Table 2.2, should include:

Auto ignition point	The temperature above which a substance will start to burn without an ignition source being necessary
Flash point	The lowest temperature at which a liquid gives off sufficient vapour to burn if an ignition source is present
Atmospheric boiling point	The temperature at which the liquid boils at atmospheric pressure, 101.325 kPa
Vapour specific gravity	Specific gravity is the ratio of a vapour's density to air at standard conditions, atmospheric pressure, 101.325 kPa and 4°C
Surface tension	The surface tension at process temperature will indicate how difficult the liquid is to seal

Table 2.2 Liquid properties reviewed during hazard assessment

The nature of the hazards will also dictate the type of pipe connections to be used; screwed, flat-face flanged, raised-face flanged, ring-type joints. Process upset conditions must be considered as part of the assessment. Upset conditions which last for more than one or two hours may have a significant impact on pump and ancillary equipment selection.

The physical location of the pump, indoor or outdoor, will decide the behaviour of the leakage once outside the pump. Will any vapour cloud quickly disperse on a breeze which always blows over the un-manned site or will a manned enclosed pump house gradually build up a dangerous concentration of vapour? Only the user can assess these questions and specify the necessary precautions.

It is the responsibility of the user to define exactly what the pump is intended to do. It is the responsibility of the pump manufacturer to supply equipment to meet the required performance.

2.1.7.5 Useful sources of information

Basic information on plant and system design, together with risk assessment, can be obtained from various sources. See Section 2.7 for more complete information. For example the British Institution of Chemical Engineers publishes several useful texts as well as the American Institute of Chemical Engineers.

Information regarding trade names and manufacturers of dangerous chemicals is available from the UK Chemical Industries Association.

BS 5908:1990 summarises the statutory requirements, storage and movement of materials, process plant piping, ventilation, fire protection etc. This publication also lists a wide range of reference literature including: UK Health Protection Agency, American Petroleum Institute (API), American Society of Mechanical Engineers, (ASME), the now defunct US Bureau of Mines, and publications of the Oil Companies' European Organisation for environment, health & safety in refining and distribution (CONCAWE).

BS 5908 classifies substances according to the operating temperature and flash point temperature, see Table 2.3, agreeing with the UK Energy Institute.

	Flash point °C	Operating temperature °C
Class I	< 21	
Class II(1)	21 to 55	below flash point
Class II(2)	21 to 55	at or above flash point
Class III(1)	55 to 100	below flash point
Class III(2)	55 to 100	at or above flash point

Table 2.3 Classification of substances

American practice is outlined in ANSI B31.3 (Chemical Plant and Petroleum Refinery Code). The UK Health & Safety Executive also produces a wide range of literature on all aspects of safety. Work with chemical substances that are classed as hazardous to health is covered in the UK by The Control of Substances Hazardous to Health (Amendment) Regulations 2003. General advice on control measures can be found in the COSHH Approved Code of Practice. (Health & Safety Executive (HSE) Books Reference number: L5). COSHH covers the following significant areas:

Regulation 6 assessment of exposure,

Regulation 7 control of exposure,

Regulation 8 & 9 use and maintenance of control measures,

Regulation 10 monitoring exposure.

Separate regulations cover the hazards of carcinogenic substances, lead and asbestos. COSHH does not apply to underground mining installations or to the hazards posed by micro-organisms.

Another useful source of information is The European Agreement concerning the International Carriage of Dangerous Goods by Road (ADR) on 30 September 1957, under the auspices of the United Nations Economic Commission for Europe, and it entered into force on 29 January 1968.

A set of new amendments came into force on 1 January 2005, and consequently, a third consolidated "restructured" version was published as document ECE/TRANS/175, Vol.I and II ("ADR 2005").

The new structure is consistent with that of the United Nations Recommendations on the Transport of Dangerous Goods, Model Regulations, the International Maritime Dangerous Goods Code (of the International Maritime Organization), the Technical Instructions for the Safe Transport of Dangerous Goods by Air (of the International Civil Aviation Organization) and the Regulations concerning the International Carriage of Dangerous Goods by Rail (of the Intergovernmental Organisation for International Carriage by Rail).

Conditions are laid down in the UK Petroleum (Consolidation) Act 1928 and in particular The Highly Flammable Liquid and Liquefied Gases Regulations SI (Statutory Instrument) 1972 No. 917, governing the storage, handling and conveyance of gases and liquids constituting a fire hazard.

According to these regulations:

- **Highly flammable liquid** means any liquid, liquid solution, emulsion or suspension which, when tested in the manner specified in Schedule 1 to the Regulations, gives off a flammable vapour at a temperature of less than 32°C and when tested in the manner specified in Schedule 2 to the Regulations, supports combustion;
- **Dangerous concentration of vapours** means a concentration greater than the lower flammable limit of the vapours.

Fire hazardous liquids are listed in the Petroleum (inflammable Liquids) Order SI 1968 No 570 under Part 1 of the Schedule and any solution or mixture containing any of those substances specified in Part 1 of the Schedule, if it gives off an inflammable vapour at a temperature below 22.8°C.

NOTE: The Fire Hazard categories specified in the Liquid Table in Section 2.6.2 are classified according to the following:

CLASS 1 Liquids having a flash point below +21°C.

CLASS 2a Liquids with a flash point above +21°C but not exceeding +30°C.

CLASS 2b

Liquids with a flash point above +30°C but not exceeding +60°C.

CLASS 3 Motor fuel and heating oil with a flash point in excess of +60°C.

Below 32°C (73°F):

Abel Apparatus: Statutory Method - BS 2000: Part 33:1982 (IP 33/59(78))

-18°C to 71.5°C (0°F to 160°F):

Abel Apparatus: Non-Statutory Method - BS 2000: Part 170:1982 (= IP 170/75(81))

-7°C to 371°C:

Pensky-Martens Apparatus - BS 6664: Part 5:1990 (= IP 34/67 and = ISO 2719)

American literature can be checked for additional information. NFPA (Fire) HAZ-01 published by the US National Fire protection Association maybe useful.

ACGHI® documentation of threshold limit values for substances in workroom air is available from American Conference of Government Industrial Hygienists.

Gas constituting a fire hazard is not classified according to a flash point, the criteria being the way in which the gas can be ignited and the flame propagated. An explosive atmosphere occurs if the percentage concentration by volume of an explosive gas or vapour in air is such that the air will burn when ignited and flame propagation occurs. The upper and lower explosive limits are referred to as the explosive range or flammability limits.

Regulations governing such installations so as to reduce the risk of explosion caused by electrical equipment are laid down and certified by National bodies, European and International agreements. This aspect of hazards and safety is fully covered in Chapter 10 but some basic information is given here.

BS 5345 sets out a code of practice for selection, installation, and maintenance of electrical apparatus for use in potentially explosive atmospheres (other than mining applications or explosive processing and manufacture), gives guidance and references for all aspects of electrical installations.

European - EN 50 014 to EN 50 020, plus EN 50 028 and EN 50 039 are the relevant European standards issued by the European Committee for Electrotechnical Standardization (CENELEC) and adopted by all EU member countries. The International Electrotechnical Committee (IEC) has issued Standard IEC 79, in 18 sections for world use.

Accredited test houses check for standard compliance and issue certificates for suitable equipment. The UK, Germany and USA have the most popular test houses, see Chapter 10.

2.2 Water

2.2.1 Demineralized water

Demineralized water is chemically pure water and can be produced by various methods :

- Multiple distillation
- Multi-bed ion exchange
- Reverse osmosis
- Electro-dialysis

Chemical impurities and dissolved gases are removed which cause the water to become chemically active. At reasonable temperatures, glass and austenitic stainless steels are completely corrosion resistant. Cast iron and bronze are both attacked resulting in contamination of the water. At higher temperatures austenitic stainless steels suffer from intergranular corrosion.

Temperature °C	Dynamic viscosity μ kg/m.s			
	0.1 MPa	1 MPa	10 MPa	100 MPa
0	$1.793 \cdot 10^{-3}$	$1.789 \cdot 10^{-3}$	$1.768 \cdot 10^{-3}$	$1.651 \cdot 10^{-3}$
25	$0.891 \cdot 10^{-3}$	$0.891 \cdot 10^{-3}$	$0.889 \cdot 10^{-3}$	$0.891 \cdot 10^{-3}$
50	$0.547 \cdot 10^{-3}$	$0.547 \cdot 10^{-3}$	$0.549 \cdot 10^{-3}$	$0.568 \cdot 10^{-3}$
75	$0.377 \cdot 10^{-3}$	$0.378 \cdot 10^{-3}$	$0.380 \cdot 10^{-3}$	$0.404 \cdot 10^{-3}$
100	-	$0.282 \cdot 10^{-3}$	$0.285 \cdot 10^{-3}$	$0.300 \cdot 10^{-3}$
150	-	$0.182 \cdot 10^{-3}$	$0.185 \cdot 10^{-3}$	$0.207 \cdot 10^{-3}$
200	-	-	$0.136 \cdot 10^{-3}$	$0.156 \cdot 10^{-3}$
250	-	-	$0.108 \cdot 10^{-3}$	$0.128 \cdot 10^{-3}$
300	-	-	$0.087 \cdot 10^{-3}$	$0.110 \cdot 10^{-3}$
350	-	-	-	$0.096 \cdot 10^{-3}$

Table 2.4 Dynamic viscosity of chemically pure water at various temperatures and pressures

The hydraulic properties of demineralized water can be determined with very great accuracy. The viscosity and vapour (saturation) pressures given in Tables 2.4 to 2.7, conform to the international standard values accepted at the ICPS-8 (Eighth International Conference on Properties of Steam, 1975). The values stated for density agree with ICPS-6 (1964).

Temperature °C	Density ρ kg/m³			
	0.1 MPa	1 MPa	10 MPa	100 MPa
0	999.8	1000.2	1004.7	1045.5
50	988.0	988.5	992.4	1027.4
100	-	958.6	962.8	1000.0
150	-	917.1	922.2	965.3
200	-	-	871.1	924.2
250	-	-	806.0	876.7
300	-	-	715.4	823.2
350	-	-	-	762.5

Table 2.5 Density of chemically pure water at various temperatures and pressures

Temperature °C	Dynamic Viscosity μ kg/m.s	Kinematic Viscosity ν m²/s	Density kg/m³	Remarks
-25	$5.842 \cdot 10^{-3}$	$5.889 \cdot 10^{-6}$	992.1	Super-cooled water
-20	$4.342 \cdot 10^{-3}$	$4.365 \cdot 10^{-6}$	994.7	
-15	$3.342 \cdot 10^{-3}$	$3.353 \cdot 10^{-6}$	996.7	
-10	$2.650 \cdot 10^{-3}$	$2.655 \cdot 10^{-6}$	998.3	
-5	$2.156 \cdot 10^{-3}$	$2.158 \cdot 10^{-6}$	999.3	
0	$1.793 \cdot 10^{-3}$	$1.793 \cdot 10^{-6}$	999.8	NB: All values are for a pressure of 0.1 MPa (1bar).
5	$1.518 \cdot 10^{-3}$	$1.518 \cdot 10^{-6}$	1000.0	
10	$1.306 \cdot 10^{-3}$	$1.306 \cdot 10^{-6}$	999.8	
15	$1.137 \cdot 10^{-3}$	$1.138 \cdot 10^{-6}$	999.2	
20	$1.003 \cdot 10^{-3}$	$1.005 \cdot 10^{-6}$	998.3	
25	$0.891 \cdot 10^{-3}$	$0.895 \cdot 10^{-6}$	997.2	
30	$0.798 \cdot 10^{-3}$	$0.801 \cdot 10^{-6}$	995.7	
35	$0.720 \cdot 10^{-3}$	$0.724 \cdot 10^{-6}$	994.1	
40	$0.654 \cdot 10^{-3}$	$0.659 \cdot 10^{-6}$	992.3	
50	$0.547 \cdot 10^{-3}$	$0.554 \cdot 10^{-6}$	988.0	
60	$0.466 \cdot 10^{-3}$	$0.474 \cdot 10^{-6}$	983.2	
70	$0.403 \cdot 10^{-3}$	$0.412 \cdot 10^{-6}$	997.7	
80	$0.354 \cdot 10^{-3}$	$0.364 \cdot 10^{-6}$	971.6	
90	$0.315 \cdot 10^{-3}$	$0.326 \cdot 10^{-6}$	965.2	
95	$0.298 \cdot 10^{-3}$	$0.310 \cdot 10^{-6}$	961.7	
99	$0.285 \cdot 10^{-3}$	$0.297 \cdot 10^{-6}$	958.9	

Table 2.6 Dynamic viscosity, kinematic viscosity and density for chemically pure water in the temperature range -25°C to 99°C and at a pressure of 0.1 Mpa (1 bar)

Temperature°C	Vapour pressure MPa	Vapour pressure, metres, column of water
0	0.00061	0.06
10	0.00127	0.13
20	0.00234	0.24
30	0.00424	0.43
40	0.00738	0.76
50	0.01234	1.27
60	0.01992	2.07
70	0.03116	3.27
80	0.04736	4.97
90	0.07011	7.41
100	0.10133	10.8
110	0.14327	15.4
120	0.19854	21.5
150	0.47600	52.9
200	1.5549	183
250	3.9776	508
300	8.5927	1229
350	16.535	2810

Table 2.7 Vapour (saturation) pressure for chemically pure water

Various characteristics of chemically pure water are shown in Table 2.8.

Melting point, ice to water	0°C at 101 kPa
Latent heat of fusion, ice to water	335 kJ/kg
Specific heat	4.182 kJ/kg K at 101 kPa and 20°C
Boiling point	100°C at 101 kPa
Latent heat of steam	2260 kJ/kg at 100°C
Coefficient of expansion g =	0.000207 per deg°C at 101 kPa and 20°C
Bulk modulus K =	2.2×10^{-9} N/m^2 at 101 kPa and 20°C

Table 2.8 Chemically pure water characteristics

2.2.2 Fresh water

2.2.2.1 General

Fresh water is the water derived from rivers, streams and wells. Generally containing less than 1% salt, sodium chloride, it can be “hard” or “soft”. Hard water is high in calcium and/or magnesium salts and results in poor lather when using soap. Dissolved gases are always present, usually oxygen and carbon dioxide. The oxygen content should be about 5 mg/l.

During treatment for drinking water chlorides are added. Unless stated otherwise, fresh water is considered to be clean; this is taken as meaning no measurable concentration of solids over 10 microns. Untreated water should have micro-organisms 0.5 to 1.0 microns.

Fresh water has many uses, among other things as raw water for drinking purposes, as cooling and process water within various industries and for irrigation.

For the purposes of pumping installations, the most interesting aspect is the corrosive action of water on the commonly used construction materials:

- steel
- grey cast iron
- bronze

The characteristics of water which contribute to corrosion are:

- pH value
- hardness and carbonic acid content
- content of various chemicals, primarily salts
- acidity

Apart from these, the temperature of the water and the following factors will influence installation corrosion and wear:

- Velocity of flow; normally several m/s in pipes and 10-40 m/s in a centrifugal pump
- NPSH available and cavitation
- Content of solid bodies, e.g. sand and sludge from various sources

Not all materials are suitable for higher velocity service. Not all materials have equal cavitation resistance. If the absolute pressure of fresh water is reduced to approximately 0.4 bara the dissolved gases will start to evolve. If this occurs in a pump suction cavitation symptoms and failures may occur. Small quantities of solids can cause high wear rates and accelerated corrosion in pumps designed and selected for clean water.

Consideration should be given to the rate of corrosion. Popular pump types can be supplied with an internal corrosion allowance providing an estimated component life. It would be very difficult here to make general rules when even an insignificant quantity of salt may magnify an attack of corrosion. Water containing chlorides can be especially troublesome. The corrosive action on steel, for example, is increased by a factor of 8 at 50 mg/litre of Cl_2, which means that steel is not practicable for this purpose.

2.2.2.2 pH value and choice of material

Natural waters usually have pH values of between 4 and 8. They are divided into two main groups according to their acid content.

- **Ground water** from deep sources:— this contains very little acid and it is thus the hydrogen ion concentration which is the decisive factor in the aggressiveness of the water. It should be pointed out that iron is attacked noticeably at pH values of 6 to 7 in this low acid water.
- **Surface water**, which is acidic:— here the pH value is no absolute measure of aggressiveness although it is important to know what it is.
- **Drinking water:** — this may be low acid ground water or high acid surface water which has, furthermore, been treated chemically and filtered. Waterworks can supply the necessary information in this respect. Special conditions apply for lime-containing water, which is dealt with in the following Section.

Since pumps used for conveying fresh water are for the most part constructed of cast-iron, some general aspects are given below about the use of this material with particular reference to the pH value of the water.

- Grey cast iron may be used without any real problems within the pH range 7 to 10. If chlorides are present, it may be that cast iron is not adequate.
- Grey cast iron can often be used within the pH range 5 to 7, but the affects of those factors arising from the lower pH values can be great. Within this pH range, cast iron is superior to steel as regards resistance to corrosion. The high carbon content of grey cast-iron (3 to 4 %) means that at moderate speeds of flow the graphite, together with corrosion products, can build up an anti-corrosive film, so-called graphitization.

For pH values at which cast-iron is not resistive, bronze, steel or stainless steel has to be used. Aluminium bronze is particularly suitable for fresh water, although nickel aluminium bronze may be preferred for longer life. (See also Chapter 7, Materials).

Steel can be used successfully under the right conditions. High velocity areas can be coated. 11-13Cr steel is popular for boiler feed applications although stainless steel is sometimes used to protect against poor water quality control. It should be emphasised that the pH value of water can always be adjusted by suitable chemical treatment before pumping. Care must be taken to ensure any chemical treatment is performed correctly and that additional corrosion problems are not introduced.

When pumping corrosive liquids, particularly liquids supporting electrolytic cells, the correct choice of material combinations for adjacent components is critical. Materials for shafts, sleeves, bushes and wear rings must also be considered. Material properties at clearances can be varied by coating or plating one or both components.

2.2.2.3 Hardness of water

The degree of hardness of water depends upon the presence of impurities, mainly calcium (Ca) and magnesium (Mg) in the form of carbonates, although non-carbonates, for example sulphates, nitrates and chlorides, also have an affect. The quantities in which these impurities occur are a measure of the hardness of the water. Hardness can be expressed in terms of the specific substance such as calcium hardness (Ca-H), magnesium hardness (Mg-H) and so on. The total hardness comprises the sum of the individual hardnesses.

Distinction can also be made between permanent and temporary hardness. Temporary hardness consists of alkali ions bound to carbonates, and permanent hardness of alkali ions bound to non-carbonates. Temporary hardness is so called because it disappears with heating. Distribution under the various headings is according to the chemically equivalent content of alkali ions. Since the various alkaline metals have different atomic weights, the unit for hardness - 1 milli-equivalent per litre (meq/l) is defined as:

$$1\ \text{meq/l} = \frac{1\text{m mol/l}}{\text{chemical valency}} \qquad \text{Equ 2.8}$$

The hardness unit 1 meq/l corresponds to the following ion contents in mg/l:

1 meq calcium hardness	=	20.04 mg/l Ca++
1 meq magnesium hardness	=	12.16 mg/l Mg++
1 meq strontium hardness	=	43.82 mg/l Sr++
1 meq barium hardness	=	68.68 mg/l Ba++

The so-called English degree of hardness, Eng°, is expressed as the equivalent content of calcium carbonate ($CaCO_3$) per Imperial gallon of water, thus:

$$1\text{Eng}° = 14.2\text{mg}\,\text{CaCo}_3\ /\text{l} \qquad \text{Equ 2.9}$$

For the various oxides of mineral alkaline metals:

$$1\text{Eng}° = 8.0\text{mg}\,\text{CaO/l}$$

$$1\text{Eng}° = 5.7\text{mg}\,\text{MgO/l}$$

$$1\text{Eng}° = 14.78\text{mg}\,\text{SrO/l}$$

$$1\text{Eng}° = 21.88\text{mg}\,\text{BaO/l}$$

and the relation in terms of meq/l is:

$$1\text{Eng}° = 0.285(\text{meq/l})$$

$$1\text{meq/l} = 3.5(\text{Eng}°)$$

It is usual in the UK to state the hardness of water in English hardness degrees. See Table 2.9.

English Hardness degrees	Classification
3	very soft
3 - 6	soft
6 - 10	medium hard
10 - 15	rather hard
15 - 24	hard
- 24	very hard

Table 2.9 English hardness degrees

Other units of concentration are used in other countries; for conversion of various hardness units, see Table 2.10.

Units	Alkali ions meq/l	German hardness degrees °dH	English hardness degrees	French hardness degrees	ppm $CACO_3$ (USA)
1 meq/l alkali ions	1.00	2.8	3.5	5.0	50.5
1 German hardness degree °dH	0.356	1.0	1.25	1.78	17.8
1 English hardness degree	0.285	0.80	1.0	1.43	14.3
1 French hardness degree	0.20	0.56	0.70	1.00	10.0
1 ppm $CACO_3$ (USA)	0.02	0.056	0.07	0.10	1.0

Table 2.10 Conversion factors for various degrees of hardness

Example:

1°dH = 1.78 French hardness degrees

1 Eng° = 0.80 German hardness degrees °dH

Soft water in general is more suitable for the majority of household tasks, than hard water. When washing, a high bicarbonate content in the water is damaging because soaps consisting of a mixture of sodium stearate and palmitate generate insoluble calcium salts of organic acids with calcium bicarbonate. Hard water always contains calcium carbonates whose solubility diminishes with increase of temperature.

That is why calcium carbonate is deposited as fur and scale in boilers, heat exchangers and other heating vessels. In order to combat the deposition of scale, which can lead to local overheating in steam boilers, soft water has to be used as a source of supply. Water intended for such applications has to be softened, or dehardened.

2.2.2.4 Carbonic acid and carbonate equilibrium

Precipitation seeping down through the ground, to become ground water, absorbs carbon dioxide (CO_2) from the air in the soil, generated there by the oxidisation of organic material or by the action of various acids on limestone. Carbon dioxide and water from carbonic acid H_2CO_3 which converts the carbonates $CaCO_3$ (limestone) and $MgCO_3$, both difficult to dissolve in water, into soluble bicarbonates $Ca(HCO_3)_2$ and $Mg(HCO_3)_2$. The latter contains some CO_2 from the original carbonate (bonded carbonic acid), and some CO_2 from the carbonic acid which converted the carbonate into bicarbonate (semi-bonded carbonic acid) (see Figure 2.8). In order to keep the bicarbonate dissolved, a certain extra amount of CO_2 is required (free attached carbonic acid).

If there is enough carbonate in the ground, and if all the CO_2 is used up in the conversion of this to bicarbonate and in keeping the bicarbonate in solution, then the water is in a state of equilibrium as regards carbonate-carbonic acid. Thus a special condition of equilibrium arises between lime and the attached free carbonic acid. If the free carbonic acid content is less than that required for equilibrium, lime is separated. If the carbonic acid content increases the lime is re-dissolved. If, on the other hand, there is a surplus of CO_2, this is called “free surplus carbonic acid” or “aggressive carbonic acid”. It is this part of the carbonic acid content which usually causes corrosion.

Water without free carbonic acid, if oxygen is present at the same time, will easily generate carbon-containing protective films which prevent corrosion on exposed pump surfaces, vessels and pipes. If there is free “aggressive carbonic acid” in the water, the build-up of the protective layer is hampered, particularly if there is a lack of oxygen. If this kind of water comes into contact with lime, concrete, treated lime or soda, the lime is dissolved until its chemical equilibrium is restored. Free carbonic acid is thus aggressive to lime.

Iron, steel and light metals are attacked by all free content of carbonic acid, so that the lower the value of pH, the greater is the speed of corrosive breakdown. Ground water often contains bicarbonates with "attached carbonic acid" and perhaps also "aggressive carbonic acid". Water containing aggressive carbonic acid can be aerated and/or dosed with a suitable alkali (lime, soda, sodium hydrate) to bond the carbonic acid and to increase the pH value. At the same time, however, it must be observed that the so-called lime-saturated pH value should not be overstepped. If that happened, troublesome precipitations might result.

For practical purposes, the pH for natural carbonated water depends almost entirely on the relationship between the "bonded" and the "free carbonic acid" and, according to Klut, is determined by the relationship:

$$pH = 6.82 + \log\left(\frac{\text{bonded } CO_2}{\text{free } CO_2}\right) \qquad \text{Equ 2.10}$$

If the equilibrium value pH is set in relation to the carbonate hardness, then relationships are obtained as shown in Figures 2.9 and 2.10. Using Figure 2.10 distinction can also be made between "attached" and "aggressive" carbonic acid.

Carbonate precipitation can cause trouble in certain assemblies in pumps, for example shaft seals and water-lubricated plain bearings.

Figures 2.9 and 2.10 can be used to measure the aggressivity of natural water when the pH value and the carbonate hardness are known. Oxygen-rich carbonated water is aggressive only when the pH value is less than the values along the equilibrium curve. In other cases, a lime protective film is built up. If the pH value is considerably below the values on the equilibrium curve, the affect of other pH reducing substances may be suspected. Corrosion is then caused by these.

In the case of waters having low oxygen content there will be no build-up of protective film. All free content of carbonic acid adds to the aggressivity of the water. The corrosion rate increases with decreasing pH value.

The oxygen content of raw water, which is so important, can vary greatly. In both spring water and ground water emanating from upper strata, the oxygen content is almost always sufficient to build up the natural lime rust-protective layer for carbonate hardnesses of 6°dH. Soft surface water, because of lack of lime, cannot build up lime rust-protective layers and is, therefore, always more or less aggressive.

2.2.3 Brackish water

Brackish water occurs at the mouths of rivers where fresh water and seawater mix. The salt and chlorides content are diluted to approximately 1 to 2.5% and 4000 ppm respectively, giving a pH range of 6 to 9. Because of the turbulent flow regime brackish water contains suspended solids, typically silt and sand. Particle sizes would range from 1 to 200 microns. The percentage of solids, and the size distribution, could only be quantified by sampling.

River water may be contaminated with industrial waste such as nitrogen compounds or caustic. Ammonia, sodium carbonate and sodium hydroxide are some of the most popular wastes. If nitrogen compounds are present, the oxygen content may be reduced due to the activity of bacteria and fungi. Small micro-organisms, as in fresh water, may be present.

Cast iron, steel and 11-13Cr steel will generally not be suitable. The presence of ammonia or caustic will cause stress corrosion cracking problems with brass. The addition of 0.25% to 4% tin to copper alloys improves corrosion resistance and resistance to stress corrosion cracking. Aluminium bronzes, with 10% aluminium, and nickel aluminium bronzes, with 4% nickel, are generally corrosion resistant. Nickel aluminium bronze has proved superior where mud or silt deposits accumulate. Austenitic stainless steels, such as AISI 316, have given adequate service on brackish water with 0.2% ammonia compounds up to 35°C. The solids content must be quantified to allow proper pump selection.

2.2.4 Seawater

Seawater is pumped continuously in two important processes; crude oil secondary recovery and desalination. Secondary recovery may require pressures up to 350 barg. Desalination, by reverse osmosis, requires only 70 barg. These two processes account for the recent developments in pump materials.

2.2.4.1 General

Seawater contains a mixture of inorganic salts. Cooking salt, sodium chloride NaCl, forms about 75% of the salt content. The salt content at great depths varies insignificantly between the oceans of the world. At a depth of 500 m the salt content is about 3.5%. At the surface, however, the salinity is affected considerably by the climatic and other factors. In the Northern Baltic, the salinity is almost 0%, and in the Red Sea it is about 4%. Evaporation, rain, polar ice, rivers and industrial/chemical effluents are all factors which affect the properties of seawater.

The temperature at the surface of the sea varies from about -2°C in the Arctic Ocean and all around Antarctica to about +37°C in the Persian Gulf. The sea bed temperature is generally between + 2°C to + 5°C. At temperatures between 10 and 15°C the oxygen content varies between 8 and 10 mg/l. Chlorides are present up to 25000 ppm. Unless polluted there is no hydrogen sulphide. Seawater properties vary throughout the year and care must be taken to establish the full range. At one location temperatures ranged from 3.3 to 27°C, pH 7.9 to 8.4, whilst oxygen varied from 72 to 100% saturation. The physical properties of seawater vary little from chemically pure water.

2.2.4.2 The qualities of seawater and corrosion

Operational experience with ships' pumps has shown that the quality of seawater is of great significance in resistance to cor-

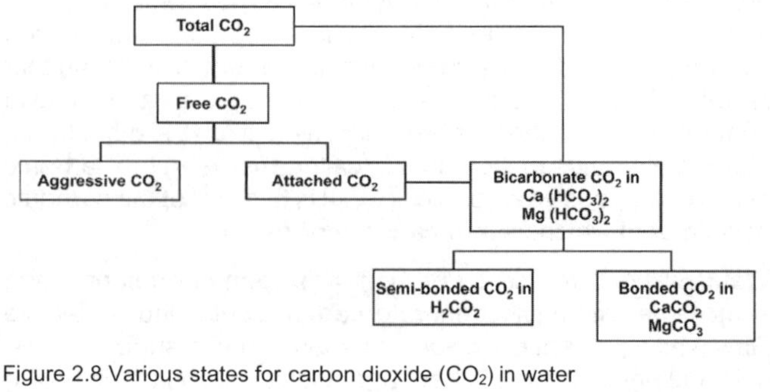

Figure 2.8 Various states for carbon dioxide (CO_2) in water

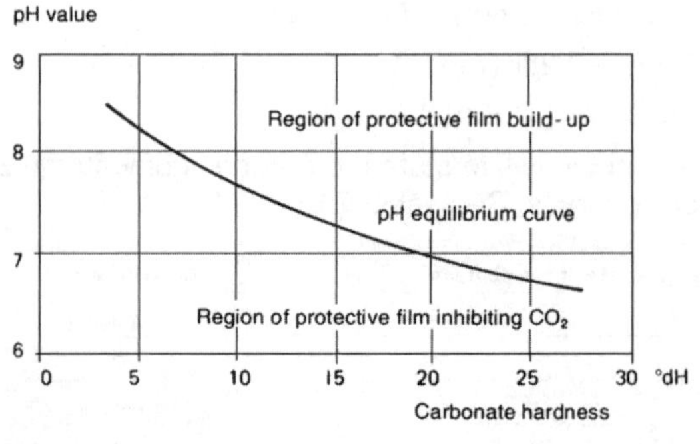

Figure 2.9 Condition of equilibrium for the creation of carbon content protective films in acidic waters

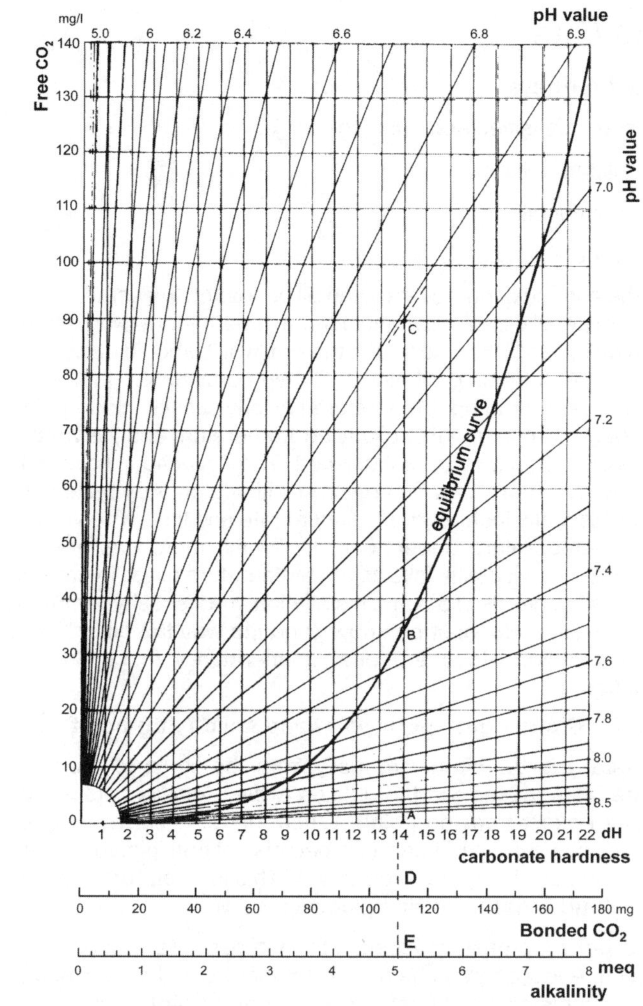

Example:

The following values are obtained after chemical analysis of raw water:

Carbonate hardness	14 °dH	(A)
Free carbonic acid	90 mg/l	(AC)
According to the nomogram:		
Free attached carbonic acid	35 mg/l	(AB)
Free aggressive carbonic acid	55 mg/l	(BC)
Bonded carbonic acid	110 mg/l	(D)
pH value	6.91	(C)
Alkalinity	5.0 meq	(E)

Figure 2.10 Carbonate hardness relative to carbonic acid content

rosion. There are examples of identical ships where the one vessel has had serious problems with its seawater pumps whilst the pumps in the sister ship have remained intact. This phenomenon has usually been explained by the fact that the ships have been plying in different types of seawater. By and large, those ships engaged in coastal and river traffic are attacked more by corrosion than are those which sail the open seas.

As regards coast-based industries and power stations, the quality of the water may vary greatly, depending upon where the installation is located. In such cases however, the quality of water to be dealt with is usually already known at the design stage, enabling the choice of pump materials to be made accordingly. This is not usually possible in the case of marine pump installations. Coastal installations and others subjected to saline atmospheres require special attention to external construction, particularly where dissimilar metals are used, to reduce the affects of galvanic corrosion. The variations in the properties of seawater and the affects these variations have on corrosion rates can be summarised as follows:

- The temperature of seawater is of relatively great importance with regard to corrosion rates. In general, higher temperatures increase the corrosion risks
- Solid contaminants damage the protective-film on the material, thus giving corrosion a chance to attack. The rate of mechanical wear can of course, also be increased
- Chemical contaminants of various types can also lead to increased rate of corrosion
- Tests have shown that oxygen contents over 1 mg/l will promote stress corrosion cracking. If pre-treatment is considered, oxygen reduction is of prime importance. The pH value of seawater is normally about 8
- The dosing of additives may perhaps be regarded as a deliberate chemical contamination of seawaters. Properly conducted, such dosing should not bring any problems from the corrosion point of view. Too strong a dose of sodium hypochlorite, sometimes used to inhibit algae, can on the other hand cause serious pitting corrosion

Summarising, it can be confirmed that clean and cold sea water with normal oxygen content gives the least corrosion problems.

2.2.4.3 Choice of materials

Care must be taken with material combinations when pumping seawater. An electrolytic cell is set up between dissimilar metals due to the high conductivity of the seawater. The less noble material should have a much larger surface area than the noble material. The problem is exacerbated by changes in the electrolytic potential caused by high flow velocities.

Steel is not very good in moving seawater; corrosion of steel increases by a factor of 4 from static to 4.5 m/s. When suitable, cast iron castings with stainless steel impellers last longer overall than nickel aluminium bronze impellers. Ni-resist and ductile ni-resist have been used for some applications; ductile ni-resist is difficult to cast and both are difficult to weld repair.

Copper alloys give good service lives in seawater and are generally easy to cast and machine. Nickel aluminium bronze can be difficult for some machining operations, like tapping. Special tools may be required. Nickel aluminium bronze and magnesium aluminium bronze are better than aluminium bronze and gunmetal. Gunmetal casings with nickel aluminium impellers have been popular. Aluminium bronze is resistant to stress corrosion cracking provided metallurgy and heat treatment are correct. Nickel aluminium bronze has exceptional resistance to cavitation damage compared to some stainless steels; BS1400 AB2 = 0.025mm; stainless steel AISI 321 = 0.305 mm in 3% NaCl. Welding of nickel aluminium bronze is possible, but heat treatment after is necessary to reduce corrosion possibilities. Aluminium and nickel aluminium bronze are limited to 18 to 25 m/s to eliminate erosion problems. Aluminium bronze can be used in combination with stainless steel and titanium without problems. Seawater piping 90/10 Cu Ni has adequate service life.

Stainless steels are the most popular materials for pumping seawater. Care must be taken when reviewing test results and when specifying materials. The chemical composition, and therefore corrosion resistance, of austenitic stainless steels varies with material production; cast AISI 316 is different to wrought AISI 316.

Ferritic steels suffer from local corrosion and pitting in seawater. Stainless steels without molybdenum are generally poor. AISI 304 is generally regarded as the lowest quality austenitic stainless steel. AISI 304 is not used for seawater. AISI 316 has proved successful in tests up to 90°C on synthetic seawater although when tested at 20°C with bubbled chlorine gas pitting occurred. The general consensus is that cast 316L casings with 316 shafts is the minimum working combination. Cast 316L has 10 to 15% ferrite and a higher chromium content than wrought 316L which has no ferrite structure.

316L castings are as good as Alloy 20 for pitting and crevice corrosion resistance, have good stress corrosion cracking resistance and are easier to cast and weld repair. The ferrite content improves corrosion resistance and aids weldability. 316 shafts have to be designed carefully to avoid sharp corners with attendant stress concentrations resulting in stress corrosion cracking. Pitting can be a problem in the stuffing box/mechanical seal area. 316 shafts and impellers have been replaced by monel at extra cost but with no real economic benefit.

Research over a number of years tended to show that molybdenum played an important part in the corrosion resistance. An early theory suggested the Cr + Mo percentage should be over 30%.

This was followed by an Uddeholm equation for Pitting Resistance Equivalent (PRE).

$$PRE = (\text{Cr weight\%}) + (3.3 \times \text{Mo weight\%}) \qquad \text{Equ 2.11}$$

Uddeholm thought that PRE should equal 28 minimum. More research and testing. Later, nitrogen was found to have an important role in the corrosion resistance. A new equation was proposed.

$$PRE = (\text{Cr weight\%}) + (3.3 \times \text{Mo weight\%}) + (16 \times N_2 \text{ weight\%}) \qquad \text{Equ 2.12}$$

Larsen of Ingersoll-Rand suggested the PRE_N value be set at 35. Some users have requested PRE_N values of over 40. Nitrogen strengthened steels had been available since the early 70s. Now more exotic varieties were compounded. Another benefit provided by nitrogen hardened stainless steels was improved cavitation damage resistance. Table 2.11 shows the range of chromium and nickel contents of current popular materials.

Designation	Composition %	
	Cr	Ni
AISI 316	16-18	10-14
ACI CF-8M	18-21	9-12
AISI 316L	16-18	10-14
ACI CF-3M	17-21	9-13
Avesta 254SMO	20	18
Avesta 254SLX	20	25
ACI CN-7M	19-22	27.5-30.5
Rex 734	21.6	9
Armco Nitronic 50	20.5-23.5	11.5-13.5
Zeron 25	24-26	5.5-7.5
Zeron 100	24-26	6-8
Ferralium 255	24-27	4.5-6.5
Ferralium 288	26-29	6-9

Table 2.11 Popular materials for use in seawater

Some exotic stainless steels have a low tensile strength combined with high ductility. Higher pressure pumps require much thicker casings resulting in heavy pumps. More costly, stronger materials may be better resulting from thinner casings and smaller corrosion allowances. Materials, such as titanium, may cost less than proprietary stainless steels.

NOTE: Water, more exactly brine, extracted from wells in oil-fields may have up to 30% salt content, mostly sodium chloride. Oxygen content can be zero but hydrogen sulphide is often present. The pH value is normally low, about 4, and the brine reacts as a weak acid causing severe corrosion. Stainless steels containing 25% Cr and 25% Ni are typically used for these applications.

See also Chapter 7 Pump materials, for further discussions of pump materials used.

2.3 Oils

2.3.1 General

Oils are classified according to their origins:

- Mineral oils
- Animal oils
- Vegetable oils

In the context of pumps they can all be treated the same. Along with mineral oils, we can include such petroleum products as solvents, petrol, kerosene and the like, which should be considered when pumping oil stocks. When pumping oil, the upper and lower operating temperature limits, viscosity, cloud point, lowest flow (pour point), solidifying temperature and the vapour pressure should all be established. The flow capability of oils follows Newton's Law. In common with water, they have constant viscosity independent of shear rate and time. The viscosity is temperature dependent, oil flowing more easily when heated. The viscosity falls as the temperature rises. In order to assess the needs of a pump installation, the viscosity-temperature relationship must be known and the way the oil behaves with variations in operational temperatures must also be clarified.

2.3.1.1 Cloud point, pour point and solidifying point

Mineral oils transform gradually from the liquid to the solid state, as opposed to other liquids (water, for example, which has an exact freezing point). When the oil is chilled, it goes cloudy at a certain temperature because of precipitation of paraffin crystals, i.e. generation of wax. This temperature is called the **cloud point** (cold filter plugging point).

The **pour point** is reached if the temperature is further reduced. A few degrees below this temperature, the oil changes to a completely solid form, the **solidifying point**.

Because of waxing, it is considered that mineral oils can only be handled by pumping at a temperature of at least 10°C above the pour point. When considering pumping high viscosity oils with solidifying temperatures near to or above the ambient temperature (for outdoor installations for example), notice must be taken of the pour point and the installation must be designed so that the pipes and pump can be heated. Pipes which are not heated must be able to be emptied in order to prevent stoppages building up if pumping is interrupted in ambient temperatures below the lowest flow temperature. Low sulphurous heating oils have higher pour point temperatures than high sulphurous oils.

Complex petroleum products, petrol for example, have a vapour pressure range which is dependent upon the most easily flowing components. This property affects the calculation of the $NPSH_a/NPIP_a$ for the pump. Mineral oils are classed as dangerous liquids in fire hazard Class 3. Light distillates such as petrol and photogene are in fire hazard Class 1, see the Liquid Table in Section 2.6.2.

2.3.2 Viscosity

2.3.2.1 Burner oils

Low sulphur oils (max 0.8% by mass - EEC Directive 25/716/EEL) are covered by classes C, and C2 have viscosities of between 1 - 2 cSt at 37°C and together with Class D oils can be stored and handled at ambient temperatures likely to be encountered in the UK. Class E - H are residual blended oils for atomising burners and normally require preheating before atomisation.

The four lines on the chart in Figure 2.11 show average viscosity/temperature relationships for fuel of Classes E to H at the maximum viscosity allowed by the specifications. Table 2.12 gives details of the class. The approximate viscosity/temperature relationship for any petroleum fuel within these classes, for which viscosity at one temperature is known, can be determined by drawing through the known viscosity/temperature intersection a line parallel to those shown. From this line can be read the temperature required for any desired viscosity e.g. that specified by burner manufacturers for proper atomisation. As the temperature is lowered towards the pour point (lowest flow temperature) there is an increasing upward deviation from the viscosity indicated on the chart. This deviation is of such a magnitude that in no case shall the chart be used within 15°C of the pour point, information on which can be obtained from the suppliers. BS 2869 : 2006 may be useful in this respect.

Class of fuel	Min. temperature for storage	Min. temperature for outflow from storage and handling
E	7°C	7°C
F	20°C	27°C
G	32°C	38°C
H		(special purpose fuels)

Table 2.12 Fuel classes

2.3.2.2 Engine oils

The SAE system was devised in America and laid down by the Society of Automotive Engineers (SAE) in 1926. The Society's aims are "to develop technical information on all forms of self-propelled vehicles including automobiles, trucks and buses, off-highway equipment, aircraft, aerospace vehicles, marine, rail, and transit systems" and is usually known by the abbreviation SAE.

Lower numbers in the SAE system mean thinner oils and the letter W after the number indicates that the oil is suitable for use in winter. The SAE system is accepted and used internationally.

Below the cloud point temperature, there is no rectilinear relationship because of waxing.

Pour point: -20°C to -30°C (lower for certain special qualities).

Lowest recommended handling temperature: 10°C to 15°C above pour point.

2.3.2.3 Gearbox oils

Below the cloud point temperature, there is no rectilinear relationship, because of waxing.

Pour point: -20°C to -30°C (lower for certain special qualities).

Lowest recommended handling temperature: 10°C to 15°C above pour point.

2.3.2.4 Industrial oils

The International Organisation for Standardisation, (ISO), has developed a system of viscosity classification for lubricants for industrial use which came into use effectively from 1977. The system consists of 18 viscosity categories stated in centi-Stokes at 40°C. Each class of viscosity is identified by an ISO VG (viscosity grade) number which in general coincides with the mean value in accordance with Figure 2.14 from BS 4231:1992, ISO 3448:1992.

ISO Viscosity Class	Kinematic viscosity at 40°C Mean value	Minimum	cSt (mm2/s) Maximum
ISO VG 2	2.2	1.98	2.42
ISO VG 3	3.2	2.88	3.52
ISO VG 5	4.6	4.14	5.06
ISO VG 7	6.8	6.12	7.48
ISO VG 10	10	9.00	11.0
ISO VG 15	15	13.5	16.5
ISO VG 22	22	19.8	24.2
ISO VG 32	32	28.8	35.2
ISO VG 46	46	41.4	50.6

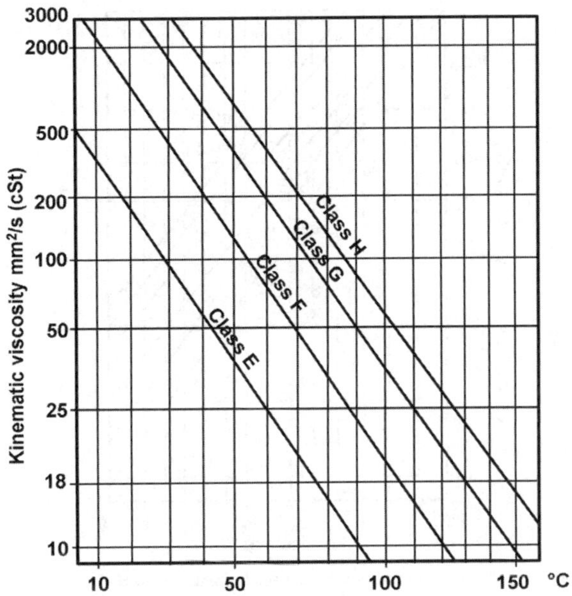

Figure 2.11 Kinematic viscosity/temperature chart

ISO Viscosity Class	Kinematic viscosity at 40°C Mean value	Minimum	cSt (mm2/s) Maximum
ISO VG 68	68	61.2	74.8
ISO VG 100	100	90.0	110.0
ISO VG 150	150	135	165
ISO VG 220	220	198	242
ISO VG 320	320	288	352
ISO VG 460	460	414	506
ISO VG 680	680	612	748
ISO VG 1000	1000	900	1100
ISO VG 1500	1500	1350	1650

Table 2.13 Viscosity classes in accordance with ISO 3448:1992, BS 4231:1992

The classification system has special advantages:

- The ISO VG number gives information on the viscosity of the oil
- ISO 3448 is fully supported by the leading national standardisation organisations such as ASTM, DIN, BSI, JIS making it easier to compare the viscosity of the oils with that specified by the machine manufacturers
- ISO 3448 is directly comparable with BS 4231:1982 for classifications at 40°C

Pour point: 20°C to 50°C (lower values for certain special qualities).

Lowest recommended handling temperature: 10°C to 15°C above pour point.

The scales in the viscosity diagrams are themselves not linear but are adjusted so that the viscosity relationships are linear. The diagram may be used to construct the viscosity relationship for other oils if the associated viscosity temperature values are known for two points. These relationships will also be linear. The Kinematic viscosity for various oils is given in Table 2.14.

Liquid	Kinematic viscosity ν mm²/s (cSt) at a given temperature	
Aviation fuel	0.7/0°C	0.6/38°C
Castor oil	300/38°C	42/54°C
Corn oil	29/38°C	9/100°C
Groundnut oil	42/38°C	23/54°C
Kerosene	2.0/20°C	1.6/38°C
Linseed oil	31/38°C	19/54°C

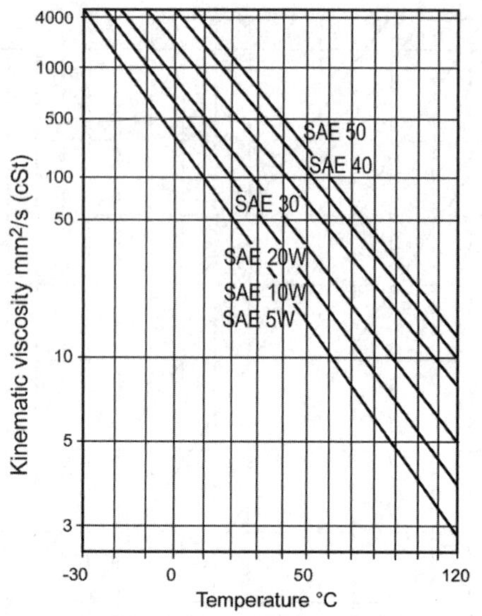

Figure 2.12 Kinematic viscosity for engine oils, SAE 5W to SAE 50

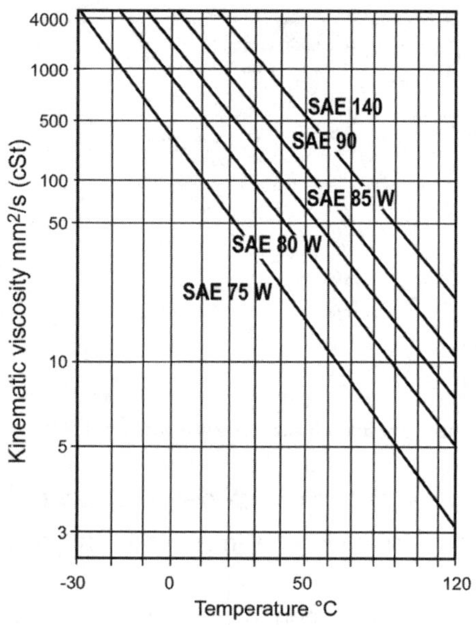

Figure 2.13 Kinematic viscosity for gearbox oils, SAE 75W to SAE 140

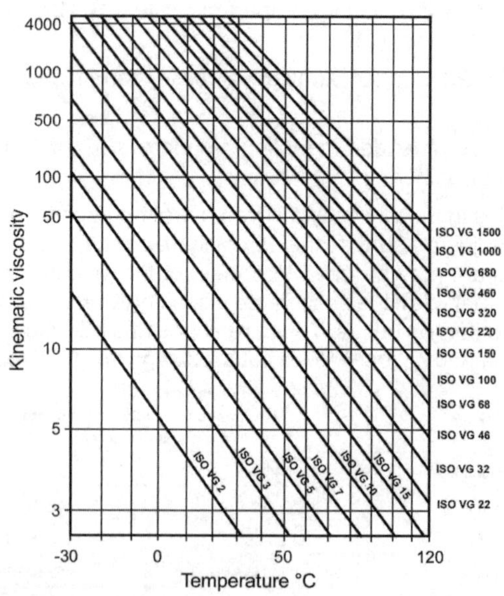

Figure 2.14 Kinematic viscosity for industrial oils in accordance with ISO 3448:1992, BS 4231:1992

Liquid	Kinematic viscosity ν mm²/s (cSt) at a given temperature	
Olive oil	43/38°C	24/54°C
Petrol	0.7/20°C	0.6/38°C
Soya oil	35/38°C	19/54°C
Turpentine	2.1/20°C	2.0/38°C
Valtran	37/38°C	22/54°C

Table 2.14 Kinematic viscosity

2.3.3.5 Other liquid data

Liquid	Density kg/m³ at 20°C	Region of boiling°C at 0.1 MPa	Vapour pressure kPa at 20°C	Fire hazard class*
Aviation fuel	720	40-150	25-35	1
Burner oils —				
BS 2869 Class D	840			3
Class E	940			3
Class F	950			3
Castor oil	960		(0)	
Corn oil	920		(0)	
Oil SAE 5W-50	900		(0)	3
Oil SAE 75W-140	910		(0)	3
Groundnut oil	910		(0)	
Oil ISO VG2-1500	880-905		(0)	3
Kerosene	780	170-250	0.1	2b
Linseed oil	930		(0)	
Liquid resin	970		(0)	
Olive oil	910		(0)	
Petrol	730	40-180	25-70	1
Soya oil	940		(0)	
Turpentine	860		(0)	2b
Valtran	920		(0)	

Table 2.15 Some liquid properties

*Refers to classification in the Liquid Table, Section 2.6.2 - See Section 2.1.7 for reference to other classifications

2.4 Liquid-solid mixtures

2.4.1 General

Particles of inorganic and organic solid matter are to be found, more or less finely distributed, suspended in liquids either as contaminants or for the purpose of transportation. Liquid-solid mixtures belong to the non-Newtonian liquids. The characteristics of the mixture depend upon:

- Pure liquid properties
- Size of the solid particles
- Density of the solid particles
- Shape, hardness and abrasiveness of the solid particles
- Deformability of the solids
- Friability of the solids
- Concentration of particles in the liquid

These parameters are of decisive importance in the choice of the pump. It is usual in pump catalogues to state the suitability of pumps for various kinds of liquids, e.g. clean, no solid contamination, slightly contaminated, contaminated, slurry, mud, transport of pulps, solids, etc. without closer definition. However, all applications for liquid-solid mixtures should be discussed in detail with pump manufacturers.

Small quantities of solids can have a serious affect on wear. Vetter, Thiel and Störk presented test results in their paper "*Reciprocating pump valve design*" for valve wear with low concentrations of 15 μm quartz with a Miller Number of 135, see Figure 2.15. These test results can be taken as representative of fine sand. A concentration by weight of only 0.25% doubled the wear rate. A concentration of 1% solids with a Miller Number of 50 in the same valves would increase valve wear by 80%.

It is helpful in pump selection to use the following parameters for the particle content of the liquid:

- Sizes <0.1, 0.1 to 1.0, 1.0 to 10 and 10 to 100 mm
- Deformable
- Abrasive
- Concentration approx. <1%, >1%

2.4.1.1 Liquid properties

The physical properties of the carrier liquid influence the properties of the liquid-solid mixture. Solids in the liquid tend to increase the viscosity. The clean liquid viscosity is necessary to evaluate the mixture viscosity. Also the abrasiveness of a mixture is affected by the liquid viscosity. Sand in oil is much less abrasive than sand in water. Solids in liquids tend to increase the relative density. Liquid relative density is required to calculate the mixture relative density.

2.4.1.2 Size of particle

Particles less than 1 μm occur in clean fresh water at very low concentrations. Solids in chemical process applications are usually small, up to 1 mm. Solids transported in liquids can vary over a wide range, dependent upon the working pressure and the choice of pump, up to 25 mm is not uncommon. Larger hard solids are encountered with dredge pumps, quarrying and building site applications.

A standard requirement is to pass a 100 mm sphere. Large soft solids, such as fruit, fish and sewage waste up to 140 mm are pumped on a regular basis. Examples of particle sizes and approximate pump capabilities are shown in Figure 2.16.

2.4.1.3 Particle density

The particle density is important for power consumption. Energy is expended moving the particles and the liquid. The power required for rotodynamic pumps is a function of the mixture density which is dependent upon the solids density.

2.4.1.4 Shape, hardness and abrasiveness

The shape of manufactured or processed solids can be defined. Naturally occurring solids may be difficult to describe. The hardness of solids can be measured and should be specified.

Because of the difficulties caused by shape, shape and hardness can be combined to give abrasiveness. Abrasiveness is quantified by testing representative samples of the mixture. The term Representative is critically important. Extrapolating data from one mixture to another is unreliable. A popular test method is the Miller Number; originally intended for reciprocating piston pumps and now standardised as ASTM G75. The test measures the weight loss of a reciprocating metal block, 28% chrome iron, due to the affect of a 50% concentration by weight in distilled water.

The Miller Number is calculated from the test results to produce a linear abrasive scale; a Miller Number of 200 produces twice as much wear as a Miller Number of 100 all other conditions being equal. Miller Numbers less than 50 are considered non abrasive; over 50 consideration should be given to wear problems. Table 2.16 indicates typical values of Miller Number.

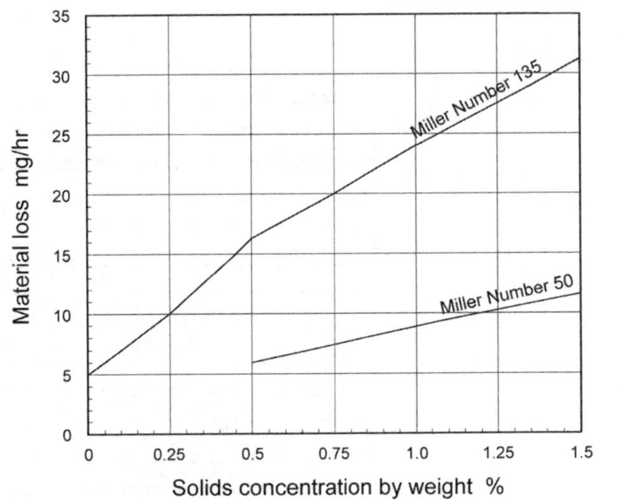

Figure 2.15 Valve material loss

Solid	Minimum values	Maximum values
Bauxite	9	134
Coal	6	57
Copper ore	20	135
Iron ore	28	157
Limestone	22	46
Magnetite	64	134
Mud, drilling	10	10
Sand	51	246
Sewage (digested)	15	15
Sewage (raw)	25	25
Shale	53	59
Tailings	24	644

Table 2.16 Typical Miller Number Values

2.4.1.5 Deformable particles

Particles of organic matter, soft, fibrous particles which can be compressed, deformed or may be subject to damage. For example, dispersions of plastic or natural rubber in water - latex. Latex can coagulate under pressure and therefore requires careful pumping. Other soft solids — fruit, vegetables and fish can be pumped with minimum damage by correct pump selection.

2.4.1.6 Friable particles

Friable means easily crushed or easily broken. Hard solids, like some coals, can be crushed or broken when trapped by certain pump components. Friable solids can be handled by gear pumps, some screw pumps and reciprocating pumps with valves.

2.4.1.7 Solids concentration

Concentration refers to the quantity of particles in suspension (dry substance = DS content) and is expressed as a percentage by weight or volume. Equations 2.13 to 2.16 show the volumetric relationships between liquid, particles and mixture. Equations 2.17 to 2.20 show the mass relationships.

$$\rho_M = \rho_L + \frac{C_V\,(\rho_s - \rho_L)}{100} \qquad \text{Equ 2.13}$$

$$C_V = \frac{100\,(\rho_M - \rho_L)}{(\rho_S - \rho_L)} \qquad \text{Equ 2.14}$$

$$\rho_L = \frac{\left(\rho_M - \frac{C_V\,\rho_S}{100}\right)}{\left(1 - \frac{C_V}{100}\right)} \qquad \text{Equ 2.15}$$

$$\rho_S = \frac{100}{C_V}\left(\rho_M - \rho_L + \frac{C_V \rho_L}{100}\right) \qquad \text{Equ 2.16}$$

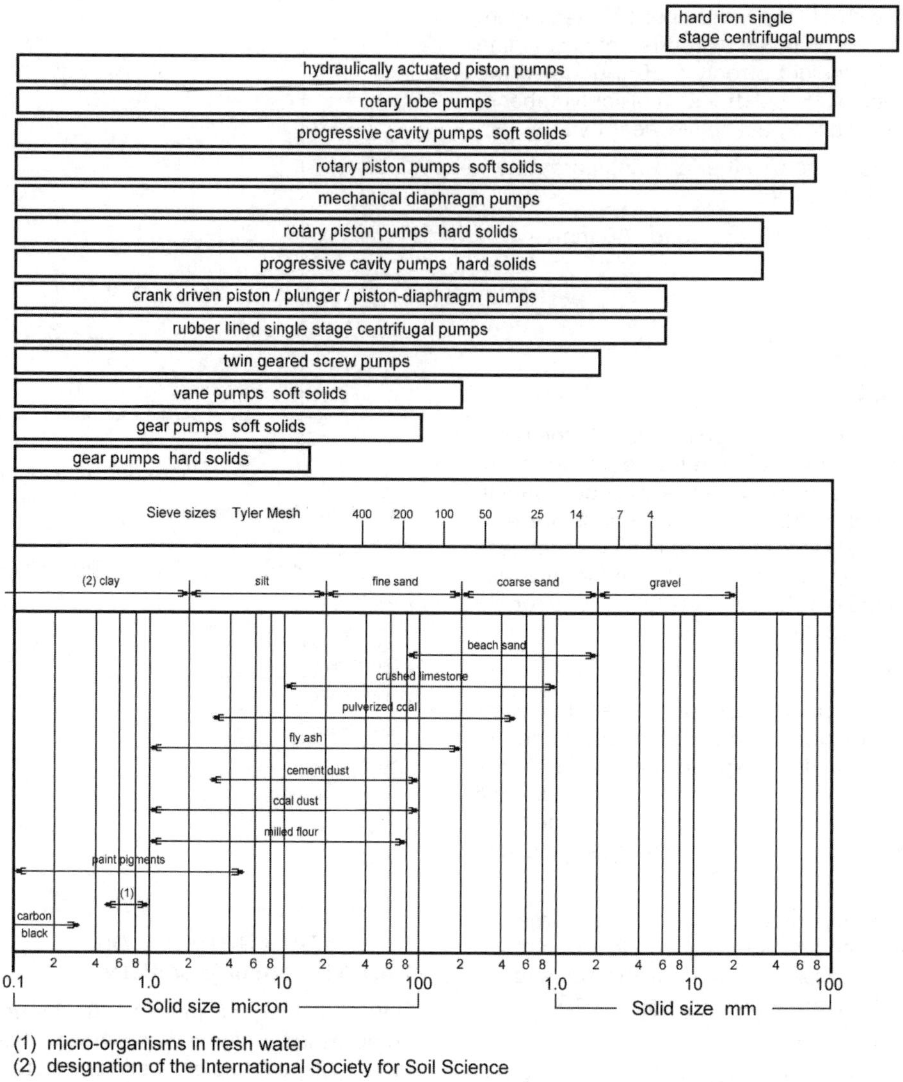

Figure 2.16 Particle sizes and pump capabilities

$$\rho_M = \frac{1}{\left(\frac{C_M}{100\rho_S} + \frac{1}{\rho_L} - \frac{C_M}{100\rho_L}\right)} \qquad \text{Equ 2.17}$$

$$C_M = \frac{\left(1 - \frac{\rho_M}{r_L}\right)}{\left(\frac{\rho_M}{100\rho_S} - \frac{\rho_M}{100\rho_L}\right)} \qquad \text{Equ 2.18}$$

$$\rho_L = \frac{\left(1 - \frac{C_M}{100}\right)}{\left(\frac{1}{\rho_M} - \frac{C_M}{100\rho_S}\right)} \qquad \text{Equ 2.19}$$

$$\frac{1}{\rho_S} = \frac{100}{C_M}\left(\frac{1}{\rho_M} - \frac{1}{\rho_L} + \frac{C_M}{100\rho_L}\right) \qquad \text{Equ 2.20}$$

where:

C_M = percentage of solids in mixture by weight

C_V = percentage of solids in mixture by volume

ρ_L = density of liquid kg/m^3

ρ_M = density of mixture kg/m^3

ρ_S = density of solids kg/m^3

The relationship between C_M and C_V is given by :

$$\frac{\rho_L C_M}{\rho_S} + \frac{C_M C_V}{50} + \frac{C_V \rho_S}{\rho_L} - \frac{C_M C_V \rho_S}{100\rho_L} - \frac{C_M C_V r_L}{100\rho_S} - C_M - C_V = 0 \qquad \text{Equ 2.21}$$

Figure 2.17 shows a nomogram for estimating the mixture density when the carrier liquid is water.

The example shows that a suspension having a particle content of 40% by weight and a particle density of 2,700 has a total density of about 1,340.

Table 2.17 indicates typical values of solids concentration found in some pump applications.

Pumped mixture	Solids concentration by weight %
Polluted municipal sewage water	0.1
Paper pulp in rotodynamic pumps	0.5 to 6
In dense pulp pumps	6.0 to 20
Centrifuged sewage	up to 35
Copper concentrate pipeline	up to 45
Coal pipeline traditional	13 to 50
Activated sewage sludge	20 to 50
Carbonation slurry	50 to 60
Limestone slurry	50 to 60
Iron ore slurry	20 to 65
Coal water fuel	up to 75

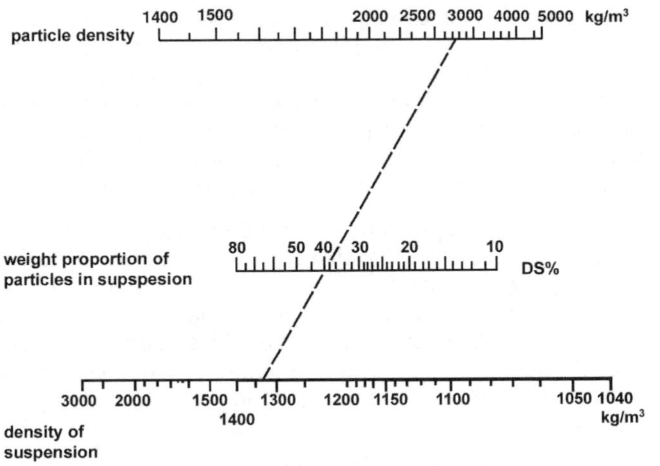

Figure 2.17 Mixture density nomogram for water

Sand	20 to 75
Clay, raw cement sludge	50 to 75
Mine backfill slime	up to 78

Table 2.17 Solids content in pumped mixtures

It is not possible to generalise the effect of solids concentration on pump selection, performance or component life. When solids are known to be present the pump selection should be fully discussed with manufacturers. Figure 2.16 indicates the maximum solid size handling capabilities of various pump types ignoring concentration. Small particles, up to 40 μm, can be pumped at 65% by weight in water and look like watery toothpaste. Solids concentrations of 80% look like toothpaste. Small solids can be pumped at high concentrations. Large solids must be pumped at low concentrations.

2.4.1.8 Velocity

One important aspect of solids handling has not been mentioned, velocity. Abrasive solids, and large solids, are pumped at much lower velocities than clean liquids.

Experience has shown that abrasive wear in pumps is related to flow velocity raised to a power between 2.5 and 5 depending upon the component. If the velocity is increased by 10% the wear rate will increase by a minimum of 27% up to 61%. The velocity-wear relationship can be used with the Miller Number to estimate the effects of duty changes.

Example:

Consider a piston pump operating at 50 rpm on drilling mud and having a valve seat life of 5000 hours. If the pump speed was increased to 75 rpm the predicted seat life would be 660 to 1800 hours. If the pump was operated at 50 rpm on mine tailings the predicted life would be 80 to 2000 hours. If the pump operated at 40 rpm on tailings then predicted seat life would be 130 to 6300 hours.

2.4.2 Sewage

Sewage is a generic term for:-

- **Soil sewage**: discharge from water-closets, urinals, slop-sinks including waste water from domestic baths etc. and even industrial effluent.
- **Surface water**: rainfall and storm water.
- **Drain water**: drainage from building sites, fields, leaks in broken pipes etc.

Waste water from households, businesses, hotels, offices and restaurants serves as a carrier for the contaminants from water closets and other sources of waste in the sewage system. The size of the waste thereafter is limited in principle to the area of the intake connected to the pipeline. It is not easy to define the limits of length of a soft and flexible objects. These objects, such as sheets of plastic, sanitary towels, etc., pass unchecked down the water closet and into the public sewer.

It is forbidden to discharge into any open drain or sewer, material which is likely to cause damage, create a hazard or which could have a prejudicial effect upon the treatment of their contents (i.e. waste matter which should be dealt with by commercial waste disposal companies). Such waste could be dirty engine oil, volatile liquids producing flammable or toxic vapour or biological waste. The legal regulations applying to local sewage works are enforced by the local water authorities supervised by government bodies such as river authorities and environmental protection agencies.

The quantity of unwanted or prohibited objects collected at rain water and sewage grills continues to increase. Especially difficult, and containing a high content of textile matter amongst others, is the sewage from densely populated city centres. The heart of the city with its restaurants, offices and department stores presents a far greater sewage problem than do the surrounding dormitory suburbs.

The demand on pumps and other means of transporting the daily sewage should be assessed according to the contaminants (permitted or not) which are contained in the sewage. The larger contaminants can be classified as deformable according to Chapter 15, thus requiring a non-clogging (chokeless) design. Different types of industry produces varying types of effluent which often contains substances which can cause damage to the sewage treatment works and the recipient water installation.

In the UK the law regarding the discharge of industrial waste into sewers is restricted by the Public Health (Drainage of Premises) Act 1937, amended in 1961. The responsibility for administering and enforcing this legislation is in the hands of the water authorities. Provisions for the restriction of pollution of tidal waters and estuaries appear in sections 2-5 of the Rivers (Prevention of Pollution) Act 1951 and amended by the 1961 Act. The Clean Rivers (Estuaries and Tidal Waters) Act 1960 also applies with limitations specified in the Schedule to the Act 1960.

The UK Control of Pollution Act 1974 sections 43 and 44 extends legislation with respect to specific pollutants. Limiting values for the discharge of substances which are toxic, corrosive, explosive and/or damaging to pump and system materials etc. are the subject of consent which must be obtained from the local water authority. (Many of these liquids appear in the Liquid tables in Section 2.6.2). Cooling water, surface water and drain water display properties as in Section 2.2. Solid precipitations from the circulating water can accumulate in cooling tower installations in the form of lime furring, sludge and concentrations of minerals giving rise to corrosion, hot spots and blockage in pumps and piping.

2.4.3 Sludge

Sludge is the name given to the residue which forms when inorganic and organic particles are separated when a liquid is cleaned. The following mechanical and chemical cleaning methods are used:

- **Sedimentation-settling**: particles fall to the bottom in a separator under their own weight;
- **Flotation**: particles are made to float by the injection of small air bubbles;
- **Centrifuging**: particles are separated by centrifugal force;
- **Filtering**: liquid is passed through a filter which allows the liquid to pass through whilst trapping the particles, the filter may consist of a grill, porous material or one or more beds of filtering media;

- **Precipitation-flocculation**: by the addition of chemical reagents, the particles form insoluble combinations -flocs- which can be separated by sedimentation or flotation.

When pumping, the dry solids (DS) content-value - or pumpability - of the sludge, is of interest as is also the size and hardness of the particles. Since sludge always occurs in the treatment of water and in industrial processes, there can be no general definition. The values obtained in the treatment of sludge in municipal sewage works can be used however as a guide. See Table 2.18.

	DS-content%
Before thickening	
Chemical sludge (scum) after precipitation or flotation	0.5 to 1
Mechanical sludge, sedimentation or flotation	2 to 3
After thickening	
Mechanically agitated sludge	6 to 10
Activated sludge	2 to 3
Bio-filtration	4 to 8
Mechanical and activated sludge	5 to 8
After straining	
In the centrifuge, filter screen press, vacuum filtering etc.	> 30

Table 2.18 DS-content after various treatment stages

In the bottom, or compression zone, of sedimentation basins, sand traps, oil tanks and other containers of stationary liquids, the bottom layer, during sedimentation is subjected to mechanical pressure due to the weight of sludge lying above it. This causes liquid to be compressed out of the bottom layer thus increasing its density. Other names given to sludge are sediment and slurry.

2.4.4 Pulps

2.4.4.1 General

Pulps are used in the cellulose and paper industries. It is a name for fibre suspensions in water. Pulps belong to the non-Newtonian liquids; group 1, the time-independent, plastic liquids.

The density of a pulp can be calculated from equations 2.13 and 2.17 with reservation for possible air content. Often the effects of the fibres and air cancel each other out. The density then coincides with that of the water.

Pulps exhibit widely varying pH values, depending on the production method and the bleaching process. Pumps made of stainless steel are usually needed.

2.4.4.2 Pulp quality

The pumping characteristics of pulp depends, in the first instance, upon the raw material used, the additives and the method of production. Examples of raw materials used are coniferous (soft) wood (fibre length 3 to 4 mm), deciduous (hard) wood (fibre length 1 to 1.5 mm) and rag (fibre length 25 to 30 mm). Basically the methods of production can be divided into chemical and mechanical methods. The chemical method is most common. Sulphate and sulphite pulps are produced both bleached and unbleached. The most important pulp to be produced mechanically is groundwood pulp.

Aside from the concentration - the dry solids (DS) content - the flow characteristics of the pulp are also affected by the fibre ratio - length/diameter and the degree of pulverisation. An illustration of the fluidity of paper pulp is shown in Figure 2.18. The percentage figures given indicate the DS-content.

2.4.4.3 Air content in pulp

Pulp differs from many other suspensions in that it consists of three phases; water, solid-fibre and air.

Air in the pulp occurs either in the form of bubbles or in a combination. Air occurs as bubbles either in a free form or attached to the fibre. Air in the combined form occurs in solution in the water or absorbed in the fibre. The content of air in the pulp depends upon; quality, concentration, additives, pulverisation, temperature and time; and also the pulp handling process. Groundwood pulp contains more air than sulphate or sulphite pulp. Better sizing increases the air content considerably as does pulverising.

An increase in temperature increases the content of free air whilst the ability of air to dissolve in water decreases. Generally speaking, the air content of pulp reduces in storage. The air content of pulp increases quickly to a certain level specific to each type of pulp when mixed with air. This happens typically when the pulp is allowed to fall freely into a vessel or cistern. Air can also be taken up by the pulp at shaft seals in pumps if there is a vacuum.

2.4.4.4 Pulp properties which complicate pumping

- **Content of solid particles**: — The pulp fibres build up a network, the density of which increases with increased concentration. It is relatively difficult to get a high concentration pulp moving, since when high energy pulses are transmitted locally in the network, it is probable that the pulp will break up and movement will take place only locally.
- **Air content**: — Air in the form of bubbles in the pulp is very inconvenient from the point of view of pumping. An air content of only 1% to 2% is enough to change the pumping characteristics.
- **Pressure drop**: — For flow at low speeds the pressure drop for pulp is much higher than for water. In general, pulp which is hot or which contains additives is easier to pump than pure, cold pulp. It can be somewhat of a problem to get pulp moving again in a pipe system after a shut-down, especially in the case of high-concentration pulp. This is because the starting resistance, “stiction”, is greater than the pressure drop at low flow speeds. Dilution with water in the pump can assist restarting.
- **Blockage characteristics**: — Flocculation of the pulp, sticks, twigs and synthetic fibre, can cause blockage in the pump. A total blockage of both pump and pipe is not an unknown occurrence.
- **The tendency to thicken in reducing sections**:— Rapidly reducing conical sections on the suction side of the pump, e.g. reduction in the suction pipe from 400 mm to 150 mm, can cause flocculation at concentrations of only 3 to 4%.

2.5 Liquid-gas mixtures

2.5.1 General

Pumping liquid-gas mixtures covers areas such as:

- Small oil and gas wells
- Boiler feed pumps
- Aircraft fuel systems
- Chemical, pharmaceutical and food processing

Small oil and gas wells: Especially offshore, where it may not be economic to lay two pipelines; one for crude oil and one for natural gas or fit pumps and compressors with associated drivers and controls.

Boiler feed pumps: Generally high speed multi-stage centrifugal pumps operating with small NPSH margins. Process upsets initiating cavitation, can cause severe costly damage, resulting in unscheduled shut-downs.

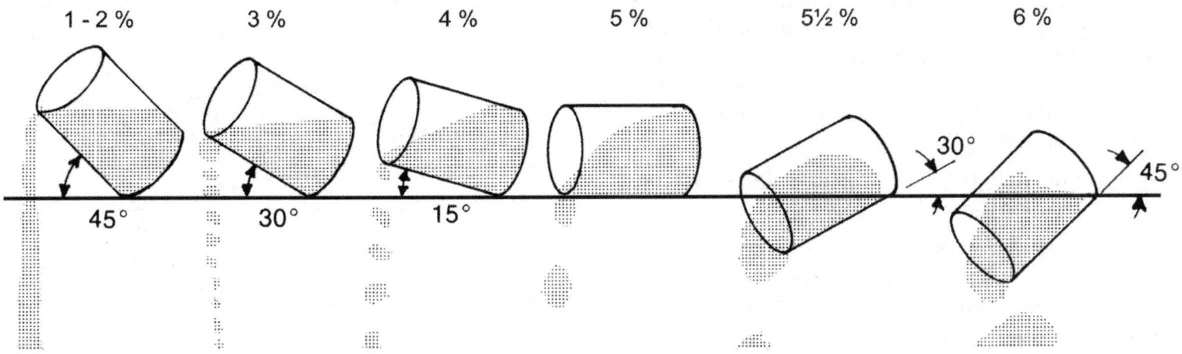

Figure 2.18 Illustration of fluidity of paper pulp

Aircraft fuel systems: Prone to air entrainment due to manoeuvring. Loss of fuel pressure and reduced flow rates can be disastrous.

Chemical, pharmaceutical and food processing operations: The handling of gas and liquids together are seen as a process simplification resulting in cost savings.

2.5.2 Useful observations

As with all pumping processes the pump manufacturer requires all the operating data to be able to assess the application. Detailed information regarding the nature of the gas, or liquid vapour, is essential. Also, details of the gas flow; steady state controlled percentage or variations from bubbles to slugs.

Users hoped it would be possible to use centrifugal pumps to handle the mixture without pre-treatment. Centrifugal pumps were the first choice equipment because of reliability and familiarity. However tests indicated that 2 to 4% gas content seriously affected pump performance and increasing gas content to 15% caused pumps to lose their prime. Development since 1989 has not significantly improved gas handling characteristics. Current configurations consist of an efficient in-line separator feeding liquid to a centrifugal pump and the gas to a compressor; both discharging into a single pipeline. This arrangement is reported to cope with 40% gas by volume. Flow controls on both machines are necessary for gas content variations.

Multi-stage axial pumps have proved successful in declining reservoir oil-field applications. Pumps have coped with 88% gas by volume with occasional slugs of 100% gas. Currently differential pressures are limited to 14 bar and inlet flows up to 500 m^3/h can be accommodated.

Progressing cavity pumps and vane pumps are both reported as running satisfactorily with liquid gas mixtures; neither suitable for 100% gas. It may be possible for diaphragm pumps or hydraulically operated piston pumps to be useful in this service.

Great success has been achieved with specially adapted geared twin screw pumps. Pumps are available which can handle gas volume fraction from 0 to 97%; dry running on 100% gas is not yet possible. Pumps are available for differential pressures of 48 bar; 69 bar in certain applications. Discharge pressures up to 153 barg are possible. Mechanical seal problems are avoided by fitting double seals with a lubricating/cooling barrier liquid.

Positive displacement pumps have the advantage in being able to cope with varying differential pressures, due to downstream friction loss changes, without substantial flow changes.

2.6 Table of liquid properties

2.6.1 Introduction

The Liquid Table in Section 2.6.2 gives useful data for pump applications. The characteristics of liquids are not given to that degree of accuracy which, in special cases, may be required in order to specify other process equipment.

Note: The most common liquids, water, oils and certain suspensions; are dealt with in more detail in Sections 2.2, 2.3 and 2.4.

Compressibility and wave speed are not included in the Table. These characteristics are stated for a number of liquids in Section 2.1.5.

The Liquid Table contains the following data:

- **Substance:** the liquid properties are arranged in alphabetical order according to the name of the substance. The trade names with synonyms are given in Table 2.19. The letter D indicates a dangerous (flammable or otherwise hazardous) liquid within the terms of the UK Petroleum (Inflammable liquids and Other Dangerous Substances) Order 1947 - Amended 1968 SI 570. The letter T indicates a high level of toxicity.
- **Chemical formula:** for identification of substance. In practice a substance unfortunately can not be assigned to a single formula because of contamination. Liquids containing contaminants which are insoluble or solid have the worst pumping characteristics, often causing troublesome wear on pumps and shaft seals.
- **Solubility in H_2O:** attention is drawn in the Liquid Table to various diagrams referring to the solubility of the substance in water. The solubility is expressed as a concentration in percentage by weight at varying temperatures. When pumping, precipitation often occurs when the concentration exceeds that stated by the solubility curve for the given temperature. Conversely, there will be precipitation if the temperature drops below the stated solubility temperature.
- **Viscosity:** is stated in mm^2/s (cSt) at +20°C. Viscosity below 5mm^2/s indicates that the liquid is easy flowing like water. Viscosity below 1mm^2/s indicates that the liquid has poor lubrication qualities and poor frictional dampening properties for acoustic and pressure waves. The Table also refers to diagrams showing dependence of viscosity on temperature. The viscosity will also be used for the determination of frictional losses in the pipework. This depends primarily on whether the flow is laminar or turbulent, see Chapter 3. In the laminar region, the pipe frictional losses are proportional to the viscosity, whereas its effect can be neglected in the turbulent region for the internal roughness of pipes which occur in practice.

  Certain liquids are non-Newtonian and require special care when determining the equivalent (apparent) viscosity.
- **Vapour pressure (absolute):** is stated in kiloPascals (kPa) at +20°C. See curves in Figure 2.21 which show the temperature dependence of vapour pressure. Note that boiling point at atmospheric pressure (101.325 kPa 760 mm Hg) gives another point on the vapour pressure curve. The

curves in Figure 2.21 are set out schematically and may deviate from the exact values. This is especially applicable to aqueous solutions - curve 13. All references to this curve apply to chemicals in solution in water. Liquid pressure near to the vapour pressure at the actual liquid temperature can cause cavitation in a pump or pipe.

- **Concentration in H_2O:** the concentration of the substance in water expressed as percentage by weight.
- **Density:** is stated in kg/m^3 at +25°C. Reference is made to various figures for other temperatures and concentrations. The density is used among other things for conversion of pressure to head and for the calculation of the power required for the pump.
- **pH region:** is an expression of acidity or alkalinity and is grouped within pH regions 0 - 4, 4 - 6, 6 - 9 and 9 - 14. This group classification assists in the choice of pump material, the pH regions coinciding with commonly used pump material resistivity to corrosion.
- **Melting point:** note that many contaminants gradually begin to degrade at temperatures immediately above the melting point which can have a harmful effect on the function of the pump.
- **Fire hazard class** stated in accordance with Section 2.1.7.

NOTE: Where there is no information in the Liquid Table, this means that the actual liquid property is not known and it does not mean that the information omitted is of no general interest in pumping.

Substance	Entered in table as
Acetic ether	Ethylacetate
Alcohol	Ethyl alcohol
Alum	Aluminium sulphate
Ammonium hydroxide	Ammoniac
Benzol	Benzene
Black liquor	Sulphate liquor
Bromite	Silver bromide
Butanol	Butyl alcohol
Carbitol	Ethyldiglycol
Caustic soda	Sodium hydroxide
Cellosolve	Sodium chloride
Cellosolve acetate	Ethylglycolacetate
Chlorhydric acid	Hydrochloric acid
Chloroacetic acid	Monochloroacetic acid
Chlorobenzene	Monochlorobenzene
Chrome alum	Potassium chrome sulphate
Chrome oxide	Chromic acid
Cooking salt	Sodium chloride
Copper vitriol	Copper sulphate
Cyancalium	Potassium cyanide
Dichloromethane	Methylene chloride
Diethylether	Ether
Dphonylether	Diphenyl oxide
Ethanol	Ethyl alcohol
Ferrichloride	Iron (III) chloride
Ferrosulphate	Iron (II) sulphate
Fixerbad	Sodium tiosulphate
Formaline	Formaldehyde
Fumaric acid	Malein acid
Glauber salt	Sodium sulphate
Hartshorn salt	Ammonium carbonate
Hydrofluoric acid	Hydrofluoric acid
Methanol	Methyl alcohol
Methylaldehyde	Formaldehyde
Naphthalene chloride	Chloronaphthalone
Natron saltpetre	Sodium nitrate
Oleic acid	Fatty acid
Palmitic acid	Fatty acid
Perhydrol	Hydrogen peroxide
Phenol	Carbolic acid
Potash	Potassium carbonate
Potash nitrate	Potassium nitrate
Potash salts	Potassium hydroxide
Pulping liquor	Calcium bisulphite liquor
Radical vinegar	Concentrated acetic acid
Sal ammoniac	Ammonium chloride
Saltlake	Sodium chloride
Slaked lime	Calcium hydroxide
Soda	Sodium carbonate
Soda lye	Sodium hydroxide
Sodium borate	Borax
Styrol	Styrene
Sublimate	Chloride of mercury
Sulphite liquor	Calcium bisulphite
Tannin	Tannic acid
Trichloroethanyl	Chloral
Urea	Carbamide
Water glass	Sodium silicate

Table 2.19 Trade names and synonyms used in the Liquid Table

2.6.2 Liquid Table

Substance		Chemical formula	Viscosity mm²/s at 20 °C	Vapour pressure kPa at 20 °C	Vapour pressure kPa at various temps. See Fig 2.21 curve no.	Conc. % in H_2O	Density at 25 °C	Density at various temps. See Fig no.	Density at various conc. See Fig no.	pH/ region	Boiling point °C at 101.3 kPa (760 mm Hg)	Melting point °C	Fire hazard class- See Section 2.1.7	Solubility in H_2O at various temps. See Fig no.	Comments
Abietinic acid		$C_{20}H_{30}O_2$										172			
Acetaldehyde	D	CH_3CHO	< 1	100	10		780/20°C	2.22			21	-125	1		
Acetic acid	D	CH_3COOH	1.2	3.5	19		1050	2.22	2.24	0 - 4	119		2b		
					19	5-100				0 - 4					
					19	50	1060			0 - 4					
Acetic acid anhydride	D	$(CH_3CO)_2O$			20		1080			0 - 4	136	- 73			
Acetone	D	CH_3COCH_3	< 1	30	14		790	2.22			57	- 95	1		E-mod and speed of sound - Table 2.1
Acetyl chloride		CH_3COCl					1100				51	- 112	1		
Alkyd solution in paint naphtha			Fig 2.19				900				150		2b		Non-Newtonian - Figure 2.19
Alkyd solution in xylene	D		Fig 2.19	0.9			1000				139		2a		Non-Newtonian - Figure 2.19
Alkyd tixotrope sol. in paint naphtha			Fig 2.19				900				150		2b		Non-Newtonian - Figure 2.19
Allyl alcohol	D	CH_2CHCH_2OH		2.3			850				97		1		
Ally chloride	D	CH_2CHCH_2CL	<1				940				45		1		
Aluminium		Al					2700					660			
Aluminium chloride		$AlCl_3$	< 5	2.2	13	5	1030		2.25	0 - 4				2.28	
				2.2		10	1090			0 - 4					
Aluminium nitrate		$Al(NO_3)_3$		2.2		10	1050		2.25	0 - 4				2.28	
Aluminium sulphate		$Al_2(SO_4)_3$		2.2	13	10	1110		2.23	0 - 4				2.28	
Ammonium alum		$NH_4Al(SO_4)_2$				10	1050			0 - 4					
Ammonium bromide		NH_4Br			13	5	1030		2.25	4 - 9					
Ammonium carbonate		$(NH_4)_2CO_3$			13										
				2.2	13	10	1030			4 - 6					
Ammonium chloride		NH_4Cl			13	10 - 50				4 - 6					
				1.8		26	1070		2.25	4 - 6					
Ammonium fluoride		NH_4F		2.3		6	1030								
				2.2		14	1060			9 - 14					
Ammonium hydroxide		NH_4OH	< 5		1			2.22		9 - 14					
			< 5	9.3		10	960			9 - 14					
			< 5	34.5		25	910								
Ammonium nitrate		NH_4NO_3			13	10 - 20			2.25	6 - 9					
				2.2	13	10	1040			6 - 9					
				0.5	13	60	1230			6 - 9					
Ammonium oxalate		$(COONH_4)_2H_2O$			13	30	1040			6 - 9					
Ammonium perchlorate		NH_4ClO_4				10	1040			6 - 9				2.28	
Ammonium persulphate		$(NH_4)_2S_2O_8$				10	1060			4 - 6					
Ammonium sulphate		$(NH_4)_2SO_4$				0 - conc			2.25	4 - 6				2.28	
				2.2		10	1060			4 - 6					
				0.7		50	1280			4 - 6					
Amyl acetate		$CH_3COOC_5H_{11}$	< 5	0.7	20		880				149	- 78	2b		
Amyl alcohol		$C_5H_{11}OH$	6				810				133	- 78	2b		
Amyl chloride		$C_5H_{11}Cl$					890				109		1		
Amyl polycaptan		$C_5H_{11}SH$					850				100				
Aniline	D	$C_6H_5NH_2$	6	0.1	24		1020	2.22			184	- 6	3		
Aniline hydrochloride		$C_6H_5NH_2HCl$				5				0 - 4					
Antimony		Sb				20	1090			0 - 4		630			
Anthracene oil		$C_{14}H_{10}$					1250				226				
Arsenic acid		H_3AsO_4					2500		2.24						
						10	1070			0 - 4					
Asphalt solution in naphtha			1500 - 7500				900				150		2b		Non-Newtonian
Aviation fuel			< 1	25			720				40		1		Tables - 2.1, 2.14, 2.15
Barium chloride		$BaCl_2$		2.2	13	10	1090		2.26					2.28	
				1.8		26	1280			6 - 9					
Beer					13		1010			4 - 6	100				
Benzene	T	C_6H_6	< 1	13	16		880	2.22			80	5.5	1		
Benzene sulphonic acid		$C_6H_5SO_3H$										525			
Benzoic acid		C_6H_5COOH					1270				250	122			

Substance		Chemical formula	Viscosity mm^2/s at 20 °C	Vapour pressure kPa at 20 °C	Vapour pressure kPa at various temps. See Fig 2.21 curve no.	Conc. % in H_2O	Density at 25 °C	Density at various temps. See Fig no.	Density at various conc. See Fig no.	pH/ region	Boiling point °C at 101.3 kPa (760 mm Hg)	Melting point °C	Fire hazard class- See Section 2.1.7	Solubility in H_2O at various temps. See Fig no.	Comments
Benzyl alcohol		C_7H_8O	5	< 0.1			1050	2.22			205		3		
Biphenyl		$C_6H_5C_6H_5$					1990				256	70			
Borax		$Na_2B_4O_7$			13		2370					741			
				2.2	13	3.5	1030								
Borax acid		H_3BO_3			14	10				4 - 6					
						50				4 - 6					
Bromine	D	Br_2	< 1/0 °C		12		3120/20 °C	2.22			59	- 7.2			See Section 2.3 and Table 2.1
Burner oils	D		Fig 2.11				840/20 °C						3		See Section 2.3 and Table 2.1
	E		Fig 2.11				940/20 °C						3		See Section 2.3 and Table 2.1
	F		Fig 2.11				950/20 °C						3		
Butane		C_4H_{10}			8	50	600/0 °C	2.22			- 0.5	138			
Butyl acetate	D	$CH_3COOC_4H_9$	< 1	2	19		880				125	- 77	2a		
Butyl alcohol	D	C_4H_9OH	< 5	0.9	18		810	2.22			118	- 90	2a		
Butyl glycol		$C_6H_{14}O_2$	< 5	< 0.1			900				171		3		
Butyric acid		C_3H_7COOH			21		960	2.22			164	6.5			
Calcium bisulphate		$Ca(HSO_3)_2$			13	25	1040			4 - 6					
Calcium chloride		$CaCl_2$			13			2.25							
Calcium hydroxide		$Ca(OH)_2$			13	5	1060		2.23	9 - 14					
Calcium nitrate		$Ca(NO_3)_2$							2.25						
Camphor		$C_{10}H_{16}O$					990				209	176			
Carbamide		$(H_2N)CO(NH_2)$					1320					132			
						50	1115			4 - 6					
Carbolic acid	D	C_6H_5OH	11				1070	2.22			43				
Carbon disulphide		CS_2	< 1	48	12		1262	2.22			46	- 112	1		
Carbon tetrachloride	T	CCl_4	< 1	20	15		1600				77	- 23			
Castor oil			Table 2.14	(0)			960								
Chloral		CCl_3CHO					1520				98	- 57			
Chloramine		$CH_3C_6H_4SO_2NCl$										183			
Chloride of lime	D	$Ca(ClO)_2$ $4H_2O$					2350								
Chloride of mercury	T	Hg Cl_2			13	0.1	1000			4 - 6					
							7150								
Chlorine	D	Cl_2			3			2.22			- 35	- 103			
Chlornaphthalene		$C_{10}H_7Cl$	2.5				1200				263	- 25			
Chloroform	D	$CHCl_3$	< 1	31	14		1490	2.22			61	2			E-mod and speed of sound - Table 2.1
Chlorosulphonic acid		$HClSO_3$	< 5		21		1280				158	- 80			
Chromic acid	D	H_2CrO_4	< 5				2700			0 - 4					
			< 5			10 - 50				0 - 4					
Citric acid		$C_6H_8O_7H_2O$			13		1550			0 - 4		153			
Copper (II) chloride		$CuCl_22H_2O$				10	3390					620		2.28	
							1090			0 - 4					
Copper cyanide		CuCN			13		2920								
Copper nitrate		$Cu(NO_3)_26H_2O$			13		2070							2.28	
		$Cu(NO_3)_23H_2O$					2320					114			
				2.2	13	10	1090								
Copper sulphate	D	$CuSO_4$					3610							2.28	
					13	10	1160			0 - 4					
				2.2	13	18	1210			0 - 4					
Corn oil			Table 2.14												
Cresol		$C_6H_4(CH_3)OH$			25		1040			0 - 4	191	30			
Cyclohexanol	D	C_6HOH	71	0.3			950	2.22			161		1		
Cyclohexanone	D	$C_6H_{10}O$	< 5	0.7			950				156		2b		
Diacetone alcohol	D		150	0.1	23		940				170	- 43	2b		
Dichloroacetic acid		$CHCl_2$-COOH	< 5				1550	2.22		0 - 4	192	- 11			
Dichloroethane	D	CH_2Cl-CH_2Cl		31.2	14		1180	2.22			59		1		
Dichloroethylene	D	$C_2H_2Cl_2$		32			1250	2.22			60	- 81	1		

2.6.2 Liquid Table

Substance		Chemical formula	Viscosity mm^2/s at 20 °C	Vapour pressure kPa at 20 °C	Vapour pressure kPa at various temps. See Fig 2.21 curve no.	Conc. % in H_2O	Density at 25 °C	Density at various temps. See Fig no.	Density at various conc. See Fig no.	pH/ region	Boiling point °C at 101.3 kPa (760 mm Hg)	Melting point °C	Fire hazard class- See Section 2.1.7	Solubility in H_2O at various temps. See Fig no.	Comments
Dichlorohydrin		$(CH_2Cl_2)_2CHOH$					1360				176				
Dichlorophenoxyacetic acid		$Cl_2C_6H_3O$								0 - 4					
		CH_2COOH													
Diesel fuel			Fig 2.11				830				210		3		
Diethyl carbonate		$(C_2H_5O)_2CO$			19		980				128	- 48	2b		
Dioctylphthalate			82				980								
Dioxan	D	$O_2(CH_2)_4$	2	3.9			1040	2.22			101	12	1		
Dipentene							850				180	- 70	3		
Diphenyloxide		$C_6H_4OC_6H_4$	4	0.03			1070				253	27	3		
Engine oil SAE 5W-50			Fig 2.12				900/20 °C						3		See Section 2.3 and Table 2.1
Ether	D	$C_4H_{10}O$	< 1	74			710	2.22			35		1		
Etherdiethylene		C_2H_4O		150			870				11	- 113			
Ethyl acetate	D	$CH_3COOC_2H_5$	< 1	13.8	16		900	2.22			77		1		
Ethyl alcohol		C_2H_5OH	< 5	8			790	2.22			78		1		
Ethyldiglycol		$HO(CH_2)_4$	< 5	0.02			990				202		3		
		$O_2C_2H_5$													
Ethyl chloride		C_2H_5Cl	< 1	133	9		890	2.22			12	- 139	1		
Ethyl glycol		$C_4H_{10}O_2$	2.3	0.5			930				135		2b		E-mod and speed of sound - Table 2.1
Ethyl glycol acetate		$C_6H_2O_3$	< 5	0.2			970				150		2b		
Ethylene diamine	D	H_2NCH_2	1.5	1.4			900			9 - 14	117	11			
		CH_2NH_2													
Ethylene glycol	D	$C_2H_6O_2$	18	< 0.13	24		1110	2.22			198		3		E-mod and speed of sound - Table 2.1
Fluorine	T	F									- 188	- 223			
Formaldehyde	D	HCHO			5		820				- 20				
Formic acid	D	HCOOH	1.5	5.3	18	20 - 50	1220	2.22			101				
Freon		CF_2Cl_2					1470				- 29	- 155			
Fruit juices					13										
Furfural	D	C_5H4O_2	1.4	0.3	22		1160	2.22			162		3		
Furfurole	D	C_5H6O_2		0.13	23		1160	2.22			171		3		
Gallic acid		$C_6H_2(OH)_3COOH$					1700			0 - 4					
Glycerine		$CH_2OHCHOH$	1200				1260	2.22			290				
		CH_2OH													
Ground nut oil			Table 2.14				910								See Tables 2.14, 2.15
Hexane		C_6H_{14}	< 1	30	14		660		2.22		69	- 90	1		
Hydrobromic acid		$Hbr \cdot H_2O$	< 1				1780		2.24	0 - 4		- 89			
				0.6		50	1520			0 - 4					
Hydrochloric acid	D	HCl		1.3	13	20	1100		2.24	0 - 4	110				
				1.3	13	36	1180			0 - 4					
Hydrofluoric acid	T	HF					990	2.22	2.24		20	- 83			
						1	1005			0 - 4					
				2.2		10	1030			0 - 4					
				1		40	1130			0 - 4					
						75	1240			0 - 4					
Hydrogen		H_2									- 253	-259			
Hydrogen peroxide		H_2O_2			21		1460				151	- 89			
				2.2	21	10	1040								
Hydrogen sulphide	D	H_2S					950/-50 °C	2.22			- 60	- 86			
Hydriodic acid		$HI.H_2O$					1700		2.24		127				
Industrial oils as per ISO -															
Cutting oil			Table 2.14	(0)			900/20 °C						3		See Section 2.3 and Table 2.1
Gear oil			Table 2.13	(0)			905/20 °C						3		

Substance		Chemical formula	Viscosity mm^2/s at 20 °C	Vapour pressure kPa		Conc. % in H_2O	Density			pH/ region	Boiling point °C at 101.3 kPa (760 mm Hg)	Melting point °C	Fire hazard class- See Section 2.1.7	Solubility in H_2O at various temps. See Fig no.	Comments
				at 20 °C	at various temps. See Fig 2.21 curve no.		at 25 °C	at various temps. See Fig no.	at various conc. See Fig no.						
Hydraulic oils			Table 2.14	(0)			880/20 °C						3		
Turbine oil			Table 2.14	(0)			885/20 °C						3		
Iodine	D	I_2					4940	2.22			184	114			
Iodoform		CHI_3					4010					120		2.28	
Iron (II) chloride		$FeCl_2$				10 - 50				0 - 4					
Iron (III) chloride		$FeCl_3$	1	2.2	13	10	1085			0 - 4					
			15	0.7	13	50	1550			0 - 4					
Iron nitrate		$Fe(NO_3)_3$		2.2		10	1080			0 - 4					
				1.8		25	1230			0 - 4					
Iron (II) sulphate		$FeSO_4\ 7H_2O$	< 5		13		1050								
Iron (III) sulphate		$Fe_2(SO_4)_3$			13	30 - 50				0 - 4					
Isobutylalcohol		$(CH_3)_2CHCH_2OH$	< 5	1.2			800				108		2a		
Isopropylalcohol		$CH_3CHOHCH_3$			16		780				82	- 89	1		
Isopropylnitrate					17		1190				100				
Kerosene			< 5	0.1			780				170		2b		See Tables 2.14, 2.15
Lacquer			200 - 500				900				150		2b		Non-Newtonian, see Section 2.4
Lactic acid		$H_6C_3O_3$			10	1020				0 - 4					
Latex			1000 - 4000			50	1050			6 - 9	100				Non-Newtonian, see Section 2.4
Lead		Pb					11350					327			
Lead acetate		$(CH_3COO)_2Pb\ 3H_2O$				10				6 - 9				2.28	
Lead nitrate		$Pb(NO_3)_2$				30				0 - 4				2.28	
Linseed oil			Table 2.14			930									
Liquid resin			20 - 300		(0)		970								
Lubricating oil SAE 75W-140			Table 2.13	(0)			910/20 °C								See Section 2.3 - Table 2.1
Magnesium carbonate		$MgCO_3$					2960								
Magnesium chloride		$MgCl_2\ GH_2O$			13		2320					708		2.28	
				2.2	13	10 - 50				4 - 6					
				2.2	13	10	1080			4 - 6					
				2.2	13	15	1130			4 - 6					
				2.2	13	25	1150			4 - 6					
				2.2	13	30	1280			4 - 6					
Magnesuim sulphate		$MgSO_4 7H_2O$			13		1680							2.28	
				2.2	13	10	1100			4 - 6					
				2.2	13	20	1300			4 - 6					
Maleic		$(HCCOOH)_2$			13		1590					130			
						50	1300			0 - 4					
Manganese chloride		$MnCl_2$			13		2980					650			
						10	1060			4 - 6					
Manganese chloride hydrate		$MnCl_2\ 4H_2O$					2010					58			
Manganese sulphate		$MnSO_4\ 7H_2O$					2090								
						30	1220			4 - 6					
Mercury	T	Hg					13600				357	- 39			E-mod and speed of sound - Table 2.1
Methyl acetate	D	$CH_3CO_2CH_3$	< 1		12		930				58		1		
Methyl alcohol	D	CH_3OH	< 1	13	14		790	2.22			65		1		E-mod and speed of sound - Table 2.1
Methylene chloride		CH_2Cl_2	< 1	72.4	11		1320	2.22			40				
Methylethylketone		C_4H_6O		15	16		810	2.22			80		1		
Methyl glycol		$C_3H_8O_2$	1.6	1.7			970				125		2b		
Methylchloride		CH_3Cl					1790				- 24		1		
Milk, fresh					13		1020								
Milk, sour					13		1020			4 - 6					
Molasses			100 - 50000												Non-Newtonian, see Section 2.4
Monochlorobenzene		C_6H_5Cl	1	1.6	20		1110	2.22			132		2a		
Monochloracetic acid		CH_2ClCO_2H	2.2 (50 °C)		24		1410			0 - 4	189	63			
Mustard			approx 5000		13										Non-Newtonian, see Section 2.4

2.6.2 Liquid Table

Substance	Chemical formula	Viscosity mm^2/s at 20 °C	Vapour pressure kPa at 20 °C	Vapour pressure kPa at various temps. See Fig 2.21 curve no.	Conc. % in H_2O	Density at 25 °C	Density at various temps. See Fig no.	Density at various conc. See Fig no.	pH/ region	Boiling point °C at 101.3 kPa (760 mm Hg)	Melting point °C	Fire hazard class- See Section 2.1.7	Solubility in H_2O at various temps. See Fig no.	Comments
Naphthalene	$C_{10}H_8$			27		1150				210	80			
Naphthalene sulphonic acid	$C_{10}H_7SO_3H$					1450					102			
					10	1030			0 - 4					
Nickel nitrate	$Ni(NO_3)_2\ 6H_2O$			13		2050				137	57			
				13	10	1050			4 - 6					
Nickel sulphate	$NiSO_4\ 7H_2O$			13		1950								
				13	10	1060			4 - 6					
Nitric acid D	HNO_3			16		1500		2.24	0 - 4	83	- 42			
				16	1	1004			0 - 4					
				16	5	1030			0 - 4					
				16	10	1050			0 - 4					
			2	16	20	1120			0 - 4					
			1.5	16	40	1250			0 - 4					
				16	65	1480			0 - 4					
				16	80	1452			0 - 4					
Nitro benzine D	$C_6H_5NO_2$	2	0.2			1200	2.22			211		3		
Nonane	C_9H_20	1	0.7			720	2.22			151	- 54	2b		
Octane	C_8H_{18}	< 1	2			700	2.22			126		1		
Octanol	$C_6H_{13}\ CHOH\ CH_3$	11				830				194	- 15			
Oleum	$H_2SO_4 + 13\%SO_3$					1910			0 - 4					
Olive oil		Table 2.14	(0)			910								
Oxalic acid	$C_2H_2O_4\ 2H_2O$			13		1650					101			
				13	10 - 50				0 - 4					
				13	0.5	1000			0 - 4					
				13	10	1020			0 - 4					
				13	15	1030			0 - 4					
Ozone	O_3									- 112	- 192			
Paraffin						770				240	45 - 55			
Pentane	C_5H_{12}	< 1	67	11		620	2.22			36	- 130	1		
Peracetic acid	CH_3COOOH					1230				105				
Petrol		< 1				730				40		1		
Phenoldicarbonate						1130				312	80			
Phenolphthalein	$C_{20}H_{14}O_4$					1280					261			
Phosphoric acid D	H_3PO_4			13		1840			0 - 4					
				13	10	1050			0 - 4					
			3	13	20	1120			0 - 4					
			1.3	13	40	1370			0 - 4					
				13	70	1530			0 - 4	181				
				13	85	1690			0 - 4					
Phthalic acid	$C_6H_4(CO_2H)_2$					1600					206			
Phthalic anhydride	$C_6H_4(CO)_2O$					1530					130			
Potassium bi-carbonate	$KHCO_3$			13		2170							2.27	
				13	10				6 - 9					
Potassium bi-chromate	$K_2Cr_2O_7$					2680								
					25	1050			0 - 4					
Potassium bi-sulphate	$KHSO_4$					2320					214			
					5	1035			0 - 4					
Potassium bromide	KBr			13		2750		2.26		1435	730			
Potassium carbonate	K_2CO_3					2420					891		2.27	
			2.2	13	20	1190			9 - 14					
Potassium chlorate D	$KClO_3$			13		2320					356		2.27	
Potassium chloride	KCl			13		1980		2.22			776		2.27	
					20	1130			6 - 9					
					10 - 30				6 - 9					
Potassium chromate D	K_2CrO_4					2730					968		2.27	

Substance		Chemical formula	Viscosity mm^2/s at 20 °C	Vapour pressure kPa		Conc. % in H_2O	Density			pH/ region	Boiling point °C at 101.3 kPa (760 mm Hg)	Melting point °C	Fire hazard class- See Section 2.1.7	Solubility in H_2O at various temps. See Fig no.	Comments
				at 20 °C	at various temps. See Fig 2.21 curve no.		at 25 °C	at various temps. See Fig no.	at various conc. See Fig no.						
Potassium chrome sulphate		$KCr(SO_4)$ $12H_2O$			13		1830								
Potassium cyanate	D	KOCN					2060								
Potassium cyanide	T	KCN			13		1520				635				
					13	5 - 10				9 - 14					
Potassium fluoride	T	KF $2H_2O$					2450				156	41			
Potassium hydroxide	D	KOH					2040		2.23		360			2.27	
				2.2	13	10 - 90				9 - 14					
				2.2	13	30	1290			9 - 14					
				2.2	13	50	1510			9 - 14					
Potassium iodide		KI					3130		2.26						
Potassium nitrate		KNO_3					2110							2.27	
				2.2		10	1080	2.25							
Potassium oxalate		$K_2C_2O_4$ H_2O					2130								
						15	1170			0 - 4					
Potassium perchlorate		$KClO_4$					2520							2.27	
Potassium permanganate		$KMnO_4$					2700								
				2.2	13	20	1040			0 - 4					
Potassium persulphate		$K_2S_2O_8$					2480								
						1 - conc				0 - 4					
Potassium silicate		K_2SiO_3										976		2.27	
Potassium sulphate		K_2SO_4			13		2660								
Propane		C_3H_8			2		580	2.22			- 42	- 190	1		
Propanol		C_3H_7OH	2.8	2.8			800	2.22			97	- 126	1		
Propionic acid		C_2H_5COOH			21		990				141	- 21			
						25	1030			0 - 4					
Pyridine	D	C_5H_5N		3.6	19		980				116	- 42			
Pryogallol	D	$C_6H_3(OH)_3$					1460				309				
						10	1030			0 - 4					
Quinine		$C_{20}H_{24}N_2O_2$										57			
Salicylic acid		$C_6H_4OHCO_2H$					1440				211	159			
						conc	1000			0 - 4					
Seawater - 4% NaCl			1	2.3	13		1020			6 - 9	100				See Section 2.2.4 - Table 2.1
Sebacic acid		$C_8H_{16}(COOH)_2$					915				240	55			
Silver bromide		AgBr					6470				432				
Silver chloride		AgCl					5560				455				
Silver nitrate	D	$AgNO_3$			13		4350		2.26		212				
Size			500 - 3000							6 - 9	100				Non-Newtonian, see Section 2.4
Sodium acetate		$NaC_2H_3O_2$ $3H_2O$			13	7	1450				123	58		2.27	
					13		1050			6 - 9					
Sodium arsenate	T	Na_2HAsO_4 $7H_2O$					1880								
Sodium bi-carbonate		$NaHCO_3$			13		2160								
					13	5	1040			9 - 14					
Sodium bi-sulphate		$NaHSO_4$ H_2O			13		2100					58			
					13	10	1080			0 - 4					
Sodium bi-sulphite		$NaHSO_3$			13		1480								
				2.2	13	10	1100								
Sodium bromide		NaBr $2H_2O$					2180								
						5 - 10				4 - 6				2.27	
Sodium carbonate		Na_2CO_3			13		2530		2.26			850		2.27	
				2.2	13	10 - 50				9 - 14					
				2.2	13	10	1150			9 - 14					
Sodium chlorate		$NaClO_3$					2480					250		2.27	
						28	1410			4 - 6					
Sodium chloride		NaCl			13		2170		2.26		1413	801		2.27	
				2.2	13	1 - conc				6 - 9					

2.6.2 Liquid Table

Substance	Chemical formula	Viscosity mm^2/s at 20 °C	Vapour pressure kPa at 20 °C	Vapour pressure kPa at various temps. See Fig 2.21 curve no.	Conc. % in H_2O	Density at 25 °C	Density at various temps. See Fig no.	Density at various conc. See Fig no.	pH/ region	Boiling point °C at 101.3 kPa (760 mm Hg)	Melting point °C	Fire hazard class- See Section 2.1.7	Solubility in H_2O at various temps. See Fig no.	Comments
Sodium di-chromate	$Na_2Cr_2O_72H_2O$			13		2520								
Sodium fluoride T	NaF					2550					988		2.27	
					5	1050			4 - 6				2.27	
Sodium fluorsilicate	Na_2SiF_6					2680								
Sodium hydrogen difluoride	$NaHF_2$					2080								
Sodium hydrosulphate	$NaHSO_4$					2440								
	$NaHSO_4$ H_2O					2100					58			
Sodium hydroxide D	NaOH			13		2130		2.23		1390	318		2.27	
			2.2	13	10 - 70				9 - 14					
			2.2	13	30	1330			9 - 14					
			2.2	13	50	1530			9 - 14					
Sodium hypochlorite D	NaOCl		2.3	13	5	1020			6 - 9					
Sodium nitrate	$NaNO_3$			13		2260		2.25			307		2.27	
				13	4	1030			6 - 9					
Sodium perchlorate	$NaClO_4H_2O$					2020					130		2.27	
					10	1070			4 - 6					
Sodium peroxide	Na_2O_2					2810								
					10	1110			6 - 9					
Sodium phosphate primary	NaH_2PO_4 $12H_2O$					1910					60			
			2.2		10	1070								
Sodium phosphate secondary	Na_2HPO_4 $12H_2O$					1520								
			2.3		50	1250			0 - 4					
Sodium phosphate tertiary	Na_3PO_4 $12H_2O$					1620								
			2.2		conc	1110			9 - 14					
Sodium silicate	Na_2SiO_3			13		2400		2.25			1088			
Sodium sulphate	Na_2SO_4 $10H_2O$			13		1460					32		2.27	
			2.2	13	1	1000			4 - 6					
			2.2	13	5	1020			4 - 6					
Sodium sulphide	Na_2S $9H_2O$			13		1420								
			2.2	13	20	1070			6 - 9					
Sodium sulphite	Na_2SO_3 $7H_2O$			13		1530							2.27	
Sodium thiosulphate	$Na_2S_2O_3$ $5H_2O$			13		1730					40			
			2.2		1-conc				4 - 6					
Soya oil		Table 2.14	(0)			920								
Stearic acid	$CH_3(CH_2)_{16}COOH$					840				380	- 70			
Styrene	C_8H_8	< 1	0.95	21		910				145		2b		
Sulphate liquor 20% DS*		Fig 2.20		13		1090								*DS - Dry substance, content by weight - see
40% DS*		Fig 2.20				1195								equations 2.13 to 2.20
50% DS*		Fig 2.20				1250								
60% DS*		Fig 2.20				1305								
Sulphur	S					2060	2.22			445	113			
Sulphur dioxide	SO_2					1380	2.22			- 10	- 73			
	SO_2 H_2O			5				2.24	0 - 4					
Sulphuric acid D	H_2SO_4			13		1030		2.24	0 - 4	338				
Sulphurous acid D	H_2SO_3			13		1840		2.24	0 - 4					
				13	10	1070			0 - 4					
				13	15	1100			0 - 4					
			2.1	13	20	1140			0 - 4					
				13	30	1220			0 - 4					
				20	40	1300			0 - 4					
				20	50	1400			0 - 4					
				20	60	1500			0 - 4					
			0.1	20	70	1610			0 - 4					
				20	80	1730			0 - 4					
				20	90	1820			0 - 4					
Sweetened juice				13										

Substance		Chemical formula	Viscosity mm^2/s at 20 °C	Vapour pressure kPa		Conc. % in H_2O	Density			pH/ region	Boiling point °C at 101.3 kPa (760 mm Hg)	Melting point °C	Fire hazard class- See Section 2.1.7	Solubility in H_2O at various temps. See Fig no.	Comments
				at 20 °C	at various temps. See Fig 2.21 curve no.		at 25 °C	at various temps. See Fig no.	at various conc. See Fig no.						
Tannic acid		$C_{76}H_{52}O_{46}$				5 - 50				0 - 4					
						10	1035			0 - 4					
Tartaric acid		$C_4H_6O_6$			13	10 - 50				0 - 4					
Tetrachloroethane		$CHCl_2CHCl_2$	1.1	0.8	20		1590	2.22			146				
Tetrachloroethylene	D	C_2Cl_4	< 1	2.4	19		1620	2.22			121				
Tetralin		$C_{10}H_{12}$		0.1			970				207		3		
Tin		Sn					5750				2260	232			
Tin (II) chloride		$SnCl_2\ 2H_2O$					2710					38			
Tin (IV) chloride		$SnCl_4$					3950				652	246			
Tincture of iodine			< 5	8			905				78		1		
Thioglycollic acid		CH_2SHCO_2H					1330				120	- 17			
Tuolene	D	$C_6H_5CO_3$	< 1	3.5	18		860	2.22			110		1		
Trichloroacetic acid	D	CCl_3CO_2H					1620			0 - 4	198	56			
Trichloroethylene	D	C_2HCl_3	< 1	7.5	16		1460	2.22			87				
Turpentine							860				156		2b		E-mod and speed of sound - Table 2.1
Unslaked lime		CaO					3250				2850	2580			
						10				9 - 14					
Valtran			Table 2.14	(0)			920								
Vinylacetate		$C_4H_6O_2$		15.5			930				73		1		
Vinegar							990			0 - 4					
Water			1	2.3	13		997			6 - 9	100				See Section 2.2 - Table 2.1
Xylene	D	$C_6H_4(CH_3)_2$	< 1	1			860	2.22			139		2a		
Zinc		Zn					7140					419			
Zinc chloride		$ZnCl_2$			13		2910					283			
				2.2	13	5	1030			4 - 6					
				2.2	13	20	1150			4 - 6					
				2.2	13	30	1220			4 - 6					
				2.2	13	40	1420			4 - 6					
				2.2	13	60	1750			4 - 6					
Zinc cyanide		$Zn(CN)_2$					1850								
Zinc sulphate		$ZnSO_4\ 7H_2O$			13		1960					100		2.28	

2.6.3 Supplementary diagrams

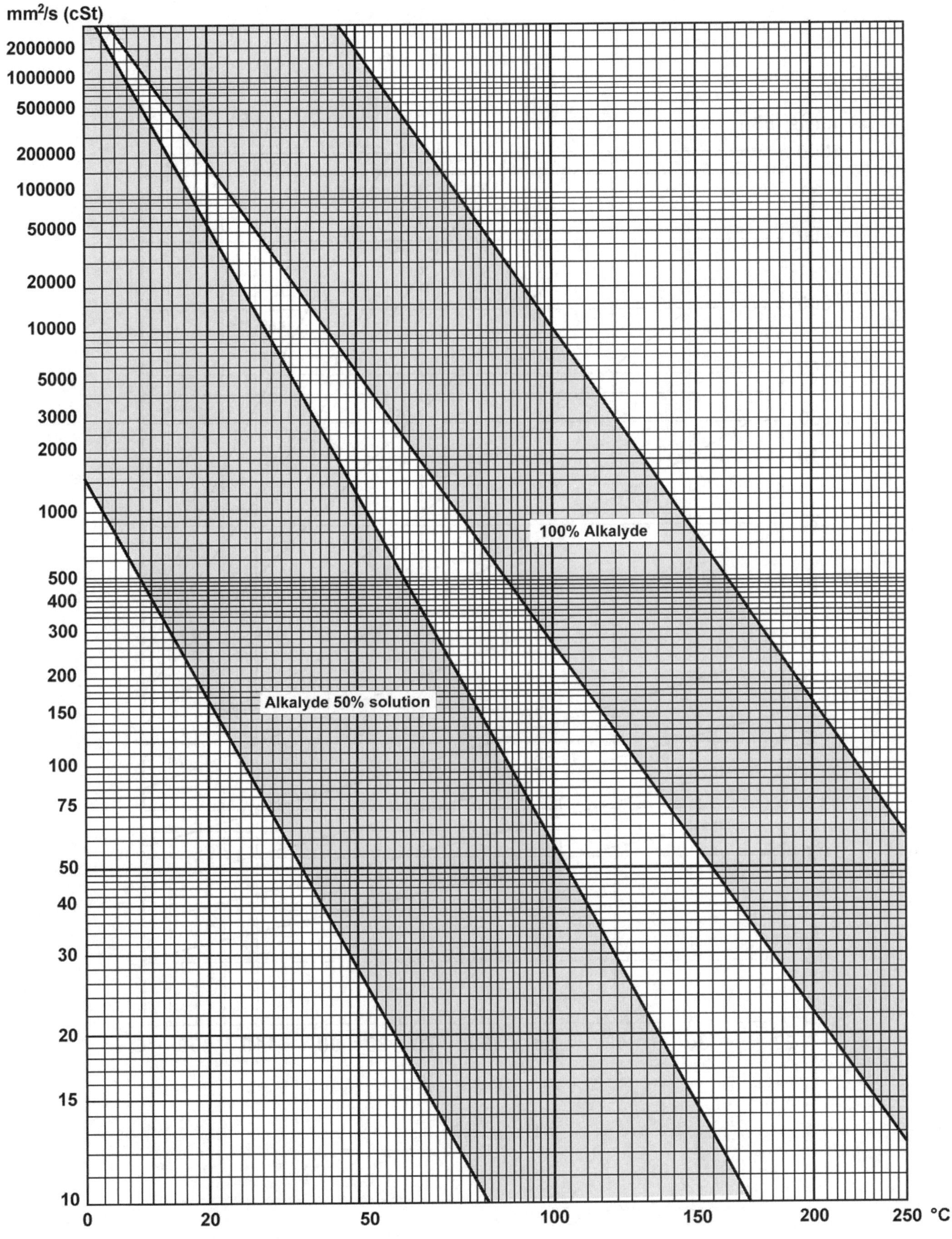

Figure 2.19 Viscosity of alkydes at various temperatures

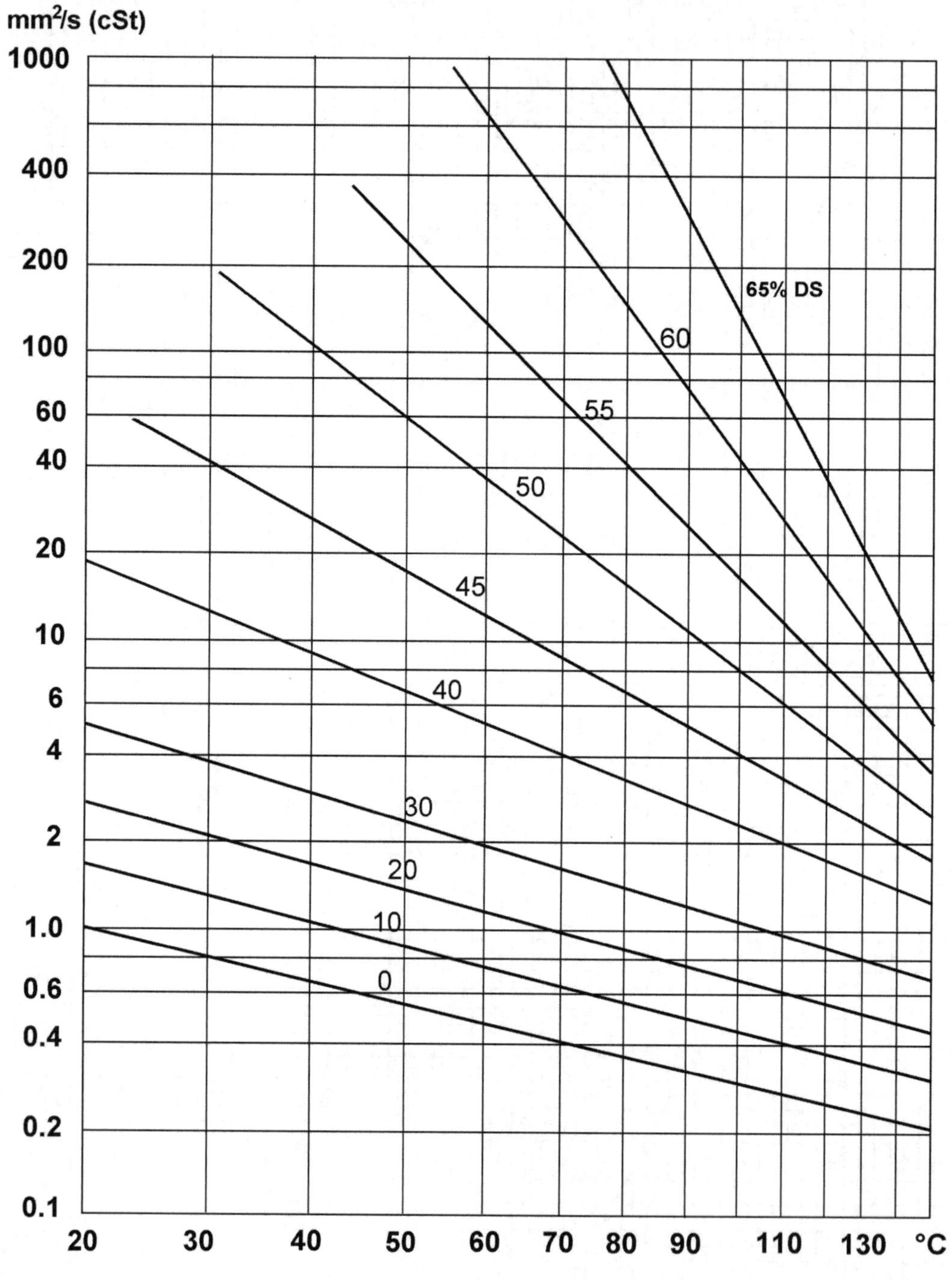

Figure 2.20 Viscosity temperature diagram for black liquor

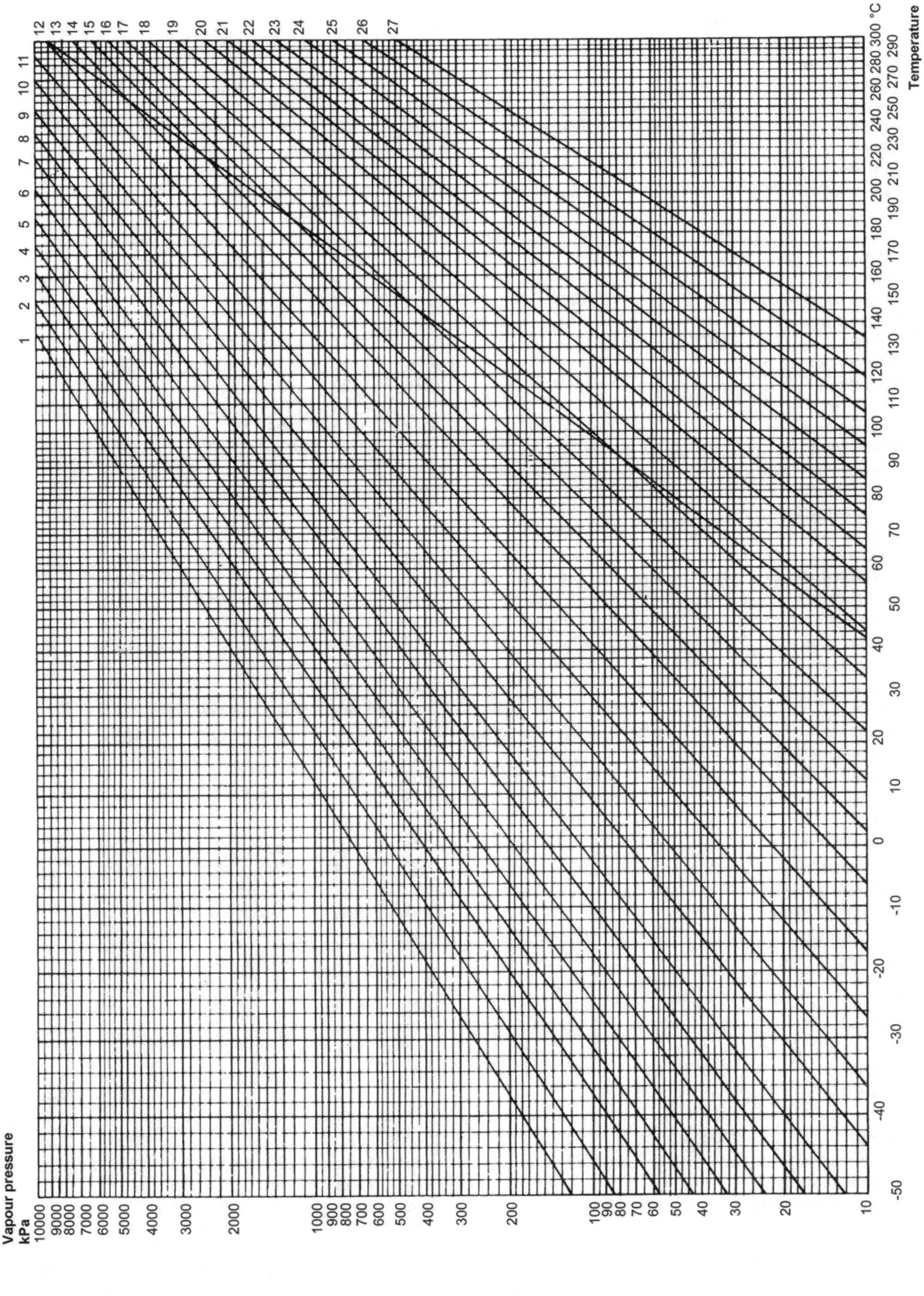

Figure 2.21 Vapour pressure diagram (see Liquid Table, Section 2.6.2 for cross references)

Substance	Temperature °C									
	-100	-75	-50	-25	0	20	50	100	150	200
Acetaldehyde						783				
Acetic acid						1049	1018	960	896	827
Acetone	920	893	868	840	812	791	756			
Ammonia			695		636	609	561	458		
Aniline					1039	1022	996	951		
Benzene					900	879	847	793	731	661
Benzylalcohol					1061	1045	1022			
Bromide					3188	3120				
Butane	698	676	652	627	601	579	542	468	296	
Butanol					825	810				
Butyric acid					977	958	927			774
Carbon bi-sulphide	1432		1362		1292	1262				
Carbon dioxide			1154	1070	925	772				
Chlorine	1717		1595		1469	1411	1314	1111	810	
Chlorobenzene					1128	1106	1074	1019	960	896
Chloroform			1618		1526	1490	1433			
Cyclohexane						779	750	700	645	578
Cyclohexanone					963	952	925			
Cyclohexene				842	830	811	780	732		
Dichloroacetic acid						1552				
Dichloroethane					1207	1176				
Dichloroethylene						1250				
Dioxane						1030	1010			
Ethanol					806	789	763	716	649	557
Ether (Diethylether)	842	816	790	764	736	714	676	611	518	
Ethyl acetate					924	901	864	797	721	621
Ethylene glycol					1128	1112				
Ethyl chloride					919	892	846			
Formic acid						1220	1184			
Furfurol					1181	1160	1128			
Glycerine					1273	1261	1242	1209		
Hexane		742	721	700	678	659	631	580	520	438
Hydrochloric acid	1235		1076		920					
Hydrofluoric acid	1660		1123		1002	987				
Hydrogen sulphide	1170		980		870					
Iodine	5060		5010		4960	4940			3780	
Methanol					810	792	765	714	650	553
Methyl acetate					959	934	894	822	734	610
Methylene chloride					1362	1326				
Methylethylketone					826	803				
Nitrobenzene					1223	1203	1174			
Nonane			769	751	733	718	694	653	609	
Octane		860	757	738	719	703	678	635	588	532
Octanol					842	829				
Pentane	737	715	693	670	646	626	596	533	460	
Phenol					1092	1071	1050			
Propane	646	619	590	560	528	501	450			
Propanol						804	779	733	674	592
Styrol						910				
Sulphur	2100		2080		2070	2060			1780	1740
Sulphur dioxide			1557		1435	1383	1296	1118	768	
Tetrachloroethane					1626	1593				
Tetrachloroethylene					1656	1621				
Toluole					885	868	839	739	737	672
Trichlorethylene						1463				
Xylene					881	864	838	795		678

Figure 2.22 Density for liquids at various temperatures

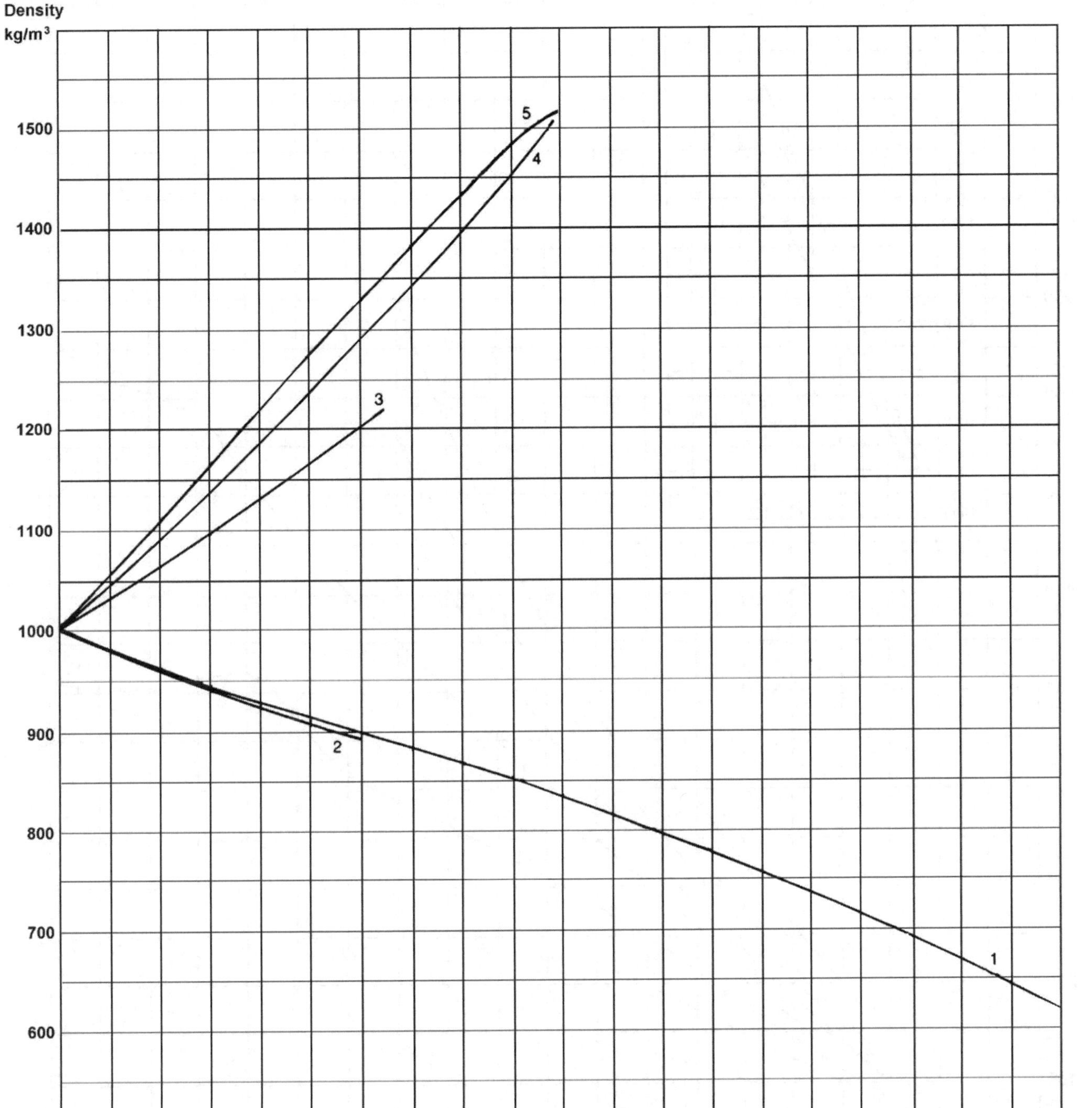

1. Ammonia solution at 15 °C
2. Ammonium hydroxide solution 20 °C
3. Calcium hydroxide solution 20 °C
4. Potassium hydroxide solution 15 °C
5. Sodium hydroxide solution 15 °C

Figure 2.23 Density against concentration for various bases

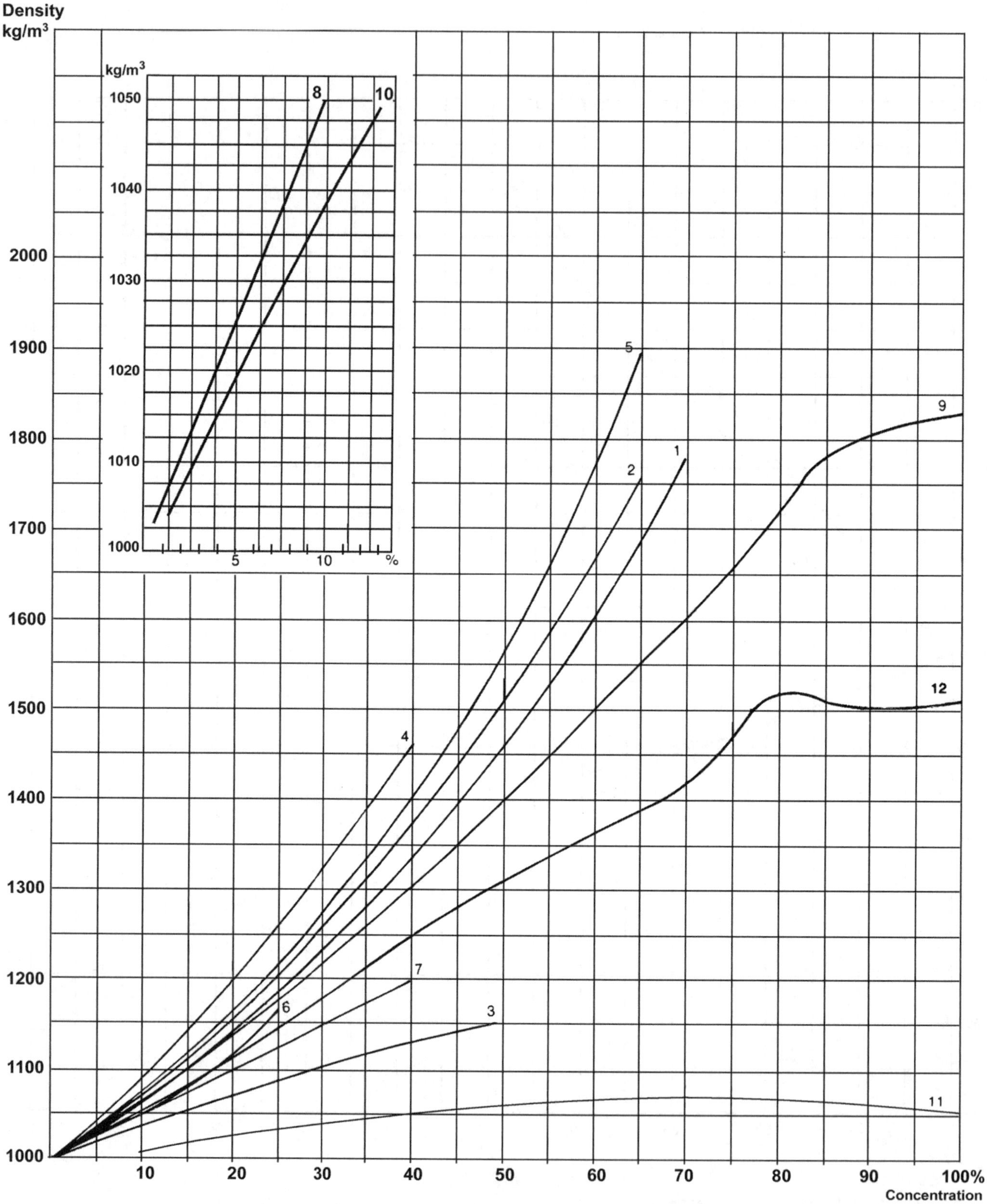

1. Arsenic acid	at	15°C	7. Hydrochloric acid	at	20°C
2. Hydrobromic acid		20°C	8. Sulphur dioxide solution		15°C
3. Hydrofluoric acid		20°C	9. Sulphuric acid		20°C
4. Iodic acid		18°C	10. Sulphurous acid		15,5°C
5. Hydroidinic acid		20°C	11. Acetic acid		20°C
6. Chloric acid		18°C	12. Nitric acid		18°C

Figure 2.24 Density against concentration of various acids

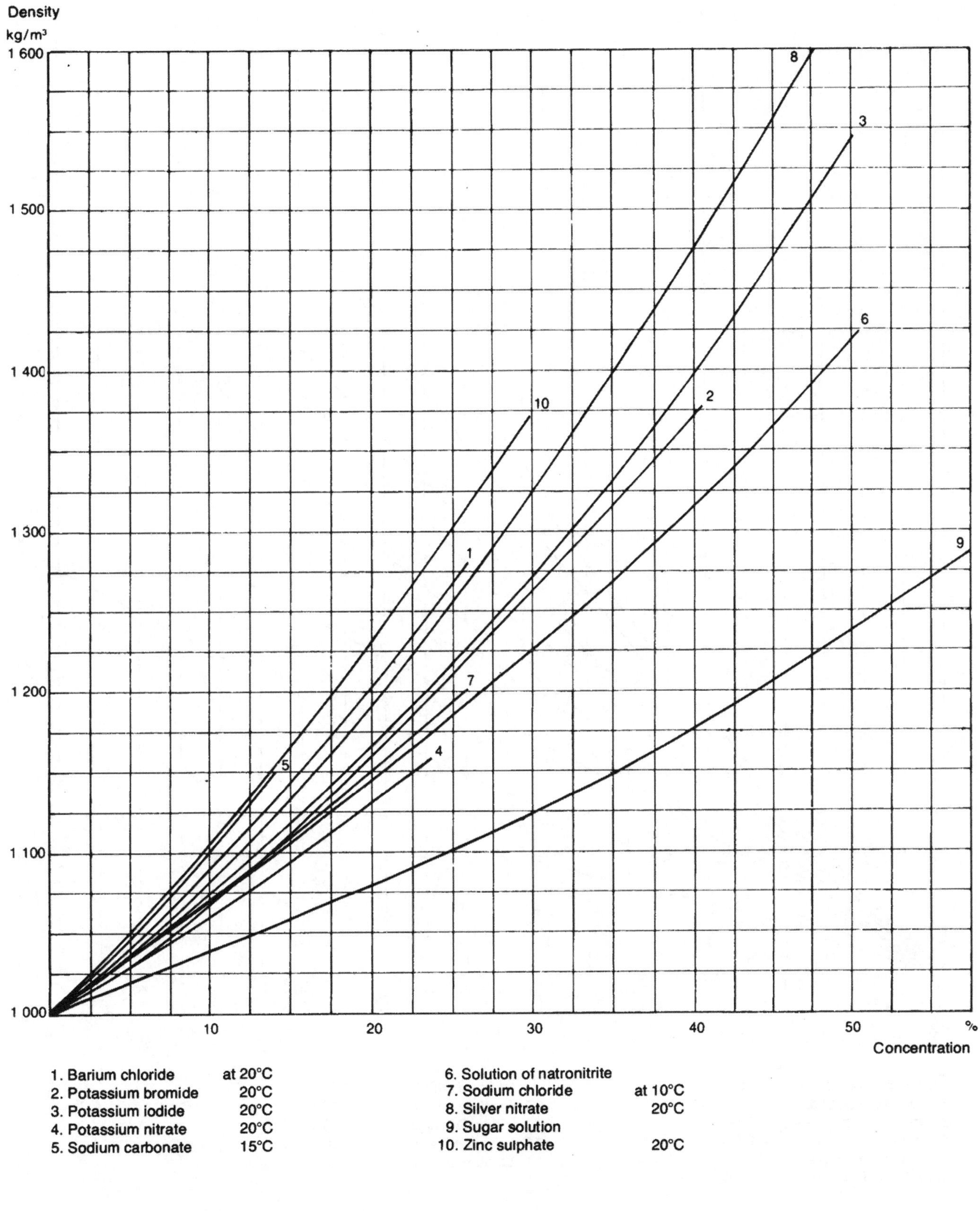

Figure 2.25 Density against concentration for various salts

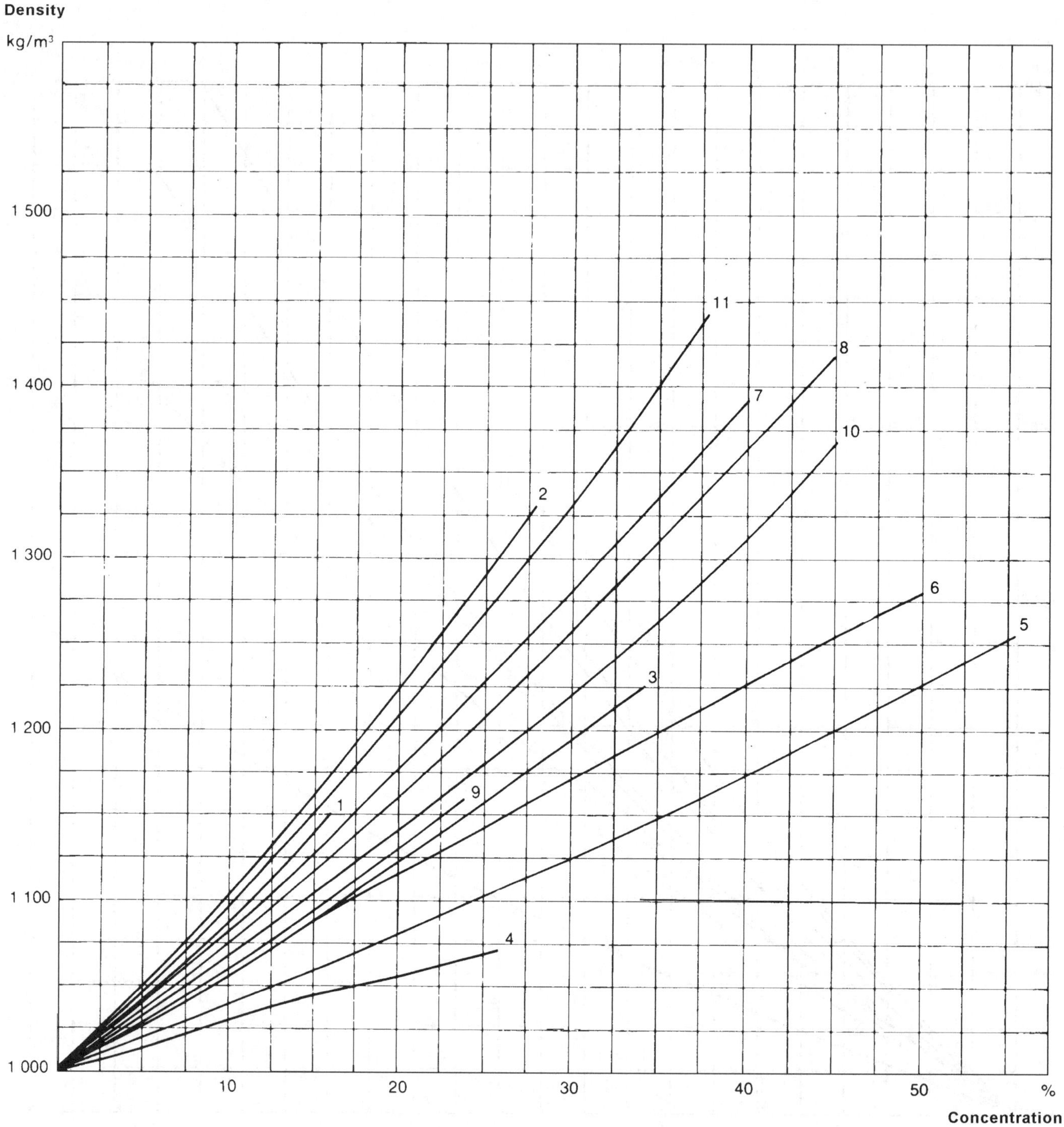

1. Aluminium chloride at 18°C
2. Aluminium sulphate 19°C
3. Ammonium bromide 18°C
4. Ammonium chloride 20°C
5. Ammonium nitrate 20°C
6. Ammonium sulphate 20°C
7. Calcium chloride at 20°C
8. Calcium nitrate 18°C
9. Potassium chloride 20°C
10. Sodium nitrate 20°C
11. Sodium silicate 20°C

Figure 2.26 Density against concentration for various salts

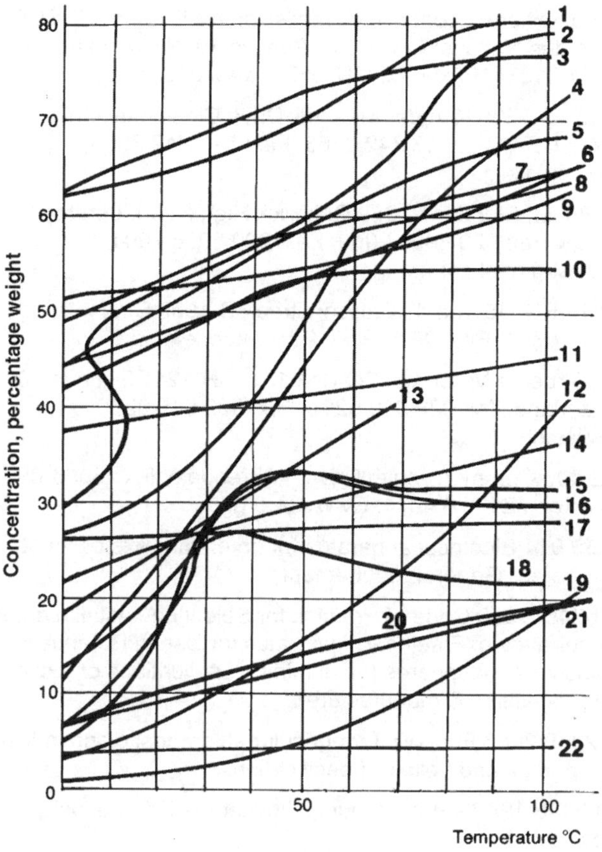

Curve no.	
Potassium bicarbonate	13
Potassium hydroxide	7
Potassium carbonate	9
Potassium chlorate	12
Potassium chloride	14
Potassium chromate	11
Potassium nitrate	4
Potassium perchlorate	19
Potassium sulphate	21
Sodium acetate	8
Sodium bicarbonate	20
Sodium bromide	10
Sodium dichromate	1
Sodium fluoride	22
Sodium hydroxide	2
Sodium carbonate	15
Sodium chlorate	5
Sodium chloride	17
Sodium nitrate	6
Sodium perchlorate	3
Sodium sulphate	16
Sodium sulphite	18

Figure 2.27 Solubility of salts in water at various temperatures

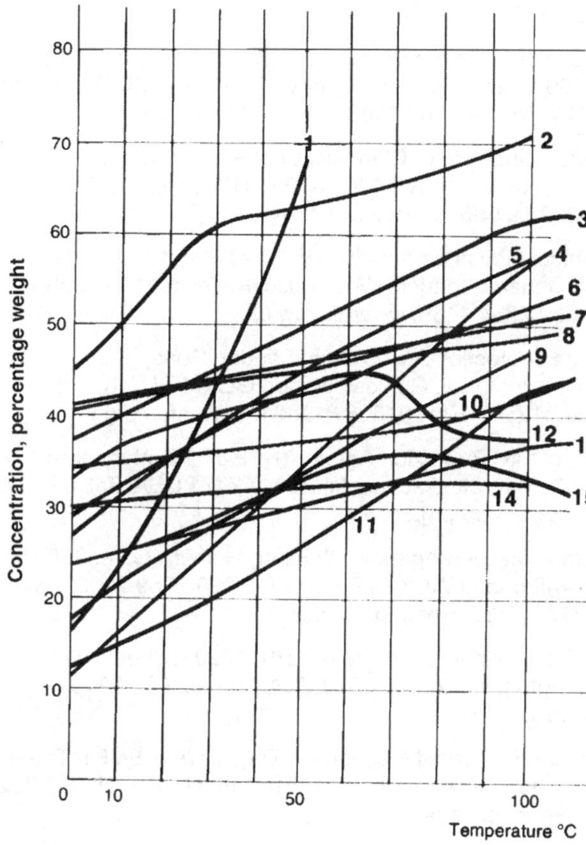

	Curve No.
Aluminium chloride	14
Aluminium nitrate	3
Aluminium sulphate	9
Ammonium perchlorate	4
Ammonium sulphate	7
Barium chloride	13
Lead acetate	1
Lead nitrate	5
Iron (II) chloride	8
Copper (II) chloride	6
Copper (II) nitrate	2
Copper (II) sulphate	11
Magnesium chloride	10
Magnesium sulphate	15
Zinc sulphate	12

Magnesium sulphate 15
Zinc sulphate 12

Figure 2.28 Solubility of salts in water at various temperatures

2.7 Useful references

Process Plant Design and Operation, D.Scott, F Crawley, Institution of Chemical Engineers, 1992, ISBN 0 85295 278 3.

Nomenclature for Hazard and Risk Assessment in the Process Industries, David Jones Institution of Chemical Engineers, ISBN: 085295297X.

Institution of Chemical Engineers (IChemE), Davis Building, 165-189 Railway Terrace, Rugby CV21 3HQ UK, Tel: 01788 578214, Fax: 01788 560833, www.icheme.org.

American Institute of Chemical Engineers (AIChE), 3 Park Avenue, New York, NY 10016-5991 USA, Tel: 212 591 8100, Fax: 212 591 8888, www.aiche.org.

Chemical Industries Association (CIA), Kings Buildings, Smith Square, London SW1P 3JJ UK, Tel: 020 7834 3399, Fax: 020 7834 4469, www.cia.org.uk.

UK Fire Protection Association, London Road, Moreton-in-Marsh, Gloucestershire GL56 0RH UK, Tel: 01608 812500, Fax: 01608 812501, www.nfpa.org.

National Fire Protection Agency (NFPA), 1 Batterymarch Park, Quincy, Massachusetts 02169-7471 USA, Tel: 617 770 3000, www.nfpa.org.

Health Protection Agency, 7th Floor, Holborn Gate, 330 High Holborn,London WC1V 7PP UK, Tel: 020 7759 2700, Fax: 020 7759 2733, www.hpa.org.uk

American Petroleum Institute (API), 1220 L Street, NW, Washington, DC 20005-4070 USA Tel: 202 682 8000, www.api.org.

American Society of Mechanical Engineers (ASME), Three Park Avenue, New York, NY 10016-5990 USA, Tel: 973 882 1167, www.asme.org.

CONCAWE, Boulevard du Souverain 165, B-1160 Brussels Belgium, Tel: 02 566 9160, Fax: 0 2 566 9181, www.concawe.be.

Energy Institute, 61 New Cavendish Street, London, W1G 7AR, UK,Tel: 020 7467 7100, Fax: 020 7255 1472, www.energyinst.org.uk.

American National Standards Institute (ANSI), 25 West 43rd Street, (between 5th and 6th Avenues), 4th floor, New York, NY 10036 USA, Tel: 212 642 4900, www.ansi.org.

Health and Safety Executive (HSE), Rose Court, 2 Southwark Bridge, London SE1 9HS UK, Tel: 0845 345 0055, www.hse.gov.uk.

United Nations Economic Commission for Europe (UNECE), Palais des Nations, CH - 1211 Geneva 10, Switzerland, Tel: 022 917 1234, Fax: 0 22 917 0505, www.unece.org.

ACGIH®, 1330 Kemper Meadow Drive Cincinnati, Ohio 45240, USA, Tel: 513 742 6163, Fax: 513 742 3355, www.acgih.org.

BSI British Standards, 389 Chiswick High Road, London W4 4AL UK, Tel: 020 8996 9000, Fax: 020 8996 7001, www.bsi-global.com.

CENELEC, 35, rue de Stassartstraat, B-1050 Brussels, Tel: 02 519 6871, Fax: 02 519 6919, www.cenelec.org

IEC, 3 rue de Varembé, PO Box 131, CH-1211 Geneva 20, Switzerland, Tel: 022 919 0211, Fax: 022 919 0300, www.iec.ch.

SAE 400 Commonwealth Drive, Warrendale, PA 15096-0001 USA, Tel: 724 776 4841, www.sae.org.

EN 50 014 Electrical apparatus for potentially explosive atmospheres- General requirements.

BS 5345 :1980 Code of Practice for Selection, Installation. & Maintenance of Electrical. Apparatus for Use in Potentially Explosive Atmospheres (save mining applications or explosive processing & manufacture).

BS 2869:2006 Fuel oils for agricultural, domestic and industrial engines and boilers. Specification.

ISO 3448:1992 Industrial liquid lubricants - ISO viscosity classification.

BS 4231:1992 Classification for viscosity grades of industrial liquid lubricants

BS 5908:1990 Code of practice for fire precautions in the chemical and allied industries.

ANSI B31.3 (Chemical Plant and Petroleum Refinery Code)

Flow of liquids

3

3.1 Fundamental equations

3.1.1 Explanation of concepts

The processes of flow occurring in nature are usually very complex and difficult to deal with. In many technical applications, however, perfectly acceptable results can be obtained from calculations based upon simplified considerations. Some of the concepts and conditions germane to this concept are discussed in this chapter.

A flow process is steady if all flow parameters - pressure, velocity, etc. - at any specific point in the flow field are independent of time. According to this definition practically all flow processes are unsteady. In practice, many can, with sufficient accuracy, be treated as steady using suitable mean time values. The fundamental equations presented in later sections are valid for steady flow.

Extra care should be taken with certain pumping installations where steady state conditions may not exist. Low head, high flow centrifugal pumps with wide impeller tip widths produce velocity variations across the flow stream. Some positive displacement pumps, with a small number of pumping elements, produce cyclic flow and pressure variations. The amplitude of these variations is dependant upon the piping system as well as the pump construction.

Flow is generally three-dimensional, i.e. the flow parameters vary with all three co-ordinates defining some point in space. In many practical cases of flow the number of dimensions studied may be reduced whilst still maintaining accuracy. A common example is pipe flow, where the parameters are assumed to vary in only one direction, i.e. the longitudinal dimension of the pipe.

One-dimensional flow assumes that flow parameters can be described by mean values across the flow through-section, see Figure 3.1. In principle, different mean values should be applied when studying continuity, momentum and energy. In the case of pipe flow, the mean velocity in the pipe is defined as the volume flow divided by the cross-sectional area of the pipe ($v = Q/A$). This mean velocity can as a rule be used with adequate accuracy for most purposes.

An important property of a flowing medium is its density and the changes in density which occur during flow. A gas is compressed, i.e. the density increases, when the pressure somewhere in the flow increases. This type of flow is called compressible flow. In a liquid system density variations are very small under great changes of pressure. Liquid flow can thus, with sufficient accuracy, be treated as incompressible. This is also the case with gases at low flow velocities when the pressure changes are small. Liquid compressibility may be important for pump selection. Hot liquids and liquefied gases may require special consideration depending upon the differential pressure. Power is required to compress liquids. Ignoring liquid compressibility can result in low flow and/or undersized pump drivers.

A streamline is a curve to which the velocity vector is tangential at any point. In the case of steady flow, the streamlines remain unchanged with time and describe the path of a liquid particle passing through the flow field. The streamlines through all

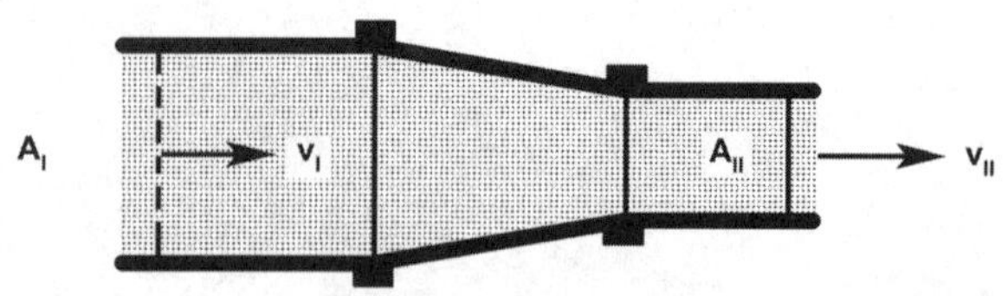

Figure 3.1 Example of one-dimensional flow

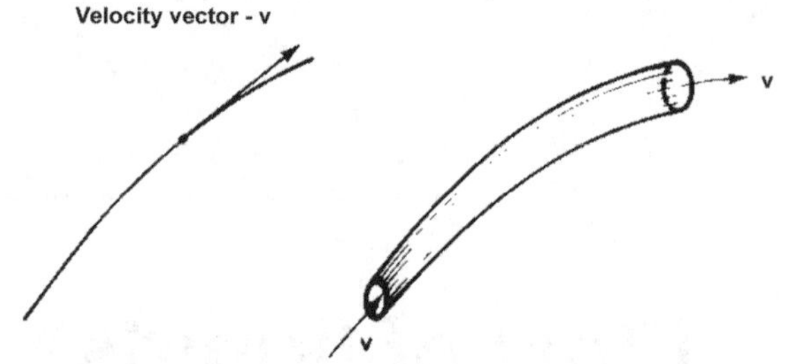

Figure 3.2 Streamline and flow tube

points on a closed curve in the flow field constitute a flow tube. See Figure 3.2. No mass will pass through the circumferential boundary surface of a flow tube. The flow tube is thus reminiscent of an ordinary pipe. In the case of a pipe, however, there are strong frictional effects associated with the wall of the pipe which do not necessarily occur in a flow tube.

3.1.2 The continuity equation

The continuity equation is a statement for the condition that mass is not created or destroyed during a flow process.

Assuming that the flow is steady, the mass flow $\dot{m}$ must be of equal magnitude everywhere along the pipe or the flow tube. In the case of one-dimensional flow in Figure 3.1.

$$\dot{m} = r_I \cdot v_I \cdot A_I = r_{II} \cdot v_{II} \cdot A_{II} \qquad \text{Equ 3.1}$$

or for an incompressible liquid flow,

$$Q = v_I \cdot A_I = v_{II} \cdot A_{II} \qquad \text{Equ 3.2}$$

where:

Q	=	volume flow (m^3/s)
v	=	Q/A = flow velocity (m/s)
A	=	cross-sectional area (m^2)

When the cross-sectional area in a pipe reduces, then, according to the continuity equation, the flow velocity increases. If the area is halved, the velocity doubles and so on.

Since there is no increase in mass at the point of branching, the mass flow entering will equal the total mass discharging per unit time. Using the symbols of Figure 3.3,

$$Q_I = Q_{II} + Q_{III}$$

or

$$v_I \cdot A_I = v_{II} \cdot A_{II} + v_{III} \cdot A_{III} \qquad \text{Equ 3.3}$$

3.1.3 Bernoulli's equation

Bernoulli's equation is an equation of motion for incompressible flow based on Newton's second law of motion adapted for fluids and Euler's equation of motion. The equation expresses the relationship between:

- static pressure
- potential energy
- kinetic energy

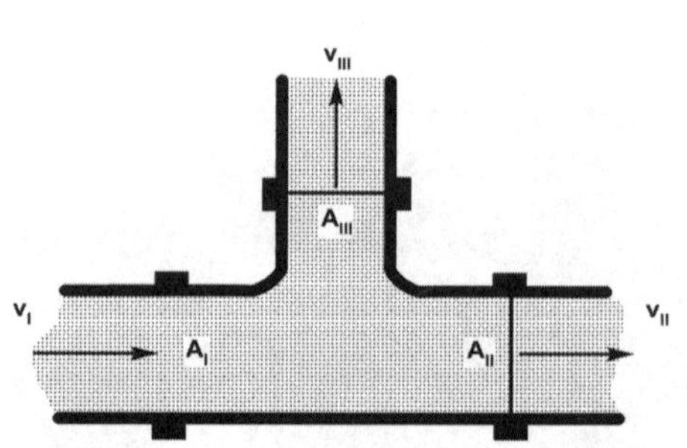

Figure 3.3 Branching

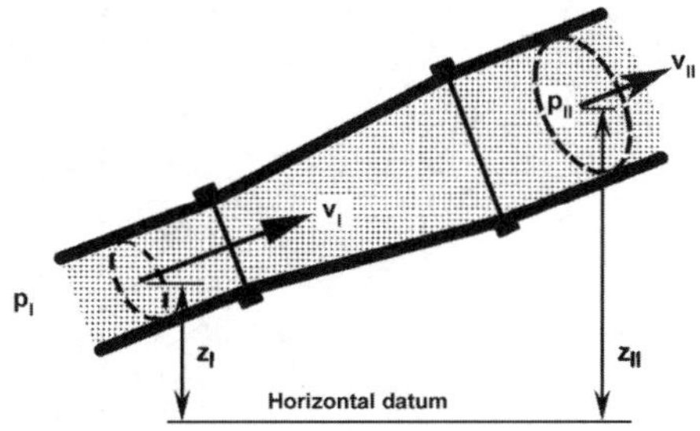

Figure 3.4 Symbols used in Bernoulli's equation

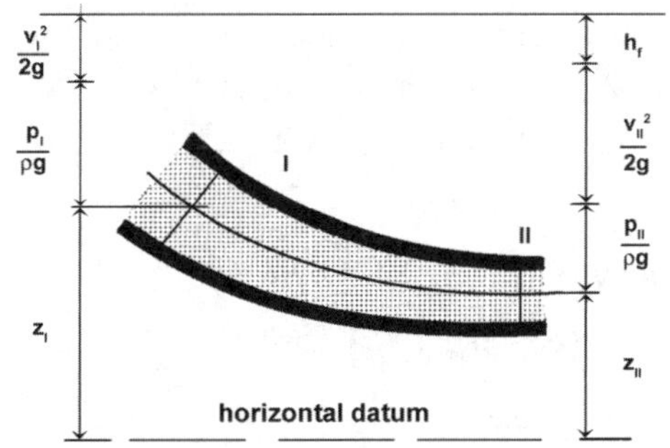

Figure 3.5 Graphic illustration Bernoulli's equation

- friction losses

Using Figure 3.4, Bernoulli's equation, for steady, one-dimensional, constant temperature and incompressible flow between stations I and II, becomes:

$$\frac{p_I}{\rho} + \frac{v_I^2}{2} + gz_I = \frac{p_{II}}{\rho} + \frac{v_{II}^2}{2} + gz_{II} + \frac{\Delta p_f}{\rho} \qquad \text{Equ 3.4}$$

where:

p	=	static pressure (gauge) (Pa)
ρ	=	density (kg/m^3)
v	=	velocity (m/s)
g	=	gravitational acceleration (m/s^2)
z	=	height above datum (m)
Δp_f	=	frictional pressure loss (Pa)

Equation 3.4 expresses the various energy terms per unit mass of fluid, J/kg.

In rotodynamic pump technology it is practical to express Bernoulli's equation in terms of "head" i.e. metres of liquid column. If all terms in Bernoulli's equation are divided through by g:

$$\frac{p_I}{w} + \frac{v_I^2}{2g} + z_I = \frac{p_{II}}{w} + \frac{v_{II}^2}{2g} + z_{II} + \frac{\Delta p_f}{w} \qquad \text{Equ 3.5}$$

where:

w	=	specific weight (N/m^3)
w	=	ρg

The various terms are then called:

$\frac{p}{rg}$ or $\frac{p}{w}$ = static head (m)

$\frac{v^2}{2g}$ = velocity head (m)

z = potential head (m)

$\frac{\Delta p_f}{w} = h_f$ = head loss (m))

Since all the terms in equation 3.5 refer to height they are easy to illustrate graphically. See Figure 3.5.

Total head is sometimes designated by H. The relationship between pressure and head is p/w = h. The three head terms can be converted to equivalent pressure terms. Static pressure plus velocity, or dynamic pressure is called total pressure or stagnation pressure.

$$\text{total pressure} = p + \frac{rv^2}{2} = wh + \frac{wv^2}{2} \qquad \text{Equ 3.6}$$

Total and static pressure are measured in different ways and illustrated in Figure 3.6. A pressure tap at right angles to the di-

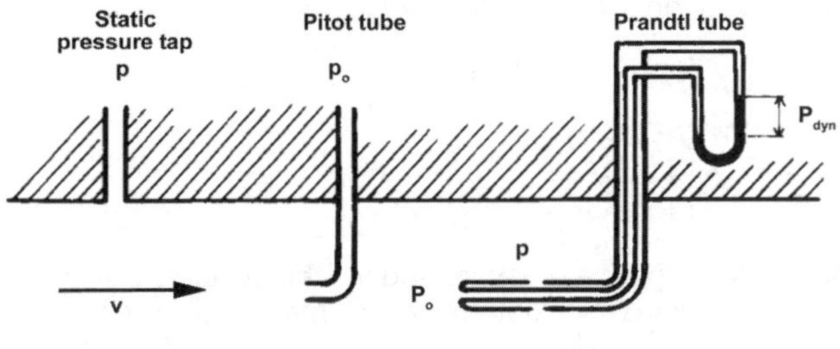

Figure 3.6 Measuring of static, total and dynamic pressure

rection of flow will sense the static pressure. Ahead of a Pitot tube the velocity is reduced to zero and the static pressure rises to stagnation pressure.

As a rule, it is easy to measure the static pressure, whereas the total pressure can easily be affected by measurement errors. A Prandtl tube or Pitot-static tube measures both total and static pressure. The differential pressure, the dynamic pressure, $p_{dyn} = wv^2/2g$ is obtained directly.

3.1.4 The momentum equation

The product of mass times flow velocity for a liquid particle is called its momentum. The momentum equation for steady flow reads:

$$F = \frac{d}{dt} \cdot \dot{m}$$

The resultant F of all external forces on a control volume is equal to the changes in momentum for the outflowing and inflowing mass per unit time, or,

$$F = \dot{m}\left(\bar{v}_2 - \bar{v}_I\right) \qquad \text{Equ 3.7}$$

where:

$\dot{m}$	=	mass flow dm/dt $\left(\frac{kg/s}{(m/s)}\right)$
$\bar{v}$	=	velocity vector

The momentum equation is illustrated in the following example. The problem is to determine the force required to hold a pipe bend in position, as illustrated in Figure 3.7.

The control volume is defined by the interior of the pipe bend.

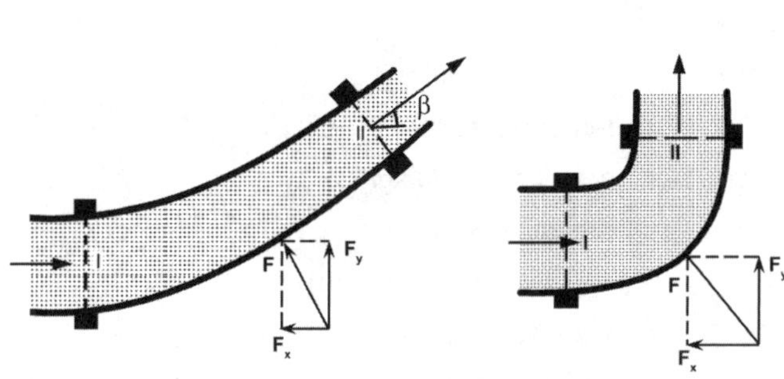

Figure 3.7 Pipe bends, arbitrary (left) and 90° bend

Then, for the arbitrary bend in the x axis:

$$-F_x + p_I \cdot A_I - p_{amb} \cdot A_I - p_{II} \cdot A_{II} \cdot \cos\beta + p_{amb} \cdot A_{II} \cdot \cos\beta = \dot{m}\left(v_{II} \cdot \cos\beta - v_I\right)$$

or:

$$F_x = \dot{m} \cdot v_I + \left(p_I - p_{amb}\right) \cdot A_I - \left[\dot{m} \cdot v_{II} + \left(p_{II} - p_{amb}\right) \cdot A_{II}\right] \cdot \cos\beta$$

Similarly, in the y axis:

$$F_y = \left[\dot{m} \cdot v_{II} + \left(p_{II} - p_{amb}\right) \cdot A_{II}\right] \cdot \sin\beta$$

For a 90° bend $\beta = 90°$, therefore:

$$F_x = \dot{m} \cdot v_I + \left(p_I - p_{amb}\right) \cdot A_I$$

$$F_y = \dot{m} \cdot v_{II} + \left(p_{II} - p_{amb}\right) \cdot A_{II}$$

$$F = \sqrt{F_x\ 2 + F_y^{\ 2}} \qquad \text{Equ 3.8}$$

NOTE: It is a gauge pressure and not the absolute pressure in the pipe which determines the magnitude of the force.

Also that the forces can be determined without detailed knowledge of the internal flow process through the bend. The momentum equation applies regardless of whether the process has losses or not.

It is often the case that the terms m and v are small compared with the others. For a 90° bend with A_I being equal to A_{II}.

$$F = \sqrt{2} \cdot \left(p - p_{amb}\right) \cdot A \qquad \text{Equ 3.9}$$

where:

p	=	pressure in pipe (absolute) (Pa)
p_{amb}	=	ambient pressure (absolute) (Pa)
A	=	pipe cross-sectional area (m^2)
F	=	fixing force (N)

The fixing force has an angle of 135° to the incoming direction of flow.

3.1.5 The energy equation

The energy equation is an extension of the energy principle which states that energy cannot be created or destroyed but can only be converted to other forms. The energy equation for steady state one-dimensional pump flow is:

$$W_{in} = \left(h_{II} - h_I\right) + \frac{\left(c_{II}^2 - c_I^2\right)}{2} + g\left(z_{II} - z_I\right) + Q_{out} \qquad \text{Equ 3.10}$$

The term h, specific enthalpy, is made up from two components; u, ($u = Tc_p$) internal energy and pv, which represents the mechanical energy due to pressure, often called flow work. The energy equation can be rewritten:

$$W_{in} = \left(T_{II}c_{p_{II}} - T_I c_{p_I}\right) + \left(p_{II}v_{II} - p_I v_I\right) + \left(\frac{c_{II}^2 - c_I^2}{2}\right) + g\left(z_{II} - z_I\right) + Q_{out} \qquad \text{Equ 3.11}$$

where:

T	=	temperature (°C)
c_p	=	specific heat (J/kg/°C)
p	=	pressure (Pa)
v	=	specific volume (m^3/kg)
c	=	velocity (m/s)
z	=	height above datum (m)
g	=	gravitational acceleration (m/s^2)

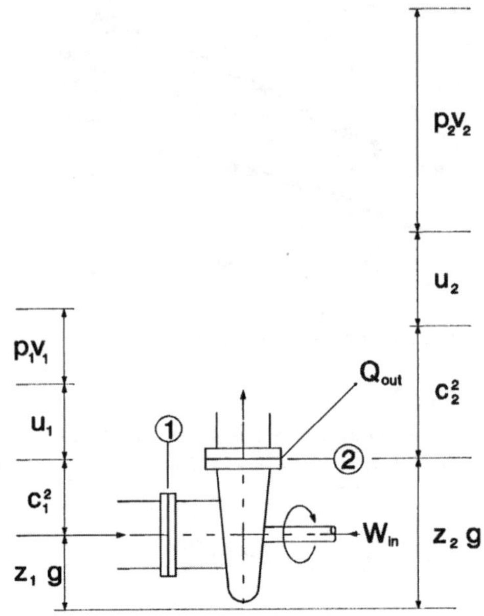

Figure 3.8 Illustration of the energy equation

W_{in}	=	work input to the shaft (J/kg)
Q_{out}	=	heat rejected (J/kg)

NOTE: In these equations the velocity is represented by c not v. Engineers frequently use the same symbol for different quantities. The context is important and should be explanatory. All the terms in the energy equation are calculated per kg of medium flowing. The terms $c^2/2$ and gz respectively represent the kinetic and potential energy of the medium per kg.

The effect of a pump on the energy equation is shown diagrammatically in Figure 3.8. The input of mechanical work is through the rotating shaft, all the variables change and some heat is rejected from the pump body. Notice particularly that a temperature rise has occurred. The change from suction conditions to discharge conditions usually takes place very quickly within the pump so the ideal process is isentropic, constant entropy, not constant temperature. The isentropic temperature rise is small for water, of the order of 0.4 °C per 100 bar pressure rise at 50°C rising to 2.6°C per 100 bar rise at 250°C.

Pump inefficiencies will generally appear as increased temperature rise and heat rejection. Noise and vibration will take their energy as portions of heat rejection. In equations 3.10 and 3.11, g is the only constant, for the location, not universally. Even c_p is a variable, from less than 4.0 to over 5.0 for water.

Rotodynamic pump designers take the centre-line of the first stage impeller eye as the datum for all calculations and parameters such as NPSHr. Most positive displacement pump designers take the centre-line of the inlet connection as the datum. A system designer will not know precisely where the impeller eye or inlet connection will be. He should base all the system calculations on the foundation level where the pump will be located or the liquid surface level for submerged pumps.

The quality of any calculated result is dependant upon the accuracy of the input data. Pumps always operate in the pressurised liquid region so that saturated liquid data will not be accurate. Steam tables, such as the UK Steam Tables, based on the 1967 IFC Formulation, contain tabulated data for pressurised water up to 1000 bara. However this data is aimed at the power generating industry and is tailored to their needs. Inconsistencies exist in the data for low temperatures at pressures close to atmospheric. This data should be examined carefully before use and "conditioned" if thought appropriate. Unconditional use of raw data can result in errors of up to 6% when compared to results obtained from calculations in Sections 3.2 and 3.4.

Computerised steam tables alleviate some of the inconsistencies.

In some cases it will be possible to ignore the effects of changes in velocity, specific volume and height, between inlet and outlet, without loss of accuracy. Pump differential head or pressure is the important criterion. No assumptions should be made on pumps which operate at low differentials, below 50m.

3.2 Pipe flow losses

3.2.1 Reynolds Number

Flow losses occur because of the effect of internal friction. Shear stresses arise as soon as velocity gradients exist.

$$\tau = \mu \cdot \frac{dv}{dy} \qquad \text{Equ 3.12}$$

where:

τ = shear stress (N/m^2)

μ = viscosity (Ns/m^2)

v = flow velocity (m/s)

y = thickness of flow stream(m)

The work done by the shear force generates heat and adds to the internal energy of the liquid. The increase in internal energy means that the static pressure becomes a little less, hence the name pressure loss or pressure drop. The level to which the liquid temperature can rise due to friction heating is dependant upon the heat transfer mechanism to the surroundings.

The flow losses are great wherever the shear stress is great, i.e. where the velocity gradients are great. High velocity gradients occur in the boundary layer, at flow around sharp corners, in strong vortex flows and so on. Viscous flow in pipes is characterised by Reynolds Number.

$$Re = \frac{v \cdot d}{\nu} \qquad \text{Equ 3.13}$$

where:

Re = Reynolds Number (dimensionless)

v = flow velocity (m/s)

d = pipe diameter (m)

ν = kinematic viscosity (m^2/s)

The following relationship exists between kinematic viscosity, ν, and dynamic viscosity, μ.

$$\nu = \frac{\mu}{\rho} \qquad \text{Equ 3.14}$$

where:

ρ = liquid density (kg/m^3)

In the case of low Reynolds Numbers, Re <2300, the flow is laminar. The liquid flows in layers with different velocities which do not mix with each other. At Re > 2300 the flow is turbulent, that is to say the liquid particles do not flow in a straight line in the direction of motion of the main flow.

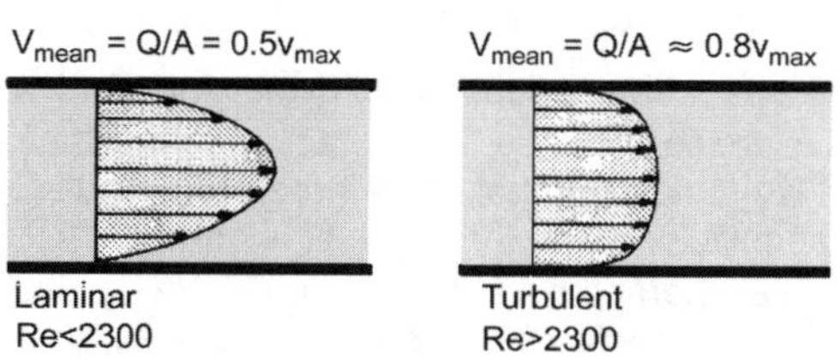

Figure 3.9 Laminar and turbulent flow

Since the velocity gradients differ, the shear stress and flow losses will also differ between the two cases. The critical value of Re Number has been stated as Re cr ≅ 2300. It should be noted however, that the flow may continue to be laminar at considerably higher values of Re if the flow is very well protected from disturbances and entrance conditions to the section are the optimum. Low values of Re Number occur at low flow velocities, in small pipe diameters or at high viscosity. See Figure 3.9.

Liquid	Kinematic Viscosity ν m^2/s	Transition to turbulent flow (for the pipe diameters given) Vcr m/s			
		d = 0.001m	d = 0.01 m	d = 0.1 m	d = 1 m
Water, 20°C	$1 \cdot 10^{-6}$	2.3	0.23	0.023	0.0023
SAE 5W 60°C	$1 \cdot 10^{-5}$	23	2.3	0.23	0.023
Burner oil Class E (BS 2869: 1970) 30°C	$1 \cdot 10^{-4}$	230	23	2.3	0.23
SAE 50 18°C	$1 \cdot 10^{-3}$	2300	230	23	2.3

Table 3.1 Transition to turbulent flow in pipes

It can be seen from the examples, see Table 3.1, that laminar flow is less usual in practice for liquids with viscosity similar to that of water, but is quite common for oils.

3.2.2 Head losses in straight pipes

The following formula applies to head losses for flow in a straight pipe.

$$h_f = \lambda \cdot \frac{l}{d} \cdot \frac{v^2}{2g} (m) \qquad \text{Equ 3.15}$$

where:

λ = loss coefficient for straight pipes (dimensionless)

l = length of pipe (m)

d = pipe diameter (m)

v = Q/A = flow velocity (m/s)

The loss coefficient is dependent upon Reynolds Number and upon the internal roughness of the pipe. This is shown in Figure 3.10.

Pipe and pipe material	Condition of pipe	Factor k
		Calculation values mm
Steel pipe	Seamless, new	0.03 - 0.05
Steel pipe	Welded, new	0.03 - 0.10
Steel pipe	In use, rusted	0.15 - 0.5
Steel pipe	Seamless, galv.	0.1 - 0.16
Steel pipe	Welded, galv.	0.1 - 0.2
Steel pipe	Heavy asphalt coating	0.5
	Mature water mains	1.2
	Stainless, acid resist.	0.045
Cast iron pipe	Bitumen coated	0.1 - 0.3
	Main conduit	0.1
	Other conduits	0.2 - 0.3
	In use, rusted	0.4 - 2.0
Copper pipe	Diameter <200 mm	0.01
Brass, glass	Technically smooth	0.002
Concrete pipe	Water main	0.1 - 0.2
	Waste, drains, new	0.2 - 0.5
	Waste, drains, old	1.0
Asbestos cement pipe	Newly laid	0.05
Earthenware pipe	Newly laid drains	0.2
	Older drains	1.0
	Mature foul sewer	3.0

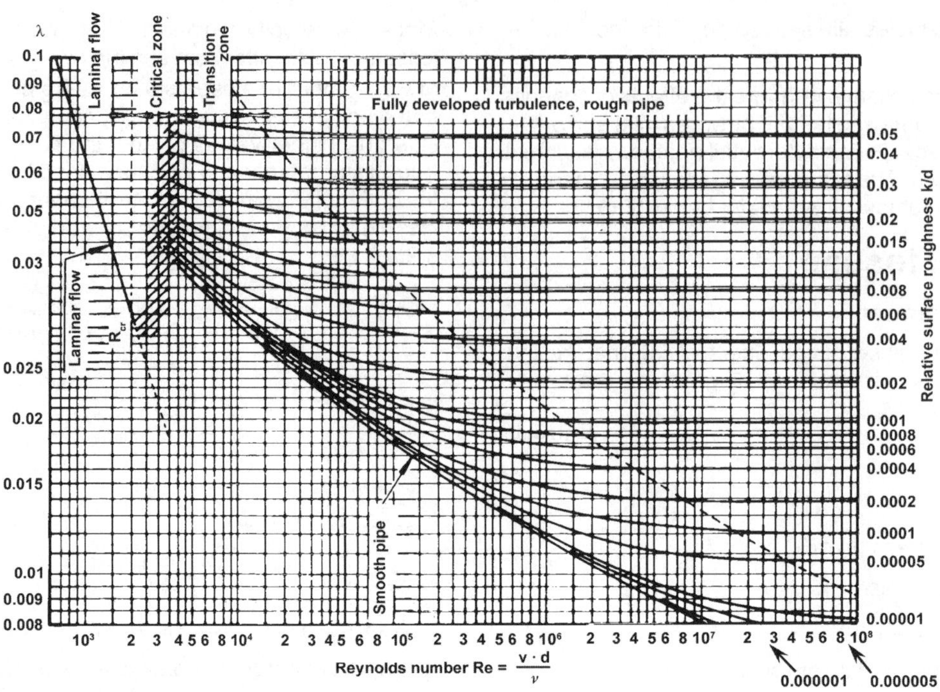

Figure 3.10 Loss coefficient for straight pipes

Pipe and pipe material	Condition of pipe	Factor k
		Calculation values mm
PE, PVC	Dia. <200> mm	0.01
PE, PVC	Dia. 200 mm	0.05
GRP	All dimensions	0.1
Wood	New	0.2 - 1.0

Table 3.2 Examples of approximate roughness in pipes

The roughness of the pipe can be assessed from the values in Table 3.2.

The loss coefficient can also be calculated from the following formulae:

Laminar flow Re < 2300

$$\lambda = \frac{64}{Re} \qquad \text{Equ 3.16}$$

Turbulent flow Re > 4000

$$\frac{1}{\sqrt{\lambda}} = -2 \cdot \log_{10}\left(\frac{2.51}{Re\sqrt{\lambda}} + \frac{k/d}{3.71}\right) \qquad \text{Equ 3.17}$$

Equation 3.17 is sometimes called the Colebrook-White equation; Colebrook in France and Prandtl-Colebrook in Germany. In the region Re = 2300 to Re = 4000 laminar and turbulent flows may alternate in various sections of the pipe. The resultant loss coefficient will then assume values between those obtained in formulae 3.16 and 3.17.

The expression given in equation 3.17 is somewhat inconvenient for manual calculation. The following formula, proposed by Miller, can be used for turbulent flow:

$$\lambda = \frac{0.25}{\left[\log_{10}\left(\frac{k}{3700d} + \frac{5.74}{Re^{0.9}}\right)\right]^2} \qquad \text{Equ 3.18}$$

Other formulae used for the calculation of straight pipe losses in the turbulent region are the Hazen-Williams and Manning formulae. These, however, do not provide any additional information to the equations presented above and have, therefore, been omitted.

3.2.3 Head losses in fittings

Head losses in bends and valves, etc. can be calculated with the help of the formula:

$$h_f = \varsigma \cdot \frac{v^2}{2\,g} \text{ (m)} \qquad \text{Equ 3.19}$$

where:

ζ = loss coefficient of fitting (dimensionless)

v = flow velocity (m/s)

The magnitude of the head losses are in principle influenced, as in the case of straight pipe flow, by surface roughness and by Reynolds Number.

Examples of approximate values of loss coefficients for losses in fittings are illustrated in Figure 3.11. All values are those for conditions of normal roughness and at high Reynolds Number, i.e. at fully developed turbulent flow. In cases where the velocity changes it must always be the highest velocity which is used for calculating h_f in accordance with equation 3.19. In the case of varying diameters, the velocity to be applied is the one resulting from the smallest diameter, and in the case of T-pieces, the velocity related to the total volume flow is applicable.

An alternative method of expressing the magnitude of head losses in fittings is the use of the concept of equivalent pipe length, l_{eq}, i.e. the equivalent length, (l_{eq}) of straight pipe giving rise to the same losses in pressure as the fitting, at the same flow velocity. By comparing equations 3.15 and 3.19:

$$l_{eq} = \frac{\zeta}{\lambda} \cdot d \qquad \text{Equ 3.20}$$

where:

ζ = loss coefficient of fitting (dimensionless)

λ = loss coefficient for straight pipes (dimensionless)

It can be seen from equation 3.20 that the equivalent pipe length is a function of λ, i.e. the surface roughness of a straight pipe and Reynolds Number. It is therefore difficult to quantify l_{eq}, as a specific property for, say, a pipe bend and is thus of an approximate character. For λ = 0.02, the equivalent pipe length of a short radius 90° bend (r = d) is l_{eq} = 20d.

For additional information and data on losses see Section 3.6.

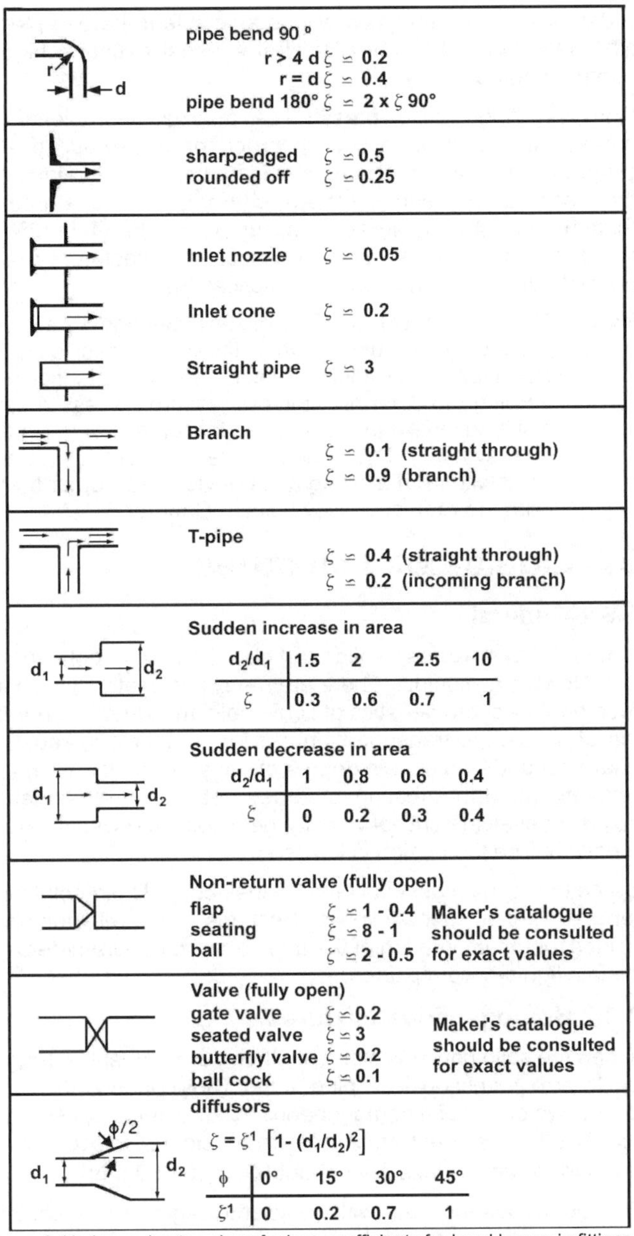

Figure 3.11 Approximate values for loss coefficients for head losses in fittings
The values for flow meters are stated in Section 3.5

3.2.4 Hydraulic diameter

For the case of flow in a partially filled pipe, in non-cylindrical ducts or in open channels, the head losses can in principle be calculated in the same way as described in earlier Sections. However, instead of using the diameter of the pipe, the hydraulic diameter, d_h, must be used in these cases.

$$d_h = \frac{4 \cdot A}{P} \qquad \text{Equ 3.21}$$

where:

d_h = hydraulic diameter (m)

A = cross-sectional area flowing (m^2)

P = wetted perimeter (m)

For a completely full cylindrical pipe:

$$d_h = \frac{4\frac{\pi \cdot d^2}{4}}{\pi \cdot d} = d$$

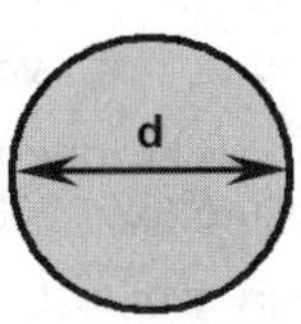

i.e. in this case the hydraulic and geometric diameters coincide.

For a half-filled cylindrical pipe:

$$d_h = \frac{4 \cdot \frac{\pi \cdot d^2}{2 \cdot 4}}{\frac{\pi \cdot d}{2}} = d$$

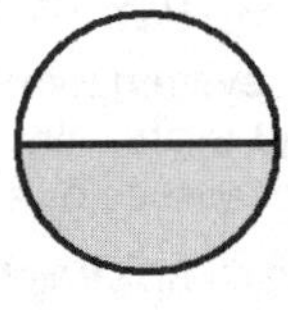

For a half-filled rectangular section:

$$d_h = \frac{4 \cdot b \cdot \frac{h}{2}}{b + 2 \cdot \frac{h}{2}} = \frac{2 \cdot b \cdot h}{b + h}$$

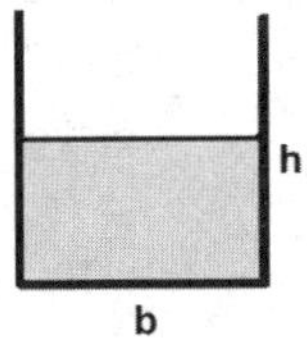

With Reynolds Number

$$Re = \frac{v \cdot d_h}{\nu} \qquad \text{Equ 3.22}$$

and relative roughness k/d_h, Figure 3.10 and equations 3.16 and 3.18 apply with unchanged numerical values. The value of Re_{cr} will change as the section proportions change and will have to be evaluated by experiment or computer simulation.

Pipes flowing entirely due to gravitational head can operate completely filled. The degree of fullness depends, among other things, on the slope of the pipe and the operating conditions. For a pipe the fullness is defined by the ratio d^1/d where d^1 is the depth of liquid in the pipe and d is the diameter of the pipe. Figure 3.12 illustrates how various flow parameters change with the degree of fullness of the pipe. Both flow velocity and hydraulic diameter have their greatest value at a filling coefficient of just below 1.

For gravity flow pipes carrying solids and liquids, care must be taken with regard to sedimentation; the avoidance of which can place restrictions on the viable configuration and operation of the pipe.

Settling liquid-solid mixtures must be transported at a velocity greater than the critical velocity. If the flow velocity is reduced below the critical velocity sedimentation occurs. Sedimentation continues until the flow velocity is increased to the critical velocity. The flow area of the pipe is reduced until the mixture flow velocity is high enough to prevent further sedimentation. The hydraulic characteristic of the pipe is modified because the shape is no longer circular.

The hydraulic diameter of the flow regime can be estimated from Figure 3.13. The reduced hydraulic diameter can be used to estimate the friction loss in the partially blocked pipe. This can only be an **estimate.** The pipe wall may be smooth and well-defined. The surface of the sediment may not be smooth and the roughness may be difficult to approximate. The surface may also be moving if the particles are small. Figure 3.13 can also be used for hydraulic diameters for partially-round pipes in special applications.

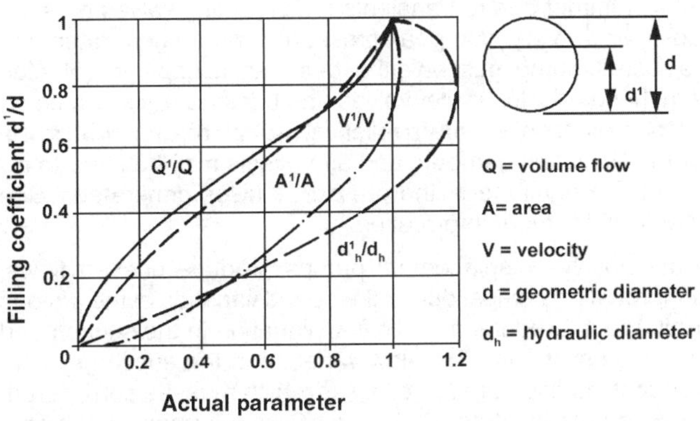

Figure 3.12 Flow in a partially-filled pipe

3.2.5 Total head losses in the pipe system

The system total losses consist of the sum of the straight pipe head losses, the head losses in fittings and valves etc. and static losses due to changes in elevation.

$$h_f = h_f(\text{straight pipe}) + h_f(\text{fittings\& valves}) + \Delta z \quad \text{Equ 3.23}$$

With the usual symbols, head loss

$$h_f = \lambda \cdot \frac{1}{d} \cdot \frac{v^2}{2g} + \Sigma\zeta \cdot \frac{v^2}{2g} + \Delta z$$

$$= \left(\lambda \cdot \frac{1}{d} + \Sigma\zeta\right) \cdot \frac{v^2}{2g} + \Delta z$$

$$= \left(\lambda \cdot \frac{1}{d} + \Sigma\zeta\right) \cdot \frac{Q^2}{\left(\frac{\pi \cdot d^2}{4}\right)^2} \cdot \frac{1}{2g} + \Delta z \quad \text{Equ 3.24}$$

where:

h_f	=	head loss (m)
λ	=	loss coefficient for straight pipe
ζ	=	loss coefficient for fittings, valves etc.
$\Sigma\zeta$	=	sum of all loss coefficients
l	=	length of pipe (m)
d	=	diameter of pipe (m)
Q	=	volume flow (m^3/s)
v	=	flow velocity (m/s)
Δz	=	(z_{II} - z_I) (m)

Another useful method for calculating head losses in fittings was proposed by William B. Hooper in *Chemical Engineering*. See Section 3.6.

The flow losses in both suction and discharge systems must be evaluated. For variable flow systems a Q-h_f curve should be produced for both systems; the suction system curve should also show NPSHa/NPIPa:

Net Positive Suction Head Available

Net Positive Inlet Pressure Available

Both functions are the liquid pressure minus the vapour pressure. It is best for the system designer to use the top of the pump foundations as the datum level; the pump manufacturer can easily adjust the data to satisfy the pump configuration. Centrifugal pumps use the centre of the impeller eye as the datum whereas reciprocating pumps use the centre of the suction connection.

In long systems, or systems where the flow velocity is high; these terms are of necessity vague; the possibility of surge or water-hammer due to transients or fast acting valves must be considered. Very high peak pressures can be generated and the system/pump must be either designed to cope or protected from these effects. Relief valves and bursting discs can be fitted for protection, standard relief valves can be damaged in the event. Pulsation dampers and air vessels may be fitted to reduce the magnitude of the pressure pulses generated. See Chapter 5 for more information.

Some positive displacement pumps produce uneven flows. Reciprocating pumps, due to the speed variation of the pistons or plungers, produce a cyclic flow variation in the suction and discharge systems. The flow variation in the suction system produces addition pressure losses which must be considered. The Hydraulic Institute (USA) and VDMA (Germany) have both produced data to allow this phenomenon to be evaluated.

In general, it is better to have more NPSHa/NPIPa than less. A wider range of pump types will be available if there is plenty NPSHa/NPIPa. Low NPSHa/NPIPa usually means bigger pumps running slowly.

Conversely high suction pressure can also cause problems; increased thrust, bearing cooling problems and reduced efficiency. The suction pressure should be considered relative to the discharge pressure. If the discharge pressure is greater than 16 barg and the suction pressure is greater than 20% of the discharge, it is advisable to consult manufacturers about possible problems and cost/size implications.

NOTE: The treatment of pipe flow losses in Section 3.2 applies to Newtonian liquids of which the commonest is water. Non-Newtonian liquids require special attention regarding the determination of pressure losses. A special type of flow loss is caused by control valves in the pipe system. The control valve pressure losses are dealt with from a sizing point of view in Chapter 5 and from the regulation viewpoint in Chapter 6.

3.3 Liquid-solid mixtures

3.3.1 General

Liquid-solid mixtures, solid particles in liquid, behave as non-Newtonian liquids. There are many parameters which affect the flow characteristics of liquid-solid mixtures. All pumping of solids, i.e. transportation of solid materials in liquid, is characterised by a certain degree of uncertainty. It is normally recommended that testing is carried out, with representative solids, to evaluate critical velocity, head loss and wear characteristic before the design is finalised.

Some of the main aspects of the behaviour of liquid-solid mixtures are set out in the following Sections, along with guideline values for the estimation of the more important parameters for certain liquid-solid mixtures.

3.3.2 Homogeneous mixtures

When the solid particles distribute themselves evenly throughout the cross-section of the pipe, without any concentration gradients, we speak of a homogeneous mixture or suspension. In practice, homogeneous mixtures are considered to occur when the particle size is less than about 50μm (0.050 mm).

Homogeneous mixtures with sufficiently high particle content behave as time-independent, plastic, non-Newtonian liquids. These kinds of liquid to some degree follow Binghams Law, see Chapter 2, Section 2.4 and also Figure 3.13.

The pressure drop for laminar flow in straight piping measured in metres head of mixture is:

$$h_f = \frac{32}{\rho g} \cdot \frac{l}{d} \cdot \left(\frac{t_0}{6} + \frac{\tan\gamma \cdot v}{d}\right) \quad \text{Equ 3.25}$$

where:

h_f	=	loss of head (m)
ρ	=	density of liquid-solid mixture (kg/m^3)
g	=	acceleration due to gravity (m/s^2)
l	=	length of pipe (m)
d	=	diameter of pipe (m)
t_0	=	viscous shear stress at boundary (N/m^2)
tan γ	=	plastic dynamic viscosity (Ns/m^2)
v	=	velocity of flow (m/s)

This is illustrated in Figure 3.14 for a homogeneous liquid-solid mixture (ρ_p = 2600 kg/m^3, d = 50 mm)

The viscous shear stress at boundary τ_0 and plastic viscosity, tan γ, can be determined with the aid of a viscometer or by mea-

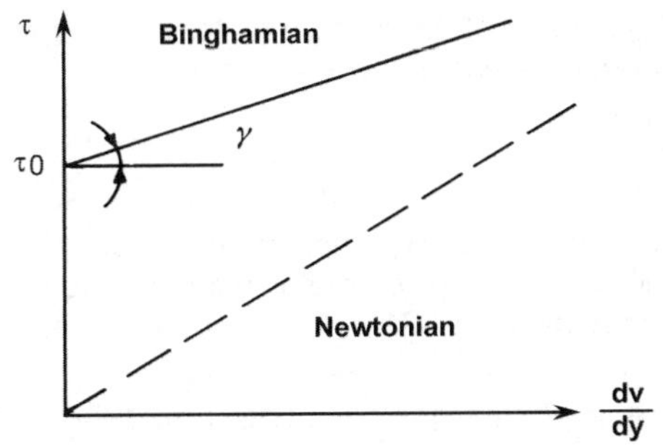

Figure 3.13 Homogenous suspensions with flow characteristics as per Binghamian and Newtonian behaviour

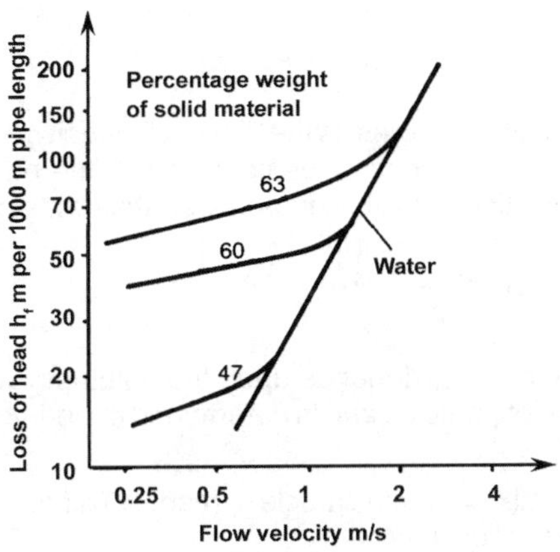

Figure 3.14 Examples of loss of head in pipe flow

suring the pressure drop in a test loop. Plastic viscosity, tan γ, is also called apparent viscosity in some texts.

The loss of head for turbulent flow in straight pipe, measured in metres of mixture is:

$$h_f = l \cdot \frac{l}{d} \cdot \frac{v^2}{2g} \qquad \text{Equ 3.26}$$

i.e. exactly the same formula as for Newtonian liquids. The loss coefficient must be determined using the apparent viscosity.

Expressed in metres liquid column, the loss of head for a homogeneous liquid-solid mixture is the same as for pure water in turbulent flow.

The transition from laminar to turbulent flow occurs, as for water, at Re ≅ 2300 but with the difference that the apparent viscosity should now be used for Reynolds Number. The velocity of flow at transition V_{cr} for Re = 2300 is:

$$V_{cr} \cong \frac{1150 \cdot \tan\gamma}{r \cdot d}\left(1 + \sqrt{\frac{1 + \tau_0 \cdot r \cdot d^2}{3600 \cdot \tan^2 g}}\right) \qquad \text{Equ 3.27}$$

where:

V_{cr}	=	critical velocity of flow at transition (m/s)
tan γ	=	plastic viscosity (Ns/m^2)
ρ	=	density of liquid-solid mixture (kg/m^3)
t_0	=	shear stress at boundary (N/m^2)
d	=	diameter of pipe (m)

Sedimentation in the pipe is avoided completely if the flow velocity in the pipe is greater than V_{cr}. Thus V_{cr} is of considerable importance when sizing pump installations for homogeneous mixtures in pipes. In short or carefully monitored pipes, $v < V_{cr}$ can sometimes be accepted.

There is an optimum solids concentration for the transportation of solid material by hydraulic methods (see Figure 3.15). If the concentration is too low, the liquid-solid mixture loses its plastic characteristic and with it the possibility of making use of the advantageous “plug” flow. Besides which an unnecessarily large amount of liquid is pumped per kilogram of solid material if the liquid is discarded. If the particle content is too high the transportation must be carried out at increased velocity due to increased transitional velocity, V_{vc}, so as to avoid sedimentation.

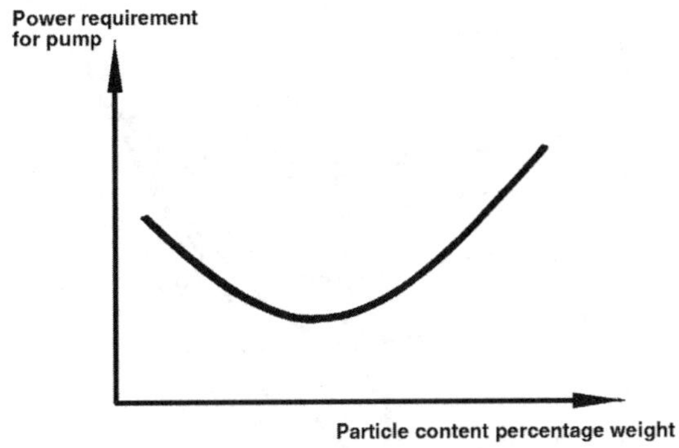

Figure 3.15 Optimum particle content for homogeneous liquid-solid mixtures for the transportation of a given quantity of solid material per unit time

If, for example, the particle concentration in the liquid-solid mixture in Figure 3.14 were to increase from 60% to 63%, the velocity of flow would have to increase from about 1.7 to 2.2 m/s. Thereby increasing the transportation capacity by a factor of 1.4, the loss of head by about 1.7 and the pump power consumption, with pipe losses only, by a factor of about 2.3. When the transportation capacity is constant but using a smaller pipe diameter the power requirement of the pump increases 1.7 times. The particle content is thus a critical economic parameter.

3.3.3 Heterogeneous mixtures

Heterogeneous mixtures or suspensions, that is to say liquid-solid mixtures with concentration gradients across the pipe section, are considered in practice to occur for particle sizes of more than about 50 μm (0.050 mm). The main difference compared with homogeneous liquid-solid mixtures is that the liquid and the particles in a heterogeneous mixture retain their identity. The flow is a two - phase flow. The viscosity of a heterogeneous mixture in principle is the same as that for the pure liquid.

The solid particles are conveyed suspended in the liquid if the particles are small and the velocity high, or by jumps if the particles are large or the speed low, as shown in Table 3.3..

Particle size mm	Normal means of transportation
0.05 to 0.2	suspended
0.20 to 1.5	combined suspended
> 1.50	jumps

Table 3.3 Normal means of transportation for various particle sizes

For all practical purposes, particle sizes in the order of 0.05 to 0.2 mm are transported suspended in the liquid. Liquid-solid mixtures with particle sizes down towards the lower limit, 0.05 to 0.1 mm, comprise the boundary with the homogeneous mixtures. The most effective way of minimizing the risk of sedimentation and blockage of the pipe is to maintain the pipe flow in the turbulent regime. The pressure losses, expressed as loss of head in metres mixture, can then be calculated in the same way as for the liquid.

Liquid-solid mixtures of particles with sizes up towards the upper limit, 0.1 to 0.2 mm, are on the boundary with the next group where the combined means of transportation prevails. The particle region 0.05 to 0.2 mm is not so well understood as other size ranges.

Mixtures with particles of 0.2 mm have been the subject of comprehensive pressure loss measurements. (See work by Durand, Condolios et al., see Section 3.6.) Figure 3.16 shows an example of a loss diagram for a horizontal pipe.

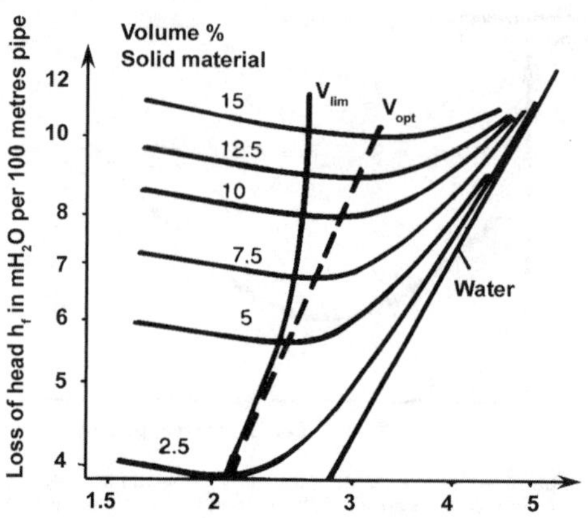

Figure 3.16 Example of loss of head in heterogeneous suspension, sand ρ_p - 2650 kg/m³, d_p mean ≅ 0.4 mm, d = 200 mm

The experimental results shown cover the region:

d	=	pipe diameter - 0 to 600 mm
d_p	=	particle size - 0.2 to 25 mm
K_{vol}	=	volumetric concentration - 2.5 to 15%
ρ_p	=	particle density liquid 0 to 5000 kg/m³ water

The loss of head expressed in metres of water (m H_2O) is greater than for pure water and exhibits a minimum at the optimum flow velocity v_{opt}. The loss of head for liquid-solid mixtures of d_p > 0.2 mm flowing in horizontal pipes can be assessed using the formula:

$$h_f = \lambda \cdot \frac{l}{d} \cdot \frac{v^2}{2g} (1 + k_f) \qquad \text{Equ 3.28}$$

where:

h_f	=	loss of head (m H_2O)
λ	=	loss coefficient for water
d	=	pipe diameter (m)
v	=	velocity of flow (m/s)
k_f	=	supplementary loss factor over and above that of pure water

The supplementary losses above those for pure water can be assessed from:

$$k_f = 85 \cdot K_{vol} \left[\frac{g \cdot d}{v^2} \cdot \left(\frac{\rho_p}{\rho_v} - 1 \right) \cdot k_p \right]^{3/2} \qquad \text{Equ 3.29}$$

where:

K_{vol}	=	proportion of solid material by volume
ρ_p	=	particle density (kg/m³)
ρ_v	=	liquid density (kg/m³)
k_p	=	factor for particle characteristics

The particle factor is assessable from:

$$k_p = \frac{V_s}{\sqrt{\frac{4}{3} \cdot \left(\frac{\rho_p}{\rho_v} - 1 \right) \cdot g \cdot d_p}} \qquad \text{Equ 3.30}$$

where:

v_s	=	sedimentation velocity for a single particle in still water (m/s)
d_p	=	particle diameter (m)

Some text books use the dimensionless "particle resistance coefficient" C_D instead of the particle factor k_p. The relationship between them is given by:

$$k_p = \frac{1}{\sqrt{C_D}} \qquad \text{Equ 3.31}$$

The optimum flow velocity v_{opt}, i.e. the flow velocity for minimum loss of head at any given volumetric concentration, can be assessed from equation 3.32.

$$v_{opt} = 3.5 \cdot K_{vol}^{0.333} \left[g \cdot d \cdot \left(\frac{\rho_p}{\rho_v} - 1 \right) \cdot k_p \right]^{1/2} \qquad \text{Equ 3.32}$$

For particles >0.2 mm and density ρ_p = 2650 kg/m³, some text books give:

$$v_{opt} = 2.9 \cdot K_{vol}^{0.15} \cdot \sqrt{g \cdot d} \qquad \text{Equ 3.33}$$

Another velocity of interest is the "limiting" velocity, below which the particles are in continuous contact with the bottom of the pipe. The limiting velocity can be evaluated, at testing, from:

$$v_{lim} = F_{lim} \sqrt{2g \cdot d \left(\frac{\rho_p}{\rho_v} - 1 \right)} \qquad \text{Equ 3.34}$$

where the factor F_{lim} depends upon the volumetric concentration and size of particles and is determined experimentally. See Figure 3.17.

Liquid-solid mixtures with particle contents of different sizes are dealt with as follows:

- Small particles which form a liquid-solid mixture with the pure liquid at the velocity used are considered to be a part of the carrier liquid. The liquid and the small particles (d_p less than perhaps 0.1 mm, certainly less than 0.05 mm) constitute a new liquid with high density and viscosity values.
- The larger particles together with the carrier liquid form a heterogeneous mixture. Where there is a variation of size within the group of larger particles, a weighted mean value is used to determine the particle factor k_p.

Small particles increase the density of the carrier liquid and reduce the ratio of densities of transported particles and carrier liquid. Because of this both the supplementary losses and the limiting velocity for minimum loss and sedimentation are reduced. Additions of small particles can thus be economically attractive in the transportation of larger particles.

The sedimentation velocity v_s used to calculate the particle factor k_p should be quantified experimentally. The effect of the density of the liquid and its viscosity, together with the geometrical shape of the particle all affect the particle factor. The measured sedimentation velocity for sand in water is illustrated in Figure 3.18. This is in accordance with Richards for quartz sand (ρ_p = 2650kg/m³) in water (10°C, ν = 1.3 x 10^{-6} m²/s) compared

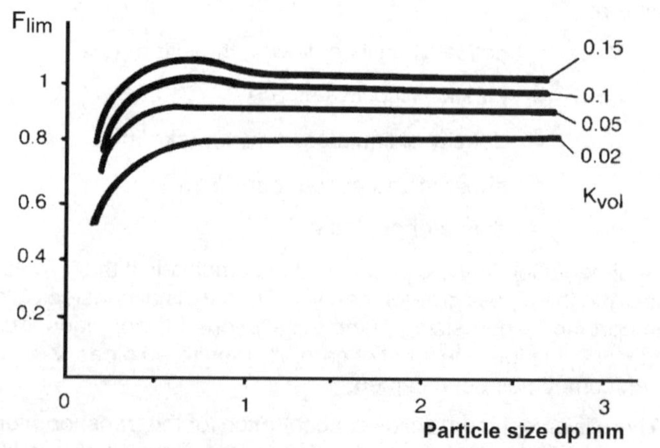

Figure 3.17 Factor F_{lim} for the determination of the limiting velocity V_{lim}

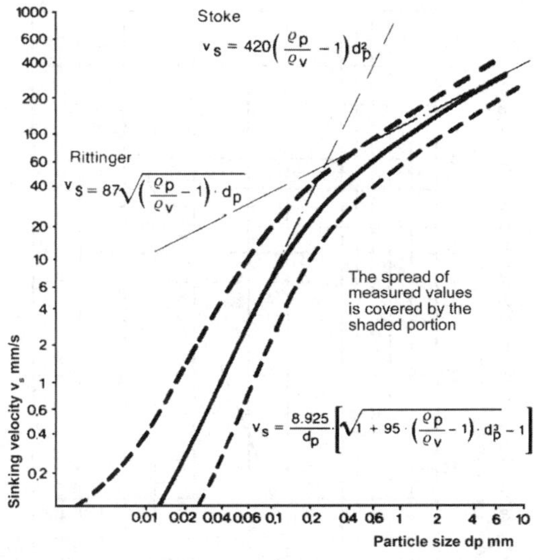

Figure 3.18 Measured sedimentation velocity

with various theoretical expressions

It is extremely difficult to give general guideline values for the most suitable flow velocity for the hydraulic transportation of solid materials. The values shown in Table 3.4 are recommended.

Solids	Mean velocity m/s
waste from ore dressing plant	1.5 - 2.1
dust from precipitators	1.5 - 1.8
fly ash	1.8 - 2.1
fine sand	2.4 - 3.0
normal sand with fine	3.4
coarse sand without fine	3.7 - 4.0
shingle, max. 13 mm	4.3

Table 3.4 Flow velocity for hydraulic transportation of solid materials

These values are based on pipe diameters between 100 and 200 mm. Note the effect of the pipe diameter on the limiting velocity in equation 3.34.

3.3.3.1 Special note

The formulae presented for the evaluation of properties for liquid-solid mixtures are the result of extensive research. Research is always being conducted and more accurate expressions may be available. However all published data is approximate. The equations are derived by finding the "best fit" to test results. It is very unlikely that any of the test results would closely approximate a particular actual operating condition and there may be no way of evaluating how far the "best fit" equation is from the closest suitable test results.

In practice, it will be the size distribution/content of the solids which determines the flow properties. When solids such as copper/iron ore, coal are crushed, variable quantities and sizes of fines are also produced. Model tests on representative crushed samples are essential to confirm calculated results. Wear tests, such as Miller Number, ASTM G75, should also be undertaken to assess what precautions, if any, are required against wear.

Discussions with pump manufacturers of several pump types, should be held as soon as approximate figures are available for solid sizes and distribution, flows and pressures. The discussions should focus on:

- **Efficiency** — how efficient will the pump be and how will the efficiency be affected by changing operating conditions and wear,
- **Wear** — which components will wear, predicted life of components and accessibility for maintenance,
- **Adjustment** — how can the pump be adjusted, at site, to optimize efficiency and component life for changing operating conditions. Multi-stage centrifugal pumps can be de-staged or have impeller diameters modified; reciprocating pumps can have different diameter pistons or plungers fitted,
- **NPSHr** — some pumps require a lot of NPSHa and this can radically effect the design of suction systems.

3.3.4 Paper pulp

3.3.4.1 General

The behaviour of pulp suspensions when flowing through a pipe is unusual from several aspects. The loss of head for chemical pulps goes through a maximum and a minimum value when the flow velocity increases from zero, and the head loss is finally less than the value for pure water.

Referring to Figure 3.19, for velocities up to point D, after the intersection of the two curves, the suspension flows like a plug surrounded by a thin boundary layer in which the whole of the viscous flow occurs. The character of the boundary layer varies according to the position of the working point on the curve:

- A to B — The surface of the plug is disturbed by intermittent contact with the pipe wall.
- B to C — The boundary layer consists mainly of pure water in laminar flow.
- C to E — The flow in the boundary layer is turbulent.

Plug flow can be regarded as ended when the stress at the edge of the plug is equal to the ultimate shear stress of the network of fibres which form the plug. When the velocity increases beyond this point the plug breaks up and finally the whole cross section flows turbulently.

The friction loss curves for mechanical pulps do not usually have any maximum or minimum. The flow characteristics are modified by the relatively low strength of the network and the high proportion of non-fibrous solid material.

3.3.4.2 Calculation of pipe flow losses

Pipe flow losses can be determined with the help of a number of diagrams shown in Figures 3.20 and 3.21.

The non-dimensional flow parameter S is defined as:

$$S = \left(\frac{V^5 \cdot \rho^2 \cdot h}{\tau_D^3} \cdot d \right)^{1/6} \qquad \text{Equ 3.35}$$

and similarly the non-dimensional loss parameter F:

$$F = \frac{r \cdot g \cdot h_f \cdot d}{4 \cdot \tau_D \cdot l} \qquad \text{Equ 3.36}$$

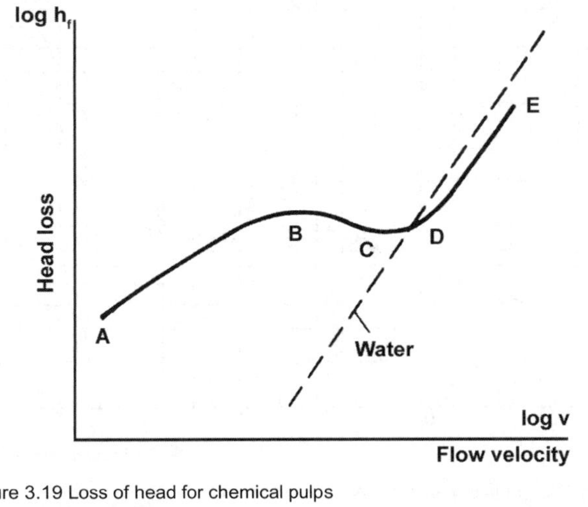

Figure 3.19 Loss of head for chemical pulps
Courtesy of ABS Group

The shear stress τ_D is a measure of the strength of the plug of plaited fibres which occurs at most flow velocities in practical operations.

$$\tau_D = \tau_D^1 \cdot f_F \cdot f_R \cdot f_M \cdot f_D \qquad \text{Equ 3.37}$$

where:

h_f	=	loss of head (m)
l	=	pipe length (m)
v	=	flow-velocity (m/s)
k	=	surface roughness of pipe (m)
d	=	pipe diameter (m)
ρ	=	density of suspension (kg/m^3)
μ	=	dynamic viscosity (Ns/m^2)
τ	=	shear stress (N/m^2)
F_k	=	length/diameter ratio of fibre

If the pulp never dries out, $f_D = 1$. If it has dried out and is absorbed again, $f_D = 0.75$.

The diagrams, especially Figure 3.20, apply to chemical pulps. The different characteristics of mechanical pulps shown in Figure 3.21 (fibres, networks, etc.) give the loss curve a different shape and generally increased losses. Correction factors should therefore be used. Approximate values can be obtained, however, using the above method provided that:

- the loss parameter F never assumes a value lower than the maximum point in Figure 3.20 for the section of the curve S >0.6.
- $F_k = 1$
- a safety factor of 1.2 should be used.

The guideline values for head losses in bends and fittings are given in Section 3.4.4 in the form of equivalent pipe lengths.

It is typical of pulp suspensions, as for all plastic non-Newtonian liquids, that a certain yield shear stress (stiction) has to be overcome before a flow can be made to start up at all. This stiction can be reduced considerably by dilution.

3.3.4.3 Calculation example

A chemical pulp, 281 m^3/h, has to be pumped through a 70 m long stainless steel pipe diameter 300 mm. The pulp has never been dried and has a fineness grading 600 C.s.f., the temperature is 30°C and the concentration 2.7% absolutely dry. The average fibre length is 2.4 mm and mean diameter 40 μm. Determine the loss of head through the pipe.

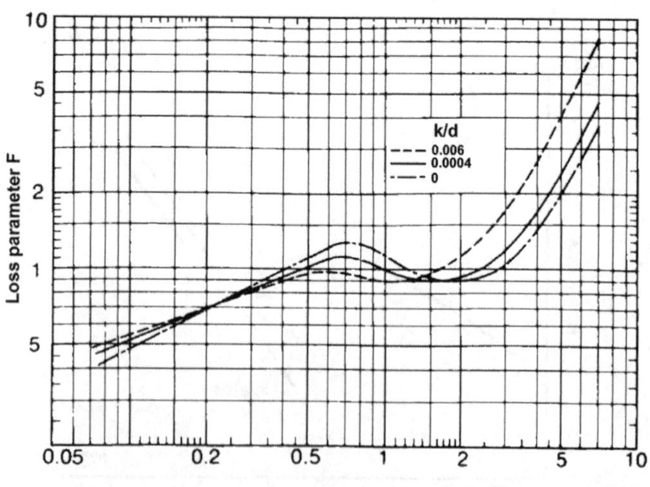

Figure 3.20 Pipe flow characteristics of chemical pulp
Courtesy of ABS Group

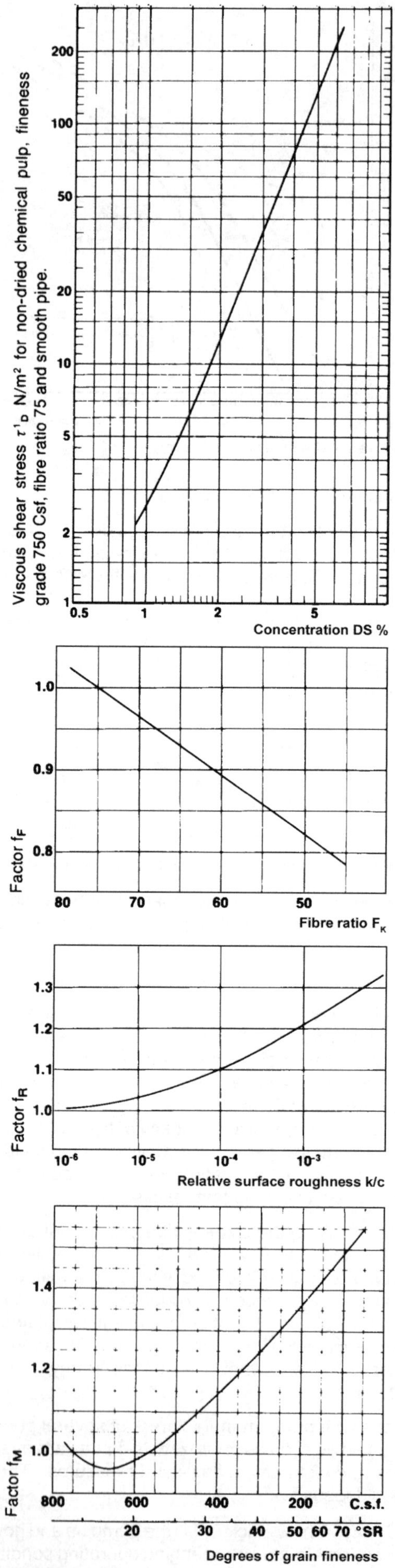

Figure 3.21 Properties of chemical pulps
Courtesy of ABS Group

Fibre ratio $F_K = 2.4/0.040 = 60 \rightarrow f_F = 0.89$

Stainless steel pipe k/d = 0.045/300 = 1.5 x 10-4 $\rightarrow f_R = 1.11$

Fineness grading 600 Csf $\rightarrow f_M = 0.98$

Undried chemical pulp $\rightarrow f_D = 1$

DS content 2.7 % $\rightarrow \tau^1_D = 27$ N/m²

where:

DS = proportion by weight of fibre (absolutely dry) (%)

Shear stress $\tau_{D \rightarrow} 27 \cdot 0.89 \cdot 1.11 \cdot 0.98 \cdot 1 = 26.1$ N/m²

Water at 30°C $\rightarrow \mu = 0.8 \cdot 10^{-3}$ Ns/m², $\rho \sim 1000$ kg/m³

Flow velocity

$$v = \frac{Q}{\frac{\pi \cdot d^2}{4}} = \frac{281 \cdot 4}{3600 \cdot \pi \cdot 0.3^2} = 1.1 \text{m/s}$$

The flow parameter

$$S = \frac{1.1^5 \cdot 1000^2 \cdot 0.8 \cdot 10^{-3}}{26.1^3 \cdot 0.300} = 0.79$$

Read off loss parameter F = 1.05 (Figure 3.20)

$$h_f = \frac{4 \cdot \tau_D \cdot l}{\rho \cdot g \cdot d} \cdot F = \frac{4 \cdot 26 \cdot 1 \cdot 70}{100 \cdot 9.81 \cdot 0.3} \cdot 1.05$$

$$h_f = 2.4 \text{m}$$

3.4 Pressure losses - nomograms and diagrams

3.4.1 General

Pre-computed pressure losses in pipes for a number of commonly applied cases of flow are set out in this section in the form of diagrams. These diagrams and nomograms can be used to calculate the losses in a system, i.e. flow losses in the pipes, valves, bends, etc. This method is quick and provides acceptable results if carried out accurately.

A number of diagrams for the determination of the associated values of velocity, frictional losses, diameter and volume flow are presented. There are also diagrams for determining the flow velocity in pipes in relation to the volume flow, mass flow and diameter together with diagrams for the determination of head loss (m) in bends fittings etc., related to the velocity of flow.

It is most practical when sizing pipes to be able, right from the outset, to assess the approximate economic flow velocity. By this means, the final economic sizing can be carried out rapidly, since fewer alternatives have to be tested. Table 3.5 gives guidelines for normal flow velocities in some common cases.

Liquid application	Flow velocity m/s
Dirty water	
Mine water lines	1.75 to 2.5
Hot water	
Condensate	1.0 to 1.5
District heating distribution	1.0 to 2.5
District heating service main	0.75 to 2.0
Domestic heating circulation	0.3 to 0.6
Feed water lines	1.2 to 4.3
Crude oil	
Oil pipelines	0.25 to 1.8
Fuel oil	
Truck loading	5 to 6
Lube oil	
Drain line (1)	0.15 to 0.75
Feed lines	1.0 to 2.3
Pump discharge line	0.6 to 1.8
Pump suction line (2)	0.45 to 1.5
Pump suction line (3)	0.3 to 1.0
Petrol	
Pump discharge	1.0 to 1.4
Pump suction	0.45 to 0.75
Refrigerant	
Condenser to receiver	≤ 1.0
Receiver to system (4)	≤ 1.5
Sewage	
Pump discharge, centrif	≥ 2.4
Pump suction, centrif	≥ 1.5
Warm water	
Economiser tubes	0.75 to 1.5
Water	
Pump discharge, centrif	8 to 15
Pump discharge, recip	3.0 to 4.5
Concrete suction bays	≤ 0.3
Pump suction headers	≤ 1.0
Pump suction, centrif	0.45 to 1.5
Pump suction, recip	≤ 0.6
Consumer supply	1.0 to 1.4
Service mains	0.6 to 1.5
Trunk main	1.0 to 3.0

(1) 1:50 fall minimum, half full, (2) flooded suction, (3) suction lift, (4) 0.115 bar / 25m maximum

Table 3.5 Guideline flow velocities for general applications

High flow velocities give small pipe dimensions giving low plant procurement costs; but cause greater pump power consumption through greater frictional losses resulting in high running costs.

Quite apart from economics, other factors can affect the choice of flow velocity:

- Noise in central heating equipment
- Erosion in copper hot water pipes
- Static electricity for liquids with low flash points
- Risk of cavitation in suction lines
- Risk of cavitation with sudden changes in velocity
- Sedimentation with liquid-solid mixtures, i.e. critical velocity
- Vibration of pipework
- Surge and pressure pulsation problems

If the NPSHa/NPIPa margin is small special attention should be given to pump suction systems. The best systems are short, straight and one pipe size larger, at least, than the pump suction connection. Any bends should be more than five pipe diameters from the pump connection. If the NPSH/NPIP margin is small it is critically important to evaluate all system losses for all flow rates. If the suction system causes pump cavitation, operational problems will reduce the maintenance interval. Typical problems include:

- Loss of pump capacity
- High rotating/reciprocating parts wear
- Sealing problems
- Fatigue failures

- Discharge pipework noise and vibration
- Pipe connection leaks

It should be remembered that thermodynamic data is for a pure substance with no contaminants. Fresh water will contain dissolved gases. Gases will evolve from solution at a much higher pressure than the vapour pressure of the water; typically at 8m absolute when cold. Gas evolution causes the same problems, and looks exactly the same, as cavitation. For liquids with dissolved gases, the NPSHa/NPIPa margin must be based on the gas evolution pressure, not liquid vapour pressure. A pump cannot distinguish between released gas and saturated vapour, the effects and the damage are the same.

Thermodynamic data must be comprehensive if used for pump applications. Data on compressed liquid state must be available as well as saturation line data.

The higher velocities listed in Table 3.5 demand better pipework designs and high quality manufacture and assembly. Poorly fitted slip-on flanges, badly aligned, will cause extra turbulence and losses. Bent pipe is preferable to fabricated bends. In most cases, a large radius should be used in place of a short one. The minimum of weld neck flanges, or connectors which use locating seal rings, such as Grayloc's, will preserve good flow patterns.

Losses in bends and valves etc. can be expressed as equivalent straight pipe lengths by using the values in Table 3.6. The values for valves are mean values and vary according to the design. By adding the equivalent length of pipe and the straight length, the total pipe length is obtained enabling the flow losses, expressed as head (m), to be calculated from the diagrams.

As pointed out in Section 3.2.3, the equivalent pipe length depends upon the straight pipe loss coefficient A, i.e. pipe roughness, Reynolds Number etc. The numerical values in Table 3.6, therefore must be regarded as approximate guideline values.

Obstacle type	Equivalent pipe length	
	d = 10 - 50 mm	d = 80 - 400 mm
Gate valve	15 - 10 x d	10 - 5 x d
Check valve (with flap)	200 - 150 x d	150 - 100 x d
Angle valve (seating)	400 - 200 x d	150 - 100 x d
Stop valve (straight seating)	1000 - 500 x d	500 - 400 x d
Pipe bend 45°	10 x d	10 x d
Pipe bend 90°	30 - 20 x d	15 - 10 x d
Tank outflow (rounded)	10 - 5 x d	10 - 5 x d
Tank outflow (sharp edged)	40 - 30x d	40 - 30 x d
T-branch through side outflow	50 - 40 x d	40 x d

Table 3.6 Conversion factors for equivalent pipe lengths

Alternatively, losses in bends, valves etc., can be separately calculated by converting the values for ζ obtained from Figure 3.11 into metres head (column of liquid) at specific flow velocities with the help of the diagrams. Note that the stated value of ζ is for turbulent flow. In the case of laminar flow conversion to equivalent pipe length is the most convenient way to calculate these losses. Figure 3.22 shows the flow velocity in pipes for volume flow (m^3h and l/min), and pipe diameter (m). Figure 3.23 shows the losses of head in bends and fittings

The applicability of the diagrams is wider than indicated. Figures 3.24 to 3.41 give frictional loss diagrams for calculating pressure losses in a wide range of pipelines. The diagrams for water can be used for all water-like Newtonian liquids, i.e. for liquids with corresponding viscosity. Note, however, that if the pressure loss is stated in Pa (Pascal) the density must also coincide with that of water.

The diagrams for oils can be used for all Newtonian liquids with higher viscosities than water, under the same conditions.

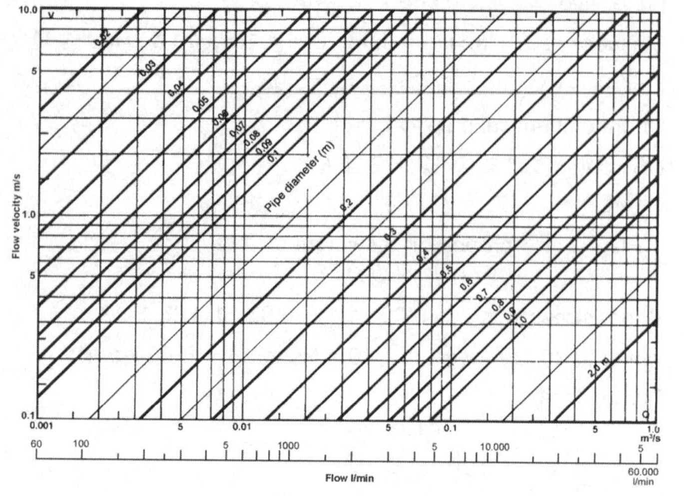

Figure 3.22 Flow velocity in pipes in for volume flow and diameter. (Interpolate between the lines for other diameters)

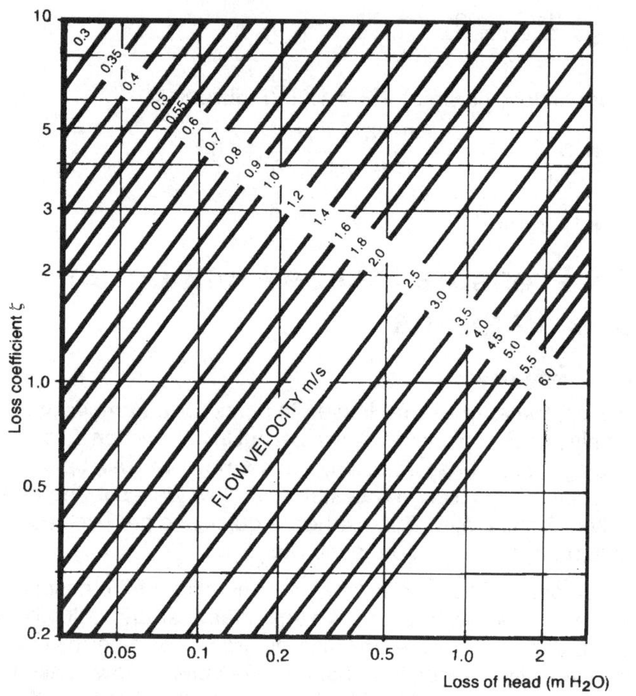

Figure 3.23 Relation between loss coefficient and loss of head in bends and fittings

3.4.1.1 Calculation example

Calculate the total loss of head in 3 x 90° short radius pipe bends in a 100 mm diameter pipe for a water flow of 700 l/m.

From Figure 3.22, velocity of flow = 1.5 m/s

From Figure 3.11, ζ = 3 x 0.4 = 1.2

From Figure 3.23, loss of head = 0.14 m

3.4.2 Water

The diagrams also apply to other Newtonian liquids having similar viscosity to that of water when the pressure loss is expressed as head (metres of liquid column).

3.4.2.1 Calculation example

A flow of 10 l/s is required through a 120 mm diameter pipe, 200 m long. Calculate the flow velocity and the loss of head. From the diagram, the velocity is 0.9 m/s and loss of head 200 x 7/1000 = 1.4 m

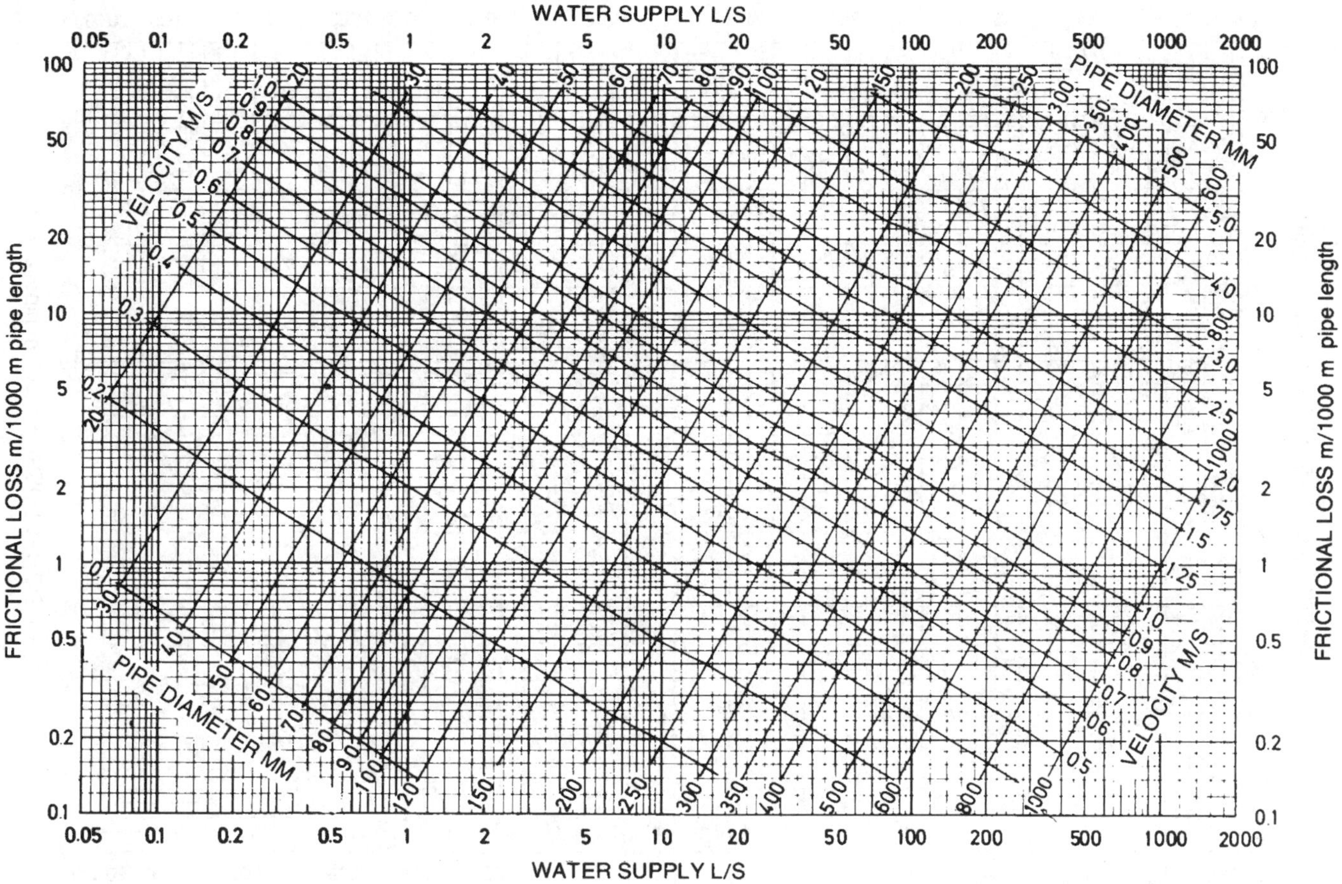

Figure 3.24 Frictional loss diagram for water at +10°C, k = 0.01 mm for diameter ≤ 200 mm, k = 0.05 mm for diameter > 200 mm. Pipe type: steel, Cu, PE, PVC

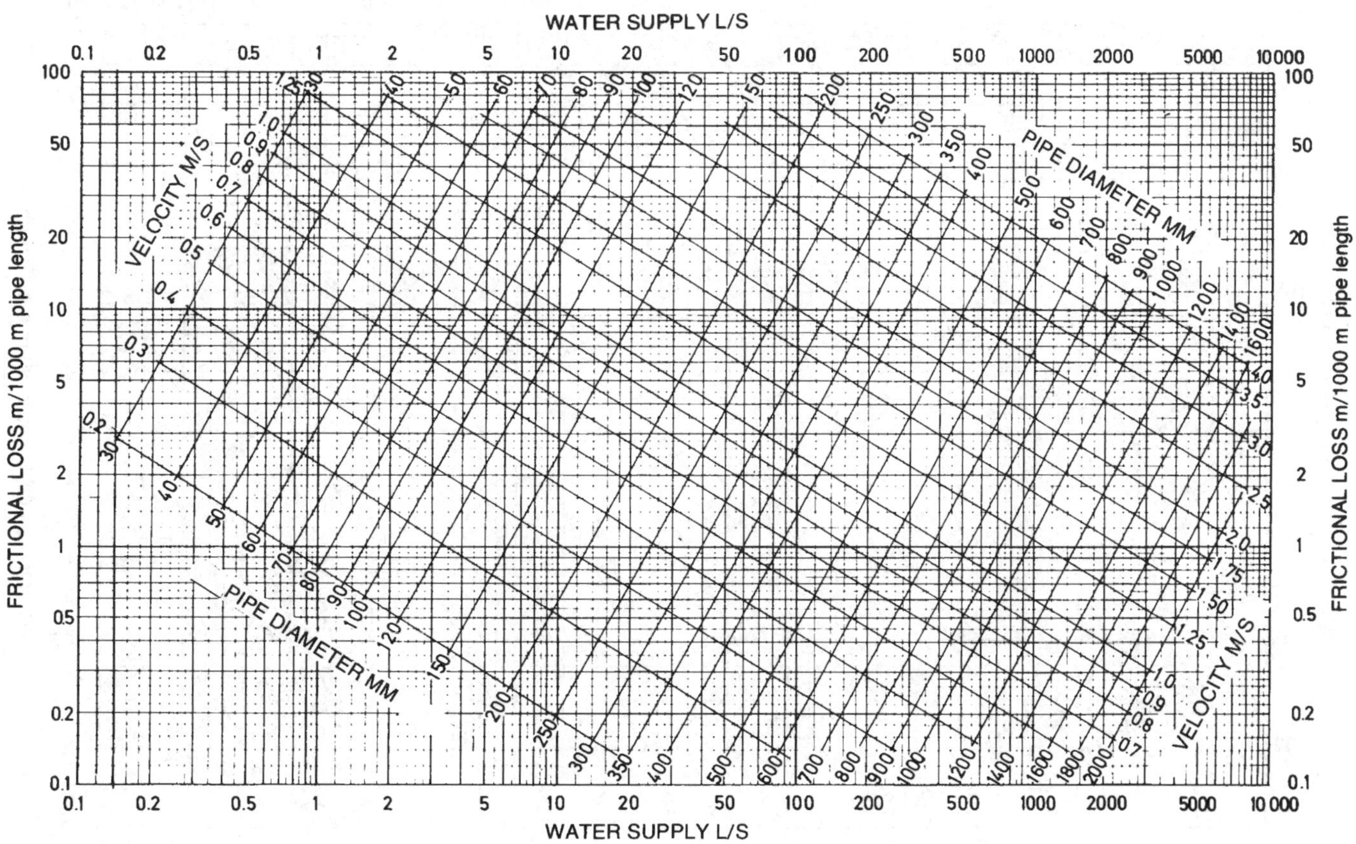

Figure 3.25 Frictional loss diagram for water at +10 ° C, k = 0.1 mm. Pipe type: steel, cast iron, GRP

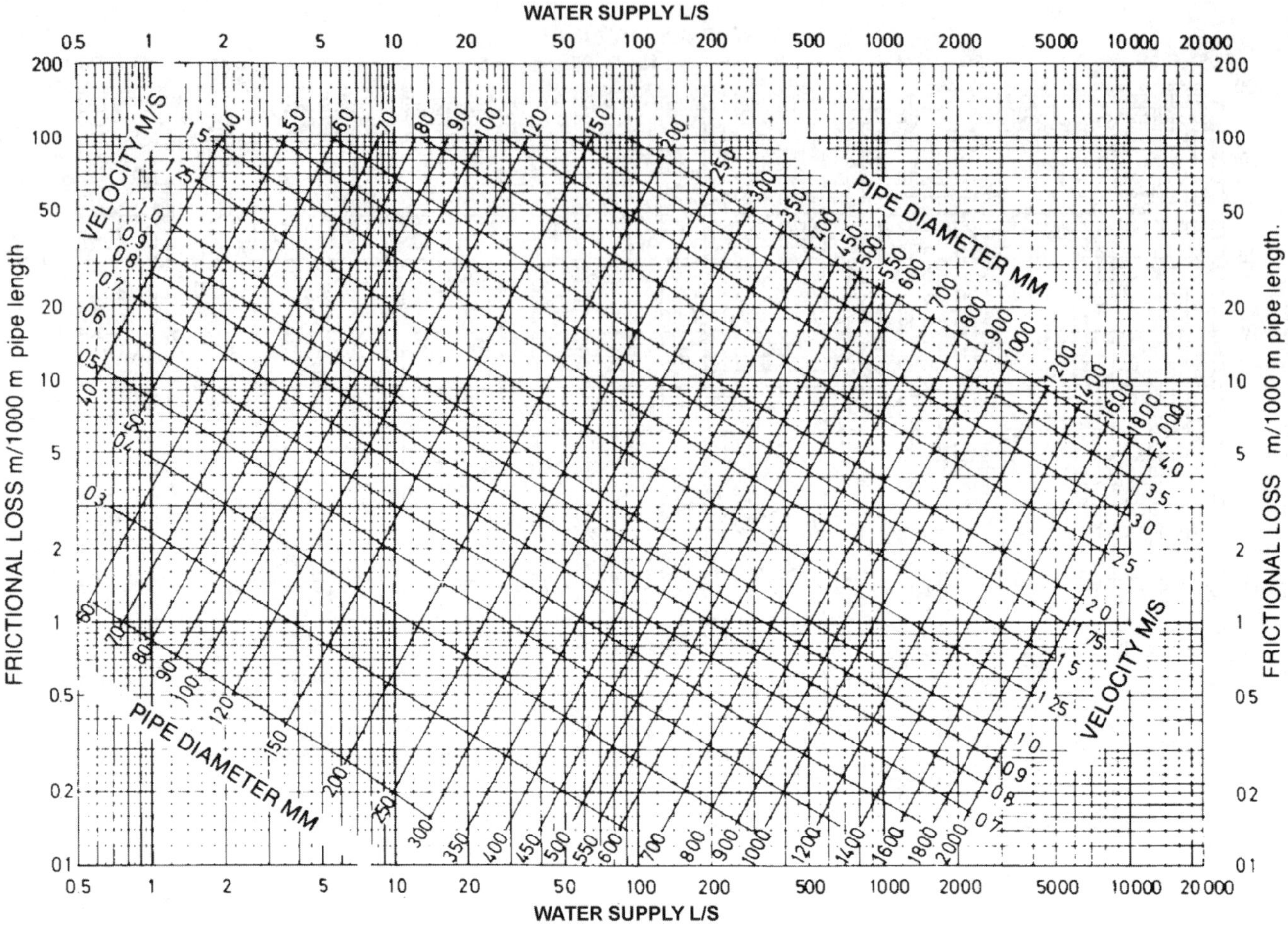

Figure 3.26 Frictional loss diagram for water at + 10°C, k = 0.2 mm. Pipe type: steel, cast iron, concrete, wood

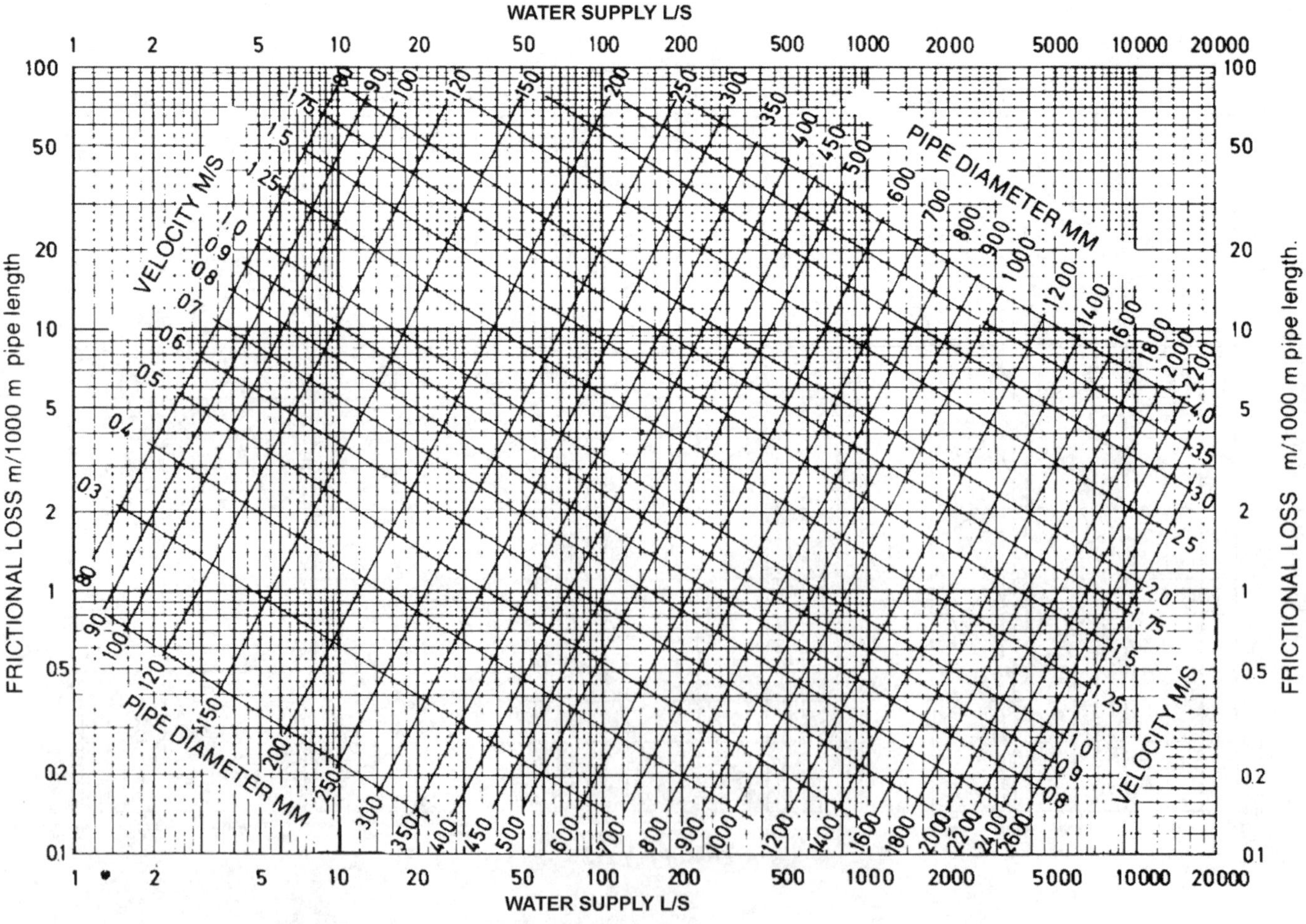

Figure 3.27 Frictional loss diagram for water at +10°C, k = 0.5 mm. Pipe type: steel, cast iron, concrete

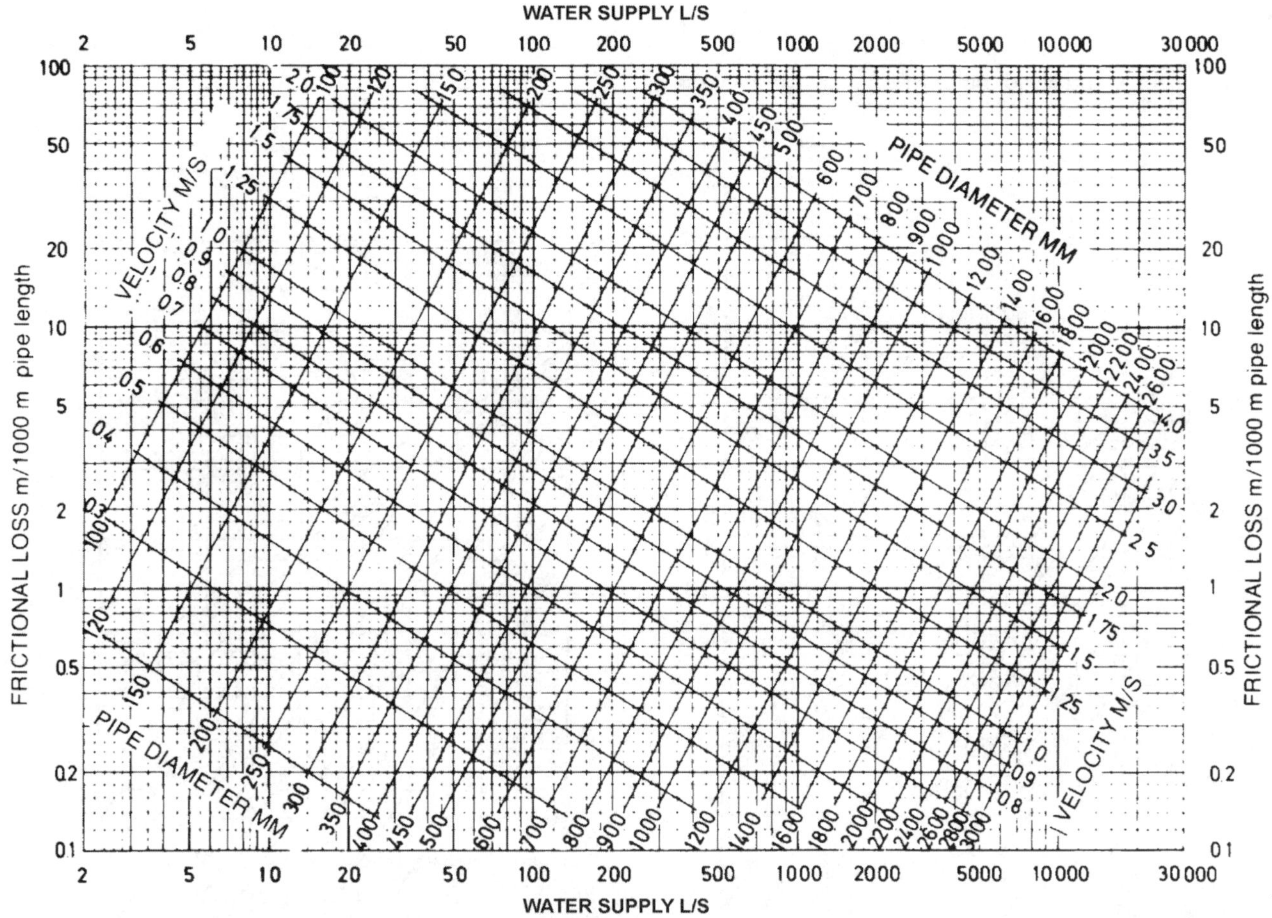

Figure 3.28 Frictional loss diagram for water +10°C, k = 1.0 mm. Pipe type: cast iron, concrete, earthenware

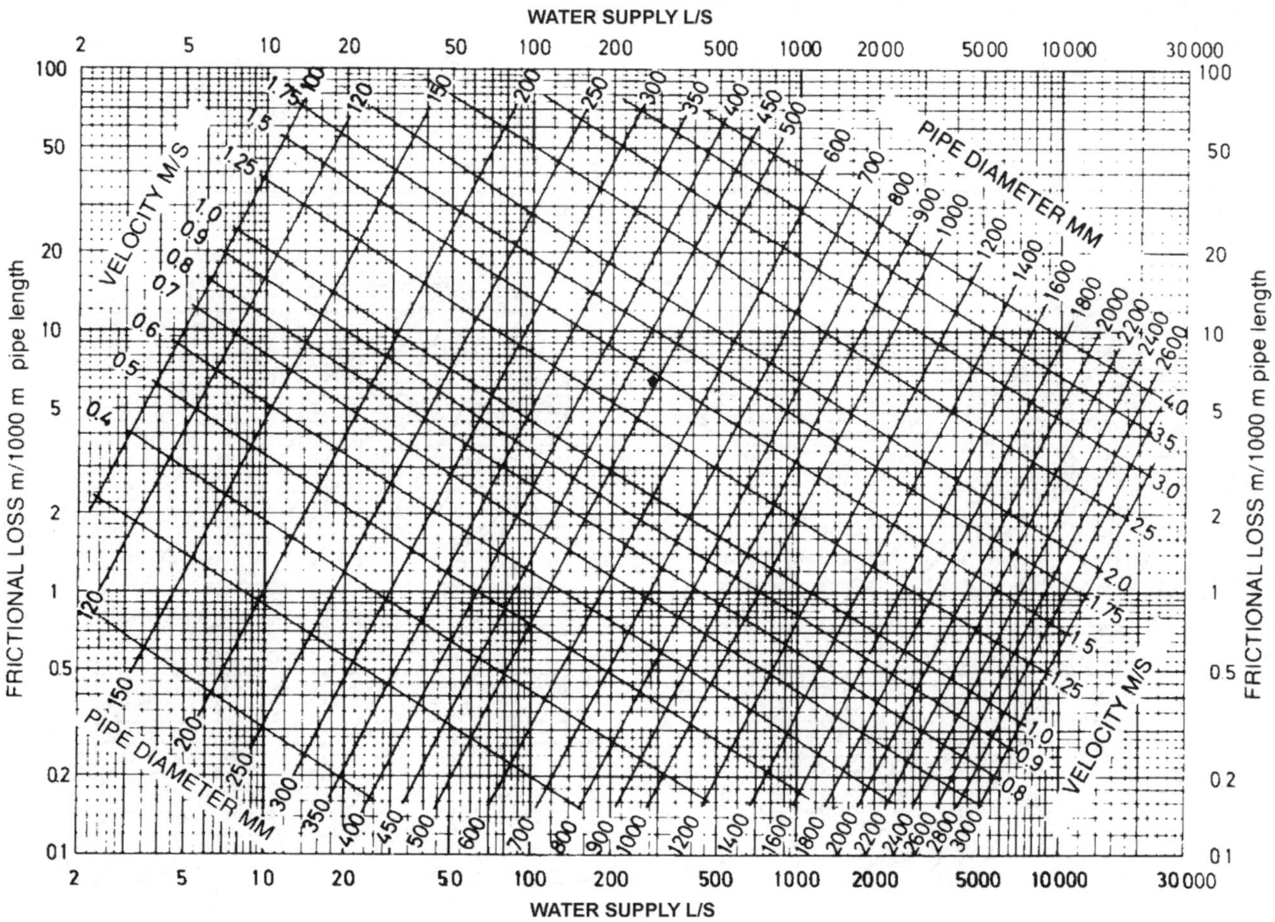

Figure 3.29 Frictional loss diagram for water at +10°C, k = 2.0 mm. Pipe type: cast iron

3.4.3 Oil

The diagrams apply equally for other Newtonian liquids at the stated viscosity.

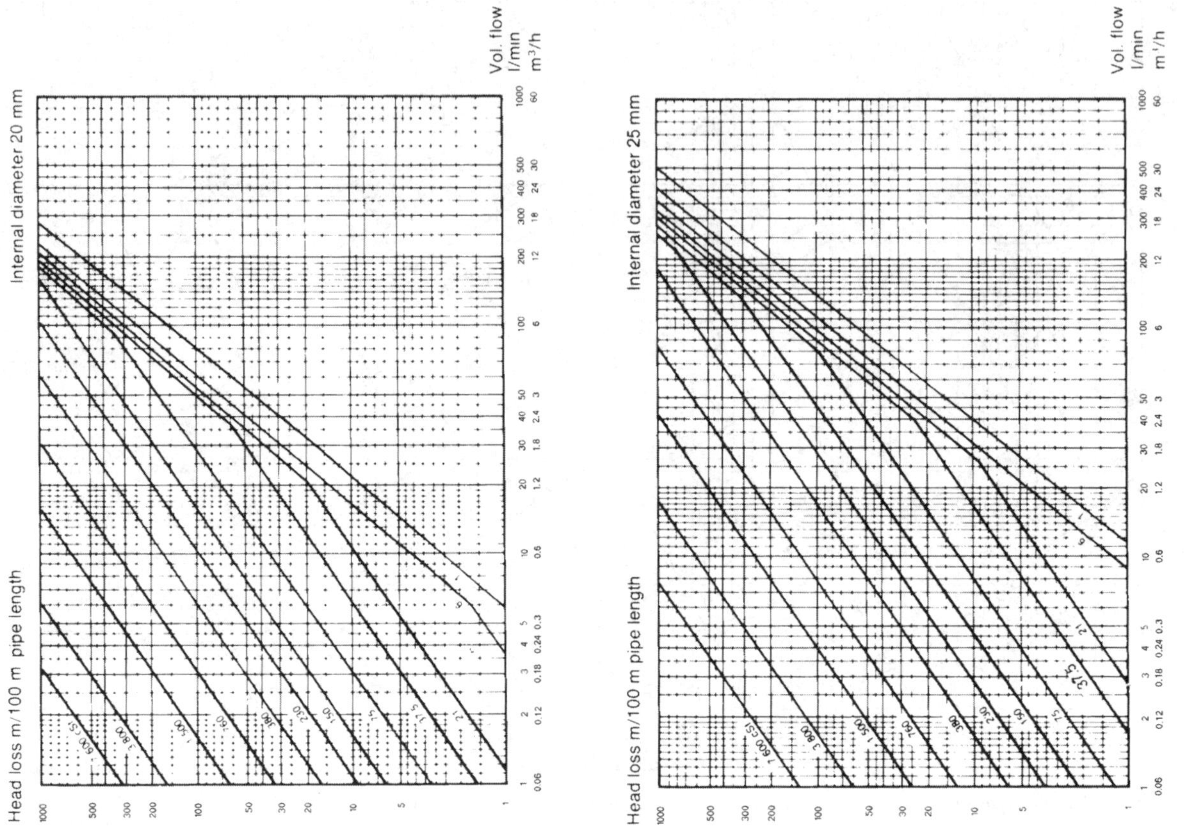

Figure 3.30 Frictional loss diagrams for oil at various viscosities, k = 0.03 mm, steel pipe

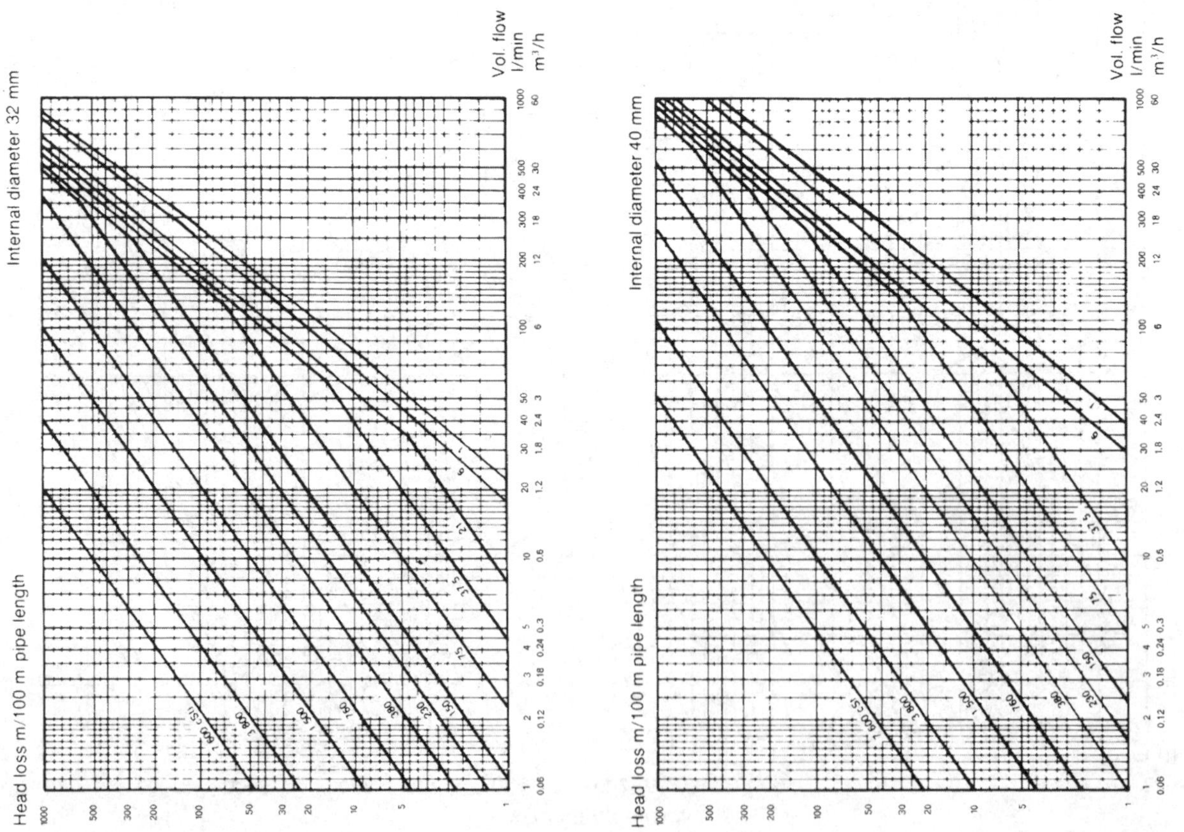

Figure 3.31 Frictional loss diagrams for oil at various viscosities, k = 0.03 mm, steel pipe

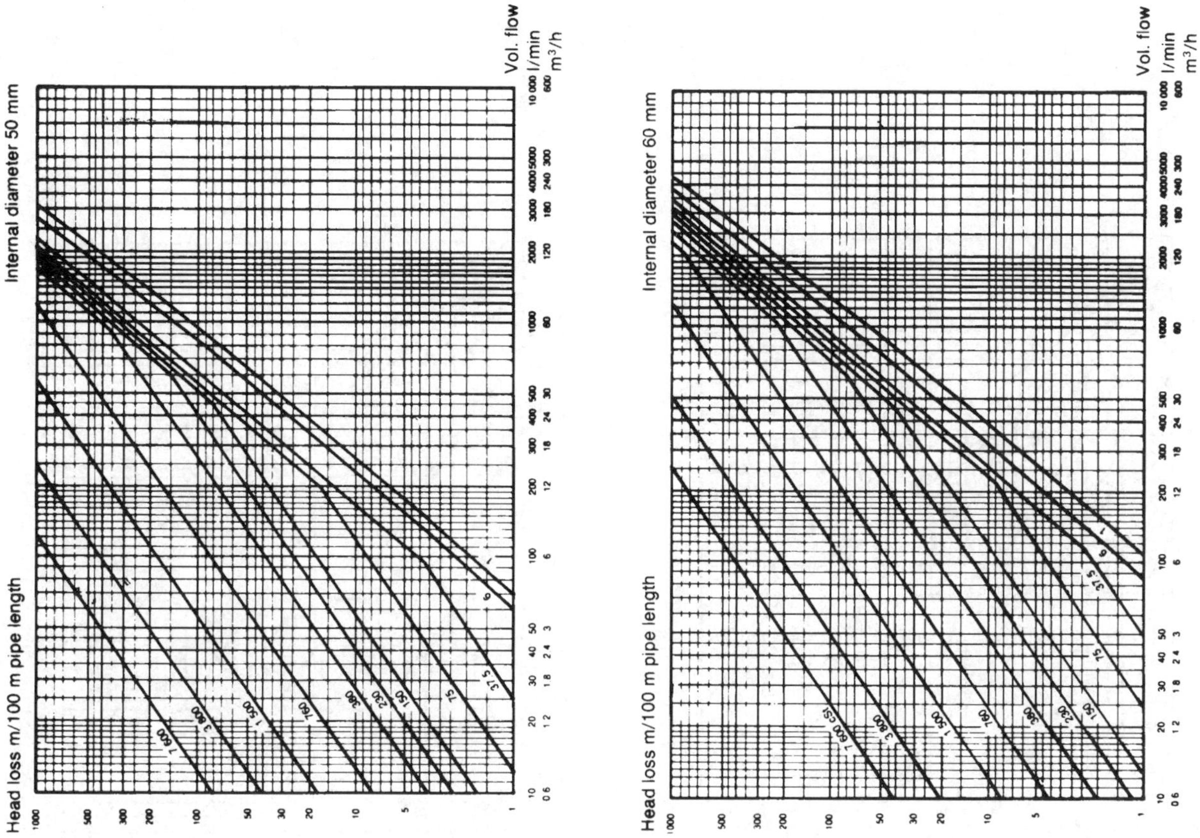

Figure 3.32 Frictional loss diagrams for oil at various viscosities, k = 0.03 mm steel pipe

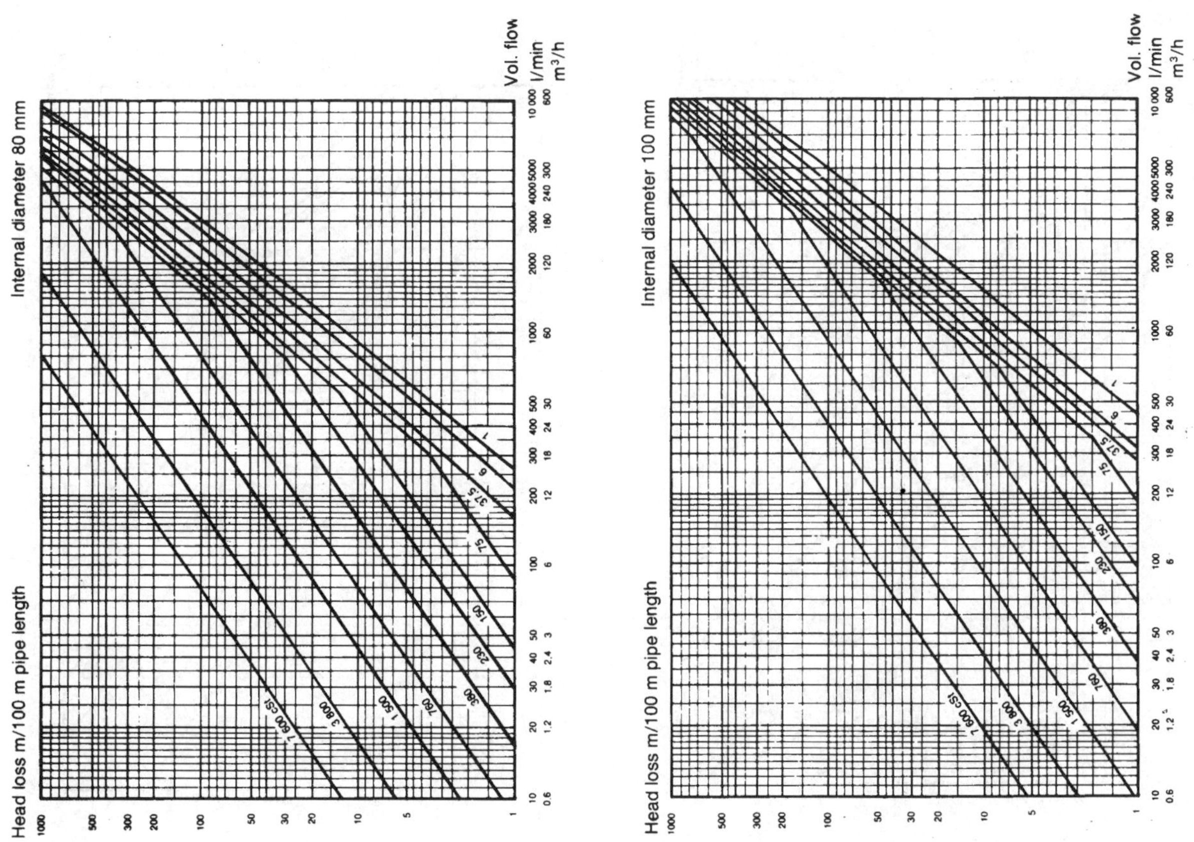

Figure 3.33 Frictional loss diagrams for oil at various viscosities, k = 0.03 mm, steel pipe

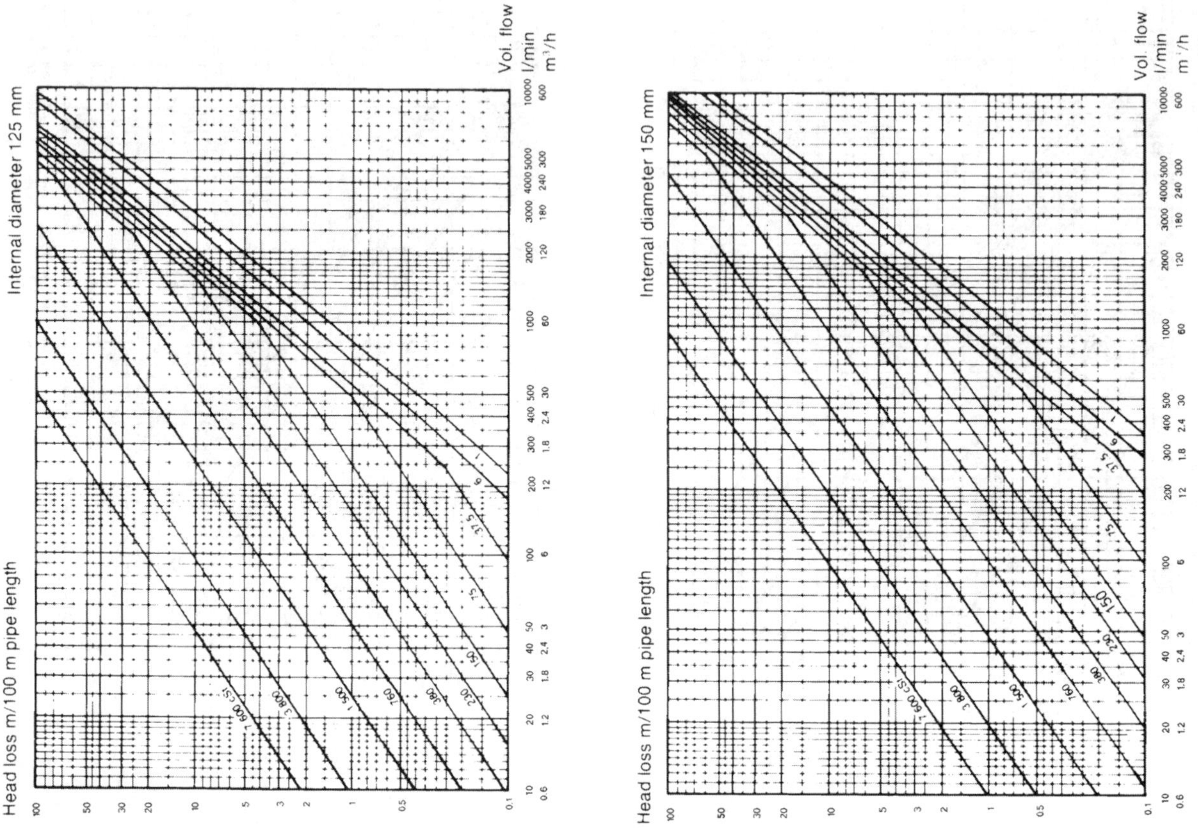

Figure 3.34 Frictional loss diagrams for oil at various viscosities, k = 0.03 mm, steel pipe

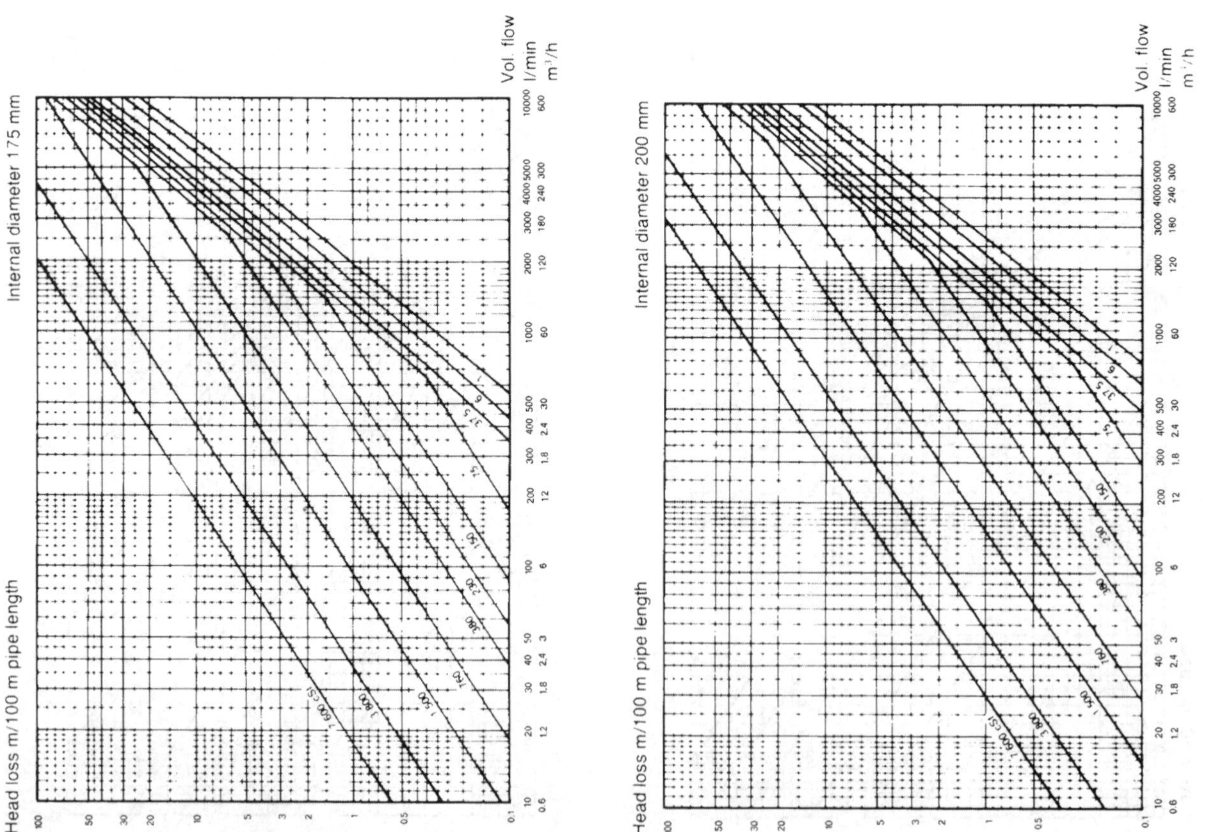

Figure 3.35 Frictional loss diagrams for oil at various viscosities, k = 0.03 mm, steel pipe

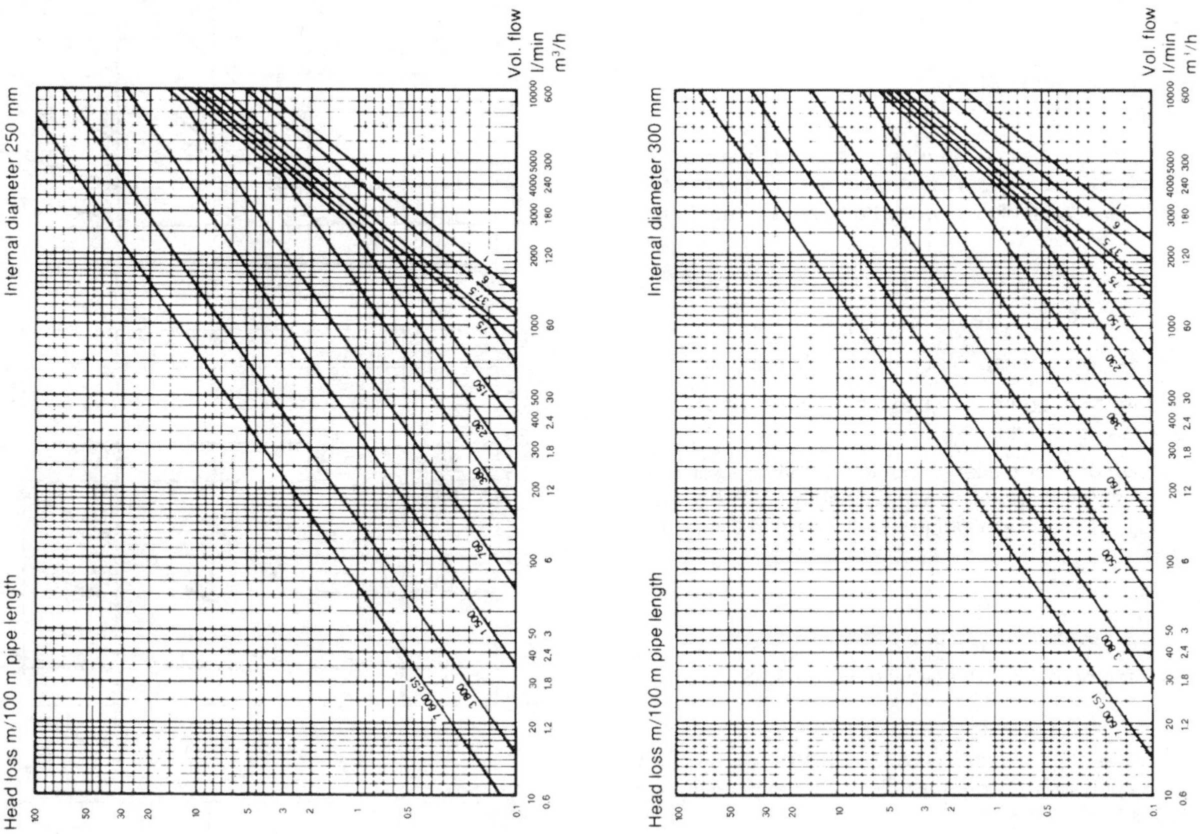

Figure 3.36 Frictional loss diagrams for oil at various viscosities, k = 0.03 mm, steel pipe

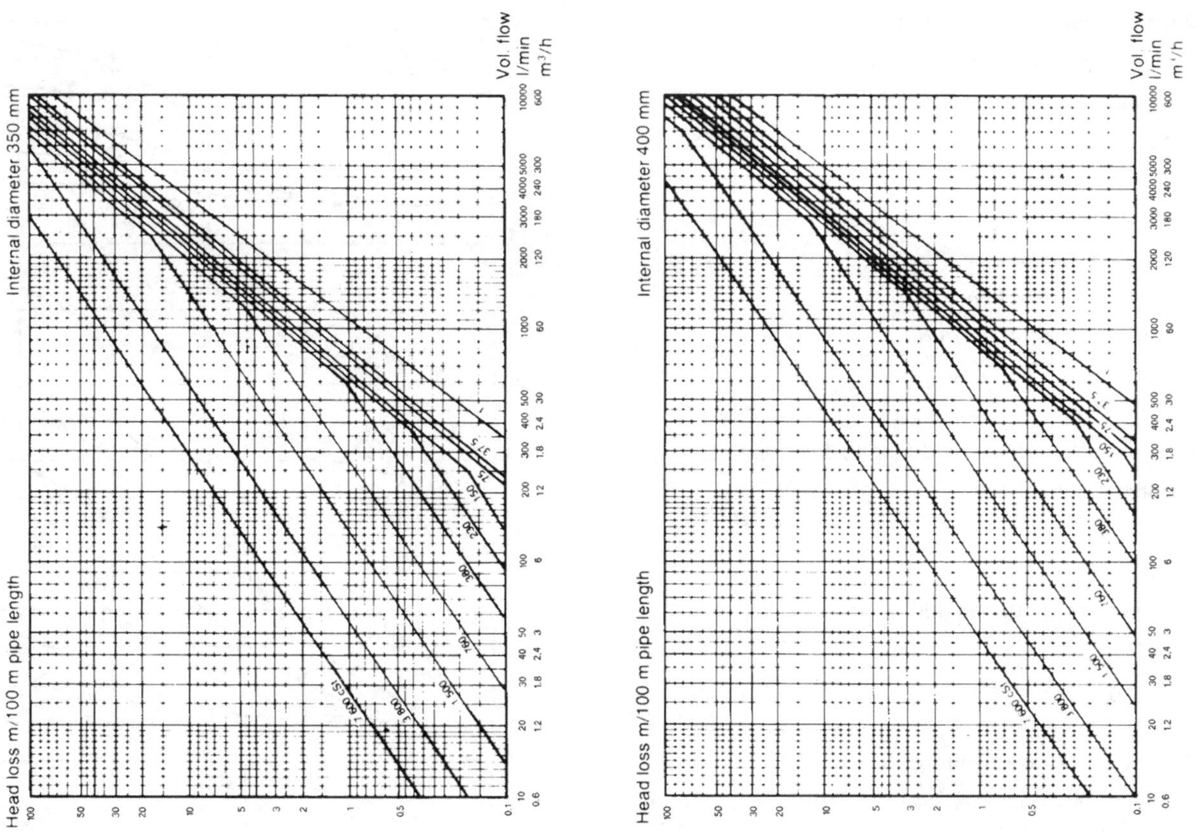

Figure 3.37 Frictional loss diagrams for oil at various viscosities, k = 0.03mm, steel pipe

3.4.4 Paper pulp

The diagrams in Figures 3.38 and 3.39 apply for paper pulp of normal quality in stainless steel pipe:

Surface roughness of pipe 0.045 mm
Undried pulp:
Fibre ratio 75
Fineness grade 750 C.s.f.
Temperature 40°C

The diagrammatic values are computed in accordance with the method set out in Section 3.3.4. If the actual pulp differs from normal pulp, the frictional losses may be assessed using the formula:

$$h_f = f_F \cdot f_M \cdot f_D \cdot h_{fN} \qquad \text{Equ 3.38}$$

h_{fN} = loss of head for normal pulp (m)

f = factors in accordance with Section 3.3.4.

It is not possible, on the other hand, to make corrections for the roughness of the pipe in this way.

The guideline values for losses in bends and fittings, are expressed by means of an equivalent pipe length - i.e. the extra length of straight pipe which would result in a corresponding pressure loss.

$$l_{eq} = L_e \cdot d \qquad \text{Equ 3.39}$$

l_{eq} = equivalent length of pipe (m)

L_e = factor in accordance with Figure 3.38

d = diameter of pipe (m)

Since the equivalent pipe length is related to straight pipe flow in principle, it is only valid for normal pulp and cannot be applied to other liquids in other pipes.

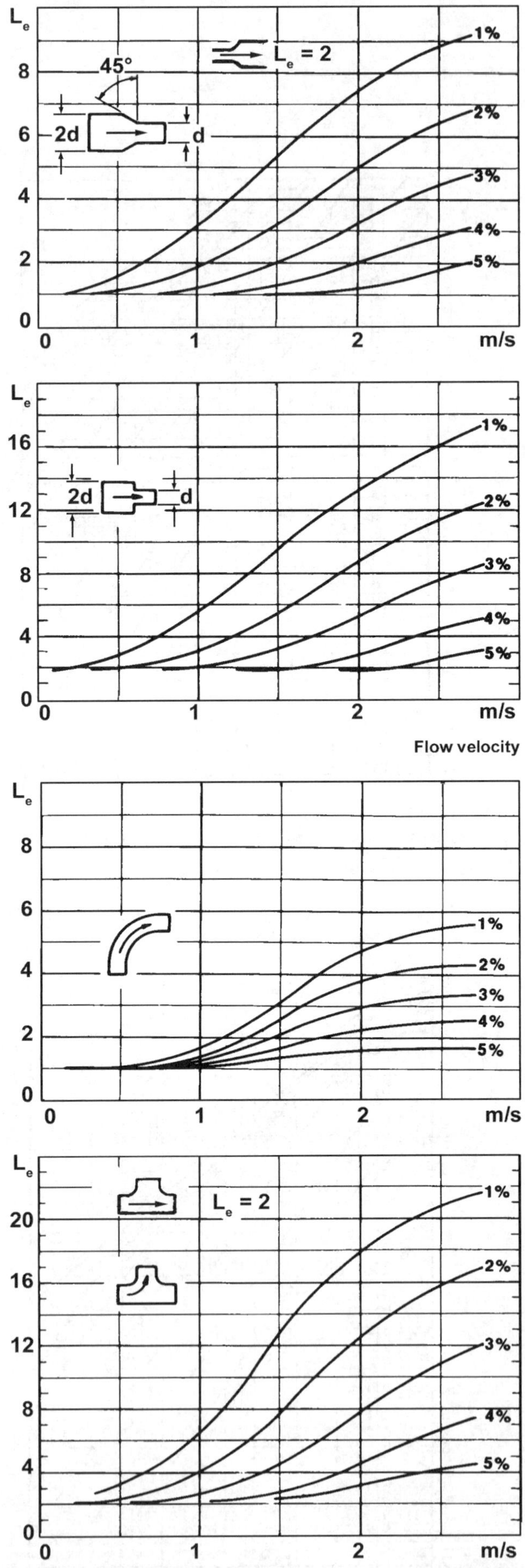

Figure 3.38 Factor for calculating equivalent pipe lengths for normal pulp in stainless steel pipe

Courtesy of ABSGroup

3.5 Flow measurement

3.5.1 General

3.5.1.1 Measurement requirements

The purpose of flow measurement may be:

- For pump performance monitoring
- For pump control
- For leakage detection
- To obtain sizing data
- To obtain costing data
- To obtain productivity data

The measuring installation may be:

- Permanent
- Temporary

The measurement requirements during testing may be:

- Continuous
- Periodic
- Intermittent

Measurement data may be needed:

- Currently
- Continuously
- Cumulatively

3.5.1.2 Methods of measurement

When choosing a suitable method of measurement, account must first be taken of the category of liquid which is to be measured:

- pure liquid compounds free of contamination,

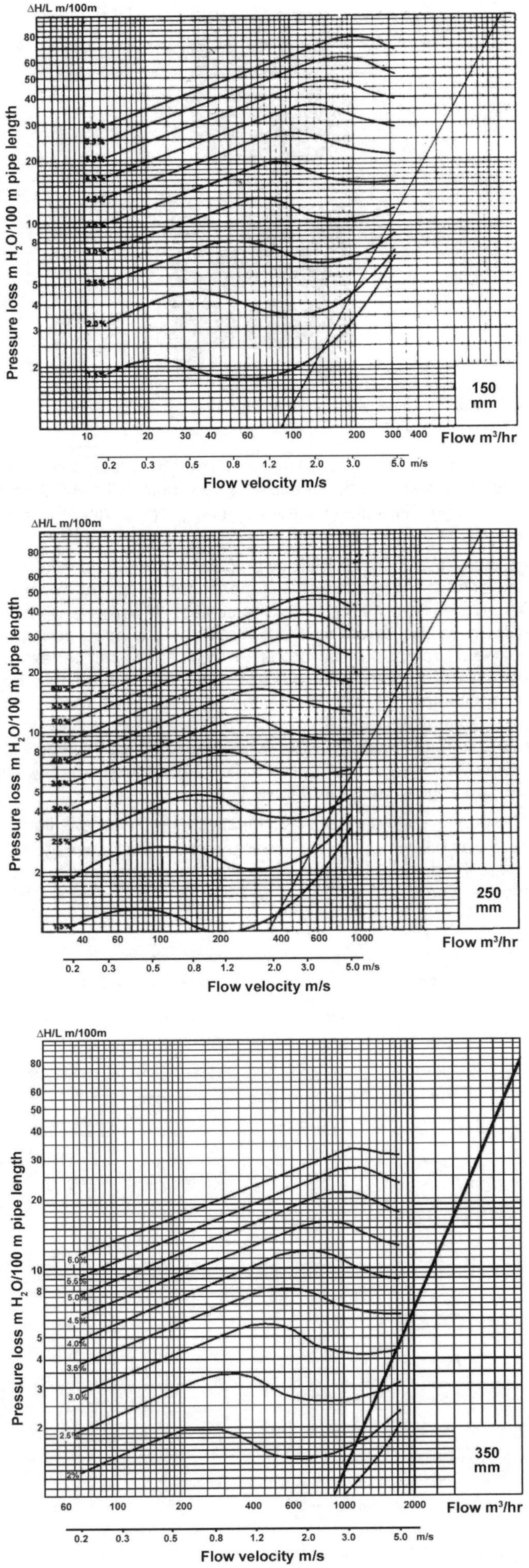

Figure 3.39 Frictional loss diagrams, normal pulp in stainless steel pipe

- contaminated liquids with a content of dissolved or undissolved; liquids, solids or gases,
- liquids having special characteristics, such as high viscosity, hot, reactive, corrosive, explosive or toxic.

Pure liquid compounds with no solid particle content or gas bubbles can be measured using most measurement equipment commercially available. The exceptions are extremely pure liquids with measuring processes which make use of contaminants normally occurring in liquids. For example, distilled water flow cannot be measured with laser or electromagnetic flow meters.

When measuring liquids contaminated with solid particles, equipment with narrow passages and sludge trapping cavities are unsuitable. When measuring liquids with special characteristics the equipment must be selected to obtain compatible performance and material combinations having due regard to the particular properties of the liquid. Laser, electromagnetic, ultrasonic and turbine devices are more suitable than others, for hazardous liquids, because of the absence of external pipe connections.

3.5.1.3 Measuring range

Instruments should be selected so that the main area of operation falls within 20% to 60% of their full-scale range.

Too large a scale may mean that a large part of the measurement occurs below 20% capacity, with resultant errors in measurement.

Too small a scale may cause the capacity of the meter to be exceeded, which will also give incorrect readings. In order to avoid measuring in unsuitable ranges, the measuring equipment in some cases may be fitted with several scales in order to minimize measurement errors.

3.5.1.4 Accuracy of measurement

Accuracy is usually stated in percentage error of full scale and applies in ideal conditions as regards installation of the equipment according to the manufacturers instructions.

It can be seen from Figure 3.40 that in the region below 10% of maximum value, the measurement accuracy is obviously poor. Within the normal measurement range, the errors extend from ± 8 to 25% of the actual measurement value, despite the claim that the meter has a measurement error within ± 5% at full-scale deflection. The absolute error is assumed to be ± 5% of maximum value within the complete measuring range.

3.5.1.5 Straight pipe length

Most flow meters require a well developed velocity profile ahead of the measuring instrument. Requirements are often placed on the minimum length of straight pipe in order to maintain the guaranteed accuracy. The necessary straight pipe lengths, in many cases, may be difficult or expensive to arrange. However, pump manufacturers will not be pleased if the

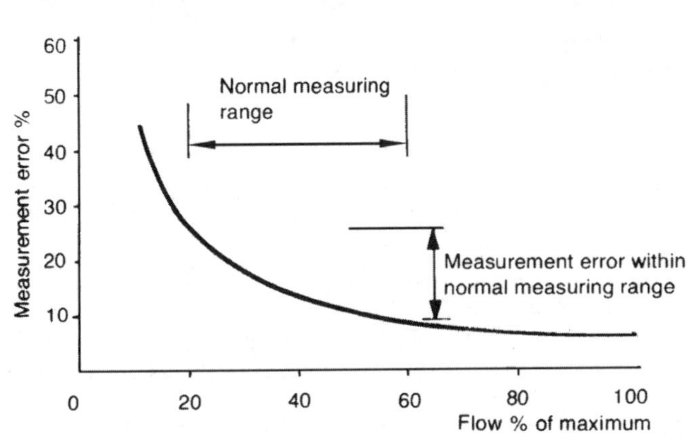

Figure 3.40 Relative measurement error with actual flow

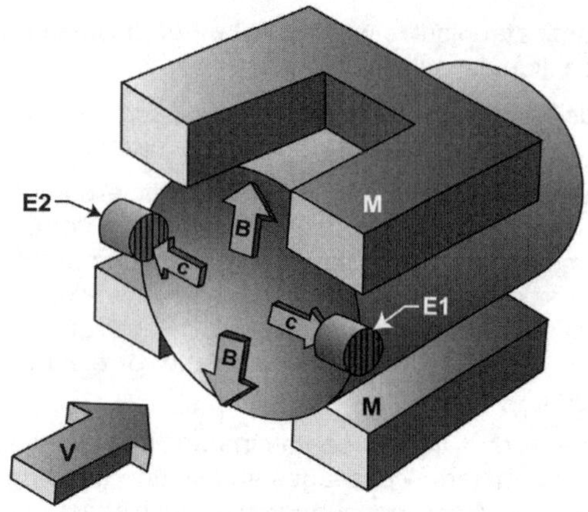

Figure 3.41 Electromagnetic flow meter principle
Courtesy of Liquid Controls

quality of their products are impugned by poorly installed flow meters. Different measuring instruments give rise to considerably varying flow losses in the installation. Loss of head associated with flow measurement can be responsible for undesirable additions to the pump energy requirement and power consumption.

3.5.2 Flow meters for pipes

This Section outlines the function and use of a number of varied flow meters. The typical properties and characteristics of these in pipes are summarised in Table 3.7; the characteristics quoted, varying of course in the case of special types.

3.5.2.1 Electromagnetic flow meters

Electromagnetic flow meters consist of a sensor and an integrally mounted or remote mounted converter. Figure 3.41 illustrates the principle and Figure 3.42 shows a typical exterior view. It consists of a lined flow tube, two electromagnetic coils (M), and two diametrically opposed electrodes (E1 & E2). Faraday's Law of Electromagnetic Induction is the basis for magnetic flow measurement. The law states that when a magnetic field is created at right angles to the flow tube through which a conductive liquid is flowing, the voltage induced in the electrodes (immersed in the fluid) is proportional to the velocity of the liquid. This induced voltage is converted to a scaled digital pulse output and, optionally, to an analog output. The analogue output signal is a linear output directly proportional to rate of flow.

Figure 3.43 Typical electromagnetic flow meter
Courtesy of Liquid Controls

As the output signal is directly proportional to the flow there is no need for linearisation. This means that it can be switched simply between different display scales. It can even be arranged so that it switches automatically to a more sensitive scale as soon as the flow drops below a certain limit, for example, 20% of full scale. The electromagnetic flow meter does not recognize the direction of flow through the meter head. This can cause problems if, for example, the meter is fitted to a pump line from a pump station which uses reverse flushing with product to clean out the pump. The reverse flush quantity will then have been measured three times.

The process liquid must be conductive to some extent; liquids as low as 0.008 μS/cm have been measured successfully. Because the device is electrical, certification is advisable for hazardous area.

Method	Liquid	Measure-ment range	Error	Straight pipe requirements	Head loss	Comments
Electromagnetic	(1)	≤ 40:1	± 0.25% possible	5D	very low	liquid must have electrical conductivity
Positive displacement	(2)	≤ 100:1	± 0.15% possible	short	moderate	various types; vane, love, rotary piston, screw
Orifice plate	(3)	4:1	2% of reading	25D	large	easily manufactured those to ISO R541 need no calibration
Nozzle	(3)	6:1	2% of reading	25D	large	those to ISO R781 need no calibration
Venturi meter	(3)	8:1	2% of reading	25D	small	
Multiple Pitot tube	(4)	10:1	± 0.5% after insitu calibration ± 1.1% uncalibrated	8D to 30D	small	
Axial turbine	(4)	20:1	0.25% FSD or 1% of reading	25D	small	
Pelton Wheel	(4)	≤ 200:1	± 0.5%	short	small	0.01 l/m to 1200 m^3/h
Ultrasonic Doppler	(5)	50:1	±2%	20D	very small	
Ultrasonic timing	clean	10:1	±0.5%	20D	very small	
Swirl	clean	10:1	1.5% FSD or 2.5% of reading	15D	moderate	
Vortex shedding	(6)	15:1	±0.65% digital ±1.35% analogue	15D	moderate	
Variable orifice	(7)	10:1	1.5% FSD	short	moderate	
Calorimetric	clean	100:1	± 5%	short	moderate	low flows

FSD = Full scale deflection

(1) Liquid can contain solid particles and gas bubbles

(2) Clean liquid is best, solids cause wear, some viscosity desirable for lubrication and to reduce slip

(3) Solids damage orifice and reduce accuracy, clean liquid preferred, range limited by Reynolds Number

(4) Clean liquid, range limited by Reynolds Number

(5) Liquid must contain impurities

(6) Slightly contaminated possible, limited by Reynolds Number

(7) Clean liquids, high viscosity possible

Table 3.7 Typical properties of flow meters in pipes

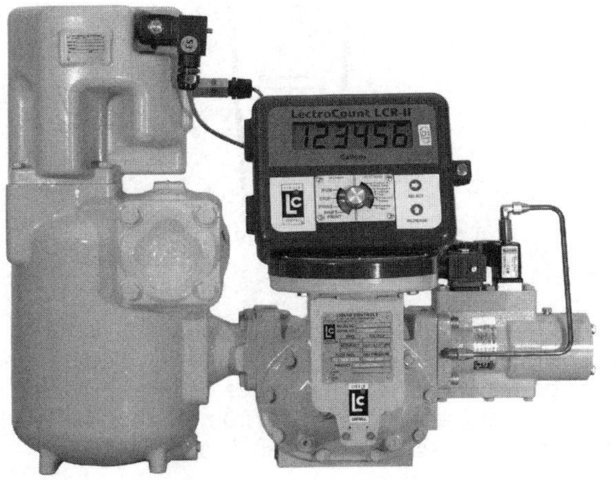

Figure 3.43 Typical positive displacement flow meter
Courtesy of Liquid controls

3.5.2.2 Positive displacement (PD) flow meters

Most positive displacement pumps, when driven as a "motor", i.e. with reversed direction of flow, work like turbines. There are as many positive types of mechanical flow meter, as there are positive displacement pumps, adapted to various liquid characteristics and capacities. A typical example is shown in Figure 3.43. This PD meter is used for refined fuels applications.

Some types, such as rotary piston flow meters, rely on the process liquid viscosity for lubrication and sealing, minimum viscosity of the order of 20 cP. Other positive displacement flow meters can cope with low viscosity non- lubricating liquids, like petrol, and very viscous liquids, over 300000 cP. Flow rates as low as 0.002 l/m can be measured and working pressures over 500 bar can be accommodated. Larger meters can measure flows of 3700 l/m at pressures up to 100 bar. PD flow meters can be completely mechanical, with no external power source required, and are eminently suitable for remote or hazardous installations.

3.5.2.3 Orifice plates and nozzles

Measuring flanges and orifices are dimensioned in accordance with ISO R541. In addition, the ISO Standard provides information about installation requirements and the calculation process for conversion of the pressure differential into flow. The guaranteed degree of accuracy is also specified. See Figure 3.44.

The range of measurement is limited at the lower end by increasing measurement errors at low Reynolds Numbers and at the top end by sharply rising head losses. Orifice plates and nozzles are simple devices with no moving parts. Problems can arise with the small bore instrument pipework connecting the manometer.

Normal straight pipe lengths are of the order of 20 x pipe diameter upstream of the orifice or nozzle and 10 x pipe diameter downstream.

As can be seen from Figure 3.45, the nozzle has lower pressure losses than the orifice plate. From this point of view, the best differential pressure flow meter is the venturi meter.

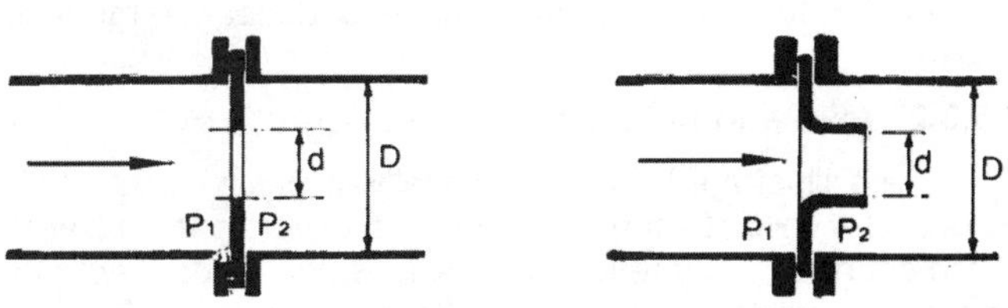

Figure 3.44 Orifice plate and nozzle. The flow is proportional to the square root of the pressure differential

Diameter ratio d/D	Loss coefficient ζ		
	Orifice	Nozzle	Venturi
0.4	85	30	6
0.5	30	10	2
0.6	12	3	1
0.7	5	1	0.5
0.8	2	0.5	0.2

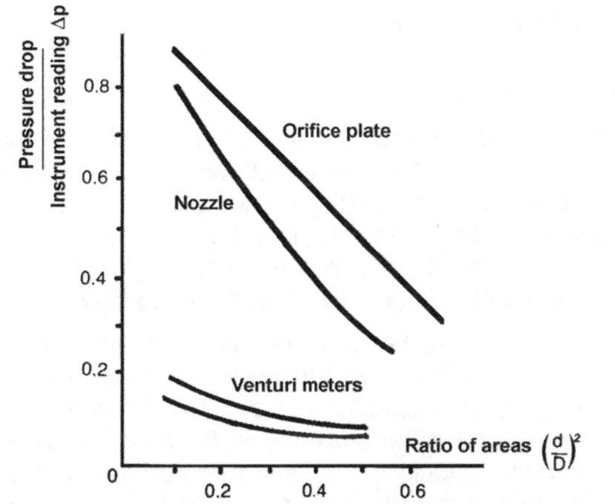

Figure 3.45 The pressure drop for a differential pressure flow meter expressed as a proportion of the instrument reading and as a loss coefficient.

3.5.2.4 Venturi meters

Venturi meters are available in long and short versions, see Figure 3.46. The lowest flow losses occur with the long version. Venturi meters are covered in ISO 5167-1 as regards dimensions, installation, calculation procedure and degree of accuracy.

3.5.2.5 Turbine meters

A turbine meter consists essentially of a turbine impeller driven by the flow of liquid, see Figure 3.47. Either the number of revolutions is measured, as for summation measurements, or the rpm (pulses per unit time) in volume flow measurements. Turbine meters are available in both axial flow and Pelton Wheel versions. Axial flow meters can be very large whereas Pelton Wheel meters are particularly suitable for very small flows with high accuracy.

Axial flow turbine meters can have aerofoil profile blades and specially shaped stators to obtain very high degrees of accuracy and are often used for calibration purposes. Simpler axial turbine meters, the so-called Woltmann meter, are not shaped for highest accuracy, however, but for general sensitivity.

3.5.2.6 Ultrasonic meters

Flow measurement using ultrasonic sound waves can be performed in two different ways and require different liquid charac-

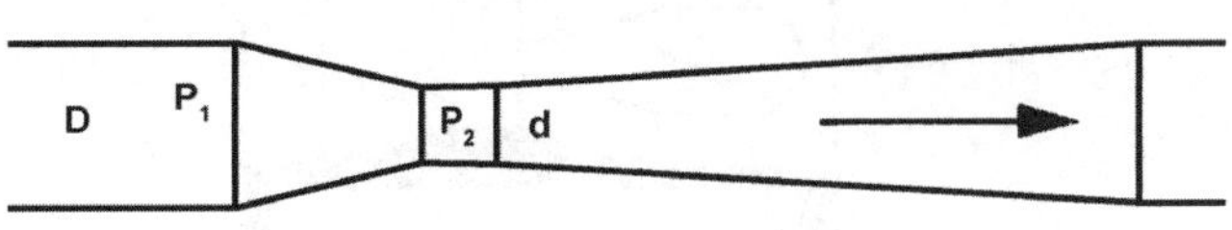

Figure 3.46 Venturi meter, long version

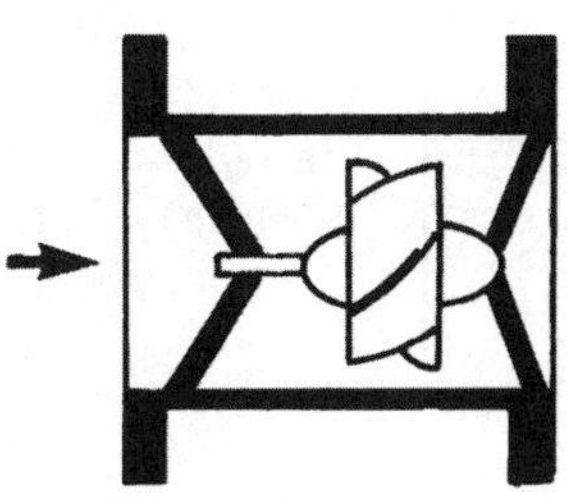

Figure 3.47 Simple turbine meter

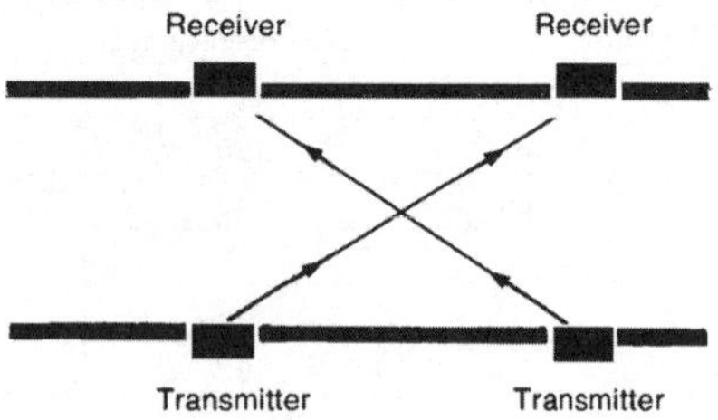

Figure 3.48 Flow measurement, ultrasonic

teristics to work correctly.

The first method requires the liquid to have some impurities, like solids or gas bubbles, to reflect the sound waves. The injected ultrasonic signal is reflected and scattered by impurities. The receiver detects the returning signals and uses Doppler techniques to convert the frequency shift into liquid velocity. The frequency shift is directly proportional to flow.

Pure clean liquids can be measured by a different method. Two transmitters inject sound waves, one in the direction of flow, and the other against the direction of flow. The difference in transit time between transmitter and receiver for these two signals is a measure of the velocity of flow and therefore of the flow itself. The difference in transit time can be converted to frequency, this being directly proportional to the flow. The frequency difference is independent of the speed of sound in the liquid. The principle is illustrated diagrammatically in Figure 3.48.

3.5.2.7 Swirl flow meters

A set of fixed vanes impart a tangential component of velocity, rotational component, at the same time as creating a secondary flow. The rotating secondary flow gives rise to pressure pulsations which are detected by a fixed sensor in the smallest section. The number of pulsations per unit time is a measure of the flow. This meter, which is shown in Figure 3.49, contains no moving parts.

3.5.2.8 Vortex shedding flow meters

When liquids flow round specially shaped bodies, there is a periodic release of vortices trailing behind the bodies. Since the boundary layer release changes from one side of the body to the other, the pressure distribution also changes around the body. The number of changes per unit time is proportional to the flow. This type of flow meter can contain no moving parts depending upon how the vortices are counted. It is shown diagrammatically in Figure 3.50.

The vortices can be counted by a thermistor detecting the cooling or heating effect changes caused by the different flow patterns. As the vortices are shed alternately from each side the

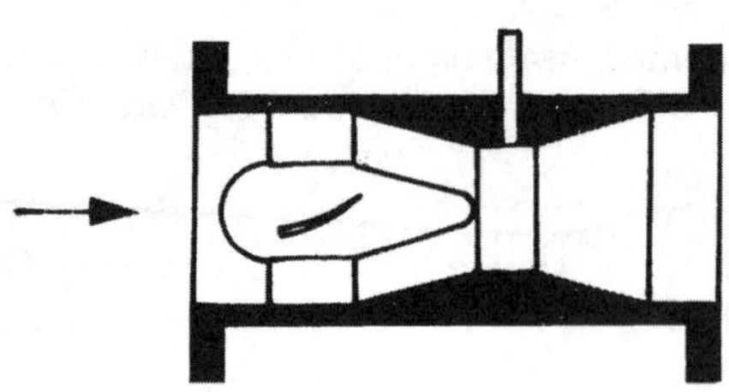

Figure 3.49 Swirl flow meter

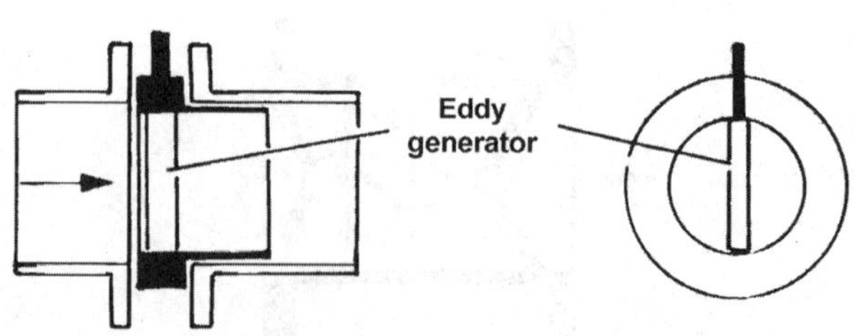

Figure 3.50 Vortex shedding flow meter

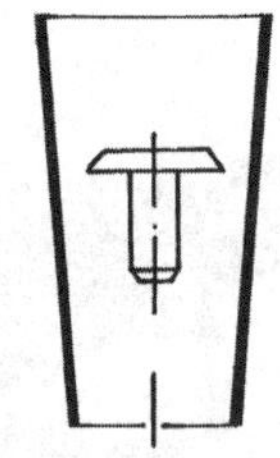

Figure 3.51 Variable orifice flow meter

pressure differential across the body changes direction. The direction of the pressure gradient can be detected by a moving sphere or disc. The vortices can be counted by a thin vane, downstream of the body, deflecting from side to side. Lastly, the vortices can be counted by ultrasonic sound waves.

3.5.2.9 Variable orifice or aperture flow meters

With this type, a spring loaded restriction, shaped like a poppet valve and mounted in a fixed orifice, moves to adjust the flow area in response to the liquid velocity. The flow can be read directly from the position of the restriction. Variable aperture flow meters are available in various sizes, and flows over 1400 m^3/h can be accommodated. See Figure 3.51.

3.5.2.10 Calorimetric flow meters

Calorimetric flow meters work on the principle of applying a known quantity of heat to the flowing liquid and measuring the consequential temperature rise. This technique is most useful in small flow rates, up to 10 l/min.

3.5.3 Flow meters for open channels

Useful data for the various methods described in this Section is given in Table 3.9.

Method of measuremen t	**Liquid**	Range of measurement	Degree of accuracy	**Remarks**
Flume	Clean to heavily contaminated	10:1	5% of max.	Several shapes available. Low head loss
Weir Notch	Clean to slightly contaminated	10:1	2% of max.	Several shapes available. Moderate loss of head. Risk of sludging
Vane	Clean	High ~ 20:1	0.5% of max.	

Table 3.9 Typical data for open channel flow meters

3.5.3.1 Measurement flumes

Measurement flumes are based on the idea that when the flow of water is so constrained a specific relationship will exist between flow and water level upstream of the flume, or, in some designs, between flow and difference in water level along the length of the flume. Flow measurement is thus converted to a level measurement.

The most common forms of measurement flume are the Parshall flume (see Figure 3.52) and the Palmer-Bowlus flume. The relationship, water level to flow, is determined empirically and therefore each flume has to be an exact copy of the master flume from which the calibration curve is derived. The degree of accuracy is stated in various sources as being in the order of ±3% to ±5% for precision made flumes and measurement devices.

3.5.3.2 Weirs and notches

These are illustrated diagrammatically in Figure 3.53. Weirs consist of a wall, of various proportions, covering the full width of the channel. Theoretical expressions can be derived, from Bernoulli's equation, for the relationship between the head above the weir to the flow rate. Theoretical expressions have proved very inaccurate and calibration of the finished installa-

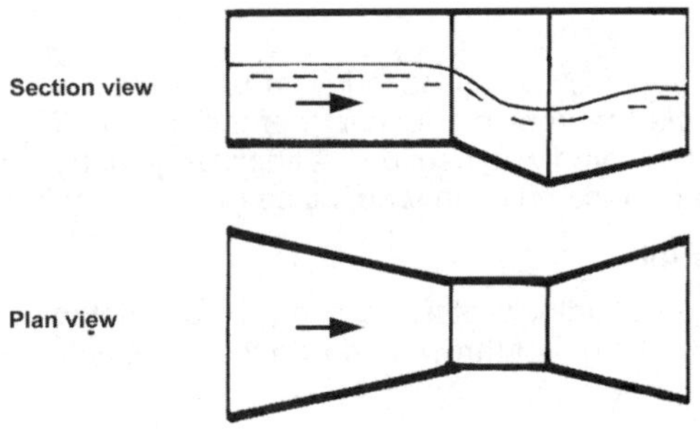

Figure 3.52 Parshall flume

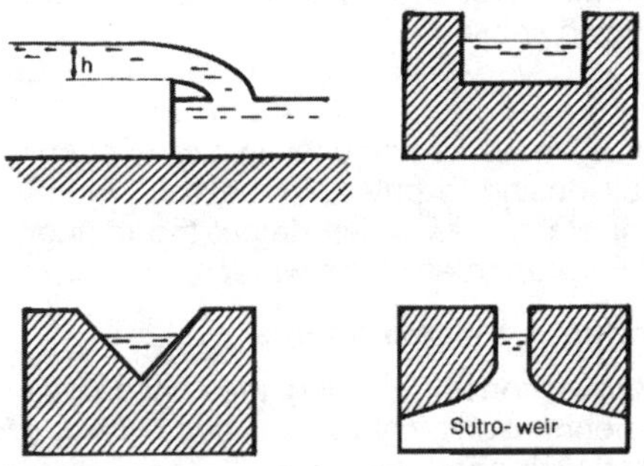

Figure 3.53 Different types of weir and notch for flow measurement

tion is essential. Empirical equations are available for designing weirs but calibration is still imperative after construction for accurate measurements. These are known as:

Francis's Formula

$$Q = 1.84(L - 0.2H)H^{1.5} \qquad \text{Equ 3.40}$$

Bazin's Formula

$$Q = \left(0.4046 + \frac{0.003607}{H}\right)\sqrt{2gL}H^{1.5} \qquad \text{Equ 3.41}$$

where:

L	=	length of weir (m)
H	=	height of liquid (m)
g	=	gravitational acceleration (m/s^2)

Notches consist of an accurately shaped sharp-edged "partial orifice" over which the liquid flows. A notch can be rectangular or triangular in shape. The flow rate is directly related to the height (H) of the upper liquid surface above the notch. The coefficient of discharge, C_d, must be evaluated for the specific installation.

Rectangular notch

$$Q = \frac{2}{3}C_d\sqrt{2g}LH^{1.5} \qquad \text{Equ 3.42}$$

Vee notch

$$Q = \frac{8}{15}C_d\sqrt{2g}\,\tan\frac{q}{2}H^{2.5} \qquad \text{Equ 3.43}$$

where:

L	=	length of weir (m)
H	=	height of liquid (m)
g	=	gravitational acceleration (m/s^2)
C_d	=	coefficient of discharge
θ	=	angle of vee notch (degree)

The Sutro-weir is not an overfall weir in the strict sense of the word. It is, however, a simple arrangement for flow measurement of heavily contaminated liquids. The Sutro-weir does not have sludge pockets and gives a linear relationship between upstream liquid level and flow.

3.5.3.3 Vanes

A very accurate flow measurement can be carried out using a measurement vane. Its use is shown diagrammatically in Figure 3.54.

The vane, which is shaped to fit the channel, moves with the liquid flow. The time is measured to travel between two predetermined positions.

3.5.4 Other instruments for flow measurements

3.5.4.1 Tanks

Measuring the time taken to fill or empty a tank of known volume is a well tried method of flow measurement. As an alternative to volume measurement, the tank can be weighed before and after filling. Properly performed, the weighing method can offer the most accurate of all known methods and is therefore often used for calibration. Both volume and weighing measurement methods are ideal when the flow is not constant, from reciprocating pumps for example.

When the weighing method is used and the results subsequently converted to volume, care should be exercised during the conversion. Ensure all parties concerned are using the same, agreed, liquid data. Even for water.

3.5.4.2 Pitot-static tube

Velocity of flow can be determined by the use of the Pitot-static tube (see Figure 3.55). Because of the velocity profile in pipe-lines and ducts, the velocity at a particular point cannot be converted to a flow. By measuring the dynamic pressure at a number of points the flow can be summated with greater accuracy. Various means of measurement with several outlet pressure points distributed around the pipe or duct section are available commercially.

3.5.4.3 Current meter

A current meter is a device used in large channels, canals and rivers, to measure the current speed at a particular position. Some instruments are mounted on traversing gantries so that a velocity distribution across the section can be averaged. The device converts the linear motion of the liquid into rotary motion which can be measured. Two designs are available. Figure 3.56 shows a propeller version, the rotary speed of which varies linearly with the velocity of flow. The other version consists of a simple Pelton Wheel device which uses hollow cones as buckets and is mounted with the axis of rotation vertical.

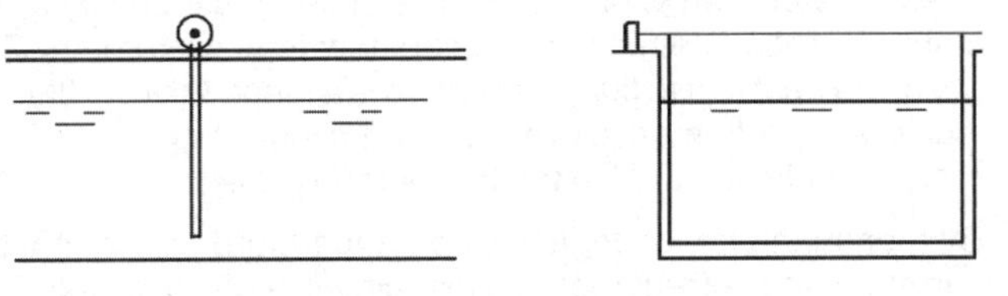

Figure 3.54 Flow measurement using vanes

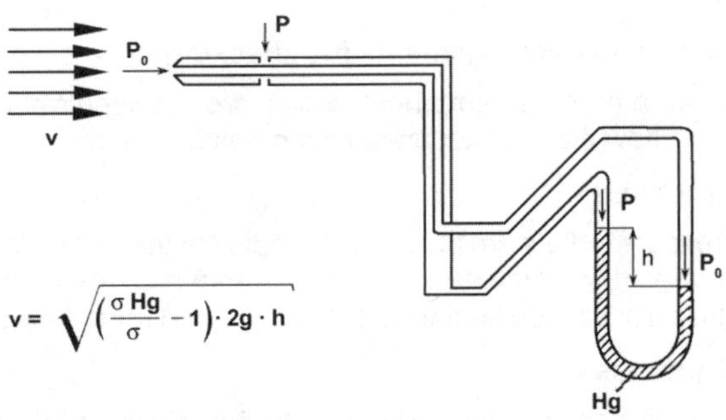

Figure 3.55 Pitot-static tube with U-tube manometer

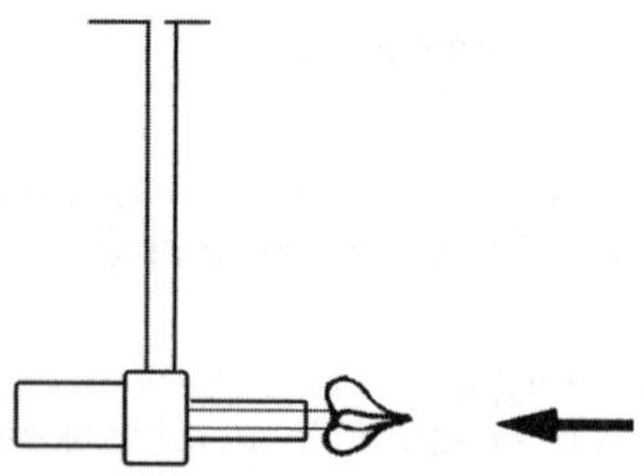

Figure 3.56 Propeller current meter

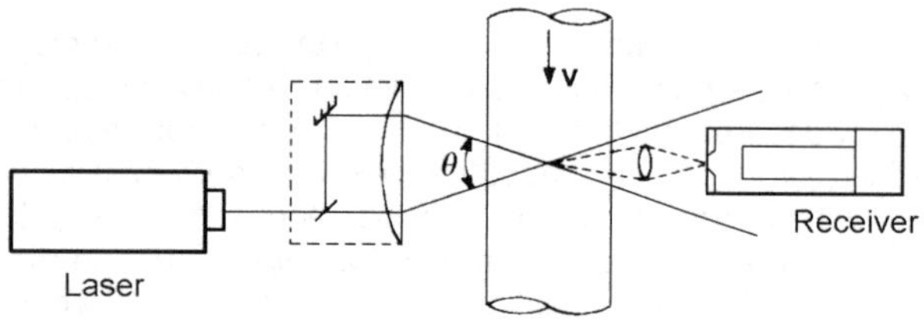

Figure 3.57 Flow measurement using lasers

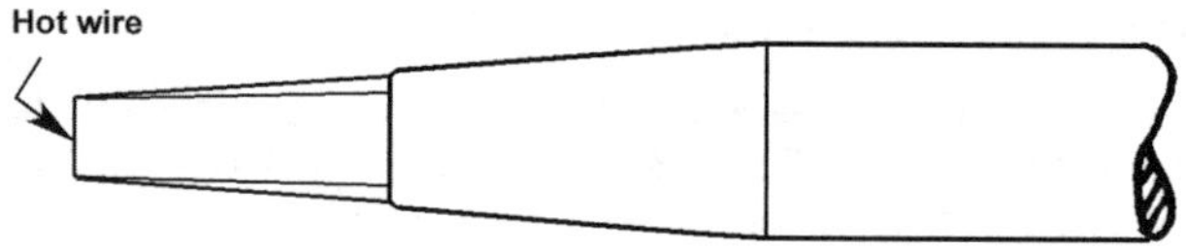

Figure 3.58 Simple probe for hot-wire anemometer

3.5.4.4 Laser-doppler

A laser beam is divided into two beams in a prism as outlined in Figure 3.57. The beams are focussed together again in the volume to be measured in the flow field, where an interference pattern is generated. A particle crossing the interference band reflects light pulses to a receiver. The frequency of the reflected light is a measure of the velocity of flow. The measuring equipment is very accurate but costly and its use is limited to the laboratory. See Figure 3.58.

3.5.4.5 Hot-wire anemometer

The hot-wire anemometer is the most well known thermal anemometer, and measures a liquid velocity by noting the heat convected away by the liquid. See Figure 3.58. The core of the anemometer is an exposed hot wire either heated up by a constant current or maintained at a constant temperature. In either case, the heat lost to fluid convection is a function of the fluid velocity. (As an alternative to wire, thin film may also be used. The meter is delicate and limited to laboratory use.)

By measuring the change in wire temperature under constant current or the current required to maintain a constant wire temperature, the heat lost can be obtained. The heat lost can then be converted into a fluid velocity in accordance with convective theory.

3.5.4.6 Various arrangements for checking flow

There are a number of methods, which are intended rather for checking flow than for accurate flow measurements:

- **Pipe bend**

  Pressure outlets are placed perpendicular to the direction of through flow at the inner and outer bend radii respectively. The error can be limited to ±10 % for a calibrated bend.

- **Sight glass**

  The position of a pivoted body in the flow (flap, paddle, etc.) can be observed through a viewing glass.

- **Pump curve**

  The flow can be assessed from pressure measurements and via the known H-Q curve of the pump. This method must be used with caution. A knowledge of how the pump performance has changed, since new, is very important.

- **Free jet**

  The outlet velocity of a free jet can be estimated by measuring the length of throw, and with the velocity the flow.

- **Floats**

  The time taken for a float, or other trace substance, to travel a given length gives a good idea of the flow in an open channel of known section.

- **Tracers**

  Trace substances can be used in pipe systems. An example is salt dosing in pulses. The salt cloud is traced by means of electrodes which detect the changed electrical conductive properties of the water.

- **Concentration measurements**

  With a known constant dosing, the concentration in the liquid is dependent on the flow. Solutions of salt in water have been used with good results.

- **Pump speed**

  For speed regulated pumps, the measurement of pump speed provides a means of checking the flow. If both the pump and system curve are known, this method can provide reasonable accuracy. Again wear in the pump and it's effect on pump performance must be considered. Positive displacement pumps produce a relatively constant output over a wide pressure range. System response is not so important in these situations.

3.5.5 Measurement standards

Methods of measurement of liquid flow are dealt with in various standards.

3.5.5.1 Measurement in pipe and duct systems

BS 1042-2.1:1983, ISO 3966:1977 Measurement of fluid flow in closed conduits. Velocity area methods. Method using Pitot static tubes.

BS 1042-2.2:1983, ISO 7145:1982 Measurement of fluid flow in closed conduits. Velocity area methods. Method of measurement of velocity at one point of a conduit of circular cross section.

BS 1042-2.3:1984, ISO 7194:1983 Measurement of fluid flow in closed conduits. Velocity area methods. Methods of flow measurement in swirling or asymmetric flow conditions in circular ducts by means of current-meters or Pitot static tubes.

BS 1042-1.4:1992 Measurement of fluid flow in closed conduits. Pressure differential devices. Guide to the use of devices specified in Sections 1.1 and 1.2. 1.

ISO TR 3313:1998 Measurement of fluid flow in closed conduits — Guidelines on the effects of flow pulsations on flow-measurement instruments.

BS 5857-1.5:1980, ISO 2975-VII:1977 Methods for measurement of fluid flow in closed conduits, using tracers. Measurement of water flow. Transit time method using radioactive tracers.

BS 5857-2.1:1980, ISO 4053-I:1977 Methods for measurement of fluid flow in closed conduits, using tracers. Measurement of gas flow. General.

BS 5857-2.4:1980, ISO 4053-IV:1978 Methods for measurement of fluid flow in closed conduits, using tracers. Measurement of gas flow. Transit time method using radioactive tracers.

BS 5857-1.4:1980, ISO 2975-VI:1977 Methods for measurement of fluid flow in closed conduits, using tracers. Measurement of water flow. Transit time method using non-radioactive tracers.

BS 5857-1.1:1980, ISO 2975-I:1974 Methods for measurement of fluid flow in closed conduits, using tracers. Measurement of water flow. General.

BS 5857-1.3:1980, ISO 2975-III:1977 Methods for measurement of fluid flow in closed conduits, using tracers. Measurement of water flow. Constant rate injection method using radioactive tracer.s

BS 6199-1.2:1991, ISO 9368-1:1990 Measurement of liquid flow in closed conduits using weighing and volumetric methods. Weighing method. Procedures for checking static weighing system.s

EN ISO 8316:1997 Method of liquid flow in closed conduits. Method by collection of the liquid in a volumetric tank. (See BS 5844, BS 5875, ISO 4185, ISO 4373).

EN 7405:1991 Guide to selection and application of flowmeters for the measurement of fluid flow in closed conduits.

EN 24006:1993, ISO 4006:1991 Measurement of fluid flow in closed conduits. Vocabulary and symbols.

EN ISO 5167-1:1997 Measurement of fluid flow by means of pressure differential devices inserted in circular cross-section conduits running full. Orifice plates, nozzles, and venturi tubes inserted in circular cross-section conduits running full.

EN ISO 5167-1:2003 Measurement of fluid flow by means of pressure differential devices inserted in circular cross-section conduits running full. General principles and requirements.

EN ISO 6817:1997 Measurement of conductive liquid flow in closed conduits. Method using electromagnetic flowmeters.

ISO TR 12764:1997 Measurement of fluid flow in closed conduits. Flowrate measurement by means of vortex shedding flowmeters inserted in circular cross-section conduits running full.

3.5.5.2 Measurement in open channels

BS 3680-10C:1996 Measurement of liquid flow in open channels. Sediment transport. Guide to methods of sampling of sand-bed and cohesive-bed materials.

BS 3680-3Q:2002 Measurement of liquid flow in open channels. Guidelines for safe working practice in river flow measurement.

BS 3680-8A:1989, ISO 2537:1988 Measurement of liquid flow in open channels. Measuring instruments and equipment. Current meters incorporating a rotating element.

BS 3680-7:2000 Measurement of liquid flow in open channels. Specification of equipment for the measurement of water level.

BS 3680-2A:1995, ISO 9555-1:1994 Measurement of liquid flow in open channels. Dilution methods. General.

BS 3680-4B:1986, ISO 4360-1984 Methods of measurement of liquid flow in open channels. Weirs and flumes. Triangular profile weirs.

BS 3680-3F:1986, ISO 1088-1985 Measurement of liquid flow in open channels. Stream flow measurement. Collection and processing of data for determination of errors in measurement.

BS 3680-11B:1992, ISO/TR 9824-2:1990 Measurement of liquid flow in open channels. Free surface flow in closed conduits. Specification for performance and installation of equipment for measurement of free surface flow in closed conduits.

BS 3680-4A:1981 Methods of measurement of liquid flow in open channels. Weirs and flumes. Method using thin-plate weirs.

BS 3680-2D:1993, ISO 9555-4:1992 Measurement of liquid flow in open channels. Dilution methods. Methods of measurement using fluorescent tracers.

BS 3680-2B:1993, ISO 9555-2:1992 Measurement of liquid flow in open channels. Dilution methods. Methods of measurement using radioactive tracers.

BS 3680-10B:1980, ISO 4363-1977 Measurement of liquid flow in open channels. Sediment transport. Measurement of suspended sediment.

BS 3680-8B:1983, ISO 3454-1983 Measurement of liquid flow in open channels. Measuring instruments and equipment. Recommendations for direct depth sounding and suspension equipment.

BS 3680-8C:1980, ISO 3455-1976 Measurement of liquid flow in open channels. Measuring instruments and equipment. Calibration of rotating-element current-meters in straight open tanks.

BS 3680-4I:1986, ISO 8333-1985 Methods of measurement of liquid flow in open channels. Weirs and flumes. V-shaped broad crested weirs.

BS 3680-11A:1992, ISO/TR 9824-1:1990 Measurement of liquid flow in open channels. Free surface flow in closed conduits. Methods of measurement.

BS 3680-4C:1981 Methods of measurement of liquid flow in open channels. Weirs and flumes. Flumes.

BS 3680-10E:1993, ISO 9195:1992 Measurement of liquid flow in open channels. Sediment transport. Sampling and analysis of gravel bed material.

BS 3680-5:1992, ISO 1070:1992 Measurement of liquid flow in open channels. Slope area method of estimation.

BS 3680-3D:1980, ISO 4369-1979 Measurement of liquid flow in open channels. Stream flow measurement. Moving-boat method.

BS 3680-4F:1990, ISO 4374:1990 Methods of measurement of liquid flow in open channels. Weirs and flumes. Round nose horizontal broad-crested weirs.

BS 3680-4E:1990, ISO 3846:1989 Methods of measurement of liquid flow in open channels. Weirs and flumes. Rectangular broad-crested weirs.

BS 3680-2C:1993, ISO 9555-3:1992 Measurement of liquid flow in open channels. Dilution methods. Methods of measurement using chemical tracers.

BS 3680-3B:1997, ISO 1100-1:1996 Measurement of liquid flow in open channels. Stream flow measurement. Guide for the establishment and operation of a gauging statio.n

EN ISO 748:2000 Measurement of liquid flow in open channels. Velocity-area methods.

3.6 Useful references

Internal Flow Systems, D S Miller, Crane Company, USA 1988. Technical Papers Nos. 409 and 410.

The Two K Method Predicts, William B. Hooper, Chemical Engineering, August 24, 1981 pp 96-99.

Calculate Head Loss Caused by Change in Pipe Size, William B. Hooper, Chemical Engineering, November 7, 1988 pp 89-92.

Hydraulic Institute, 9 Sylvan Way, Parsippany NJ, 07054 USA. Tel 973 267 9700, Fax 973 2679055, www.pumps.org.

VDMA (German Engineering Federation), Lyoner Straße 18, D-60528 Frankfurt, Germany. Tel 069 66 03-0, www.vdma.org.

Transport hydraulique et décantation des matériaux solides. Durand, R. & Condolios, E. (1952). Deuxième Journée de l'Hydraulique, 27-55, France.

Capillary conduction of liquids through porous mediums, L.A. Richards, 1931. Physics 1:318–333.

Measurement of fluid flow by means of pressure differential devices, Part 1: Orifice plates, nozzles, and Venturi tubes inserted in circular cross-section conduits running full, ISO 5167-1.1991(E).

Fluid meters: Their theory and application, edited by H. S. Bean, 6ed Report of ASME Research Committee on Fluid Meters. American Society of Mechanical Engineers (ASME),1971.

Pump theory

4

4.1 Introduction

Use is made of pumps everywhere there is a need for liquid transportation. Sometimes it is clean liquid which is pumped, sometimes the liquid is used as a medium for carrying thermal energy or solid suspended material etc. Due to widely varying operational requirements there are a large number of different pump designs available. A classification into four main categories has been made in an attempt to provide a clear picture of the types of pumps in use today and forms the basis of the pump types described in detail in Chapter 1. The pump types in Chapter 1 should be studied individually to assist in the selection of the appropriate type.

- Rotodynamic pumps
- Special rotodynamic pumps
- Positive displacement pumps
- Other pumps

The construction of rotodynamic pumps is characterised by one or more impellers equipped with vanes which rotate in a pump casing. Those forces which come into being when liquid flows around the vanes are utilised in the conversion of energy. The rotodynamic pumps are called radial pumps or axial pumps depending on the main direction in which the liquid flows through the impellers. Alternatively, the terms "centrifugal pumps" for radial pumps, and "propeller" for axial pumps are used. Of course, intermediate forms between purely radial and purely axial pumps are also found and as such, are termed "mixed-flow" pumps. Figure 4.1 shows the most usual and most important of the rotodynamic pumps.

Certain aspects of the special rotodynamic pumps are closely related to the radial rotodynamic described above. These pumps are not true rotodynamic pumps because their performance does not meet the expectations predicted by the rotodynamic equations. The kinetic energy of the liquid is transformed into static head but the actual pump performance must be confirmed by testing. The peripheral pump uses impellers with vanes to increase the liquid kinetic energy whereas the Pitot tube pump utilises a rotating drum. See Figure 4.2. Both pumps are very useful because the differential head increase is more than a centrifugal pump of the same speed.

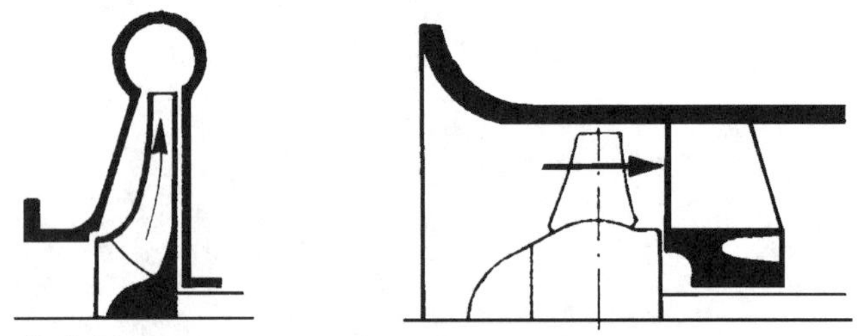

Figure 4.1 Examples of rotodynamic pumps

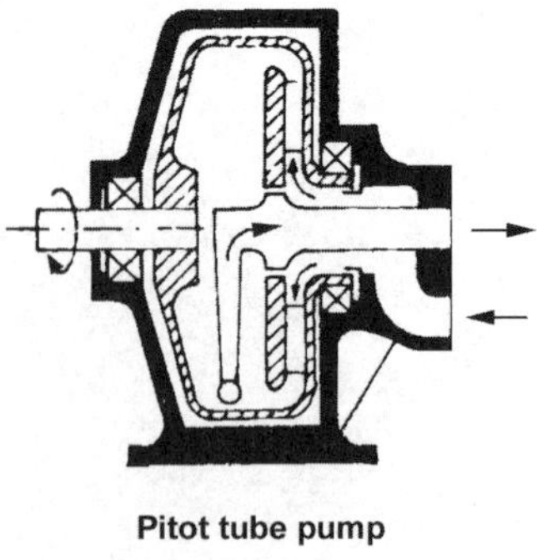

Figure 4.2 Example of special rotodynamic pumps

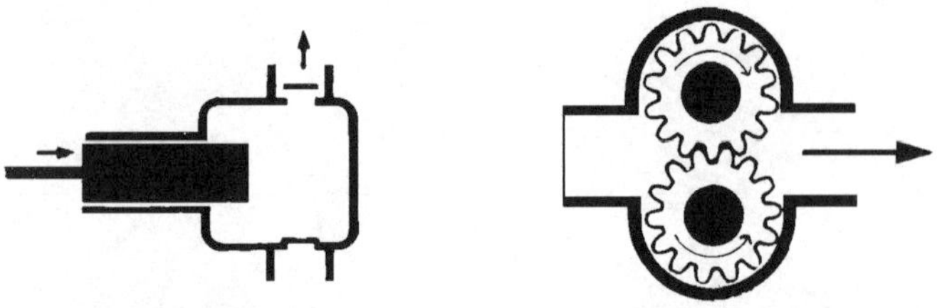

Figure 4.3 Examples of positive displacement pumps

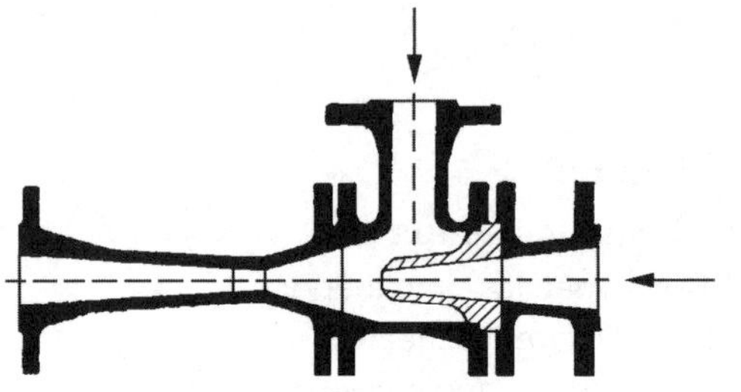

Figure 4.4 Example from the category Other pumps

Positive displacement pumps operate with enclosed liquid volumes which are forced forward in the direction of pumping, or squeezed and expelled into the pump outlet, see Figure 4.3. The oldest and perhaps the most typical positive displacement pump is the piston pump. A certain volume of liquid is pressed out of the cylinder for each stroke of the piston. The plunger pump is a variant of the piston pump. The piston pump belongs to the sub-category among the positive displacement pumps which operate by means of a reciprocating motion. The other sub-category comprises rotating or rotary positive displacement pumps. These include, for example, gear pumps where the liquid is transported in gaps between the gear teeth and where the difference in pressure is maintained by means of, among others, the seal afforded by the contact surface of the meshing gears. Other examples of rotary displacement pumps are screw pumps, vane pumps, hose pumps, etc.

Ejectors and air-lift pumps are typical examples of the main category termed "Other pump types". The static pressure in the nozzle of the jet pump falls when the speed of the motive fluid increases. The pumped liquid is sucked in and mixed with the motive fluid. The overall speed of the fluid flow is reduced and the static pressure is increased in the jet pump venturi (diffuser). Because jet pumps can handle liquids and gases they are useful for priming other pumps, see Figure 4.4. Jet pumps powered by steam or compressed air are also called ejectors.

Air-lift pumps function by reducing the effective density of the liquid by entraining large volumes of gas, usually compressed air. The liquid may be converted to foam if the surface tension is adequate. Atmospheric or gas pressure acting on the bulk liquid surface lifts the reduced density mixture above the stationary liquid. Performance is very variable and testing and tuning are usually required to achieve the desired results.

The performance of pumps, and above all rotodynamic pumps, is usually shown in the form of a curve. The main interest is the relationship between the volume of liquid transported per unit of time, the so-called volume flow or flow rate or just "flow" and the pressure increase or head rise of the pump. The different main categories have, in this respect, drastically different characteristics, see Figure 4.5.

The volume flow in positive displacement pumps is virtually independent of the pressure difference, i.e. the performance curve is almost a vertical line. The performance curve for the centrifugal pump shows increasing pressure rise with decreas-

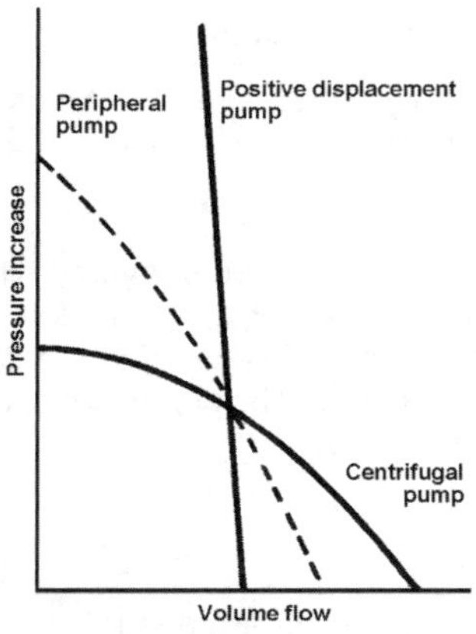

Figure 4.5 Pump curves for different pump types

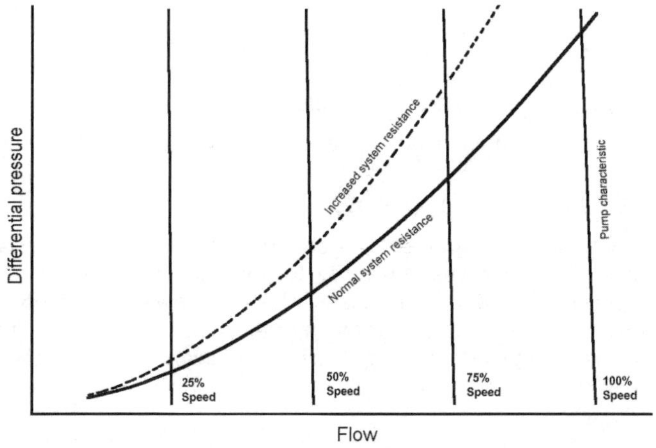

Figure 4.6 Positive displacement pump and system characteristics

ing volume flow. Peripheral pumps have a curve lying between these two groups of pumps, (see Chapter 1, Section 1.4.1). The flow in a pump is determined by the relationship of the specified pump pressure increase to the back-pressure, i.e. the resistance in the system, which is required in order to drive forward the flow.

The piping system has a characteristic curve, flow-differential pressure or head, which must be derived by the system designer. A pump can only operate where the pump characteristic intersects the system characteristic, see Figure 4.6. The pump characteristic can be modified by adjusting the pump speed. The system characteristic can be changed by opening or closing valves or by changing the levels in tanks. If everything remains constant the pump flow will be constant. However, the pump could wear and the pipework become partially blocked due to deposits or corrosion products. Operating conditions do tend to change with the passage of time.

If the liquid pressure in a pump suction is reduced to the vapour pressure of the liquid, the liquid evaporates. So-called cavitation occurs when the vapour bubbles return to liquid form in areas of higher pressure, this phenomenon causes intense local pressure spikes. Cavitation is not desirable since it reduces both pump performance and the operating life of the pump. The concept of NPSH/NPIP is employed to describe the risk of cavitation and states the necessary pressure or head measured above the vapour pressure of the liquid such that the effects of cavitation are limited or eliminated. For further details see Section 4.4.1.

4.2 Rotodynamic pumps

4.2.1 Rotodynamic pump theory

4.2.1.1 Differential head

In the past, pumps were used almost exclusively for transporting water from a lower to higher level, e.g. in a mine, water distribution or in an irrigation installation. Then it was practical to use the concept of delivery head (delivery height) as a criterion of the pumps performance ability. Despite the fact that today pumps are used to a large extent for purposes where the alteration in the elevation of the medium is of subsidiary importance, this concept of head still remains. It is useful because the ideal pump differential head is related to the speed of the vanes, velocity head $c^2/2$. In this Section the pump differential head will always be shown as ΔH and called differential head. In some publications it may be shown as $H_{t2\text{-}1}$ indicating the difference in total heads between suction and discharge. In pumping circles it is standard practice only to refer to "head" on the understanding that everyone knows that it is really differential head.

The process which takes place when a liquid is passing through the pump is isentropic, i.e. the exchange of heat between the pumped liquid and the environment is so small that it can be neglected and is more than compensated by extra heat generated by inefficiencies. An equal amount of mass flows through per unit of time in and out through the pump, the external leakage normally being negligible in comparison with the mass flow through the pump. Some pump designs have internal recirculation for thrust balancing which complicates the theory a little. Such a process is described by the Energy Equation (see also Chapter 3, Section 3.1.5) for steady flow calculated per unit of mass of pumped liquid and using the terms as given in Figure 4.7. Absolute flow velocity is designated by c in this Chapter in line with current practice for rotodynamic machinery.

$$l_t = u_3 - u_1 + \frac{p_3 - p_1}{\rho} + \frac{c_3^2 - c_1^2}{2} + g(h_3 - h_1) \qquad \text{Equ 4.1}$$

Put into words, the Energy Equation states the following: The mechanical work l_t, which is supplied via the pump shaft, causes: an increase of the internal energy (u), an increase in the static pressure (p/ρ), an increase in kinetic energy ($c^2/2$) and an increase in potential energy (gh), in the liquid.

It is desirable that as large as possible a part of the shaft work supplied results in an increase of static pressure in the liquid. In principle kinetic and potential energy can be completely converted into static pressure. It is more difficult to make use of the increase in the internal energy. It corresponds to an unnecessary and undesirable increase in temperature of the liquid and is considered as a loss. It is against this background that the pump's differential head ΔH is defined as that useful part of the change in state of the liquid measured in metres of liquid column.

$$\Delta H = \frac{p_3 - p_1}{\rho g} + \frac{c_3^2 - c_1^2}{2\,g} + h_3 - h_1 \qquad \text{Equ 4.2}$$

where:

p = static pressure (N/m^2)

c = absolute velocity (m/s)

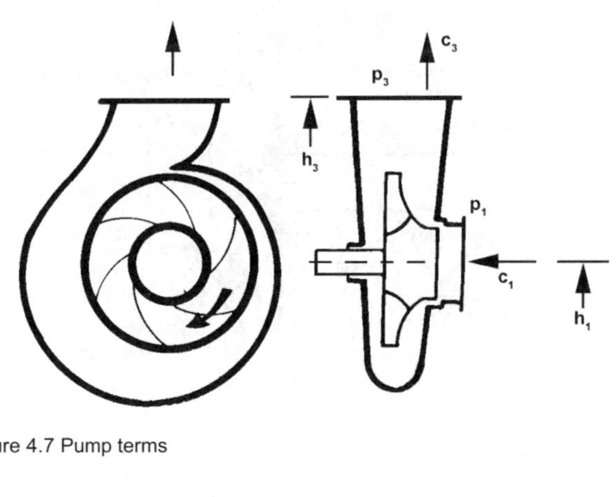

Figure 4.7 Pump terms

h = potential head (m)

r = density (kg/m³)

g = gravitational acceleration (m/s²)

Note: The potential head of the inlet flow is taken at the centre-line of the impeller.

4.2.1.2 Losses and efficiency

Varying liquid properties can change the proportions of the power flow. Increased viscosity, for example, leads to higher impeller friction and blade losses. The shaft power P supplied is defined as the product of rotary moments and angular velocity at the pump's shaft coupling. The pump efficiency is designated by η.

$$\eta = \frac{P_u}{P} \qquad \text{Equ 4.3}$$

where:

P_u = useful work output (W)

P = power input (W)

The useful power (P_u) is thus the shaft power which remains from that supplied after all losses have been overcome. The useful power is applied to the pumped liquid and causes it to change state. The useful power is associated with the previously mentioned mechanical work according to the following relationship. (See Figure 4.8.)

$$P = \dot{m} \cdot l_t \qquad \text{Equ 4.4}$$

where:

$\dot{m}$ = the mass flow through the pump

$$P_u = \dot{m} \cdot g \cdot \Delta H = \rho \cdot Q \cdot g \cdot \Delta H \qquad \text{Equ 4.5}$$

where:

$\dot{m}$ = mass flow (kg/s)

ρ = density (kg/m³)

Q = volume flow (m³/s)

g = gravitational acceleration (m/s²)

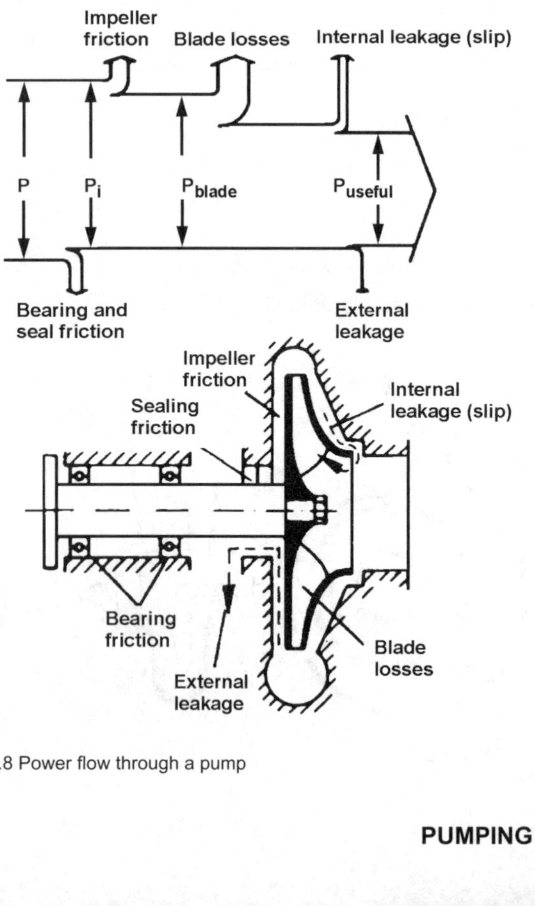

Figure 4.8 Power flow through a pump

ΔH = differential head (m)

The pump efficiency can be shown to be the product of other efficiencies concerned with specific losses within the pump.

$$\eta = \eta_m \cdot \eta_i \cdot \eta_c \cdot \eta_v \qquad \text{Equ 4.6}$$

A part of the shaft power is used in overcoming friction resistance in bearings and seals, P_{jm}. This loss of power does not generally affect the liquid but is transformed into heat which is given off to the environment. The pump mechanical efficiency is given by:

$$\eta_m = \frac{P - P_{jm}}{P} \qquad \text{Equ 4.7}$$

The impeller efficiency, η_i, is a measure of how much of the power available, (P - P_{jm}), is converted to kinetic energy in the liquid and how much power is lost in blade losses and disc friction.

The casing efficiency, η_c, is a measure of how much kinetic energy produced by the impeller, is recovered as static pressure rise compared to losses which appear as increased internal energy, temperature increase.

The volumetric efficiency, η_v, evaluates how much flow is delivered from the pump discharge compared to the flow through the impeller. External leakage must go through the impeller, and be worked on, before it can escape from the seal. Internal leakage, slip, allows a small quantity of liquid to recirculate around the impeller, reducing the effective pump flow.

The individual losses mentioned above are not constant but vary depending upon where the pump operates on its curve. At low flows/high heads, bearing friction can increase due to higher axial thrust and radial loads. The power conversion to kinetic energy in the impeller is influenced by the velocity triangles, see Figure 4.11. For a constant speed pump, the velocity triangles can only be correct for one flow rate, Best Efficiency Point (BEP). At other flows mismatches in the blade and liquid angles occur causing additional losses.

How the casing efficiency varies is dependant upon the type of casing design. Kinetic energy can be recovered as static pressure rise in two types of diffuser; vaned and vaneless. A vaned diffuser will suffer the same velocity triangle problems as the impeller with off-design flows. The vaneless diffuser or volute does not have blade angles to cause extra losses but does operate with flow paths away from the optimum in the specially shaped passages, again increasing losses. Internal leakage is affected by the pump differential head and the measures employed by the manufacturer to reduce it, such as wear rings, and wear in the pump.

A very important efficiency to consider, in today's increasingly energy conscious climate, is the overall efficiency, η_{gr}, often called the "wire-to-water" efficiency. This should be evaluated from the actual power consumed by the pump driver compared to the pump useful power. Overall efficiency takes into account any drive train systems between the driver and the pump and is the only valid efficiency to be used in Life Cycle Costs. Extra care should be exercised with electric motors controlled by variable frequency invertors. Invertors can inject harmonics into the mains supply which may cause serious errors in electrical power measurement.

When considering "wire-to-water" efficiency and Life Cycle Costs it is essential to consider the pump unit in a proper context. In the case of boiler feed pumps, applications where heat is added to the liquid after pressurisation, the normal pump efficiency is inappropriate. The heat added by the pump is not a loss but is a useful energy addition; the "wire-to-water" efficiency should be increased to include the pump heating effect. "Wire-to-water" efficiency can be maximised by matching the pump unit to the system requirements. Oversized pumps, throt-

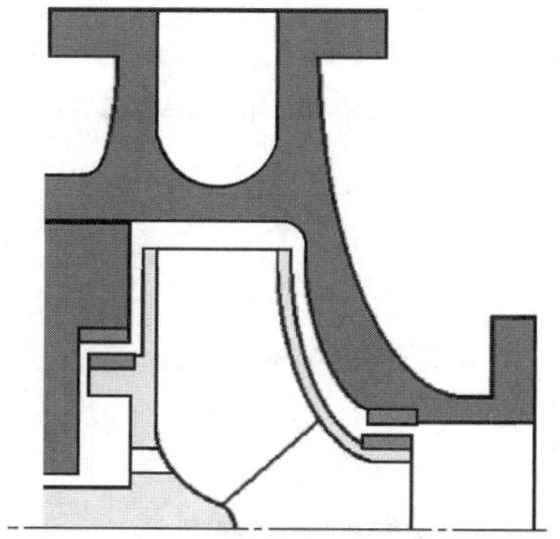

Figure 4.9 Wear rings on a centrifugal pump

tled to produce the correct flow, are very inefficient. As the power consumed by a pump costs much more than the initial purchase cost a small improvement in efficiency can result in a considerable reduction in Life Cycle Costs.

As mentioned earlier, wear rings are employed to reduce internal leakage slip. Figure 4.9 shows where rings are fitted on a centrifugal pump impeller and its casing. Sometimes rings are only fitted to the front of the impeller. Wear rings can be of various designs, see Figure 4.10. Wear rings are fitted to enclosed impellers, one on the front and one on the back. It is possible to avoid wear rings and just provide a tight clearance between the casing and the impeller, but adjustment for wear can be a problem. Wear rings are intended to be a replaceable spare part. Liquid leakage from higher pressure zones, after the impeller has imparted energy, to suction zones is reduced by imposing tight clearances in the leakage path. The actual radial clearance is strictly controlled.

Plain rings, shown on the left in Figure 4.10, are by far the most popular style. They are the easiest to manufacture and fit. Also, a plain ring does not restrict axial float. The rotor can move without rubbing on a ring. Stepped rings, shown in the centre, are fitted to some pumps, quite often at the insistence of the purchaser or pump user. Some pump users find a stepped ring provides better performance and longer life. The stepped ring concept can be taken further by introducing a radial flow portion to the ring combination. Stepped rings inhibit rotor axial movement and axial setting must be controlled accurately. Stepped rings also limit assembly possibilities.

Wear rings are primarily for internal leakage reduction, but do perform another important, secondary, function. Wear rings act as hydrodynamic bearings, bearings lubricated by the process liquid. The bearing function is quite useful in multistage pumps, when the wear rings effectively 'stiffen' the shaft; they reduce radial movement of vibration, for example. The material combination can be very important if rubbing is likely.

Rubbing is most likely to occur during start-up or shutdown, or during low speed operation when the hydrodynamic bearing effects are reduced. Wear ring material combinations are the result of 'trial and error' over the years, see Table 4.1. Newer materials, more expensive materials, can be adopted to solve problems such as very low viscosity or short term dry running. Alloys containing 75% or more nickel have proved effective. Modern engineering plastics, like PEEK, are proving very useful. The cost can seem exorbitant when compared to 'old-fashioned' metal.

Casing wear ring	Impeller wear ring
Cast iron	Cast iron
Ni-resist	Ni-resist
Bronze	Bronze
13Cr steel	13Cr steel
Stainless steel/Colmonoy 6	Stainless steel/Colmonoy 6

Table 4.1 Standard wear ring material combinations

Metal wear rings can be through-hardened, case-hardened or surface hardened. Laser hardening is used by some pump manufacturers. Most wear rings are ground to provide a very smooth bearing surface. Occasionally, wear rings are knurled to provide a rough surface. Micro-machining is a relatively new technique to improve bearing performance.

4.2.1.3 Euler's equation

Figure 4.11 shows a pump impeller with associated velocity triangle diagrams. The velocities u_1 and u_2 are the peripheral velocities of the impeller blade at inlet and outlet. The outlet peripheral velocity is commonly called the tip speed. Liquid enters the impeller at a velocity c_1 at a radius r_1. The liquid leaves the impeller at a radius r_2 with a velocity c_2. The design of the blade forces the relative velocity, i.e. the velocity which is experienced by an observer travelling with the movement of the blade, to alter its magnitude and direction from w_1 to w_2. Due to this the absolute velocity c_2 at the impeller outlet will also deviate from the flow velocity c_1 at entry. The tangential component of absolute velocity is designated by c_u, i.e. in the direction of the peripheral speed.

By applying the Law of Conservation of Momentum from fluid mechanics to the flow through the impeller, we obtain in the tangential direction the following

$$M_t = \dot{m}(r_2 \cdot c_{2u} - c_{1u}) \qquad \text{Equ 4.8}$$

Here M_t, is the rotary moment which the impeller must impart to the liquid, in order that the flow according to Figure 4.8 may be produced. The flow velocities in equation 4.8 are intended to be representative mean values for the liquid which is flowing through the impeller.

During the time Δt the pump shaft rotates through the angle $\Delta\phi$ at the same time as the mass Δm passes in and an equal mass Δm passes out through the impeller tip. If both sides of equation 4.8 are multiplied by $\Delta\phi$ and at the same time $\dot{m} = \Delta m/\Delta t$ is introduced, then

$$M_t \cdot \Delta\phi = \frac{\Delta\phi}{\Delta} \cdot \Delta m \, (r_2 \cdot c_{2u} - r_1 \cdot c_{1u}) \qquad \text{Equ 4.9}$$

but $M_t \cdot \Delta\phi/\Delta m$ is the work per unit mass which the rotary moment carries out whilst the impeller rotates through angle $\Delta\phi$. This work per unit mass is called blade work and is designated

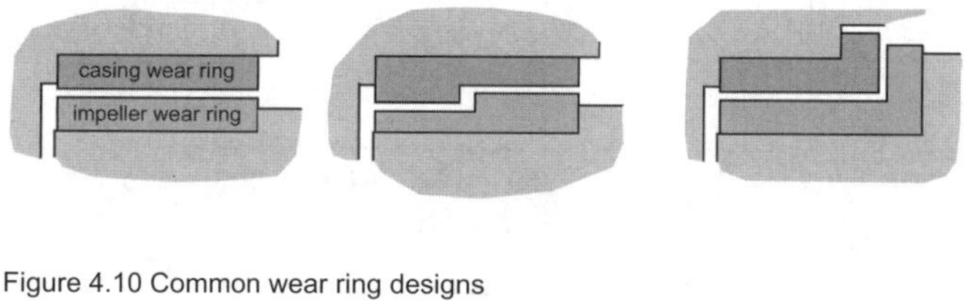

Figure 4.10 Common wear ring designs

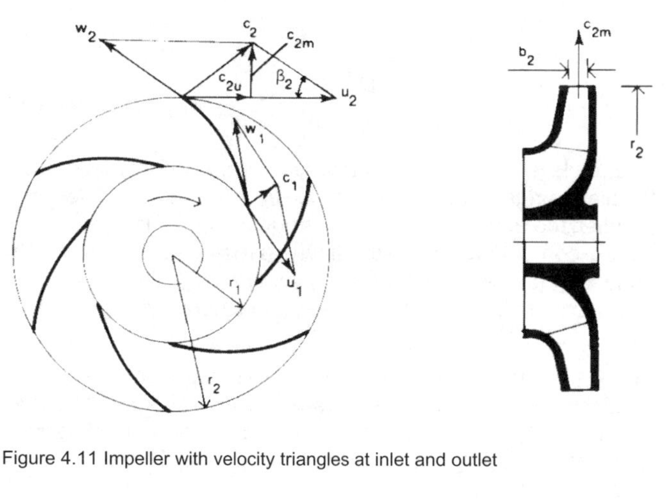

Figure 4.11 Impeller with velocity triangles at inlet and outlet

l_b. Furthermore, the quotient $\Delta\phi/\Delta t$ is equal to the constant angular velocity ω of the pump impeller.

Thus:

$$l_b = \omega(r_2 \cdot c_{2u} - \phi \cdot c_{1u}) \qquad \text{Equ 4.10}$$

That part of the blade work which corresponds to the blade losses results in an increase of the internal energy of the liquid whilst the rest ($\eta_h \cdot l_b$) provides a useful change of the state of the liquid ($g \cdot H$). If then the peripheral speed $u = r_2 \cdot \omega$ is applied, we finally obtain:

$$\frac{g \cdot \Delta H}{\eta_h} = u_2 \cdot c_{2u} - u_1 \cdot c_{1u} \qquad \text{Equ 4.11}$$

where:

g = gravitational acceleration (m/s^2)

ΔH = differential head (m)

η_h = hydraulic efficiency

u = peripheral velocity (m/s)

c_u = tangential component of the absolute velocity (m/s)

Equation 4.11 is Euler's equation as it is normally written for pumps.

4.2.1.4 Pump curves

In accordance with Euler's equation differential head is dependent upon the size and direction of the velocity vectors and the size of the hydraulic losses. Both these factors are affected by, among others, the flow rate Q, which passes through the pump.

In many cases the tangential component of the absolute velocity entering the impeller is small, i.e. $c_{1u} \approx 0$. In such cases Euler's equation is simplified to:

$$\frac{g \cdot \Delta H}{\eta_h} = u_2 \cdot c_{2u} \qquad \text{Equ 4.12}$$

If the flow through the pump were free of losses $(\eta_h = l)$, the ideal differential head, ΔH_i, would be given by the following equation:

$$\Delta H_i = \frac{u_2 \cdot c_{2u}}{g} \qquad \text{Equ 4.13}$$

ΔH_i is often called the Euler head. With reference to Figure 4.11

$$\Delta H_i = \frac{u_2}{g}\left(u_2 - \frac{c_{2m}}{\tan\beta_2}\right)$$

and

$$c_{2m} = \frac{Q}{2\pi \cdot r_2 \cdot b_2}$$

or

$$DH_i = \frac{u_2}{g}\left(u_2 - \frac{Q}{2p \cdot r_2 \cdot b_2 \cdot \tan\beta_2}\right) \qquad \text{Equ 4.14}$$

The angle β_2, in the velocity triangle is somewhat smaller than the blade angle at the impeller blade tip. This angular difference is called the deviation angle and is due to the blade's inability to control completely the relative flow of a real liquid as opposed to an ideal liquid with zero viscosity and density. The size of the deviation angle is primarily dependent upon the number of blades.

For a given pump, operating at a certain constant speed, the ideal differential head H_i decreases linearly according to equation 4.14 with increasing flow Q. The actual differential head H differs from H_i due to the hydraulic losses h_f. These, as in all

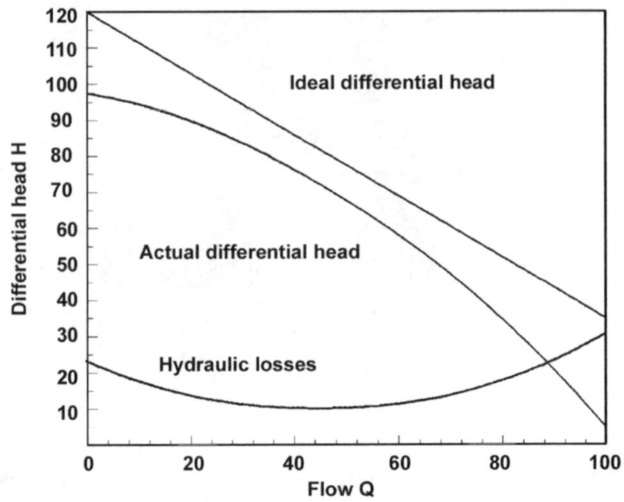

Figure 4.12 Centrifugal pump H-Q curve

other cases of flow, are dependent upon the inlet flow direction towards the body around which the liquid flows.

In the case of pumps, the inlet flow direction towards for example the blades, varies with the flow rate. A certain inlet flow angle gives the most favourable flow and thus the smallest losses. Both higher and lower values of Q result in an increase of h_f. By subtracting h_f from H_i, the actual H-Q curve of the pump at constant speed is obtained, see Figure 4.12. The shape of the H-Q curve varies from one pump type to another and between different designs of the same pump types. Detailed design particulars, such as: blade profiles, blade angles, number of blades, area ratios and diffuser type, all affect the curve shape.

In addition to the pump's H-Q curve, the required shaft power P, the pump's efficiency η and NPSHR, as shown in Figure 4.13 are also usually given as a function of the flow in a pump characteristic curve. The shaft power is based on water, i.e. specific gravity = 1.0.

Variations in specific gravity will affect the power, and the discharge pressure even though the differential head is constant. The pump can, in principle, operate at any point whatsoever along the H-Q curve. The position of the operating point in an actual case is determined by the characteristics of the system, suction plus discharge, to which the pump is connected. The art of good pump selection is to operate as close as possible to the Best Efficiency Point (BEP); also some users specify the duty point must be to the left of BEP.

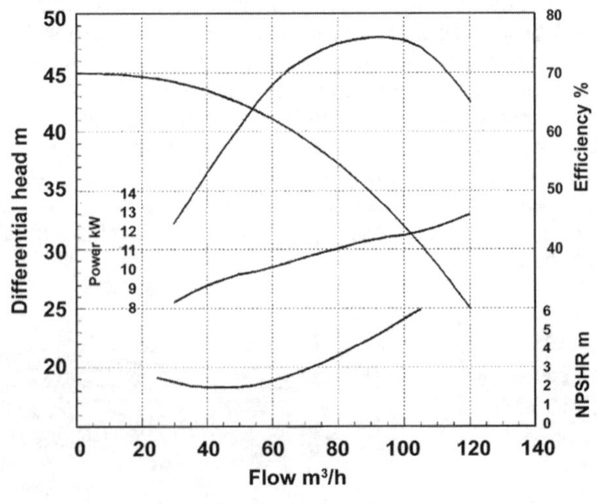

Figure 4.13 Constant speed, fixed diameter pump curve

4.2.1.5 Specific speed

An important parameter which is frequently used to describe different pump types and to characterise a duty point is specific speed N_s, defined by the following relationship:

$$N_s = n \cdot \frac{Q_{BEP}^{0.5}}{H_{BEP}^{0.75}} \qquad \text{Equ 4.15}$$

where:

N_s = specific speed (dimensionless)

n = shaft speed (rpm)

Q_{BEP} = pump flow at best efficiency point (m^3/h)

H_{BEP} = pump ΔH at best efficiency point (m)

Note that N_s is not dimensionless and therefore has different numerical values in different systems of units. N_s can be defined in words as the mechanical speed of a pump identical with the one under consideration, and which with identical velocity vector diagrams gives a flow of 1 m^3/h at a differential head of 1m. All identically shaped pumps have therefore the same specific speed independent of their size.

The specific speed is used, among others, for characterising the shape of rotodynamic pump impellers, see Figure 4.14. Drawing on experience gained in this field, it is generally possible to estimate which impeller shape will, under normal circumstances, give the best results for a given speed, flow and differential head. Figure 4.13 illustrates this relationship. The axial flow pump has shown itself to be the most suitable type in the case of, for example, a large flow and a small differential head, i.e. in the case of high specific speed. Axial flow pumps are therefore said to have high specific speeds.

There is another form of specific speed in popular use, Suction specific speed, N_{ss}. Suction specific speed is evaluated using the same equation as specific speed but NPSHR is used in place of differential head. N_{ss} is an indication of how good the suction performance of a pump is. Some users specify a limiting range, 175 to 215, of N_{ss} in an attempt to limit cavitation problems.

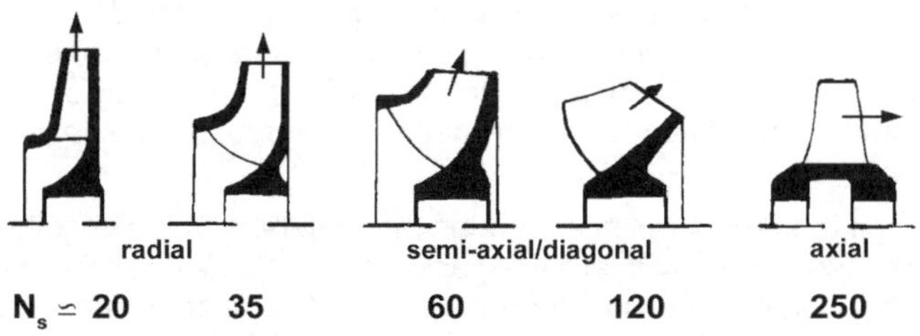

Figure 4.14 Pump impellers having different specific speeds

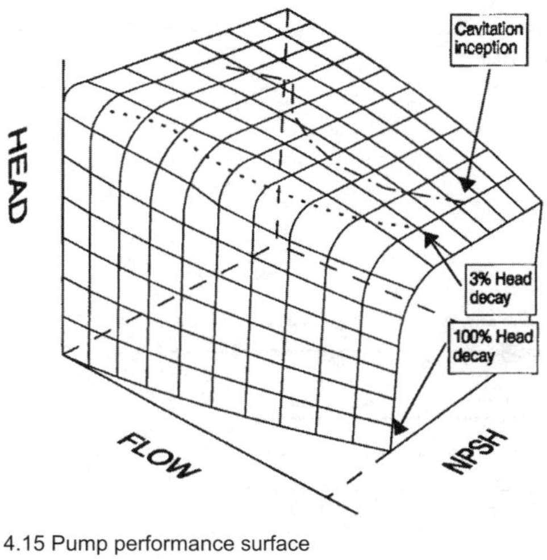

Figure 4.15 Pump performance surface

Cavitation, especially when operating well away from BEP, is a very complex phenomenon and specifying any N_{ss} value will not necessarily solve or help the problem. Other problems associated with rotodynamic pumps operating well away from the BEP are surge and internal recirculation. Both of these problems are associated with low flows. Manufacturers take great care to avoid surge and recirculation problems by specifying minimum flows for pumps. It is up to the user to correctly assess the full operating regime and communicate this to the manufacturer.

A very interesting and educational paper was presented by R. Palgrave, some years ago, on the relationship and effects of both specific speed parameters. The paper, indicates the alternatives available to the pump designer in response to changing operating conditions and customer restraints. The paper shows pump performance as a 3D surface, which connects flow, head and NPSH, rather than a conventional 2D curve and is shown in Figure 4.15.

4.2.1.6 Similarity

Under certain conditions, the performance of a pump can be recalculated in a specially simple manner which is valid for all sizes of pump at different operational speeds. These conditions are as follows:

- The pumps compared should be identical or geometrically similar
- The pumps compared should operate at similar operational points, i.e. with geometrically similar velocity diagrams
- Any difference in efficiency should be negligible

In a comparison between two pumps I and II, which fulfil these conditions, the following relationships between their respective speeds n_I and n_{II}, (dimensions expressed as the impeller diameters D_I and D_{II}) and their performance, will apply:

- For the flows

$$\frac{Q_I}{Q_{II}} = \frac{n_I \cdot D_I}{n_{II} \cdot D_{II}}$$

- For the differential heads

$$\frac{H_I}{H_{II}} = \frac{n_I^2 \cdot D_I^2}{n_{II}^2 \cdot D_{II}^2}$$

- For the power requirements

$$\frac{P_I}{P_{II}} = \frac{n_I^3 \cdot D_I^3}{n_{II}^3 \cdot D_{II}^3} \qquad \text{Equ 4.16}$$

The Affinity Laws which apply to a pump with a fixed impeller diameter, form a special case of the general similarity relationships. When $D_I = D_{II}$, then

$$Q \propto n$$

$$H \propto n^2$$

$$P \propto n^3 \qquad \text{Equ 4.17}$$

When the principles of similarity are used in practice, the variation in efficiency between the pumps compared is ignored except when calculating power. Similar pumps have the same blade angles. Geometrically similar velocity diagrams imply the same flow angles. The inlet flow direction, which is so important for the hydraulic losses, is therefore unaltered in the cases compared. Against this background a constant hydraulic efficiency would seem to be a reasonable approximation. In practice the efficiency will improve slightly as speed increases. In the case of very large pumps which are tested by means of model tests in the laboratory, a somewhat higher efficiency can be expected for the full-scale pump than that which is measured during the model test. Parasitic losses tend to increase as size reduces.

Similarity is very useful to pump manufacturers. Similarity allows a pump curve to be adjusted for changes in pump speed and small changes of impeller diameter. Similarity is not quite so useful for pump users. Similarity does not predict changes in operating conditions. The pump curve can be adjusted then the system curve must be added. A pump can only operate at the intersection of the pump characteristic and the system characteristic. The similarity equations apply to pumps — not pumps in systems!

4.2.1.7 Axial thrust and radial forces

The pump impeller works best if the impeller tip width is aligned centrally with the throat of the diffuser. The pump shaft is adjusted during assembly to accomplish this. The unbalanced axial thrust (axial force) on the pump shaft must be absorbed by a thrust bearing or balanced out with a special arrangement. A special balance disc or drum, however, always requires a certain amount of driving power and internal leakage and therefore reduces the overall efficiency of the pump. It is thus advantageous if the axial force can be limited directly by means of the pump impeller design. This is not always possible and some large pumps have a balance disc/drum to reduce the axial thrust and a thrust bearing to locate the shaft. Balance discs and drums are always designed to leave a residual thrust which must always be in the same direction to avoid thrust bearing problems. (See Figure 4.16.)

The application of the Momentum equation, equation 3.7, in Chapter 3, to a control volume, precisely that which encloses the impeller, gives the following, using the terms as shown in Figure 4.17.

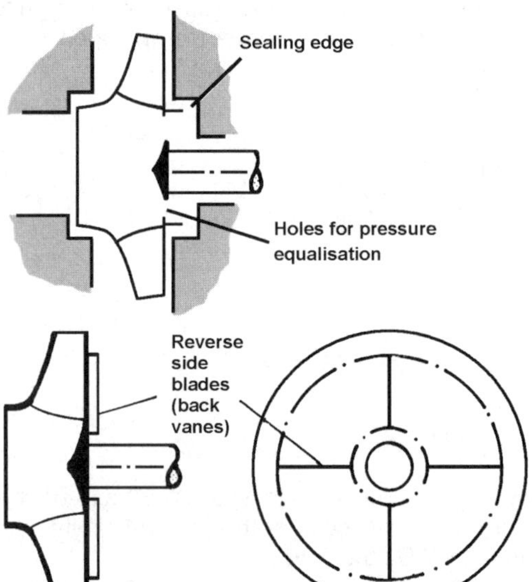

Figure 4.16 Methods for limiting the axial thrust

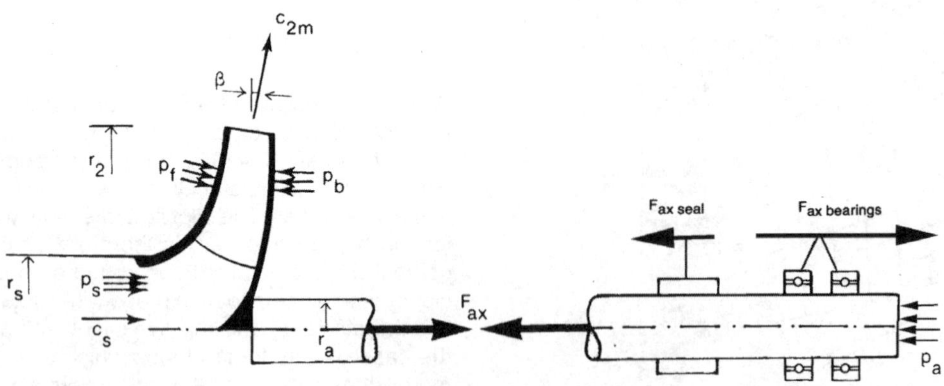

Figure 4.17 Illustration of axial thrust

$$F_{ax} = \int_{r_a}^{r_s} p_b \cdot 2\pi \cdot r \cdot dr + \int_{r_s}^{r_2} p_b \quad 2\pi \cdot r \cdot dr - \int_{r_s}^{r_2} p_f \cdot 2\pi \cdot r \cdot dr - p_s \cdot \quad p \cdot r_s^2 + \dot{m} \cdot c_{2m} \cdot \sin\beta - \dot{m} \cdot c_s \qquad \text{Equ 4.18}$$

If the pressure on the front and back face of the impeller for $r < r_s$ is assumed to be equal ($p_b = p_f$) and if the angle β is small ($\sin \beta = 0$) the above expression can be simplified to

$$F_{ax} = \int_{r_a}^{r_s} p_b \cdot 2\pi \cdot r \cdot dr - p_s \cdot \pi \cdot r_s^2 - \dot{m} \cdot c_s \qquad \text{Equ 4.19}$$

In order to limit F_{ax} it is normal procedure to provide the back wall of the impeller with an extra sealing edge at the same diameter as r_s, back wear ring, and at the same time to bore holes through the wall close to the shaft for pressure equalisation. Then

$$F_{ax} \quad - p_s \cdot \pi \cdot r_a^2 - \dot{m} \cdot c_s \qquad \text{Equ 4.20}$$

As well as the axial force from the impeller, the shaft is also subjected to forces from seals and bearings as well as the atmospheric pressure at the end of the shaft. The bearing force is

$$F_{ax\,bearing} = (p_a - p_s) \cdot \pi \cdot r_a^2 - \dot{m} \cdot c_s + F_{shaft\,seal} \qquad \text{Equ 4.21}$$

In general the forces generated by seals and atmospheric pressure on the open shaft end are very small. Another way of limiting the axial thrust is to provide the impeller with blades on the reverse side. This causes the rotation of the liquid on the reverse side to be greater than that achieved by friction on the front side. Again see Figure 4.16. The increased rotation therefore causes lower pressure for $r < r_2$ and also therefore a reduction in the actual force. In the case of multi-stage pumps or double suction impellers, low axial thrust can be obtained by having impellers facing in opposite directions.

Considerable radial forces occur during partial loading in all pumps equipped with volute casings. The design of such pumps seeks to achieve an even pressure distribution around the impeller. This condition determines the shape of the volute casing. During part load, however, the pressure around the periphery of the impeller will vary. The radial forces on the impeller are taken up by the bearing via the shaft. The diameter of the shaft and the distance from the bearings determines the deflection.

Since the pressure varies around the periphery of the impeller the flow in the individual blade passages will also vary as the impeller rotates. In simple theory the liquid flow is considered as constant velocity across the peripheral section of a blade passage. In reality the flow may be in two or three distinct zones with different velocity profiles.

The size of the radial force can be estimated by means of the following formula:

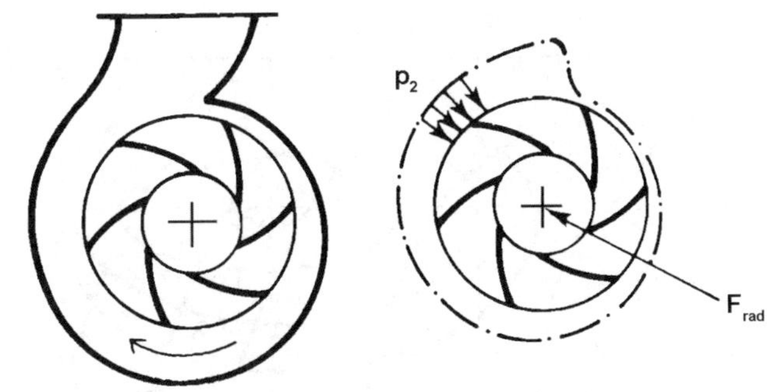

Figure 4.18 Illustration of radial forces on the impeller in the case of reduced flow ($Q < Q_{BEP}$)

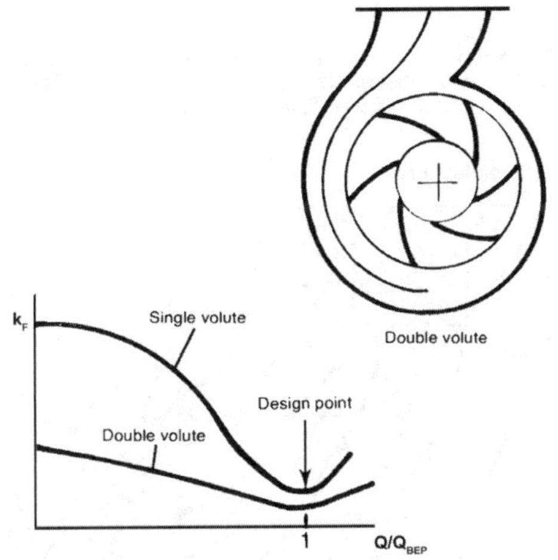

Figure 4.19 Examples of the effect of the position of the operating point and the type of the volute casing on the parameter k_F

$$F_{rad} = k_F \cdot D_2 \cdot b_2 \cdot \rho \cdot g \cdot \Delta H \qquad \text{Equ 4.22}$$

where:

F_{rad} = radial force (N)

k_F = empirical constant (dimensionless)

D_2 = diameter of impeller (m)

b_2 = tip width (m)

ρ = density of liquid (kg/m^3)

g = gravitational acceleration (m/s^2)

ΔH = differential head m)

The parameter k_F assumes different values for different designs of pumps and also varies greatly with the volume flow through the pump. Figure 4.18 illustrates this.

For a centrifugal pump with a single volute, k_F can have a value up to 0.4 at zero discharge (Q=0).

An effective method of limiting the size of the radial force is to provide a double volute, see Figure 4.19.

The radial force is greatest at Q=0, the shut-off head. The radial force causes a deflection of the pump shaft and subjects the shaft to rotary fatigue. The condition that the maximum shaft deflection at the zero discharge should be less than 0.05 mm at the shaft seal is often used as the guiding value in deciding the dimensions of the shaft.

The size of the axial thrust and radial force is a determining factor in the design and construction of the pump (clearances, bearings, bearing arrangement, etc.). They are also often the primary cause of breakdowns. Note that both the axial thrust and the radial force increase with partial load and are greatest at zero discharge. Pumps with front and back wear rings on each impeller have much stiffer shafts than simple theory predicts. Each wear ring acts as a hydrodynamic bearing, to some extent, and provides support for the shaft.

Large pumps, and high differential head pumps, often have restrictions regarding low flow operation. The manufacturer may have stated the pump should not be operated continuously at less than 25% flow. Low flow operation is restricted to reduce heat build up in the pump; it obviously helps with axial thrust and radial load.

4.2.2 Rotodynamic pump curves

4.2.2.1 Presentation of data

In addition to the relationship between head and flow (H-Q) the complete data set for the pump also comprises power input, efficiency and NPSHr. Figure 4.20 shows an example of a complete pump characteristic. This diagram also shows performance curves for different diameters of impeller, i.e. the different performances which can be attained for one and the same pump by machining the impeller. With moderate adjustments of the impeller diameter the NPSH curve is not affected and the efficiency curve is only affected to an insignificant degree.

Rotodynamic pumps are usually designed to be driven directly by standard squirrel-cage electric motors. This philosophy means that most pumps run at one speed out of a choice of eight.

Number of poles	50 Hz	60 Hz
2 pole	2900 rpm	3500 rpm
4 pole	1450 rpm	1750 rpm
6 pole	960 rpm	1160 rpm
8 pole	730 rpm	880 rpm

Table 4.2 Approximate full load speeds of electric motors

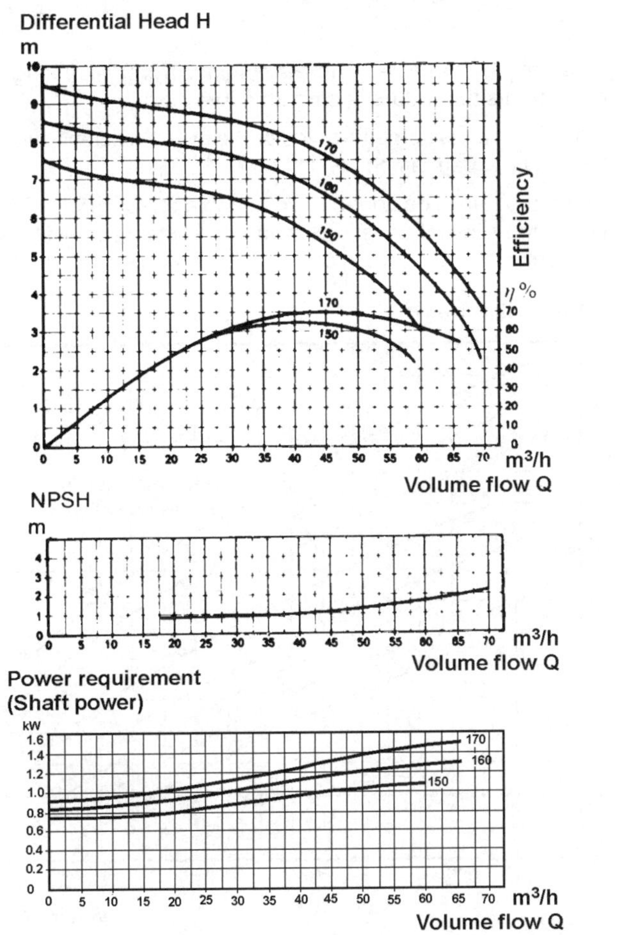

Figure 4.20 Example of performance curve for a centrifugal pump with a speed of 1450 rpm on water (Numbers 150-170 refer to the impeller diameter in mm)

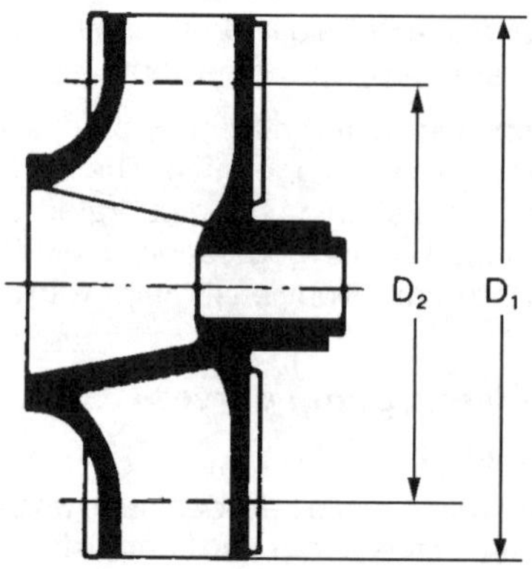

Figure 4.21 Impeller with diameter reduced from D_1 to D_2

Motors are available with more than eight poles but these are not popular and are costly. The actual full load speed of the motor is dependant upon its size. Large motors are more efficient and have less slip and therefore run closer to the synchronous speed. From Table 4.2 it can be seen that there is a speed ratio of about two between 2-pole motors and 4-pole motors.

For a pump designed at 2900 rpm to run at 1450 rpm it would produce 50% flow and 25% differential head. Pump manufacturers needed a cost-effective mechanism to increase the operating range of a pump without changing motors or speeds with gearboxes or V-belts. Reducing the impeller diameter, from the maximum possible, is a simple and very effective solution. The impeller can be reduced to about 70% on most designs.

In the absence of pump performance data for different diameters, the modification in performance in the case of reduction in the diameter from D_1 to D_2, according to Figure 4.21, can be estimated according to the following rule:

$$\frac{H_2}{H_1} = \frac{D_2^2}{D_1^2} \quad \text{and} \quad \frac{Q_2}{Q_1} = \frac{D_2}{D_1} \qquad \text{Equ 4.23}$$

A new approximate H-Q curve is obtained by recalculating sufficient points to produce a smooth curve.

The efficiency is sometimes indicated by iso-efficiency lines on the H-Q diagram as shown in Figure 4.22.

When a pump is designed to be capable of accepting different types and sizes of driver, the pump characteristic will show the efficiency of the pump alone. If the characteristic shows an overall efficiency, including driver and any transmission losses, it will be stated clearly.

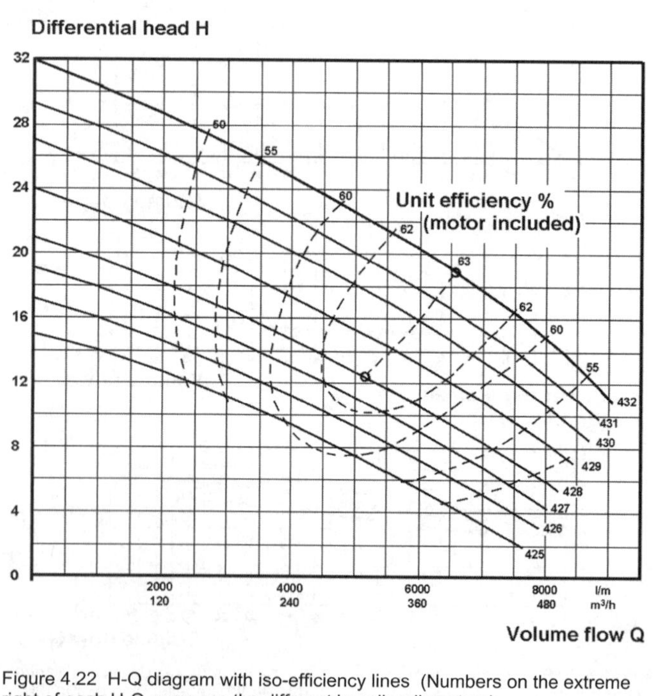

Figure 4.22 H-Q diagram with iso-efficiency lines (Numbers on the extreme right of each H-Q curve are the different impeller diameters)

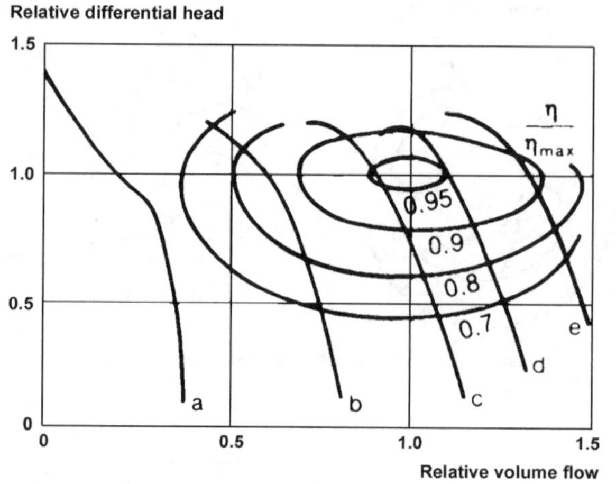

Figure 4.23 Pump curves for an axial flow pump with adjustable impeller blades. The curves a to e designate different blade settings. The performance is expressed relative to Best Efficiency Point (η_{max})

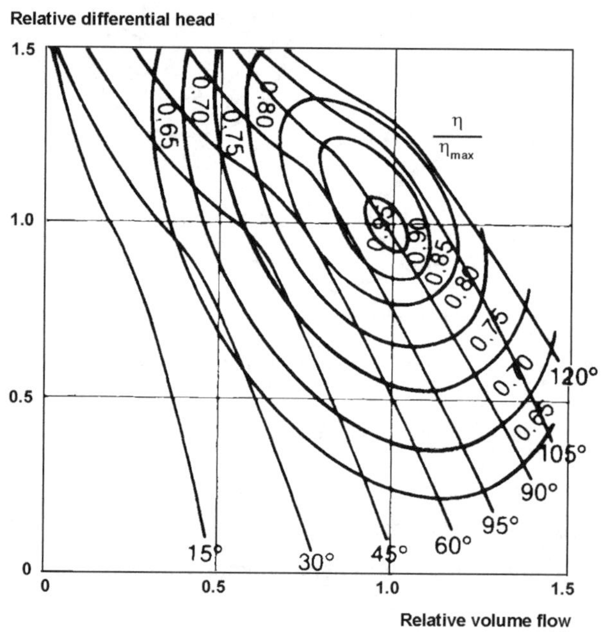

Figure 4.24 Pump curves for mixed flow pump with adjustable inlet guide vanes (The angular values relate to the angle setting of the guide vanes)

In the case of axial or mixed flow pumps, different characteristics can be achieved through adjustment of the angular position of the impeller blades, see Figure 4.23. Similarly the performance can be modified by adjusting stationary inlet guide vanes, see Figure 4.24.

If the pump speed is varied then performance changes can also be expressed as a series of curves in the H-Q and the P-Q diagrams. For example, an existing pump curve with H-Q and P-Q diagrams can be recalculated from a maximum speed to some other speed by recalculating 4 or 5 points on the maximum curves, one after the other, to the new speed; and with the help of the affinity laws and the transmission efficiencies for the actual speed converter. Figure 4.25 shows an example of such recalculated pump performances. A more detailed treatment of speed regulated pumps is given in Chapter 6.

The pump curves described so far have related to one and the same size pump. In order to be able to display data for many sizes of pump simultaneously, range charts are used. Logarithmic scales may be used on either or both axes to show the individual performance range with some accuracy, see Figure 4.26. Diagrams of this type are used both in order to be able to find a suitable size of pump quickly, and to systematise classifi-

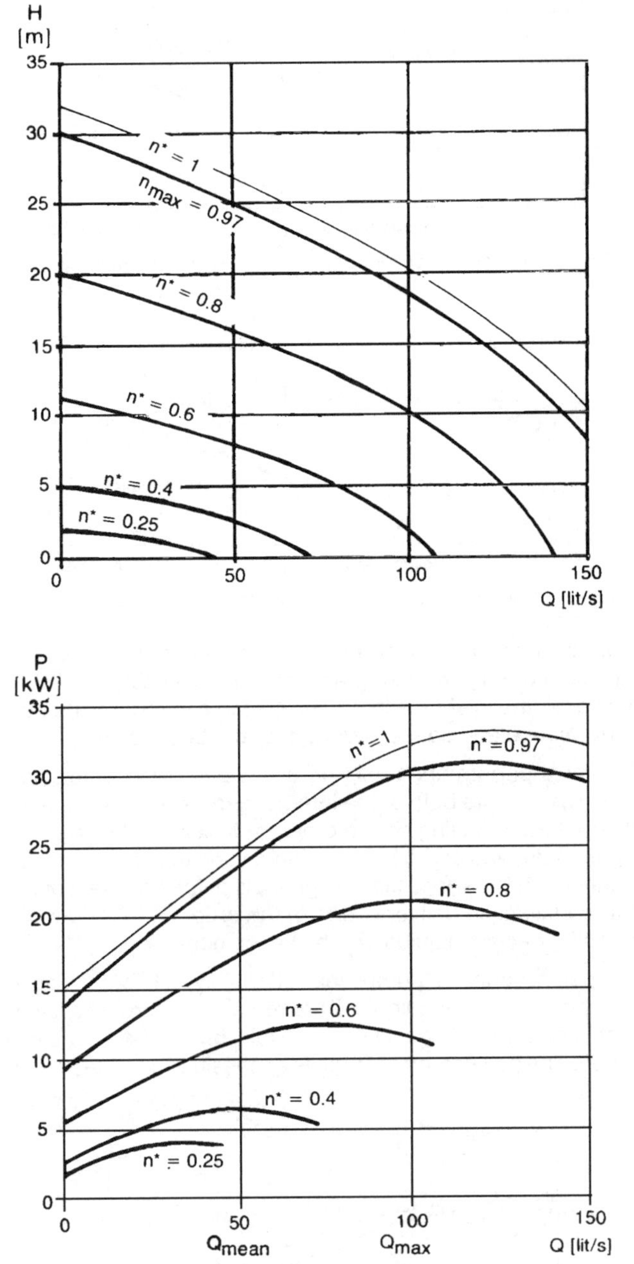

Figure 4.25 Example of performance curves for a variable speed pump. The power requirement relates to a pump with fluid coupling. The speed is expressed relative to the full speed of a direct driven pump. Due to slip in the hydraulic coupling at the full load, n^*_{max} = 0.97 which gives somewhat lower performance than for corresponding direct drive pumps

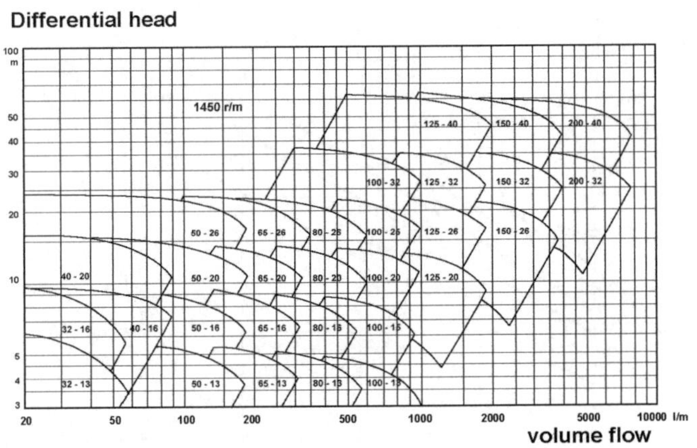

Figure 4.26 Range chart for small single stage pumps. (The first number of each designation relates to the discharge dimension in mm and the second number to the maximum impeller diameter in cm)

cation into different sizes. Due to the use of different pump impeller diameters or blade angles, one pump size covers a "window" in the diagram. Each pump covers a flow range of approximately 2:1 and a head range of 1:6. In order to cover varying needs for volume flow and differential head, 20 to 60 different sizes of pump can be required in a range to provide good coverage. It is very useful to indicate the BEP as a point on the maximum and minimum impeller curves. This shows very quickly how good a selection will be.

Good range design for a specific type of pump is very complicated and many factors must be considered. Problems arise because small pumps cost more to manufacture than large pumps. Good range coverage is achieved with centrifugal pumps when pumps are spaced on a geometric power scale. The step size should lie between 1.2 and 1.75. Experience has shown that because of the small pump cost problem, small pump design should not be compromised by interchangeability considerations but designed specifically for that size.

A rotodynamic pump's H-Q curve is stated as being stable or unstable depending upon whether the differential head is constantly rising or not when flow decreases, Figure 4.27. The differential head when Q = 0 is called "shut-off head" or "closed valve head" and in the case of a centrifugal pump with a stable pump curve is the highest pressure the pump can produce. The unstable part of the curve can cause difficulties due to the fact that the point of interception with the system curve is not clearly definable in some cases. Unstable pump curves are therefore undesirable and are usually avoided when the friction losses in the system are small and when a number of pumps are operating in parallel.

Depending upon the slope of the H-Q curve, a distinction is made in theoretical considerations between steep and flat curves. As a measure of this slope, the equation H_{max}/H_O, is evaluated for the differential head at optimum BEP and closed valve, as shown in Figure 4.28.

When drawing a H-Q curve the choice of scales can cause the same pump curve to appear to be flat or steep. Choice of an operating point to the right or left of the flow at best efficiency, Q_o in Figure 4.28, determines in practice if the curve, together with a piping system will function as a steep or a flat pump curve.

Parallel pump operation can be very difficult, especially when the system curve is mostly static pressure and there is little pipe friction, such as in boiler feed pumps. Experienced users specify a minimum value, percentage head, for the slope of the H-Q curve between the duty point and closed valve.

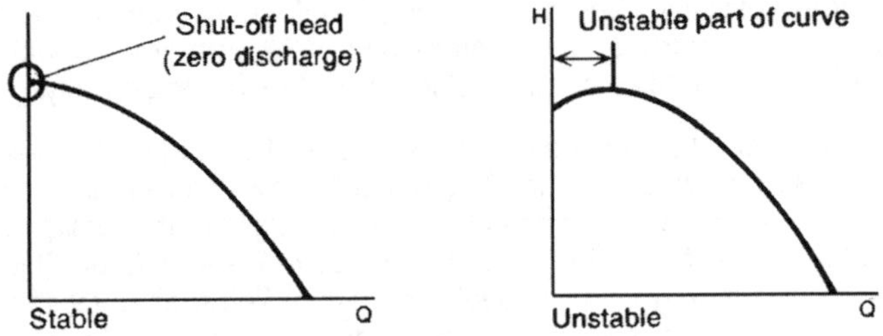

Figure 4.27 Stable and unstable pump curves

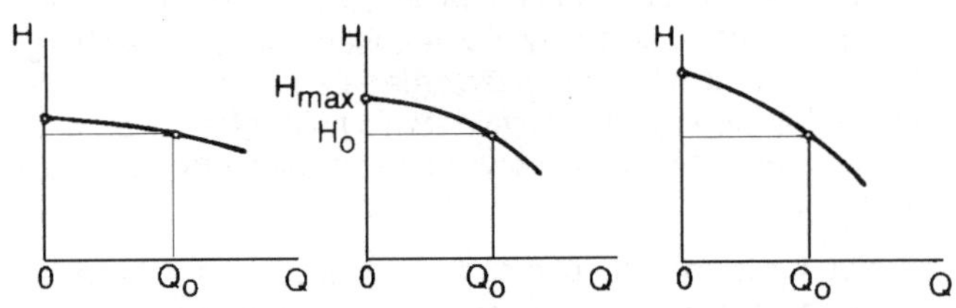

Figure 4.28 Flat and steep pump curves for centrifugal pumps

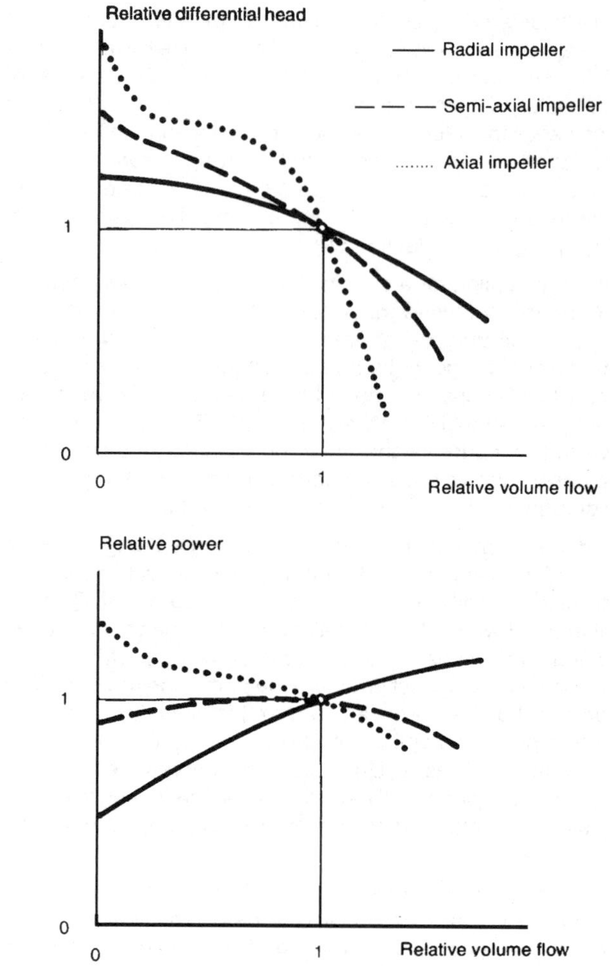

Figure 4.29 Rotodynamic pump curves relative to the BEP

The many varieties of rotodynamic pump — centrifugal pumps, mixed-flow pumps, axial pumps; have performance curves which differ greatly in appearance as shown in Figure 4.29. An increased specific speed gives an increasingly steep H-Q curve whilst the power curve alters from rising to decreasing with the flow.

The shape of the power curve is, together with variations in the flow, the determining factor for the size of the driver. In the case of axial pumps, see Figure 4.29, the greatest power requirement is when the flow is zero, which can mean that the pump's starting conditions must be adapted to this fact. The power requirements which are given in pump catalogues, unless otherwise stated, refer to liquids comparable with cold water, i.e. with a density of 1000 kg/m^3 and a viscosity of 1 cP approximately.

In the case of liquids having different densities and viscosities from that of water, the use of the unit "m" for differential head can cause misunderstandings. "m" stands for "metres of liquid column". If reference is made to "metres of water column" it is usually written "m H_20" to avoid confusion. The reason for this confusion is that pumps give the same differential head in meters liquid column (m), irrespective of the density of the liquid. The pump's power requirement, and discharge pressure, is on the other hand proportional to the density. For densities different from that of water the power requirements and discharge pressure quoted in the purchase order always relate to the liquid stated. When pumping liquids which differ from water, the manufacturer should state how the pump will perform, on test, with water.

The standards for acceptance testing of rotodynamic pump performances is covered by various current national and international standards, for further details see Chapter 17, Section 17.1.9 and Chapter 12. The manufacturers' guarantees relate

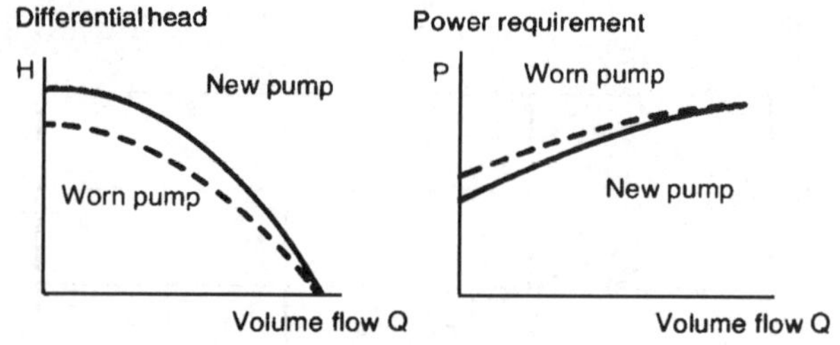

Figure 4.30 The effect on performance of wear in seals and clearances

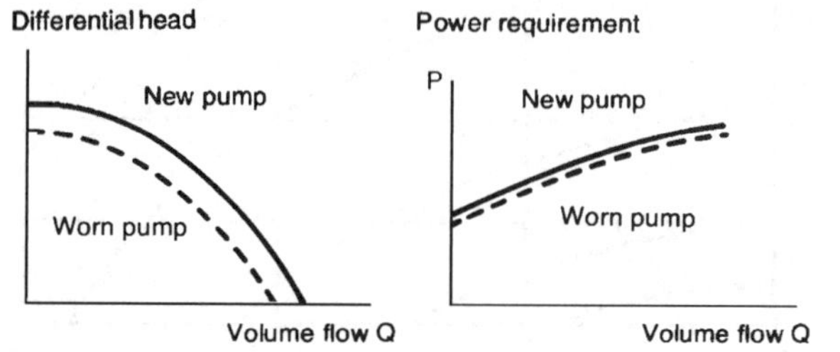

Figure 4.31 The effect on performance of wear on pump blades

to an installation designed and built to recognised standards and sound engineering practice. An unsuitable design of, above all, the inlet pipe in an installation can reduce the pump performance and cause severe operational problems.

A pump's performance falls off with wear of the pump parts. This wear occurs both at the internal and external seals and at clearances, affecting the internal and external leakage, see Figure 4.30; and at the blades, decreasing the blade work, see Figure 4.31. It is very difficult to give a limit for the extent of wear but reductions in delivery head of the order of 10% during the pumping of clean liquids are, however, possible.

The performance of pumps falls off rapidly with increasing viscosity of the pumped liquid. This reduction expresses itself in such a manner that the H-Q curve falls but shut-off head is retained. The power requirement rises considerably due primarily

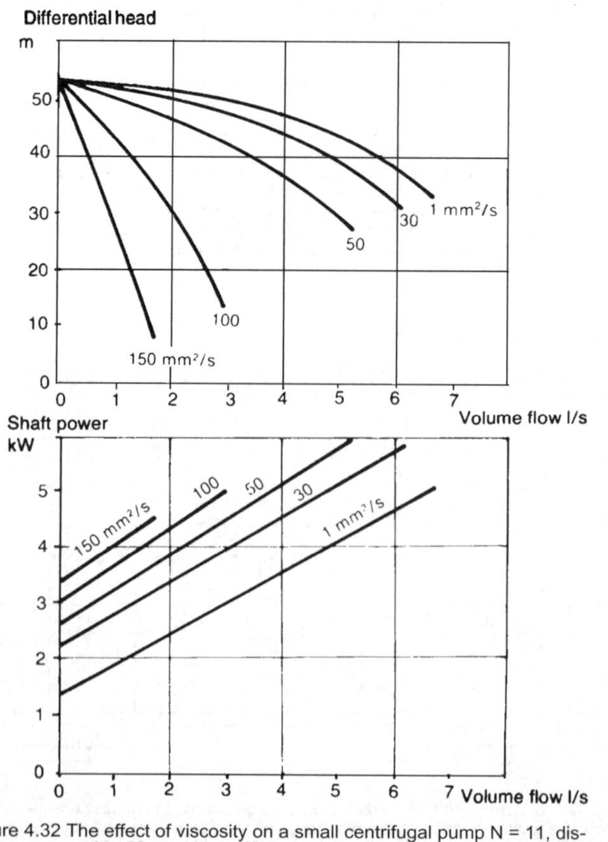

Figure 4.32 The effect of viscosity on a small centrifugal pump N = 11, discharge connection 50 mm. Max efficiency falls from approx 50% to approx 5%

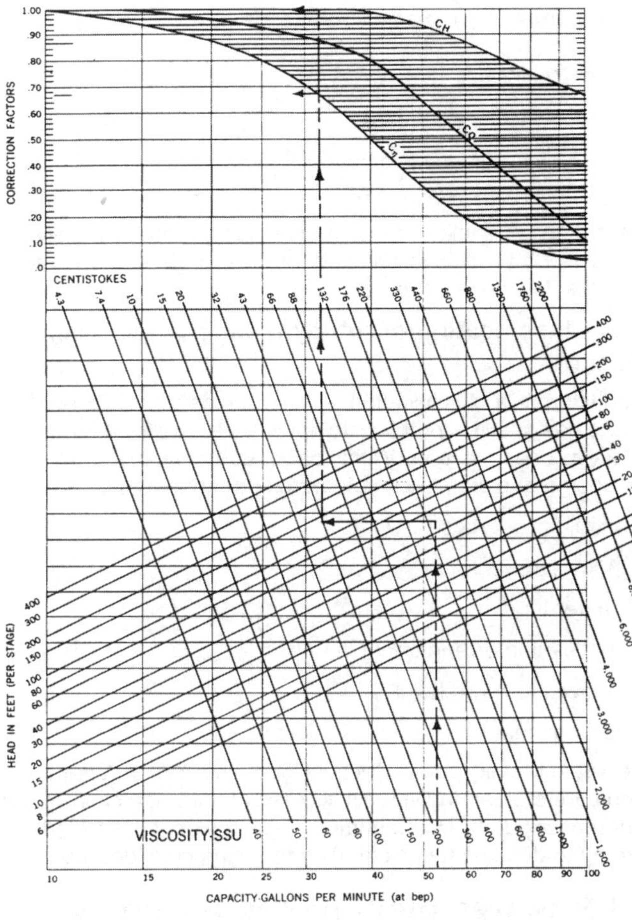

Figure 4.33 Viscosity corrections for small centrifugal pumps

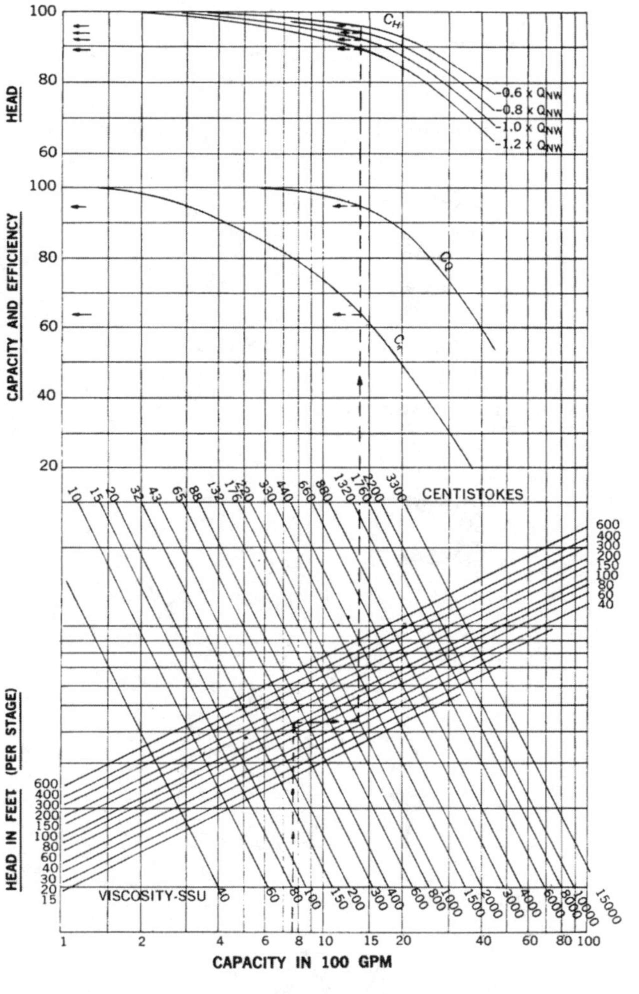

Figure 4.34 Viscosity corrections for large centrifugal pumps

to the increase of impeller friction. Figure 4.32 shows typical examples of the reduction in performance for smaller centrifugal pumps.

Figures 4.33 and 4.34 show correction factors, in USgpm and feet, for large and small centrifugal pumps. These graphs are based on testing carried out by Stepanoff of the former Ingersoll-Rand and are reproduced by permission of Flowserve Corporation. For further details see Chapter 15 on pump selection. PD ISO/TR 17766:2005 uses the Stepanoff data to produce a mathematical correction factor method. PD ISO/TR 17766:2005 gives performance corrections for all worldwide designs of centrifugal pumps of conventional design, in the normal operating range, with open or closed impellers, single or double suction, pumping Newtonian fluids.

4.2.3 Classification of rotodynamic pumps by hydraulic design

4.2.3.1 Impeller

The impeller, together with the casing, constitute the most critical parts in a rotodynamic pump. The shape of the impeller is determined by a combination of factors: flow, differential head and speed. These three parameters are used to define the specific speed in accordance with equation 4.15, in Section 4.2.1.5. In principle all pumps with the same specific speed can be geometrically similar. The variation, in practice, is due to different design techniques and manufacturing methods for small and large impellers and also the effect of surface roughness.

Experience teaches the hydraulic designer how the impeller should be shaped in order to obtain the greatest efficiency and suitable curve shape. The pump is named according to the flow path through the impeller, such as: centrifugal, mixed flow and axial, see Figure 4.35. There are an unlimited number of intermediate impeller shapes to those shown in order to meet every performance requirement.

Two impellers can be manufactured as one for parallel operation, two single stage impellers back-to-back, a double suction impeller as shown in Figure 4.35. A double suction pump should have a lower NPSHr than an equivalent single suction pump. For liquids contaminated with solids or sewage and where production must be simplified due to manufacturing problems of the tough materials used, the design principle of "greatest possible efficiency" is discarded and the impeller is given a much simpler shape. A distinction is made in theory between closed, semi-open and open impellers as well as special shapes for contaminated liquids. See also "non-clogging" pumps.

4.2.3.2 Multi-stage pumps

The efficiency of pumps is dependent upon the specific speed, as shown in Figure 4.36. Single stage pumps, required to produce high differential heads or handle low flows will have low specific speeds when calculated by equation 4.15. High differential heads can be produced by using several low head impellers in series. The specific speed will now be higher because the flow is the same as before but the differential head per im-

Closed centrifugal impeller. The illustration on the right shows the impeller with the front shroud removed

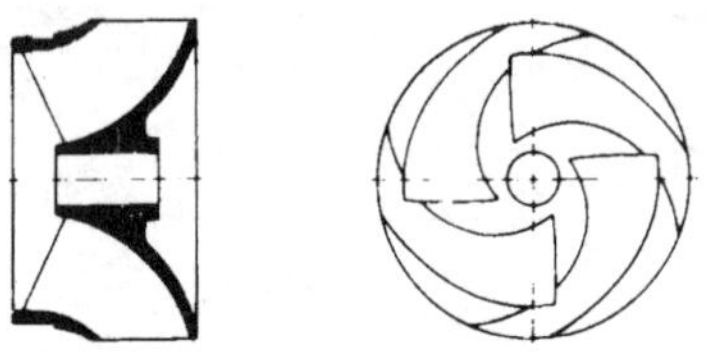

Closed mixed flow impeller. The illustration on the right shows the impeller with the front shroud removed

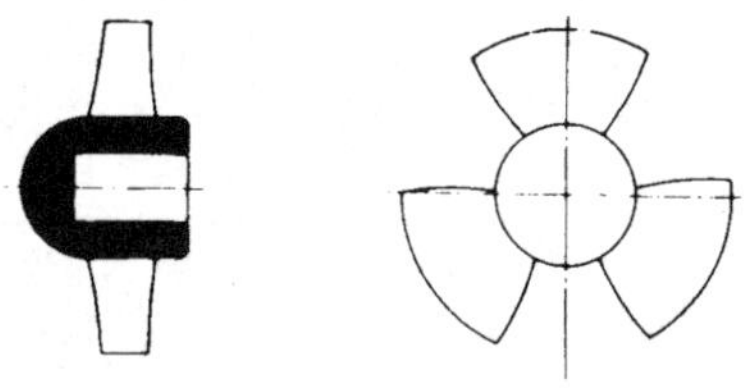

Axial impeller (propeller)

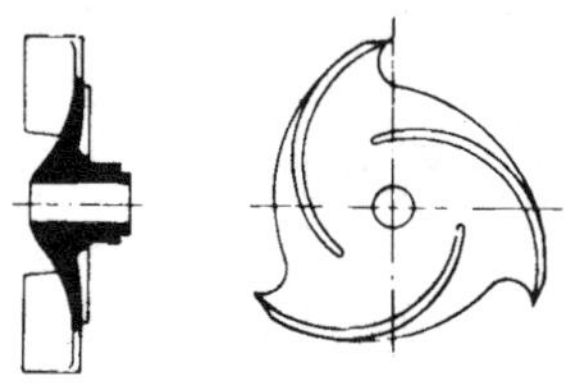

Open centrifugal impeller

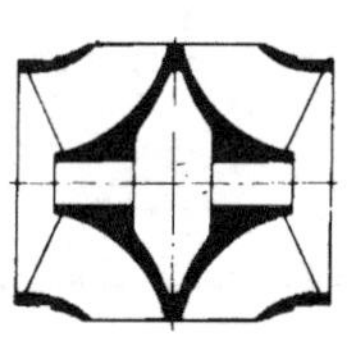

Double suction mixed flow impeller

Figure 4.35 Examples of different impeller types

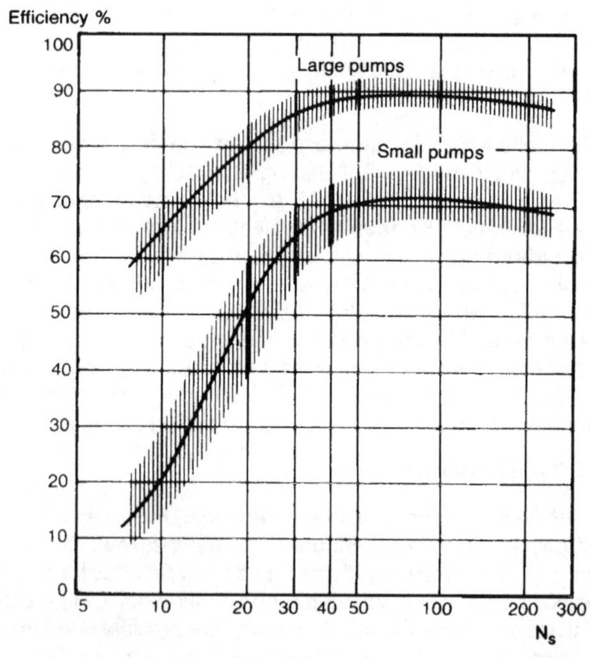

Figure 4.36 Pump efficiency as a function of specific speed

peller is smaller. A multi-stage pump consists of several similar impellers working on the same liquid consecutively. Doubling the number of impellers increases the specific speed by $2^{0.75}$ = 1.68. The following efficiencies, for example, can be read off from Figure 4.36 for a small pump:

No of stages	N_s	Efficiency %
1	10.0	21
2	16.8	42
4	28.2	63

The range of application of multi-stage pumps for delivery heads over 50 and up to 100 m, is, as a general rule, dependent upon the flow and speed.

4.2.4 Classification of rotodynamic pumps by application

In addition to a classification in accordance with different hydraulic designs, it is practical to apply a classification for different areas of application. Within each such area, there are special design and construction features dependent upon the:

- Liquid properties
- Allowable leakage
- Driver type
- Installation arrangement
- Allowable noise level
- Operational safety

The rotodynamic pump Sections 1.3 and 1.4 of Chapter 1 should be studied individually to assist in the selection of the appropriate pump. These pump types have been classified according to detailed technical and marketing considerations.

4.3 Positive displacement pumps

4.3.1 Rotary positive displacement pump theory

Basic principles

One of the simplest rotary pumps, and probably the easiest to understand, is the external gear pump, see Figure 4.37. One gear is driven by an external power source, the other meshes with the driven gear. The gears are enclosed in a close fitting casing. The radial clearance around the gear teeth is controlled by machining; the axial clearance down the sides of the gears is usually controlled by shimming during assembly. Clearance is necessary to avoid seizure during operation. The axial clearance is also necessary to allow lubrication.

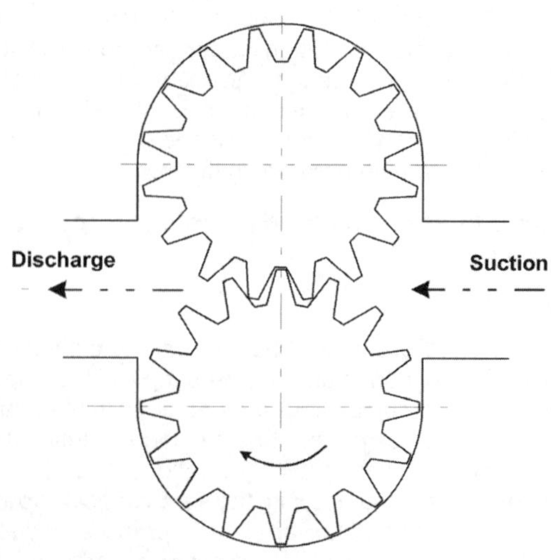

Figure 4.37 Diagrammatic external gear pump

Liquid is carried around the periphery of the casing in the voids between the teeth. The theoretical flow is given by:

$$Q_{th} = 0.001 \times a \times w \times (2 \times N) \times rpm/60 \qquad \text{Equ 4.24}$$

where:

Q_{th}	=	theoretical flow (cc/s)
a	=	area between teeth (mm^2)
w	=	tooth width (mm)
N	=	number of teeth in each gear
rpm	=	speed (1/minute)

Equation 4.24 applies to gear pumps where identical gears are used; this is not essential for gear pump designs but it is the most usual. Theoretical flow cannot be realised in practice. Leakage occurs directly from discharge to suction where the gears mesh; this leakage can be very small and is dependant upon the tooth form and the clearance. Most leakage occurs around the periphery between the tips of the teeth and the casing. This leakage is complex to analyse but approximates as follows:

$\propto$	differential pressure
$\propto$	gear width
$\sim\propto$	tip clearance^3
$\propto$	$1/\mu$
$\propto$	1/N
$\propto$	1/gear tip length
$\sim\propto$	1/function of tip speed

Gear pumps rely on viscosity to reduce leakage. As viscosity reduces the leakage becomes a greater proportion of the theoretical flow. The gear pump characteristic shown in Figure 4.48 in Section 4.3.3 indicates the effect of speed, viscosity and differential pressure. Leakage is reduced at higher speeds for constant viscosity and differential pressure.

A popular light oil grade for centrifugal pumps with rolling bearings is ISO 32. At 20 °C these oils would have a viscosity of 85 mm^2/s. Water has a viscosity about 1 mm^2/s. It can easily be seen in Figure 4.48 how the leakage losses escalate as less viscous liquids, more like water, are pumped.

The mechanical efficiency of gear pumps is controlled by several mechanisms and assumptions should not be made when calculating power required. At low viscosities the efficiency tends to improve with increasing speed. At high viscosities the efficiency may be constant or reduce with increasing speed. The distinction between high and low viscosity is a function of pump size. Mechanical efficiency usually increases as the differential pressure increases. Small pumps can have efficiencies of 25% and large pumps 70%.

Because gear pumps have multiple pumping elements, for example, the teeth, the flow pattern may have a slight ripple which can cause pressure pulsations. Pumps with few teeth are worst. Tests by Professor Edge, of Bath University, have indicated peak-to-peak pressure pulsation levels of 3.25 bar with a mean suction pressure of 3 bara.

Wear in rotary displacement pumps

The internal leakage varies as the cube of the clearance between the rotating components and the casing. Displacement pumps are thus susceptible to increased clearances due to wear. Displacement pumps used for "abrasive" liquids are constructed from wear-resistant materials and operated at lower speeds. Exceptions to these guidelines can be made by utilising principles of construction which eliminate leakage by the use of an elastic element, peristaltic pumps, for example.

Considering elementary principles, wear is a function of:

- Surface pressure between components
- Relative rubbing speeds between components
- A dependance upon the coefficient of friction

or with more usual wear theory, the combination of surface pressure and rubbing speed, PV.

The heat produced per unit of surface area is dependant upon PV and the coefficient of friction and can be a significant factor.

The technique of creating wear-resistant rotary pumps is concerned with reducing the product PV. Low values of velocity can result in large pumps, the only solution then is to reduce the surface pressure.

The driving force for rotary pumps can be transmitted to the various components in different ways. In the case of gear pumps the external drive is normally applied to only one of the gears, while the other is driven by engagement with the gear teeth of the first. However, if both the gears are driven externally, the load relationship within the process liquid is completely different. It is necessary to synchronise the shafts by means of a separate synchronising gear so that the teeth pass within each other when rotating without contact. The surface pressure between the interactive gear components is thus reduced to nil. The shape of the interactive gear surfaces can now be chosen with regard to the pumped fluid. Instead of gear teeth, the pump elements can now be given smoother shapes or radically different shapes, as in the case of lobe pumps and rotary piston pumps, (see Chapter 1, Section 1.5.8).

Within the rotary pump group there is a special "family" of pumps whose interactive components are guided by external devices as in the case of the lobe pump mentioned above. Even pumps where the rubbing pressure is eliminated in other ways also belong to this group. Examples of such pumps are rotary piston pumps and twin screw pumps. (See Chapter 1, Section 1.5.4.)

With the rubbing pressure eliminated or greatly reduced the duty range can be increased. Depending upon other design features, this range of application can be extended to contaminated liquids, low viscosity, some dry-running capability or high speed.

Rotary pumps suitable for abrasive liquids should be constructed according to the following:

- The active components should be positively located by external means or have greatly reduced sliding pressure.
- The forces generated on the pumping elements by the liquid pressure should be absorbed by bearings which are not in contact with the liquid.

4.3.2 Reciprocating positive displacement pump theory

The theoretical principles of positive displacement pumps are basically very simple. For each working cycle, each stroke, a certain volume of liquid is drawn into and expelled from the cylinder virtually irrespective of the differential pressure. The volume of liquid depends upon on the displacement (swept volume) and the volume of the cylinder. The maximum attainable pressure is primarily dependent only upon the mechanical construction and the driving power available. The maximum pressure can be limited by means of a pressure relief valve installed in the system or integrated in the pump. Because reciprocating pumps can develop such high pressures they must always be operated with over-pressure protection. A relief valve or bursting disc must be fitted. In hydraulic power applications it has become standard practice to incorporate the over-pressure protection with other functions, such as unloading, by using a common pilot-operated valve. In process applications this approach would not be tolerated; a dedicated device must be fitted solely for protection.

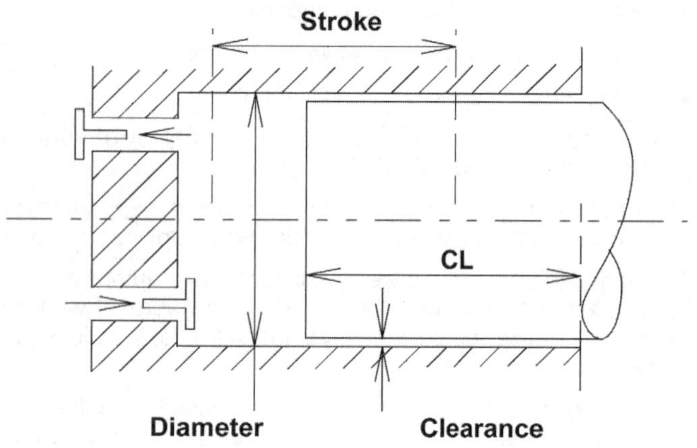

Figure 4.38 Basic reciprocating pump

Consider the single cylinder pump shown in Figure 4.38. Ignoring leakage and compressibility, the ideal flow delivered is calculated from:

$$Q_{th} = 0.001 \times \text{displacement} \times n/60$$

and

$$\text{displacement} = s \times \pi d^2/4 \qquad \text{Equ 4.25}$$

where:

Q_{th}	=	theoretical flow (cc/s)
n	=	strokes per minute (1/minute)
d	=	piston diameter (mm)
s	=	stroke (mm)

The displacement of the piston establishes the name of the pump type, positive displacement. Displacement is also called swept volume as it is the volume swept out by the face of the piston.

The ideal power required for the pump is given by:

$$P_{th} = Q_{th} \times (p_d - p_s) \qquad \text{Equ 4.26}$$

where:

P_{th}	=	ideal power (W)
Q_{th}	=	ideal flow (m^3/s)
p_d	=	discharge pressure (gauge) (Pa)
p_s	=	suction pressure (gauge) (Pa)

Equation 4.26 is not very useful for engineering applications. The Pascal is an extremely small unit of pressure, the kPa is very small. In order to avoid confusion at site, where suction pressure gauges may be calibrated in kPa and discharge gauges in MPa, the pump industry prefers to use the "bar". 1bar = 100kPa. The following equations are more practical.

$$kW = \frac{m^3/h \times \Delta bar}{36} \qquad \text{Equ 4.27}$$

$$kW = \frac{L/s \times \Delta bar}{10} \qquad \text{Equ 4.28}$$

The ideal power is based on ideal flow. This Section on theory commenced by excluding leakage and compressibility. Obviously real pumps leak and real liquids are compressible. Compressibility is the easiest to tackle.

It can be seen from Figure 4.38 that the piston stroke does not extend right to the end of the cylinder. The volume of liquid trapped in the cylinder, together with liquid compressibility, affects the volumetric efficiency. The volumetric efficiency is the ratio of the liquid volume pumped to the displacement of the piston. We have already seen that the discharge volume may be smaller than the suction volume, this leads to two definitions of volumetric efficiency — suction and discharge.

$$ve_S = 100 \times \left(\frac{\left(1-\Delta p \times \chi \times \left(1 + \frac{V_{int}}{V_{sw}}\right)\right)}{(1-\Delta p \times \chi)} \right) \qquad \text{Equ 4.29}$$

$$ve_D = 100 \times \left(1 - \Delta p \times \chi \times \left(1 - \frac{V_{int}}{V_{sw}}\right)\right) \qquad \text{Equ 4.30}$$

$$ve_S = 100 \times \left(1 - \frac{V_{int}}{V_{sw}} \times \left(1 - \frac{\rho_D}{\rho_S}\right)\right) \qquad \text{Equ 4.31}$$

$$ve_D = 100 \times \left(1 - \left(\left(\frac{V_{int} + V_{sw}}{V_{sw}}\right) \times \left(1 - \frac{\rho_S}{\rho_D}\right)\right)\right) \qquad \text{Equ 4.32}$$

where:

ve_S	=	suction volumetric efficiency (percentage)
ve_D	=	discharge volumetric efficiency (percentage)
Δp	=	differential pressure (bar)
χ	=	compressibility (*)
V_{int}	=	clearance volume (cc)
V_{sw}	=	swept volume (cc)
ρ_D	=	discharge density of liquid (kg/m^3)
ρ_S	=	suction density of liquid (kg/m^3)

(*) the units for compressibility are "contraction per unit volume/bar differential pressure".

Compressibility is the inverse of the bulk modulus which has units of N/m^2. The bulk modulus of water at 20 °C from atmospheric pressure to 100 barg is approximately 2.1 x 10^4 bar. The presence of entrained and dissolved gas or air in the liquid increases the compressibility and reduces the volumetric efficiency.

The formulae for volumetric efficiency indicate the importance of keeping the clearance volume as small as possible. Also the volumetric efficiency reduces as compressibility increases. At low differential pressures the compressibility of water can be used for most other liquids without creating significant errors. Some liquids are much more compressible than water, liquid gases for example can be six times as compressible, and using water compressibility will create problems of undersizing and low flow.

Leakage further reduces the actual liquid volume delivered by the pump. The term "slip" is used to collect together all the liquid losses which can be evaluated and those which must be compensated by an experimental allowance. Slip is evaluated as a percentage and subtracted from the volumetric efficiency.

Because there is a clearance around the periphery of the piston, leakage can occur due to the pressure rise within the cylinder; this leakage can be approximated from:

$$Q_L = 0.00156 \times \left(\frac{d}{CL}\right) \times \frac{h^3 \times \Delta p}{\mu} \qquad \text{Equ 4.33}$$

where:

Q_L	=	leakage flow (cc/min)
d	=	cylinder bore (mm)
CL	=	contact length (mm)
h	=	clearance (mm)

Δp = differential pressure (bar)

μ = viscosity (cP)

As the piston and cylinder wear the leakage rate rises dramatically. The liquid viscosity may be lower in the clearance due to frictional heating. This leakage is fairly constant and independent of pump speed, strokes per minute.

The valves shown diagrammatically are spring-loaded automatic valves. The differential pressure across the valve, in the direction of flow, opens the valve and liquid flow forces hold the valve open. As the flow reduces the valve starts to close due to the spring force. The pressure losses and leakage through valves of this generic type are impossible to generalise due to the infinite range of proportions and shapes. However, the following relationships can be used for guidance:

- For fixed valve lift

  $\Delta p \propto Q$

  $\propto \mu$

- For fixed flow

  $\Delta p \propto m$

  $\propto l/\text{lift}^3$

One valve loss, particularly troublesome at high speed, is valve backflow. Automatic valves are closed by the spring; the spring applies a force to the mass of the valve and the valve accelerates. At higher speeds, there is insufficient time for the valve to close, unless a stronger spring is fitted. The valve remains slightly open when the piston travel has reversed allowing some liquid to escape. After inception, this loss varies with speed and is almost linear.

Some pumps are fitted with actuated valves, or valve timing is incorporated in the rotating cylinder block of hydraulic pumps. These valves are usually rotary and driven directly by the main shaft. The valve timing is optimised for the design conditions. Operating at other conditions causes backflow or over-compression.

Valve flow losses are complicated because the liquid flow is approximately sinusoidal. The reciprocating pumps described here are usually used for hydraulic power applications. The liquid is oil or water-oil emulsion providing good lubrication for the close fitting piston. The pumps are axial or radial piston. The pistons are driven by a mechanism which produces simple harmonic motion. The piston dynamics, sine and cosine functions, are shown in Figure 4.39.

The liquid is unable to follow the piston dynamics exactly due to inertia. The liquid inertia causes the liquid to accelerate and decelerate faster than the piston, causing pressure pulsations. Liquid flow variations and pressure pulsations caused by reciprocating pumps can not be approximated easily as they are a function of the system as well as the pump. In critical applications, and with low viscosity liquids, an acoustic analysis should be performed which will evaluate pressure pulsations, flow variations and acoustic resonance.

Two variations of reciprocating pump design used for process applications where the liquid may not have good lubricating properties, are shown diagrammatically in Figure 4.40. The plunger pump is single acting whereas the piston pump is double acting.

The reciprocating motion is produced by a crankshaft with connecting rods and crossheads, the slider-crank mechanism. There is no side thrust applied to the piston or plunger from the driving mechanism, this is applied by the crosshead to the crankcase. The length of the connecting rod compared to the throw of the crankshaft is important because it affects the piston/plunger dynamics and the crosshead side thrust. The sine and cosine functions of simple harmonic motion are modified by connecting rod proportions as shown.

$$x = r(1-\cos\theta) + l\left(1-\sqrt{1-\lambda^2\sin^2\theta}\right) \qquad \text{Equ 4.34}$$

$$v = \omega r\left(\sin\theta + \frac{\lambda\sin\theta\cos\theta}{\sqrt{1-\lambda^2\sin^2\theta}}\right) \qquad \text{Equ 4.35}$$

$$a = \omega^2 r\left(\cos\theta + \frac{\lambda\left(1-\sin^2\theta\left(2-\lambda^2\sin^2\theta\right)\right)}{\left(1-\lambda^2\sin^2\theta\right)^{3/2}}\right) \qquad \text{Equ 4.36}$$

where:

x = travel (m)

v = velocity (m/s)

a = acceleration (m/s/s)

r = crank throw (m)

l = connecting rod centres (m)

ω = angular velocity (rad/s)

λ = r/l (non-dimesional)

θ = crank angle (degree)

The three parameters are plotted in Figure 4.41, for λ=0.25, alongside the simple harmonic motion curves. Notice that the:

- Curve shapes are modified
- Maximum velocity is not at 90 °

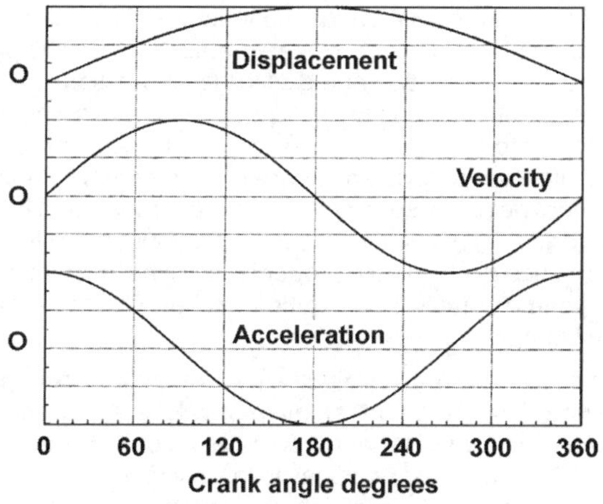

Figure 4.39 Simple harmonic motion piston dynamics

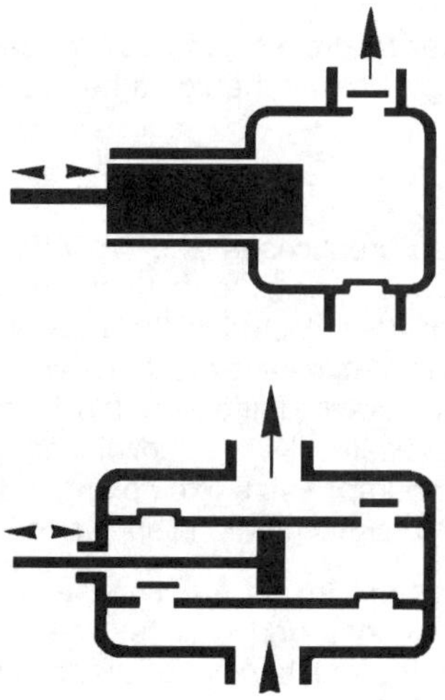

Figure 4.40 Process plunger and piston pumps

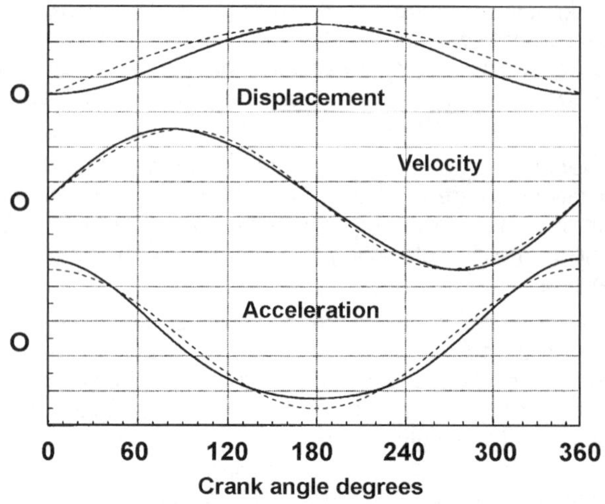

Figure 4.41 Travel, velocity and acceleration of slider-crank mechanism

- Forward and return acceleration is different
- Maximum acceleration is greater

Having an infinitely long connecting rod produces simple harmonic motion. In practice, connecting rods are 4 to 7 times the crankshaft throw.

A plunger pump is very similar to the piston pumps described previously for hydraulic fluid power except a stationary sealing stuffing box prevents most leakage. The piston pump shown is double acting and has moving seals on the piston as well as a stationary sealing stuffing box. The prime distinction between piston and plunger pumps is the absence of moving seals in the plunger pump. Plunger pumps are not a type of piston pump.

Piston pumps are used for pressures up to about 150 barg and plunger pumps for higher pressures. There is no absolute limit on plunger pumps, 10000 barg is possible, but popular pumps operate up to 1200 barg. In Chapter 1, Figure 1.90 shows a piston pump for about 25 barg and Figure 1.91 shows a plunger pump for about 350 barg . Figure 1.92 shows a plunger pump built as vertical pump.

The stuffing box prevents leakage of process liquid to the atmosphere. Simple seal designs are used for non-hazardous liquids such as water and ethylene glycol, Figure 4.42. Hazardous liquids; petrol, methanol, LPG; or liquids contaminated with hazardous gases like hydrogen sulphide, can use very complicated stuffing boxes depending upon the location, Figure 4.43.

Auxiliary systems can be fitted to stuffing boxes, see Section 8.5, in Chapter 8.

Reciprocating pumps for process applications can be fitted with various valve designs to suit the liquid handled, Figure 1.96, Chapter 1.

Mechanical losses

Reciprocating pump crankcases and actuating mechanisms are designed to exert a specific force on the piston or plunger, the rod load. Pumps are selected so that they operate as close to the design rod load to minimise mechanical losses. Different piston/plunger diameters are fitted to match the operating pressure. Figure 4.44 indicates the effect of discharge pressure on a 3" stroke plunger pump, with a fixed plunger diameter and a simple single packed stuffing box as shown in Figure 4.42.

At pressures below the design pressure, friction becomes a more significant influence and reduces the mechanical efficiency. In general, the mechanical efficiency of reciprocating pumps is not affected by pump speed or viscosity. Larger pumps may be more efficient; pumps of 1000 kW can have a mechanical efficiency of 94% at design pressure. Complicated,

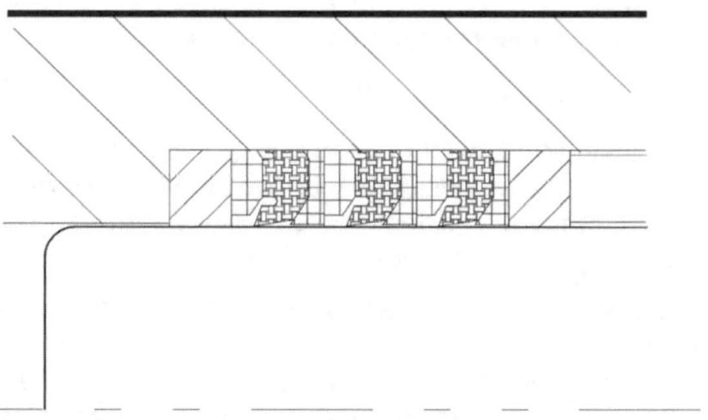

Figure 4.42 Non-adjustable stuffing box for water

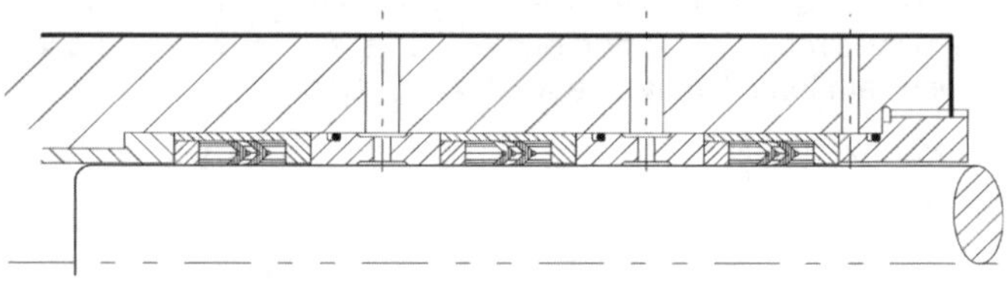

Figure 4.43 Self-adjusting stuffing box for hazardous liquids

multi-seal stuffing boxes increase friction and reduce the mechanical efficiency.

Wear in reciprocating pumps

The main wearing parts in reciprocating pumps are piston seals, stuffing box packing and valves. Wear in stuffing box packing produces an external leak which operators or maintenance staff can recognise and act upon. Wear in piston packing and valves produces internal leaks, increased slip, and can only be detected as a reduction in pump flow.

Reciprocating pumps can handle liquids contaminated with solids. The Miller Number should be evaluated. Piston pumps in standard materials are adequate for Miller Numbers up to 50. Above 50, severe wear problems may occur. Plunger pumps can handle the most abrasive liquid/solid combinations by running at reduced speeds, using ball valves and flushing the stuffing box to protect the packing.

Both piston and plunger pumps can be modified for solids handling by introducing a diaphragm to completely isolate the reciprocating driving element from the process liquid. Diaphragm pumps without pistons or plungers are also possible. Diaphragms can be driven directly by another fluid, liquid or gas, which has timed control valves to regulate the motion.

Flow and pressure pulsation

For many positive displacement pumps the flow varies during the period of the working cycle. Extreme examples of flow variation occur in single and twin cylinder reciprocating pumps, see Figure 4.45.

During a working cycle the instantaneous flow varies, to a greater or lesser extent, for most positive displacement pump types. In order to reduce variations in flow it is possible to increase the number of cylinders or elements or to provide pulsation dampers at the suction and discharge connections. These can be filled with air/gas or equipped with air cushions or springs separated from the liquid by an elastic membrane. The size of the flow variation is an important feature, which unfortunately cannot be easily specified by the manufacturer. The flow variation, and the resulting pressure pulsations, are a function of the pump design, the liquid properties and the installation design. The same pump will have different characteristics in a different installation.

Flow variations and pressure pulsations measured during factory testing will not be representative of values experienced at site, unless the pipework and liquid conditions are identical. Flow resistance in a test rig consists largely of throttling losses in control valves i.e. with a resistance varying as the square of the flow. Dampening in rubber hoses for example, also has

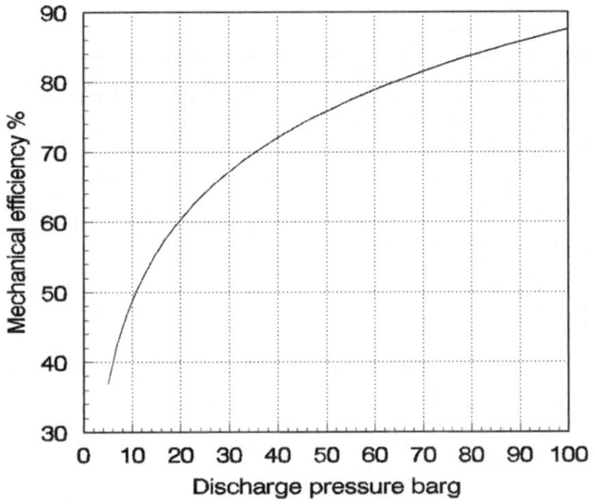

Figure 4.44 Discharge pressure and mechanical efficiency

considerable influence. If flow variations and pressure pulsations are critically important for an installation, then a computer simulation, analogue or digital, must be performed of the complete system to verify instantaneous flow and pressure values together with sources of acoustic resonance.

NOTE: The flow variations shown in Figure 4.45 are based on an incompressible liquid, an ideal pump and no suction or discharge systems. These conditions cannot be reproduced in practice. Flow diagrams of this type are purely indicative and very approximate. No practical data can be derived from such diagrams and no useful calculations can be performed for flow variations, pressure pulsations or pulsation damper sizing.

The minimum flow for single and twin cylinder pumps is zero so the total flow variation is 100%. Screw pumps have flow variations of a few percent. Pressure pulsations caused by flow variations are the primary cause of noise and vibration in positive displacement pump piping. In the case of gear pumps for example, it is very easy for a small volume of liquid to become trapped in a pocket at the moment of gear contact, the volume of which changes during the course of rotation. This gives rise to harsh running and high noise levels. The problem can be considerably reduced by introducing a slight helix angle on the gear; by providing relief grooves in the side plates or by increasing the side clearance.

4.3.3 Positive displacement pump curves

Figure 4.46 shows a positive displacement pump curve drawn in the same manner as a rotodynamic pump curve. Little detailed information can be gained from the curve and two curves must be drawn separately to show the power relationship.

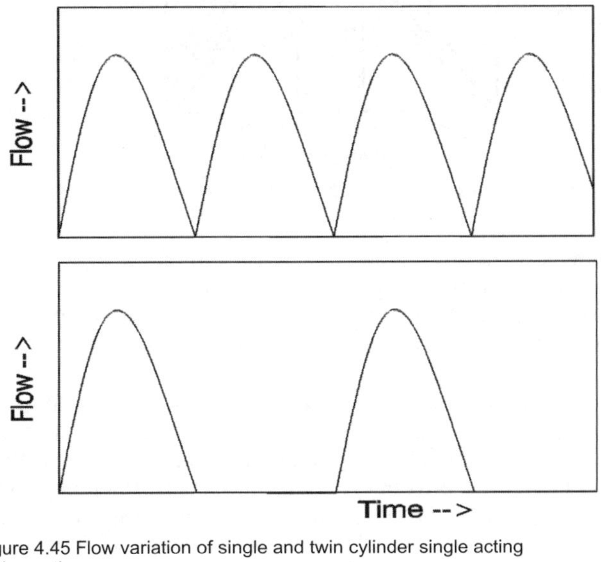

Figure 4.45 Flow variation of single and twin cylinder single acting reciprocating pumps

If curves are drawn it is better to plot the discharge pressure along the abscissa. This then conforms to mathematical standards; the abscissa being the independent variable. Flow, power and mechanical efficiency can be plotted on the same curve, see Figure 4.47. For sensible flow information to be retrieved an expanded scale must be used and zero flow is not necessarily shown. The characteristic curve of flow v pressure may not be linear. Compressibility and slip effects may be more pronounced at higher pressures. Curves for different speeds may not be the same shape. When various liquids are pumped, or operating conditions vary, additional curves may be required. For fixed speed pumps with a single duty point, tabulated data is preferred.

Figure 4.48 shows the change in fixed speed pump flow resulting from changes in differential pressure and liquid viscosity for an external gear pump. It can be seen that the reduction in flow is very small for relatively large changes in differential pressure and is affected by the liquid viscosity.

As mentioned earlier, viscosity and speed can play an important role in the performance of rotary pumps such as gear pumps. Figure 4.49 indicates the speed, viscosity and differential pressure effects on a gear pump.

To make pump selection easier performance curves can be drawn differently.

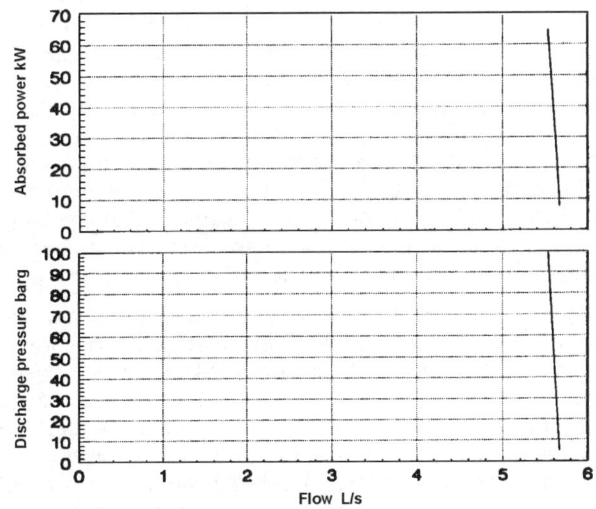

Figure 4.47 Performance curve for positive displacement pump

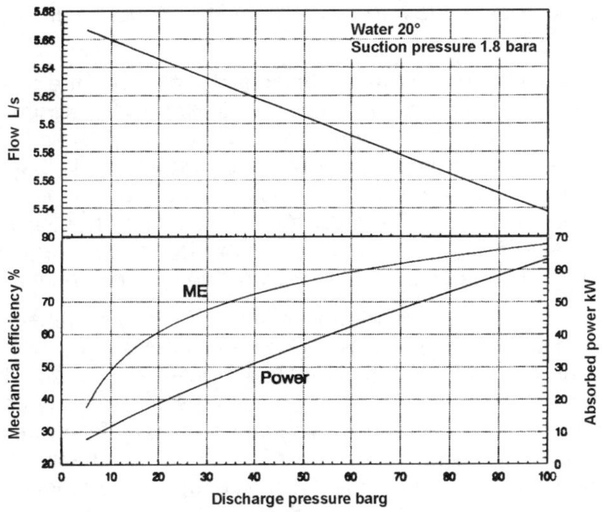

Figure 4.47 Performance curve for positive displacement pump

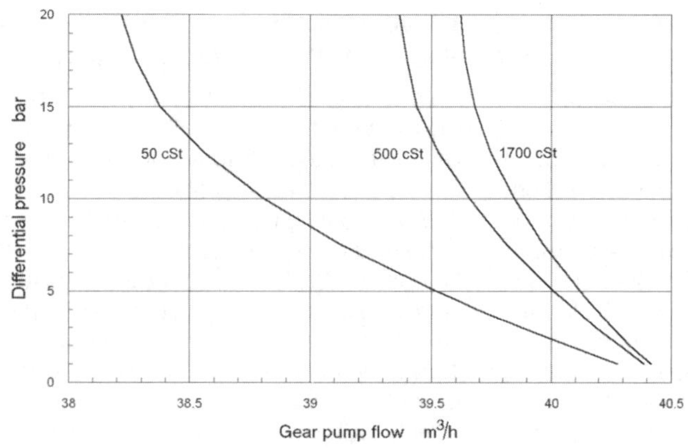

Figure 4.48 The effects of slip on gear pump flow

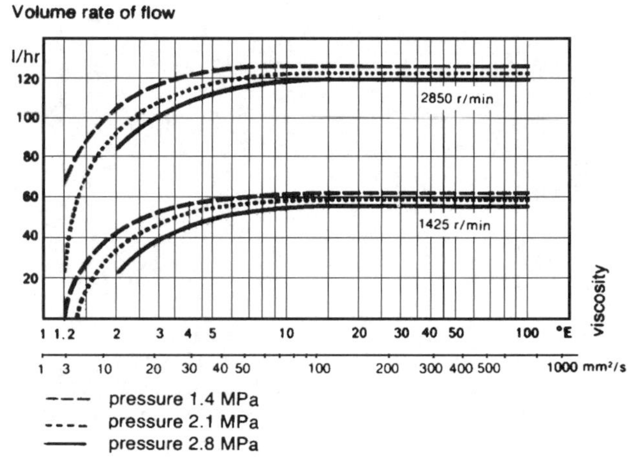

Figure 4.49 The effect of viscosity, speed and dp on a small gear pump

Figures 4.50 and 4.51show the curves for a lobe pump for two different viscosities.

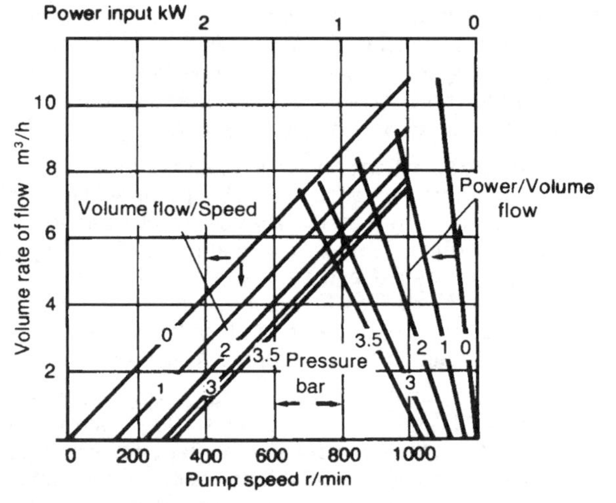

Figure 4.50 Performance curve for a lobe pump in water

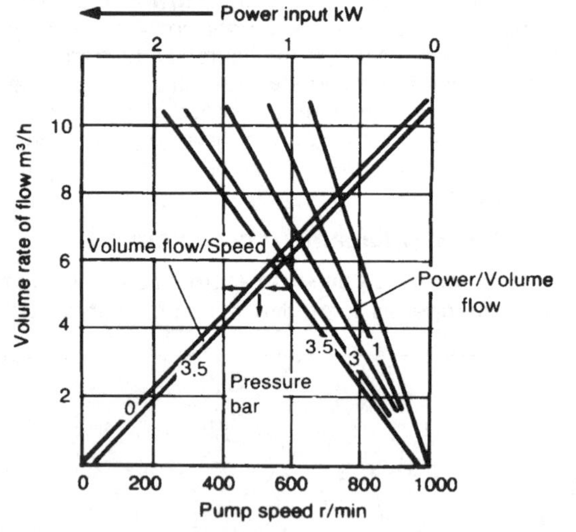

Figure 4.51 Performance curve for the lobe pump in Figure 1.70, Chapter 1, on liquid at 100 cSt

4.3.4 Types of positive displacement pumps

Rotary positive displacement pumps are generally self-priming i.e. they can begin pumping with a "dry" suction pipeline and pump casing. Reciprocating positive displacement pumps are self-priming, in theory, although they tend to be poor compressors. In general, reciprocating pumps should be vented and primed prior to operation unless the manufacturer specifically states that self-priming is allowable.

All positive displacement pumps can pump a wide variety of liquids of varying characteristics, from very low viscosity liquids to very viscous pastes. The construction, application and pumping characteristics of different types of displacement pumps varies considerably however. It is therefore necessary to understand thoroughly the specific characteristics of each type of pump in order to be able to determine the suitability of a particular pump for a particular application. Gear pumps are not suitable for abrasive liquids for example. A progressive cavity pump with its flexible stator is more suitable. If the discharge pressure is relatively high a vane pump with flexible vanes is unsuitable, a gear pump or piston pump being a better choice.

The specified volume flow per working cycle is relatively constant and reduces insignificantly for moderate pressure increases. Liquids with high compressibilities, liquid gases, should always be evaluated for density changes. Control of the flow should be carried out by increasing or decreasing the pump speed. The flow can also be controlled by arranging a by-pass between the discharge and suction source. Do not, however, attempt to throttle the discharge in order to reduce the flow, as in the case of centrifugal pumps. This results in increased pressure, increased power consumption and increased loads on the working components, without significantly reducing the flow.

Serious damage and injury to personnel can occur if the pump discharge should become completely blocked, by the closing of a valve for example. The pressure will increase rapidly and can become so great that it causes failure of pipelines and other vital components. A dedicated pressure relief device should always be fitted as a safety precaution. The opening pressure of which should be selected so as to provide a good margin of safety for the protection of the system and its components.

A certain amount of internal leakage (slip) is always present in displacement pumps, this is due to the clearance between the working elements and inefficiencies of internal valves. The clearance can be of the order of 0.01 to 0.9 mm and is necessary in order to counteract galling and to reduce friction within the pump. Leakage occurs from the discharge side back to the suction side, i.e. against the direction of flow. The amount of slip, which reduces volumetric efficiency, varies greatly according to the type of pump construction. A progressive cavity pump with its long rotor and effective sealing rubber stator has less slip than a conventional gear pump having only line contact between the gears. The pumped liquid, if it is sufficiently viscous, can help to "seal" the clearances and reduce internal leakage.

Positive displacement pumps should not be installed to make use of their maximum suction lift capacity. Pump wear is unavoidable and this means that in time, the maximum suction performance will be reduced. This means that the pump loses its priming ability with subsequent loss of its pumping function. Unnecessary service costs are thus incurred in order to constantly maintain the pump in an "as new" condition.

4.3.5 Classification of positive displacement pumps by application

As discussed in Section 4.3.4, there are many types of positive displacement pumps. The rotodynamic pump, Sections 1.3 and 1.4 of Chapter 1, should be studied individually to assist in the selection of the appropriate pump. These pump types have been classified in accordance with their mode of operation.

NOTE: Reciprocating pumps for hydraulic power applications usually have no seals and rely on the pumped liquid for all lubrication. Reciprocating pumps for process applications do not rely on the pumped liquid for any lubrication. Process pumps can be used for hydraulic power applications.

4.3.6 Classification of other pump types by application

As mentioned in Sections 4.2.4 and 4.3.5, the 'other pump' types in Section 1.6 of Chapter 1, should be studied individually to assist in the selection of the appropriate pump. These pump types have also been classified in accordance with their mode of operation.

4.4 Suction performance

4.4.1 Cavitation

The word cavitation originates from the Latin meaning "hollow". Cavitation occurs when the static pressure somewhere within the pump falls to or below the vapour pressure of the liquid. Some of the liquid then vaporises giving rise to the formation of vapour bubbles. These vapour bubbles are entrained in the liquid and transported until they reach an area where the static pressure is greater than the vapour pressure.

At this point the vapour bubbles implode since they can no longer exist as vapour. Each implosion causes a severe pressure wave. The continuous repetition of this sequence of events occurring many times and at high frequency can cause mechanical damage to the pump materials. Furthermore, the hydraulic performance of the pump is reduced by the onset of cavitation. Cavitation is therefore an undesirable phenomenon and should be avoided.

The severity of cavitation damage is related to the thermodynamic properties of the liquid and the pump operating conditions. The important factors to consider are:

- Latent heat of vaporisation
- Differential pressure
- Rate of pressure rise
- Volume of liquid in component
- Volume of vapour
- Material of component
- Time of exposure

Water tends to be the worst liquid for cavitation damage because the latent heat of vaporisation is large compared to other liquids. The time of exposure is very important. If a pump cavitates for a few seconds during start-up the reliability and performance may not suffer noticeably. If a pump operates with cavitation for hours the likelihood of damage is increased. Figure 4.52 shows mild cavitation damage on a centrifugal pump impeller vane. Figure 4.53 shows severe cavitation damage to a centrifugal pump impeller.

Figure 4.54 shows cavitation damage on a reciprocating pump valve. The damage is very unusual because it is on the outside of the seat taper. The valve was probably dislodged by the forces generated by the cavitation and then the damage occurred due to the high velocity liquid flowing around the seat.

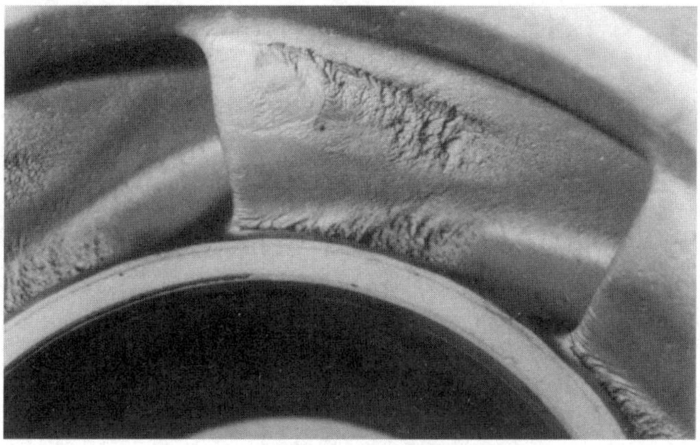

Figure 4.52 Mild cavitation damage on a centrifugal pump impeller vane

Figure 4.53 Severe cavitation damage to a centrifugal pump impeller

Figure 4.54 Cavitation damage on a reciprocating pump valve

The risk of cavitation is greatest where the pressure within the pump is lowest. The lowest pressure in a centrifugal pump occurs on the suction side of the impeller blades, slightly downstream of the front edges where the liquid has started to accelerate, see Figure 4.55. At the pump's inlet connection at the same elevation as the shaft the pressure is somewhat greater, by Δp. At this point, which lies at the static suction head h_s above the liquid free surface, the flow velocity is c_s. By applying Bernoulli's equation, see equations 3.4 and 3.5, in Chapter 3, the flow through the suction pipe is:

$$p_s = p_a - \rho h_s - \rho g h_{fs} - \rho \frac{c_s^2}{2} \qquad \text{Equ 4.37}$$

$$p_{min} = p_s - \Delta p \qquad \text{Equ 4.38}$$

where:

p_{min} = lowest pressure within the pump (N/m²)

p_s = suction pressure (N/m²)

Δp = local pressure reduction (N/m²)

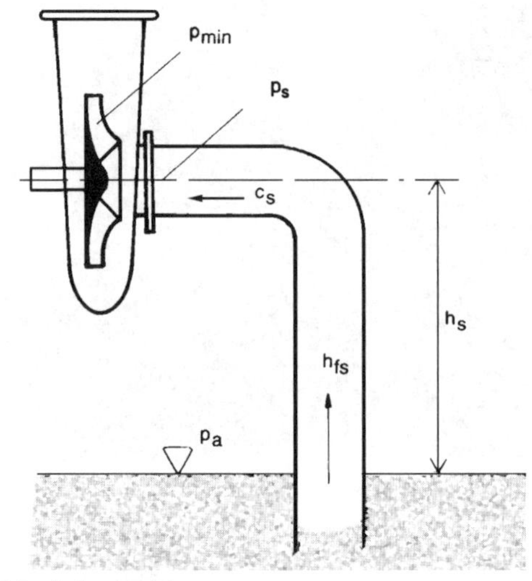

Figure 4.55 Cavitation model

c_s = Flow velocity in inlet pipe (m/s)

h_s = Static suction head (m)

h_{fs} = Friction losses in suction pipe (m)

p_a = Pressure on liquid surface (N/m^2)

Δp covers any losses within the pump, after the suction connection, such as velocity increases or friction losses.

Positive displacement pumps require a corresponding pressure differential, Δp above vapour pressure, in order to avoid cavitation within the pump. The lowest pressure in a pump is determined partly by external factors such as suction pressure and partly by factors concerned with construction and design. The latter consists of the local pressure drop Δp and the flow velocity at the inlet connection c_s. In order to avoid cavitation it is necessary to ensure that p_{min} is greater than the vapour pressure of the liquid p_v.

4.4.2 Net Positive Suction Head and Net Positive Inlet Pressure

The concept of NPSH, Net Positive Suction Head, is encountered in connection with cavitation in rotodynamic pumps . The definition of NPSH is as follows:

NPSH = suction pressure - vapour pressure

$$\text{NSPH} = \frac{p_s - p_v}{\rho g} \qquad \text{Equ 4.39}$$

In this context the suction pressure is the pressure at the NPSH datum. For rotodynamic pumps the NPSH datum is the shaft centre-line, except for vertical pumps when it is the front face of the impeller, see Figure 4.56.

In some cases when purchasing a new pump, the user will not know where the NPSH datum is going to be. In these cases, the NPSHa and the suction pressure should be referred to the top of the foundations. NPSH is a differential head, therefore the pressures are converted to metres of liquid when required.

Some textbooks state that NPSH refers to the suction total pressure, the suction stagnation pressure. This makes life very complicated. The pump suction pressure is the static pressure; head has been deducted to produce the velocity head of the flowing liquid. This is the value to which the pump differential head is added. It does not make sense to calculate a different suction pressure, the stagnation pressure, for NPSH purposes.

NPSH has two forms:

- The system form
- The pump form

The system form is designated NPSHa, Net Positive Suction Head Available. NPSHa is calculated from the suction system configuration and is the suction pressure minus the vapour pressure at the NPSH datum. NPSHa tells the pump manufacturer how much Δp is available for losses in the pump suction.

The pump form is designated NPSHr, Net Positive Suction Head Required. NPSHr is a characteristic of the pump and specifies how much Δp is necessary for losses in the pump suction.

NPSHa and NPSHr are the difference between two pressures or heads; there is no requirement to specify gauge or absolute.

Both forms of NPSH are variables, not constants. Both are a function of flow. In a fixed suction system configuration, NPSHa will increase as flow is reduced because friction losses and the velocity head requirement are reduced. Conversely, NPSHa will reduce as flow increases because the same losses increase. NPSHr for a pump is predicted during the design phase and confirmed by testing. NPSHr increases with flow and, at some point, increases as flow is reduced.

Rotodynamic pumps are optimised for a specific operating condition; flow and differential head; the Best Efficiency Point (BEP). When operating at other conditions parasitic losses increase and blade/liquid angles are mismatched. As flow is reduced below BEP angular mismatch increases. At a certain point, the losses created by mismatch become larger than the gains produced by reduced friction losses and NPSHr increases, see Figures 4.12 and 4.13.

In an ideal world, the NPSHr value for a pump would signify the initiation of cavitation. However it is extremely difficult, if not impossible, to detect the first indications of cavitation. Because of this problem various criteria have evolved as bench-marks for defining cavitation. The most popular definition is for 3% head loss. The pump is tested at constant flow and the suction pressure is reduced until the differential head has decayed by 3%, see Figure 4.57. The NPSHr is calculated from the suction pressure and the vapour pressure.

A complication to the 3% head loss criterion arises with multi-stage pumps. 3% head loss of what ? Some manufacturers use 3% of the pump differential head. In fact it should be 3% of the first stage differential head as this is the only impeller

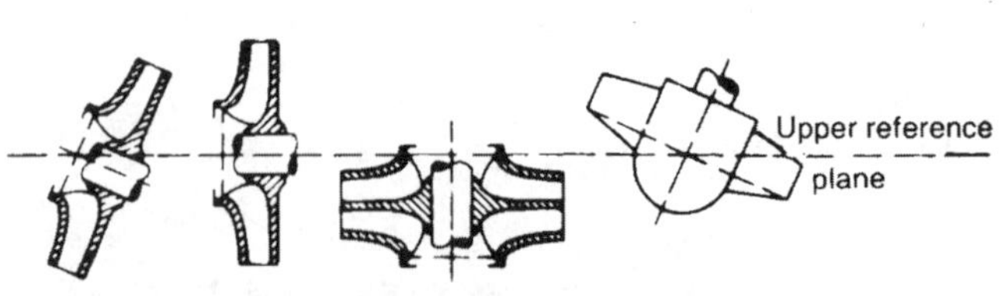

Figure 4.56 NSPH datum for rotodynamic pumps

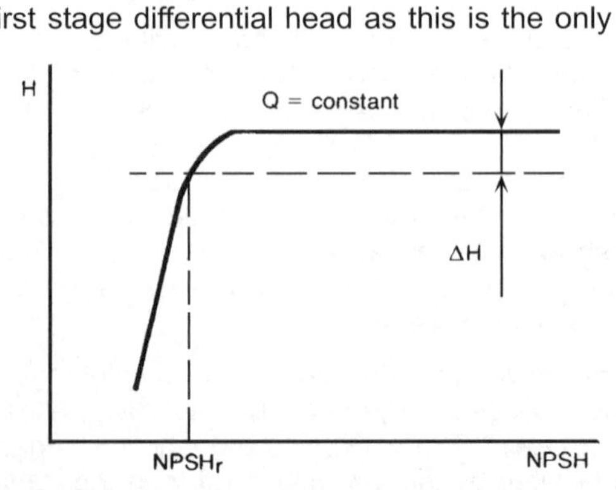

Figure 4.57 Test determination of required NPSH, NPSHr

which is cavitating. How strictly this requirement needs to be enforced depends upon the size of the pump and how much margin exists between NPSHa and NPSHr.

As mentioned earlier, other NPSHr definitions exist. Cavitation causes damage due to the implosion of the vapour bubbles resulting in a loss of performance and increased power consumption. The concept of damage limitation was studied and a 40000 hour NPSHr criterion has evolved. If effect, the pump manufacturer guarantees that the pump will suffer no loss of performance or require parts replacement due to cavitation damage for 40000 hours of operation. This value of NPSHr is somewhat greater than the 3% criterion and tends to be applied to large pumps and pumps built of costly materials. Obviously, the 40000 hour criterion can not be applied to pumps which will suffer extensive corrosion damage or be eroded by abrasive solids.

Acoustic measuring techniques have advanced during the past few years and the general opinion is that cavitation produces a distinctive noise at about 400 kHz to 500 kHz. This is obviously outside the range of human hearing and detectors are required. Devices are available which include detectors and suitable circuits to initiate alarms or trips. It is possible in the future that a definition will be based on the strength of the cavitation signal.

NPSHr is a function of pump speed. The Affinity Laws can be used for small speed changes. The effect of impeller diameter changes should be referred to the manufacturer.

4.4.3 Permissible suction lift

The permissible suction lift of a pump is controlled by:

- Liquid vapour pressure
- Liquid density
- Pressure on the liquid surface
- Velocity in the pipe
- Friction losses in the pipe
- NPSHr of the pump

Equations 4.38 and 4.39 indicate the mathematical relationship of the variables. P_{min} must never fall below p_v if cavitation is to be avoided. Equation 4.37 can be rewritten in terms of p_v and NPSHr and rearranged to evaluate the limiting value of h_s:

$$h_s = \frac{p_a}{r} - gh_{fs} - \frac{c_s^2}{2} - \frac{p_v}{r} - \text{NPSHr} \qquad \text{Equ 4.40}$$

where p_a and p_v are pressures and NPSHr is in metres.

If the pump is drawing liquid from a vessel which is open to the atmosphere then the maximum suction lift will vary with atmospheric pressure. Atmospheric pressure varies, for example, due to the height above sea level. This is illustrated in Table 4.4. The height of the barometer at sea level is assumed to be 760 mm Hg. Atmospheric pressure also varies according to climatic conditions. At sea level the atmospheric pressure does not normally fall below 720 mm Hg, 960 mbar, 9.8 m H_2O.

Height above sea-level m	Atmospheric pressure Pascal Pa	Millibar mbar	Torr mm Hg	Water column m H_2O
0	$1.013 \cdot 10^5$	1013	760	10.3
1000	$0.899 \cdot 10^5$	899	674	9.2
2000	$0.795 \cdot 10^5$	795	596	8.1
3000	$0.695 \cdot 10^5$	695	521	7.1

Table 4.4 The variations of atmospheric pressure due to height above sea level (760 mm Hg at sea level)

Table 4.4 The variations of atmospheric pressure due to height above sea level (760 mm Hg at sea level)

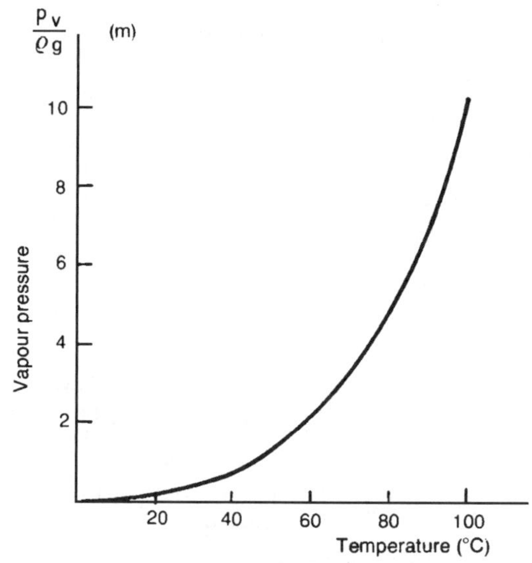

Figure 4.58 Vapour pressure for water

The vapour pressure of a liquid is dependent upon temperature, and Figure 4.58 shows the vapour pressure for water at various temperatures. When pumping warm water or hydrocarbons for example p_v can be large. When pumping boiling liquids the pump must be placed below the liquid free surface so that static head will be greater than the losses in the pipework and the NPSHr. Vapour pressures for a wide selection of liquids appear in Chapter 2, Section 2.6.2.

It should be noted that neither ISO or CEN have approved steam tables for the properties of water. Care must be taken to ensure all interested parties work to the same figures.

The losses in the suction pipe should be kept as low as is practically possible, since they have the effect of reducing the permissible suction lift for cavitation-free operation. The suction line should be of large diameter, short and without unnecessary bends, valves etc. Suction pipe losses and NPSHr vary with flow. It is therefore important to know the minimum and maximum flow at which the pump is to operate and check for adequate NPSHa.

Cavitation and NPSH have so far been explained in the context of rotodynamic pumps. Positive displacement pumps have similar phenomena and criteria although not identical.

Cavitation in positive displacement pumps causes a reduction in flow rather than head or pressure. Any vapour bubbles which form fill part of the displaced volume and reduce the liquid capacity. The bubbles collapse when the pressure increases and flow is reduced by the vapour bubble volume. The imploding bubbles cause material damage as in rotodynamic pumps. Additionally, in pump types such as reciprocating, severe shock waves are produced which can cause fatigue failure in components and bearing failure.

The concept of NPSH becomes NPIP, Net Positive Inlet Pressure, again in two forms, NPIPa and NPIPr. Pressure is used rather than head because all the other pump parameters are expressed as pressure. In positive displacement pump types which are used for process applications handling a wide range of liquids, NPIP is considered to be an equivalent to NPSH; NPIP is quoted in a completely analogous manner. However, some positive displacement pump types are used mostly on one type of liquid and a different approach has become accepted.

Gear pumps and screw pumps are used mostly on lubricating oil. Lube oil has a very low vapour pressure and NPIP testing is difficult due to the high vacuums which must be achieved and attendant air leaks. Also, air and gases evolving from solution tend to mask the true NPIPr value. It has become the practice

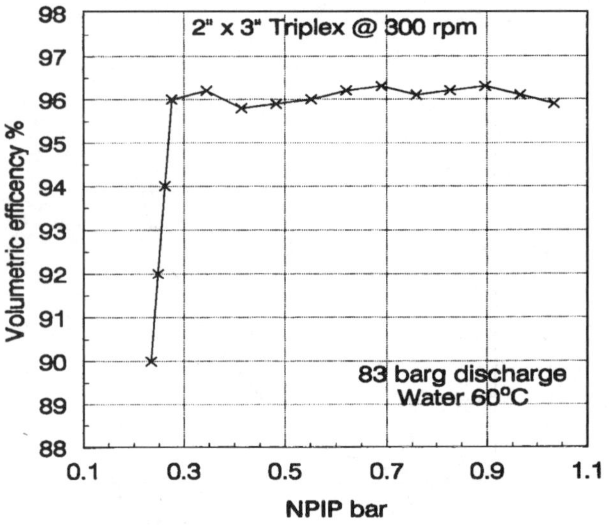

Figure 4.59 Plunger pump NPIP test result

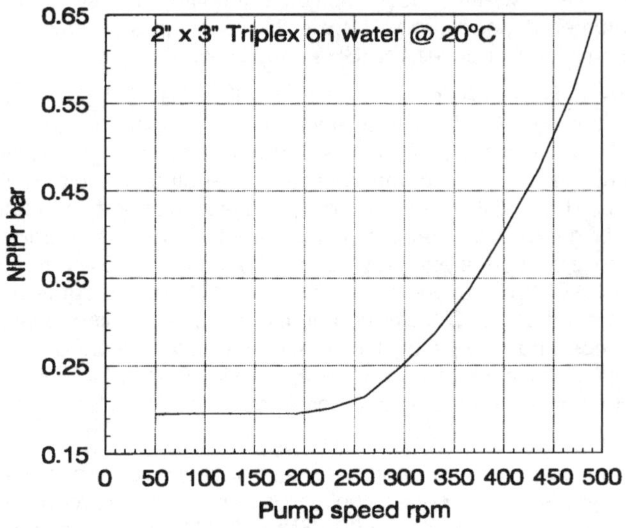

Figure 4.60 Plunger pump NPIP test result

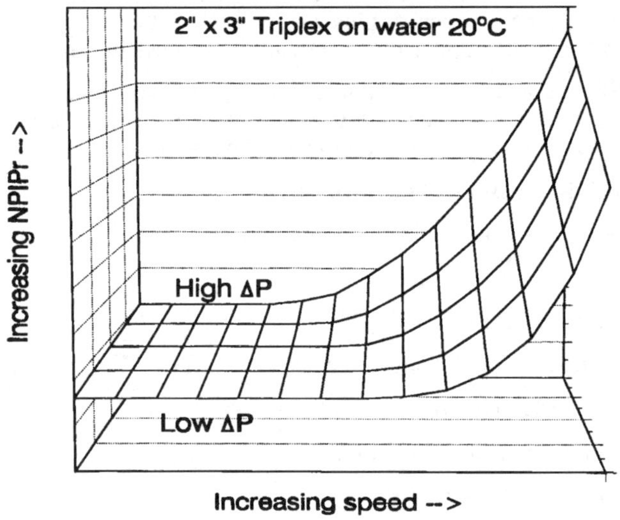

Figure 4.61 NPIPr surface

for these pump manufacturers to specify a minimum suction pressure. Unfortunately some still call this NPIPr and quote values in bara. The picture is further complicated by some positive displacement pump manufacturers who do not fully understand NPSH or NPIP and quote values of NPIPr in bara. It is worthwhile double checking with manufacturers to be sure exactly what figure is being quoted.

Plunger pumps are used extensively for process applications and data is readily available. Pumps are tested at constant discharge pressure while the suction pressure is gradually reduced. When cavitation is initiated, the flow per revolution and the volumetric efficiency are reduced. It is possible that the pump flow may increase slightly due to a very slight increase in speed caused by reduced power consumption. Pumps driven by fixed frequency squirrel-cage motors will run at slightly different speeds depending upon the absorbed power. The flow per rev or volumetric efficiency can be plotted against NPIP, see Figure 4.59.

Reciprocating pumps with plate or plug valves have very distinctive NPIPr characteristics as can be seen from Figure 4.59. Manufacturers of API 674 and 676, reciprocating and rotary positive displacement pumps advise that NPIPr should be specified at 3% flow loss. This is obviously a direct translation of the rotodynamic pump criterion of 3% head loss. 3% flow loss can not be applied universally to reciprocating pumps. Most reciprocating pumps are built with plate valves. The NPIPr characteristic is very clear. NPIPr values should be quoted for 0% flow loss.

Cavitation in reciprocating pumps can produce catastrophic failures due to fatigue. Pumps should not be operated routinely while cavitating. The effects of cavitation on rotary positive displacement pumps should be discussed in detail with the manufacturer, on a case-by-case basis, when deciding on the operating conditions and the purchasing specification.

Reciprocating pump NPIPr characteristics are plotted for constant differential pressure against speed. Some curves may show more than one viscosity or more than one plunger diameter. Increasing viscosity increases NPIPr although the relationship is not easily defined and varies considerably from pump to pump. Increasing viscosity has a more pronounced effect on small pumps than large pumps. Viscosity can also affect the valve dynamics resulting in reduced speed operation. The NPIPr curve does not approach zero at zero speed, see Figure 4.60. A minimum value, depending upon the pump and valve design, is required to avoid cavitation. It may be possible for the NPIPr value to increase at very low speeds, similar to the effect in centrifugal pumps. No test results have been reviewed which show this effect but the possibility should not be ignored if pumps are required to operate below 10% speed.

As explained in Section 4.2.1.5, and shown in Figure 4.15, NPSH can be considered as a 3D surface. The same concept can be applied to reciprocating pumps showing the relationship between speed, differential pressure and NPIPr, see Figure 4.61. The surface shown can be considered as the relationship for a specific viscosity. A higher viscosity would produce another surface, not necessarily of the same shape, above the surface shown. The effect indicated where NPIPr reduces as differential pressure reduces can be completely negated by poor suction valve design.

Rotary positive displacement pumps tend to have a linear NPIPr relationship with speed. The slope of the curve increases as viscosity increases. Usually two or three viscosity curves are shown on one graph to allow interpolation for the required value.

4.4.4 Cavitation effects on pump operation

Cavitation in rotodynamic pumps, if allowed to continue for prolonged periods of time, can cause mechanical damage to the pump. Surface material is removed by the implosion of the vapour bubbles and characteristic damage is visible when inspected. Clearances can be increased causing permanent reduction in performance. Mechanical seals are prone to physical damage. The hydraulic performance is also reduced by the onset of cavitation. The performance is affected in different ways

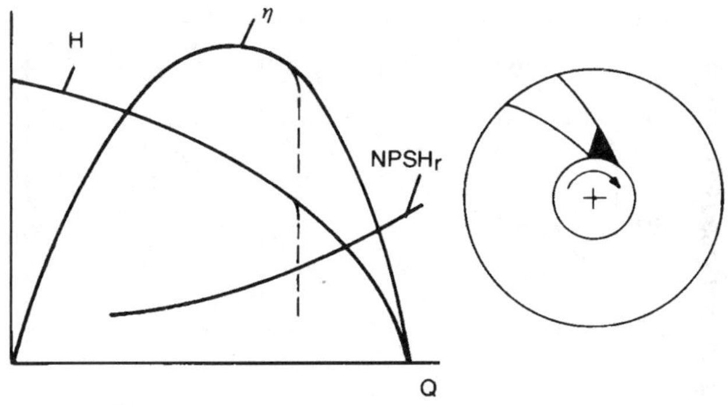

Figure 4.62 Cavitation in a centrifugal pump

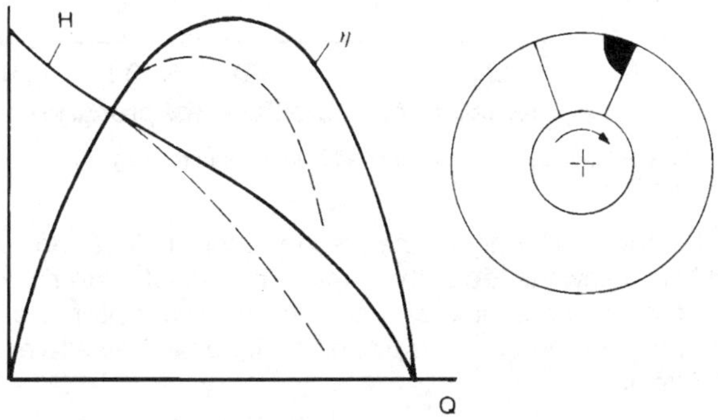

Figure 4.63 Cavitation in an axial flow pump of N_s = 200

depending upon the type of pump. In general, there is an accompanying increase in noise and vibration.

The full lines shown in Figure 4.62 represent the pump's differential head and efficiency for cavitation-free operation. The pump's cavitation sensitivity in the form of NPSHr is also shown. The pump has a low specific speed, N_s = 20. The pump is intended for a system having a certain static suction head. When the flow increases, the NPSHr also increases, while p_{min} decreases. When the positive suction static head is reduced eventually the limits of cavitation-free operation are reached and the pump performance declines rapidly (dashed lines). The rapid decline in performance is associated with the narrow vane passages which are characteristic of pumps having low specific speed. The complete length of the front edges of the vanes are at approximately the same diameter and are subjected to the same velocity. With the onset of cavitation it requires only a slight increase in flow to completely fill the vane passages with vapour. The pump's delivery head can thus be reduced to zero.

The relationship for axial flow pumps is different. Here cavitation occurs on the suction side of the profiles at the tips of the blades where the relative velocity, Δp is greatest. Despite cavitation at the tips of the blades there still remains a large usable amount of blade passage, Figure 4.63. The decline in performance is therefore a more gradual process in the case of axial pumps.

The onset of cavitation in positive displacement pumps causes a reduction in flow rate since part of the displacement becomes filled with vapour. When the vapour bubbles collapse, severe pressure waves can be created which result in fatigue failures. In reciprocating pumps, classic cavitation damage includes:

- Cylinder cracks
- Broken bolts and studs
- Broken valve plates and springs
- High bearing wear

Externally, signs of cavitation include a characteristic “pinking” noise, increased pump vibration and increased pipework vibration due to increased levels of pressure pulsations in both the suction and discharge.

For reciprocating pumps, and some rotary pumps with only a few lobes or teeth, an additional loss, to those already explained, is experienced in the suction pipework. The flow variation caused by the small number of pumping elements requires some additional head to be available to accelerate the flow to keep up with the changing velocity of the pumping elements. This phenomenon has been studied by the Hydraulic Institute in the USA and the German Engineering Federation (VDMA). The extra head required is a function of the:

- Pump type
- Number of pumping elements
- Pump speed
- Velocity in the suction pipe
- Length of the suction pipe
- Nature of the liquid

This extra loss is called Acceleration Head Loss and can be approximated for water, amine, glycol, from the following equation.

$$H_{acc} = 4LUnC \qquad \text{Equ 4.41}$$

where:

H_{acc} = Acceleration Head Loss (m)

L = suction pipe length (m)

U = average velocity in pipe (m/s)

n = pump speed (s^{-1})

C = pump characteristic non-dimensional

The pump characteristic values for single acting reciprocating pumps are:

1 cyl	C = 0.4
2 cyl	C = 0.20
3 cyl	C = 0.07
4 cyl	C = 0.11
5 cyl	C = 0.04
6 cyl	C = 0.06
7 cyl	C = 0.03
8 cyl	C = 0.045
9 cyl	C = 0.025

For other liquids multiply the water value by:

- urea, carbamate, liquids with traces of entrained gases 1.5
- de-aerated water 1.1
- hydrocarbons 0.75
- hot oil 0.6

The approximation given in equation 4.41 must be used with caution. It is, of necessity, an extremely simplistic evaluation of a very complex phenomenon. It was proposed for use on relatively short pipe runs, 15 to 20 diameters. On long runs it tends to be pessimistic, on short runs optimistic. The formula tends to zero as the variables reduce. In practice, Acceleration Head Loss tends to a finite minimum depending upon the pump detail design. If acceleration pressure loss will cause pipe vibration or reduce NPIPa below NPIPr then pulsation dampers may solve the problem. A correctly selected, installed and maintained pulsation damper can attenuate pressure pulsations by 90%.

Positive displacement pumps should always operate with a safety margin between NPIPa and NPIPr. A fixed value of safety margin would be too simplistic and a calculated value is

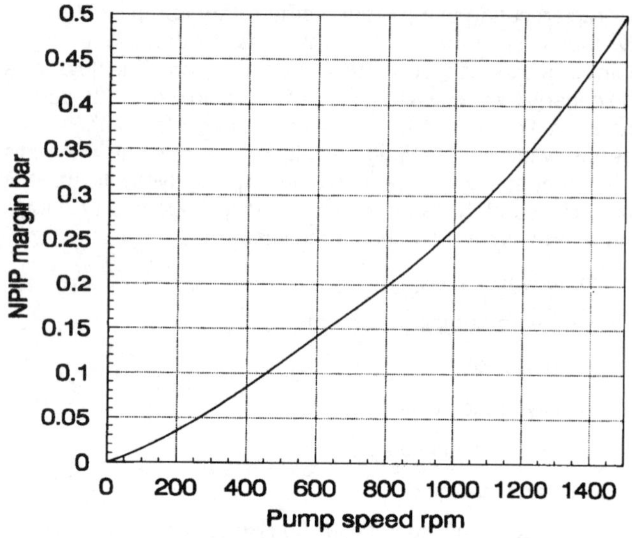

Figure 4.64 Recommended margin of NPIPa over NPIPr for pump speed

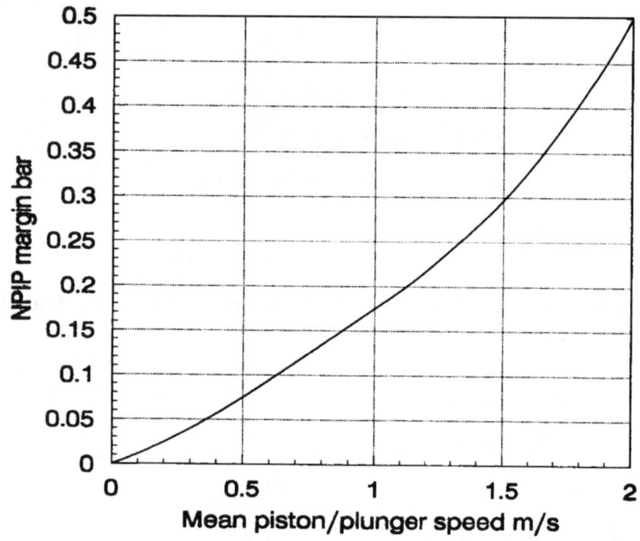

Figure 4.65 Recommended margin of NPIPa over PNPIPr for mean piston/plunger speed

very difficult. Figures 4.64 - 4.66 show guidelines for reciprocating pumps from the British Pump Manufacturers Association.

Values of margins are obtained from both Figures 4.64 and 4.65. A multiplier is obtained from Figure 4.66 depending upon how highly loaded the pump is at the rated conditions and applied to the highest value obtained from Figures 4.64 or 4.65.

4.4.5 Self-priming

A rotodynamic pump operating with air rather than water produces a head rise which is about 1/1000. If the pump is placed above the level of the upper liquid surface and the suction line is filled with air the pump is thus incapable of removing this air by itself when starting. This is referred to as a centrifugal pump being unable to evacuate its own suction line. Before pumping can start it is therefore necessary to fill the pump with liquid, venting and priming. This can be arranged in different ways, see Figure 4.67. Placing the pump below the level of the upper liquid surface, flooded suction, eliminates the problem of evacuation and reduces the risk of cavitation. The NPSH can be achieved for example, by immersing the pump or both the pump and the drive motor, so-called submersible pumps.

Rotodynamic pumps are available in a wide range of designs and configurations. **Not all pumps are self-venting.** Pumps installed as indicated in Figure 4.65 as "dry" may still require venting and priming. Venting and priming after installation or

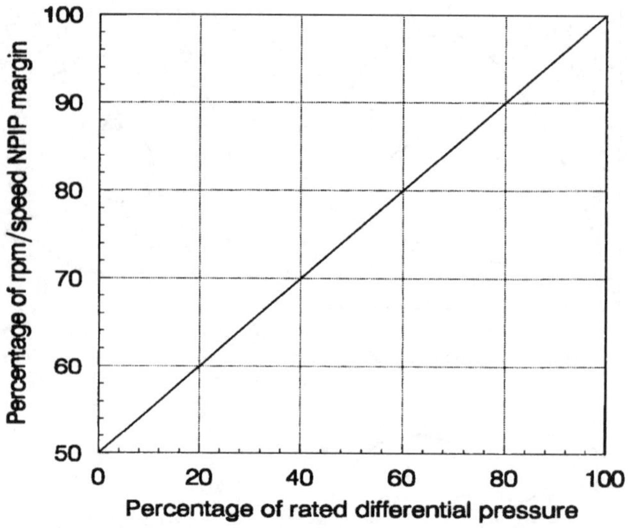

Figure 4.66 Percentage of recommended margin for percentage of rated differential pressure

maintenance, will only be necessary once, unless the liquid level falls below the top of the pump. The manufacturer's operating instructions should be checked to ensure proper pump operation. Submerged pumps are designed to be self-venting and priming.

An alternative method of keeping the pump filled with liquid is shown in Figure 4.68. The non-return (foot) valve prevents liquid draining from the suction line when the pump is stopped. The foot valve is however, always subject to a small amount of leakage and also has the disadvantage that it causes large pressure losses within the suction line when pumping. Provision must be made to fill the suction line initially. The volume of the evacuating tank must be several times greater than the volume of the suction line. It is also necessary to check that cavitation does not occur in the suction line.

The air in the suction line and pump casing can be evacuated with the help of a smaller self-priming pump, Figure 4.69. The priming pump must be connected to the highest point on the pump casing. A wide variety of pump types can be used as evacuating pumps. The most usual are vacuum pumps, side-channel and jet pumps.

All positive displacement pumps are in principle self-priming providing that the inner seals are sufficiently effective and that "dry" running is acceptable during starting. It is therefore advisable to check with the manufacturer with regard to "dry" running in the case of applications having long suction lines. Reciprocating pumps should always be vented and primed prior to starting.

In the case of specially designed self-priming centrifugal pumps the pump is built into a tank containing liquid. The evacuating process is then carried out automatically if the pump is first filled with liquid. The suction line must be dimensioned with regard to both flow losses and evacuation time and must rise continuously towards the pump. Too large a suction line makes evacuation more difficult. A useful rule is to dimension for a suction head loss of 1 to 2 m. If the volume of the suction line above the surface level of the liquid is very large, for example, for suction pipe lengths of several hundred metres, a foot valve must, without exception, be placed below the liquid surface.

4.4.6 Effects of dissolved and entrained gases

The pumped liquid may contain certain quantities of dissolved gas or entrained gas. Lubricating oil, for example, at atmospheric pressure and room temperature can contain up to 10% air by volume, petrol can contain up to 20% air by volume. The ability of a liquid to dissolve gas reduces with reduced pressure and with increased temperature. Gas can therefore evolve in

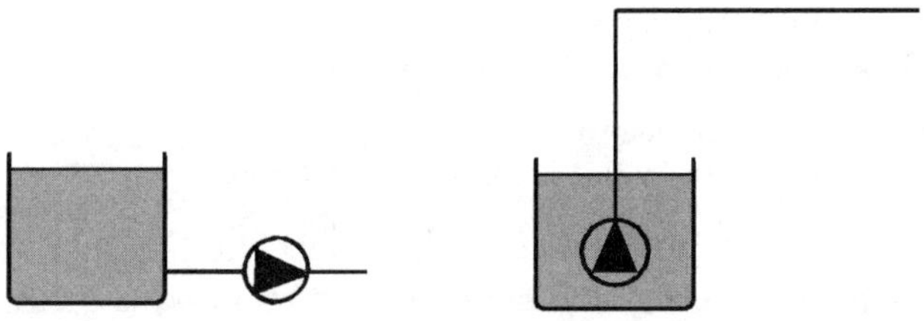

Figure 4.67 Pump installation without suction line evacuation

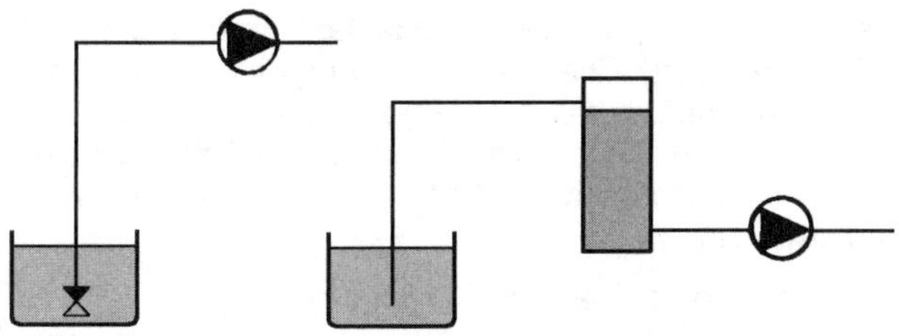

Figure 4.68 Foot valve and evacuating tank

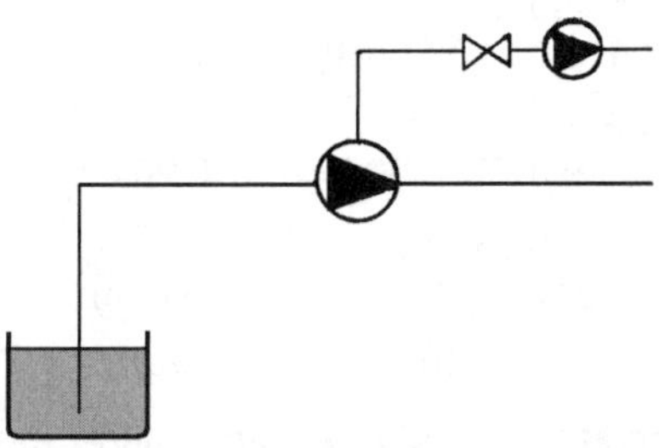

Figure 4.69 Evacuating pump

areas of localised low pressure and re-dissolve when the pressure increases again. This process is similar in some ways to the formation and implosion of vapour bubbles, or cavitation, but is considerably less violent and causes no mechanical damage unless the pressure changes are large, as in positive displacement pumps. When gas separation and cavitation occur simultaneously in rotodynamic pumps the gas has a dampening effect on the implosion process and in this way helps to limit the extent of damage caused by cavitation. Small amounts of air have been introduced into the pump suction on purpose to alleviate certain site problems.

The pumped liquid can also contain entrained gas, usually air. This entrained gas has considerable influence on pump performance. In rotodynamic pumps both the delivery head and the efficiency are noticeably reduced by increased quantities of entrained gas. In positive displacement pumps entrained gas reduces the liquid flow rate and creates pressure pulsations which can cause premature component failure and severe pipe vibration.

As shown in Figure 4.70, the originally stable H-Q curve becomes unstable due to the influence of entrained air. For the system curve shown there are, for a given concentration of entrained air, two alternative intersection points and there is risk of unstable operating conditions.

Certain conditions can cause air to collect in the pump impeller, which can lead to loss of prime. The risk of air collecting is greatest for flows which are less than the design flow. It is normally possible to pump without other difficulties arising, other than reduced efficiency, for air concentrations of up to 2 to 4% by volume measured at the pump intake connection.

Positive displacement pumps usually cope with larger concentrations of air better than centrifugal pumps. For positive displacement pumps operating with a suction lift, the theoretical volume flow is reduced considerably, up to 50%, if the liquid

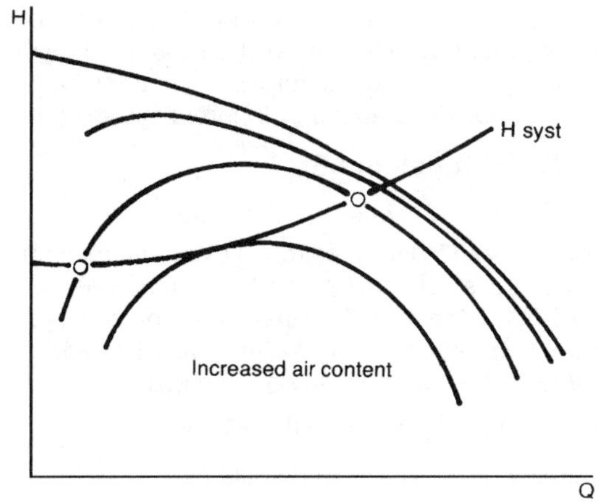

Figure 4.70 The influence of entrained air in water on a centrifugal pump's H-Q curve

contains entrained or dissolved gases. Entrained gas will expand and dissolved gas will evolve and partially fill the displacement volume. Severe pressure pulsations, causing vibration and fatigue damage, may result. Consult pump manufacturers and provide them with full operating conditions, including details of gas contamination, before deciding on a pump type and selection.

Pumps handling cold fresh water from rivers, reservoirs and tanks can suffer damage and short component lives due to entrained and dissolved air. Experience has shown that an allowance must be made for air evolution. The vapour pressure of the water should be increased by 0.24 bar.

Developments in rotodynamic pump technology has resulted in increased gas handling capabilities. Production centrifugal pumps are limited to about 15% by volume before the pump loses prime. Special centrifugal pumps have been able to handle 40% by volume **under controlled test conditions**.

More recent developments in multi-stage axial flow pumps can handle 98% gas volumes for pressures up to 20barg. Special versions of positive displacement pumps have been developed to cope with entrained/dissolved gas continuously. Twin geared-screw pumps can handle 97% gas at pressures up to 40barg. These pumps are also capable of handling small entrained solids, such as sand.

NOTE: It cannot be stressed too often or too strongly that successful pump installations are based on good communications between the user and the pump manufacturer. The pump manufacturer issues guarantees on the basis of operational data provided. If important data is withheld, the guarantee will be invalidated and the manufacturer will not supply replacement parts or service engineers free-of-charge.

The user should not judge which information is important and which is trivial. Forward all information to the pump manufacturer and let the manufacturer decide. This course of action will provide the user with a "water-tight" guarantee. If certain aspects of operation are impossible to quantify, discussions with manufacturers will indicate whether the problem is important or insignificant and if there could be an effect on performance or reliability. It is in the best interests of all concerned to discuss points which need clarification. In general, data sheets for pumps are poor. Full operational conditions are not described in any detail. Filling in an API 610 or API 674 data sheet will not assure a "water-tight" guarantee.

The performance and reliability of many modern pump installations is hampered by poor pipework design. Pumps are not designed to cope with any pipework which the piping designer can fit in. Pumps are designed to operate successfully within an appropriate system. Pump purchasers should ensure the system designer is fully experienced with the types of pumps being considered. Pump purchasers do not like buying prototype pumps; but every new pump installation has a prototype piping system. The pump manufacturer will not accept the responsibility for pump failures caused by inherent system defects.

4.4.7 Calculation examples

Example 1

A pump is required to lift water from a river. The minimum barometric pressure is 735 mm Hg and the maximum water temperature is 10 °C. At the rated flow the suction pipe losses, including velocity head and entrance losses, are 1.5 m and the pump is 2 m. What is the maximum suction lift possible?

Water vapour pressure = 0.01227 bara

Water density = 999.8 kg/m^3

Mercury density = 13600 kg/m^3

Atmospheric pressure ≡
735 x 13600 / 999.8 / 1000 = 9.997 m

Water vapour pressure ≡
0.01227 x 100000 / 999.8 / 9.80665 = 0.125 m

Allowance for air evolution ≡
0.24 x 100000 / 999.8 / 9.80665 = 2.448 m

NPSHa = 9.997 - 1.5 - 2 - 0.125 - 2.448 = 3.924 m

Therefore, the maximum suction lift is 3.924 m without any safety margin between NPSHa and NPSHr. If a self-priming pump was used, then the margin for air evolution would not be required and the maximum suction lift would increase to 6.372m.

The density used for water is compressed water data, not saturated liquid data. The water is not at saturated pressure therefore it is compressed.

Example 2

Water at a maximum temperature of 60 °C is to be pumped from an open tank. The minimum barometric pressure is 735 mm Hg and the pump suction will be 3 m below the minimum liquid surface level. If the suction pipe losses, including velocity head and entrance losses, are 4 m at the rated flow what is the NPSHa for the pump?

Water vapour pressure = 0.1992 bara

Water density = 983.2 kg/m^3

Mercury density = 13600 kg/m^3

Atmospheric pressure ≡
735 x 13600 / 983.2 / 1000 = 10.167 m

Water vapour pressure ≡
0.1992 x 100000 / 983.2 / 9.80665 = 2.065 m

NPSHa = 10.167 + 3 - 4 - 2.065 = 7.102 m

Notice that the head equivalent of atmospheric pressure has changed even though the barometric pressure is the same. The water density is reduced because of the increased temperature. Allowance for gas evolution is not necessary because the temperature is high enough to drive off any gas.

Example 3

Water at a temperature of 100 °C is to be pumped from an open tank. The minimum barometric pressure is 735 mm Hg and the suction pipe losses, including velocity head and entrance losses, are 4 m at the rated flow. If the pump has an NPSHr of 5 m at the rated flow, how much static suction head is required to prevent cavitation?

Water vapour pressure = 1.01325 bara

Water density = 958.1 kg/m^3

Mercury density = 13600 kg/m^3

Atmospheric pressure ≡
735 x 13600 / 958.1 / 1000 = 10.433 m

Water vapour pressure ≡
1.01325 x 100000 / 958.1 / 9.80665 = 10.784 m

NPSHa = 10.433 - 4 - 5 - 10.784 = -9.351 m

So the minimum static suction head required is 9.351 m without any safety margin. The head equivalent of atmospheric pressure is increased due to the reduced water density.

Example 4

Water is to be pumped from a closed tank which is pressurised to 4 barg 0.1 barg. The minimum barometric pressure is 735 mm Hg. The maximum water temperature is 150 °C. At rated flow the suction pipe losses are 3 m and the NPSHr of the pump is 3 m, how much positive static head is required to prevent the pump cavitating?

Water vapour pressure = 4.76 bara

Water density = 916.7 kg/m^3

Mercury density = 13600 kg/m^3

Atmospheric pressure ≡
735 x 13600 / 916.7 / 1000 = 10.904 m

Pressure in tank =
(3.99 x 100000 / 916.7 / 9.80665) + 10.904 = 55.287 m

Water vapour pressure ≡
4.76 x 100000 / 916.7 / 9.80665 = 52.949 m

NPSHa = 55.287 - 52.949 - 3 - 3 = -3.662 m

The minimum static head required is 3.662 m without any margin.

Example 5

A three cylinder single acting pump, running at 600 rpm, is to be supplied with water at 60 °C from an open tank. The minimum barometric pressure is 735 mm Hg. The minimum water level above the pump suction is 5 m. The suction line velocity is 0.75 m/s and the line length is 10 m. The suction line losses total 2 m and the pump NPIPr is 0.7 bar. Will the system operate successfully ?

Water vapour pressure = 0.1992 bara

Water density = 983.2 kg/m^3

Mercury density = 13600 kg/m^3

Atmospheric pressure ≡
735 x 13600 x 9.80665 / 1000 / 100000 = 0.980 bara

Static pressure =
5 x 983.2 x 9.80665 / 100000 = 0.482 bar

Friction pressure drop =
2 x 983.2 x 9.80665 / 100000 = 0.193 bar

H_{acc} = 4 x 10 x 0.75 x 10 x 0.07 = 21

H_{acc} ≡ 21 x 983.2 x 9.80665 / 100000 = 2.025 bar

NPIPa = 0.98 + 0.482 - 0.1992 - 0.193 - 2.025 = - 0.955 bar

The system will not work as originally described. The acceleration head loss is too great. A larger suction pipe could be used; this would reduce the friction loss and the acceleration head loss. This would be the best solution for really critical installations. Or a pulsation damper could be fitted; the acceleration head loss could be reduced to 0.2025 bar which would provide an NPIPa/NPIPr margin of 0.867 bar. Figure 4.64 recommends a margin of about 0.15 bar for pumps at 600 rpm. So 90% attenuation of H_{acc} is not necessary for NPIPr purposes. Check the effect of pressure pulsations on pipework vibration before deciding on attenuation required.

4.5 Useful references

American National Standard for Centrifugal Pumps, Std. No. ANSI/HI 1.1-1.6.

The relationship and effect of both specific speed parameters, Paper presented by R. Palgrave, to the Institution of Mechanical Engineers, September 1984, at the conference on Deaerators and Boiler Feed Pumps.

PD ISO/TR 17766:2005 Performance corrections for all worldwide designs of centrifugal pumps of conventional design, in the normal operating range, with open or closed impellers, single or double suction, pumping Newtonian fluids.

Using Centrifugal Pumps for High Viscosity Liquids, John D. Constance, Plant Engineering, September 16, 1976, pp 163-166.

Investigation of the Performance of Centrifugal Pumps When Pumping Oils, Robert L. Daugherty, Bulletin 130, Goulds Pumps Inc, Seneca Falls, N.Y., 1925.

A Further Investigation of The Performance of Centrifugal Pumps When Pumping Oils, Robert L. Daugherty, Bulletin 130, Goulds Pumps Inc, Seneca Falls, N.Y., 1926.

Effect of Viscosity on the Hydraulic Performance of a 2x1LF-10 Centrifugal Pump, R.B. Erickson, Duriron Lab. and DuPont Jackson Lab. Development Report, 5 May 1995.

Flowserve (formerly Durco) Pump Engineering Manual, Dayton, 1980, pp 100-103.

Pumping Highly Viscous Fluids with Centrifugal Pumps, J.F. Gülich, World Pumps, 1999, No 8 & 9.

Kreiselpumpen. Ein Handbuch für Entwicklung, Anlagenplanug und Betrieb, J.F. Gülich, Springer, ISBN 3-540-56987-1, Berlin 1999, pp 70-72, 107, 538-550.

Prediction of Viscosity Effects in Centrifugal Pumps by Consideration of Individual Losses, C.P. Hamkins, H.O. Jeske, P.H. Hergt, (from a lecture at the Third European Congress Fluid Machinery for the Oil, Petrochemical, and Related Industries, The Hague, Netherlands, 18-20 May 1987).

Verlustanalyse an einer Kreiselpumpe auf der Basis von Messungen bei hoher Viskosität der Fördermediums, P. Hergt, B. Stoffel and H. Lauer, VDI Report No. 424, 1981, pp 29-38.

Pumping of Liquids, F.A. Holland and F.S. Chapman, Reinhold, N.Y. 1966, pp 249-256.

The Influence of Viscosity on Centrifugal Pump Performance, Arthur T. Ippen, ASME Paper No A-45-57, (Annual Meeting of the American Society of Mechanical Engineers, New York, N.Y. 27 November 1945).

Aporte al Estudio Sobre la Influencia de la Viscosidad, en la Caracteristica de Bombas Centrifugas, O.J. Maggiolo Campos, Boletin de la Facultad de Ingenieria de Montevideo, Año XVI, Vol VI, No 4, October 1952, pp 487-510.

Influence of the Viscosity of the Liquid to be Handled on the Operating Reaction of Centrifugal Pumps with Different Speeds, G Mollenkoff, (in German) Pumpentagung, Karlsruhe '78, 28 September 1978, Section K10.

Influence de la viscosité et des jeux sur le fonctionnement d'une pompe centrifuge, R. Ouziaux, Student thesis, C.N.A.M., France, 12 December 1969, pp 80-86.

Evaluation of Performance Correction Factors for High Power Centrifugal Pipeline Pumps for Higher Oil Viscosity, S.V. Saxena, J. Kuhlman and H. Renger, (in German), Fachgemeinschaft Pumpen im VDMA, Pumpentagung, Karlsruhe, 30 September - 2 October, 1969, Section C7.

Centrifugal and Azial Flow Pumps Theory, Design and Application, J.J. Stepanoff, John Wiley, N.Y. 1948, pp 310-318.

Centrifugal Pump Operation on Viscous Liquids, D.Y. Sukhanov, (in russian) MASHGIZ, Moscow, 1952.

Performance of Centrifugal Pumps at Low Reynolds Number, (1st Report, Experimental Study), K. Tanaka and H. Ohashi, (in Japanese), Transactions of JSME Ed. 50 No 449, Doc No 83-007, January 1984, pp 279-285.

Optimum Design of Centrifugal Pumps for Highly Viscous Liquids, K. Tanaka and H. Ohashi, Proceedings of the 13th AIHH Symposium at Montreal, Canada 1986, 9 No 35.

Equations Correct Centrifugal Pump Curves for Viscosity, Z. Tunzo, G. Takacs and J. Zsuga, Oil & Gas Journal, 29 May 2000, pp 57-61.

Umrechnung der Kennlinien von Spiralhäusepumpen bei Betrieb mit zähen Flüssigkeiten, KSB Worksheet, No 38.1, 15 April 1983.

The "Sudden-Rising Head" Effect in Centrifugal Oil Pumps, Li Wen-Guang, World Pumps, 2000, No 10.

American National Standard for Centrifugal Pumps for Design and Application (ANSI/HI 1.1-1.2-2000, 1.3-2000).

Resistance of Flow through an Annulus with an Inner Rotating Cylinder, Y. Yamada, Bulletin JSME, Vol 5, No 17, 1962, pp 302-310.

Roughness on Frictional Resistance of Enclosed Rotating Disc, J.W. Daily, R.E. Nece, Transactions of ASME, Journal of Basic Engineering, 1960, No 82, pp 553-560.

Torque Resistance of a Flow Rotating Co-axial Cylinders Having Axial Flow, Y. Yamada, Bulletin JSME, Vol 5, No 20, 1962, pp 634-641.

Standard Practice for Conversion of KinematicViscosity to Saybolt Universal Viscosity of to Saybolt Furol Viscosity, ASTM Designation D2101-93 (Reapproved 1999) with editorial corrections in August 2000.

How Centrifugals Perform When Pumping Viscous Oils, A.J. Stepanoff, Power, June 1949.

Reynolds Number in the Design of Centrifugal Pumps for Viscous Liquids, R.J. MacMeekin, Ingersoll-Rand Co Internal report, September 1949.

Pumps and piping systems 5

5.1 System curves

The relationship between fixed speed pumps and piping systems is covered by the following sections. Variable speed pumps produce an infinite number of pump characteristics, which can be derived using the affinity laws, and make it possible for a pump to operate virtually anywhere on any system curve.

5.1.1 Pump nominal duty point

When connected in a pipe system a pump will operate at a point of equilibrium between the pump and the pipe system. At the point of equilibrium the energy supplied by the pump is equal to the losses due to resistance in the system. The pump characteristics in this respect are usually represented by the H-Q curve. The equivalent curve for the pipe system is called the system curve and is designated H_{syst}. H_{syst} must include suction **and** discharge losses. Remember both curves indicate differential head.

The flow at which the pump and system curves intersect each other, is the flow which will pass through the pipe system. In order to determine correctly the size and dimensions of the pump and pipe system a knowledge of the characteristics of their respective curves is necessary. This section deals with the characteristics of the pipework system.

Figure 5.1 shows typical curves for a centrifugal pump with a low specific speed, when pumping a water-like liquid. The curves vary in shape when pumping other liquids and when using other types of pumps. Generally, however, the flow is always determined by the balance between the pump and system.

5.1.2 Single pipe system

The system differential head is usually divided into a static component H_{stat} and a loss component h_f.

$$H_{syst} = H_{stat} + h_f \qquad \text{Equ 5.1}$$

The static component, which is generally independent of flow, comprises the difference in static pressures and the level differences between the boundaries of the system. Using the designation as in Figure 5.2:

$$H_{stat} = \frac{p_B - p_A}{r\,g} + h \qquad \text{Equ 5.2}$$

where:

p	=	static pressure (N/m²)
ρ	=	density of the liquid (kg/m³)
g	=	acceleration due to gravity (m/s²)
h	=	difference in elevation (m)

The losses include friction losses in straight pipe and losses in valves and fittings etc.

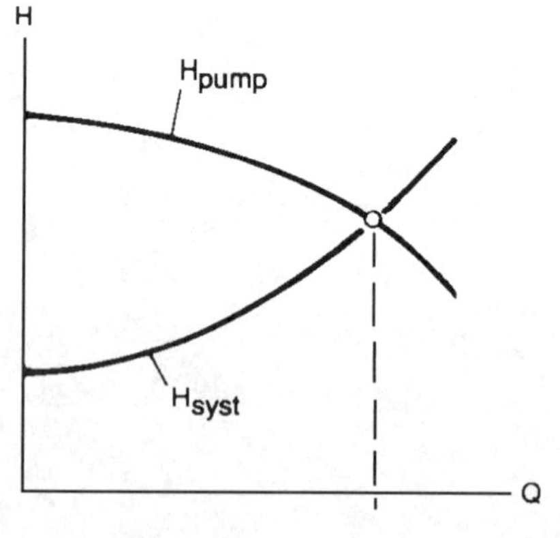

Figure 5.1 Pump nominal duty point

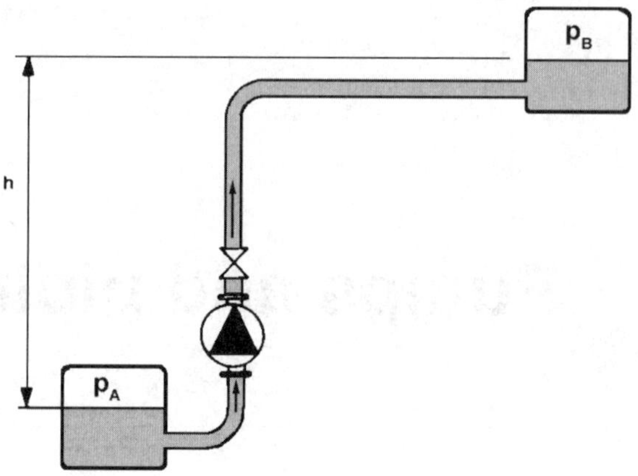

Figure 5.2 Example of single pipe system

$$h_f = h_{(straightpipe)} + h_{(fittingsandvalves)} \qquad \text{Equ 5.3}$$

Using the terms as defined in Chapter 3

$$h_f = \lambda \cdot \frac{l}{d} \cdot \frac{v^2}{2g} + \Sigma\zeta \cdot \frac{v^2}{2g} = (\lambda \cdot \frac{l}{d} + \Sigma\zeta) \cdot \frac{v^2}{2g}$$

$$= (\lambda \cdot \frac{l}{d} + \Sigma\zeta) \cdot \frac{Q^2}{\left(\frac{\pi \cdot d^2}{4}\right)^2} \frac{1}{2g} \qquad \text{Equ 5.4}$$

where:

λ	=	loss coefficient for straight pipe
ζ	=	loss coefficient for valves and fittings
Σζ	=	sum of all loss coefficients
l	=	pipe length (m)
d	=	pipe diameter (m)
Q	=	flow (m³/s)
v	=	flow velocity (m/s)

For a given pipework system of l and d, with water-like liquids the loss coefficients λ and ζ are often independent of Q for large values of Reynolds Number. The loss of head then becomes approximately:

$$h_f = \text{constant} \cdot Q^2 \qquad \text{Equ 5.5}$$

If the pressures at the system boundaries are such that PA = PB = atmospheric pressure, then H_{stat} equals the elevation difference h, the system differential head then becomes:

$$H_{syst} = H_{stat} + h_f = h + \text{constant} \cdot Q^2 \qquad \text{Equ 5.6}$$

or shown graphically in Figure 5.3.

Note that the system curve represented in Figure 5.3 is based on p_A, p_B and h being independent of the flow Q. It is also as-

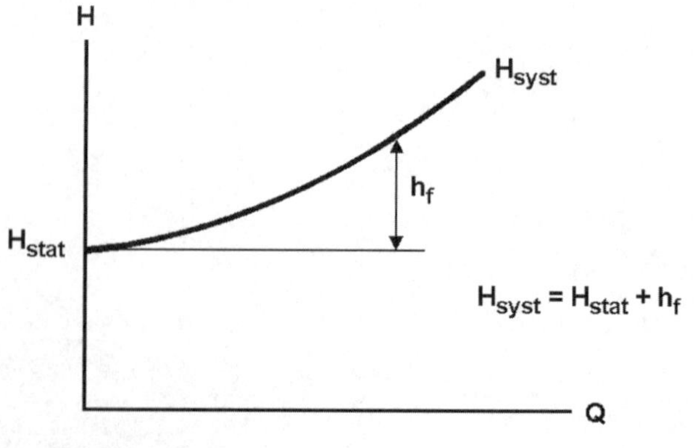

Figure 5.3 System curve

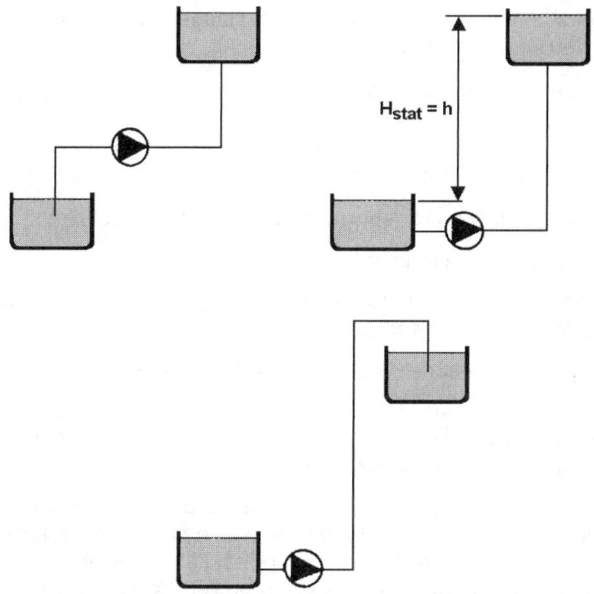

Figure 5.4 Pipework systems with the same static head

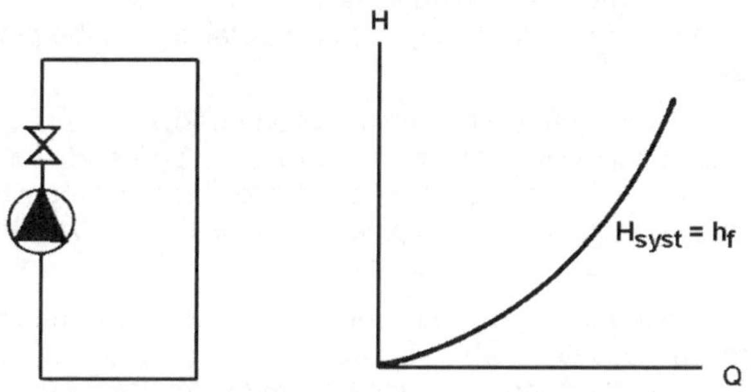

Figure 5.5 Piping system where H_{stat} = 0

sumed that λ and ζ are independent of Reynolds Number. These conditions are often, but not always, fulfilled.

Figure 5.4 shows three different pipework systems having the same level differences and therefore the same differential static head. Notice that, from a practical pumping perspective, the systems are different. The first system involves a suction lift whilst the other two have flooded suctions. The physical location of the pump has a significant impact on system design and pump type selection. Non-return valves, essential to prevent reverse flow, are not indicated. In some cases automatic closing isolating valves may be necessary to prevent undesirable forward flow. For certain piping systems, circulating systems for example, H_{stat} = 0 and the head of the system consists entirely of pipe flow losses.

For other piping systems with short pipes and considerable pressure or level differences the flow losses are negligible and $H_{syst} = H_{stat}$.

Systems similar to Figure 5.5 are called "pipeline" systems because all the losses are friction losses. Systems similar to Figure 5.6 are called "chemical" systems because it is typical of the arrangements found in chemical and petrochemical installations. When developing the system curve the flow losses are calculated according to Chapter 3.

5.1.3 Variable system curves

Certain conditions can cause the system curve to vary according to the operating situation. Some examples of such situations are shown in Figures 5.7 - 5.11.

When starting up, the discharge pipes are filled with air above the liquid level of the lower tank. The pump must lift the liquid to the highest point in the line. The static delivery head reduces when the line is completely filled with liquid. If the pump's differential head at the zero discharge point, called the "shut-off head", does not exceed H_{stat} at the start then pumping cannot begin.

For installations similar to Figure 5.8 the static head varies as the level difference between the liquid surfaces in the vessel changes. Similar system curve changes occur in enclosed vessels, in which the pressure is dependent upon the operational situation.

The system curve's loss component can also vary according to the operating situation. The most usual case being when rust or other deposits begin to build up in a pipeline after a period of operation. This causes an increase of flow resistance in the pipe-

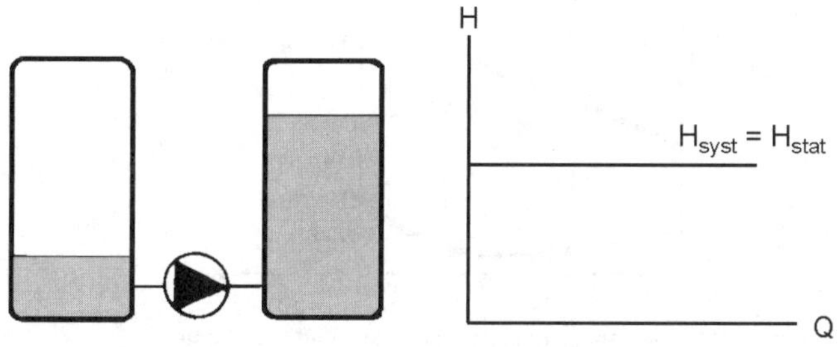

Figure 5.6 Piping system where h_f = 0

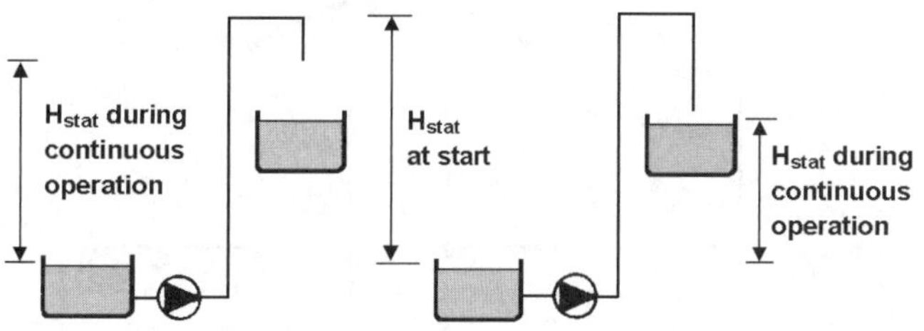

Figure 5.7 Static head at start and during continuous operation

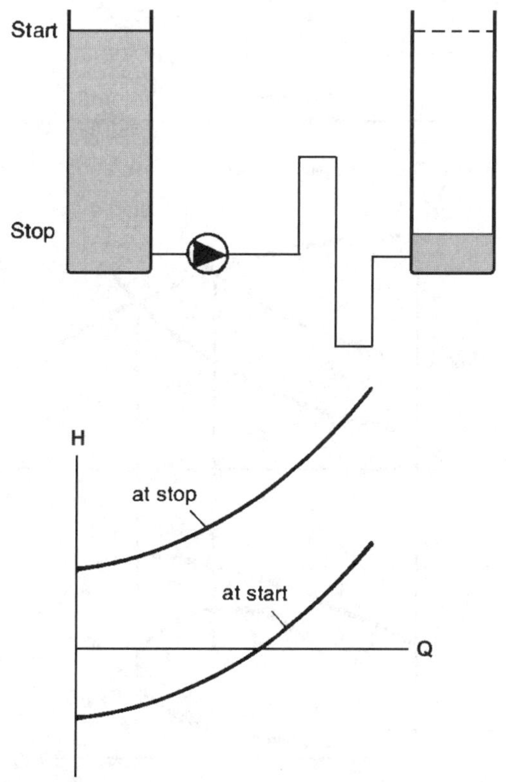

Figure 5.8 System curve with varying static head

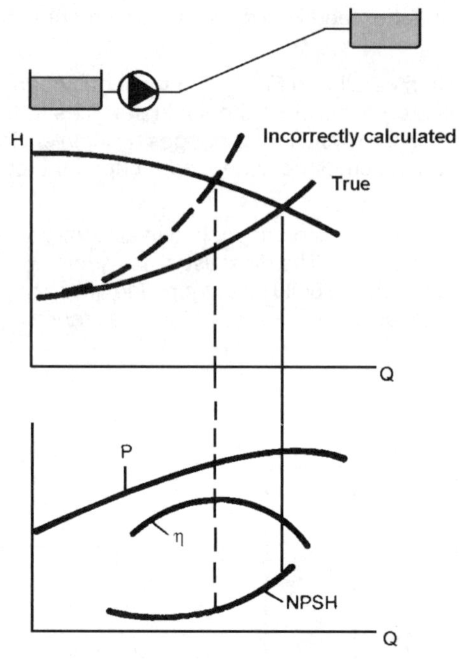

Figure 5.9 Consequences of incorrectly calculated pipe losses

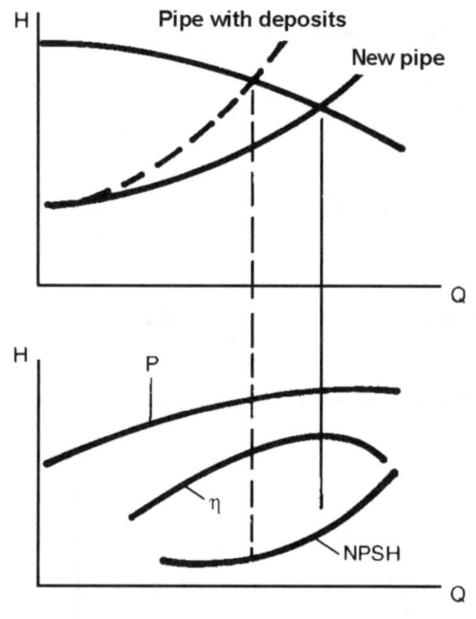

Figure 5.10 The effect of deposits on pipes

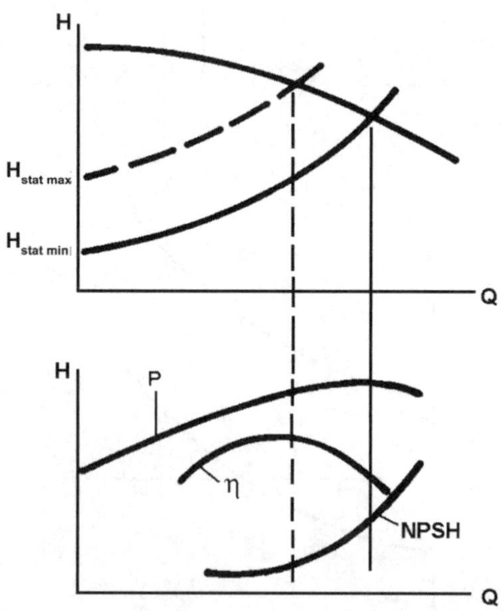

Figure 5.11 The effect of varying static head

line which results in a steeper system curve. Other examples of varying loss curves are:

- Temperature variations
- Viscosity
- Reynolds Number
- Concentration variations in the case of liquid-solid mixtures
- Dosage variations etc.

An intentional change to the system curve is brought about by throttle-regulation, and discussed in Chapter 6.

The pump H-Q curve is also affected by viscosity and wear for example, see Chapter 1.

The relevance of the system curve to operating situations can be illustrated by some examples:

The pump installation shown in Figure 5.8 has been designed to be "on the safe side", the pipe losses having been estimated too high. The actual system curve results in a larger flow than required, which means greater pump power requirement and reduced efficiency, see Figure 5.9. This situation gives rise to the risk of pump overload and cavitation. The desired flow can be maintained by introducing a flow resistance into the piping system.

Figure 5.10 illustrates the reduction of pump flow caused by the build up of deposits (scale, rust, etc.) in the pipeline. Since the size of pump has been chosen in order to obtain optimum efficiency for operation with new pipes, the subsequent flow reduction results in reduced pump efficiency.

The pump in Figure 5.11 has been chosen to suit a particular operational situation, which is equivalent to $H_{stat,\ max}$ i.e. when the liquid levels in the lower and upper vessels are lowest and highest respectively.

In the alternative case $H_{stat,\ min}$ a larger flow is obtained. When plotting pump and system curves, it is not sufficient to plot only the differential head. The NPSHa and the NPSHr should be plotted for the various flows considered. If allowances are added to the system losses, for eventual fouling for example, the operating conditions for the new pipe must be checked if balancing valves are not fitted.

If the plotted characteristics indicate that there is no pump available to suit the system NPSHa, the "system" must be modified. In this context, the system includes some operating conditions. In order to provide more NPSHa the suction pressure must be increased or the difference between the suction pressure and the vapour pressure must be increased. The following courses of action should be considered:

- Re-route the suction pipe to make it shorter
- Move the pump closer to the suction source and reduce the length of the suction pipe
- Lift the suction source up to increase H_{stat} on the suction side
- Increase the suction pipe diameter
- Reduce the liquid temperature to reduce the vapour pressure

5.1.4 Branched pipe systems

The first example of a branched pipe system is a circulation system, see Figure 5.12. In such a case $H_{stat} = 0$. The flow Q_p, which passes through the pump, is divided at the branch point. Continuity dictates that

$$Q_p = Q_A + Q_B \qquad \text{Equ 5.7}$$

Both branches A and B have their own system curves, the sum of which determines the resultant system curve. The intersec-

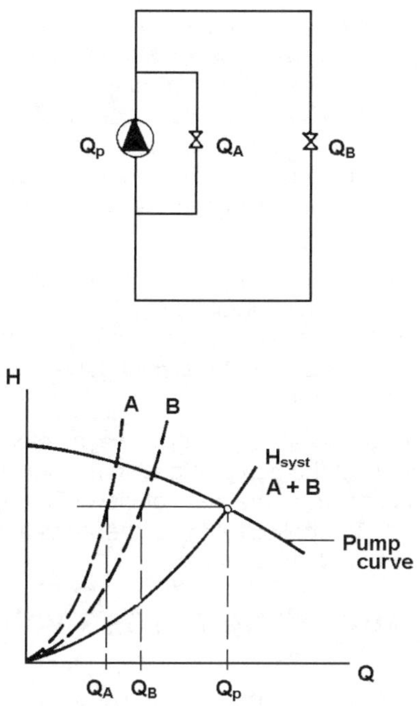

Figure 5.12 Branched circulation system

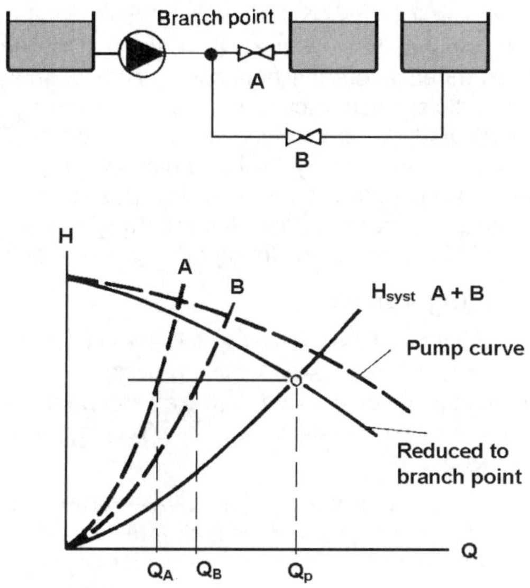

Figure 5.13 Branched pipe system $H_{stat} = 0$

tion point of the resultant system curve and the pump curve determines the pump's operating point.

Figure 5.12 also shows how large a portion of the pump flow flows through each respective branch. The division of flow depends upon the magnitude of the losses in each respective branch. In this example it is assumed that the branch point is near to the pump i.e. that flow losses between the pump and the branch point can be neglected.

The next example, Figure 5.13, also takes into consideration the losses in the main pipeline up to the branch point. This is done by reducing the pump curve by the magnitude of the losses up to the branch point. The "reduced" curve is then matched to the remaining system curve as before.

In the third example as shown in Figure 5.14, the pump curve is first reduced to the branch point. Thereafter the system curves A + B are added to the resultant system curve H_{syst} A + B. The intersection point between the reduced pump curve and the resultant system curve determines the pump's operational point.

In the fourth example as shown in Figure 5.15, there is a level difference, namely a supply head between the supply vessel on the suction side and the branch point. The system curve for the

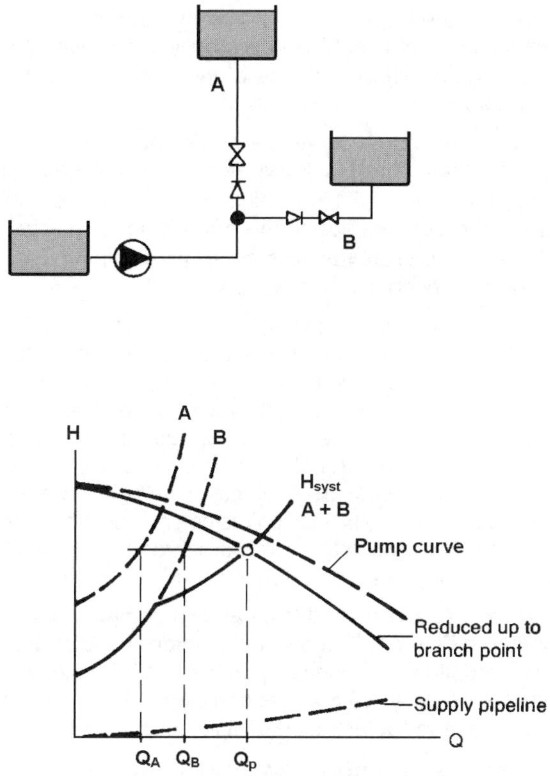

Figure 5.14 Branched pipe system with static differential head

supply pipeline exhibits therefore, a static head, $H_{stat} < 0$. The method is the same as in the previous examples. First reduce the pump curve with the supply line curve up to the branch point. Then determine the system curves for the branches A + B from the branch point. The resultant branch curve is then matched to the pump's reduced curve.

In principle, this method can be applied to calculate any pipework system, however complicated and containing any number of branch points. The first step is to determine the resultant sys-

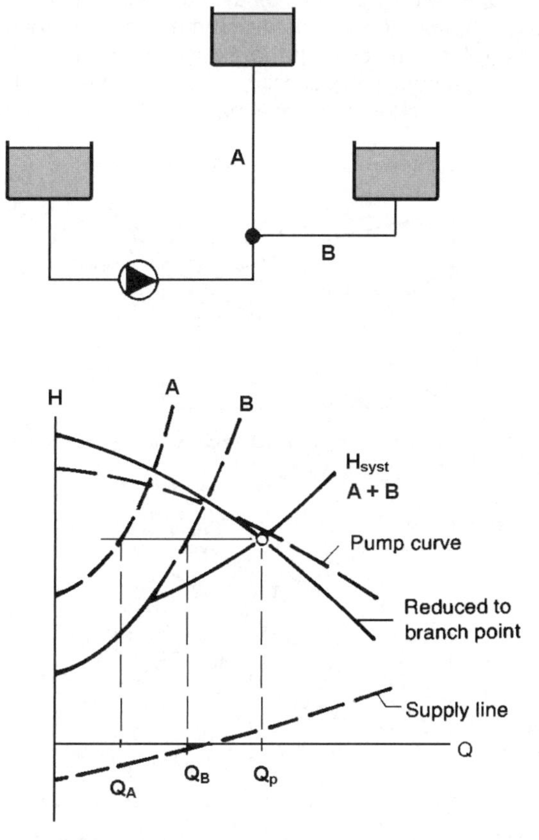

Figure 5.15 Branched pipe system with static supply line

tem curve for the branch furthest from the pump in relation to the furthest branch point. The next step is to deal with the next branch point and so on until the branch point nearest to the pump is reached.

Remember the system curve is a combination of suction and discharge losses. The H for the vertical axis in Figures 5.1, 5.3, 5.5, 5.6, 5.8 and 5.9 to 5.15 is really ΔH, the losses from the beginning of the suction pipe to the end of the discharge pipe. Rotodynamic pump curves indicate ΔH across the suction and discharge connections of the pumps.

A common problem is the sizing of a piping network with many branches and many tapping points in such a way as to guarantee a predetermined flow and pressure at each tapping point. Sizing is carried out so that the pipeline requiring the most pressure, usually the pipeline to the furthest tapping point, is dimensioned first. The branch lines are then calculated on the basis of the excess pressure available at the branch points. The excess pressure at the branch points can be such that it is necessary to equip a branch line with extra throttling to prevent excessive flow and unbalancing the system.

Throttle valves can be fitted to balance a complex system. As was shown in Figure 3.25, some applications have well defined velocity restrictions depending upon the quality of the pipework construction. Also, throttle valves can be used to correct imbalance caused by irregular fouling of different parts of the system.

5.1.5 Viscous and non-Newtonian liquids

In the earlier figures, the pump and system curves have been representative of the relationships for centrifugal pumps and water-like liquids. For viscous and non-Newtonian liquids the curves have different shapes.

For viscous liquids the laminar flow regime is maintained at higher flow velocities. The system curve thus becomes modified. The centrifugal pump H-Q curve falls off more quickly with increased flow for higher viscosities.

High viscosity liquids are normally pumped by positive displacement pumps. Rotodynamic pumps can handle higher viscosity liquids by derating the pump, usually reducing the speed to reduce the torque required. A common bench-mark used to separate rotodynamic pumps, low viscosity, from positive displacement pumps, high viscosity, is 300 cSt. This is not an absolute number; it depends upon the size of the pump and can

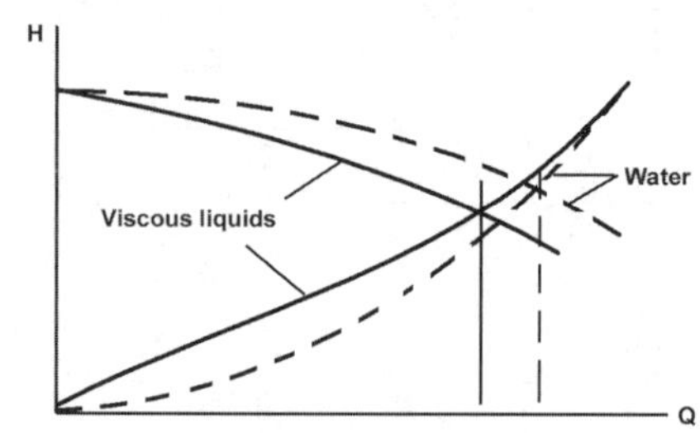

Figure 5.16 Rotodynamic pump and system curves for more viscous liquids compared to water

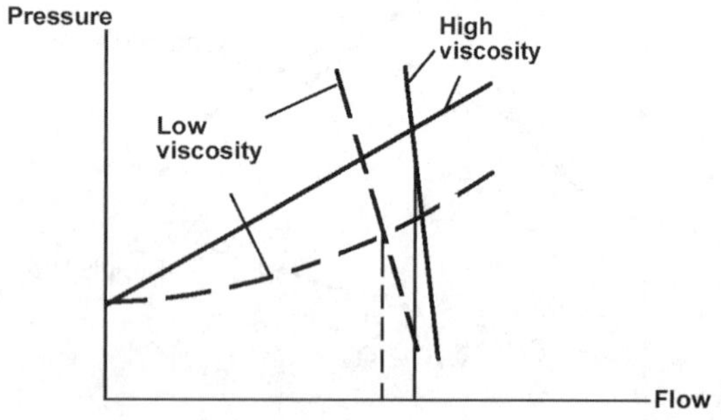

Figure 5.17 Rotary positive displacement pump and system curves for different viscosities

be exceeded when circumstances necessitate using a rotodynamic pump. The high viscosity tends to reduce the internal leakage, slip, of rotary positive displacement pumps making them more efficient. High viscosity causes higher pipe losses. The system curve becomes completely linear if laminar flow is maintained at the duty conditions. The increased flow and increased pressure means greater power requirements. See Figures 5.16 and 5.17.

The pumping of chemical paper pulp is an example of system curves for non-Newtonian liquids, see Figure 5.18. The liquid characteristics of pulp suspensions is reflected in the shape of the system curve. The centrifugal H-Q curve also changes in relation to that for water.

Since the pump's operating point is always determined by the equilibrium between the pump and system, the shape of these curves is significant from the point of view of pump installation performance. This applies also to the sensitivity to disturbances and part-load characteristics.

5.2 Pressure drop across valves

5.2.1 General

For pump installations which use valves as a means of controlling the flow, the pressure drop across the valves may constitute a considerable proportion of the piping system head losses. The system curve which is the basis for the determination of the required differential head for the pump, cannot be defined with sufficient accuracy if the pressure drop across the valves, particularly control valves, is not considered. The following considerations apply to the choice of valves and the pressure drop across the valves for the purpose of determining the required pump performance. A more detailed treatment of flow control is to be found in Chapter 6, Flow regulation.

5.2.2 Isolating valves

Isolating valves are those valves which are intended to be either fully open or fully closed. Isolating or stop valves are installed in an installation in order to make components accessible for service; to direct flow in another direction and for other similar functions.

The key requirements of an isolating valve are low pressure drop in the open position and good leak-free sealing in the shut position. Information regarding the calculation of pressure drop in the fully open position is given in Chapter 3. The most reliable method, however, is to obtain the value of the loss coefficient λ or flow coefficient K_v from the valve manufacturer.

5.2.3 Non-return valves

Non-return or check valves are installed in a pipe system in order to prevent the reverse flow of liquids. There are many different types of non-return valves available, and it is normal to select the one which gives least pressure drop. In some cases it is necessary to have exceptionally fast-closing check valves. These are biased, by means of spring loading for example, and give rise to large pressure drops. These valves can also give rise to tremendous pressure surges due to the water hammer effects they create. Very slow closing valves are available specially designed to eliminate water hammer. Since the effective flow area varies with the flow through the valves, the loss coefficient ζ also varies with flow in non-return valves. Pressure drop calculations are best made by using data supplied by the valve manufacturer.

5.2.4 Control valves

The purpose of control valves, in contrast to the other valve groups, is to create head losses and thereby regulate the flow in the pipework system. The choice of valve is therefore made on the basis of entirely different criteria than for isolating or non-return valves.

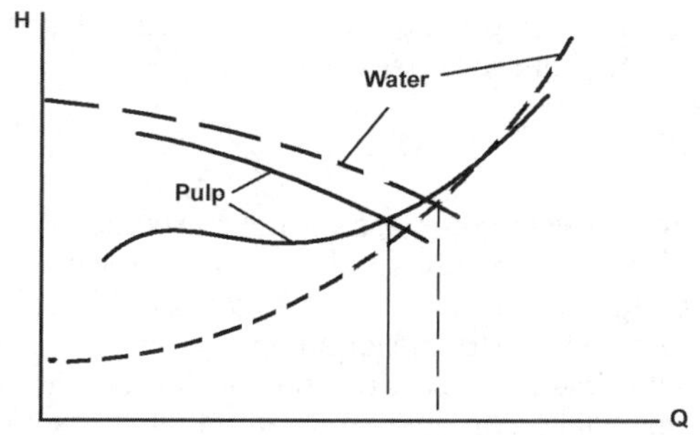

Figure 5.18 Rotodynamic pump and system curves for water and paper pulp

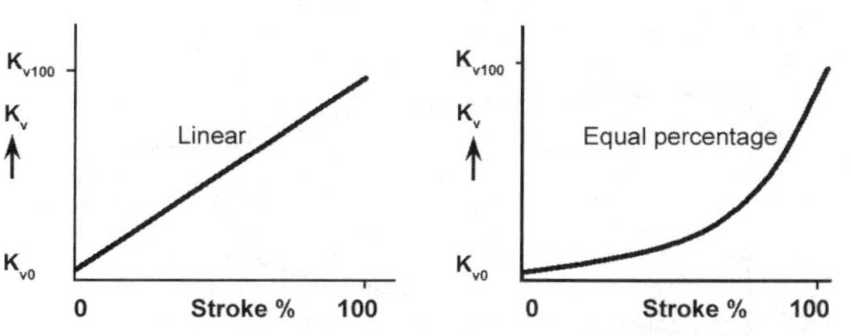

Figure 5.19 Linear and logarithmic control valve characteristics

The principal requirements of a control valve are based upon its regulating characteristics. These characteristics are also dependent upon the interaction between the pump, control valves, pipe system and other control equipment. The valve can only regulate flow by increasing the liquid velocity locally to create a high pressure drop. The capability of the valve to regulate ceases when the pressure drop across the valve approaches zero. If it is required to sustain regulation capability even at Q_{max}, maximum flow in the proposed process, the valve must have an available pressure drop at that flow which is somewhat greater than that for the fully open valve position. Control valves are sized so that they are never fully open or closed; a 10% margin of travel at each end of the stroke is a minimum safety margin.

The design of other control equipment is simplified if the flow through the pipe system has fairly linear characteristics in relation to the position of the valve. This requirement also makes a large pressure drop across the valve desirable at Q_{max}.

The flow through a valve is calculated using the formula:

$$Q = K_v \sqrt{\frac{\Delta p}{\rho}} \qquad \text{Equ 5.8}$$

where:

Q = flow (m^3/s)

K_v = valve flow coefficient (m^2)

Δp = pressure drop across the valve (Pa)

ρ = density of the liquid (kg/m^3)

A measure of the valve capacity and size is its flow coefficient A_{v100} at the fully open position, 100% stroke. To make a rapid assessment of a suitable valve size the following rule of thumb is used:

$$K_{v100} \geq \frac{Q_{max}}{\sqrt{g\,(0.1 \cdot H_{stat} + 0.3 \cdot h_{fpipe})}} \qquad \text{Equ 5.9}$$

K_{v100} = flow coefficient at fully open position 100% stroke (m^2)

Q_{max} = maximum flow with regulation requirement (m^3/s)

g = acceleration due to gravity (m/s^2)

H_{stat} = system static head (m)

h_{fpipe} = total head loss for pipe system at Q_{max} excluding valve losses (m)

If the two sides of the equation are equal then this indicates that the valve at full stroke would just precisely allow the flow Q_{max} to pass through and constitutes a theoretical lower limit of valve capacity.

The given pressure drop at Q_{max} is a compromise between, achieving good regulating characteristics for the system without using too large a valve on the one hand, and not necessitating too high an additional pump differential head with associated pump and energy costs on the other. Control valves are not sized by the pipe size. A control valve will frequently be two sizes smaller than the pipe.

The next step is to select the valve characteristic. The two most common are linear and logarithmic, equal percentage. In principle a valve can be given any characteristic by specially shaping the internal components, the valve trim i.e. plug etc. The valve is described as having a linear characteristic if its flow coefficient increases in direct proportion to the stroke. If K_v increases logarithmically in relation to the stroke then the valve is referred to as having an equal percentage characteristic.

Between these two types are valves with quadratic, V-port characteristics. Many valves display characteristics which differ to a greater or lesser extent from these mathematically defined forms.

The following rule of thumb is used as a guide when selecting a suitable valve characteristic:

If the relationship between the valve pressure drop at Q_{min}, minimum flow which is to be regulated, and at Q_{max} is less than 3, try a valve having a linear characteristic. If $\Delta pQ_{min}/\Delta pQ_{max} > 3$, choose a logarithmic characteristic.

With a lower limit of K_{v100} and with a preferred valve characteristic, a preliminary choice of control valve can be made. Alternatives to A_v are used in practice:

$$K_v = \frac{Q}{\sqrt{Dp \cdot \frac{rH_2O}{r}}} = \left[\frac{m^3/h}{bar^{1/2}}\right] = \frac{K_v \cdot 10^6}{28} \qquad \text{Equ 5.10}$$

$$C_v = \frac{Q}{\sqrt{Dp \cdot \frac{rH_2O}{r}}} = \left[\frac{US\ gpm}{psi^{1/2}}\right] = \frac{K_v \cdot 10^6}{24} \qquad \text{Equ 5.11}$$

In catalogues, the specified value for K_v and C_v are derived by testing with water at 5 to 30°C, which must be considered when selecting a valve for high viscosity or non-Newtonian liquids.

Having made a preliminary selection of pump and control valve and after calculating the system curve excluding the valve, it is possible to check the pump installation regulating characteristics.

5.2.4.1 Calculation example

For a particular pump the system curve, excluding the valve, is determined, using Figure 5.20 upper illustration. The flow is to be valve regulated within the range 0.03 to 0.1m^3/s. Select the valve and pump size and check the regulating characteristics of the pump installation.

At maximum flow with regulating requirements:

$Q_{max} = 0.1\ m^3/s$, $H_{stat} = 18$ m, $h_{fpipe} = 20$ m

According to the rule of thumb, Equ 5.9, at Q_{max}

$h_v = (0.1 \cdot 18) + (0.3 \cdot 20) = 7.8$ m

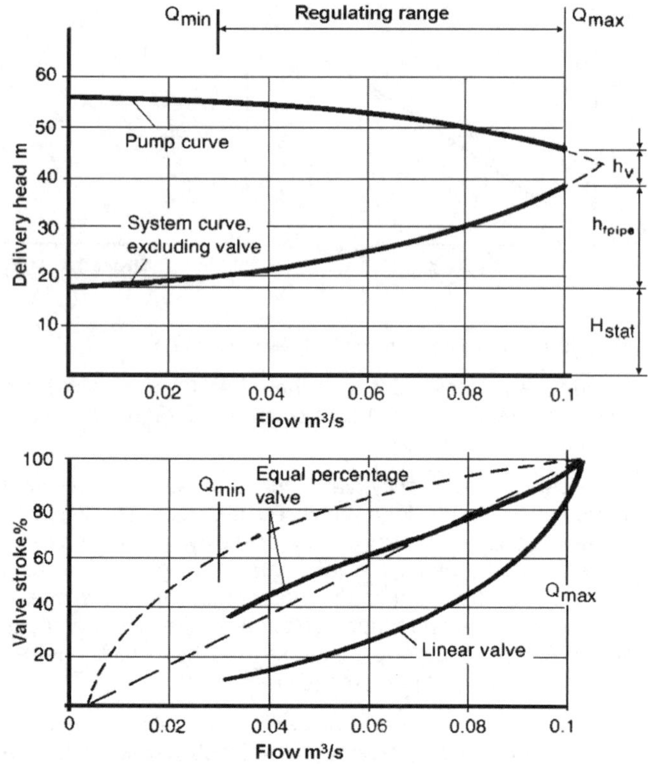

Figure 5.20 Evaluating the regulating characteristics of the pump installation

Select a pump whose H-Q curve passes through the point, H = 18 + 20 + 7.8 = 45.8 m at Q = 0.1 m³/s.

According to the rule of thumb, Equ 5.9,

$$K_{v100} \geq \frac{0.1}{\sqrt{9.81 \cdot 7.8}} = 0.0114$$

From the pump and system curves the largest and smallest pressure drop across the valves within the regulating range can be read. The relationship between these is 35/7.8 = 4.5 which is > 3 and indicates a logarithmic valve characteristic. Choose a regulating valve with K_{v100} = 0.013 having a logarithmic characteristic. Calculate the valve setting, percentage stroke, at various flows by using the formula:

$$Q = K_v \cdot \sqrt{g \cdot h_v}$$

$$K_v = K_{v0} \cdot e^{\frac{s}{100} \cdot \ln\left(\frac{K_{v100}}{K_{v0}}\right)} \qquad \text{Equ 5.12}$$

where:

s = actual setting (stroke) (%)

K_{v0} = flow coefficient at s = 0 (m²)

The calculated result is shown by the lower illustration in Figure 5.20.

In the example K_{v0}= 0.04 · K_{v100}, i.e. the valve regulating range is 1:25. For comparison the pump installation regulating characteristics have also been calculated for a valve having linear characteristics. For the linear valve:

$$K_v = K_{v0} + (K_{v100} - K_{v0}) \cdot \frac{s}{100} \qquad \text{Equ 5.13}$$

The largest flow which can pass through the pump installation is 0.102 m³/s and corresponds to the valve fully open condition. Within the regulating range the valve stroke varies between 35 and 96% of full stroke for the logarithmic valve. Corresponding figures for the linear valve are 87 and 9%. The logarithmic valve gives the installation an approximately linear characteristic.

The chain-dotted curves in the lower illustration in Figure 5.20 show the installation's regulating characteristics at constant pressure drop across the valve, i.e. without considering the effects of the pump and the shape of the system curve.

Note that the pump data selected in the example neglects tolerances of the installation components, which is not to be recommended in practice.

5.2.4.2 Summary

- The pressure drop across control valves can be considerable and cannot be neglected when determining pump operating characteristics.
- The regulating characteristics of the installation determines the pressure drop required across a control valve.
- The smaller the valve pressure drop in relation to the pipe system losses, the more distorted the valve characteristic.
- Valve pressure drop must be "paid for" with increased pump power and can generate considerable energy costs.
- Rules of thumb can never be applied universally.
- All components in the pump installation are subject to certain hydraulic tolerances. Deviations from the nominal performance for pumps, control valves and other fittings, together with inaccuracies when calculating pipe losses etc. are unavoidable and must be given careful consideration. In valve regulated systems the control valve is the component which is used to compensate for the component tolerances.

5.3 Multiple pump systems

5.3.1 Series pump operation

Series pump operation is used when high differential heads are required. Series operation is especially applicable for high differential heads in combination with comparatively small flows. By dividing up the differential head between more than one pump, a higher specific speed, N_s, is maintained for each individual pump, thereby increasing efficiency.

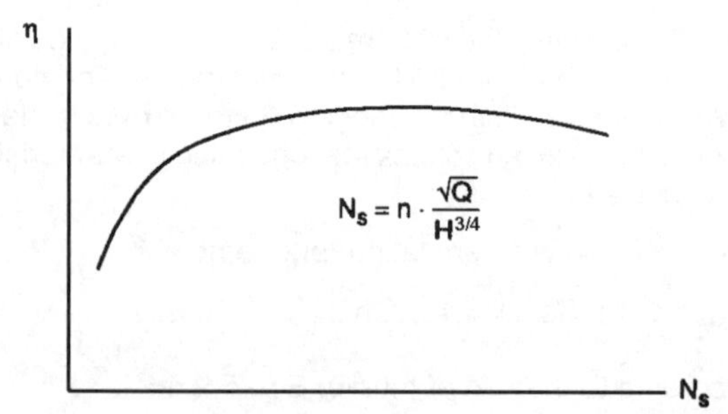

Figure 5.21 Maximum efficiency for centrifugal pumps at various specific speeds

Multi-stage pumps are a compact, integral form of series connection. Systems incorporating a number of pumps, where the pumps are situated in close proximity to one another are simply dealt with by calculating a common performance curve for the complete "pump package". The common pump curve is then compared with the system curve.

For series operation the resultant H-Q curve is obtained by adding the actual differential heads for each value of flow, Figure 5.22.

The resultant efficiency at various flows is calculated from the relationship:

$$h = \frac{r \cdot g \cdot Q\,(H_1 + H_2)}{P_1 + P_2} \qquad \text{Equ 5.14}$$

When two similar pumps are connected in series the same efficiency curve applies for the "package" as for each separate pump unless the liquid is very compressible. For very compressible liquids, liquified gases are the worst, the flow throw

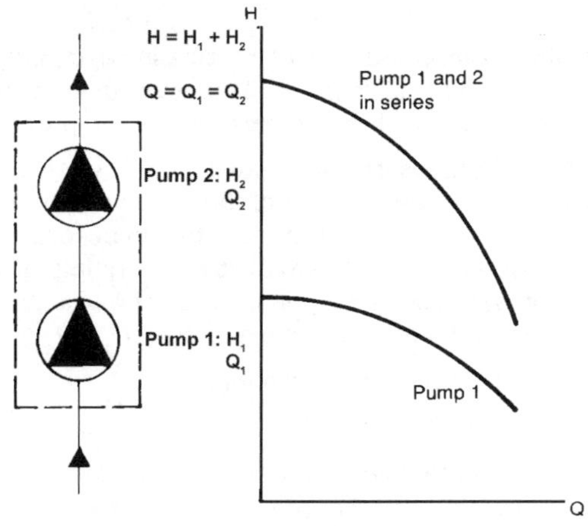

Figure 5.22 H-Q curve for series pump operation

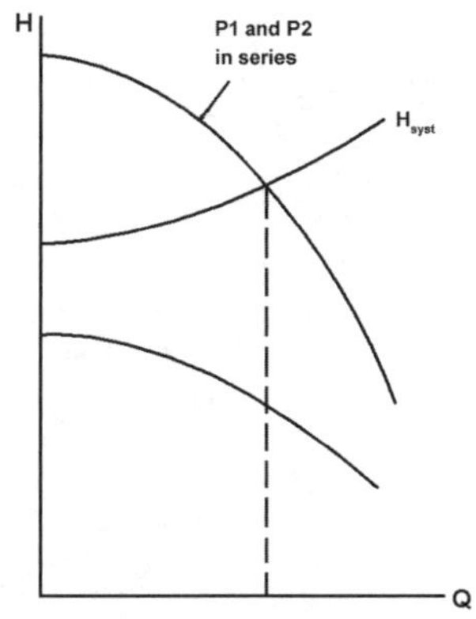

Figure 5.23 Pump nominal duty point for series operation

the second pump may be considerably smaller than the first if the pump differential head is high.

The pumps P_1 and P_2 shown in Figure 5.23 cannot operate individually on the system because one pump does not develop sufficient differential to overcome H_{stat}. Depending upon the pump type and size, precautions may be needed to prevent one pump running without the other. The second pump may cavitate if the first pump is not running and both pumps may overheat. Instrumentation and control logic should be fitted to protect the pumps. It is possible to pump through a rotodynamic pump. The head loss through a pump is proportional to the specific speed. As the liquid flow through a pump increases it will begin to rotate in the normal direction. The liquid and impeller angles will match and the head loss will not rise as steeply. The pump should not overspeed.

For most liquids, the same flow passes through both pumps. The inlet pressure for P_2 is the pump outlet pressure of P_1. The pump pressure rating and shaft seals can therefore be affected. The axial thrusts, which are dependent on the differences between the internal and atmospheric pressure, will also be different for the two pumps. For low differential pressures, identical pumps may be used. As the differential pressure increases, the second pump may require significant modifications to accommodate the increased inlet pressure.

One very common occurrence of series operation is used to overcome the problem of low NPSHa discussed in Section 5.1.3. Some rotodynamic pumps are required to produce very large differential heads. Even with multi-stage pumps, 12 stages are common, the ΔH at normal motor speeds is not sufficient. The pumps are driven at higher speeds by step-up gearboxes or by steam/gas turbines. The NPSHr of the pump increases.

To overcome the problem, and to avoid very costly solutions such as raising the suction source, a low power booster pump is operated in series with the main pump. The function of the booster is to supply NPSHa only. The differential head is therefore very small compared to the main pump and the power absorbed is low; efficiency is not really a consideration.

The booster pump would generally be a single stage pump, perhaps double entry, running slowly to match the NPSHa. If the main pump is geared up from an electric motor, the booster pump may be geared down from the same motor. If the main pump is turbine driven, the booster will probably be motor driven. In this type of series pump installation the pumps will be completely different, reflecting their respective functions.

Booster pumps are also used on positive displacement pumps. Depending upon the nature of the liquid the booster pump may be another positive displacement pump or a rotodynamic pump. High pressure positive displacement pumps working

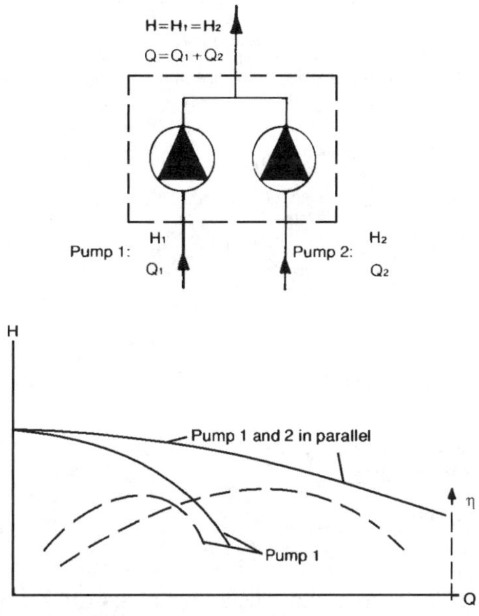

Figure 5.24 Parallel pump operation

with water-like liquids may use a centrifugal pump booster to provide adequate NPIPa.

Pumps handling more viscous liquids would use a low pressure positive displacement pump. When two positive displacement pumps are used, the booster should be oversized slightly to allow for wear. A back-pressure regulator and a relief valve should be fitted to the inter-connecting pipework to prevent excessive pressures. It is also a good idea to fit a pulsation damper, with extra gas volume, to reduce any pulsation levels and function as an accumulator.

5.3.2 Parallel pump operation

For parallel pump operation, the resultant H-Q curve is obtained by adding the actual flows for each head value, Figure 5.24.

Parallel pump operation is used to achieve large flows and deal with greatly varying flows by operating a varying number of pumps. The effect of parallel operation is dependent largely upon the shape of the system curve.

For steep, friction dominated system curves only a small increase in flow is achieved by parallel operation. A much better effect is obtained if the system curve is dominated by static head.

The pump's specification, Q and H, is based upon the pump's nominal duty point, B in Figure 5.25. If only one pump operates, the duty point will then be at point A. This can cause conditions where there is risk of cavitation, because of the lower NPSHa value at point A than at B, and risk overloading of the motor.

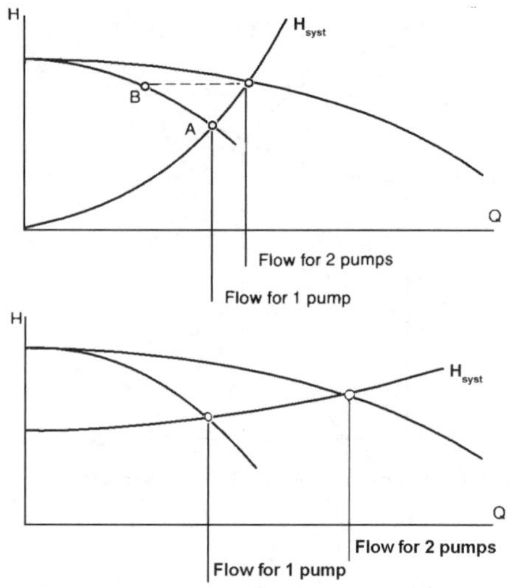

Figure 5.25 Parallel operation of 2 similar pumps with different system curves

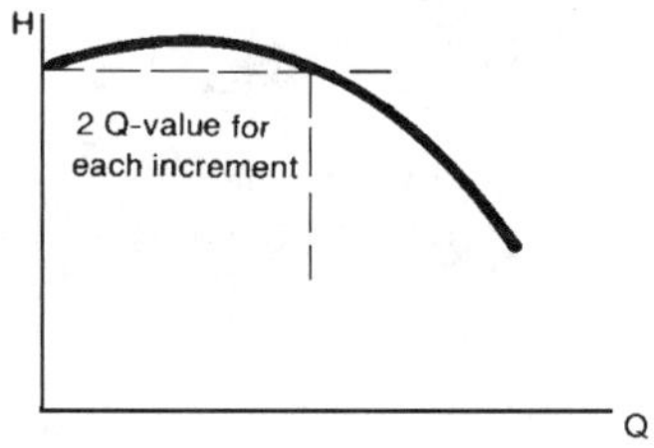

Figure 5.26 Unstable pump curve

The resultant efficiency curve for the "package" is calculated from the relationship:

$$\eta = \frac{r \cdot g \cdot H\,(Q_1 + Q_2)}{P_1 + P_2} \qquad \text{Equ 5.15}$$

For two similar pumps the efficiency values are the same for each separate pump except that the Q values are displaced by a factor of two, Figure 5.25.

Unstable pump and shallow pump curves should be avoided when pumping in parallel. In any case the intended operational range should not fall within the unstable region without taking special precautions. Special minimum flow valves or orifices can be fitted to prevent the pump reaching the peak of the curve. See Figure 5.26.

Typical problems which can occur include irregular flow due to varying operational points and the inability of a pump to open it's non-return valve when starting. Non-return valves are fitted to prevent reverse flow through stationary or standby pumps. Characteristic curves for positive displacement pumps are generally so steep that parallel operation causes no problems. Different types of pump can be operated in parallel to achieve the best operating efficiency at all flows.

5.3.3 Pressure boosting

Pressure boosting pumps are primarily employed in long pipelines and large circulation systems. A common problem is that of pressure in a distribution network being insufficient for certain higher locations, for example, the water supply to high-rise apartment blocks; or when an additional tapping point is connected to an existing network.

Pressure boosting is a form of series connection where the pumps are located at considerable distances from each other. The advantages of pressure boosting are:

- A better distribution of the specified pressure in the pump installation is achieved
- Pump and energy costs can be reduced, (applies to branched pipework systems), see Figures 5.27 and 5.28

In Figures 5.27 and 5.28, higher pressure rating may be required for the upstream section of the pipeline and pump for Alternative I, or for the second pump in the event that the pressure boosting for Alternative I is also divided into two pumps.

Branched pipework systems are usually sized by calculating for the pressure requirements of the furthest tapping point. The required pump specification is therefore based according to this pressure requirement and the total flow consumption. The application of pressure boosting principles can provide an alternative, as illustrated by the following example.

Pump data for Alternative I becomes:

Q = 1, H = 0.5 + 0.5 = 1, P = 1

Pump data for Alternative II becomes:

Q1 = 1, H1 = 0.5 + 0.1 = 0.6, P1 = 0.6

Q2 = 0.5, H2 = 0.5 - 0.1 = 0.4, P2 = 0.20

The total installed pump power P = 0.80

The smaller power requirement for Alternative II is explained by the fact that the surplus pressure in pipe BC is eliminated. In alternative I this excess pressure must be throttled to prevent tapping point C from overflowing and reducing the flow at D.

A better illustration of the pump's operating situation is obtained from drawing the pump and system curves. In Figure 5.29 the respective pump and system curves are shown both separately and in relation to each other. The curves are compiled in accordance with Section 5.1.3. The points marked indicate the situation at the specified flow condition.

The H-Q diagram shows, for example, that pressure boosting is not required at all for total flows Q << 0.46, even if the complete flow is tapped off at D. The largest flow which can pass through the pressure boosting pump is Q ≅ 0.67. This occurs when C is completely closed and is sized for the pressure boosting pump motor. The largest flow at C, D completely closed, is ≅ 0.87. For Q > 0.8, H_B < 0.25, then D is receives no flow unless the pressure boosting pump is started.

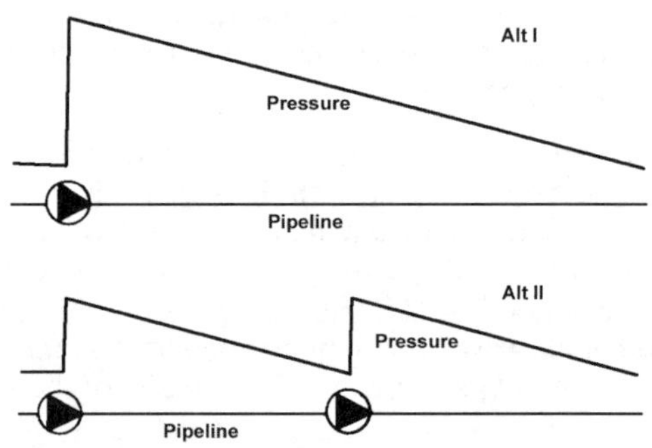

Figure 5.27 Schematic pressure distribution in a pipeline

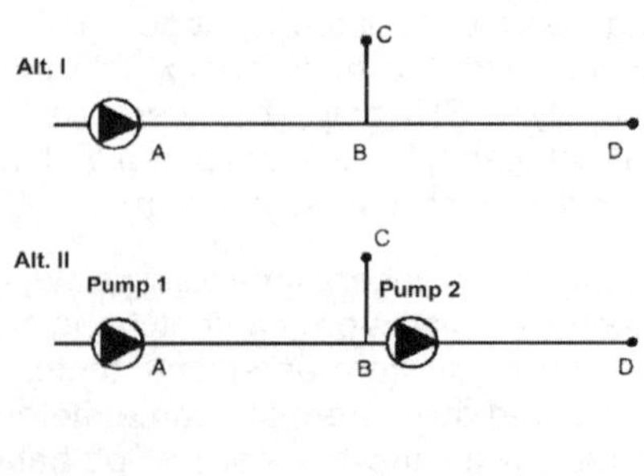

Data supplied		
Length	Flow	H_{syst}
AB	1	0.5
BC	0.5	0.1
BD	0.5	0.5

Figure 5.28 Example of conditions for pump sizing in a branched pipe system

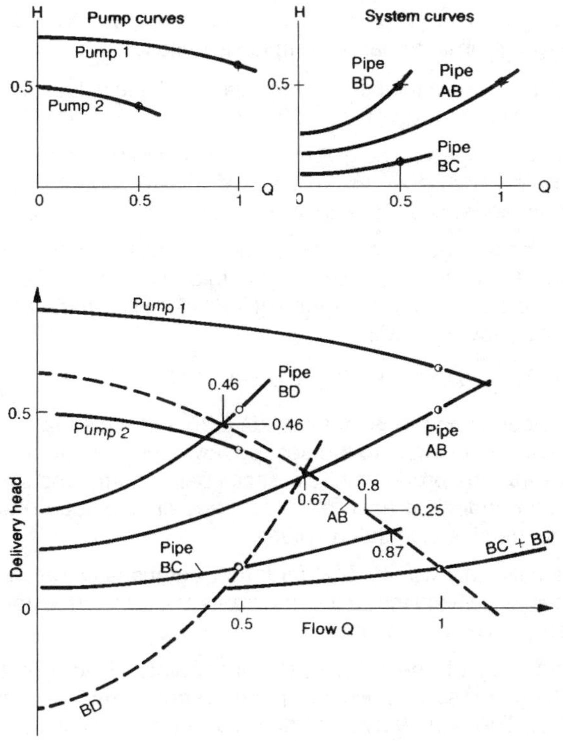

Figure 5.29 Operating conditions for pressure boosting example in Figure 5.28

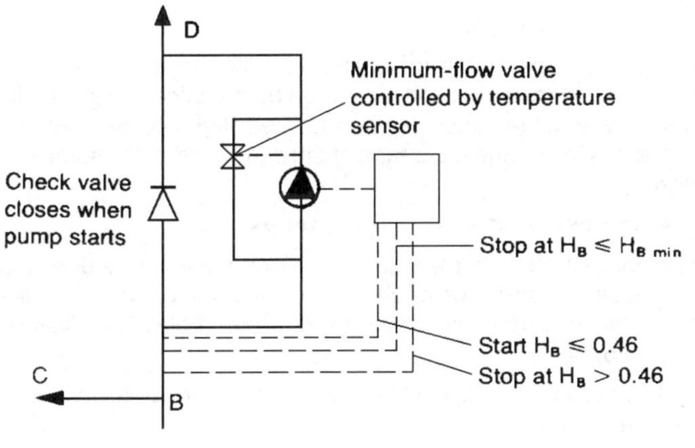

Figure 5.30 Installation of pressure boosting pump

The pressure boosting point can be controlled in the following way and is illustrated in Figure 5.30. To guarantee full flow at D, $Q_D = 0.5$, independent of consumption at C, the pressure boosting pump is started by a signal from a pressure sensor at the branch point when H_B falls below 0.46 and is stopped when H_B exceeds 0.46 by a certain margin. Within the range $0.19 \leq HB \leq 0.46$, the total flow $0.46 \leq Q \leq 0.87$, a situation where flow $Q_D = 0$ can occur.

The input power is transmitted to the liquid, the temperature rises and the risk of seal leakage and damage due to cavitation increases. Excessive temperatures can be avoided by means of a temperature sensor which transmits a signal to a valve which opens a by-pass line. To protect the pressure boosting pump against cavitation in the event of low pressure on the inlet side, which could occur if Pump 1 became damaged for example, a pressure sensor is installed which stops the pressure boosting pump when HB falls below the required cavitation limit, H_{Bmin}.

The most energy conserving method of regulation would be to control the speed of the pressure boosting pump according to the constant pressure criteria in D.

5.3.4 Pressure maintenance

Pressure maintenance, as the name implies, is intended to sustain a certain pressure in a pump installation or circulation system. The liquid being pumped is often hot and, because of the high temperature, has high vapour pressure. The pressure maintenance prevents cavitation at high points in the pipework and pumps, in other words cavitation is avoided. Pressure maintenance also compensates for unintentional leakages and makes up variations in the volume of liquid and pipework due to changes in temperature and pressure.

Typical cases, where pressure maintenance is necessary, are district heating networks and hot water circulation systems in buildings. Pressure maintenance is carried out at a point in the pipe network where it will be most effective, which will vary from case to case. If the pressure maintenance pump is placed on the suction side of the main pump then a constant inlet pressure is maintained, whilst the outlet pressure varies with change in flow and vice-versa.

The simplest method of providing pressure maintenance is by means of a high-level tank. A higher pressure can be provided if required by the introduction of a pressure maintenance pump, Figure 5.31.

The by-pass line ensures a minimum flow to the pressure maintenance pump, even though no liquid is supplied to the main pipeline, thus preventing unacceptable temperature rises in the pump. The by-pass line can also be used to maintain constant pressure for varying flows to the main pipeline.

The main line pressure is equal to the pressure increase due to the pump plus the supply pressure. Constant throttling in the by-pass line causes the pressure level in the main line to vary by an amount ΔH dependent upon the flow in the main line. Vari-

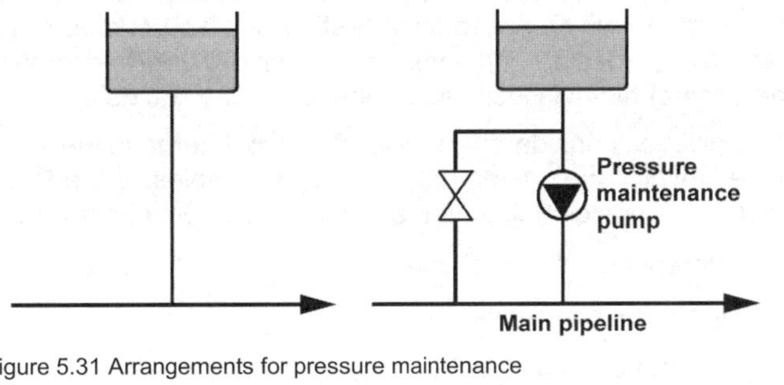

Figure 5.31 Arrangements for pressure maintenance

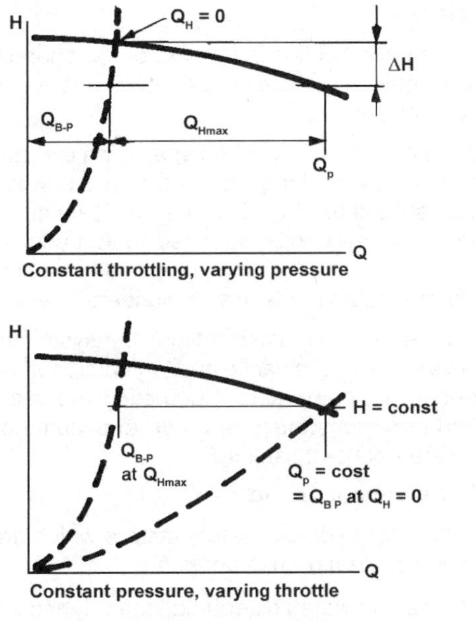

Fig. 5.32 Pressure maintenance pumps for different control methods

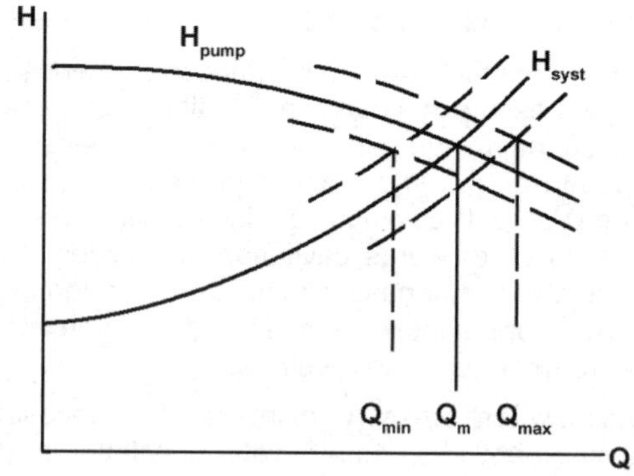

Figure 5.33 Illustration of hydraulic tolerances
——— calculated value — — — — tolerance limits

able throttling enables constant pressure to be maintained, Figure 5.32.

Regulation of pressure maintenance can also be carried out in other ways, for example, by throttling the delivery pipe or by regulating pump speed.

5.4 Pump hydraulic data

5.4.1 Hydraulic tolerances

5.4.1.1 General

All components in the pump installation have tolerances in respect of hydraulic performance. To ensure achievement of a definite flow through the installation it is necessary to consider the hydraulic tolerances when determining pump data, Figure 5.33.

Pump hydraulic tolerances are carefully set out in the pump testing standards. The maximum permissible deviations from the guarantee data are thus defined by specifying the Standard according to which acceptance testing and approval is to be carried out. Special conditions must however be agreed upon in the case of high viscosity and non-Newtonian liquids.

It is however, considerably more difficult to determine the tolerance limits of the system curve, H_{syst}. Examples of the tolerances which accumulate in the complete system curve are:

- Tolerances for pipe diameter
- Pipe bore roughness
- Valves, bends and fittings
- Flow meters and instrumentation
- Variable filter Δp

together with uncertainties in respect of the characteristics of the process liquid at various temperatures and pressures, and the methods of calculation.

Another type of uncertainty, when specifying the pump data, is the future flow and head requirement. This is however, a question which should be clarified before selecting a pump. Rotodynamic pumps can be supplied so that additional impellers can be fitted. Reciprocating pumps can be built to accommodate various sizes of pistons or plungers.

Because of the difficulty in determining the system curve tolerance limits a general allowance is often applied when calculating the necessary pump data. Such allowances are a safety margin and provisions must be made to accommodate them when they are not used usefully.

5.4.1.2 Rotodynamic pumps

The magnitude of hydraulic safety factors which are normally applied to the system requirements are:

- +5% of the calculated differential head due to losses other than H_{stat} for pumps, P > 30 kW
- +2% of the required flow to allow for leakage (may require a larger margin for large complex installations)
- +10% of the calculated differential head due to losses other than H_{stat} for pumps, P < 30 kW

The percentage values given apply to Newtonian liquids. The uncertainty increases in the case of non-Newtonian liquids. The consequences of the allowances are illustrated as follows:

For a mass-produced centrifugal pump, class C, purchased according to a catalogue curve, the tolerances for differential head are ± 6%. For a tolerance of ± 8% for H_{syst} at the desired flow, the allowance will be:

$$\sqrt{0.06^2 + 0.08^2} = 0.10 \text{ i.e. given as } 10\%$$

The calculation means that there is a 95% chance that the actual flow will be equal to the desired flow. The formula also assumes that the production tolerances of the pump and system are independent of each other and equally distributed about their respective calculated values.

The smaller allowance, 5%, for larger pumps is influenced by the stricter delivery tolerance requirements and more extensive system curve calculation.

Another way of approaching the uncertainty of pump/system matching is to specify two duty points. A rated duty and a nominal duty. The rated duty is the flow plus a leakage allowance at the head required to cover all the highest losses. The pump characteristic must pass through this point. The nominal duty is the basic design flow without any leakage at the head required for a new clean system. This point must be on or below the pump characteristic.

Pumps built specifically for a project can have their characteristics "fine tuned" to suit. Impellers can be machined to a specific diameter and the vanes can be modified slightly in the eye and at the tips to change the shape of the characteristic to some degree.

5.4.1.3 Positive displacement pumps

For positive displacement pumps, which have very steep H-Q curves, the system curve tolerances have little effect on the actual flow. The pump flow tolerances are completely definitive, Figure 5.34.

The required flow will most certainly be achieved if the following rules are applied:

- Specify the pump flow as -0% +3% of the design flow.
- Specify the pump suction pressure as the system minimum.
- Specify the pump discharge pressure as the system maximum.

Allowances must be made in the discharge system for overpressure when the relief device operates.

The three rules may not work effectively if the liquid properties, viscosity and compressibility, and the differential pressure vary

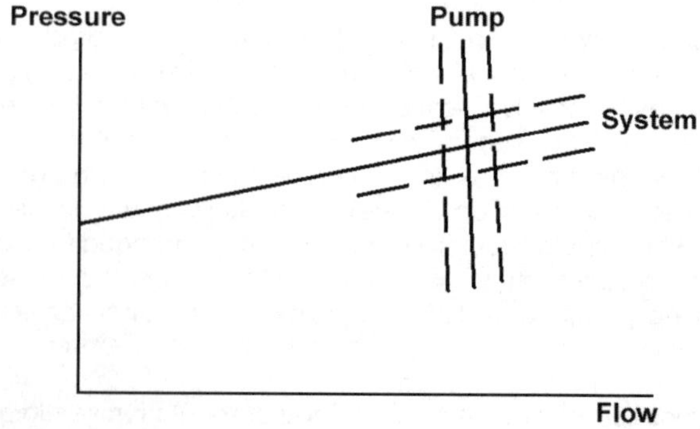

Figure 5.34 Illustration of hydraulic tolerances for positive displacement pumps

over a wide range. The pump volumetric efficiency could change dramatically resulting in a significant variation in flow. Liquids such as lubricating oils, when the operating temperature range is wide, and liquefied gases, are likely to create problems. Failure to specify the full working pressure range, suction or discharge, could result in the incorrect selection of ancillary products such as pulsation dampers.

To operate any pump at the best efficiency, whether or not the system characteristic is difficult to predict accurately, select a pump which can be adjusted, on site, when the actual equilibrium point can be measured during initial operation. Impeller diameters can be reduced, piston/plungers increased or reduced, or the pump speed can be adjusted. Vee-belts and timing belts are very useful for implementing small changes in speed; the smallest pulley can be specially manufactured to give the desired ratio. The energy cost saving in a single year may easily pay for the extra components. Oversized pumps waste a lot of energy.

5.4.2 Pump H and Q at maximum efficiency η_{max}

On the basis of the installation liquid requirements the mean flow Q_m and maximum flow Q_{max} are determined. Then:

$$H_{Qmax} = H_{stat} + h_{fpipe} + h_v + h_{tol} \qquad \text{Equ 5.16}$$

where:

H_{Qmax} = the pump head at Q_{max} (m)

H_{stat} = static head (m)

h_{fpipe} = head losses in the pipe system excluding the control valve at Q_{max} (m)

h_v = pressure drop across control valve at Q_{max} (m)

h_{tol} = allowance for hydraulic tolerances at Q_{max} (m)

In this way a point Q_{max}/HQ_{max} is established on the intended pump H-Q curve, see Figure 5.35. Except in the case of on-off control systems however the pump will very rarely, if ever, be required to operate at this point. The pump operating point, design point, η_{max}, should preferably occur at a lower flow, as near to Q_m as possible. If the difference between Q_{max} and Q_m is great, more than 15%, and if Q_m is chosen to coincide with the pump's η_{max} then there is risk of over loading the pump at Q_{max}. Normally, however, a flow which is 20% greater than the design flow is not considered to present any problem for rotodynamic pumps providing the manufacturer is consulted and power is available.

Using the above, the following rules of thumb are given for the determining of pump hydraulic specification data for continuously regulated pump installations.

- Determine Q_m and Q_{max} according to the process requirements.
- Calculate H_{Qmax}.

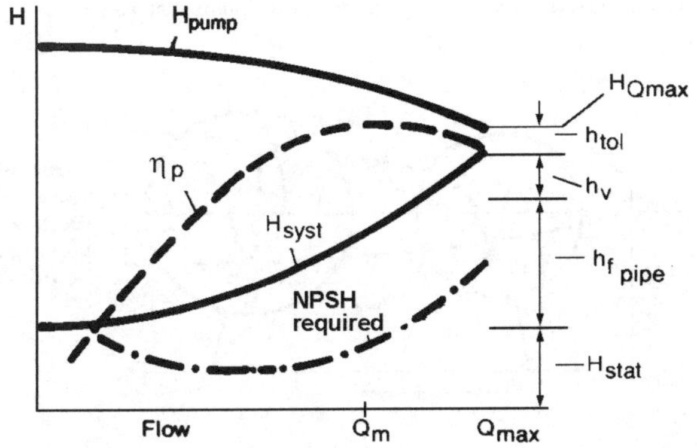

Figure Illustration of recommended pump data

- Select a pump which can cope with Q_{max}, H_{Qmax}. The pump's η_{max} should lie as near to Q_m as possible.
- Check the risk of cavitation, $NPSHr_{Qmax}$, and the required shaft power at Q_{max}. Check with the pump manufacturer if there are any other reasons which may prevent operation at Q_{max}.

The matching of pump flow to Q_{max} when the production tolerances are known can be accomplished by:

- Remachining the pump impeller(s)
- Constant throttling or by-pass
- Speed matching (belt or gear drive)

5.4.3 Minimum pump flow

There are many reasons why it is necessary to ensure a minimum flow for a rotodynamic pump:

- Heating of the pumped liquid.
- Increased cavitation risk when Q approaches zero.
- Increased vibration levels when Q approaches zero.
- Axial thrust and radial force on the pump impeller are greatest at Q = 0.
- The power required for an axial pumps is greatest at Q = 0.

Pump efficiency is low in the vicinity of zero discharge, Q = 0. Most of the power is converted to heat, which must be carried away by the pumped liquid, Figure 5.36. The increase in temperature of the pumped liquid can be calculated from the relationship:

$$\Delta T = \frac{g \cdot H}{c_{sp}}\left(\frac{1}{h} - 1\right) = \frac{P(1-h)}{\rho \cdot c_{sp} \cdot Q}$$

$$\cong \frac{P_{Q=0}}{\rho \cdot c_{sp} \cdot Q} \qquad \text{Equ 5.17}$$

where:

H = differential head (m)

g = acceleration due to gravity (m/s²)

c_{sp} = specific heat (J/kg °K)

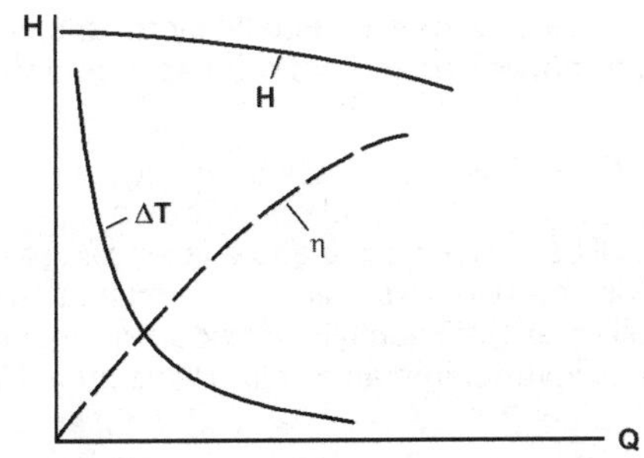

Figure 5.36 Temperature rise of liquid when pumping near to the point of zero discharge

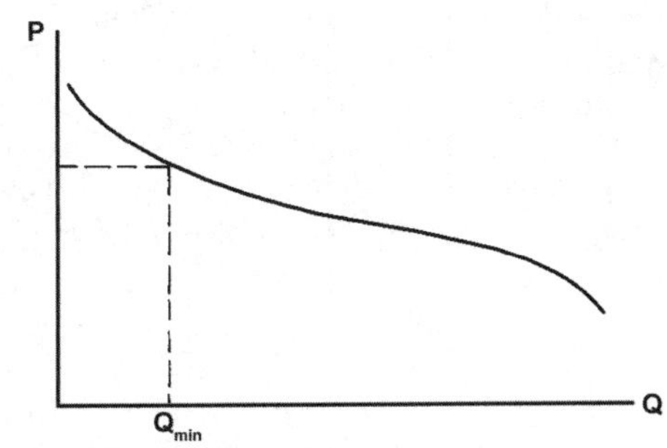

Figure 5.37 Power curve for axial pump

η = pump efficiency

Q = flow through pump (m^3/s)

ρ = density of liquid (kg/m^3)

P = supplied shaft power (W)

ΔT increases rapidly as the flow through the pump approaches zero ($Q \to 0$, $\eta \to 0$). Apart from increased material stresses due to temperature rise, the vapour pressure of the liquid also increases with corresponding cavitation risk and the shaft seal liquid film is also in danger, so that a breakdown of sealing function may be imminent.

For axial pumps, propeller pumps and pumps with high specific speed, the required shaft power increases as flow decreases at constant speed, Figure 5.37.

By having a specified minimum flow through the pump a smaller motor can be used without risk of overloading.

5.4.3.1 Devices for obtaining a minimum flow

A minimum pump flow at Q = 0 in the system can be maintained by introducing a by-pass line into the system, (see Figure 5.38).

The return line should return the liquid to the pump sump in order that a large liquid volume should be available if circulation pumping is protracted. The by-pass flow can be controlled in various ways:

- Constant throttling
- Loaded check valve
- Pressure sensor after the pump
- Flow sensor in the main line
- Temperature sensor after the pump

In cases employing constant throttling, such as an orifice plate or locked needle valve, a certain flow will pass through the by-pass pipe even at Q_{max} in the main line and should then be added to Q_{max} when defining the pump size. A loaded non-return valve or pressure sensor requires a stable pump curve whose differential head falls continuously as pump flow increases.

In the case of positive displacement pumps a relief valve or other overpressure protection device must always be provided as protection against blockage in the discharge line. Flow regulation cannot be carried out by throttling in the main line. Regulation of fixed speed pumps is usually by means of a by-pass line. Other alternatives are speed regulation and variable displacement.

NOTE: Speed regulation using variable frequency inverters to supply standard squirrel cage motors is popular due to ease of use, wide choice and efficiency. Variable speed pumps can overcome all system mismatch problems, usually very efficiently. Introducing a variable frequency inverter into an existing fixed speed installation may cause problems. Inverters can produce unexpected electrical side effects. The best solution is to ask the motor supplier to supply an inverter. Then the supplier is responsible for the inverter and the motor with no divided warranty problems resulting.

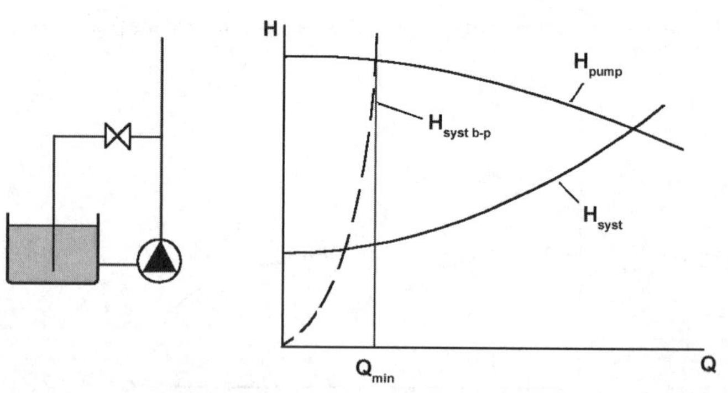

Figure 5.38 By-pass line for a minimum flow

5.5 Water hammer

5.5.1 Hydraulic gradient

In order to obtain a better understanding of the distribution of pressure throughout a pump installation, use is made of the hydraulic gradient. By definition, for a pressure head H_t:

$$H_t = \frac{p^1}{rg} + z \qquad \text{Equ 5.18}$$

where:

H_t = pressure head above liquid surface (m)

p^1 = static pressure above atmospheric (Pa)

ρ = density of the liquid (kg/m^3)

g = acceleration due to gravity (m/s^2)

z = elevation above liquid surface (m)

In this method of representation the pipeline inner positive pressure is the difference between the H_t-line and the elevation of the pipeline, see Figure 5.39. The Figure also shows a high point where there is negative pressure, that is below atmospheric pressure. Parameters which should be observed are the prevailing inner positive and negative pressures as regards sizing for stress and margins of safety against the risk of cavitation. The cavitation risk is greatest in pumps, valves and other installation components having restricted sections together with high points in the system.

The hydraulic gradients for steady flow take on different appearances for different operating points. The extreme value for the H_t-line often occurs, however, in conjunction with load changes associated with a transient, a time dependent process.

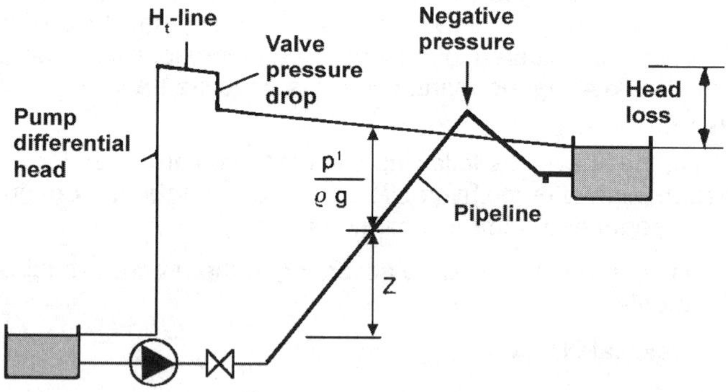

Figure 5.39 Illustration of hydraulic gradient for a pump installation

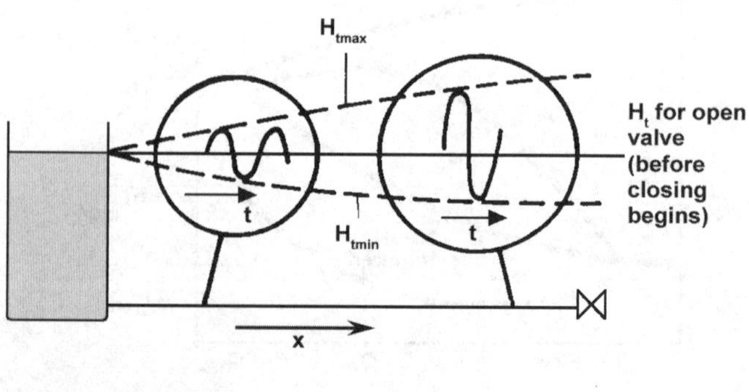

Figure 5.40 Pressure head lines for closing valve

When closing a valve the pressure increases in front of the valve, see Figure 5.40. Increasing pressure gives rise to a pressure wave, which travels through the pipeline at the speed of wave propagation "a". The magnitude of the wave speed "a" is a function of the liquid and the pipe. Pressure waves would normally travel at the speed of acoustic waves in the liquid, which is a function of the liquid properties only.

$$c = \sqrt{\frac{K}{\rho}} \qquad \text{Equ 5.19}$$

where:

c = acoustic velocity (m/s)

K = liquid bulk modulus (Pa)

ρ = liquid density (kg/m^3)

Water has an acoustic velocity of approximately 1450 m/s. Constraining the liquid within a pipe reduces the acoustic velocity because of the flexibility of the pipe wall, see equation 5.20.

$$a = \sqrt{\frac{g/w}{\frac{1}{K} + \frac{1}{8E}(d_o^4 - d_i^4)}} \qquad \text{Equ 5.20}$$

where:

a = wave speed (m/s)

g = acceleration due to gravity (m/s^2)

w = specific weight (N/m^3)

E = Young's Modulus of Elasticity of the pipe (N/m^2)

d_o = pipe outside diameter (m)

d_i = pipe inside diameter (m)

The elasticity of the pipe wall has a significant effect on the wave speed, but air or gas bubbles have a much more severe effect and can reduce it to 100 m/s. The pressure wave is reflected at the end of the pipe as a negative pressure wave, returns to the valve, is reflected again and so on. In this way a process of pressure oscillation is set up in the pipeline and continues, even after the valve is completely closed, until the oscillations are dampened out by friction. The hydraulic gradient lines H_{tmax} and H_{tmin} connect the highest and lowest pressures which occur at different points along the pipeline during the oscillation process. The greatest pressure variations are to be found in the valve and occur when the valve is closed rapidly. The greatest amplitude of pressure surge is:

$$\Delta p_{max} = \rho \cdot a \cdot v \qquad \text{Equ 5.21}$$

where:

Δp_{max} = max. pressure variation (Pa)

ρ = density of the liquid (kg/m^3)

a = speed of wave propagation (m/s)

v = velocity of flow before valve begins to close (m/s)

For water in steel pipes $\rho \cong 1000$ kg/m^3, $a \cong 1100$ m/s, and a flow velocity of 1 m/s creates, for rapid valve closing, a water hammer pressure of 1.1 × 106 Pa = 11 bar ≅ 110 m. The pressure process at the valve cannot be affected before the pressure wave has been reflected from the end of the pipe and returned to the valve. This time is referred to as the pipeline reflection time and is designated t_r.

$$t_r = 2 \cdot \frac{L}{a} \qquad \text{Equ 5.22}$$

where:

t_r = pipeline reflection time (s)

L = the pipe length (m)

a = wave speed (m/s)

If for example the length of pipe is 1100 m, then $t_r \cong 2$ s. All valve closing times ≤ 2 s thus give maximum water hammer pressure.

For a plastic pipe $\Delta p_{max} = 3 \times 10^5$ Pa ≅ 30 m is obtained for valve closing times ≤ 7.36 s. Increasing the closing time reduces the magnitude of water hammer and very slow closing valves result in no line shock at all.

5.5.2 Causes of water hammer

Water hammer in pump installations is caused by changes in flow and especially by rapid flow changes. Rapid flow changes occur during:

- Rapid valve movement
- Starting and stopping of pumps
- Special situations

Since the shock of water hammer can reach considerable magnitude there is risk of damage to the pump installation. Apart from the pressure oscillations which occur with normal variations, the prediction of unintentional operational conditions must also be considered. Accidental pump stoppage caused by a power failure or power cut is one such situation which a pump installation must contend with sooner or later. An estimation of the magnitude of the line shock must therefore be carried out for every installation.

Water hammer caused by cavitation in the pipeline is difficult to estimate. If the absolute pressure somewhere in the pipeline is allowed to fall below the vapour pressure of the liquid, vapour is formed in the pipeline. The original column of liquid divides itself into two which, with different speeds and separated by a growing vapour bubble, flows on through the pipeline. Eventually the two columns of liquid begin to approach each other causing the vapour bubble to implode as the liquid columns reunite, resulting in a large pressure surge. The magnitude of the water hammer shock depends upon the instantaneous difference in velocity of the two columns of liquid at the moment when they reunite, which, in the worst case, can be greater than the steady flow velocity in the pipeline.

Cavitation in pipelines must, therefore, be avoided which also underlines the need to control H_{tmin}. The risk of unacceptable negative pressures are greatest at high points, downstream of valves and pumps and in long suction lines.

The following situations are especially susceptible to water hammer:

- Cavitation due to heating or insufficient cooling in thermal installations
- Trapped air-pockets; pump starts against closed or nearly closed valves in an insufficiently vented system
- Automatic re-start of pump, which rotates backwards after a power failure or after reverse flushing
- Acoustic resonance effects due to periodical disturbances caused by valve oscillations, pump vane passing frequency, positive displacement pump flow variations

5.5.3 Pump behaviour after power loss

5.5.3.1 Pump curves

The pump is sometimes compelled to operate at abnormal operating points especially in the case of pump installations without a non-return valve. Multiple rotodynamic pump installations with parallel operation always have non-return valves fitted to prevent reverse flow back to suction through any stationary

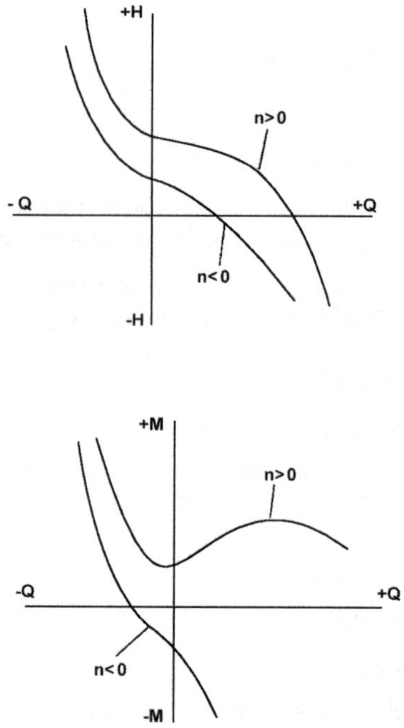

Figure 5.41 Complete performance curves for a centrifugal pump (H = differential head, Q = flow, n = speed, M = shaft torque)

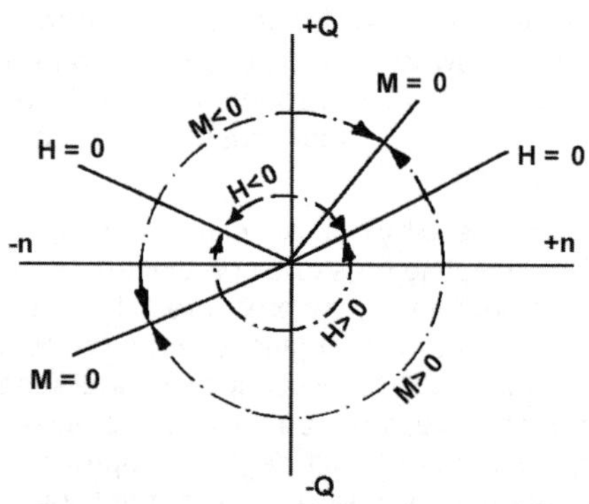

Figure 5.42 Possible centrifugal pump operating conditions

pumps. Single rotodynamic pump installations must have a non-return valve fitted when there is positive static head in the discharge system. Reciprocating pumps do not usually need non-return valves unless a by-pass is used for starting.

Other types of positive displacement pumps may need non-return valves, so consult the pump manufacturer when system characteristics are known. Reverse flow can occur in the event of power failure for example. Inasmuch as reversal of flow or direction of rotation are occurrences which must be considered, a knowledge of pump curve behaviour under these conditions is advisable.

Figure 5.41 shows that a normal centrifugal pump possesses limited pumping capacity even when rotating in the reverse direction and that it functions exceptionally well as a turbine. Values of negative Q indicate reverse flow through the pump. Values of speed, $n < 0$, indicate reverse rotation of the pump. Reverse rotation can be incompatible with shaft driven oil pumps resulting in considerable damage to mechanical seals and bearings; overspeed may also damage the pump. Pumps with low specific speed tend not to overspeed; overspeeding can be a serious problem with high specific speed pumps.

An important consideration from the point of view of water hammer is the rapidly increasing pressure differential across the pump which occurs with reverse flow and pump speed. Another way of illustrating a possible pump operating condition is shown in Figure 5.42.

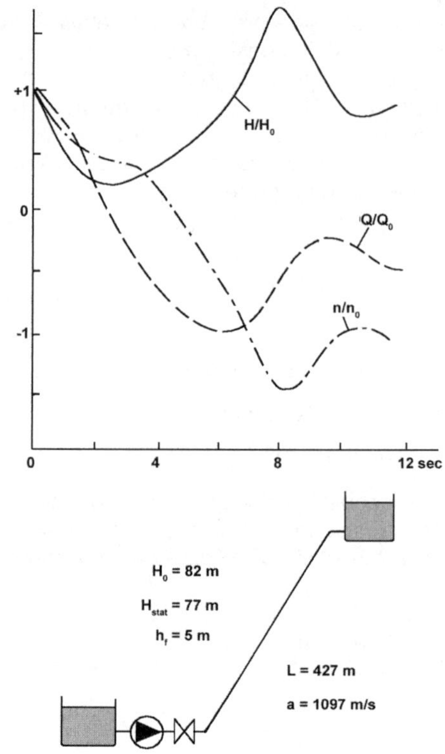

Figure 5.43 Transient process during power failure conditions at pump

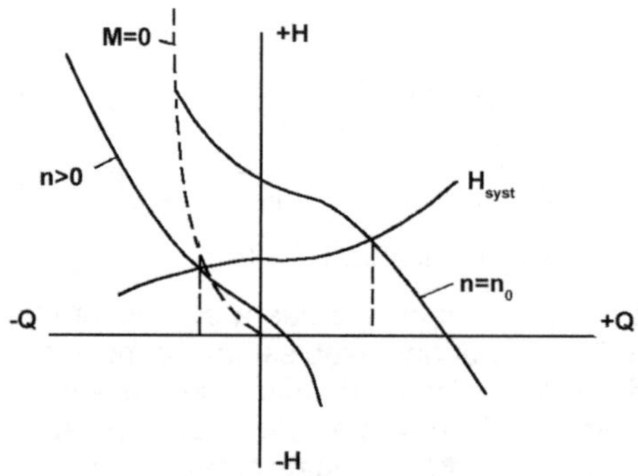

Figure 5.44 Steady, initial and final conditions during power failure

5.5.3.2 Pump stops without non-return valve

The transient process in the event of power failure is illustrated by an extreme example, see Figures 5.43 and 5.44. The installation is equipped with a slow-closing valve which remains open during the process.

When the driving torque fails, the impeller is braked by the hydraulic torque. The decreasing speed also reduces Q and H for the pump. The flow through the pump changes direction whilst the speed is still positive. The pressure in the pump reaches its maximum value. The direction of rotation also changes a few seconds later. Very high pressures occur in the pump when retarding reversed flow at high reversed speed.

The pressure oscillations are eventually dampened out and a steady flow condition is obtained. The pump reaches its so-called runaway speed.

5.5.4 Protection against water hammer

5.5.4.1 Protection equipment

Protection equipment is necessary when the pump installation would be damaged by the resultant:

- Excess pressure
- Risk of cavitation
- Negative pressure in the line

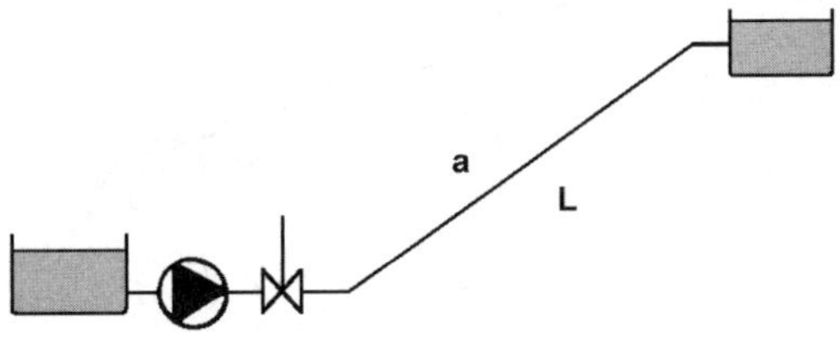

Figure 5.45 Pump installation with slow-closing valve

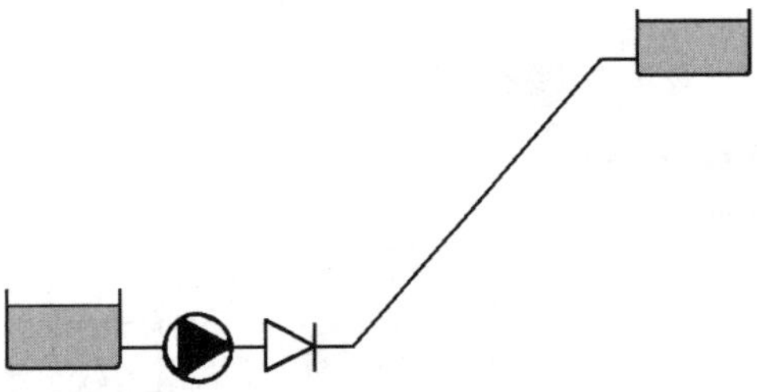

Figure 5.46 Pump installation with non-return valve

- Reverse running and overspeed

Rotodynamic pumps should usually be fitted with non-return valves; this is not considered protection in this context. Installations subjected to repeated water hammer must be designed to withstand fatigue.

Detailed water hammer calculations are complicated and costly. Specialist consultants can be employed to investigate the effects and make recommendations for protection. This raises the question as to when a detailed investigation is necessary. For certain installations simple estimations can give an indication as to the magnitude of water hammer.

5.5.4.2 Valve closing time T

In Figure 5.45 the pump starts and stops against a closed valve. The pressure changes at the valve in connection with closing can be approximated by:

$$\pm\Delta H = \frac{2\cdot L\cdot v}{g\cdot T} \quad \text{for } T > 2\cdot\frac{L}{a} >$$

$$\pm\Delta H = \frac{a\cdot v}{g} \quad \text{for } T \leq 2\,\frac{L}{a} \qquad \text{Equ 5.23}$$

where:

L	=	pipe length (m)
v	=	initial flow velocity (m/s)
g	=	acceleration due to gravity (m/s^2)
T	=	valve closing time (s)
a	=	wave speed (m/s)

The above formula assumes a linear valve closing characteristic, i.e. linear reduction in flow in relation to time.

5.5.4.3 Pump stops with non-return valve

In Figure 5.46 the pressure changes with non-return valves in conjunction with the pump stopping are approximated by:

$$\pm\Delta H = \frac{2}{\frac{g\cdot C}{v\cdot L} + \frac{K}{H}} \qquad \text{Equ 5.24}$$

Equation 5.24 does not apply to special non-return valves which have delayed action and/or anti-slam features.

Calculation example:

The following data applies to a pump installation with a non-return valve; pipe length L = 1700 m, delivery head H = 22 m and flow velocity v = 1.3 m/s in the pipeline. Estimate the magnitude of pressure surge when the pump stops.

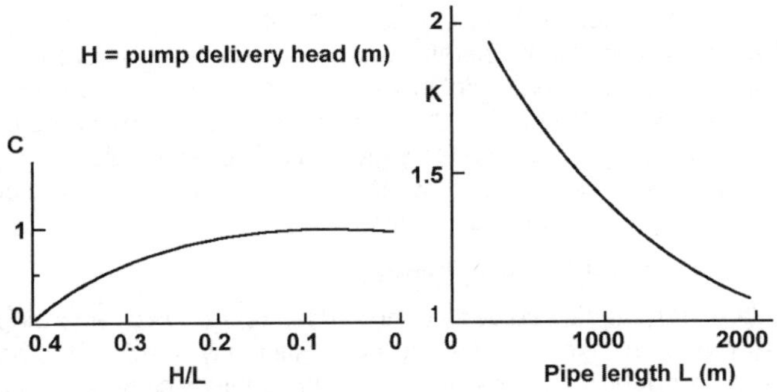

Figure 5.47 Values for C and K in Equ 5.24

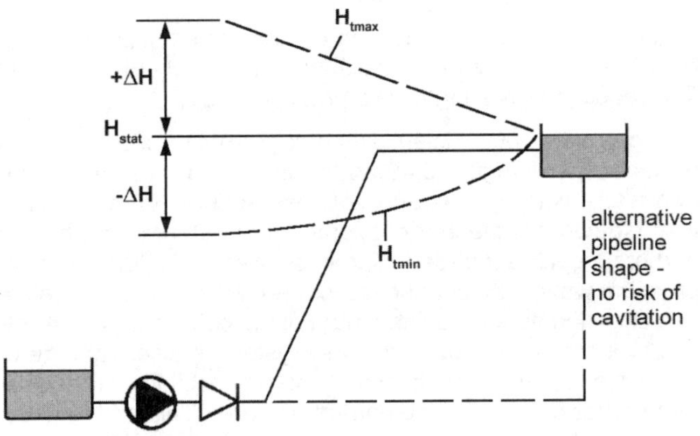

Figure 5.48 Reviewing hydraulic gradients

L = 1700 m → K = 1.1, H/L = 22/1700 → C = 1

Then according to equation 5.24

$$\pm\Delta H = \frac{2}{\frac{g\cdot 1}{1.3\cdot 1700} + \frac{1.1}{22}} = 37 \text{ m}$$

5.5.4.4 Pump stops without non-return valve

The installation is the same as in Figure 5.45. The pump stops because of power failure. The slow-closing valve remains open.

$$\text{If } \frac{h_f}{H} > 0.2 \qquad \text{Equ 5.25}$$

then

$$H_{tmax} \leq H$$

where:

h_f	=	pipe system head loss (m)
H	=	pump delivery head (m)

The size of the pressure pulse, -ΔH, is estimated from Equ 5.24 using the values from figure 5.47.

5.5.4.5 Hydraulic gradients

The rules of thumb take no account of the shape of the hydraulic gradient along the pipeline. The gradients shown in Figure 5.48 are based on estimates derived from experience. It is, however, possible to conclude that there are definite risks of cavitation in the case of the pipeline shape as represented by the continuous line. The chain-dotted line routing eliminates this risk. It then remains to be shown whether or not H_{tmax} at the valve is acceptable.

By means of these simple estimations it is possible to judge many non-critical cases, usually having short pipe lengths and low flow velocities. In cases of doubt it is advisable, as the next stage of investigation, to refer to the sizing diagrams which are published in technical literature etc. These cover more thorough calculations for pump stops without non-return valves for certain simple pump installations and also quote figures for H_{tmax} and H_{tmin} at mid-point in the pipeline. (See work by Kinno and Kennedy.) If there is still cause for doubt then the only solution is to carry out a detailed digital computer simulation using the Method of Characteristics. See also work by Parmakian, or an analogue computer simulation.

5.5.4.6 Protection equipment

Water hammer is avoided during normal operation by ensuring that load changes are carried out sufficiently slowly. Starting and stopping of pumps is carried out against closed valves and/or by using by-passes. Extra protection equipment is therefore usually installed in order to protect the pump installation against power failure.

A large proportion of protection equipment attempts to eliminate the basic cause of water hammer, namely the too rapid flow reduction resulting from a power failure.

A surge column of sufficient diameter and height provides good protection, see Figure 5.49. In the event of power failure the non-return valve closes and isolates the pump from the pipeline. Liquid from the surge column flows out into the pipeline and prolongs the retardation process, weakening the negative pressure wave. The column refills when the flow in the pipeline has changed direction. The mouth of the column can be throttled to increase dampening in the system. A disadvantage of the surge column is that it must be at least as high as the maximum pressure head at the non-return valve. This problem can be avoided by means of an air chamber, Figure 5.50.

During normal operation the air pressure in the air chamber is equal to the pump delivery head. When pumping stops liquid flows out of the air chamber and the air pressure reduces. The greater the volume of air the smoother the process. The system can be suitably dampened by throttling the chamber at the flow inlet. A disadvantage with the air chamber is the need to provide a compressor to replace the air which is dissolved in the liquid and carried away in the pipeline. This may still be considerably cheaper than building a much taller water column.

The problem of air dissolution in the water can be greatly reduced by using a proprietary surge arrester which uses a nitrogen charge stored within an elastomer envelope. Direct contact of gas and water is eliminated. The elastomer envelope is not 100% gas tight and a very small gas loss is experienced. The gas volume can be topped up as necessary during routine maintenance.

The retardation of the pump rotor is also affected by the moment of inertia of the rotating parts. By fitting the pump with a flywheel (Figure 5.51) it is possible to prolong the run-down time. The method is most effective for high speed pumps and moderate pipe sizes. A disadvantage is increased starting difficulties. The starting problems can be alleviated by using modern soft starters. Variable speed pumps should not have starting problems.

With low or negative static heads there is a risk of negative pressure on the pressure side of the pump. The pressure drop across the pump is reduced by means of a by-pass line to augment forward flow when the pump has stopped, Figure 5.52.

The other main group of protection equipment is intended to eliminate water hammer at a somewhat later stage of the transient process.

Negative pressure, with the associated cavitation risks can be avoided by placing a surge tank at critical points (Figure 5.53). In the event of negative pressure, liquid flows from the tank into the pipeline. The tank is refilled during normal operation via a separate level controlled filling line.

An alternative method is to let air into the pipeline when the pressure falls below atmospheric pressure. The pipeline must be vented, however, prior to re-starting. In installations having large static delivery heads it is often the excess pressures which occur upon reversal of the liquid column and in bringing it to rest, which are most critical. These pressures can be reduced with the aid of control valves, (Figures 5.54 to 5.56).

In order not to amplify the initial pressure wave, the negative pressure wave, the controlled by-pass valve remains closed, in principle, until the flow at the pump is reversed. It then opens

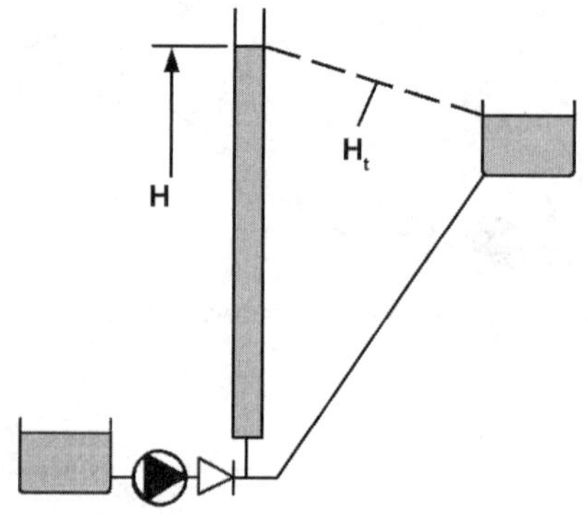

Figure 5.49 Surge column

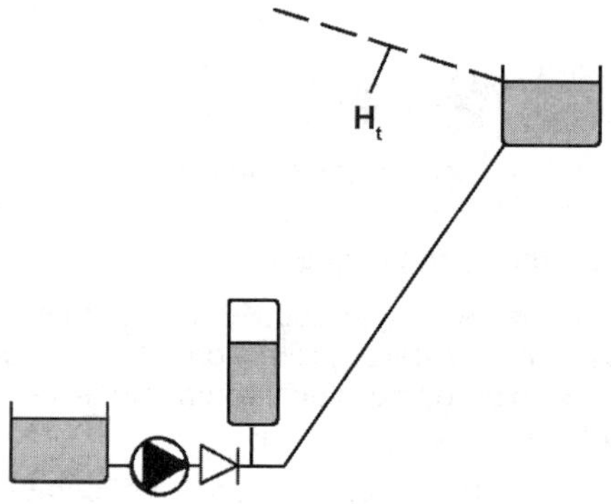

Figure 5.50 Air chamber

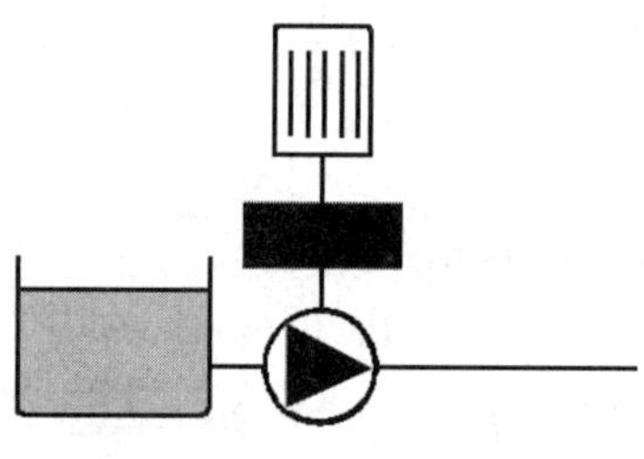

Figure 5.51 Pump with flywheel

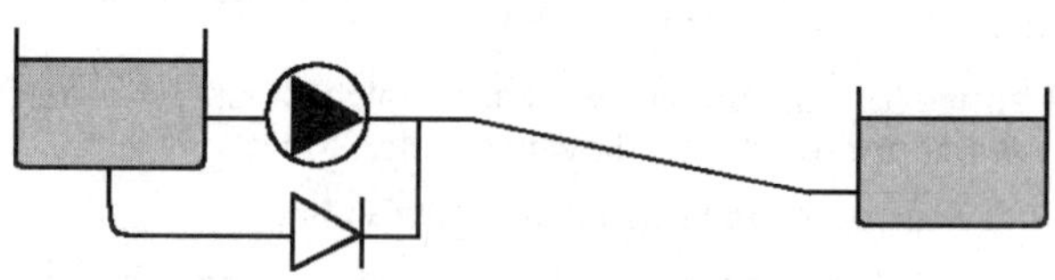

Figure 5.52 Pump with reverse by-pass

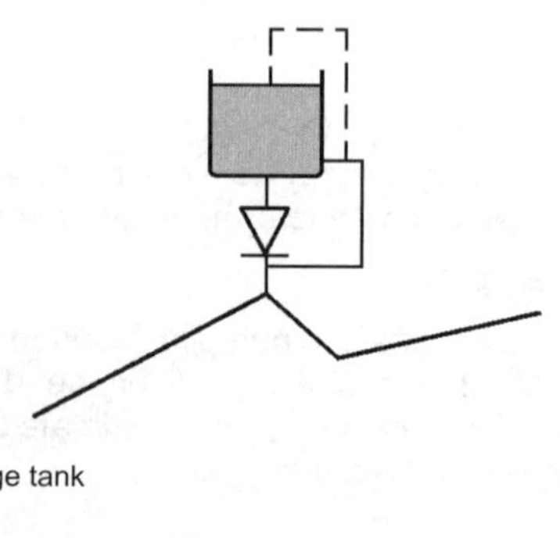

Figure 5.53 Surge tank

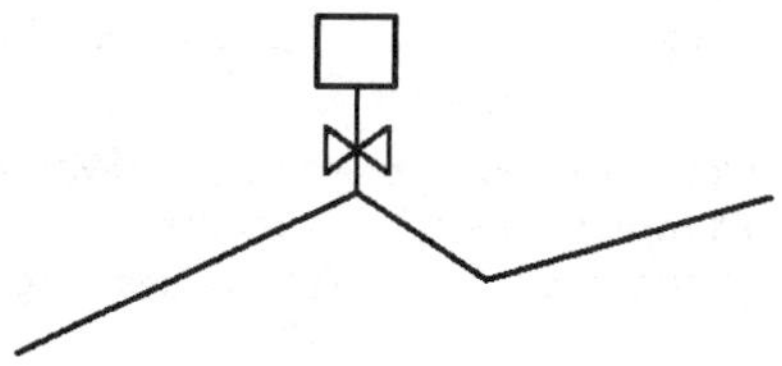

Figure 5.54 Automatic air bleed valve

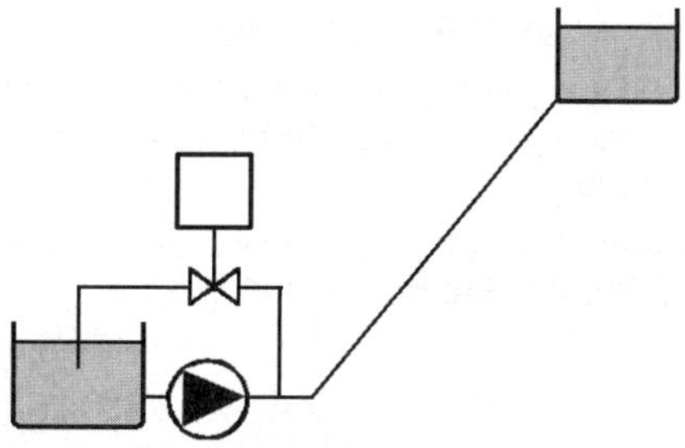

Figure 5.55 Controlled valve in by-pass line

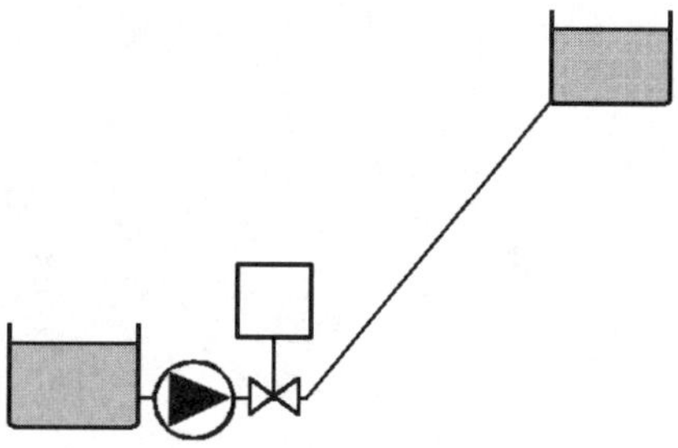

Figure 5.56 Controlled valve in main line

rapidly, thereby relieving pressure in the pipeline and finally it closes slowly.

The controlled valve in the main line should remain open until the flow at the pump reverses. It then closes rapidly to its optimum position, ideally to retard the reversed flow with optimum distribution of pressure drop between the pump and valve.

Pump reverse rotation can be prevented by means of a reverse running clutch, Figure 5.57. The reversed pump flow is then retarded more smoothly and the pressure surge is dampened. The risk of reverse overspeed is eliminated.

Apart from the pure pressure surge dampening characteristics, the choice of protection equipment is also affected by other factors; for example, liquid characteristics, construction, maintenance and procurement costs. Each case must be judged on its own merits as to the most suitable choice.

5.6 Pressure pulsations

As shown in Section 5.5, pressure surges can occur due to isolated operating conditions such as start-up, shut-down and valve operations. These are not the only conditions under which rapid pressure changes occur. Some systems run continuously with pressure pulsations.

5.6.1 Rotodynamic pumps

It is generally thought, and often stated in literature, that rotodynamic pumps produce a continuous steady flow pattern without pressure pulsations. Axial flow pumps probably come closest of all rotodynamic pumps to fulfilling this requirement; mainly because of the low differential head produced.

Mixed flow pumps tend to be used for low differential head applications with high flow rates and consequently have wide impeller tip widths. Wide impeller passages, in the axial direction, produce uneven velocity streams with the liquid. The liquid leaving the pump discharge connection is not a steady stream but a mixture of several flow streams. As the liquid flow pattern stabilises in the discharge pipework, pressure pulsations and

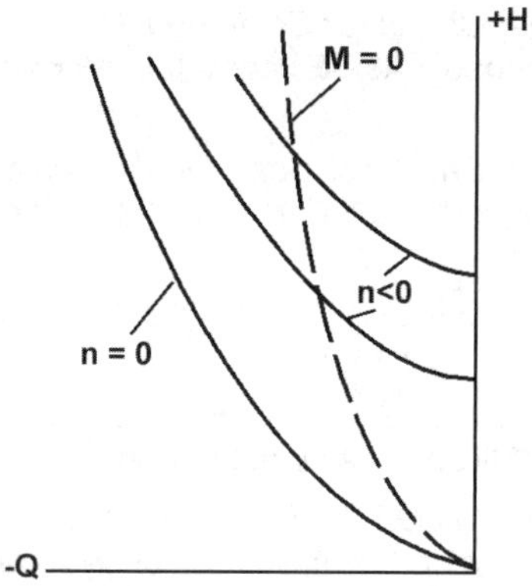

Figure 5.57 Effects of reverse running clutch

flow variations can be measured. If flow measuring devices are to be used close to the pump discharge some form of flow straightening may be necessary.

Centrifugal pumps operate at a wide range of speeds. As speed is increased the circumferential flow pattern between adjacent vanes is modified. Liquid tends to be forced closer to the advancing vane and dragged away from the receding vane. It has been suggested that the flow between a pair of vanes will segregate into three distinct flow streams moving at different velocities. The action of one or two cutwaters produces additional pulsations. It has been proposed in new test standards that fluctuations in flow readings of ± 3%, smallest tolerances, are acceptable. Obviously flow variations of this magnitude will have associated pressure pulsations.

Centrifugal pumps can have additional problems. When operating at low flows some pumps surge in the suction pipework due to recirculation in the impeller eye. Pump manufacturers should be able to quantify the onset and magnitude of suction surge.

5.6.2 Positive displacement pumps

The flow variations of reciprocating pumps are well documented, even if incorrectly. These flow variations are of sufficient magnitude to produce pressure pulsations which create pipework vibration which do not require special instrumentation to detect. Some other positive displacement pumps produce pressure pulsations. Lobe pumps and gear pumps with few teeth deserve special attention.

Newer test standards recognize the problems caused by pressure pulsations and include test requirements. However, testing on any test rig will not reproduce site conditions or site pulsation levels. Pressure pulsations can be amplified by pipework resonance and cause severe problems. Problems with site pipework can only be detected by computer simulation or site operation.

Most pressure pulsation problems can be controlled by pulsation dampers, see Chapter 11, Section 11.6, or by simple pipework modifications to eliminate resonance. The proprietary gas-charged surge arrester, mentioned in Section 5.5.4.6, can also be used to attenuate pressure pulsations created by continuous cyclic flow variations. Surge arresters can be sized to provide an attenuation up to 90%. Correct sizing is dependent upon a fairly accurate prediction of all operating pressures.

5.7 Useful references

Water Hammer Control in Centrifugal Pump Systems, H. Kinno, Journal of the Hydraulics Division, May 1968, pp 619-639.

Water hammer analysis, J. Parmakian, 1963, Dover Publications, New York.

Water – Hammer Charts for Centrifugal Pump Systems, H. Kinno & J.F. Kennedy, ASCE Journal of the Hydraulics Division, 1965.

Water – Hammer Charts for Centrifugal Pump Systems, (H. Kinno & J.F. Kennedy, ASCE Journal of the Hydraulics Division, 1965.

Fluid transients in systems, E. B. Wylie, V. L. Streeter,. (1993), Prentice Hall, New York.

Hydraulic analysis of unsteady flow in pipe networks, J.A. Fox, 1987, MacMillan Press Ltd, London.

Simple Guide for Design of Air Vessels for Water Hammer Protection of Pumping Lines, Journal of Hydraulics Engineering, D. Stephenson, (Director, Water Systems Research Group, Univ. of the Witwatersrand, P.O. Box 277, WITS 2050, South Africa), Volume 128, Issue 8, pp. 792-797 (August 2002).

Simple Guide for Design of Air Vessels for Water Hammer Protection of Pumping Lines, D. Stephenson, Giuseppe De Martino et al., Journal of Hydraulics Engineering, 130, 273 (2004).

Simple Guide for Design of Air Vessels for Water Hammer Protection of Pumping Lines, D. Stephenson, Journal of Hydraulics Engineering, 130, 275 (2004).

Practical designs against pump pulsations, M. Corbo, C. F. Stearns, Proceedings of 22nd International Pump Users Symposium 2005

Contractual and physical risks from water hammer, Geoffrey Stone, World Pumps, February 2006.

Flow regulation and control 6

6.1 Introduction

When dealing with liquid transportation, a pump installation is sized to handle a maximum flow which, in practice, might never occur. This principle of sizing is nevertheless correct, since inadequate pumping capacity on isolated occasions can lead to the most damaging consequences. It is also obvious that sensible ways of varying the flow are important, not only for the functioning of the pump installation itself but also for the process for which the pump is included as a major or minor component.

Plant procurement costs generally amount to a small proportion of the total pump investment cost of a plant. But not always! Some pumps cost over $2M each and the remainder of the installation is just pipework. In most installations the functional quality of the pump may be the critical factor for the overall operation and its associated running costs. Advanced methods of flow control, e.g. speed regulation of pumps, are nowadays the best methods of varying the flow, but with the disadvantage that initial investment costs for the pump equipment are increased. The advantages are reduced running costs overall, smaller requirements for storage tanks and sumps, smaller electric cabling, and a reduction in starting current.

With existing installations, the method of flow regulation affects, most of all, the energy costs. As a result of increasing energy costs, the optimum methods of pump control have changed greatly from those with low initial costs and high energy consumption to those with relatively high investment costs and good power economy. For pumps with a power requirement greater than 10 kW, it will even now be beneficial to convert to more power-saving methods of regulation. See also Chapter 14.

Pump flow can be controlled in steps, discrete flow regulation or smoothly, continuous flow regulation or infinitely variable flow regulation. Combinations of the two methods are also used. The choice of method is determined by the overall requirement of the pump installation, variation of flow, timescale, operating economics, initial cost and technical possibilities.

In this Chapter, pump, may also mean pumps, for example multiple pumps of the same size operated in parallel. When the detailed description of the installation is explained it will become obvious whether a single pump or multiple pumps are involved.

The operating point of the pump at a particular speed is the intersection of the system characteristic and the pump curve as in Figure 6.1. At this point, the energy dissipation of the system is equal to the energy supplied by the pump.

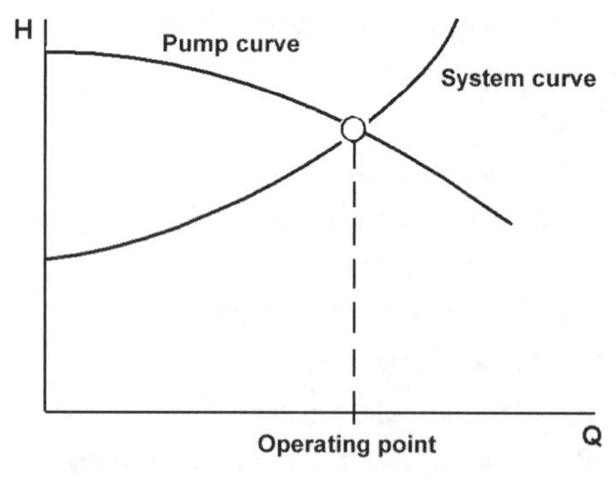

Figure 6.1 The operating point of pump and system

6.2 Variable flow requirements

Whether or not a pump needs flow regulation is dependent upon the pump type and the nature of the process involved. The pump flow may not match process demands because of:

- Safety margins and hydraulic tolerances
- Failure to utilise full production capacity
- Variations in requirement caused by climatic variations, e.g. summer, winter, rain, drought
- Variations in production processes

Flow variation with time can be described either by a running curve, Figure 6.2 or by a constancy curve, Figure 6.3.

In the constancy curve, time has been summated for all periods of flow having the same magnitude. A point on the constancy curve gives the time during which the flow at least amounts to a particular value.

Constancy curves can be made up for various periods of time ranging from a full 24 hours to the lifetime of the installation. Constancy curves are usually made up for a period of 1 year.

Unfortunately, constancy curves are not that easy to compile, but a simple analysis of the actual liquid transportation often gives adequate guidance for the determination of maximum, minimum and average flow. The mean flow, for example can be determined from the annual production capacity and maximum flow from peak consumption.

The pump size is obviously determined by the maximum flow figure, whereas pump economy is determined by the duration of flow. Economic assessments are easily made if the volume flow is known at 100%, 50% and 0% constancy. It would be even better, of course, if the values at 25% and 75% are also known.

The following simple equation is obtained by the use of Simpson's Rule for integration:

$$Q_{mean} = \frac{1}{6} \cdot Q_{100} + \frac{2}{3} \cdot Q_{50} + \frac{1}{6} \cdot Q_0 \qquad \text{Equ 6.1}$$

where:

Q_{mean} = mean flow (m^3/s)

Q_{100} = flow at 100% constancy, i.e. the flow taking place 100% of operational time ($Q_{100} = Q_{min}$) (m^3/s)

Q_{50} = flow at 50% constancy (m^3/s)

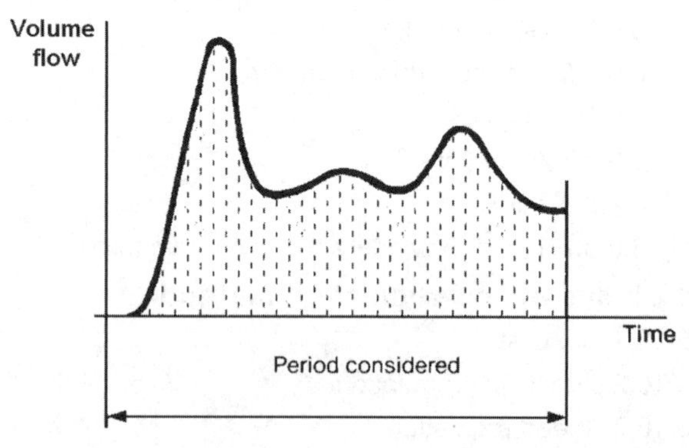

Figure 6.2 Flow demand as a function of time, running curve

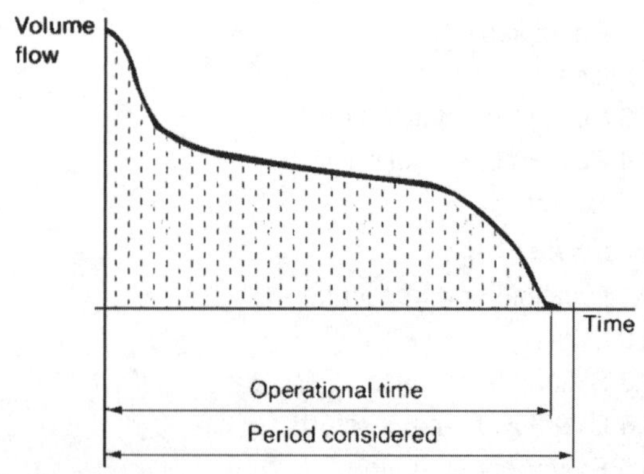

Figure 6.3 Constancy curve for flow

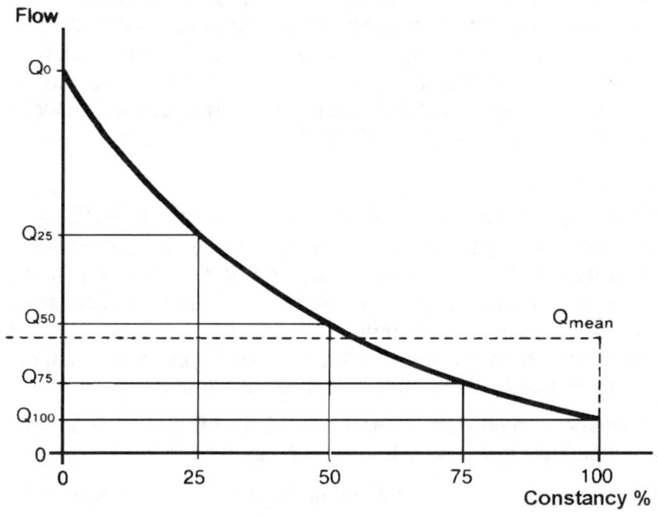

Figure 6.4 Graphical representation of Equations 6.1 and 6.2

Q_0 = flow at 0% constancy ($Q_0 = Q_{max}$) (m^3/s)

or if more points on the curve, see Figure 6.4, are known:

$$Q_{mean} = \frac{1}{12} \cdot Q_{100} + \frac{1}{3} \cdot Q_{75} + \frac{1}{6} \cdot Q_{50} + \frac{1}{3} \cdot Q_{25} \qquad \text{Equ 6.2}$$

where Q_{75} and Q_{25} are expressions for the flow at 75% and 25% constancy.

6.3 Flow regulation

6.3.1 Methods

As Table 6.1 shows, flow can be regulated in many ways. Throttle regulation, by-pass regulation and overflow by-pass regulation all mean that the control takes place outside the actual pump itself. Some of the flow produced by the pump is thrown away, wasting energy.

Method	Advantages	Disadvantages
	Discrete variations	
Start-stop control	Cheap, easy to operate	Flow surges, water hammer, wear and tear on switchgear
On-off by-pass return	Relatively cheap, easy to operate	Flow surge problems eliminated with slow acting valve
Pole-changing motors	Relatively cheap, easy to operate	Must stop to change speed, limited number of speed options, flow surges, water hammer
Multi-speed gearbox	Easy to operate	Must stop to change speed, flow surges, water hammer, may need maintenance staff intervention (1)
Vee belts and pulleys	Cheap	Must stop to change speed, flow surges, water hammer, needs maintenance staff intervention
	Continuously variable	
Throttling	Cheap, easy to operate	Wastes energy, not for solids handling applications
Modulating by-pass return	Easy to operate	Wastes energy, not for solids handling applications
Overflow by-pass return	Easy to operate	Wastes energy
Variable speed	Energy efficient	Can be costly to install
Adjustable blade angle	Energy efficient	Axial pumps only, low differential head
Adjustable inlet guide vanes	Energy efficient	Limited flow range, can be costly to modify pump
Part loading	Energy efficient	Archimedean screw only
Adjustable stroke	Energy efficient	Reciprocating pumps only, very costly in larger pumps

(1) Portable engine driven pumps frequently use semi-automatic transmissions, no need to stop to change gear.

Table 6.1 Pump flow regulation methods

Other methods change the actual performance of the pump including power consumption, either within the pump unit or by means of variable filling, e.g. Archimedean screw pumps and displacement regulation. As can be seen from Table 6.1, the only universally applicable method is speed regulation, i.e. it can be used for all types of pumps, from the smallest dosing pump with power requirements of 10 to 100 W to the largest possible pump units in power stations with power requirements up to 60 MW.

There are no rules against combining various methods, e.g. the parallel operation of several units with continuous regulation by means of throttling or speed regulation. The choice of optimum methods requires great experience and often extensive economic and process technical analyses. Sometimes there are also special problems to be considered, such as noise levels from control valves when throttling.

As mentioned earlier, it is possible to combine control methods. If a modulating by-pass was fitted to most of the discrete variation methods the problems of surge and water hammer could be eliminated.

Series and parallel operation are dealt with in Chapter 5, Sections 5.3.1 and 5.3.2 respectively. Examples of pump performance for pumps with adjustable impeller blades, or with adjustable inlet guide vanes, pre-rotation, are shown in Figures 4.21 and 4.22 respectively. Archimedes rigid screws flow regulation is dealt with in Section 1.5.3, Chapter 1. Metering pump flow regulation is dealt with in Metering pumps in Section 1.5.19, Chapter 1.

Where changes in flow requirements are needed over longer periods, more than about six months, say, it may be economic to change the diameter of a centrifugal pump impeller or piston/plunger or modify the blade angles on a propeller. See also Chapter 4, Section 4.2.2.

6.3.2 Control parameters

Control parameters may be flow, pressure, or some quantity derived from these such as level, temperature, concentration and so on. Some examples of these are shown in Table 6.2.

Control parameter	Sensor type	Example of application
Discrete level control	Level switch	Fresh water reservoirs Emptying separators
Proportional level control	Level transmitter	Process control on reactors and vessels
Constant flow	Flow meter	Process control for variable pressure systems
Discrete pressure control	Pressure switch	Charging vessels and accumulators. Pressure maintenance
Constant pressure	Pressure transmitter	Process control of vessels and accumulators
Proportion control for mixing	Flow meter in each circuit	Chemical processing Water treatment
Proportion control for mixing	Flow meter in mixer outlet	Chemical processing Water treatment

Table 6.2 Some control parameters

The principles of regulation can also be shown by the H-Q diagram, see Figure 6.5 for the simplest cases.

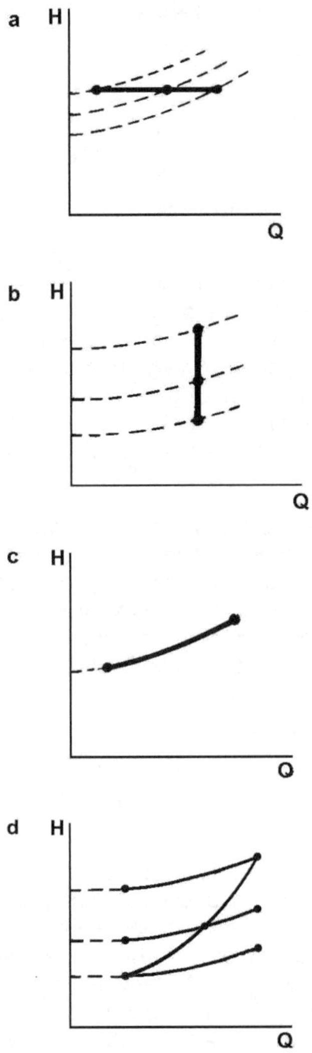

Figure 6.5 Pump regulation for system configurations
(a) constant pressure for variable system
(b) constant flow for variable system
(c) variable flow for a fixed system
(d) variable flow for variable system

6.4 On-off control of constant speed pump

6.4.1 Principles

On-off control means that one or more pumps are started and stopped, or loaded and unloaded, systematically, according to the capacity requirement. Loading and unloading is accomplished by an isolating valve in a by-pass. When the flow requirement does not coincide with the pump capacities the pumps will work intermittently at a starting frequency of 1 to 15 starts per hour for small pumps, reducing to 1 to 4 starts per hour for larger pumps. To account for the difference between instantaneous flow demand and the supply flow, some form of liquid storage or reservoir must be provided.

On-off control and load/unload control is arranged automatically so that pumps are started at a high level in a storage vessel on the suction side of the pump or at low level in a storage vessel on the discharge side. When pumping between a suction tank and a discharge tank there is no need for other vessels or tanks to store the imbalance between supply and demand. The storage capacity will be included in the suction/discharge tanks.

For systems using pressure rather than level control, gas charged storage vessels, called accumulators, are used to store the imbalance. Pumping begins with a low pressure signal from the accumulator and stops at a high pressure signal. The accumulator(s) must be sized so that motors do not start more frequently than their specification allows. A combination of on-off and load-unload can be very effective with larger pumps to eliminate starting problems and also increases the efficiency of load-unload used in isolation.

6.4.2 Costs

Of the discrete flow control methods, the on-off and load-unload methods are the most usual. The on-off control method is the cheapest of all methods to install. The additional costs for full automatic control may be limited to a single level switch, extra control contacts in the motor starters and some wiring. The starters for the pump motors are required for every installation, control contacts are standard with some models.

However, consideration must be given to the following when additions are necessary to the pump installation:

- Cost of extra storage in which to hold the liquid during the regulation cycle
- Additional costs for heavy duty starting equipment since this has to be selected to cope with a very large number of regular starts
- Additional costs for equipment to assist in the alleviation of water hammer when starting/stopping and loading/ unloading
- Additional costs for pipework and fittings if these have to be designed to withstand the fatigue stresses and vibration induced by water hammer

6.4.3 Starting/loading problems

The number of permitted starts per hour is a critical factor in total initial cost for the pump installation. The type of starting used, direct-on-line, DOL, or star-delta, S-D, or a modern soft start system affects the frequency of starting. Small motors can be started 20 times per hour without problems, depending upon the operating conditions. Larger motors, 125 kW and above, have strict restrictions on starting, 4 starts per hour maximum is very common. Experience shows that the number of starts chosen can be 5 to 10 starts per hour. Thus there will be 20,000 to 50,000 starts per year and up to the time of financial write-off, the number of starts will amount to several millions, thus some components have to be sized to withstand fatigue if precautions are not taken.

Even if the starting time of the pump is short, about 0.25 s with DOL and about 1 s to 1.5 s for S-D, this can put a heavy strain on the mains supply.

Loading and unloading by automatic by-pass removes the electrical problems associated with starting and stopping. This approach should be seriously considered as motor sizes increase. A slow acting by-pass valve can eliminate the surge and water hammer problems. Fatigue may have to be considered if parts of the discharge pipework system cycle from low pressure to high pressure with every loading cycle. Pressure system code compliance, without fatigue considerations, is usually limited to 7000 pressure cycles. Much larger cycle counts may reduce the allowable stress levels to 50% resulting in design difficulties or routine replacement of piping sections.

The selection of the by-pass valve is of critical importance for the reliability of the whole system. Water systems can impose unexpected difficulties due to dissolved air and the deposition of salts. By-pass valves must have a throttling capability for the smooth transition from full flow to zero flow.

Reciprocating pumps can avoid the starting and by-pass valve problems by using suction valve unloaders. During periods of zero flow a pneumatic or hydraulic control system lifts the suction valves off their seats. No discharge flow is possible and "slugs" of liquid recycle in the suction manifold. When flow is required the suction valves are lowered, in sequence, to commence the normal pumping operation. If gas charged pulsation dampers are fitted they can be slightly oversized to provide some surge protection.

6.4.4 Stopping/unloading problems

A liquid flowing in a lengthy pipeline possesses considerable kinetic energy. When the liquid decelerates with the stopping of the pump, large pressure variations occur. This can be calculated and is, for abrupt stops:

$$\Delta H = \frac{a \cdot v_o}{g} \qquad \text{Equ 6.3}$$

where:

ΔH = change of pressure head (m)

a = wave speed (m/s)

v_o = initial liquid velocity (m/s)

g = gravitational acceleration (m/s^2)

In order to reduce the variation of head the flow has to be decelerated over a period of time, t_o (s), which can be roughly assessed in a pipeline of length L (m) as:

$$t_o = (15 \text{ to } 60) \cdot \frac{L}{a} \qquad \text{Equ 6.4}$$

In the case of pipelines with high points or with low resistance to fatigue, the stopping time may have to be doubled. The normal arrangements used to increase the stopping time or to reduce the stresses are:

- Slow-closing valves
- Air-chambers or pulsation dampers
- Pumps with flywheels
- Surge columns
- Overpressure and underpressure valves

See Chapter 5, Section 5.5.4 for full details.

In all cases the flow continues for a time t_o. To avoid sucking air into the pump and pipeline, a sufficient supply of liquid must be available on the suction side of the pump, whilst that same volume of liquid must not cause trouble on the delivery side, through surge for example.

For air chambers, the volume of liquid is contained in the air chamber and in other cases, the storage tank or pump sump has to supply the corresponding increase in volume. When the flow requirement or inflow is negligible, which is the most unfavourable condition, the mean flow during the stop cycle can be estimated to be about ⅔ of the maximum flow. The partial storage volume V_o (m^3), between stop signal and stop flow will, for the decelerating time t_o s, be as before:

$$V_o = (10 \text{ to } 4) \cdot \frac{L}{a} \cdot Q_p \qquad \text{Equ. 6.5}$$

where:

Q_p = pump flow (m^3/s)

With longer pipelines, especially for installations with slow closing valves, a large part of the operational period may be taken up by the start and stop cycles, see Figure 6.6. During these transients, throttle regulation or modulating by-pass takes place, see also Sections 6.7 and 6.8. Power consumption is determined by the relationship between the times for pure on-off operation and those for valve operation. Throttle regulation and modulating by-pass can increase energy consumption by anything up to 50 %.

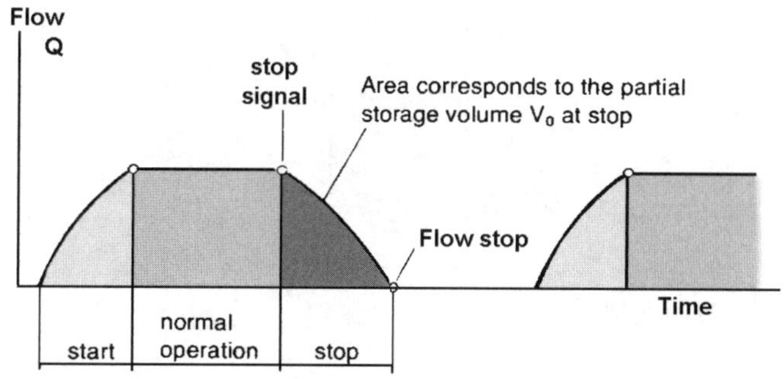

Figure 6.6 Flow variation for on-off and load/unload control with long pipeline

6.4.5 Operational sequences

There is a choice of several different operational sequences for switching in and out parallel operation units. Critical factors to be considered are:

- Number of permitted starts per pump per hour
- Number of permitted starts during the lifetime of the installation
- Permitted variations in flow
- Costs of storage tanks, sumps and accumulators

The differences between various possible sequences, designated A, B and C, are shown in the following three schematic examples, Figures 6.7, 6.8 and 6.9 respectively, applied to an installation with two identical parallel operating pumps P1 and P2. It becomes clearest if the condition is examined when the supply flow is equal to 1.5 times the flow of one pump. The other operational possibilities— when the supply flow is less than the flow of one pump, and when the supply flow is greater than the flow of two pumps, must also be considered for a complete control strategy. In these examples, on and off can be synonymous with load and unload.

In **Operational Sequence A**, P1 operates continuously and P2 operates on and off in response to a level switch signal. The cycle time is obtained when sizing P2's storage or partial sump volume. In this operational sequence there are moderate flow variations.

In **Operational Sequence B**, P1 starts first, and then, at a higher level, P2 starts. Stop occurs simultaneously for both pumps at sump minimum level. The flow variation is very wide, but the storage or partial sump volume for P2 is reduced to almost half of that for Sequence A. Because of the common stop,

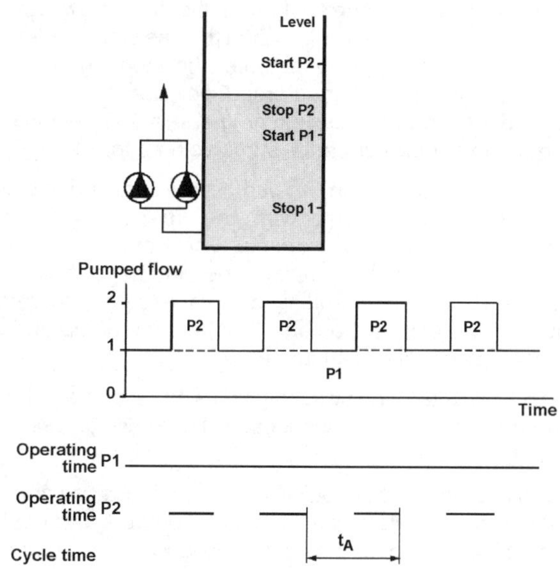

Figure 6.7 Flow variation and operating times for operational sequence A

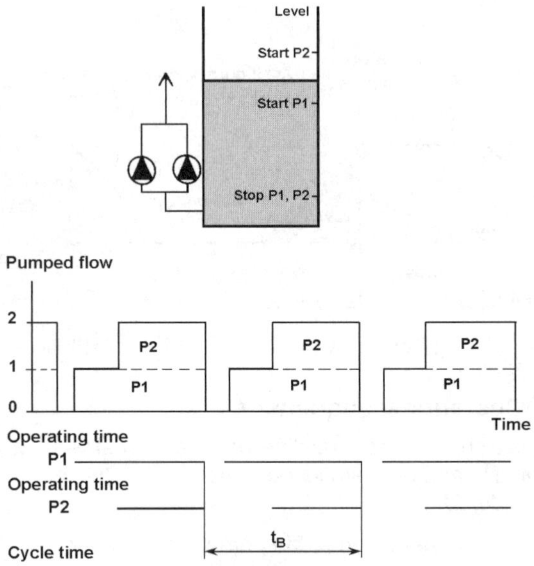

Figure 6.8 Flow variations and operating times for operational sequence B

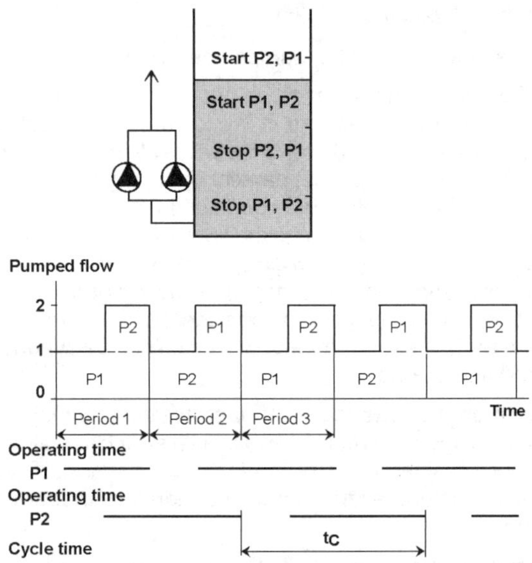

Figure 6.9 Flow variation and operating times for operating sequence C

one sensor signal is saved. Operational Sequence B has been applied to about 90% of all sewage pump installations up to the present time.

In **Operational Sequence C**, there is the same flow variation as in Operational Sequence A, except that the stop sequence is the same as the start sequence, i.e. the stop signal always stops the pump which started first. Operational Sequence C means a doubling of the cycle time, i.e. $t_C = 2_{tA}$, meaning the halving of the storage or partial sump volume for P2.

Where there are more than two pumps, the same principles as above are used, as also with the speed changing of pole-changing motors, for example. For a number of pumps working in parallel, the partial storage volume is then determined for each pump depending upon its respective incremental flow Q1, Q2, Q3, etc., as obtained from the pump and system curves, according to Figure 6.10.

In practice, to determine switching times and storage volumes, supply flows very close to pump capacities would be used, e.g. 99%, 101%, 199%, 201%, etc.

Powerful Programmable Logic Controllers, PLCs are currently available with multiple digital inputs and outputs and cost approximately $260; about the same as a good pressure switch. These devices allow on-off control to be completely optimised. For example, the PLC could keep a running total of the "hours

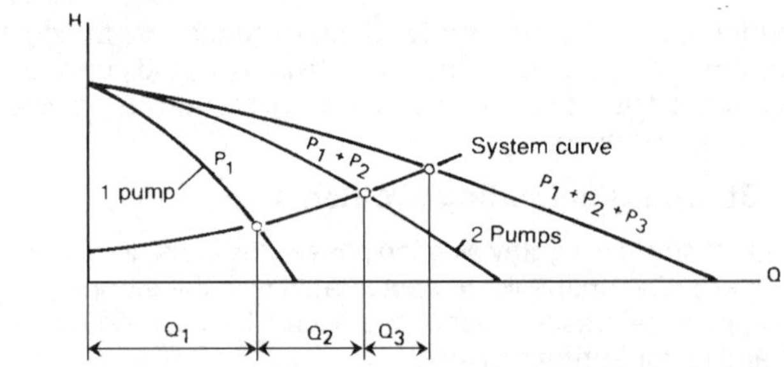

Figure 6.10 Flow increments Q1, Q2, and Q3 obtained by switching in various numbers of parallel operating pumps

run" for each pump and start the pump with the least hours first and switch it off last. A simple programme would try to balance the operating time across all the pumps. Operating times could also be biased so that one pump had few hours and could be considered as a stand-by.

When considering how many pumps to install, the possibility of failure and allowances for unavailability due to routine maintenance must be considered. The need for stand-by capacity must be weighed against the consequences of unavoidable reduced capacity.

6.4.6 Storage volumes

Storage and pump tank/sump volumes, see Figure 6.11, are most easily calculated by the use of a filling time, time constant, for the various partial volumes V, and V_2, and so on, defined as:

$$t_2 = \frac{V_2}{Q_2} \text{ etc.} \qquad \text{Equ. 6.6}$$

Time t_1 thus corresponds to the time taken to fill volume V_1 with flow Q_1.

In the case of an accumulator, Q varies because of counter-pressure from the pre-compressed volume of air or nitrogen. Boyle's Law at constant temperature: $p_1V_1 = p_2V_2$, should be used to calculate the change in gas volume with pressure. Time t_1 for the pump switched on first depends solely on the number of permitted starts per unit time, depending on whether starting is applied to several pumps alternately.

Time t_1 is determined by

$$t_1 = \frac{3600}{4 \cdot m} \text{ (s)} \qquad \text{Equ. 6.7}$$

where:

m = number of starts per hour per pump if the pumps are not alternated

or

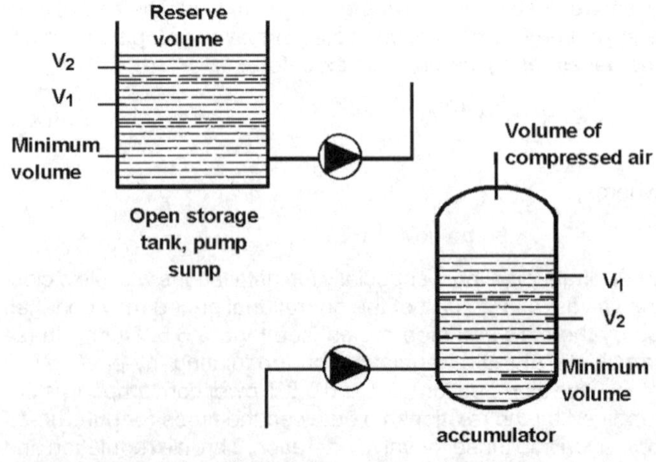

Figure 6.11 Storage with associated minimum volumes, partial volumes V_1 and V_2 and the volume of air for various arrangements

m = number of starts per hour for the whole installation, if the pumps are alternated

Here the number of starts per hour is a maximum when the mean flow per hour is equal to half the maximum flow. The times t_2, t_3, t_4, etc., are heavily dependent upon the operational Sequences A, B and C.

Table 6.3 shows various values for up to four pumps operating in parallel. In the table it is assumed that no more pumps are included in each operational sequence range than are needed to satisfy the flow requirements. An example also follows.

Number of starts/hr per pump	5			10			15		
Sequence	A	B	C	A	B	C	A	B	C
Filling times for:									
pump 1: t_1	180	180	180	90	90	90	60	60	60
pump 2: t_2	180	72	90	90	36	45	60	24	30
pump 2: t_3	180	48	60	90	24	30	60	16	20
pump 2: t_4	180	40	45	90	20	22	60	13	15

Table 6.3 Table for determination of filling time for partial volumes in pump tanks/sumps

The air volume in an accumulator is calculated using Boyle's Law for isothermal compression, Figure 6.12.

At the absolute pressures p_{off} and p_{on}, there are respectively the volumes of air V_{Aoff} and V_{Aon}:

$$p_{off} \cdot V_{Aoff} = P_{on} \cdot V_{Aon} = P_i \cdot V_{Ai} \qquad \text{Equ 6.8}$$

where:

p_i = initial absolute pressure (kPa)

V_{Ai} = initial volume of air (m^3)

Using the effective volume of water $V_1 = V_{Aon} - V_{Aoff}$:

$$V_{Ai} = \frac{P_{on}}{P_i} \cdot \frac{P_{off}}{P_{off} - P_{on}} \cdot V_1 \qquad \text{Equ 6.9}$$

Combining equations 6.7 and 6.9, if only one pump of flow Q, delivers to the accumulator:

$$V_{Ai} = \frac{P_{on}}{P_i} \cdot \frac{P_{off}}{P_{off} - P_{on}} \cdot Q_1 \cdot \frac{3600}{4 \cdot m} \qquad \text{Equ 6.10}$$

where:

V_{Ai} = total volume of accumulator if not pre-charged (m^3)

V_{Ai} = volume of air in the accumulator when pre-charged (m^3)

p_{on} = absolute switch-on pressure (kPa)

p_{off} = absolute switch-off pressure (kPa)

p_i = absolute initial pressure (kPa)

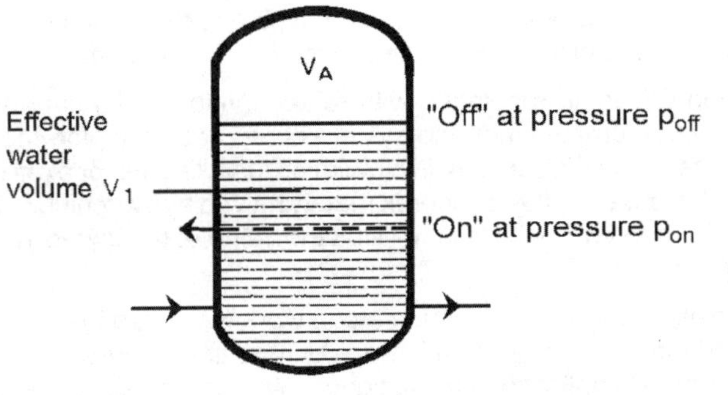

Figure 6.12 Accumulator with designations for working volumes

Q_1 = flow from pump, mean value of flow at p_{on} and p_{off} (m^3/s)

m = maximum number of starts per hour per pump (n/hr)

If several pumps are working on the same accumulator, the effective water volumes V_1, V_2, V_3, etc. are determined for each one as above. The various on and off switching pressures are then bound entirely by Boyle's Law, equation 6.8, which necessitates careful evaluation of the various quantities.

When liquid is supplied to processes which themselves are controlled by on-off valves the process control valves can create surges and water hammer effects. Large on-off process demands can have local accumulators to reduce peak loads on the supply system. The cost of local accumulators is offset by the reduction in supply pipe size necessary to feed the demand.

6.4.7 Power consumption

An on-off controlled pump operating at constant differential head has constant running power consumption, whilst the operational time varies with volume flow. The relative operational times, or factor of intermittency i_Q when transporting flow Q with a pump of flow Q_p will be:

$$i_Q = \frac{Q}{Q_p} \qquad \text{Equ 6.11}$$

The energy consumption E, i.e. the power requirement multiplied by time will be, for time t:

$$E = \frac{Q}{Q_p} \cdot P_{Qp} \cdot t \qquad \text{Equ 6.12}$$

In the case of varying flow the power consumption is summated for various intervals of flow. Simpson's Rule integration formula provides the total power consumption in simple terms thus:

$$E = P_{mean} \cdot t_o \qquad \text{Equ 6.13}$$

$$P_{mean} = \left(\frac{1}{6} \cdot \frac{Q_{100}}{Q_p} + \frac{2}{3} \cdot \frac{Q_{50}}{Q_p} + \frac{1}{6} \cdot \frac{Q_0}{Q_p}\right) \cdot P_{Qp} \qquad \text{Equ 6.14}$$

or at 5 points taken from the flow constancy curve:

$$P_{mean} = \left(\frac{1}{12} \cdot \frac{Q_{100}}{Q_p} + \frac{1}{3} \cdot \frac{Q_{75}}{Q_p} + \frac{1}{6} \cdot \frac{Q_{50}}{Q_p} + \frac{1}{3} \cdot \frac{Q_{25}}{Q_p} + \frac{1}{12} \cdot \frac{Q_0}{Q_p}\right) \cdot P_{Qp} \qquad \text{Equ 6.15}$$

where:

E = energy consumption (kWh/yr)

P_{mean} = mean power throughout the year (kW)

t_0 = operational time (hr/yr)

Q_0, Q_{50} = flow at 0%, 50% etc. constancy (m^3/s)

Q_p = flow at pump operating point (m^3/s)

P_{Qp} = pump power @ flow Q_p (kW)

For pumps operating at variable discharge pressure, with accumulators for example, consider the power consumption to be the mean of the two values at low and high pressure. Power can also be "wasted" due to throttling or modulated by-pass during starting and stopping. Pumps which run continuously but are loaded and unloaded will consume some power when unloaded. This energy consumption should be added to the power calculated from equation 6.15.

On-off control has good power economy when there is a high proportion of static differential head. The power economy may be expressed in terms of regulation efficiency η_r comprising the ratio between head requirement at each part of the system

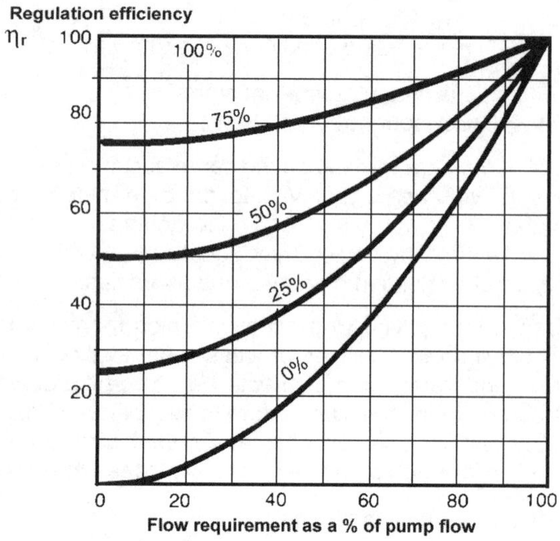

Figure 6.13 Regulation efficiency

curve and the pump head at flow Q_P. The regulation efficiency is shown in Figure 6.13 for on-off control of a single centrifugal pump working with water or liquid with characteristics like water for various ratios of H_{stat}/H_o. Ratio of the static delivery head as a percentage of total differential head H_o at the operational point of the pump. The total efficiency factor for liquid transportation is obtained by multiplying η_r by pump and motor efficiencies. See also Chapter 14, Section 14.3.1.

6.4.8 Examples

On-off control is used extensively in pumps for water, sewage and domestic drinking water installations. It is also commonplace in large hydraulic power packs used in steelworks and chemical processing. Some installations use 10 or 12 pumps to reduce the supply step size and also to limit the size of the liquid store. On-off control has many practical applications.

Certain special cases omit the liquid store, for example booster pumps, see Chapter 5, Section 5.3.3, working in "dead" pipeline networks. In order to avoid too frequent starting, an electrical interlocking device with, say, a time relay must be fitted. In certain flow ranges, pressure variations due to starting and stopping of pumps is inevitable.

In the case of circulatory pumping in totally enclosed systems, there is no requirement for a flow storage reservoir, but there is often a requirement for storage of some other parameter, e.g. heat in a thermal store.

Calculating storage volumes

The pump flows in Table 6.4 and storage volumes in Table 6.5, have been determined from the pump and system curves for a pump installation using three parallel connected pumps, see also Figure 6.10.

Number of pumps	Total Flow	Flow increment
1	100 L/s	100 L/s
2	180 L/s	80 L/s
3	240 L/s	60 L/s

Table 6.4 Pump flows

Number of pumps	Filling time	Partial storage volume
1	t_1 = 90 s	V_1 = 100 · 90 = 9000 L
2	t_2 = 45 s	V_2 = 80 · 45 = 3600 L
3	t_3 = 30 s	V_3 = 60 · 30 = 1800 L

Table 6.5 Storage volumes

From Table 6.3, there are 10 starts per hour using operational Sequence C;

The total storage volume will be
9000 + 3600 + 1800 = 14,400 litres ≡ 14.4 m^3.

Added to this is a minimum volume and, for closed storage, also a volume of air, see Figure 6.11, as well as a storage volume dependent upon pressure surge, see Sections 6.4.3 and 6.4.4. The sensor signals for start-stop are arranged at levels or pressures corresponding to the various partial volumes. If operational Sequence A had been selected the storage volume would have increased to (100 + 80 + 60) - 90 = 21,600 litres ≡ 21.6 m^3.

A storage volume for operational Sequence B = 100 - 90 + 80 - 36 + 60 - 24 = 13,320 litres ≡ 13.3 m^3, is obtained in the same way.

If only the number of starts for each pump and not the total number of starts for the installation has been considered, there can be reductions in storage volume. With complete rotation using 1, 2 and 3 pumps respectively in operation and applying the principle of operational Sequence C, $t_1 = t_2 = t_3$ can be reduced to 30 seconds, giving a total storage volume of 7.2m^3.

Furthermore, in many cases the partial storage volumes can, to some extent, be absorbed within each other, in which case, the minimum total storage can be estimated to be approximately 4m^3.

It should be noted that a part of the cost saving for storage will then contribute towards more complex automation, although these costs are dropping dramatically with modern developments.

6.5 Pole-changing induction motors

6.5.1 Principle and application

Special versions of induction motors with arrangements for changing the number of poles can be used for pump control. Pump speed is inversely proportional to the number of poles, see Chapter 10.

Two distinct versions of motor are available with similar, but not identical, properties. Pole-changing motors achieve two or more speeds with one set of windings connected in different configurations. A dual wound motor has more than one set of windings; one for each speed. Small motors can have a switch built into the motor for local manual speed control. Usually all the winding terminations are brought out through the terminal box to the starter enabling remote automatic speed changes. Speed changing under load is normal. Star-delta (S-D) starting on all windings of dual wound motors is possible. With pole-changing motors, S-D starting is only possible on the lowest speed. Soft starts can be incorporated but the wiring becomes complicated. Pole-changing motors, and dual wound, can be used in those situations when a limited range of speeds; 2:1 or 4:3, 4:2; can be useful.

In some cases motor power is not always proportional to speed. Rotodynamic pump performance changes in accordance with the Affinity Laws, see Section 4.2.1.6, equation 4.16, in Chapter 4. Positive displacement pumps react differently to speed changes. The pump flow is approximately proportional to speed, depending upon the pump type and the viscosity of the liquid; the ability to generate pressure is unaffected.

Pole-changing regulation with rotodynamic pumps is particularly favourable when the system curve has only pipe friction losses as in Figure 6.14, in circulation installations for example. In this case the flow is approximately halved by switching to 750 r/min. Systems with H_{stat} will limit the value of the lowest useful speed.

Positive displacement pumps can work with systems with low and high H_{stat}. This is why these pumps are favoured in many chemical applications using constant pressure vessels and reactors. Gear, lobe, screw and progressing cavity pumps are

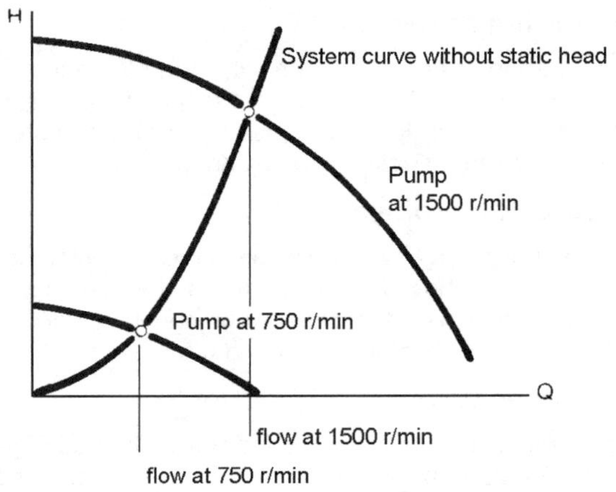

Figure 6.14 Example of change in duty point due to pole-changing 750/1500 r/min

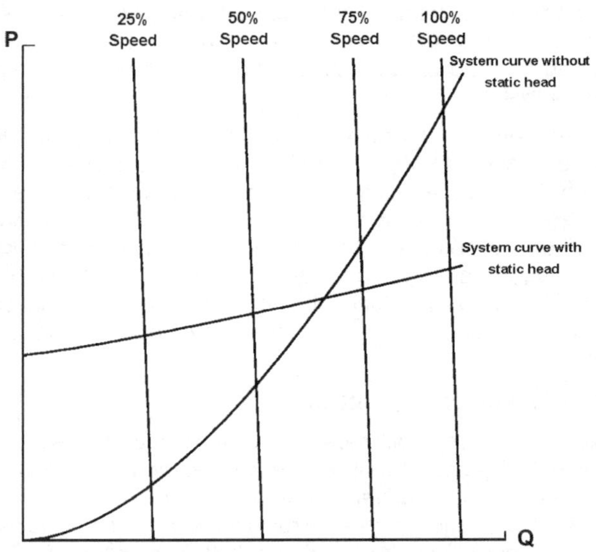

Figure 6.15 Positive displacement pumps at various speeds

used when viscosity is suitable; reciprocating pumps can be used for any viscosity. Figure 6.15 indicates the effect on the duty point of varying the speed of a PD pump when working with friction only systems and systems having a high H_{stat} value.

The power required by the pump is dependent upon the flow, the pressure and the corresponding pump and motor efficiencies at the duty point. Irrespective of the pressure, the flow varies very much in proportion to the speed change. The change in pressure is controlled by the system characteristic.

6.5.2 Costs

From the cost point of view, a variation factor of 2 is the most favourable since this gives pump speeds of, say, 750/1500 or 1500/3000 r/min, although other speed ranges are also possible. Some motors with four speeds are available, 1500/1000/750/500 r/min at 50 Hz. Pole-changing motors are cheaper than a motor plus a frequency inverter. If continuously variable speed is not necessary, and all-electric controls is an asset, the pole-changing motor should be considered. At 20 kW a two speed motor is 25% more costly than a standard motor, at 37 kW it is 50% more. At 135 kW a three or four speed motor is double the cost of a standard motor.

6.5.3 Starting problems

Starting problems are not as bad as those involved with fixed speed pumps with on-off or load-unload control because the pump can be started at low speed. Once the system has stabilised at low speed the motor can be switched to the next speed. The period of time required for stabilisation will depend upon the pump size and the system size.

6.5.4 Stopping problems

As with starting, stopping problems can be reduced by changing speed down in steps, with stabilisation periods, before switching off at the lowest speed. Power failure, when operating at the fastest speed, will cause the same problems as other systems. The reliability of the power supply, together with the likely problems, must be assessed when deciding if precautions are warranted, see Section 6.4.4.

6.5.5 Storage volumes

Since pole-changing regulation is a discrete method of regulation, liquid storage is necessary. Sizing is carried out in the same way as for on-off control, Section 6.4.6, with extra steps to allow for the speed increments. The permitted number of pole changes is usually between 1 and 15 times per hour but should be confirmed with the motor manufacturer once specific running conditions are known.

6.5.6 Power consumption

Rotodynamic pump and motor efficiencies are for practical purposes the same at the various speeds, which leads to good operational economy, see Figure 6.16. The efficiency of positive displacement pumps is dependent upon the discharge pressure, which is in turn, dependent upon the system characteristic. The cost of pole-changing motors is of the order of 20% to 50% higher than for normal, squirrel-cage induction motors. A limiting factor, to some degree, is the reduced availability of this type of motor from manufacturers.

6.5.7 Examples

Consider the pumps and system curves shown in Figure 6.10. It was shown in Section 6.4.8 that the nominal storage volume was 14.4 m^3. If the pump curves shown referred to a 4-pole pump, 1480 r/min, then it may be possible to consider pole-changing. Using the Affinity Laws, the pump curves are re-plotted at 6 pole speeds, 986 r/min. The closed valve head is greater than H_{stat} so the pumps will operate with the system curve successfully. The new values of Q_1, Q_2 and Q_3 become 60, 45 and 35 l/s.

Using the same switching per hour and operational Sequence C, the volumes V_1, V_2 and V_3 become 5400, 2025 and 1050 litres giving a total volume of 8475 litres. Pole switching produces a reduction in basic volume of 41%. Further reductions in volume, considering combining partial volumes as in Section 6.4.8, will lead to further reductions.

The flow range of the pump package has been increased from 240/100, 2.4:1, to 240/60, 4:1. Without considering mixed

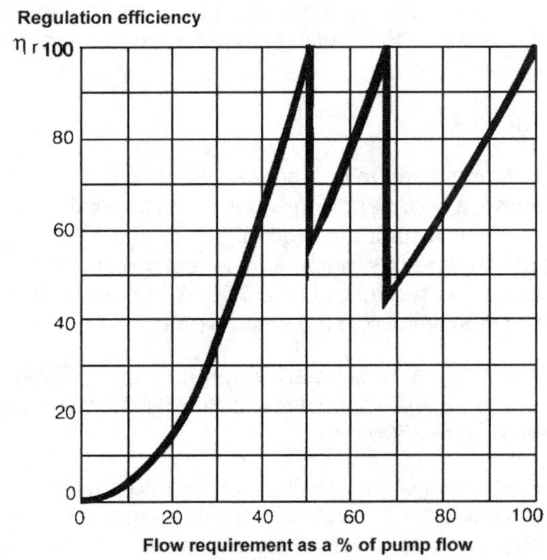

Figure 6.16 Regulation efficiency of a centrifugal pump using a pole-changing motor with 4, 6 and 8 poles. There is no static delivery head.

speed operation, the operating capacities available are 25%, 42%, 44%, 58%, 75% and 100%.

The example from Figure 6.10 could not use pole-changing if the pumps had been running a 2-pole speeds. The next lower speed would be 4-pole and the pumps would not produce sufficient differential to overcome H_{stat}.

Pole-changing motors can offer increased flexibility and reduced storage volumes. It may be possible to run slow rotodynamic pumps in parallel with fast rotodynamic pumps; the relative shapes of the pump and system curves, together with the driver power installed, will determine whether this is feasible. Of course there are no problems running slow positive displacement pumps with fast positive displacement pumps.

6.6 Multi-speed gearboxes

6.6.1 Principles

The principle and application of multi-speed gearboxes is similar to that of pole-changing motors except there are no strict limitations on the speeds available. Manual change multi-speed gearboxes are nearly always built specially "to order" so the gear ratios can be optimised for the application. If a two pump system is considered, speeds of 100% and 50% could be chosen. This would provide pump package capacities of 25%, 50%, 75% and 100%. Again considering two pumps, gear ratios of 33%, 66% and 100% would produce package capacities of 17%, 33%, 50%, 67%, 83% and 100%. Two and three speed gearboxes are relatively common; more ratios are rare. Manual change gearboxes usually have internal dog clutches, the gears are always in mesh, for selecting the ratio. One or two levers are moved to change a ratio, while the drive is stationary. It may be necessary to jog or inch the motor to line up the dog clutches before engagement. On units up to about 250 kW, the high speed coupling can be turned by hand. On larger units facilities must be provided for barring, jogging or inching.

Semi-automatic gearboxes are frequently used on engine-driven portable pumps. Versions of these gearboxes are available for motor drives. Just like their automotive counterparts, gears can be changed onload without stopping. The gearbox is fitted with a torque converter or fluid flywheel to smooth the transition from one speed to the next. Remote and automatic control of gear changes is simple. The drive can be disengaged to give on-off control as well as variable speed. Larger units are fitted with a flywheel lock to save the 2 to 4% speed and energy loss due to flywheel slip. Gearboxes can be fitted with auxiliary power take-off to drive associated equipment. Standard gearboxes are available with up to ten gears. Choice of ratios is limited, but because of the multiple speed increments available this is not a problem.

6.6.2 Costs

For small manual change gearboxes, about 40 kW for a reciprocating pump, a two speed gearbox will cost approximately 4.5 times as much as a standard fixed ratio gearbox. A three speed box will cost 10 times as much. As the gearbox size increases the cost impact is reduced. For a 700 kW gearbox, two and three speed cost twice and 3.5 times as much.

Semi-automatic gearboxes start at approximately 28000 kW for an air operated four speed unit with fluid flywheel suitable for input speeds up to 3500 r/min.

The increased cost of the power transmission should be considered with the reduced costs of storage volumes. The overall scale of the pumping project will determine the relative cost proportions. The reduction in drive efficiency of semi-automatic gearboxes without fluid flywheel locks must be considered.

6.6.3 Starting problems

Manual change gearboxes are similar to pole-changing motors, see Section 6.5.3. Semi-automatic gearboxes, with more ratios, can have really low speed starts and smaller increments between the ratios effectively removing the problem.

6.6.4 Stopping problems

Again, manual change gearboxes are similar to pole-changing motors, see Sections 6.4.4 and 6.5.4. Semi-automatics do not experience problems with normal stops but power failures can be a problem, see Sections 6.4.4 and 6.5.4.

6.6.5 Storage volumes

Speed changing by fixed ratios is a discrete method of regulation and requires some form of liquid storage to compensate for the imbalance between supply and demand. Manual speed changing should only be considered when demand varies periodically rather than continuously. Manual speed changing can be viable if the demand only varies significantly say once or twice a day or less. At any given speed the pump should be considered as fixed speed and volumes calculated as shown in Section 6.4.6.

Semi-automatic gearboxes have no restrictions on gear changes and remote automatic control is relatively simple which includes disengaging the drive. The flows possible are in discrete steps, however small, and even with the additional on-off control some small storage capacity may be necessary. However, the storage capacity will be based on the smallest pump flow thereby reducing the storage required dramatically compared to manual speed change, on-off or load-unload options.

6.6.6 Power consumption

Power consumption will depend upon the curve shapes and the intersection points. Pump efficiency depends upon flow and head or pressure, driver efficiency depends upon the load. Manual change gearboxes are relatively simple so the mechanical efficiency should be about 95 to 97%. The basic efficiency of semi-automatic gearboxes will be slightly lower because of the greater complexity. Also the fluid coupling/fluid flywheel may lose up to 4% when not locked up.

Manual gear changing is very efficient at transmitting power but poor at changing gear. Semi-automatic gearboxes are good at changing gear at the expense of reduced transmission efficiency. The type and size of the pump package, limitations on storage volumes and cost, considered with the operational requirements will show which gearboxes are suitable.

6.6.7 Examples

Referring to the example shown in Figure 6.10, the minimum pump speed possible, allowing a 10% margin for wear and increased losses, is 60% of the speed shown. This would give a slowest running speed of 888 rpm based on an original 4 pole motor speed. Unusually, in this case $Q_1 = Q_2 = Q_3 = 40$ l/s. The basic storage volume would be 6.6 m^3. The flow capacity increments would be 17%, 33%, 42%, 50%, 75% and 100%. Using this pump on this system would produce a minimum safe capacity of 17%.

Positive displacement pumps do not have a closed valve head problem and any minimum speed could be selected to suit operations. Minimum speeds for reciprocating pumps can be as low as 5% of design.

6.7 Throttling by control valve

6.7.1 Principles

In the case of throttle regulation for rotodynamic pumps, Figure 6.17, the operating point of the pump changes because the system curve is modified. As the flow area of the control valve is reduced the pressure drop increases, the flow through the pump reduces from Q_1 to Q_2.

Throttle regulation creates an extra flow loss $h_{f\,throttle}$ in the system. To overcome this, power must be delivered to the pump, thus:

$$P_{throttle} = \frac{\rho \cdot Q_2 \cdot h_{f\,throttle}}{1000 \cdot \eta_2} \quad (kW) \qquad \text{Equ 6.16}$$

In a completely zero loss regulation system, the power supplied would be less by $P_{throttle}$.

As can be seen from Figure 6.17, $h_{f\,throttle}$ and therefore $P_{throttle}$ are greater when the system curve is steeper for the same change of flow from Q_1 to Q_2.

Throttle regulation and by-pass regulation are the most common of the continuously variable methods of regulation, but throttle regulation cannot be used for positive displacement pumps. In view of rising energy costs, however, the method is becoming uneconomic despite the relatively low procurement cost of a control valve.

Throttle regulation is a continuously variable flow method. Operation is possible at any point on the system curve where the pump curve is higher. If the pump can be operated at closed valve without problems there is no requirement for storage volumes. The problem is efficiency; all the throttling energy is wasted. The choice of throttling is dependent upon the size of the pump, the cost of energy, how often throttling will be used and for how long. If the pump is small and/or energy is cheap throttling may be attractive. If throttling is only necessary for start-up and shut-down it may be attractive. Most pumps are installed with isolating valves. The throttle valve can be fitted as a by-pass around the isolating valve. After throttling for start-up the isolating valve can be opened fully for normal operation.

6.7.2 Costs

Installation costs are dependent upon the size of valve required and the mode of operation desired. Local manual control would be sufficient for start-up/shut-down that took one hour say every month. If continuous throttling was necessary a fully automatic control valve would be required. A 1 inch manual throttle valve, specifically designed for continuous operation would cost about $1500. A 2 inch low pressure constant flow control valve in cast iron suitable for water would be about $1000.

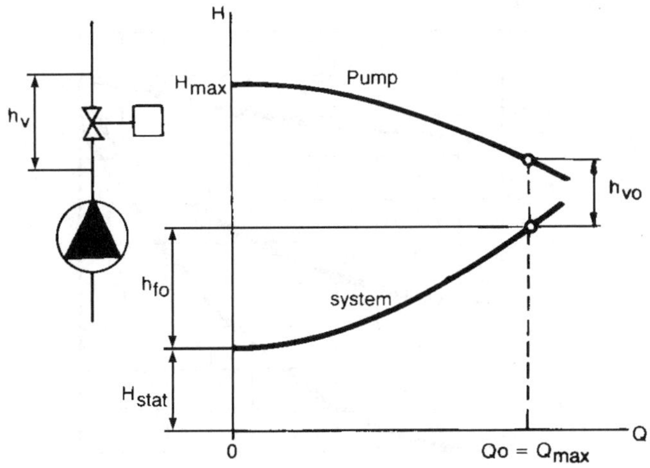

Figure 6.18 Relationship between pump and system characteristics for successful throttling

6.7.3 Sizing the control valve

The control valve is sized so that the control function itself has the best possible regulation characteristics. This means that the parameters to be controlled, i.e. the flow or the head, should be as linear as possible in relation to the movement of the valve. It is a requirement that a loss of head should always exist across the valve even at maximum flow Q_0, Figure 6.18.

When regulation is necessary in a system which has a simple smooth characteristic, as shown in Figure 6.18, the following approximation may be used for the minimum head loss at maximum flow:

$$h_{vo} = 0.1 \cdot h_{stat} + 0.3 \cdot h_{fo} \quad (m) \qquad \text{Equ 6.17}$$

The sizing of the control valve and its linearity for this case has already been dealt with in Chapter 5, Section 5.2.4.

When the system curve is variable, throttling may be used to obtain a constant flow. The loss of head due to the valve must be selected at a considerably greater value than that given by the rule of thumb in equation 6.17. An analysis of regulation characteristics should also be carried out in the manner shown in the example in Section 5.2.4, of Chapter 5.

In throttle regulation, the power requirements of the pump are determined entirely by its power curve, regardless of what the

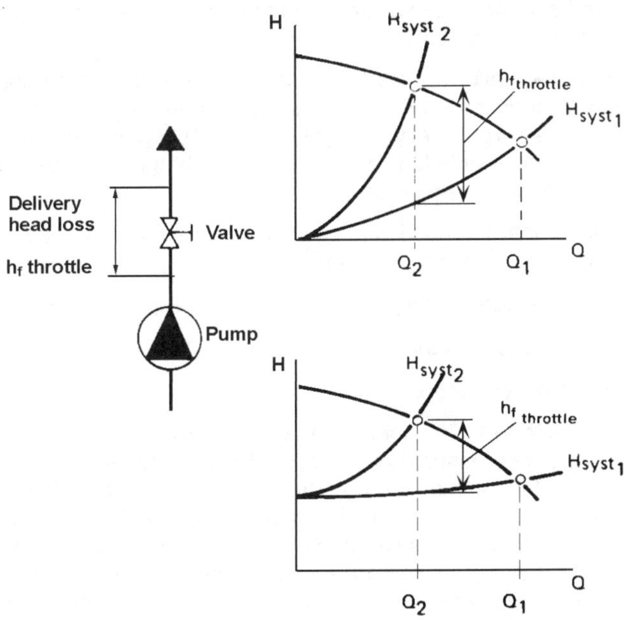

Figure 6.17 Throttling by control valve, schematic and curves

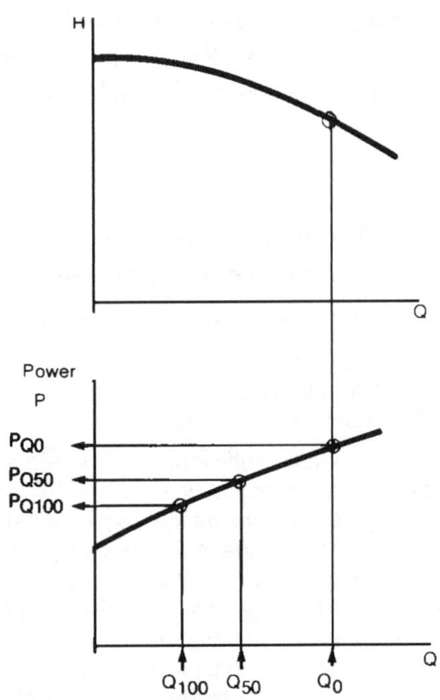

Figure 6.19 Pump power for throttle regulation

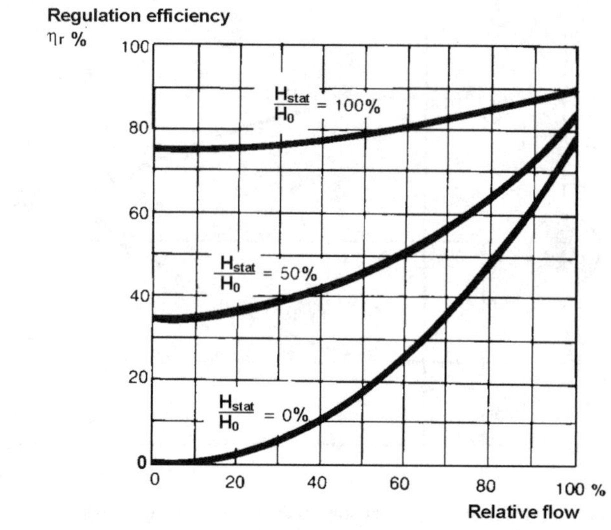

Figure 6.20 Regulation efficiency for throttling of a centrifugal pump with a shut-off head about 20% higher than the operating point at maximum flow

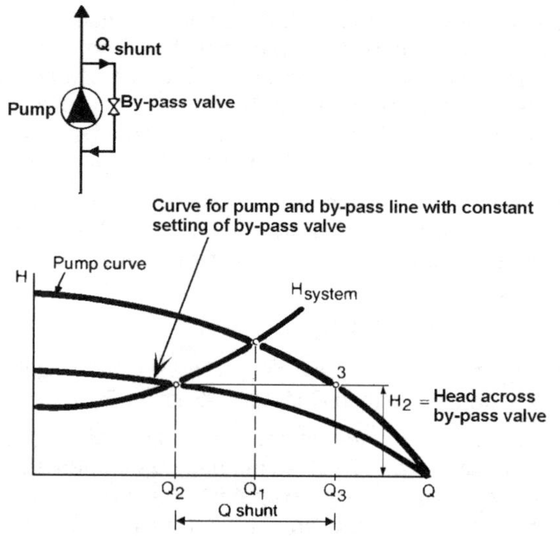

Figure 6.21 By-pass regulation, schematic arrangement and curves for centrifugal pump and by-pass line are reduced to a combined curve

system curve looks like, see Figure 6.19, i.e. the power requirement can be read directly from the pump power curve. In Figure 6.19 P_{Q0}, P_{Q50} and P_{Q100} are the power values corresponding to flow Q_0, Q_{50} and Q_{100} respectively, see Figure 6.22. These flows apply to mean flows at 100%, 50% and 10% constancy.

Energy consumption E is power multiplied by time, or

$$E = P_{mean} \cdot t_0 \qquad \text{Equ. 6.18}$$

where:

E	=	annual energy consumption (kWh/year)
t_0	=	operational time per year (hr/year)
P_{mean}	=	mean power during operation (kW)

The mean power is most simply determined by the use of Simpson's Rule integration formula in the same manner as flow in Section 6.4, equations 6.1 and 6.2.

For throttling, the regulation efficiency is the ratio between the differential head of the system curve compared to the pump differential head. Figure 6.20 shows the regulation efficiency for the special case of a centrifugal pump working with water or a liquid with water-like qualities. The head H_0 applies to the head loss of the valve, i.e. $H_0 = H_{stat} + h_{fo}$, see also Figure 6.21. The loss of head due to the valve has been chosen in accordance with equation 6.17 and the pump curve is assumed to have its closed valve head some 20% higher than the operating point for maximum flow. In the case of throttle regulation, pump efficiency varies with flow and the motor efficiency varies with absorbed power.

6.7.4 Power consumption

Throttling increases the power consumption. The head loss through the control valve is completely destroyed but must be "paid for" with energy from the pump. If throttling must be used continuously it will be a continuous waste of energy. If throttling is only required for start-up or shut-down it can be accomplished in a by-pass around the isolating valve and once the isolating valve is opened the throttling loss will be zero.

6.7.5 Example

Consider the pumps and system shown in Figure 6.10. The process has been redesigned and only one pump is required. The storage volume has been moved to another part of the process. During start-up and shut-down, which last 2 hours each every month the flow through the system must be reduced to 50 l/s. The head drop through the valve must be 60 m. What size of valve is required ?

Using Equation 5.10 (in Chapter 5) with

$Q = 180\ m^3/h$ and $\Delta p = 60 \cdot 1000 \cdot 9.81 / 100000 = 5.89$ bar

a minimum K_V of 74.2 is obtained. Reviewing manufacturers' data reveals the smallest valve suitable to be 80 mm, which would be at 80% open. This is considerably smaller than the pump discharge which is 150 mm. With the valve wide open K_V = 110, trying to pass 100 l/s would result in a pressure drop of 14.3 bar, 146 m. This pressure drop is higher than the closed valve of the pump so the pump would be unable to run at the normal duty point. This is an ideal case for putting the throttle valve in a by-pass around the discharge isolating valve.

6.8 By-pass return

6.8.1 Principle and application

By-pass regulation is sometimes called shunt regulation because of the similarity with shunts in electrotechnology. In practice, it is simply called by-pass. In by-pass regulation, as shown in Figure 6.21, liquid is fed back from the delivery side of the pump into the suction side of the pump. In practice, as with relief valves, the by-pass should be piped back to the suction source not to the suction pipe adjacent to the pump.

Operating points are most easily determined by combining the curves for a rotodynamic pump and the by-pass line into a single curve.

By-pass regulation is used for positive displacement pumps and only rarely for rotodynamic pumps. The energy consumption for these is generally greater with by-pass regulation than it is with throttle regulation. (See for example the pump curves in Figure 4.26, Chapter 4).

By-pass control is used with positive displacement pumps in three main modes of operation:

- Low pressure starting
- Load/unload control
- Small flow changes

Some positive displacement pumps operate in systems where the discharge pressure is almost constant, consequently starting takes place at high torque. This type of starting can generally only be accomplished with direct-on-line (DOL) starting. Not all electric supplies can cope with the current demands of DOL and star-delta is preferred, especially for motors over 15 kW. Star-delta starting does not provide 100% motor torque so that starting the pump on load is not possible. A by-pass, returning to suction source, provides a low pressure, low torque starting option.

The pump starts with the by-pass wide open, the discharge pressure generated is only sufficient to overcome friction losses of the by-pass line and valve. Once the pump is running at full speed the by-pass valve is gradually closed, the discharge pressure increases until the non-return valve opens and the pump is online. The by-pass valve is completely closed and incurs no energy losses. The pump may be stopped by gradually opening the by-pass valve to reduce water hammer and pressure surges.

Step capacity control can be achieved on positive displacement pumps by loading and unloading the pump rather than starting and stopping by on-off. A by-pass valve is opened and closed and the pump flow varies from 100% to zero. The by-pass valve can be fitted with dampers to ensure long opening/closing times to eliminate water hammer and pressure surges.

When only small flow reductions are necessary a by-pass valve can be economically attractive compared to other methods. As only a small flow passes through the by-pass the amount of energy wasted compared to the pump power is small and only a small valve is needed.

Positive displacement pumps working with non-Newtonian liquids rarely have sufficient information to accurately predict system curves to allow the control characteristics to be analysed. Large safety margins are advisable. It should be noted that the control valve flow area may not be adequate for liquids containing larger particles. The power requirement is determined by the flow, $Q_{max} + Q_{shunt}$, and the discharge pressure according to the system curve at the actual system flow. Because of this, the regulation efficiency will be relatively high, Figure 6.22.

It is generally not necessary to have 20% extra capacity with positive displacement pumps. If the flow rate is critical then an allowance should be made for wear in the pump between overhauls. Depending upon the pump type, and how corrosive/abrasive the liquid is, this should be between 2% and 5%.

6.8.2 Costs

Costs for by-pass valves are similar to throttle valves. A 1 inch manual valve specifically designed to cope with the throttling and high speed liquid would cost about $1500. Valves are available up to 2.5 inches; a 1 inch valve has a C_v of 27, a 2.5 inch is 79.

6.8.3 Sizing the by-pass

The by-pass valve is sized in a similar manner to a throttle valve. Once the flow and pressure drop are known equations 5.10 or 5.11 (in Chapter 5) can be used to evaluate K_V or C_V. A slightly larger valve is not a problem; it just requires closing a little. A slightly smaller valve will stop the pump from operating down the system where it was intended.

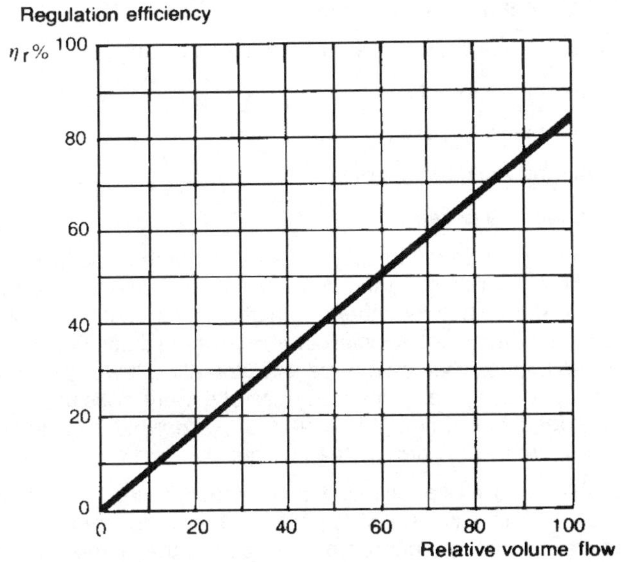

Figure 6.22 Regulation efficiency for by-pass regulated positive displacement pump with 20% by-pass at maximum required flow

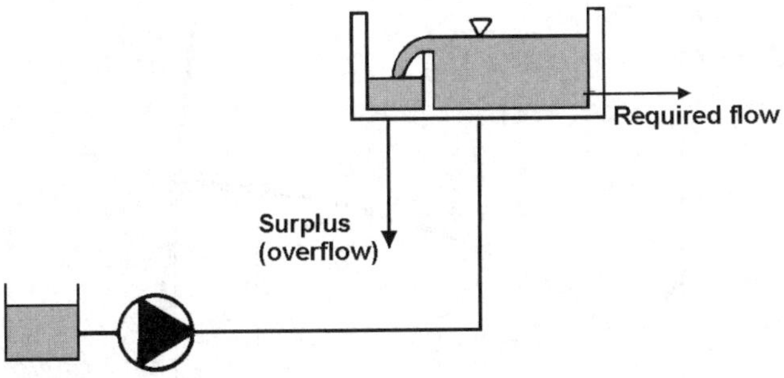

Figure 6.23 Overflow control

6.8.4 Overflow by-pass return

With overflow control the liquid is pumped into a reservoir at almost constant level. The required system flow is taken from the reservoir. The surplus liquid flows over a weir, which creates the constant head, and goes back to the suction side of the pump or escapes in to a drain pipe, Figure 6.23. The pump in such cases operates at constant flow and constant power requirement.

Overflow control presents a simple and reliable method but uses a relatively large amount of energy. The regulation efficiency depends upon the proportion of static delivery head, Figure 6.24. The figure makes the assumption that there are no unwanted level differences.

6.8.5 Power consumption

For rotodynamic pumps, energy consumption is determined as shown in Section 6.7.4, i.e. operating points on the H-Q curve are established, thereafter the power requirement of the pump is derived from its power curve. The regulation efficiency for rotodynamic pumps is generally lower with by-pass regulation than it is with throttle regulation, Figure 6.19. In spite of this, by-pass regulation of a rotodynamic pump may be justified because of the lower loss of head across the control valve with consequent lower noise levels and reduced risk of cavitation in the control valve. By-pass valves can also be closed completely so there is no energy wasted at all. The operating points of the rotodynamic pump should, however, always be checked for cavitation and abnormal vibration levels.

When a by-pass valve is closed there are no energy losses. When it is open the pump must provide the system flow and the by-pass flow, both at system pressure. The wasted power loss with by-pass regulation is:

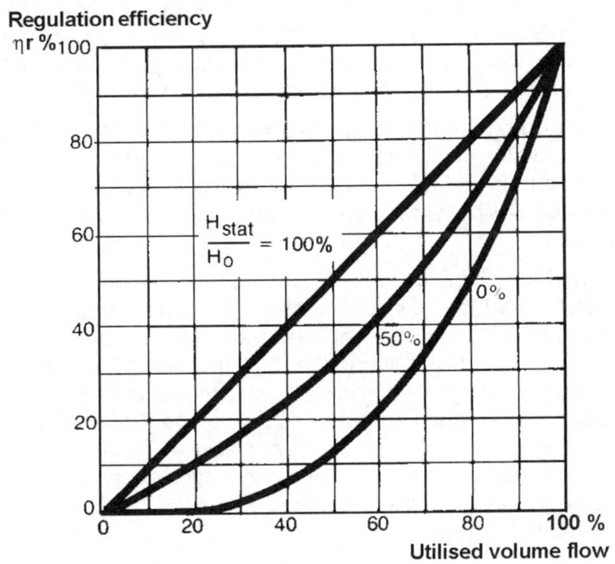

Figure 6.24 Regulation efficiency with overflow control, $H_0 = H_{stat} + h_{f0}$

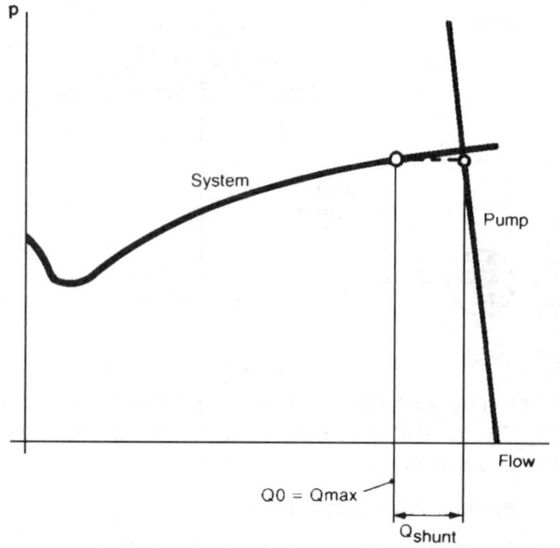

Figure 6.25 By-pass regulation of a positive displacement pump. The system curve shown applies to a non-Newtonian liquid

$$P_{shunt} = \frac{\rho \cdot g \cdot Q_{shunt} \cdot H_2}{100 \cdot \eta_3} \text{ (kW)} \qquad \text{Equ 6.19}$$

where η_3 is for flow Q_3 (in Figure 6.21).

For positive displacement pumps using by-pass regulation, the power consumption is determined by the back pressure, i.e. by the system curve. In the same way as for throttle regulation, the valve must be given priority in the system control loop. This is achieved by increasing the required flow, Figure 6.25.

6.8.6 Example

A fixed speed reciprocating pump operates in a system with a high static pressure. The nominal duty point is 190 m³/h at 184 barg, the suction pressure is 1 barg. For starting purposes the system flow must be reduced to 100 m³/h, which corresponds to a discharge pressure of 53 barg, what size of by-pass valve is required ?

Q_{normal} = 190 m³/h, p_{normal} = 184 barg, Q_{start} = 100 m³/h

The flow to be by-passed is 90 m³/h and the differential pressure available is 131 barg. From equation 5.11 (in Chapter 5) the K_V of a suitable valve would be 7.86. Reviewing manufacturers' data, a 25 mm valve has the required K_V at approximately 85% opening. The extra capacity of the valve will allow for other losses in the by-pass system.

6.9 Infinitely variable speed

Infinitely variable speed provides infinitely variable flow. There are some fixed speed pumps which can provide infinitely variable flow. Reciprocating metering and dosing pumps are usually variable stroke and flow can be varied from zero to 100%.

For small flows metering pumps should be considered. Metering and dosing pumps can be surprisingly large. Plunger pumps can be built in variable stroke versions. These pumps were first developed as boiler feed pumps and pumps up to 75 kW were available.

For hydraulic applications, liquids with very good lubricating qualities, variable displacement vane pumps and variable stroke piston pumps are available. Some variable stroke piston pumps can actually reverse the flow direction, -100% to 0 to +100% flow, while running at constant speed. For hydraulic applications consider vane and piston pumps. Variable speed pumps will generally not offer zero flow. A minimum pump speed is required to assure proper lubrication of bearings. Variable speed pumps will tend to offer 5% to 100% or 10% to 100% maximum flow ranges.

The need to have infinitely variable speed to provide infinitely variable flow will ultimately depend upon:

- The liquid
- The size of the pump
- The type of pump
- The flow range required

6.9.1 Principles

Many pumping applications require the flow to be adjusted to several different values which sometimes cannot be accurately predicted. In these situations it is advisable to make the flow infinitely adjustable over a range. Changes in operating conditions due to fouling or corrosion in pipework, variations in process conditions or wear in the pump can be accommodated without compromising operational requirements.

With rotodynamic pumps, throttling and by-pass regulation could satisfy most of the needs. With positive displacement pumps, by-pass regulation could satisfy all of the needs. However, both throttling and by-passing are inefficient. Energy supplied by the pump to the liquid is destroyed in a valve, and may not be economically viable in larger installations or when used continuously.

Pump speed variation allows the pump capacity to be matched to the process demand. In Sections 6.5 and 6.6 pump speed was shown to be adjusted in discrete steps; pump capacity may or may not match the demand but it will be much closer than it would have been without speed adjustment. The ability to vary the pump speed infinitely allows pump capacity to match process demands exactly.

Pump speed can be varied in many different ways. Two main approaches are adopted; a fixed speed driver with a variable ratio transmission and a variable speed driver. The two approaches can also be combined.

6.9.1.1 Variable ratio transmissions

Some pumps, typically larger pumps, require some form of power transmission because the pump speed does not coincide with the driver speed. In these instances variable ratio can sometimes be incorporated in the speed reduction. The popular methods for varying speed ratios are:

- Mechanical variators
- Hydrostatic couplings
- Fluid couplings
- Torque converters
- Eddy current couplings

6.9.1.2 Mechanical variators

Mechanical variators are a sub-group comprising:

- Variable ratio belts
- Variable ratio chains
- Variable ratio friction drives
- Variable ratio gearboxes

Belt and chain drives usually have fixed diameter pulleys and sprockets. By adjusting the diameter of the pulleys/sprockets the speed ratio between the two shafts can be varied. Variable speed belt drives use special cone pulleys to adjust the diameters. Adjustment can be manual or automatic. Power transmitted is up to 25 kW and speed ranges of 9:1 are possible. Variable speed belt drives are mounted like normal belt drives, inside a guard provided by the purchaser.

Variable ratio chain drives are constructed like gearboxes, housed within a casing which acts as the oil reservoir. Fixed gear ratios can be incorporated to adjust the output speed. Some variable ratio chain drives are designed to be constant power transmissions rather than constant torque devices; this feature can be useful in some applications and must be considered in the cost evaluation. Small units are capable of 6:1 speed ranges, the larger units, 600 kW with a service factor of 1, only 3:1.

Variable ratio friction drives use discs, rollers or ball bearings clamped between two flat or specially shaped plates to transmit the power. One plate is mounted on the input shaft, the other on the output shaft. The variable ratio is created by adjusting the contact circles on the plates. Units up to 15 kW are popular; output speed ranges are about 6:1.

The variable ratio gearboxes described should not be confused with multi-speed gearboxes in Section 6.6. Variable ratio gearboxes are epicyclic gearboxes with small control drives attached to the outer gear annulus. The basic epicyclic gear is designed for the required full speed ratio. Then, by driving the outer gear annulus with a low power source, the gear ratio can be made infinitely variable. They are usually built specially "to order" so any configuration is possible.

6.9.1.3 Hydrostatic couplings

Hydrostatic couplings consist of a variable displacement pump and a fixed or variable displacement hydraulic motor mounted in one housing which acts as the oil reservoir. The fixed speed driver provides the power to the variable displacement pump; variable stroke piston pumps are most usual. The high pressure oil drives the output hydraulic motor. Hydrostatic drives are constant torque devices, full load torque is available at all speeds. Up to 200% instantaneous torque can be used at start-up. Speeds down to 0 r/min are not usual, most units have a minimum continuous speed limitation dependent upon the input speed to the unit. With 1450 r/min motors, the minimum speeds would be about 50 r/min. Hydrostatic couplings are very stable and have a linear characteristic at constant torque. Over-running protection is inherent in the design as is controlled deceleration. Output speeds up to the input speed are standard, 1900 r/min maximum, speed range 27:1 with powers up to 22 kW. Speed control can be adapted to suit the process requirements. Cooling is by natural convection around the housing, high ambient temperatures would restrict coupling capacity.

Hydrostatic couplings can be built as separate units. A power pack, with reservoir, motor, pump and cooler, can supply high pressure oil to a remote hydraulic motor. This format can be useful in hazardous areas. All the electrical equipment can be in a safe area with only the hydraulic motor in the hazardous area. Standard packages of this style are available for powers up to 150 kW. Specials can be built to order to cover virtually any speed and power requirements.

6.9.1.4 Fluid couplings

Fluid couplings are one of the two types of coupling which use oil to transfer power from the drive to the pump. Because these are "torque" transmitters they are always fitted in the drive train at the highest speed. Fluid couplings would be fitted in the input side of reduction drives and the output side of speed increasing drives.

The torque which is transmitted, and consequently the power, is dependent upon how much oil is circulating. A scoop in the oil reservoir adjusts the proportions of oil in the coupling to oil in the reservoir. When the coupling is full of oil 100% torque is transmitted. When used with centrifugal pumps on a pure friction system curve with no static head the fluid coupling is capable of speed ranges of about 8:1.

For constant torque loads, due to oil cooling limitations, the speed range is only about 4:1. At speeds less than 100%, the excess power is given up to the circulating oil and must be extracted by suitable cooling. The fluid coupling must always operate with some slip, consequently the maximum output speed is about 98.5% of the input speed. Typical input speeds are up to 1800 r/min. Couplings over 650 kW are available as standard. Larger units, up to 1.5 MW, use a pump rather than a scoop to control the oil flow. Some of the larger units have built-in gearboxes. Controls can be supplied to suit any system.

6.9.1.5 Torque convertors

Torque convertors are the rotodynamic equivalent of hydrostatic couplings. A centrifugal pump on the inlet shaft transfers energy to the oil. Inlet guide vanes direct the oil through an inward flow radial turbine on the outlet shaft. Torque convertors transmit torque and should be in the highest speed shaft. Torque convertors are almost constant power devices, rather than constant torque. At low output speeds output torque is increased. The output speed can also be increased above the input speed. One problem with torque convertors on manual control is that they can overspeed if the output load is suddenly reduced. Oil cooling is by external heat exchanger. Couplings have been supplied for over 10000 r/min and powers over 10MW are not uncommon. Some units have integral gearboxes. Controls can be fitted to suit any system.

A special hybrid consisting of a fluid coupling and a torque convertor is available in certain sizes. The internal controls select the appropriate drive mechanism to optimise efficiency.

Liquid couplings have an important attribute which is not related to their variable speed capabilities. All liquid couplings effectively isolate the two sections of the drive train. If a torsional analysis is necessary, the total drive train can be treated as two separate systems divided at the coupling. A liquid coupling will not transmit torsional vibrations and it will act as a dampener for cyclic torque and speed variations.

6.9.1.6 Eddy current couplings

Eddy current couplings are the electrical version of fluid couplings but can transmit more than 100% motor torque for starting. A small control current, typically less than 0.5 kW, adjusts the speed and torque relationship between the driven and the driving halves. Couplings are manufactured in standard sizes; the torque and speed range being a function of the driver/driven power to the coupling rating. Couplings are available up to 5MW.

6.9.2 Variable speed drivers

Variable speed drivers are a very diverse group with very different characteristics. All of the following are currently used:

- Petrol engines
- Diesel engines
- Gas engines
- Crude oil burning engines
- Air motors
- Steam turbines
- Gas turbines
- Turbo-expanders
- Hydraulic turbines
- AC squirrel cage induction motors
- AC slip ring motors
- Switched reluctance motors
- DC motors

The choice of driver is dependent partially on the size required, the speed range and the energy source available.

6.9.2.1 Petrol engines

Petrol engines are used mainly for portable applications, comprising both small light packages which can be man-handled and trailer mounted units. Speed adjustment is by throttle control of the engine. The useful range of speed available is dependent upon the torque characteristic of the pump. Rotodynamic pump torque, for pumps operating on purely friction systems, varies approximately as the square of the pump speed, so allowance must be made for changing efficiencies. As pump speed is reduced from the rated condition, torque reduces quite rapidly. In these situations it may be possible to achieve a 4:1 speed range. For rotodynamic pumps operating with high static heads compared to friction heads the engine may struggle to cover a 2:1 speed range depending upon the pump characteristic.

Positive displacement pumps with frictional system losses may approach a 4:1 speed range; with constant discharge pressure systems the engine may have to be oversized to achieve any speed range if stalling is to be avoided. Small engines can operate up to 5000/6000 r/min, larger engines to 3000 r/min. A wide variety of sizes are available, up to 150 kW being popular.

6.9.2.2 Diesel engines

Diesel engines are used for fixed installations as well as portable machines. They are used on sites where electricity is not available or is not capable of supporting the pump load. All the comments on petrol engines regarding speed ranges apply to diesel engines. The speed available from diesel engines is lower; small engines about 3000 r/min, large engines down to 600 r/min. Small diesel engines of 5 kW are available, large fixed installations use engines up to 5 MW.

6.9.2.3 Gas engines

Gas engines are popular in fixed installations where natural gas is readily available. The natural gas can be used locally, usefully, rather than incurring additional cost by transportation or wasting by flaring-off. Full details of the gas specification must be available to the engine manufacturer to allow engine outputs to be predicted. Gas engines are more similar to diesel engines than petrol engines. All the comments on diesel engines apply to gas engines.

6.9.2.4 Crude oil burning engines

Like natural gas engines, crude oil burning engines are popular in fixed installations where there is a good source of supply and other energy supplies are limited or non-existent. Crude oil burning engines are similar to diesel engines. Some types of crude oil may require considerable pre-treatment before being suitable for use in the engine. Some engines may require starting and warming-up on diesel or gas before the crude oil can be used. Decisions and costing for crude oil engines can only be made when all the details of the actual crude oil are available. Extra costs incurred for pre-treatment or starting on other fuels may indicate other energy sources to be more cost effective. The comments for speed and speed ranges given for diesel engines apply to crude oil engines except that small engines are not available.

Dual fuel engines

As indicated in crude oil burning engines it is possible to run engines on more than one fuel, not generally at the same time. The most popular combinations are diesel and natural gas, or crude oil and natural gas. Full details of both fuels must be given to the engine manufacturer.

Turbo-charged engines

The details given in previous engine categories apply to naturally aspirated engines; that is engines which draw the air supply through an air filter from the atmosphere. Turbo-charged engines are becoming very popular. Turbo-charging significantly improves the engine efficiency and increases the power output without much change in the space required. In some cases turbo-charging seriously modifies the shape of the engine torque curve so that torque reduces rapidly either side of the optimum operating point. This change in engine characteristic can considerable reduce the operational speed range of any pump with any system. This feature must be considered, with all the other points, when deciding upon the type and size of an engine.

NOTE: The use of the waste heat, from the engine exhaust, can significantly improve the overall operating efficiency of the pump installation. The heat can be utilised for water or process heating or steam raising. In general, the extra cost of heat exchangers will be quickly recovered due to the large amount of heat available.

Engines for hazardous areas

Internal combustion engines can be modified to be suitable for operation in some hazardous areas. This requirement must be specified at the initial inquiry stage. Not all engines can be modified; not all manufacturers are interested in modifications. The surface temperature requirements, together with any relevant approvals, must be specified. Hot areas of the engine are cooled or insulated, electrical auxiliary equipment is uprated. Special attention is paid to air filters and silencers for arresting sparks.

Problems with engines in hazardous areas can be alleviated, in some cases, by installing the engine in a safe area separated from the hazardous area by a fire wall. Power transmission is via a spacer coupling which has a special mechanical seal, set in the fire wall, to prevent direct leakage of the hazardous substances.

NOTE: All engines can be used with multi-speed gearboxes to increase the overall useful speed range. The engine provides the infinitely variable speed capability and the gearbox can be used to provide several operational speed ranges. This approach should be considered when wide speed ranges are required.

6.9.2.5 Air motors

Pumps can be driven by compressed air. Air motors are available as standard assemblies. Torque characteristics should be almost constant so wide speed ranges are possible with all pumps. Sizes are limited to about 5 kW, speeds up to 9600 r/min are possible.

6.9.2.6 Steam turbines

Steam turbines are popular for driving larger pumps, but also small pumps in situations, such as refining and chemical processing, where steam is plentiful. Continuous processes are best so that time waiting for warm-up is periodic rather than regular. Steam turbines can be condensing or pass-out, that is to say the turbine expands the steam to a lower pressure which is then used in another part of the process. The torque characteristic is fairly constant, wide speed ranges are possible. Steam turbines are built from about 25 kW; there is no practical limitation on the largest size. Small turbines tend to run quite fast, 6000 r/min and faster, but there are no fixed speeds, the governor can be adjusted for a range of speeds.

6.9.2.7 Gas turbines

Gas turbines are useful for producing plenty of power from a small space; they also tend to produce much noise. The turbine itself can be small, but has a tendency to be overshadowed by the air filters and exhaust ducts. Gas turbines are ideal for driving fast pumps, speeds of 10000 to 15000 r/min are common. Slower pumps will require reduction gearboxes. Warming up is generally not a problem; full load can be available in 1 or 2 minutes. Gas turbines can utilise a variety of liquid and gaseous fuels. Crude oil burning is an option from two manufacturers.

NOTE: The usefulness as a variable speed driver is dependent upon the torque characteristic which in turn is dependent upon the gas turbine design. A single shaft machine, one in which the compressor and the turbine are mounted on one shaft, or two separate shafts but geared together at a fixed ratio, does not have an ideal torque curve. Below the design speed, torque reduces at a power slightly higher than a square law. The point of zero drive torque occurs at about 20% speed. This type of gas turbine will be suitable for some rotodynamic pumps whose system characteristics have no static head.

The other type of gas turbine is the free power turbine type; the compressor and its turbine are on one shaft, the turbine for external drives is mounted on a separate shaft and can rotate at virtually any speed required. The torque curve of this gas turbine is ideal. The compressor can be allowed to run up to full speed, if necessary, before the pump is started. Starting torque can be over double the normal running torque, reducing almost linearly to full load torque at full speed. Any type of pump or system can be operated with a free power turbine gas turbine. Speed ranges over 5:1 are possible. Turbine cooling may be a problem for low speed continuous operation with constant torque pumps.

Gas turbines start at about 25 kW. Overall efficiency can be improved by utilising the waste heat in the exhaust gases. It is more cost effective to design the pump/system around a standard gas turbine than to ask the turbine manufacturer to modify the turbine to suit the pump/system. The world supply of gas turbines, in some sizes, is extremely limited. When discussing operational requirements for gas turbine drivers also ensure that machines will be available for delivery to suit the project timescale. Some projects have been postponed for two years waiting for turbine availability. Several manufacturers operate allocation systems and it is essential to reserve a place in the queue.

6.9.2.8 Turbo-expanders

Turbo-expanders are effectively the power turbines of gas turbines without the gas generator. They can be used where a supply of gas is available which can be expanded. The gas utilised may be process gas which is expanded between process stages or waste gas which is expanded to atmospheric conditions. Speed and torque curves can be adjusted by control and by-pass valves; most pump requirements can be accommodated. They are built in standard sizes which can be adapted to suit individual conditions. Turbo-expanders are also built specially "to order" when a sufficient number of machines is required.

6.9.2.9 Hydraulic turbines

Pumps can be driven by hydraulic turbines when a source of high pressure liquid is accessible. The turbine can be a Pelton Wheel, a Francis turbine or a pump running in reverse. The pump can be axial flow or radial depending upon the head and flow available. All hydraulic turbines can be controlled to produce variable speed outputs. The choice of turbine type will depend upon the characteristics of the driven pump system. Pumps running in reverse have become popular over the last 10 to 15 years. Many parts can be made common to the pump and only one manufacturer is involved. Size and speed is virtually unlimited.

6.9.2.10 AC squirrel cage induction motors

A standard, off-the-shelf motor, driven by a variable frequency inverter can produce speed ranges of 10:1, and over, without problems. This method of speed control is the most popular of all methods for motors up 125 kW, perhaps even 250 kW.

Full load torque, plus a margin, is available for starting without transmitting large current demands into the supply system. Full load torque is available over the whole speed range. Some motors may need derating by up to 10% to work with some types of inverters. There is a need to match the frequency inverter to the motor in the larger sizes. Care should be taken with continuous operation at low speeds. Some motors may need a separate motor driven cooling fan.

Discussions should be held with the motor and inverter manufacturers regarding operating conditions. Special attention may be needed with starting torque or overload capabilities for operating at relief valve settings. If accurate speed control is required, for metering pumps for example, some feedback from the motor to the inverter may be necessary. Some small inverters can accept single phase power at 240 V and drive three phase motors at 380 V. Some inverters can produce output frequencies up to 120 Hz.

The motor manufacturer must be consulted if motors are to be overspeeded on a regular basis. Virtually any control system can be interfaced to an inverter. Large inverters can transmit significant harmonic signals into the mains supply. Generating authorities should be consulted when contemplating drives over 500 kW. All inverter drives must comply with the statutory requirements for electromagnetic pollution in all countries.

6.9.2.11 AC slip ring motors

Slip ring motors are also called AC wound rotor motors. Slip ring motors were the standard before squirrel cage motors were invented. The lack of slip rings, plus simpler starting, made squirrel cage motors much more attractive for most fixed speed applications. However, slip ring motors continued to be used for limited variable speed applications and those applications which involved large inertias requiring long run up times.

The external resistance, connected in series with the rotor windings, allows the speed/torque characteristic of the motor to be modified. At low resistance, when the slip rings are shorted, the torque characteristic is very steep giving a wide range of torque values for small variations in speed. As the rotor resistance is increased the speed/torque characteristic becomes shallower giving corresponding torque values at lower speeds. Full load torque can be made available at any speed. This method of speed variation can be applied to most practical pump installations. However, it is not efficient. At reduced speeds, energy is dissipated in the rotor resistances and external cooling, motor driven fans, may be necessary. The motor efficiency reduces approximately in proportion to the motor speed.

The sliprings and brushes of wound rotor motors require maintenance and this is a disadvantage compared to squirrel cage motors. Problems can arise with hazardous area requirements. AC slip ring motors can be useful for large installations, with heavy inertias creating starting problems, when speed variation is only required periodically. If a pump was also required to run as a turbine, then the slip ring motor could become an alternator and produce power.

6.9.2.12 Switched reluctance motors

The switched reluctance motor, srm, is not a new motor design but an old motor design concept which has become viable with the advances of modern electronic power switches. The srm is similar in some respects to the squirrel cage motor. The motor rotor does not require copper or aluminium conductors and can be made from a single steel forging. Very high motor speeds are possible. The power electronics are used to switch the supply current to different stator windings. A srm installation is similar to a variable frequency inverter supplying a squirrel cage motor; the srm "inverter"' is much simpler than the squirrel cage version.

The srm package is very useful because of the extra features possible compared to the inverter/squirrel cage package:

- 100% torque at zero speed
- 100% torque holding at zero speed
- No motor overheating at low speed
- No motor overheating due to starting inrush currents
- Precise speed control

- Very high speeds possible

The srm does not have a synchronous speed and cannot be operated without the power electronic controls. The switched reluctance motor can be expected to produce significant changes in the ways variable speed rotodynamic and positive displacement pumps are applied. The very low speeds possible should result in very small accumulators being fitted rather than large storage tanks. Specialist motor manufacturers are concentrating on motors up to 300kW.

6.9.2.13 DC motors

DC motors were the principal electrical variable speed drivers before variable frequency inverters became economically viable. Most applications which used to have DC motors can now be driven by variable frequency AC motors. DC motors have two main advantages over AC motors; they can be designed to run at any particular speed and have a faster response to transient conditions.

DC motors do not have synchronous speeds like AC motors. A DC motor could be designed to run at 4500 r/min, for example, to eliminate the need for a gearbox. If rapid speed changes are required as part of the normal operating conditions then DC motors may be better than AC. However, consideration should be given to the flow variations and pressure pulsations caused by any rapid speed changes.

The speed of a DC motor, up to its base speed, is controlled by the armature voltage. When the armature current equals the rated current, full load torque is available. These motors can provide full load torque from zero speed up to base speed. Speeds above base speed, produced by field weakening, are at reduced torque. Above the base speed, DC motors are constant power, not constant torque. Motor efficiency is high over the whole speed range. Modern motors are powered via electronic controllers in a similar manner to variable frequency AC motors. DC motors have commutators and brushes which require maintenance attention similar to sliprings. Hazardous area requirements can cause problems. DC motors are not as popular as AC motors and are not mass produced in the same quantities.

6.9.3 Costs

Procurement costs for various speed regulation equipment depends upon the shaft power, maximum speed and the environment in which the drive is going to be located. To some degree, extra allowances must be made for ancillary cooling equipment and in the case of electrical methods, of the effects of feedback in the mains from thyristors, and protection against incoming mains transients. The method of speed regulation selected may affect the procurement costs by a factor of 5. Figure 6.26 indicates the cost of standard three phase squirrel cage AC motors. Figure 6.27 indicates values for additional cost for speed regulation in relation to the costs for standard constant speed squirrel cage induction motors. For choice of equipment see Section 6.10.

6.9.4 Conversion of pump curves to various speeds

For rotodynamic pumps the Affinity Laws govern performance at speeds n_1 and n_2. If equation 4.16, in Chapter 4, is rewritten for constant diameter impellers then:

$$\frac{Q_2}{Q_1} = \frac{n_2}{n_1}$$

$$\frac{H_2}{H_1} = \left(\frac{n_2}{n_1}\right)^2$$

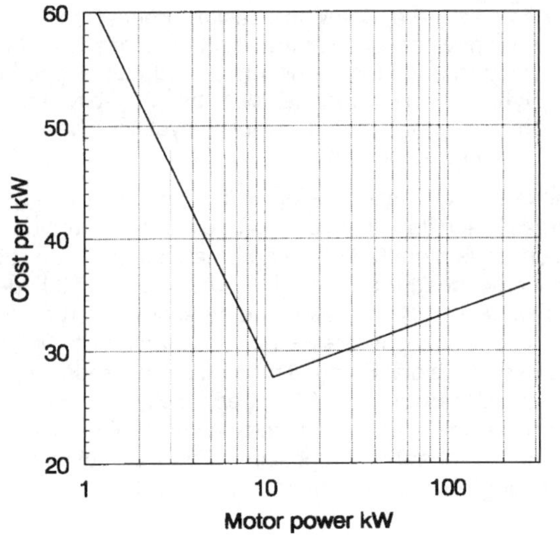

Figure 6.26 Costs of standard three phase squirrel cage AC motors

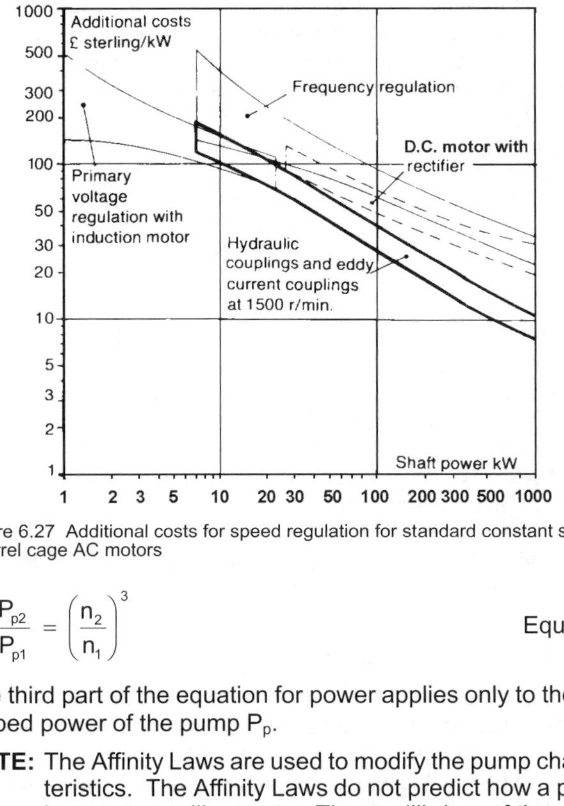

Figure 6.27 Additional costs for speed regulation for standard constant speed squirrel cage AC motors

$$\frac{P_{p2}}{P_{p1}} = \left(\frac{n_2}{n_1}\right)^3 \qquad \text{Equ 6.20}$$

The third part of the equation for power applies only to the absorbed power of the pump P_p.

NOTE: The Affinity Laws are used to modify the pump characteristics. The Affinity Laws do not predict how a pump in a system will operate. The equilibrium of the pump and system characteristics must be found in order to calculate the pump power and efficiency.

For any given pump in any given system, in accordance with Figure 6.28, the minimum required speed, n_{min}, can be calculated from the following equation:

$$n_{min} = n_{max} \cdot \sqrt{\frac{H_{stat}}{H_{max}}} \qquad \text{Equ 6.21}$$

The lowest required speed has to be known when selecting speed regulation equipment.

The performance of speed regulated pumps is usually described as a family of curves for H-Q and Q-P as shown in Chapter 5, Figure 5.23. These curves were determined by means of the Affinity Law, equation 6.20. The power requirement in accordance with equation 6.20 refers to the power requirement at the pump shaft. The power requirement for the electric motor is obtained by dividing by the efficiencies of both the motor and the transmission, see also Figure 6.29. Frequency inverters and thyristor controllers for DC motors must

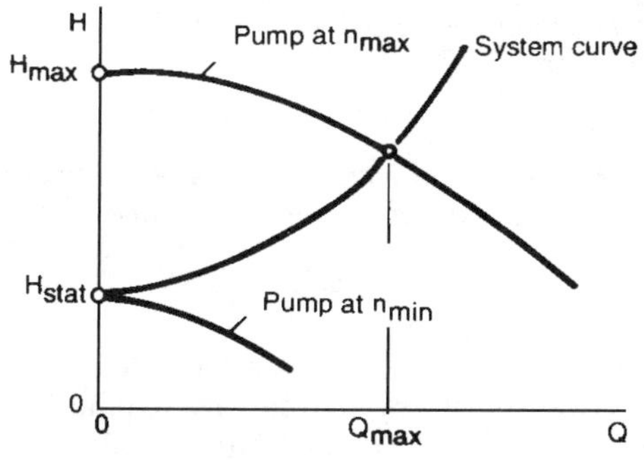

Figure 6.28 Relationship for lowest required speed, n_{min}

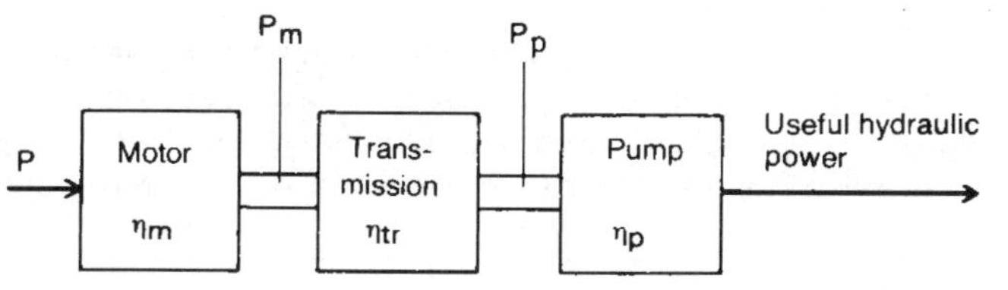

Figure 6.29 Pump unit with power transmission

be included. The overall efficiency is also called the "wire to water" efficiency.

For positive displacement pumps the flow is approximately proportional to speed, Q $\propto$ n. The first part of equation 6.20 can be used but some allowances may be necessary for variations in slip. Slip will vary with speed and differential pressure. The differential pressure must be evaluated from the system characteristics, see Figure 6.15.

Pump power is the product of flow and differential pressure divided by the efficiency. Just as with rotodynamic pumps, pump efficiency varies with operating conditions.

Thus for rotodynamic pumps, drive power P:

$$P = \frac{P_m}{h_m} = \frac{P_p}{h_m \cdot h_{tr}} = \frac{r \cdot Q \cdot DH \cdot g}{h_m \cdot h_{tr} \cdot h_p \cdot 1000} (kW) \qquad \text{Equ 6.22}$$

For positive displacement pumps with flow in m^3/h and differential pressure in bar, drive power P:

$$P = \frac{P_m}{h_m} = \frac{P_p}{h_m \cdot h_{tr}} = \frac{Q \cdot DP}{h_m \cdot h_{tr} \cdot h_p \cdot 36} (kW) \qquad \text{Equ 6.23}$$

6.9.5 Efficiencies of various methods

If the Affinity Laws are applied to a centrifugal pump then, when the pump speed is regulated down to half speed, the volume flow also reduces by a half, the differential head to a quarter and the power required at the pump shaft to an eighth based on a system without static head. Similarly, when regulating down to 20% of maximum speed the pump shaft power goes down to 0.8% of the power at maximum speed based on zero static head. The variable speed pump characteristics must be plotted against the system curve to evaluate the speed, torque, efficiency and power for each duty point. At low speed, losses within the pump, such as bearing friction and seal friction, may distort the power curve so that the Affinity Laws underestimate the power required.

For positive displacement pumps the procedure is exactly the same. The pump characteristic shapes will be different as will the variation in pump efficiency. The pump torque and power for any duty point will be dependent upon the intersection with the system curve.

The characteristics of variable speed devices differ considerably. Not only is it essential to know the speed, torque and power of each duty point, but it is essential to know the timescale of operation at each duty point. It does not make sense to purchase a costly system which provides high efficiency at 10% flow if the only time low flow is necessary is during start-up and shut-down which occurs twice a year for a couple of hours.

Figures 6.30 to 6.34 indicate the general characteristics of the popular methods used for speed variation. The actual speed range possible is dependent upon the actual torque required at any speed. Efficiencies for drivers do not include the efficiency of any speed reduction necessary. Gearboxes should have transmission efficiencies of 95 to 98%. Efficiencies for variable ratio transmissions do not include the driver efficiency. Electric motors, the most popular driver, will have efficiencies between 80 and 96% depending upon the size.

The two curves in Figure 6.34 represent the variation in efficiency due to size. The dotted lines indicate a possible reduc-

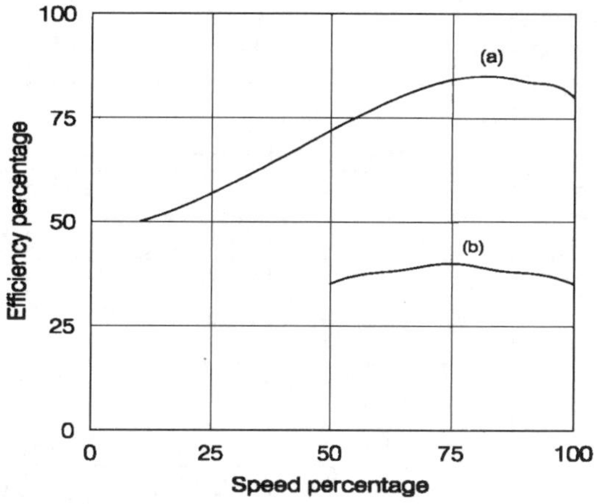

Figure 6.30 Speed range and efficiency —
(a) torque convertor (b) engine

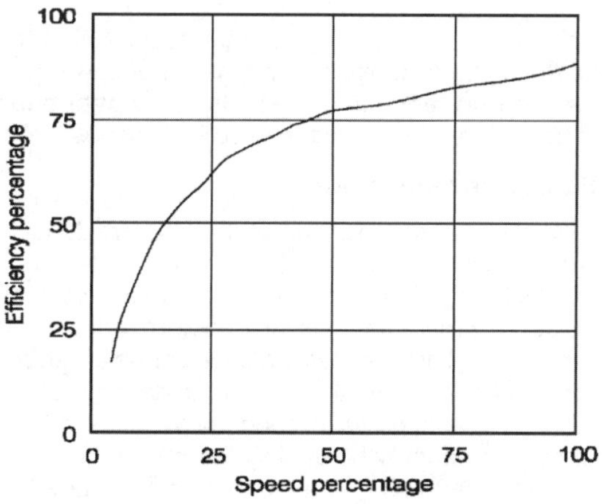

Figure 6.31 Speed range and efficiency of hydrostatic transmission

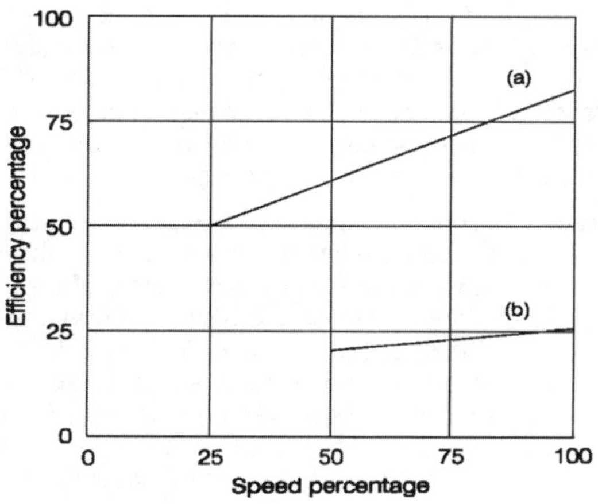

Figure 6.32 Speed range and efficiency —
(a) steam turbine (b) gas turbine with free power turbine

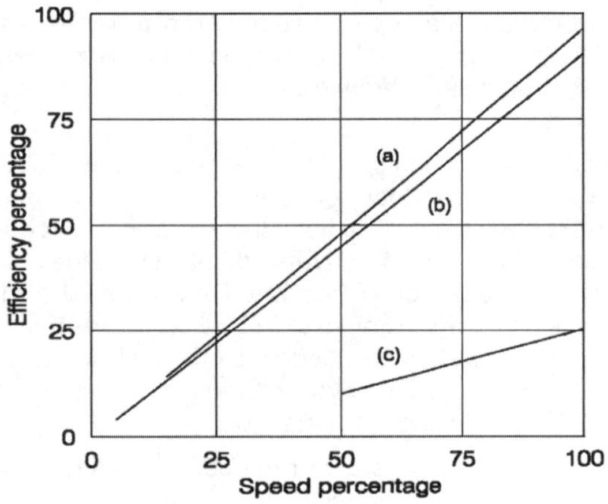

Figure 6.33 Speed range and efficiency —

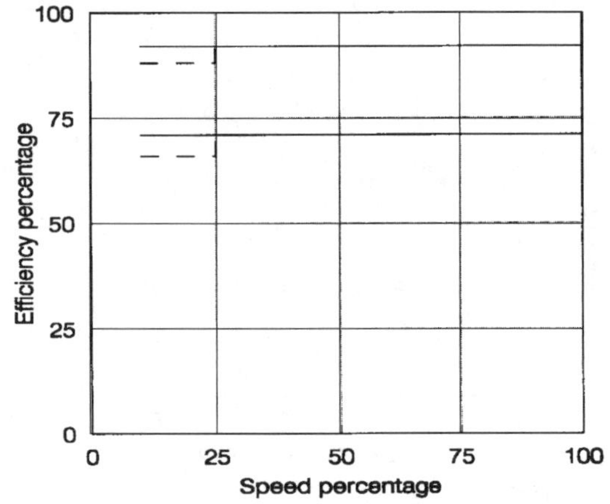

Figure 6.34 Speed range and efficiency of electric motors

tion in efficiency at low speed due to the requirement to start a separate motor driven fan to cool the main motor. Any power required for cooling fans fitted to electronic control equipment must be considered in the overall efficiency assessment.

6.9.6 Power consumption

In recent years, users whose operating costs are dominated by pump energy consumption have pioneered the concept of 'wire to water' efficiency. The pump and drive train efficiencies are not considered in isolation but are combined to produce an overall efficiency. This process includes variable frequency inverters for AC motors and thyristor controllers for DC motors. Some users have expressed concern over the accuracy of power measurements, when using the pump manufacturer's standard instrumentation, for electronic motor speed controllers. The controllers distort the power supply by injecting harmonics. Some users have insisted that the pump manufacturer needs modern, costly, analysis equipment to accurately measure the power consumption. This may in fact be strictly correct and theoretically possible. However, the pump user will probably be charged for electricity based on measurements taken by an old-fashioned kWh meter. This then, is the way to measure efficiency when the pump is being tested.

In operation, of course, the performance of a speed-regulated pump follows the requirement of the system curve in the H-Q diagram. By transferring the intersection points of the system curve with the various pump curves on the H-Q diagram to the power diagram the dotted line curve in Figure 6.35 is obtained. This curve indicates the power requirement when operating along the system curve. The power requirement is for a pump with fluid coupling. The speed is expressed relative to maximum speed for a direct drive pump no = 1. Because of slip in the fluid coupling a slightly lower maximum speed pump curve is produced compared to an equivalent direct drive pump.

The energy consumption E in kWh/year will be the power multiplied by time, or

$$E = P_{mean} \cdot t_0 \qquad \text{Equ 6.24}$$

where t_o = operational time per year (hrs) and where the mean power is determined by using Simpson's Rule:

$$P_{mean} = \frac{1}{6} \cdot P_{Q100} + \frac{2}{3} \cdot P_{Q50} + \frac{1}{6} \cdot P_{Q0} \quad \text{(kW)} \qquad \text{Equ 6.25}$$

The values P_{Q100}, P_{Q50} and P_{Q0} are the values for power requirement taken from the system curve at volume flow Q_{100}, Q_{50} and Q_0 respectively, i.e. the flow at 100%, 50% and 0% constancy. See also Figure 6.36. Taken at various values of Q provides the basis for determination of mean power in accordance with equation 6.25.

In speed regulation, the regulation efficiency is always equal to 1 because supply always equals demand. The losses in speed regulation are instead described by a transmission efficiency η_{tr}, see also Figures 6.29 to 6.34. The variations in pump speed

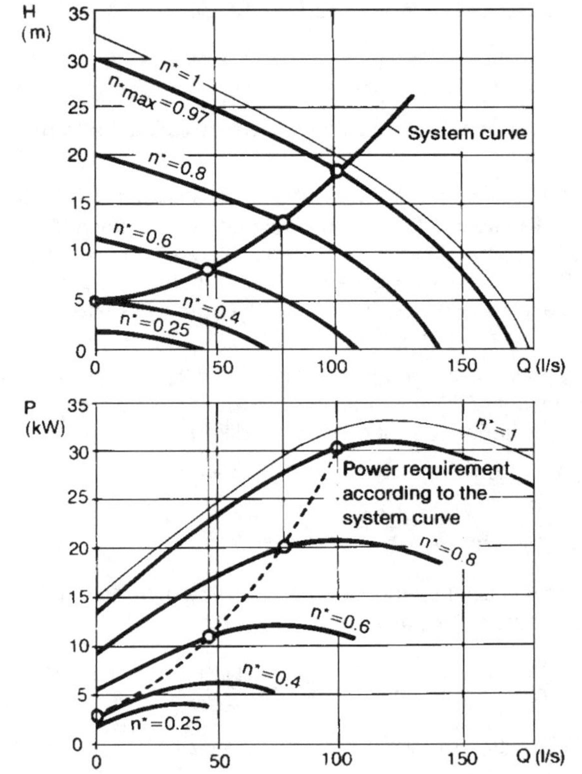

Figure 6.35 Example of performance curves for a speed regulated pump

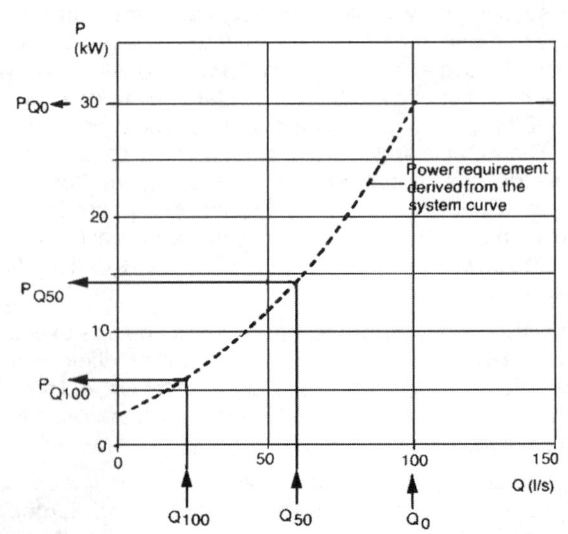

Figure 6.36 Power requirement for speed regulated centrifugal pump in accordance with Figure 6.35

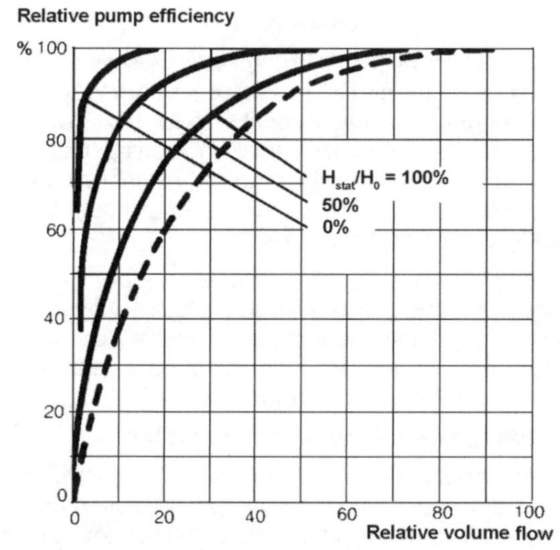

Figure 6.37 Examples of pump efficiency variation for a speed regulated centrifugal pump

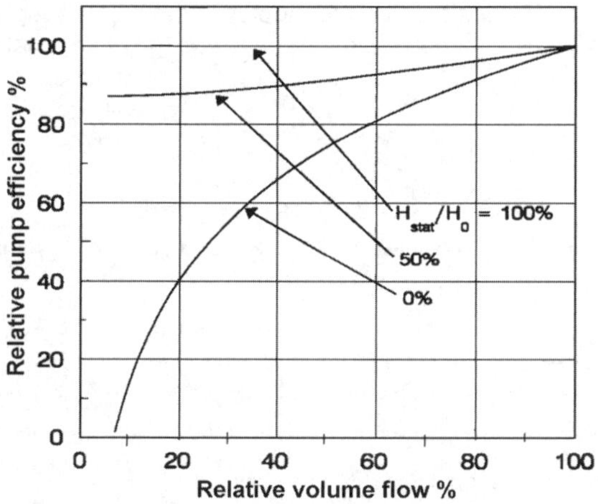

Figure 6.38 Examples of pump efficiency variation for a speed regulated reciprocating pump

and also transmission efficiency will depend upon the shape of the system curve. In some drive arrangements, AC motors with low speed pumps for example, transmission efficiency must include everything but the pump.

$$h_{tr} = h_i \cdot h_m \cdot h_g \quad \text{Equ 6.26}$$

where:

η_i = inverter efficiency (decimal)

η_m = motor efficiency (decimal)

η_g = gearbox efficiency (decimal)

The transmission efficiency for a large drive train would be

η_{tr} = 0.97 · 0.96 · 0.97 = 0.903

Rotodynamic pump efficiency varies more favourably when speed regulated than with other methods of regulation, Figure 6.37. The dotted curve represents the normal efficiency change for a constant speed pump when 100% flow is at BEP. With frictional losses only, (H_{stat} = 0), pump efficiency remains constant over the larger part of the flow regulating range. It should be noted, that if H_{stat} is not equal to zero, then better part-load efficiencies are obtained if the operational point of the pump at full load lies to the right of best efficiency point. Some process users will not relish operating passed BEP and detailed calculations will be required to confirm the financial benefits.

Reciprocating pumps, not surprisingly, react differently to system variations. As static head increases in a friction system a reciprocating pump is able to operate closer to its optimum rod load even as flow is reduced. The best system is one where the discharge pressure is constant irrespective of flow rate, see Figure 6.38.

6.9.7 Regulation system schematics

A schematic showing the principle of operation for the complete regulation process of a speed regulated pump, by variable ratio transmission, is shown in Figure 6.39. The control system can be local or more usually a sub-system in the plant control system. Linearity between signals and required pump function is of the utmost importance for functional stable regulation. This condition is emphasised particularly in public water supplies where very great variations of flow are common. In this application, unfortunately, a number of installations using speed regulation are unusable for the simple reason that the importance of the regulation function was not fully appreciated. In many cases it is necessary to measure additional parameters, e.g. pump speed, n, or flow, Q, (dotted signals in Figure 6.39) and to use these as feedback into the regulation system, if stability is to be achieved.

Linearity is exploited through comparison between the required control parameter, generally flow, Q, or speed, n. If both are measured, the control system can compensate for wear and report on pump condition. When H_{stat} is not equal to zero, Q and n are no longer linear, see Figure 6.40. For rotodynamic pumps operating at high differential heads Q and n will never be linear. As can be seen from Table 6.6 it must be possible to regulate the speed at small values of flow with very great precision. The Table has been compiled with the help of additional pump curves and by interpolation.

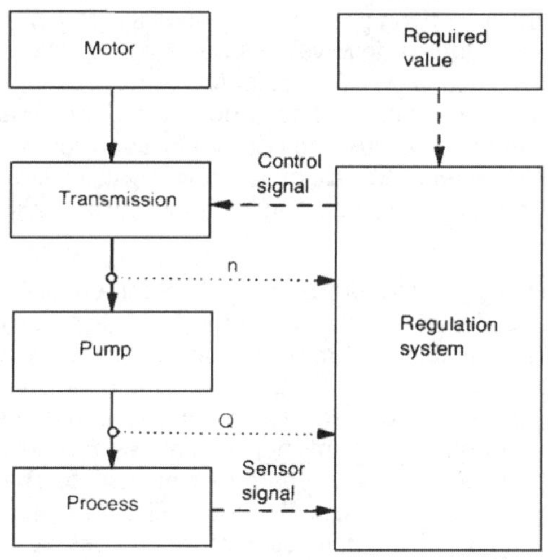

Figure 6.39 Schematic diagram for speed regulated pump

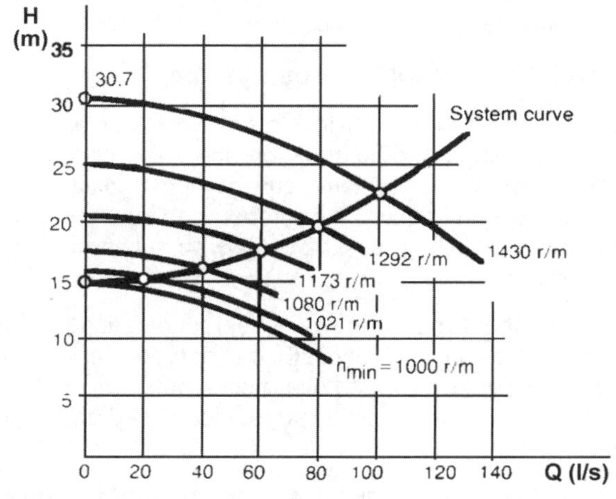

Figure 6.40 Examples of pump curves for determining the relationship between flow Q and speed n

Ql/s	0	5	10	15	20	40	100
n r/min	1000.0	1001.3	1005.2	1011.7	1020.7	1080	1430

Table 6.6 Data compiled from Figure 6.40

Alternative methods are to linearise the speed by means of so-called indirect flow measurement or for example to eliminate speed completely from the control loop and to replace it by flow. When precise flow control is required, the method adopted for flow measurement must be selected with great care. It must be emphasised that rotodynamic pumps produce some cyclic flow variations and pressure pulsations; centrifugal pumps at high speed can be quite unacceptable. Flow measurement by orifice plates, for example, can be impaired by noise in the signal and may require averaging or smoothing.

Positive displacement pumps can suffer from complex relationships for speed and flow. Slip tends to be a variable rather than a constant. Changes in speed, differential pressure, and in some designs temperature, make it advisable to use flow measurement as the primary control signal. Metering pumps usually run slow enough for slip to be constant; direct speed control will produce proportional flow control.

The dynamic interaction between a speed regulated pump, reservoir or pump sump and pipeline can also be a cause of instability. In such a case the problem has to be studied with regard to the response time of the speed regulation, time constant, delay time, filling time of the reservoirs and transient changes in the flow requirement due to the opening or closing of valves, among other things. Relatively slow changes in pump speed are always preferred.

To build up an installation comprising the various component parts, without carrying out a stability analysis from the point of view of regulation can involve considerable risks. When buying packaged solutions one should, for the same reason, try to obtain functional guarantees based upon the actual case. The process parameters must then be clearly specified as regards pipeline characteristics, size of pumps, reservoirs, storage and operational conditions, including variations in the liquid properties.

Local packaged systems are available commercially as an alternative to the user selecting component parts in order to make up a complete system. These include pump, drive equipment with speed regulation and control system. In some cases several pumps are included, of which at least one is speed regulated. The control system then comprises facilities for the switching in and out as well as providing for interchange between the various pumps, so that the method of operation can be chosen to suit the total function of the process. An interface to plant control systems can be fitted by using one of the standard data communication interface protocols; such as 4-20 mA, RS 232, RS 423, IEEE 488, and 60488 or HART and others; depending upon the nature and complexity of the data.

6.9.8 Schematics for various systems

Schematics are an important tool in system design and control logic. A large amount of information about the various piped and electric/electronic systems can be communicated using simple symbols and boxes. The symbol used for a pump, in the examples shown in Figures 6.42 to 6.48, complies with several national standards.

This symbol should not be used in actual schematics because it does not convey anything about the pump type. If engineers of different disciplines view and use a schematic, recommendations and improvements can be suggested based on the pump types shown.

The symbols shown in Figure 6.41 are proposed for consideration.

6.9.8.1 Variable ratio coupling

A variable ratio coupling can be hydraulic or electric, see Figure 6.42. Fluid couplings and hydrostatic drives are torque devices, torque convertors are almost constant power devices and eddy current couplings are slightly better than constant torque. Fluid couplings, hydrostatic drives and torque convertors can be controlled electrically or pneumatically, eddy current couplings are usually controlled electrically.

Semi-automatic multi-speed gearboxes can be controlled in a similar manner. The gearbox would be "pulsed" by a digital signal, which triggers the pneumatics, to change up or down in response to the demand. Mechanical variators can usually be controlled electrically or pneumatically.

6.9.8.2 Voltage regulated induction motor

For an AC induction motor the torque is proportional to the square of the applied voltage. Speed regulation can be obtained by varying this. A limiting condition, however, is that the motor should be equipped with a rotor having increased resistance, a so-called high resistance rotor, in order for the losses to be reasonable. The variable voltage is obtained by means of a transformer or a thyristor connected in the motor supply, Figure 6.43. The minimum speed depends on the motor character-

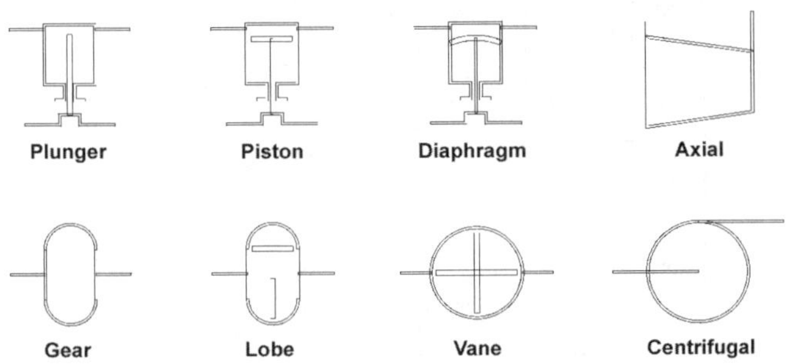

Figure 6.41 Typical symbols for different pump types

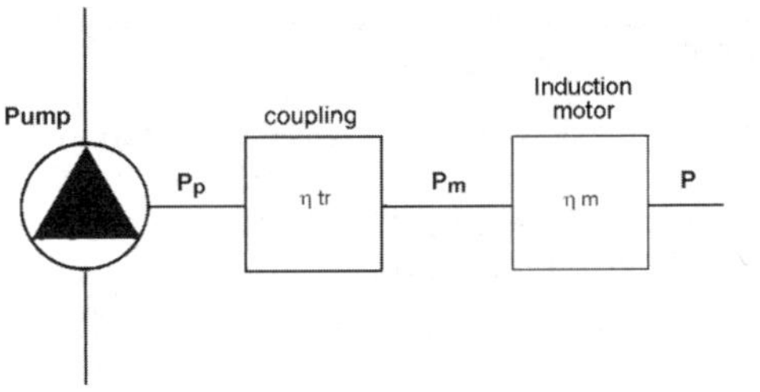

Figure 6.42 Schematic for speed regulation using variable ratio coupling

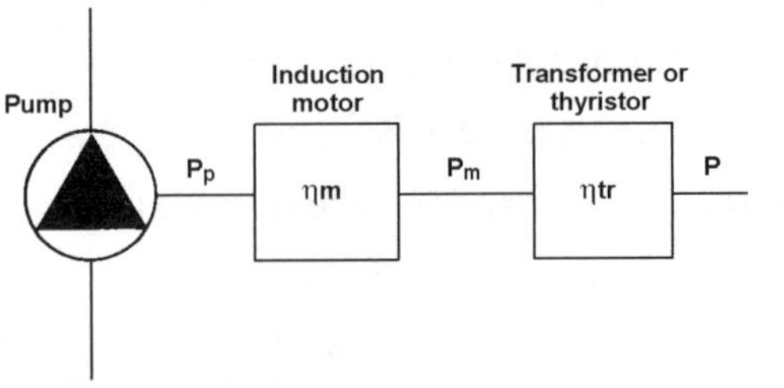

Figure 6.43 Schematic for voltage regulated AC motor

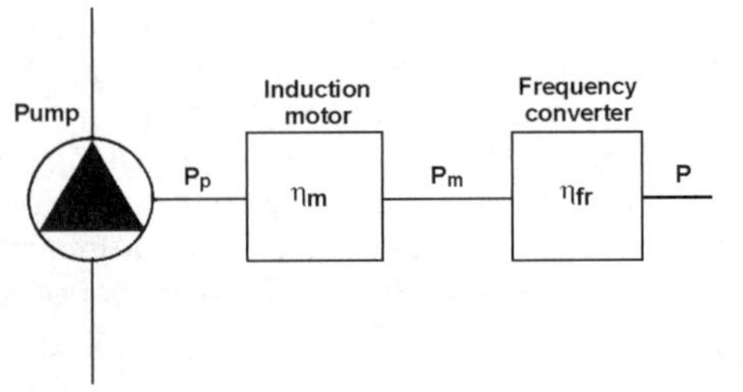

Figure 6.44 Schematic for frequency regulation

istics, mainly the magnitude of the rotor resistance. Control would normally be electric.

6.9.8.3 Frequency regulated induction motor

The speed of an AC induction motor is directly proportional to the frequency. In frequency regulation the frequency and also the voltage to the motor is varied simultaneously. The frequency converter, or inverter, is connected in the motor supply, Figure 6.44. In the frequency converter, the mains power is first rectified and then the DC current is chopped up to form AC at the required frequency and voltage. Because the frequency converter is electronic, electric control is the ideal; all the protocols listed in Section 6.9.6 are used depending upon the control source.

6.9.8.4 DC motor drive

For a DC motor with constant field, the speed is approximately proportional to the armature voltage. The armature voltage can be controlled quite simply for any required value by using a thyristor rectifier to convert AC to DC, Figure 6.45. Electric control signals similar to frequency converters.

6.9.8.5 Slipring induction motor

The speed of an AC slipring or wound rotor induction motor can be controlled by taking power from the rotor via the sliprings, Figure 6.46. The power can be dissipated in an external regulating resistance or fed back into the mains via a thyristor rectifier. In the latter case the rotor voltage is rectified and then chopped back into AC and is fed back into the mains via a transformer or sub-synchronous cascade rectifier. Electric control similar to frequency converters is standard.

6.9.8.6 Engines

Industrialised engines are fitted with governors which control the speed. The governor is set to a specific speed and adjusts the fuel system to maintain the speed as the load varies. External speed control systems send signals to the engine governor to adjust the set speed. Most governors have a "tick-over" or "idling" setting which allows the engine to run slowly during warming-up or when the pump torque is low, on by-pass or unloaded. Governors are available to match any control signal. Figure 6.47 shows a typical engine arrangement.

6.9.8.7 Air motors

The speed of air motors is regulated by a flow control valve in the air supply. Control valves may be actuated by air, liquid or directly by an electric positioner. The control signal can be pilot air or liquid or digital or analogue electric. The pump flow should be measured and the speed adjusted to produce the correct flow, see Figure 6.48.

6.9.8.8 Turbines

Steam and gas turbines are fitted with governors in a similar manner to engines. The governor is set to the desired speed. In steam turbines, the governor actuates steam valves which control how much steam and how many nozzles are in operation to provide the power to operate at the set speed. Gas turbine control is identical to engines, the governor controlling the fuel supply, see Figure 6.47.

6.9.8.9 Turbo-expanders and hydraulic turbines

Both turbo-expanders and hydraulic turbines derive their power supply from flowing pressurised fluid streams. Neither device has any control over the fluid source so a flow control valve must be used to regulate the proportion used. Additional controls may be required in the primary supply to divert or by-pass fluid which is not used because of load conditions. The pump user and the manufacturer must agree who supplies the valves. In principle, control is exactly the same as for air motors, see Figure 6.48.

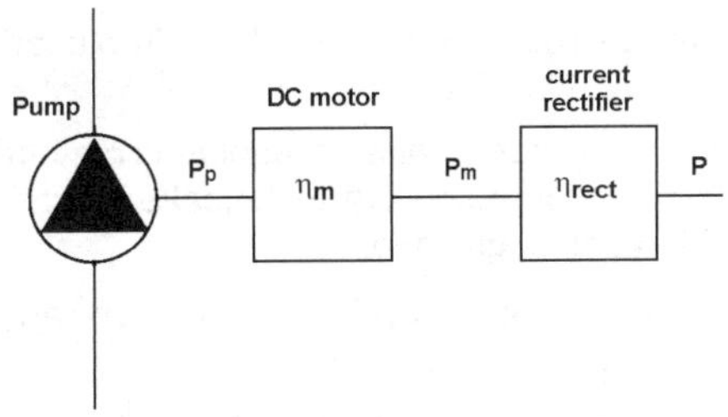

Figure 6.45 Schematic for DC motor drive

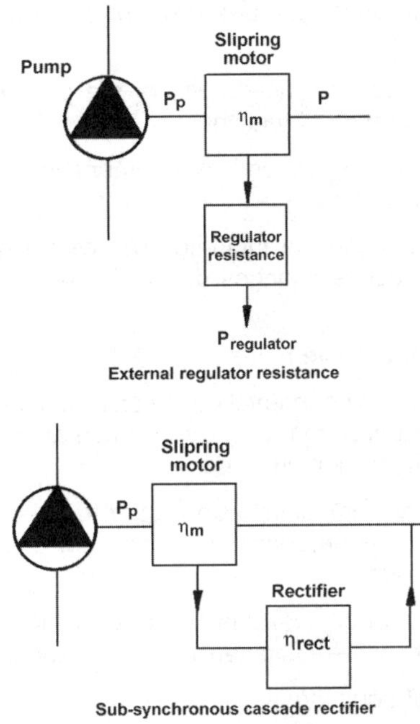

Figure 6.46 Schematic for speed regulation using a slipring motor

6.10 Choice of flow regulation method

6.10.1 General

The choice of flow regulation in pumping has traditionally been made largely on the basis of the initial investment required for the equipment. This method of assessment can be blamed on design/procure/build contractors who are not responsible for the ultimate running costs of the plant. As a result of increased energy costs, however, it has become essential to take into account the total costs during the whole life of the equipment, "life cycle cost", see also Chapter 14. This approach brings with it a drastic change of established practice, meaning among other

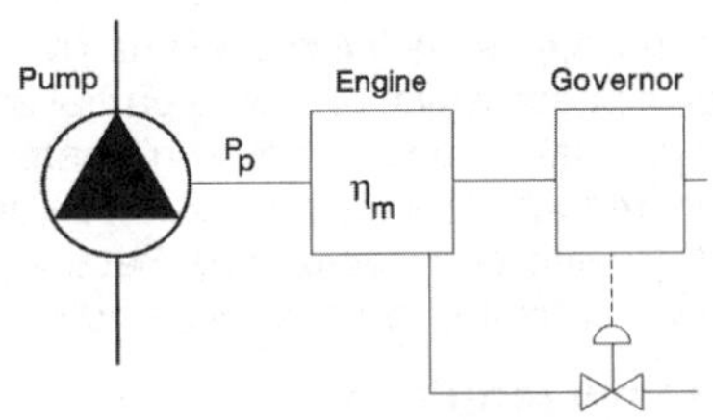

Figure 6.47 Schematic for engine speed regulation

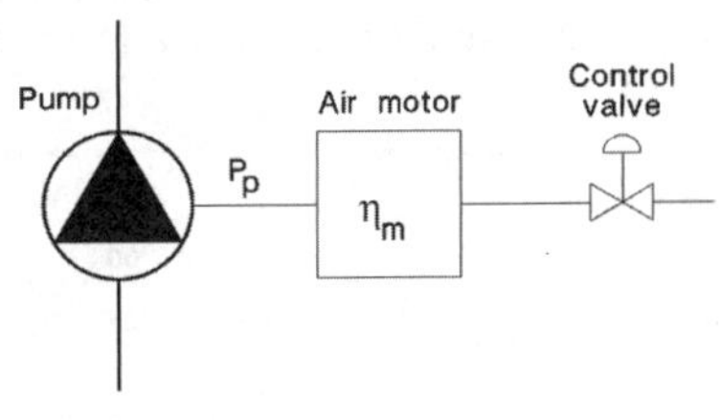

Figure 6.48 Schematic for air motor control

things, that throttle regulation is often being replaced by speed regulation.

In order to make a correct choice of regulation method, it is necessary to carry out an extensive technical/economic analysis where consideration is given to:

- System requirements for cyclic flow variations and pressure pulsations and flow range
- Constancy of flow throughout the life of the equipment
- Shape of the system curve and the variation in the system curve due to start-up/shut-down, corrosion, fouling, seasonal production
- Criticality of service, how much of the process depends upon the operation of this unit
- Reliability, i.e. the general requirement for back-up and stand-by equipment
- The ability to modify the equipment slightly or significantly to cope with process changes due to future refinements or modifications
- Primary control system
- Sensitivity to environmental conditions, particularly humidity, dust, temperature range, gases such as methane, ammonia and hydrogen sulphide
- Serviceability, particularly with regard to the various skills required of local personnel involved, e.g. electrical, electronic and mechanical
- Energy costs of all available sources, including variations during the life of the equipment and/or seasonal variations
- Initial cost of equipment
- Life, cost and availability of spare parts
- Service life of the equipment and estimated profits

The flow/head required and constancy of flow decide whether the flow shall be distributed among several pump units, of which in its turn only one may need to be flow regulated.

The shape of the system curve and constancy of flow will guide the selection of pump type and flow regulation method.

The need for a stand-by unit, and if the stand-by units requires regulation, is dependent upon the criticality of the service. Reliability of the unit may be improved by imposing limitations on the normal operation; reduced speed, reduced power, reduced operating pressure. The requirement to have a stand-by unit affects the investment calculation significantly; additional space, piping, valves, cables, control equipment; since the cost savings from maintaining production have to cover the investment and operational costs of the stand-by unit.

As can be seen in Chapter 14, the cost of energy is the dominating operational cost, for which reason a partial optimization of investment for regulation equipment and energy savings will give satisfactory results. The energy costs can be determined by use of the total efficiency factor, see Section 14.3.1, or by use of total power consumption at various flows.

6.10.2 Direct flow regulation

Table 6.7 compares the power requirements for flow control of a centrifugal pump using different speed methods and with flow regulation using throttle and on-off regulation to the hydraulic power.

	Flow				
	20	40	60	80	100
Hydraulic power kW	0.8	6	22	51	100
Variable frequency AC (large)	1	6.5	24	55	109
motor or DC motor (small)	1.5	8.5	31	72	140
Mechanical variators	8	16	33	61	110
Fluid coupling	10	23	44	74	115
Torque convertor	8	18	37	61	125
Hydraulic coupling	10	19	38	72	129
Throttling	84	90	97	106	110
On-off control (see Figure 6.6)	24	51	76	97	110

Table 6.7 Comparison of power requirement with various methods of flow regulation of a centrifugal pump with zero static differential head

The comparison applies to one specific case in which the system's static differential head is zero, i.e. the only losses are pipe frictional ones. For other operational cases, particularly with increasing proportion of static differential head, the power savings are less for speed regulation. When the differential head consists only of static head, there is in certain cases a 10% to 20% greater power consumption for speed regulation than for on-off control. In these circumstances a positive displacement pump may give better results, see Figure 6.38.

The loss of head in the control valve in accordance with equation 6.17 is not included. This corresponds in this case to a power of about 35 kW.

From Table 6.7 it can easily be seen that throttling is inefficient. Throttling, or by-passing, should only be undertaken when the flow reduction necessary is small and/or the operating time at reduced capacity is short. On-off control is inefficient for flow regulation on purely friction systems. If a cheap reliable source of energy can be obtained the cost of inefficiency can be negated and the choice of method will be based on initial installation costs.

6.10.3 Speed regulation

To determine energy consumption it is necessary to summate power consumption over a period of time using a constancy diagram. If there is no constancy diagram, a rough estimate can be made using mean flow, or better still that flow which has 50% constancy. When studying the constancy of flow together with the performance of the pump already chosen, it is usually difficult to differentiate between the extreme flow requirement and the approximate safety margin. The normal maximum flow is generally less than 80% of the pump flow. The flow with 50% constancy is therefore usually quite low and can be judged to be:

- About 60% of maximum pump flow in process industries
- About 30% of maximum pump flow in public water supplies
- About 40% of maximum pump flow in heat transport (central heating and district heating)

Reducing safety margins and operating pumps closer to their BEP will lower operating costs.

A comparison of the power requirements in Table 6.7 at 60% of mean flow shows a difference between throttle regulation and speed regulation by fluid coupling of 97- 44= 53 kW. For example, for a capitalised cost for 1 kW of about $525, the difference in the capitalised energy costs will be about $28,000. Since the additional costs for speed regulation by fluid couplings could be about $7,000, the economic superiority of speed regulation is indisputable, see Figure 6.27.

This example illustrates that in many cases speed regulation is well worth considering. The cost of energy is critical. If energy is virtually free, e.g. waste steam or high pressure gas/liquid, there is no need to do any calculations. The energy source must be available for the lifetime of the pump though. As a result of general studies, it can be said that speed regulation of rotodynamic pumps for a continuous flow requirement is the best method of regulation at virtually all powers, if the system

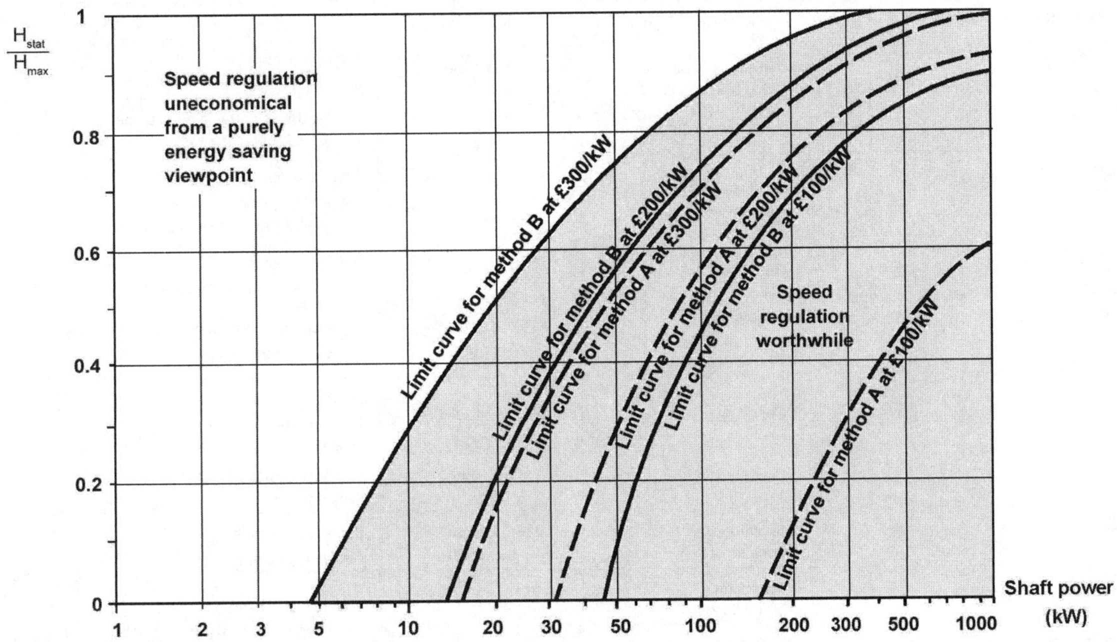

Figure 6.49 Approximate regions of viability for speed regulation. Method A applies to AC frequency regulation and B to fluid couplings
Figures in £/kW comprise capitalised energy costs based on maximum shaft power, see Table 14.8, in Chapter 14

curve comprises pipeline losses only. A review of other cases for rotodynamic pumps is shown in Figure 6.49. It can be seen that the capitalised value of energy costs per kW and the shape of the system curve, H_{stat}/H_{max} as in Figure 6.50, has considerable effect.

The limit of economic viability for speed regulation according to Figure 6.49 gives a first guide regarding the choice of method of regulation for rotodynamic pumps. As H_{stat}/H_{max} increases, positive displacement pumps should be considered. As a second step in the assessment, the investment costs for the actual installation can be compared with the possible energy savings using speed regulation. Intensive studies of normal pump installations actually show that possible energy savings using speed regulation compared to throttle regulation amount to 20% to 60% and depend heavily upon the shapes of the system and pump curves.

The guideline values shown in Figure 6.51 for energy savings can serve as a guide at the initial project stage and apply most closely to the process industries. In this illustration, Method A includes mechanical variators, frequency controlled AC motors and thyristor controlled DC motors; Method B includes fluid couplings and eddy current couplings. For water mains, central heating and domestic purposes, the energy savings are greater because of low mean flow. When the flow constancy is known, it is then of course possible to determine energy consumption for various regulation methods for the actual pump installation. The calculations can be simplified to a great extent by using Simpson's Rule as shown in previous Sections.

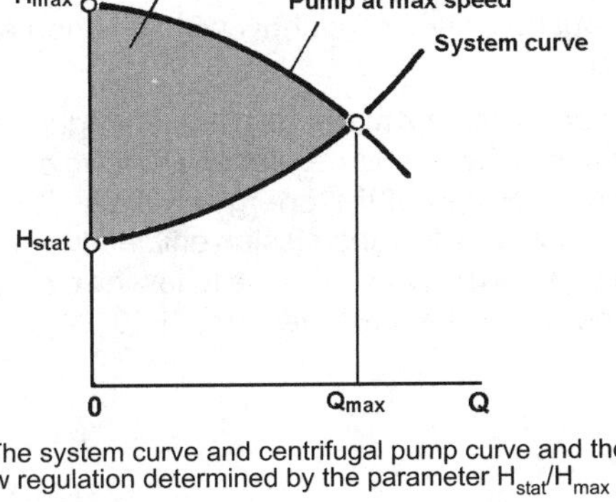

Figure 6.50 The system curve and centrifugal pump curve and their effect on method of flow regulation determined by the parameter H_{stat}/H_{max}

Safety factors applied to flow and differential heads carry with them dramatic increases in costs of pump installations. A flow safety margin of just 25% can double both the investment and the operational costs for a pump installation. The consequence of the increased power consumption due to the safety margin is eased to some extent with speed regulation as can be seen in Table 6.8, a summary of costs per kW of shaft power:

	Throttle reg £/kW	Speed reg £/kW
Increase in pumps costs	15-60	15-60
Increase in motor costs	20	20
Increase in costs for reg equipment, approx.	10	50
Capitalised energy costs, approx.	300	50
Total of additional costs, approx.	345/390	135/180

Table 6.8 Summary of costs per kW of shaft power

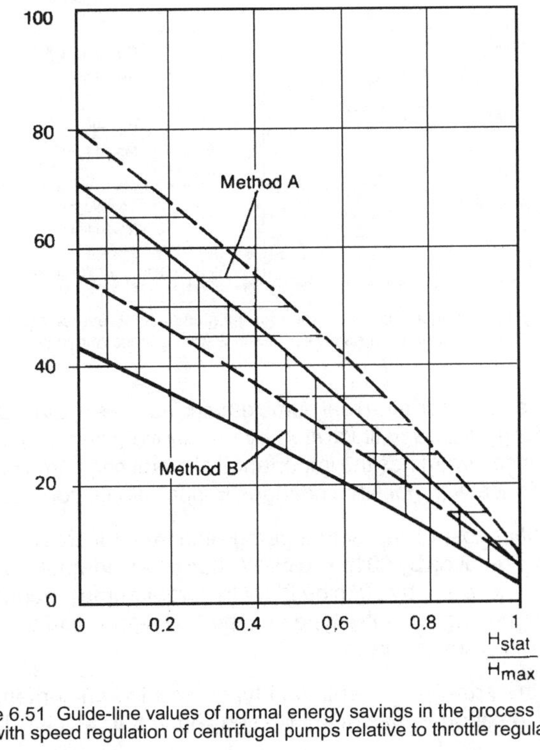

Figure 6.51 Guide-line values of normal energy savings in the process industries with speed regulation of centrifugal pumps relative to throttle regulation

Method	Pumps	Power kW	Application
Multi-speed gearbox	Positive displacement pumps	10 - 2000	Pipeline pumps Transfer pumps
Semi-automatic gearbox	Positive displacement pumps	100 - 400	Engine driven portable test pumps Oil-well fracturing
Mechanical variator	Metering pumps Positive displacement pumps	1 - 10 10 - 200	Chemical dosing General purpose Variator can include speed reduction
Hydrostatic transmission	Centrifugal	5 - 100	Ship bilge pumps One power pack drives several remote motors
Hydrostatic coupling	Centrifugal Positive displacement	5 - 30 5 - 30	Drive prevents over-running and turbining Constant torque drive for high H_{stat}
Fluid coupling	Centrifugal Centrifugal	5 - 100 10 - 200 50 - 5000	Sewage pumps Liquor & pulp pumps Hostile environment Circulating pumps for district heating Very reliable
Torque converter	Centrifugal Centrifugal	50 - 20000 50 - 5000	Boiler feed pumps Circulating pumps
Eddy current coupling (*)	Centrifugal Mixed flow	5 - 300	Water works & sewage
Multi-pole AC motor	Centrifugal Mixed flow Axial	0.05 - 25	Short pipeline pumps Transfer pumps Circulating pumps
Frequency controlled AC motor (*)	All pump types	0.1 & up	Most applications
Variable speed Switched reluctance motor	Centrifugal Positive displacement (rotary) Positive displacement (reciprocating)	up to 300 up to 150 up to 300	High speed pumps General process and hydraulic power packs
Primary voltage controlled AC motor	Centrifugal Mixed flow Axial	0.01 - 20	Short pipeline pumps Circulating pumps Booster pumps
Slipring AC motor	All pump types	20 - 500	Water works Sewage
DC motor (*)	All pump types	1 - 1000	Applications requiring precise speed control and rapid response
Petrol engine Diesel engine	Centrifugal Mixed flow Positive displacement	1 - 300	Portable pumps for site work, drainage Portable hydrotest pumps
Gas engine Crude oil engine	Centrifugal Positive displacement	100 - 3000	Pipeline pumps
Air motor	Centrifugal Mixed flow Axial	1 - 25	High speed pumps up to 9600 r/min from std compressed air
Steam turbine	Centrifugal Mixed flow Axial	25 & up	Boiler feed pumps Used in refineries as main pump, electric stand-by
Gas turbine	Centrifugal Mixed flow Axial	25 & up	Used where steam & electricity not available, popular off-shore
Turbo-expander	Centrifugal Mixed flow Axial	10 & up	Used in continuous processing, refineries with hp gas streams
Hydraulic turbine	Centrifugal Mixed flow Axial Positive displacement	10 & up	Used in continuous processing, refineries, RO, with hp liquid stream

Table 6.9 Typical applications for speed regulation RO ≡ reverse osmosis fresh water production
(*) Control equipment requires cooling air free from excess humidity

The need to determine real pump data as early as the initial project stage, is clearly illustrated by the above costing example which also underlines the importance of modifying pump data in accordance with possible changes in operational conditions.

The initial cost of the actual pump equipment increases with speed regulation by 80 to 150 £/kW shaft power in small pumps, about 10 kW, and by 20 to 80 £/kW for larger pumps, about 100 kW, depending upon the type of speed converter and ancillary equipment, see Figure 6.27.

All of the previously mentioned types of drive equipment are suited to various types of application. Table 6.9, which is by no means exhaustive, shows just how widely varied applications and sizes are.

When comparing various methods of speed regulation it should be noted that it is the speed regulation itself which is responsible for the greater part of the energy savings. The method of speed regulation and its transmission efficiency only have marginal effects, as can be seen from the following examples of typical investment calculations, see Table 6.10.

	DC motor drive AC variable frequency	Fluid coupling Eddy current
Capital investment costs	£15000	£8000
Savings in energy costs £ p.a.	£5000	£4000
Pay back time	3 years	2 years

Marginal investment	£15000 - £8000 = £7000
Marginal savings	£5000 - £4000 = £1000 p.a.
Pay back time	7 years

Table 6.10 Typical investment calculations

On the basis of these calculations both groups seem to be entirely acceptable. If the marginal investment is now studied, we get: The pay back time of 7 years is matched roughly by the write-off time of 30 years at the interest factor 15% and means that variable speed motors in this case are uneconomical compared with variable speed couplings.

In the same way as in the marginal calculation one may find that variable speed motors offer the same economic viability as variable speed couplings only when the procurement costs for motors do not exceed couplings by more than 20% to 30% i.e. the same value as corresponds to the greater energy savings of motors.

6.11 Useful references

HART Communication Foundation, 9390 Research Boulevard, Suite I-350, Austin TX 78759 USA. Tel: 512-794-0369, Fax: 512-794-3904, www.hartcomm.org.

IEEE488.2 -1992 (R2004) IEEE Standard Codes, Formats, Protocols, and Common Commands for Use with IEEE Standard. 488.1-1987, IEEE Standard Digital Interface for Programmable Instrumentation, (Now replaced by IEC/IEEE 60488 -2-2004).

Materials for pumps

7

7.1 Introduction

Material losses from corrosion costs industry a fortune every year. This is equivalent to saying that the output from every fifth steelworks emerges in the form of rust. Steel can be considered as an unstable material. After millions of years existing as an iron oxide it spends most of its time trying to revert to its original state. The problem of pump materials is quite different from that of simple common rusting, because of the complex effects brought about by the pumped liquids themselves.

Since pumps are used in most processes, on-going process development creates new problems for which solutions based on experience do not yet exist. Examples of these are:

- The catalytic manufacture of sulphuric acid. The production rate has been increased by removing various cooling operations. Through this, the temperature rises by 50°C, which increases the material corrosion problems of the pump by a factor of three.
- Adding fluoric acid to chromic acid baths to improve the shine on chromium plated surfaces. The new acid mixture is much more corrosive than before.

Individual materials and combination of materials for pumps are determined by a number of interactive factors, see Figure 7.1. The end result is usually an inspired compromise between all of these, whilst at the same time, unfortunately, some of the factors may be more or less unknown quantities.

In Figure 7.1, "Manufacturing techniques" is a broad heading and includes some aspects which affect the useful life of a pump and the ability of the user to repair a pump. It is preferable to use materials which are easy to machine. If a material is so difficult to machine that "as cast" parts must be used then pump performance will normally be reduced. It is also better to use materials which can be welded.

Intricate components tend to be cast. Casting is not 100% successful. Provisions must be made to allow for approved repair methods. Welding is the most useful repair technique. Components made in materials which cannot be welded may have to be scrapped, increasing the overall cost. Pumps manufactured from materials which are difficult or impossible to weld will prove difficult or impossible to repair when faults occur in service. Some materials which can be welded should not be repaired by welding. This is controlled by the Quality Control (QC) Plan, see Chapter 12. Stationary materials in close proximity to moving materials must be compatible.

Attention must be paid to the "bearing" qualities of some materials, which may be affected by the liquid properties. In some cases, non-metallic materials in particular, process upset conditions can change the physical properties of the material. In these cases, pump design conditions should protect the integrity.

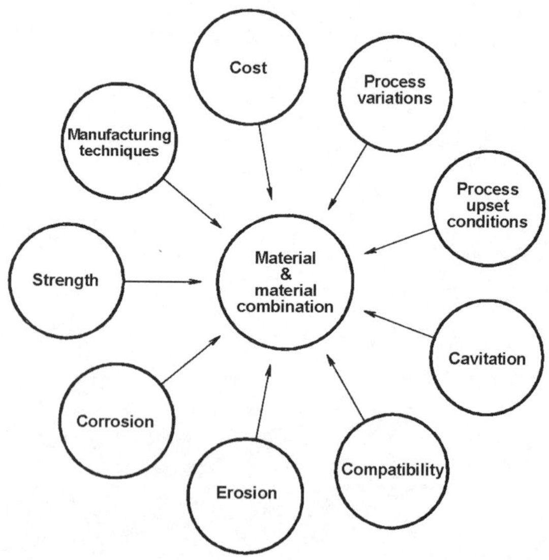

Figure 7.1 Factors affecting material suitability for pumps

The strength of materials can be judged by the allowable stress values given in pressure vessel codes. Local laws may specify a particular code to follow. ASME VIII and BS 5500 are popular and are specified as the limitation in many rotodynamic pump specifications. Pressure vessel stress levels are useless if components must be designed for fatigue resistance, as is the case with reciprocating pump cylinders for example. The fatigue endurance limit of the particular material, in contact with the particular liquid, will dictate the allowable stress levels.

When reviewing different specifications for materials it is important to consider materials which are "equivalent". General purpose materials may not be equivalent to materials specifically designated for pressure containment. Manufacturing methods can impart integrity which is not easily indicated by physical properties or chemical composition.

7.2 Typical materials

The majority of materials used currently in pumps can be classified within the following main groups:

- Material with iron as its main constituent
- Material with significant proportions of chrome and nickel
- Material with copper or aluminium as its main component
- Other metallic materials
- Non-metallic materials

The mechanical properties are defined in the usual way by the tensile strength and elongation. Sometimes there are other physical properties which are of interest such as impact resistance and hardness. The pump pressure rating PN, see also Section 7.3, is largely determined by the strength of materials and the temperature. For plastics at room temperatures as for metals at higher temperatures, the resistance to deformation must be considered. Creep resistance describes the effect of time under stress on permanent deformation.

Temperature plays an important part in the properties of solid materials. In general, as temperature increases materials become more elastic and deform more easily. This effect is reflected in the allowable stress which should be used for design

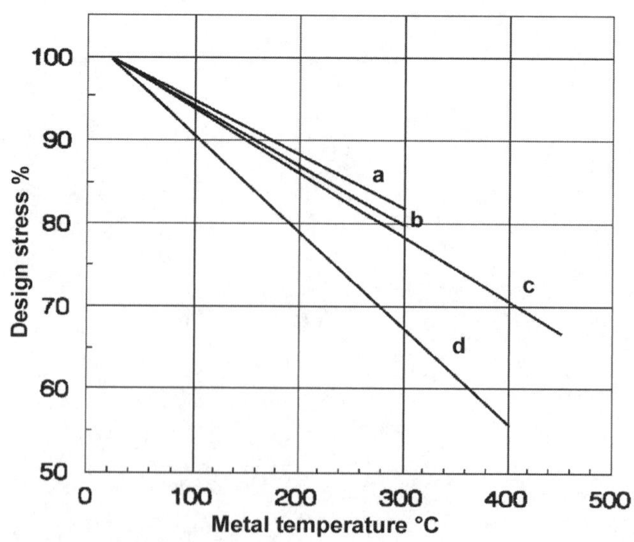

a = spheroidal graphite iron
b = grey cast iron
c = austenitic chrome iron
d = cast steel

Figure 7.2 Variation in material design stress with temperature

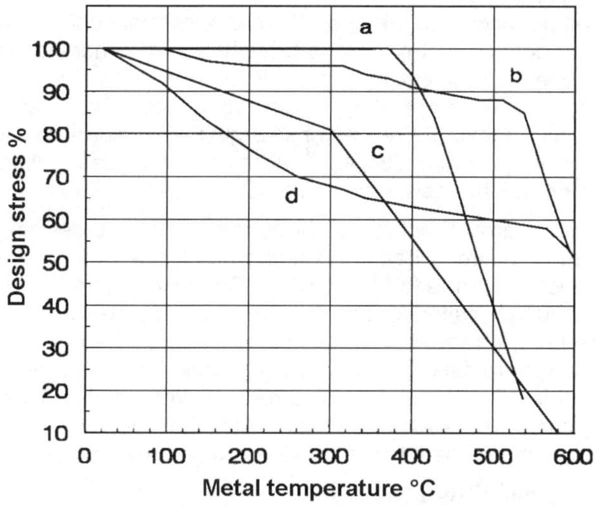

a = high tensile steel bolting (B7)
b = cast 18/8/2 austenitic stainless steel, large deformation
c = cast 11 - 13 Cr steel
d = cast 18/8/2 austenitic stainless steel, small deformation

Figure 7.3 Variation in material design stress with temperature

purposes. Well-respected pressure vessel standards are a source of good information. Pumps are not pressure vessels; the pump casing or cylinder must do much more than just retain internal pressure. However the data in pressure vessel standards can be used for guidance on the temperature effects. Figures 7.2 and 7.3 are based on data published in ASME VIII and BS 5500.

Low temperature can be a problem. The definition of low temperature varies with material types. Most materials become brittle and become susceptible to shocks. Both extremes of the temperature range need effective consideration. Creep can be a problem with metals at high temperature. Creep can be a problem with non-metallic materials at relatively low temperatures, see Figure 7.4.

The tables presented in the subsequent sections, in some cases, show several material designations in one category. National and international standards are generally not identical for requirements of production and quality control. The designations quoted are extracted from other publications, such as ISO 7005, and must be treated with some caution. The decision as to the suitability of any material for a specific application rests with the design engineer responsible for product liability.

7.2.1 Grey cast iron

General purpose cast iron is manufactured in various strengths, BS 1452, Grades 180-300, the grades referring to the tensile strength. Cast iron specifically for pressure containing parts, ASTM A278, has similar grades, see Table 7.1. In general, cast iron is not defined by a chemical composition but only by physical properties. Some specifications limit proportions of certain elements. The tensile strength depends primarily upon the content of carbon and reduces as the carbon content increases. The carbon occurs as graphite flakes enclosed in a steel-like matrix. The flake formation means that the value of elastic limit is low. The impact strength of cast iron is affected by the flakes but cast iron can be used down to -50°C, and maximum temperatures of about 200/250°C for general purpose cast irons. Because of the relatively low tensile strength, grey cast iron impellers are restricted to "low" peripheral speed pumps. Note also that the strength of cast iron is related to the thickness; specifications must be reviewed during the design process.

Grey cast iron is the most commonly used material in centrifugal pumps. From the corrosion point of view, it has a very wide field of application. Grey cast iron can be used without any great risk in water, or aqueous solutions, with pH values between 6

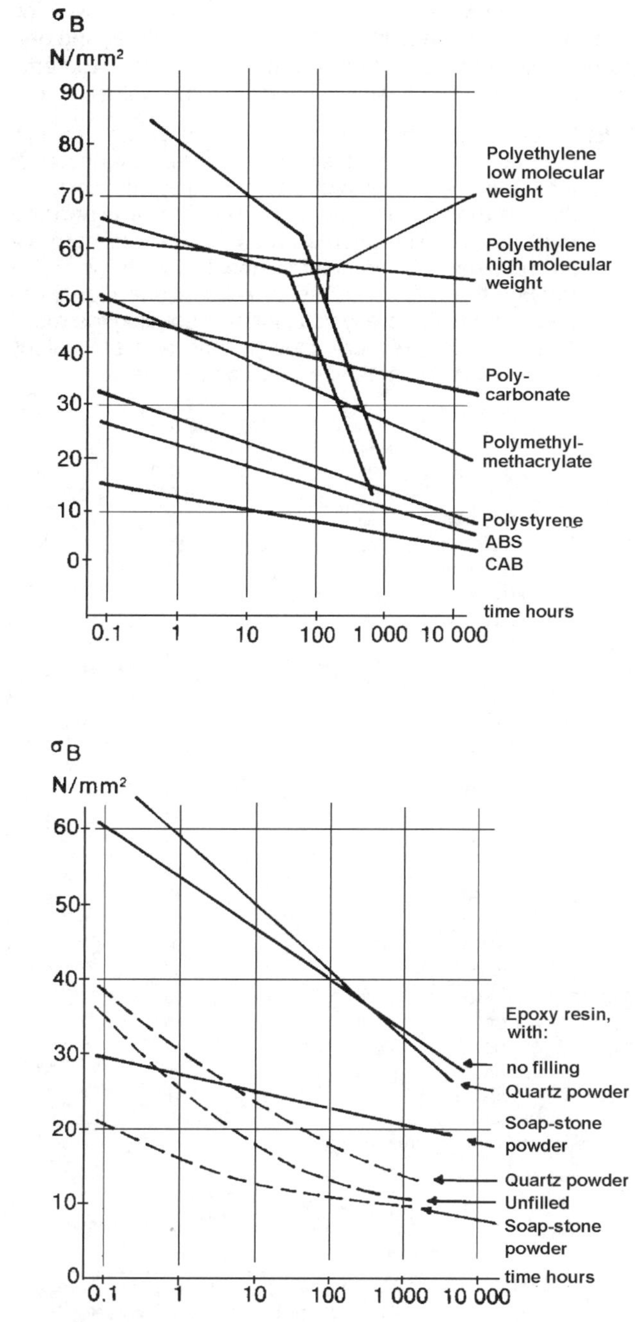

Figure 7.4 Creep of non-metallic materials
——— 23°C - - - - - - - - 60°C

and 10. Grey cast iron can be used in more acidic media, but it is dependent to a high degree upon what has caused the lower pH value. Temperature dependence is also more accentuated in this region. Because of its high carbon content, grey cast iron quickly acquires a protective graphite layer, which means, from the strength point of view, that poorer quality cast iron with higher carbon content has better protection against corrosion. Grey cast iron is also superior to normal steel for the same reason.

Grey cast iron is used for bearing housings because of its compressive strength and damping qualities. It is used as standard material for crankcases for all types of reciprocating pumps.

Meehanite® is a proprietary foundry process for producing cast iron under a more rigorously controlled manufacturing cycle. It tends to be stronger and can be cast in thinner sections and should be considered when ever cast iron is appropriate. (Meehanite® is an organisation comprising a world-wide network of licensed foundries sharing co-ordinated development and interchange of technology.)

Cast iron can be welded or repaired by welding. However, the weld material will have a different chemical structure, and perhaps physical properties, from the parent material. Extreme caution should be exercised if cast iron is to be welded.

NOTE: When designing pressure vessels or pump casings, the maximum stress due to internal pressure occurs at the inner wall. For vessels, casings, without holes through the wall, the inner tensile stress will be approximately equal to the internal pressure. The internal pressure can never be greater than the design stress. Holes through the wall will create stress concentrations, values of 1.5 to 3.5 are typical, further reducing the maximum internal pressure. Design tensile stress values given are for room temperature conditions.

Standard	Grade	Tensile strength min MPa	Max temp °C	Approx. max tensile design stress MPa	Remarks
EN 1561 See DIN 1691 ASTM A48	EN JL 1020	150	≈ 300	38	As cast design values
	EN JL 1030	200	≈ 300	50	
	EN JL 1040	250	≈ 300	63	
	EN JL 1050	300	≈ 300	75	
	EN JL 1060	350	≈ 300	88	
ISO 185	200	200	300	50	As cast
	250	250	300	62	
ASTM A126	A	145	230	36	As cast
	B	214	230	53	
ASTM A278	C120	138	230	35	
	C125	172		43	
	C130	207		52	
	C135	241		60	
ASTM A278	C140	276	345	70	Restricted carbon content P ≤ 0.25% S ≤ 012%
	C145	310		77	
	C150	345		86	
	C155	379		94	
	C160	414		103	
Meehanite	GF150	150	≈ 350	40	As cast design value
	GE200	200	≈ 350	50	
	GD250	250	≈ 350	62	
	GB300	300	≈ 350	75	
	GM400	400	≈ 350	100	

Table 7.1 Specifications for grey cast iron

Standard	Grade	Tensile strength min MPa	Max temp °C	Approx max tensile design stress MPa	Remarks
EN 1563 See DIN 1693	EN JS 1015	350	≈ 350	102	Elongation 22%
	EN JS 1025	400	≈ 350	116	Elongation 18%
	EN JS 1040	450	≈ 350	130	Elongation 10%
	EN JS 1050	500	≈ 350	145	Elongation 7%
	EN JS 1060	600	≈ 350	156	Elongation 3%
	EN JS 1090	900	≈ 350	225	Elongation 2%
ISO 1083	350-22	350	350	87	Elongation 22%
	400-15	400	350	100	Elongation 15%
	500-7	500	350	125	Elongation 7%
	600-3	600	120	150	Elongation 3%
ISO 2531	400-5	400	350	100	Elongation 5%
ASTM A536	60-40-18	413	≈ 350	102	Elongation 18%
	65-45-12	448	≈ 350	112	Elongation 12%
	0-55-06	551	≈ 350	137	Elongation 6%
	100-70-03	689	≈ 350	172	Elongation 3%
	120-90-02	827	≈ 350	206	Elongation 2%
Meehanite	SFF350	350	350	87	Elongation 24%
	SFF400	400	350	100	Elongation 20%
	SF400	400	350	100	Elongation 17%
	SFP500	500	350	125	Elongation 8%

Table 7.2 Specifications for spheroidal and nodular cast iron

7.2.2 Spheroidal and nodular cast iron

Spheroidal or nodular cast iron is called SG iron and ductile iron. See Table 7.2. By analysis, this grade of cast iron is very like ordinary cast iron with regard to carbon and silicon content. In contrast to ordinary cast iron, however, the graphite is not in flakes but occurs in the form of small spheroidal nodules in a mainly pearlite matrix. This is brought about by adding small quantities of magnesium to the melt before pouring. Because of the spheroidal form of the graphite, spheroidal or nodular cast iron has a substantially greater strength than ordinary cast iron and is much more ductile. Furthermore, the elastic limit is greater and the resistance to impact is better.

Further improvement in the elastic limit and impact resistance can be obtained by heat treatment, whereby the basic matrix is changed towards a ferritic structure. Spheroidal or nodular cast iron is ideally suited for pressure vessels subjected to internal pressures, where these were formerly always assigned to cast steel. Spheroidal or nodular cast iron also has some welding properties. The resistance to corrosion is very similar to that of grey cast iron at low flow velocities. At higher flow velocities this material often has a lower resistance than ordinary cast iron.

7.2.3 Low alloy steel

A steel can be classified as "low alloy" if none of the alloying elements exceeds 4.5%. Typically alloying elements are rarely greater than 3%. Low alloy steel castings and wrought material only differ slightly in chemical analysis because of certain additives intended to improve castability. Some low alloy steels have small additions of nickel and chromium in order to improve their strength at elevated temperatures. Steel castings are used mainly for high pressure and hot liquid pumps but are being replaced more and more by spheroidal or nodular cast iron and higher alloyed steel.

From the corrosion aspect, steel castings are not as good as grey cast iron or spheroidal or nodular cast iron, because, by preference, the carbon content should not exceed 1.5%, which is insufficient to provide a protective graphite layer. The carbon content affects the heat treatment of the steel and the weldability. Trace quantities of chrome and nickel improve hardening qualities, ductility and impact resistance. High strength bolting materials have about 0.4% carbon and 1% chrome. Low alloy steel plate and forgings are used for various components, see Table 7.3.

7.2.4 Alloyed cast iron

The properties of flaked and spheroidal cast iron can be altered to a large extent by the addition of alloying elements which affect strength, workability, and resistance to corrosion and abrasion. The following elements can be used:

7.2.4.1 Copper, (for flaked graphite cast iron only)

Copper additives give the basic matrix of cast iron a more uniform pearlite structure with less ferrite and a better form and distribution of the graphite. The tensile strength increases by 10% to 20% with additions of 1% and 2%, with corresponding increases in hardness. Corrosion resistance to very concentrated sulphuric acid and solutions of sulphuric acid, for example, is markedly improved.

7.2.4.2 Nickel and chromium

Small additions of nickel and/or chromium cause the structure to be more even and finely distributed giving advantages similar to those for copper. Furthermore, casting material integrity is improved.

The additions of 20% Ni (Ni-Resist) causes the matrix to become austenitic and non-magnetic. This quality has shown itself to be very useful for seawater and other chloride containing liquids, even at high temperatures. The strength is similar to the corresponding grades of grey cast iron and spheroidal or nodular cast iron.

Ni-Resist, with alloying proportions of about 6% Ni and 9% Cr produces a martensitic matrix with chrome carbides, resulting in hardness figures between 500 and 600 BHN. The hardness can be further increased by heat treatment or by increasing the

Standard	Grade	Major constituents	Tensile strength min MPa	Max temp°C	Remarks
EN 10213-2	GP240GR	0.21C 0.6Si 0.6Mn 0.03P 0.02S	420	≈ 200	Cast RT min(1)
DIN 1681 Werkstoff ISO 4991 ASTM A216	GS-45 1.0435 C23-45B WCA	0.25C 0.25Cr 0.15Mo 0.4Ni	430	400	Cast RT min (**1**)
EN10213-2	GP280GH	0.21C 0.6Si 0.5Mn 0.03P 0.02S	480		Cast RT min
DIN 1681 ISO 4991 ASTM A216	GS-52 C26-52H WCB	0.30C 0.25Cr 0.15Mo 0.4Ni	480	390/480	Cast RT min
BS 1503	164-490	0.25C 0.25Cr 0.1Mo 0.4Ni 1Mn	490	480	Forged
EN10213-2	G12MoCrV5-2	0.12C 0.45Si 0.55Mn 0.5Mo 0.4Cr 0.03P 0.02S	610	≈ 400	Cast RT min
BS 1503	221-550	0.35C 0.25Cr 0.1Mo 1.1Mn 0.4Ni	550	350	Forged
BS 3100 ISO 4991 ASTM A352	AL1 C23-46BL LCB	0.2C 1.1Mn		≈ 425	Cast -46°C min
BS 970 Werkstoff	070M20 'N' 1.0402	0.2C 0.7Mn	400	≈ 200	Wrought
BS 970	080M30 'N'	0.3C 0.75Mn	460	≈ 200	Wrought
BS 970 Werkstoff	080M40 'N' 1.0503	0.4C 0.8Mn	510	≈ 200	Wrought
BS 970 Werkstoff	708M40 'S' 1.7225	0.4C 0.85Mn 1Cr 0.2Mo	770	≈ 500	Wrought (**2**)
AISI Werkstoff	1040 1.1186	0.4C 0.75Mn	551	≈ 200	Wrought
AISI Werkstoff	4140 1.7225	0.43C 0.8Mn 1Cr 0.2Mo	860	≈ 500	Wrought (**2**)
EN 10113-2 Werkstoff	S355N 1.0483	0.2C1.2Mn + N_2	470		Structural plate
EN 10113-2 Werkstoff	S420NL 1.0570	0.2C 1.35Mn 0.15Cr + N_2	500		Low temp structural plate -30°C
BS 4882 ASTM A193 DIN 17200	B7 B7 42Cr.MO.4	0.4C 0.85Mn 1Cr 0.2Mo	860	400/500	Bolt/stud -45°C min
BS 4882 ASTM A194 DIN 17200	2H 2H C45	0.4C	PLT (**3**)	450	Nut 0°C min

Table 7.3 Properties of low alloy steel — (1) RT ≡ room temperature — (2) reduced allowable stress over 350°C — (3) PLT ≡ proof load tested

Standard	Grade	Major constituents	Tensile strength min MPa	Max temp°C	Remarks
ASTM A436	Type 1	2.1Cr 15.5Ni 1Mn 6.5Cu 1.9Si	172		Wear resistant, best anti-galling properties
BS 3468 DIN 1694 See ISO 2892	L-NiCr 20 2 GGG-NiCr20 2	1.8Cr 20Ni 1Mn 0.3Cu 1.9Si	170	650	Good for alkalis -50°C min
ASTM A436	Type 2	2.1Cr 20Ni 1Mn 0.3Cu 1.9Si	172	705	Good for corrosive environments
BS 3468 DIN 1694 See ISO 2892	S-NiCr 20 2 GGG-NiCr20 3	1.8Cr 20Ni 1Mn 0.3Cu 2.2Si	170	650	
ASTM A439	Type D-2	2.25Cr 20Ni 1Mn 2.25Si	400	815	8% elongation, wear/galling resistant
BS 3468 See ISO 2892	S-NiMn 23 4	0.2Cr 23Ni 4.2Mn 2Si	440		-196°C min
ASTM A439	Type D-2M	0.2Cr 22.5Ni 4Mn 2.2Si	417		-196°C min wear/galling resistant
BS 3468 ASTM A436 See ISO 2892	L-NiCr 30 3 Type 3	3Cr 30Ni 1Mn 0.3Cu 1.5Si	172	815	Erosion resistant, good corrosion resistance
ASTM A439	Type D-3	3Cr 30Ni 1Mn 1.9Si	379	815	6% elongation, superior erosion resistance
ASTM A439	Type D-4	5Cr 30Ni 1Mn 5.5Si	414	815	Wear/galling resistant
ASTM A436	Type 4	5Cr 30.5Ni 1Mn 0.3Cu 5.5Si	172		Best grade for erosion, corrosion, oxidation resistance
ASTM A436	Type 5	0.05Cr 35Ni 1Mn 0.3Cu 1.5Si	138	425	Dimensionally stable
ASTM A439	Type D-5	0.05Cr 35Ni 1Mn 1.9Si	379		Elongation 20%

Table 7.4 Typical alloy cast irons

chrome content up to 30%. Thus a material can be produced with very good resistance to abrasion and erosion. Because of the considerable hardness, the material can only be worked by means of grinding so that special designs are required for pumps.

7.2.4.3 Silicon

Silicon alloyed cast iron, silicon iron, is highly resistive to acids if the silicon content is greater than 13%. This material is resistant to sulphuric acid at all concentrations and also to many other non-organic and organic acids, though not to hydrofluoric acid, hot concentrated hydrochloric acid, sulphurous acids, sulphites and hot alkalis. An addition of 3% molybdenum makes the material resistive also to hydrochloric acid and solutions containing chlorine.

Silicon iron is very difficult to cast and displays poor integrity under pressure. For this reason, it should never be used in pumps with an internal pressure greater than 4 barg. As the material is hard and very brittle, special design principles are normally

Standard	Grade	Major constituents	Tensile strength min MPa	Max temp°C	Remarks
EN 10213-2 ASTM A296 Werkstoff DIN 17445	GX8CrNi12 CA-15 1.4027 G-X 20 Cr 14	0.1C 0.55Mn 12Cr 1Ni	540	600	Castings
BS 970 Pt4 Werkstoff DIN 17440	420S45 1.4028 1.4034 X 30 Cr 13 X 46 Cr 13	0.32C 1Mn 13Cr 1Ni	690	400	Bar
BS 1503 ASTM A182 Werkstoff	410S21 F6b 1.4001	0.12C 1Mn 12.5Cr 1Ni	590	400	Forgings
BS 4882 ASTM A193 Werkstoff DIN 17440	B6 B6 1.4006 X 10 Cr 13	0.15C 12.5Cr 1Mn	758	425	Fasteners -28°C min
EN 10213-3	GX3CrNi13-4	0.05C 1mn 0.7Mo 12.7Cr 4.2Ni	760		Castings -120 °C min

Table 7.5 11-13% Cr materials

used for pumps, e.g. liners of silicon iron in a steel casing to take the pressure stresses. Table 7.4 lists the important alloys.

7.2.5 11-13% Cr steel

Rotodynamic pump manufacturers use 11-13% Cr steel as an improvement on low alloy steel when full stainless steels are not required. Material integrity is greatly improved because of carbide formation due to the addition of chrome. The structure is martensitic. Strength and hardness are improved, as are heat resistance and scaling properties. It can be used up to 600°C but strength falls off rapidly over 300°C. Weldability suffers slightly as the chrome content is increased. Traces of nickel improve hardenability and also impact properties. Hardness values between 180 and 450 BHN can be obtained. It is used for pump casings, impellers, shafts and wear rings and can be a good material choice for boiler feed water. Alloys with 4% Ni have improved corrosion resistance and better low temperature properties. 11-13 Cr steel is not generally used by reciprocating pump manufacturers. Materials are shown in Table 7.5.

7.2.6 Stainless steel

It is a general rule for steels that the content of alloying elements should not exceed 50% of the total. A characteristic of all stainless steels is that they contain a minimum of 15% chromium. In the presence of oxygen this forms a thin invisible film of chromic oxide which is chemically resistant and inhibits any direct contact between the surrounding medium and the steel. It is, however, necessary that oxygen is present in order to maintain the film. The steel is said to be in a passive state. In a reducing atmosphere on the other hand, there can be no build-up of the chromic oxide layer and the steel is then said to be in an active state.

The designation "stainless" steel is not altogether correct, as these steels can easily be subject to corrosive attack depending upon external circumstances relative to alloying content, degree of heat treatment, welding and so on. In general, stainless steels exposed to normal atmospheric conditions will not corrode or discolour.

The corrosion resistance of stainless steels is dependent upon the rate at which the passive oxide layer is dissolved into solution. This rate is slowed down mainly by alloying with metals which themselves have high resistance to corrosive media. Such alloying elements are, in the main, nickel, molybdenum and copper. The effects of chromium are then less important from the point of view of chemical resistance.

The presence of molybdenum in significant quantities greatly enhances the corrosion resistance. The structure of the iron matrix is changed by the various alloying additives. The type of structure is used to divide stainless steel and acid resistant steels into the following groups, see also Figure 7.5:

- Ferritic, which can be hardened by heat treatment
- Martensitic, which can be hardened by heat treatment
- Austenitic, which are totally non-magnetic and cannot be hardened by heat treatment
- Ferritic-austenitic, which are paramagnetic.

The austenitic stainless steels have lower tensile strength and greater corrosion resistance and toughness than the other groups. This makes them easier to manufacture into the various forms, such as sheet metal and piping. Because pumps are largely produced from castings, the chemical analysis can be chosen with greater freedom.

Using the same main analysis there are a number of standardised variants of materials. The main reason for this is the modification of chemical structure around a weld, usually carbide precipitation. Carbide precipitation is eliminated in the various standard variants by:

- Low or extra-low carbon content
- Addition of titanium or niobium in proportions of 5 to 10 times the carbon content to stabilise the carbides.

Carbide precipitation can be corrected by a post weld heat treatment. In general, heat treatment is not preferred. Components which have welded attachments, or may require weld repairs, should be made from low carbon varieties.

Cast and wrought versions of the same variety may have slightly different chemical compositions to accommodate the different manufacturing processes

Austenitic stainless steels have low strength. The strength can be improved by inoculating the steel with nitrogen or by cold working.

Most stainless steels are recognised by their designations proposed by the American Iron and Steel Institute, AISI, and the Alloy Casting Institute of America, ACI, or Steel Founders Society of America, (SFSA). The most well known are the AISI designa-

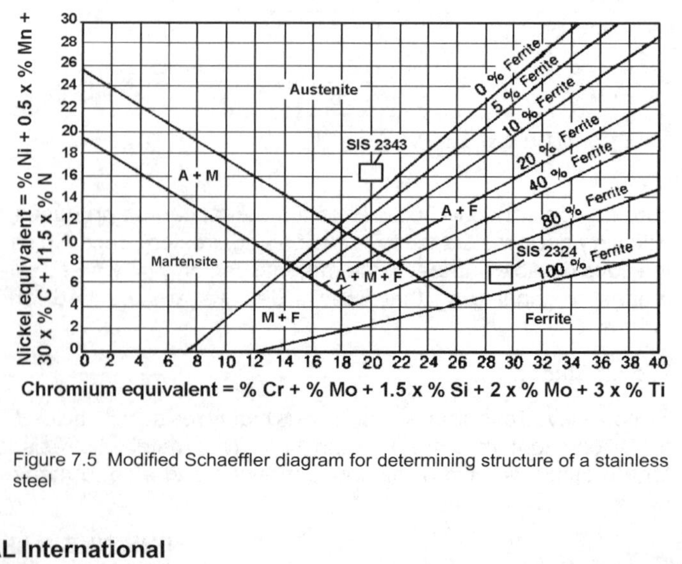

Figure 7.5 Modified Schaeffler diagram for determining structure of a stainless steel

tions for wrought products. The majority of the popular stainless steels belong to the 300 series. The inclusion of "L" means low carbon version; "N" means nitrogen strengthened. The AISI designations are generic, a full product specification is required for purchasing.

AISI	ACI	Nominal composition Cr/Ni/Mo
303	CF-16F	18/9/0.6
304	CF-8	19/9
304L	CF-3	19/10
316	CF-8M	17/12/2
316L	CF-3M	17/12/2
317	CG-12	19/13/3
347	CF-8C	18/11

Table 7.6 Popular austenitic stainless steels

Table 7.6 indicates the relationship between the popular AISI and ACI grades and Table 7.7 summarises their various compositions.

7.2.7 Super stainless steel

Materials are always required with better corrosion resistance. The expansion of offshore oil production, together with secondary recovery by seawater, led to a demand for materials with superior pitting resistance. Molybdenum and nitrogen content were found to be of critical importance. These super stainless steels are the result of much research by oil companies, pump manufacturers and steel makers. Super stainless steels can be welded by most popular techniques; resistance to stress- corrosion cracking is not impaired, although some loss of toughness and resistance to intercrystalline corrosion may be apparent. See Table 7.8.

7.2.8 Copper alloys

Copper alloy components are used largely for salt water and weak cold chloride solutions; for corrosive properties of seawater and choice of materials for use with seawater see Section 2.2.4, in Chapter 2. The alloys have varying casting characteristics associated with solidifying temperatures, which in turn are

Standard	Grade	Major constituents	Tensile strength min MPa	Max temp°C	Remarks
Werkstoff DIN 17445 ISO 4991	1.4308 G-X 6 CrNi 18 9 C47	0.12C 17Cr 7Ni 2Mn	460	700	Castings
BS 1503 Werkstoff DIN 17440	304S11 1.4306 X 2 CrNi 18 9	0.03C 18Cr 10.5Ni 2Mn 0.3Mo	480	450	Forgings
ASTM A351 ISO 4991	CF-3M C61LC	0.03C 18Cr 9Ni 2Mo	480	454	Castings
EN 10213-4 Werkstoff DIN 17445	GX5CrNiNb19-11 1.4552 G-X7 CrNiNNb 18 9	0.07C 19Cr 9Ni 1.5Mn	480	680	Castings
BS 1503 Werkstoff DIN 17440	347S31 1.4550 X 10 CrNiNb 18 9	0.08C 18Cr 10.5Ni 2Mn 0.3Mo	510	450	Forgings
EN 10213-4 Werkstoff DIN 17445 ISO 4991	GX5CrNiMo 19-11-2 1.4408 G-X6 CrNiMo 18 10 C60 C61	0.08C 19Cr 10Ni 2.5Mo 1.5Mn	480	700	Castings
BS 1503 Werkstoff DIN 17440 ISO 2604-1	316S31 1.4401/36 X 5 CrNiMo 17 12 2 F62	0.07C 17.5Cr 12Ni 2Mn 2.2Mo	610	700	Forgings
ASTM A351 DIN 17445 ISO 4991	CF-8M G-X3 CrNiMoN 17 13 5 C57	0.08C 19.5Cr 10.5Ni 1.5Mn 2.5Mo	483	815	Castings
SIS Werkstoff DIN 17440	2343 1.4436 X 5 CrNiMo 18 12	0.05C 17.5Cr 12Ni 2.8Mo	450	400	Castings
ASTM A351	CG-6MMN	22Cr 13Ni 5Mo + N_2	517	565	Castings
ASTM A182 ISO 2604-1	F316L F59	0.03C 16Cr 12Ni 2Mo	448	454	Forgings
BS 1503	318S13	0.03C 22Cr 5Ni 2Mn 3Mo	640	300	Forgings
ASTM A182	XM-19	22Cr 13Ni 5Mn 2Mo + N_2	689	565	Forgings
SIS Werkstoff	2324 1.4460	0.05C 25Cr 5.7Ni 1.5Mo 2Mn	600	315	Cast
BS 1506	304S71	0.07C 18Cr 9.5Ni 2Mn			Fasteners B8N
BS 1506	316S51	0.07C 17.5Cr 12Ni 2.2Mo 2Mn			Fasteners B8MH
BS 4882 ASTM A193	B8 B8	0.08C 19Cr 9Ni 2Mn	517	425/575	Fasteners -254°C
BS 4882 ASTM A193	B8M B8M	0.08C 17Cr 12Ni 2.5Mo 2Mn	517	425/575	Fasteners -196°C
ASTM A193	B8R	0.06C 22Cr 12.5Ni 5Mn 2Mo + N_2	689		Fasteners

Table 7.7 Austenitic stainless steels

Standard	Grade	Major constituents	Tensile strength min MPa	Max temp°C	Remarks
EN 10213-4	GX2CrNiMoN22-5-3	0.03C 22Cr 5.5Ni 3Mo 2Mn N_2	600	≈ 300	Castings
	FMN	0.05C 25Cr 5Ni 2Mo 0.8Mn N_2	695	≈ 300	Castings
	Ferralium 255	0.08C 25.5Cr 5.5Ni 3Mo 2Mn N_2	830	≈ 300	Castings
	Amazon 256-Cu	0.04C 25.5Cr 5.5Ni 3Mo 1.5Mn N_2	760	≈ 300	Wrought
	Zeron 25	0.03C 25Cr 6.5Ni 2.5Mo 1.5Mn N_2	650	≈ 300	Castings
	Zeron 100	0.03C 25Cr 7Ni 3.5Mo 1Mn N_2	756	≈ 300	Castings
	Ferralium 288	0.08C 27.5Cr 7.5Ni 2.5Mo 2Mn N_2	800	≈ 300	Castings
	254SMO	0.02C 20Cr 18Ni 6.1Mo 0.5Mn N_2	651		Castings

Figure 7.8 Super stainless steels

determined by the alloying elements. For tin-bronze with about 10% tin, the temperature range is 200 to 250°C. The prime result of this is that a certain amount of microscopic porosity cannot be avoided in the solidifying alloy. This porosity need not mean that the casting is not pressure-tight. The wide solidifying temperature range makes the alloy relatively easy to cast.

Brass with 30% to 35% zinc and aluminium-bronzes with about 10% aluminium have solidifying temperature ranges between 200 and 210°C. The structure is fairly non-porous although large cavities due to coring may occur which can be eliminated by careful metal feeding. These alloys with narrow solidifying temperature ranges are more dependent upon the shape of the casting and involve higher costs.

Gunmetal made up of about 5% each of tin, lead and zinc, is easily cast. The pressure-tightness quality increases with lead content up to about 7%.

From the corrosion point of view, copper and its alloys give widely varying results. Here, as with other materials, it is a solid oxide film which provides the protection against further corrosion. The rate of build-up of this protective film varies and is affected by flow velocities. Table 7.9 lists some commonly used alloys.

Standard	Grade	Major constituents	Tensile strength min MPa	Max temp°C	Remarks
BS 1400 ISO 1338 DIN 1714 ASTM B148	AB1 CuAl10Fe3 CuAl10Fe C95200	9.5Al 2.5Fe 1Ni	500		Castings
BS 1400 ISO 1338 DIN 1714 ASTM B148	AB2 CuAl10Fe5Ni5 CuAl10Ni C95500	9.5Al 4.5Fe 4.5Ni	640	» 300	Castings » -150°C
BS 1400 ISO 1338 DIN 1705 ASTM B145	LG2 CuPb5Sn5Zn5 CuSn5ZnPb Gr836	5Sn 5Zn 5Pb	200	» 300	Castings -196°C min
BS 1400	PB4	9.7Sn	190		Castings
BS 2872 ISO	CA 107 CuAl7Si2	6Al 2.2Si	520		Forgings
BS 2872 ISO DIN 17665 ASTM B150	CA 104 CuAl10Ni5Fe4 CuAl10Ni C63200	9.5Al 4.5Fe 4.5Ni	700	» 500	Forgings
BS 2874	CA104	9.5Al 4.5Fe 4.5Ni	700	400	Fasteners, suitable for cryogenics

Table 7.9 Common copper alloys

7.2.9 Aluminium alloys

Aluminium alloys, Table 7.10, are not used to any great extent in pumps, because of their low resistance to corrosion and low hardness. The exceptions are those areas where low weight is a prime factor, for example in building site ground drainage, and also where manufacturing costs are important as in domestic pumps. The range of application can be increased by inhibiting electrolytic corrosion or actually making use of it, for example by:

- Insulating from more noble materials such as steel
- Connecting to less noble materials, so-called sacrificial anodes of zinc or magnesium

Standard	Grade	Major constituents	Tensile strength min MPa	Max temp °C	Remarks
BS 1490 ISO 3522	LM 4 Al-Si5Cu3	5Si 3Cu	230		Castings
BS 1490 ISO 3522 DIN 1725 ASTM B85	LM 9 Al-Si10Mg AlSi10Mg SG100A	11.5Si 0.5Mn 0.4Mg	240	150	Castings
BS 4300/11 ISO	5454 AlMg3Mn	2.7Mg 0.7Mn	275		Forgings
BS 1472 ISO	2014A AlCu4SiMg	4.5Cu 0.8Mn 0.5Mg 0.7Si	370		Forgings
	AGS 6060		300		Fasteners
	AZ5GU-707 5	5.5Zn 2.3Mg 1.5Cu	550		Fasteners

Table 7.10 Common aluminium alloys

7.2.10 Nickel alloys

Nickel base alloys, as also pure nickel itself, come as a natural extension at a point where stainless steels are no longer satisfactory from the corrosion point of view. They are not easy to cast as a rule, and require high annealing temperatures, about 1200°C, in order to maintain the correct structure and grain size which are necessary for good corrosion resistance. Nickel alloys are dominated by a group usually called by the trade name of Hastelloy. Similar alloys are manufactured under different trade names. See Table 7.11 for a list of the chemical composition of the commonest nickel alloys.

Hastelloy B has high corrosion resistance in strong hydrochloric and sulphuric acids. Unfortunately, the alloy is sensitive to high flow velocities and is rarely used in centrifugal pumps.

Hastelloy C is the most common of the nickel alloys because of its resistance in both oxidising and reducing atmospheres. It is used mainly for chloride solutions, e.g. liquids containing free active chlorine, bleaching fluids or hypochlorite and chlorine dioxide solutions, where the chlorine content should not exceed 15 g/l at room temperature. It also has a good resistance to acids at temperatures below 70°C.

Hastelloy D possesses good resistance to erosion, due to its high hardness, approximately 400 BHN. It has its best application in sulphuric acid, against which it has high resistance at all concentrations and high temperatures, and within certain limits for boiling acids also. Maximum corrosion resistance is achieved only after about six weeks in operation, which is the time it takes for a protective sulphate film to build up on the surface.

Monel is more resistant than nickel under reducing conditions and more resistant than copper under oxidising conditions. In summary, it can be said that monel is more resistant to corrosion than are its main components. Re-precipitation of dissolved copper cannot occur, since nickel, copper and monel lie very close to each other in the electrolytic series, galvanic series. Monel however, is somewhat difficult to cast and often results in porous castings. It is therefore used mainly for non pressurised components, such as impellers, or in wrought form for shafts. Monel has very good resistance to salt water especially at higher flow velocities. In stationary seawater, biological growth may cause local concentrations of oxygen which lead to localised attacks.

7.2.11 Other metallic materials

A number of metallic materials are available for special purpose use. Some of these are discussed in the following Sections:

7.2.11.1 Lead and lead alloys

Because of its poor mechanical properties, pure lead is not used as a material for pumps. A 10% to 25% addition of antimony, Sb, improves its strength characteristics considerably.

Standard	Grade	Major constituents	Tensile strength min MPa	Max temp°C	Remarks
	Hastelloy B	28Mo 5Fe 1.2Co 0.5Cr	≈ 950	760	Cast, Very ductile
	Hastelloy C	17Mo 16.5Cr 5.7Fe 4.3W	≈ 490	980	Cast, Very ductile
	Hastelloy C	17Mo 16.5Cr 5.7Fe 4.3W	≈ 690	980	Wrought, Very ductile
	Alloy 20	36Fe 20Cr 3.5Cu 2.5Mo 2Mn 1Si	670		Cast, Very ductile
	Langalloy 7R	23Cr 6Mo 6Cu 5Fe 2W	420		Cast, acid resistant
BS 3076	NA 18	30Cu 2.7Al 1Fe 0.8Mn	1000		Fasteners
BS 3076	NA 14	15Cr 8Fe	830	600	Wrought
	Inconel 625	22Cr 9Mo 5Fe + No + Ta	725/900	≈ 800	Fasteners & wrought
	Inconel X750	15.5Cr 7Fe 2.5Ti	1130	≈ 650	Wrought
	Monel K-500	30Cu 2.7Al 1Fe 0.8Mn	900		Wrought
	Incoloy 800	32Ni 21Cr 1.5Mn	517/690	815	Wrought

Table 7.11 Common nickel based materials

Standard	Grade	Major constituents	Tensile strength min MPa	Max temp°C	Remarks
ASTM Gr 2	CP		380	300	Wrought & cast
ASTM Gr 5	Ti-6Al-4V	6Al 4V	900	325	Wrought & cast
	TiMetal 550	4Al 4Mo 2Sn	1100	400	Wrought
	TiMetal 551	4Al 4Mo 4Sn	1200	400	Wrought
ASTM Gr Beta C	Beta	8V 6Cr 4zr 4Mo 3.5 Al	1240	450	Wrought

Figure 7.12 Common titanium alloys

The alloy is called hard lead. For higher pressure applications, a structural pressure retaining casing could be lined with lead.

Lead has a very good resistance to acids, with the one exception of nitric acid. In sulphuric acid, lead sulphate is built up on the surface of the lead which is difficult to dissolve and acts as a protective film. Lead can be used for sulphuric acid at a concentration of 30% and temperature 100°C, but temperatures have to be reduced at higher concentrations. In hydrochloric acid a film of lead chloride builds up, which protects the underlying metal to some extent. If the acid is in motion however, the surface layer wears away easily, so that lead pumps are not recommended for hydrochloric acid.

7.2.11.2 Titanium

Titanium has low density, 4500 kg/m^3, and very good tensile strength. Titanium alloy castings can have strengths up to 890 MPa, forgings up to 1200 MPa. Titanium is relatively difficult to work however. It is cast centrifugally in graphite moulds in vacuum chambers in the presence of inert gas. Titanium can be wrought and is used because of its low density for dynamic components such as valve plates.

Titanium is being used more and more in the chemical industry mostly because if its good corrosion resistance to solutions containing free active chlorine. Corrosion is completely excluded in such solutions as chlorine dioxide in water, hypochlorites, chlorates and metal chlorides.

In some cases titanium can be a more economic solution, than the proprietary super stainless steels listed in Table 7.8. Because of its very good strength-to-weight ratio titanium can be used to advantage in corrosive applications where weight is important.

7.2.11.3 Tantalum

Such exotic metals as tantalum, hafnium, niobium and germanium have lately come more into their own for modern industrial products, where very high corrosion resistance is required for reliability. Tantalum has a corrosion resistance very nearly equal to that of glass. It is resistive to boiling sulphuric and hydrochloric acids. The density of tantalum is 16,600 kg/m^3. The ability to cast and weld is extremely limited, so that parts with complex shapes do not lend themselves to economic production. For rotodynamic pumps tantalum is generally used to provide an internal lining. Some reciprocating pumps can be easily manufactured solely from wrought products without any welding.

7.2.11.4 Tungsten carbide

Tungsten carbide is a sintered material originally produced as a high quality cutting tool material. Its hardness, abrasion resistance and improved manufacturing techniques have led to its adoption in several key areas of pump construction. Initially produced only with a cobalt binder it is now available with nickel as an alternative. The main area of use is in seal faces for mechanical seals. It is also found in valve seats of reciprocating pumps operating on abrasive liquid-solid mixtures. Tungsten carbide can be sprayed as a surface coating, see Section 7.2.13.

7.2.12 Non-metallic materials

Non-metallic materials are extremely important in pump construction. Most pumps use some non-metallic materials even though major components are metallic. Most pump designs embody gaskets and seals which rely on the very different properties of non-metallic materials. Non-metallic materials can be divided into five main categories:

- Thermoplastics
- Thermosetting plastics
- Elastomers
- Ceramics
- Minerals

Thermoplastics, thermosetting plastics and elastomers are often confused; in some cases the physical properties are very similar and all three groups of materials are used for the same function. Gaskets, seals and "O" rings are examples of components which are produced in all three groups of material. These materials also tend to absorb liquids when in contact; this property must be considered during the design and application stages.

Ageing, and the effects of ultraviolet radiation must also be studied. Some materials are only suitable for relatively low temperatures, up to 50°C for example. "High" temperature tends to age some non-metallic very quickly. A temperature increase from 20°C to 60°C could reduce the useful life by 80%. Care must be exercised to ensure that no operating conditions occur outside the temperature range. In some processes the system commissioning requires short-term exposure to higher than normal temperatures; these instances should be discussed fully with the pump manufacturer so the long-term effects can be evaluated.

Plastics can be filled with various materials, either for bulk or to improve the strength. Glass, carbon and KEVLAR® fibres are used for strengthening. Thermal expansion, especially in the axial direction of the plastic pump, has to be taken into account when routing the pipework. Similarly, the pump must be relieved of pipework forces and moments to the fullest possible extent.

Many companies have registered trade names for materials. The same material may be known by many different names depending upon the company which produced it. Just as with metals, some materials may be slightly different and possess remarkably different physical and chemical properties. Great care should be exercised when specifying specific materials.

7.2.12.1 Thermoplastics

The following popular materials are thermoplastics:

Acetal
Acrylic
acrylonitrile butadiene styrene (ABS)
aromatic nylons (ARAMID)
fluorocarbons
fluorinated ethylene propylene (FEP)
Nylon
polyester ether ether ketone (PEEK)
polyacetal
polybutylene (PB)
polycarbonate
polychlorotrifluoroethylene (PCTFE)
polyether
polyethylene (PE)
polyethylene chlorotrifluoroethylene (PECTFE)
polyethylene terephthalate (PETP)
polyethylene tetrafluorethylene (PETFE)
polyfluoroalkoxytetrafluoroethylene (PFA)
polyimides
polymethyl pentene (PMP)
polymide
polypropylene (PP)
polysulphone (PS)
polytretrafluoroethylene (PTFE)
polytrichloroethylene (PTFCE)
polyvinylfluoride (PVF)
polyvinylidene fluoride (PVDF)
polyphenylene oxide PPO
polyvinyl chloride (PVC)

Thermoplastics are characterised by their ability to be repeatedly softened by heating and then hardened by cooling; their shape can be changed many times. Notice that PTFE is a thermoplastic and not an elastomer, even though it is used in many applications where elastomeric properties are required.

A sub group within thermoplastics are the thermoplastic elastomers. These are man-made materials which have rubber-like properties with enhanced chemical resistance. The thermoplastic elastomers can have their chemical compositions adjusted to bestow particular properties to the material.

Some thermoplastics have excellent impact, abrasion and tear resistance. Compounds of this type would be used for internal liners on some solids handling applications.

The fluorocarbons are a sub group containing FEP, PCTFE, PECTFE, PETFE, PFA, PTFE and PVDF.

Acetal is a high melting point plastic with high strength and rigidity. Acetal can be extruded, cast and moulded. It is used for a wide variety of purposes including impellers, casings, bearing housings and bearings.

Aromatic nylons are usually formed as fibres such as ARAMID®, KEVLAR® or NOMEX®. These fibres are used to reinforce and strengthen moulded or cast materials. The fibres can also be used for braiding in the case of soft packing.

Popular thermoplastic materials are filled or alloyed with other materials to improve specific properties. Nylon is a favoured bearing material; it can be reinforced with glass to increase strength and stiffness; it can also be alloyed with molybdenum disulphide to improve mechanical properties, increase high temperature capabilities and improve bearing performance.

PTFE is probably the most well-known thermoplastic after Nylon. PTFE has many trade names, TEFLON® for example. Renown for its corrosion resistance and chemical inertness, PTFE has many uses in pumps. Pump casings, impellers, seal faces, gaskets and "O" rings. Its wide temperature range, -250 to 250°C, makes it suitable for an extensive range of applica-

	ABS	cPVC	PB	HDPE	PP	uPVC	PVDF
Strength	high	high	very low	low	low	high	high
Impact resistance	high	low	high	high	high	low	high
Ductility	low	high	high	high	high	high	high
Temperature range	-40 to 60°C	0 to 80°C	-15 to 95°C	-40 to 60°C	-10 to 80°C	0 to 60°C	-40 to 140 °C
Fabrication	adhesive	adhesive	fusion welding	fusion welding	fusion welding	adhesive fusion welding	fusion welding
Suitable for:	hygienic high purity water potable water	alcohols mineral acids paraffinic HCs	hygienic potable water	acids alkalis inorganic solvents organic solvents direct sunlight	dilute acids alkalis organic solvents waste water water	acids alkalis hygienic potable water	aliphatic HCs, aromatic H's alcohols high purity water inorganic acids organic acids direct sunlight
Unsuitable for:	acetic acid alcohol petrol vegetable oil direct sunlight	aromatic HCs chlorinated HCs ketones organic solvents	direct sunlight	strong oxidising acids	aromatic HCs concentrated acids chlorinated HCs direct sunlight	aromatic HCs chlorinated HCs ketones	alkaline amines alkalis

Table 7.13 Typical properties of popular thermoplastics

tions. Alloying, or sintering, with other compounds can extend the properties and applications of PTFE.

Typical additions include glass fibre, bronze, carbon, graphite, molybdenum disulphide and stainless steel. PTFE is also machined for the construction of pistons, bellows and other components which need to be corrosion resistant. It has the advantage of having a very low coefficient of friction and is generally not recommended for pumps designed for handling liquids containing abrasive solid particles.

PEEK is attracting attention because of its chemical resistance and physical properties. Some manufacturers are using it for wear rings, stuffing box bushings and valve plates.

Thermoplastics such as polyvinyl chloride (PVC), polypropylene (PP), polyethylene (PE) and PTFE are used as blow-moulded parts and also in sheet or tubular form for coating.

PTFE/FEP/PFA are chemically resistant to most liquids with the exception of a few chlorinated compounds, molten potassium and molten sodium.

Table 7.13 indicates some of the general properties of popular thermoplastics and limitations on use.

7.2.12.2 Thermosetting plastics

The following materials are typical thermosetting plastics:

Alkyd
Amino
epoxides, epoxy resins
furanes
Melamine
phenolic resins
polyester
polyurethane (PU)
silicones
urea
vinyl ester resins

Thermosetting plastics are formed to the final shape during manufacture and the shape remains fixed until the material is destroyed.

As with thermoplastics, thermosetting plastics can be filled or reinforced with other materials. Asbestos, and other mineral fibres, together with cotton are often used. Great care must be taken when machining composites to avoid exposing the fibres to the process liquid. Selected parts of components can be treated and some surfaces can be inoculated with PTFE or graphite to reduce friction and wear.

Pump bodies and impellers are produced in GRP, glass reinforced plastic. A matrix of glass fibres is coated with epoxy resin, vinyl ester resin or phenolic. A chemical reaction between the two components of the resin, filler and hardener, causes the mixture to solidify. Many resin formulations are available including casting versions. Resins can be selected for broad chemical resistance. Epoxy resins can be used up to 98°C; vinyl ester up to 205°C.

Non-metallic materials are used because of their excellent corrosion resistance properties. However, the physical properties tend to be very inferior. Non-metallic materials cannot be considered as a direct replacement for metals. Figure 7.6 indicates the approximate design stresses for some popular materials compared to cast iron. Cast iron is a low strength metallic material. Novatron ® is a nylon alloy.

The low strength problem can be accommodated in several ways. For short production runs and special designs the major pump components can be machined from the solid. Figure 7.7 shows the construction of a small centrifugal pump. Most of the machining is straightforward; impellers which do not have radial vanes can present difficulties. The flanges are screwed on to the casing stub connections and then locked in position with adhesive. Machining from the solid is costly and time consuming. Non-metallic materials can be moulded in a similar way to casting metals.

Figure 7.8 shows the style of construction of a moulded pump. It is a style of construction which is used for many single-stage back-pullout centrifugal pumps. The bearing bracket would be made from cast iron and the shaft from steel. Neither are in direct contact with the process liquid. The moulded casing is supported by a cast iron housing. The housing performs two important structural functions. Casing distortion due to internal pressure is greatly reduced; alignment is improved and con-

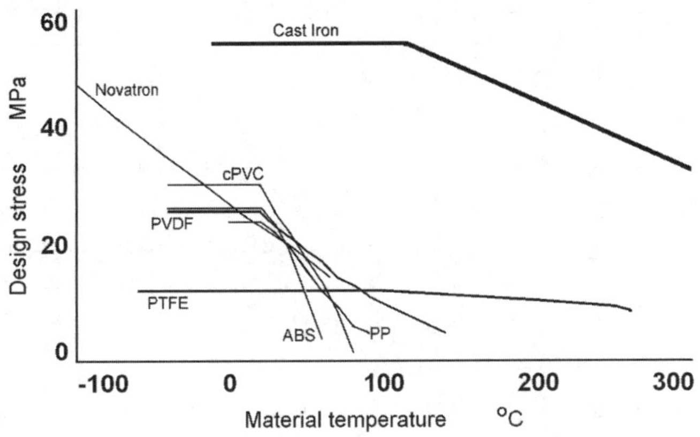

Figure 7.6 Temperature effects on material design stresses

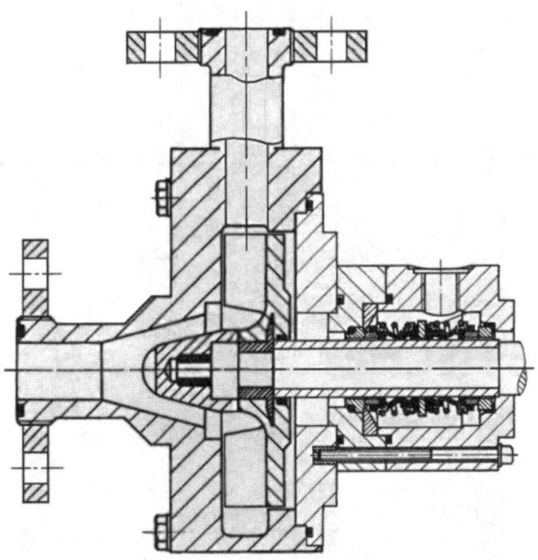

Figure 7.7 A typical small non-metallic centrifugal pump

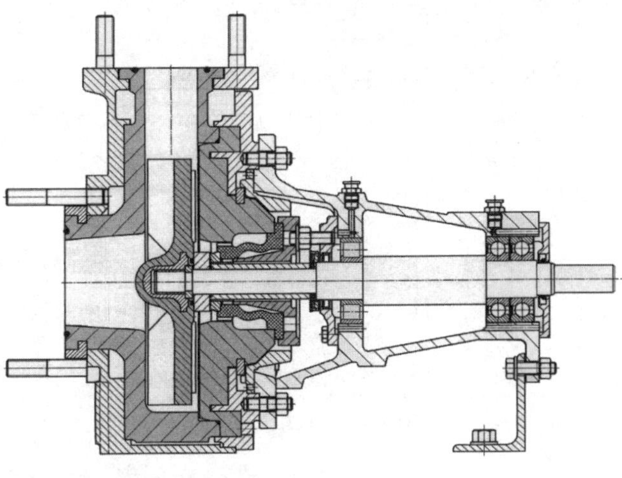

Figure 7.8 A typical moulded centrifugal pump

	PVC	NR IR	SBR	AU EU	IIR	PP	CR	XNBR	CSM	NBR	EPDM	EPM	HNBR	ETFE	ACM	FKM	VMQ	PTFE	FFKM	PFA	PI	PEEK
Low temperature °C	-35	-60	-50	-50	-30	0	-40	-30	-20	-45	-55	-45	-30	-100	-20	-15	-60	-250	-29	-250	-240	-18
High temperature °C	70	80	90	100	100	104	110	110	120	130	140	140	140	150	160	175	200	230	232/316	260	480	260
Abrasion resistance		E	G	E	F	G	G	E	E	G	G	E	E	G	F	G	P	P	P	P	E	
Tear resistance		E	P/F	E	G		G	E	E	G	G	G	E	F	F	F	P		F			
Gas permeability		H	L	M	L		L	L	L	L	L		L	L	L	L	L	L	L		M/H	
Compression set		E	G	G	P		E	G	F	G	G	F	E	G	P	G	G		G			
Resilience		E	F	G	P		G	F	P	F		P	F		F	F	P	P		P	P	
Coefficient of friction		H		H				L		L	L		L			H		L	L	L	L	L/M
Radiation resist		G	F	G	P		F	F	F	F	F		G		F	P	P	P	G	P	E	E
Weather resist		P	F	E	E		P	P	E	P	E	E	G	E	E	E	E	E	E	E	E	
Ozone resist	G	F	F	E	E		G	P	E	P	E	E	G	E	E	E	E	E	E	E		G
Sunlight ageing		P	P	E	E	P	E	P	E	P	E		G			G	G	E		E		
Mineral oil (LAP*)	P	F	P	G	U	P	F	E	G	E	P	P	E	G	E	E	P	E	E	G	G	G
Mineral oil (HAP**)	P	F	P	P	U	P	G	E	G	E	P	P	E	G	E	E	G	E	E	G	G	G
Silicone fluids	F	G	F	F	F	G	F	F	F	F	G	G	F		G	E	P	P/G	E	P/G		G
Diester fluids		P	P	G	G		P	P	P	P		P	P			G	F	G	G	G		
Phosphate esters		P	P	G	E	P	P	P	P	P	E/G	E/G	P	G	P	P	F	G	E	G		
Silicate esters		P	P	F	P	P	F	F	F	F	P	P	F		P	P	F	G	E	G	G	
Aliphatic hydrocarbons	E	P	P	E	P	G	F	G	G	G	P	P	G	G	P	E	P	E	E	E	G	G
Aromatic hydrocarbons	U	P	P	G	P	F	P	F	P	F	P	P	F	G	P	E	P	E	E	E	G	G
Halogenated solvents	U	P	P	P	P	F	P	P	P	P	P	P	P	E			P	E		E	P	G
Ketones	U	G/F	P	P	P	G	G	P	P	P	G	F	P	P	P	P	P	G	G	E	G	G
Aromatic petroleum	G	P	P	G	P	P	P	G	P	G	P	P	G	F	F	E	P	G	E	G	G	G
Non-aromatic petroleum	G	P	P	G	P	P	G	G	F	G	P		G	F	G	E	G	G	E	G		G
Dilute acid	E	G	G	G	G	E	P	P	E	G	G	E	G		P	G	F	E	E	E	P	G
Concentrated acid	E	F	P	P	F	F	P	P	G	P	G	E	G	G	P	G	P	E	G	E	F	G
Dilute alkali		G	G	P	G	E	G	F	G	G	E	E	G		P	G	F	E	E	E	P	G
Concentrated alkali		F	F	P	G	G	P	P	G	P	G	E	G			P	P	E		E	P	G
Cold water	G	E	E	G	G	G	F	P	G	G	E	E	E	E	U	G	F	E	E	E		G
Hot water	U	F		U	F	P	P	P	F	G	E	E	E	E	U	G	F	E	E	E	P	G
Steam	U	U	U	U	U	U	P	P	P/F	P	E	E	E	E	U	U	P	G	E	E		G

Table 7.14 Comparative properties of elastomers

E = excellent, G = good, F = fair, U = unsuitable, H = high, M = medium, L = low (*LAP = low aniline point, **HAP = high aniline point)

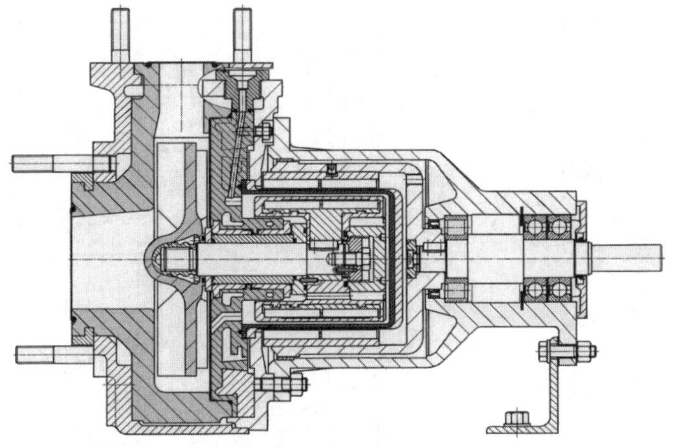

Figure 7.9 A typical magnetic drive non-metallic pump

trolled. Piping loads are transmitted through the housing to the baseplate. The corrosion resistant casing does not have to be designed to cope with external forces and moments. The low strength of the non-metallic casing is augmented by an external structural support. This concept can be used for the highest integrity centrifugal pumps, see Figure 7.9. Current advances in material technology allow the pressure containment "can", around the magnet assembly, to be made of ceramic.

7.2.12.3 Elastomers

Some elastomers occur naturally, natural rubber is the most obvious, others are manufactured for specific physical and chemical properties. The most popular elastomers are:

- ethylene propylene (EP)(EPDM)(EPT)
- fluoroelastomer (FPM)
- natural rubber (NR)
- NEOPRENE® (CR)
- nitrile rubber (NBR)
- perfluoroelastomer (FFKM)
- silicone (Si)
- styrene butadiene rubber (SBR)
- urethane (AU)(EU)

Natural rubber is compounded with fillers and other materials and then vulcanised, heated with sulphur compounds, to produce a solid elastic material for processing. The hardness and chemical resistance being defined by the vulcanisation process. Modern materials have been developed for increased resistance to oils, sunlight and high temperatures.

Modern compounds are generally known by the registered trade name of the manufacturing company. Attempts have been made to standardise the nomenclature of generic groupings, ISO and ASTM abbreviations are used. The list below indicates the general progression to higher temperature operation and increased chemical resistance. Each individual case must be evaluated on its own merits.

- NR, NBR, SBR Perbunan®, Buna N®, Buna S®
- EPDM Dutral®, Vistalon®
- FPM Viton®, Fluorel®
- PTFE (thermoplastic) Teflon®, Fluon®
- FFKM Kalrez®, Chemraz®

Some of the latest man-made compounds are very costly. Elastomer seals and gaskets can sometimes be replaced by metal parts. Silver plated stainless steel "O" rings are much more cost-effective than exotic elastomer compounds. An overview of elastomer properties is given in Table 7.14.

7.2.12.4 Ceramics

Ceramics are used in three forms in pumps – cast sintered and sprayed, see Section 7.2.13 for sprayed applications. Cast ceramic is manufactured just like pottery. A wet clay mix is poured into moulds and then fired. The resulting component is very hard and almost completely chemically inert. Industrial ceramics are not glazed like pottery and some grades exhibit slight porosity; this can be remedied by suitable sealants. Some ceramic compounds are sintered from granules to produce the desired shape. Machining to finished sizes is by grinding and lapping.

Silicon carbide, silicon nitride and alumina, aluminium oxide, are the most common ceramics. In rotodynamic pumps, silicon carbide is alloyed with silicon and carbon to produce mechanical seal faces and bearings in magnetic drive and canned pumps. In reciprocating pumps, alumina is used for plungers, bushings and ball valves.

7.2.12.5 Minerals

Asbestos used to be one of the most widely used minerals; an excellent gasket material for high temperature applications. Due to its unacceptable health hazards, asbestos is rarely used in its free state. Asbestos is used as a filler or reinforcing in some grades of thermosetting plastics and elastomers. Typical applications include dry running and product-lubricated plain bearings.

Carbon and graphite, because of their resistance to chemicals and low coefficients of friction, are used for mechanical seal components and product-lubricated bearings.

7.2.13 Coatings

Various pump components are coated to achieve different objectives. Coatings are applied to achieve corrosion/abrasion/wear resistance and running compatibility. Coatings can be applied by various techniques; brushed, dipped, physical vapour deposition, sprayed, weld deposit, vacuum fused and electroplated.

Like elastomers, coatings are best known by their trade names rather than by chemical composition or standard designations. Metallic, non-metallic and ceramic compounds are used for different applications. Table 7.15 lists sprayed coatings and indicates the major constituent.

Trade name	Major constituent
Stellite 6	cobalt
Stellite 12	cobalt
Colmonoy 5	nickel
Colmonoy 6	nickel
Metco 130SF	aluminium oxide/titanium dioxide
Colmonox	chrome oxide (dipped)
UCAR LW-15	tungsten carbide

Table 7.15 Common coating materials used in pumps

Colmonox is a dipped coating which is built up in several layers. Chemical treatment between dip cycles eliminates any porosity.

Pure elements, such as nickel and chromium, are deposited electrolessly for corrosion and wear resistance. Titanium nitride is applied by physical vapour deposition in a vacuum chamber. A very thin hard coating is applied to finished components without dimensional change. Thermoplastic materials can be applied to metal components to provide corrosion resistance and compatible bearing surfaces.

Coatings such as epoxy resins can be applied by brush. An extension of this idea is to incorporate solids within the resin. Glass beads have been used with great success.

Coatings are used extensively on new components such as:

- shafts

- sleeves
- wear rings
- balance drums
- bushings
- plungers
- piston rods
- valve seats

Coatings can also be used to reclaim damaged or worn components. Coatings with glass beads have been used to repair pump casings. Worn shaft bearing lands can be reclaimed with hard chrome plate. Special nickel alloys are available to coat cast iron where wear has occurred.

7.3 Material strength and integrity

There are no international standard for hazards caused by liquids. National standards exist, see Section 4.7 in Chapter 4, but there is little agreement or correlation of test methods and permissible doses.

In Germany, for example, it is not permissible to pump petrol using cast iron pressure-containing components. In France all road petrol tankers have cast iron pumps. In general, most proprietary specifications do not allow the use of cast iron for flammable liquids. The strength of material used related to the working pressure and the level of integrity required for the material are both related to the perceived hazards and the cost of failure.

Ratings of pumps follow pipe flange ratings. ISO 7005 has flanges which are suitable for metric and inch sizes. The following nominal pressure ratings, barg, are covered:

2.5
6
10
16
20
25
40
50
110
150
260
420

Figures 7.2, 7.3 and 7.25 show that the strength of materials is dependent upon temperature. In ISO 7005, PN 50 refers to a set of dimensions, not a guaranteed pressure rating. With some specified materials, PN 50 flanges are only suitable for 42.5 barg at room temperature. PN 420 flanges are suitable for 431 barg in certain materials at room temperature. Some users and manufacturers will think 431 barg a very high pressure and of only academic interest. This is not so. Oilfield applications regularly operate at pressures over 400 barg and standard reciprocating pumps extend to 1000 barg.

High pressure pumps require high strength materials. If a pump casing was infinitely thick, the highest tensile stress would be in the bore and equal to the liquid pressure. If cast iron was used, the theoretical maximum allowable pressure, not considering fatigue, would be about 200 barg. As the wall thickness was reduced to economical proportions the maximum stress would increase. Figure 7.10 shows the stress values for a pump casing of 250 mm bore subjected to an internal pressure of 200 barg.

If it was necessary to have a wall thickness of only 10 mm then a much stronger material would be required, 13Cr alloy steel for example. If the casing was to be cast in 18/8/2 austenitic stainless steel, then the minimum wall thickness would be about 17 mm.

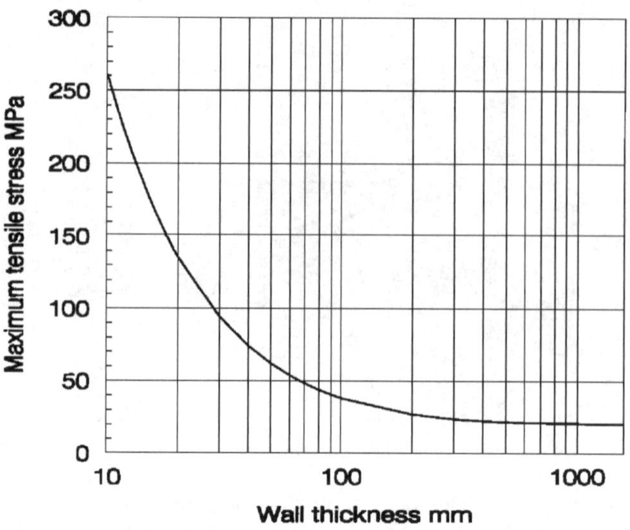

Figure 7.10 Casing tensile stress for various wall thicknesses

The design and manufacturing processes are complicated because of the vast range of materials available. Manufacture at constant thickness and vary the pressure rating; manufacture at various thicknesses for constant pressure rating. Obviously the cost of the material plays an important role as does the batch size.

For any given application the pump designer must ensure that the pressure retaining materials are thick enough and strong enough to withstand the loads and stresses imposed on the pump. The design process is further complicated by the imposition of corrosion allowances. The pump user may require material to be added in the bore to allow for corrosion and erosion, 3 mm is a typical value. When the pump is new, it is stronger than necessary. In the corroded condition, the casing will have reduced to the design thickness.

Integrity of pressure retaining parts is checked during production by hydrotesting i.e. pressure testing with water. Castings are only machined sufficiently to provide good seals and then hydrotested.

This first test is used to find casting weaknesses and porosity. The length of time a component must be held under pressure is a function of the material thickness. Thin casting can be checked in 10 minutes, heavier castings are usually tested for 30 minutes. Castings which fail this first test are evaluated to see if reclamation is cost effective. Depending upon the material, thickness, position and extent of fault, the casting will be reclaimed or scrapped.

For mass-produced standard pumps, the pump user has no control over this aspect of production. The pump manufacturer will have standard procedures to be adopted for various failure modes. All repairs implemented will be compatible with the published range of operations for that product. Repair techniques may include welding, impregnating, caulking, plating or plugging. Purchasers of standard products accept the manufacturer's warranty of suitability for service.

With pumps built specifically to order the situation is different. The purchaser may chose to limit the type of repair techniques used and also the extent of repairs. Pumps for hazardous liquids are generally not repaired by impregnating, caulking or plating. Plugging may be allowable if the thread in the bore can be seal welded. The purchaser may define repairs as "minor" and "major", depending upon the position and nature of the fault and the extent.

Minor faults may be repaired by agreed techniques, where the purchaser is informed of the location but production is not impeded. The agreed techniques include:

- The inspection methods used to determine the extent of the fault and the qualifications of the operators
- The repair techniques to be used, approved procedures, including post repair heat treatment, and qualifications of staff
- Post repair inspection to prove the integrity of the repair

When a major fault is detected production stops, the purchaser is informed and nothing happens until the fault has been inspected by a representative of the purchaser. Then discussions between the two parties take place to decide if, and how, the fault can be repaired. One important part of any such discussion is the effect on delivery.

Castings will be hydrotested again in the fully machined condition, usually when the pump is fully assembled. This final pressure test assures that all material is sound and that bolted and screwed joints do not leak. Mechanical seals may or may not be fitted as agreed. Leakage through mechanical seals may or may not be permissible. Leakage through soft packed stuffing boxes must be permissible.

Although up until now have been discussed castings here but the same principles apply to fabrications and forgings. Forgings, however, should only be repaired with great reluctance, a "major" repair, and never in highly stressed regions. Cosmetic repairs can be made in any low stress zone which is not in contact with the process liquid.

The pressure at which hydrotests are conducted is subject to agreement and depends upon the design and construction and operating temperature. Small pumps tend to have suction and discharge flanges with the same pressure rating, so the complete casing may be tested at one pressure. Larger pumps may have a low pressure suction section and a high pressure discharge section here, the casing would be divided by a suitable diaphragm and tested at two pressures. It is not unusual for pumps to be fitted with high pressure suction connections when the suction pressure is low. The user and the manufacturer must agree whether the hydrotest pressure will be based on the suction pressure or the flange rating.

Some flange specifications, ISO 7005 and ANSI B16.5 for example, specify hydrotest pressures as part of the flange specification. In general the hydrotest pressure is 1.5 times the rated pressure. Components with standard flanges attached cannot be hydrotested at higher pressures. Other components, mechanical seals say, may limit the hydrotest pressure to a much lower value. Hydrotest pressures and pressure ratings must be clarified during the quotation stage. Individual components may be hydrotested at higher pressures than the assembled pump. There is no theoretical basis for having lower test pressures as the rated pressure increases. If the perceived hazard is as great, or greater because of the higher pressure, producing increased leakage through similar faults, the safety margins should be maintained across the pressure range. Mass-produced pumps may not be hydrotested in the final assembled condition; clarification from the manufacturer should be sort if problems with leaks occur at site.

Integrity can be interpreted in another manner, namely, design integrity. Design integrity, or proof testing, is when the manufacturer shows the basic design and construction to be sound. This problem arises particularly when corrosion allowances are added to pump casings. The proves the new casting is good, it does not show how the casting will react when it is 3 mm thinner.

A British standard BS 7929 addresses this theoretical problem of design proving. Calculation and physical testing are allowed. The manufacturer would show by calculation, computer model or finite element analysis, that stresses and strains in the "thin" component would not impair its ability for pressure containment and other essential functions.

Alternatively, the new "thick" component could be subjected to higher test pressures to create the stresses and/or strains which would exist in the "thin" component. For prototype pumps, and extensively modified standard pumps, discussions during the quotation stage should clarify the need, or otherwise, of design proving. Consideration must also be given to standard and prototype pumps which are designed and constructed to long established rules. There would be no point in forcing a manufacturer to produce a finite element analysis of a pump which had been operating successfully for twenty years.

7.4 Corrosion and erosion

7.4.1 Liquid corrosion of metals

The corrosion of metals in liquids is largely an electrochemical process, i.e. there is a positive electrode, the anode, and a negative cathode, between which an electric current will flow. If a ferrous component is immersed in acidic water, the anode-cathode process will occur over its entire surface. The cathode in this case comprises local elements of more noble metals in the galvanic series segregated in the alloying process, or non-metallic inclusions or by local reaction with oxides, sulphides and so on.

Metal is dissolved by the generation of metal ions at the anode. Oxygen dissolved in the liquid is reduced by the generation of hydroxyl ions, OH-, at the cathode, or in the absence of oxygen, hydrogen ions, H+, are used up in the generation of hydrogen gas. Where metallic and hydroxyl ions come into contact, corrosion products are generated, usually in several stages of chemical reaction. The final corrosion product usually comprises metal oxides, which are difficult to dissolve, or metal hydroxides. Corrosion products generate a more or less dense film on the surface.

The incidence of protective films resulting from corrosion, or those generated in other ways, are necessary as a rule for most metals in use in order that further corrosion should be inhibited, or contained within reasonable limits. Only the most noble metals, like gold and platinum, are immune within the greater part of the stable region of water.

Protective films can be removed by fast-flowing liquid. If a metal surface has a protective layer of corrosion products, this may be one of two types:

7.4.1.1 Porous, relatively thick films

These are the film on copper alloys and graphite on cast iron. Thick protective films often have complex constructions, because contaminants in the liquid, such as lime, may be included as a significant component. For this reason it is difficult to predict whether any effective film will be generated at all. At speeds normally found in centrifugal pumps, these films often have a limit, which is a step function, above which they get carried away by the flow of liquid. These films can be very successful in positive displacement pumps which operate at much lower liquid velocities. Positive displacement pumps use aluminium bronze or nickel aluminium bronze for seawater applications which require super-duplex stainless steels in centrifugal pumps. Thick films are relatively sensitive to solid particles, cavitation and even air bubbles in the flowing liquid. Even if such disturbance is sporadic, the damage may be great because the films take a relatively long time to regenerate, days or weeks even.

7.4.1.2 Dense, thin films

Examples are those passive films normally found on stainless steel, titanium and so on. The very dense and thin passive films usually consist of one single metal oxide only, for example chromic oxide on stainless steel. These films have very good adhesive qualities and are not removed even by very high liquid velocities.

On the contrary, a flowing liquid tends to keep the film intact by preventing crevice corrosion and pitting. It is only when the flowing liquid contains solid particles or when flow conditions are such that cavitation occurs that a passive film will be damaged. Passive films usually regenerate quickly after damage and occasional penetrations by solid particles or cavitation do not usually lead to further corrosion. If, for any reason, a passive film does not exist on stainless steel, titanium and so on, then heavy corrosion attack can be expected.

Chemical reactions increase exponentially with temperature. As temperature increases, however, secondary effects can come into play, resulting in reduced corrosion. The oxygen content of water at atmospheric pressure decreases as temperature increases, thus reducing the corrosive effect on, for example, iron. If there is positive pressure, dissolved gases such as O_2 and H_2 will be present in the water even at high temperatures and will still influence corrosion.

When a centrifugal pump operates at closed valve, the absorbed motor power will be converted to heat in the pump, which rapidly causes high temperatures. Especially in the case of corrosive liquids, this can lead to heavy attacks because a passive film can be destroyed. Heating can also mean mechanical damage resulting in deteriorating material strength and properties. Plastic pumps are especially prone to softening with heat and should therefore be fitted with temperature sensors. Metallic pumps often use elastomer gaskets; overheating can result in leaks. When considering chemical attack and temperature ratings all pump components must be evaluated.

7.4.2 Corrosion rate

For a material to be considered in practice as being completely resistive, the material lost from the surface should not exceed about 0.1 mm per year. If the material losses are of the order of magnitude of 1 mm per year, the material can still be used but must obviously be checked at regular intervals.

Other chemical reactions occur at the cathode with certain other liquid-metal combinations. The chemical rule of thumb that a reaction speed doubles with a temperature increase of 10°C also applies to the corrosion process. Corrosion rates are expressed either as penetration depths or loss of weight per unit area per unit time. The conversion factors for some commonly used units are tabulated in Table 7.16. For depth of penetration, the conversion applies only to steel or substances of the same density.

Unit	Conversion factors		
	$g/m^2/h$	$g/m^2/day$	mm/year
1 $g/m^2/h$	1	24.0	1.12
1 $mg/dm^2/day$	0.004	0.10	0.0046
1 $mg/cm^2/day$	0.417	10.00	0.465
1 mm/year	0.890	21.4	1
1 mm/month	10.800	258.0	12
1 inch/year	23.0	554.0	25.4
1 mil/year	0.023	0.554	0.0254
1 $oz/ft^2/year$	0.035	0.84	0.0394

Table 7.16 Conversion of various units for corrosion rate (1 mil = 0.001 inch)

7.4.3 Types of corrosion

Units for rate of corrosion assume an even attrition over the whole surface, i.e. uniform corrosion. This often does not happen and other forms of corrosion can occur:

- Electrolytic (galvanic) corrosion, when different metals come into contact, or very close porximity, in the presence of a conducting medium.
- Crevice corrosion, in crevices and small confined spaces, where quantities of liquid may become more or less isolated from the main body of the liquid.
- Pitting, meaning that the corrosion is concentrated at certain points, pits.
- Intergranular or intercrystalline corrosion usually associated with redistribution of alloying elements during heat treatment or welding.
- Selective corrosion, where one single alloying element is dissolved out, e.g. dezincification in brass.
- Erosion-corrosion, which occurs in flowing media and is the commonest complication in pumps, see also Section 7.4.3.2.
- Stress-corrosion, where a combination of mechanical stress and attack by the liquid may lead to fracture.

Crevice corrosion and pitting are related phenomena, and are most troublesome for stainless steel in a chloride-containing environment. These attacks occur in crevices or holes where the conditions for setting up a protective film are absent because of screening from the surrounding solution. Special care must be taken in pumps to prevent crevice corrosion at rotor fixings or on shafts rotating within seals. The risk of crevice corrosion is reduced in flowing liquids. Crevice corrosion will not usually occur so long as the pump is working. During extended stationary intervals in chloride-containing liquids, however, the pump should be emptied and flushed out with clean water.

Corrosion fatigue is also important in pumps and especially for shafts. Even seemingly insignificant corrosion may reduce the fatigue strength of a material by a factor of 3 to 10.

7.4.3.1 Galvanic corrosion

Where there is a combination of different metals, galvanic corrosion may occur. This means that the less noble metal, the anode, has a higher corrosion rate than the more noble metal, the cathode, which is protected. An indication of the risks of galvanic corrosion can be obtained by studying the galvanic or electrolytic series, see Table 7.17. In practice, metals with a potential difference of 0.2V can be connected without any trouble.

CORRODED END - ANODE (Least noble)	Electrode potential V
Magnesium alloys	- 1.51
Zinc	- 1.09
Cadmium	- 0.73
Aluminium alloys	- 0.75 to - 0.64
Aluminium alloys, cast	- 0.70
Iron and steel	-0.55
Stainlesss steel, all types	-0.53
Lead-tin solder	-0.52
Lead	-0.51
Tin	-0.47
Nickel	-0.25
Inconel	-0.25
Titanium active	-0.25
Brass	-0.22
Copper	-0.22
Bronze	-0.22
Nickel-copper alloys	-0.20
Stainless steel, all types, passive	-0.15
Monel	-0.10
Silver solder	-0.07
Silver	- 0.05
Gold	+0.18
Titanium, passive	+0.19
Platinum	+0.33
PROTECTED END - CATHODE (Most noble)	

Table 7.17 Galvanic series for metals with potential values measured in seawater

Galvanic corrosion is of great significance in pumps because of the principle of using different materials for various parts, see Section 7.7.2. However, the effects of erosion-corrosion seem

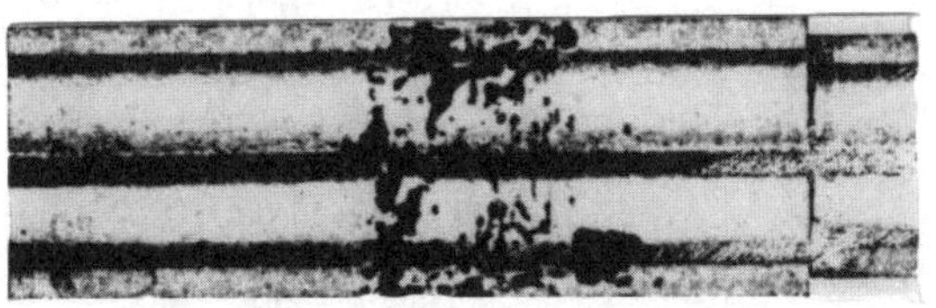

Figure 7.11 Extreme galvanic corrosion on 13Cr steel pump shaft

Figure 7.12 Erosion-corrosion on a 13Cr + Mo steel impeller

to create relatively wide departures from good practice from the galvanic corrosion point of view.

Where the anode surfaces are large and cathode surface is small, in the case of pump casings and impellers for example, the potential differences for certain favourable combinations can be twice as large.

Figure 7.11 shows a drastic case of galvanic corrosion. The shaft had been lying in stock fitted with graphite containing soft packing, the corrosion being caused by condensation from humid atmosphere. The attack has been accelerated by crevice corrosion.

7.4.3.2 Erosion-corrosion

Erosion-corrosion is particularly common in rotodynamic pumps where the flow velocity nearly always exceeds 20 m/s. Characteristic of erosion-corrosion are the horseshoe-shaped patterns scavenged in the direction of flow and bright surfaces free of corrosion products, see Figure 7.12. In this illustration the impeller was used for sulphate liquor. Note that the rivets are completely intact. These were made of SIS 2343. For this application the impeller should have been made of SIS 2343 or SIS 2324.

When metal comes under attack, oxidisation media O_2 or H+ for example, are consumed at the surface of the metal and metal ions are generated. Thus the concentration of metal ions will be greater and the concentration of oxidisation media will be lower at the metal surface than in the main body of the liquid, see Figure 7.13. The difference in concentration at the metal surfaces is dependent upon the difficulties in transportation of the participating compounds which is necessary for the continuation of the corrosion process.

Obviously the transportation will be aided and, at the same time the rate of corrosion will be accelerated, if the liquid is in motion, which is generally the case in pumps, (see Figure 7.17). Care should therefore be taken when using corrosion data from tests in stationary liquids for the selection of pump materials, since corrosive attack inside a pump may be ten times worse than it is in a still liquid. See also Figure 7.14, where at (Re_d)critical, the change over from laminar to turbulent flow occurs.

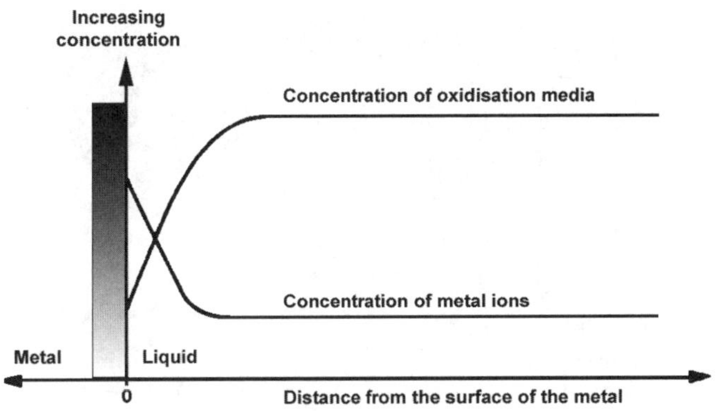

Figure 7.13 Concentration differences of general corrosion at a metal surface

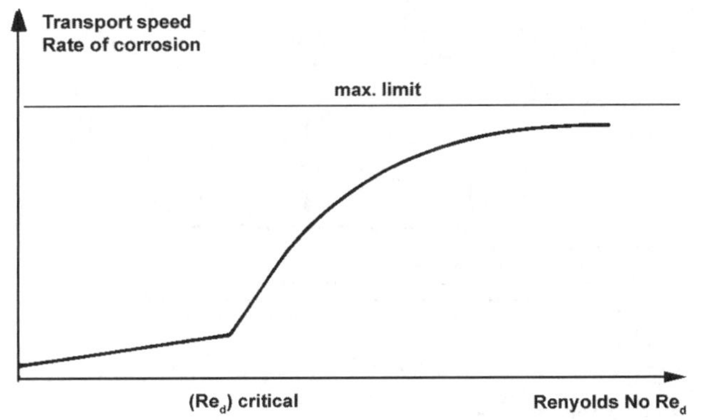

Figure 7.14 The principle of increased rate of general corrosion with liquid velocity.

7.4.4 Corrosion testing

Methods of corrosion testing in flowing liquids include:

- Pipe flow
- Rotating cylinder, freely rotating or enclosed
- Rotating disc
- Spraying a sample piece with a liquid spray.

If the purpose of the test is to determine the probable corrosion rate then the test rig should as far as possible resemble the condition intended to be tested. This is so that the test takes place at the same Reynolds Number, i.e. similar flow. Otherwise the result will only be a comparison between different materials.

When the corrosion test is intended for pumps, it would be advantageous to use an enclosed rotary disc at a speed such that the Reynolds Number is about the same as that for the pump impeller.

$$Re_d = \frac{u_2 d_2}{\mu} \qquad \text{Equ 7.1}$$

where:

u_2 = impeller peripheral velocity (m/s)

d_2 = impeller tip diameter (m)

μ = liquid viscosity (m^2/s)

Instead of rotating a complete disc of test material, small electrically connected samples can be set into the fixed wall opposite the disc, see Figure 7.15, which are then exposed to the same velocity as the rotating disc. This positioning of the sample pieces has the advantage that electrical measurements can be carried out simply, see Figure 7.16. Beside which, surfaces subjected to various liquid velocities can at least be of the same size. In a solid rotating disc, the larger part of the surface area is subjected to the highest liquid velocities, whereas in the pump

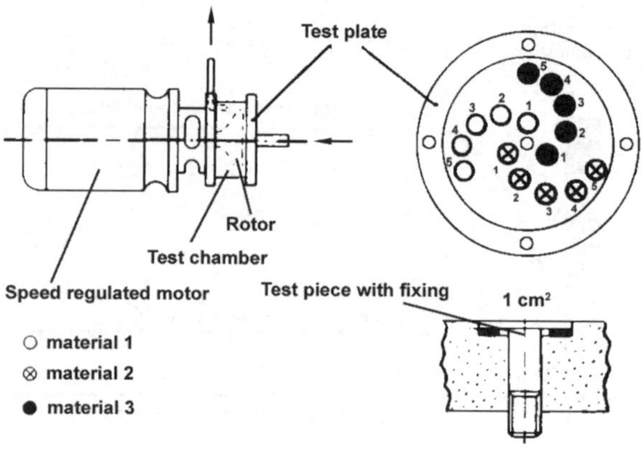

Figure 7.15 Corrosion test set

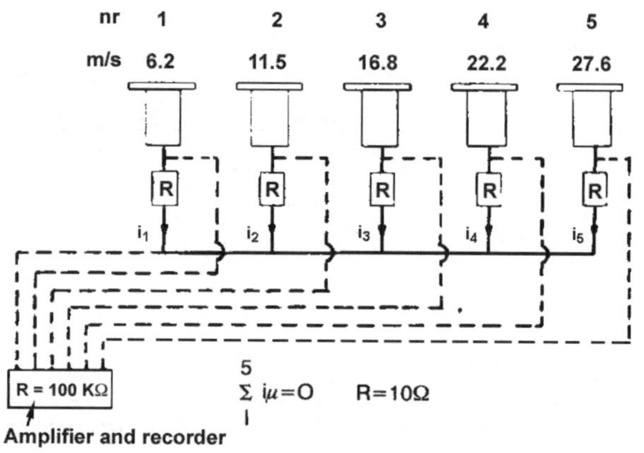

Figure 7.16 Electrical connection of test pieces subjected to various liquid velocities

the opposite is the case, because of the large surface area of the pump casing.

The test arrangements shown in Figure 7.15 and Figure 7.16 will resemble the corrosion process in a pump, since all the relevant liquid velocities are present simultaneously, and redistribution of metal losses due to galvanic currents will occur between parts of the same metal subjected to various liquid velocities.

The galvanic currents are among the items measured in corrosion testing, apart from the usual weighings, potential measurements and so on. If metal corrodes, there are actual electrolytic currents flowing between parts of the same metal surface subjected to differing liquid velocities. Thus, by recording the currents, it is possible instantaneously to check the occurrence of corrosion, and at the same time qualitatively to follow changes in the rate of corrosion, build-up of protective film of corrosion products etc. Unintentional changes in the rate of corrosion can also easily be recorded.

Materials generating thick protective films require relatively long test times, and worthwhile laboratory tests can be difficult to carry out since it is not easy to maintain control over the composition of the liquid over long periods of time. In these cases it is better to move the laboratory to the liquid rather than the other way round. When there are decidedly passive films generated normally on the test samples by contact with the air, it is best to remove this film, i.e. activate, in order to establish whether the passivity is restored in the actual liquid. After activation, passivity will be restored after only a short time, within hours, if it is going to be restored at all. If the general corrosion is uniform with time after a certain stabilising period, this is confirmed by the fact that the electrolytic currents also are uniform. The test time is then regulated so that stable conditions apply during as great a part of the time as possible.

Figure 7.17 shows some examples of test results using the set up shown in Figure 7.15.

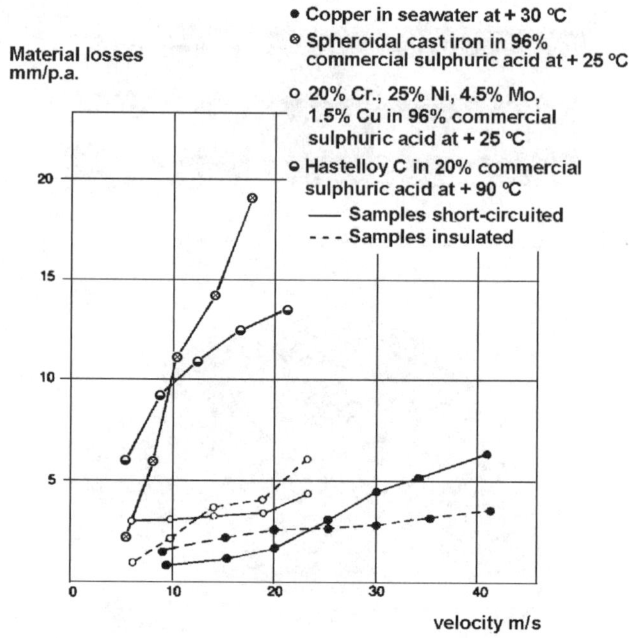

Figure 7.17 Onset of corrosion as a function of flow velocity

7.5 Abrasion resistant materials

7.5.1 Pump selection

Liquids containing solid particles can present considerable problems. To some extent the problems can be eased by choosing a pump design suitable for the abrasive conditions or reducing the velocities involved. Surfaces subject to wear can also be fitted with replaceable liners or designed with a good margin for wear. Examples of this are those pumps which are used for transporting abrasive solid materials such as ash, sand, asphalt, etc. In these pumps, not only are the wear margins large but even the pump shaft systems are designed to cope with the imbalance resulting from such wear.

Two main groups of material are used for abrasive liquids, hard metals and various types of rubber cladding. Wherever rubber cladding can be used, depending upon particle size and liquid temperature, it is superior to the hard metal materials. Hard metals can be used as coatings over normal metals. In this respect, hard coatings can be replaced to extend the lives of worn components. Ceramics are also very good but, so far, plastics have not given very good results.

Wear increases greatly with flow velocity for all qualities of material. Wear increases to a power of between 2.5 and 5 with velocity. This relationship infers that wear for a centrifugal pump with differential head H increases in the proportion $H^{2.5/2}$ to $H^{5/2}$. Delivery head per pump stage is limited therefore and is a maximum of about 50 m for rubber-clad pumps and about 100 m for pumps made of wear-resistant metals. Positive displacement pumps can produce pressure without recourse to high velocities. Positive displacement pumps are ideally suited for liquids with high concentrations of solids because these liquids assume an apparent high viscosity. High viscosity reduces the efficiency of rotodynamic pumps.

7.5.2 Hard metallic materials

The following applies to hard materials, mainly steel with a high carbon and chromium content:

- Material losses increase greatly when the hardness of the particles is greater than the material in the pump or liner.
- Material losses increase with the size and mass of the particles.

- Material losses increase if the particles have sharp edges and the particle material has sufficient strength to maintain the edges.
- Material losses increase as the particle concentration increases.
- Wear-resistance is improved by increasing the hardness of the pump material, but this effect can only be significant if the pump material hardness figure exceeds about 300 BHN. (If Miller Number test results are corrected for soft materials, normal carbon steel for example, a multiplier of over 1000 is required.)
- Combination of general hardness in the pump material with wear-resistance in individual elements in the pump material is a good thing. Examples of this are Ni-Hard, a martensitic chromium steel, where the wear-resistance is obtained from chromium carbide granules embedded in a hard matrix.

7.5.3 Rubber cladding

The wear-resistance of rubber cladding is due to the fact that solid particles more or less bounce on the rubber surface without actually damaging it. It should be pointed out that:

- Soft rubber is better than hard rubber.
- Rubber gives favourable results if the particles strike the surface at right-angles, angle of incidence 90°, or if they move along the surface of the rubber. If the angle of incidence lies between 5° and 30° the results will be much less favourable.
- The thickness of rubber should be 2 or 3 times the size of the particle in order to exploit the bouncing effect.
- Sharp particles can cut the surface of the rubber to pieces.

Natural rubber is suitable for temperatures up to 70°C, whilst nitrile rubber can be more suitable when mineral oil contaminants are present, although it is not so wear-resistant. Urethane rubber has become popular for parts of pumps used for abrasive pumping for temperatures up to about 50°C, see Figure 7.18.

Urethane rubber has excellent wear properties and good resistance to oils, but polyester-based compounds are sensitive to hydrolysis. In this respect, polyester based urethane rubber, Adiprene® is better. Typically, urethane rubbers are considerably higher in price than natural rubber; however, this is compensated for by a simpler manufacturing process involving forming in open moulds.

Glass beads set in resin have been applied to various metal casing parts to protect against corrosion and erosion.

7.6 Materials resistant to cavitation damage

When cavitation occurs inside a pump, the vapour bubbles implode resulting in local pressure waves up to about 10000 MPa. Implosions close to a metal surface in a pump impose repeated loads on the surface with the result that fragments of material are loosened. Cavitation damage thus has a characteristic appearance, see Figure 7.19. The surface becomes rough and pitted in appearance, and is limited to those zones in a pump where low pressure can occur. If cavitation is allowed to persist then severe material loss can occur, see Figure 7.20.

In reciprocating pumps the continuous action of the repeated implosion loads cause fatigue failures in important components. The classic failures include:

- Suction valve plates
- Liquid end studs

Figure 7.18 Pump impeller made of urethane coated steel and pump casing of solid urethane rubber

- Cylinders

If cavitation damage has been caused in a pump, the first thing to do is to improve the NPSHa/NPIPa of the installation by:

- Increasing the available suction head or pressure or reducing the suction lift
- Lowering the temperature of the liquid in the pump intake either by direct or indirect cooling

If neither of these actions is feasible the pump speed should be reduced with production consequences. If this is not feasible or if the cavitation process is purposely used for regulation, as is the case for certain condensate pumps, then a cavitation-resistant material must be used.

A choice can be made by studying the results of comprehensive experiments, see Figure 7.21, which shows that it takes a certain amount of time before cavitation damage appears. The process is similar to the incubation time for some infectious diseases.

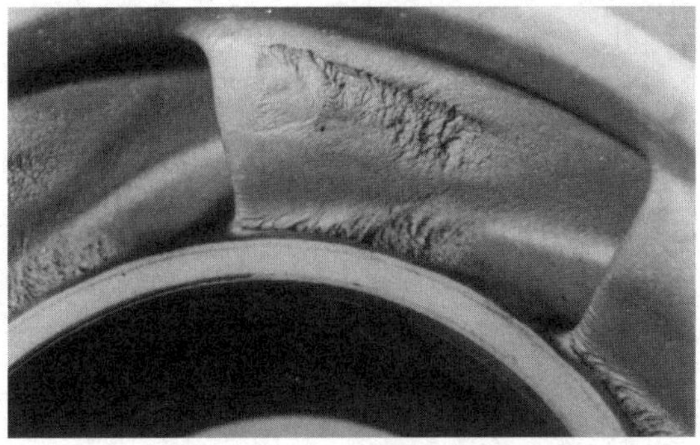

Figure 7.19 Typical cavitation damage of pump impeller

Figure 7.20 Severely damaged pump impeller

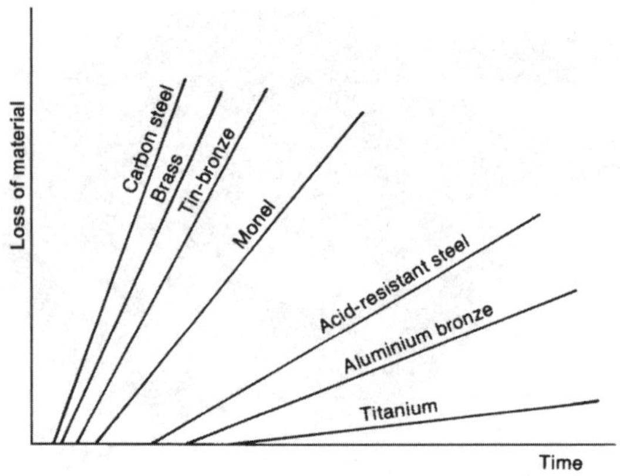

Figure 7.21 The resistance of various materials to cavitation damage

7.7 Material selection

7.7.1 Basic information

A review of preliminary material choice can be found partly in Section 2.2, Chapter 2, and partly in Section 7.7.3. This information enables the selection of materials suitable for a wide range of liquids. In more specific cases, it is recommended that a study should be made of special literature such as that referred to in Section 7.8. It is, however, inevitable that even this background will not be adequate in many special cases, so that judgement has to be made based on practical experience of similar cases. The information required for choice of material has been collected in Table 7.18 in the form of a check list.

It is the users'/purchasers' responsibility to provide all relevant data to allow pump type and material selection to be accomplished. Successful pump operation depends upon the initial decisions made. Decisions based on incorrect or incomplete data will tend to be suspect.

The pump manufacturer cannot be held responsible for wear or corrosion resulting from inappropriate decisions based on incomplete or incorrect information. The manufacturers warranty will be invalidated.

Liquid
Full description
trade name, chemical formula, concentration
hazardous/flammable/toxic
controlled by legislation
abrasive/corrosive
all constituents including any trace elements
vapour pressure
viscosity
viscosity characteristic with shear
compressibility
solidify temperature
crystallisation temperature
SG/Cp/pH
Liquid contaminants
dissolved/entrained gas release pressure
solids
size/concentration distribution
hardness
rigid/deformable/friable
abrasiveness (Miller Number)
Operating conditions
max/min temperature
max rate of temperature change
max/min pressures
max/min flow
respective NPSHa's/NPIPa's
continuous or intermittent
time between routine maintenance
stand-by cycles
full of product/flushed
hazards during stand-by
dry running
Special requirements
hygienic
sterile
allowable contamination of liquid
cleaning of pump
chemical/steam
Environment
Site conditions
indoor/outdoor/onshore/offshore
attended/unattended operation
min/max ambient temperature
max black bulb temperature
min/max relative humidity
altitude
hazardous area classification
atmospheric pollution
allowable product leakage rate
conditions during transport if different to site
Material
User/purchaser material recommendations
recommended materials
prohibited materials
Physical properties
strength
ductility
impact resistance
creep resistance
erosion resistance
thermal shock resistance
Chemical properties
resistance to corrosion
corrosion allowance
predicted life

Table 7.18 Check list for material choice

7.7.2 Material combinations

In a pump some components more than others are subjected to stress, therefore it is natural that these should be made of better materials. Some are subject to compressive stress and others tensile. Some components are pressure retaining while others are completely surrounded by liquid or are external to the pressure boundary. Some components are subject to dynamic stressing with obvious considerations to fatigue.

It may also be necessary for parts which are in permanent or occasional contact to have good bearing properties. This is the case in displacement pumps, plungers and bushings or pistons and liners, or with wear rings and balance drums in centrifugal pumps. Generally speaking, the materials should differ completely, but if they are the same, their hardness should differ by at least 50 BHN.

There is a certain established practice for combinations of materials in centrifugal pumps. This requires that the pump impeller and sundry small parts should be made of better material than the casing. In the case of shafts, attention must be paid to corrosion fatigue, which occurs with relatively small corrosive attacks and, in time, causes the shaft to fail. An alternative is to seal off the shaft completely from the liquid.

There are several different impeller materials for use with the simplest pump casing material grey cast iron, as shown in Table 7.19.

Table 7.19 also shows the relative costs of the various materials. This shows an approximate picture only of price levels, since the cost ratios depend on size of pump, basic prices for the various metals and so on. If the shaft is protected against the liquid, it can be made of lower grade metal.

Wear ring material combinations are listed in Table 7.20. When the impeller is made of grey cast iron, there is a risk of rusting after a test run, or whenever the pump is stationary. A better choice of material for the impeller would be gunmetal, plastic or stainless steel. If the pH value of the liquid is more than about 8 to 9 then gunmetal is not sufficiently resistant.

Casing	Impeller	Shaft	Approx. relative cost
Grey cast iron	Grey cast iron	Stainless steel	0.97
Grey cast iron	Gunmetal	Stainless steel	1.0
Grey cast iron	Plastic	Stainless steel	0.95
Grey cast iron	Stainless steel	Stainless steel	1.08
SG iron	Stainless steel	Stainless steel	1.25
Cast steel	Grey cast iron	Steel	1.1
Cast steel	Cast steel	Steel	1.2
Cast steel	Bronze	Steel	1.25
Cast steel	13Cr steel	Steel	1.3
Cast steel	Stainless steel	Stainless steel	1.5
Cast steel	Super stainless steel	Super stainless steel	2.0
13Cr steel	13Cr steel	13Cr steel	1.4
13Cr steel	13Cr steel	Stainless steel	1.4
Stainless steel	Stainless steel	Stainless steel	1.8
Super stainless steel	Super stainless steel	Super stainless steel	3.0
Hastelloy C	Hastelloy C	Hastelloy C	4.0
Titanium	Titanium	Titanium	10

Table 7.19 Some commonly used material combinations for centrifugal pumps

Casing wear ring	Impeller wear ring
Cast iron	Cast iron
Ni-resist	Ni-resist
Bronze	Bronze
13Cr steel	13Cr steel
Stainless steel/Colmonoy 6	Stainless steel/Colmonoy 5

Table 7.20 Standard wear ring material combinations

Rotary positive displacement pump materials are so varied it is not wise to try to generalise. Reciprocating pumps have evolved to a point where manufacturers offer very similar material groups. Table 7.21 lists cylinder materials in ascending order of corrosion resistance.

Low pressure	High pressure
Carbon steel	High tensile steel
Cast iron	
28Cr iron	28Cr iron
Aluminium bronze	Nickel aluminium bronze
AISI 304/CF-8 stainless steel	N_2 strengthened austenitic stainless steel
AISI 316/CF-8M stainless steel	N_2 strengthened austenitic stainless steel
Super stainless steel	Super stainless steel

Table 7.21 Reciprocating pump cylinder materials

7.7.3 Corrosion resistance

From the corrosion point of view the choice of materials is made from two entirely different viewpoints:

- The material selected must last the intended life of the pump.
- The material should also be such that the liquid handled by the pump is not contaminated. Examples of the effects on processes can be any thing from rusty water in a drinking water installation, to "poisoning" of catalysts in a chemical process because of corrosion products from the pump. Hence the process effect can rule out some materials in certain cases, even though they, in themselves, have an adequate resistance to corrosion.

When it comes to corrosion from chemicals in aqueous solutions, it is the pH value and the liquid temperature which provide a preliminary guideline for choice of materials, see Table 7.22.

pH	Material
0 - 4	Stainless steel
0 - 4	Nickel alloy
0 - 4	Silicon iron
0 - 4	Plastic
4 - 6	Bronze
4 - 6	Stainless steel
6 - 9	Grey cast iron
6 - 9	Non-alloyed steel
6 - 9	Bronze
9 - 14	Stainless steel
9 - 14	Nickel alloys
9 - 14	Plastic
9 - 14	Grey cast iron

Table 7.22 pH value and choice of material for water 0°C to 50°C

At very low and very high pH values, i.e. with strong acids or strong alkalis, the resistance of the various materials must be studied more closely, e.g. by the use of corrosion tables and diagrams, and also the literature listed in Section 7.8.

Table 7.23 reviews the typical fields of application for various materials. See also the respective descriptions of materials in Section 7.2. Within the pH range 6 to 9, grey cast iron is the most commonly used material. Exceptions are certain salt solutions, mainly chlorides, seawater for example, where bronze is used with acceptable results. For the use of grey cast iron for ordinary cold water see also Section 2.2, Chapter 2.

When a material is not completely resistant to a liquid, the effect of dissolved or solid contaminants, from erosion and cavitation and occasional temperature increases, can be important. This effect can mean that practical performance may vary from installation to installation.

Materials	Applications
Steel	De-aerated hot water, boiler feed pumps
SG iron Nodular iron	As above. Better resistance to corrosion at high liquid velocities than cast steel but not as good as cast iron
Grey cast iron	Has good resistance to many liquids assuming the graphite film is not damaged by high velocities, aeration, cavitation or solids. 120°C max, pH >5.5
13Cr steel	De-aerated hot water, boiler feed pumps 350°C depending upon O_2 treatment A good replacement for steel for temperatures over 200°C, better thermal stability, higher pressure capabilities
Silicon iron » 14Si	Resistant to acid corrosion, HCl, H_2SO_4, and strong salt solutions. Not suitable for thermal shock
Ni-resist 20Ni 3Cr	Hot , seawater
Gunmetal, bronze	Salt water, seawater, brine and other moderately corrosive water solutions. Al bronze and NiAl bronze better than tin bronze in seawater. NiAl bronze for higher pressures
Monel	seawater, brines
Austenitic stainless steels	Boiler feed water, general corrosive applications, poor in chloride solutions
SIS 2324, SIS 2343	Good for high velocity applications. Not recommended for some acids and chloride solutions
Alloy 20	Better resistance to acids and chlorides
Hastelloy C	Very resistant to corrosion. For H_2SO_4 at all concentrations and moderate temperatures. Dilute HCl, strong chloride solutions at higher temperatures
Titanium	Chloride solutions, chlorine dioxide and other liquids for bleach manufacture
Non-metallic materials	Acids, alkalis, corrosive reagents and solvents at moderate temperatures. Some also resistant to solid erosion. Temperature and pressure limitations

Table 7.23 Materials for centrifugal pumps and their main areas of application

7.7.4 Corrosive effects of unknown materials

In some cases it is practically impossible to specify the liquid to be pumped. A typical example would be the various residues and effluent from the plating industry. In this case, the only known conditions may be that the liquid contains troublesome constituents of chlorides, acids and perhaps even hydrofluoric acid, and that the corrosion resistance of materials available at reasonable cost levels will be inadequate. The problem is further complicated if the liquid also contains solid abrasive particles.

Problems of this type have to be resolved by the use of common sense and a certain amount of experimentation during the early operational period. When the corrosive effects of liquids are uncertain, it is not only the pumps which will give problems. Pumps do not work without pipework. Pipework may have flanged or screwed connections. An early decision on the connection type and other hazards may help solve this dilemma and will prove very educational.

Many installations require tanks and these are often not easy to change once production has started. The pumps may prove to be a minor problem compared to the other parts of the installation. Here are some factors worth noting:

- Any particles should be separated if possible before the pump. If this is feasible, then various plastics for example would make a good choice in this instance. But if solid contaminants are present, then a metallic material should be chosen, such as silicon iron, chrome iron or the highest grade of high alloy stainless steel that is considered to be economically viable. Perhaps a rubber lined pump may be suitable.
- If possible, corrosion tests should be made on small material samples. The samples should be suspended where they can be retrieved and in such a way that galvanic corrosion does not take place.
- Choice of pump supplier should be made so that several alternative materials can be obtained for the component parts. In this way, individual parts or even the whole pump, can be replaced with better materials in the event of unexpected corrosion damage. Pump internals can usually be replaced easily. A long lasting casing can have many sets of internals. Internal pump corrosion does not lead to external leaks. The best casing and seal possible should be used.
- Since some experimental error may occur, the pump installation should be set up with both pump servicing and the consequences of breakdown firmly in mind.
- When practical operational results are available later, it will then be easier to optimise the economic choice of materials, balancing useful life and procurement costs of various materials.

7.8 Conclusions

Successful pump applications rely on good communications with the pump manufacturer. The system designer should hold preliminary discussions with several manufacturers about the most suitable pump type and the possible alternatives for materials. Pump manufacturers have plenty of experience with many liquids and are able to offer advice on many processes. As can be seen from Chapter 1, there are many different pump types. Materials that work successfully in one pump type may not be successful in a different pump type.

Pumps are never installed in isolation; there are always pipes and valves. Valve manufacturers also have much experience with many liquids. But a valve is not a pump and valve materials may not be the best choice. Accurate operating condition definitions are the keys to success. The sources in Section 7.9 are suggested for additional guidance on material selection.

7.9 Useful references

Meehanite Worldwide Corporation, IMMCO House, 38 Albert Road North, Reigate, Surrey, RH2 9EH UK, Tel: 01737 24478, Fax: 01737 226644, Email: immco@meehanite.fsnet.co.uk, www.meehanite.co.uk

ASME International, Three Park Avenue, New York, NY 10016-5990 USA, Tel: 973 882 1167, Email: infocentral@asme.org.

American Institute of Chemical Engineers (AIChE),3 Park Avenue, New York, NY 10016-5991 USA, Tel: 212 591 8100, Fax: 212 591 8888, www.aiche.org

American Petroleum Institute (API), 1220 L Street, NW, Washington DC 20005-4070 USA, Tel: 202 682 8000, www.api.org

American Iron and Steel Institute AISI, 1140 Connecticut Avenue, NW Suite 705, Washington, D.C. 20036 USA, Tel: 202 452 7100, www.steel.org.

Steel Founders' Society of America (SFSA), 780 McArdle Drive, Unit G, Crystal Lake, IL 60014 USA Tel:815 455 8240, www.sfsa.org.

The Institute of Materials, Minerals and Mining (IOM3), 1 Carlton House Terrace, London, SW1Y 5DB, UK, Tel: 020 7451 7300, Fax: 020 7839 1702, www.iom3.org.

Energy Institute, 61 New Cavendish Street, London, W1G 7AR, UK, Tel:020 7467 7100 Fax:020 7255 1472, www.energyinst.org.uk.

Institution of Chemical Engineers, Davis Building, 165-189 Railway Terrace, Rugby CV21 3HQ UK, Tel: 01788 578214, Fax: 01788 560833, www.icheme.org

The Plastics Institute of America Inc., University of Massachusetts, Lowell , 333 Aiken Street, Lowell, MA 01854-3686 USA, Tel:978 934 3130, Fax: 978 458 4141 Email:info@plasticsinstitute.org, www.plasticsinstitute.org.

The British Plastics Federation, 6 Bath Place,Rivington Street, London EC2A 3JE UK, Tel: 020 7457 5000, Fax: 020 7457 5045, Email: bpf@bpf.co.uk, www.bpf.co.uk.

NACE International (Formerly The National Association of Corrosion Engineers) 1440 South Creek Drive, Houston, Texas 77084-4906 USA , Tel: 281 228 6200, Fax: 281 228 6300, www.nace.org.

Rapra Technology Ltd (Independent plastics and rubber specialist organization), Shawbury, Shrewsbury, Shropshire SY4 4NR UK, Tel: 01939 252413, www.rapra.net.

Royal Society of Chemistry, Burlington House, Piccadilly, London W1J 0BA UK, Tel:020 7437 8656, Fax: 020 7437 8883, www.rsc.org.

American Chemical Society, 1155 Sixteenth Street, NW, Washington, DC 20036 USA, Tel: 202 872 4600, Fax: 202 776 8258, www.chemistry.org.

National Metals Technology Centre Ltd, Swinden House, Moorgate Road, Rotherham, Yorkshire S60 3AR UK, Tel:01709 724990, Email: info@namtec.co.uk, www.namtec.co.uk

Stainless Steel Advisory Service, Broomgrove, 59 Clarkehouse Road, Sheffield S10 2LE UK, Tel: 0114 267 1260, Fax: 0114 266 1252, Email: enquiry@bssa.org.uk, www.bssa.org.uk.

ASM International (The Materials Information Society), 9639 Kinsman Road , Materials Park, OH 44073-0002 USA, Tel: 440 338 5151, Fax: 440 3384634, Email: customerservice@asminternational.org, www.asminternational.org.

TEMA (Tubular Exchanger Manufacturers Association), 25 North Broadway, Tarrytown, NY 10591 USA, Tel: 914-332-0040, Fax: 914-332-1541, Email: tema@tema.org, www.tema.org.

National Centre of Tribology, ESR Technology Ltd, 410 Whittle House, Birchwood Park, Warrington, Cheshire, WA3 6FW UK, Tel: 01925 843410, Fax: 01925 843500, Email: nct.tribology@esrtechnology.com, www.esrtechnology.com.

Corrosion Resistance of Metals and Alloys, Monograph 158, F. LaQue, H. R. Copson, Reinhold Publishing Co., New York, 1963.

Uhlig's Corrosion Handbook, 2nd Edition, R. Winston Revie (editor), John Wiley & Sons, December 2005, ISBN 0-471-78494-X.

An Introduction to Corrosion and Protection of Metals, Gösta Wranglen, 1985, Frank Amato Publications, ISBN 0412260409.

Corrosion Data Survey — Metals Section, 6th Edition, Edited by D. L. Graver, 1985, NACE, USA, ISBN 0-915567-07-5.

DD 38:1974 Recommendations for cast materials for use in the pump industry, BSI. (Withdrawn but worth reviewing).

BS PD 5500:2006 Specification for unfired fusion welded pressure vessels.

BS 7929:1999 Guide for the verification of a pump casing design procedure.

EN 13445 Unfired pressure vessels.

EN 603-2:1997 Aluminium and aluminium alloys. Wrought forging stock. Mechanical properties.

EN 603-3:2000 Aluminium and aluminium alloys. Wrought forging stock. Tolerances on dimensions and form.

Process seals and sealing 8

8.1 Introduction

The seal is the most vulnerable element in a pump and it is well worth devoting a great deal of attention to its choice, style, installation and maintenance. By this means the cost for spare parts and repairs as well as leakage and down-time costs can be kept to a minimum. As well as the initial purchase cost, consideration must be given to the amount and type of leakage which **will** pass through the process seal while operating. The leakage, liquid or vapour, may pose a variety of hazards; toxicity, flammability, health, radioactivity, environmental; or may simply be unpleasant. If vapour is released it is essential to know whether the vapour is heavier or lighter than air.

The effects on personnel, local environment and global environment must be considered thoroughly. Many hazardous liquids are controlled by strict legislation. The penalty for illegal pollution may be very severe. The purchaser should be aware of all the implications of handling each specific liquid so that meaningful discussions can be held with the relevant safety authorities, the pump manufacturer and the seal manufacturer. The purchaser must inform the suppliers of any peculiar local regulations which are applicable.

The methods normally used for sealing shafts and reciprocating rods are:

- Lip seals
- Soft packings
- Mechanical seals, (rotary only)

In the special case when the liquid is so hazardous that leakage cannot be tolerated under any circumstances and the risk of seal failure is unacceptable, the following pump classifications in Chapter 1 should be consulted:

- Magnetic drive and canned motor rotodynamic pumps
- Peristaltic and rotary peristaltic pumps
- Diaphragm and air-operated double diaphragm pumps
- Metering pumps
- Direct-acting reciprocating pumps
- Ejectors

8.2 Rotary shafts

8.2.1 Non-contacting seals

A simple non-contacting seal can be constructed by passing the shaft through a tight clearance bore. The narrow passage acts as a long orifice and creates a pressure drop as liquid flows through. Some leakage must occur to create the pressure drop.

Ignoring the rotation of the shaft, at constant differential pressure the leakage rate is proportional to the cube of the clearance and inversely proportional to the viscosity and the length. Eccentricity of the shaft tends to increase the leakage. If the flow in the seal is laminar, eccentricity can more than double the leakage rate. Figure 8.1 shows how a non-contacting seal could be applied to a vertical pump. If the leakage can be returned to the pump suction, this type of sealing can be cost effective and durable.

The plain bore as shown is used as a back-up seal on some mechanical seal designs. In the event of mechanical seal failure, the throttle bush restricts the leakage rate by creating a high pressure drop.

The effectiveness of non-contacting seals can be greatly improved by using multiple seals connected by "sudden enlargements". This type of non-contacting seal is called a "labyrinth seal" and is used extensively in compressors and turbines. Having steps in the shaft, as well as the casing, further improves performance. When the shaft diameter is increased considerably, a flinger is created which prevents the axial motion of liquid by centrifugal force. Flingers are used in bearing housings to prevent oil leakage.

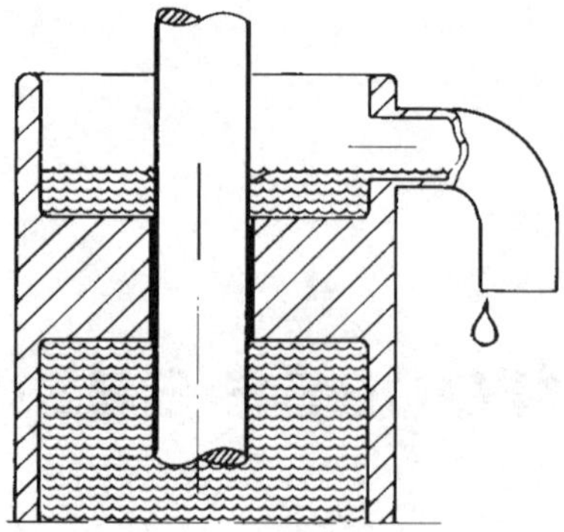

Figure 8.1 Shaft seal with radial clearance

8.2.2 Lip seals

Figure 8.2 shows an elastomeric lip seal which can be used at 0.2 to 0.5 bar differential at surface speeds up to 14 m/s. The lip is held in contact with the shaft by a garter spring. Designs are available which incorporate two lips on one seal. Seals can be manufactured without springs, which rely instead on the elasticity of the elastomer. Lip seals can be mounted in series for increased pressure capabilities; typically gear pumps for lubricating oil.

Most applications are as a bearing seal and secondary seal for outer quenching of a mechanical seal. However, new designs are capable of sealing 10 bar differentials and surface speeds of 40 m/s are possible. Many compounds are available, including PTFE, which is not a true elastomer. Some designs can be fitted with back-up rings, for additional support, and the back-up ring can include a non-contact throttle bush.

8.2.3 Hover seals

An elastomeric hover seal, see Figure 8.3, can be used for moderate pressures and where dilution by the sealing liquid can be allowed in the process. It tolerates relatively large radial movements and can be used for solid-liquid mixtures.

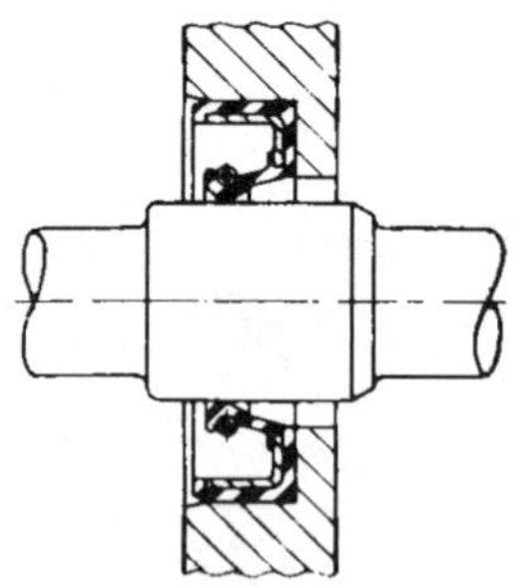

Figure 8.2 Elastomeric lip seal

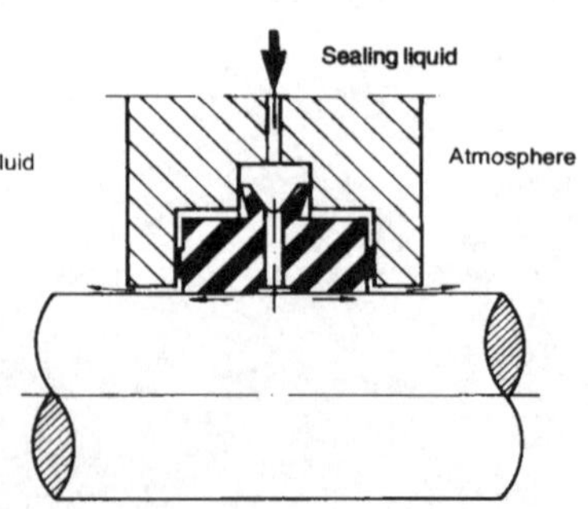

Figure 8.3 Elastomeric hover seal

8.2.4 Auxiliary pumps

An auxiliary pump can be constructed by fitting vanes on the back of centrifugal pump impellers, see Figure 8.4; only the impeller next to the seal or stuffing box need be modified. These vanes are sometimes called 'pump out vanes'. The pumping action of the vanes reduces the pressure at the shaft and discourages flow down the back of the impeller. Some form of back-up seal, on the shaft itself, is necessary for when the pump is stationary. An auxiliary pump is useful when handling liquids difficult to seal and for solid-liquid mixtures.

8.2.5 Soft packing

A stuffing box with a soft packing material is the traditional seal for pumps, the preload being applied by means of an axial compressive force, Figure 8.5. Stuffing box designs for rotating shafts and reciprocating rods/plungers are very similar but material selections can be significantly different. Another variant, Figure 8.6, applies pressure radially and produces a better loading pattern.

8.2.6 Mechanical seals

Up till about 30 years ago, most rotodynamic pumps were fitted with soft packed stuffing boxes. Mechanical seals were available, but not mass-produced and consequently costly. Manufacturing techniques and materials improved and simultaneously pump users were convinced of the operational advantages of reduced routine maintenance. Currently, probably 95% of all rotodynamic pumps built for process applications are fitted with mechanical seals.

In a mechanical seal, sealing takes place between a stationary and a rotating axial face, see Figure 8.7. The basic concept is simple but extreme accuracy is required for the seal to function. The important components are simple and relatively small and can be used in multiple sets to build complicated seal units for hazardous liquids.

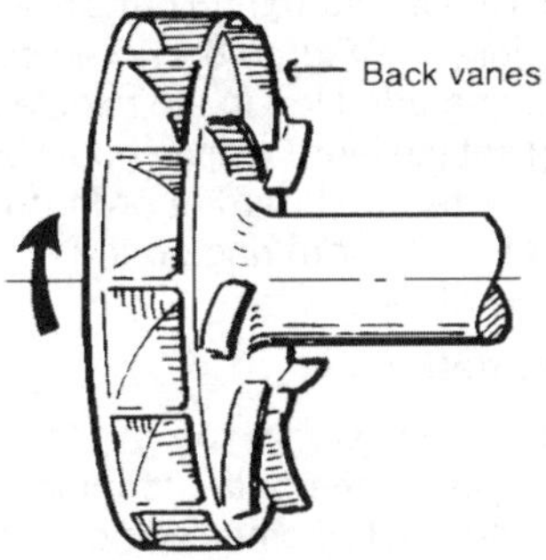

Figure 8.4 Auxiliary pump on centrifugal impeller

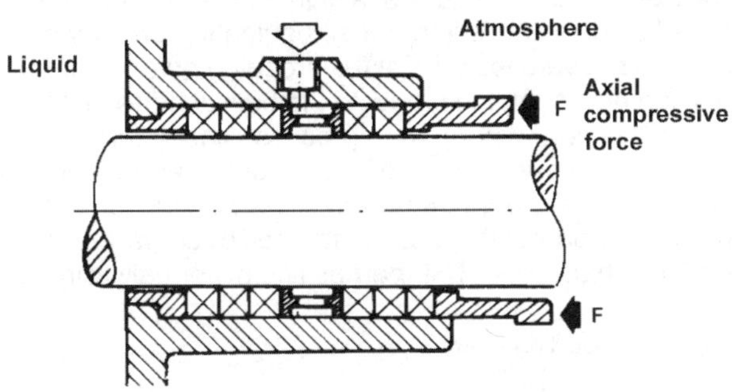

Figure 8.5 Traditional soft packed stuffing box

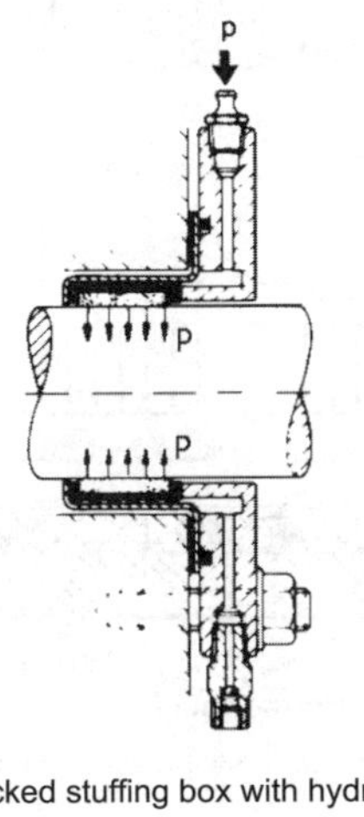

Figure 8.6 Modern soft packed stuffing box with hydraulic radial loading

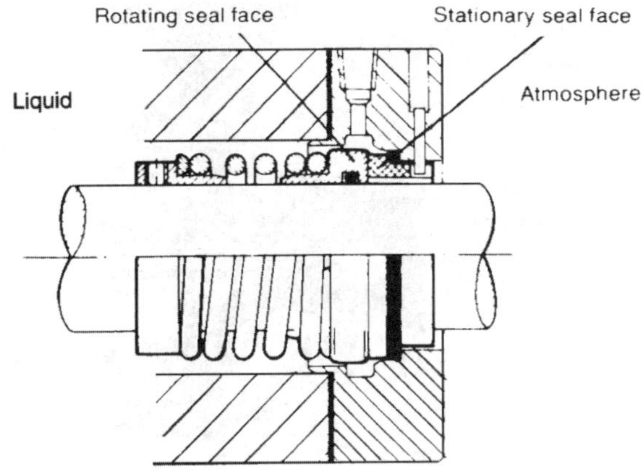

Figure 8.7 Mechanical seal with single driving spring

8.3 Process liquid seals for rotary shafts

A common feature of different sealing methods is the condition in the seal area or stuffing box, see Figure 8.8, which plays a critical part in the design of the seal. The operating conditions are described by:

- The process liquid with possible solid contaminants and gas content
- Temperature, which can deviate from that of the process liquid by use of various cooling/heating or flushing arrangements
- The shaft peripheral speed
- Liquid pressure at the shaft seal

The pressure at the shaft seal p_s, for a single stage pump, can be calculated in the following manner:

$$p_s = p_i + p_d \qquad \text{Equ. 8.1}$$

where:

p_i = inlet pressure to the pump

p_d = pressure differential within the pump between inlet and seal area, which varies according to the type of pump.

In the case of single stage, single entry centrifugal pumps having normal axial thrust balancing by means of a back clearance with relief holes or back vanes, p_d is approximately 30% of the pump's pressure rise.

For multi-stage pumps, the pressure at the seals is dependent upon the pump construction, i.e. the arrangement of the impellers and the type of axial thrust balancing employed. It is preferable not to have a high pressure seal and a low pressure seal. Identical seals reduce the spares inventory and reduce possible errors due to fitting similar parts in the wrong seal.

The pressure in the seal area p_s and the peripheral speed v are usually combined into a pv value, in the same manner as in plane bearing calculations. At the same time, the pv value pro-

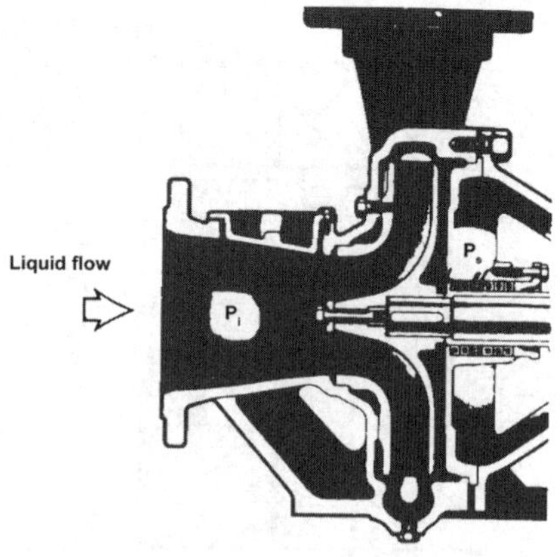

Figure 8.8 Pressure conditions in a centrifugal pump

vides a measurement for the frictional heat developed in the seal. In addition to the frictional heat generation, the temperatures at the surfaces of the seal itself are also dependent on the liquid temperature and the heat dissipation.

8.3.1 Soft packing

8.3.1.1 Operating principles

A soft packing set usually consists of two to six rings which are compressed by the axial tightening of a gland against the outside ring, see Figure 8.9.

A radial reaction pressure is exerted on the shaft and the bore of the stuffing box as a result of this compressive force. Due to the friction from, above all, the stuffing box, the radial reaction pressure is greatest at the gland and decreases towards the bottom ring.

Figure 8.9 shows a typical simple soft-packed stuffing box with a metal throat bush in the bottom of the stuffing box. In good stuffing box designs, a metal follower ring is fitted between the packing and the gland to promote even pressure distribution, circumferentially, around the shaft. The gland plate is usually attached by four studs. Adjusting packing evenly, via the four nuts, is a skilled task and it is very easy to produce an unbalanced axial load. It is also very easy to over tighten packing to reduce leakage to impractical levels.

The liquid under pressure will penetrate between the packing rings and the shaft to form small fluid reservoirs. There are therefore two different pressures to be considered, i.e. the liquid pressure in the reservoirs themselves, and the radial abutment pressure between the recesses. In the case of a soft packing charged with liquid pressure, the radial pressure varies as shown in Figure 8.10. The liquid pressure reduces gradually through the soft packings to atmospheric pressure at the gland.

A direct consequence of the distribution of the radial pressure along the length of the soft packing is that wear on the shaft is greatest at the gland, at A in Figure 8.10. If wear occurs at B then it is due to the action of abrasive particles in the liquid or shaft eccentricity.

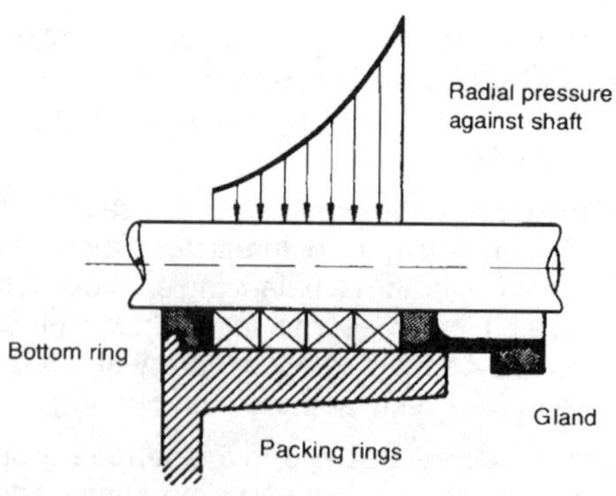

Figure 8.9 Soft packing with radial pressure distribution due to axial load

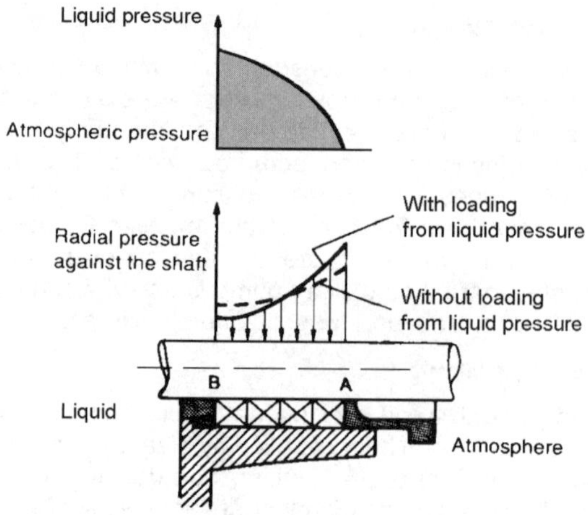

Figure 8.10 Pressure distribution in working soft packing

The extent of leakage depends on the radial pressure which in its turn is regulated by means of the axial compressive force. Leakage must always be present in order to remove frictional heat and to make up the liquid lost due to evaporation. The leakage rate in the case of working soft packing which is properly adjusted is 10 to 60 drops per minute.

Soft packing sets with a lantern ring have a somewhat different pressure distribution, see Figure 8.11. The barrier/buffer liquid or lubricant should be admitted in the middle of the box. In the case of a five ring set the lantern ring should be placed with two rings inboard and three outboard in order to avoid excessive axial movement of the lantern ring during adjustment, thereby blocking the supply of liquid.

8.3.1.2 Design variations

Before looking at different stuffing box arrangements in detail, it is worthwhile clarifying some of the terminology which is used when stuffing box design becomes complex. It is essential that all parties concerned describe the same function or component by the same terms.

- **Barrier liquid** — a liquid, at a higher pressure than the process liquid, which is introduced behind the primary seal to prevent outward leakage of the process liquid. A barrier liquid can also cool and lubricate the seal and prevent the ingress of air when the process liquid pressure is sub-atmospheric. As no process liquid can leak into the seal, a barrier liquid effectively protects against solids ingress. The barrier liquid is constrained by some form of secondary shaft seal. The barrier liquid normally circulates

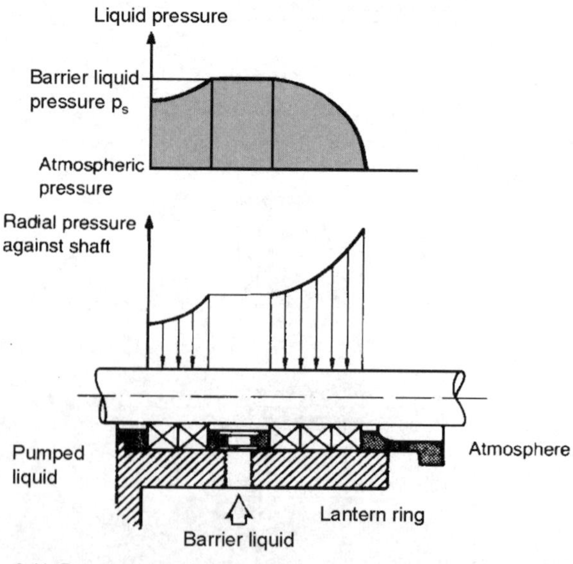

Figure 8.11 Pressure drop and radial pressure for soft packing with lantern ring

through an external piping system driven by a pumping ring on the pump shaft.

Pumping rings are only used with mechanical seals. Variable speed pumps may require an external motor driven pump for circulation to ensure adequate flow at low pump speeds. Soft packed stuffing boxes will require an external pump unless thermo-syphon circulation can be used.

- **Buffer liquid** — a liquid, at a lower pressure than the process liquid, which is introduced behind the primary seal to lubricate and/or cool the seal, and/or dilute any leakage from the primary seal. The buffer liquid is constrained by some form of secondary shaft seal. The buffer liquid normally circulates through an external piping system driven by a pumping ring on the pump shaft.
- **Flush liquid** — a flow of clean liquid, which is introduced in front of the primary seal, to cool and/or lubricate the seal or to prevent the ingress of solids which may be contained in the process liquid. The flow rate of the flush liquid is dependent upon the differential pressure available.
- **Lubrication** — a small flow of suitable lubricating liquid introduced behind the primary seal. The lubricant may mix with the process liquid and contaminate it. Some lubricant will leak past the secondary seal together with process liquid. All the lubricant is "lost".
- **Quench liquid** — a flow of cool liquid or steam, which is introduced on the atmospheric side of the last seal, to condense any escaping vapour or prevent the formation of crystals or solids which would interfere with the seal function. Alternatively, a flow of warm liquid or steam to prevent ice formation when the process liquid temperature is below 0 °C. A flow of cold water can be used to provide an atmospheric seal to prevent air ingress when the process liquid pressure is sub-atmospheric.

Plain stuffing box

The process liquid forms a fluid film and must therefore be clean in order to avoid unacceptable wear. This design, see Figure 8.12, can cope only with moderate temperatures where the dissipation of heat takes place by means of the leakage and conduction through the shaft and stuffing box. There is a risk of air being sucked in at low inlet pressures. Because process liquid leaks out of the packing, this type of packing arrangement is not suitable for any liquid which damages the environment. It may be suitable for some hazardous liquids depending upon the nature of the immediate surroundings.

The stuffing box arrangement as shown can be improved by the addition of a throat bush and a follower ring, see Figure 8.14. If a liquid quench was fitted, the problem of air entrainment at low pressure would be eliminated

Stuffing box with lantern ring

The arrangement in Figure 8.13 is also suitable for clean liquids. The lantern ring can perform several functions, such as barrier or buffer liquid or lubrication. The barrier or buffer liquid could be the process liquid taken from a tapping on the pump. Both of these options would solve air entrainment at low operating pressures. If the process liquid was not oil, a separate oil supply would be required for lubrication.

If the barrier liquid was "safe" this stuffing box arrangement could be used for hazardous liquids. The liquid supplied must be selected taking into account the process liquid and should maintain a pressure which exceeds the pressure at the seal by 1 to 1.5 bar. The quantity of barrier liquid which leaks into the process liquid is very small, a fraction of a litre/minute.

Flushed stuffing box

In the case of abrasive solid-liquid mixtures, the throat bush can be modified to allow the introduction of a clean flush liquid, see

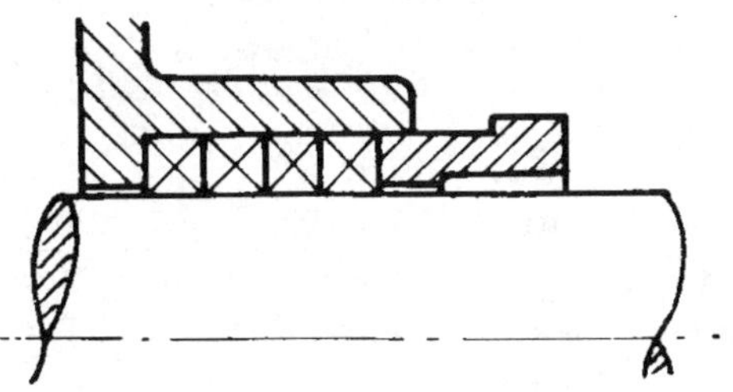

Figure 8.12 Plain stuffing box

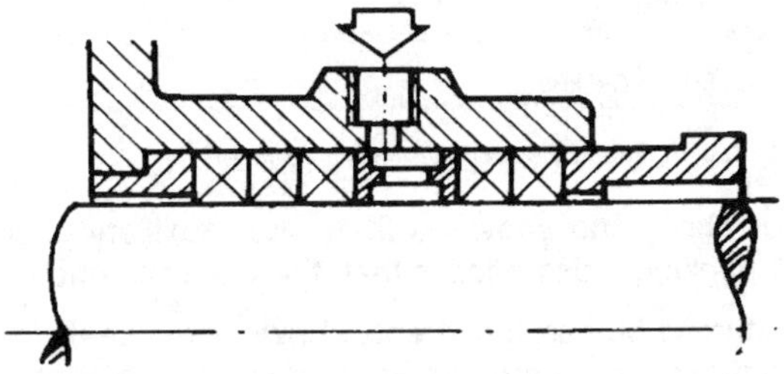

Figure 8.13 Stuffing box with lantern ring

Figure 8.14, in order to prevent abrasive particles from entering into the box. In this case the quantity of flush liquid which leaks into the process liquid can be considerable. The minimum flush required can be approximated by considering a velocity of 0.1 m/s through the throat bush clearance when worn.

Stuffing boxes can be fitted with lantern rings as well as flushed throat bushes. Contaminated hazardous liquids can be pumped or contaminated liquids requiring lubrication.

Stuffing box with cooling

With liquid temperatures in excess of 80 to 120 °C, depending upon the packing material, cooling of the stuffing box should be considered. Cooling of the stuffing box and shaft can be accomplished by the barrier or buffer liquid pumped through the lantern ring. However this method does not cool the bottom of the box or the front rings of packing. External cooling of the packing rings is relatively ineffective because of the poor thermal conductivity of the packing material.

The cooling chamber should therefore extend in front of the packing rings in the case of highest temperatures, see Figure 8.15. Cooling of the gland should also be provided if there is a possibility of vapour forming in the packing. Any vapour created in the packing will shorten packing life dramatically. Gland cooling prevents heat being conducted through the shaft to the front

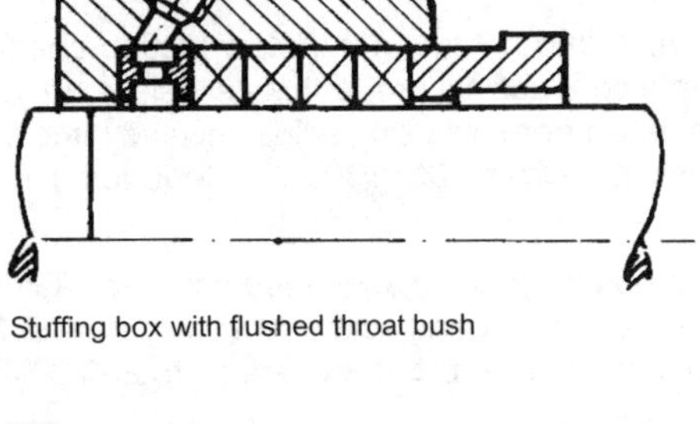

Figure 8.14 Stuffing box with flushed throat bush

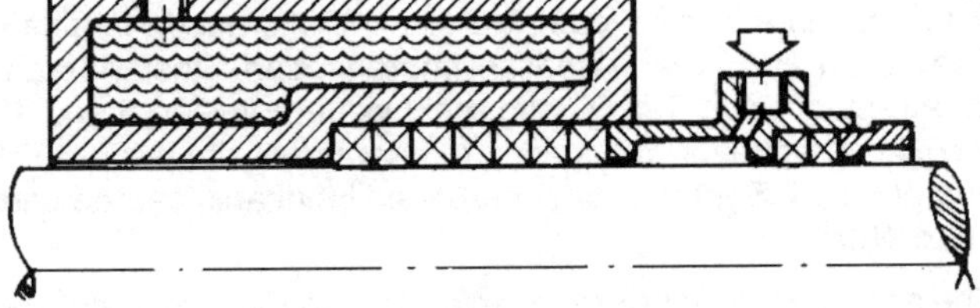

Figure 8.15 Cooled stuffing box with cooled gland

Packing	Fibre	Filler/coating	pH	Press bar	Speed m/s	Max °C	Min °C	Applications
Solid plaited	Flax	graphite	5 - 9	20	8	120		water, seawater
Cross plaited	Ramiex	PTFE	4 - 11		15	120	-30	water, brine, water-oil emulsions
Cross plaited	PTFE	PTFE/PTFE	0 - 14	10	4	250	-100	weak and strong acids, weak and strong alkalis, solvents
Cross plaited	PTFE	PTFE/PTFE/lubricant	0 - 14	25	10	250	-100	weak and strong acids, weak and strong alkalis, solvents
Cross plaited	Aramid	PTFE/lubricants	2 - 13		20	250		water, oils, solvents, medium strength acids, food
Cross plaited	Aramid/-Graphite	PTFE/lubricants	2 - 13	20	20	260		Water, chemicals, mild acids, alkalis, sewage, abrasive slurries
Cross plaited	Aramid	PTFE/graphite/lubricants	1 - 14	20	25	290		weak acids, weak alkalis, oils, abrasive slurries
Cross plaited	BCX	PTFE/lubricants	3 - 13	100	10	300		acids, alkalis, oils, solvents, water
Compressed foil	Graphite		0 - 14	20	30	500+	-200	acids, alkalis, solvents, hot oil and water

Table 8.1 Packing materials and operating parameters

bearing. The gland shown is fitted with auxiliary packing, a smaller packing cross-section than the main packing.

All the stuffing box arrangements shown, indicate the packing rubbing directly on the pump shaft. Wear of the shaft can be avoided by using a sleeve. The sleeve can be made of a different material to the shaft, a material with better wear and corrosion resistance properties. A pump with a carbon steel shaft could be fitted with a sleeve of hardened 12Cr steel or solid Stellite®, for example. When wear occurs, the sleeve is replaced, not the complete shaft.

8.3.1.3 Packing material

The most common packing material is plaited or braided fibre, but packing compounds and moulded rings are also used. Table 8.1 provides a survey of packing materials.

A braided packing consists of:

- A fibre, the basic material
- An optional coating on each fibre
- An optional impregnation or filler applied during braiding

More than one type of fibre may be used. Modern designs sometimes incorporate a strong central core and/or reinforced corners. Fibres currently in use include:

Vegetable fibres: hemp, flax, linen, cotton; used for temperatures up to 100 °C. Poor chemical resistance.

White asbestos and **blue asbestos** have a wide operating range and can be used for temperatures up to 500 °C. The use of asbestos is now forbidden by legislation in certain countries and asbestos has largely been replaced by other materials.

Synthetic fibres: Aramid®, Kevlar®, PTFE; have excellent sliding properties and are resistant to a very wide range of liquids. Maximum temperature about 260 °C.

Graphite fibre and foil has a very low friction coefficient and good thermal conductivity properties as well as good chemical resistance, apart from oxidising acids such as nitric acid. Maximum temperature from 500 to 3000 °C depending upon the liquid.

Glass fibre has largely replaced blue asbestos. Glass fibre is not resistant to hydrofluoric acid and strong alkalis. Maximum temperature about 480 °C but will reduce to 260 °C if PTFE lubricants are used.

In order to reduce friction the packing is impregnated with a suitable animal, mineral, vegetable or synthetic-based lubricant. The natural and mineral lubricant impregnation withstands temperatures of up to 125 °C and a pH value between 5 and 10. PTFE impregnation withstands temperatures of up to 260 °C and pH 0-14. Synthetic silicone-based lubricants can be used up to 480 °C.

The style of stuffing box required and the packing material necessary are both functions of all the operating conditions. The pump selection can be greatly influenced by operating conditions. No part of the pump application or design should be considered in isolation.

8.3.1.4 External systems

As stuffing boxes can have liquid or steam introduced from external sources, these obviously involve external systems. VDMA 24297 and API 610/EN ISO 21049 have standard arrangements designated for most applications.

The most popular stuffing box system for soft-packed boxes is actually an internal system with no external components. An internal drilling leads clean liquid from the pump discharge branch to the throat bush, a clean flush. This is classified as "A" in VDMA and Plan 01 in API.

If the process liquid was contaminated, or was a solid-liquid mixture, then feeding raw medium to the throat bush would be counterproductive. In these instances, a completely internal system is not viable. An externally piped system, including a cyclone separator to remove the solids, would be fitted; VDMA "B" or API Plan 31. In its simplest form an external system can consist of a piece of pipe and an orifice, Plan 11. This would be used when an internal drilling was impossible. VDMA 24297 does not differentiate between internal and external systems, only functions.

Most modern stuffing box systems are designed for mechanical seals. See Section 8.3.2.7 for a description of popular schemes. Pumping rings are usually not possible in soft-packed boxes, external pumps must be fitted if required.

8.3.1.5 Maintenance

Mechanical seals have replaced soft packing, in many applications, because of reduced routine maintenance. Once fitted, mechanical seals cannot be adjusted. See Section 8.3.2.7 about comments on strainers. Soft packing, on the other hand, must be inspected regularly and adjusted if required. Adjusting packing is a skilled job. Different applications and different packing types require different techniques.

Correct installation and bedding-in, after selecting the correct seal arrangement and materials, is a prerequisite of good packing life. It is difficult to assess the useful operating life of packing. In some hot applications, a good packing life will be three months. In some water pump applications, using large slow pumps, good packing life can be three years. In order to obtain the longest life possible the following procedure should be adopted:

- Read the pump manufacturer's instructions first.
  Some packing rings must be soaked in oil prior to installation. Failure to comply will inevitably shorten packing life.
- The stuffing box must be completely empty and perfectly clean before repacking.
- Inspect the old packing rings as removed, check for indications of overheating, extrusion and chemical attack. Note

the order of component removal from the box and check that the box was assembled correctly the last time.

- Check the throat bush and gland bores for wear. Replace if the clearance is greater than the manufacturer's recommendations.
- Check the surface finish of the shaft and the stuffing box bore. The shaft should be better than 0.4 μm R_a, 16 micro inches CLA, but not too good. Not better than 0.2 μm R_a, 8 micro inches CLA. The stuffing box bore should be better than 1.6 μm R_a, 64 micro inches CLA. Both surfaces should be free of scratches
- When replacing the throat bush, whenever possible lubricate the bore. Special lubricants may be required for hot applications, and definitely for food applications.
- Whenever possible, lubricate the shaft and stuffing box bore prior to fitting the rings.
- Fit each ring individually, tap into place with a piece of split pipe of approximately the same proportions as the rings. Lubricate lantern rings when fitted. Stagger ring joints.
- Grease or lubricate the gland threads before the gland is tapped into place. Tighten gland nuts only hand tight.
- Vent and prime pump if necessary, start pump and run for about 15 minutes before tightening gland. Tighten evenly by 1/6th turn about every 10 to 30 minutes until leakage is reduced. If a tapping to the lantern ring is available, oil while running.

Hard packing will bed-in much slower than soft packing. Hot applications will probably need the packing adjusted as the pump warms up. Hot pumps will probably need the gland loosened as the pump cools down for shut-down. In some cases the packing bedding-in will take up to a week.

8.3.1.6 Trouble-shooting

If operating problems occur, there is one very important action which must be done immediately — check the operating conditions. If the pump is not running at the correct speed, suction head, discharge head and temperature, data sheet conditions, then it may not be surprising that the seal gives problems. Check the process liquid. Is it the same, exactly the same, as the data sheet liquid? If all the operating conditions do not agree with the data sheet values then it is an upgrade problem not a trouble-shooting problem. If operating conditions are correct, Table 8.2 can be used for guidance.

NOTE: Only use spare parts supplied by the original equipment manufacturer. It is not possible to completely 're-verse engineer' a component from the finished part. It is easy to check the chemical composition and physical properties. It is impossible to detect if the material was vacuum remelted, or of X-ray or ultrasonic quality, or has the full heat treatment history. Dimensional inspection will not reveal whether the part is on top or bottom tolerance. Fitting parts, made by others, will invalidate any manufacturer's warranty.

The correct functioning of any external system must be checked regularly and liquid supplies for barrier, buffer and lubrication must be maintained.

Problem	Probable cause	Remedy
Packing begins to leak after a short period of operation.	Debris in process liquid or operating condition transients during start-up.	When process conditions have stabilised, disassemble and repack box.
Packing begins to leak after a short period of operation	Packing materials not suitable for operating conditions.	Check operating conditions against specification. Consult pump and packing manufacturer.
Packing begins to leak after a short period of operation	Packing materials not suitable for cleaning or stationary conditions.	Consult pump and packing manufacturer.
Packing extrudes into throat bush and/or gland .	Large clearance.	Check clearance, if too big replace throat bush and/or gland. Check life of component, is it short or adequate.
Packing extrudes into throat bush and/or gland .	Packing too soft.	Fit hard end rings to packing set. Glass reinforced PTFE is common. Metal rings with tight clearance can be used but must be a very good bearing material; phosphor bronze or cast iron.
Leakage from outside of gland.	Packing not fitted correctly.	Strip stuffing box and repack in accordance with manufacturer's instructions.
Leakage from outside of gland.	Packing wrong cross-section.	Fit correct packing size.
Leakage from outside of gland.	Stuffing box or gland worn.	Strip stuffing box and check dimensions and surface finish. Contact manufacturer for replacement parts if badly worn.
Packing scored on outside surface.	Packing wrong cross-section, rotating with shaft.	Fit correct packing size.
	Stuffing box worn.	Check stuffing box dimensions. If badly worn, oversize packing may be possible. Fit 1/4" instead of 6 mm.
Packing rings extrude or flow into adjacent rings.	Packing rings cut too short.	Fit die formed rings prepared by packing manufacturer.
	Packing too soft, or overheating slightly.	Fit hard spacers between rings.
	Slight attack by liquid.	Upgrade packing material.
Packing burnt or charred in the bore.	Dry running.	Prevent pump operating without process liquid.
	Lubrication failure.	Change packing type to grade with more or different lubrication impregnated. Fit lantern ring and lubricate or supply buffer liquid.
Packing seriously attacked, reduced in volume.	Packing incompatible with process liquid, or barrier, buffer, lubricant, when used.	Check process liquid quality and barrier, buffer, lubricants. Change packing grade.
Excessive leakage.	Packing attacked by liquid(s).	Check liquid(s). Change packing grade.
	Leakage through ring joints, rings too short.	Fit die formed rings prepared by packing manufacturer.
	Stuffing box incorrectly packed or reassembled.	Strip stuffing box and repack in accordance with manufacturer's instructions.
	Shaft eccentricity.	Check shaft run-out. Check pump bearings.

Table 8.2 Trouble-shooting guide for soft packing

8.3.2 Mechanical seals

8.3.2.1 Operating principles

The sealing of a rotating shaft passing through a pump casing is achieved by means of a rotating, optically flat seat, rubbing against a stationary, optically flat, seat with a fluid film in between them, see Figure 8.16. Sealing between the rotating seat and shaft or sleeve, and the stationary seat and gland, is usually performed by means of an elastomer such as an "O" ring, "V" ring, etc. The primary purpose of the spring is to provide an initial pressure between the sealing seats. One of the great advantages of this type of sealing is that it does not cause wear to the shaft or the sleeve.

The theory as to what happens between the seat faces is explained schematically in Figure 8.17. The seal should be considered as drip-free but not leak-free. The very low leakage rate, known as diffusion, will escape in the form of vapour on the atmospheric side of the seal.

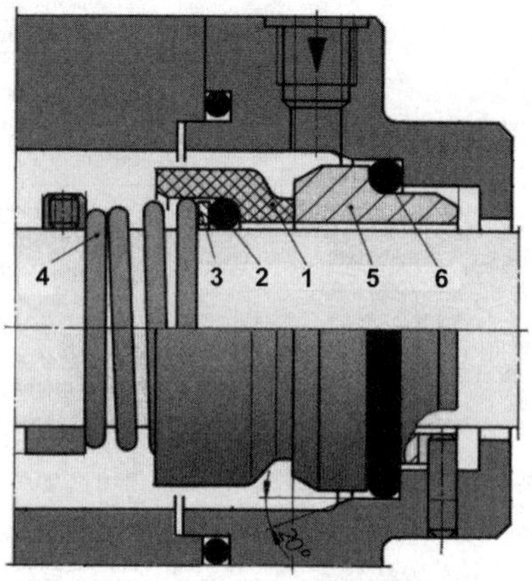

1 = Seal face (rotating) 4 = Helical spring
2 = Rotating "O" ring 5 = Stationary seat
3 = Thrust washer 6 = Stationary "O" ring

Figure 8.16 A simple unbalanced seal arrangement for a pump
Courtesy of Burgmann Industries GmbH & Co KG

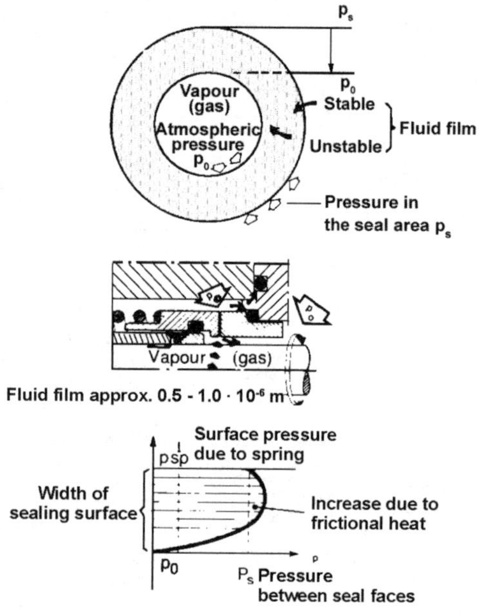

Figure 8.17 Fluid distribution across seat faces and operating pressures

Leakage through the seal faces in the case of mechanical seals amounts to 0.05 to 400 cm^3/hour depending on operating conditions, dimensions and choice of seal face materials. Manufacturers have formulae or graphs to evaluate leakage. No dripping, however, should be observed, only the formation of vapour which can lead to a certain amount of crystal formation.

A general and very important rule is that a mechanical seal should never be run dry, there must always be a fluid film between the faces. Dry running can occur even when the pump is full of liquid if the temperature is so high that the liquid vaporises in the stuffing box. Only specially designed seals can be run on the vapour phase of a liquid. Dry running seals are available for specific applications. The seal operating conditions must be fully described and include details of the vapour properties and the length of time the seal will operate without liquid lubrication or cooling.

At high pressures and high speeds the sliding surfaces must be relieved so that the fluid film is kept stable. The frictional heat must be limited at temperatures approaching the vapour temperature. In such cases, the face pressure is reduced hydraulically. This is done by balancing the ratio between the sealing diameter "S" and the inner diameter of the stationary seat. The pressure between the seal faces is thus varied according to the liquid pressure. Dependent upon the relationship between the

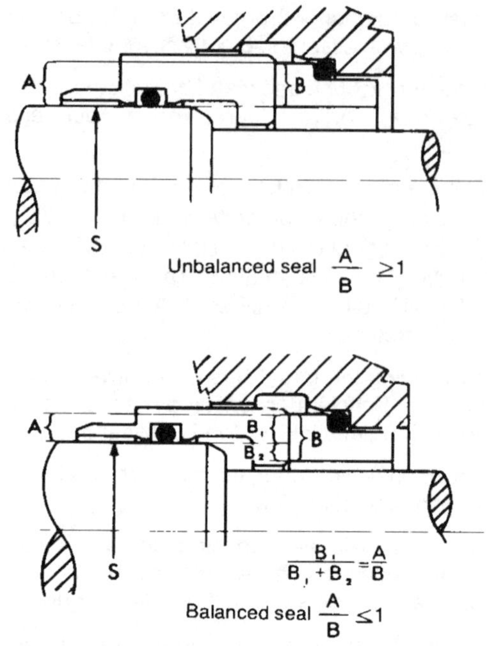

Figure 8.18 Hydraulic balancing of mechanical seals

two surfaces "A" and "B" in Figure 8.18, the seal is termed, balanced or unbalanced.

The limit between balanced and unbalanced seals is determined by the pv factor and the operating temperature. Balanced standard seals can be used for pressures up to about 16 bar at speeds of 20 m/s. In the case of higher pv values, hydrodynamic seals having a controlled leakage must be used. See also Section 8.7, Useful references.

8.3.2.2 Surface flatness

The flatness of the seat faces must be of a very high accuracy. The leakage is approximately proportional to the third power of the thickness of the fluid film which normally is in the order of 0.001 mm. Flatness is normally expressed in light bands, since checking normally takes place using an optical flat and monochromatic light, usually sodium light. A deviation of one light band corresponds to an out of flatness of 0.25 μm.

Normal standards of flatness are better than 1 to 2 light bands with variations up to approximately 5 light bands in the case of shaft diameters over 100 mm or in the case of special face materials. In addition to the number of light bands the light pattern must also be considered, see Figure 8.19. Large differences in seal tightness occur, for example, between a spherical surface, light pattern 3, and a wavy surface, light pattern 5. A check of seat faces is recommended as a standard measure prior to the installation of a mechanical seal.

Because of the accuracy of mechanical seal components, any work done on seals must be carried out in clean conditions. All components must be thoroughly cleaned prior to assembly.

8.3.2.3 Design principles

The generation of the initial sealing pressure, i.e. the initial face load is provided by using different types of springs. Single springs, multiple springs, Bellville washers and wave springs are the most common, see Figure 8.20. The drive, i.e. transmission of the torque to the rotating seat, can when single springs are being used, be carried out by the spring itself, separate driving elements used with other spring types. At least two driving points are necessary in order to guarantee a satisfactory loading of the seal faces.

The top illustration in Figure 8.20 shows a seal utilising two Bellville washers to provide the axial load. The rotating seat is positively driven by the shroud extending from the drive collar. The next picture shows the same type of seal but the axial load

Light pattern 1
The surface is flat within 1 light band - 0.25 μm. The distance between the bands does not tell us anything about flatness only about the size of the air gap.

Light pattern 2
The surface is almost flat but the edges have been rounded, probably in connection with hand polishing.

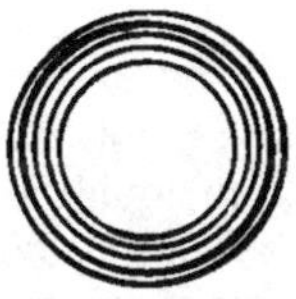

Light pattern 3
The surface is concave or convex to 3 light bands. This can be decided by pressing on the centre of the glass. If the rings then move outwards the surface is convex.

Light pattern 4
The surface is not completely flat. The tangent is drawn to the wavy bands. Half the number of light bands which are bisected by the tangent is a measurement of the flatness.

Light pattern 5
The surface goes in waves. It is difficult to achieve tightness. The cause can be residual tension after tightening in a chuck. The flatness fault is evaluated by comparison with light pattern 1.

Light pattern 6
The surface goes in waves. The same phenomena as light pattern 5 but 4 high points.

Figure 8.19 Seat flatness checks using light and an optical flat

is provided by multiple helical compression springs which are located in pockets in the drive collar. For this particular seal the manufacturer recommends multiple springs should be only considered for seals larger than 100mm.

The next illustration shows a rubber bellows style seal. The axial load is provided by a single helical compression spring which also provides the driving torque. In the event of spring problems some, or all, of the torque may be transmitted by the bellows and considerably shorten the life. The bottom illustration shows a metal bellows seal. Both the axial load and the driving torque are provided by the bellows.

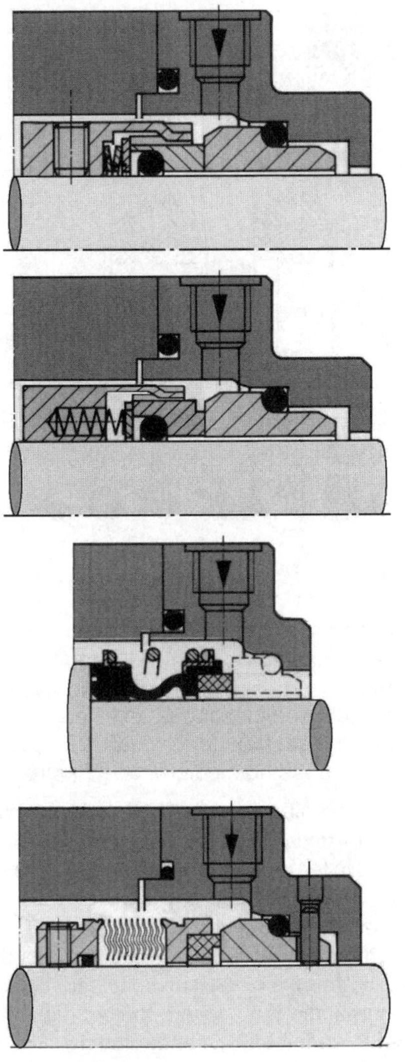

Figure 8.20 Examples of face loading and drive
Courtesy of Burgmann Industries GmbH & Co KG

Single springs should generally be chosen in preference to multiple spring types. The reasons for this are that driving and face loading can be combined in one element, that the single spring is robust and withstands corrosion, as well as the fact that the risk for blocking is less than in the case of small springs. In addition, the single spring is easier to handle.

The single helical spring has one major disadvantage; the spring handing must be changed if the direction of rotation is changed. For example, a double suction between bearings on a pump, with two stuffing boxes, would have a different seal at each end, i.e. different springs. To overcome this problem, many standard mechanical seals are fitted with Bellville washers. The springs are usually inside the stuffing box in the process liquid. Special mechanical seals can have the springs outside the liquid on the atmospheric side.

The shape of the rotating seat varies with the method of driving, engagement and the loading, as well as the materials of which the seal faces are made. In order to take up any vibration in the shaft, parallel deviations in the "O" ring grooves as well as any effects of the driving arrangement, the stationary seat should be mounted as flexible as possible. Figure 8.21 shows various methods of attachment. "a" is DIN standard construction and provides the greatest flexibility.

As shown in Figure 8.22 static seals are produced in a great many varieties. "O" rings dominate. "O" rings, type "a" are standard fittings, type "m" for temperatures up to 250 °C, wedge shaped seal rings are often used for high temperatures, type "c".

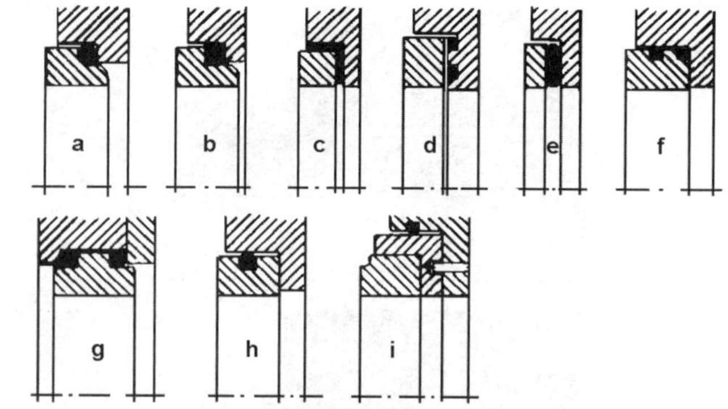

Figure 8.21 Examples of how a stationary seat can be fitted and sealed

Figure 8.22 Examples of sections of static seal rings

8.3.2.4 Temperature considerations

To achieve acceptable temperature conditions at the seal faces, frictional heat must be dissipated to avoid dry running due to loss of liquid between the faces. Heat dissipation takes place by means of heat transfer through the seats and by convection to the surrounding medium. See Figure 8.23.

Each seal and combination of materials has what is called a minimum ΔT in order to be able to function, see Figure 8.24. ΔT is the difference in temperature between the boiling point of the liquid at the seal cavity pressure and the actual temperature of the liquid. Usually ΔT should be at least 20 °C. The seal manufacturer can provide information for each type of duty and seal arrangement. By means of various installation options, internal and external circulation or cooling, the available ΔT can be increased, see the definitions at the beginning of Section 8.3.1.2. The required ΔT of a seal can also be reduced by using materials with good thermal conductivity or with low friction, etc. In such special cases the ΔT required can be reduced to approximately 5 °C. In Figure 8.24, the operation range of the seal is to the left of each respective speed boundary. The liquid is water. The distance between the vapour curve and the seal's limit is the required ΔT for each speed

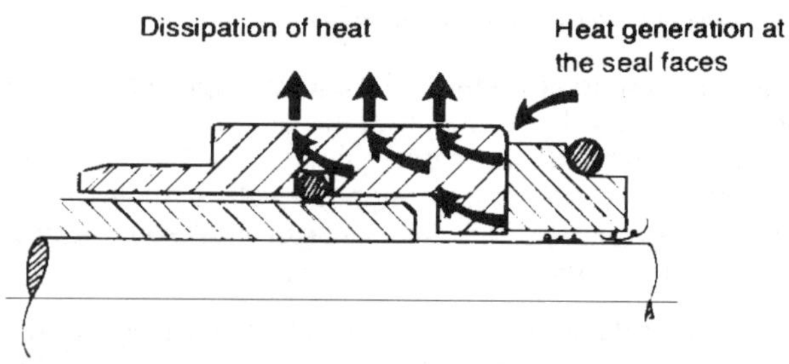

Figure 8.23 Heat dissipation in a mechanical seal

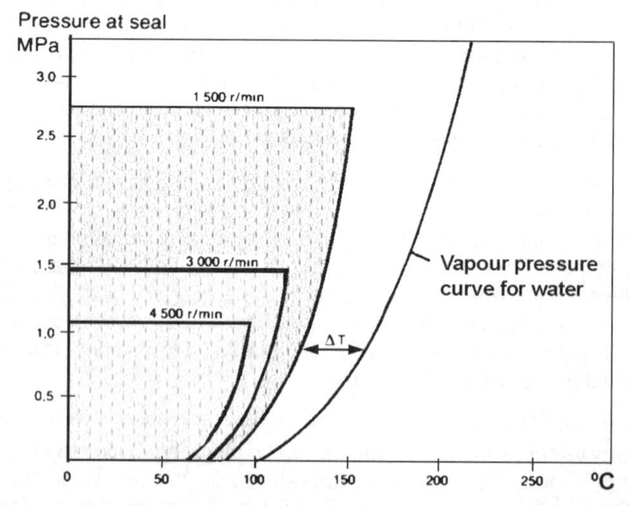

Figure 8.24 Pressure and temperature relationships at three different speeds for a balanced seal

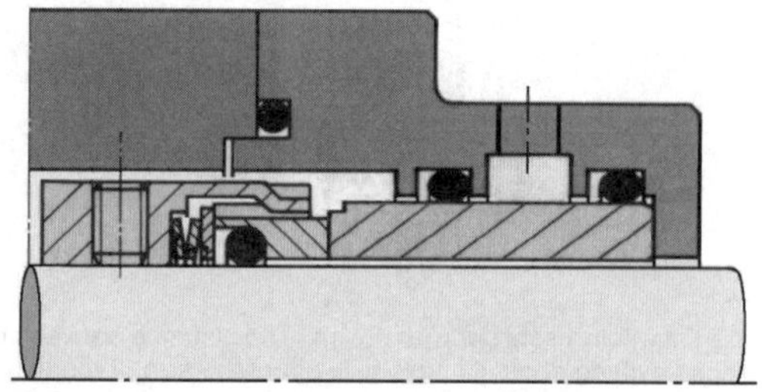

Figure 8.25 Seal without circulation but with external heating/cooling
Courtesy of Burgmann Industries GmbH & Co KG

8.3.2.5 Design variations

Single seal without circulation

A seal without circulation, see Figure 8.25, can be used in favourable cases such as clean liquid and moderate temperatures. Cooling or heating can take place in direct contact with the stationary seat, if ports are provided in the gland plate.

Single seal with circulation of pumped liquid

The purpose of the circulation is:

- To remove the frictional heat from the seats,
- To remove very fine particles and crystals from the area around the seal faces,
- To cool or to heat the seal cavity in relationship to the pump in general.

This is the most common seal arrangement and by varying the circulation system and the choice of material most liquids can be handled, both cold and hot, clean and slightly contaminated, see Figure 8.26.

Figure 8.35 also shows suitable external systems; Plan 11, Plan 12, Plan 21, Plan 22, Plan 31, Plan 52 and Plan 62. There are certain exceptions and they are described later.

Single seal with flushing

Plan 31 in Figure 8.35 is for heavily contaminated liquids and suspensions, but the cleaned process liquid is piped to a modified throat bush, not the seal cavity. The flow of liquid through the throat bush discourages solids from entering the seal cavity. An external clean liquid may be used when dilution in the process is permissible. See also Figure 8.14 which shows a flushing throat bush in a soft packed stuffing box.

Single seal for slurries without flushing

If dilution cannot be allowed due to process, technical or economic reasons, a balanced reversed seal can be used with hard-wearing surfaces, e.g. tungsten carbide versus tungsten

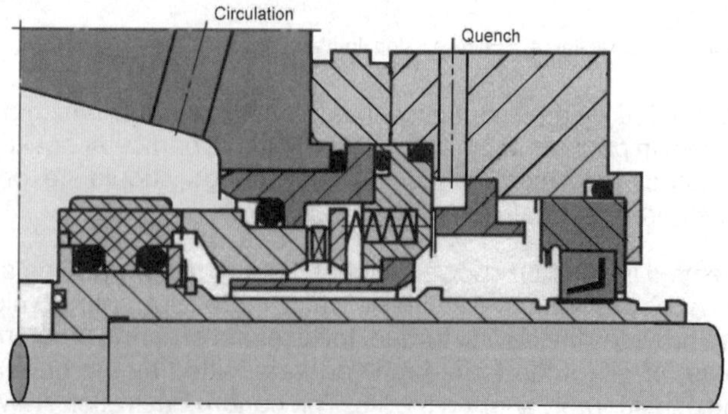

Figure 8.26 Seal with circulation or quenching
Courtesy of Burgmann Industries GmbH & Co KG

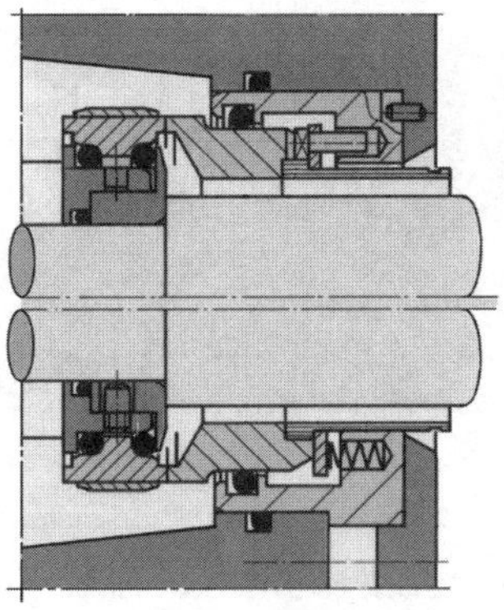

Figure 8.27 Balanced reversed seal for slurries
Courtesy of Burgmann Industries GmbH & Co KG

carbide, see Figure 8.27. The spring is located outside the slurry and all the parts in contact with the slurry are so designed that clogging is prevented. A Plan 31 system would still be beneficial. Quench for cooling and keeping the atmospheric side of the seal clean is also recommended.

Figure 8.27 shows two versions of the same seal design. The upper view shows a seal which is assembled from inside the pump when the impeller is removed. This style allows for a very simple seal cavity design without external bolting. The lower view shows the same seal for assembly from the low pressure side of the seal cavity; a gland plate is required. In both styles the rotating seat is mounted on a carrier which is clamped between the back of the impeller and a step in the shaft.

Single seal with quench

The quench, see Figure 8.26, can be used to:

- Remove dangerous liquids or gases which escape,
- Cool or heat the seal from the atmospheric side,
- Rinse away any wear powder and crystals,
- Lubrication for dry-running protection.

See also Section 8.3.1.2 and Plan 62 in Figure 8.35.

Double seals

Figure 8.28 shows a double seal, consisting of two seals usually mounting back-to-back and having a selected barrier or buffer liquid in between. The pressure of the liquid placed between the seals is dependent upon its precise function, see Section 8.3.1.2. A double seal involves extra equipment in order to pressurise, circulate and possibly cool or heat the liquid. Double seals are therefore only recommended where regarded as being necessary for protection of the pump, personnel or the environment.

Typical installations where double seals are required:

- suspensions and slurries in which the liquid is hazardous
- liquids which are toxic, radioactive or liable to explode
- liquids which have dissolved or entrained gases which are toxic or flammable

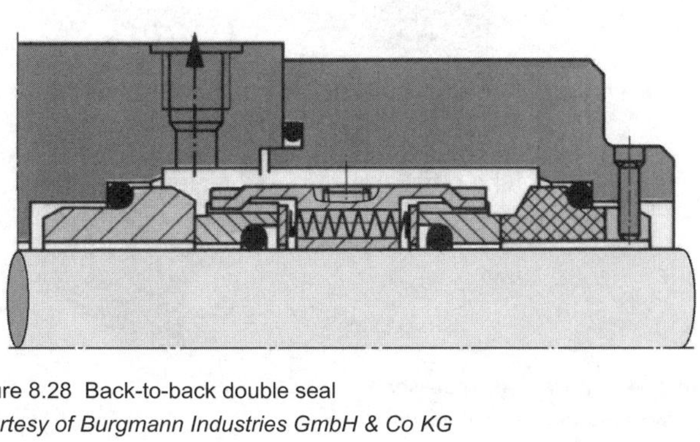

Figure 8.28 Back-to-back double seal
Courtesy of Burgmann Industries GmbH & Co KG

Figure 8.29 Typical back-to-back unbalanced double seal
Courtesy of Burgmann Industries GmbH & Co KG

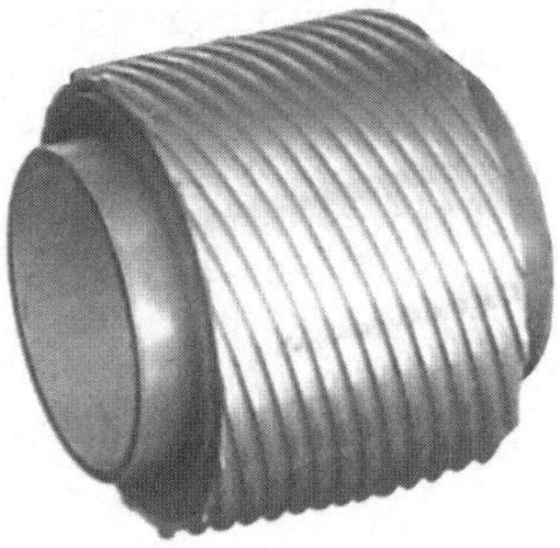

Figure 8.30 Typical pumping ring
Courtesy of Burgmann Industries GmbH & Co KG

- when there is a risk of severe crystal formation
- when the liquid is very corrosive to pump external components and the surroundings

Figure 8.29 shows the double seal components. The driving slots, where the drive collar transmits the torque to the rotating seats, are easily seen. When it is necessary to circulate a barrier or buffer liquid through the seal cavity the outer diameter of the drive collar can be covered with screw helices to form a screw pump, see Figure 8.30. The screw pump formed is generally known as a "pumping ring" or a "pumping screw". The action of the pumping ring is dependent upon the direction of rotation. For double-ended pumps the seals may be identical except for the drive collar and pumping ring.

Double seals can be fitted facing the same way i.e. tandem seals. In extreme cases, three seals can be fitted. For all hazardous liquid applications, the advice and experience of the seal and pump manufacturer should be obtained.

Single seal of bellows type

Used for very corrosive liquids and hot or cold liquids. The parts in contact with the liquid are made of ceramic, elastomer or ceramic and CrNi steel, see Figure 8.31. Elastomer bellows can be made from natural rubber (NBR), EPDM, Viton® or PTFE. Pressures up to 12 or 16 barg can be accommodated while operating between -20 and 120, or sometimes 140 °C. Metallic bellows seals can operate over a much wider temperature range, typically, somewhere between -100 and 400 °C. Pressures are generally limited to 25barg.

For PTFE bellows, pressure can be up to 6 bar with a temperature range of between -20 °C and 120 °C. For metal bellows seals, pressures go up to 16 bar with a temperature range from -100 °C to 350 °C. Typical applications include asphalt, bitumen and proprietary heat transfer fluids.

Cartridge seals

All the mechanical seal designs reviewed so far require "setting" or "adjusting" during assembly. Seal assembly is performed on the pump which may be outdoors under unfavourable weather conditions or local pollution.

Also it is usual for the pump manufacturer to supply the gland plate. Communications between the pump and seal manufacturers must be good. However, seal assembly problems first

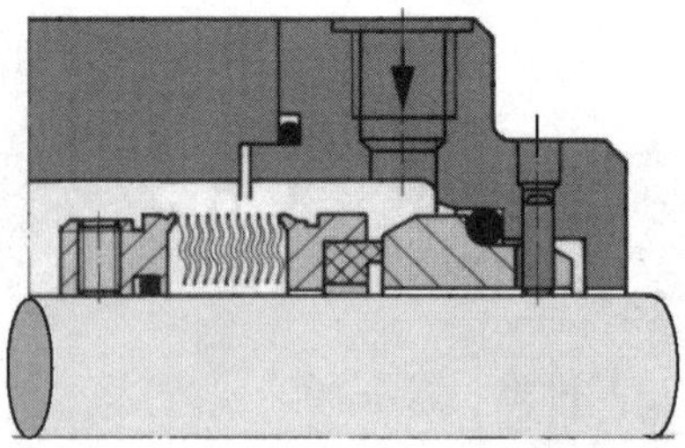

Figure 8.31 Chemical bellows seal
Courtesy of Burgmann Industries GmbH & Co KG

appear in the pump manufacturer's works where any problems can be quickly rectified. Many pump users, including most oil and gas users, thought the use of cartridge seals could significantly improve pump reliability. For an indication of general process equipment reliability look at Tables 13.2 to 13.6. Removing the necessity for seal adjustments was seen as a major improvement. When viewed in the correct perspective seal adjustment is a very small part of a much bigger operation which may include dismantling most of the pump. The reliability of a pump operating at site is controlled by the actions of:

- The purchaser
- The system designer
- The piping designer
- The pump manufacturer
- The seal manufacturer
- The installation team
- Process operators and control room staff
- Maintenance staff

The use of cartridge seals started to become widespread when ISO 21049/API 682 specified them as a minimum requirement. A seal, supplied as a fully assembled subsystem, can be tested before fitting to the pump and after repair. Seal inspection and repair can be performed in the workshop under controlled cleanliness conditions. The cartridge design is very good for complex seal arrangements. Site staff do not have to work outdoors with a number of components and little springs, trying to assemble a seal in the bottom of a hole. (Rather like trying to assemble a watch when you have both hands and feet inside the case!) ISO 21049/API 682 only permits the use of cartridge seals which can be fitted as complete assemblies. Figure 8.32 shows a balanced double cartridge seal suitable for oil and gas applications up to 150 barg and 150 °C.

The seal arrangement shown is very different from all of the seal styles shown previously. Most of the seal is not in the "pump" but within a seal cavity which is bolted on to the pump stuffing box. After the seal assembly is located correctly and bolted securely to the stuffing box the seal sleeve can be attached to the pump shaft. The seal settings are maintained by "transport lugs" (the grey strips bolted to the gland plate which engage in the slot in the sleeve). The seal shown uses a clamping ring to compress the sleeve on to the shaft. When the sleeve is attached the transport lugs can be disengaged and the seal installation is complete. The double seal shown has a throttle bush to reduce leakage from the drain connection.

Split seals

One of the problems with seal and soft packed stuffing box maintenance is dismantling the pump until a bare shaft end is exposed. It may be necessary to remove a bearing housing and the coupling. Seals which can be mounted "across" a shaft rather than "along" it could save much time. Soft packing can be split so that the wearing part, the packing rings, can be replaced without major dismantling. Mechanical seals can be "semi-split"; that is the normal wearing components can be split for rapid replacement. Sufficient axial clearance must be available, when the gland plate is withdrawn, to allow the radial displacement of the longest component; probably the stationary seat. For a 50mm seal the stationary seat is 32mm long; this length is maintained up to 175mm seals.

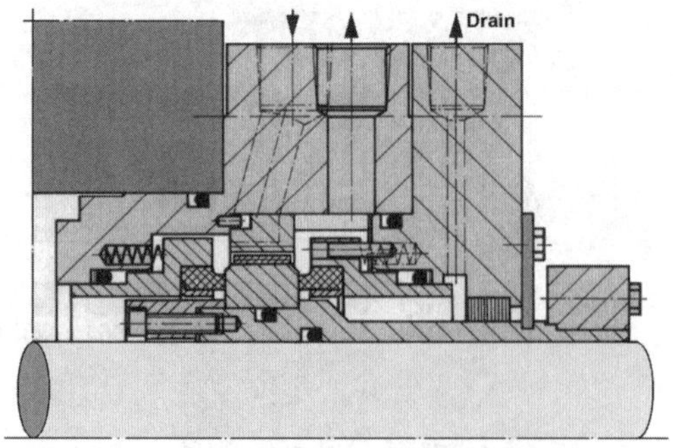

Figure 8.32 A balanced double cartridge seal with pumping ring
Courtesy of Burgmann Industries GmbH & Co KG

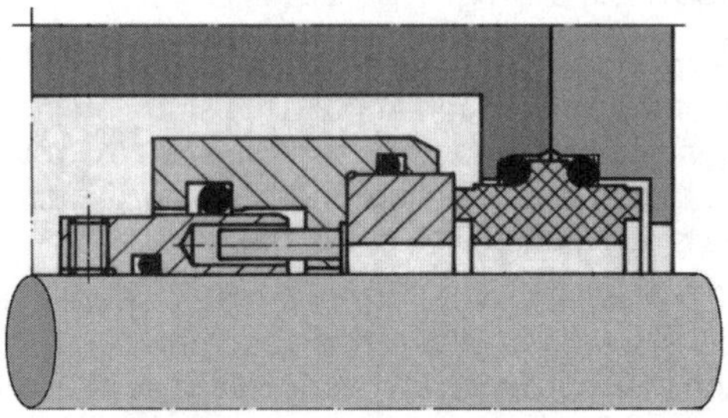

Figure 8.33 A typical semi-split balanced single seal
Courtesy of Burgmann Industries GmbH & Co KG

Figure 8.33 shows a typical single balanced seal with split rotating and stationary seats. Split "O" rings are used on both seats. The stationary seat is double sided. Semi-split seals can be used up to 40 barg and 150 °C. Completely split single seals are available. The gland plate and all the internals are split. Gland plates cannot be attached to the pump stuffing box by studs; bolts must be used. Split single seals can handle 6.9 barg at up to 120 °C, see Figure 8.34.

Gas seals

Gas seals for pumps are available. A gas seal can be fitted as the second seal when a double seal is used. This arrangement is feasible when the pressurised liquid within the pump will change to vapour when the pressure is reduced. A throttle bush is fitted between the seals to restrict the flow of product to the second seal. Under normal operating conditions the small vapour leak from a pair of liquid faces is sufficient to allow a gas seal to operate satisfactorily. The interspace pressure, between the seals, can be monitored. If the primary liquid seal begins to fail the increased leakage will create a higher interspace pressure; this can be used to trigger an alarm. If the interspace pressure rises sufficiently the interspace will fill with liquid and

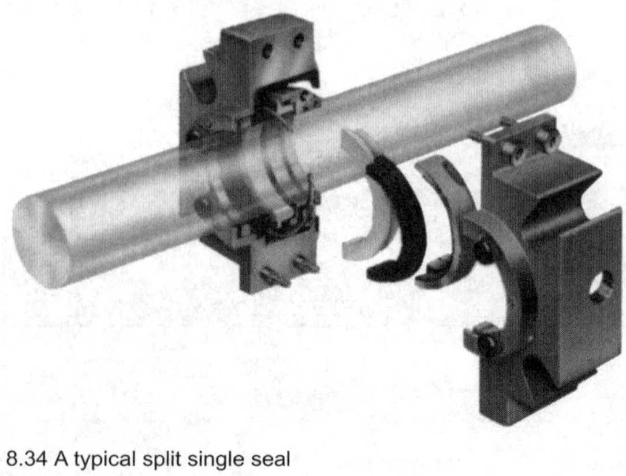

Figure 8.34 A typical split single seal
Courtesy of Flowserve Corporation

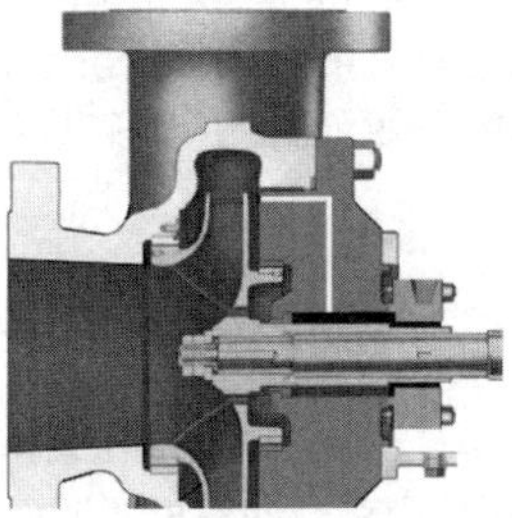

API 682 Plan 01 / ISO 5199 Plan 01 / ISO 13709 Plan 01

Internal recirculation from pump discharge to seal cavity.

Recommended for clean liquids only.

Internal drillings should have an orifice facility included to adjust recirculation flow to maintain stable seal face conditions for the specific operating conditions.

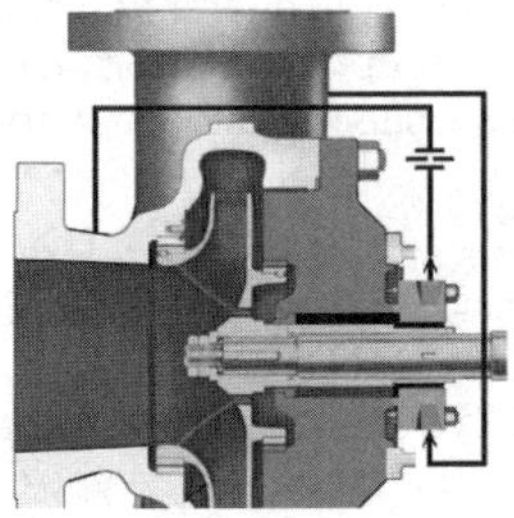

ISO 5199 Plan 03 / ISO 13709 Plan 14

Recirculation from pump discharge through seal cavity and return to pump suction.

ISO 5199 does not show orifice plates.

ISO 13709 is uncertain whether one or two orifices are necessary.

An orifice in the suction return line would be best.

Recommended for clean liquids.

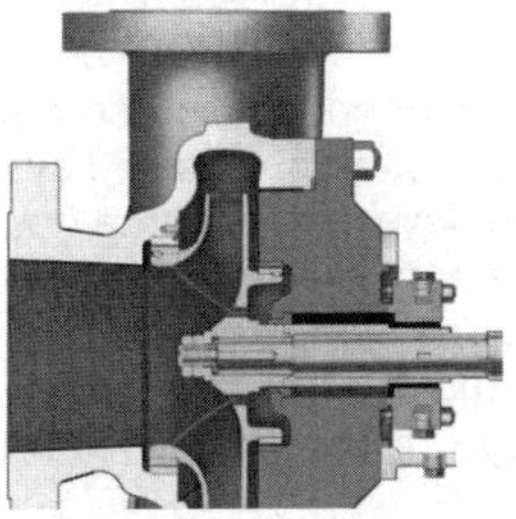

API 682 Plan 02 / ISO 13709 Plan 02

Seal housing or gland plate with two plugged connections for use with possible future circulation system.

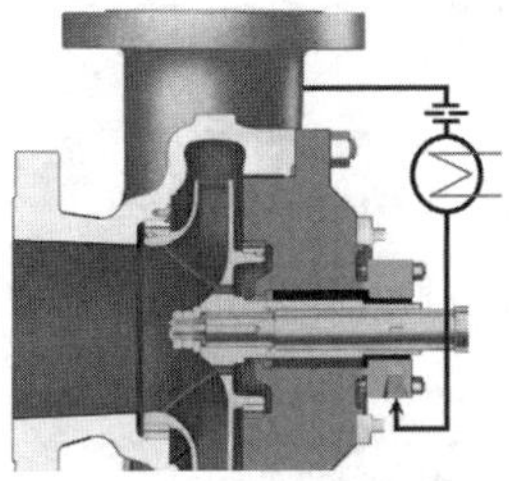

API 682 Plan 21 / ISO 13709 Plan 21 / VDMA 24297 Plan 6.3

Recirculation from pump discharge through a flow control orifice and a cooler to the seal housing or the gland plate.

ISO 5199 Plan 06 does not show any flow control device.

API 682 Plan 22 / VDMA Plan 6.4 or 6.8 are better because filtration is included.

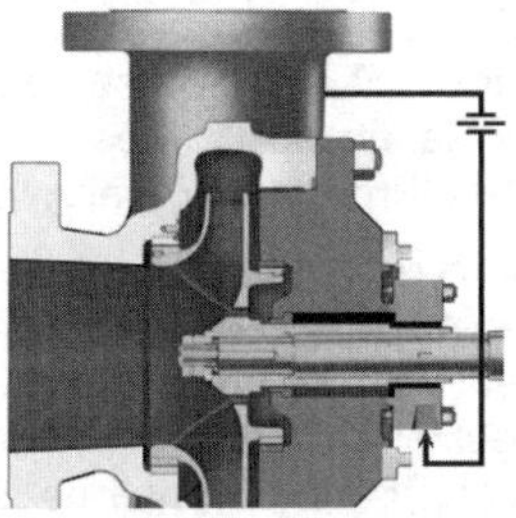

API 682 Plan 11 / ISO 5199 Plan 11 / ISO 13709 Plan 11 / VDMA 24297 Plan 6.1

Recirculation from pump discharge through a flow control orifice or device to the seal housing or gland plate.

ISO 5199 does not stipulate the orifice.

VDMA 24297 stipulates a flow control device.

The flow into the seal area is intended to enter close to the seal faces and then flow along the shaft into the pump.

Recommended for clean liquids.

Orifices are good flow control devices when the operating conditions are well defined (there is nothing for operators to adjust).

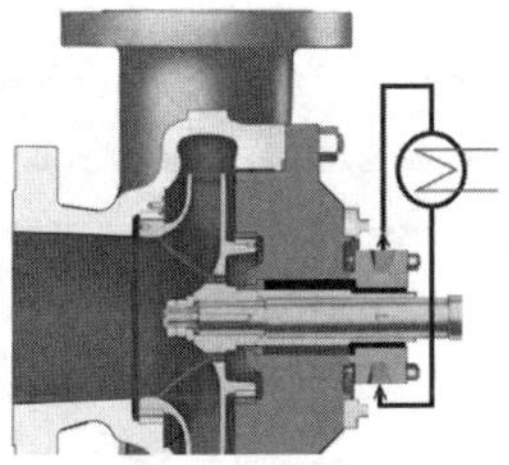

API 682 Plan 23 / ISO 5199 Plan 06 / ISO 13709 Plan 23 / VDMA 24297 Plan 6.5

Recirculation of seal cavity liquid by pumping ring through external cooler.

Filtration should be included to prevent any solids from accumulating in the cooler.

In some cases a thermo-siphon circulation may be possible removing the need for a pumping ring. Thermo-siphon systems won't be able to include filtration.

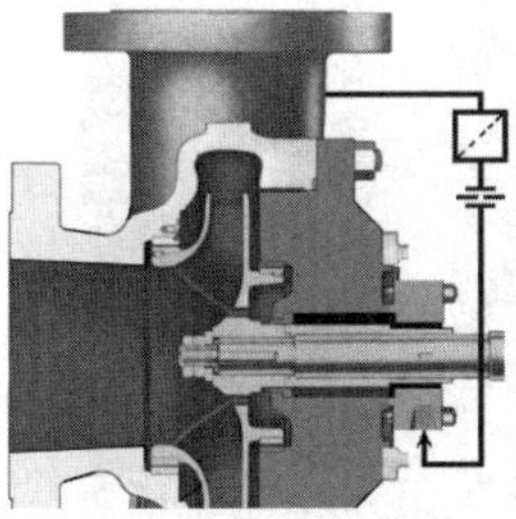

API 682 Plan 12 / ISO 13709 Plan 12 / VDMA 24297 Plan 6.2

Recirculation from pump discharge through strainer and flow control orifice or device to the seal housing or gland plate.

Recommended for clean liquids which do not normally contain solids.

Use API 682 Plan 31 / ISO 5199 Plan 05 / ISO 13709 Plan 31 which include a cyclone separator instead of a strainer for liquids which always contain solids. In this case the pump should be protected by a strainer or filter to remove oversize solids. The reject line from the cyclone separator should be piped into the suction before the pump strainer/filter. API 682 Plan 41 / ISO 13709 Plan 41 / VDMA Plan 6.4 include a cooler.

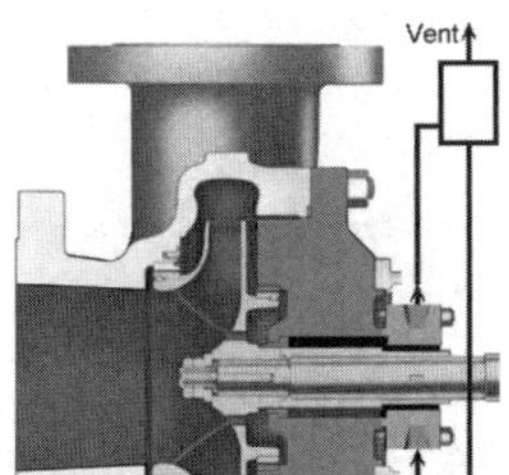

API 682 Plan 52 / ISO 5199 Plan 10 / ISO 13709 Plan 52

Pumped circulation of a buffer liquid contained in an unpressurised external reservoir.

Used with double seals and an unpressurised seal cavity.

The reservoir would normally be vented to a closed drains system or a flare stack.

ISO 5199 Plan 12 and VDMA 24297 Plan 6.9 use a vessel pressurised by the pump discharge; API 682 Plan 53 / ISO 13709 Plan 53 use an external pressure source; to create a supply of barrier liquid.

Note: No instrumentation or valving is shown. Only the connections for the piping plan described are shown. Piping plans can be combined for more complex systems.

Figure 8.35 Mechanical seal piping arrangements

the backup seal can operate as a normal liquid seal. The pump can be allowed to run until it is convenient to examine the primary seal.

The advantage of this arrangement is the lack of barrier or buffer liquid systems. This type of sealing can only be applied when the seal leakage is not considered as pollution or hazardous to the local environment. Gas seal faces are machined differently to normal liquid seal faces and are not interchangeable.

Secondary gas seals can be used with a buffer gas supply to prevent harmful vapour leakage. During normal operation it is buffer gas, usually nitrogen, which leaks through the secondary seal. If the primary seal fails some product vapour will escape. This may be allowable as a transient condition until the seal can be repaired. If high pressure gas is available the interspace can be pressurised with a barrier gas which prevents all product leakage. The pump user should check the site facilities for nitrogen supplies at the correct pressure.

The advantages of gas secondary seals should be considered carefully. It is difficult or impossible to check the condition of the secondary seal while the primary seal is functioning correctly. When the secondary seal is required there is no certainty the seal will function adequately. If the liquid is subject to any type of legislative control for emissions the regulatory authorities are unlikely to be impressed by secondary gas seals.

8.3.2.6 Materials

The material of the two seat faces can be the same or different. The combination of materials is determined, amongst other factors, by wear, strength, friction conditions, lubrication, chemical resistance and thermal conductivity.

Seal face materials	Combination qualities	Typical applications
Cr Mo steel alloy - carbon/graphite	Short-term dry running possibilities	Clean liquids, moderate pressures, moderate speeds
Ceramic - carbon/graphite	Short-term dry running possibilities	Clean and slightly contaminated water, water solutions, sea water, chemicals
Tungsten carbide - carbon/graphite	Short-term dry running possibilities, suitable for high temperature/pressure	Clean and contaminated liquids
Ceramic - PTFE	Very broad corrosion resistance	Clean liquids, not too hot
Silicon carbide - carbon	Short-term dry running possibilities, suitable for high temperature/pressure	Clean and contaminated liquids
Silicon carbide - silicon carbide	Very broad corrosion resistance, abrasion resistant	Clean and contaminated liquids
Silicon carbide - tungsten carbide	Very broad corrosion resistance, abrasion resistant	Clean and contaminated liquids
Tungsten carbide - tungsten carbide	Very broad corrosion resistance, abrasion resistant	Clean and contaminated liquids

Table 8.3 Properties of seal face material combinations

Table 8.3 shows some common combinations for the two seal faces. Tungsten carbide and ceramics can either be used in solid form or as a surface coating on, for example, stainless steel.

Material	Temperature of pumped liquid °C	Sealing element	Typically resistant to
Nitrile rubber a Low nitrate b Medium nitrate c High nitrate	 - 40 to + 80 - 35 to + 90 - 30 to + 100	O-rings, bellows, sleeves, etc	Water, mineral oils and other petroleum products
Ethylene-propylene rubber	- 40 to + 150	ditto	Animal and vegetable oil, chemicals. Not suitable for petroleum products
Silicone rubber	- 60 to + 200	O-rings	Weak and oxidising chemicals
Fluorocarbon rubber	- 5 to + 225	O-rings, bellows, sleeves, etc.	Chemically resistant with certain exceptions - e.g. ketones
PTFE Polytetrafluoro-ethylene	- 100 to + 250	ditto	Chemically resistant
Fibre reinforced PTFE	+ 150 to + 250	Sleeves, O-rings, etc.	Hot oils, asphalt, etc.
Perfluorelast (Kalrez)	- 30 to + 400	O-rings	Chemically resistant

Table 8.4 Elastomers for mechanical seals

Elastomers of different types, see Table 8.4, are used as flexible elements and for static seals. The material names used in Table 8.3 are generic. Each material may have up to five variants to optimise specific properties. The seal manufacturer can easily select the best variant(s) if accurate operating conditions are provided.

8.3.2.7 External systems

Mechanical seals can be fitted with various external systems to enable the seal to cope with the operating conditions. External systems are designed to protect the seal faces from adverse conditions and, in some cases, to assist in the sealing function.

Some popular system plans EN ISO 21049/API 682, ISO 5199, ISO 13709 and VDMA 24297 are listed in with illustrations to indicate the piping arrangements, Figure 8.35.

Seal piping arrangements, or seal harnesses as they are also known, can be quite complicated when the process liquid is hazardous. Figure 8.36 shows a multi-stage radially-split pump during manufacture. The pump is double-ended and so has two seals. The pressure vessel reservoirs for the Plan 53 systems are easy to see. To maintain the drive-end seal the coupling and the bearing housing must be removed. To maintain the non-drive-end seal the shaft-driven oil pump and the bearing housing must be removed. For a multi-stage pump of this style the seals are identical in theory but are manufactured slightly differently; the seals are handed.

To provide a sensible piping arrangement all the seal piping is confined to one side of the pump. One seal assembly is left-hand the other right-hand. The seal internals are identical except for the pumping ring; it is handed as well. Seal assemblies for double-ended pumps must be clearly marked as DE and NDE to avoid assembly mistakes. Figure 8.37 shows a typical Process and Instrumentation Diagram (P & ID) for the seal assemblies. The sealing systems are identical and separate. Notice that cooling is implemented by a coil inside the reservoir.

None of the standards and specifications reviewed include details of vent and drain pipework; this pipework is assumed to be fitted. Seal cavities and gland plates, supplied by the seal manufacturer usually have screwed connections. Clause 2.3.16 of API 682 states "Gland plates and seal chambers shall have provision for only those connections required by the seal flush plan." None of the popular seal systems mention vents or

Figure 8.36 A typical multi-stage pump seal installation
Courtesy of Flowserve Corporation

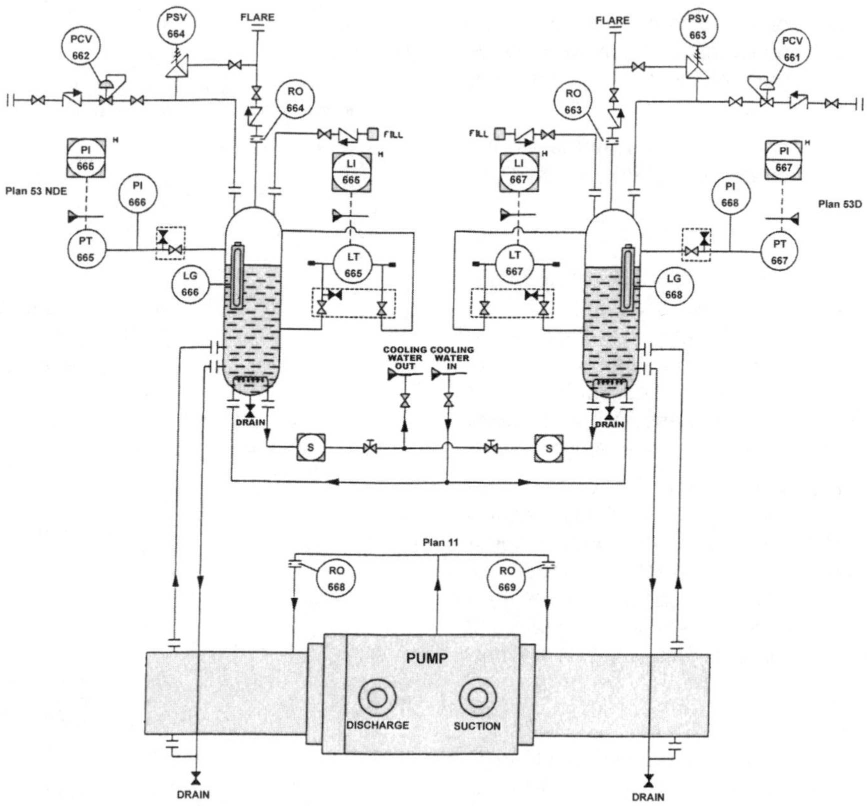

Figure 8.37 A typical double plan process and instrumentation diagram
Courtesy of Flowserve Corporation

drains for the seal cavity. However, the clause continues "If additional tapped connection points are specified..........". This is one of API's illogical clauses. This point illustrates why API committees find it difficult to adopt a logical "standards" approach to requirements.

The American pollution and safety requirements are not necessarily suitable for all locations and hazardous liquids. Seal piping is generally screwed into the seal chamber or gland plate. The piping specification may require the remainder of the piping to be butt-welded and flanged. The weakness of the screwed connection is overlooked. It is very rare for the seal chamber or gland plate connections to be seal welded. Seal welding of threaded connections is not recommended for corrosive applications. If the liquid/vapour is extremely hazardous all the pipe connections should be of very high integrity. Flanging to the circumference of a seal chamber or gland plate is not impossible.

Plans 52 and 53 require that the liquid must have some desirable properties, in areas where the process liquid is deficient. Some necessary qualities include:

- Safe, i.e. not hazardous
- Clean
- A better temperature, hotter or colder
- Good lubricant
- Good thermal conductivity

Buffer liquids are generally a hydraulic oil or water. These two have successfully coped with most applications in the past. Both are obviously safe at reasonable temperatures.

Table 8.5 indicates popular barrier liquids which have been used extensively in the past.

Liquid	Temperature range
Ethylene chloride	-130 to -15 °C
Propanol	-120 to + 0 °C
Ethanol	-105 to +50 °C
Methanol	-90 to +40 °C
Butanol	-80 to +90 °C
Paraffin	-30 to +160 °C
Hydraulic oils	-30 to +80 °C
Mineral oils	-10 to +200 °C
Water	+5 to +80 °C
Ethylene glycol	0 to +175 °C
Vegetable oils	+10 to +130 °C
Glycerine	+100 to +260 °C
Heat transfer oils	+100 to +350 °C

Table 8.5 Traditional barrier liquids

As mentioned earlier one of the criteria was for the liquid to be safe. This requirement is necessary in case the liquid leaks out of the second seal. These are some of the liquids "standardised" in past installations:

- Ethylene chloride; ethyl chloride, chloroethane — highly flammable and toxic
- Propanol — flammable and slightly toxic
- Ethanol, ethyl alcohol — flammable, differing opinions about its health risk
- Methanol, methyl alcohol — flammable and toxic
- Butanol, butyl alcohol — flammable and toxic

These liquids are covered by both UK and USA safety legislation, and by the ADR transportation regulations, see Section 17.1.2 in Chapter 17. Obviously the use of these liquids as barriers or buffers needs very careful consideration of the immediate pump surroundings, especially if the pump is situated in an enclosed pump house.

Alternative safe barrier liquids should be considered for all applications. Some liquids which have been tested by seal manufacturers include water + soluble oil, water + antifreeze, polyalkylene glycol and synthetic oil.

Special attention should be paid to the external systems with regard to pressure and the consequences of unintentional shutdown or breakdown.

8.3.2.8 Maintenance

The operating life of any seal is primarily dependent upon selecting the right seal and the right material, i.e. the optimum with regard to the type of duty. The service life as well as the leakage rate are further influenced by the condition of the pump and its design and by the manner in which it is operated, continuous or intermittent, etc. Depending upon the speed, a pair of seats should last for at least one year in a clean liquid environment.

Regular maintenance of a mechanical seal is limited to checking for leakage once a week. If the pump is handling a hazardous liquid controlled by safety legislation, the normal vapour leakage may have to be measured and logged. From the time when leakage appears as drops, the seal can usually be operated for a week or two before the seats need to be replaced. It will also be necessary to check the function of any external barrier or buffer liquid system.

8.3.2.9 Trouble-shooting

Leakage is the only indication of a faulty mechanical seal. But to be able to evaluate if and when the seal must be dealt with, it is not sufficient just to notice leakage. The occurrence of leakage does not provide sufficient information as to the course of action to be taken. It is also necessary to evaluate the circumstances, when and where, and the appearance of the leaking liquid.

Table 8.6 reviews some typical problems and offers remedies. In more difficult cases the pump and seal manufacturers should be contacted and should also be present when the seal is disassembled since the cause can easily be destroyed during disassembly. Always read the manual before carrying out any work on a pump or seal.

Problem	Probable cause	Remedy
Newly installed seal leaks badly when stationary.	Incorrect installation. Rotating seat hang-up.	Check manual, disassemble completely. Check cleanliness. Rebuild.
Newly installed seal leaks moderately when stationary and when running.	Damaged 'O' rings.	Strip, check cleanliness and rebuild.
	Seat faces not flat.	Depending upon the materials of the seat faces the seal may bed-in.
Newly installed seal is pressure tight when tested but leaks when running.	Single spring seal running in wrong direction.	Disassemble and check spring handing.
	Incorrectly fitted stationary seat.	Read manual, disassemble and rebuild.
	Shaft run-out too great.	Check run-out, check bearings. Replace parts as necessary.
	Eccentricity between shaft and sleeve.	Check run-out of sleeve to shaft. Remove sleeve and check sleeve bore to sleeve OD.
	Vibration.	Check for shaft or sleeve run-out.
Seal begins to leak after a short period of operation.	Debris in process liquid or operating condition transients during start-up.	When process conditions have stabilised, disassemble seal and rebuild.
	Seat or elastomer materials not suitable for operating conditions.	Check operating conditions against specification. Consult pump and seal manufacturer.
	Seat or elastomer materials not suitable for cleaning or stationary conditions.	Consult pump and seal manufacturer.
Seal was working correctly then suddenly starts to leak badly.	Cracked seat(s).	Thermal shock or dry running. Investigate cause and prevent repetition.
	Bearing failure.	Inspect bearings, replace as necessary.
	Loss of barrier/buffer liquid.	Add instrumentation to external system.

Table 8.6 Trouble-shooting guide for mechanical seals

8.4 Reciprocating rods

There are two main differences between rotary shafts and reciprocating rods:

- A reciprocating rod tends to carry liquid through the seal,
- A reciprocating rod stops moving twice every cycle.

Because of these differences, the technology used for rotating shafts is not always applicable to reciprocating applications. Reciprocating seals can also be placed in one of two categories; stationary seals and moving seals. Moving seals are usually associated with piston seals.

Modern plunger pumps operate at mean linear velocities of up to 2 m/s, and piston pumps up to 1 m/s. The maximum instantaneous velocity is about 20% greater. A wide variety of seals are available, however some are for hydraulic power applications, e.g. piston/cylinders for hydraulic actuation. Hydraulic seals are generally suitable for 0.5 m/s and are intended to work with hydraulic oil, water-oil emulsions or water. Hydraulic seals may not be suitable for continuous operation in process pumps and seal materials may not be resistant to the wide range of liquids generally encountered.

The need for rod seals can sometimes be completely eliminated by the use of diaphragms or bellows. Short stroke pumps, typically metering and dosing pumps, have successfully used diaphragms and bellows to eliminate the leakage along rods. Precautions must be taken in the event of fatigue failures and extra instrumentation can be fitted for failure detection. Diaphragms and bellows cannot be applied to longer stroke machines because of the space required to accommodate the stroke.

8.4.1 Lip seals

Lip seals, similar in design to rotary seals, are used for preventing oil escaping along crosshead extensions where they pass through the crankcase wall. Similar seals are mounted in reverse to prevent the ingress of dirt and moisture. Seals for this type of application are limited to about 100 °C.

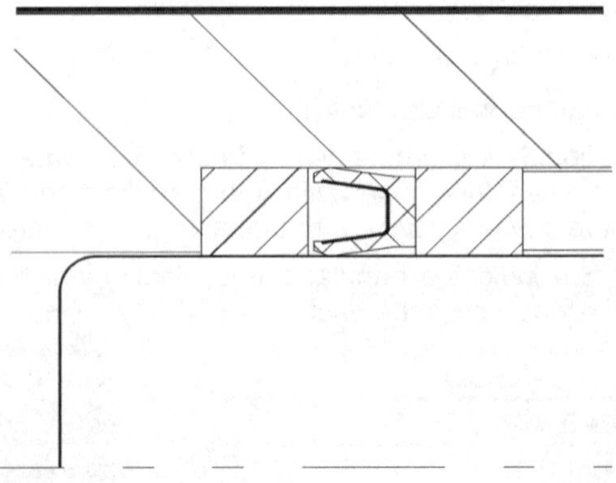

Figure 8.38 "U" ring for process seals

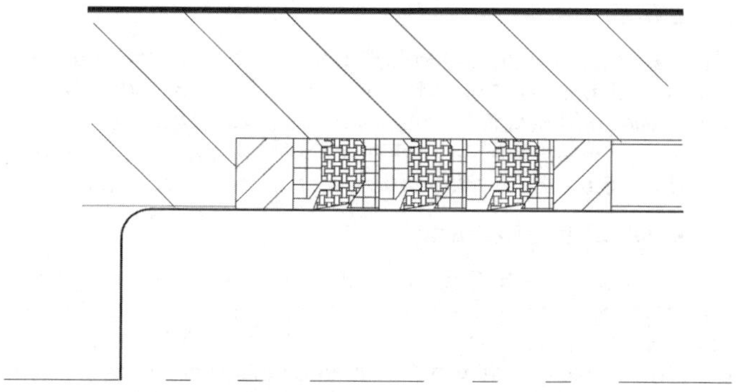

Figure 8.39 Semi-chevron packing set

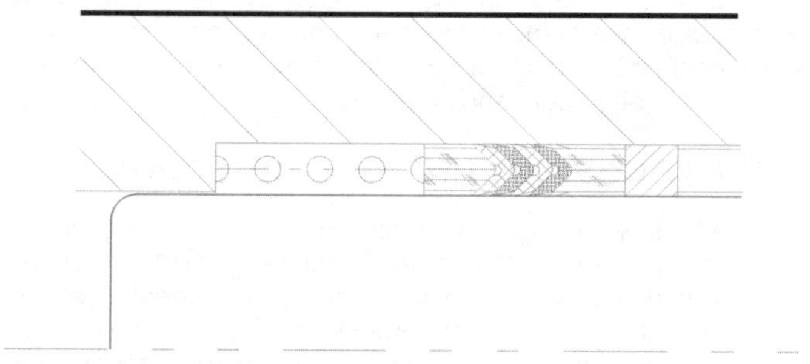

Figure 8.40 Spring-loaded chevron packing

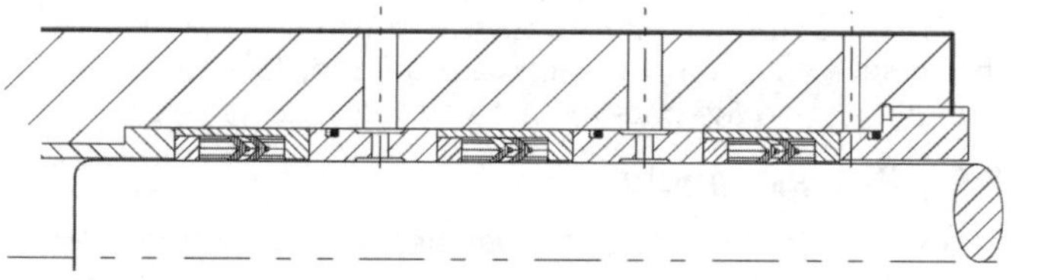

Figure 8.41 Triple packed cartridge chevron stuffing box

Special lip seals in the form of "U" rings can be used for process sealing, see Figure 8.38. Generally made of FKM elastomers or PTFE and energised by pressure plus a stainless steel spring, these seals can be capable of sealing up to 300 bar. Like soft packing rings, "U" rings can be stacked in series. The "U" ring is not compressed axially and must be fitted with a gland which locks to provide a fixed installation gap. No adjustment is possible or necessary.

Another style of lip seal which has proved successful is the semi-chevron; the chevron style is discussed later. The semi-chevron is a combination of styles; soft packing and a chevron ring, see Figure 8.39. The outside diameter of the packing is almost flush; the axial gland load compresses the OD against the bore of the stuffing box. The bore of the packing is recessed to form pockets for the sealing lip. The sealing force on the lip is initiated by interference and supplemented by liquid pressure. The packing is not adjustable and requires no maintenance. The gland is tightened to provide an initial compression on the packing but no further adjustment is necessary.

Semi-chevrons are available for over 200 bar at temperatures up to 105 °C. A packing set normally consists of three units with three sealing lips. They can also be arranged in series. A triple packed box has been operated successfully. Lantern rings between the sets can be used for buffer liquids or lubrication.

Soft packing can be used for any pressure. At high pressures, over 300 bar say, the routine maintenance, adjustment, can become tiresome. Chevron packing was designed to remove the necessity for adjustment. Chevron packing has internal and external sealing lips. Packing sets can be compressed manually, but this requires skill and experience and is not recommended. Most sets are spring-loaded and the gland is locked solid, see Figure 8.40. Spring loading can be by a single large spring or multiple small springs.

Chevron sets are available in various compounds and allow operation over 2000 bar at temperatures up to 290 °C. Pure PTFE is available but is not suitable for higher speed pumps.

Chevron packing can be fitted as a cartridge assembly. Offshore pumps, handling hazardous liquids, have been equipped with three chevron sets in series with buffer and lubrication systems, see Figure 8.41. For most high pressure applications, lubrication can greatly extend packing life.

In the past only pump manufacturers produced cartridge packing sets. The chevron packing was purchased from the packing manufacturer in the normal way but the pump manufacturer designed and manufactured the cartridge to suit a specific application. At least one packing manufacturer is now offering cartridge assemblies. The cartridges are pre-compressed by a circlip on the low pressure side of the packing. This method of construction reduces the strength of the cartridge, which is already a fairly light assembly compared to the rest of the stuffing box. The structural stability of thin wall cartridges should be fully investigated before designing or approving the rest of the stuffing box. Cartridge designs, which use the full width available, plunger/rod od to stuffing box id, to transmit the axial forces are much stronger in practice.

8.4.2 Soft packing

Soft packing for reciprocating rods is very similar to the soft packing used for rotary shafts. Because of the different running conditions the design limits are modified, Table 8.7 indicates the scale of the modifications.

Rotary		Reciprocating	
Speed m/s	Pressure bar	Speed m/s	Pressure bar
2	5	1	150
10	10	1	100
25	20	2	250

Table 8.7 Comparison of soft packing capabilities

8.5 Process liquid seals for reciprocating rods

Seals on reciprocating rods, unlike rotodynamic pump shaft seals, operate alternatively at suction and discharge conditions. If multiple seals are fitted, the seals behind the primary seal may operate at reasonably constant conditions.

8.5.1 Operating principles

Soft packing seals for reciprocating rods operate on the same principles as rotary seals. Allowances must be made for the rod attempting to drag the liquid through the packing. Most stuffing boxes have between three and seven rings of packing. Fitting too many rings is as bad as fitting too few. With too few rings it is difficult to create adequate pressure drop within the stuffing box and leakage is always great. With too many rings the outer rings receive too little lubrication and burn up.

Proprietary pump specifications which specify a fixed number of rings in the stuffing box are misguided. Pressures higher than the pump operating pressure can be generated in the stuffing box and the packing rating must take this into account.

8.5.2 Design variations

The design of stuffing boxes for reciprocating rods generally follows the same pattern as rotary shafts. The major exception is that mechanical seals are not available as an option. Small pumps may have a diaphragm or bellows to eliminate the leak-

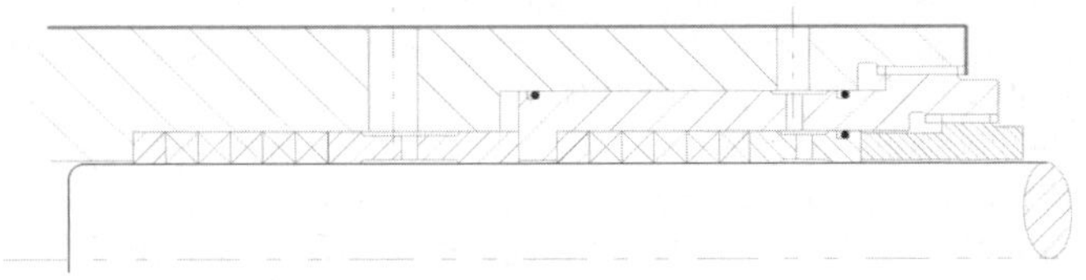

Figure 8.42 Stuffing box with adjustable primary and secondary packing

age problem. With packing, more sets are added as the liquid becomes more hazardous or problematic to seal.

Section 8.4.1 indicated that lip seals can be stacked in series as in Figure 8.41. Soft packing sets can also be stacked in series; the stuffing box design becomes complicated, correct adjustment is dubious and the rod sizes permissible within a fixed centre distance are greatly reduced.

Figure 8.42 shows a stuffing box with two sets of independently adjustable soft packing, both rated at 100%. Lubrication or a buffer liquid can be supplied to the middle lantern ring. A quench could be applied across the atmospheric side of the second packing. Additional rings could be fitted after the second lantern ring but some liquid supply would definitely be required.

Just as with rotary sealing systems, reciprocating rod seals can be multiple units; it is important to specify or know whether secondary/tertiary seals are rated for full pressure. When handling difficult liquids the use, if possible, of non-adjustable packing styles will achieve better packing life and pump availability.

8.5.3 Packing material

Packing materials suitable for rotary shafts, see Table 8.1, in general, are suitable for reciprocating applications. Reciprocating rods do not have sleeves, the rod may be hardened or coated. Packing with Aramid® or Kevlar® fibres tends to be abrasive. Very hard coatings, such as chrome oxide or tungsten carbide, should be used.

Braided and plait packing is normally cut to length from a roll. Reciprocating rods also use moulded packing rings which are solid and must be assembled from the end of the rod. Rings can also be machined from solid material. Special profiles and diameters can easily be accommodated when the rings are machined.

8.5.4 External systems

External systems are similar to their rotating counterparts except barrier liquids are not used and the seal cavity cannot be flushed. Lubrication is the most common system. The lubricant is pumped by a special single cylinder reciprocating pump to each stuffing box. The lubricant mixes with the product and some is lost.

Low pressure quench is the second most popular system. The quench liquid is constrained by a close fitting bush in the gland or an auxiliary packing set.

Heating and cooling of the packing can be accomplished by circulating liquid through a lantern ring. Because the rod is reciprocating, liquid is carried to the packing at both sides of the ring. The rod is heated or cooled by direct contact with the liquid.

Flushing through the throat bush is popular on solids handling pumps and for liquids which may crystallise in the packing. Flushing can be low pressure on the suction stroke when supplied by a low pressure external source. High pressure flushing is like lubrication, a single cylinder from another pump is piped to each stuffing box. The flush injection is synchronised to flush on the back stroke when the rod would tend to draw solids into the box.

8.5.5 Maintenance

Maintenance of adjustable soft packing is a skilled job. The pump and packing manufacturers' instructions must be followed for good results. If site labour is poor or the site is inaccessible, non-adjustable packing should be specified. See Section 8.3.1.5 for the maintenance section for rotary shafts.

8.5.6 Trouble-shooting

See Section 8.3.1.6. Except for the reference to bearing wear and packing rotating, all comments are valid. If spring-loaded packing runs too hot and tries to stick to the rod it is possible to drag the packing axially in the stuffing box. This can cause longitudinal scratches on the packing OD. It is possible to have rods running eccentrically in the stuffing box. Periodically, the rod should be checked to ensure it is central. Some rod connections allow for radial misalignment, others do not. As the pump wears, crossheads may drop in their guides, effectively lowering the centre of the rod. If eccentric packing wear is a problem the cylinder alignment should be confirmed.

8.6 Process liquid seal selection

As can be seen from the preceding Sections, sealing technology is complicated and there are a wide range of choices open to the pump designer and pump user. Environmental controls, together with increased personnel safety requirements, have imposed limitations on the allowable leakage rates of some liquids and vapours.

A logical approach must be adopted for seal selection to achieve a sound, practical, operational solution. The pump and seal manufacturer must know everything the user knows about the liquid. If any approximations are needed, the user must provide these; and the data must be distinguished as such.

8.6.1 Process liquid

The seal selection process must start with the process liquid itself. If the process liquid is legally registered as hazardous then limits on exposure and vapour concentrations will be published. From the design of the immediate pump surroundings, an allowable concentration can be converted to an allowable leakage rate. If a rotodynamic pump is being considered and the liquid is so hazardous that leakage cannot be tolerated, then Section 8.1 should be read again and a sealless pump selected. If a reciprocating pump is considered for very hazardous liquids, try to use small pumps with diaphragms or bellows. If the liquid is only flammable, then flammability limits will be published and an allowable leakage can be calculated.

Consideration must be given to what happens to the liquid when it does leak out of the seal:

- Will it remain liquid or will it vaporise?
- If it remains liquid, will it cause damage to the pump or the immediate surroundings?
- Can it be piped to normal drains or must it be piped in closed drains for treatment?
- If the escaped liquid vaporises; is the vapour heavier or lighter than air?
- Where will it accumulate? What damage will it cause?

Obviously the nature of the process liquid affects more than just the seal selection. However, the pump seal may be one of the few places where the liquid has a chance to escape.

8.6.2 Size, speed and pressure

The rate of leakage from different types of seal depends on the pressure at the seal, shaft diameter, speed and, in the case of soft packings, also to a very large extent upon maintenance. The following are given as guide values:

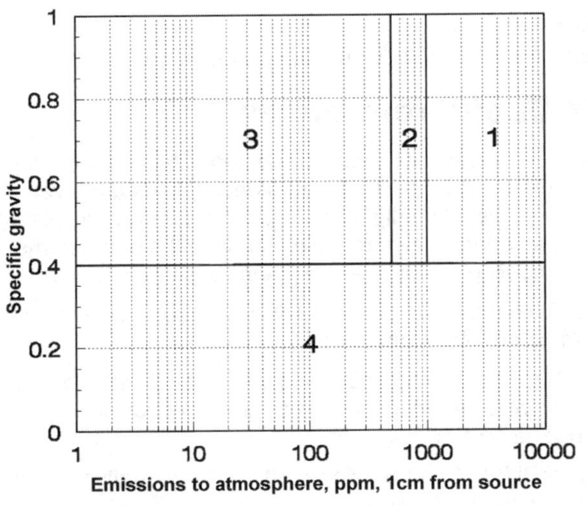

Chart area	Acceptable technology
1	Single, double or tandem seals acceptable
2	Single seals may be acceptable. Depending on actual operating conditions, seal size, pressure and temperature, tandem or double seals may be required to meet emission regulations.
3	Tandem or double seals acceptable.
4	Double seals required.

Figure 8.43 Mechanical seal selection guidelines

- Mechanical seal, single, 0.2 to 20 cm^3/hour
- Soft packing, single, functioning well and properly maintained, 5 to 50 cm^3/hour

The values refer to seals working correctly. When a seal begins to fail, the increase in leakage is dependent upon the whole stuffing box design. Leakage should also be considered as a proportion of the pump flow. Soft packing leakage rates can be as low as 0.01%.

Figure 8.43 shows guidelines to assist with mechanical seal design choices. The graph applies to mechanical seals up to 152 mm diameter, pressures up to 40 bar and speeds up to 3600 rev/min.

Leakage should be a part of the design philosophy. Not “What shall we do if the seal leaks?”, but “What shall we do with the leakage when the seal leaks?”. When treated correctly, leakage can be recycled and is not necessarily a waste disposal problem.

8.6.3 Local environment

The local environment surrounding the pump plays a important role in the consideration of leakage. If the pump is in an enclosed pump house, leakage will tend to accumulate within the it depending upon the arrangements made for ventilation. If the pump house is unmanned and personnel visits are once a week, the requirement for ventilation will be low. Vapour may accumulate to toxic concentrations for personnel or form an explosive mixture. Mechanical ventilation may be necessary, or the outlet from the pump house may have to be ducted to a flare stack for disposal.

If the pump house is open, leakage will tend to disperse naturally. If the liquid has an obnoxious smell, this may not be acceptable. If the pump site is close to habitation, natural dispersion may cause problems in the community. Remote pump sites with open pump installations avoid local dispersion problems. The concentration is very dilute, perhaps undetectable, before reaching any community.

One remote type of installation which has special problems is the offshore installation. Offshore installations have limited escape possibilities if anything does go wrong. Although surrounded by water, the water is not always safe. Offshore in the Gulf of Mexico the water is relatively safe, whereas in the North Sea, especially in the winter, the water is definitely not safe. Extra safety precautions must be taken to protect personnel.

If leakage is damaging to the environment then not only the local environment must be considered. The global environment may be at risk as well.

8.6.4 Cost

The initial cost for soft packing materials is relatively low compared with the initial cost for a mechanical seal. The initial cost is, however, not of prime importance and the whole-lifecost is largely determined by a number of secondary costs associated with maintenance, leakage and external systems.

The level of routine maintenance required is dependent, not only on the seal, soft packing or mechanical seal, but also on the operating conditions. High pressure and high speed require more maintenance. The possibility of changing the working pressure is not usually within the options available.

Changing the pump speed is a viable option. The increased cost of a larger pump must be weighed against the reduced maintenance costs and the extended operating time between routine shut-downs. Individual construction and operating costs are required to properly evaluate this option. The user should always have an operating strategy when pumps are purchased and the pump manufacturer should be aware of the target operating time between shut-downs. If the level of maintenance skills at the site are low, it may be best to fit cartridge seals where possible. Complete cartridges could be refurbished by the seal manufacturer ready for replacement.

The cost of liquid lost due to leakage may be considerable. Additional initial cost should be considered to reduce leakage or recycle leakage. The feasibility of either course will be dependent upon the difficulties of sealing the liquid. Suitable barrier or buffer liquids may not be available or cause as much trouble as the process liquid. Loss of any liquid in some closed loop processes may cause tremendous problems; a seal-less pump may be necessary even if the liquid is safe.

External systems may require extra instrumentation or maintenance. Some stuffing box lubricators may require the oil reservoir to be topped up every 24 hours. Extra storage space will be required for special lubricants and barrier/buffer liquids.

The true cost of sealing solutions is difficult to analyse. Very good data must be accessible and the data trends must be stable. Possible changes in economic conditions, over the life of the pump, may necessitate changes in sealing philosophy. Can the pump selected be adapted to other sealing systems ?

8.6.5 Standardisation

EN ISO 21049 specifies requirements and gives recommendations for sealing systems for centrifugal and rotary pumps used in the petroleum, natural gas and chemical industries. It is applicable to hazardous, flammable and/or toxic services where a greater degree of reliability is required for the improvement of equipment availability and the reduction of both emissions to the atmosphere and life-cycle sealing costs. It covers seals for pump shaft diameters from 20 mm (0.75 in) to 110 mm (4.3 in).

This Standard is also applicable to seal spare parts and can be referred to for the upgrading of existing equipment. A classification system for the seal configurations covered by this International Standard into categories, types, arrangements and orientations is provided.

The 2004 edition of ISO 21409 is identical to API 386 3rd edition.The most widely used Standard for seals is probably DIN 24960; parts of this were extracted for ISO 3069. DIN 24960 specifies dimensions for mechanical seals, seal cavities, seal designations and materials. DIN 24960 was referenced in the original DIN standards for standard water and chemical pumps, now ISO 2858 and ISO 5199. 24960 applies to shafts from 10 mm to 100 mm. Single seals, double and tandem seals are specified.

The Swedish pulp and paper industries technical cooperative has published a standards for soft packings, SSG 1300, 1320 and 1321 which comprises dimensions, installation and materials.

A recent change to sealing standardisation is the introduction of ANSI/API RP682, Shaft sealing systems for centrifugal and rotary pumps. Even though the title states "Shaft sealing systems", the standard only deals with mechanical seals, there is no mention of soft packed boxes. API 682 has replaced the seal section of API 610.

It has already been stated that API is controlled by pump users. Pump manufacturers have made it known to API that they would like to stop redesigning their pumps every 4 or 5 years to comply with the users latest fashion. API 682 cannot be applied universally to all existing pumps because of the enlarged seal cavities required. While it is recognised by most people that larger seal cavities, not extended soft packed stuffing boxes, can extend mechanical seal life, it is not possible to impose such drastic changes in a short time span.

EN ISO 21049 is only applicable to seals between 30 mm and 120 mm, temperatures from -40 °C and 260 °C, and pressures from 0 to 34.5 barg. A list of applicable liquids includes; water, sour water, caustics, amines, some acids and most hydrocarbons. "Some acids" obviously needs clarification. Only cartridge seals are acceptable to API 682. Hook sleeves are not acceptable. The complete seal, including sleeve, must be removable so that the complete assembly can be tested. The minimum requirement is a single balanced seal.

EN ISO 21049 has standardised "qualifying" tests for seals based on four test liquids; water, propane, 20% NaOH and mineral oil. The results of the qualifying test must be submitted as part of any seal quotation. 21049 does allow pump testing with alternative face materials when the contact face materials are unsuitable for the test liquid, water.

EN ISO 21049 has a comprehensive seal selection procedure comprising 10 pages. Seal type/style are covered, as are material recommendations, piping plans and guidance for selecting barrier/buffer liquids.

EN ISO 21049 can be used as a reference source for other industries; it contains much useful information. However, the contents should not be considered in isolation. Discussions with seal manufacturers may provide better, alternative solutions to difficult applications.

Pump users who have large numbers of pumps usually standardise on one seal manufacturer. This type of standardisation is intended to reduce inventory cost and space, and also reduce the chances of fitting incorrect parts to a seal. The pump user can also benefit by increasing his bargaining power when purchasing a large number of spares from one source. Seal manufacturers with a wide product range profit from this philosophy.

In some applications, very special seals are required, and manufacturers must be selected for a specific product.

8.7 Useful references

API 610 Centrifugal Pumps for General Refinery Services. (In Appendix D and E to API 610 descriptions are given of mechanical seal configurations, circulation systems (circulation of a separate medium, circulation of the pumped medium), cooling systems and a coding system for seals.)

DIN ISO 5199 and VDMA 24297 are identical with the meaning of API 610 in many decisive points.

NF EN 12756 Mechanical seals - Principal dimensions, designation and material codes, April 2001, Association Francaise de Normalisation.

ANSI/API 682 Shaft Sealing Systems for Centrifugal and Rotary Pumps, 3rd Edition.

EN ISO 21049:2004 Pumps - Shaft sealing systems for centrifugal and rotary pumps.

ISO 3069:2000 End-suction centrifugal pumps - Dimensions of cavities for mechanical seals and for soft packing.

DIN 24960 End Suction Centrifugal Pumps - Dimensions of cavities for mechanical seals and for soft packing.

VDMA (German Engineering Federation), PO Box 71 08 64, 60498 Frankfurt/Main, Germany, Tel 069 66030,
Fax: 069 6603 1511, Email: Kommunikation@vdma.org, www.vdma.com

Association Française de Normalisation (AFNOR), 11, rue Francis de Pressensé F-93571 La Plaine Saint-Denis Cedex France, Tel:01 41 62 80 00, Fax: 01 49 17 90 00, www.afnor.fr.

SSG Teknik AB (Swedish pulp and paper industries' technical co-operative organization), Köpmangatan 1, SE-852 31 Sundsvall, Sweden, Tel: 060 123850, Fax: 060 150710, Email: info@ssg.se, www.ssg.se.

Seals, A G Fern with BS Nau, Oxford University Press, ISBN 0-19-859148-9 / 0198591489.

Seal Users Handbook, R M Austin, B S Nau, British Hydromechanics Research Association, ISBN: 0900983337.

Axiale Gleitringdichtungen Gleitringdichtung, E Mayer, 1982, VDI-Verlag GmbH.

The British Pump Manufacturers' Association (BPMA),The McLaren Building, 35 Dale End, Birmingham B4 7LN UK, Tel: 0121 200 1299, Fax: 0121 200 1306, enquiry@bpma.org.uk, www.bpma.org.uk.

Hydraulic Institute, 9 Sylvan Way, Parsippany, NJ 07054, USA, Tel: 973 267 9700, Email: gromanyshyn@pumps.org, www.pumps.org.

BHR Group Ltd, The Fluid Engineering Centre, Cranfield, Bedfordshire MK43 OAJ UK, Tel:01234 750422 , Fax: 01234 750074 , Email: solutions1@ bhrgroup.com, www.bhrgroup.com.

Shaft couplings

9

9.1 Introduction

A shaft coupling transfers torque between two in-line, or nearly in-line, rotating shafts. The magnitude of the torque in the shafts is equal, although slipping and disengagement can cause speed variations. In its simplest, and perhaps oldest form, the coupling acts as a means of joining shafts. Another function is to join two shafts which are not necessarily in perfect alignment with each other. The coupling in this case must be capable of absorbing such misalignment. Modern couplings, between pump and driver, must be capable of rapid disassembly.

9.2 Types of coupling

Shaft couplings can perform many different functions and have varying characteristics. They are usually divided into three main groups with sub-divisions, namely:

Non-disengaging couplings

- Solid
- Torsionally rigid
- Torsionally flexible

Disengaging couplings

- Clutch with manual over-ride mechanism
- Free-wheeling clutches

Limited torque couplings

- Non-controlled
- Controlled and variable

Some of the requirements for flexible couplings, including definitions, performance and operating conditions, dimensions of bores, reference to components as well as an appendix on alignment are to be found in BS 3170. Friction clutches and power-take-off assemblies for engines, and their requirements are included in BS 3092. Process pumps to API 610 standard have spacer couplings in accordance with API 671.

For pump applications it is usual to use a coupling from the first group above, although special installations make use of disengaging clutches and limited torque couplings. Examples being centrifugal clutches to reduce starting loads when using a direct-on-line starting induction motor. Hydrodynamic clutches can be used for reducing starting loads and speed regulation. Combinations of brakes and reverse locks can be used to prevent reverse pump rotation for example.

Power recovery hydraulic turbines are often coupled to the non-drive end of pump motors so that the turbine can unload the motor. The coupling used is a free-wheel type with manual over-ride so that the pump/motor can start-up before the turbine. Once the turbine runs up, as it tries to rotate faster than the motor, the clutch locks automatically and power is transmitted.

Non-disengaging couplings maintain, after assembly, a more or less flexible but continuous transmission of the rotational movement. The connection is only broken for disassembly, repair, etc. Flexible couplings of one form or another, which are capable of absorbing residual misalignment, are most common, although solid couplings do have their areas of use, see Figure 9.1.

One example is the split muff coupling, the main advantage being its ease of assembly. It is best used for low speed applications due to the difficulties in balancing. The sleeve coupling is mounted and removed by oil-injection; being almost symmetrical, balancing is easy.

Another example is in the case of long-shaft pumps and stirrers, Figure 9.2, where, because of space requirements when disassembling, it is necessary to split the shaft. Production technicalities also necessitate the joining of long shafts. The most usual couplings in these cases are disc or flanged couplings.

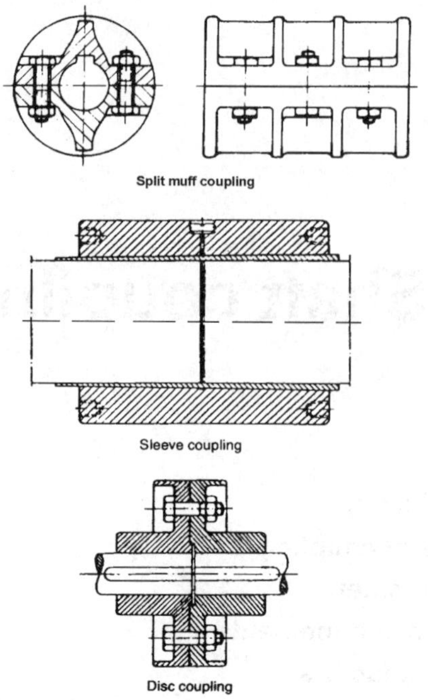

Figure 9.1 Examples of solid shaft couplings

Torsionally rigid flexible couplings consist of various types of diaphragm and gear couplings, see Figure 9.3. Couplings with a single functional element have the ability to take up angular and axial misalignment. Couplings with two functioning elements separated by a fixed "spacer", are also able to cope with radial misalignment, whereby the magnitude of the radial misalignment is determined by the angular misalignment multiplied by the distance between the coupling elements.

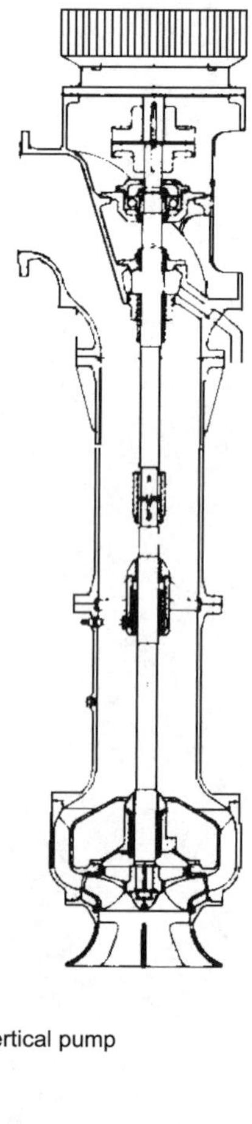

Figure 9.2 Long-shaft vertical pump

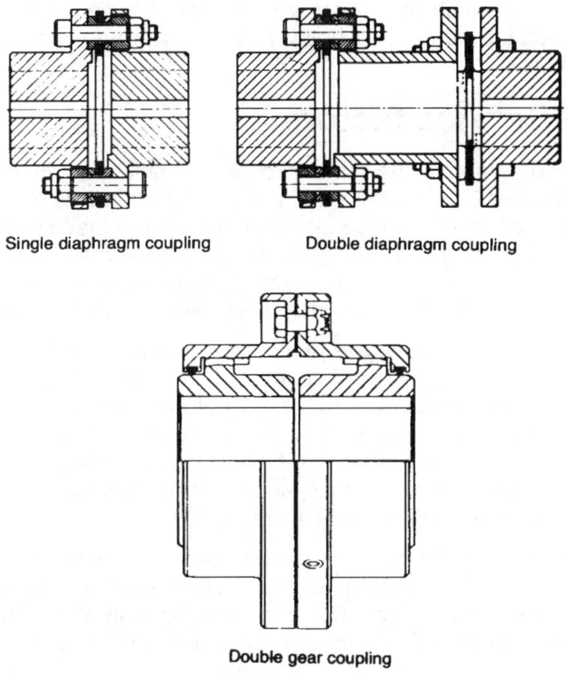

Figure 9.3 Examples of torsionally rigid flexible couplings

Torsionally flexible shaft couplings usually consist of flexible rubber, plastic or even steel elements, Figure 9.4. The first mentioned coupling elements require somewhat larger coupling diameters because of their lower load carrying capacity. Single element couplings can accommodate radial misalignment as well as angular and axial. The flexible spring coupling is interesting because it is designed to have a variable torque/deflection characteristic. Together with dampening provided by the grease lubricant, the variable torque/deflection characteristic provides a powerful torsional vibration dampener.

The torsionally flexible couplings shown can be built with two working elements and a spacer to allow additional radial misalignment. In order to simplify disassembly and service of some machines, spacer couplings are preferred, Figure 9.5. Removal of the spacer enables the rotating elements to be serviced without necessitating the removal of the whole machine. API 610 specifies 5" spacer lengths as a minimum. EN 22858, ISO 2858, for 16 bar back pull-out pumps, specifies spacer lengths from 100 mm to 180 mm depending upon pump size.

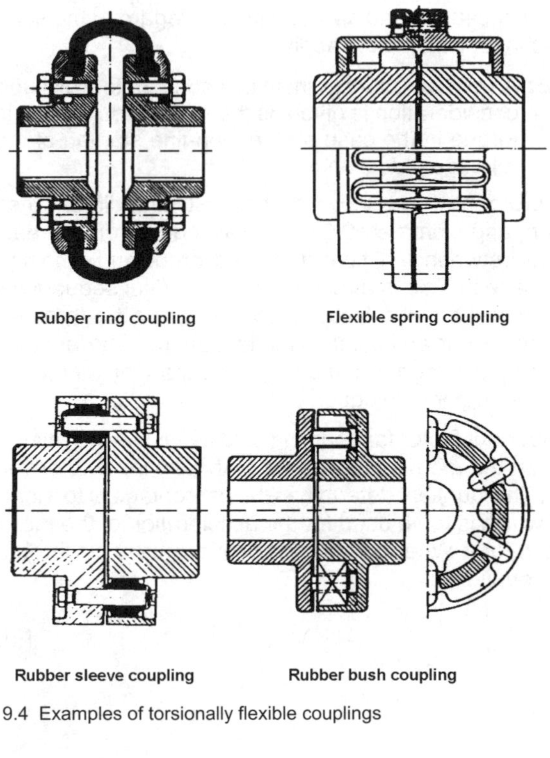

Figure 9.4 Examples of torsionally flexible couplings

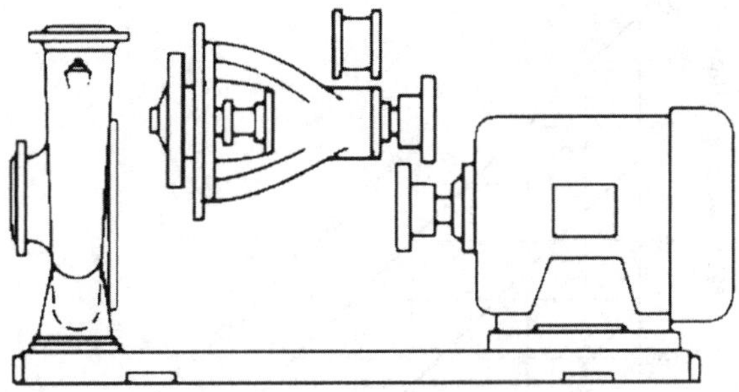

Figure 9.5 Simplified disassembly using a spacer coupling

9.3 Misalignment

Three types of movement or deviation can occur between two shafts, see Figure 9.6, namely:

- Radial misalignment, where the shafts are parallel although not lying on a common centre line.
- Axial misalignment, end float, where the shaft centre lines are in alignment although the axial position is incorrect and axial movement may be possible.
- Angular misalignment, where the centre lines of the respective shafts are not parallel.

The deviations can occur singly or in combinations. Also the individual deviations can change with operating conditions.

A typical changing condition is from cold to running temperature conditions. Thermal growth causes machine centre heights to increase slightly as they warm up. API 610 process pumps, are centre-line mounted to avoid thermal growth of the pump casing affecting the pipework. However, the motor driving the centre-line mounted pump is foot mounted and does have thermal growth. In this situation, motors are mounted low so that the thermal growth will expand the motor to near perfect alignment.

In large machines changes in ambient temperature or sunshine can affect the alignment. The thermal growth phenomenon can be further complicated when the drive and non-drive ends of a machine expand at differing rates. Not only does the radial alignment change, but also the angular alignment. Accurate on-line measurement is necessary to check for this condition.

Suppliers of couplings provide information relating to the maximum permissible deviations, usually stated for each individual

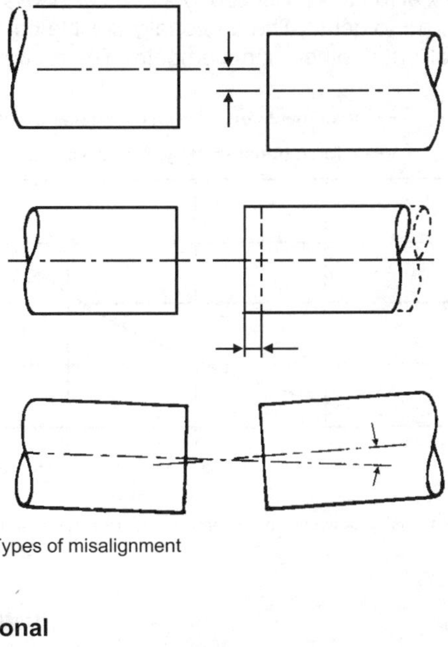

Figure 9.6 Types of misalignment

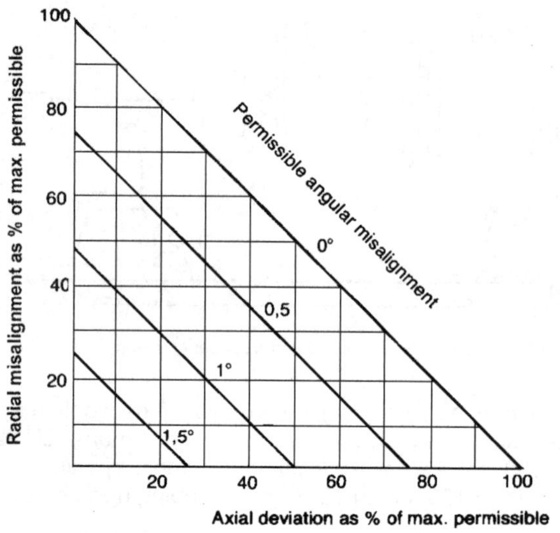

Figure 9.7 Permissible angular misalignment as function of axial deviation and radial misalignment

type of deviation. It is important to know the maximum permissible values of combined misalignment and how the maximum permitted deviations are influenced by speed and the torque transmitted. Figure 9.7 shows this for a particular size of double-diaphragm spacer coupling.

The service life of both couplings and machines, normally machine bearings, are influenced by misalignment. Just how much the life of the machine is affected can only be judged when information regarding the precise magnitude of the torque and forces transmitted due to misalignment is known. It is usual to refer only to the amount of misalignment permitted for a specific coupling type. But it is the amount of misalignment tolerated by the machine, Figure 9.8, which should really be investigated.

9.4 Forces and moments

A solid coupling is only designed and constructed to be subjected to torsional power transmission torques and axial forces. Flexible couplings can be subjected to bending moments as well as axial and radial forces. The solid coupling does not allow the shafts to move independently of each other. Torque and axial movement are transmitted directly from one shaft to the other. Diaphragm and gear couplings transmit torque directly but react differently to axial and radial movement. A diaphragm coupling allows the shafts to move both axially and radially, the diaphragms are deformed, and an axial force and a radial moment are generated.

The double gear coupling also allows axial and radial movement. No axial force is produced, but a radial load is produced rather than a moment. The torsionally-flexible coupling produces radial loads rather than moments. The rubber ring coupling will produce an axial force in response to axial movement, whereas the other couplings will slide to accommodate axial movement.

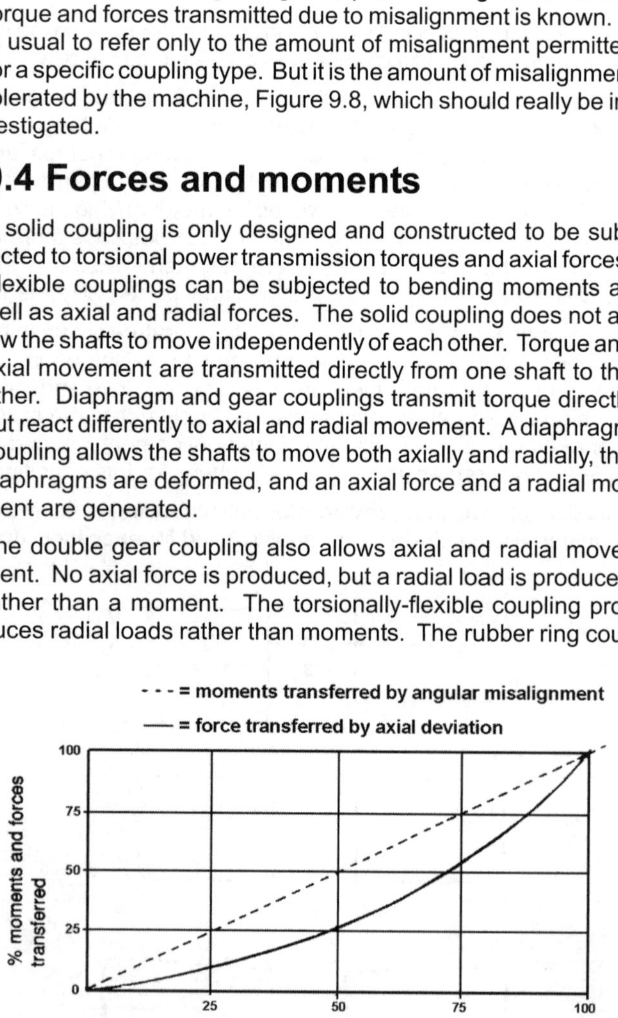

Figure 9.8 Relationship between misalignment and transmitted forces/ moments

9.5 Service factors

When determining the size of flexible and solid couplings, it is usual to evaluate a service factor or safety factor. This primarily takes account of the type of driving and driven machines, the mode of starting and duty cycle. Most coupling manufacturers publish nominal ratings for each coupling type and size together with lists of service factors for various duty conditions.

Drives with AC squirrel-cage motors and rotodynamic pumps generally have a service factor of 1.0. Arduous duty cycles, such as high pressure descaling, would increase the service factor to 1.25. If the pump driver was an engine, the service factor would increase between +0.5 and +1.0 depending upon the number of cylinders. Service factors for an AC motor driving a reciprocating pump can be as high as 3.0.

To compare different couplings objectively a new method has been developed which takes into consideration the frequency of starting, temperature, the moments of inertia of the driving and driven machine, normal torque and maximum torque.

This method has been presented in the German coupling Standard DIN 740, which, apart from the method of calculation, also contains dimensional standards. It is often used for pumps manufactured according to ISO 2858, BS 5257, and can form a basis for internal standardisation to simplify handling, storage, etc. There are, however, two additional service factors which should be considered.

The first is the effect that shaft misalignment can have on the coupling. A factor based on the extent of allowable misalignment expressed as a percentage of the maximum permissible deviation, should also be given.

The second factor should take into consideration the level of vibration of both machines, at least vibration velocities above 1.5 to 2 mm/s RMS. Note that for rotodynamic pumps, vibrations of up to 5 mm/sec RMS can be permissible, which is approximately equivalent to balancing degree Q16 according to VDI 2060 or quality class B for class IV machines according to BS 4675. The size of the various factors and their influence on coupling speed varies with different types, which is why the calculations and values given in DIN 740 must be used with a certain amount of caution and always with due regard to the suppliers' instructions, which must apply.

NOTE: A very important point in this context, to which too little consideration is given, is the magnitude of the starting torque in the case of direct-on-line starting of a squirrel-cage induction motor.

Measurements have shown that almost immediately after connection, approximately 0.04 s, a maximum torque is reach- ed which is between 6-10 times the rated torque and even higher in some cases. This is a result of the electrical sequence in the actual motor and the fact that connection of the three phases does not occur absolutely simultaneously. The actual maximum torque is therefore much greater than the starting torque quoted in motor catalogues.

An important factor for coupling calculations is the relationship between the moments of inertia of the driving and driven machine. This quotient determines the percentage of torsional moment which is to be used for the acceleration of the motor and pump rotors. When starting, the torque passing through the shaft coupling is:

$$M_k = M_i\left(1 - \frac{J_{mo}}{J_{ma} + J_{mo}}\right) = M_i\left(1 - \frac{t_o}{t}\right) \qquad \text{Equ. 9.1}$$

where:

M_k = coupling torque at start (Nm)

M_i = internal motor torque (air-gap torque) at start (Nm)

J_{mo} = moment of inertia of motor (kgm^2)

J_{ma} = moment of inertia of driven machine (kgm^2)

t_o = motor starting time without load (s)

t = motor starting time with load (s)

The moment of inertia of a centrifugal pump can be given in relation to that of the motor; the following figures can be used as a guide:

2-pole motor $J_{ma} = 0.04$ to $0.2\ J_{mo}$

4-pole motor $J_{ma} = 0.25$ to $0.6\ J_{mo}$

6-pole motor $J_{ma} = 0.60$ to $3.2\ J_{mo}$

By inserting these figures in equation 9.1 and assuming that M_i is 6 to 10 times the rated torque, values for coupling torques at starting of up to 3.8 times the rated torque for 4-pole motors and 7.6 times for 6-pole motors are obtained. Therefore care must be taken when sizing couplings which are subjected to direct-on-line starting, especially when the driven machine has a large inertia.

9.6 Speed

Centrifugal forces increase with speed squared. The material of the coupling and the permissible peripheral velocities must be calculated. The maximum peripheral velocity for grey iron, for example, is 35 m/sec. To avoid vibrational damage it is necessary, for couplings which are not fully machined, to carry out both static and dynamic balancing at much lower speeds than those which are fully machined.

The mass of the coupling is often quite small in relation to the rotating masses in the driving and driven machines. For a pump unit the relationship of coupling/total rotor weight is approximately 0.02 to 0.08. It therefore follows that out-of-balance in the coupling normally has less effect on bearings and vibration than out-of-balance in the actual main components. However the actual position of the coupling relative to the bearings may change this.

The following relationship applies:

$$F = m \cdot e \cdot \omega^2 \cdot 10^{-3} \qquad \text{Equ 9.2}$$

where:

F = out-of-balance force (N)

m = out-of-balance mass (kg)

e = distance from centre of rotation to centre of gravity of out-of-balance mass (mm)

ω = angular velocity (rad/s)

For highly resilient rubber element couplings with a spacer, the out-of-balance can be further increased by whirling. It is also important that balancing is carried out using whole keys, half keys or without keys, depending upon the method of balancing the attached component.

9.6.1 Example

A fully-machined coupling can be assumed to have an inherent degree of balancing, without dynamic balancing, equivalent to VDI 2060 Ql6 to Q40, i.e. approximately 0.08 mm permissible centre line deviation at 3000 rev/min.

If the concentricity tolerance for the shaft bore in the hub is 0.05 mm, the maximum centre-line deviation can therefore be 0.13 mm. This is not abnormal. In many cases the tolerance alone reaches this value. This centre-line deviation generates an out-of-balance force of about 12 N per kg coupling weight at 3000 rev/min. A coupling for 50 kW can weigh 10 to 15 kg, which thus generates a rotational out-of-balance force of 120 to 180 N.

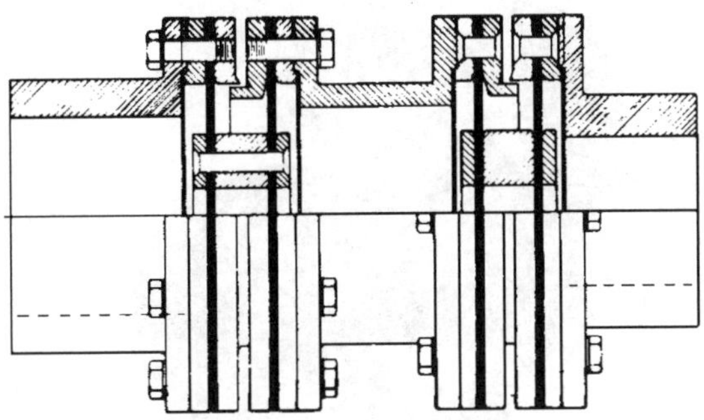

Figure 9.9 Non-sparking diaphragm coupling

Most couplings have no components which can move radially to create out-of-balance forces. The gear coupling is different. The teeth on the hubs and the spacer must have clearance at the top and bottom; this allows the spacer to move radially. In theory, the angle of the teeth flanks should provide a centralising force to counteract any tendency for the spacer to run eccentrically. Problems have been experienced with gear couplings and special attention should be paid to radial clearances and spacer weight.

The flexible spring coupling has a spring which in theory could move and run eccentrically. These couplings are usually used on positive displacement pumps at speeds which are low enough not to have balance problems.

9.7 Size and weight

The importance of small size and low weight to achieve as small a moment of inertia as possible, as well as reducing the out-of-balance forces, has been mentioned previously.

In certain extreme cases light-alloy metal spacers and diaphragms are used to reduce weight. Apart from the need to maintain a small size/transmitted torque ratio, it is also important, from the cost and standardisation point of view that the coupling should be able to accommodate large variations in shaft diameter.

9.8 Environment

Corrosive and abrasive environments affect the service life of the coupling by causing abnormal wear to the component elements. Extremes of heat and cold affect the strength and elasticity of the component materials. Oils, chemicals, sunlight and ozone can completely destroy a rubber element. A coupling made entirely of metal such as a diaphragm or flexible spring coupling, for example, is usually the only solution in such cases.

The process industries offer a very poor environment. In the petrochemical industry for instance, in refineries as well as oil and gas tankers, for example, it is necessary to use non-sparking couplings.

A non-sparking diaphragm coupling can be manufactured by making the diaphragm of monel and the remaining components of carbon steel or bronze. See Figure 9.9. Non-sparking types are usually used in conjunction with flameproof electric motors in environments where there is risk of explosion, either continuously or normally during operation. Statutory regulations must be observed, see Section 9.2.8.

A flexible spring coupling has the important elements housed in a seal cover and coated with lubricant, in the form of grease. Environmental changes have little effect on the coupling. Instances of spring breakage are rare, but any parts which could create a spark are fully enclosed, see Figure 9.10.

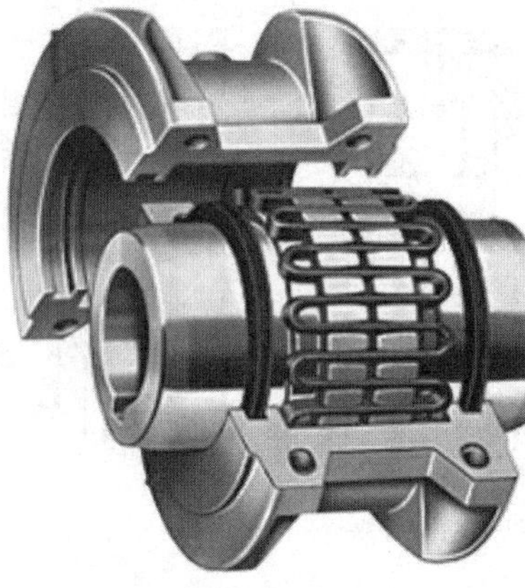

Figure 9.10 Flexible spring coupling

Thread diameter mm								Shaft journal diameter
d_1	d_2	d_3	d_4	t_1 $^{+2}_{0}$	t_2 min	t_3	t_4	d_6
M3	2.5	3.2	5.3	9	13	2.6	1.8	7-10
M4	3.3	4.3	6.7	10	14	3.2	2.1	11-13
M5	4.2	5.3	8.1	12.5	17	4	2.4	14-16
M6	5	6.4	9.6	16	21	5	2.8	17-21
M8	6.8	8.4	12.2	19	25	6	3.3	22-24
M10	8.5	10.5	14.9	22	30	7.5	3.8	25-30
M12	10.2	13	18.1	28	37.5	9.5	4.4	31-38
M16	14	17	23	36	45	12	5.2	39-50
M20	17.5	21	28.4	42	53	15	6.4	51-85
M24	21	25	34.2	50	63	18	8	86-130

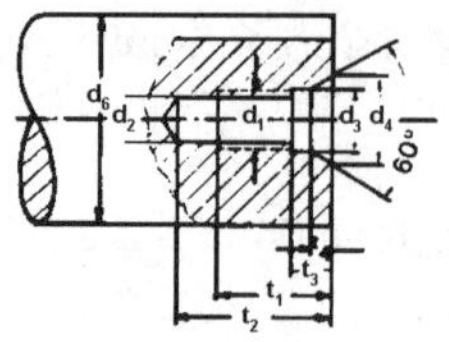

Figure 9.11 Tapped assembly hole in electric motor shaft

Another method of overcoming explosion risks, especially on board ship and with engine drivers, is by means of gas-tight bulkheads and bulkhead fittings consisting of two mechanical seals with barrier fluid between them, together with bellows which absorb misalignments. This type of fitting must be equipped with non-sparking shaft couplings.

9.9 Installation and disassembly

To maintain maximum operational reliability and to simplify assembly and service it is important that the machines connected are securely mounted, preferably on a common foundation and baseplate. Guards must be fitted to rotating parts according to safety requirements, see Section 9.13.

9.9.1 Alignment

Alignment of couplings or, more correctly, alignment of the shafts which the coupling is to connect, should be carried out as accurately as possible. For pumps packaged on baseplates with their driver and other equipment, provisional alignment should be achieved by "chocking" the baseplate during leveling. After grouting, the alignment should be set correctly by adjusting the shims. A perfect alignment should be considered as an economic possibility, since alignment can considerably affect both service life and maintenance costs. See Section 9.11 for methods of shaft alignment.

9.9.2 Baseplates

It is normal practice to bolt the pump directly to the baseplate. Other drive train equipment is shimmed to achieve correct alignment. In the case of cardan shafts the angular deviation should be equally distributed between the two joints to avoid unequal rotational velocities. Furthermore, a universal coupling should always rotate with a slight amount of angular misalignment to promote lubrication.

9.9.3 Attachment

The attachment of a coupling half to a shaft usually presents a dilemma. The hub should be securely attached and preferably absorb part of the torque, to reduce the load on the key, as well as being easy to detach. The practice of hub attachment is similar to that for motor shafts where the fit is usually H7/k6, light push fit up to 48 mm diameter. A push fit H7/m6 is preferred for diameters above 55 mm. Some pump manufacturers prefer a positive interference fit, typically 0.001 mm per mm of shaft diameter. These couplings are heated for mounting and dismounting. Large couplings become unwieldy. Oil injection on shallow taper shafts, without keys, can be very successful.

The tighter fit is brought about by the fact that the height of the key is reduced from 12.5% of the diameter at 24 mm diameter to only 6% at 100 mm shaft diameter. This reduction should also be compensated for by increasing the length of the hub. In

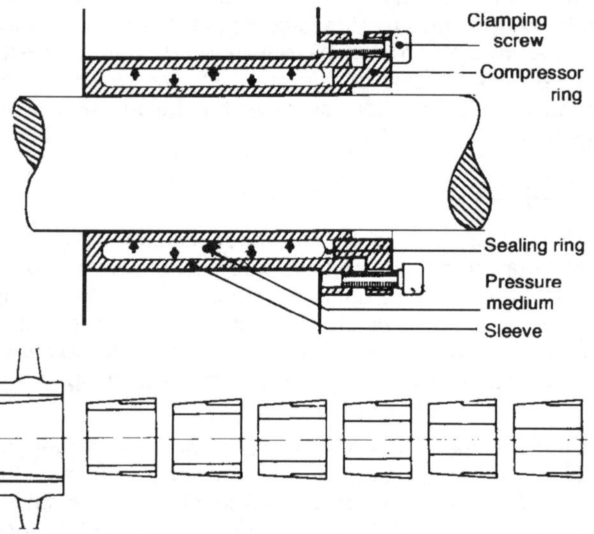

Figure 9.12 Examples of clamping sleeves

the case of electric motors the key does not normally extend right to the end of the shaft, which also increases the strain on the key. This must also be compensated for by increased hub length.

Assembly and disassembly of the coupling halves must be carried out carefully to avoid damage to the shaft ends and bearings. This operation could be simplified considerably if motor, pump and coupling suppliers fitted their equipment with suitable lugs, etc., to assist the attachment of pullers. For electric motors a tapped hole in the end of the shaft, as shown in Figure 9.11 can be supplied at extra cost, and ought to be standardised on all equipment.

Other methods of attaching the coupling halves are shrink fits, bolted joints or some form of clamping sleeve, Figure 9.12. Taper bushes are used primarily for chain wheels and pulleys, but can be a useful alternative for couplings where space permits. Some manufacturers offer taper bushes as an alternative to parallel bores. The hydraulically loaded clamping sleeve shown is a relatively new innovation and is not used extensively in pumps.

The resilient elements in the shaft coupling must be easy to purchase, replace or repair. That it must be possible to replace without disturbing the machines or coupling hubs, goes without saying.

9.10 Service life

The life of the coupling is influenced by many factors, which vary according to the style of construction. One which above all

Method	Shaft coupling type	Measuring device and location	Zero setting and notation rules**	Parallel misalignment mm	Inclination* mm per 100 mm measured length	Remarks
I	Short shaft coupling Machined outer diameter Machined on insides	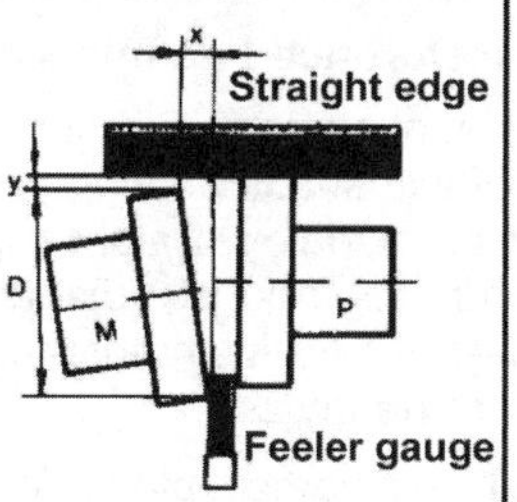	Misalignment according to the figure is positive i.e. the difference is measured above on the motor side	Measured directly as dimension y	$L = \frac{100 \cdot x}{D}$	Make due allowance for bearing end float in the machines
II	Short shaft coupling Requires at least a good surface at measuring pointer Machined on insides	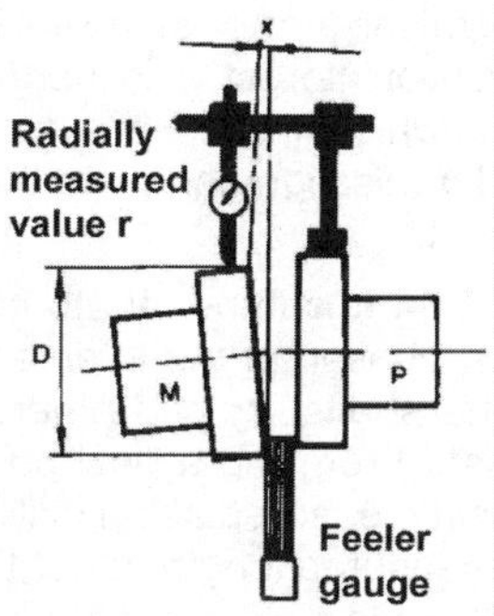	For vertical location zero set the dial above Measured value is read after rotating one half turn	$y = \frac{r}{2}$	$L = \frac{100 \cdot x}{D}$	Make due allowance for bearing end float in the machines (Zero set the dial indicator underneath if the pointer is resting on the pump half)
III	Short shaft coupling Good surfaces at the measuring pointers	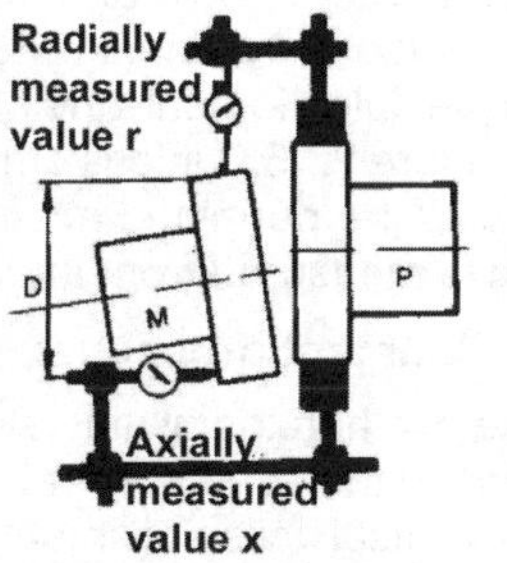	For vertical location zero set both dial indicators in the position shown i.e. for radial deviation above and axial deviation underneath The dials are read after rotating one half turn	$y = \frac{r}{2}$	$L = \frac{100 \cdot x}{D}$	Make due allowance for bearing end float in the machines If both dial gauges are placed with their pointers on the pump half, then zerosetting should be carried out from underneath
IV	Long shaft couplings, i.e. couplings with a distance between the coupling halves Good surfaces at the measuring pointers	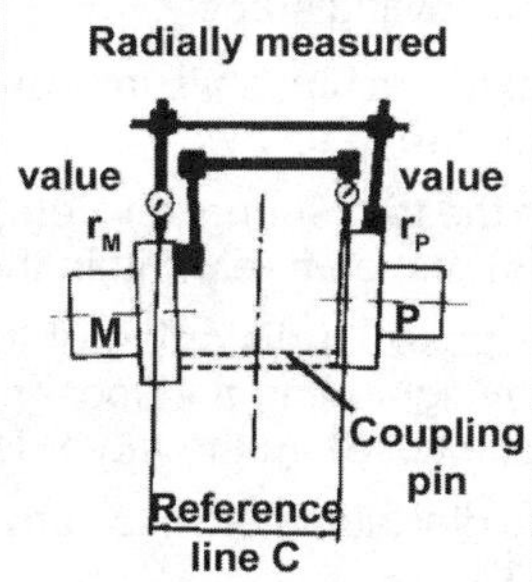	Zero set both dial gauges from above The measurements r_M and r_P are read after rotating one half turn	$y = \frac{r_M - r_P}{4}$***	$L = \frac{r_M + r_P}{2 \cdot C} \cdot 100$	N.B. Measurement can also be carried out on "smooth" shaft ends.
V	Long shaft couplings, i.e. couplings with distance between the coupling halves Good surfaces at the measuring pointers	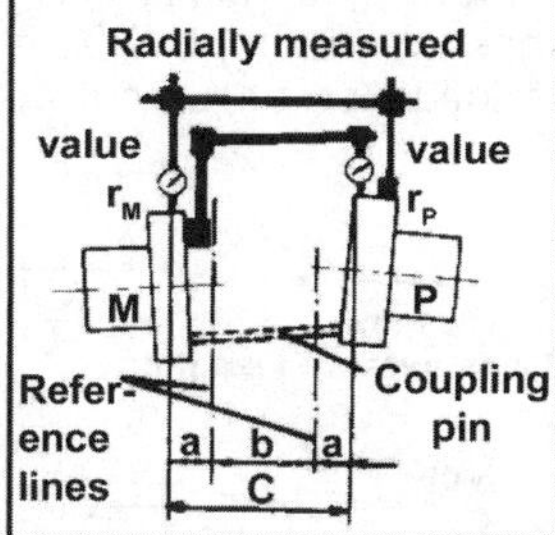	Zero set both dial gauges from above The measurements r_M and r_P are read after rotating one half turn	$y_M = \frac{r_M}{2}$*** $y_P = \frac{r_P}{2}$	$L = \frac{r_M + r_P}{2 \cdot C} \cdot 100$	Similar to method IV Notice the position of the reference lines for calculating angular misalignment

* Angular misalignment in degrees = $\frac{180}{p} \cdot \frac{L}{100} = 0.57L$, ** Dial gauge readings are reckoned to be positive when the pointer moves clockwise,

*** Parallel misalignment refers to a reference line as show in the figure

Figure 9.13 Shaft alignment measuring methods

affects couplings with rubber elements, is the surrounding environment. The service life of a gear coupling is largely dependent upon regular lubrication using the correct type of lubricant according to the ambient temperature, etc. Flexible spring couplings are available with special grease which lasts five years, require almost no routine maintenance, and have no effects on the environment. Alignment affects the service life of all couplings irrespective of type or manufacturer.

For certain types of installations it can be desirable to use a coupling that allows a certain amount of emergency drive even in the event of failure of the flexible element. For other installations it may be necessary to use a limited torque coupling with overload protection.

It is important to carry out regular service and alignment checks according to the manufacturer's instructions, and equally important that these instructions are placed in the hands of the personnel concerned. Unfortunately methods or regulations for assessing the degree of wear are often lacking.

9.11 Shaft alignment

9.11.1 General

Flexible shaft couplings are normally used to transfer torque between rotating shafts where the shafts are not necessarily in perfect alignment. It should be noted that a flexible coupling is not an excuse for poor alignment. Careful alignment is important for the purpose of achieving maximum operational reliability whilst reducing service and maintenance.

When carrying out alignment, consideration must be given to relative movements of the respective machines due to thermal expansion and deformation caused by pipe forces/moments and setting of baseplates on foundations, etc. In certain cases, such as electric motors with plain bearings, notice must be taken of the electric motor's magnetic centre. Alignment should be carried out at various stages during installation.

When alignment is carried out at cold temperatures, it is necessary to make allowances to compensate for the thermal expansion caused by the difference in temperature to that of the normal operating temperature of a pump, pipeline and driver. When possible, a final check should be made at operating conditions after a few weeks in service. Alignment checks should then be carried out at regular intervals. Misalignment, apart from being caused by any of the previously mentioned loads and deformations, can depend upon worn bearings and loose holding down bolts. An increase in vibration levels can often be caused by a change in alignment.

Within the petrochemical industry and refineries, reports are frequently made with respect to alignment. The reports note the alignment prior to and after operation, before removing the pump or dismantling for repairs. The same procedure is carried out to check alignment of hot pumps after warm running.

Correct alignment can be achieved in many ways depending upon the type of equipment and degree of accuracy required. Information regarding alignment requirements is usually to be found in the pump manufacturer's instructions.

Never use the limiting values for the coupling as given by the coupling manufacturer since they greatly exceed the values for machines if smooth running and long service life are to be achieved.

As a guide, a final alignment check should not produce greater parallel misalignment than 0.05 to 0.1 mm or an angular misalignment exceeding 0.05 to 0.1 mm per 100 mm measured length. For the definition of misalignment see Section 9.11.2.

Alignment is adjusted by means of brass or stainless steel shims, usually placed beneath the machine supports. Baseplates are generally machined so that a minimum number of shims are always required under the motor. Horizontal adjustment is performed by moving the machine sideways on its mountings. Large machines must have horizontal jacking screws fitted. Sometimes the pump and driver are fixed after final adjustment by means of parallel or tapered dowels.

9.11.2 Alignment measurement

9.11.2.1 Reference line misalignment

In principle, alignment is based upon the determination of the position of two shafts at two points. Measurement or assessment can be made by straight edges, feeler gauges and dial indicators for the various radial and axial distances or run-out, see Figure 9.13. Adjustment is continued until these deviations are zero, or nearly zero.

Two shafts in a vertical plane, for example, can display two deviations from their common centre line, namely parallel misalignment and angular misalignment, see Figure 9.14. The amount of misalignment at the flexible section of the coupling is that which is of interest. It is therefore appropriate to use a reference line which passes through the flexible section. Parallel and angular misalignments are then referred to this reference line, Figure 9.15.

In Figure 9.14 it is important to note that if the reference line were to be chosen at the intersection point of the two centre lines of the shafts, point A, then only angular misalignment would exist. From a practical point of view angular misalignment is best measured as an inclination expressed as mm per 100 mm measured length rather than as an angular measurement in degrees.

The position of the reference line depends upon the type of coupling and naturally should always be located in relation to the flexible section of the coupling. For couplings with spacers and one or two flexible elements the position of the reference line is shown in Figure 9.15. Unless otherwise stated by the coupling manufacturer the permitted misalignment is considered to be that which is measured from the reference line.

9.11.2.2 Alignment procedure

In the case of a horizontal unit, alignment is best carried out by first aligning in the vertical plane, followed by transverse alignment. For vertical units alignment is measured in two directions at 90° to each other. For a horizontal unit, alignment is carried out in the following steps:

1. Align the machines visually and check that the coupling is not crushed in any way.
2. Attach the measuring device(s) and check that the dial indicator(s) moves freely within the area to be measured.
3. Check possible distortion of the motor mounting or baseplate by tightening and loosening each, holding down bolt individually. Shim the motor feet if distortion is present.
4. Set the dial indicator(s) to zero in the position shown in Figure 9.13.
5. For methods II, III, and IV in Figure 9.13, rotate both shafts simultaneously through 180°, half revolution, thus eliminating the influence of run-out between shaft bores and the

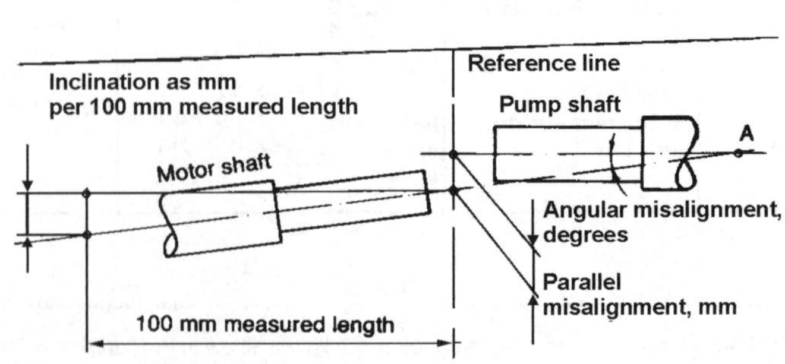

Figure 9.14 Misalignment of two shafts in a common plane

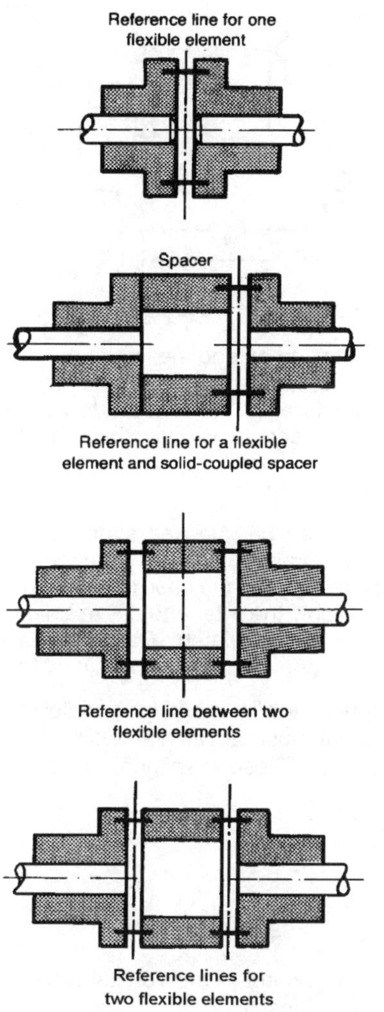

Figure 9.15 Location of reference lines for various types of coupling

outer diameter of a coupling half. The coupling halves need not then be cylindrical. Determine the measured values according to Figure 9.13. Note the measured values with plus or minus signs, see Figure 9.13 for notation. Determine parallel and angular misalignments.

6. Determine shim thickness according to Section 9.11.3 or 9.11.4 and adjust.
7. Carry out checks according to steps 4 and 5.
8. Carry out transverse alignment in the same manner as in the vertical plane.
9. Perform final alignment checks in both vertical and transverse directions and record for future reference remaining parallel or angular misalignments in both vertical and transverse directions. Also make note of operational conditions at the time of alignment, for example, cold motor with warm pump.

9.11.2.3 Choice of method

Figure 9.13 shows the five most common measuring methods. From the point of view of accuracy it is difficult to compensate for manufacturing tolerances between the two halves of the coupling by using a straight edge and feeler gauge, method I.

The difference in accuracy between method III and method IV is determined by the differences in the dimensions D and C respectively. Accuracy increases in both cases as each respective dimension increases, whereby method III is chosen if D is larger than C and method IV or V is chosen if C is larger than D. The choice of method is also determined, apart from accuracy, by the available measuring surface and by attachment facilities and space requirements of the measuring devices.

The difference between methods IV and V lies in the location of the reference lines. Method IV is universally applicable and suitable for smooth shafts or where it is sufficient to measure the total parallel misalignment and inclination. In the case of a coupling with two flexible elements, method V is suitable if the angular misalignment for each element is first calculated individually.

Optical methods are now available. Light sources and mirrors are attached to each coupling half. The units are connected to a small dedicated portable computer which, when supplied with information regarding the feet position, will calculate the respective feet adjustments. Similar optical devices can be attached to machine casings to detect differential expansion when warming up.

9.11.3 Determination of shim thickness

Using the measured parallel and angular misalignment, the necessary shim thickness can be calculated directly. The misalignment is expressed as positive or negative, + or -, according to Figure 9.16, which shows positive misalignment.

The shim thicknesses are calculated from the simple relationship:

$$U_1 = y + L \cdot \frac{F_1}{100} \qquad \text{Equ 9.3}$$

$$U_2 = y + L \cdot \frac{F_2}{100} \qquad \text{Equ 9.4}$$

where:

U_1	=	shim thickness at foot 1 (mm)
U_2	=	shim thickness at foot 2 (mm)
y	=	signed parallel misalignment (mm)
L	=	inclination expressed as mm per 100 mm measured length
F_1 & F_2	=	distance from coupling reference line to each respective foot (mm)

In Figure 9.16 the coupling reference line usually passes through the middle of the coupling.

9.11.3.1 Example

An indicator reading shows parallel misalignment y = +0.28 mm and inclination L = -0.06 mm/100 mm.

The distances to the feet are F_1 = 300 mm and F_2 = 500 mm.

The shim thicknesses required are:

$$U_1 = +0.28 - 0.06 \cdot \frac{300}{100} = 0.28 - 0.18 = 0.10\text{mm}$$

$$U_2 = +0.28 - 0.06 \cdot \frac{500}{100} = 0.28 - 0.30 = 0.02\text{mm}$$

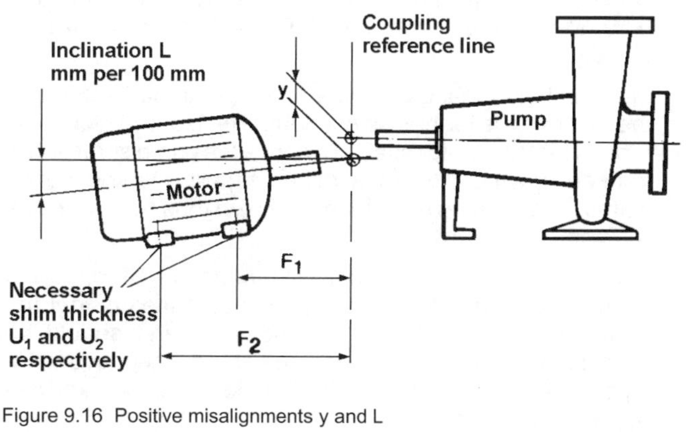

Figure 9.16 Positive misalignments y and L

Shims of thickness 0.1 mm are placed under foot 1. The calculated value of U_2 = -0.02 mm means that 0.02 mm should be removed from foot 2, but can probably be accepted as permissible misalignment.

Equations 9.3 and 9.4 can also be combined so that parallel and angular misalignments can be determined in cases where it is not possible to fit the calculated shim thickness. In which case:

$$y = \frac{U_1 + U_2}{2} \qquad \text{Equ 9.5}$$

$$L = \frac{U_2 - U_1}{\frac{F_2}{100} - \frac{F_1}{100}} \qquad \text{Equ 9.6}$$

where: y and L are residual misalignments

U_1 and U_2 respectively (with sign notation) are shim thickness deviations.

For the previous example, when the proposed correction has been carried out, the residual misalignment is:

$$y = \frac{0 - 0.02}{2} = -0.01 \text{ mm}$$

$$L = \frac{-0.02 - 0}{\frac{500}{100} - \frac{300}{100}} = -0.01 \text{ mm}/100 \text{ mm}$$

9.11.4 Graphical determination shim thickness

The required shim thickness can also be determined graphically by drawing the position of the shaft in respect of the measured values using a greatly enlarged vertical scale, 100:1 for example, and a reduced horizontal scale, 1:5 or 1:10 for example.

The method is illustrated by the following example carried out according to measuring method IV or V in Figure 9.13 with the various stages:

1. Fit the measuring device according to method IV or V and take readings r_P and r_M on the dial gauge.

Example:

dial reading at pump half gives r_P = -1.40 mm

dial reading at motor half gives r_M = +1.20 mm

2. Determine the dimensions C, F_1 and F_2. Note that the reference line in this example has been chosen to pass through the measuring pointer as shown in Figure 9.17

Example:

Measured results

C = 180 mm

F_1 = 470 mm

F_2 = 890 mm

3. Draw up a diagram on squared paper as shown in Figure 9.19. Mark in the dimensions C, F_1 and F_2 on the horizontal scale.
4. Mark half the measured value at the pump half, 0.5 r_P, on the vertical axis furthest to the right. The positive sign for r_P means that the motor shaft lies above the pump shaft and is marked upwards, whilst a minus sign is marked downwards. The reading r_P = -1.4 mm should thus be marked as -0.7 mm, i.e. downwards.
5. Mark half the measured value at the motor half, 0.5 r_M, at distance C. The reading's positive value means that the motor shaft lies below the pump shaft and should be marked as a minus value and vice versa for negative readings. The reading r_M = + 1.2 mm should thus be marked as -0.6 mm, i.e. downwards.

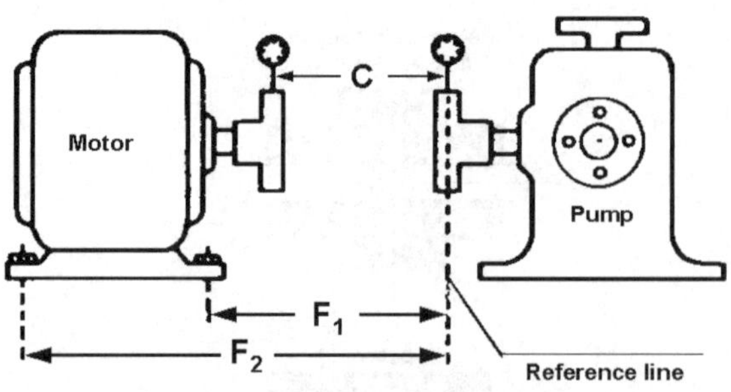

Figure 9.17 Length measurements and location of reference line

6. Join both points and extend the line to the motor feet locations F_1 and F_2 respectively. The motor shaft shown in the example lies 0.44 and 0.21 mm too low at the respective foot locations and should be raised by shims of corresponding thickness, after which transverse alignment is carried out in the same manner.
7. The alignment can be checked simply by using the two measured values and r_M and the distance "b" between the two flexible elements. In the case of couplings with two flexible elements, only the total angular misalignment of each element should be calculated. Parallel misalignments are experienced as angular misalignments by the coupling.

To calculate angular misalignment, the parallel misalignment at the flexible element must be calculated first, i.e. calculated at both reference lines. These misalignments are:

$$h_P = \frac{r_P}{2} - \frac{a \cdot L}{100} \qquad \text{Equ 9.7}$$

$$h_M = \frac{r_M}{2} - \frac{a \cdot L}{100} \qquad \text{Equ 9.8}$$

The angular misalignment in the vertical plane is then determined from the relationship:

$$\propto_M = \frac{h_P}{b} \text{ (radians)} = 57.3 \cdot \frac{h_P}{b} \text{ (deg)} \qquad \text{Equ 9.9}$$

$$\propto_P = \frac{h_M}{b} \text{ (radians)} = 57.3 \cdot \frac{h_M}{b} \text{ (deg)} \qquad \text{Equ 9.10}$$

The angular misalignments in the horizontal plane β_M and β_P are calculated in the same way.

Thereafter, the total angular misalignment, θ, per flexible element is calculated from the relationships:

$$\theta_M^2 = \propto_M^2 + \beta_M^2 \qquad \text{Equ 9.11}$$

or

$$\theta_P^2 = \propto_P^2 + \beta_P^2 \qquad \text{Equ 9.12}$$

9.11.5 Optical alignment

Recent advances in micro-electronics and laser technology have allowed optical alignment techniques to become portable and cost effective. A laser source is mounted on one shaft and a mirror is mounted on the other. The source module includes a detector which measures the position of the returned beam. The shafts are rotated incrementally through 90 ° and readings stored. A small control unit, sometimes small enough to be hand held, which is programmed with the drive geometry calculates the shim adjustment necessary to achieve good alignment. Figure 9.19 shows a typical set up for a small pump. Laser alignment can be used for shafts which are 10 m apart.

Similar equipment can be attached to pump casings, gearboxes, motor stators or baseplate pads to monitor movement or deflection under operating conditions.

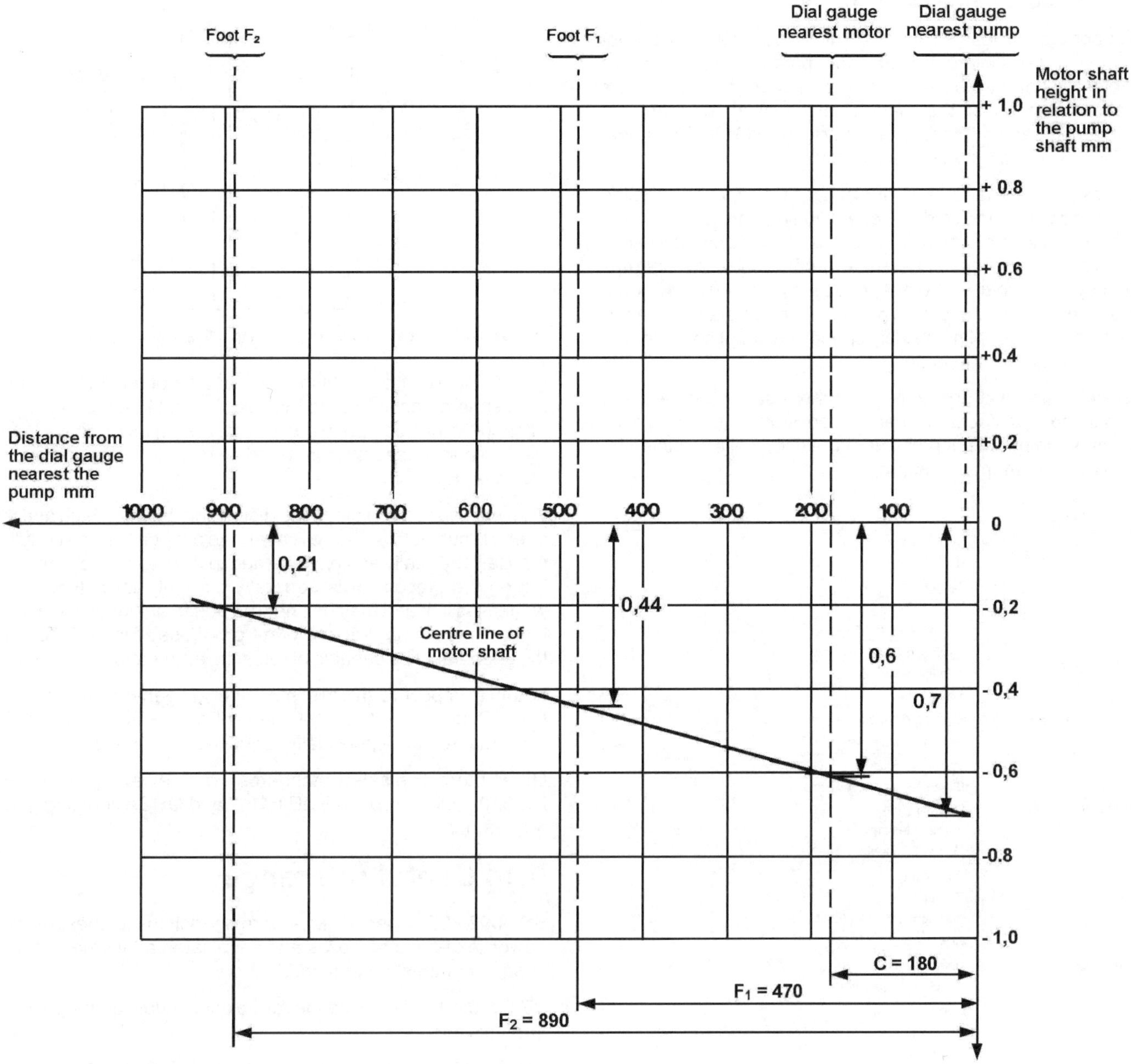

Figure 9.18 Diagram for the example in Section 9.11.4

9.12 Choice of coupling

9.12.1 Costs

In general the cost per kW of a coupling is only a fraction of that of a pump or motor. A pump usually costing at least 40 times that of a coupling and a 4 pole electric motor at least 20 times.

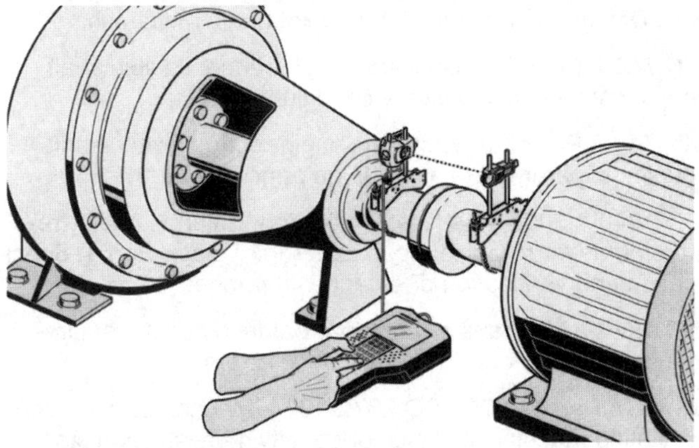

Figure 9.19 A typical laser alignment set up
Courtesy of Prüftechnik Alignment Systems GmbH

The cost varies according to the size and the type. The market for couplings is very competitive, the cost difference between manufacturers is usually small.

Gear couplings are the most costly. If a spray oil lubrication system is required this obviously increases the total cost considerably. Diaphragm and flexible spring couplings, together with the rubber buffer couplings, are about the same cost. Some of the rubber ring couplings are surprisingly expensive.

A good way to compare the cost of couplings is to set the price in relation to the torque and range of shaft end sizes to which the coupling can be fitted. The same pump shaft can, for example, be used for a torque range of 1:20 which occasionally means that the shaft end dimension and not the torque is used when selecting the size of a coupling.

Furthermore, the motor shaft is often larger than the corresponding pump shaft. The motor shaft is dimensioned for bending stress to a greater degree than the pump shaft; for example a motor is often used for belt drive. This can also mean using a larger size coupling.

9.12.2 Factors influencing choice

It is important, not least of all from an initial cost point of view but also cost and space required for spare parts, to establish a viable internal standard by which a small number of type or style variations can cover the majority of coupling requirements within a company or plant. The factors reviewed in the checklist, Table 9.1 should be considered.

For most centrifugal pumps, the diaphragm spacer coupling has become the standard. These couplings are very reliable and can easily cope with the loads and speeds encountered in most situations. For higher speed applications, e.g. pumps driven by gas turbines, the gear coupling is preferred by some users. Positive displacement pumps operate better with a torsionally flexible coupling; flexible spring and couplings with rubber cushioning are favourites.

Users who have a large number of pumps usually choose a single coupling manufacturer whenever possible. This philosophy increases the purchasing power of the user while reducing inventory requirements for spares.

Factor	Influencing parameters
Type of coupling	Non-disengaging Disengaging Torque limitations Torsionally rigid Torsionally flexible
Type of movement	Radial and axial deviation Angular deviation
Forces and moments	Torsional moment Bending moment Axial and radial forces
Operational factors	Starts per hour Electricity supply frequency Operating time Ambient temperature Moment of inertia
Speed	Balancing Strength Burst protection (safety flange)
Size, weight	Shaft bore Space requirements Spacer for disassembly
Environment	Corrosive Abrasive Temperature Hazardous (non-sparking required)
Installation and disassembly	Horizontal and vertical shafts Alignment Fit
Others	Attachment facilities etc. for alignment measuring device Replaceable wear elements Service life Routine maintenance Internal standard Costs Coupling safeguards

Table 9.1 Factors to be considered in shaft coupling selection

9.13 Machinery guards

The pump manufacturer is normally responsible for machinery guards. In the case of standard pumps, a distributor may package the pump with its driver and other equipment and it would become the distributor's responsibility to supply and fit guards.

Standard guards are generally made of painted steel. Sometimes is used because it is easier to bend and may not need painting. When pumps are to be installed in a potentially hazardous environment special motors are used to reduce the chances of the motor igniting any gas present. A steel coupling rubbing on a steel guard could cause a spark and is not appropriate. Onshore, in these situations, an aluminium or bronze guard would be fitted. Offshore pumps in potentially hazardous atmospheres have bronze guards; the salt laden atmosphere offshore is not compatible with most aluminium alloys. Aluminium and bronze guards would be described as "non-sparking" guards.

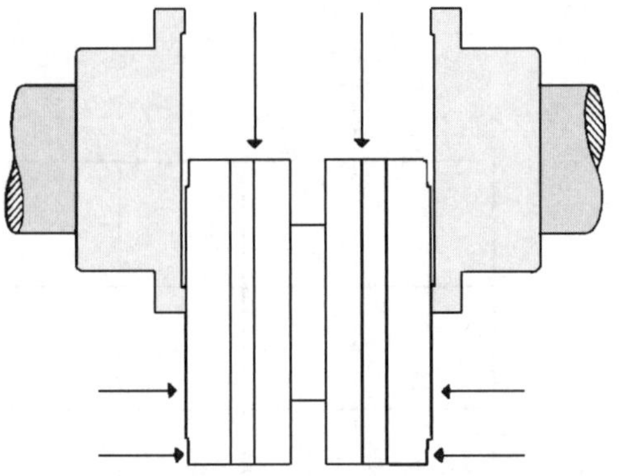

Figure 9.20 Burst-proof diaphragm coupling with spigotted spacer

With high speed couplings the distinction between high and low speed is subjective. There is a remote chance that the coupling may fail physically and explode due to the centrifugal force acting on the pieces. It is generally thought that bolting is the weakness link and may be sheared due to an unforeseen overload. If the coupling is not "burst-proof", see Figure 9.20, then the guard must be capable of retaining any scattered material.

Within Europe, the safety of machinery in general is covered by EN ISO 12100 : 2003. The safety of pumps is covered by EN 809. Guards are specifically regulated by EN 953.

Other interesting safety standards worth reviewing include BS PD 5304, DIN 31001, ASME B15.1 and OHSA coupling guard requirements.

9.14 Useful references

BS 3092:1973 Specification for main friction clutches, main power-take-off assemblies and associated attachments for internal combustion engines.

BS 3170:1972 Specification for flexible couplings for power transmission.

API 610 Centrifugal Pumps for General Refinery Services.

ANSI/API 671 Special-Purpose Couplings for Petroleum, Chemical, and Gas Industry Services, 3rd Edition1998, American Petroleum Institute.

NF EN 22858 End-suction centrifugal pumps (rating 16 bar). Designation, nominal duty point and dimensions, June 1993, Association Francaise de Normalisation.

ISO 2858:1975 End-suction centrifugal pumps (rating 16 bar). Designation, nominal duty point and dimensions.

DIN 740-1 Power transmission engineering; flexible shaft couplings; technical delivery conditions, 1986.

DIN 740-2 Power transmission engineering; flexible shaft couplings; parameters and design principles, 1986.

ISO 1940-1:2003 Mechanical vibration. Balance quality requirements for rotors in a constant (rigid) state, Part 1: Specification and verification of balance tolerances.

ISO 1940-2 Mechanical vibration. Balance quality requirements of rigid rotors. Part 2: Balance errors.

BS 4675:Part 1:1976, ISO 2372-1974 Mechanical vibration in rotating machinery. Basis for specifying evaluation standards for rotating machines with operating speeds from 10 to 200 revolutions per second.

EN ISO 12100-1:2003 Safety of machinery. Basic concepts, general principles for design. Basic terminology, methodology.

EN 809 Pumps and pump units for liquids - Common safety requirements 1998.

EN 953: 1998, Safety of machinery - Guards - General requirements for the design and construction of fixed and movable guards.

BS PD 5304:2005 Guidance on safe use of machinery.

DIN 31001-1 Safety design of technical products; Safety devices; Concepts, safety distances for adults and children.

ASME-B15.1, 1992 Safety standard for power transmission apparatus.

OHSA 1910.219 - Mechanical power-transmission apparatus.

ISO 8821:1989 Mechanical vibration - Balancing - Shaft and fitment key convention.

ISO 11342:1998 Mechanical vibration - Methods and criteria for the mechanical balancing of flexible rotors.

BS 6861-2:1997 Mechanical vibration. Balance quality requirements of rigid rotors. Balance errors.

AGMA 515 Balance classification of flexible counplings, American Gear Manufacturers Association.

ISO 2491:1974 Thin parallel keys and their corresponding keyways (Dimensions in millimetres).

ISO 2492:1974 Thin taper keys with or without gib head and their corresponding keyways (Dimensions in millimetres).

ISO 3117:1977 Tangential keys and keyway.

ISO 3912:1977 Woodruff keys and keyways.

OHSA (Occupational Safety & Health Administration), 200 Constitution Avenue, NW, Washington, DC 20210 USA, www.osha.gov.

AGMA (American Gear Manufacturers Association), 500 Montgomery Street, Suite 350, Alexandria, VA 22314-1581 USA, Tel: 703 684 0211, Fax: 703 684 0242, www.agma.org

Drivers for pumps

10

10.1 Basic electrical theory and principles

10.1.1 Units

Electric current is available in two distinct types; alternating current, AC, and direct current, DC. AC is the most popular because it is much easier to distribute and control. AC is the type of electricity that power stations generate and that is used every day in the home.

DC is the type of electricity produced by batteries. The electrical system in cars is usually 12 V DC, larger commercial vehicles are 24 V DC. Both AC and DC can be generated by rotating machines. Also, one can be converted to the other electronically.

The most common electrical units in use are given in Table 10.1.

Unit	Symbol	Designation
Current	I	Ampere (A)
Voltage	U	Volt (V)
Resistance	R	Ohm (Ω)
Active power	P	Watt (W) or kilowatt (kW) 1 kW = 1000 W 1 W = 1 J/s (see Energy)
Apparent power	S	Volt-ampere (VA) or kilo volt-ampere (kVA) 1 kVA = 1000 VA
Reactive power	Q	Volt-ampere reactive power (VAr) or (kVAr)
Power factor	cos φ	Dimensionless, cos φ = p/S
Energy		Joule (J) or wattsecond (Ws) Kilowatt hour (kWh) is usually used for convenience 1 kWh = 3,600,000 J

Table 10.1 Electrical units

10.1.2 Systems

Electric current is supplied to the consumer as either single phase or 3-phase AC with a frequency of 50 or 60 Hz, i.e. 50 or 60 oscillations per second. Single phase current is supplied to domestic consumers when the total load is less than 50 A. 3-phase current is supplied to all consumers when the load is higher. The supply frequency of 50 Hz is used throughout Europe but not worldwide. America uses 60 Hz, as do some Middle East countries. Some offshore oil platforms also use 60Hz supplies.

When describing the electric supply, a full description must be given, the following are typical of 24 V DC motor supplies:

110 V	1 ph	50 Hz
240 V	1 ph	50 Hz
380 V	3 ph	50 Hz
415 V	3 ph	50 Hz
660 V	3 ph	50 Hz
3.3 kV	3 ph	50 Hz
6.6 kV	3 ph	50 Hz

The 110 V supply described is used for site work where increased electrical safety is assured by using a centre tapped transformer; the maximum voltage to earth is 55 V.

When pump packages are supplied with instrumentation, the user specifies electrical supplies for various other functions. Control supplies may be 24 V DC or 110 V 1ph AC. For critical installations, emergency power supplies will be available to allow equipment to operate during power cuts. Emergency supplies can be AC or DC depending upon the source.

Drivers for pumps such as engine driven pumps may have electric auxiliaries and the engine may have electric starting. 24 V DC is the most popular.

When a user purchases a pump package with controls, the electric supplies for various functions will be specified as shown:

Electric motors, > 250 kW	3.3 kV 3 ph 50 Hz
Electric motors, ≤ 250 kW	380 V 3 ph 50 Hz
Motor heaters	220 V 1 ph 50 Hz
Controls	110 V 1 ph 50 Hz

AC electricity is generated as 3-phase in the power station. It is distributed at high voltage as a 3-phase supply. Before it arrives at the consumer its voltage is reduced to a suitable level. A small commercial consumer will have electricity delivered at 380 V or 415 V 3-phase. The consumer can then install equipment to produce a single phase supply.

Figure 10.1 schematically illustrates a 3-phase system. Between the system voltage U_s, also referred to as mains voltage, and the phase voltage, U_f, the following relationship applies:

$$U_f = \frac{U_s}{\sqrt{3}} \qquad \sqrt{3} = 1.732$$

The 3-phase supply shown in has the neutral grounded. This is the normal situation on land. Offshore electrical systems may not have the neutral grounded and the neutral must be conducted individually to each device. This type of supply is called "4 wire 3-phase" and must be included in any specification.

In a 3-phase supply with system or mains voltage of 415 V a suitable electric motor can be connected to all phase cables. The motor is designated as a 3-phase motor. Smaller electric motors and other loads, up to several kW, can be constructed for connection between a single phase cable and the system's neutral cable, a normal wall socket. These types of motors are designated as single-phase motors.

For the units given in Table 10.1 the following relationships apply:

For direct current DC

$$U = I \cdot R \qquad \text{Equ 10.1}$$

$$P = U \cdot I = I^2 \cdot R = \frac{U^2}{R} \qquad \text{Equ 10.2}$$

Direct current motors are used for variable speed pumps and emergency pumps, see Section 10.3.3 and 6.9.

For single phase AC current

$$U = \frac{I \cdot R}{\cos\varphi} \qquad \text{Equ 10.3}$$

$$P = U \cdot I \cdot \cos\varphi \qquad \text{Equ 10.4}$$

For 3-phase AC current

$$P = U \cdot I \cdot \sqrt{3} \cdot \cos\varphi \qquad \text{Equ 10.5}$$

If the required pump shaft power is P_p kW and the efficiency of the intended 3-phase motor is ηm, then the current required in amps is:

$$I = \frac{P_p \cdot 1000}{U \cdot \sqrt{3} \cdot \cos\varphi \cdot \eta m} \qquad \text{Equ. 10.6}$$

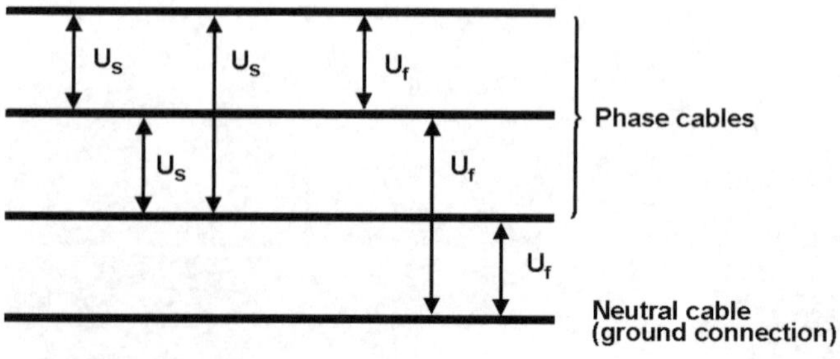

Figure 10.1 3-phase electric system

See also example calculations in Section 10.7

10.1.3 Power

An electric motor for AC consumes not only active power, P, which is converted to mechanical work, but also a reactive power, Q, required to build up the necessary magnetic field in the motor. This reactive power does not perform any work and is not converted into heat but is nevertheless a load on the electricity supply. Electric power suppliers make a charge therefore for consumed reactive power if it exceeds a certain value.

10.1.3.1 Power factor

This relationship is illustrated in Figure 10.2. The active and reactive powers combine to form the apparent power, S. This trigonometrical relationship between P and S constitutes the power factor cos φ.

The power factor can be determined by measuring an electric motor's power input, current and voltage, at its rated power. The tolerances are:

$$\frac{1 - \cos\ \varphi}{6}$$

For electric motors the power factor usually lies between 0.7 and 0.9. It is lower for small motors, see Figure 10.3. The power factor is dependent upon the motor load the table in Figure 10.3 gives approximate values for 4 pole motors of various size at various loads.

The power required by the pump determines the power which the electric motor supplies. This means that if a pump at certain operating conditions requires 11 kW then the motor supplies this power irrespective of whether the rated motor power is 15 or 7.5 kW. This will result in overloading in the case of the 7.5 kW motor, which can cause damage to the motor winding due to overheating. Some form of motor protection device would therefore be necessary, whereby, after a certain period of time under overload conditions, the motor is automatically disconnected from the supply, causing the pump to stop, see Section 10.4.

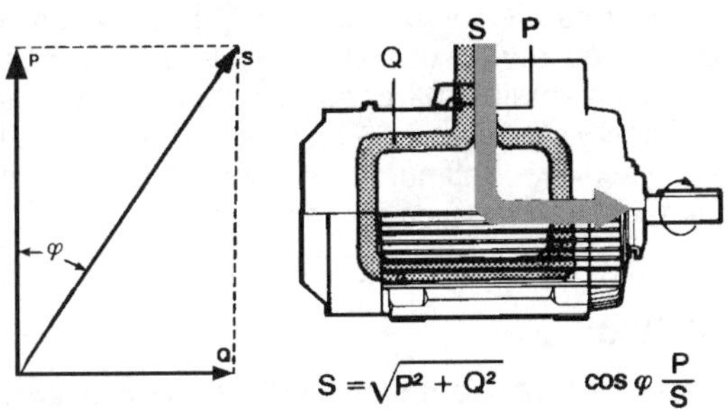

Figure 10.2 Power factor

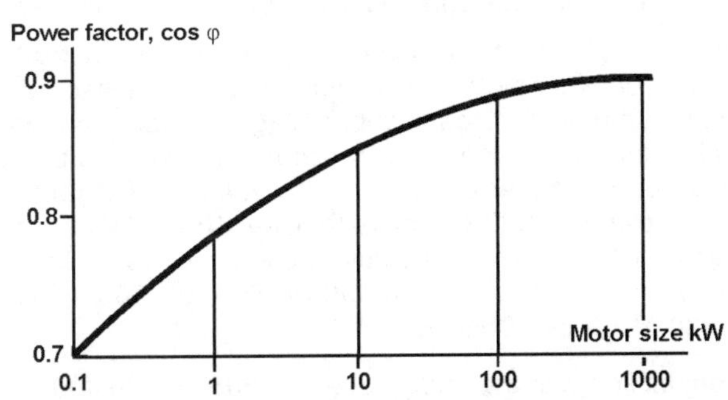

Motor size kW	Degree of loading (Shaft output)			
	¼	½	¾	4/4
< 0.75	0.5	0.6	0.68	0.75
0.75 - 7.5	0.55	0.65	0.73	0.8
7.5 - 75	0.6	0.7	0.78	0.85
> 75	0.65	0.75	0.83	0.90

Figure 10.3 Variation in power factor with motor size and load

A pump motor must, of course, be capable of providing the power which the pump requires. A small margin, depending upon the pump size, is added to the calculated power required to account for small discrepancies in the calculations and minor overload situations which may occur. The motor size may be controlled by starting requirements rather than running requirements. Too large a motor should not, however, be chosen. The procurement costs are unnecessarily high, the efficiency and power factors are lower, whilst the starting loads imposed on the supply are also unnecessarily high.

10.1.4 Phase compensation

If an installation contains many electric motors, large quantities of reactive power are consumed and the overall power factor is lowered. The problem can be aggravated by flow regulation systems such as load-unload when motors can be running at very low loads. A solution to this is to install capacitors, which generate reactive power and thereby increase the power factor. This procedure is usually called "phase compensation" or "power factor correction".

The capacitors can be applied to motors individually or to groups of motors. Extensive calculations must be performed for sizing phase compensation capacitors. The economics must be fully investigated to justify the extra initial investment. Modern systems include automatic compensation over a range of power factor values by switching various capacitors in and out of circuit.

Improvement of the installation's power factor results in a conservation of energy. It is carried out in order to:

- Avoid excess consumption of reactive power involving extra cost
- Reduce loads on transformers and electric cables
- Maintain voltage potential within the system

10.1.5 Motor speed

The speed of an AC electric motor is determined by the frequency of the supply and the number of poles in the motor stator according to the following relationship:

$$n = \frac{2 \cdot f \cdot 60}{p} \qquad \text{Equ 10.7}$$

where:

n	=	speed (r/min)
f	=	supply frequency (Hz)
p	=	number of stator poles

From equation 10.7 it can be seen that a change in frequency causes a change in speed. Electric motors can thus be speed regulated by means of varying the frequency, see Section 10.3.2.2.

In Europe the frequency is 50 Hz, therefore the speed of electric motors is 6000 divided by the number of poles. At least two poles are required which produces a maximum speed of 3000 r/min; 4 poles gives 1500; 6 gives 1000; 8 gives 750 r/min, etc. In the US and some Middle East countries as mentioned, the frequency is 60 Hz.

Electric motors which are constructed so that they exactly follow the stipulated frequency and number of poles are referred to as synchronous motors. For large sizes it is necessary, however, to have a special exciting winding, supplied with direct current, together with special starting equipment. This is not necessary for induction motors or squirrel-cage motors, which are characterised by the fact that there is a speed differential between the frequency of the electrical supply, rotating magnetic field of the stator, and the motor speed.

An induction motor therefore has a 1 to 5% lower speed than a corresponding synchronous motor. This slip increases with increased motor loading and decreases with motor size. Because of its simple construction the induction motor is most popular for pump operation.

Most standard 3-phase squirrel-cage motors are designed to operate satisfactorily when the supply frequency is within -5% and +2% of the nominal frequency.

AC electric motors can be equipped to enable the number of poles to be varied in operation, thereby obtaining changes in speed. Such motors are called "pole-changing motors" and to a certain extent can be used for speed regulation of pumps, see Section 10.3.2.1.

In most AC motors the rotor is short-circuited and so has no connections with the stator winding or any external current supply. Such motors are called short-circuited or squirrel-cage motors. In order to limit the starting current and to regulate the speed a variable resistance can be connected to the rotor whereby the rotor must be equipped with sliprings. This type of motor is referred to as a slipring or wound-rotor motor. Usually once the motor is up to running speed the sliprings are shorted to produce a squirrel-cage motor.

The speed of a DC motor is a function of the motor design and the voltage applied. The DC motor stator does not have a rotating magnetic field which is dependent upon the supply frequency. The DC motor is somewhat similar to a slipring motor in that it has a wound rotor, an armature. But instead of sliprings, which supply current to all the windings all the time, the armature has a commutator which supplies current to some of the windings some of the time. The stator and armature need not necessarily be supplied at the same voltage. The motor speed is controlled by the armature voltage up to the full power design speed. Further speed increase is possible, at the expense of torque, by reducing the stator voltage, field weakening.

10.1.6 Torque

The torque of an electric motor is a measure of the motor's ability to rotate a shaft against resistance. The following relationship exists between torque, power and speed:

$$T = \frac{30000 \cdot P}{\pi \cdot n} \qquad \text{Equ 10.8}$$

where:

T = torque (Nm)

P = power (kW)

n = speed (r/min)

When starting an electric motor it is necessary to develop a greater torque than for the normal operating torque, so that the motor can accelerate to operating speed without overheating.

Figure 10.4 illustrates a typical 3-phase induction motor speed/

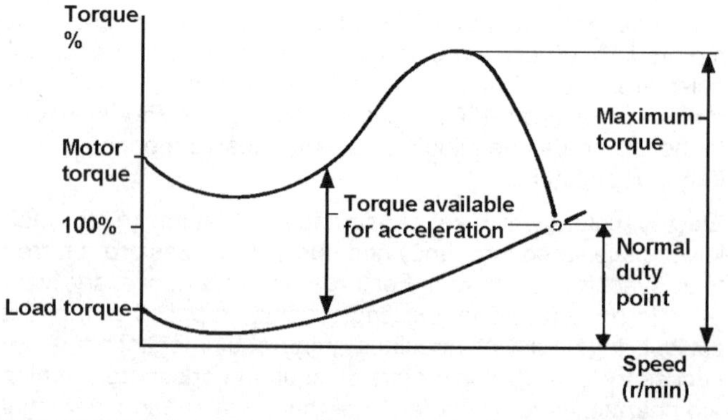

Figure 10.4 Induction motor speed/torque curve for direct-on-line starting

torque curve. When starting "direct-on-line", the motor has a relatively high torque, usually 1.5 to 2 times the full load torque. As speed increases the torque first reduces, followed by an increase to maximum torque and finally rapidly reduces to the point at which the motor torque is equal to the load torque; equilibrium being reached at the designed duty point.

Electric motor catalogues usually state the starting torque in relation to the full load torque and full load current. They also assume full supply voltage, but this is not always a valid assumption.

Maximum torque is a measurement of the electric motor's momentary overload capacity. This is usually required to be at least 1.6 times the full load torque, although most electric motors have values which are considerably higher.

The various critical torques, relevant to specific motor designs, have special names:

- **Locked rotor torque** — The torque at the instant the current is applied before the rotor begins to rotate.
- **Pull-up torque** — The minimum torque the motor will supply during run up to full load speed
- **Pull-out torque** — The maximum torque the motor will supply during run up to full load speed.

Some motor designs have a pull-up torque which is less than the full load torque. Depending upon the pump type and the starting conditions of the pump, this design feature can be unacceptable.

Single phase motors produce no torque when starting and require an auxiliary winding to facilitate starting and to ensure rotation in the correct direction. The following categories of motors are supplied:

- Motors with a resistance starter winding, which is disconnected automatically after starting
- Motors with auxiliary winding in conjunction with a capacitor. For low starting torque requirements, the capacitor can be connected during operation, whilst for larger starting torque requirements the capacitor must be disconnected automatically after start-up. Other combinations are necessary to achieve optimum conditions for large starting torques and good maximum performance
- Motors with commutators, "universal current motors"

10.1.7 Voltage

Generally small 3-phase motors, with the exception of pole changing motors, can be used for 380 V and 220 V. The motor in this case is marked Y/Δ 380/220 V and, depending upon the method of connection, can be used for either voltage.

Normally 380 V or 415 V is used for 3-phase motors up to about 250 kW. For larger motors, up to approx 450 kW or installations with large motor loadings, higher voltage may be required for both technical and economical reasons. According to present international standards for standard voltage, IEC 60038, a system voltage of 660 V should be used. This voltage has the advantage that it enables squirrel-cage motors to be constructed for connection to either 660 or 380 V since $\sqrt{3} \cdot 380 = 660$, see Section 10.1.2.

In conjunction with the 1983 issue IEC 60038, 230/400 V has replaced 220/380 V as the international value for public low voltage networks. According to this standard, the system's nominal voltage should not have exceeded 230/400 V + 6% - 10%, i.e. the range between 244 V and 207 V, during the transitional period up to 2003.

As specified in DIN IEC 60038, this ensures that all equipment that is rated for 220/380 V can be further operated until the end of its service life without its safety being impaired. The above details also apply to the 380/660 V or 400/690 V system's nominal voltages.

In the case of larger motors local voltage drops are often achieved by using transformers. For motors connected to high voltage supplies of 11 kV or 22 kV, transformers are normally used for a nominal motor voltage of 3.3 or 6.6 kV. Motors larger than 1000 kW can be connected directly to 11 kV supplies.

The choice of motor voltage must be made from technical and economical considerations. The determination of voltage involves not only the motor but the complete electrical equipment such as cabling, transformer requirements, switchgear, etc. Generally it can be said that for larger motor loads it is advantageous to use a higher voltage than 380 V (415 V).

If the supply voltage reduces then the motor torque is also reduced, in theory. Standard low voltage motors are designed to operate correctly on the nominal voltage ± 6%. Low voltage producing reduced torque applies when starting as well as during normal operation. The torque reduces in proportion to the square of the voltage. It is therefore important to ensure that electric cables to the motor are suitably sized to avoid excessive voltage drop. The maximum acceptable figures usually being 10% when starting and 5% during normal operation. The voltage drop during starting is higher because of the increased current flowing, typically between 4 and 7 times full load current depending upon the starting method. To ensure that starting is not a problem, some user specifications require the motor to be able to start at 80% nominal supply voltage.

10.1.8 Starting

Electric motors need to start and stop. The type of equipment required is dependent up on the type of motor, the starting torque required and the capabilities of the supply system. 3-phase motors are manufactured in power ratings down to 0.06 kW, single phase motors are rarely used in sizes over 7.5 kW.

Motors are started by contactors, sometimes called circuit breakers. Under normal circumstances a contactor appears to be a switch. When the start button is pressed the motor starts. When the stop button is pressed the motor stops. The difference is seen when there is a power cut or an overload. When the power is restored after a power cut, motors which were running before do not start. When an overload occurs, the motor appears to switch itself off, after the overload condition has cleared the motor does not start again.

A contactor reacts this way because a normal switch could be very dangerous, allowing motors to start unexpectedly. When the start button is pressed, what actually happens depends upon the size and voltage of the motor. For small, low voltage motors the button may press the switch contacts together, making the circuit. Once the circuit is made, a small electric coil holds the contacts "in". On high voltage motors, pressing the start button energises a low voltage coil which closes the contacts, again a coil holds the contacts "in". Pressing the stop button breaks the circuit of the "hold-in" coil and the contacts open or come "out".

When a power cut occurs there is no power for the motor so it stops; there is also no power for the "hold-in" coil so the contacts open. If power is restored the contactor is in the off position; the motor cannot start. When an overload occurs the power supply to the "hold-in" coil is broken so the motor stops as if the stop button had been pressed.

Motor control is a little more complicated than this. 3-phase motors can be started by various methods, see Section 10.4. Single phase motors are started by a simple two-pole contactor. Single phase motors, other than commutator motors, have the complicated starting control performed automatically by a centrifugal switch inside the motor.

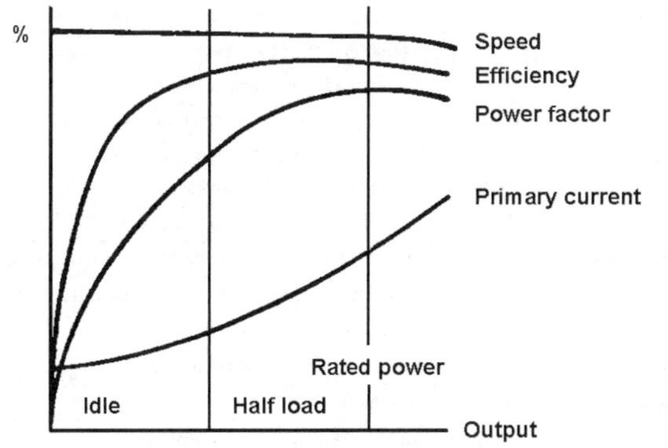

Figure 10.5 Variation in efficiency of electric motors with load

Controlling motors by electrically operated contactors has the advantage of allowing more than one motor to be started by the same signal. Also automatic control of motors becomes easy. The control system produces a simple low voltage digital signal which can open or close the contactors. Control systems can be constructed from separate relays and timers which are wired to perform the necessary logic functions. Alternatively, PLCs could be used. PLCs have digital outputs which can be rated at 110 V or 220 V to power the contactor hold-in coils.

Contactors can be extended to operate several circuits simultaneously. Not all circuits must close, and it is possible to open some while closing others. The motor contactor can generate digital logic signals for the control system as well as operating indicator lamps.

When considering starting motors, the regularity of starting must be evaluated. Small motors, 0.06 kW, can be rated for 2300 to 5000 starts per hour. Larger motors, 132 kW, are rated for only 30 starts per hour. When the pump duty cycle requires frequent motor starting, the motor manufacturer should be consulted with all available data. The type of starting method and the run up time, together with environmental conditions, are crucial.

DC motors are not usually used as fixed speed drives, see Section 10.3.3 about starting variable speed motors.

10.1.9 Efficiency

The efficiency of medium size electric motors, 10 to 100 kW, is usually 85 to 90% at full load. Better values can be obtained for larger motors, whilst 75 to 85% is usual for smaller motors. Losses in motors consist of bearing and friction losses together with electrical and magnetic losses in the windings. Some manufacturers produce energy efficient motors, which are 1.5 to 3% more efficient than their standard motors. Whether the increased efficiency can be cost effective is dependent upon local energy costs and the constancy of the application.

Figure 10.5 shows speed, efficiency, power factor and current consumption of an electric motor. It can be seen that efficiency varies with motor load.

10.2 Regulations and standards

10.2.1 Controlling authorities

The same regulations apply to the installation of electric motors as for other electrical installations and equipment.

Principal national standards authorities are listed in Table 10.2. A comprehensive list of countries and their national standards bodies who are members of ISO, International Standards Organisation are given in Section 10.11.

Country		National standards authority
Belgium	IBN	Institut Belge de Normalisation
	BIN	Belgisch Instituut voor Normalisatie
Canada	SCC	Standard Councils of Canada

Country		National standards authority
Denmark	DS	Dansk Standardiseringsrad
Finland	SFS	Suomen Standardisoimisliitto r.y.
France	AFNOR	Association Francaise de Normalisation
Germany	DIN VDE	Deutsches Institut für Normung Verband Deutscher Elektrotechniker
Greece	ELOT	Hellenic Organization for Standardization
Luxembourg	ITM	Inspection du Travail et des Mines
Ireland	NSAI	The National Standards Authority of Ireland
Italy	UNI	Ente Nazionale Italiano di Unificazione
Netherlands	NNI	Nederlands Normalisatie Instituut
Norway	NSF	Norges Standardiseringsforbund
Portugal	IPQ	Instituto Portugues da Qualidade
Spain	AENOR	Asociacion Espanola de Normalizacion y Certificacion
Sweden	SIS	Standardiseringskommissionen i Sverige
United Kingdom	BSI	BSI, British Standards
United States	ANSI	American National Standards Institute

Table 10.2 National Standards Authorities

The European Community has agreed to work together to establish European Standards. General standards are produced by CEN, European Committee for Standardisation. Electrical standards are issued by CENELEC, European Commission for Electrotechnical Standardization. CENELEC issues two types of standard. ENs, which are identical in principle to ENs issued by CEN; each member country must publish the EN and withdrawn any national standards which conflict.

CENELEC's other standards are HDs, Harmonising Documents. Each national standard body must publicise the existence of the HD but it does not constitute a standard which requires compliance. The member countries agree to drift towards HD compliance at a future date when it will become an EN.

Other major industrial countries have national electrical standards. Compliance with these may be necessary when pump packages are built for export.

International organisations have established standards for motors and other electrical equipment with the hope that eventually all electrical equipment will be interchangeable. This is perhaps an ideal which will never be realised. The more parties involved in negotiating standards the weaker the standard tends to become and the longer it takes to finalise. The International Electrotechnical Commission is the authoritative worldwide body responsible for developing consensus global standards in the electrotechnical field. IEC is dedicated to the harmonization and voluntary adoption of these standards, supporting the transfer of electrotechnology, assisting certification and promoting international trade.

There are 15 countries covered by CENELEC standards, IEC standards cover about 44 countries.

Area of design	BS	CEN/CENELEC	IEC / ISO
Balance, (residual vibration)	4999 Pt 142	≡ HD 347	≠ 34-14 ≡ ISO 2373 Gr N
Cooling arrangements	EN 60034-6	EN 60034 Pt 6	= 34-6
Dimensions	4999 Pt 141		≠ 72 & 72A
Hazardous area classification	5501 Pt 1		≠ 79-10
Insulation	2757		≡ 34-7
Mounting arrangements	EN 60034 Pt 7	EN 60034 Pt 7	≡ 85
Noise	4999 Pt 109	≡ HD 53.9	≠ 34-9
Performance	EN 60034-1 & -9	≡ HD 53.1	≡ 34-1
Physical protection	EN 60034 Pt 5	EN 60034 Pt 5	≠ 34-5 = 529
Starting	EN 60034-12	≡ HD 53.12	≠ 34-12
Temperature limitations	EN 50014 & 50018	EN 50014 & 50018	≠ 79-0 & 79-1
Terminals and rotation	4999 Pt 108	≡ HD 53.8	≠ 34-8
Thermal protection	EN 60730-2-2	EN 607302-2	≠ 730-2-2

Table 10.3 British, European and international motor design standards

The important areas of electrical equipment design which require standardization and regulation are shown in Table 10.3. The relationship between British, European and international standards are shown.

Equipment specifically for hazardous areas can be designed to reduce the attendant risks by using different philosophies. Table 10.4 indicates the relationship between standards. BS, EN and IEC/ISO have been rationalised.

Construction		BS	CEN/CENELEC	IEC / ISO
General requirements		EN 60079	EN 60079	IEC 60079
Oil immersed	"o"	EN 60079-6	EN 60079-6	IEC 60079-6
Pressurised	"p"	EN 60079-2	EN 60079-2	IEC 60079-2
Powder filled	"q"	EN 60079-2	EN 60079-2	IEC 60079-2
Flameproof	'D' "d"	EN 60079-1	EN 60079-1	IEC 60079-1
Increased safety	"e"	EN 60079-7	EN 60079-7	IEC 60079-7
Intrinsically safe	"i"	EN 60079-11	EN 60079-11	IEC 60079-11
Encapsulated	"m"	EN 60079-18	EN 60079-18	IEC 60079-18
Non-sparking	"N"	EN 60079-15	En 60079-15	IEC 60079-15

NOTE: "i" can also be "ia" and "ib", similarly "N" can be "n"

Table 10.4 Standards for hazardous area equipment

BS 4683 Part 2 and IEC 79-1 regulate the construction and testing of flameproof apparatus.

BS 5345, Code of practice for selection, installation and maintenance of electrical apparatus for use in potentially explosive atmospheres (other than mining applications or explosive processing and manufacture is worth reviewing. There is no alternative, or close equivalent in European standards. Parts 1 and 2 are similar to some standards, IEC 79-10 and IEC 79-14, but the remainder of 5345 is unique. Figure 10.6 indicates the acceptable types of construction which can be used in the various zones, see Section 10.2.8.

Commercial organisations also have standards which regulate motors in specific applications, take shipping for example. The following bodies have specific motor regulations for marine applications. Off-shore oil installations may be included.

Lloyd's Register

Germanischer Lloyd

Det Norske Veritas

Bureau Veritas

OCMA, Oil Companies Materials Association, has specifications relating to motor protection. EEMUA, Engineering Equipment and Materials Users Association and EEMAC, Electrical and Electronics Manufacturing Association of Canada, both have motor specifications.

NOTE: In Europe, it is important to remember the ATEX Regulations, **REGULATIONS** !! not specifications or standards, came into full effect on the 30th June 2006. These regulations will be implemented with some vigour after the recent serious accidents.

10.2.1.1 Specification compliance

In order to ensure motor manufacturers comply with all the relevant standards independent test houses have been established to validate specific motor designs. Test houses are mostly employed for equipment destined for hazardous areas, usually potentially explosive atmospheres. Motors are tested on load under laboratory controlled conditions to ensure that construction tolerances limit air gap sizes and that surface temperatures do not exceed prescribed values.

Zone	Type of protection
0	Ex 'ia' Ex 's' (specifically certified for use in Zone 0)
1	Any type of protection suitable for Zone 0 and Ex 'd' Ex 'ib' Ex 'p' Ex 'e' Ex 's'
2	Any type of protection suitable for Zone 0 or 1 and Ex 'N' or 'n' Ex 'o' Ex 'q'

Figure 10.6 Electrical apparatus construction for various zones

Test houses are located in several countries and each issues their own certification. The areas of expertise are not identical, some test houses are preferred to others for certain equipment. The test houses listed in Table 10.5 are in alphabetical order and not in any order of superiority.

Country		Test house
Australia	NSW	New South Wales Mines Approval
Austria	TÜV-A	
Belgium	ISSEP	
Canada	CSA	Canadian Standards Association
	EMR	Energy, Mines and Resources Canada
Denmark	DEMKO	Dansk Elektrische Materialkontrol
France	INERIS	Institut National de l'Environment Industriel et des Risques
	ISSeP	Institut Scientifique de Service Public
	LCIE	Laboratoire Central des Industries Electriques
Germany	DMT-BVS	Bergbau-Versuchsstrecke Dortmund-Derne
	PTB	Physikalisch-Technischen Bundesanstalt
Italy	CESI	Centro Elettrotecnico Sperimentale Italiano
Netherlands	N.V. KEMA	Keuring van Electrotechnische Materialen
Norway	NEMKO	Det Norske Veritas
Spain	LOM	Laboratorio Oficial J M Madariaga
Sweden	SP	Swedish National Testing & Research Institute
United Kingdom	EECS	Electrical Equipment Certification Service, incorporating:
	BASEEFA	British Approvals Service for Electrical Equipment in Flammable Atmospheres
	MECS	Mining Equipment Certification Service
		SCS
		SIRA Certification Service
USA	UL	Underwriters Laboratory
	FM	Factory Mutual

Table 10.5 Test houses

10.2.2 Physical protection

IEC 60034 Part 5 classifies the types of enclosure for rotating electrical machines in respect of the degree of protection afforded to personnel against contact with live parts, and against ingress of solid foreign particles and liquids. The various types of classifications are designated by using the letters IP followed by two numbers. The same method is also applied in IEC 34-5 and IEC 529 but the design requirements are not identical. The table in Table 10.6 reviews the standard designations for various protection categories.

First digit		Second digit	
0	No protection	0	No protection
1	Protection against solids ≥ 50 mm	1	Protected against vertical drips
2	Protection against solids ≥ 12 mm	2	Protected against drips 15° from vertical
3	Protection against solids ≥ 2.5 mm	3	Protected against spray 60° from vertical
4	Protection against solids ≥ 1.0 mm	4	Protected against splashing water
5	Protection against dust	5	Protected against water jets
5	Protection against dust	6	Protected against heavy seas
5	Protection against dust	7	Protected against intermittent immersion
5	Protection against dust	8	Protected against continuous immersion

Table 10.6 Physical protection categories for electric motors

The dust used for testing is talc, 75 μm nominal size at a concentration of 2 kg/m^3. The test for intermittent immersion is conducted with water for 30 minutes. The minimum immersion depth is 150 mm for the top on the motor, the minimum immersion for the bottom of the motor is 1 m.

Motors which are weather-proofed are designated by the addition of the letter "W" between "IP" and the number.

In certain cases electric motors must be equipped with condensation drainage facilities, usually by drilling a 6 to 8 mm dia- meter hole at the lowest point. IP 22 used to be the standard for motor protection, most motors now are IP 44 minimum. Motors for offshore platforms are usually IP W 55. Motors for hot, humid atmospheres are often required to be "tropicalised"; this treatment has not be standardised.

European and international standards recognize the effect of vibration on personnel. Vibration from motors can be transmitted through support structures to the ground. To limit personnel discomfort, and ensure proper bearing life, European Standard HD 347 has been incorporated in BS 4999 Part 142. IEC 34-14 has identical requirements to ISO 2373 Grade N, which are similar to European requirements.

Noise continues to be an increasing problem. Motor noise is specified in HD 53.9 and BS 4999 Part 109. IEC 34-9 is similar.

10.2.3 Cooling categories

In IEC 34-6 the categories for air cooling are designated by means of the letters IC followed by two figures. The designations used in BS 4999: Part 21 utilises the same two figures but omits the letters IC. The first character refers to the type of cooling and the second to the method of circulation. The most common methods of cooling electric motors and their designations are shown in Table 10.7. The IEC Standard also contains a more detailed code for more complicated cooling systems. Standard air cooling designs are based on an ambient temperature of 40°C and an altitude of 1000 m maximum.

BSI Standard	IEC Standard	Cooling Category
0/1	IC01	Free circulation, self circulation
0/6	IC06	Free circulation, independent components mounted on machine
1/1	IC11	Inlet duct ventilated, self circulation
1/7	IC17	Inlet duct ventilated, independent and separate device or coolant system pressure
2/1	IC21	Outlet duct ventilated, self circulation
2/7	IC27	Outlet duct ventilated, independent and separate device or coolant system pressure
3/1	IC31	Inlet and outlet duct ventilated, self circulation
3/7	IC37	Inlet and outlet duct ventilated, independent and separate device or coolant system
4/1	IC41	Frame surface cooled, self circulation
5/1	IC51	Integral heat exchanger (using surrounding air), self circulation

Table 10.7 Motor air cooling categories

Motors for pumps in boreholes and similarly limited spaces usually have product lubricated bearings. Cooling is achieved by using the pumped product. The rotor chamber is filled with liquid and protected against freezing. The motor is connected to the pump by means of an intermediate section, which also

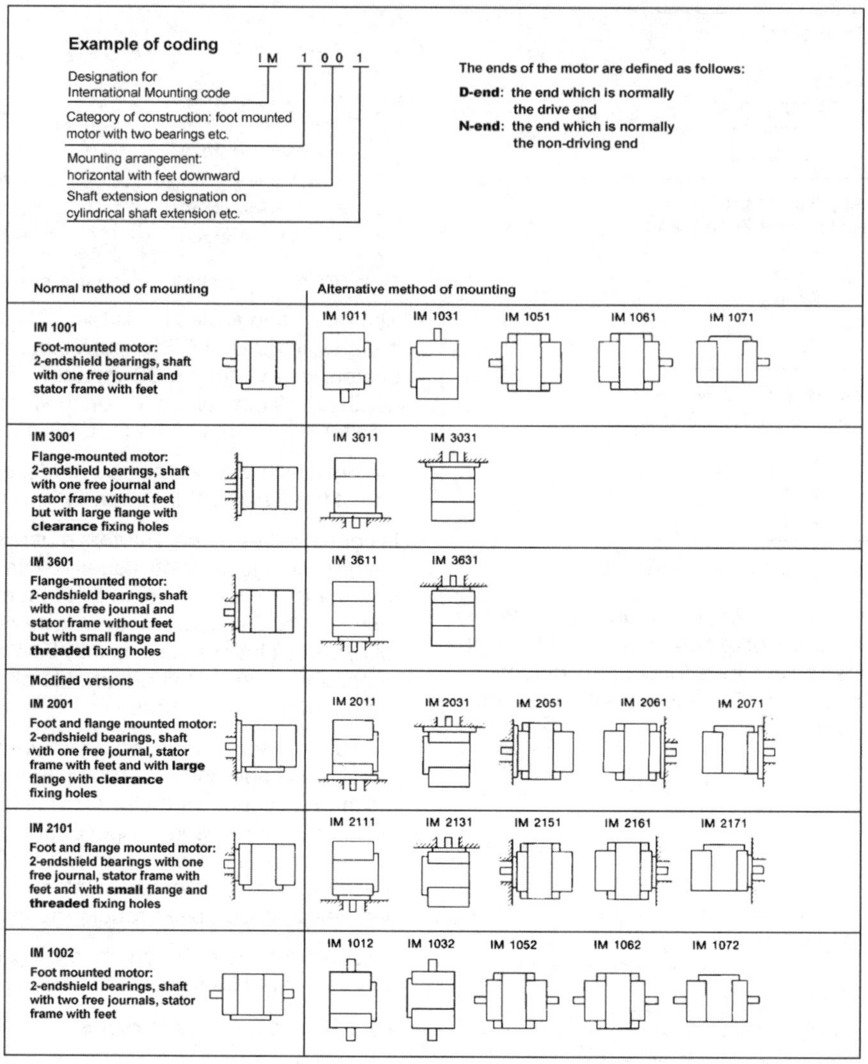

Figure 10.7 Mounting arrangements for electric motors

serves as the pump inlet. Because of the limited space both the pump and motor must be specially designed.

Motors for pumps immersed in larger vessels can be connected directly to the pump casing. Shaft sealing is by means of two mechanical seals in an oil bath. The motor is normally cooled by means of the pumped liquid, which is effective and requires less motor power than for an equivalent air-cooled motor. The motor should be fitted with temperature detectors, thermistors, for 120°C, integrated in the stator windings to protect against dry running. The detectors control a contactor which can be incorporated in the terminal box.

Typical cooling systems include:

- **Direct cooling** of the motor outer casing, which can be corrugated to increase the surface area, is used when the motor is permanently immersed in liquid which does not exceed 40°C
- **Casing cooling** using the pumped liquid. Part or all of the process liquid passes between the outer casing and the motor casing. The pump can be either "wet" or "dry", although dry-running is not permitted. At part flow cooling, the liquid must not contain solid particles. The temperature of the liquid must not exceed 40°C
- **Oil cooling**. Oil from the pump's oil case is circulated in the outer casing during operation
- Casing cooling using external liquid which passes between the outer casing and the motor casing. Temperature of the pumped liquid can normally be up to 70°C. This system is used in potentially explosive atmospheres. The motor should stop in the absence of positive pressure
- A "wet motor" consists of a 3-phase or single phase cage induction motor, hermetically attached to the pump casing. The pump impeller is connected directly to the motor shaft. The rotor and bearing in the motor are lubricated and cooled by the pumped liquid. This construction constitutes a leak-free pump suitable for transportation of hazardous liquids and for use in potentially explosive atmospheres
- In a wet motor the pumped liquid completely fills the motor casing. Both the stator and the rotor are immersed in the liquid. For wet rotor motors the stator winding is kept dry by means of a thin stainless steel shroud which separates the stator from the liquid. This type of construction is used for particularly troublesome media and also very small pumps, such as domestic central heating circulators. Wet motors and wet rotor motors are also used for submersible pumps
- Motor cooling can be monitored by embedding temperature detectors in the motor windings. Thermistors are normally used but resistance temperature detectors, RTDs, can be used at extra cost. One detector per phase is the minimum requirement; two per phase for critical applications. See BS 4999 Part 111, EN 60730-2-2, IEC 34-11 and IEC 730-2-2

10.2.4 Mounting arrangements

The design of electric motors with respect to bearings, shaft extension and methods of fixing is covered by BS 4999: Part 107, EN 60034 Part 7 and IEC 34-7. The motor construction allows for operation in all mounting positions, a motor which is normally intended for a particular mounting position is also shown in alternative positions. Figure 10.7 shows examples of various types of mounting arrangement of motors together with their international standard designations. For practical reasons larger manufacturers of electric motors often have their own methods of designation.

10.2.5 Terminal markings and direction of rotation

Terminal markings and direction of rotation are covered by BS 4999: Part 108 which is directly equivalent to CENELEC HD 53.8. IEC 34-8 is a similar specification. According to these standards the phases in a 3-phase motor should be designated by the letters U, V, W and the external earth or neutral connection designated by the letter N. The normal direction of rotation in electric motors is clockwise, viewed on the shaft drive end, for phase sequence U, V, W.

3-phase electric motors can be reversed by changing the connection positions of any two phases. This operation is easily carried out by a contactor. **Not all** motors are suitable for reverse operation but most standard motors are satisfactory. Motors, specifically designated as "quiet", are probably not suitable. One method of reducing motor noise is to fit a special cooling fan which is designed for one specific direction of rotation. Care must be exercised when reverse running is required.

Some pumps are not suitable for reverse running. Shaft driven oil pumps may not function causing bearing lubrication problems. Pumping rings in mechanical seals can be another problem. The direction of motor rotation must be checked during installation/commissioning. Start the motor when the coupling spacer is removed. Change two phases if necessary. Figure 10.8 shows fixing dimensions for flange-mounted motors with mounting arrangement IM 3001 and 3002.

Designation for main dimensions

A	=	Distance between fixing hole centres (end view)
B	=	Distance between fixing hole centres (side view)
C	=	Distance from fixing hole centres at drive end of motor to shoulder of shaft
D	=	Diameter of shaft extension
E	=	Length of shaft extension from the shoulder
H	=	Shaft centre height
K	=	Diameter of holes in feet or mounting pads of machine

Dimensioning for foot mounted motors with mounting arrangement IM 1001 and 1002

M	=	Pitch diameter of fixing holes
N	=	Spigot diameter
P	=	Outside diameter of flange
S	=	Fixing hole diameter
T	=	Depth of spigot

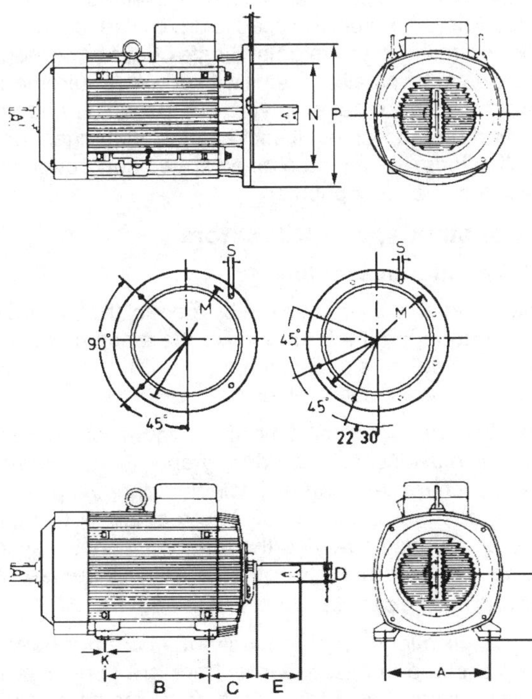

Figure 10.8 Standard dimensions for electric motors

10.2.6 Rated power, centre height and outline dimensions

The IEC Standards 72 and 72 A contain recommendations concerning power, centre heights, foot mounting dimensions, shaft end dimensions and flange dimensions for electric motors. The preferred series of electric motor output (kW) are as follows; 0.37, 0.55, 0.75, 1.1, 1.5, 2.2, 3, 4, 5.5, 7.5, 11, 15, 18.5, 22, 30, 37, 45, 55, 75, 95, 110, 132. Ratings and performance within Europe are covered by HD 53.1 which has been incorporated in BS 4999 Part 101. Motor output is structured in HD 231.

According to the relevant electric motor standards, dimensioning should be carried out in the manner shown in Figure 10.8 using standardised dimensions. Interchangeability is then possible between different manufacturers providing the non-standardised dimensions are reasonably similar. It should be pointed out that this standardisation is not yet fully implemented, so that it is prudent to carry out checks when purchasing replacement motors.

For induction motors, which are the most common type of electric motor, there is a special standard, BS 4999 Pt 141 : 2004, which is very similar to IEC 72/72A.

10.2.7 Temperature classification

Winding insulation thickness is normally designed to cope with the hottest point for an ambient temperature of +40°C at an altitude of up to 1000 m. If the motor is exposed to excessively high temperatures then the service life of the insulation, and thus the motor, is considerably reduced. It is therefore necessary to use suitably designed insulation if the motor is to be operated at high temperatures.

Motors which will operate at ambient temperatures below 40°C and a sea level are capable of safely producing more than the standard nameplate rating without problems. The motor manufacturer should be contacted to assess the possibility of a motor upgrade.

EN 60085 : 2004, which is related to IEC 85, categorises insulation material according to temperature and insulation classes which are designated by means of a number and a letter. These represent the upper limit of use for normal operating conditions and acceptable service life for the particular insulating material, see Figure 10.9. For example, the 120 (E) indicates that the insulation can be used up to a maximum of +120°C, ambient temperature of 40 + temperature rise 75 + safety margin 5.

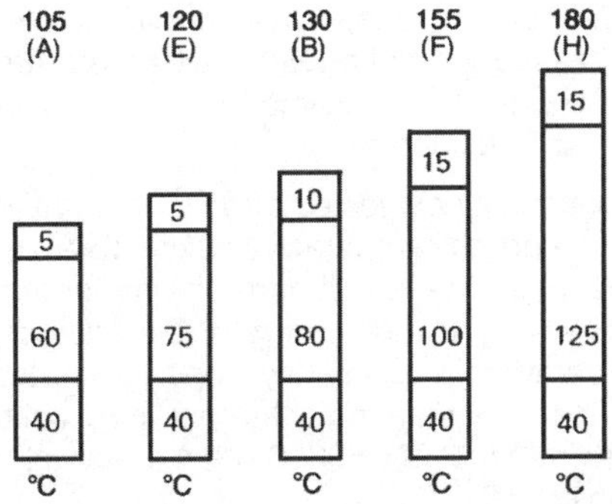

Figure 10.9 Temperature classification for insulating material

The temperature rise of an electric motor can be determined by means of direct temperature readings or by measuring the resistance of the motor windings when cold and hot, respectively, as follows:

$$\text{Temperature rise } ^\circ\text{C} = 225 - \left(\frac{\text{cold resistance}}{\text{hot resistance}} - 1\right) + 5 \pm 15$$

For extra safety, and increased insulation life, some purchasers will specify Class “F” insulation with the temperature rise limited to Class “B” specification. This means the insulation runs 20 to 25°C cooler than designed.

Motors for potentially explosive atmospheres have additional restrictions placed on temperature rises. To avoid ignition of any gas present in the atmosphere, the motor temperature is restricted according to a grouping for the gas(es) likely to be present. All flammable gases have been allocated one of six categories. The maximum motor temperature, according to BS 4683, EN 50014 and EN 50018 for flameproof motors, are shown in Table 10.8.

Temperature class	Maximum surface temperature°C
T1	450
T2	300
T3	200
T4	135
T5	100
T6	85

Table 10.8 Maximum surface temperatures for flame-proof motors

10.2.8 Potentially explosive atmospheres

Special regulations apply for electrical equipment and installations. For example where there is a potential fire or explosion hazard because of the generation, handling or processing of flammable liquids, vapours, mists, etc. These regulations also apply in areas where there are other kinds of explosive materials, such as dust. Fire hazardous liquids, vapours and gases are defined and classified according to their flash point, see Section 2.1.7.

Electrical equipment should not be installed in potentially explosive locations if this can possibly be avoided. If this is unavoidable then the apparatus must conform to the relative standards and regulations.

BS 5345 offers guidance in the selection and installation of electrical apparatus. The code corresponds to the IEC recommendations for electrical apparatus for use in explosive atmospheres, although the numbering systems are different. BS 4683: Part 2 = IEC 79-1, Electrical apparatus for explosive atmospheres, prescribes features for flameproof enclosures and groups them according to the type of atmosphere for which they are suitable and gives dimensions and permitted gaps.

Hazardous areas are classified in zones according to internationally accepted concepts, BS 5345 Part 2 and IEC 79-10, as follows:

Zone 0 in which an explosive gas-air mixture is continuously present, or for long periods.

Zone 1 in which an explosive gas-air mixture is likely to occur in normal operation.

Zone 2 in which an explosive gas-air mixture is not likely to occur in normal operation, and if it occurs it will exist only for a short time.

In practice, physically defining a zone or zones around a piece of equipment may be a problem. American API 500 and NFPA 30 produces very small zones. RoSPA and the HSE CS5 regulations produce much larger zones. There are no agreed European or international regulationsbecause of the different approaches adopted by chemical and electrical engineers.

As a rule standard squirrel-cage three-phase motors, mechanically and electrically adapted, or derated, are used for hazardous environments. To maintain effective protection it is important that both installation and servicing are carried out thoroughly and carefully.

Potentially explosive gases are divided into groups. Group I is reserved for underground mining applications. Normal industrial and processing applications are covered by Group II, which is subdivided into IIA, IIB and IIC. BS 5345 Part 1 and IEC 79-12 list gases and vapours which have been classified, indicating their temperature class and group. A few gases and vapours cause problems and have not been classified; acetylene is an example. A motor environment should be described fully: Zone 1 Group IIA T3.

Potentially explosive atmospheres can be created by dust clouds. Layers of dust on equipment can be ignited by hot surfaces as well as by flame or spark, see BS 6467 and BS 7535. Some dust clouds can be ignited by temperatures as low as 280°C and dust layers at 225°C.

Zones for dust are designated Z — dust present under normal operating conditions, and Y — dust present only under abnormal operating conditions. Physical protection testing specifically for dust, IP rating, is conducted using French chalk with particle sizes down to 1 μm. IP 6x designates complete immunity to dust ingress. The minimum physical protection recommended for Zone Y or Z is IP 65.

10.2.9 Certification

Electrical equipment manufacturers design in accordance with the requirements for the standards relating to their equipment. To be certain that their equipment does comply with all the requirements, the equipment is submitted to an independent test authority, see Section 10.2.1, who checks the equipment and certifies compliance.

When equipment has been certified the type of certification is shown on the equipment nameplate. The following information appearing:

- applicable standards
- test house identification
- test house certificate number

As part of a project quality assurance programme, the purchaser may ask for a copy of the actual test certificate. In extreme cases, say a T5 or T6 temperature limitation, the purchaser may send an inspector to review the original test house certificate and confirm the photocopy to be a true copy.

10.3 Motor types

The most common type of electric motor used for driving pumps is probably the multi-speed, single phase canned motor with a wet rotor; this is the motor type used to drive water circulators in domestic central heating systems. Single phase cage induction motors are used for smaller pumps with power requirements up to several kW, e.g. barrel emptying pumps which need to be portable and be fed from wall sockets. For industrial and process applications, the 3-phase, fixed speed squirrel-cage motor is definitely the most popular.

10.3.1 Constant speed AC motors

10.3.1.1 Synchronous motor

This type of motor is characterised by the fact that speed is totally dependent upon the frequency of the electric supply and the number of poles, see Section 10.1.5. The motor runs at constant speed irrespective of load.

The synchronous motor must be fitted with a special starter winding, which is disconnected during operation. Also necessary is a DC energised magnetic winding; the DC being supplied by a generator in the motor. The reactive power can be regulated by means of varying the magnetisation. A synchronous motor can, therefore, contribute to an increase of the power factor cos φ, see Sections 10.1.3 and 10.1.4.

For pump applications synchronous motors are normally used in larger sizes and in cases where there are special requirements. The reason being that this type of motor is relatively complicated, costly and sensitive to frequency variations in the supply system.

10.3.1.2 Squirrel-cage motors

Squirrel-cage seems an odd name for a motor type. It is an old name. If the design was invented today it would probably be called something like "Hamster wheel". The important electrical parts of the rotor resemble the exercise wheel found in hamster cages. Squirrel cage motors are also called simply "cage" motors.

Three-phase squirrel-cage induction motors are the predominant type for industrial and process pump applications. The construction is simple, requiring a minimum of maintenance and spare parts. Their starting qualities are generally good. There are usually two types of rotor to choose between: simplex and double-cage. The former is used for direct-on-line starting if the starting torque does not need to be high. The simplex rotor is the most common and is especially suited when frequent momentary overloading occurs.

Motors for submersible pumps are usually 3-phase motors. Although single phase motors are not unknown for this purpose.

Squirrel-cage motors are constructed on the principle that a speed differential exists between the rotating stator magnetic field and its rotor, thereby maintaining induction in the rotor in a simple and reliable manner. The speed of an induction motor is therefore somewhat less than that of a corresponding synchronous motor with the same number of poles, see Section 10.1.5. The speed of an induction motor reduces with increased load, see the relationship between torque, power and speed in Section 10.1.6. The direction of rotation can be changed by means of simple switching.

Squirrel-cage motors are popular because the simple design can be adapted to suit many situations. Slipring and DC motors, because of their internal rubbing connections, are not as easy to protect for hazardous area situations. Also the internal rubbing connections are subject to wear and require maintenance. Motors of over 100MW have been supplied.

10.3.1.3 Slipring motors

In this type of motor the stator construction is identical to that of the squirrel-cage motor. The rotor windings are connected to rotating sliprings on the rotor. By connecting the rotor windings to an external power supply via resistances during starting, considerable limitation of the starting current can be achieved. The resistance can be adapted to suit the desired starting torque.

During starting the resistance can be successively reduced as required as speed increases, and this can be arranged as an automatic function. The resistance is completely disconnected at full load speed and the sliprings are shorted to create, in effect, a cage rotor, the principle of which is shown in Figure 10.10. The resistance is successively reduced as speed increases. Figure 10.11 shows the effect of a starting sequence of a slipring motor with five resistance steps.

The connection of starting resistance to the rotor windings is via brushes rubbing on the sliprings. Sliprings and brushes have been standardised in BS 4999 Part 147 which is identical to IEC 136. Slipring motors are used when there is a requirement for low starting current and in cases where the starting torque is relatively high or the run up time is long. This type of motor can also be used when speed regulation of the pump is required.

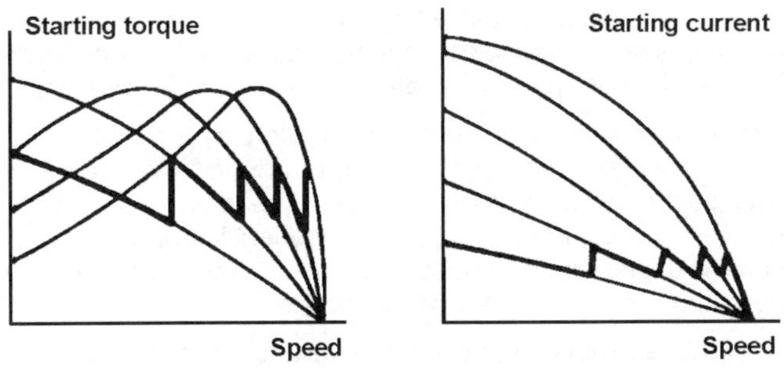

Figure 10.11 Starting torque and current for a slipring motor

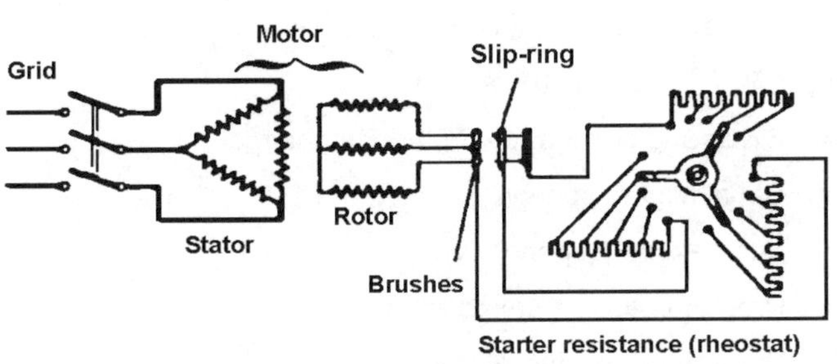

Figure 10.10 Wiring diagram of a slipring motor

10.3.2 Variable speed AC motors

10.3.2.1 Pole-changing motors

By using pole-changing motors, see Section 10.1.5, discrete speed regulation can be achieved. For pump operation it is common to use two-speed motors whereby the combination of pole numbers determines the speed. The combination 2/4 gives speeds of 3000/1500 r/min, the combination 2/8 gives 3000/750 r/min. The speeds quoted are ideal without consideration of slip. Lower efficiencies are normally expected at low speeds. Three and four-speed motors are available.

For the usual type of two-speed motor the stator is fitted with separate windings for each set of poles. The output of the motor will thus be relatively low when only half of the stator capacity is used for each set of poles. The same winding can, however, be used for both sets of poles by means of a special connection method, the Dahlander connection, see Figure 10.12. Motors with this type of winding cannot be Star-Delta started. (See Section 10.4.3.)

Pole-changing motors are constructed as cage electric motors. Starting can be achieved at either the low or high speed. In the latter case, however, it does not always follow that the starting current will be correspondingly lowered.

10.3.2.2 Variable frequency motors

Section 10.1.5 showed that AC motor speed is a function of the supply frequency and the number of poles in the motor stator. Pole-changing motors achieve discrete speeds by changing the number of stator poles. Up until about 30 years ago it was very difficult and costly to change the frequency of AC power supplies. Then the development of power electronics made it possible to convert AC to DC and then DC to any frequency AC. The DC current was switched on and off at high speed to generate a sinusoidal waveform.

The new solid state power electronic components allowed high currents to be switched at voltages up to about 600. This was ideal for low voltage 3-phase systems and standard 3-phase squirrel-cage motors. As frequency inverters developed from the original current and voltage source devices, additional features were added. The motor supply current was compensated,

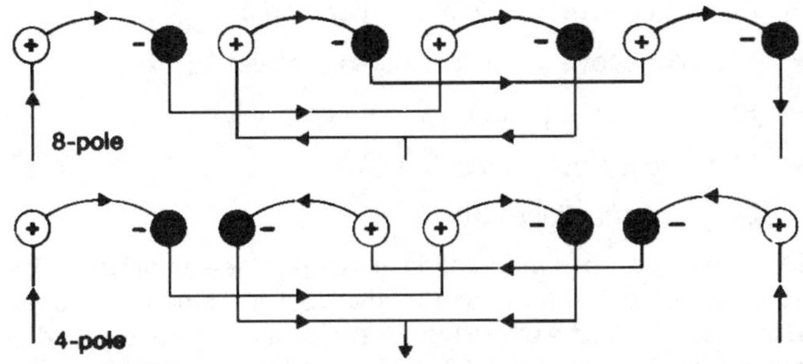

Figure 10.12 Principle of pole-changing stator winding for two-speed motors

at low frequencies, to allow full load torque to be developed. The high speed switching technique was refined to allow varying "switched on" periods, known as pulse width modulation.

Modern inverters can drive multiple motors simultaneously. In theory, inverters can offer a speed range of 1% to 100%, in practice the minimum allowable pump speed will be 5 to 10%. inverters are capable of supplying frequencies up to 400 Hz, which is useful if a high speed motor is required. Contactors for starting are built into the inverter cubicle.

The 3-phase supply produced by inverters is not as smooth as the standard current available from the grid. The distortion of the waveform can cause heating in the motor. Some motors may require derating to operate with an inverter. Others may make more noise when operating with an inverter. Motors which are required to run at low speeds, at full torque, for extended periods may need to be fitted with a separated motor driven cooling fan. The inverter cubicle may have to be mounted "close" to the motor; the maximum cable length should be confirmed.

Some inverter packages have poor low frequency performance resulting in cyclic motor speed variations, or "cogging". Inverter packages emit radio frequency interference, RFI. The interference is transmitted from the inverter into the electric supply system. Small inverters produce little interference. Large inverters can produce significant interference, enough to cause problems for electricity supply companies. If an inverter drive over 500 kW is planned, consult the local electricity supplier. The small problems which can occur are more than compensated for by the increased versatility of continuously variable speed over a wide range. In order to ensure compatibility between the motor and the inverter, it is advisable to buy both from one vendor.

The use of an inverter to supply motor power can invalidate the hazardous area certification of a motor. This problem can be overcome by certifying the motor and the inverter together. The German test house PTB specialises in this type of testing. Certification problems, and overheating problems, can sometimes be remedied by fitting thermistors in the motor windings which are connected to the contactor "hold-in" coil. Overheating problems with motor shafts have occurred; winding temperature detectors may not solve this particular difficulty.

Modern inverter circuits are sophisticated and can offer many useful features; the following are pertinent to pump applications:

- Short circuit protection
- Earth fault protection
- Set minimum speed
- Set maximum speed
- Programmable acceleration
- Programmable deceleration
- Torque limit setting for starting
- Torque limit setting for normal running
- Automatic reset to minimum speed on trip
- Smooth recovery from momentary power failure
- Slip compensation (for accurate flow only)
- Motor voltage/current indication
- Diagnostic fault indication

Modern pulse width modulation drives produce an almost constant power factor, close to 1.0, irrespective of the motor duty. Depending upon the pump type, the inverter may be required to supply 150% torque for 10 or 20 seconds for starting. The method of speed control must be compatible with the process control system; manual, 0 to 10 V DC; 4 to 20 mA DC, RS 232, RS 422, RS 485, IEEE 488. Remote control and telemetry, via radio or microwave links, is possible for outlying installations.

Variable frequency supplies can be applied to synchronous motors to produce variable speed motors. Slip compensation is not required because the motor is locked to the supply frequency. Salient pole rotor machines are ideal for high speeds. These drive packages would be used for larger power requirements, 1 MW and larger. Designs are available for 5 MW at 8000 r/min. Motors are installed for 3.5 MW at 7720 r/min, 5.3 MW and 9.15 MW. Brushless motors are possible by using standard induction techniques for the rotor excitation current. The lagging power factor, which varies with speed, may need correction on large powers. The increased cost of these motor/inverter packages can be weighed against the increased efficiency. The modified inverter design can be up to 7% more efficient than conventional squirrel-cage inverters.

The switched reluctance motor attracts much attention because of three useful features. The motor rotor can be manufactured from a single steel forging and does not require windings or a "cage". The rotor is suitable for very high speeds. The power electronic controls are simpler and cheaper than those required for squirrel-cage motors. The motor does not tend to overheat at low speeds when delivering high torques. The variable speed switched reluctance motor could replace a variable speed squirrel-cage motor and a gearbox. High motor speeds allow the use of smaller pumps for some applications. Direct drive positive displacement pumps are a possibility.

Some inverters have been supplied for motors over 100MW. Always purchase the motor and the inverter controls from the same supplier. By following this advice there will not be any problems fixing either or both if incompatibility appears to be affecting performance.

10.3.3 Variable speed DC motors

Before the advent of power electronics, variable speed drives which required accurate speed and torque control would have been DC. The DC power supply would have been supplied by a motor-generator set; a squirrel-cage motor driving two dynamos. One dynamo would supply the field current, the other for the armature, commonly known as Ward-Leonard. By using two dynamos, the two current supplies could be controlled independently. Also control was achieved through the low power excitation currents; heavy current power cables were wired direct from the dynamos to the motor.

Power electronics changed DC motor control as significantly as it affected AC motor control. DC could now be derived electronically without motor-generators. DC motor control equipment includes two sets of rectifiers to allow independent supplies to the motor field and armature. DC motors can be designed for any voltage, up to about 600 V, and any speed. Small motors, up to 7.5 kW, have field voltages of 150 to 360 V and armature voltages of 180 to 300 V. Larger motors have armature voltages up to 500 V. Motor speeds are generally between 1500 and 3000 r/min.

Common specifications for DC systems include:

- 20:1 speed range at constant torque
- 150% torque for 10 seconds
- 2% speed change for 100% load change
- Set minimum speed
- Set maximum speed
- Programmable acceleration
- Programmable deceleration
- Set current limit

For accurate speed control tacho feedback can be added, and the speed change reduced to 0.5%. Small standard units are available with 0 to 10 V DC or 4 to 20 mA DC control signals. Larger units can have the same control facilities as AC inverters. For units over 7.5 kW, speed stability can be as good as 0.1% and overload current rating extended to 30 seconds.

Power electronic controllers for DC motors produce a lagging power factor which varies with speed. This will not be a problem on small drives. Small variable speed DC systems are very economical but the motors can be larger than standard AC; AC variable frequency systems become more attractive as the drive power increases.

There is no simple guide to the economic changeover. The particular application requirements will determine which drives are suitable. AC variable frequency drives can now do almost everything the DC drive was intended to do. DC motors can be a problem for hazardous areas. The commutator and brushes, like sliprings and brushes, require extra maintenance over simple squirrel-cage motors.

10.4 Motor starters

Single phase starting was covered in Section 10.1.8 and 3-phase starting is now discussed in some detail.

10.4.1 Protection

Figure 10.13 shows how a typical 3-phase motor wiring diagram. All electric installations must be safe and include protective devices for personnel, the wiring as well as the motor and any motor control equipment. IEC 60947-4 covers motor starters and contactors. It has been reported that over 60% of motor failures were caused by:

- Insulation failure
- Overloading
- Stalled rotors
- Single phasing

The various protective devices are described in the following Sections.

10.4.1.1 Short circuit protection

The purpose of the short circuit protective device is to quickly disconnect the motor, starter/contactor and any control equipment from the supply in the event of a short circuit or earth fault. Protection can be provided by fuses or circuit breakers. Fuses are old-fashioned and slow to replace, but extremely reliable and fail-safe. Miniature plastic-cased circuit breakers, using magnetic sensing, are available for the smallest loads. Resetting is very quick. Short circuits can be caused by broken cables and damaged insulation.

10.4.1.2 Overload protection

This is built into the starting contactor by means of a current detection device, an overload relay. Two popular methods are available; thermal and magnetic.

Thermal protection is accomplished by passing the motor current through a bimetallic strip. Due the difference in expansion of the two metals the strip bends as it warms up. The temperature rise of the strip is related to the current passing through. When the strip bends passed a set limit, the "hold-in" coil circuit is broken and the contactor opens.

Magnetic protection is implemented by using the variation in magnetic strength of a coil as the flowing current varies. When a set magnetic pull is exceeded the "hold-in" coil circuit is broken. Dampening of thermal bimetallic strips is inherent in the mass of the strip. Dampening of the magnetic devices is provided by adjustable dash-pots.

Motor stopping with overload symptoms can be caused by:

- Pump running out on curve
- Liquid SG too high
- Pump running at relief valve accumulation
- Poor cable insulation
- Single phasing
- Bearing seizure
- Solids trapped in wear rings

Single phasing is caused when the connection between one of the phases is broken while a motor is running. The motor will continue to run but at a slightly lower speed. The current drawn by the two working phases will increase by about 75% causing overheating in the windings. A motor with a broken phase connection cannot be started.

10.4.1.3 High temperature protection

Motor winding temperature detection has already been mentioned. Thermistors normally, or resistance temperature detectors, RTDs, are optionally embedded within the motor field windings. One per phase is standard, two per phase for critical applications. As the windings warm up during normal operation the temperature can be measured and indicated. A high temperature switch can be included in the circuit to alarm and/or trip at set temperatures.

The motor will overheat under the following conditions:

- Developed power too high (overload)
- Too frequent starting
- Single phasing
- Earth fault in a winding
- Ambient air too hot
- Cooling system fault

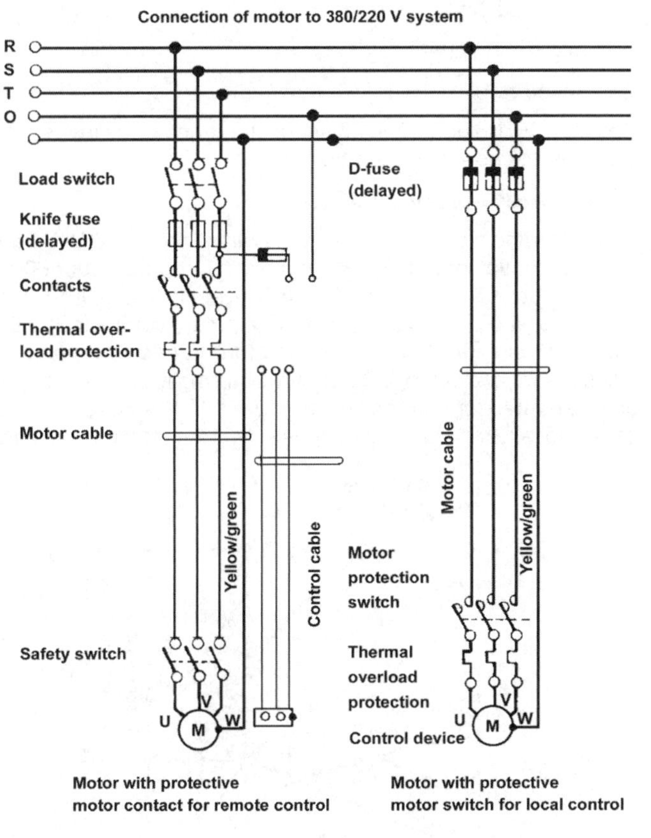

Figure 10.13 3-phase motor wiring diagram

- Fan problems
- Motor covered in dust
- Motor mechanical problem

Thermistors and RTDs can work with the starter overload protection and allow current protection to be set a little higher for better starting performance.

10.4.1.4 Safety switch

For safety reasons, when the motor controls are distant from the motor, in a control room for instance, it is necessary to fit some local control next to the motor. In the event of an emergency, personnel close to the motor can stop it before too much damage is done. The local safety switch must have a "stayput" stop button to prevent the motor being restarted remotely. Also, it is a good idea to have a lockable isolator for maintenance purposes.

10.4.1.5 Ammeter

An ammeter is used for measuring the motor's operating current. It is connected to one phase only on the assumption that the phases are balanced. In the event of single phasing, the current will increase significantly or drop to zero. Regular readings from the ammeter should be logged, these can indicate trends in the motor or pump performance.

Manufacturers produce motor protection devices which are completely independent of the starter/contactor. The devices measure parameters in all three motor phases. If a fault is detected the starter/contactor is tripped using the "hold-in" coil.

The following conditions can be monitored:

- Unbalanced load
- Single phasing
- Short circuit
- Earth fault
- Thermal overload
- Repeated start
- Long run up time
- Locked rotor

Sections 10.1.8 and 10.3.1.3 should be read before investigating squirrel-cage motor starting in detail.

10.4.2 Direct-on-line (DOL)

When starting a squirrel-cage motor direct-on-line, DOL, there is a substantial inrush of current from the supply, about 600% full load current, FLC, which results in a local voltage drop. This affects the motor as well as other equipment connected to the supply. Suppliers of electricity therefore lay down certain restrictions in relation to DOL starting. Normally DOL starting is only permitted for motor sizes up to 7.5 kW. It is therefore necessary to take suitable precautions to limit the starting current of larger motors. Figure 10.14 shows the difference in torque and current for DOL and star-delta, Y-Δ starting for a squirrel- cage motor.

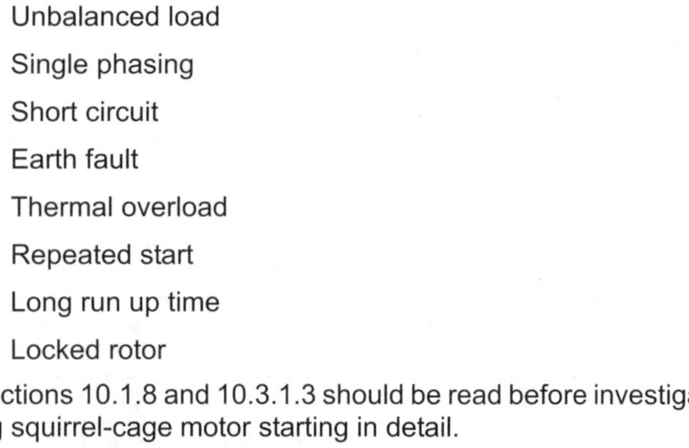

Figure 10.14 Squirrel-cage motor starting torque and current

DOL starting will start most pumps. Most will have a locked rotor torque, (see Section 10.1.6), of 130% and a pull-up torque of over 110%. With the safety margin between required power and installed power, most motors will accelerate the pump fairly quickly. However, some motors have very poor starting characteristics. Locked rotor torque can be as low as 80% and pull-up torques down to 70%. These motors may have been derated for hazardous area operation. Pumps with high inertias, long drive trains, or positive displacement pumps starting on load, may have serious problems when the starting torque reduces to less than 100%. In these situations a sealed fluid coupling may help.

10.4.3 Star-Delta (Y-Δ)

This is the usual method for starting large motors. In principle, this means that when starting, the motor's windings are connected between the phase cables and the neutral cable, Y connection. Two windings are connected between each pair of phases. Each winding then receives a lower voltage than that for which they are intended.

A 3-phase motor for 380 V only receives 220 V, see Section 10.1.2, and the motor develops only a small part of its starting torque, which of course means that the acceleration time is extended. At approximately 90 to 95% of normal operating speed the stator winding connections are changed over automatically to the system phase cables, Δ connection, which then receives full voltage. Figure 10.14 again illustrates the comparison between DOL and Y-Δ respectively for a squirrel-cage motor. Star-Delta starting requires about 200% FLC.

If Star-Delta starting is to be used, care must be taken to ensure that the run up time is not so long as to allow the heat generated by the motor, prior to achieving full speed, to reach damaging proportions. Star-Delta creates more starting heat than DOL. It is also necessary to check that the motor torque will be sufficient to aid starting i.e. it must be greater than the load torque at all times during run up.

It should also be noted that changing from Y to Δ can cause a substantial torque and current pulse if performed too early . For this reason it is desirable to achieve a speed which is as near to maximum speed as possible using the Y connection. Y-Δ starting is often unsuitable for pump motors because of the rapid increase in load torque with speed. Starting positive dis- placement pumps direct-on-line is usually impossible. If DOL is not possible and if the starting torque is too low when using Y-Δ starting then a soft start, variable frequency drive or slipring motor may be necessary. A sealed fluid coupling may help.

10.4.4 Soft starts

Soft starts are an electronic method of controlling starting current. The soft starter is wired between the motor and a DOL contactor. By using a thyristor to control the supply voltage to the motor, the starting current can be controlled. The supply voltage can be programmed to increase over a fixed period of time or the supply current can be limited during the starting sequence.

Soft starters allow a very smooth acceleration of the motor unlike DOL and Y-Δ. Some sophisticated units allow the motor torque to be held at a preset value during starting. Other versions have energy conservation optimisers which reduce the power consumption when the motor is running light. A soft start will probably still peak at a minimum of 200% FLC.

Because the soft start is electronic, a wide range of extra features can be built in without a large increase in cost. The following features are available on some units:

- Adjustable thermal overload protection
- Unbalanced phase detection
- Single phasing
- Inch or jog function
- Low speed-low load operation
- Internal diagnostics
- Adjustable soft stop

Soft starters cost about twice as much as Y-Δ starters. The extra features and protection, plus possible savings on cabling, maximum demand charges and reduced surge problems, may make them worthwhile.

10.4.5 Resistance

Star-Delta starting reduces the starting current by reducing the starting voltage. For a 380 V system, the starting voltage would be reduced to 220 V. Starting current and torque is a function of the square of the applied voltage. By applying 58% voltage, the current and torque are reduced to 33% full load values.

By using resistance connected in series with the windings, any suitable value of voltage can be applied. By decreasing the resistance value in steps, similar to a slipring motor rotor resistance, the motor torque and current can be gradually increased. This type of starting is similar to soft start but without the electronics. Because of the reduced starting torque, and the lack of diagnostics associated with electronic methods, resistance starting must be monitored carefully to ensure the motor accelerates successfully.

10.4.6 Reactor and transformer starters

Other methods of limiting the starting current are reactor and transformer starters. Both these methods involve reduction of both starting current and starting torque. It is therefore necessary to check that sufficient starting torque is available.

10.5 Noise

The quiet running of electric motors is increasingly important and special construction modifications can be made to achieve this. However, the noise level requirements of the motor should not be more stringent than for other components, which can often generate quite high noise levels. The noise level of the motor should not be taken in isolation. The noise level of the pump unit should be considered, and if too high the various options available should be reviewed.

Noise from motors is generated by the cooling system, by the bearings and also by the magnetic circuit. The dominant source of noise depends upon the size of the motor, its speed, design and construction. Bearing and magnetic noise cause vibration in the motor and this can be transmitted through the baseplate/skid to other locations.

Small motors are generally quiet; as motor speed and size increases the noise level increases. Table 10.9 illustrates the trends for squirrel-cage motors with IP 54 enclosures.

		Sound pressure level dB(A)	
Motor power kW	Motor speed r/min	Normal design	Quiet option
0.75	2850	63	
0.75	1400	48	
0.75	920	65	
0.75	700	48	
132	2965	93	84
132	1480	84	81
132	985	80	
132	735	74	

Table 10.9 Electric motor noise at 1 m

A standard requirement is for equipment packages to have noise levels below 84 dB(A). As pump packages increase in power, this noise restriction creates problems. Also a complex package may have multiple motors; oil pumps, circulation pumps, cooling water pumps. A partial solution is to locate equipment in areas which are designated "Ear Protection Zones". Personnel in these areas must wear ear protection. This approach is much more cost effective, when possible, than fitting acoustic enclosures. Acoustic enclosures are very costly and can lower the overall efficiency of the pump unit if ventilating fans are necessary. Acoustic enclosures are disliked by maintenance departments because access to the machine is restricted.

Electric motors are very reliable and routine maintenance, weekly or monthly, is low. Acoustic enclosures around motors cause less problems than enclosures around pumps.

There is no international agreement on restricting noise levels. National regulations or EU regulations apply.

10.6 Maintenance

The electric motor requires regular attention, as does the pump itself, in order to give good performance and reliability. The basic prerequisites for long service life of an electric motor are, apart from selecting the correct motor, careful alignment, mounting and proper lubrication. Regular balancing checks to prevent out-of-balance running should also be made. Vibration can be caused by faulty bearings and windings or irregular air gaps and also by out-of-balance of the couplings. Motors fitted with accelerometers can be monitored continuously for vibration problems.

Most electric motors use rolling contact bearings and it is important to ensure that they are correctly lubricated. The service intervals between lubrication depends upon the type of bearing and the environment in which it operates. Lubrication instructions should form part of the motor documentation and must be followed carefully. Removal of bearings for inspection should be avoided whenever possible since they can be damaged when being withdrawn and reassembled. Checks should also be made by listening to bearing noise and observing the lubricant. See Chapter 12 for further information.

Motor control equipment must also be maintained. Starters which operate regularly must have their contacts examined and replaced if necessary. Poor connections at the starter can lead to single phasing and high resistance causing low motor volts.

10.7 Engines

This Chapter has been devoted to electric motors. Engines are used to drive pumps in certain circumstances. The most popular occasions are for portable pumps and for fixed installations where electricity has insufficient power or is unreliable. In some cases a very good supply of fuel is much cheaper to use than electricity. See Chapter 6, Section 6.9.2 for information on the different engine types available.

Engines can be treated like electric motors in most respects but there is one major difference which cannot be ignored. Motors supply the driving torque to the pump very smoothly, once started. Engines, on the other hand, do not have such a smooth torque characteristic. Engines produce a cyclic speed irregularity due to the firing of the individual cylinders. The magnitude of the irregularity is dependent upon the number of cylinders. Engines with six or more cylinders are fairly smooth. A large number of pumps, particularly small contractors site pumps, are driven by single or two cylinder engines.

The nature of the power supply must be remembered when choosing and sizing couplings and gearboxes. Service factors must be increased. The torsional stressing of the pump shaft may need reconsidering depending upon the ability of the chosen coupling to dampen the cyclic speed changes.

The cyclic speed changes of engines can excite torsional vibrations in the drive train. In general, this is not a problem for rotodynamic pumps with few stages. It can however be a severe problem for reciprocating pumps. Whenever a reciprocating pump is driven by an engine, a torsional vibration analysis must be performed. If the analysis highlights excessive amplitudes or stresses, or nodes at gear meshes, then corrective action must be taken. Changing coupling stiffness is the first modification.

Torsional vibrations caused by engines can almost certainly be eliminated completely by using a fluid coupling at the engine crankshaft.

10.8 Turbines

Turbines are used to drive large pumps or small pumps when a free steam supply is available, see Chapter 6, Section 6.9.2 for the different types available. Gas turbines are used in situations where a large amount of power is required in a small space or high speed is required.

API 611 and ANSI/API 612 are steam turbine specifications for refinery applications. API 616 is a specification for gas turbines for refinery applications. Turbine driven pumps generally run at 3000 r/min and faster. Fast pumps can experience problems with lateral shaft vibrations and critical speeds.

Rotodynamic pumps are categorised as "stiff shaft" or "flexible shaft". If the pump always operates below the first critical speed then it is "stiff shaft". If, during run up, the pump passes through the first critical speed then it is "flexible shaft". The problem is complicated because of the variables involved.

The following aspects of pump design affect the critical speed:

- Length of shaft between bearings
- Number of stages
- Type of bearings, plain or rolling
- Bearing housing support
- Type of coupling
- Design of driver
- Other drive train items, gearbox, etc

The critical speed is further complicated by differences of opinion as to whether the critical speed should be "dry" or "wet". The critical speed can be calculated for all the rotating masses running in air; this is easy. When the pump is operating the rotor is surrounded by liquid. Liquid in the wear rings, centre bush and throat bush(es) creates a bearing effect so that the "wet" critical speed is higher than the "dry". The "wet" can be difficult to calculate accurately as the bearing effects of some of the components cannot be predicted precisely.

In theory, a pump handling a liquid with a viscosity of 100 cSt will derive more bearing stiffness from internal close clearances than a pump handling a 5 cSt liquid. It is worthwhile considering critical speed calculations and monitoring during test when the wear ring material combinations are marginal for the liquid, and there is high speed operation with many stages.

10.9 Power recovery turbines

Some pumps operate with power recovery turbines as the sole driver. Processes which operate continuously and have pressurised liquid streams which must be expanded to a lower pressure are ideal. However in most cases the turbine is used to augment the main driver and unload it. A main driver is usually required for start-up until the process has stabilised. Oil refining and reverse osmosis water purification plants are popular applications. Adding a power recovery turbine to a pump which has a main driver considerably lengthens the shaft system.

Adding a power recovery turbine to a motor driven pump package is not the most efficient use of the fluid energy supply. Power recovery turbines assisting an electric motor cause the motor to operate at low load and hence low efficiency and low power factor. It would be more efficient to use the power recovery turbine to drive an alternator which could produce electricity for use anywhere on the site.

Section 10.8 should also be read in relation to critical speed, as well as Section 6.9.2, in Chapter 6, for details of power recovery turbine designs.

10.10 Motor selection illustration

A clean fixed speed water pump for a flow of 25 l/s at 65 m differential head is to be purchased and connected to a 3-phase system with 415V supply, and the electric current requirement is to be calculated. The manufacturer's literature shows the pump efficiency to be 70%. The following method may then be used:

First determine the pump's power requirement, see Chapter 4, combine equations 4.3 and 4.5.

$$P = \frac{r \cdot Q \cdot g \cdot \Delta H}{h}$$

$$P = \frac{1000 \cdot 0.025 \cdot 9.81 \cdot 65}{0.7}$$

$$P = 22773 \text{ W} = 22.8 \text{ kW}$$

As the pump is fixed speed, a 3-phase squirrel-cage motor is appropriate. A safety margin of 10% is considered necessary so the minimum motor size acceptable would be $22.8 \cdot 1.1 = 25$kW.

Select a 30 kW motor. The motor has a full load efficiency of 89%, power factor of 0.89 and current of 53 amps. At normal operating conditions the motor will deliver 76% power, when the efficiency will be 88% and the power factor 0.87.

Using equation 10.6 from Section 10.1.2 the motor current can be calculated:

$$I = \frac{22.8 \cdot 1000}{415 \cdot 1.732 \cdot 0.87 \cdot 0.88}$$

$$I = 41.4\text{A}$$

During normal operation of the pump unit the current consumption should, therefore, be approximately 41 A.

An ammeter connected to the pump motor gives a good indication of normal operation. Any increase or decrease in the current consumption indicates that there is a fault somewhere or the pump duty has changed and that an investigation is necessary.

The power supply to which the pump is connected must be fitted with fuses or a circuit breaker. The motor also requires a starter. The fuses or circuit breaker, and the motor starter must be rated for the motor full load current, not the running current. When the motor is started, star-delta probably, the motor will absorb 2 to 3 times the full load current, not 2 to 3 times the operating current, see Figure 10.14. The fuses must therefore be of the delayed action type to prevent spurious tripping. The cable sizing can be based on the normal running current providing starting does not occur too often.

10.11 National Standards bodies

The following is a list of countries and their national standards bodies who are members of ISO, International Standards Organisation.

Algeria - IANOR - Institut algarien de normalisation
Argentina - IRAM - Instituto Argentino de Normalizacian

Armenia - SARM - National Institute of Standards and Quality
Australia - SAI - Standards Australia International
Austria - ON - Austrian Standards Institute
Bangladesh - BSTI - Bangladesh Standards and Bangladesh Standards and Testing Institutio,
Belarus - BELST - Committee for Standardization, Metrology and Certification of Belarus
Belgium - IBN - The Belgian Institution for Standardization
Bolivia - IBNORCA - Instituto Boliviano de Normalizacian y Calidad
Bosnia and Herzegovina - BASMP - Institute for Standards, Metrology and Intellectual Property of Bosnia and Herzegovina
Brazil - ABNT - Associao Brasileira de Normas Tacnicas
Brunei Darussalam - CPRU - Construction Planning and Research Unit, Ministry of Development
Bulgaria - BDS - Bulgarian Institute for Standardization
Canada - SCC - Standards Council of Canada
Chile - INN - Instituto Nacional de Normalizacion
China - SAC - Standardization Administration of China
China - CSSN - China Standards Information Center
Colombia - ICONTEC - Instituto Colombiano de Normas Tacnicas y Certificacian
Costa Rica - INTECO - Instituto de Normas Tacnicas de Costa Rica
Croatia - DZNM - State Office for Standardization and Metrology
Cuba - NC - Oficina Nacional de Normalizacian
Czech Republic - CSNI - Czech Standards Institute
Denmark - DS - Dansk Standard
Ecuador - INEN - Instituto Ecuatoriano de Normalizacian
Egypt - EO - Egyptian Organization for Standardization and Quality Control
El Salvador - CONACYT - Consejo Nacional de Ciencia y Tecnologa
Estonia - EVS - Eesti Standardikeskus
Ethiopia - QSAE - Quality and Standards Authority of Ethiopia
Finland - SFS - Finnish Standards Association
France - AFNOR - Association franaise de normalisation
Germany - DIN - Deutsches Institut fur Normung
Greece - ELOT - Hellenic Organization for Standardization
Grenada - GDBS - Grenada Bureau of Standards
Guatemala - COGUANOR - Comisian Guatemalteca de Normas
Guyana - GNBS - Guyana National Bureau of Standards
Hong Kong - ITCHKSAR - Innovation and Technology Commission
Hungary - MSZT - Magyar Szabva¡nyagyi Testalet
Iceland - IST - Icelandic Council for Standardization
India - BIS - Bureau of Indian Standards
Indonesia - BSN - Badan Standardisasi Nasional
Iran - ISIRI - Institute of Standards and Industrial Research of Iran
Ireland - NSAI - National Standards Authority of Ireland
Israel - SII - The Standards Institution of Israel
Italy - UNI - Ente Nazionale Italiano di Unificazione
Jamaica - JBS - Bureau of Standards, Jamaica
Japan - JISC - Japan Industrial Standards Committee
Jordan - JISM - Jordan Institution for Standards and Metrology
Kazakstan - KAZMEMST - Committee for Standardization, Metrology and Certification
Kenya - KEBS - Kenya Bureau of Standards
Republic of Korea - KATS - Korean Agency for Technology and Standards
Kuwait - KOWSMD - Public Authority for Industry, Standards and Industrial Services Affairs
Kyrgyzstan - KYRGYZST - State Inspection for Standardization and Metrology
Latvia - LVS - Latvian Standard
Lebanon - LIBNOR - Lebanese Standards Institution
Lithuania - LST - Lithuanian Standards Board
Luxembourg - SEE - Service de l'Energie de l'Etat, Organisme Luxembourgeois de Normalisation
Malaysia - Department of Standards Malaysia
Malta - MSA - Malta Standards Authority
Mauritius - MSB - Mauritius Standards Bureau
Mexico - DGN - Direccian General de Normas
Moldova - MOLDST - Department of Standardization and Metrology
Morocco - SNIMA - Service de Normalisation Industrielle Marocaine
Netherlands - NEN - Nederlandse Norm, maintained by the Nederlands Normalisatie Instituut (NNI)
New Zealand - SNZ - Standards New Zealand
Nicaragua - DTNM - Direccian de Tecnologa, Normalizacian y Metrologa
Nigeria - SON - Standards Organisation of Nigeria
Norway - SN - Standard Norge (Standards Norway)
Oman - DGSM - Directorate General for Specifications and Measurements
Pakistan - PSQCA - Pakistan Standards and Quality Control Authority
Palestine - PSI - Palestine Standards Institution
Panama - COPANIT - Comisian Panamea de Normas Industriales y Tacnicas
Papua New Guinea - NISIT - National Institute of Standards and Industrial Technology
Peru - INDECOPI - Instituto Nacional de Defensa de la Competencia y de la Proteccian de la Propiedad Intelectual
Philippines - BPS - Bureau of Product Standards
Poland - PKN - Polish Committee for Standardization
Portugal - IPQ - Instituto Portuguas da Qualidade
Romania - ASRO - Asociatia de Standardizare din Romania
Russian Federation - Rostekhregulirovaniye - Federal Agency for Technical Regulation and Metrology
Saint Lucia - SLBS - Saint Lucia Bureau of Standards
Saudi Arabia - SASO - Saudi Arabian Standards Organization
Serbia and Montenegro - ISSM -Institution for Standardization of Serbia and Montenegro
Seychelles - SBS - Seychelles Bureau of Standards
Singapore - SPRING SG - Standards, Productivity and Innovation Board
Slovakia - SUTN - Slovak Standards Institute
Slovenia - SIST - Slovenian Institute for Standardization
South Africa - SABS - South African Bureau of Standards
Spain - AENOR - Asociacian Espaola de Normalizacian y Certificacian
Sri Lanka - SLSI - Sri Lanka Standards Institution
Sweden - SIS - Swedish Standards Institute
Switzerland - SNV - Swiss Association for Standardization
Syrian Arab Republic - SASMO - The Syrian Arab Organiza-

tion for Standardization and Metrology

Taiwan (Republic of China) - BSMI - The Bureau of Standards, Metrology and Inspection

Tanzania - TBS - Tanzania Bureau of Standards

Thailand - TISI - Thai Industrial Standards Institute

Trinidad and Tobago - TTBS - Trinidad and Tobago Bureau of Standards

Turkey - TSE - Tark Standardlari Enstita

Uganda - UNBS - Uganda National Bureau of Standards

Ukraine - DSSU - State Committee on Technical Regulation and Consumer Policy of Ukraine

United Kingdom - BSI - British Standards Institution

United States of America - ANSI - American National Standards Institute

Uruguay - UNIT - Instituto Uruguayo de Normas Tacnicas

Venezuela - FONDONORMA - Fondo para la Normalizacian y Certificacian de la Calidad

Vietnam - TCVN - Directorate for Standards and Quality

10.12 Useful references

IEC 60038 Ed 6.2b, IEC Standard voltages.

EN 60079-1:2004 Electrical apparatus for explosive gas atmospheres. Flameproof enclosures 'd'.

EN 60079-14 Electrical apparatus for explosive gas atmospheres - Part 14: Electrical installations in hazardous areas (other than mines) (IEC 60079-14:2002); German version EN 60079-14:2003 DIN-adopted European Standard.

IEC 60034-5 Ed. 4.0 b:2000 Rotating electrical machines - Part 5: Degrees of protection provided by the integral design of rotating electrical machines (IP code) – Classification.

BS 4999-141:2004 General requirements for rotating electrical machines. Specification for standard dimensions.

EN 60085:2004 Electrical insulation. Thermal classification.

NFPA (Fire) 30-00 Flammable and Combustible Liquids Code.

BS 4999-147:1988 General requirements for rotating electrical machines. Specification for dimensions of brushes and brush-holders for electrical machinery.

IEEE 488.1-2003 Higher Performance Protocol for the Standard Digital Interface for Programmable Instrumentation.

IEC 60947-4-1 Amd.1 Ed. 2.0 b:2002 Amendment 1 - Low-voltage switchgear and controlgear - Part 4-1: Contactors and motor-starters - Electromechanical contactors and motor-starters.

Physikalisch-Technische Bundesanstalt (PTB), Bundesallee 100, D-38116 Braunschweig Germany, Tel: 0531 592 3006, Fax: 0531 592 3008, Email: presse@ptb.de, www.ptb.de.

ANSI/API 611 Fourth Edition 1997 General-Purpose Steam Turbines for Petroleum, Chemical and Gas Industry Services.

ANSI/API 612 Sixth Edition 2005 Petroleum, Petrochemical and Natural Gas Industries-Steam Turbines-Special-purpose Applications.

API 616 Fourth Edition 1998 Gas Turbines for the Petroleum, Chemical, and Gas Industry Services.

Fire & Explosion ATEX Regulations Part 1 - The Regulations implement a European Directive called, the ATEX Directive No.1992/92.

IEC 60529, Degrees of protection provided by enclosures (IP code).

IEC 1000-3-2, Electromagnetic compatibility (EMC) - Part 3: Limits - Section 2: Limits for harmonic current emissions (equipment input current ~16 A per phase).

IEC 1000-3-3, Electromagnetic compatibility (EMC) - Part 3: Limits - Section 3: Limitation of voltage fluctuations and flicker in low-voltage supply systems for equipment with rated current ~16 A.

Ancillary equipment

11

11.1 Introduction

Ancillary equipment is frequently supplied as part of a complete package but which is not regulated by easily accessible standards. In this context it is worth explaining two definitions which form the basis of European pump standards.

A "pump" is a single piece of equipment which terminates at its suction and discharge connections and its shaft end. Pump feet are not mentioned but these are another terminal point.

A "pump unit" consists of a pump and any other associated equipment which is defined as being included in an assembly. Whenever the term "pump unit" is applied, it must be defined to show the content intended. Equipment supplied loose, for mounting in the user's process pipework, would not constitute part of a pump unit. In practice, the smallest pump unit would be a pump plus a baseplate; when the purchaser was supplying the motor and coupling. A large pump unit could consist of several pumps, couplings and motors, pipework and controls all mounted on one baseplate.

Small pumps can be installed without any additional equipment. This applies particularly to in-line pumps mounted in and supported by the pipework. Larger pumps tend to have other pieces of equipment to assist with the prime function, pumping liquid.

11.2 Mountings

Pump units are generally mounted on a structure to support and align the various components. The structure itself must perform additional functions depending upon its duty.

11.2.1 Baseplates

Pump units are mounted on baseplates to allow the unit to be bolted down. Unless otherwise stated, baseplates are designed for mounting on concrete foundations and to be supported over the whole length by grouting. The pump supplier must be aware if a baseplate is intended for mounting on structural steelwork with intermittent supports. The design of the baseplate side members may need adjusting to preserve alignment. If the baseplate is to be grouted in with epoxy grout rather than concrete, a special paint finish will be required on the contact surfaces. If the baseplate is to be welded down to structural steelwork the pump supplier should be informed.

The baseplate should cover the perimeter of the pump unit. Only in exceptional circumstances should equipment extend beyond the baseplate outline. By allowing the baseplate to form a perimeter, delicate and sensitive equipment can be protected by a substantial static structure. If equipment must extend over the baseplate edge, provision should be made to protect the equipment by enlarging the concrete plinth. Problems with motor terminal boxes overhanging the side of the baseplate can be cured by top-mounted terminal boxes.

When a pump unit is built "to order" the purchaser is in control and provides lists of national or international standards, and includes proprietary specifications for different aspects of the pump/pump unit. Baseplates are however a problem. There are no good standards. Consequently, the purchaser has to use inappropriate ones such as "Structural steelwork for buildings".

Baseplates are not designed to the same sort of criteria as buildings. Deflections are much smaller and stressing is much lower. The pump supplier will normally have plenty of experience designing and fabricating baseplates and will simply require a few little reminders about good practice.

For example, all structural welds must be continuous. Main longitudinal members must be one continuous piece with no joints. Welds for main cross-members to main longitudinals must be on all sides of the cross member. Pads for equipment must be machined and thicker and larger than the equipment foot. Pads for pumps and drive train must be fully supported. Drive train pads must be machined low, about 3 mm, so that shims are always fitted. At least two pads, one at each end, should be wide enough to accommodate a spirit level without disturbing equipment. Baseplates, with a fully assembled weight of over 2500 kg must have levelling bolts at each foundation bolt.

Individual pieces of equipment over 500 kg must have provision for horizontal positioning bolts; these bolts can be detachable. Provision must be provided for a four point lift with equal length slings. If a lifting beam is necessary the pump supplier must state so and quote a price for its supply. Lifting arrangements must comply with insurance company regulations.

The simplest lifting arrangement is to incorporate full width tubes across the baseplate. The user inserts a longer bar through the tubes to accept the eyes of the slings.

Baseplates for larger pumps are designed to cope with transportation and installation without suffering permanent distortion. This is not always true for smaller pumps. Some small pumps may be delivered with temporary bracing or reinforcement to prevent permanent distortion.

Baseplates can be constructed with or without various desirable features, which must be specified at the inquiry stage.

These options are:

Drain rim — The baseplate incorporates a full length sloping trip tray with one drain connection.

Full depth cross-members — The structural cross-members extend to the bottom of the main longitudinal members.

Shallow cross-members — The cross-members do not extend closer to the bottom of the main longitudinal members than a specified distance.

Lifting lugs — Four lugs, shaped to accept standard shackles, are welded on the outside of the outer longitudinal structural members.

Top covers — All accessible areas of the baseplate top are fitted with covers; solid, grid, permanent, removable.

Laser mounts — A baseplate can be fitted with specific brackets to allow optical alignment equipment to be fitted to assist levelling.

In some instances, options may not be possible. Shallow cross-members may not be viable if the supplier has agreed that the pump can accept very large nozzle forces and moments.

When a user purchases a special pump from a manufacturer it is important to have a general arrangement drawing showing the overall sizes, positions of nozzles and foundation bolts. The pump manufacturer may include a preliminary drawing with the quotation but the user needs a certified drawing to finalise pipework and foundations.

Chapter 10, Section 10.2.6 shows that some dimensions are standardised on standard motors. Special motors, and other types of drivers, need a drawing to confirm all the relevant dimensions. Manufacturers of motors and engines only issue certified drawings after a purchase order has been placed. The pump manufacturer cannot issue a certified pump general arrangement drawing until receipt of the certified drawings for the drive train. At the beginning of a pump order, time is often taken up with issuing purchase orders and finalising certified drawings.

11.2.2 Skids

A skid is very much like a baseplate but it is not intended to be fastened down to concrete or structural steelwork. It is designed to rest on a reasonably flat surface. Consequently the longitudinal structural members are slightly deeper, and perhaps heavier, than a similar baseplate.

Most of the comments for baseplates are valid for skids. The obvious exceptions are those requirements for baseplate levelling screws and an option for permanent distortion during transportation or installation. Skids are very rugged. A skid package will operate satisfactorily with four points of support.

Skids are constructed with angled or rounded ends on the main longitudinal members to assist movement. Towing points, to accept slings, are provided at both ends. Full depth cross-members are not an option.

11.2.3 Trailers

Trailers are a slightly more complicated version of a skid. When purchasing a trailer-mounted pump one of the two major options must be specified before any other decisions can be made; off-road or roadworthy.

Off-road trailers are designed specifically for rough terrain and cannot be towed on public roads in Europe. Trailers can have two or four wheels and the wheels may or may not have pneumatic rubber tyres. Four wheel trailers may have a steering axle; brakes are an option not mandatory. Depending upon the size and design of the trailer, stabilising jacks may be essential. There are no standards specifically for off-road trailers. Safety requirements inherent in the Machinery Directive must be met, but there are no specific design requirements.

Roadworthy trailers must comply with the current relevant legislation. These trailers must be complete with pneumatic rubber tyres, suspension, brakes and lights. It may be worthwhile specifying a maximum speed for towing.

ISO 1102, 1285, 3731, 4141, 8703 and 8785 may prove useful.

Some pumps are packaged on a vehicle; the size of the vehicle being dependent upon the size of the pump and the nature of any accessories. The facilities required on such vehicles must be fully specified.

11.2.4 Lifting frames

Lifting frames can be considered as a special version of a skid. Pumps mounted in lifting frames are not intended to be bolted down but are not "ruggedised" in the same way as skids. Pumps mounted in lifting frames are intended to be manhandled and must be as light as possible, see Chapter 1. The purchaser must know the "environmental" conditions of use and how many personnel will be available for manhandling. There may be strict limitations on size and shape if pumps are to be carried through doors or hatches.

Lifting frames are best constructed from tube or rectangular hollow sections. The size should be suitable for a good grip; designated lifting points are not always accessible. Thin wall, high strength alloy tube can be used to reduce weight. The lifting frame should completely surround the pump and no protuberances are allowed. It is essential to know how many operating orientations can be used; upside-down or on the side.

Lifting frames are used for site pumps, fire fighting pumps and general purpose marine pumps. Petrol engines provide the best power to weight ratio but electric pumps could be used if a supply was available. The manufacturer requires a detailed specification of the complete "operating environment" to package a pump unit successfully.

11.3 Belt drives

Belt drives are an economical method of speed increasing, transmitting and reduction. The speed does not necessarily have to change. Belt drives, 1:1, are used as an alternative to couplings. The driver and driven shafts can be moved to better locations not possible when using a coupling.

Belt drives should be considered when the following conditions exist:

- Power rating up to 200 kW
- Speed ratio up to 7:1
- Belt linear speed up to 40 m/s
- Infrequent speed changes
- Ambient temperature -30 to 65 °C
- Fairly clean ambient air
- Shaft phasing unimportant, (Vee-belts)

Belt drives are available for higher powers but these generally involve special pulleys incurring extra costs and causing spares availability problems. Belts can be used at higher speeds but this necessitates using a specially designed drive. Pulleys with quick mounting hubs are standard for inch and metric sizes. Pulley sizes can be changed fairly quickly for irregular speed modifications.

Belts are available certified for underground mining applications and for potentially explosive atmospheres. Cast iron metric pulleys generally comply with ISO 4183 and belts ISO 4184.

Vee-belts with a service factor of 1.0, for use with rotodynamic pumps, would have an efficiency of about 96%. Three cylinder reciprocating pumps would normally have a service factor around 1.6, efficiency 94 to 96%.

Vee-belts do slip slightly and require occasional tension checking. These shortcomings can be overcome by using toothed belts. Higher powers can be transmitted in the same space requirements at slightly higher efficiencies. Toothed belts are not popular because of the noise they produce. Toothed belt drives with 1500 r/min motors are very noisy, 105 dB(A) being possible. The silencing of belt guards causes problems with air circulation and belt cooling.

Belt drives cannot be fitted with continuous monitoring instrumentation. At present, to inspect the condition of a belt drive the drive must be stopped and the guard removed. Belt guards should be treated in a similar manner to coupling guards, see Chapter 9, Section 9.13.

11.4 Gearboxes

Gearboxes can be applied to any shaft speed and any ratio is possible. By selecting the correct service factor, working life can be long with minimal maintenance.

The most popular style of gearbox is the parallel shaft helical gearbox. Table 11.1 shows the range of ratios possible with varying numbers of gear meshes or gear pairs.

Number of meshes	Range of ratios
1	1.0 to 6.4
2	6.6 to 21
3	18 to 100
4	90 to 525

Table 11.1 Gear ratios of parallel shaft helical gearboxes

A pump unit can become quite long by the addition of a gearbox and an extra coupling. To reduce the overall length, a flange-mounted motor could be bolted directly to the side of the gearbox. Another method would be to replace one of the helical ratios with a bevel ratio. The gearbox input shaft could be located vertically on top of the gearbox. Epicyclic gearboxes and in-line helical gearboxes, both with flange mounted motors are popular for sizes up to 100 kW to minimise package dimensions.

Gearboxes are available with more than one ratio. Parallel shaft helical gearboxes are built with up to four ratios. These gearboxes are manual change and require the pump to be stopped before a change can be made. Special provision, "inching" or "jogging" of the drive motor, may be necessary in order to engage the dog clutches. Small units can be turned by hand from the high speed coupling. This type of gearbox is suitable when discrete speed changes are acceptable and changes do not occur too often. Maintenance personnel may be required to change gear if operators are unhappy or lack confidence for technical intervention. Efficiency is high, and because the gearboxes are made to order, ratios are exactly as required.

Semi-automatic multi-speed gearboxes, similar to automotive transmissions, are available from a few manufacturers. These gearboxes, designed for commercial vehicles, rail locomotives and marine applications, are normally used on portable engine driven pump packages. The number of gears varies from four to ten. A limited choice of ratios is possible but the multiplicity of ratios more than compensates for the lack of an optimum ratio. Speed changing can be manual or automatic via a process control system.

Epicyclic gearboxes are popular for larger and high speed applications; steam and gas turbine drivers. Very costly in comparison to parallel shaft helical boxes; but maybe more efficient. Epicyclic gearboxes can be used as a variable ratio transmission by adding a small control drive.

Gearboxes tend to rely on natural convection cooling. When necessary, fans can be added to the high speed shaft(s) to encourage forced convection. When fans are insufficient, an external cooling circuit can be added for the lubricating oil.

Most gearboxes have rolling contact bearings. Special versions can be built with plain bearings. ANSI/API 613 for high speeds and API 677 for special purpose gearboxes for refinery applications both require plain bearings as standard. Gearboxes with rolling contact bearings can manage with a simple splash lubrication system. Gearboxes with plain bearings may need an external pressurised lubrication system. API 614 specifies the requirements for refinery lubrication systems. API 614 systems tend to be "over-engineered" and require a lot of space; individual applications should be evaluated and designed accordingly.

Gearboxes are sized by using service factors. The service factor is dependant upon the smoothness of the drive and the smoothness of the load. The service factor adjusts the working tooth stresses to suit the application. Rotodynamic pumps handling clean liquids driven by an electric motor would have a service factor between 0.8 and 1.25. If the driver was a four cylinder engine, the service factor would increase to 1.0 to 1.5. For a reciprocating pumps the equivalent service factors would be 1.5 to 2.0 and 1.75 to 2.25.

There are no European standards for service factors. National standards exist in UK, Germany and USA. The American Gear Manufacturers Association (AGMA)is the most prolific for issuing standards. ANSI/AGMA 2001 covers rating factors for spur and helical gears. DIN 3990 covers service factors. The British Standard is considered to be the most conservative for through hardened gear designs. Most modern spur and helical gears tend to be case hardened; AGMA is considered to be the most conservative. DIN 3990 is thought to produce the most realistic designs for case hardened gears.

ISO TR 13593 was going to be a proper standard rather than a technical report. However, there are so many rating options, to keep everybody happy there is no standardisation. Service factors have become application factors. 13593 was not intended to confuse gearbox users, but that is the result.

Application factors will not deal with the life problem. The best way to specify a gearbox is to pick an old, well-tried standard that has been used successfully for many years. AGMA 913 is an old standard for spur and helical gears. Rotodynamic pumps driven by standard squirrel-cage motors for continuous operation, not frequent starting would be AGMA 913 SF= 1. For a three cylinder reciprocating pump driven by a squirrel-cage motor the factor would be AGMA 913 SF= 2. For pumps with five or more cylinders the service factor can be reduced to 1.75. To be absolutely sure of what is purchased it would be advisable to specify the minimum acceptable life of teeth and bearings, i.e.

- Minimum tooth life at rated conditions 85000 hours
- Minimum bearing life at rated conditions 30000 hours

If pumps are subject to cyclic operation, including duty changes, on-off or load-unload flow regulation, then the full operating conditions should be discussed with the gear manufacturer for suitable recommendations.

Unlike belt drives, gearboxes can be fitted with instrumentation to monitor the "health" of the teeth and the bearings. Accelerometers can be fitted to each shaft and the itself. Harmonic analysis of the vibration signals can show how wear is developing and if bearings are working correctly. Analysis of oil samples can indicate the onset of oil degradation and the level of metal particles.

11.5 Relief valves

Relief valves are fitted to piping systems and vessels to prevent excessive internal pressures damaging the equipment. When it is likely that a pump will produce pressures in excess of the system or vessel design pressure a relief valve should be fitted.

In the case of positive displacement pumps a relief valve must always be fitted to the discharge. Relief valves should be fitted as dedicated pieces of equipment. Modern designs allow relief valves to be combined with load-unload valves. Although this saves costs it is not a preferred method of construction. Relief valves fitted to pump discharges must be sized to pass the total pump flow. They are sometimes fitted in pump suction systems to prevent the pump experiencing excessive suction pressure leading to high axial thrust and subsequent bearing wear. In reciprocating pumps, high suction pressure can result in rapid wear of the crosshead pin bearings.

Some useful terms used in conjunction with relief valves are:

Set pressure — The pressure at which the relief valve will begin to open.

Overpressure — The pressure which is required for the relief valve to pass 100% flow. Sometimes called accumulation. For liquid relief valves purchased without any special requirements the overpressure will be 25% of the set pressure. Overpressure can be set as low as 8% in some cases.

Reseat pressure — The pressure at which the valve will close and reseal. Normally between 80 and 97% of the set pressure.

Back pressure — The pressure at the relief valve outlet connection when piped into a system.

Modulating action — The lift of the valve is proportional to the flow.

Pop action — Once the valve opens the lift is 100% until it closes.

Relief valves are connected to a short branch on the process pipework. The branch must be short so that there is no significant pressure drop between the process pipe and the valve seat. When the valve opens the flow is piped to a low pressure system. If the pump differential pressure is less than 20 bar, the relief valve outlet may be piped into the pump suction although it is always preferable to pipe the valve back to the suction source. If the relief valve outlet is not at atmospheric pressure, a back pressure is present, the valve may require bellows compensation and/or the size increased.

A pressure surge is created when a pop action valve opens; this may induce a pipe kick. Also, there may be a significant momentum reaction at the valve body. Check the magnitude of the reaction and reinforce or brace as appropriate.

A special application of relief valves, not directly involving pumps, is thermal relief. When a piping system is built it is possible to have a length of pipe with an isolating valve at each end. If the pipe is full of liquid when both valves are closed high pressures can be created when the liquid expands due to temperature changes. Small spring-loaded relief valves are fitted to relieve the excess pressure.

Relief valves are built in three different styles to suit different applications and to provide different facilities. Two additional designs are available for special applications.

11.5.1 Spring-loaded

Spring-loaded valves are of the most basic and simplest of designs. The valve is held closed by a spring load. The liquid pressure under the valve seat has to overcome the spring load before the valve can open. As more liquid flows through the valve the lift increases. When the liquid stops flowing the valve reseats and seals. Manufacturers have proprietary nozzle and poppet designs, some with adjustable lift characteristics, to suit various applications. Spring-loaded relief valves are suitable for gases and liquids; pumps require a valve with a liquid trim. High viscosity liquids, such as lubricating oils and crude oil, may require modification to prevent the poppet sticking. Spring-loaded relief valves are the first choice for most applications.

Variable flow applications can create problems for spring valves. If the valve is required to pass flows less than 30%, stability and chattering can cause high wear rates and damage the seat resulting in poor sealing.

Spring-loaded valves are generally set at +10% with 10% overpressure. Running at overpressure is considered as intermittent. If longer running at overpressure conditions is necessary, this must be specified in the operating conditions. In some cases it is possible to use 25% overpressure to advantage; a smaller valve can be fitted. Low pressure applications and when the installed motor power is much larger than required are typical cases

11.5.2 Pilot-operated

Pilot-operated relief valves use liquid pressure to hold the valve closed. When the set pressure is reached the pilot releases all, or some, of the holding pressure and allows the valve to open. When the process pressure falls, the pilot re-energises and the valve is firmly closed. Pilot-operated valves are more costly than spring-loaded valves but they are more versatile. A pilot-operated valve can be set to have zero overpressure — a pop action. Once the set pressure has been reached, the valve opens 100%. A modulating pilot is also available.

The set pressure can be fixed closer to the operating pressure because the valve is normally clamped closed hydraulically. The reseat pressure can be adjusted to be as close as 97% of the set pressure. A set pressure of +5% and a 5% overpressure is a good compromise for system operation. A pop action valve can create pressure pulsations within the system, for this reason a modulating valve is preferable. Pilot-operated valves do not suffer instability at low flows.

11.5.3 Air-assisted

An air-assisted valve is a modified spring-loaded valve with pneumatic control. The poppet, as well as the spring load, is attached to a pneumatic diaphragm.

The valve is held closed by the spring force and the air pressure above the diaphragm. A control valve senses the pressure in the process pipe. When the liquid pressure reaches the set pressure the control valve releases the air above the diaphragm. The valve opens as a normal modulating spring-loaded valve. When the liquid pressure reduces to the reseat pressure the air is reapplied to the diaphragm, closing the valve.

The addition of the air assistance allows the set pressure, overpressure and reseat pressure to be brought closer together. As with pilot-operated valves, the operating pressures can be + 5%, 5% and 97%. A slightly higher reseat is possible.

The benefit of an air-assisted valve is in its back-up capability; if the pneumatics fail it will operate as a normal spring-loaded valve. The drawback is the cost and need for an air or gas supply. The valve does not use gas continuously; only when the valve operates, bottled gas is adequate.

11.5.4 Shear pin and buckling pin relief valves

These operate slightly differently to other valves. A shear pin valve is held closed by a relatively weak pin restraining a plunger. For the plunger to move, and allow the valve to open, the pin must be broken by direct shear at two points.

A buckling pin relief valve is held closed by a slender pin under direct compression at its ends. When the relief valve reaches set pressure, the axial load on the pin exceeds the critical load, the pin buckles and the valve opens.

In both designs the valve does not reset after opening. These styles of relief valve are often used for liquid/solid mixtures.

11.5.5 Bursting discs

Bursting discs or rupture discs can be used for over pressure protection. Both these devices consist of a thin membrane which is highly stressed. Under over pressure conditions the ultimate stress of the membrane is exceeded and the membrane fails. The system must be shut down to allow the disc to be replaced. Discs are very useful for hygienic, sterile or extremely hazardous applications. Disc assemblies can be built which are very easy to clean and sterilise. Most discs relieve to atmosphere and low pressure pipework is not fitted.

However, disc assemblies can easily be clamped between two flanges to allow the low pressure liquid to be recovered. A "clamped" disc version can be used in series with a normal relief valve for very hazardous applications. The disc performs as the primary safety device; during normal operating conditions there is no leakage. Over pressure conditions cause the disc to fail allowing liquid to flow through a normal relief valve. When operating conditions return to normal the relief valve closes and permits the process to operate. Normal operation can continue until it is convenient to replace the bursting disc.

11.6 Non-return valves

11.6.1 General

Non-return valves are incorporated in the pipe system to prevent, undesirable reverse flow. Normally the valve cannot be actuated by external sources, it is self-actuating, i.e. the flow alone determines its function. A non-return valve should be able to operate under varying conditions with respect to media, pressure, temperature, flow and viscosity. Non-return valves selected for clean liquid operation should not be expected to function as well when handling abrasive solids. A desire common to both suppliers and consumers is that it should be possible to use the same type of non-return valve under different operating conditions, i.e. its application should be universal.

In many pipe systems non-return valves have performed indifferently. A chattering non-return valve is extremely irritating, if not to say, hazardous. The valve is subject to considerable wear. The chattering releases forces which work on the pipes and supports, pumps and the valve itself. The risk of a breakdown of the system is therefore very great. Chattering valves indicate a change in flow velocity and perhaps the use of an incorrect valve type. For increased system reliability the true cause should be investigated.

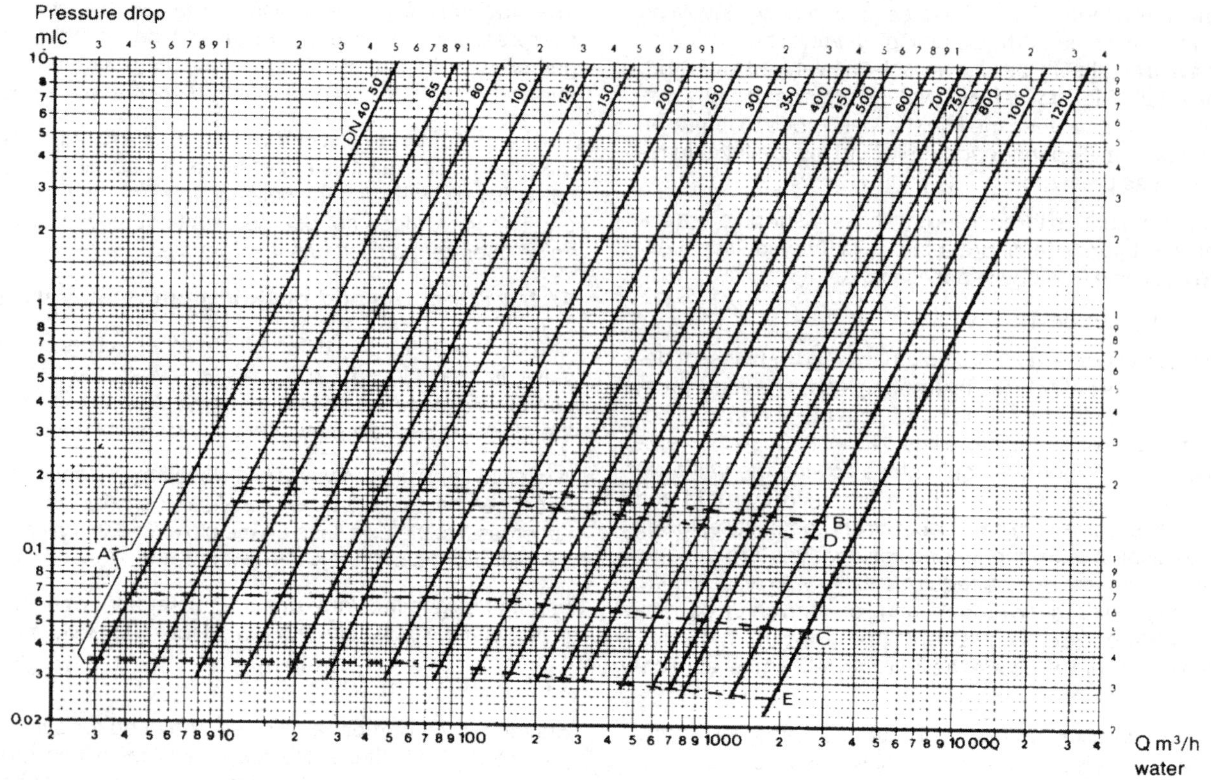

A. Before the valve is fully open the pressure drop is greater than indicated on the curve, see Figure 11.2

B. Valve is fully open - applies to designs with springs and horizontal pipes.

C. Valve fully open- applies to designs without springs and horizontal pipes.

D. Valve fully open - applies to designs with springs and vertical pipes.

E. Valve fully open - applies to designs without springs and vertical pipes

Figure 11.1 Butterfly check valves for water

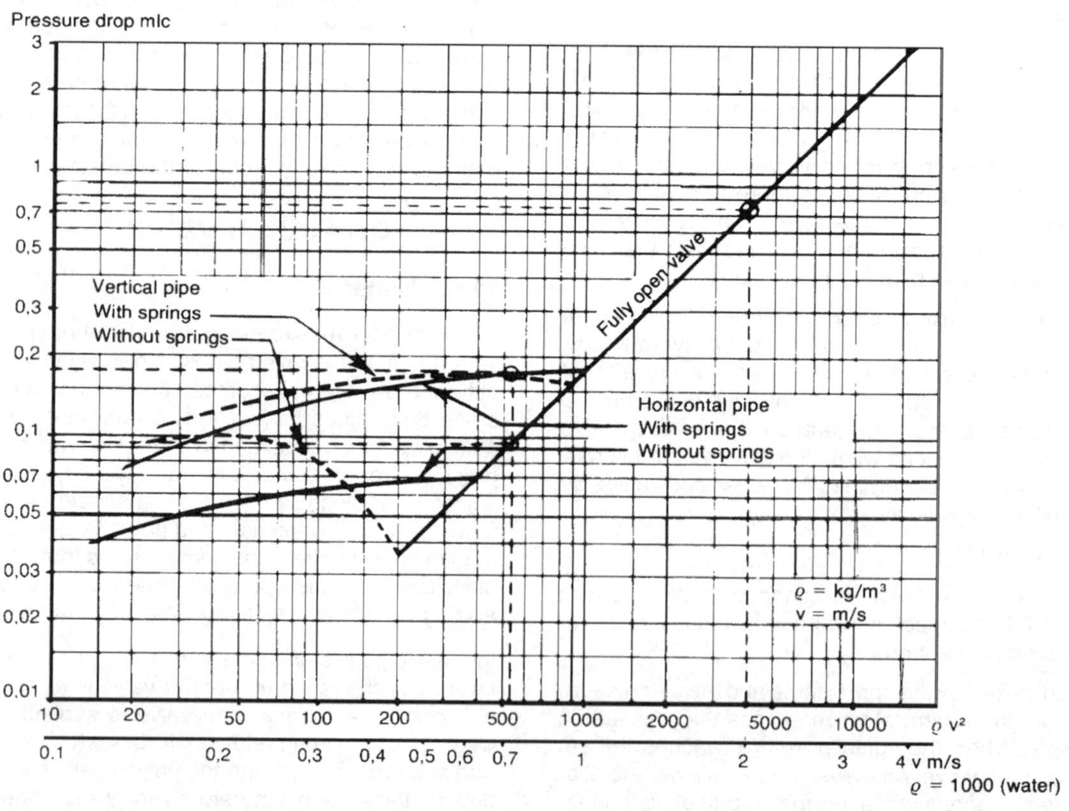

Figure 11.2 Butterfly check valves for liquids other than water

11.6.2 Water hammer and pressure surge in pipes

Following a pipe fracture or shutdown of a pump in a pipe system, the forward flow is reduced and after some time it may be succeeded by a reverse flow. If this remains in operation for some time, the result can be that the pumps reverse their direction of rotation and attain high speeds. This usually entails damage to pumps, couplings and motors. Efforts are made to prevent reverse flow with the aid of automatic non-return valves.

Automatic non-return valves rely on forward flow to hold the valve open. The amount of valve opening or lift is dependant upon the flow velocity. When the flow velocity reduces to zero the valve closes. Due to the inertia of the moving components within the valve there can be a short delay between the flow stopping and the valve closing. The delay period can allow reverse flow to start. When the reverse flow is arrested, pressure surge takes place in the pipe system both upstream and downstream of the non-return valve.

11.6.3 Choice of type of non-return valve

For non-return valves the most important requirements are:

• Controlled closure in the event of reverse flow	reduced risk of line shock
• Low opening pressure	reduced risk of line shock
• Low flow resistance	low energy consumption
• Low maintenance requirements	low operating costs
• Reliable	low operating costs

The range of non-return valves is considerable. Table 11.2 lists popular types of non-return valve with their respective merits.

Type	Flow resistance	Risk of water hammer	Maintenance	Temperature restrictions
Swing-check valve	low	yes	low	no
Poppet valve	high	slight	medium	no
Disc valve	high	yes	medium	no
Piston valve	high	slight	medium	no
Ball check valve	medium/high	slight	low	no
Rubber tube valve	high	yes	medium	as low as 80 °C
Butterfly check valve	high	yes	low	no

Table 11.2 Types of non-return valves

11.6.4 Sizing non-return valves

It is important that the valve is sized so that it is fully open under normal steady state running conditions. If there is a flow variation the valve must be selected very carefully. At maximum flow the valve should not be wide open and up against the "stops". Minimum flow should not correspond to less than 50% opening. The inertia of the moving parts must be minimised and the movement damped.

In the case of non-return butterfly valves, operating under steady-state conditions, sizing is carried out in accordance with the following procedures and Figures 11.1 and 11.2.

For water

Use Figure 11.1:

1 Enter the figure along the "x" axis at the minimum normal flow.

2 Move vertically until the pipe diameter line is intersected.

3 Check valve operation by noting relative positions of curves A, B, C, D and E.

4 If valve not fully open, select a smaller diameter.

5 Check pressure drops at all flows.

For liquids of similar viscosity to water but differing densities

Use Figure 11.2:

1 Calculate the flow velocity in the pipe for the minimum normal flow.

2 Calculate ρv^2.

3 Enter the figure along the "x" axis at the ρv^2 value.

4 Move vertically until the valve lines are intersected.

5 If valve not fully open, select a smaller diameter at higher velocity.

6 Check pressure drops at all flows.

Examples

Using Figure 11.2

Example 1:

Water flow velocity 2m/s

pressure drop 0.73 mlc

Choose valve with spring

Example 2:

Air horizontal pipe

density r = 1.3 kg/m^3

flow velocity 20 m/s

$\rho \cdot v^2 = 1.3 \cdot 20^2 = 1.3 \cdot 400 = 520$

where:

v	=	velocity (m/s)
ρ	=	density (kg/m^3)
Q	=	water flow (m^3/s)

a) Without springs:

pressure drop 0.09 mlc

The valve is fully open

b) With springs:

pressure drop 0.17 mlc

There is a risk of clattering. Therefore choose a valve without springs.

11.6.5 Pressure drop

The pressure drop across the valve according to Figure 11.2 is based on the resistance figure Z = 3.6 for valves up to and including DN 250. For larger valves the drop in pressure is lower and the pressure drop values in the diagram should therefore be reduced by the factors stated below.

DN	300 - 350	factor 0.89
	400 - 700	factor 0.83
	750 - 1000	factor 0.78
	1200	factor 0.70

The pressure drop is dependent on density and flow velocity.

11.6.6 Pitfalls

11.6.6.1 Centrifugal pumps operating in parallel

Pumps for parallel operation are usually installed as shown in Figure 11.3. To avoid problems in cases when a second pump is to be started while the first is in operation, it is necessary for the pumps to have stable characteristics, see Chapter 5, Figure 5.24. The characteristics should be steep rather than flat, see Chapter 5, Figure 5.25.

In Figure 11.3:

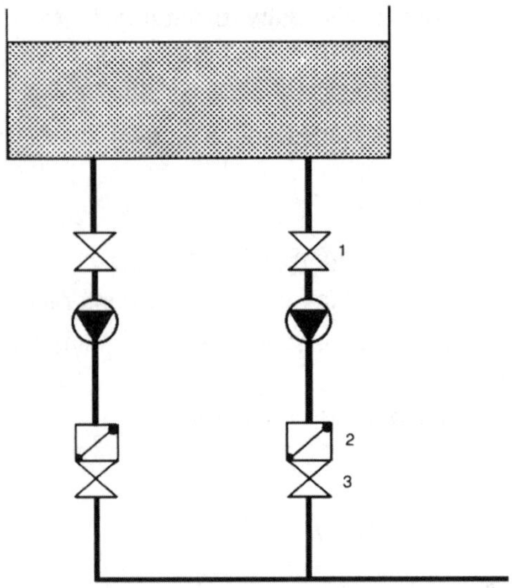

Figure 11.3 Pumps connected in parallel

1. Suction isolating valve for pump. Select a full-bore valve with a low pressure drop to avoid loss of NPSHa and the risk of cavitation. Full-bore ball valve or wedge gate valve.
2. A non-return valve with low pressure drop is selected to avoid losses of energy and with low risk of water hammer, i.e. poppet or ball check valve or in the case of small sizes a piston valve.
3. Discharge isolating valve after the pump with low pressure drop to avoid great losses of energy. Preferably a ball valve or wedge gate valve.

11.6.6.2 The suction side of centrifugal pumps

A centrifugal pump requires a certain pressure above the vapour pressure on the suction side, NPSHr, in order to operate correctly and to avoid cavitation. The non-return valve should therefore be located on the delivery side.

11.7 Accumulators

Accumulators are used for pressurised liquid storage in systems which require variable flow rates or intermittent flow requirements. Chapter 6, Sections 6.4, 6.5 and 6.6 show that storage volumes are necessary when the pump flow does not match the flow demand. Accumulators are suitable for pressurised applications up to 345 barg.

The accumulator consists of a pressure vessel which houses a bladder, rather like a balloon. The bladder is attached to the top of the pressure vessel by its gas valve. The bottom of the pressure vessel is piped to the process system. The bladder is pre-charged with nitrogen. When the accumulator is charged with process liquid, the nitrogen in the bladder is compressed to a higher pressure as its volume is reduced.

During the working cycle, the system pressure will fall when the pump is unable to match the demand. As the pressure falls, the nitrogen in the bladder expands and pushes some of the liquid out into the system. The next time the pump flow is greater than the system flow the accumulator will be recharged. The volume of liquid which can be stored is a function of the volume of gas.

As stated in Chapter 6, Section 6.4.6, it is the application of Boyle's Law for gases at constant temperature which controls the pressure and volume relationships. If more liquid is required, more gas must be provided. If a smaller pressure change is required, more gas must be provided.

Guidelines are available, which indicate the range of working pressures possible, and there are design charts. Working pressure ranges of up to 3.5 are possible, but this is far too large for normal pump applications, see the run-down of accumulators at the end of this Section.

Accumulators are manufactured with gas volumes of 200 l as standard. If these are too small, accumulators can be collected in racks and piped together. The pressure/volume characteristic can be enhanced by adding more gas volume. Standard nitrogen bottles can be piped into the gas side of the bladders. The liquid volume per unit volume of pressure drop is thus increased.

Bladders are not completely gas tight. As with tyres on a car, the gas pressure must be checked regularly and topped up when necessary.

Another application for accumulators is in lube and seal systems on large machines. Large machines, particularly high speed machines, do not always stop when the power is switched off. Such machines are normally stopped by control logic so that after the power has been switched off to the main driver the auxiliaries — lube oil pumps, seal oil pumps, cooling water pumps — carry on running for 30 seconds or a minute to allow the main pump to decelerate. After a preset time the control logic switches off the auxiliaries.

Controlled run-down such as this works well in normal circumstances but can cause problems in emergencies. If the main motor is high voltage and the auxiliaries are low voltage, it is possible to have a fault in the low voltage supply so that all auxiliaries are lost but the main motor is still running. Interconnection between the HV and LV systems could switch the main motor off as soon as the LV supply is lost. However, there are no auxiliary supplies for run-down.

The problem is solved by fitting and charging run-down accumulators. The auxiliaries would normally be started and allowed to stabilise before the main motor was started. This period can be used to charge the accumulators. Once charged, the system acts as though they are not there. When auxiliary power is lost the accumulators discharge their liquid into the auxiliary systems during the run-down. If the main motor was required to continue operating under these conditions it would be necessary to fit stand-by pumps to all the auxiliary services powered from an alternative supply, probably 24 V DC.

11.8 Pulsation dampers

Reciprocating pumps, cause flow variations and pressure pulsations in the system pipework because of the nature of the piston/plunger motion. Pulsation dampers are fitted to the suction and discharge systems of these pumps in order to reduce the severity of the flow disturbances. Pulsation dampers attempt to disconnect the pump from the system in terms of inertia. The flow variations produced by a pump cannot be changed. If the density of the liquid were zero, then there would be no pressure pulsations, no pipe vibration and the effect in the suction system, "Acceleration Head Loss", would disappear. The pulsation damper absorbs excess liquid and provides liquid when the pump is deficient. This is very similar to the action of accumulators but it happens much faster.

Flow variations are difficult to measure. Pressure pulsations are easy to measure and their effect is easy to observe. Pressure pulsations can be measured with a pressure transducer and displayed on an oscilloscope or captured for analysis by computer.

The peak-to-peak pressure value is easily evaluated. The effect of pressure pulsations is vibration of pipework. The product of the peak-to-peak pressure and the pipe bore area is a force; this force acts alternatively along the pipe. If the force is large in comparison to the mass of the pipe and the stiffness of the pipe supports, the pipe will vibrate. By reducing the peak-to-peak pressure, the severity of the vibration will be reduced. Pressure pulsations in suction pipework can create a serious problem for pumps. The negative half of the pressure pulse reduces the static pressure of the liquid in the pipework. If the static pressure is reduced below the vapour pressure the liquid will boil and form vapour bubbles. If the vapour bubbles enter the pump the pump will cavitate. Cavitation can drastically reduce pump reliability and significantly increase expenditure on spare parts. Dampers reduce the peak-to-peak pressure by reducing the flow variation.

Peristaltic pumps can suffer from significant flow variation induced pressure pulsations. Pulsation dampers can be used to reduce problems and are manufactured in three basic types:

- Gas-charged
- Inertial
- Acoustic

11.8.1 Gas-charged dampers

A gas-charged damper is similar in many respects to the accumulators described in Section 11.6. It relies on the increased compressibility of a volume of gas to absorb some of the flow variation.

Consider the discharge stroke of a single cylinder pump. When the piston/plunger accelerates towards maximum velocity the pressure at the pump increases due to liquid inertial loads and increased friction losses in the pipe system. If a pulsation damper was placed close to the pump, some of the liquid could flow into the damper, and increase the gas pressure, rather than flow up the discharge pipe. When the piston/plunger decelerated and the pressure at the pump decreased, the liquid would be forced out of the damper due to the gas pressure. The flow variation into the discharge pipe would be reduced thereby reducing the pressure pulsations. The suction system would react in a similar manner.

The volume of liquid which flowed into the damper would depend upon the gas volume and the nature of the gas. Nitrogen is nearly always used so the gas characteristics become constant. The volume of gas used in a damper is the important sizing criterion. The pump user or manufacturer specifies the residual pressure pulsations acceptable and the damper manufacturer calculates the gas volume required.

Gas-charged dampers are constructed in various styles. A popular style for suction systems is a vertical vessel which contains approximately 75% liquid. The remainder of the vessel is charged with nitrogen. No bladder or diaphragm is used to separate liquid and gas. This type of damper, also called a suction stabiliser, is very successful for low pressure and hot applications.

At higher pressures the nitrogen is absorbed more quickly into the liquid. This style of damper is a "flow-through" type; since all the liquid passes through the damper. Flow through dampers are also constructed with bladders or diaphragms to contain the gas charge, see Figure 11.4. Some dampers are constructed in a similar way to accumulators and are attached to a branch on the process pipe, Figure 11.5.

Gas-charged dampers must be pre-charged before use. The gas charge is usually 60 to 80% of the working pressure. Dampers without bladders or diaphragms can suffer problems with gas loss when the gas expands during shut-downs. In general, these dampers do not create a pressure loss and are the most used type of damper.

In some installations, the spares inventory is reduced by using high pressure pulsation dampers on low pressure suction systems. High pressure dampers are constructed with fairly rugged bladders and diaphragms; some over 20 mm thick. Heavy, thick diaphragms do not work very well with low differential pressures trying to move them. Consequently high pressure dampers do not achieve the predicted pulsating pressure attenuation. High pressure dampers are suitable for high pressures. Low pressure dampers are essential for good performance on low pressure suction systems.

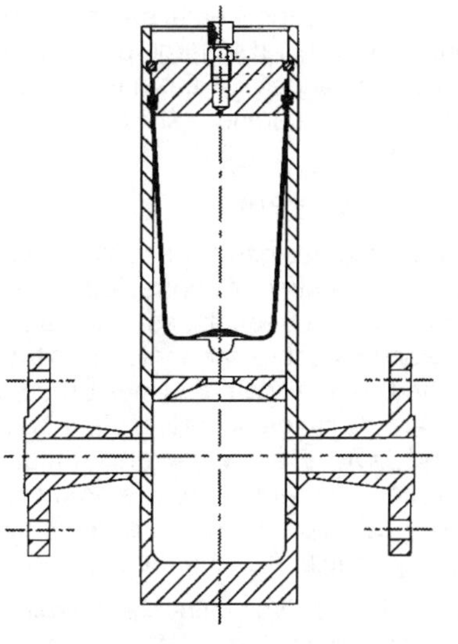

Figure 11.4 Flow through gas charged bladder damper

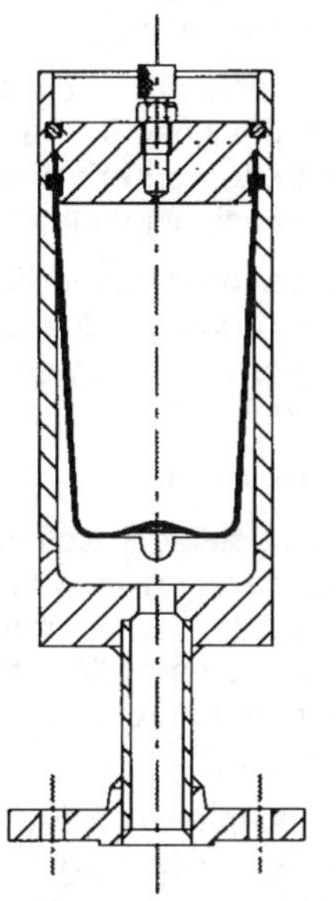

Figure 11.5 Single port gas charged bladder damper

11.8.2 Inertial dampers

Inertial dampers rely on a moving mass of liquid to limit the velocity changes in the discharge system. A mass of liquid inside a pressure vessel, is made to rotate as the liquid flows through. The mass of liquid acts as a flywheel, in mechanical terms, and reduces the flow variations. Inertial dampers are used for high temperature applications where elastomers are unsuitable, see Figure 11.6.

Figure 11.6 A typical inertial damper

Courtesy of Flo-Dyne Ltd

There is no pressure limit on inertial dampers and they are unaffected by pump speed or discharge pressure. Inertial dampers do create a pressure loss and this must be considered in the pump selection. Inertial dampers are only used as discharge dampers.

11.8.3 Acoustic dampers

Acoustic dampers are designed specially for each application. Pressure vessels consisting of connected volumes and choke tubes attenuate discrete frequencies. The dampers are tuned to remove the worst pressure harmonics from the flowing liquid. Gas-charged and inertial dampers are broad-band devices; both attenuate a wide range of frequencies. Acoustic dampers are narrow-band devices; a few discrete frequencies are attenuated. Like inertial dampers, there are no moving parts other than the liquid. Acoustic dampers are unaffected by working pressure but very sensitive to changes of speed.

They would be of little use on infinitely variable speed pumps. All of the pump flow passes through the damper and a pressure loss is induced. Acoustic dampers can be used on the suction and discharge sides of the pump.

There are two major obstacles to popular use of acoustic dampers. Because they are designed for each specific application the cost is relatively high. To work effectively they must be large. Acoustic dampers are much bigger than any other damper.

When choosing a pulsation damper the initial cost must be weighed against running and maintenance costs. Gas-charged dampers may require gas every two weeks. If the gas charge is lost, can the process continue ?

11.8.4 Pressure pulsations

At the beginning of this Section the effect of the peak-to-peak pressure pulsation on the pipework was discussed. The pressure pulsation creates an axial force which shakes the pipework. Some shaking force is allowable as the inertia of the pipework plus the restraining forces of the supports restrict the pipe movement.

Figure 11.7 indicates acceptable peak-to-peak pulsation levels as a function of the mean process pressure. These values are based on pipework mechanics only. The process or instrumentation connected to the pipework may require lower pulsation levels. Pressure pulsations in the pump suction pipework may also require lower levels. Pressure pulsations in the pump suction pipework reduce the NPSHa/NPIPa, this must be considered when deciding upon acceptable levels.

The pressure pulsation radiated by a reciprocating pump is a complex signal which includes all the integer harmonics from 1 · r/sec and greater. The power of the signal is in the lower harmonics, usually starting at r/sec x number of plungers. The

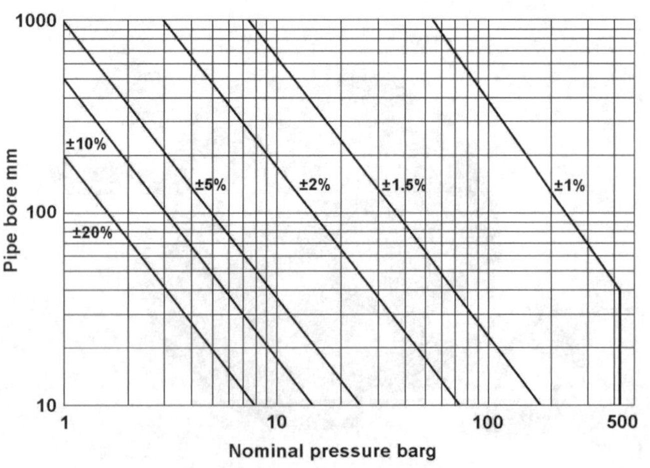

Figure 11.7 Acceptable peak-to-peak pressure pulsations for pipework

pressure pulsation can cause lateral vibration of the pipework, and resonance if the natural frequency of the pipe span is very close to one of the integer harmonics.

It is also possible to achieve acoustic resonance. A pipework system can be tuned to various frequencies due to the pipe length and the arrangement of branches and other connections. If one of the pipework acoustic frequencies is very close to a pump integer harmonic very serious vibration problems may arise. The magnitude of the problem is dependent upon the operating conditions and the potential resonant frequency.

11.8.5 Computer simulation

System mechanical and acoustic resonance can be predicted by computer simulation. The complete system must be analysed. The discharge system cannot be analysed without the suction system and vice versa. Complete and accurate piping isometrics must be available for analysis. There is little point in analysing one system and building another. Analysis can either be analogue or digital. Very few companies had analogue computers. Those who did, accrued a huge amount of practical experience, firstly on compressors, then pumps.

The analogue method has largely been replaced by digital computer studies. There are a few more companies in this field now. Not all of these have much experience! or expertise!! Not all companies perform the simulations the same way. Insist on very detailed descriptions of simulations proposed. All companies will show thick finished reports including masses of data which is incomprehensible. Look for one which can explain things in understandable terms.

The following points should be raised with any prospective simulation contractor:

- Can the simulation model show pump run-up and run-down conditions?
- Can the model process transient conditions?
- Can all types of pulsation damper be modelled?
- How easy is it to change parameters to evaluate modifications?
- Can the interaction between multiple pumps be modelled?

Simulations for a single pump system are very expensive. The analysis cost must be judged against the pump unit cost and the criticality of the service.

It is not possible to give generalised advice on when simulations should be performed but it is possible to state some obvious facts.

- Pressure pulsations are largely attenuated by liquid viscous effects.
- Pulsations will radiate further in low viscosity systems than in high viscosity.
- Pumps running fast, both r/min and m/s, will produce worse pulsations than pumps running slow.
- Pipework with poor supports will vibrate more easily than properly supported pipework.
- It is much better to have pipework for reciprocating pumps at ground level securely attached to concrete.
- Piping systems running at high liquid velocities will tend to give more problems than low velocity systems.

If it is difficult to decide whether a simulation should be performed seek independent advice. Good advice can only given after reviewing all the information; this is the information that would be supplied to allow the simulation to be performed.

11.9 Instrumentation

11.9.1 Integrated control and diagnostics

Many pump suppliers have developed standard packages, complete with instrumentation and starting equipment, for popular applications. These include:

- Twin circulating pump packages for commercial and industrial heating
- Pressure maintaining packages for closed systems
- Trailer-mounted engine driven site pumps
- Dosing pump packages complete with suction tank

The ultimate in packaging is to be found among high pressure cleaner suppliers. Pumps are packaged in floor-mounted coin-operated booths for garages, on two and four wheel trailers, and truck-mounted for complete portability.

Pump packages which are built to the user's specifications can incorporate almost anything. As pump motor sizes increase the motor starters are generally not included in the package but are housed remotely in motor control centres. The pump package would include control logic and local operating functions.

Typical equipment found in local control panels is:

- Pump start/stop buttons
- Hand/off/auto switches
- Lag/lead switches
- Indicator lamps
- Hours run counters
- Ammeters
- Fault/trip annunciators

11.9.2 Annunciators

Annunciators provide visual indications which reveal specific messages. Not all the messages relate to faults or trips; it is possible to indicate normal functions. The annunciator is fitted with a test button to check all indicators are functioning. Most annunciators have connections to allow remote signalling of fault and trip conditions. In automatic packages , a wide range of annunciator message information is displayed, see Table 11.3.

Annunciator message
Pump 1 running
Pump 2 running
Pump 1 tripped
Pump 2 tripped
Low suction pressure
Low low suction pressure
High discharge pressure
High high discharge pressure

Table 11.3 Typical annunciator messages

When a pump motor trips the annunciator does not know the reason; the annunciator receives a signal from the motor starter to notify that the contactor has opened when it should have closed. The operator must go to the motor control centre to look at the motor starter. Motor control centres are now very sophisticated and a group of motor starters may have something equivalent to an annunciator which indicates a wider range of motor faults.

Operating conditions designated "low" or "high" are fault or alarm conditions. The annunciator flashes the relevant lamp and sounds an audible signal. The alarm function is to warn operators that some operating condition is outside normal limits and should be investigated and corrective action taken. The pump continues to operate. "Low low" and "high high" are trip conditions, indicating something has switched off. If, after the alarm signal, corrective action was not or could not be taken and the operating condition deteriorates, the control system takes over and shuts down the equipment. At the alarm condition, the control system can be designed to start stand-by equipment when this is fitted.

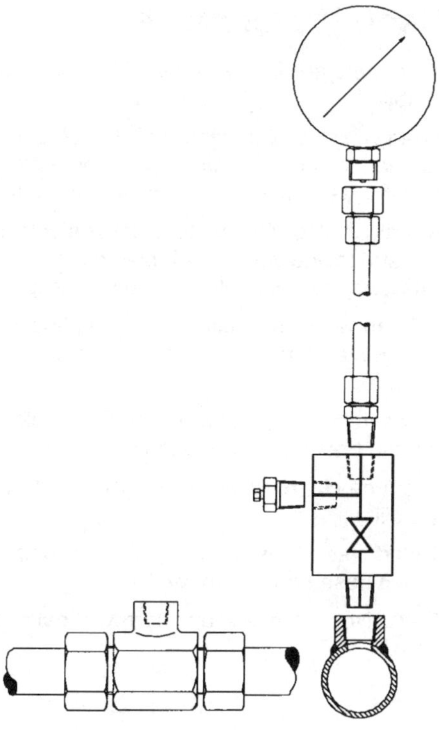

Figure 11.8 Pressure gauge for hazardous liquid

Latest control systems, using PLCs, can manage very complicated systems. The use of PLCs allows transducers to replace switches. A transducer can monitor the rate of change of a parameter and take pre-emptive action rather than wait for the parameter to reach a prescribed limit.

11.9.3 Connection to process pipework

One important aspect of instrumentation which is sometimes not specified is how instruments should be connected to process pipework. This part of detail design becomes more important when the process liquid is hazardous. In some cases the user should specify that no screwed connections are permissible for sealing process liquids. Compression fittings, flanges and other connectors must be used. Valves are used to isolate instruments from the process to allow servicing and calibration. Bleed, vent and drain valves may also be required. For liquids such as water and lube oil, bleed and vent plugs could be acceptable, but not for propane.

For example, Figure 11.8 shows how a pressure gauge should be connected to a hazardous liquid system. The user must be fully aware of the hazards posed by the process liquid. These hazards must be discussed with the pump supplier to allow appropriate designs to be considered.

When pressure switches are fitted, a pressure gauge should also be included so that local operators can see the normal pressure and report on trends. If multiple switches are fitted, for separate alarm and trip functions, they must be connected to the same point on the process pipe.

Although the latest electronic instruments allow much more sophisticated control of pumps and equipment; qualified, experienced maintenance staff must be on hand to look after the equipment if high costs are not to be incurred using the supplier's personnel on site.

11.10 Useful references

ISO 1102 Commercial road vehicles - 50 mm drawbar eye - Interchangeability

ISO 1185 Road vehicles - Connectors for the electrical connection of towing and towed vehicles - 7-pole connector type 24 N (normal) for vehicles with 24 V nominal supply voltage

ISO 3731 Road vehicles - Connectors for the electrical connection of towing and towed vehicles - 7-pole connector type 24 S (supplementary) for vehicles with 24 V nominal supply voltage.

ISO 4141-1 Road vehicles - Multi-core connecting cables - Part 1: Test methods and requirements for basic performance sheathed cables.

ISO 8703 Road vehicles - Caravans and light trailers - Control device for inertia braking - Calculation of forces

ISO 8755 Commercial road vehicles - 40 mm drawbar eye – Interchangeability.

ISO 4183 Belt drives - Classical and narrow V-belts - Grooved pulleys (system based on datum width).

ISO 4184 Belt drives - Classical and narrow V-belts - Lengths in datum system.

ANSI/API 613 Special Purpose Gear Units for Petroleum, Chemical and Gas Industry Services.

API 677 General Purpose Gear Units for Petroleum, Chemical and Gas Industry Services.

API 614 Lubrication Shaft-Sealing and Control-Oil Systems for Special-Purpose Applications.

ANSI/AGMA 2001-D04 Fundamental Rating Factors and Calculation Methods for Involute Spur and Helical Gear Teeth.

DIN 3990-21 Calculation of load capacity of cylindrical gears; application standard for high speed gears and gears of similar requirement.

ISO/TR 13593:1999 Enclosed gear drives for industrial applications. See also ISO/TR 13593:1999/Cor 1:2005.

AGMA 913-A98 Method for Specifying the Geometry of Spur and Helical Gears.

AGMA, 500 Montgomery Street, Suite 350, Alexandria, VA 22314-1581 USA, Tel: 703 6840211, Fax: 703 684 0242, www.agma.org.

Quality, inspection and testing 12

12.1 Introduction

Most pumps have the important components manufactured from metal, of which cast iron, carbon steel and stainless steel are the most common. Cast iron forms the basis of all steel products; even though some new steels are made from scrap steel. Iron ore can be converted to iron by three methods:

- blast furnace
- sintering or pelletised/blast furnace
- direct reduction

In a blast furnace, iron ore reacts with hot coke to produce pig iron. The sintering or pelletising process prior to the blast furnace operation is added to allow blending of iron ores and also to control the size of the blast furnace feed. Sintering or pelletising improves the blast furnace operation and reduces energy consumption. Direct reduction produces sponge iron from iron ore pellets by using natural gas. Most iron is produced from sintered iron ore and coke. The steel maker controls the sintering process to produce a consistent iron quality.

Modern blast furnaces are capable of providing, together with computer modelling, in-process control. Iron is taken from the blast furnace as finished material for iron foundries. Iron is transferred to the oxygen steel process for conversion to various grades of steel. Iron from direct reduction plants is mixed with scrap steel in an electric arc furnace to produce various grades of steel.

The pump design engineer is not generally concerned which production method is used. For some alloy steels, used for some arduous applications, the designer may specify vacuum remelted steel for reduced impurities or may require additives to promote fine grain structure. Principally the design engineer looks at the physical properties of the material to judge suitability.

Standard tests are applied, solely to assess compliance with the published specifications. Some materials are characterized only by their physical properties or chemical composition, others by both. Grey cast iron is specified by its physical properties. Some low grades of carbon steel are specified by their chemical composition, no physical properties are necessary. Most materials are described by both.

For the following physical properties defined in Section 12.1.1, standard test pieces are stretched in a machine which simultaneously measures the increase in length and the applied load. There are several different test piece sizes which give slightly different results. One standard test piece is very small, this fits a machine called a Hounsfield Tensometer. Very small test pieces are useful when samples must be taken from castings or finished parts.

12.1.1 Physical properties

The mechanical properties referred to in Sections 12.1.1.1 to 12.1.1.6, are a function of the grain direction of the material. Unless specified otherwise all properties relate to the longitudinal direction. Properties in the transverse direction and through direction may be lower depending upon the physical treatment of the material and its grain structure.

12.1.1.1 Ultimate strength

This is the strength of the material when it fractures. See Chapter 7, Section 7.2 for typical values.

12.1.1.2 Limit of proportionality

This is the strength of the material when the relationship between stress and strain ceases to be linear. In low carbon steel this is classified as the yield point, the onset of plastic deformation, the material does not return to its original length when the load is removed. Most designs do not stress materials beyond the limit of proportionality.

12.1.1.3 Elongation

How much the material has increased in length when it fractured, known as elongation. Different test pieces have different gauge lengths, each gauge length gives a slightly different result. Good elongation properties, 15 to 20%, are required for complex components which are highly stressed. Good elongation indicates ductility. Ductility is necessary so that components can deform very slightly to spread the load. A good cast iron may be 4%.

12.1.1.4 Reduction in area

Ductile materials "thin" slightly as they are stretched. When the material fractures, the cross-sectional area of the fracture is less than the original test piece. Reduction in area is reported in most American standards but not used very much in Europe.

12.1.1.5 Hardness

Hardness is the ability of the material to withstand surface indentation. No special test piece is required, raw material and finished parts can be tested. Several scales of hardness are used; Brinell Hardness Number, Vickers Pyramid Hardness and Rockwell. Approximate conversions are available between scales. In carbon steels, the hardness is directly related to the strength.

12.1.1.6 Impact strength

The ability of the material to withstand shock or impact is important. A special test piece is required to fit the test machine. Most materials lose impact strength as the temperature reduces. Depending upon the material, impact properties should be checked when operating below 0 °C. Two different tests are used which give different results, very approximate conversions are available. Charpy is the most popular, Izod is little used. A popular bench-mark for offshore equipment is 27 J at the design temperature. It is normal to check three test pieces.

12.1.1.7 Fatigue strength

All the tests defined so far can be performed fairly quickly; "test the pieces today, get the results tomorrow". Fatigue is different. A special test piece is either subjected to repeated tensile loads or repeated bending loads.

For repeated tensile loads, the test piece experiences cyclic loads from 0 to + value. A bending test piece is loaded from -value to +value. To find the endurance limit the test piece must not fail. A test piece will appear satisfactory if it lasts five million cycles, 5,000,000. If the machine runs at 3000 r/min this will take 1667 minutes, i.e. 28 hours. Of course, it will not be possible to guess the correct stress so several tests must be run.

Testing for fatigue in air is simple, testing for fatigue in contact with a specific liquid is more complicated. It is not usual for the fatigue strength of materials to be checked on a contract basis. For critical pumps, fatigue testing of finished components may be necessary. Most pump designs are not based on fatigue. Exceptions include components for reciprocating pumps. The manufacturer should state if the life of any component is limit by fatigue when running at the rated conditions.

Fatigue can be induced in components by the operating environment. Pumps which operate with start-stop or load-unload control can be prone to fatigue failures. If the control system does not provide a gradual change in flow/pressure/torque some components are likely to have very short lives.

12.1.1.8 Creep resistance

Creep is the permanent distortion of the material after being subjected to a stress for a long period of time. This is not normally a problem in pumps, although those built of GRP or PTFE may suffer at temperatures below 250 °C. It is however usually considered in steam and gas turbines, i.e. hot components. Creep testing is similar to fatigue testing but creep tests can last for years. Published research data is therefore often used when necessary.

12.1.2 Heat treatment

Many materials require heat treatment to achieve the correct condition or strength. Carbon steels are hardened and tempered to achieve high strength, usually at the expense of ductility. Austenitic stainless steels are stress-relieved, softened or solution-annealed to modify the physical or chemical properties. The final condition is usually confirmed by taking hardness readings. When components are heat treated to achieve specific physical properties a test piece is heat treated as well. The necessary physical tests are conducted on the test piece.

12.1.3 Chemical composition

When a metallic material is produced as a raw material, its chemical composition is checked. When cast iron is converted to carbon steel in the oxygen process, all the relevant elements are weighed before being put into the converter. Before the steel is poured, the chemical composition is checked. When the steel is poured a sample is cast. The sample is analysed and its chemical properties are the properties of the melt. Certificates will show the name of the steelmaker and the melt, cast or heat number.

The chemical composition may show elements which are not required by the specification. Low carbon steels may show traces of nickel, chromium and molybdenum. The trace elements are a welcome addition because they tend to enhance the physical properties of the material. Impurities such as sulphur and phosphorous, will be shown very accurately. The chemical composition of specific components, when necessary, can be traced back to the original melt.

On rare occasions, a sample will be taken from a component and analysed. Latest techniques only require very small samples. It is possible to analyse material without destroying it. Only a few techniques are available which can analyse material without removal from the component. No methods can detect carbon. However sufficient accuracy is present to differentiate between 304 and 316 stainless steel.

12.1.4 Corrosion resistance

Corrosion resistance of materials is judged by studying published research. Pump manufacturers carry out long term research on corrosion and develop materials to cope with specific problems. If a pump user wishes to pump a new liquid, of which previously no pump manufacturer has had experience, the user should conduct basic corrosion testing.

12.1.5 Non-destructive testing

Raw material, raw castings and completely finished components can be examined physically to determine the quality of certain aspects of the material. This type of examination falls into two categories:

- Surface inspection
- Interinal inspection

Surface inspection looks for discontinuities in the surface which could be detrimental to the service life of the component. Cracks in the surface create stress raisers which can lead to fatigue failures. Pinholes in the surface may indicate porosity.

Internal inspection can show the integrity of the material and identify any impurities, inclusions or voids in critical locations. Impurities, inclusions and voids detract from the cross-sectional area available for stressing and create stress raisers. Porosity can lead to problems of leakage.

When flaws are detected it has to be decided whether the flaw is serious, if it can be repaired or whether it **should** be repaired. Some national standards, particularly pressure vessel standards, have categories for defects. The manufacturer's requirements may be more or less stringent than published standards. If the flaw is in raw material, a casting or piece of plate, it may be more cost-effective to scrap it rather than expend more time and money on repairs. If the flaw is in a semi-finished piece there may be more incentive to repair. If the flaw is in a finished component there will no doubt be compelling financial reasons for a repair.

12.1.5.1 Radiographic inspection

Radiography, X-ray, is accepted as the highest grade of internal inspection and the most costly. Radiography is good because a permanent record of the inspection is available at any time for review. It is used mostly for welds but also for critical areas in castings. Components can be taken to fixed x-ray equipment. Large components or assemblies are radiographed using radioactive isotopes. Strict safety precautions must be enforced.

12.1.5.2 Ultrasonic inspection

Ultrasonic inspection is popular because it can be conducted with portable equipment. The visual display indicates the position of flaws in respect to the thickness of the material. The size of the flaw must be assessed by an experienced operator. This technique is very good for inspecting large flat plates prior to fabrication or forgings.

12.1.5.3 Dye penetrant inspection

Dye penetration, or "dye pen" is a surface inspection method which is ideal for finding pinholes and hairline cracks. The surface is first sprayed with dye which is allowed to soak into any surface defects. After a specified time, the dye is washed off and the component cleaned. Chalk is finally sprayed onto the surface. If surface defects exist, the dye trapped in the defect is drawn into the chalk by capillary action. Defects are outlined by dye indications in the chalk. The technique is generally used on finished machined surfaces. A skilled operator can judge the depth of the defect by the size of the "bleed-out".

Dye penetration is particularly useful on welds and surface coatings. Weld integrity is degraded significantly by the presence of surface imperfections. It can indicate if a surface coating has achieved complete coverage without porosity, pinholes or cracks and works on non-metallic coatings as well as metallic.

12.1.5.4 Magnetic particle inspection

Magnetic particle, or "mag particle" is a surface inspection method which is popular for cast materials. This inspection can only be conducted on materials which can be magnetised by an electric current. The surface is coated with a liquid bearing small magnetic particles. If the surface contains flaws, the magnetic flux is concentrated around them, drawing the magnetic particles towards the flaw.

12.1.6 Repairs

When repairs are undertaken, usually by welding, the repair must be inspected. Initially the faulty material must be removed by any suitable means and then checked to ensure that all faulty material has been removed; this is normally by dye penetrant or magnetic particle. The repair is then carried out. The component must then be inspected by the same method which found the original flaw. Pumps built specifically for a purchaser may have repairs categorised into "major" and "minor".

12.1.7 Welding

Welding is a skilled occupation. There are many types of welding and the welder undertaking any particular job must be proficient at that type of welding. Welders are graded by the types of components they weld, the positions of the welds and by the types of equipment used. The welding of pipework is thought to be the most difficult type of welding, in particular welding pipework to pump casings.

Welders for pressure vessels have to be qualified and is usually assessed by an insurance company. Designing weld methods for exotic materials requires very special materials knowledge. All welding for process liquid parts should be supported by certification.

12.1.8 Inspection

Inspection of all components is carried out as a matter of course. The degree of inspection is dependent upon the nature and function of the component and the batch size. Mass-produced parts are not scrutinised as closely or individually, as a small number of specially made parts. General dimensional inspection of parts does not normally form part of the purchaser's involvement. In special cases, the purchaser may wish to witness the measurement of specific dimensions.

12.1.9 Assembly

The key process of pump assembly is the final dimensional check. If components do not have the correct fit or clearance it will be very apparent. Selective assembly of components is frowned upon unless the components are always supplied as a sub-assembly. Spare parts cannot be selectively assembled on site. Assemblies may have fits, clearances and alignment measured and recorded before testing or dispatch. This record is sometimes called a "History Sheet".

12.1.10 Packaging

Packaging is the building of the pump and any associated equipment into a pump unit.

In Chapter 11 it was mentioned that a simple pump unit may be a pump and a baseplate and the purchaser supplies and mounts his own driver. At the other extreme, a pump unit may consist of several pumps and drivers with pipework and control equipment mounted on one baseplate or skid. When control equipment is supplied, the pump manufacturer is often required to wire all the control elements together.

If a control panel is mounted on the package then switches and transducers can be wired to the panel. If the control panel is mounted separately, the pump manufacturer will probably wire the package instrumentation to a terminal box. The purchaser's inspector or instrument engineer may wish to witness the wiring process. Some cable glands are complicated to assemble and wiring routes can be better assessed with a full size 3D package; wiring practices inside instruments and terminal boxes may be considered critical.

12.1.11 Pressure testing

Pressure testing of components and assemblies confirms the material integrity and the basic strength of the design. Porosity of castings can be detected easily without lengthy ultrasonic or costly radiographic inspections. Pressure testing does not always indicate the presence of thin walls. Whenever possible components should be pressure tested in a manner which closely represents the pressure effects during normal working. Clamping components in fixtures modifies their elastic behaviour and does not highlight problems such as distortion of mating flanges.

Pressure testing does not automatically mean hydrotesting with water. Some tests may be conducted with oil, others with a detergent and water mixture. Some components may be pressurised with compressed air and submerged in a water tank. Critical components for very hazardous liquids may be tested with helium.

The pressure to which components may be subjected can be limited by standard flange designs. The purchaser and the manufacturer must ensure that each fully understands the requirements and the limitations of pressure testing.

12.1.12 Running tests

Running tests are thought by some to be the ultimate quality control test. This may or may not be the case. Many types of test are possible. The following list has been proposed for positive displacement pumps.

It is only for fully-assembled pumps. Pump rotors fitted to test casings would not comply with the testing requirements.

(ISO has not prepared a suitable test standard for positive displacement pumps. It seems unlikely a good standard will be available in the near future.)

Confirmation test: A running test conducted to confirm the pump unit functions as required. No test measurements are made. The pump unit actually performs the specified functions. This type of test is limited to fully packaged equipment such as engine driven, trailer mounted pumps complete with suction and discharge hoses.

Performance test: A running test conducted to confirm the mechanical and volumetric efficiency and the mechanical soundness. The following test readings should be taken and recorded:

> Inlet pressure, inlet temperature, outlet pressure, outlet flow, pump speed and pump power input or torque.

The number of sets of readings to be agreed by the manufacturer.

NOTE: Power input may be measured electrically at the input to a calibrated test motor or the contract motor when data on efficiency is supplied by the motor manufacturer. Driving torque may be measured directly or indirectly as a torque reaction when appropriate.

NPIP test: A running test conducted to confirm the NPIPr is less than the NPIPa. The purchaser and the supplier should agree the flow loss criterion to be used. All the readings required for a performance test should be recorded.

NOTE: An NPIP test is not part of a performance test and must be specified separately. Test rig modifications may be necessary which have cost and delivery implications.

MIP test: A running test conducted to confirm the Minimum Inlet Pressure is sufficient to support the rated operating conditions.

NOTE: MIP tests are conducted when NPIPr tests are inappropriate. NPIPr is based on the liquid vapour pressure. If the liquid vapour pressure is extremely low or the process liquid is a variable mixture of compounds NPIPr becomes an invalid criterion. The pump manufacturer will specify a suitable Minimum Inlet Pressure to ensure proper pump performance.

Linearity test: A running test conducted on dosing pumps to show that the variation in flow with stroke or speed adjustment conforms to a straight line relationship. All the readings required for a performance test should be recorded.

Repeatability test: A running test conducted on dosing pumps to show that the pump delivery is consistent for specific stroke or speed settings. All the readings required for a performance test should be recorded.

Steady state accuracy test: A running test performed on dosing pumps to show the precision of each dose under steady state operating conditions. All the readings required for a performance test should be recorded.

Gas handling test: A performance test conducted with specific volumes of gas entrained in the inlet pipework. All the readings required for a performance test should be recorded.

NOTE: This test should only be conducted after the normal performance test. This test can only produce useful results when the test is conducted at data sheet conditions with the data sheet liquid or a close approximation.

Self-priming test: Conducted to show that the pump can evacuate gas from the inlet pipe and draw liquid into the pump and operate normally without external intervention. The pump casing should be filled, according to the suppliers instructions, before the test.

Mass-produced self-priming pumps should be fitted with a standard inlet arrangement described by the supplier. Pumps tested for contract should be fitted with pipework to provide a similar gas volume to the site arrangement. When necessary, the static lift can be increased by applying a partial vacuum. All the readings required for a performance test should be recorded plus the time taken to evacuate the inlet.

Product data for mass-produced pumps should indicate all the test parameters including outlet pressure, liquid viscosity and temperature.

Mechanical run: A running test, at specified conditions, conducted to confirm mechanical soundness of the pump and any other auxiliary equipment. The following test data should be recorded:

inlet pressure

outlet pressure

pump speed

Site simulation test: An extended performance test where the pipework local to the pump replicates the site pipework. An NPIP test, if required, must be specified separately. All the readings required for a performance test should be recorded. The purchaser should not specify the requirement to run a site simulation test until details of site process pipework are available for the supplier's review.

Strip inspection: Involves a partial disassembly of agreed portions of the pump to confirm correct operation of components. Retest after strip inspection is not required unless a fault is detected.

String testing: A running test which involves all, or part, of the contract drive train and any other equipment within the suppliers scope. The full extent of the test, including all equipment, operating conditions, test readings and whether mounted on the baseplate/skid/trailer, must be agreed.

Endurance run: A running test which involves all, or part, of the contract equipment at specified operating conditions and is intended to confirm the reliability of the equipment. The purchaser and the supplier should agree the readings and frequency of readings to be taken and the total running time.

Relief valve operation testing: Involves increasing the pump outlet pressure until the relief starts to open, then further increasing the pressure until all the pump flow is relieved. The pressure is then reduced until the relief valve closes and seals. All the readings required for a performance test should be recorded. This test may be applied to integral and external relief valves.

NOTE: This test is only valid when the test liquid closely approximates the contract liquid.

Vibration testing: Involves measuring the vibration severity, rms velocity, at agreed locations under agreed operating conditions.

NOTE: With some pump types, particularly reciprocating pumps, vibration levels experienced on the test bed, when baseplates are not grouted in or pumps are supported by temporary mounts, will be higher than those under normal operation at site. Under these circumstances, vibration data can only be considered as informative and not representative of the final installation.

Pulsation measurement testing: Involves using pressure transducers and electronic instrumentation to measure the pressure pulsations, in inlet and/or outlet piping or within the pump cylinder or casing, over one or more cycles, during running tests. Pulsation measurement readings should be taken at agreed operating conditions.

NOTE: When pulsation measurements are made on the manufacturer's test bed using test pipework the results can only be considered as informative. Such test results cannot confirm or guarantee the pulsations at site.

Noise testing is not included in this list because noise testing is sometimes very difficult and is the subject of a complete proposed standard in its own right. Most pump manufacturers do not have quiet rooms and noise testing is conducted on the normal testbed, sometimes adjacent to the machine shop. Background noise tends to be a problem. If the pump is tested with a slave motor, rather than the contract motor, the slave motor noise must be isolated because it cannot be treated as part of the package noise. Most manufacturers can offer a noise survey; a noise test not complying with any standard but reporting the background noise at each test point with the test reading.

12.1.13 Painting

Painting is the last operation prior to dispatch. Some paint finishes, especially for offshore requirements, are quite complicated and have many intermediate stages. The purchaser's inspector may wish to witness some operations.

Paint specifications may require the raw material to be blast cleaned to a bright finish. The surface has obviously to be dry and free from loose fragments. Blast cleaning may have to be performed in a controlled atmosphere. Inspectors may need to check the environment and to see the surface before the primer is applied. Individual coats of some paint finishes have specified thickness and the inspector may wish to measure the paint thickness as coats are applied. The total paint thickness and the finish of the top coat will be the final inspection.

Some components are machined on all surfaces. Pump manufacturers may not wish to blast clean a surface which is machined for design purposes. Paint systems are available for applying to bright machined surfaces, the only pre-treatment being a thorough de-grease.

12.1.14 Purchased equipment

When the pump manufacturer packages a pump to form a pump unit, specifications and purchase orders will be issued for the purchased equipment. If the manufacturer is building his standard unit, all the specifications issued will be the pump manufacturer's. If the pump unit is built especially for a user, the user may impose requirements on the bought out equipment. It is possible that conflicts may arise between the two sets of specifications. The pump manufacturer is responsible for supplying a complete working package, irrespective of what the purchaser's specifications require.

12.1.15 Functional testing

Pump testing is complicated. Current standards do not address the problem of package testing, that is to say how much of the contract equipment is used for the test. Another part of testing which is not covered by specifications is the functional testing of packaged instrumentation and control equipment.

The following inspection functions are normal:

- Wiring continuity
- Earth and leakage tests
- Circuit verification
- Calibration of instruments

- Verification of set points
- Simulation of control logic

If the purchaser wishes to exercise control over these activities then proprietary specifications must be issued.

12.1.16 Witnessing

The purchaser, or his representative or the certifying authority, may wish to watch certain activities. The certifying authority is usually an insurance company who wants to confirm that statutory requirements have been fulfilled.

The viewing of activities is approached in two ways:

- **Activities which are interesting or complicated may be observed**. The manufacturer informs the purchaser when the activity will take place, usually five to ten days notice is normal. The activity proceeds whether the purchaser is present or not.
- **Critical activities are witnessed**. The manufacturer informs the purchaser/certifying authority when the activity will take place. The activity cannot be performed if the inspector is not present. A hold is placed on that aspect of the contract. Both observing and witnessing is costly and takes time. The purchaser has no automatic rights to either. The full extent of external participation in inspection functions must be agreed during the pre-contract negotiations. Changes to external inspection, or the unavailability of inspectors at specific times can lead to extended deliveries and increased costs.

The purchaser may wish to observe/witness activities at the pump manufacturer's sub-suppliers. This can only be done with prior agreement. Also it is important for the purchaser and the pump manufacturer to agree who has to pay for any third party inspection or the certifying authority.

12.1.17 Clarification of specifications

Pump users who have a large population of pumps and who purchase many pumps every year, will have proprietary specifications covering most aspects of design, operation and maintenance.

The general trend is to take a national or international standard and amend certain clauses. Some clauses are deleted as being irrelevant, others are edited to reduce options and some options are modified to become mandatory requirements. When the published standard is thought to be weak, the proprietary version includes new and rewritten clauses.

In Chapter 11 it was stated that pump users can specify standards which are inappropriate for the application. A similar problem can arise with proprietary specifications. When this arises it is essential that the pump manufacturer has access to the purchaser's engineers who understand why specific clauses were written. The manufacturer must have the opportunity for technical discussions with engineers who can change or delete parts of the specification.

Many engineers involved in purchasing pumps do not fully understand the working principles of the pumps and associated equipment. They are therefore totally reliant on published standards and specifications. Such engineers cannot engage in meaningful technical discussions. Discussions are avoided by selecting equipment which fully complies with the nominated documentation. Under these circumstances the best equipment for the process may not be purchased. The purchaser is at the mercy of unscrupulous suppliers.

With modern business practices there is a great risk that the purchaser may buy the pump he specifies rather than buy the pump he needs. One way to visualise the consequences of this problem is the following edict:

"Trouble-shooting is making the pump do what you specified. Making the pump do what you want is up-grading!"

The difference between troubleshooting and up-grading is — who pays for the work!

12.1.18 Certification

It is clear from the range of testing and inspection operations discussed that the scope for certification is immense. The cost and delivery implications cannot be overlooked. Table 12.1 lists, in the order of popularity, purchasers' favourite documentation.

Component/Assembly	Certification
Pump	Hydrotest Performance Material conformance Material physical Material chemical Material impact Welder qualification Weld procedures Vibration Noise
Pipework	Hydrotest Material physical Material chemical Welder qualifications Weld procedures Cleanliness
Motor	Hazardous area Type test Noise Performance Vibration

Table 12.1 Popular certificates

A material conformance certificate is a declaration from the manufacturer that the materials used comply with the specifications. Certificates of conformance which begin "I .... certify that to the best of my knowledge......" are not acceptable.

When pumps are destined for cold climates, impacted-tested material becomes very important. Even baseplate material is certified. Pipework does not really suffer from impact certification problems. It is easier to use 304 or 316 stainless steel, for which many welders are qualified, rather than use a specialist nickel alloy.

As liquids become more hazardous and materials more exotic the requirements for non-destructive testing and operator qualifications increase.

Although the quality of the material has been discussed, what about the quality of the certification? When a material certificate specifies physical properties or chemical composition, how accurate is it, how valid is it and which authority certified it? The first standard to grade certification was DIN 50049. However, this was superseded by the DIN-adopted European Standard, EN 10204.

Certification is classified as follows:

2.1 **Certificate of Conformance**. The manufacturer certifies the materials conform to the order specification.

2.2 **Works Report**. The manufacturer certifies, on the basis of tests performed on the batch, the materials conform to the order specification. No actual certificates are supplied.

2.3 **Works Certificate**. The manufacturer certifies, on the basis of tests performed on the contract material, the materials conform to the order specification. No actual certificates are supplied.

3.1.A **Certificate issued by the manufacturer's Quality Control or Inspection Departments**; not someone from manufacturing, stating the test results of the actual material.

3.1.B **Certificate issued by an independent test house**, employed by the manufacturer, stating the test results of the actual material.

3.1.C **Certificate issued by an independent test house**, employed by the purchaser, stating the test results of the actual material.

Certification to 3.1.C is clearly the best from a pump purchaser's viewpoint. It is also the most costly and has the longest impact on delivery. On all but the most critical applications, certification to 3.1.B has become the standard for large pumps built to order. The purchaser's inspector can insist on positive material identification on random components to validate certification.

The pump manufacturer retains quality records for a considerable length of time, 20 years is not uncommon.

Manufacturer's quality could be considered as an extension of certification quality. But how much control does the manufacturer have over the manufacturing process ? This aspect of Quality Assurance is standardised in BS 5750, EN 29000, ISO 9000 and the United States ANSI/ASQC Q9000 series all describe the same set of quality principles.

The basis of manufacturing control is recognised as lying within strict operational instructions. Every employee has a written job description. Routine inspection and manufacturing procedures must be written down and strictly controlled.

All of these documents should form part of the manufacturer's Quality Assurance Manual, QAM. The manuals are issued to all relevant departments. Personnel are not allowed to keep individual copies of often-used procedures since these can become out-of-date and result in incorrect operations. Lines of communications are defined with critical decisions. Special emphasis is given to error reporting and feedback to the designer. Companies who advertise compliance with the standards have been audited by a third party. Routine inspections are performed at regular intervals to ensure standard compliance.

Compliance with the three standards mentioned is not essential for good manufacturing control. The basic principles can be incorporated into a manufacturing system without strict compliance with the standards. The critical items which are essential for effective control are a quality plan, for components or complete assemblies, and the power invested in the quality control department. Inspection at every stage of manufacturing is essential.

Total Quality Management is a phrase which is used by management to boost the image of companies. If quality plans are not used, if an independent department does not check components/assemblies at each stage of production, what does Total Quality Management have to offer the purchaser?

Quality Assurance is not intended as a means of achieving higher or better quality levels. Quality Assurance ensures goods are manufactured to the quality level specified and is intended to produce consistency. Well-regulated manufacturing systems will refine designs and manufacturing techniques as the fault reporting system highlights component or assembly failures; this is an essential cost reduction feature. Improvement in product quality is usually driven by the purchaser, directly or indirectly.

12.1.19 Documentation

All pumps and pump units will be supplied with some documentation. The extent of the documentation will depend upon the nature of the pump and the method of purchase. However, some documentation is controlled by legislation and the purchaser must receive this documentation irrespective of the method of purchase.

In Europe safety is controlled by the Machinery Directive which is implemented by compliance with standards. It is the duty of the pump designer to design safe pumps. If complete safety cannot be incorporated in the design, documents must be provided which detail the areas of concern and what precautions must be taken by the user. In general these will deal with the fitting of guards, protection against pressurised liquids and isolation of electric supplies prior to working on equipment. Other documentation the purchaser would reasonably expect to receive is discussed later.

Documentation can also apply to manufacturers' published literature. Typically manufacturers show performance curves and list materials of construction either by generic term e.g. carbon steel, stainless steel, bronze; or standard specification. Information that should be in publications to endorse the quality and reliability of the product is also discussed later.

12.2 Mass-produced pumps

Mass-produced pumps can be bought by two methods; "over-the-counter" and by enquiry. Many standard pumps are purchased "over-the-counter" from distributors. The use of distributors enables a manufacturer to have his products widely available at many outlets. Some distributors have technical knowledge and skills and can remove some of the skilled back-up required from the manufacturer. A purchaser can review published data, select a pump for an application and purchase a pump or unit over-the-counter. Alternatively, the purchaser can send the pump manufacturer or distributor an enquiry with the duty conditions. From the subsequent quotation, accompanied by standard literature, the purchaser can decide which pump to buy.

Before purchasing a pump or unit the following questions should be answered, either from reading the published data or direct questions to the distributor or manufacturer:

- Is the pump designed to a standard?
- Is the pump manufactured to a standard?
- Will my assembled pump be subjected to any running tests?

Some manufacturers do not test all pumps. If not all pumps are tested, how many pumps in a batch are tested and what is a typical batch size. What liquid is used ? Some manufacturers test rotating assembles in test casings rather than the contract casing. This does not fulfil the prime requirement of running tests, final assembly verification.

- Will my assembled pump be pressure tested?

If an assembled pump is pressure tested it is important to know the pressure and how the stuffing box was sealed.

- To what standard are running tests conducted?
- What is the tolerance band on the published data for head, flow, NPSHr/ NPIPr and efficiency?

Some unscrupulous manufacturers publish data which is the best possible attainable. Tolerances will all be negative. Some pump designs relying on moulded non-metallic materials may have unexpectedly wide tolerance bands on flow.

- What Quality Assurance documentation will be supplied with the pump or unit?

The minimum documentation would be a hydrotest certificate and a final inspection release certificate. For a standard pump constructed of 316 stainless steel or a more exotic alloy, it would be reasonable to expect a Works Report, 2.2 as in Section 12.1.18.

Other standard documentation would include:

- Parts lists and assembly drawings

- Installation instructions and guide-lines
- Maintenance and safety instructions
- Trouble-shooting guide
- Telephone, fax number, e-mail and web site of service department

12.3 Custom-built pumps

When a pump is built to the purchaser's specification, the pump order will have been placed on the basis of quotations received from various manufacturers. The purchaser will have issued an enquiry detailing the operating conditions and any other requirements. The purchaser will have evaluated the quotations based on:

- Pump duty point and efficiency
- Compliance with specifications
- Manufacturing schedule and delivery
- Manufacturer's experience
- Confidence in manufacturer
- Documentation
- Cost

Cost is a difficult variable. If the quotation is evaluated by a contractor, initial cost plus a cursory look at the efficiency may suffice. A user will evaluate life-cycle costs. Documentation plays an important role in business and technical documentation plays a critical role in modern plants. The manufacturer should have listed all the documents which would form part of the technical contract. A Vendor Documentation Requirement form or VDR, should have been completed and returned. This form lists all the technical documents and at what time they will be submitted. A preliminary Quality Control Plan, QP, would have been submitted with the quotation to indicate the inspection included.

An important part of any quotation is a good installation list; this together with a good quality plan and a good VDR can instil confidence. A quality plan indicates the important inspection functions the manufacturer will conduct to ensure an acceptable quality level. The QP also shows the areas where control is taken from the Quality Control department, QC, and assumed by the manufacturer's Engineering department or the purchaser's inspector or engineer. In some contracts a Certifying Authority may also have inspection rights; the CA would normally be an insurance company. The QC department is always in overall control of any contract.

Some problems may be due to design rather than to manufacturing. If a pump casing fails hydrotest, and inspection confirms the machining to be correct, the problem will be referred to the Engineering department. Some components are critical, and any problems are also referred directly to Engineering. If the purchaser and the manufacturer have agreed to differentiate between minor and major repairs, major repairs will result in a purchaser's hold point. If the pump has difficulty in achieving the desired performance on test, both sets of engineers will be involved.

If the preliminary QP is not acceptable to the purchaser, the modifications must be agreed during the pre-contract negotiations. Extra certification and hold points not only affect the cost but also the delivery.

The Energy Industries Council, one of the leading UK trade association for UK companies supplying capital goods and services to the energy industries worldwide, considered the problem of QP's and issued a proposal, *Guideline for the generation of Quality Plans, QAG 01*. The proposal advocated a style of QP, on A4 paper, which could be word processed. The basic idea was to have each component on a separate sheet, organised in a specific format. Complicated components could have their quality requirements extending over several pages.

Table 12.2 indicates a slightly modified format which has been used very successfully. The references in the "Q.C. Procedure Reference" column — ES and WIM — refer to the manufacturer's standard inspection and workshop instructions. Inspection to national, European or international standards can be performed when appropriate. Components which are not listed in a QP are subject to the manufacturer's standard quality requirements. If the purchaser is considering placing an order, he may ask for a copy of the manufacturer's Quality Control Manual. This would normally be an "Uncontrolled Document". The prospective purchaser would not receive modified or new procedures as they were released.

When a contract is subject to such structured quality requirements, it is usual for the purchaser and the manufacturer to meet very early after the purchase order placement to clarify any ambiguous points and present a contract manufacturing schedule. This meeting, sometimes called a "Kick-off meeting" or a "Vendor co-ordination meeting" takes place between the manufacturer's engineers and contract management and the purchaser's engineers and inspectors. If a Certifying Authority is imposed, they must be represented so that their involvement and hold points are correctly interpreted. This meeting confirms the quality requirements and the documentation requirements shown on the VDR. The major and minor repair categories may be finalised on a component-by-component basis. The manu-

CYLINDER											
REF NO.	PROCESS DESCRIPTION OR Q.C. ACTIVITY	Q.C. PROC. REF	ACCEPTANCE CRITERIA			VERIFYING DOCUMENT	INSPECTION				
			IN HOUSE	CLIENT	ACTUAL		IN HOUSE		CLIENT		CERTIFYING AUTHORITY
							ENG	QC	INSP	ENG	
1	VERIFICATION OF MATERIAL COMPLIANCE	NITRONIC 50	CONFORMS TO SPEC.		EN 10204 3.1.B	CERT. OF MECH. & CHEMICAL PROPS.		H	X		
2	ULTRASONIC INSPECTION	ES 509-005	CONFORMS TO SPEC.			CERT. OF U.T. INSPECTION		H	X		
3	GOODS INWARD INSPECTION	ES 509-018	CONFORMS TO P.O. & SPEC.			INSPECTION CARD		R			
4	INSPECTION AFTER MACHINING	TO DRAWING	CONFORMS TO DRAWING			INSPECTION CARD		R			
5	INSPECTION/Q.A. REVIEW & RELEASE FOR ASSY	WIM 9-200-81	INSPECTION/QC REQMTS. COMPLETE			INSPECTION CARD		R			

H ≡ hold point or witness point R ≡ document retained at manufacturer's works X ≡ document included in purchaser's data pack

Table 12.2 Example of Quality Plan format for a component

facturer may agree to submit progress reports during the contract.

As the contract progresses the manufacturer's contract engineer and the purchaser's inspector can work through the plan and ensure all inspections are completed and that documentation is available.

One important document, which is only used occasionally, is a design review. Pumps for very critical applications, very hazardous liquids or subject to legislative control, usually through a certifying authority, have important areas of design reviewed by a third party. This review is considered as an additional safety measure and would evaluate important areas of construction such as:

- Pump casing pressure stressing
- Effect of nozzle loads on casing stressing
- Shaft torsional and lateral vibration analysis
- Baseplate stressing

It is quite normal for small pressure vessels, seal pots and pulsation dampers, to have their pressure vessel calculations examined and approved.

The pump test is one of the most important aspects of the quality assurance programme. The preliminary Quality Plan with the quotation should have outlined the tests proposed and how they related to the rated operating conditions. The vendor co-ordination meeting should also have clarified any missing data.

Manufacturers' test facilities do have limitations and all rated operating conditions cannot be accommodated. Typical limitations are suction pressure, temperature and power. High suction pressure is difficult to provide, especially on large pumps. Rotodynamic pumps can be tested at the contract differential head at low suction pressure; the absorbed power will be very similar. Positive displacement pumps can be tested at the contract differential pressure or the contract discharge pressure if extra power is provided. Discussions with the manufacturer are essential.

Pumps routinely operate at temperatures over 250 °C. Manufacturers of large pumps will have difficulty sustaining such high temperatures. A compromise of a temperature around 100 °C may be possible. Modern pumps can be quite large, over 50 MW. Most manufacturers do not have power supplies to test such large pumps. Pumps up to 5 MW should not be a problem.

The choice of test liquid can cause formidable problems. Most popular rotodynamic pump test specifications call for water tests. Many process and chemical pumps operate with liquids of viscosities of up to 300 cSt. Centrifugal pumps operate at higher viscosities when running at low speeds. The liquid viscosity can have a positive effect on vibration levels and the criti-

12.4 Guidelines for documentation

12.4.1 Rotodynamic pumps

Speed r/min	Discharge press barg	Stages	Materials	Pump documentation									
				Pump assy hydro	Performance test	String test	Cert of conform	Casing material	Impeller material	Shaft material	Rubbing parts	Torsional analysis	Lateral analysis
≤ 3600	≤ 40	1	Cast iron casing/cast iron impeller Carbon steel casing/carbon steel impeller Bronze casing/bronze impeller	■	□		2.1						
≤ 3600	≤ 100	1	Carbon steel casing/carbon steel impeller Bronze casing/bronze impeller	■	■		2.2						
≤ 3600	≤ 100	1	Stainless steel casing/stainless steel impeller	■	■			3.1.A	3.1.A		3.1.A		
> 3600	≤ 40	1	Any materials	■	W	W		3.1.A	3.1.A	3.1.A	3.1.A	n	n
> 3600	> 40	1	Any materials	W	W	W		3.1.B	3.1.B	3.1.B	3.1.A	n	n

n ≡ required r ≡ required if pump modified, i.e. impeller diameter changed W ≡ witnessed

Table 12.3 Documentation for single stage overhung pumps

Speed r/min	Discharge press barg	Stages	Materials	Pump documentation									
				Pump assy hydro	Performance test	String test	Cert of conform	Casing matrial	Impeller material	Shaft material	Rubbing parts material	Torsional analysis	Lateral analysis
≤ 3600	≤ 100	≤ 6	Cast iron casing/cast iron impeller Carbon steel casing/carbon steel impeller Bronze casing/bronze impeller	■	■		2.2						
≤ 3600	≤ 100	≤ 6	Stainless steel casing/stainless steel impeller	■	■		2.3						
3600 to 6000	≤ 100	≤ 6	Carbon steel casing/carbon steel impeller	■	W	W	2.3					■	■
3600 to 6000	≤ 100	≤ 6	Stainless steel casing/stainless steel impeller	■	W	W		3.1.B	3.1.B	3.1.B	2.3	■	■
> 6000	≤ 100	≤ 6	Any materials	■	W	W		3.1.B	3.1.B	3.1.B	3.1.B	■	■
≤ 3600	≤ 100	> 6	Carbon steel casing/cast iron impeller Carbon steel casing/carbon steel impeller	■	■		2.2			3.1.B	3.1.B		■
≤ 3600	> 100	> 6	Any materials	W	W			3.1.B	3.1.B	3.1.B	3.1.B	■	■
> 3600	> 100	> 6	Any materials	W	W	W		3.1.B	3.1.B	3.1.B	3.1.B	■	■

Table 12.4 Documentation for multi-stage pumps

cal speed. Testing on water can provide unrepresentative results and lead to a pump failing test. Differences between site performance and test performance must be clarified before any test procedures are agreed.

A good performance test is one which closely approximates the operating conditions. Some positive displacement pumps rely on viscosity to operate correctly so there would be no point in specifying water as the test liquid. The test liquid must be agreed with the manufacturer, but one closely approximating the properties of the process liquid would obviously be better than an arbitrary choice based on dogma.

Depending upon the equipment being supplied by the manufacturer, the actual test set up should be clarified. Will the pump be tested on the baseplate with the contract motor? Will the pump be tested on temporary supports and driven by a calibrated slave motor? Will the pump performance test be conducted as part of a string test with all the contract equipment available? This whole aspect of testing is not covered by current standards and not by the proposed ISO standard for rotodynamic or positive displacement pumps.

Larger, more complex pump units would normally have two volumes of documentation supplied with the equipment, an installation and maintenance manual, and a data book. The maintenance manual would contain all the information listed for standard pumps. A typical data book would contain:

Quality plan

Material certificates

12.4.2 Positive displacement pumps

Speed r/min	Discharge press barg	Power KW	Materials	Pump documentation: Pump assy hydro	Performance test	String test	Cert of conform	Casing material	Rotor material			Torsional analysis	Lateral analysis
≤ 250	≤ 50	≤ 10	Cast iron casing/cast iron rotor Carbon steel casing/cast iron rotor Carbon steel casing/carbon steel rotor Bronze casing/bronze rotor	n			2.1						
≤ 250	≤ 50	≤ 10	Stainless steel casing/stainless steel rotor	n			2.2						
≤ 250	≤ 160	≤ 10	Cast iron casing/cast iron rotor Carbon steel casing/cast iron rotor Carbon steel casing/carbon steel rotor Bronze casing/bronze rotor	n			2.2						
≤ 250	≤ 160	≤ 10	Stainless steel casing/stainless steel rotor	n			2.3						
≤ 250	> 160	≤ 10	Any materials	n	n			3.1.A					
≤ 250	≤ 50	≤ 100	Cast iron casing/cast iron rotor Carbon steel casing/cast iron rotor Carbon steel casing/carbon steel rotor Bronze casing/bronze rotor	n	n		2.2						
≤ 250	≤ 50	≤ 100	Stainless steel casing/stainless steel rotor	n	n			3.1.A					
≤ 250	≤ 160	≤ 100	Any materials	n	n			3.1.B	3.1.B				
≤ 250	> 160	≤ 100	Any materials	W	W			3.1.B	3.1.B				
≤ 250	> 160	> 100	Any materials	W	W	W		3.1.B	3.1.B				

Table 12.5 Documentation for rotary pumps
Torsional and lateral vibration analysis should be considered for powers over 30 kW when engines are used as drivers.

12.4.3 Reciprocating positive displacement pumps

Mean piston or plunger speed r/min	Discharge press barg	Power KW	Materials	Pump documentation: Pump assy hydro	Performance test	String test	Cert of conform	Crankshaft material	Liquid end material	Rubbing parts material		Torsional analysis	Lateral analysis
≤ 0.5	≤ 160	≤ 25	Any materials except stainless steel	n			2.1					(*)	
≤ 0.5	≤ 160	≤ 25	Stainless steel liquid end	n			2.2					(*)	
≤ 0.5	≤ 435	≤ 25	Any materials except stainless steel	n	■		2.1					(*)	
≤ 0.5	≤ 435	≤ 25	Stainless steel liquid end	n	■				3.1.A			(*)	
≤ 0.5	> 435	≤ 25	Any materials	n	■				3.1.B			(*)	
0.5 to 1.0	≤ 160	≤ 100	Any materials except stainless steel	n	■				3.1.A			(*)	
0.5 to 1.0	≤ 160	≤ 100	Stainless steel liquid end	n	■				3.1.A	2.1		(*)	
0.5 to 1.0	≤ 435	≤ 100	Any materials	n	■			3.1.A	3.1.B	3.1.A		(*)	
0.5 to 1.0	> 435	≤ 100	Any materials	W	■			3.1.A	3.1.B	3.1.B		(*)	
1.0 to 1.5	≤ 160	≤ 200	Any materials	W	■			3.1.A	3.1.B	3.1.B		(*)	
1.0 to 1.5	> 160	≤ 200	Any materials	W	W	W		3.1.A	3.1.B	3.1.B		(*)	
> 1.5	> 160	> 200	Any materials	W	W	W		3.1.B	3.1.B	3.1.B		(*)	

(*) A torsional analysis should be performed on all pumps driven by a reciprocating engine or reciprocating air motor except if a fluid coupling is used.

Table 12.6 Documentation for reciprocating pumps

Weld procedures

Welder qualifications

Radiographs

Non-destructive test (NDT) reports

NDT operator qualifications

Hydrotest certificates

Pump test procedures

Pump test results

Motor certification

Motor test results

Instrument certification

Wiring check results

Functional test results

Paint procedure

Paint inspection certificate

Overall dimension and weight certificate

Inspector's release certificate

If the purchaser's inspector carries out random positive material identification, the material chemical composition certificate would be marked as such next to the inspector's stamp.

12.5 Useful references

Marchwood Laboratory, Dept. of Materials Engineering, Applied Science Building, University of Wales Swansea, Singleton Park, Swansea SA2 8PP Wales UK, Tel:01792 205864, www.swan.ac.uk/mateng/test/marchwood.

School of Engineering, Solid Mechanics Group, University of Surrey, Guildford, Surrey GU2 7XH, UK, Tel: 01483 259264, Fax: 01483 306039, www.surrey.ac.uk/MME/Research.

McMaster Steel Research Centre, McMaster University, Steel Research Laboratory, 1280 Main Street West, Hamilton, Ontario L8S 4L8 Canada, Tel: 905 525 9140, www.mcmasteel.mcmaster.ca.

Energy Industries Council, (EIC), Newcombe House, 45 Notting Hill Gate, London W11 3LQ UK, Tel: 020 7221 2043, Fax: 020 7221 8813, www.the-eic.com.

DIN 50049 Inspection Documents for the Delivery of Metallic Products – (Withdrawn, - Replaced by EN 10204).

BS 5750-8:1991 Quality systems. Guide to quality management and quality systems elements for services.

ISO 9000 Compendium 9th Edition 2001, International Standards for Quality Management.

ANSI/ASQC Q9000-1-1994 Quality Management and Quality Assurance Standards - Guidelines for Selection and Use.

ANSI/API STD 610 Centrifugal Pumps for Petroleum, Petrochemical and Natural Gas Industries, Tenth Edition 2004.

EN 1151:1999 Pumps. Rotodynamic pumps. Circulation pumps having an electrical effect not exceeding 200 W for heating installations and domestic hot water installations. Requirements, testing, marking.

BS 4617:1983 Methods for determining the performance of pumps and motors for hydraulic fluid power transmission.

ISO 4409:1986 Hydraulic fluid power - Positive displacement pumps, motors and integral transmissions - Determination of steady-state performance.

EN ISO9906:2000 Rotodynamic pumps. Hydraulic performance acceptance tests. Grades 1 and 2.

EN ISO5198:1999 Centrifugal, mixed flow and axial pumps. Code for hydraulic performance tests. Precision class.

EN ISO9001:2000 Quality management systems. Requirements.

ISO 4412-1:1991 Hydraulic fluid power — Test code for determination of airborne noise levels — Part 1: Pumps.

ISO 4412-2:1991 Hydraulic fluid power — Test code for determination of airborne noise levels — Part 2: Motors.

BS 5944-5:1985 Measurement of airborne noise from hydraulic fluid power systems and components. Simplified method of determining sound power levels from pumps using an anechoic chamber.

BS 5944-1:1992, ISO 4412-1:1991 Measurement of airborne noise from hydraulic fluid power systems and components. Method of test for pumps.

BS 5944-2:1992, ISO 4412-2:1991 Measurement of airborne noise from hydraulic fluid power systems and components. Method of test for motors.

BS 5944-6:1992, ISO 4412-3:1991 Measurement of airborne noise from hydraulic fluid power systems and components. Method of test for pumps using a parallel piped microphone array.

BS 6001-2:1993, ISO 2859-2:1985 Sampling procedures for inspection by attributes. Specification for sampling plans indexed by limiting quality (LQ) for isolated lot inspection.

BS 6002-4.1:1994, ISO 8423:1991 Sampling procedures for inspection by variables. Specification for sequential sampling plans for percent nonconforming. Known standard deviation.

BS 7201-1:1989 Hydraulic fluid power. Gas loaded accumulators. Specification for seamless steel accumulator bodies above 0.5 L water capacity.

BS 7250:1989, ISO 8426:1988 Methods of determining the derived capacity of hydraulic fluid power positive displacement pumps and motors.

ISO 8278:1986 Aerospace — Hydraulic, pressure compensated, variable delivery pumps — General requirements.

EN 287-1:2004 Qualification test of welders. Fusion welding. Steel.

EN ISO 9606-2:2004 Qualification test of welders. Fusion welding. Aluminium and aluminium alloys.

EN ISO 15607:2003 Specification and qualification of welding procedures for metallic materials. General rules.

EN ISO 15609-1:2004 Specification and qualification of welding procedures for metallic materials. Welding procedure specification. Arc welding.

EN ISO 15614-1:2004 Specification and qualification of welding procedures for metallic materials. Welding procedure test. Arc and gas welding of steels and arc welding of nickel and nickel alloys.

EN ISO 15614-2:2005 Specification and qualification of welding procedures for metallic materials. Welding procedure test. Arc welding of aluminium and its alloys.

EN 10045-1:1990 Charpy impact test on metallic materials. Test method (V- and U-notches).

EN 10204:2004 Metallic materials. Types of inspection documents.

ANSI/HI 1.6 (M104) Centrifugal Tests 2000, (For centrifugal, sealless centrifugal and regenerative turbine pumps of all industrial types except vertical multistage diffuser type and submersible pumps. Includes procedures on setup and conduct of hydrostatic and performance tests).

ANSI/HI 6.1-6.5 (M113) American National Standard for Reciprocating Power Pumps for Nomenclature, Definitions, Applications, and Operation 2000.

ISO/CD 148-3 Metallic materials — Charpy pendulum impact test — Part 3: Preparation and characterization of Charpy V-notch test pieces for indirect verification of pendulum impact machines.

ISO 148-3:1998 Metallic materials — Charpy pendulum impact test — Part 3: Preparation and characterization of Charpy V reference test pieces for verification of test machines.

ISO 9906:1999 Rotodynamic pumps — Hydraulic performance acceptance tests — Grades 1 and 2.

ISO 5198:1987 Centrifugal, mixed flow and axial pumps — Code for hydraulic performance tests — Precision grade.

EN 626-2, EN 12162.

EN 12639, Liquid pumps and pump units - Noise test code - Grade 2 and 3 of accuracy.

EN 12296 Biotechnology - Equipment - Guidance on testing procedures for cleanability.

EN 12297 Biotechnology - Equipment - Guidance on testing procedures for sterilizability

EN 12298 Biotechnology - Equipment - Guidance on testing procedures for leak tightness.

IEC 1000-4-1, IEC 1000-4-2, IEC 1000-4-4, IEC 1000-4-5, IEC 1000-4-7, IEC 1000-4-8, IEC 1000-4-9, IEC 1000-4-10, IEC 1000-4-11, IEC 1000-4-12.

IEC 60034-2 Recommendations for rotating electrical machinery (excluding machines for traction vehicles) - Part 2: Determination of efficiency of electrical machinery.

IEC 60051 Recommendations for direct acting electrical measuring instruments and their accessories.

ISO 4288 Rules and procedures for the measurement of surface roughness using stylus instruments.

ISO 3274 Instruments for the measurement of surface roughness by the profile method - Contact (stylus) instruments of consecutive profile transformation - Contact profile meters.

EN ISO 5199 Centrifugal pumps - Technical requirements - Class II.

EN ISO 9908 Technical specifications for centrifugal pumps - Class III.

EN ISO 15783 Technical specification for sealless rotodynamic pumps - Class II.

IEC 1000-4-1, Electromagnetic compatibility (EMC) - Part 4: Testing and measurement techniques Section 1: Overview of immunity tests - Basic EMC publication.

IEC 1000-4-2, Electromagnetic compatibility (EMC) - Part 4: Testing and measurement techniques Section 2: Electrostatic discharge immunity test - Basic EMC publication.

IEC 1000-4-4, Electromagnetic compatibility (EMC) - Part 4: Testing and measurement techniques Section 4: Electrical fast transient/burst immunity test - Basic EMC publication.

IEC 1000-4-5, Electromagnetic compatibility (EMC) - Part 4: Testing and measurement techniques Section 5: Surge immunity test.

IEC 1000-4-7, Electromagnetic compatibility (EMC) - Part 4: Testing and measurement techniques Section 7: General guide on harmonics and interharmonics measurements and instrumentation, for power supply systems and equipment connected thereto.

IEC 1000-4-8, Electromagnetic compatibility (EMC) - Part 4: Testing and measurement techniques Section 8: Power frequency magnetic field immunity test - Basic EMC publication.

IEC 1000-4-9, Electromagnetic compatibility (EMC) - Part 4: Testing and measurement techniques Section 9: Pulse magnetic field immunity test - Basic EMC publication.

IEC 1000-4-10, Electromagnetic compatibility (EMC) - Part 4: Testing and measurement techniques Section 10: Damped oscillatory magnetic field immunity test - Basic EMC publication.

IEC 1000-4-11, Electromagnetic compatibility (EMC) - Part 4: Testing and measurement techniques Section 11: Voltage dips, short interruptions and voltage variations immunity tests.

IEC 1000-4-12, Electromagnetic compatibility (EMC) - Part 4: Testing and measurement techniques Section 12: Oscillatory waves immunity tests - Basic EMC publication.

MSS SP-55-2001 Quality Standard for Steel Castings for Valves, Flanges and Fittings and Other Piping Components - Visual Method for Evaluation of Surface Irregularities.

MSS SP-53-1999 (R2002) Quality Standard for Steel Castings and Forgings for Valves, Flanges and Fittings and Other Piping Components-Magnetic Particle Exam Method.

MSS SP-54-1999 (R2002) Quality Standard for Steel Castings for Valves, Flanges and Fittings and Other Piping Components - Radiographic Examination Method.

MSS SP-93-1999 (R2004) Quality Standard for Steel Castings and Forgings for Valves, Flanges, and Fittings and Other Piping Components - Liquid Penetrant Examination Method.

MSS SP-94-1999 (R2004) Quality Standard for Ferritic and Martensitic Steel Castings for Valves, Flanges, and Fittings and Other Piping Components - Ultrasonic Examination Method.

MSS SP-112-1999 R04 Quality Standard for Evaluation of Cast Surface Finishes - Visual and Tactile Method.

Installation and maintenance 13

13.1 Installation

13.1.1 General

The installation of pumps can be carried out by the pump manufacturer, the driver manufacturer or the contractor building the installation. Large pumps are usually installed as early as possible while access at site is at its best. Small pumps can usually be fitted in at any time. The pump should be ordered so that its delivery occurs at the correct time in the overall site programme.

The certified pump drawing must be available to allow foundations or structural steelwork to be designed and manufactured before the pump is delivered.

The space allocated for the pump must be sufficient to permit installation, care, disassembly and maintenance. Transport routes and lifting facilities must be available. Sufficient space must be available around the installed pump for the access of personnel. Operators may need to make frequent running adjustments. Maintenance personnel will require space to remove components and assemblies. Some pump designs require withdrawal space at the non-drive end. The motor may need space for the withdrawal of the rotor. The building or plot must have drainage facilities for leakage and priming and for liquids used for flushing sealing devices as well as water if used for quench. The risks of flooding the pump unit, in addition to the electrical and control systems, which can be very costly and time consuming to dry out, must also be taken into account.

The pump should be sited as low as possible in relation to the liquid level in the suction system and in a place permitting the shortest possible suction pipe with the minimum number of bends.

13.1.2 Ventilation

The ventilation of the pump site is very important. The electric motor of the pump must receive the necessary cooling. If hazardous liquids, harmful to the environment or highly flammable, are to be pumped, special ventilation requirements must be observed. It should be noted that the heat generated by the pump's electric motor can be considerable. The electrical and control systems for the pump must also be protected against damage.

13.1.3 Noise and vibration

Consideration must also be given to the noise caused by the pump, its motor and drive train. Considerable attention is being devoted to noise from pump units and their pipe systems, and special measures may have to be taken. To reduce vibration transmission, which may be further conveyed through the building structure, it may be necessary for the pump foundation and parts of the piping system to use isolation mountings.

Mounting the pump unit on isolation mountings can create piping and bearing problems. Piping close to the pump nozzle can fracture if the pump vibrates but the piping does not. This problem can be solved by using flexible pipes adjacent to the pump. Short bearing life can be a problem if the pump does move appreciably on its isolation mountings. It is better to use an isolated foundation block.

Acoustic enclosures may have to be erected around the pump unit. Permissible noise levels are often controlled by legal, safety and environmental requirements at places of work. In the case of noise exceeding 75 to 85 dB(A), specific measures must usually be taken. Staff can be protected by declaring a "Ear protection zone".

13.1.4 Siting

If the pump unit has been on site for some time, it may have been necessary to carry out some preventative maintenance. Shafts may have been turned to maintain an oil presence in the bearings or to prevent rolling bearings from "Brinelling". It is the user's responsibility to ensure the pump unit is stored in suitable conditions and that the factory preservation instructions have been followed.

On unpacking the pump unit it must be checked thoroughly for any damage which may have occurred in transit. Also check the packing list to ensure all parts are present. Some delicate instruments may have been removed and packed separately. If there is damage or shortages, the manufacturer must be informed immediately. If delicate parts have been packed separately, repack them temporarily. Remove any temporary bracing or locking clamps; replace parts as instructed by the manufacturer.

13.1.5 Baseplate and alignment

Before placing the pump on its foundation, thoroughly clean the top of the foundation, remove any thin ridges and roughen the top to provide a good key for the grout. Prepare enough shims to level the baseplate, two sets for each foundation bolt. The shims should be longer than the width of the bottom flange. A packer should be placed either side of each foundation bolt, about 20 to 35 mm thick. Remove coupling spacers or driving pins. Lift the pump over the foundation block and fit the foundation bolts into the baseplate holes before lowering onto the packers. Level the baseplate, by adding shims, using an accurate spirit level on the machined pads. Long baseplates may be fitted with targets for laser or optical alignment.

NOTE: It is the upper surfaces of the machined pads on the baseplate which are important for levelling. The underside of the baseplate is usually not machined and will not be exactly straight.

Horizontal pumps and motors are usually supplied complete and aligned on a common baseplate. (See Section 13.1.6 for separate pumps and motors.) Whether the pump has been tested or not, the pump and motor will have been accurately aligned prior to dispatch from the manufacturer's works. Check the coupling alignment, see Chapter 9, Section 9.11. Adjust the shimming until the coupling alignment is close, and consult the manual for thermal growth correction. Protect the tops of the foundation bolts and fit shuttering around the block in preparation for grouting. A good mixture for grout is one part of cement to two parts of sharp sand. The final consistency should be able to flow easily.

Pour the grout and ensure the top of the foundation is covered evenly. If full depth cross members are fitted, ensure complete support. If a cast iron baseplate is used, fill it completely. Remove the shuttering after two days, but allow the grout to harden for at least 10 days. Do not allow the grout to dry out too quickly, and protect it from direct sunlight. In hot environments it may be necessary to cover the grout with damp sacks. Also, protect the grout from frost if the temperature is low. Sacks, covered with polythene sheets will be adequate unless the temperature is very low.

Tighten up the foundation bolts and smear with grease. Check the coupling alignment, remember thermal growth corrections and adjust the shimming under the motor feet if necessary. Record all the alignment settings. Spacer couplings are not an excuse for poor alignment and diaphragm coupling life will be short with poor alignment. Pumps with Cardan shafts should have some radial misalignment, and the manufacturer's instructions should be followed.

The suction and discharge process pipework should be finalised after the grout has hardened and final alignment has been completed. The pipework must line up naturally with the pump connections. Do not force pipework into alignment. Remove the piping sections adjacent to the pump and clean out the pipework, as best as possible. Fit a temporary suction strainer when pipework is rebuilt. Recheck coupling alignment with recorded figures and modify the pipework if necessary.

Jog the driver, to check for correct rotation direction, before replacing coupling spacers or drive pins. It is important that coupling bolts are tightened with the correct torque and that the appropriate quality of bolt is used. Information on the subject should be obtained from the manufacturer's instruction book. Fit any loose equipment which was removed for transportation and wire up the unit, "locking-off" all local isolators.

13.1.6 Separate pumps and motors

Pumps driven by electric motors over 5 MW, or steam or gas turbines, will probably be delivered in two sections, pump and driver. Start with the largest section; level and grout in as described above. After the grout has hardened, align, level and grout in the second section. When the second section grout has hardened, then proceed with final alignment and assembly.

13.2 Foundations

With the exception of pumps directly incorporated into the piping, as well as some submersible pumps and site pumps fitted with skids or trailers, a pump requires a solid foundation. This may be of concrete or a steel structure. Standard pump baseplates are designed to be grouted to a continuous concrete foundation block. If the pump is to be installed on a steel structure, the pump manufacturer should be informed of the size and position of the steelwork. Baseplate modifications may be necessary.

It is the user's responsibility to provide adequate support for the pump unit. The size of a foundation block will depend upon the nature of the sub-soil and the magnitude of the vibrations produced by the pump. Increasing the mass of the foundation block will reduce the amplitude of radiated vibration. A block isolated by proprietary elastomer mats may be necessary. Structural steelwork will be much more flexible than a concrete foundation block. Consideration must be given to the natural frequency of the support structure.

Further, the foundation should be sufficiently high to enable the connection of pipes and to ensure adequate space for drain lines, oil drains from reservoirs and for the fitting and maintenance of filters in the pipework. The foundation block should be longer and wider than the baseplate to provide extra physical protection, from such items as forklift trucks and barrows.

Horizontal pumps are usually supplied with pump and motor mounted on a common baseplate of cast iron or fabricated steel, tested and aligned. Large pumps may be delivered with the driver separate from the pump package. It may be necessary to provide a split level foundation.

Vertical pumps with flange-mounted motors are supplied with a base or mounting plate, also tested and aligned. In the sizing of foundations attention must be paid to the relationship between the height of the pump and the surface of the base or mounting plate. For example, if the pump unit is divided, in such a way that the electric motor can be mounted above an overflow level, separate foundations must be designed for both the motor and the pump. Large vertical pumps can suffer from structural vibration problems. Sound foundation designs may help to alleviate such difficulties.

13.3 Tanks and sumps

13.3.1 General

Many pump installations have an open vessel, a tank or pump sump, from which the pump receives the liquid to be transported. Figure 13.1 illustrates different tank arrangements. The pump performance stated by the manufacturer is based on an uninterrupted flow to the pump inlet. Defective design of the pump sump or tank and suction pipe will result in inferior pump performance, possible damage and increased running costs.

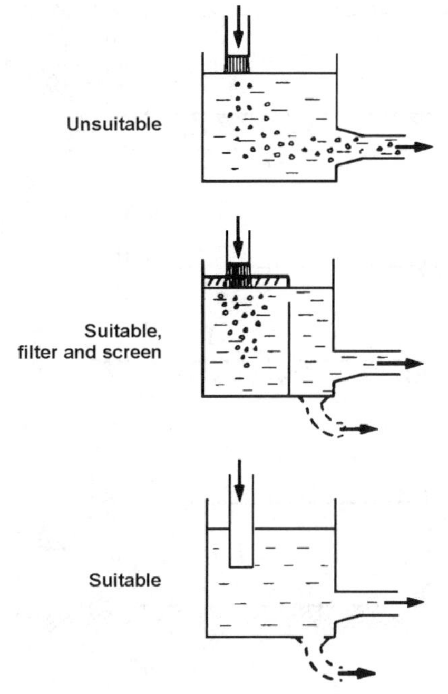

Figure 13.1 Pump suction tanks

The most common faults affecting the function of a pump due to defective suction design are:

- Entrainment of air
- Foam formation
- Uneven velocity distribution in the pump inlet
- Formation of undesirable turbulence
- Suction pressure too low, (insufficient NPSHa/NPIPa)

All these faults are dependent upon the properties of the process liquid and lead to different degrees of reduced pump performance and damage for various types of pumps. Viscous liquids and liquid-solid mixtures demand, as a rule, the utmost care in the design of the suction system of the pump installation.

To prevent the entrainment of air the end of any pipe filling a tank or sump should extend below the surface of the liquid. In

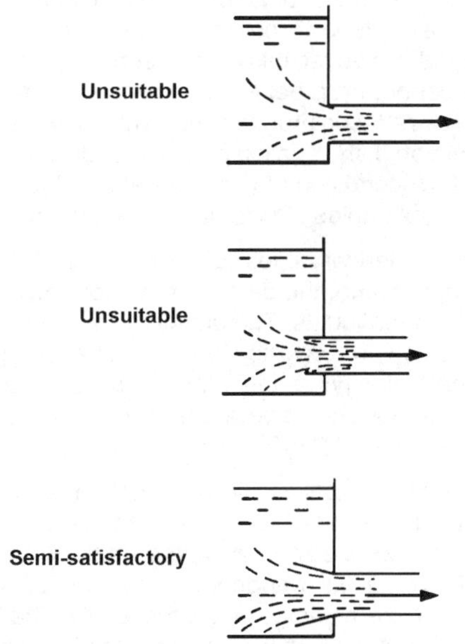

Figure 13.2 Unsuitable connections for suction pipes

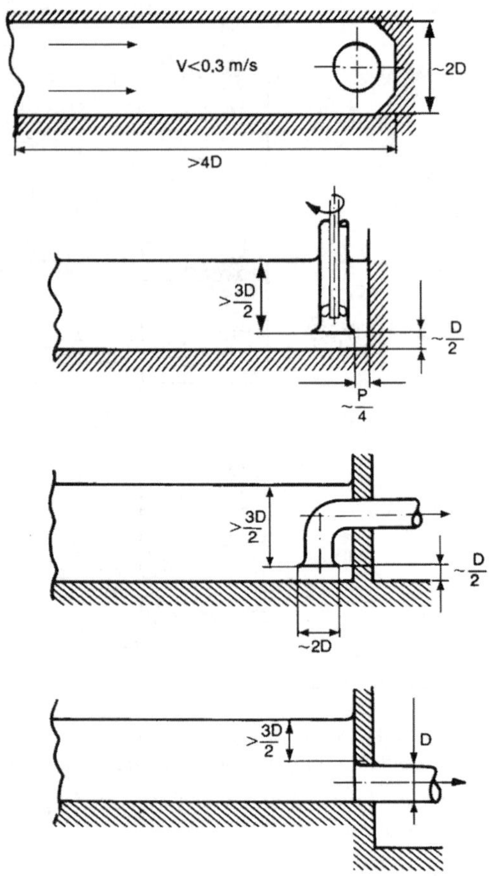

Figure 13.3 Recommended designs for the connection of suction pipes

certain cases, however, the pipe end must be located above the liquid surface.

The volume of the tank or pump sump represents a compromise between, on the one hand, economic factors associated with low initial cost for the plant and, on the other hand, technical criteria such as maximum permissible number of starts per hour in the case of on-off regulation or adequate retention time for degassing. An example of the effect of liquid properties in this respect, is the different rates of ascent of air-bubbles in oil and water.

The design of the pump sump or tank should be such as to ensure an even low flow velocity of the liquid. Local high velocities increase the risk of the formation of turbulence, vortexes and the associated problems of air entrainment. Areas of very low velocity have a tendency to collect particles which may be present in the liquid. However, this is not a serious problem. The local velocity cannot drop below the critical velocity so the flow area cannot be reduced progressively to zero. A sensitive place is the connection of the suction pipe to the tank. The risk of unfavourable flow conditions is considerable. Figures 13.2 and 13.3 illustrate examples of unsuitable and suitable designs.

Although the numerical values stated in Figure 13.3 refer to water and a single pump, the design of the connections are also applicable to other liquids. The unsuitable connections shown in Figure 13.2 can be converted to acceptable arrangements by reducing the velocity at the pipe entrance. Velocities for rotodynamic pumps should be less than 1.5 m/s and reciprocating pumps less than 0.3 m/s.

Low pressure at the pump suction connection, with consequent risk of cavitation damage, can be avoided by means of small losses in the suction pipe and restricted suction lift, see Chapter 4, Section 4.4 on suction performance. Whenever possible, the pump is placed to advantage below the level of the liquid in the suction tank, thereby permitting the liquid to flow down into the pump with positive head.

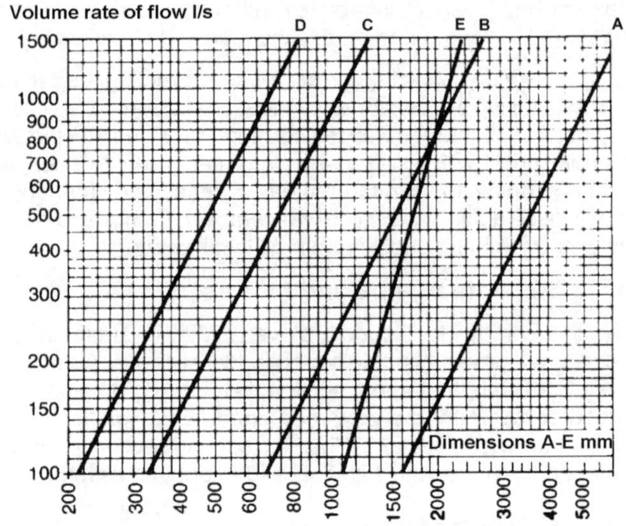

Figure 13.4 Nomogram for determining the dimensions of a pump sump

13.3.2 Submersible pumps

The following guidelines are applicable to pump sumps for sewage pumps with flows exceeding 100 to 200 l/s, i.e. for medium to large pumps. To a considerable extent these also apply to pump sumps for stormwater and rainwater. These pump sumps however, do not have problems with sewage sediment but rather with sand and soil.

A pump sump must have a sufficiently large volume to avoid too frequent starting and stopping of the pumps, otherwise there is a risk of the pump motors overheating. The problem of air being entrained into the water, the "waterfall" effect, arises if the inlet to the sump is located above the surface of the liquid. In such cases it is necessary to ensure that the pump sump is designed so that the water is degassed before it reaches the pumps. Otherwise there is a risk that the pumps will run unevenly, vibrate and even lose prime if the amount of air in the water is too great.

Vortexes are a problem which can also cause pumps to operate unevenly and create vibration. Often these eddies originate at the pump intakes and can be reinforced into a vortex if the pump sump is incorrectly designed. In the case of totally calm liquid surfaces it is necessary to have greater submersion depths to avoid eddies other than those required when the liquid surface is disturbed.

The inlet to the pump sump must be located in such a way that the flow of liquid to the pumps may be evenly distributed. If the flow in to the sump discharges immediately in front of a pump, there is considerable risk of disturbance to the pump intake. This can increase or reduce the power consumption of the motor at the same time as the capacity of the pump is changed. In extreme cases this can cause the motor protection system to be activated.

The formation of sludge and scum must also be borne in mind when designing a pump sump. It must be designed to prevent the formation of dead zones with stagnant water. Similarly it must be possible to pump away scum at regular intervals.

Pump sump volumes for on-off flow regulation are dealt with in detail in Chapter 6.

Four alternative solutions are given for pump sumps with different locations for inlets. The dimensions of the pump sump are calculated according to the flow per pump. The dimensions A to E for the different designs are determined by using the nomogram shown in Figure 13.4.

If the available height is restricted, the volume required is most easily obtained by increasing dimension A.

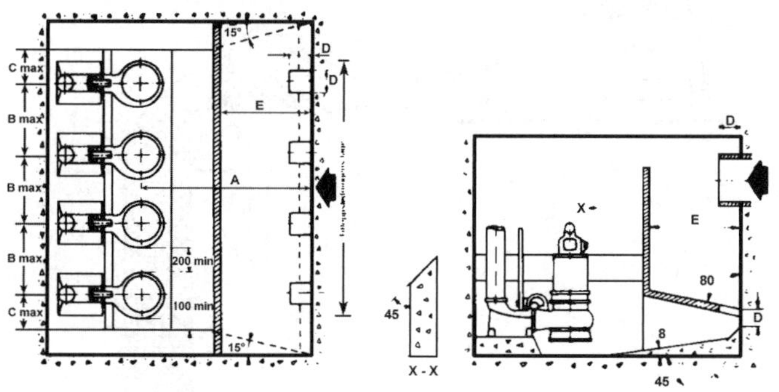

Figure 13.5 Alternative 1 — central, inline inlet at high level

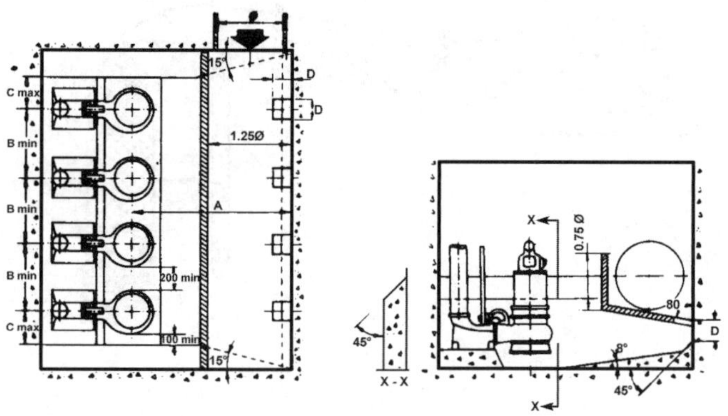

Figure 13.6 Alternative 2 — side inlet with inlet trough

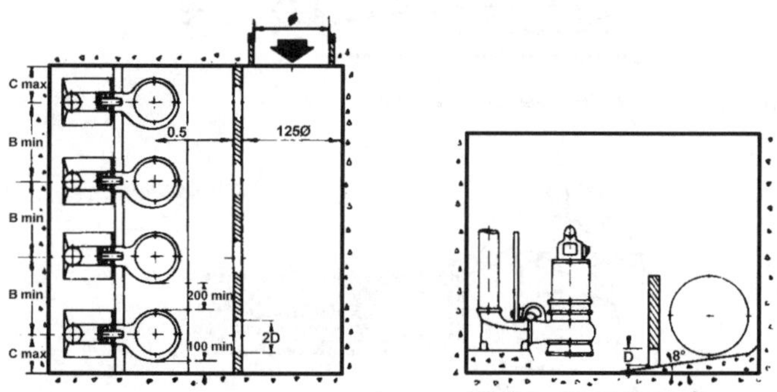

Figure 13.7 Alternative 3 — side bottom, inlet

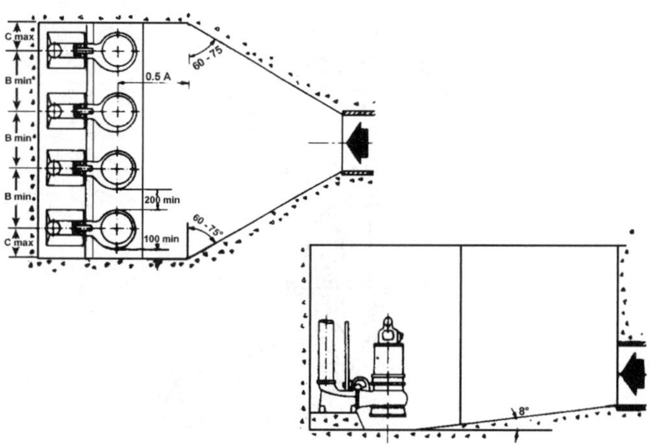

Figure 13.8 Alternative 4 — central inline inlet at low level

Note: Dimension B: The clearance between two pump casings should not be less than 200 mm.

Dimension C: The clearance between a side wall and pump casing must not be less than 150 mm.

Alternative 1

Central, inline inlet at high level, see Figure 13.5.

Due to the high location of the inlet, plenty of air is entrained into the incoming water. The sizing of the inlet trough allows time for the water to degas before it is led into the pump chamber through holes in the bottom of the trough. The side wall, section X-X, which is extended at least half way up the motor housing, prevents the formation of eddies between the outer pump and the side wall. The continual movement of the water means that the risk of sedimentation is minimal, provided that the minimum dimensions are not exceeded too greatly.

The trough wall immediately opposite the inlet must be high enough to catch the inflowing water. A weir can be arranged at the side so that floating waste can be removed by pumping.

Alternative 2

Side inlet with inlet trough, see Figure 13.6.

This is a variant of Alternative 1. The square openings in the inlet trough are necessary to spread the flow evenly through the entire sump and to allow time for degassing any air which may still become whipped into the incoming water. Compared to Alternative 1, the flow conditions in the pump chamber are somewhat more uneven and turbulent. The breadth of the inlet trough and the height of the baffle are calculated with respect to the diameter of the inlet, the former 1.25 x the diameter and the latter 0.75 x the diameter.

Alternative 3

Side bottom, inlet, see Figure 13.7.

A partition wall separates the pump sump from the inlet and forms a type of inlet chamber. From this chamber water flows through square openings with measurements 2D x D located immediately in front of every pump. The openings are dimensioned so that the flow velocity through them does not exceed 1 m/s. In this way the water is led without risk of eddies forming or of sedimentation at the pump intakes.

The partition wall should be at least high enough to reach a point on a level with half way up the pump motor housing. At the same time its height should not exceed a level situated a little below the highest starting level, thereby making possible the removal of floating waste by pumping. The breadth of the intake chamber should be 1.25 x the diameter of the inlet pipe, provided the maximum flow velocity in the inlet pipe does not exceed 3 m/s. The distance between the pump centre lines and the wall can be as little as 0.5 x A.

Alternative 4

Central inline inlet at low level, see Figure 13.8.

This is a simple and functional arrangement which gives a smooth turbulent-free flow to the pumps. The tapered inlet section should have an included angle of 30 to 60°. The distance between the pump centre lines and the end of the inlet section should not exceed 0.5 x A.

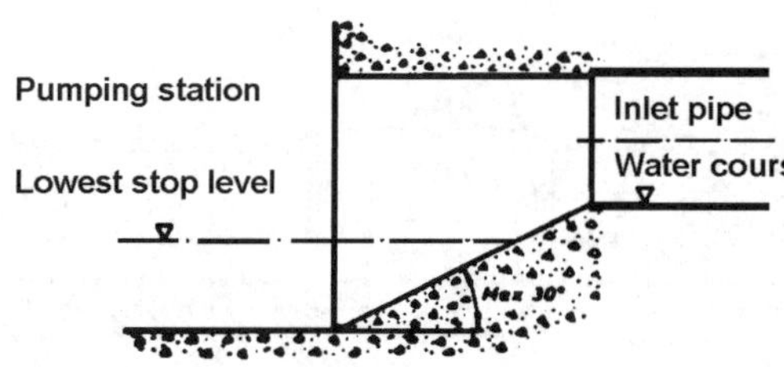

Figure 13.9 Recommended arrangement of low level inlet pipe

Sometimes it is necessary to empty the inlet pipe completely. Alternatives 1 and 2 meet this requirement. In the case of Alternatives 3 and 4, connection may be made according to Figure 13.9. In the case of major flows, if this arrangement is not followed and the inlet pipe is allowed to discharge directly above the lowest stop level in the pump sump, air entrained due to the "waterfall" effect becomes a problem.

13.4 Pipework and pipe systems

Pumps with back-pull-out features, axially-split and radially-split pumps are designed so that rotating elements can be removed without disturbing the pipework. Not all pumps have these facilities. The pipework must be designed to accommodate the style of pump used and enable the pump unit to be dismantled without involving extensive work on the pipes. This may mean that spool pieces have to be incorporated in the pipe adjacent to the pump. It should be noted that it is very difficult to fit in a pipe section with flanges as well as gaskets unless it is possible to displace neighbouring sections. This problem is eliminated if a bend is removed.

The need for flexible and expansion joints to compensate for pipe movement and to accommodate pipe expansion in the event of changes in temperature must also be borne in mind. Necessary supports and mountings for the pipe must also be provided. The pump should not be considered as a pipe support or anchor. Considerable forces are released by flow surges and pressure pulsations in the case of rapid valve operation or during the starting and stopping of pumps, see Chapter 5. Figure 13.10 illustrates an example of normal mounting requirements of pipes adjacent to a pump. Notice that the suction pipe is larger than the pump suction connection. Pipes should be provided with rigid clamps at the supports.

When attaching pipes to the pump, care must be taken to ensure that they align correctly and do not impose stresses in the casing. Stresses of this kind can cause cracks in the pump casing. Also distortion can cause galling between the impeller and wear rings. Checking the coupling alignment before and after the pipes are fitted can quantify distortion produced by the pipes.

The pump suction pipe should be designed to rise or fall evenly with few bends. If a valve must be incorporated, it should be of a suitable type, see Section 13.4.1. If the pump is lifting on the suction then the suction pipe must be completely free of air leaks. Also the suction pipe must be designed without changes of section which could trap air pockets. Examples of correctly and incorrectly aligned suction pipes are shown in Figure 13.11.

Bends should not be fitted directly to the suction connection of the pump. This can lead to uneven loads resulting in both a deterioration in performance as well as abnormal axial loads. Figure 13.12 shows the worst case of this problem —a bend fitted close to the suction of a single stage double suction impeller. The bend disturbs the flow pattern which results in uneven flow distribution to the two sides of the impeller. This type of installation leads to short bearing life and vibration problems. With this type of pump bends close to the pump can only be in the vertical plane.

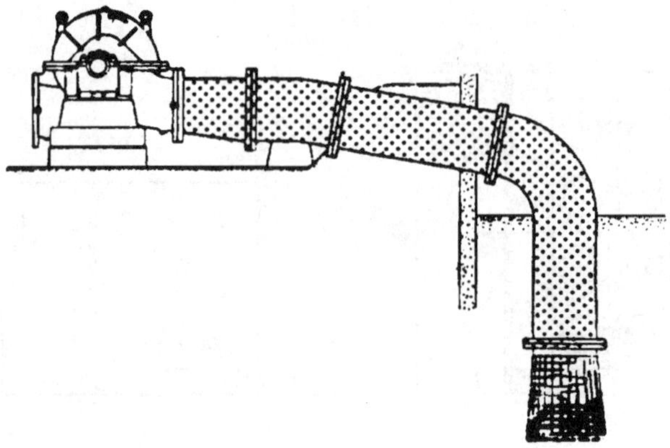

Correct routing of suction pipe

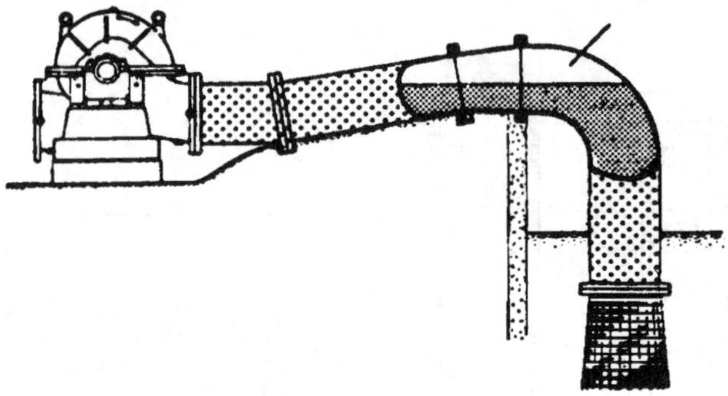

Incorrect routing of suction pipe

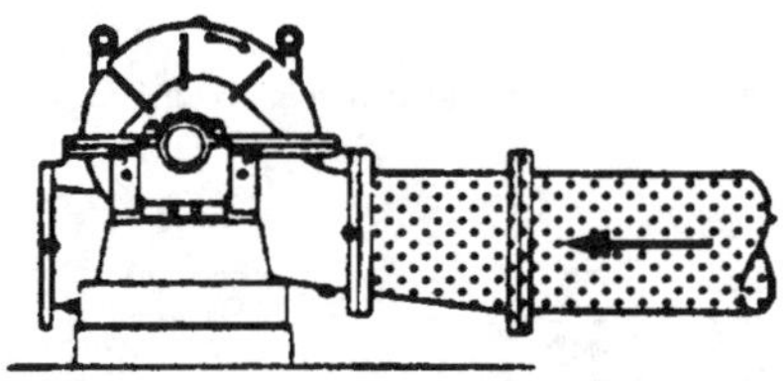

Correct installation with eccentric reducer

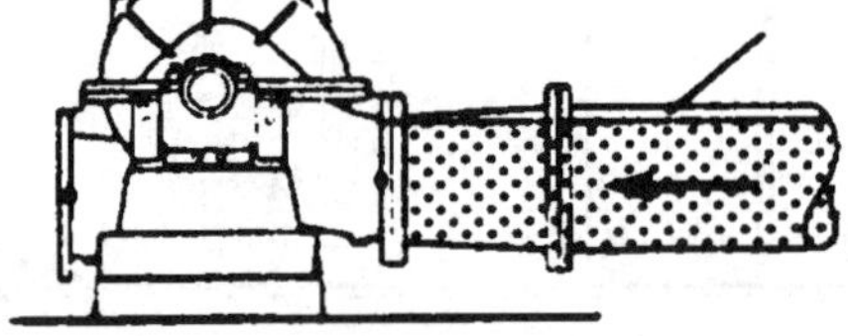

Formation of air pocket due to use of concentric reducer

Figure 13.11 Examples of correctly and incorrectly designed suction pipes

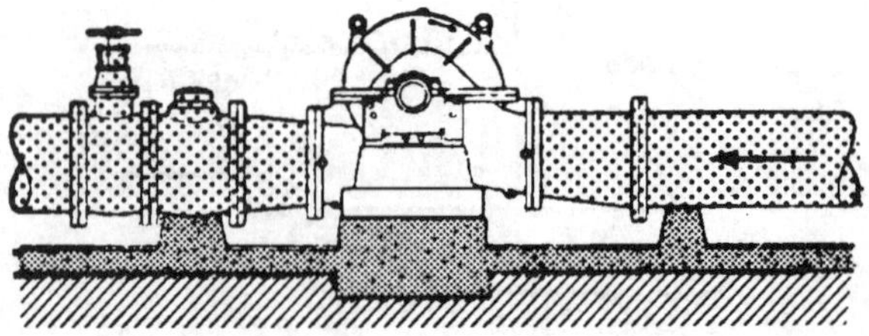

Figure 13.10 Correct mounting of suction and delivery pipes

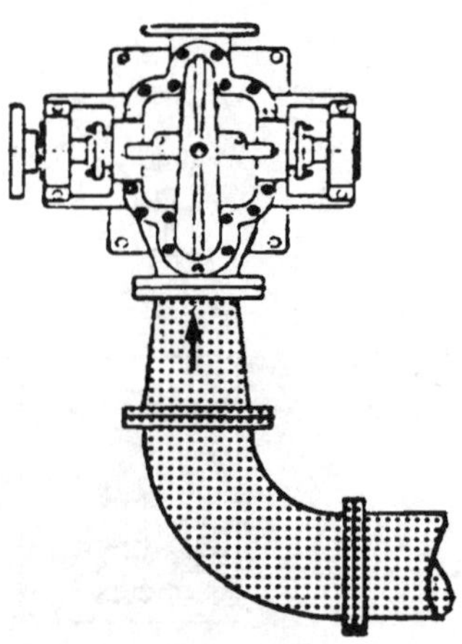

Figure 13.12 Incorrect installation of a bend close to a pump

Chapter 3, Table 3.5 lists typical values for velocities in pipes. In multiple pump installations, pumps for one duty may be supplied by a header. The header should be larger in diameter than the branches supplying each pump and the header diameter maintained constant over the full length. Do not attempt to save costs by reducing the header diameter after pump branches. In theory this may be attractive, in practice it makes future plant modifications and expansion difficult.

Square tees in suction headers are difficult to assess for pressure losses. A full size swept tee, followed by an eccentric reducer, is preferable. In suction systems, "pulled" bends are more suitable than welded pressed bends. For reciprocating pumps, a full size swept 45° branch, followed by 3 x D of straight pipe and then a 45° set followed by the eccentric reducer has provided the best results. For reciprocating pumps, bends should be avoided in suction and discharge; pairs of 45° sets separated by 3 pipe diameters give much less pipework vibration.

Some textbooks suggest the fitting of bends to attenuate noise and vibration transmission by the liquid. This is a fallacy. Consider what the bends in a trumpet or French horn do to the noise! From a vibration point of view the best pipework design is straight; it is the bends which cause most problems.

Methods for calculating flow losses in pipe systems are given in Chapter 3. For the requirements of standards on pipework and installations see Chapter 17, Sections 17.1.3 to 17.1.8.

13.4.1 Valves in pipework

The pipework arrangement is very important for the reliability of continuous processes. Some equipment may be essential for start-up and shutdown. Consider how most pumps will be prepared for start-up. Valves must be strategically placed for venting and priming. Vent valves on the pump casing may be essential. If the liquid has to be lifted into the pump for priming, an ejector on the casing may be necessary rather than a vent valve. After shutdown the pump may have to be drained. Perhaps the process liquid is very corrosive and the pump should be flushed with a suitable liquid to reduce material loss. Perhaps the liquid is hazardous and the pump must be flushed before maintenance can be performed. Extra connections on the pump and/or pipework will be necessary. Plan for all modes of operation; not just the normal process operation.

Valves of various types will generally be installed in the pipe system associated with the pump. When choosing valves, allowance must be made for the pressure drop which always takes place. The difference in this respect between various types of valves intended for the same application can be considerable. Information on this subject should be requested from possible suppliers and evaluated, see also Chapter 5. Manually operated isolating valves with position indicators, e.g. with rising screw, are suitable for use when it is necessary to establish rapidly the position of the valve. In the case of larger sizes, geared valve types are necessary.

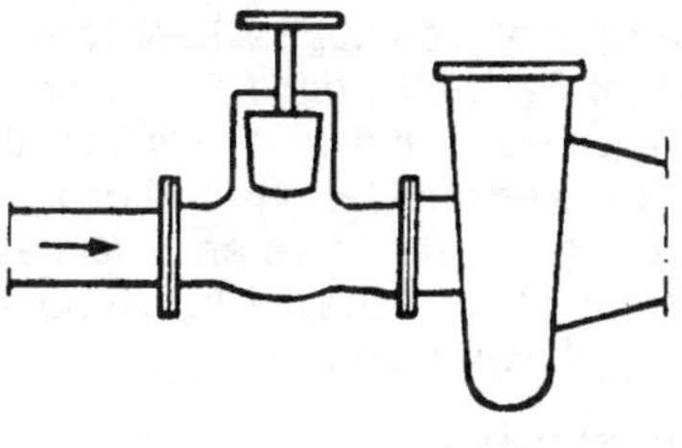

Gate valve, straight unimpeded flow

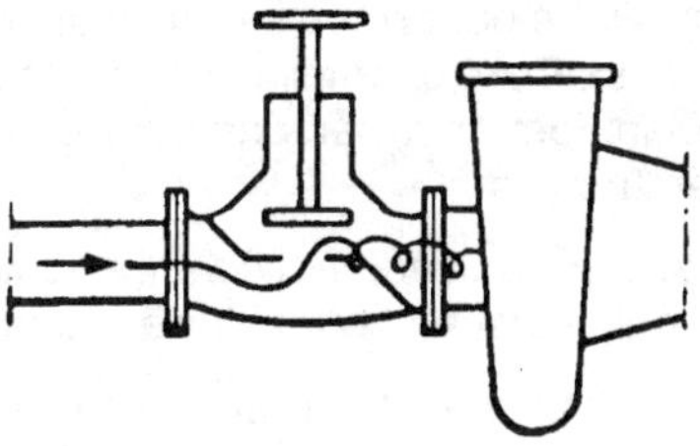

Globe valve, turbulence

Figure 13.13 Valves in suction pipes

Non-return valves of the flap type with external levers are often of practical use. In this way it is possible to achieve functional control as well as enabling the valve to open manually. The disadvantage is difficulty in efficiently sealing the shaft bushing to the external lever. Non-return valves should be provided with inspection covers so that the operation can be tested and flaps and balls replaced without dismantling the valve. If a non-return valve is required for the system, it should be located between the pump discharge connection and an isolating valve, so that it is not necessary to drain the delivery pipe when repairing the valve. When pumps are operating in parallel, the delivery pipe of every pump must have its own non-return valve so that one pump cannot pump backwards through another. In long piping systems it may be necessary to fit damped non-return valves, which cannot slam shut, to prevent or reduce pressure and flow surges.

When pneumatically or hydraulically operated valves are used, a special operating and control system is required. This can be relatively complicated. As a rule small electrically-controlled solenoid valves are used in such systems. It is usually necessary to insist on comparatively clean operating media in such cases to avoid the risk of blockages and subsequent functional defects. If pneumatic or hydraulic valves must close in the event of pump stoppage caused by an electrical power failure, the solenoid valves must fail "closed" and the operating medium must be independent of the electric circuit. This generally means the incorporation of an accumulator. Pneumatically and hydraulically operated valves must include adjustable orifices or needle valves in the fluid power circuits so the speed of valve closure can be adjusted.

Valves operated by electric motors are quite expensive, but allow very good regulation. Naturally they do not function in the event of a power failure.

If throttle regulation is applied by means of a valve, the throttle valve must always be located in the delivery pipe. Valves in the suction pipe must only be used for isolation. They should be of a type permitting unimpeded flow to ensure the minimum losses and undisturbed flow into the pump. Globe valves must not be installed in a suction pipe immediately before a pump. See Figure 13.13.

Best results for low pressure drop and minimum flow disturbance are achieved by full bore ball valves. Relief valves are fitted in positive displacement pump systems to prevent harmful increases in pressure, see Chapter 11, Section 11.5.

In pump suction systems, strainers or filters of various kinds may be necessary to provide protection against objects which can cause damage within the pump. These must be accessible for cleaning or replacement. The pressure drop across strainers or filters must be measured as this increases with the degree of blockage. It will be necessary to install a differential pressure indicator to check conditions. Standard screens are designated by width of mesh or diameter of holes as well as number of meshes or holes per unit of surface. For this purpose there are national Standards, see Table 13.1.

DIN 4188	Former Designation Meshes		Width of Mesh		USA Standard	Tyler	British Standard	Afnor
Meshes mm	Every cm	Every cm^3	Micr.	Inch	Meshes per inch			Module
			45	.0018	325	(325)	(350)	
0.063	100	10000	63	.0025	230	(250)	240	19

DIN 4188	Former Designation Meshes		Width of Mesh		USA Standard	Tyler	British Standard	Afnor
Meshes mm	Every cm	Every cm^3	Micr.	Inch	Meshes per inch			Module
0.09	70	4900	90	.0036	170	170	170	
0.125	50	2500	125	.0049	120	115	120	22
			180	.0071	80	80	85	
0.25	34	576	250	.0098	60	60	60	25
			355	.0139	45	42	44	
0.5	12	144	500	.0197	35	32	30	28
			710	.0280	25	24	22	
1	6	36	1000	.0394	18	16	16	31
			1400	.0552	14	12	12	
2	3	9	2000	.0787	10	9	8	34
			2800	.1102	7	7	6	
4	(1.5)	(2.25)	4000	.1570	5	5	34	37
			5600	.2200	3.5	3.5		

Table 13.1 Comparison of national Standards for screen mesh

Pump sumps for water containing sand should be equipped with a sand and gravel trap, in which the rate of flow is reduced to about 0.3 m/s so that sedimentation may take place. Sand traps of this type must be emptied at suitable intervals if they are to function efficiently.

Pumping stations for sewage are sometimes provided with grills to trap rags, pieces of wood, etc. and thereby prevent blockage of the pumps. The cleaning of these grills is a complicated process and is usually carried out by mechanical means in large-scale systems. Reverse flushing through the pump during every shut-down period can reduce the risk of blockage considerably. If the delivery pipe is very short, its entire volume can be used for reverse flushing; in other cases an arrangement of valves according to Figure 13.14 is also effective.

In Type A the ball in the air valve should be of a suitable weight to achieve good functional effect. When the pump is started a small amount of liquid passes through the air intake, which should therefore be located in the pump sump or other drained well.

In Type B the non-return valve is a pneumatically or hydraulically operated shut-off valve.

The volume of liquid required for reverse flushing is 50 to 150 litres, depending on the size of the pump. A check should be made that the pump is able to tolerate a certain degree of reverse rotation. This does not, however, need to involve any high speed operation. Care must be taken to ensure that reverse flushing of this kind does not cause damage to the pump. It may be necessary to install devices for filling and tapping, air separation and sampling, etc. in the pump piping system.

Accurate flow measurement is generally required for metering pumps to check the amount of dosage. Such pumps are often supplied with graduated vessels which can be sealed off with valves as shown in Figure 13.15. In dosage systems it is always necessary to take into account the risk of over dosage due to defects in the control system, ejector effects, etc. It may therefore be necessary to incorporate special devices to counteract such risks.

13.5 Care and maintenance

Modern pump units are often supplied with large quantities of paperwork and manuals. It is not unusual for 10 sets to be provided; 20 on occasions. However, when a service engineer from the manufacturer visits the site he often cannot find a manual for reference. Paperwork for pump units is costly to produce but it is full essential information. Process operators and site maintenance staff are the people who need the information on a day-to-day basis.

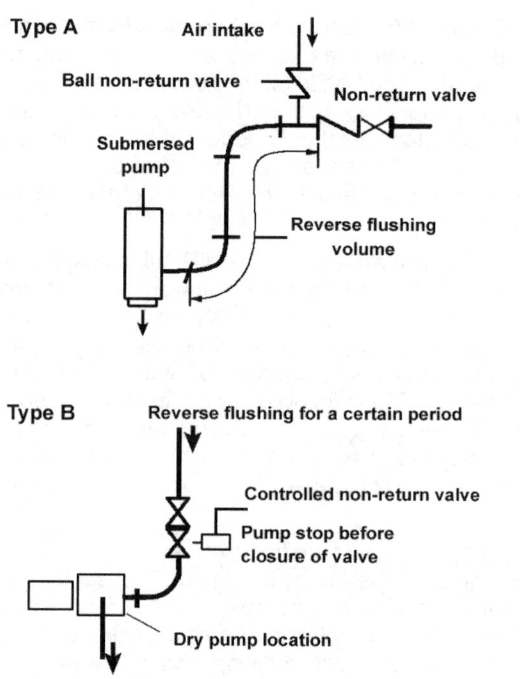

Figure 13.14 Reverse flushing systems for sewage pumps

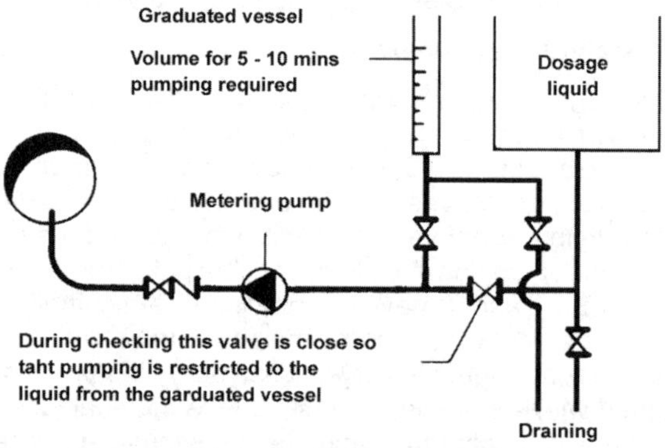

Figure 13.15 Graduated vessel for flow measurement of metering pump

It is essential that site staff are given original data produced by the pump manufacturer. Some contractors extract what they think is "essential" data from the pump manual and produce process manuals. Costly errors have been made which were avoidable.

13.5.1 General considerations

The first prerequisite for long service life and reliable operation of a pump unit is naturally, as applies to all types of machinery, correct installation. Secondly, once the pump is operating it should not be subjected to working conditions other than those for which it was specified and designed. The third is that the unit should receive the necessary care and maintenance. This is not generally carried out to preserve the economic value but to maintain reliability of operation.

13.5.2 Commissioning

The transition from installation to operation is "commissioning". Commissioning can be performed by site staff or the manufacturer's staff. If site staff are unfamiliar with some aspects of the pump unit, the manufacturer's personnel can combine commissioning with staff training.

The following list can be used as a commissioning guide. The manufacturer's help must be sought if clarification is required.

1 Read carefully the manufacturer's instructions regarding preparation for starting.

Then read all of this list.

The manufacturer's instructions take priority.

If in doubt, consult the manufacturer.

2 Fit temporary suction and discharge pressure gauges if necessary. Fit extra gauges to measure pressure drop across temporary suction strainer.

3 Check minimum flow by-pass is fitted correctly on rotodynamic pumps.

Check by-pass is fitted to positive displacement pumps.

4 Check if a mechanical seal or packing is fitted. If soft packing is fitted ensure packing is not too tight and has plenty of lubrication for bedding-in.

5 Check all external nuts and bolts are tightened correctly.

6 Check relief valve, if fitted, for correct setting. The relief valve manufacturer should have sealed and tagged the valve with contract settings.

7 Check gas precharge on pulsation dampers and accumulators for run-down systems. If the pump is to be commissioned at lower pressures than normal adjust damper precharge pressures. Precharge to 70% of nominal liquid pressure unless the manufacturer quotes alternative values.

8 Check external cooling, heating and quench systems are piped up correctly and supplies are available. Heating and cooling may be connected to the pump casing as well as the stuffing box(es). Check thoroughly.

9 Flush out all equipment which has been treated with rust preventatives.

10 Check specifications for lubricants and other liquids which must be used in reservoirs. Fill reservoirs to correct levels.

Do not use close approximations of liquids as a temporary measure.

11 Turn all rotating equipment by hand. If not free, investigate before proceeding further.

12 Check all power supplies are available to control system. Test all indicator lamps.

Ensure all systems are set to "OFF" except the system to be started which should be set to "HAND".

Check the resistance of all motor winding insulation with a Megger. Log results, switch on space heaters if condensation appears to be a problem.

Note: See IEEE 43-2000 IEEE Recommended Practice for Testing Insulation Resistance of Rotating Machinery, Institute of Electrical and Electronics Engineers, 2000, ISBN: 0738119245.

13 Spin motors uncoupled very briefly to check direction of rotation, adjust if necessary.

14 Start the smallest auxiliary system first. Check if any pumps on the system require venting or priming; do this before switching on. Some systems are equipped with hand pumps for priming.

Check and record operating conditions, make allowances for system being cold, check with manufacturer's figures.

Inspect all pipe runs and connections for leaks.

If the pump does not develop correct discharge pressure or flow seems low, stop immediately and investigate. See troubleshooting chart in Section 13.5.5.

NOTE: Venting is removing all the air from the pump casing. Some pumps are self-venting so that when the suction pipework is vented, and the discharge pipe up to the isolating valve, the pump will vent. Priming is filling the pump casing with liquid. If the pump is not self-venting there will be one or more vent valves on top of the casing.

NOTE: Do not start motors too frequently if problems occur. Do not restart motors while the shaft is rotating.

15 Simulate fault conditions and check instrument set points.

Check operation of annunciator and control logic for starting stand-by equipment.

16 Allow system to run for at least one hour, then stop and inspect filters.

If more than one pump is fitted in a system, start the other pump(s) and run each for one hour.

17 If pumps have stand-bys, set controls to "AUTO" and check correct operation.

18 Start all the auxiliary systems in turn by repeating steps 14 to 17.

19 Complicated drivers, such as engines and gas turbines, will have their own auxiliary systems, so start these first. Remove coupling spacers or driving pins and start the engine or turbine, according to the manufacturer's instructions, and run without the pump. When the driver is running properly, commission the pump.

20 Vent and prime the pump if necessary. If the process liquid is much hotter or colder than ambient conditions, the pump temperature must be adjusted slowly without thermal shock. The pump casing temperature should not change faster than 4°C per minute. Large high pressure pumps may need much slower changes. If the casing is heated or cooled by auxiliary systems, ensure even temperature distribution through the whole casing and shaft.

21 Check the manual for the pump starting instructions. Small rotodynamic pumps should be started with the discharge valve closed. As pump size increases, the pump may be started with the discharge valve partially open. Pumps fitted with minimum flow by-passes will probably start with the discharge valve closed.

Positive displacement pumps must be started with the discharge valve open unless a starting by-pass is fitted. If one is fitted, close the discharge valve and open the by-pass valve.

22 Turn on external quench and external heating or cooling circuits.

23 Start the pump. Rotodynamic pumps should have the discharge valve opened slowly as soon as the pump has reached rated speed. The valve does not have to be opened fully but sufficiently to allow a reasonable flow through the pump.

Positive displacement pumps can be run at low discharge pressure until the pump unit has been checked for correct operation. After successful operation at low pressure the discharge pressure can be increased in stages by throttling the by-pass valve. Over a period of about one hour the pressure should be increased to rated discharge pressure. If all systems operate correctly, the discharge valve should be opened and the by-pass valve closed completely.

24 Pumps fitted with torque convertors may experience difficulties running at very low loads. The discharge pressure or flow may have to be increased until the convertor can stabilise.

25 While the pump is running check the stuffing box(es) and bearings for overheating. Check the leakage at the stuffing box(es), adjust soft packing carefully, see Chapter 8.

Record operational values from all gauges fitted. Monitor progress of cooling water and heating medium temperatures. Monitor the condition of all filters, including the temporary suction strainer.

Stop the pump immediately if the discharge pressure falls unexpectedly.

26 When the pump is seen to be operating correctly, simulate auxiliary pump failures and monitor the automatic change-over to stand-by pumps.

27 After 12 hours running remove and clean any gearbox filters.

If pump operation is satisfactory, take readings and calculate overall pump unit efficiency. Take vibration readings, at marked locations on all bearing housings and suction/discharge pipework, and analyse harmonics. For reciprocating pumps, take pressure pulsation readings and analyse harmonics.

After 750 running hours, remove all oil filters and wash out oil filter housings. Drain all lubricating oil and replace. Fit new oil filters.

Remove coupling spacers or drive pins, check coupling alignment.

28 After an appropriate period of successful running remove all the temporary equipment, including the suction strainer. The system may then be handed over to the production department.

Give production operators log sheets to record operating conditions.

NOTE: If the equipment is not put into service immediately, then run at least once a month until warmed up.

NOTE: In the event of operational problems, the log sheets are a good guide to the changing condition of the pump. The pump manufacturer's service engineers will ask to see the log sheets if they are ever called to site.

13.5.3 Care of equipment

In the selection of pumps and the planning of installations, it is essential to take account of all the maintenance aspects. The assessment of a particular pump should be based on technical grounds connected with maintenance; ease of dismantling, availability of spare parts, trade skills required. Further, steps must be taken to ensure that adjacent pipe spools can be loosened easily, that working space, lifting facilities and adequate access routes are available. See Section 13.1.

The distinction between servicing and care, on the one hand, and preventive maintenance, on the other, is difficult to draw. The aim of care and maintenance is, of course, to prevent breakdowns as far as possible, with costly loss of production and subsequent process plant operational problems.

The necessary amount of care and maintenance is difficult to estimate. Factors such as the need for reliability of operation, the properties of the liquid pumped, the working environment of the pump, etc., exercise a marked influence. In general it can be recommended that care, as well as maintenance, should be incorporated into regular routines and that some form of database of faults noted and measures taken should be initiated and up-dated. Modern computers allow the storage and retrieval of vast quantities of information. Software is available which permits the storage of machinery data and each machine's service history.

With only a cursory glance, computer records do not appear to offer any advantage over card files or machine wallets. Consider some typical questions which are easily answered by a computerised system but virtually impossible for a manual system, unless you examine every machine wallet:

- How many machines use bearing 32010X?
- How many machines use ISO 320 oil?
- How many times has "Joe" checked pump alignments?
- How many machines went off-line when No.2 boiler blew a tube?

A properly organised computerised system can offer information unimaginable when only using cards and wallets. It also provides the basis for good cost control. It is estimated that the annual maintenance costs of a pump unit including spare parts amount to between 10% and 20% of the purchase cost of the unit.

Before pump failures are discussed, it is useful to consider how often pumps cause major problems as opposed to other equipment. Data published in 1988 in *Hydrocarbon Processing* revealed the following facts regarding the 100 worst incidents world-wide over a 30 year period. Table 13.2 lists the type of plant involved and Table 13.3 lists the type of equipment involved. The worst incident in the 30 year period was caused by a piping problem and cost $114,000,000 in property damage.

The worst incident in Europe, since 1945, took place in the UK in 2005, at a fuel storage depot. There was a large explosion followed by a very large fire. This incident has been traced to instrument failure; two level instruments on a petrol storage tank failed, resulting in, approximately, 300 tonnes of spilt petrol. One has to wonder what the operators were doing while the tank was overfilled. There was so much vapour in the air that even vehicle engines off site in the local area, failed to respond to normal throttle control! There must have been an 'operator' failure as well. At the very least, the operators had no 'experience' or 'interest' in what they were doing.

Type of plant	Number of incidents
Refineries	42
Petrochemical plants	16
Plastic-rubber plants	11
Chemical plants	10
Gas processing	7
Bulk terminals	7
Pipelines	2
Underground storage	2
Special	2
Gas oil production	1
Drilling rigs	0

Table 13.2 Types of plant involved in major incidents

Type of equipment	Number of incidents
Piping systems	29
Miscellaneous or unknown	22
Storage tanks	17
Reactors	10
Process holding tanks	5
Heat exchangers	4
Valves	4
Process towers	3
Compressors	2
Pumps	2
Gauges	2

Table 13.3 Types of equipment involved in major incidents

From Table 13.3 it can be seen that pumps created 2% of the problems overall and only 7% of the problems caused by pipework. No one would deny that pumps have problems, but those problems must be placed in context.

Tables 13.4, 13.5 and 13.6 review some practical cases of pump faults and their frequency of occurrence. These give some idea as to the types of functional troubles to be anticipated.

Component	Relative frequency of breakdown %
Centrifugal pump	88
Coupling	8
Electric Motor	4

Table 13.4 Frequency of breakdown of a horizontal centrifugal pump

Sub-component	Number of items	Relative frequency of breakdown %
Sundry small items	96	41
Soft packings	-	16
Mechanical seals	3	10
Control valves for sealing water	2	7
Roller bearings	2	8
Gland halves	4	3

Sub-component	Number of items	Relative frequency of breakdown %
Shaft	1	10
Impeller	1	4
Pump casing	1	1

Table 13.5 Frequency of breakdown of an axially-split pump with a capacity of 5,200 l/min @ 17.8 m

Type of fault	Frequency %
Defective bearings	24
Blockage by foreign matter and formation of plugs	14
Gland packings, shaft sleeves	12
Packing	11
Worn out and damaged impellers	11
Broken shafts	7
Corrosion of components	4
Motor defects	4
Erosion and wear	4

Table 13.6 List of faults experienced by a large number of pumps of different types used in the cellulose industry

A general conclusion to be drawn from Tables 13.4 to 13.6 may be that faults in the pump are much more common than in couplings or motors. This should not be surprising. Faults affecting bearings and packings are of relatively high frequency in the pump, as are also blockages and damage to impellers when troublesome liquids are pumped. A control study of a sewage disposal plant with about 250 pumps in operation in pumping stations and purification plants yielded much the same results.

The second most common fault, "Blockage by foreign matter and formation of plug" is not really a pump problem. It is, more correctly, a pump application problem or a system failure. The term "foreign matter" suggests that material which should not have been there is blocking the pump; this is not the fault of the pump. Formation of plugs suggests the pump was running too fast, an application problem, or the process liquid was "off-spec", a system problem.

There are two causes of unreliability and failure which do not appear in popular surveys; inappropriate piping design and inexperienced operators. Pump manufacturers expect a pump to be installed with 'sensible' piping. The suction piping is the most important. If a pump has a poor liquid supply then a wide variety of problems will ensue.

REMEMBER! a pump does not **suck**. It is a waste of time trying to pull liquids; they have incredibly low tensile strengths; the liquid column will break. Liquids can only be **pushed**. The suction piping must provide an efficient route from source to pump. Liquid falling into the pump is by far the most reliable arrangement. Suction piping which is self-venting and self-priming is much better than all other styles; it can save a lot of man-hours during start-up.

Very few engineers seem capable of distinguishing between 'good' and 'bad' pump piping. Operators must understand the equipment installed in the plant. Many simple failures, during commissioning, show that operators do not understand the basic fundamentals of pump use. A much higher level of 'useful' education is required; not degrees in pump design; practical knowledge about what pumps 'can' and 'can't' do!

13.5.4 Preventative maintenance

The care of a pump unit should be carried out by routine inspection performed according to a definitive schedule. The timing intervals required are dependent upon the working conditions and environment of the pump and the demands of operational reliability. Operational reliability should have been a major part of the pump specification and evaluation criteria.

Good maintenance practices cannot reverse deficiencies of pump selection. Some pumps may require attention every day; topping up oil reservoirs for example. Periods exceeding one week are inadvisable, especially in the case of pumps in distant pump-houses and lacking alarm systems. Observations such as listening and feeling for vibrations, and checks on pressure, flow and power consumption should be performed with every inspection in addition to the checking of packings and mechanical seals. More detailed investigation should, of course, be undertaken if deviations from normal operation are noted. Section 13.5.5 includes troubleshooting guides which may be of assistance.

If several pumps are included in the system, the starting sequence should be adjusted with the "lag-lead" switch so that the running time is divided between all the units. Automatic and alarm devices should also be regularly tested.

Dosing pumps often require special checks for flow measurement, necessitating the use of graduated vessels or other measuring equipment, and also the testing of control equipment.

The concept of care must include the external cleaning of pump and motor in addition to the cleaning of screens and filters on the suction side and in pump sumps. The necessary cleaning of pump casings and impellers may also be classed under the same heading. This work may be extensive, depending upon the type of pump and the nature of the product. The possibility of regular reverse flushing can considerably reduce trouble caused by blockage, see Section 13.4.

Cleaning of the casing and impeller means that the unit must be able to tolerate a certain amount of low speed reverse rotation, which should be checked with the supplier, and that due care should be exercised.

Reciprocating positive displacement pumps used in sewage and solid handling applications frequently have easy access to valves for inspection and cleaning. These should be inspected a regular intervals.

Pumps used in hygienic applications such as food processing, dairy products, etc., must be cleaned and sterilised in accordance with the prevailing regulations.

Maintenance in this context refers to preventive maintenance intended to reduce the number of breakdowns and the resulting unscheduled shut-downs. Attempts have been made to calculate optimum maintenance statistics for pumps, but the results are uncertain and the systems are difficult to handle.

Generalisation of pump maintenance is a waste of time. Pumps operate in widely differing circumstances with a vast array of materials. The best policy is to initiate a strict, very regular, routine inspection of the equipment and continue this until a pattern of equipment behaviour is apparent. At this stage, it may be possible to relax the inspection routine in some areas.

It is important to remember that the pump cannot be held responsible for the effects of process upsets. If the pump requires an overhaul after a serious process upset this should not affect the pump's normal inspection regime. The cause of the upset must be traced and rectified. This is the area where the normal inspection routine should be modified.

Timing	Activity
Daily	
	Check and record process operating conditions
	Check and record values of all indicators on the unit
	Check oil levels; add oil as required (1)
	Check levels in any external reservoirs; add liquid as required
	Check liquid level in gas charged suction stabilisers; add gas as required
	Check water in radiators of engines
	Check bearings and stuffing boxes, by hand, for temperature
	Check for leakage from mechanical seals
	Assess leakage from soft packed stuffing boxes
	Check vibration levels by hand
	Test annunciator lamps and audible alarm
	Listen for unusual noises or change in noise

Timing	Activity
	Weekly
	Pumps with valve covers, inspect valves
	Pumps with soft packing, back-off gland slightly, oil and reset
	Switch over gearbox filters and clean used filter
	Switch over fuel filters and clean used filter
	Vent water side of coolers
	Vent filters
	Check for leaks on pipework
	Check for leaks at shaft seals
	Weekly
	Check glands on control valves, back-off and reset
	Check air filters on engines and gas turbines
	Check water in batteries
	Assess pressure pulsations from pressure indicators
	Check gas pre-charge pressure in dampeners on stand-by units
	Adjust "lag-lead" switches so that stand-by equipment runs 1 week in 4
	Monthly
	Vent all gauge connections
	Drain off water from all oil reservoirs
	Smear grease on important external nuts and exposed stud threads
	Take temperature readings on bearings and stuffing boxes
	Take vibration readings on pump and driver and at marked locations on the process pipework
	Quarterly
	Clean all air breathers
	Clean all cooling fins on housings
	Clean motor fan and any air filters
	Check vee-belt tension
	Check coupling bolts
	Check battery connections, smear with grease
	Take oil samples and analyse
	Take vibration readings and analyse harmonics
	Biannually
	Check torque settings of important fasteners + holding down bolts
	Check clearances of sliding bearings, wear rings, balance disc
	Check stuffing box clearances
	Check for general wear and corrosion
	Check seals on all drive shafts
	Check battery cell voltages individually
	Take readings and calculate overall pump unit efficiency
	Annually
	Drain all oil reservoirs
	Clean reservoirs internally
	Clean any internal oil passages
	Check and clean any orifices
	Check and clean any strainers
	Check oil pumps internally for wear
	Check condition of any rubber hoses
	Check seals on auxiliary pumps
	Check reciprocating pump valves, change springs
	Clean water side of coolers
	Refill with new oil
	Grease motor bearings (2)
	Check motor insulation resistance
	Inspect relief valve seats, poppets and springs
	Check calibration and set points of instruments
	2 years
	Thoroughly inspect all pressure vessels
	Re-hydrotest all pressure vessels (3)

(1) Do not mix different grades of oil. Do not mix different brands of oil.

(2) Check motor manual, regreasing period is based on motor size.

(3) Local regulations may be more stringent

Table 13.7 Typical routine maintenance schedule

Table 13.7 lists the type of typical routine inspections and maintenance functions which should be considered. Not all pumps will have all the systems mentioned. The timings of these activities may be relaxed if the pump unit performance is stable and acceptable. Oil changes may need to be more frequent if operating conditions are severe. Oil must be changed at least every 18 months irrespective of operational time or apparent condition. Checking the gas pre-charge pressure is the best indicator of a healthy bladder in a damper or an accumulator. If the bladder is chemically attacked it will lose gas faster.

Small handheld instruments are available which are capable of measuring; temperature, speed, vibration, bearing condition and cavitation assessment. Devices such as these greatly improve the necessary data collection to improve equipment reliability.

Overhaul implies taking the pump to pieces and examining all parts. In the case of pumps working with abrasive liquid-solid mixtures or other difficult liquids, overhauls must be carried out relatively frequently, possibly every year. Routine inspection will indicate the need for a complete or partial overhaul. Standard process pumps and sewage pumps usually require overhauls at intervals of three to five years. For pumps handling clean water the intervals can be still further extended. Boiler feed pumps, operated at steady-state stable conditions, have been known to last 10 years between shutdowns. Stable operating conditions promote a low-wear running condition. During overhauls the pump is cleaned internally and externally.

Bearings and other items subject to wear are checked and where necessary replaced. Wear ring clearance should be checked. This should not exceed the manufacturer's dimension, as the mechanical efficiency of the pump is adversely affected if the clearance is large. The impeller and the pump casing are checked for deposits, cavitation and corrosion damage. If the impeller is damaged it must be repaired or replaced. An impeller in poor condition means a drastic reduction of pump performance.

The clearance of rotary positive displacement pumps should be checked; enlarged clearances produce higher slip losses, lower flow and lower efficiency. The valves of reciprocating pumps should be inspected. Worn seats form poor seals and reduce pump flow. Many valve types can be ground-in to improve the seal. Valve springs should be replaced regularly; usually every year.

Pump overhauls are expensive and it is important to consider whether they should be carried out more than a few times. In the case of small and simple pumps it is usually more profitable to replace the pump than to perform a complete overhaul.

13.5.5 Trouble-shooting guides

Trouble-shooting pumps is somewhat taken for granted; it may not be what you think. Trouble-shooting may be defined as: making the pump do what it was specified to do.

The first step is to check the actual operating conditions against the datasheet conditions. If the conditions are different then it is perhaps not surprising that the pump does not perform correctly. If the plant operating conditions have changed since the pump was specified and purchased then the pump may require "up-grading".

A review of Tables 13.8, 13.9 and 13.10 maybe helpful.

Symptom	Possible cause	Remedial action
Pump won't start		
	Power failure	Check all power supplies
	Blown fuses	Check all fuses
	Motor single phasing	Check motor wiring
	Control switch contacts open	Check all instrument switches in motor contactor circuit
No flow		
	Suction valve closed	Check all valves in pump pipework for correct position
	Discharge valve closed	Vent and prime pump
	Pump not primed	Check suction pipework, replace gaskets
	Air leaks in suction pipes	Check stuffing box
	Excessive suction lift	Check calculations, check liquid temperature

Symptom	Possible cause	Remedial action
	Suction pipe blocked or closed valve	Check suction valve, check suction pipework
	Impeller blocked	Pump vibrates badly; strip pump
	Differential head too high	Check pressure gauges, check calculations
	Pump speed low	Check pump speed, check motor frequency, check motor voltage
	Wrong direction of rotation	Check rotation
Low flow		
	Differential head too high	Check pressure gauges, check calculations
	Pump speed low	Check pump speed, check motor frequency, check motor voltage
	Impeller partially blocked	Pump vibrates badly, strip pump
	Insufficient NPSHa	Noisy pump, check calculations Small air leak through stuffing box Gas evolving from liquid
	Pump wear or damage	Excessive wear ring clearance Excessive balance drum/disc clearance Damaged impeller
Low head		
	Pump running out on curve	Check flow, throttle discharge
	Pump speed low	Check pump speed, check motor frequency, check motor voltage
	Impeller too small	Check flow, check calculations, check pump curve
	Pump wear	Excessive wear ring clearances Excessive balance drum/disc clearance
High flow		
	Differential head too low	Check pressure gauges, check calculations; throttle discharge valve
	Liquid viscosity too low	Check data sheet conditions, check calculations; throttle discharge valve
	Impeller diameter too big	Check data sheet conditions, measure impeller; throttle discharge valve
	Pump speed high	Check pump speed, check motor frequency; correct pump drive
High power consumption		
	Pump power high	Liquid SG high, check liquid Liquid viscosity high, check liquid Differential head low, flow high, check operating conditions Pump bearings/stuffing boxes binding Pump speed high, check speed
	Electrical faults	Voltage and frequency low, check supply Motor problem, check motor
Pump vibration		
	Poor alignment	Check alignment Check coupling assembly Check baseplate grouting Check process pipework
	Insufficient NPSHa	Check operating conditions, check gauges
	Damaged impeller	Check vibration harmonics, strip pump
	Bearing problem	Check bearing clearance Check vibration harmonics
Motor vibration		
	Poor alignment	Check alignment Check coupling assembly Check baseplate grouting
	Bearing problem	Check bearing clearance Check vibration harmonics
	Motor rotor fault	Loose rotor bars

Table 13.8 Trouble-shooting guide for rotodynamic pumps

Symptom	Possible cause	Remedial action
Pump won't start		
	Power failure	Check all power supplies
	Blown fuses	Check all fuses
	Motor single phasing	Check motor wiring
	Control switch contacts open	Check all instruments switches in motor contactor circuit
No flow		
	Suction valve closed	Check suction pipework valves
	Discharge valve closed	Check for flow through relief valve
	Pump not primed	Vent and prime pump
	Air leaks in suction pipes	Check suction pipework, replace gaskets, check stuffing box
	Excessive suction lift	Check calculations, check liquid temperature
	Wrong direction of rotation	Check rotation
Low flow		
	Differential pressure too great	Check:- pressure gauges, calculations, relief valve flow
	Bypass valve partially open	Check bypass valve
	Pump speed low	Check:- pump speed, motor frequency, motor voltage
	Excessive wear	Check:- pump clearances, relief valve seat; replace worn parts
	Low viscosity	Check:- liquid temperature, liquid properties
	Insufficient NPIPa	Check:- pressure gauges, liquid temperatures
	Gas/vapour entrained in liquid	Check:- condition of liquid supply
Low discharge pressure		
	Pump speed low	Check:- pump speed, motor frequency, motor voltage
	Low viscosity	Check:- liquid temperature, liquid properties
	Excessive water	Check:- pump clearances, relief valve seat; replace worn parts
	Bypass valve partially open	Check:- bypass valve
High discharge pressure		
	Relief valve stuck	Check:- relief valve
	Relief valve incorrectly set	Check:- relief valve settings
	High viscosity	Check:- liquid temperature, liquid properties
High power consumption		
	Viscosity high	Check:- liquid temperature, liquid properties
	Discharge pressure high	Check:- pressure gauges, operating conditions
Pump vibration		
	Poor alignment	Check:- alignment, coupling assembly, pump holding down bolts, baseplate foundation bolts, baseplate grouting
Pump vibration		
	Insufficient $NPIP_{at}$	Check:- operating conditions, instrumentation, liquid properties
	Entrained gas/vapour in liquid	Check:- condition of supply liquid

Table 13.9 Trouble-shooting guide for rotary pumps

Symptom	Possible cause	Remedial action
Pump won't start		
	Power failure	Check:- all power supplies
	Blown fuses	Check:- all fuses
	Motor single phasing	Check:- motor wiring
	Control switch contacts open	Check:- all instrument switches in motor contactor circuit
No flow		
	Suction valve closed	Check:- suction pipework valves
	Discharge valve closed	Check:- flow through relief valve
	Pump not primed	Vent and prime pump
	Air leaks in suction pipes	Check:- suction pipework, stuffing box; replace gaskets
Low flow		
	Differential pressure too great	Check:- pressure gauges, calculations, relief valve flow
	Bypass valve partially open	Check:- bypass valve
	Pump speed low	Check:- pump speed, motor frequency, motor voltage
	Excessive wear	Check:- pump valves, relief valve seat; replace worn parts
	Insufficient NPIPa	Check:- presssure gauges, liquid temperature
	High viscosity	Check:- liquid properties; increase liquid temperature
	Gas/vapour entrained in liquid	Check:- condition of liquid supply
Very low flow		
	Broken valve spring(s)	Check:- suction/discharge valves; replace spring(s)
Low discharge pressure		
	Pump speed low	Check:- pump speed, motor frequency, motor voltage

Symptom	Possible cause	Remedial action
	Excessive wear	Check:- pump valves, relief valve seat; replace worn parts
High discharge pressure		
	Bypass valve partially open	Check:- bypass valve
	Relief valve stuck	Check:- relief valve
	Relief valve incorrectly set	Check:- relief valve settings
High power consumption		
	Discharge pressure high	Check:- pressure gauges, operating conditions
Pump vibration		
	Poor alignment	Check:- alignment, coupling assembly, pump holding down bolts, baseplate foundation bolts, baseplate grouting
	Insufficient NPIPa	Check:- operating conditions, instrumentation, liquid properties
	Entrained gas/vapour in liquid	Check:- condition of liquid supply

Table 13.10 Trouble-shooting guide for reciprocating pumps

Harmonics have been mentioned during commissioning as well as troubleshooting. Vibration and pressure pulsation harmonics can be analysed by specialists and the cause(s) of the problem diagnosed. Benchmark readings must be taken immediately after commissioning for comparison later if problems arise.

For soft packed stuffing box problems see Chapter 8, Sections 8.3.1.6 and 8.5.6, for mechanical seals see Chapter 8, Section 8.3.2.9.

13.5.6 Stocking spare parts

When buying pumps it is important to check the pump supplier's ability to maintain a stock of spare parts. The supplier should be able to provide a quotation for:

- Commissioning spares
- Spares for 1 year
- Spares for 2 years
- Insurance spares

Spares quotations will include delivery times. The spares quoted will be the parts which the supplier knows from experience will wear or become damaged. Insurance spares may be required if the pump is on a critical service and space is not available for stand-bys. Spares are costly. Some manufacturers may offer a discount if the spares are purchased with the complete pump. If the quotation does not mention discounts, ask.

The assessment of spares requirement is dependent upon many factors; the process liquid, the operating conditions, the number of hours of operation per year, delivery times for spare parts from the manufacturer, storage costs, etc. Table 13.11 shows a proposal for stocking spares in relation to the number of pumps installed for centrifugal pumps based on the German VDMA Standard 24296. The VDMA German Engineering Federation is one of the key association service providers in Europe and offers one of the largest engineering industry networks in Europe.

The pump supplier should also be asked about the supply of spare parts if a pump model is made obsolete and current production ceases.

When failure of the pump involves risk of damage, for example the possibilities of flooding or vital cooling systems being put out of action, a stand-by unit with an automatic starting system should always be installed if space allows. When a number of pumps of the same size share a service, it is prudent to have one complete unit in reserve for change-over in the event of a breakdown or in the case of a planned overhaul. Naturally the pipe connections, electrical installation, etc. should permit really rapid change-over with the minimum of inconvenience and effort.

Essential processes which run hot or cold may require the pump to be preheated or precooled before normal operation can proceed. The allowable step change in temperature is a function of pump size and is unlikely to be larger than 50°C. Quick-start stand-by equipment will require extra pipework for the circulation of heating/cooling mediums. "Hot" stand-bys must be checked periodically; pumps with long shafts should be rotated 90° regularly.

If overhauls and repairs are to be performed by the user's staff, it is also important that the tools required for the purpose should be obtained. Pump suppliers can make the necessary recommendations. If special tools are necessary, the pump supplier should quote costs for the tools when quoting for the pump. The use of unsuitable tools frequently results in more serious damage than that for which measures were to be taken.

Spare parts	Number of pumps					
	2	3	4	5	6	8
	Number of spare parts					
Impellers	1	1	1	2	2	3
Wear rings	1	1	1	2	2	3
Sealing rings	2	2	2	3	3	4
Shafts with impeller attachments	1	1	2	2	2	3
Shaft sleeves	2	2	2	3	3	4
Bearings	1	1	2	2	3	4
Mechanical seals, sets	4	6	8	8	9	12
Other seals, sets	4	6	8	8	9	12
Stuffing box packings	16	16	24	24	24	32
Sealing rings	2	3	4	4	4	6
Seats	2	3	4	4	4	6
O-rings	2	3	6	8	8	10
Seat packings	2	3	6	8	8	10
Springs	1	1	1	1	2	2

Table 13.11 Proposal for stocking spare parts for centrifugal pumps

13.6 Useful references

Troubleshooting centrifugal pumps and their systems, Ron Palgrave, Elsevier, 2003, ISBN 1856173197.

Know and understand centrifugal pumps, Bachus, Elsevier, 2003, ISBN 1856174039.

Predictive maintenance of pumps using condition monitoring, Beebe, Elsevier, 2004, ISBN 1856174085.

Pump efficiency and economics 14

14.1 Economic optimisation

14.1.1 General

An important characteristic of pumping plant, from an economic point of view, is its efficiency factor, i.e. the efficiency with which it converts the absorbed energy, usually electrical energy, into hydraulic power. This is especially important, since efficiency factors normally have extremely low values. Reasons for pump efficiency and economic factors are discussed and suggestions for improvement are presented. The secondary costs for the main process, of which the pump unit is a part, can be considerable if factors such as process adaptation and available capacity are not considered when designing the plant. The most precious form of energy in pumping is hydraulic energy, which warrants attention because of the parameters which not only determine the hydraulic power requirements but also the energy consumption.

The basic requirement of every pump installation is to transport liquid. The object of economic optimisation is to enable the liquid conveyance to be carried out at the lowest possible cost. The actual liquid transportation, however, often constitutes a small part of a much larger process. An evaluation of the liquid transport costs must, therefore, also include the costs of adaptation to the main process, or if adaptation is not carried out, the resultant costs of a less effective main process.

Since the required liquid transportation cannot be performed without the contribution of all components in the pumping plant, the complete liquid transportation system should be considered when carrying out economic optimisation. Only when the remaining costs are not affected, can partial optimisations give completely accurate results. Partial optimisations do, however, have the advantage that they are easier to evaluate and are therefore applied extensively, despite the risk of certain errors.

14.1.2 Whole-life costs

The cost which should be minimised is the plant whole-life cost, i.e. the summation of all the costs which occur during the total economic life span of the plant. This begins with planning and drafting and ends when the plant is written off or otherwise disposed of. Whole-life costs are traditionally divided into investment and operating costs.

Investment costs consists of the sum of the following:

- planning
- design and drafting
- equipment
- installation and commissioning
- process adaptation
- staff training
- disposal, writing off

Operating costs consist of the sum of:

- maintenance, parts and labour
- energy consumed

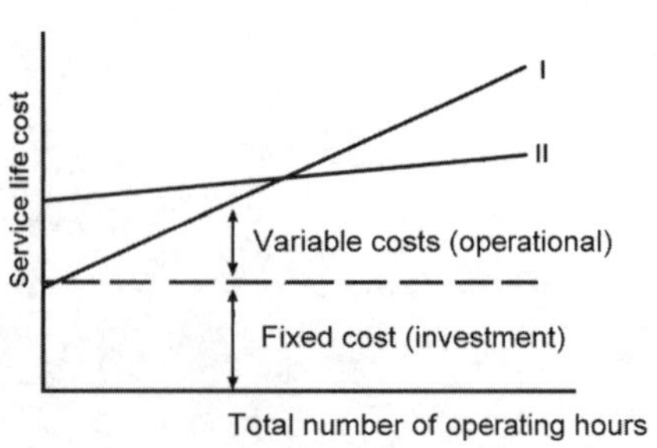

Figure 14.1 Graphical representation of whole-life costs

If the pumping plant is critical to a process and space is not available for stand-by capacity, the cost of lost production can be considered as a running cost.

The investment costs are basically fixed price in character, i.e. in principle they are independent of the extent to which the plant is used. Whereas the operating costs increase with the number of hours of the operation.

The two alternatives in Figure 14.1 show the influence which the number of operating hours has in respect of plant whole-life costs. Plant I, which is characterised by its simple construction with low investment costs, gives the lowest total cost despite its higher operating costs, low system efficiency, when the total number of operating hours is low.

For longer operating times, however, the relatively expensive plant II, with its high efficiency, shows the lowest whole-life cost. In practice, the operating costs will not be linear. As the plant ages, efficiency will be reduced and maintenance costs and spares usage will increase.

The distribution of costs for the various items of expense are naturally different for each case. An example of the items, together with their cost distribution, is shown in Table 14.1. These figures are derived from estimates made within the cellulose and paper industry.

Expense item	Cost %	Cost %
Pumps including mounting	4	
Temporary liquid storage	8	
Electrical equipment	12	
Instrumentation	8	
Pipework	20	
Administration	8	Investment 60 %
Maintenance including operational administration	11	
Energy	29	Operating 40 %
Total	100	100

Table 14.1 Distribution of whole-life costs for a liquid transportation system

According to Table 14.1 therefore, the cost for pumps represents the smallest part of the fixed costs, whilst energy and maintenance are the largest part of the total costs. A doubling of the pump investment costs would only increase the total cost by 4%. A reduction of the pump energy consumption by 25%, which in many cases is quite feasible, would reduce the whole-life costs by as much as 7%. It can also be seen from the Table that energy costs, 29%, together with the investment costs for the pipework, 20%, are the two dominant costs and account for half of the total cost of the system.

Even if the values in Table 14.1 refer to a special branch of industry, similar, but not identical, distribution trends are to be found in many pump installations. It must be remembered that the cellulose and paper industry is characterised by a predominance of stainless steel and super stainless steel pumps. And also that this industry handles very difficult products which result in very robust pumps that suffer particular operational problems. Analysis of other industries has shown the whole-life electrically energy consumption to be between 10 and 20 times the pump purchase cost.

Very small improvements in operating efficiency can return a substantial reduction on overall energy expenditure. Within sewage systems, water supply, heating, ventilation, sanitation and district heating, the cost of pipes and fittings usually show higher values but, together with energy costs, they still constitute the largest items. On the other hand, large engines, steam or gas turbines, will be much more costly than equivalent electric motors and inflate the purchase cost of the pumps. Engines may be much less efficient if waste heat recovery is not implemented. However low efficiency may not be a problem if the fuel cost is very low or "free". For small, simple pumping plant, however, the costs associated with labour and administration can often predominate. In such cases it is important that the plant is of a simple construction being easy to install and requiring little maintenance.

14.1.3 New and existing plant

The design and layout of new plant offers a greater freedom of choice with respect to economic optimisation than is the case for existing plant. For new plant the conditions are more favourable for total optimisation, since all the plant components can be chosen freely.

14.1.3.1 Process adaptation

Normally, economic optimisation is carried out for a number of technically feasible alternatives. The costs for the various items are determined and summated as in the previous Section. The alternative selected is that which fulfils the liquid transportation requirements whilst incurring the lowest whole-life costs. The opportunities presented by a new plant should be seized upon to achieve good adaptation to the process. Since the major costs are often associated with the main process, of which the liquid transportation is only a part, considerable cost reductions can be achieved by suitably designing the pumping plant.

The following examples are given in order to explain what is meant by the term "process adaptation".

- The transfer of process liquid between two process operations without synchronised flow requirements; the process availability is increased if temporary storage is included in the liquid transportation system.
- The pumping of sewage water to a sewage works regularly; constant flow assists the purification function of the sewage works.
- The pumping of drinking water to a water tower; the matching of supply to consumption results in a smaller water tower.
- The pumping of hazardous liquids; careful selection of the pump size and design of the shaft seals or use of glandless pumps reduces the costs of environmental protection.
- The gentle pumping of food products; reduces product damage and costs.
- The correct choice of low speed pumps for abrasive liquid-solid mixtures; reduces wear, costs of spare parts and labour costs.
- The scheduled adjustment of pump configuration, adding a stage, changing impeller diameter, changing piston/plunger diameter; matching flow/pressure requirements to system changes to reduce energy consumption.

Optimisation of existing plant should also, of course, consider the costs of process adaptation. Here, however, many parameters are already established which imposes limitations. Optimisation of existing plant assumes therefore, a partial optimisation function. The question is then: whether an extra investment can reduce the operating costs sufficiently to make the improvement profitable?

Examples of modifications proposed for existing plant are:

- Replacement of pump; replacement of impeller with larger diameter; reduction of impeller diameter
- Change method of regulation or control
- The addition of temporary liquid storage
- Pipework or system modifications to reduce pressure losses
- The introduction of improved shaft sealing arrangements

All modifications are introduced with the intention of reducing future maintenance and energy costs.

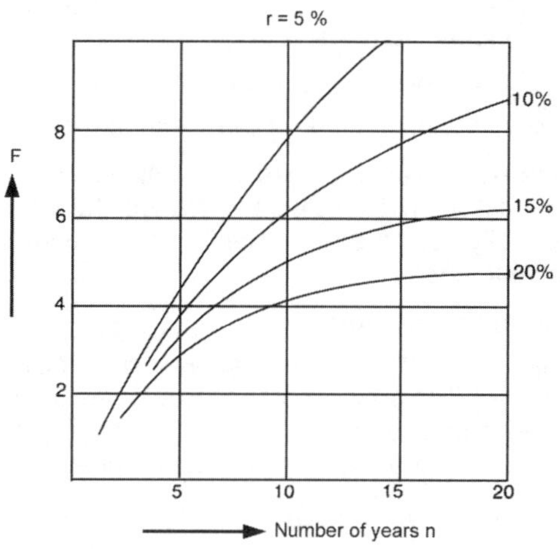

Figure 14.2 Present capitalised value factor

Another type of optimisation occurs when the transportation requirements can be performed by alternative existing pumping units. Such situations appear, for example, in large pipe networks, which are supplied by several pumping stations.

Here since the fixed costs for the plant cannot be changed, the rule is to choose the pump unit which results in the lowest operational costs. The operational costs in these instances are complicated by differing pump efficiencies and varying system pressure losses due to distance and pipe size.

14.2 Economic assessment criteria

14.2.1 Investment calculation - new plant

14.2.1.1 Present capitalised value method

The pumping plant whole-life cost K_L can be calculated as the sum of the investment cost K_I and the capitalised operational costs K_{Dcap}. The operational costs must be paid annually and therefore be summated to a capitalised present value. This is carried out with the aid of a present capitalised value factor F. The whole-life cost becomes:

$$K_L = K_I + K_{Dcap} = K_I + F \cdot K_D \qquad \text{Equ 14.1}$$

where:

K_L	=	whole-life cost (currency)
K_I	=	investment cost (currency)
K_{Dcap}	=	capitalised operational cost (currency)
F	=	present capitalised value factor
K_D	=	annual operating costs (currency/annum)

The present capitalised value factor is calculated from the relationship:

$$F = \frac{1-(1+r)^{-n}}{r} \qquad \text{Equ 14.2}$$

where:

F	=	present capitalised value factor
r	=	estimated annual profits (decimal)
n	=	number of years

Example:

Two alternative designs of pumping plant have been proposed. Their cost review is as follows:

Investment I = £10,000 annual operational costs = £1,000
Investment II= £7,500 annual operational costs = £1,500

Which plant is preferable from an economic point of view, if the estimated profits are 15% and the service life is 10 years?

If r = 15% and n = 10 years, the present capitalised value factor F = 5, from Figure 14.2. Thus:

I $K_L = £10000 + 5 \cdot 1000 = £15000$

II $K_L = £7500 + 5 \cdot 1500 = £15000$

Both plants are similar in this particular instance. If it is probable that the service life will not exceed 10 years, then investment II would normally be preferred since it requires a smaller initial capital investment.

14.2.1.2 Annuity method

For the annuity method the annual plant costs are calculated and minimised. The annual costs consist of the sum of the annuity interest and repayments for a proposed loan which covers the investment cost and the annual operational costs.

$$K = A + K_D = a_F \cdot K_I + K_D \qquad \text{Equ 14.3}$$

where:

K	=	annual cost (currency/annum)
A	=	annuity (currency/annum)
K_D	=	annual operating costs (currency/annum)
a_F	=	annuity factor (decimal)
K_I	=	investment (currency)

The annuity factor is identical to the reciprocal of the present capitalised value factor

$$a_F = \frac{I}{F} \qquad \text{Equ 14.4}$$

and is determined by service life and estimated profit. Applying this method, for the two investments in the previous example:

$$\text{I} \quad K = \frac{10000}{5} + 1000 + £3000$$

$$\text{II} \quad K = \frac{7500}{5} + 1500 = £3000$$

The two alternatives are equivalent, as before.

14.2.2 Investment calculation - existing plant

14.2.2.1 Present capitalised value method

For existing plant the question is often asked if an improvement of the plant can reduce the operational costs in such a way as to reduce the whole-life costs. According to the present capitalised value method the summation of the savings during the service life will be expressed at today's current value.

$$B_L = K_{Dcap} - K_I = F \cdot K_D - K_I \qquad \text{Equ 14.5}$$

where:

B_L	=	total savings during working life (currency)
K_{Dcap}	=	capitalised reduction of operating costs (currency)
F	=	present capitalised value factor
K_D	=	annual operating cost reduction (currency/annum)
K_I	=	investment cost (currency)

Example:

By investing £3,000 it is possible to reduce the annual costs for liquid transportation by £1,000? The plant is designed to operate for 15 years and the estimated profits to be a minimum of 15%. How great are the savings during the service life of the plant at today's currency value ?

According to Figure 14.2 the present capitalised value factor F = 5.9 for n = 15 years and r = 15%.

$$B_L = 5.9 \cdot 1{,}000 - 3{,}000 = £2{,}900$$

Over and above the repayment of investment, including estimated profits, an extra £2,900 is obtained.

If several alternatives exist to achieve a similar technical improvement, then it is normal to choose the alternative which produces the greatest savings.

14.2.2.2 Annuity method

The annuity method answers the question: How great the first year's saving will be at today's currency value?

$$B = K_D - A = K_D - A_F \cdot K_I = K_D - \frac{K_I}{F} \qquad \text{Equ 14.6}$$

where:

B	=	annual saving (currency/annum)
K_D	=	annual operating cost reduction (currency/annum)
A	=	annuity (currency/annum)
a_F	=	annuity factor
K_I	=	investment cost
F	=	present capitalised value factor (currency)

Using the values from the previous example,

$$B = 1000 - \frac{3000}{5.9} = £492$$

For otherwise similar alternatives, the size of the annual saving is decisive when making the selection.

14.2.2.3 Pay-off method

The pay-off method uses an imaginary "repayment time" defined by the relationship:

$$T_p = \frac{K_I}{K_D} \qquad \text{Equ 14.7}$$

where:

T_p	=	pay-off time (decimal)
K_I	=	investment cost (currency)
K_D	=	annual operating cost reduction (currency/annum)

The shorter the pay-off time the more profitable the investment. By comparing with equation 14.5, it is found that

for $B_L = 0$ then $T_p = F$.

i.e. the pay-off time and present capitalised value factor have the same numerical value when the total saving is equal to zero.

Using the values from the earlier example

$$T_p = \frac{3000}{1000} = 3 \text{ years}$$

With an estimated profit requirement of 15% and F = 3, the actual repayment time, using Figure 14.2, is approximately 4.3 years.

14.2.2.4 Investment grant

Grants for energy saving investments may apply. For up-to-date information it is best to seek the advice of the relevant local or central government authority.

In the UK for example, the most relevant schemes at the moment would seem to be:

Enhanced Capital Allowances (ECA). ECAs enable a business to claim 100% first-year capital allowances on their spending on qualifying plant and machinery. There are three schemes for ECAs:

- Energy-saving plant and machinery

- Low carbon dioxide emission cars and natural gas and hydrogen refuelling infrastructure
- Water conservation plant and machinery

Businesses can write off the whole of the capital cost of their investment in these technologies against their taxable profits of the period during which they make the investment.

This can deliver a helpful cash flow boost and a shortened payback period.

The Market Transformation Programme (MTP) is a DEFRA initiative that develops policy strategies for improving the resource efficiency of traded goods and services in the UK.

The MTP quantifies current thinking on how the daily use of products, systems and services impacts on the environment. MTP uses market projections and policy scenarios to explore alternative future developments.

The Energy Saving Trust (EST) was set up by the UK Government following the 1992 Rio Earth Summit and is one of the UK's leading organisations addressing the damaging effects of climate change. The Energy Saving Trust's goal is to achieve the sustainable and efficient use of energy, and to cut carbon dioxide emissions, one of the key contributors to climate change. The Energy Saving Trust is a non-profit organisation funded by the Government and the private sector.

14.2.3 Estimated profits and service life

14.2.3.1 Estimated profits

Profit estimates should, in principle, correspond to the interest on capital which would otherwise be realised from an alternative investment. It is a measure of a company's profitability and is higher than the current bank rate. The profit estimate increases with reduced capital resources, since it is necessary to be more particular when investment capital is limited. Profit estimates are rarely considered at less than 15%.

The methods of calculation previously reviewed assume that the annual operating cost reductions are of the same magnitude from year to year. The majority and largest savings on the operational side are achieved by reducing energy consumption. Energy costs, currency per kWh, can also be expected to rise more quickly than other costs, which means that energy savings will become more profitable with time.

One way of considering energy cost increase when making economic calculations is to use a corrected profit estimation.

$$r_k = r - e + i \qquad \text{Equ 14.8}$$

where:

r_k	=	corrected profit estimate (percentage)
r	=	uncorrected profit estimate (percentage)
e	=	rate of increase of energy costs (percentage)
I	=	general inflation rate (percentage)

A more rapid rate of increase of energy costs can in this way be transferred to a reduced profit estimation requirement for energy saving investments.

14.2.3.2 Service life

The economic service life is determined by factors such as write-off rules, the technical service life of components and the planned period of use of the plant. As with other parameters for pumping plant, the economic service life is dependent upon the size and type of industry.

As an approximation with the exception of small plant the following applies:

Buildings	40 years
Pipelines, underground	50 years
Other pipelines	20 years
Machines	15 years
Control equipment	10 years
Instrumentation	10 years

The service life of control equipment and instrumentation may be taken as 10 years for financial planning. The rate of change within the electronics industry is very rapid. Control systems and instrumentation should be reviewed every 5 years using the "Present capitalised value" method, to see if improved equipment or improved control strategies could reduce existing operating costs and hence reduce whole-life costs.

14.2.4 Energy costs

Energy costs depend upon the amount of energy consumed, the prevailing energy price scale and fixed costs for the supply installation. Premiums may be levied if the "maximum demand" is exceeded. In the most usual cases of electric motor operation, energy prices are determined by the relevant electricity tariffs.

14.2.4.1 Tariffs

The basis of most forms of tariff is a fixed charge dependent upon the "maximum demand" taken by the consumer and designed to cover:

- The costs dependent upon maximum load, e.g. interest and depreciation of generating plant, rates, taxes, insurance, salaries
- The costs incurred for each consumer, e.g. transformers, meters, meter reading labour, service cabling. And a running charge depending on the energy supplied, e.g. fuel, losses and maintenance of the supply plant, equipment, etc.

For industrial consumption two-part tariffs are common with the fixed charge proportional to maximum kW or maximum kVA demand; and a running cost, kWh, which may be dependent upon the time of day and/or year, i.e. peak and off-peak periods, and also include, for example, a fuel cost variation clause. A kVA maximum demand is preferable since it takes into account the effect of low power factor. It involves, however, more expensive metering equipment. The cost of metering maximum demand makes it uneconomic, in any case, for loads of less than 20 to 50 kW.

If supplies are taken at a high voltage instead of the usual lower voltage for distribution, the maximum demand charge may be less since the consumer then has the option to provide his own transformer.

14.2.4.2 Annual energy costs

The energy costs are a product of energy consumption and cost per unit. The annual energy cost will thus be:

$$K_E = k_e \cdot E \qquad \text{Equ 14.9}$$

where:

K_E	=	annual energy cost (currency)
k_e	=	energy cost (currency per unit)
E	=	annual energy consumption (units)

or capitalised for the service life of the pumping plant

$$K_{Ecap} = F \cdot k_e \cdot E \qquad \text{Equ 14.10}$$

Chapters 6 and 10 discussed power recovery turbines and engines and gas turbines using local fuel supplies. Every energy supply has a cost. In the case of high pressure liquid, which is supplied to power recovery turbines, the pipework and the control valves will have fixed and running costs. The same type of costs will apply to waste process steam. Local fuel supplies, such as crude oil or LNG, will have a value either as refining stock or sale to a consumer. These costs must be considered when evaluating the potential of site energy supplies.

14.3 Important system characteristics

14.3.1 Pumping efficiency

The cost of energy is the dominant cost item in most pumping plants. Energy, for which the pump user must pay, is the quantity measured by the electrical supplier or equivalent. The annual energy consumption consists of the product of the power used and the period of actual power consumption. See equation 14.9 for constant load applications and equation 14.11 for variable load applications.

$$E = \int_0^{t_0} P \cdot dt = \sum_0^{t_0} P \cdot \Delta t = P_m \cdot t_0 \qquad \text{Equ 14.11}$$

where:

E = annual energy consumption (kWh/annum)

P = actual power used (kW)

t = time (hours)

P_m = mean annual power used (kW)

t_0 = operational hours (hours/annum)

The power used depends partly upon the useful hydraulic power required to maintain the necessary flow in the pipe and upon the efficiency of converting the electrical power, or equivalent, into useful hydraulic power. This efficiency is called the pumping efficiency factor and is defined as:

$$\eta_E = \frac{P_h}{P} = \eta_r \cdot \eta_p \cdot \eta_{tr} \cdot \eta_m \cdot \eta_{others} \qquad \text{Equ 14.12}$$

where:

η_E = pumping efficiency factor (decimal)

P_h = required hydraulic power (kW)

P = power absorbed (kW)

η_r = flow regulation efficiency (decimal)

η_p = pump efficiency (decimal)

η_{tr} = drive train efficiency (decimal)

η_m = motor efficiency (decimal)

η_{others} = other component efficiencies (decimal)

The flow regulation efficiency illustrated in Figure 14.3 takes into account the power losses which are caused as a direct result of the method of flow regulation. Such power losses are caused by throttle regulation in the discharge pipe, by-pass control and on-off or load-unload control.

The expression for regulation efficiency for on-off control consists of the quotient of energy requirements. Whereas the others are represented by the quotient of power requirements.

The pumping efficiency will obviously be determined for the actual operating point and not the rated duty point efficiency, maximum flow point or any other arbitrary operational point. The power absorbed by auxiliary systems; lube oil, seal oil, cooling water; must be included in the overall pump efficiency.

The drive train efficiency represents losses in gears, couplings and speed variators. It is important to consider the transmission efficiency when speed regulation is used. Losses in rectifiers, frequency converters, regulation resistances, additional motor losses, etc., i.e. losses when utilising electrical methods of speed regulation, are traditionally calculated as transmission losses.

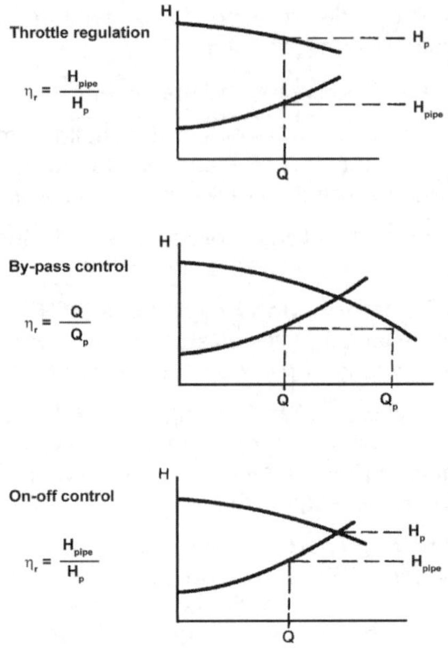

Figure 14.3 Flow regulation efficiency

The motor efficiency should be taken at the actual operating point. A common misconception is to assume that the motor efficiency is still high even at part load. At part load the percentage of reactive power is increased, this may be costed as a separate item. Do not forget about the power absorbed by separately driven cooling fans; this is an integral part of the motor efficiency.

The efficiency of other components takes into consideration losses for phase compensation and protection against supply disturbances, power requirements for supplementary ventilation and other environmental power consuming arrangements which can be directly associated with energy conversion in the pumping plant when working.

It is extremely important to remember that it is the pumping efficiency factor and not the efficiency of isolated components which is the characteristic unit for energy conversion efficiency. If all the component efficiencies are high at the actual operating point then a high pumping efficiency factor will obviously be obtained. However, it only requires one of the component efficiencies to be low to cause a poor total pumping efficiency.

Example:

Determine the pumping efficiency factor for a flow of 70% of maximum flow for the installation shown in Figure 14.4, for (I) speed regulation and (II) throttle regulation.

Using values from Table 14.2:

Case I		Case II	
$\eta_r = 1$	(from Figure 14.3)	$\eta_r = 0.44$	(from Figure 14.3)
$\eta_p = 0.75$	(estimated)	$\eta_p = 0.75$	(estimated)
$\eta_{tr} = 0.75$	(estimated)	$\eta_{tr} = 1$	(estimated)
$\eta_m = 0.90$	(estimated)	$\eta_m = 0.90$	(estimated)
$\eta_{others} = 1$	(not included)	$\eta_{others} = 1$	(not included)
$\eta_E = 0.51$		$\eta_E = 0.30$	

Table 14.2 Estimated component efficiency values

The efficiency factor for on-off control becomes evident, if it is assumed, for example, that the pump operates at full flow, 100% flow, for 75 % of the time. During 30% of the time the pump is then at test. In this case higher pipe losses are obtained during the time the pump operates and the efficiency factor becomes:

$$\eta_E = 0.75 \times 1 \times 0.90 \times 0.60 = 0.41$$

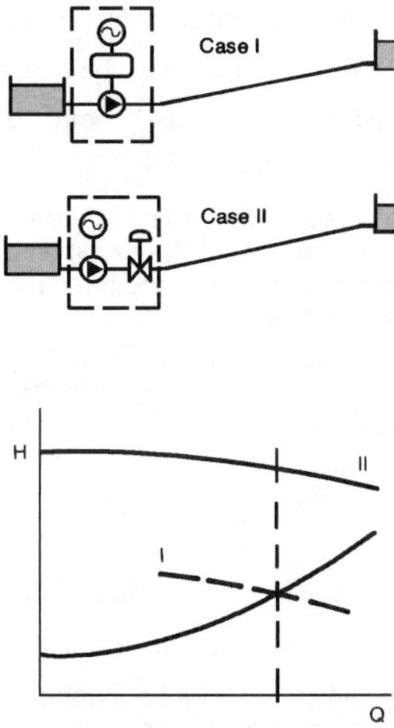

Figure 14.4 Illustration of pumping efficiency factor

For multi-pump systems the efficiency factor can be determined in a similar way. Here, as always, it is only the total efficiency for the actual flow and not the maximum efficiency of individual components which is characteristic for energy conversion efficiency. For on-off control of multi-pump systems there are a number of pertinent operating points. The efficiencies for these flows are greatly influenced by the sizes of the pumps selected.

An example of this is shown in Figure 14.5. In the case of three identical pumps having maximum efficiency at Q_{max}, poor part load efficiencies, shown as circles on the graph, are obtained. By choosing larger pumps, with η_{max} at a somewhat higher Q value, better part load efficiencies can be obtained, shown as crosses on the graph. Only by choosing three different sizes of pump can the highest efficiency be achieved for all three operating points. Three different pumps, however, involve the stocking of additional spare parts and more expensive maintenance. Also the pumps cannot be used as stand-bys for each other.

The basic pumping efficiency is very important. Pump efficiency is greatly influenced by size and choice of pump. Table 14.3 shows typical efficiencies for single stage overhung impeller centrifugal pumps, the most popular rotodynamic pump. Table 14.4 shows typical efficiencies for reciprocating plunger pumps.

Absorbed power kW	Efficiency %
3	37 to 59
7.5	38 to 74
15	50 to 78
30	50 to 79
55	55 to 83
110	60 to 85

Table 14.3 Typical best efficiencies for single stage centrifugal pumps

Absorbed power kW	100% Pressure Efficiency %	50% Pressure Efficiency %	25% Pressure Efficiency %
18.5 to 45	88	76	65
45 to 200	90	78	67
200 to 600	92	80	68
600+	94	84	72

Table 14.4 Typical efficiencies for reciprocating plunger pumps

The efficiency of plunger pumps is closely related to discharge pressure, not speed. Suction pressure does influence the efficiency when high. It can be seen from Tables 14.3 and 14.4 that

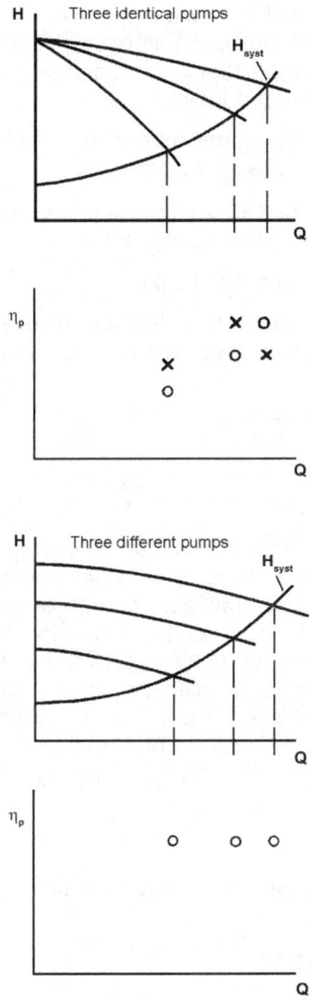

Figure 14.5 Example of pumping efficiency for three pumps connected in parallel

plunger pumps are much more efficient than centrifugal pumps. Reciprocating pumps are more costly than centrifugal pumps from an initial investment point of view, but the running costs can be much lower. Reciprocating pumps can be operated at reduced speeds, without loss of efficiency, to extend the life of parts such as packing and valves.

14.3.2 Demand variations

The primary cause of a low pumping efficiency factor is due to the variation, with respect to time, of the desired flow through the pipes, insufficient consideration having been given to this aspect at the design stage. Normally the mean flow per annum — total supplied volume of liquid divided by a calendar year — is about 10% to 15% of the installed flow capacity Q_{max}. The efficiency factor calculated as a mean efficiency over the year — theoretical hydraulic energy consumption/actual electrical energy consumption — usually gives a value in the region 5% to 40%, which should then be compared with the normal maximum momentary value of 55% to 75%.

The reasons for production or demand variations and the low mean flow are many and some brief examples are as follows:

- Seasonal variations in industry production
- Variations in supply of sewage water
- Variations in quantities of rainfall/snow
- Variations in water consumption
- Variations in heating requirements for central heating
- Component and calculation tolerances
- Margins of safety over and above the normal tolerances

A survey of flow variations with respect to time is a necessary basic requirement for all economic optimisations of liquid transportation. It is desirable that both progressive curves and constancy diagrams are produced.

Modifications for the improvement of the efficiency factor and hence the costs can be of two types:

- To improve the part load efficiency by means of suitable flow control and regulation equipment.
- To reduce extreme flow peaks.

The latter modification involves the addition of a liquid reservoir of some kind in the liquid transport system. Examples of liquid reservoirs are:

- Sumps and reservoirs
- Water towers
- Accumulators

Sumps and reservoirs offer the largest storage volumes at atmospheric pressure. Water towers have limited capacity although they can be substantial. Accumulators are small in comparison, they can be mounted in parallel, but they are the only option for pressures up to 350 bar. Unfortunately, there is no general method for assessing the optimal effect of various modifications and it is necessary to make estimates for each individual case. There is no doubt, however, that considerable cost reductions can be made by improving the degree of demand variation.

14.3.3 Availability

Availability is generally defined by the relationship:

$$A = \frac{MTBF}{MTBF + MTTR} \qquad \text{Equ 14.13}$$

where:

A = availability (decimal)

MTBF= Mean Time Between Failures (hours)

MTTR= Mean Time To Restore (hours)

The MTBF should be the summation of operating hours per annum. The MTTR should be the summation of all repair times per annum. MTTR should include all those hours of routine maintenance or adjustment when the pump must be stopped. Another concept frequently overlooked, MTBA, Mean Time Between Adjustments, includes maintenance work performed while the pump is running. MTBA does not affect availability but it does affect running costs.

Centrifugal pumps used within the process industry, for well-tried, normal operating conditions, achieve values of availability in the order of A = 0.99 to 0.9999, i.e. pump shut-down due to pump malfunction is between 1 and 80 hours per annum for continuous operation. These values also include the squirrel cage induction motor but not other electrical equipment. At the other extreme some high pressure water pumps, used for portable cleaning applications, have an availability of 0.88 on a daily basis. Some pumps must have the seals replaced every day.

Availability calculations are applied mainly to multi-pump systems. The percentage of operating time which a given number of pumps in a pumping station can be expected to be available for operation, is calculated from the formula:

$$P = \frac{N!}{K!(N-K)!}(1-A)^K \cdot A^{(N-K)} \qquad \text{Equ 14.14}$$

where:

P = system availability (decimal)

N = number of pumps installed (integer)

K = number of pumps unavailable (integer)

A = availability of individual pumps (decimal)

! = factorial (4!=24, 0!=1)

All the pumps are thus assumed to have the same availability values. By using equation 14.14 the resultant availability of a pumping station can be calculated and the effect of a stand-by pump can be demonstrated.

Without reserve pump		With reserve pump	
Function requirement	Resultant availability	Function requirement	Resultant availability
1 of 1	0.9900	1 of 2	0.9999
2 of 2	0.9801	2 of 3	0.9997
3 of 3	0.9703	3 of 4	0.9994
4 of 4	0.9606	4 of 5	0.9990
5 of 5	0.9510	5 of 6	0.9985

Table 14.5 Availability of pumping station with and without stand-by pump

From Table 14.5 it can be seen that a pumping station's availability is considerably improved by the installation of a stand-by pump. Function requirement 3 of 4 means that 3 pumps out of 4 will be available for operation. The individual pump availability has been taken as 0.99.

The user must define the availability required, at the inquiry stage, as this parameter has a crucial effect on pump selection. If the user wished to operate for 50 weeks every year the availability would have to be 0.959. If the user wished to operate for 154 weeks and allow two weeks for repair the availability would have to be 0.987. For this type of availability calculation to apply in practice it is necessary to maintain the stand-by pump to ensure that it will operate when required. In cases where it is possible to calculate the cost of a non-functioning pump, for example, in the form of lost production, availability calculations offer a direct economic optimisation possibility. Stand-by pumps usually run a limited number of hours regularly, one week in four say, to ensure the pump is functioning correctly.

14.3.3.1 Liquid transport system

The liquid transport system constitutes a link in a more extensive main process. The design of the system affects the availability of the main process, see Table 13.3 in Chapter 13.. The pipework may be 14 times more likely to create serious problems than a pump. The type of problem posed can be illustrated by a typical practical example. A process consists of two sub-processes for liquid treatment connected by a liquid transport system, see Figure 14.6.

The liquid transport system comprises the connection between the two liquid processes and in Figure 14.6 it can be characterised as a rigid connection. If the sub-process 1 must be shut down because of an internal malfunction, then the liquid transportation and sub-process 2 must also be shut down. The resultant availability for the process will thus be:

$$A = A_1 \cdot A_p \cdot A_2 \qquad \text{Equ 14.15}$$

The sub-process can consist of a large number of complicated and sophisticated components and so have a considerably lower availability than the liquid transport system. For this reason it is often unsuitable to make the sub-processes rigidly dependent upon each other.

By introducing a tank into the liquid transport system as in Figure 14.7, it is possible to loosen the connection between the two sub-processes. If the tank is sufficiently large, then the two sub-processes can become totally independent of each other. Sub-process 1 can deliver liquid to the tank while sub-process 2 is being repaired and vice versa. If $A_1 = A_2$ and $A_{p1} = A_{p2}$ the resultant process availability becomes:

$$A = A_1 \cdot A_{p1} = A_2 \cdot A_{p2} \qquad \text{Equ 14.16}$$

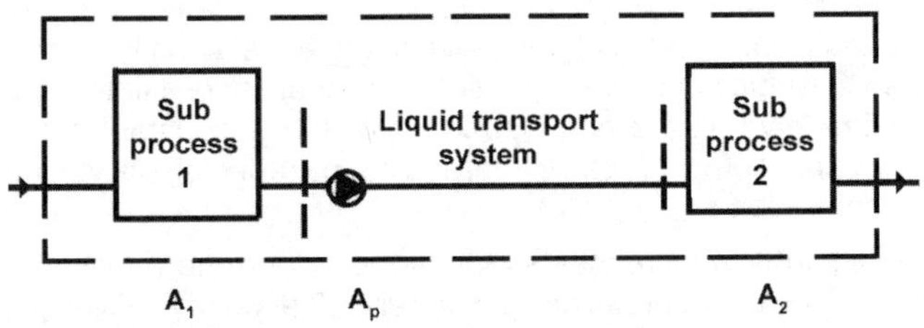

Figure 14.6 Schematic arrangement of a simple liquid system

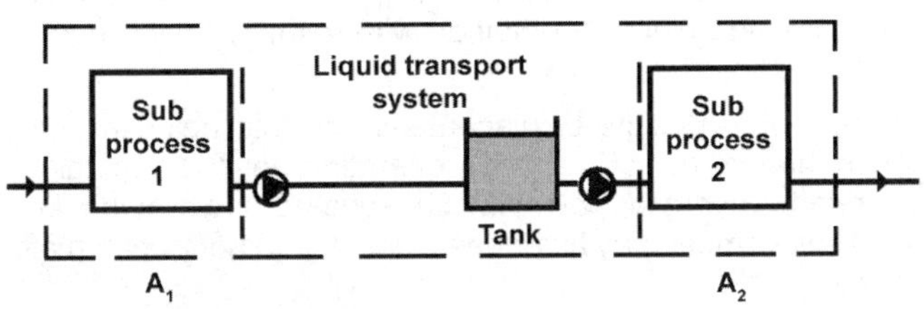

Figure 14.7 Schematic arrangement of a flexible liquid system

The higher value of availability results in a lowering of the mean capacity during operation for the same production.

The required installed capacities of all components in the sub-processes and the liquid transport system are therefore reduced. Lower mean flow also results in lower power and energy consumption if the efficiency of the smaller components is as high as the larger components. These possible cost reductions should be weighed against the investment costs of the tank.

For numerical values of $A_1 = A_2 = 0.90$ and $A_{p1} = A_{p2} = 0.99$ for mean capacity and mean flow during operation, the following results are obtained:

without buffer $A = 0.9 \times 0.99 \times 0.9 = 0.802$

with buffer $A = 0.9 \times 0.99 = 0.891$

With a tank large enough to completely isolate sub-process 1 from sub-process 2 the availability is increased by 11%.

14.3.4 Hydraulic power

14.3.4.1 General

The useful hydraulic power from the pump unit is used to satisfy the pipe system differential head requirements at the actual flow.

$$P_h = \rho \cdot Q \cdot g\,(H_{stat} + H_{fpipe}) \cdot 10^{-3}\ \text{(kW)} \qquad \text{Equ 14.17}$$

To achieve an optimal economic result, it is important that the hydraulic power is kept as low as possible. A limitation of the required hydraulic power is worthwhile, especially with low pumping efficiency factors of normally 5% to 40%. A reduction of P_h by 1 kW reduces the electric power consumption by 2.5 to 20 kW because of the low efficiency of converting electricity to hydraulic power.

14.3.4.2 Static delivery head

The static delivery head is determined when planning the plant and is often difficult to influence. It should be remembered that for installations where liquid is first pumped up to a tank in order to be subsequently gravity fed to a consumer, the energy required is always greater than if the liquid were pumped direct to the consumer. Such cases should be redesigned, if possible, at the planning stage.

14.3.4.3 Pipe flow losses

The approximate magnitude of pipe friction losses in straight pipes is described by the relationship (assuming a constant friction factor):

$$h_{fpipe} = \lambda \cdot \frac{l}{d} \cdot \frac{Q^2}{\left[\dfrac{\pi \cdot d^2}{4}\right]^2} \cdot \frac{1}{2g} \qquad \text{Equ 14.18}$$

$$h_{fpipe} = \frac{\text{constant}}{d^5} \qquad \text{Equ 14.19}$$

See also Chapter 3, Section 3.2. The pipe friction losses for a given flow in a 200 mm pipeline are 3.05 times greater than for the same flow in a 250 mm pipeline, i.e. (250/200)5 = 3.05. This calculated example illustrates the importance of correct pipe sizing. Pipe losses are not only friction losses. Pipe fittings, flanges and bends also make a contribution. The selected pipe velocity must take into account the quality of the assembled pipe system and any problems caused by turbulence at poorly aligned connections.

The corresponding expression for losses in valves and fittings, etc., is:

$$h_{fvalves\ and\ fittings} = \frac{\text{constant}}{d^4} \qquad \text{Equ 14.20}$$

See also Chapter 3, Section 3.2. Here also there is a large dependence upon size.

For laminar flow:

$$h_{fpipe} = \text{constant} \times d^{-4}$$

and

$$h_{fvalvesandfittings} \times \text{constant } d^{-3}$$

Demand and production variations lead not only to a low pumping efficiency factor, but also increase the hydraulic power requirement and hydraulic energy consumption. Hydraulic energy consumption is relative to the conversion from pressure to heat, pressure drop, caused by the effects of pipe friction.

For pump installations where $H_{stat} = 0$, which is assumed here solely for the purpose of simplifying the illustration of the importance of flow variations, then:

$$p_h = r \cdot Q \cdot g \cdot h_{fpipe} = \text{constant} \cdot Q^3 \qquad \text{Equ 14.21}$$

The installed hydraulic power, which is required in a system with mean flow Q_m and demand/production variation $\pm\Delta Q$ is:

$$F_h = \text{constant} \cdot (Q_m + \Delta Q)^3$$

$$= \text{constant} \cdot Q_m^3 \left(1 + \frac{\Delta Q}{Q_M}\right)^3$$

$$= P_{hm} \cdot \left(1 + \frac{\Delta Q}{Q_m}\right)^3 \qquad \text{Equ 14.22}$$

The hydraulic energy consumption becomes dependent upon the shape of the flow demand/supply curve and is illustrated below for three simple special cases.

The hydraulic energy consumption becomes:

$$E_h = \left[1 + 3\left(\frac{\Delta Q}{Q_m}\right)^2\right] \cdot P_{hm} \cdot t \quad \text{Square}$$

$$E_h = \left[1 + \left(\frac{DQ}{Q_m}\right)^2\right] \cdot P_{hm} \cdot t \quad \text{Saw tooth}$$

$$E_h = \left[1 + \frac{3}{2}\left(\frac{\Delta Q}{Q_m}\right)^2\right] \cdot P_{hm} \cdot t \quad \text{Sinusoidal} \qquad \text{Equ 14.23}$$

Figure 14.8 and Table 14.6 show that the demand/supply variations have a considerable influence upon both the installed hydraulic power and the energy consumption. From the economic

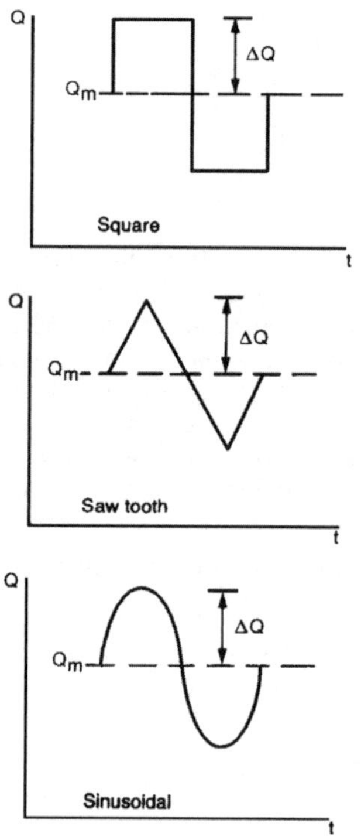

Figure 14.8 Three special cases of demand/supply variations

$\Delta Q/Q_m$	P_h/P_{hm}	$E_h/P_{hm} \cdot t$		
		Square	Saw tooth	Sinusoidal
0	1	1	1	1
0.1	1.33	1.03	1.01	1.02
0.2	1.73	1.12	1.04	1.06
0.5	3.37	1.75	1.25	1.38

Table 14.6 Hydraulic installed power and hydraulic energy consumption for different types of demand/supply variation

point of view, reducing the flow peaks and troughs can give good returns.

14.3.4.4 Suspensions

In the case of suspensions and liquid-solid mixtures, yet another variable must be considered, namely the solids concentration. The adaptation of the quantity of solid material transported per unit of time can be made by varying the flow, changing the concentration or a combination of both. Valve regulation is not usually suitable because of the risk of blockage and high wear rates. Other feasible alternatives are illustrated by the example in Figure 14.9.

Referring to Figure 14.9, at nominal capacity, operating point A at 65% centrifugal pump flow, a certain quantity of solid material

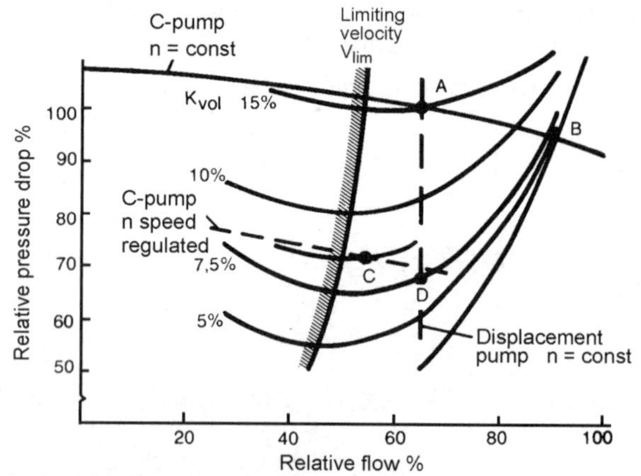

Figure 14.9 Approximate effects of pumping heterogeneous suspensions with nominal capacity, A, and at half capacity solid flow at B, C and D

is transported through the pipeline. The relative required hydraulic power is set at point A as being equal to unity. It is desired to reduce the quantity of solid material per unit time to approximately half the nominal capacity. For a constant speed centrifugal pump this can be carried out by reducing the volume concentration K_{vol}.

The liquid flow increases simultaneously and the relative hydraulic power becomes approximately 1.3 at point B. The liquid velocity increases because the pressure loss through the pipe reduces as the solids concentration is reduced, see Chapter 3, Section 3.3. In the case of a speed regulated centrifugal pump the operating point C is obtained with relative power requirement of 0.6.

Alternatively, a positive displacement pump can be used for the transportation. At half capacity, operating point D is obtained with relative hydraulic power 0.7. The original power of the positive displacement may have been less due to better pump efficiency.

The risk of unstable operation and blockage of the pipe is also illustrated by Figure 14.9. The centrifugal pump curve can maintain equilibrium with the pipe loss curve at two points, at which the operating point at the lower flow leads to blockage in the pipeline. Since the curves intersect each other at nearly the same angle then the risk of instability is relatively great. Corresponding risks do not occur when using a positive displacement pump. The actual power consumption can only be confirmed by evaluating complete pump data.

Considering positive displacement pumps for long pipeline applications introduces yet another variable, the number of pumping stations. Positive displacement pumps can produce high pressures without high speeds or high liquid velocities. The pipe friction losses can consequently be higher before the mixture must be re-pressurised. Positive displacement pumps can handle solid mixtures up to 70% by weight. Pump efficiency is not necessarily a function of solids concentration.

14.4 Partial optimisation

14.4.1 Economic pipe diameter

The two major costs for many pumping plants are the costs of pipelines on the investment side and the energy costs on the operational side. The investment costs increase with increased pipe diameter whilst at the same time operational costs decrease, see Figure 14.10.

The total costs curve represents a minimum for a certain pipe diameter. As indicated in the Figure the shape of the total costs curve is such that it rises more rapidly with reduced diameter, to the left of the economic diameter, than in cases of increased diameter, to the right of the optimum value. In cases of doubt, choose the one having the largest pipe diameter from the feasible alternatives.

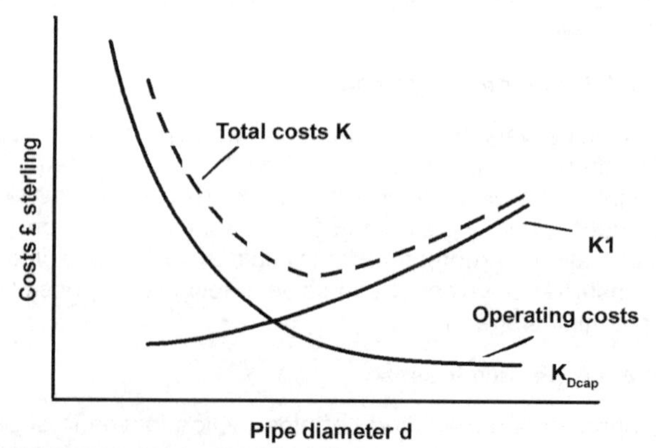

Figure 14.10 Effects on costs of pipe diameter

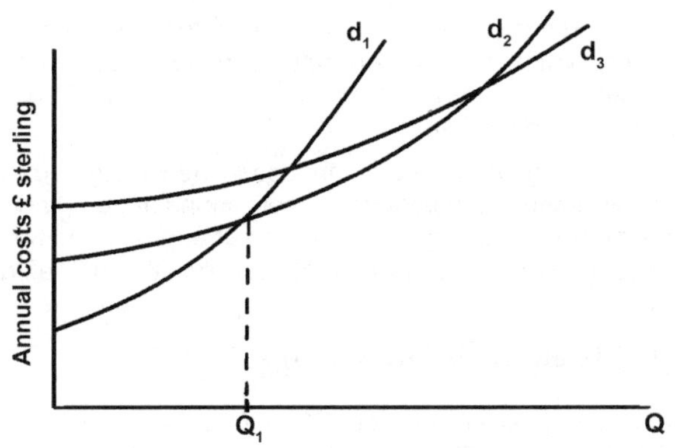

Figure 14.11 Annual costs for a pumping plant using various pipe diameters

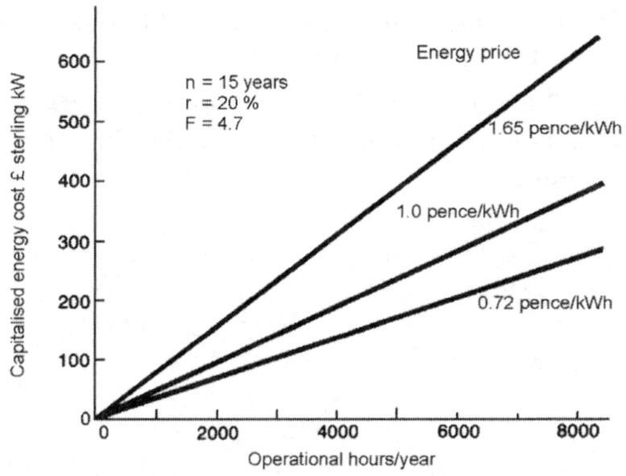

Figure 14.12 Capitalised value in £ sterling/kWh for a reduction of the mean annual electric power used

By calculating the total costs for a number of different pipe diameters it is possible to establish the most economic pipe diameter, see Figure 14.11. The economic diameter varies for different situations. Short operating periods and costly pipe sections, for example, stainless steel, tend to reduce the economic pipe diameter and to increase the economic flow velocity.

By carrying out a partial optimisation, i.e. if it is assumed that all costs are independent of the pipe diameter except for the costs of the pipeline and energy, the following relationship is obtained:

$$Q_1 = \sqrt[3]{\frac{k_2 - k_1}{1/d_1^5 - 1/d_2^5} \cdot \frac{\eta_E \cdot \pi^2 \cdot a_F \cdot 10^3}{8 \cdot \rho \cdot \lambda \cdot t \cdot k_e}} \qquad \text{Equ 14.24}$$

where:

Q_1	=	the mean flow for which pipe diameters d_1 and d_2 give the same annual cost (m^3/h)
k	=	cost/m of pipeline (currency/m)
d	=	pipe diameter (m)
η_E	=	pumping efficiency factor (decimal)
a_F	=	annuity factor (decimal)
ρ	=	liquid density (kg/m^3)
λ	=	pipeline loss co-efficient
t	=	operating time per year (hour/annum)
k_e	=	cost of energy (currency/kWh)

For this Q_1 value the total energy costs are the same. If the mean annual flow for the plant is greater than Q_1 then it is economically feasible to choose pipe diameter d_2. If Q_m is much greater than Q_1 then the procedure must be repeated using diameters d_2 and d_3.

The procedure for valves and other fittings is carried out in a similar manner:

$$Q_1 = \sqrt[3]{\frac{k_2 - k_1}{1/d_1^4 - 1/d_2^4} \cdot \frac{\eta_E \cdot \pi^2 \cdot a_F \cdot 10^3}{8 \cdot \rho \cdot \zeta \cdot t \cdot k_e}} \qquad \text{Equ 14.25}$$

where:

k	=	valve cost including connectors for pipe diameter d

It should be noted that an improved pumping efficiency factor tends to reduce the economic pipe diameter.

14.4.2 Component efficiency

When purchasing new equipment, it is often possible to choose between a cheap pump having low efficiency and an expensive pump with high efficiency. The comparative costs for the two alternatives are:

$$P_{1m} \cdot t \cdot k_e \cdot F + K_{I1} \Leftrightarrow P_{2m} \cdot t \cdot k_e \cdot F + K_{I2} \qquad \text{Equ 14.26}$$

where:

P_m	=	consumed mean electrical power (kW)
t	=	operating hours per year (hours/annum)
k_e	=	energy cost (currency/kWh)
F	=	present capitalised value factor
K_I	=	investment cost (currency)

It is important to remember that the energy costs are debited according to readings taken from the electricity meter and not from the pump shaft. Because of cable losses, the electric power used is always greater than the pump shaft power. The capitalised value of 1 kW mean electric power is $t \cdot k_e \cdot F$. See Figure 14.12.

The efficiency of other components can be tested in the same way. Conversion to electrical power used must always be made, however. The closer the power consumption lies to the useful hydraulic power in the conversion chain, the greater the energy saving and the greater the motivation for additional investment.

14.4.3 Existing plant

In the case of existing plant there are certain limitations to energy saving possibilities. Changes to the layout and pipework are often difficult to make. Possible improvements are replacements or the addition of supplementary components which do not require too great a disturbance to the plant.

All reductions of the hydraulic power, by means of improvements which reduce the static differential head or pressure drop, in the case of throttle-regulated plant, only result in a greater pressure drop across the control valve and are thus worthless from an energy conservation point of view. To realise a saving of energy some form of improvement must also be carried out on the pump side. For speed-regulated plant all pressure reductions are automatically used.

When planning a project, at the time of determining pump specifications, it is often the case that not all of the plant details are known. It is not therefore unnatural to choose pump sizes with a certain safety margin "to be on the safe side". For this reason many pumps are oversized and result, if adjustments are not made, in unnecessarily high energy costs.

In the case of basic oversizing, the following corrections to the pump may be considered:

- Machining the pump impeller
- Fitting a smaller or larger piston/plunger
- De-staging a multi-stage pump
- Fitting a smaller motor

- Changing the gear ratio of a gearbox

It may also be advantageous to review the connected process(es) to optimise or de-bottleneck and use the excess flow or pressure.

A more common reason for unnecessarily high energy costs is, however, demand/supply variations. In which case the above improvements cannot be applied.

Attention should instead be directed to corrections such as:

- Energy efficient methods of control, the most usual being speed regulation
- Energy efficient methods of control together with reduction of pressure drop and reduced flow peaks and troughs.

The fact that energy conservation proposals and suggestions cannot be realised in practice unless adaptation is carried out on the pumping side, is a strong argument for speed regulation or some other form of energy efficient method of control. Since the investment costs for the pump unit are usually a small proportion of the energy costs, the addition of energy conservation equipment is nearly always profitable. The secondary costs of an oversized pump are of a very limited nature when using energy conservation methods of control and the excess pump capacity can be considered as a reserve in the event of possible future production increases.

Insufficient engineering hours are used when designing a project to fully assess the economic implications of pump and associated equipment selection. The use of approximate operating data to select equipment always results in inefficient utilisation.

14.5 Useful references

Enhanced Capital Allowances (ECA), DEFRA, 6/H15, Ashdown House, 123 Victoria Street, London SW1E 6DE UK, Tel: 020 7082 8709, Fax: 020 7082 8708.

The Market Transformation Programme (MTP), Future Energy Solutions, PO Box 222, Didcot, OX11 0WZ UK, Tel: 0845 600 895, www.mtprog.com.

The Energy Saving Trust (EST), 21 Dartmouth Street, London SW1H 9BP UK Tel: 020 7222 0101, Fax: 020 7654 2460, www.est.org.uk.

Pump selection

15

15.1 General operating conditions

Everything about a pumping application is important. Prospective purchasers should not make value judgements on what is important and what is unimportant. All the important data which is missing or is incorrect will devalue the manufacturer's warranty. The manufacturer's warranty is given for specified operating conditions; undocumented operating conditions are not guaranteed. Operating conditions and corrosion may be experienced while the pump is stationary. Standby equipment may experience a more severe "environment" than the operating equipment.

Engineers completing or filling in data sheets sometimes read too much into the question. Data sheets request "Discharge pressure"; the data sheet does not say "All pressures up to and including". The range of operating variables is critically important. Concentrating on "maximums" is a mistake; the full range of values is important.

15.1.1 Liquid properties and operating conditions

For pump selection for an application, the following information must be known:

- Flow
- Temperature
- Viscosity
- Suction pressure
- NPSHa/NPIPa
- Differential head or pressure
- Discharge pressure
- Constituents in the liquid
- The properties of the liquid

Constituents in the liquid, in this context, mean solids, gases and other liquids. Pumps selected for clean liquids may only last 20% of the time predicted if abrasive solids are present. Entrained gases may reduce the effective NPSHa/NPIPa by over 3.5 m; pumps do not always have a long life. Liquids as trace elements can have a serious impact on the corrosion properties of the bulk liquid.

The properties of the liquid are of critical importance for selecting the correct type of pump. The first obvious property is the vapour pressure. The liquid temperature plus the vapour pressure plus the suction system arrangement provide the NPSHa and the NPIPa. The liquid density varies with temperature; this will affect the conversion to pressure in rotodynamic pumps. The liquid viscosity varies considerably with temperature. Perhaps the liquid is non-Newtonian, a pulp or emulsion. If the liquid is extremely hazardous, then a pump which cannot leak may be required. The corrosive properties of the liquid may be important for the type of pump. Normally, liquid corrosion would decide the choice of material for the pump. If the pump must be manufactured from a material which cannot be cast or welded, the choice of pump types is severely restricted.

Any variation in the operating conditions must be quantified. For any specific variation, the other conditions must be stated. The duration of expected running must also be specified. Changes in operating conditions are always assumed to be relatively slow; slow enough not to cause instability or surges. Rapid changes in temperature can create distortion; for instance it is possible to crack thick casings due to thermal shock. Any rapid changes in operating conditions must be identified. For pumps running continuously, it is usual to specify the Mean Time Between Failures, MTBF.

Chemical processing and oil refining plants can operate for 25 or 50 weeks without a shut-down. Table 15.1 shows typical operating descriptions which might be useful.

Operation	Description
Continuous	Over 8 hours running in any 24 hour period
Light	3 to 8 hours running in any 24 hour period
Intermittent	Up to 3 hours running in any 24 hour period
Irregular	The pump operates for differing times with various periods of extended rest between operation
Cyclic	The pump operates with a set pattern of rest or unloaded running followed by a period on-load

Table 15.1 Plant operating descriptions

The definitions given in Table 15.1 are very generalised and appear in some standards and specifications. The important factor is whether the equipment warms up fully. If equipment runs long enough for all temperatures to stabilise then it is considered to run "continuously". Small equipment can warm up very quickly. Large equipment may take much longer than 8 hours for all temperatures to stabilise.

More costly, slower equipment can, through higher reliability and availability, pay for itself very quickly from increased plant output.

If starting or stopping the pump results in significant changes to the liquid properties or system conditions, these effects must be clarified. High suction pressure during start-up for example, can create tremendous bearing wear problems. Pumps starting cold, at high viscosity, may not achieve the desired performance. Routine maintenance operations on the pump, such as steam or chemical cleaning, may result in modified material selections.

Variations in flow determine whether the application should be shared by several pumps. The system curve, flow/pressure characteristic, then determines whether the division should be between units of equal or different sizes.

Great variations in the viscosity of the liquid may mean, for example, that a centrifugal pump should be used alternatively with a positive displacement pump in cases of low and high viscosity respectively. A positive displacement pump may be able to cope with the complete viscosity range. Viscosity is extremely important. Some pumps rely on viscosity for lubrication. Other pumps rely on viscosity for sealing. Some pumps are unaffected by viscosity.

Variations in flow demand and the consequential change, or otherwise, of differential head or pressure determine the type of flow regulation to be used. In this context the working time of the pump is of considerable importance. The efficiency of the pump unit, at all its operating conditions must be used to calculate the total energy requirement, kWh/annum. The energy costs for pumping can be as great as the purchase cost of the pump after only a few months of operation. The cost of the energy supply must be considered carefully. Can alternative energy sources by used? Can the energy total be reduced by power recovery?

Flow should also be considered both from the point of view of long term changes and also from short term fluctuations. Different regulation methods, with different initial costs and running efficiencies can be applied depending upon the frequency of the changes.

The effect of system pressure changes on the pump flow may be important. A positive displacement pump can offer almost constant flow over a wide range of pressures. This feature may be important and can simplify flow control methods.

Operating conditions can change as the installation wears, corrodes or fouls. Changes to the pump unit, from as new to progressive ageing may be used to optimise performance at different duty points.

The location of the pump within the site also has a marked influence on the choice. The location determines the NPSHa and NPIPa and evacuation and also whether the pump is of the dry pit, wet pit or submersible type. Requirements for space can also lead to a choice between horizontal and vertical designs. The environment at the installation site may also raise special demands with regard to the driver for the pump.

The size, nature and concentration of any particles are also, of course, important factors. Hard abrasive solids will generally have a much more serious effect on pump life and efficiency than soft, deformable solids like wax. The abrasive properties of hard solids should be quantified by testing. The Miller Number test is used extensively for reciprocating pumps. Wax, however, may coat pipe walls and impeller vanes and progressively choke a pump and its system.

When using positive displacement pumps it is important to know the liquid compressibility. As differential pressures increase, the compressibility becomes more essential to evaluate the volumetric efficiency. Few liquids are less compressible than water. Liquified gases, such as CO_2, can be up to six times more compressible than water with consequential reductions in volumetric efficiency.

The hazards posed by the liquid are an important factor. Hazards to personnel, the site and the environment, and the time period involved, must be considered carefully. The location of the site may influence decisions. An oil production platform in the North Sea must be treated differently to a remote site in the heart of the French countryside. Local installation and statutory regulations may also require specific actions to be implemented. The pump may be situated in a hazardous area caused by potentially explosive gases. Surface temperatures imposed on electrical equipment will apply to mechanical equipment, even if not stated. Pneumatically or hydraulically operated pumps may be necessary.

When selecting pumps for a range of applications for one site, select the pump which is best for the application providing whole-life costs are considered. Selecting the correct pump will optimise efficiency and reduce running costs. The cost of extra spare parts for different machines will be outweighed by the reduced energy consumption and extended running times of appropriate equipment.

The skills of local personnel must also be considered. Most pumps are relatively simple and can be maintained with standard tools using established techniques. Some pumps, and/or drivers, may have complicated systems requiring specialist knowledge. In these cases the costs of factory staff for maintenance must be evaluated.

For choice of pump, the auxiliary systems necessary for the pump should also be taken into consideration. The pump itself must often be provided with protection against dry running, against overheating when running against a closed valve and, in the case of positive displacement pumps, against unintentional throttling in the suction or discharge. In many cases shaft seals require sealing or cooling liquids at the correct pressure and with a guaranteed flow.

Sometimes the properties of the liquid may be uncertain. When this problem arises it is best to discuss the effects of variations on pump performance, and perhaps life and reliability, with pump manufacturers or consultants. Values must be placed on uncertain parameters. If this is left to pump manufacturers, they will chose values which are best for their particular pump. The user must analyse the problem and select safe values for the limits of operation.

Table 15.2 lists some typical important ranges commonly encountered. The Compressibility units in this table show the reduction in unit volume per bar differential pressure.

Parameter	Range
Density	500 to 1700 kg/m^3
Solids content	up to 80% by wt
Solids size	up to 100 mm
Viscosity	0.15 to 20000 cSt
Vapour pressure	negligible to 130 barg
Compressibility	1.96E-7 to 1.95E-5

Table 15.2 Typical liquid, liquid-solid mixture properties

As with liquid properties, operating conditions can be uncertain. Table 15.3 lists typical ranges of operating conditions.

Condition	Range
Temperature	-100 to 350 °C
Suction pressure	0.15 to 140 barg
NPSHa	0.75 m and greater
Discharge pressure	0.25 to 4000 barg
Flow (can be zero in some cases)	0.00000001 to 15000 m^3/h
Power	0.05 kW to 5 MW

Table 15.3 Typical pump operating conditions

15.2 Selection of pump according to duty and capabilities

Decisive factors for the choice of pump are:

- The properties of the process liquid
- The desired pump performance
- Flow and differential head or pressure
- Certain specific requirements in connection with the pump installation

A suitable pump choice is made when all features and properties most suited to the application are assessed. In some cases, however, it is necessary to accept a choice of pump which is not ideal with respect to all features. Then the choice must be made entirely on the basis of the requirements of the specific case, i.e. a compromise must be reached between technical features such as wear capacity or resistance to corrosion. It is important also to consider the economic effects or the difference in characteristics between rotodynamic pumps and positive displacement pumps.

Centrifugal pumps are the most suitable types for about 80% of all pump requirements. Positive displacement pumps are commonly mistaken for high pressure, small flow and for viscous liquid applications. Some positive displacement pumps can be used to advantage for arduous low pressure applications; pumping sewage and viscous tank bottom sludge with sediment. Table 14.4 in Chapter 14 showed for example, that small plunger pumps running at 25% rated pressure can be more efficient than centrifugal pumps.

Liquid properties such as viscosity, resistance to chemical attack, variations in the rate of the chemical reaction with temperature and contamination are especially important not only for choosing the type of pump but also for the specifications intended to serve as a basis for purchasing. Some chemicals will not react at one temperature but will cause substantial damage if the temperature rises above a certain point. Variations in pH can make what is sometimes a harmless solution into a highly dangerous and corrosive solution. Some pumps may be resistant to chemical compounds over a short period but may be attacked over a long period. This factor is important in cleaning, where a corrosive chemical may be used for a short period to remove foreign matter but would damage the pump if it was left in contact for a long period.

15.3 Selection of pump according to hydraulic performance

When considering viscosity, the limitations given for any pump design are complicated by pump size. Small pumps have higher parasitic losses, lower efficiency, due to friction losses in the small passages. As pump size increases these losses become smaller and less important. Increased viscosity, in rotodynamic pumps, increases the parasitic losses and reduces pump performance and efficiency. As a general rule, a viscosity of 300 cSt is a good bench mark for evaluating viscosity effects. The debilitating effect of viscosity on rotodynamic pumps can be mitigated by running pumps at reduced speed.

In the following Sections it is important to remember that viscosity limitations are not "cast in stone" but indicate areas of concern where the transition from low viscosity to viscous and viscous to highly viscous are blurred.

15.3.1 Pumps for low viscosity liquids

Suitable ranges of operation for flows and differential heads can be indicated on the basis of popular rotodynamic pump availability for different types of pumps for low viscosity liquids according to Figure 15.1.

Double suction axially-split pumps and axial propeller pumps can have flows over 50000 m^3/h. Figure 15.2 indicates the flow and discharge pressure capabilities of positive displacement pumps. In some cases it is necessary for the liquid to have certain lubrication properties.

Plunger pumps are shown with a maximum pressure of 2000 barg; pumps are available for pressures up to 10000 barg. Very few applications use pressures over 1000 barg. Performance requirements, purchasing cost, running costs, estimated service life and reliability of operation must be considered when choosing pumps. Rotodynamic pumps of various designs cover nearly 80% of all pump requirements. However, looking at Figures 15.1 and 15.2, for flows from 1 to 100 m^3/h, and pressures up to 200 m or 20 barg, there are plenty of pump types to choose from.

In some instances the choice of pump is very simple, based on the duty of a single pump:

- Flow over 2271 m^3/h ∴ rotodynamic
- Flow less than 0.001 m^3/h ∴ positive displacement
- Differential head over 7000 m ∴ plunger pump

In some special applications there are dedicated designs which almost without exception are the most suitable. Some examples of these are heating, water and sanitation pumps.

With rotodynamic pumps, in certain borderline cases, it is possible to choose between single-stage and multi-stage pumps. An analysis of the efficiency and running costs on the one hand and of operational reliability on the other will provide sufficient grounds for a decision.

The speed of the pump is a factor which is often discussed. In principle the choice should be in favour of a speed sufficiently high to ensure that the NPSHa in the plant, less a suitable margin, is used. Possible fears of a reduction in operational reliability in the case of higher speeds should not be allowed to result in the choice of a lower speed, but rather in better quality and testing requirements when purchasing. Requirements of this kind can refer to critical speeds or bearing life or sealing problems, for example. It should, however, be observed that a higher speed may result in higher noise levels requiring acoustic treatment.

In the borderline cases between rotodynamic and positive displacement pumps the form of the pump characteristics and method of regulation often influence the choice. Both types of pump have their distinct advantages, see Chapter 4 with regard to different pumps designs and characteristics.

A difficult problem for all pumps is the presence of air or gas in the liquid. The effect on the effective vapour pressure must not be overlooked. Rotodynamic pumps suffer from a reduction of efficiency when the gas content on the suction side is in excess

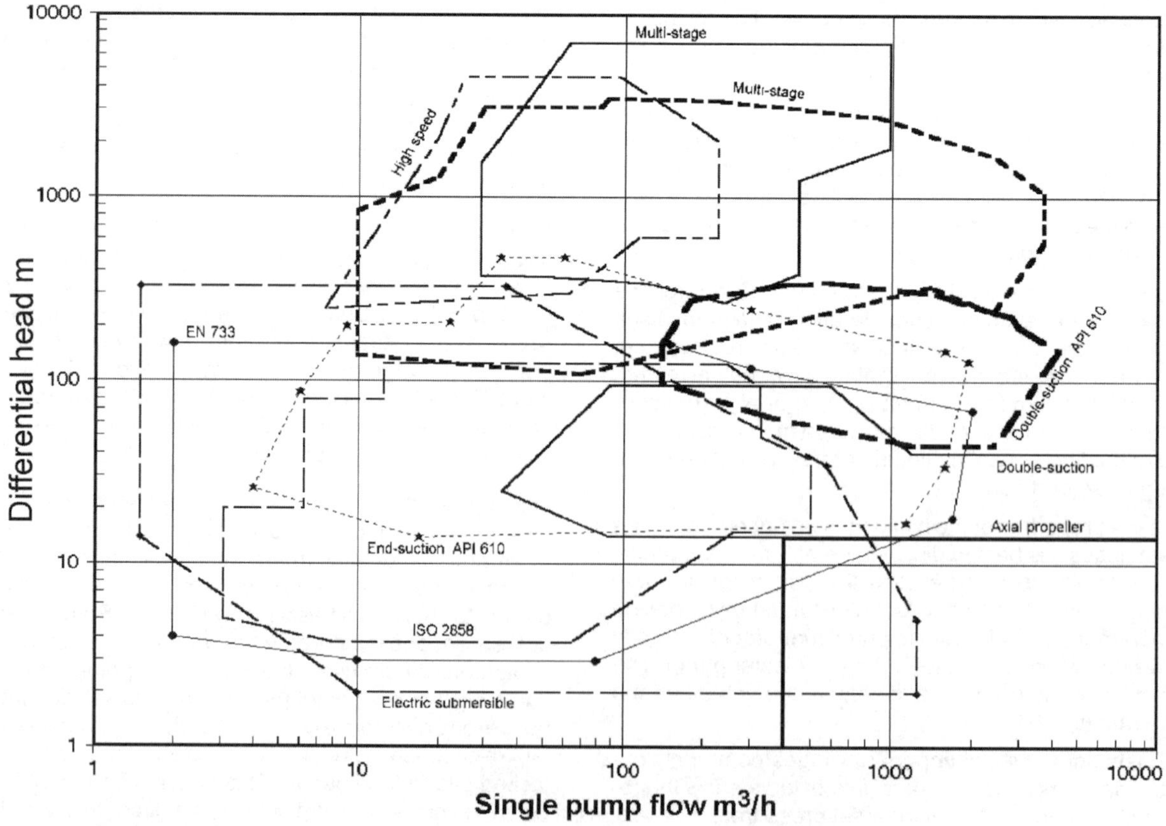

Figure 15.1 Operating ranges of popular rotodynamic pump types

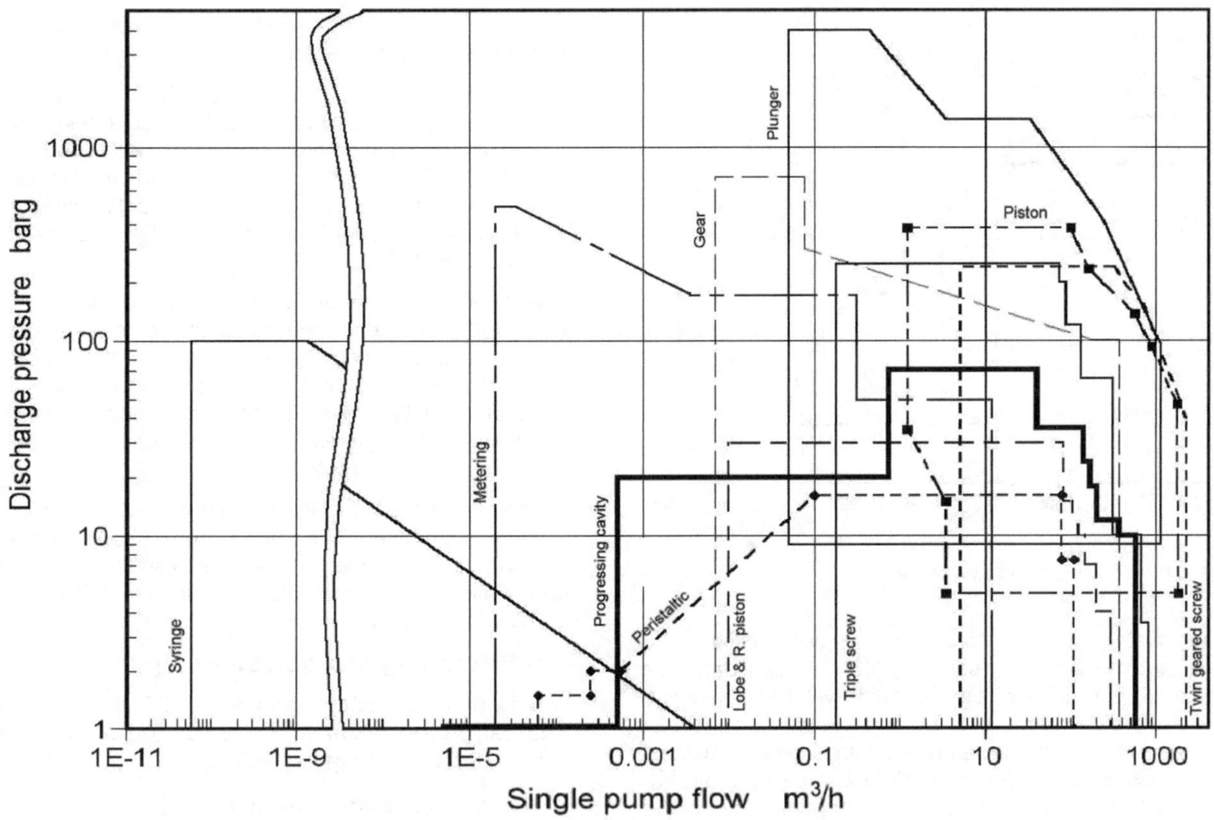

Figure 15.2 Operating ranges of positive displacement pump types

of about 1% by volume. Some rotodynamic pumps, centrifugals for example, can cope with up to 15% without losing prime. Some positive displacement pumps, particularly geared twin screw, can cope with much larger volumes. Dry running must be specified as an operating condition when required. The ability of the pump to handle slugs of gas or liquid must be evaluated.

15.3.2 Pumps for viscous liquids

The viscosity of the liquid can greatly affect the choice of pump. The effect is dependent on the flow and differential head and therefore the size of the pump. The effect of viscosity on the efficiency of centrifugal pumps is shown in Figure 15.3.

The five pumps from A to E are taken from the range outlined by ISO 2858 for end suction centrifugal pumps; F indicates the effects on a very large pump. This data was prepared by using the viscosity correction graphs shown in Figures 4.31 and 4.32. Reduced efficiency leads to increased power consumption which results in higher shaft stresses. It may be the pump shaft stressing which limits viscous operation.

In rotary positive displacement pumps the increase in viscosity reduces slip and improves volumetric efficiency, hence greater flow per revolution. But increased viscosity can also lead to increased fluid friction losses. In triple screw pumps for example, the differential pressure is limited by the viscosity. On one range of pumps, used mainly for lubricating oil duties, the differential pressure capabilities vary from 21/40 bar at 2 cSt to 160 bar at 40 to 200 cSt. The volumetric efficiency improves by about 10% which leads to a slight improvement in mechanical efficiency. Minimum inlet pressure will tend to increase with viscosity.

Depending upon the valve design, reciprocating pumps are not affected very much by medium viscosity operation. An increase in NPIPr will be apparent.

Intelligent economic pump selections for viscous liquid operations should be based on whole-life costing. We know that energy consumption is likely to be more important than initial pump cost. Review pump efficiencies at operating conditions to assess actual energy consumption.

15.3.3 Pumps for highly viscous liquids

As has been stated the problems caused by viscosity are related to pump size. In general, for high viscosities only positive displacement pumps are suitable. The choice is then to select the most suitable type.

At high viscosities the NPIPr and MIP will increase. The speed of the positive displacement pump may need to be reduced. When choosing pumps for highly viscous liquids the design of

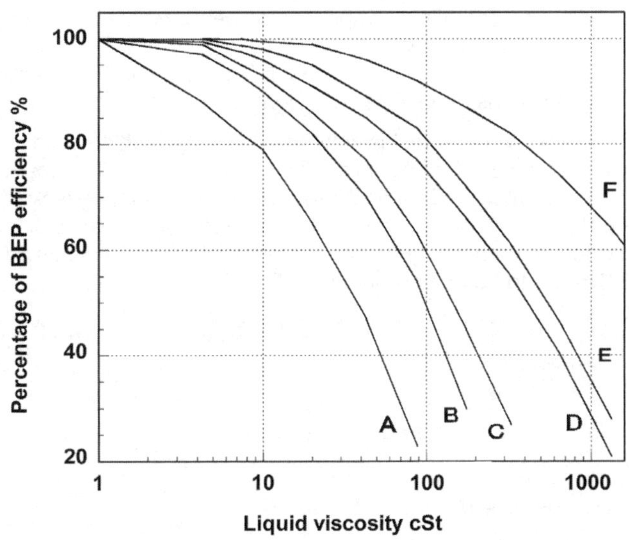

(A) BEP duty 2 m³/h at 11 m differential
(B) BEP duty 11 m³/h at 30 m differential
(C) BEP duty 25 m³/h at 30 m differential
(D) BEP duty 110 m³/h at 45 m differential
(E) BEP duty 225 m³/h at 45 m differential
(F) BEP duty 2500 m³/h at 185 m differential

Figure 15.3 The viscous effects on efficiency of centrifugal pumps

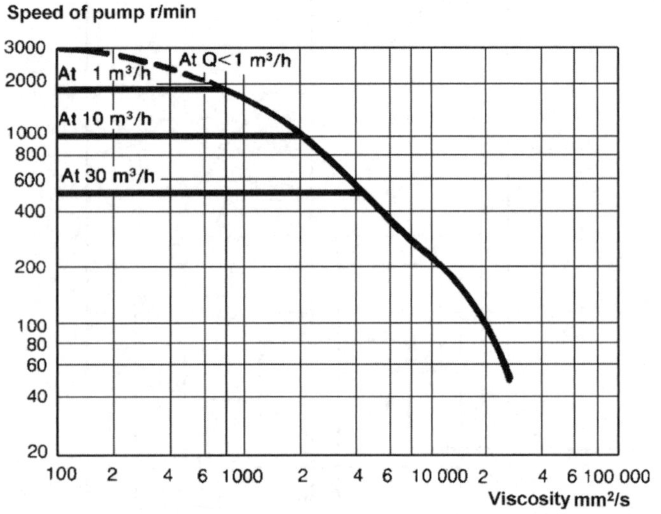

Figure 15.4 Speed of gear pumps for viscous liquids

the suction system must be studied in great detail. NPIPa and MIP must be evaluated very carefully. If the liquid cannot be supplied at the correct flow rate the pump will never function correctly. Severe problems may be experienced on the discharge side of the pump. Large pumps with large connections and low speeds are to be preferred. For extremely viscous liquids, and solids such as wax or lard, feed hoppers and augers may be necessary. Those positive displacement pumps which can handle the maximum pumpable viscosities, approximately 10000 Poise or Stoke, are progressing cavity pumps, lobe and rotary piston pumps and reciprocating plunger/piston pumps.

Figure 15.4 indicates the recommended speed of gear pumps for viscous operation.

Again, ultimate pump choice should be made based on energy consumption and whole-life costs.

15.4 Pumps for liquid-solid mixtures

When pumping solids the distinction must be made between abrasive and non-abrasive solids. The Miller Number test is used for reciprocating pumps. Values below 50 are considered non-abrasive. Review Figure 3.13 in Chapter 3 for the effects of very low concentrations of solids. It is possible to pump solids concentrations up to 70% by weight.

The Miller Number test is conducted using a high chrome cast iron test block. Most pump applications will use a different material. Stauffer, of Escher Wyss, conducted metal tests with a sand-water mixture. The bench mark material was a case-hardened carbon steel. Some interesting data, adjusted for 27Cr iron to be unity is shown in Table 15.4.

Material	Relative abrasion resistance
Hard chrome plate	2.06 / 2.28
Tungsten carbide	1.39 / 4.12
Stellite 6	0.83 / 3.31
Ni-Hard	0.89 / 1.11
27 Cr cast iron	1
13 Cr steel, 441 BHN	0.32
316L stainless steel	0.26
Carbon steel, 195 BHN	0.22
Al bronze	0.13
Ni Al bronze	0.12
Cast iron	0.09

Table 15.4 Relative abrasion resistance to sand in water

There is no consistent method of selecting pumps for solids handling applications. This is because all the pump types which are used rely on different techniques and manufacturers adopt different philosophies for derating and viscous corrections depending upon their experience.

Centrifugal pumps designed specifically to handle solids react differently to centrifugal pumps designed for clean liquids. This problem is due to the different design methods used. The rheology of the mixture can be a problem; assuming Newtonian behaviour can lead to poor pump selections. Rheology has very serious implications for the pipeline design and predicting the pressure drop.

However, rheology is not a problem to all pump types. Plunger and piston pumps can have the plunger/piston diameter optimised at site after commissioning without too much trouble. The purchaser would have to admit to being uncertain about the Q-H characteristic. The solids concentration can be adjusted to modify the mixture rheology. Chemical additives are also available for the same purpose.

Efficiency and predicted life of wearing parts are the essentials for whole-life costing. The manufacturer should be able to advise the life of any rapidly wearing parts and the Mean Time Between Failures. The time taken to replace rapidly wearing parts is also very important.

15.4.1 Pumping non-abrasive solids

Already, at solids concentrations of 0.5%, the service life of a pump can be severely affected. Liquid-solid mixtures which can be considered as non-abrasive are:

- aragonite (calcium carbonate)
- bauxite
- chalk
- clay
- coal
- gilsonite (natural asphalt)
- gypsum (selenite)
- lignite
- limestone
- mud, drilling
- potash
- rutile (titanium dioxide)
- sewage

The critical element of pump selection is the size of the particles and whether any stringy or fibrous material will be present.

One particular application of non-abrasive pumping is liquid coal fuels. Crushed coal is mixed with water and some trace chemicals to produce a liquid-solid mixture which can be handled by piston pumps. The mixture acts as an emulsion and can be stored for reasonable periods without separation. The liquid coal fuel is pumped from the mine direct to the boiler, usually at a power station to produce electricity, and is injected into the boiler. The solid-liquid mixture is not de-watered. The liquid-solid mixture is a complete fuel.

15.4.2 Pumping abrasive solids

Liquid-solid mixtures which can be considered as abrasive are:

- alundum (Al_2O_3)
- copper concentrate
- fly ash
- iron ore
- limonite (Fe_2O_3)
- microsphorite

- magnetite
- phosphate
- pyrites (FeS_2, $CuFeS_2$)
- sand
- shale
- serpentine (magnesium silicates)
- tailings
- tar sand

Again the particle size and concentration is critical in pump selection.

Some users have manufactured their own jet pumps. The basic jet pump principle has been adapted in various ways to reduce the effects of wear from the solids. Mine operators seem to be the most versatile in this respect.

15.4.3 "Gentle" pumping

"Gentle" pumping can be redefined as low shear or delicate handling pumping. The process liquid is not subject to surfaces which travel at high velocities. "Gentle" pumping is used when the solids to be transported by the liquid are to be damaged as little as possible or the liquids are not to be emulsified. For this purpose the "products" can be anything from delicate bacteria cultures to foodstuffs, live fish and root crops. (See Chapter 16, Section 16.10 for an example of the pumping of delicate solids.)

The basis for a systematic choice of pump is seldom available. In the case of rotodynamic pumps, the differential head per stage is a measurement of the stirring of the liquid. Front and back shrouds of an impeller create shear in an almost stationary liquid. In order to avoid damage to the transported particles the passages through the pump should be large. This means that among rotodynamic pumps, submersible non-clogging and submersible pumps with standard motor should be used, and among positive displacement pumps in the first instance lobe and rotary piston pumps are of special interest. For higher pressures plunger and piston pumps can be used.

Figure 15.5 shows a comparison between damage to products caused by two different types of non-clogging pump. The figure demonstrates that in this case the channel impeller type pump is superior to the free-flow pump. On the other hand if bacteria cultures are pumped as active sludge in a purification plant other test results indicate that the free-flow pump is clearly superior to the channel impeller type. These examples illustrate the difficulties in a general assessment of the "gentle handling" property.

15.4.4 Pumping waste water, sewage

Low pressure vertical plunger pumps

The properties of the pump must be formulated in terms of the waste water to be pumped. A modern waste water pump must be designed to pump waste water of the consistency found today and likely to be found during the next 10 years. It is possible to grade the water to some extent. The extreme cases range from very rural areas to large city centres. Whatever the scope of a definition of waste water, the following requirements apply:

- Minimum risk of blockage (non-clogging)
- High degree of operational reliability (mechanical)
- Low running costs (energy consumption and service)
- Simple, quick maintenance in the event of breakdown

In recent years the definition self-cleaning, non-clogging, has been used less and less to describe the ideal pump. This was a pump equipped with a through-flow impeller of channel type. Previously this type of pump could be considered, with the waste water of that time, to have satisfied the special require-

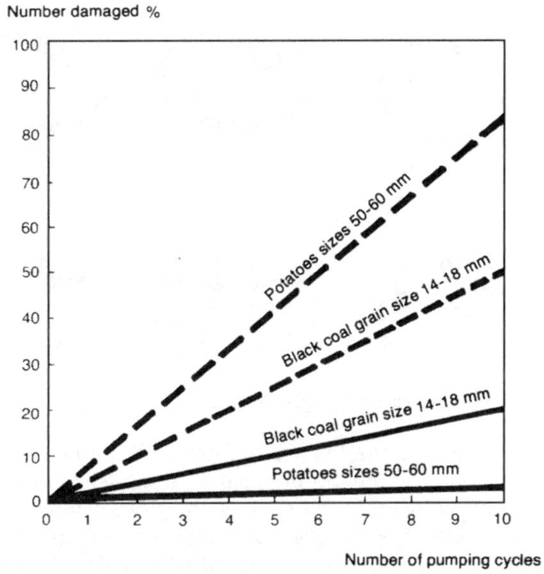

Figure 15.5 Results of pumping tests with a free-flow pump - - - - and a channel impeller pump ———. For black coal the damage criterion was when the size of grain was reduced to less than 12 mm

ments for which it was designed. However, the waste water produced today and still more likely, that to be produced in the future, makes it increasingly difficult for pumps with through-flow impellers, channel impellers, to meet the non-clogging requirements.

The dimensions and design of the channels in relation to the type of contaminant found in the liquid are of critical importance to the non-clogging capability of the pump. The design of the cross-sectional area of a channel impeller varies from rectangular to circular. It is not known for certain whether the design of the channel has any real effect on the non-clogging properties of the impeller. A circular through-flow area may be assumed to offer superior characteristics for the pumping of spherical objects than an impeller with a rectangular cross-section. In the case of textile contaminants a rectangular design may be preferable. Available statistics concerning breakdowns due to the blockage of channel impellers give little guidance to the assessment of different designs of impeller, see Figure 15.6.

Impellers for waste water pumps are usually designed with one or two channels. Apart from reasons connected with the hydraulic design, which can dictate different numbers of channels, there is also a lack of reliable assessment of the respective through-flow capacities of one and two channel impellers. In general it may be said that two outlets from the impeller are better than one, provided that the accessible area of each channel is of an acceptable size. The acceptable size of a channel area is of course dependent on the size and character of the contaminants in the waste water.

Certain guide values have established themselves in purchase specifications. Often the area is indicated as corresponding to a spherical through-flow of 75 mm for small pumps and of 100 or 125 mm for larger pumps. In some cases there is a requirement that the through-flow area should be as large as that of the discharge pipe. In such cases it may be advantageous to have suction and discharge connections the same size. The specification of these values probably has its basis in the fact that the magnitude of the H-Q curve in relation to the motor speeds in current use does not allow scope for larger impellers.

The speed of the impeller has a definitive effect on the flow of the through-flow impeller and thereby on the size of the through-flow area. With the aid of the Laws of Affinity, it can be demonstrated that a given pump at a certain speed will, at double the speed, double its flow, the delivery head will increase four times while the power requirement is increased eight times, neglecting all system effects. For pumps with small flow

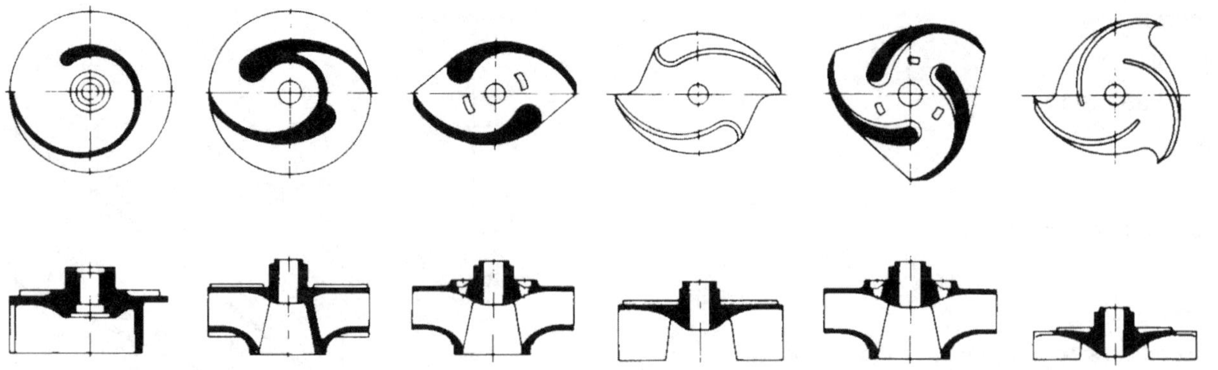

Figure 15.6 Various designs of channel impeller

and moderate pressure the speed must be determined on the basis of the minimum permissible through-flow area. The speed greatly influences the dimensioning of the through-flow area of a through-flow impeller. On the other hand the speed in itself is without significance with regard to the ability of the pump to handle contaminants.

Thanks to the hydraulic design of the interior passages, the flow through a free-flow pump is not dependent on speed. The only function of the impeller is to supply energy to the liquid and it can be dimensioned without regard to the through-flow area. The low symmetrical impeller creates negligible deflection to the shaft. This results in increased service life for shaft seals and bearings.

For normal waste water the free-flow pump may be regarded as non-clogging. The choice of a non-clogging pump represents a compromise between non-clogging properties and energy consumption. On the grounds of experience it may be claimed that free-flow pumps are the best choice up to capacities of about 5 kW and up to about 10 kW at remotely located pumping stations. Travelling costs for cleaning and maintenance greatly affect the optimum. Waste water pump schemes favour submersible high-speed free-flow pumps with capacities of up to 30 kW. In the case of larger pumps, through-flow types will probably continue to dominate the field.

For duties up to 45 m on systems with a large H_{stat} component, a low pressure vertical plunger pump may be better. These pumps run at low speeds and combine "gentle" pumping with solids handling. Built for quick, simple maintenance, the low speed ensures long operating cycles. Positive displacement principles means easy, accurate flow control.

For waste water pumps the manner of installation deserves attention. Submersible pumps are easy to install and replace for maintenance. Dry pit pumps are usual in the case of high power capacities and generally operate at a higher efficiency. Their speed can be easily regulated and they can be more reliably supervised. They are, however, more costly to install. Due to maintenance problems and industrial safety requirements, the current trend is away from submersible to dry pit pumps.

15.5 Check lists for pump purchase specification

Many problems and unnecessary costs can be avoided if the following check lists are used. Irrespective of the method of purchase, "off the shelf" or by enquiry, the following data should be collated and reviewed by all interested parties. Although more than 100 check items are included under more than 15 main groups, they cannot be regarded as entirely exhaustive, because in certain special fields attention must be paid to some additional factors. Above all, check lists should be used when planning and purchasing pumping equipment for unusual, critical or difficult liquids.

For normal applications in a particular branch of industry, e.g. a dairy, or for a certain type of pump, e.g. submersible sewage pumps, a special check list or specification can easily be drawn up. Terms described as "rated" relate to the pump's normal operating conditions; the conditions to be reproduced during testing. Items designated with (pd) apply particularly to positive displacement pumps. It must be remembered in the context of constant speed positive displacement pumps that there can only be one rated flow. If maximum and minimum flows are specified the method of achieving the flow range must also be specified.

The data collated must be accurate. When accuracy is not possible, the data presented must be suitably qualified. There must be no ambiguity. The data collected applies to the pump, not the system. Some system data may be necessary, H-Q characteristic for variable flow systems in some cases, but this must be described as "system data". It is important that the data does not require interpretation. A pump manufacturer or distributor must not be required to "read-between-the-lines" to extract important information.

When users or contractors work on large installations some decisions taken early in the project are forgotten. All relevant data should be reviewed when preparing the final pump unit specification.

Liquid properties

See Table 7.18, Chapter 7.

Solids properties

See Table 7.18, Chapter 7.

Pump capacity

Suction head, pressure — minimum, rated, maximum

Suction temperature — minimum, rated, maximum

Rate of change of suction temperature

Flow — minimum, rated, maximum

Flow duration — constancy

Differential head — at minimum, rated, maximum flow

Suction and discharge pressure — at minimum, rated, maximum flow (pd)

Suction and discharge conditions at starting

Differential head for filling pipe system with consideration to possible syphon effects

Differential head affected by possible starting inertia of liquid

Discharge pressures for relief valve operation (pd)

Pump characteristic — rotodynamic or positive displacement

NPSHa, NPIPa — minimum, rated, maximum

Suction system — self priming, evacuation time

Flow regulation — throttling,
by-pass,

on-off,
load-unload,
adjustable guide vanes,
infinitely variable speed,
discrete speed steps,
manual or automatic control.
Needs system characteristic !
Possibility of liquid reverse flow ?
Also review Table 7.18, Chapter 7

Pump, mechanical requirements

Shaft deflection at seals at rated conditions
Service life of bearings at rated conditions
Suction pressure rating
Discharge pressure rating
Suction hydrotest pressure
Discharge hydrotest pressure
Suction connection allowable forces and moments
Discharge connection allowable forces and moments
Thermal growth of pump
Mean time between failures
Mean time to restore
Starting torque requirements
Permissible number of starts per hour
Permissible number of starts per year
Internal clearances
Changes in efficiency resulting from clearance variations
Dry running capabilities
Heat tracing or cooling
Warm-up time or cool down time
Hot stand-by
Extent of inspection without removing pipework
Extent of maintenance without removing from pipework
Access for cleaning out residue, if required
Effect of pump running backwards
Balancing
Vibration
Noise level
Possibility of modifying pump for duty changes; increased impeller, plunger, piston diameter, de-staging

Shaft seal

See Table 8.1, Figures 8.30, 8.31 and 8.37, Chapter 8.

Material

See Table 7.18, Chapter 7.

Pipes and pipe connections

Type of connection — screwed, flanged, special
Special surface finish on flanges
Pressure ratings
Pipe movement and thermal strain
Flexible pipe connections
Pipe spools adjacent to pump
Suction pipe routing (NPSHa/NPIPa)
Location of pump process connections
Minimum distance from pump to bends or valves
Location of pump auxiliary service connections
Isolating valves
Non-return valves
Relief valve (pd)
Tapping points for measuring pressure and/or temperature
Possibility of water hammer during normal start
Possibility of water hammer during normal stop
Possibility of water hammer if power failure
Allowable axial shaking forces for pressure pulsations (pd)
Insulation against heat and/or cold
Heat tracing
Design and facilities for mechanical cleaning or inspection by "pig"
Will pump manufacturer review and advise on pipework (pd)?
Acoustic and mechanical analysis of pipework (pd)

Drivers

Electric motor:

type,
speed(s),
AC / DC,
speed regulation,
direction of rotation,
voltage, phases, frequency,
method of starting,
starting torque characteristics,
number of starts per hour,
area classification,
certification,
type of physical protection,
cooling system,
power, minimum, rated, maximum,
efficiency and power factor,
bearings and lubrication,
space heaters,
location of terminal box,
electrical protection/monitoring equipment,
thermal growth,
balancing,
vibration,
noise level.

See Chapter 10.

Engine:

speed,
speed/torque characteristic,
naturally aspirated, turbo-charged,
direction of rotation,
governor,
fuel, dual fuel, calorific value,
fuel consumption,
efficiency,
method of staring,
cooling system,
time to warm-up,
controls and instruments,
auxiliary equipment,
waste heat boiler,

hazardous area modifications or mount in safe area,
thermal growth,
balancing,
vibration,
noise level.

See Chapter 6, Sections 6.9.2 and the end of Chapter 10.

Turbine:

speed,
speed/torque characteristic,
direction of rotation,
condensing/pass-out,
steam conditions,
governor,
over-speed trip,
steam consumption,
efficiency,
bearings and lubrication,
barring system,
time to warm-up,
steam valves,
controls and instrumentation,
thermal growth,
balancing,
vibration,
noise level.

See Chapter 6, Section 6.9.2 and the end of Chapter 10.

Drive train

Couplings:

service factor,
maximum speed,
spacer length,
long spacer for fire wall,
capacity for misalignment,
mounting,
balancing,
lubrication.

See Chapter 9.

Gearbox:

service factor,
service life of gears and bearings at rated conditions,
exact ratio,
efficiency,
cooling,
bearings and lubrication,
minimum speed,
thermal growth,
balancing,
vibration,
noise level.

See Section 11.4, Chapter 11.

Vee-belts:

service factor,
ratio,
efficiency,
cooling,
pulley balancing,
noise level.

See Section 11.3, Chapter 11.

Ancillary equipment

Baseplate, skid, trailer, lifting frame — see Chapter 11, Section 11.2.1

Relief valves — see Section 11.5 (pd)

Accumulators — see Section 11.7

Pulsation dampers — see Section 11.8 (pd)

Instrumentation — see Section 11.9

Environment

Stand-by facilities

Electricity supplies

Auxiliary services supplies

Industrial safety standards

Overall pump package noise level

Allowable transmitted vibration level

Leakage in the form of liquid

Leakage in the form of vapour

Safety with respect to hazardous gasses

Radiation of heat

External condensation on cold surfaces, possible formation of ice on pump

Permissible gas concentration in premises (ppm) and necessary ventilation

Atmospheric pollution

Biological attack

Altitude

Weather variations

Indoor, outdoor, onshore, offshore, sheltered, sunshade, attended, unattended

Erection and installation

Horizontal or vertical shaft

Submerged pump

Space and transport routes for erection and maintenance

Lifting facilities for installation

Lifting facilities for maintenance

Foundation; concrete, structural steelwork

Dynamic loads on foundation

Static loads imposed by pipework

Space for future developments

Drainage for safe leakage

Drainage for hazardous leakage

Drainage of the pump contents

Drainage of lubricants and cooling liquid

Vents for hazardous vapour

Vents for steam

Maintenance

Parts lists and cross-sectional drawings

Lubrication schedule

Routine maintenance instructions

Repair instructions

Trouble-shooting guide for all equipment

List of rapidly wearing parts with predicted life

Quotation for commissioning spares

Quotation for 1 years operations spares

Quotation for 2 years operations spares

Quotation for insurance spares

Quotation for special tools

Pump curves

Locations of measuring points for preventive maintenance (vibration, temperature, electric current, etc.)

Hours-run meter

Staff training

Location of manufacturer's nearest service centre

Location of manufacturer's nearest spares stockist

NOTE: Manufacturers will not supply drawings of spare parts. Drawings contain proprietary information which is confidential. Using spare parts manufactured by others will invalidate any guarantee or warranty. The pump manufacturer has no responsibility for the damage caused or production lost due to "pirate" parts.

Quality Assurance

Pumps designed and built to a standard

Pumps manufactured to a standard

Fully assembled pump subjected to a running test

Fully assembled pump subjected to a pressure test

Typical batch size of mass-produced pumps

Number of pumps in a batch tested

Tolerance band on published pump performance data

Fully assembled pump unit tested

Which standard is used for testing?

What quality assurance documentation supplied?

Purchase conditions

Quotation validity

Length of guarantee for workmanship and materials

Delivery time

Point of delivery

Price

Currency

Drawings

Manuals

Certification

Erection at site

Commissioning

NOTE: Manufacturers are unlikely to give a guarantee regarding the performance of the pump at site or how the performance will degrade with time. The pump performance and life are greatly affected by the quality of the installation and particularly by the pipework design. Pump manufacturers are not normally involved in the foundation design or the pipework design. Bearing life can be shortened dramatically by poor foundations, misalignment and pipe strains. Pump performance can be severely impaired by poor pipework design. The pump manufacturer shows how the pump can operate by testing in the factory. The factory tests should be conducted as close to the site operating conditions as possible. A site simulation test may be possible. Except in the cases of fully packaged equipment such as portable high pressure cleaners, factory testing cannot reproduce site conditions. Problems such as acoustic resonance or pipework vibration can only be detected by computer simulation. The pump manufacturer should be prepared to work at site to analyse and isolate problems. The pump user must be prepared to pay the pump manufacturer for solving problems outside his sphere of control.

15.6 Purchasing

Chapter 4 deals with the criteria and standards which are of prime importance in connection with specification. Chapter 12 deals with the implementation of testing and quality controls. These are concerned, amongst other things, with guarantees of performance and quality. Criteria for service life and reliability are lacking however. Over and above the technical requirements, suitable general terms of delivery should be drawn up.

In the processing industry and at refineries API 610, 674, 675 and 676 are usually applied as requirement criteria as they specify in detail how a pump should be designed, manufactured and tested. Reference may be made where applicable to API standards for guidance for pumps to meet these requirements. However, it must be remembered that an inspector looking at a pump cannot certify that it will be capable of operating without interruption for three years.

The purchase documents should contain references to legislative regulations and requirements affecting design, to the properties of the liquid as well as to operating conditions and function of the pump in order that the manufacturer may be able to determine the necessary material and the design of the pump. There is no point in referencing obscure regulations. Manufacturers do not have easy access to all standards and legislation. If particular, local regulations must be observed, a copy should be supplied with the enquiry.

It is important that the specifications detail the pump operations. Safety margins should be shown separately as such.

Remember the site cost-saving edict:

"Trouble-shooting is making the pump do what you specify. Making the pump do what you want is up-grading !"

Pump applications & solutions 16

16.1 Modern pumps for hydrocarbon-based and synthetic heat-transfer oils

Introduction

As soon as the temperature in a heat-transfer system becomes too high for water, a plant operator must choose between hydrocarbon-based or synthetic oils as the heat-transfer liquid. This usually happens once process temperatures in excess of 200 °C become necessary. Until recently, hydrocarbon-based heat-transfer oils were often the first choice simply because they cost less. But synthetic oils are now beginning to assume a more prominent role.

The advantages of synthetic oils are appreciated over the long term. Synthetic oils pay for themselves in just a few years, with the exact time depending on the application, the system, and of course the temperature. This is especially important since as temperature rises, a greater amount of decomposition products with varying boiling point properties will form in the oil. These contaminants influence the properties of the heat-transfer liquid itself – usually in a negative way. The consequences extend beyond just having to replace the thermal transfer fill more often. More serious are deposits that form in the system, since these are difficult or impossible to remove. An operator's choice of oil will determine whether new oil is needed every year, every three years, or longer. In fact, when modern synthetic oils are used, systems with process temperatures up to about 300 °C can stay in service for their entire service life with one fill. So clearly, synthetic oils are worth the significantly higher price.

Furthermore, once process temperatures rise above 300 °C, only high quality synthetic oils will be appropriate. At these temperatures, service life will be dependant on the chemistry of the thermal transfer liquid; one has to look at more than just the maximum permissible process temperature to make a judgment. Also important, is the fact that even with these products, any increase in temperature will affect service life. Ten degrees Centigrade one way or the other will influence how long the oils can stay in service: ten degrees higher usually doubles the product's rate of decomposition. In any case, more costly maintenance of the system will be required. However, when compared to hydrocarbon-based heat-transfer oils, the amount of maintenance associated with synthetic oils is significantly lower.

Pump performance

The quality of the oil also directly influences pump performance. If a large number of low-boiling point compounds form pockets of vapour in the pump, it will lose prime and stall. Under these conditions, the pumps no longer run at design conditions. They are subjected to higher stress and have lower pumping performance. This becomes a vicious circle where low-boiling point compounds reduce pump flow, which increases stress on heat exchanger heating surfaces (higher metal temperatures), accelerating decomposition of the thermal transfer oil (higher localised oil temperatures), and ultimately forming more vapour bubbles and surface fouling. Nevertheless, synthetic heat transfer oils do place special demands on pumps. Compared to hydrocarbon-based oils, their lubrication qualities are much poorer.

Pump design

Hot-oil pumps have been specially designed that are specifically adapted to the properties of synthetic oils. Two factors are critical: poorer lubrication and often lower viscosity, which are essential because these properties improve the transfer of heat. These two factors make traditional pump designs inadequate. Starting the pumps is difficult, since low viscosity liquids do not generate an adequate lubricating film at low surface speeds, and high wear is to be expected – and therefore higher maintenance costs – when they are in service.

Seals and bearings

In particular, the presence of low-boiling point compounds in the synthetic oil places great stress on the mechanical seals. Special design solutions are needed to compensate for this stress. A large sealing area and the ability to collect vapours in a special section of the pump (where they can be easily removed) are critical for extending the pump's service life and increasing its reliability. When used for hot water applications, dissolved gases which evolve during heating can be collected here. In addition, the bearing and the seal must be adapted precisely to each other, the low viscosity, and the overall application.

Carbon bearings combined with non-relieved mechanical seals and silicon carbide bearings combined with relieved seals have proved invaluable. These combinations also make the pumps insensitive to sludge and other types of solid contamination in the oil. Finally, quench liquid buffers further increase the service life of the mechanical seal and provide greater security against leaks. A quench further reduces temperature at the mechanical seals and eliminates the possibility that fluid leaking out of the seal will oxidise and cause damage to the seal. See Figure 16.1.

This principle also makes it possible to use non-hermetic pumps with nasty liquids (hazardous, flammable, toxic and noxious) since the use of a non-critical quench liquid greatly thins the pumped liquid in a way that is comparable to the effects of a corresponding double-action mechanical seal. From the operation and efficiency point of view, non-hermetic pumps have the advantage that they consume less power since there will be lower frictional losses from the pump rotating parts and no eddy current losses during the transmission of power.

The pump units

The low cost pump units handle both low viscosity, synthetic heat-transfer oils up to 350 °C as well as hot water up to 207 °C with the identical material configuration. They make traditional water pumps with special seals obsolete.

The pump units employ a flexible modular system. Depending on the specific application, they are available in mono-block, bearing bracket and in-line versions. They are economical to produce and simple to maintain and a uniform combination of

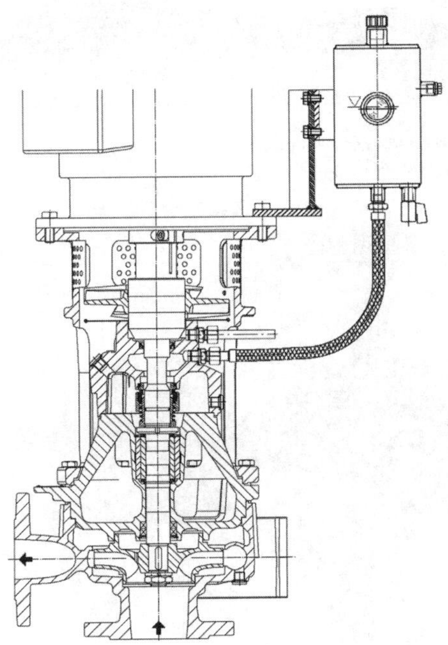

Figure 16.1 A vertical mono-block pump with quench tank
Courtesy of Allweiler AG

materials - cast iron, nodular cast iron and 13% chrome - covers all pumped liquids.

There are 12 possible combinations – ready for the most diverse applications – using a few basic components: the three design types, two casing designs, PN 16 with dimensions according to EN 733; and PN 25 with dimensions according to EN ISO 22858) and two bearing/seal combinations:

Silicon carbide bearing combined with a balanced mechanical seal for hot water up to 207 °C

Carbon bearing combined with a non-balanced mechanical seal for hot water up to 183 °C and most applications involving heat-transfer oils.

Temperature considerations

When synthetic oils are used, the application usually involves a high-temperature system where large temperature differences are a major factor. These make expansion movements within the pump unavoidable due to external forces that affect the pumps through the system and the piping. That is why pumps for synthetic oils in particular must be mechanically very durable and rigid, but without transferring too much heat to the bearing and shaft seal.

The range of pumps can handle this situation, since the process liquid lubricated bearing is no longer part of the bearing housing, but instead is now part of the casing cover. It is therefore possible to move the separating interface between the casing cover and the bearing housing even further toward the drive end. At the same time, the diameter of the sealing face/spigot and therefore the centre between the two parts has been enlarged. The result is an optimised, skeleton-frame-type structure whose mechanical stiffness equals that of bearing housings found on traditional centrifugal pumps. Making the inboard bearing part of the casing cover also increases the length of the thermal barrier protecting the mechanical seal and outboard bearing. In this way, a low temperature level is achieved without the need for cooling ribs. The position of the interface also enables additional savings, since it is possible to forgo a second 'seal housing' for the mechanical seal. See Figures 16.2 and 16.3.

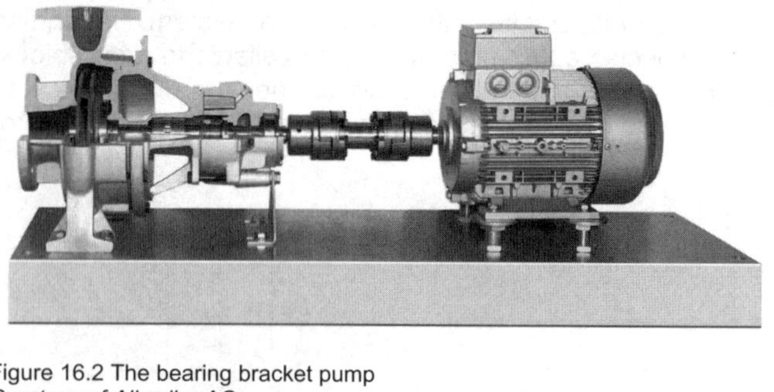

Figure 16.2 The bearing bracket pump
Courtesy of Allweiler AG

Figure 16.3 Section through the pump
Courtesy of Allweiler AG

Since the plain bearing is located behind the heat barrier where temperature is lower, there is no possibility of dry-running in the bearing. In designs where the bearing is located directly behind the impeller, by contrast, dry-running can occur as soon as the pumped liquid gets close to its vapour point – even when the pump is filled! Thanks to its lower temperature, liquid in the plain bearing of the pump is always sufficiently pressurised. Furthermore, optimisation calculations have shown that, assuming the same bearing geometry behind the heat barrier, this design provides an ideal lubricating film thickness in the bearing, maximising its effectiveness. The higher viscosity of the liquid more than compensates for the less ideal impeller cantilever length.

The carbon bearing

Practical tests have shown that the specific configuration and design of the carbon bearing generally result in a stable lubrication film. As long as the pumped liquid is free of particularly abrasive, fine-grain constituents, the carbon bearings will be subject to virtually no wear. Even normal contamination like corroded flakes from the pipes of hot water systems are not a problem, since a safety throttle keeps most of it away from the bearing. The throttle is located in front of the heat barrier and directly behind the impeller where it protects particularly against elevated leakage if the main seal fails. As a result, the bearing will achieve the stated long service times even when the pump is operated close to the minimum volume limit. The two bearing types – carbon and silicon carbide – are mutually exchangeable by changing the inner unit, which consists of the shaft, bearing and shaft seal.

The mechanical seal has been moved away from the impeller in the sealing area. The large-volume seal housing is designed to strip gas bubbles from the mechanical seal, collect the gas/vapour, and vent the pump. This protection against dry-running greatly lengthens the service life of the seal when pumping liquids with low lubricity that also tend to emit gas, even when the pump is in a vertical position. The bearing housing can also be used for the mono-block and in-line pumps, which are often installed vertically without loss of functionality. See Figures 16.4 and 16.5.

Bearing geometry

For heat transfer pumps the bearings are not pressed in, but instead allowed to tilt. This has two benefits:

First, if a bearing bracket is not aligned and centred precisely after maintenance, for example, it will automatically find its proper position,

Second, the bearing will move naturally if there is pressure from the shaft. This avoids point strains and lengthens service life.

A spacer coupling utilising two Cardan joints is used for all high-temperature applications and especially for systems that

Figure 16.4 Section through the mono-block pump
Courtesy of Allweiler AG

Figure 16.5 Section through the in-line pump
Courtesy of Allweiler AG

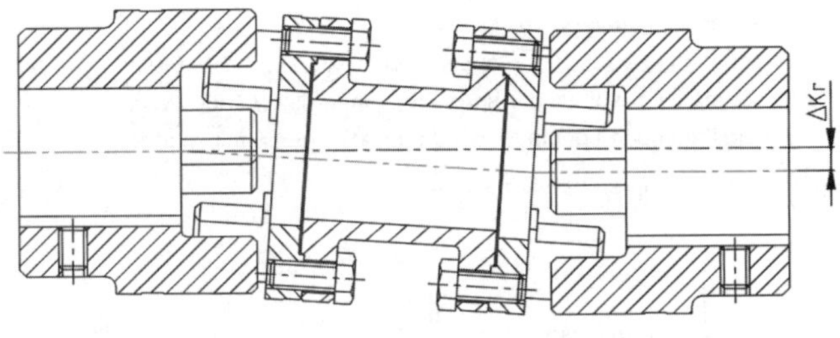

Figure 16.6 Spacer coupling
Courtesy of Allweiler AG

exhibit large thermal growths. These couplings safely accommodate even very large misalignments. See Figure 16.6.

Special heat-transfer pumps also provide a high level of investment security. If system is to be converted to modern synthetic oils, existing pumps can be adapted to the elevated temperature requirements by replacing the insert units. Such versatility from using a single pump type in a variety of heat-transfer liquids greatly simplifies processes for the operator and plant designer.

Application courtesy of Allweiler AG

16.2 Pumping fresh water to a high reservoir

Operating conditions

The clean water collected in a low reservoir, is to be pumped up to a high reservoir. The high and low water levels in the two reservoirs and the altitude of the site for the location of the pump are shown in Figure 16.7. The pipe linking the reservoirs is long, about 20 km, and during much of the day water is drawn from the pipe by consumers with consequent loss of pressure.

The daily consumption of water varies considerably. Furthermore, other water resources are operated simultaneously. For these reasons and also with regard to economy it must be possible to adjust the pumping from the water works in question from between about 50 to 115 litres per second. In cases of emergency, it should be able to pump about 170 litres per second.

System curves

The system curves are shown in Figure 16.8. It proved possible to install the pumps 1 metre below the lowest water level in the low reservoir so that a positive head of 1 to 3 metres was obtained on the suction side. At the high reservoir the difference in height between high water and low water is 5 metres. The

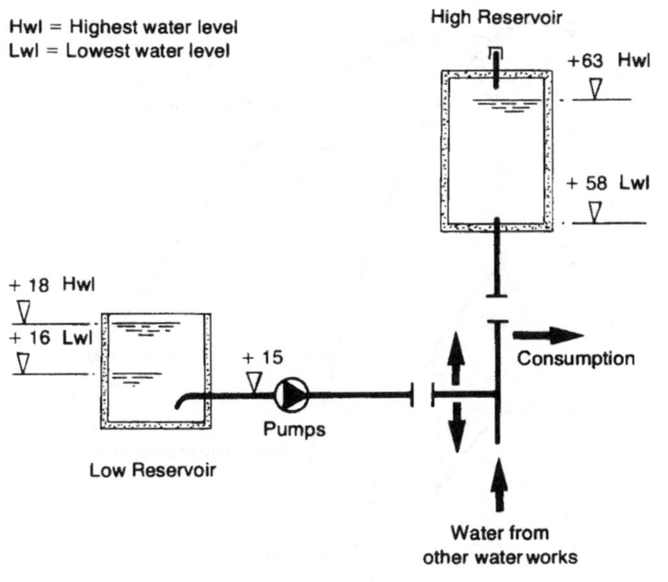

Figure 16.7 The system

static delivery head in the system therefore varies between 47 m (+63 - +16) and 40 m (+58 - +18).

The losses in the pipe system are shown by curves A and B in Figure 16.8. From time to time the high reservoir may be disconnected for cleaning purposes or as the result of a breakdown. Pumping must then be directed to another higher reservoir and this results in an increase of static losses by about 3 m. Curve C illustrates the upper limit of the head requirement in this situation. The normal operating zone for the pumps is clearly indicated in Figure 16.8. The highest operating efficiency should therefore lie within this zone.

Choice of pumps

On the basis of the great variations in flow demand, and respective head requirements found in this case, a combination of pumps was deemed necessary. However, it was scarcely possible to find space for more than three. High reliability of operation required that a stand-by unit should, at least to some extent be available. The necessary delivery heads implied that virtually only centrifugal pumps could be considered. From the running and maintenance points of view it was desirable that all the pumping units should be identical.

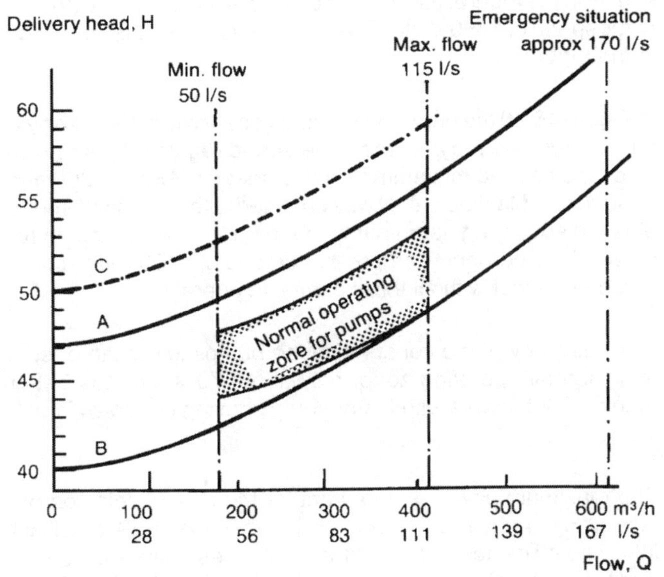

Figure 16.8 System curves

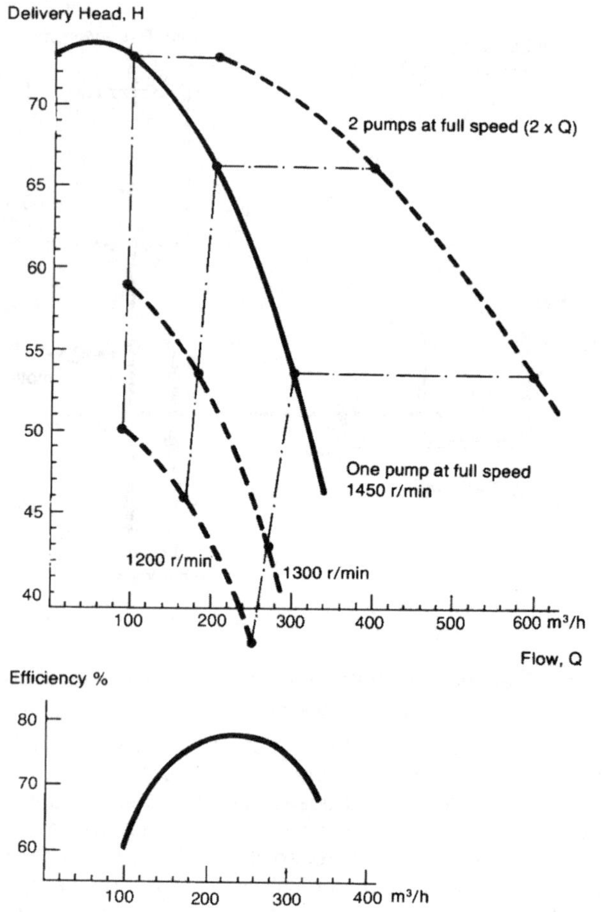

Figure 16.9 Pump curves

For the best possible reliability, company policy was to use pumps running at 4 pole motor speeds. Standard, double suction axially-split pumps would only fit the duty when running at 2 pole speeds. The use of three pumps would mean a motor power of 55 to 75 kW depending upon the pump efficiency. It had been decided to use a variable speed DC motor on one pump for supply adaptation. Mass-produced DC motors of the size necessary only ran up to 2200 rpm. Eventually a two stage axially-split pump was found; the first stage being double suction for good suction performance.

The pump curve is shown in Figure 16.9. Curves calculated according to the Affinity Laws for 1200 and 1300 r/min are also shown in the Figure, as well as a combined curve for two pumps at full speed of 1450 r/min. The efficiency curve is also included in the Figure.

In Figure 16.10 the pump curves have been overlaid on the system curves. As may be seen, the speed regulated pump provides the desired minimum flow at a speed between 1200 and 1300 r/min. Maximum flow was obtained with a constant speed plus the speed regulated pump. One unit was then kept in reserve. An emergency temporary demand of 170 litres per second can be met with all three pumps in operation.

The efficiency of the constant speed pumps is not the best in their normal operating zone, only about 70%, but it is better than that obtainable with faster pumps having cut-down impellers.

The maximum power requirement for the pumps selected occurs at H = 46 m and Q = 340 m³/h, see Figure 16.9, and is 61 kW. The efficiency of the electric motor is estimated to be 90% and the pumping units are supplied with 75 kW standard 4 pole squirrel cage motors.

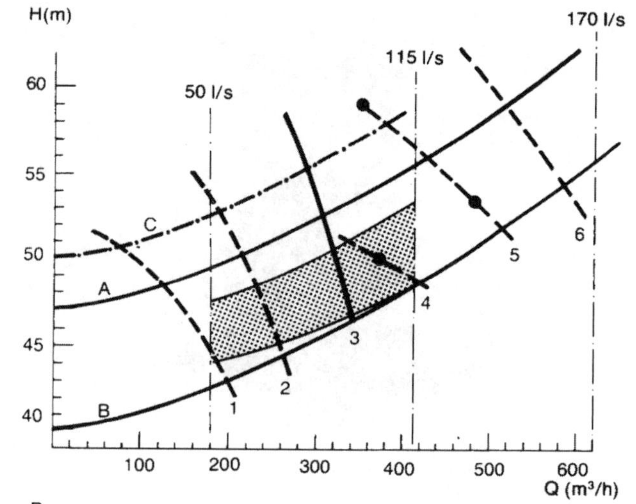

Pump curves:
1: Speed regulated pump, 1200 r/min
2: Speed regulated pump, 1450 r/min (full speed)
3: Speed regulated pump, 1300 r/min
4: Constant speed pump + speed regulated pump 1200 r/min
5: Constant speed pump + speed regulated pump 1300 r/min
6: Constant speed pump + speed regulated pump full speed

Figure 16.10 Pump and system curves

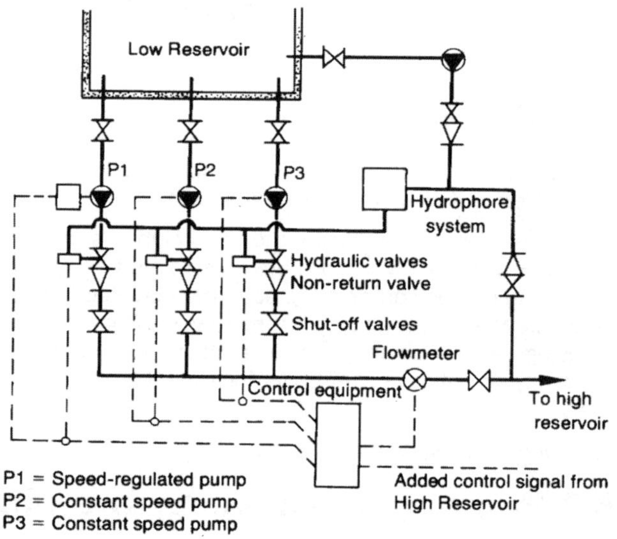

Figure 16.11 Schematic arrangement of pump installation

Installation of pumps

Figure 16.11 indicates schematically how the pumps were installed. The pumps work in parallel with the speed regulated pump as the lead unit. When this reaches full speed, one of the two constant speed pumps is started; the constant speed pumps operate alternately, controlled by switched sequencing, and the speed regulated pump reduces to its minimum speed. In the case of increasing flow requirements it increases speed again. The procedure is the reverse in the case of decreasing water requirements.

Control of the variable speed as well as the starting and stopping of the constant speed pumps is regulated by a theoretical daily demand which in turn is constantly corrected by input from the water level in the high reservoir.

The delivery pipes of the pumps are fitted with hydraulically controlled, slow operating valves to reduce water hammer when starting and stopping. Pumps are therefore always started with the valve closed and when stopping the valve closes first. For safety reasons the pumps are also supplied with non-return valves.

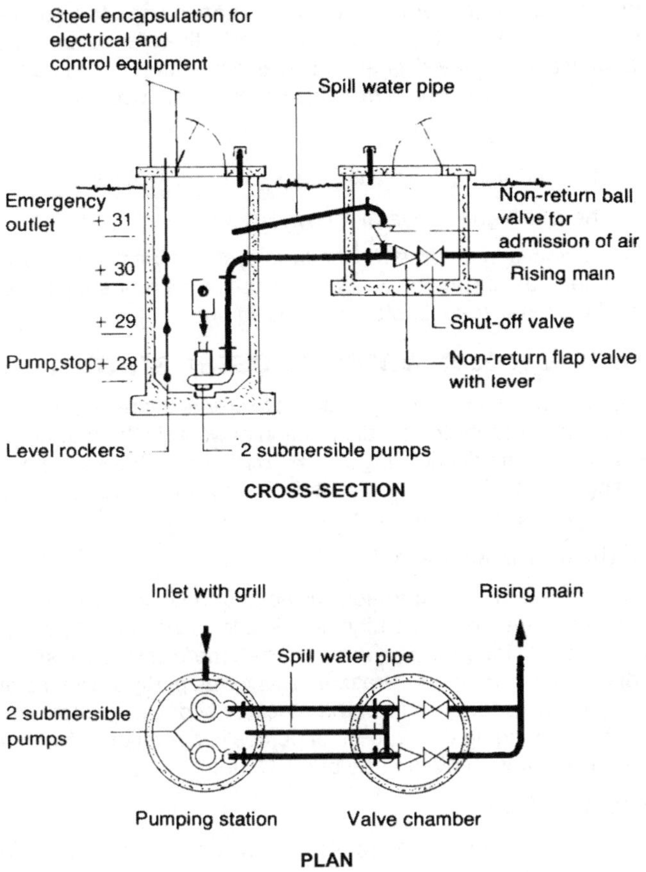

Figure 16.12 Arrangement of a small sewage pumping station

16.3 A small sewage pumping station

Introduction

The pumping station serves a small housing estate. The supply of sewage during the initial period is a minimum of approximately 1 litre/s and a maximum of approximately 5 litre/s. It is estimated that gradually the minimum flow will increase to 2 litre/s and the maximum will increase to about 8 litre/s. The positive elevations at the pumping station are shown in Figure 16.12. The rising main from the station is about 700 m long and discharges into a well at level +55 m. The diameter of the pipe is 150 mm and the volume is thus about 12.4 m^3.

As may be seen in the Figure, pump stop occurs at + 28 m. The station has an emergency outlet, or overflow, at + 31 m. The static delivery head will be maximum 55-28 = 27 m and minimum 55-31 = 24 m. The stop level of the pumps has been placed as low as possible, approximately 250 mm above the floor in the pump sump, in order to avoid the formation of sludge deposits.

The station consists of two units, the pumping station itself and a valve chamber. The pumping station is constructed of pre-cast concrete rings with rubber joints. The internal diameter is 2 m and the volume per metre of height is about 3.14 m^3. The station is equipped with two identical submersible centrifugal pumps with quick release pipe couplings to allow quick pump removal. The pumps are started and stopped via level rockers, one of which is also used to transmit a signal to the control centre when the water level in the station is too high. Level rockers are proprietary level switches which change attitude when floating on the liquid surface or are submerged.

In the valve chamber, which is also constructed of pre-cast concrete rings, there is a system of valves which ensures reverse-flushing of the pumps at every stop to prevent clogging.

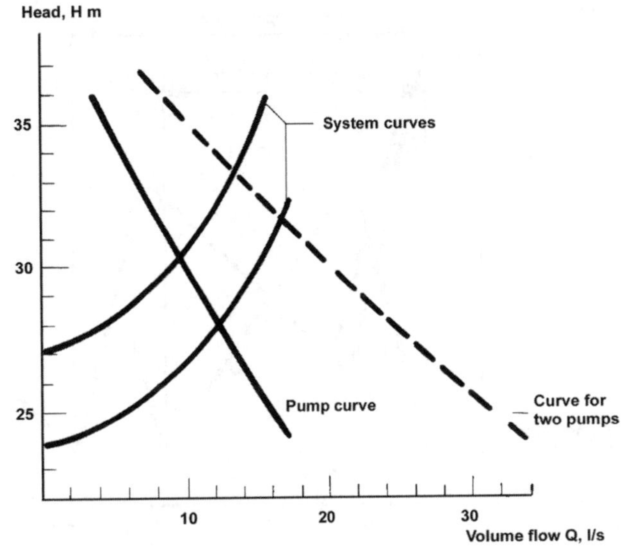

Figure 16.13 Pump and system curves

During pump shut-down, air is introduced into the system via the non-return ball valve, the ball should have a specific gravity of 0.9 to function efficiently and the water in the rising main in front of the flap non-return valve flows back to the pumping station and flushes the pump clean. The amount which flows back in this particular case is about 90 litres. Thus for a short time the pump rotates backwards.

When the pump starts up it is impossible to prevent a slight jet of liquid being expelled through the non-return ball valve as this cannot be closed instantaneously. A spill water pipe returns the liquid to the pumping station.

Pump and system curves

These are illustrated in Figure 16.13. As stated, the static rising delivery head can vary between 27 m and 24 m. The losses in the rising main and its valves are shown by the system curves. The size of pump chosen has flows from 9 to 12 litre/s, depending on the differential head. Figure 16.13 also shows the curve for both pumps in operation. The efficiency of the pumps in the operating zone is about 70%. The size of the motor is 9.6 kW, the normal output being about 6 kW, corresponding to a current of about 12 A. The pumps are started direct-on-line.

Operating conditions

As mentioned, the flow of sewage to the station in the initial period is 1 to 5 litre/s. The volume of the pump sump is about 3140 litre/metre height. Pumping should occur at least once every

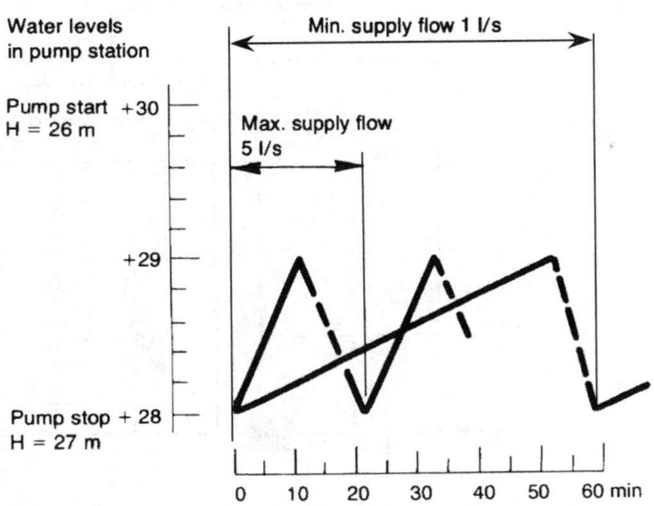

Figure 16.14 Operating sequence with 1 m working level

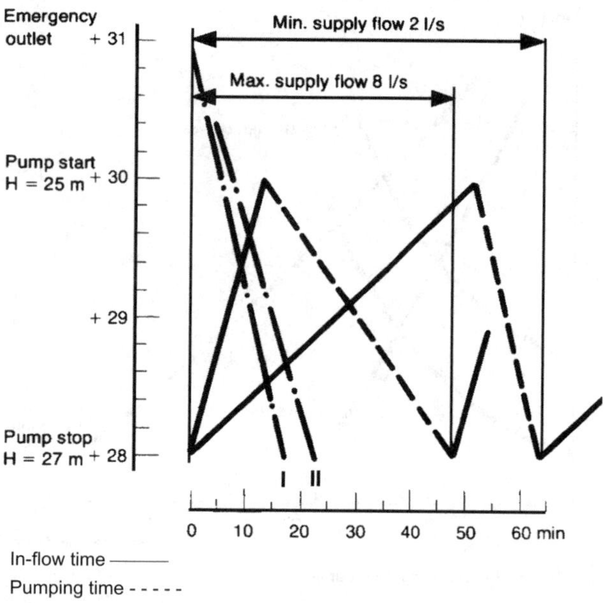

Figure 16.15 Operating sequence with 2 m working level

hour to ensure water replacement. The start level of the first pump has therefore been provisionally adjusted to level +29 m, see Figure 16.12, i.e. when 1 m of the sump is utilised. The operating sequences will then be as shown in Figure 16.14. At minimum supply flow the sump used is filled in about 52 minutes, 3140/(1 · 60) and pumping takes place for about 6 minutes. Output is approximately 10 litre/s at H_{mean} = 26.5 m, see Figure 16.14, minus inflow, i.e. $3140/((10^{-1}) \cdot 60)$.

As already mentioned, the volume of the rising main is 12.4 m^3 and after 4 hours the water in the pipe has been replaced. At maximum inflow the sump level rises sufficiently in 10.5 minutes and pumping out takes equally long, pump start occurring every 21 minutes. The pump motor is rated at 15 starts per hour.

Gradually, as the minimum and maximum rates of flow are increased, the sump volume will also be increased. Figure 16.15 illustrates operating conditions at full supply flow and when 2 m from the sump, a volume = 2 x 3140, is used. This Figure also includes two curves, I and II, showing the time required for the emptying of the station from the emergency outlet level, e.g. after a power failure, with one pump in operation and minimum in-flow, I, and both pumps in operation and maximum inflow, II.

In calculating these times no allowance has been made for the reverse-flushing flow volumes. These, however, would be a marginal effect only.

As may be seen from this, normally only one pump is required to operate. Changeover takes place at every pump start so that both units are used. The second pump therefore acts as a standby and is started when the level in the sump exceeds the permissible value, +30.5 m. This situation can occur when:

- The lead pump fails
- The lead pump is partially blocked
- There is high rain water inflow

The station is monitored by an alarm system via telemetry and is inspected only once a week, when the sump is flushed clean and functional checks, etc., are carried out.

16.4 A fresh water booster station

The function of the booster station is to create the necessary increase in pressure for the distribution of water in the higher areas of an urban district. Figure 16.16 illustrates the schematic arrangement of the booster station and its control system, which has been in operation for 20 years or so.

Method of operation

Normally, the pump operation will be automatically adjusted to water consumption in the high area in such a way that the pressure in the distribution pipe is maintained more or less constant. Adjustment to the actual pressure is affected partly by means of changes in speed of one of the electrically driven pumps and partly by changing the number of pumps in operation. Variable speed is obtained by the use of a DC motor.

Pump curves

The effects of running the pumps in combination are shown in Figure 16.17.

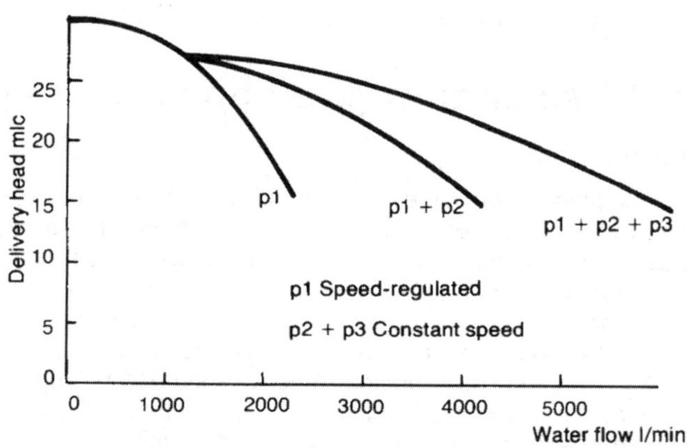

Figure 16.17 Combined pump curves

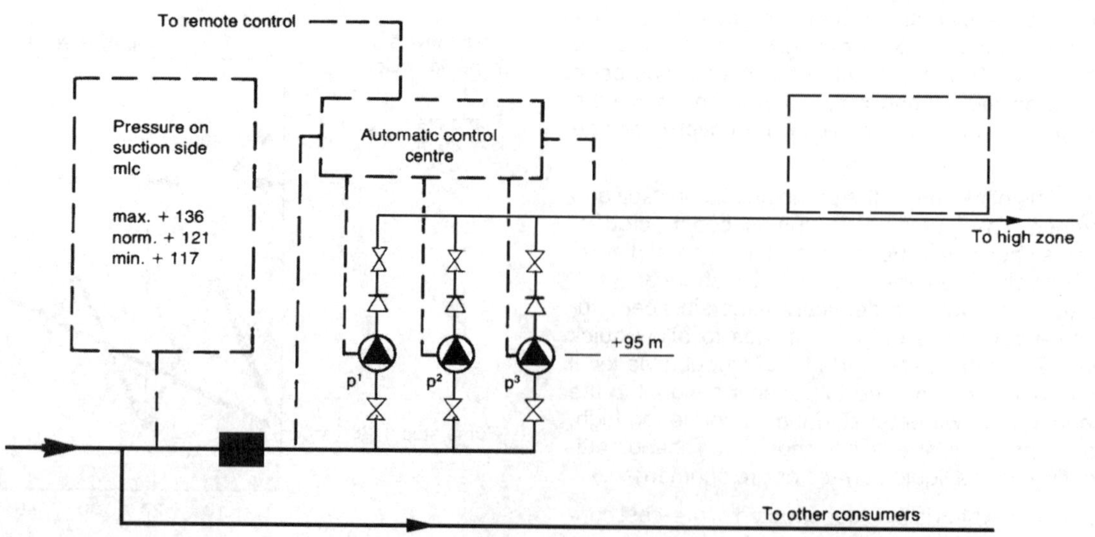

Figure 16.16 Schematic arrangement of booster station and control system

Valves

Each pump has a manual isolation valve located in the suction pipe to the common suction header. Each pump is fitted with a non-return valve and an isolating valve in the discharge connection to the common header. The gate valves are of the full bore rising stem design. A mechanical indicating, totalising flow meter is fitted in the booster station supply pipe.

Control

A pressure transducer, fitted to the distribution pipe, measures the changes in pressure caused by variations of consumption and transmits signals to the logic controller. This initiates the necessary changes in speed and number of pumps operating so that the pressure is restored to the predetermined value.

The pumping station operates without local personnel supervision. Control is performed by monitoring the water pressure on the delivery side of the pumping station with a local logic controller. In addition, certain fault condition signals are transmitted to the central control station.

Operating sequence

When water consumption is low, the speed regulated pump, p^1, operates alone. Speed is increased as consumption increases. When consumption surpasses the capacity of p^1, either p^2 or p^3 is brought into operation. Which of the two pumps, p^2 or p^3, is to be started up is decided manually by a lag-lead switch or automatically by a timer. With the intention of reducing variations in pressure experienced when starting a constant speed pump, the speed of pump p^1 is reduced as another pump comes into operation. A similar procedure is followed when a constant speed pump is withdrawn from operation. At peak consumption all three pumps are in operation.

Limitations of operation

Pumping station operation is suspended in the case of insufficient pressure on the suction side or excessive pressure on the delivery side. Not until the pumping station has been inspected is it possible to start up the pumps again. The speed regulated pump can only be started at minimum speed.

16.5 Magnetic drive pumps for Chlor-alkali manufacturing — mercury cell processing

Introduction

Chlor-alkali electrolysis is one of the most important processes in the chemical industry and provides the basic chemicals chlorine (Cl_2), caustic soda ($NaOH$) and hydrogen (H_2). Chlorine and sodium hydroxide are among the top ten chemicals produced in the world.

Important products that contain chlorine or where chlorine is involved in the manufacture are medicines, pigments, pesticides and plastics such as polyurethane, polycarbonate, epoxy resins, silicone and PVC. Chlorine likewise plays an important role in the manufacture of high purity silicon for use in electronics, and in the production of TiO_2 for the extraction of white pigments.

Caustic soda is used in large quantities for the treatment of effluent and in the breakdown of bauxite in the manufacture of aluminium. Caustic soda is also needed for the extraction of cellulose, in the production of phenol resin and in the processing of cotton. Another important use of caustic soda is in the manufacture of detergents and soaps.

The hydrogen gained during the electrolysis is of secondary importance.

The mercury cell process

Both chemicals are produced through electrolysis of sodium chloride (salt, NaCl). The mercury cell is one of several industrial processes which use electrolysis to split sodium chloride. Mercury is of course a very poisonous substance.

Salt brine is introduced into a container (cell) with two electrodes from a DC power source. The carbon anode (negative electrode) is suspended from the top. Mercury flows along the floor of this chamber in the opposite direction. The mercury goes into a chamber underneath, where water flows in the opposite direction. Chlorine is collected from the carbon anode, and sodium is recovered from the cathode. A pump returns the mercury to the top chamber.

A typical cell is approximately 2m wide and 10m long. A plant consists of many cells operating in parallel (see Figure 16.18).

The mercury cell cathode consists of a slowly flowing layer of mercury across the cell bottom. Sodium/mercury amalgam is produced at the cathode and flows to a separate reactor, known as the decomposer. The amalgam is reacted with water to produce hydrogen gas and caustic soda solution. A 50 % caustic solution is produced directly from the cell.

The pump

When mercury functions as an electrode, it is not consumed in this closed-loop process. Mercury is circulated from the decomposer back to the cell using a centrifugal pump. Several different pump types are used. The pump seal is an obvious leak point for mercury. Early sealless pumps had previously been tried and failed in this severe environment. The flow and head are fairly low, but the fluid weighs 13560 kg /m³. The first magnetic drive pumps were initially installed in this application

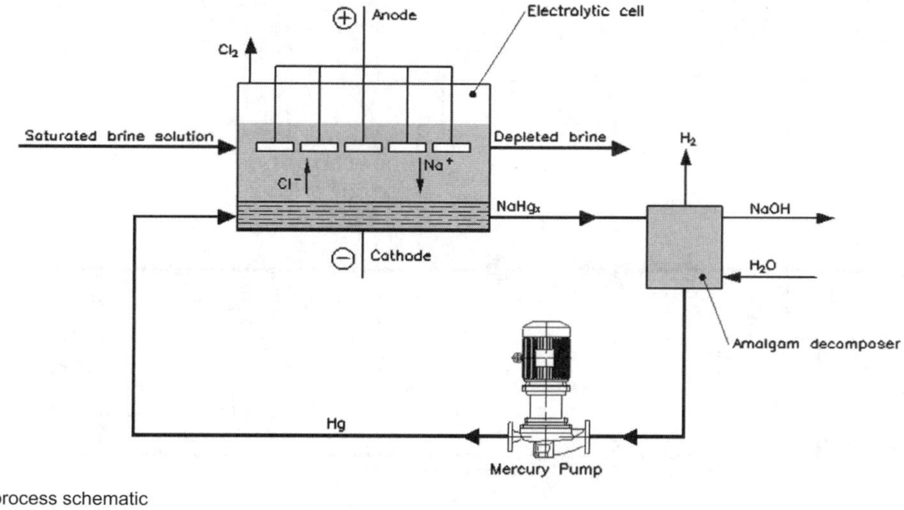

Figure 16.18 Mercury cell process schematic
Courtesy of Courtesy of Dickow Pumpen KG

Figure 16.19 Section through the magnetic drive pump
Courtesy of Courtesy of Dickow Pumpen KG

in early 1998 and have subsequently overcome many of these problems successfully, (See Figure 16.19).

Caustic soda produced by this method is of high purity. On the negative side, the toxic nature of mercury is associated with the greatest potential for adverse environmental and health effects. Strict safety procedures and process controls are followed to prevent workplace exposure and to minimize mercury emissions.

Mercury emissions

Mercury emissions from Chlor-alkali producers have declined significantly over the years. The current contribution of this industry to total natural and anthropogenic mercury emissions is less than 0.1 percent. Western European mercury process plants reduced mercury emissions from 26.60 g/ metric tonne in 1977 to 1.25 g/ metric tonne in 2000. US producers made a voluntary commitment to the US Environmental Protection Agency to reduce mercury emissions by 50%.

Mercury cell use has declined since 1972, and now represents less than 20 percent of all worldwide production. It is claimed that in the future there will be fewer and fewer mercury cells operating as the older plants are shut down or converted to membrane cells. It is also claimed that European and North American producers are committed to not building any new mercury cell facilities.

Magnetic drive pumps are currently used in many mercury cell plants operating in North America.

Application courtesy of Dickow Pumpen KG

16.6 Domestic heating systems and pumps

Introduction

The piping and instrumentation around the boiler of a typical domestic central heating system are shown in Figure 16.20. The functions of the constituent components of the system are described.

Boiler

The outlet water temperature from the boiler, set at 90°C, is controlled by an integral thermostat which switches the burner on and off.

Expansion vessel

The expansion vessel accommodates changes of volume of the water and maintains a suitable static head in the system in order to avoid cavitation.

Pipes

Pipes are sized in accordance with the pressure drop and the length of the runs. Pipes are normally copper but steel and stainless steel are also used on occasions. Figure 16.20 shows a double pipe system. Single pipe systems are also used but are more difficult to control.

Radiators

These transfer heat from the hot water to the air in the rooms. Suitable distribution of water flow to the radiators is ensured by means of adjustable throttle valves. Fitting thermostatic control valves to the radiators removes the need for preset throttle valves.

Hot water for taps

An independent water circuit is shown for heating domestic hot water for taps. In some designs, the hot water heating is controlled by an electrically driven divertor valve which changes the priority between hot water heating and room heating.

Mixing valve

Regulates the system water temperature by admitting return water. Often located close to, or built into, the boiler.

Clock-thermostat

Reduces the system water temperature at night or at other desired times when a lower room temperature is acceptable. This can be used for energy conservation through the day when the house is not occupied or as frost protection.

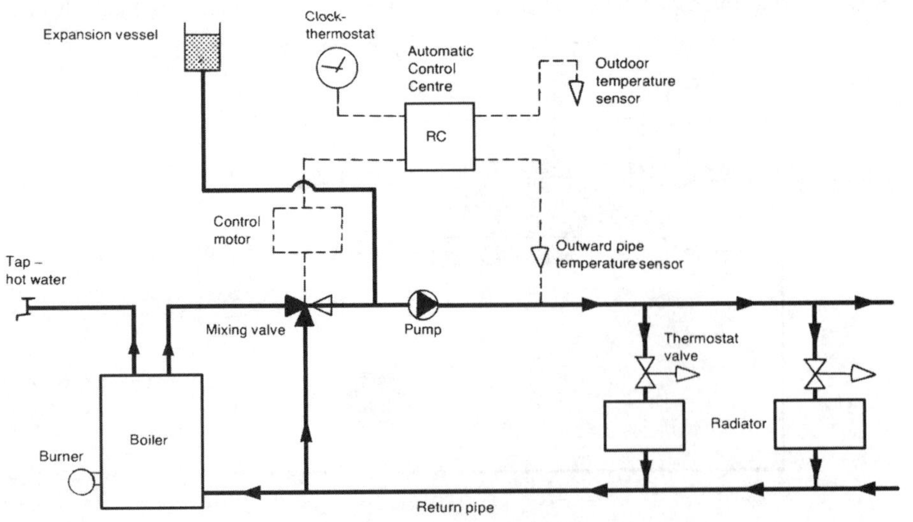

Figure 16.20 Boiler and controls for domestic central heating

Automatic control centre

Receives signals from the system water temperature and outdoor temperature sensors. Adjusts the thermal load via the mixing valve by a suitable outward pipeline temperature as a function of the outdoor temperature. In this way a more even room temperature is obtained and a lowering of the room temperature at night is made possible.

Thermostat valves

Valves throttle the flow to radiators depending upon the local air temperature, thereby permitting individual room temperature adjustment. They also make allowance for other heat sources, from the sun, electric sources, etc.

Design temperature

Forward pipe temperature approximately 80 °C, return pipe approximately 60°, $\Delta t = 20$ °C. The heat required is adjusted in accordance with outdoor temperature.

Pump

The pump circulates a flow which is nearly constant depending upon the pressure drop created by the thermostatic valves and the mixing valve. Pumps can be too large and can cause flow noise. This is avoided by using pumps with built-in flow adjustment by throttling or by-pass or speed control. Wet rotor pumps are often chosen to avoid shaft seals.

Pitfalls

Check the rotation of three phase pumps when installing. The ingress of contaminants from pipes during installation and corrosion products from boiler and pipes causes wear to the pump bearings. The system must be heated up and drained before finally put into operation. The system should be cleaned out regularly with chemical dosing. In the summer when the system is not used, the pump should be run regularly to prevent any sediment collecting and causing possible damage to the pump.

16.7 Adjustable jet pump

Introduction

Many thousands of adjustable jet pumps have been installed in heating plants, mainly in Europe.

Description

Figure 16.21 shows an infinitely variable jet pump with corresponding pressure and velocity distributions. Regulation is by means of an adjustment needle valve. In the Figure:

G1 = motive mass flow with inlet pressure P1

G2 = pumped mass flow with inlet pressure P2

G3 = G1 + G2 with outlet pressure P3

Two central heating schemes are shown in Figure 16.22. When comparing the upper conventional system with the lower using the adjustable jet pump, it can be seen that the jet pump replaces both the mixing valve and the circulation pump. In order to drive the jet pump the necessary pressure must be generated elsewhere. Both plants can be controlled by the outdoor temperature and can be fitted with a timer to provide for reduced indoor temperatures at night.

Regulation

For varying heating requirements caused by variations in outdoor temperature, the setting of the adjustment needle valve is changed. This results in changes in the mixing ratio G2/G1, the temperatures t2 and t3 and the flow G3 through the radiator circuit. In a comparison with the conventional system at reduced heating requirements the result is:

- Lower return pipe temperature

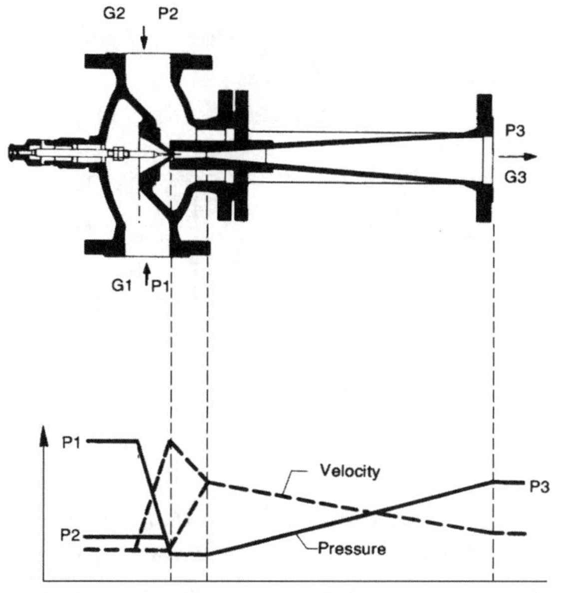

Figure 16.21 Cross-sectional arrangement and pressure/velocity distribution for a jet pump

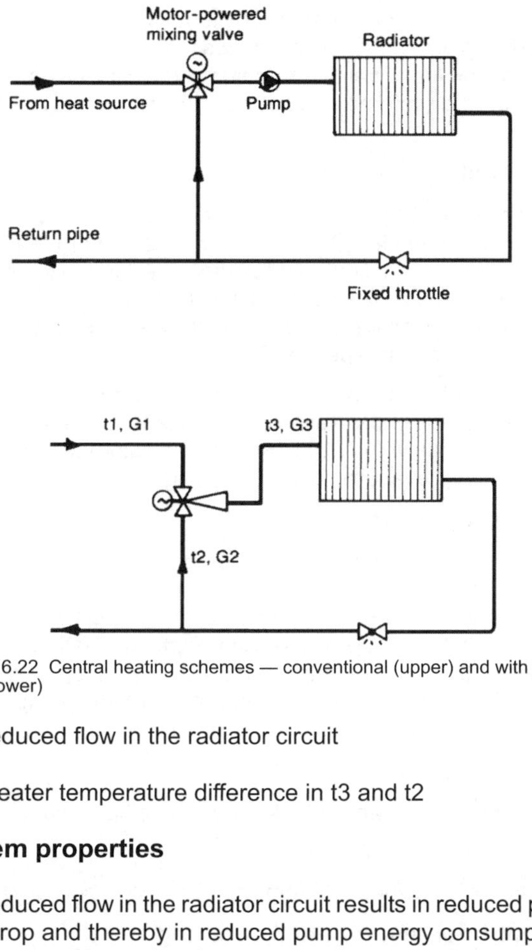

Figure 16.22 Central heating schemes — conventional (upper) and with jet pump (lower)

- Reduced flow in the radiator circuit
- Greater temperature difference in t3 and t2

System properties

The reduced flow in the radiator circuit results in reduced pressure drop and thereby in reduced pump energy consumption. In the case of larger heating systems the necessary pressure for the jet pump is supplied by a centrally located carefully sized centrifugal pump. The adjustable jet pump can in theory replace the circulation pump and the mixing valve in all existing sub-circuits of the heating system.

16.8 Updating of water supply for industrial use

Figure 16.23 shows a schematic arrangement of a water extraction plant feeding a booster station after modification to variable speed control.

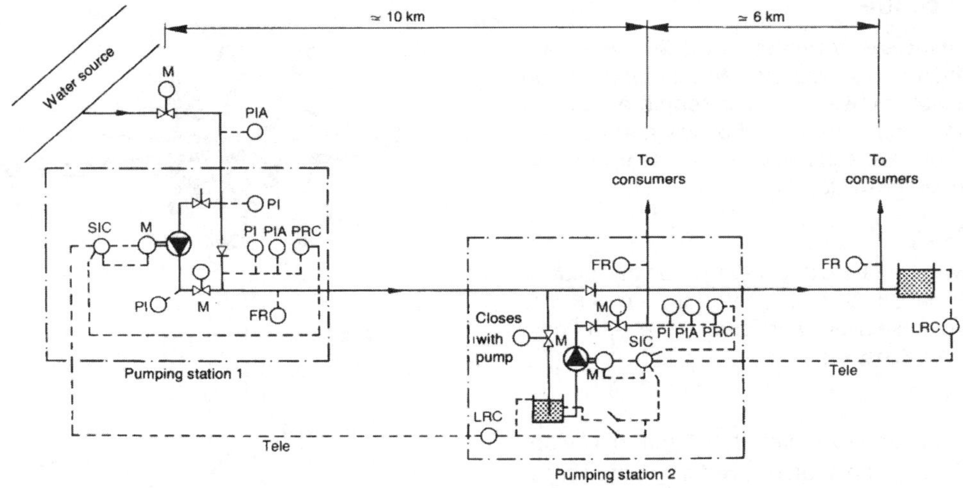

A = Alarm C = Control F = Flow I = Indicating L= Level M = Motor P = Pressure R = Recording S = Speed Tele = Telemetry

Figure 16.23 Schematic arrangement of modernised water supply system

Control before updating

Before modernisation both pumps were driven at a constant speed. The flow was controlled by a manual throttle valve, set on the basis of experience, at various times. Variations in water consumption led to varying throttle valve positions and energy wastage. The system could not react to unexpected changes in demand.

Control after updating

After modernisation both pumps were driven at variable speeds by thyristor controlled DC motors. The speed of the pump in Pumping station 1 is controlled by the water level in the sump at Pumping station 2. The speed of the pump in Pumping station 2 is controlled by the water level in a storage tower at the end of the main distribution pipeline. Throttling is no longer required.

The speed of the pumps cannot be regulated to zero. In the case of extremely low water demand the pumps are switched off. The storage volumes in the sump and storage tower being adequate to prevent rapid on-off cycling.

Alternative control

In the event of faulty transmission of remote control signals it is possible to control the speed of the pumps in accordance with the pressure on the discharge side of the respective pumps. If Pumping Station 1 is out of service, Pumping Station 2 can be controlled by the level in the sump at the station. If Pumping Station 2 is out of service, Pumping Station 1 can pump directly to the storage tower at slightly lower flow rates.

System details

Installed motor capacity 960 kW per pumping station.

Flow approximately 3400 m^3/h

Diameter of pipelines 800 mm

Operational results after updating

Modernisation has resulted in a reduction in water pumping for the industrial plant of about 40%, i.e. water savings of about 2160 m^3/h. The consumption of electricity has been reduced by about 60% corresponding to energy savings of about 10 GWh per year.

16.9 Economic aspects of energy utilisation

Choice of pumps

When selecting circulation pumps for heating systems, due consideration must be given not only to the relative merits of a particular type of pump, its efficiency and serviceability etc., but also to the overall economy of the heating system as a whole.

Depending upon the size and type of heating system and the relative costs of heating fuel and electrical power sources, it can be shown that, in spite of its lower efficiency, a wet rotor circulation pump can be more economical in terms of total energy costs, than a dry motor pump.

In most instances the final choice of circulation pump depends upon factors other than optimum energy utilisation. In small and medium plants it is advantageous to use wet rotor circulation pumps, Figure 16.24, since they can, to a large extent, operate without service or maintenance. In large plants, which are often supervised by operating personnel, it is usual to utilise

Figure 16.24 Wet rotor circulation pump, direct coupled in-line version

Figure 16.25 Circulation pump with standard motor and seal, single and twin head version

pumps driven by standard motors, dry motor pumps, Figure 16.25, because of their higher efficiency.

In borderline cases between small and large plants it may be necessary to carry out a comparative evaluation of the two types. In this comparison the question of total energy utilisation should directly influence the choice of pump. The dry pump motor losses are radiated and convected to the surrounding air, whilst the wet pump motor losses are almost entirely given up to the heating system media.

The following describes methods of evaluating the economic aspects related to the use of wet or dry motor pumps respectively — two extreme cases.

Definitions and assumptions

When comparing wet rotor and dry motor pumps, it is usual to refer to pump efficiency, η_P, which is defined as the relationship between the input and output power, equation 16.1, the difference in the two values being equal to the pump losses, equation 16.2.

For pumps, the input energy is defined as the required shaft input power, P, or as the input electrical power supplied from the electric supply system for the pump unit as a whole, the motor losses being also taken into account. The output is usually defined as work done, Pq, equation 16.3.

The motor efficiency is defined as mechanical power output divided by electrical power output, the difference between the two being mechanical and electrical losses, see equation 16.4. In close coupled pumps the pump power input and the motor power output must be equal.

The overall efficiency of the pump unit, η_U, is obtained by multiplying the motor efficiency, η_M by the pump efficiency, η_P, equation 16.5.

All the losses in a pump unit, and even pump work is converted into heat and can be utilised by the heating system. The pump output power, P_q, is dissipated by friction which results in heat. The pump losses, P_{VP}, are converted into heat within the pump. If the pump unit is mounted within the heated space, the motor losses, P_{VM}, are given up directly to the surroundings.

In order to make a proper comparison between various types of pump unit, account must be taken of the economic relationship of the heating fuel and the electrical energy supplied to the pump motor. In the following analysis the term "economic efficiency", η_E, is used to represent the relationship between the output and input energy, with due consideration to the specific energy costs of the losses.

The cost per heat unit extracted from the fuel, allowing for the efficiency of the heating system, η_H, is defined by "a" in equation 16.6 and the cost for electrical energy supplied by "b" in equation 16.7, both values in currency/kJ.

Standard motor pump units

Figure 16.26 illustrates the energy flow for a pump unit with a standard motor and seal. The motor takes energy from the electric supply system and supplies the pump with mechanical energy via the shaft coupling. The motor energy losses are given up to the surroundings, see equation 16.4.

The mechanical power is converted into hydraulic energy, equation 16.3, and the pump losses, equation 16.8, are converted into heat and transferred to the liquid.

When calculating the "economic efficiency" of a heating system using a "dry" pump unit, then the unit efficiency must include the pump losses, equation 16.9. The energy costs related to the pump losses being the only losses which can be usefully utilised by the heating system.

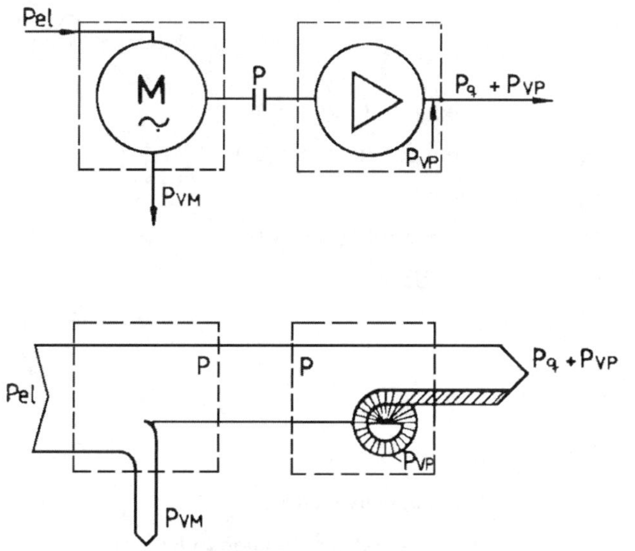

Pel = Electrical energy supplied
P = Mechanical shaft power
P_{VM} = Motor energy losses
P_q = Hydraulic pump output
p_{VP} = Pump energy losses

See equations for symbol explanations

Figure 16.26 Energy flow diagram for a standard motor pump unit

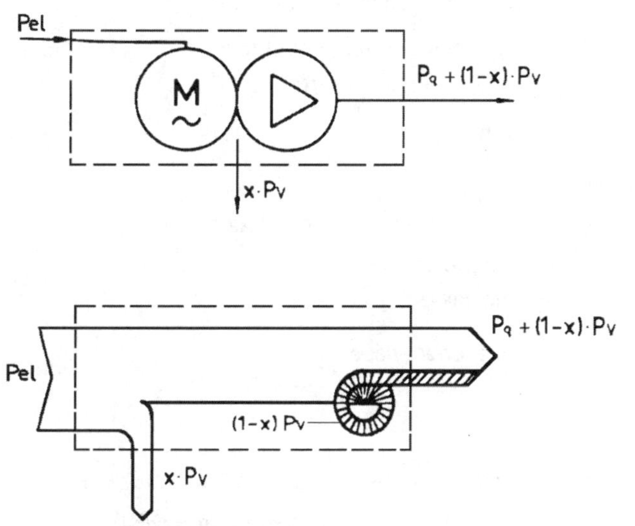

Pel = Supplied electrical power
P_q = Hydraulic pump output
P_V = Pump aggregate's energy losses
X = Losses to surroundings

See equations for symbol explanations

Figure 16.27 Energy flow diagram for a wet rotor motor pump unit

Wet rotor motor pump units

Figure 16.27 illustrates the energy flow for a pump unit with a wet rotor motor. A large proportion of the energy consumed by the motor is utilised by the heating system.

The greater part of the motor heat losses and the pump friction losses are utilised by the plant as additional heat. The heat losses dissipated to the surroundings are negligible and need not, therefore, be included in the calculations. The "economic efficiency" of a wet rotor motor pump unit is obtained by considering the pump unit efficiency and the value of the additional heat supplied, equation 16.10.

Equations

$$\eta_P = \frac{\text{pump output power}}{\text{pump input power}} = \frac{P_q}{P} \qquad \text{Equ 16.1}$$

$$P_{VP} = \text{pump power input} - \text{pump power output} ↵$$

$= P - P_q$ Equ 16.2

where:

η_P = pump efficiency (decimal)

P_q = pump output power (W)

P_{VP} = pump losses (W)

P = pump input power (W)

P_q = work done

$P_q = Q \cdot \Delta H \cdot \rho \cdot g$ Equ 16.3

where:

Q = flow (m^3/s)

ΔH = differential head (m)

ρ = liquid density (kg/m^3)

g = gravitational acceleration (9.81 m/s^2)

Motor energy losses η_M:

$$\eta_M = \frac{P_{el} - P_{VM}}{P_{el}} \quad \text{Equ 16.4}$$

where:

μ_M = motor efficiency (decimal)

P_{el} = electric power input (W)

P_{VM} = motor losses (W)

$\eta_U = \eta_P \cdot \eta_M$ Equ 16.5

where:

η_U = unit overall efficiency (decimal)

$$a = \frac{\text{heating fuel cost}}{\text{calorific value} \cdot \eta_H} \quad \text{Equ 16.6}$$

$$b = \frac{\text{electrical energy cost}}{3600} \quad \text{Equ 16.7}$$

where:

a = heating fuel energy cost (currency/kJ)

b = electrical energy cost (currency/kJ)

η_H = heating system efficiency (decimal)

Pump losses P_{VP} are:

$P_{VP} = Q \cdot \rho \cdot C_p \cdot \Delta t$ Equ 16.8

where:

C_p = specific heat of liquid (kJ/kg/K)

Δt = temperature rise across pump (°C)

$$\eta_{EDry} = \frac{P_q + \left(\left(P_{el} - P_q - P_{VM}\right) \cdot a/b\right)}{P_{el}}$$

substituting ($P_{el} - P$) for P_{VM}

we obtain

$$\eta_{EDry} = \frac{P_q + \left(\left(P_{el} - P_q - (P_{el} - P)\right)\right) \cdot a/b)}{P_{el}}$$

which can be simplified to

$$\eta_{EDry} = \frac{P_q + \left(\left(P - P_q\right) \cdot a/b\right)}{P_{el}}$$

which can be written as

$$\eta_{EDr}y = \eta_{UDry} + \left(\left(\eta_{MDry} - \eta_{UDry}\right) \cdot a/b\right) \quad \text{Equ 16.9}$$

where:

η_{EDry} = economic efficiency for a pump unit with a standard motor (decimal)

$$\eta_{EWet} = \frac{P_{el} + \left(\left(1 - x\right) \cdot P_v \cdot a/b\right)}{P_{el}}$$

which can be rewritten as

$$\eta_{EWet} = \eta_{UWet} + \left(\left(1 - x - \eta_{UWet} + x \cdot \eta_{UWet}\right) \cdot a/b\right)$$

realising

$x \cdot \eta_{UWet} \approx 0.03$

it is considered neglible

$$\therefore \eta_{EWet} = \eta_{UWet} + \left(\left(1 - \eta_{UWet} - x\right) \cdot a/b\right) \quad \text{Equ 16.10}$$

η_{EWet} = economic efficiency for a pump unit with a wet rotor motor (decimal)

$$a/b_{crit} = \frac{\eta_{UDry} - \eta_{UWet}}{1 - \eta_{UWet} - x - \eta_{MDry} + \eta_{UDry}} \quad \text{Equ 16.11}$$

$$\text{oil cost} = \text{calorific value} \cdot \rho \cdot a/b_{crit} \cdot b \cdot \eta_H \quad \text{Equ 16.12}$$

where:

oil cost (currency/m^3)

calorific value (kJ/kg)

ρ = density (kg/m^3)

a/b_{crit} = value from equation 16.11 (decimal)

b = electrical energy cost (currency/kJ)

η_H = heating system efficiency (decimal)

Comparison

Two extreme cases have been chosen to demonstrate the relative differences in total economy when considering similar heating systems.

1. Very low fuel costs and high electricity costs, i.e. the ratio a/b approaching zero,
2. The cost of fuel and the cost of electricity is approximately the same, i.e. the ratio a/b approaching unity.

The expression "economic efficiency", defined previously, is used here to calculate the effect of energy prices and should not be considered as a value of efficiency in the normal sense.

Case 1. In for example, the Netherlands, where gas is relatively cheap and electricity is relatively expensive, the situation is very similar to 1. above.

According to equations 16.9 and 16.10 and with a/b = 0 the "economic efficiency" for both dry and wet rotor motor pump units will be equal to the unit efficiency, η_U.

Since the unit efficiency for a dry motor pump is always higher than for a wet rotor motor pump, then it follows that the "economic efficiency" for a dry motor pump is always greater than for a wet rotor motor pump, i.e.

$\eta_{EDry} > \eta_{EWet}$

Case 2. In Sweden, for example, electrical energy has been relatively cheap by international standards. The relationship between fuel and electrical energy costs, a/b, is almost unity, if the normal heating system efficiency for larger installations of

0.75 to 0.80 is taken into account. This is very similar to 2. above.

If a/b is 1, then from equation 16.9 for dry motor pump units, the "economic efficiency" is equal to the motor efficiency. For wet rotor motor pump units operating under the same conditions, a/b = 1-x, where "x" is the losses to the surroundings, then the "economic efficiency" is determined by applying equation 16.10.

Measurements have shown that the losses "x" to the surroundings are normally between 4 and 7%, for wet rotor pumps with motors up to 2.2 kW, which is less than the corresponding dry motor pump losses of about 20%.

It can therefore be concluded that, for heating systems in these circumstances, better running economy will be achieved using wet motor circulation pumps than by using dry motor pumps, i.e.

$$\eta_{EWet} > \eta_{EDry}$$

Conclusions

The heat output of a heating plant is provided by the heat extracted from the fuel and the power supplied to the circulating pump.

The electrical energy contribution, although being relatively small, is important when considering the optimum economy of the complete heating system.

A comparison of heating system energy costs for similar systems in different countries shows that in countries like Sweden where electrical energy is relatively inexpensive, the ratio a/b approaches 1 thus making wet rotor motor circulation pumps, despite their lower efficiency, more economical in service than dry motor pumps.

In countries such as the Netherlands where electricity is more expensive, the ratio a/b approaches 0 and dry motor circulation pumps are more "economically efficient".

16.10 Pumping of delicate solids – food and live fish

Introduction

Screw centrifugal impeller pumps are being used to transport even the most perishable of foods, for example oranges, apples, cranberries, bean sprouts, beans, peas and mushrooms around food processing plants (See Figure 16.28).

For many years also they have been used in fish farms for the transfer of live fish, grading or tanker-loading duties; helping to reduce and eliminate the damage and high labour costs incurred with the more traditional manual methods previously employed. See Figure 16.29.

Figure 16.28 Food processing of mushrooms
Courtesy of Hidrostal Ltd

Figure 16.29 Handling live trout
Courtesy of Hidrostal Ltd

Pumping food

The pump and its associated pipework system is particularly beneficial where space is limited or where there is a need to transport large volumes over long distances, or on complicated routes, where the alternative is transferring produce between numbers of different conveyor belts. See Figure 16.30.

One of the most important requirements of equipment used in food processing is of course, an ability to handle produce gently and this is where screw centrifugal impeller pumps have an advantage, being originally designed to pump live fish. As a consequence, the pump is renowned for its ability to transport even the most delicate of materials without damage.

The pump is characterised by several unique features including an ability to handle delicate solids without damage and viscous materials at a low shear rate. This allows the transportation of even the most fragile materials over a wide range of pumping rates. It has a very smooth hydraulic path, large free passages and high efficiencies which produces low re-circulation of the liquid within the pump. This in turn results in the use of smaller electric motors, lower power costs and reduced maintenance costs.

Cleaning the pump provides minimal disruption to the factory processes; this is achieved by flushing with clean water or pumping a cleaning agent through the pipelines. Because the pump system is easier to clean it is more hygienic. As well as promoting rapid and effective cleaning, the fact that the pipework is enclosed reduces the danger of dirt falling into the

Figure 16.30 Typical food processing plant
Courtesy of Hidrostal Ltd

Figure 16.31 The pumping of live fish
Courtesy of Hidrostal Ltd

system or indeed any produce falling out. This has the double benefit of increased productivity and promotes a safer, cleaner and more pleasant working environment.

Pumping live fish

In this application, pumps are supplied as trolley or skid mounted units, with a variety of drive options. For duties where a suction lift is required the unit is easily primed by use of a fitted high capacity, hand-operated diaphragm pump.

A variable speed drive arrangement is incorporated to ensure that the pump speed/duty can be tailored to enable optimum performance for gentle, damage free handling to be achieved. Units can be available in a range of pump sizes and configurations suitable for handling a wide range of fish sizes from fingerlings up to full grown salmon. See Figure 16.31.

Application courtesy of Hidrostal Ltd

16.11 Submersible sump pumps for construction projects

Introduction

Electrically driven sump pumps are readily available from stock in various sizes from 1 kW up to 20 kW both for sale and leasing. Standard pumps of up to 60 kW can usually be supplied on short delivery from the manufacturer. Pumping of this kind is primarily intended for temporary drainage work for building and construction activities on site. In the case of major construction projects, the installations may be of a permanent nature. Pumps constructed of aluminium may be restricted to certain water grades.

Types of drainage

The nature of the drainage is characterised by the type of civil engineering activity. Industrial and housing construction does not normally require involvement with rivers or lakes and drainage is therefore mainly concerned with groundwater, storm and rainwater etc. The drainage period is short in relation to the construction period and the pumping requirement is often of a limited nature. Sizes 1 to 5 kW easily meet the requirements.

Harbour and bridge construction works involve their own type of drainage characterised by large flows and moderate to low delivery heads. Discharge pipes can often be short, further limiting the total delivery head. Typical of this type is drainage and drying out of coffer dams. This type of draining is often simple to plan and supervise. Pumps of low head of 10 to 30 kW are required to yield flows of 180 to 600 m^3/h.

The greatest pump investment, in percentage terms of the total costs, is often required by hydro-electric power plant construction works and mining. The investment is particularly large in the case of long deep tunnel sections. Sophisticated dewatering techniques involving severe demands on planning and maintenance are mainly concerned in the case of tunnel driving.

A typical four year power plant construction project illustrates the implications for pump choice.

General plant information

Headrace tunnel 4200 m long, area 24 m^2

Two transport tunnels, each 400 m, area 15 m^2

Tailrace tunnel 5600 m, area 30 m^2

Three transport tunnels, each 350 m, area 15 m^2

Maximum head 62 m

Differences in level in the transport tunnels between ground level, open excavation, and lowest level, i.e. at entrance to main tunnel.

Tunnel 1	41 m
Tunnel 2	40 m
Tunnel 3	34 m
Tunnel 4; 5	36 m

The transport tunnels are driven at a gradient of 1:10.

Transport tunnel 1 may be used as typical for drainage techniques for tunnels and rock chambers.

Pump sizing data

Total estimated flow of water Q_b comprises:

- Estimated leakage of water due to poor rock
- Estimated penetration of melt and rain water through open excavation
- Estimated amount of drilling water

Total difference in elevation levels between lowest water course and discharge pipe outlet = H_{stat}

Total length of pump pipeline = L_{tot},

Pipe details: quick-coupling pipes 100 mm diameter and 150 mm diameter.

Two drilling units are used with a total water requirement of 8.4 m^3/h when in operation. Leakage from the rock is estimated to be 6 m^3/h and melt/rain water to be 12 m^3/h, Q_b = 26.4 m^3/h. A safety margin added to ensure pumps could always cope with water flow brought the flow to Q_{max} = 36 m^3/h.

An estimate of the pressure loss in 100 mm and 150 mm pipes respectively with a design flow of 36m^3/h indicates 100 mm pipe to be acceptable. Pressure loss H_f = 9 m. Total differential head:

$$H_{stat} + H_f = 41 + 9 = 50 \text{ m}$$

The drainage is divided into two phases.

Driving of the tunnel from tunnel face and open excavation to level at -41 m at the main tunnel. During this phase, use is made of pumps capable of pumping full flow but about half the differential head, i.e. 36 m^3/h x 25 m. The intention is that the pump should be light and well-matched to the actual duty. When about half of the length of the tunnel has been driven, i.e. about 200 m and H_{stat} = 20.5 m, the first pumping station can be constructed and made permanent for the entire construction phase.

After the transport tunnel has been put to use, water flows from the main tunnel now under construction. Admittedly, the 8.4 m^3/h drilling water from the transport tunnel itself has now been eliminated, but the four units in the main tunnel produce 16.8 m^3/h and this must be added to the leakage from the main tunnel.

A rough estimate indicates that Q_b through the transport tunnel will be 42 m³/h.

This means that:

Q_b = 42 m³/h

H_{stat} = 41 m

Q_{max} estimated to be 60 m³/h

Pressure loss H_f then becomes 20 m for 100 mm pipe, which is too high a loss. If the pipe size is changed to 150 mm, H_f at 60 m³/h will be 4 m. Thus:

$H_{stat} + H_f = 41 + 4 = 45$ m

This station may be constructed to two designs: a pump chamber blasted out of the rock or a steel pump tank.

The advantage of the steel tank, see Figure 16.32, is that it can be easily moved and can therefore follow the advance of the tunnel downwards. This enables the waste pump on the drilling car to work constantly with the same length of hose and delivery head.

On reaching level -20 m a complete pumping station is established at that level and an additional similarly equipped station follows the working face down to the beginning of the main tunnel at level -41 m.

When the transport tunnel joins up with the main tunnel two options for the continued drainage of the transport tunnel may be considered, Figure 16.33.

Option 1 — Retention of two pumping stations each with a total delivery head of 22.5 m and maximum flow of 60 m³/h.

Option 2 — Installation of one central pumping station at level -41 m, the pumping duty then being 60 m³/h x 45 m.

Option 1 is preferable because the pump size used permits greater range of application and reduces the risk of water hammer in the case of power failure. The advantage of Option 2 is that inspection can be limited to a single station.

The total pumping requirement for the transport tunnel until the commencement of Phase 2 is shown in Figure 16.34.

General installation instructions for drainage

Determination of size may be found by:

- Difference in elevation
- Length of pump pipe
- Flow

The difference in elevation between the liquid surface and the pipe outlet is always known. The length of the pump pipe can always be determined.

The flow of water to be pumped can usually only be estimated and this must be done on the basis of experience. When this has been done, the manufacturer's instructions should be consulted, to calculate the pipe friction loss and this added to the difference in elevation.

$H_{stat} + H_f = H_{tot}$

Flexible pipes

When using hose as a pump pipe, a 20% addition should be made for unavoidable kinks. Use the shortest possible route.

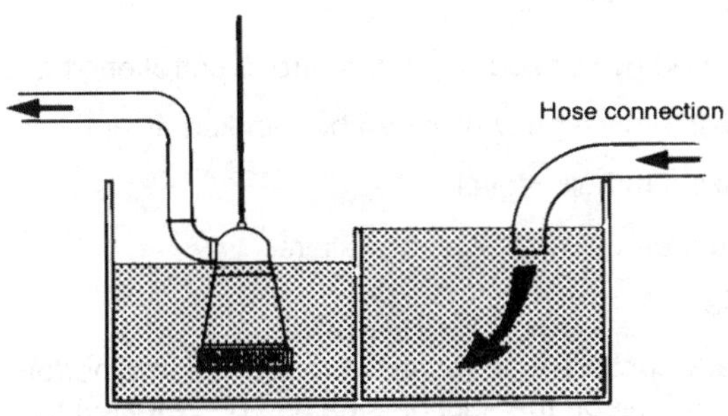

Figure 16.32 Pump station — steel tank option

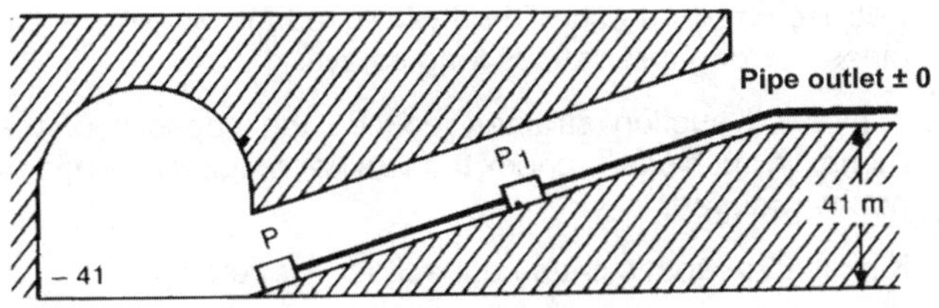

Figure 16.33 Drainage with two pumping stations

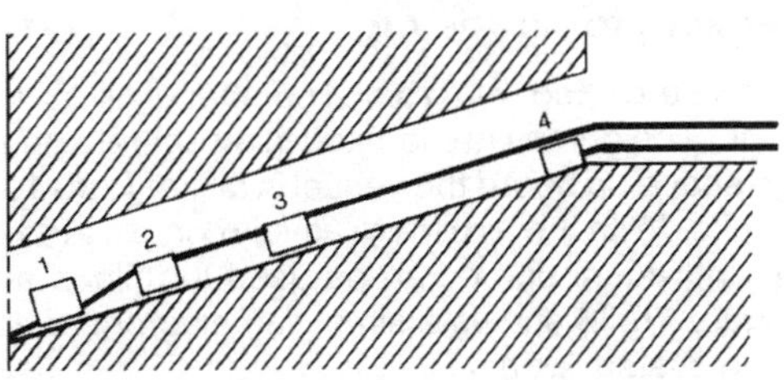

1. Drilling car with air-powered diaphragm pump
2. Pumping station with submersible electric pump 60 m³/h x 22.5 m
3. Pumping station with submersible electric pump 60 m³/h x 22.5 m
4. Submersible electric pump for rain and melt water with its own pipe. Size about 30 m³/h x 5 m

Figure 16.34 Drainage with four pumping stations

Figure 16.35 Correct routing of flexible pipes

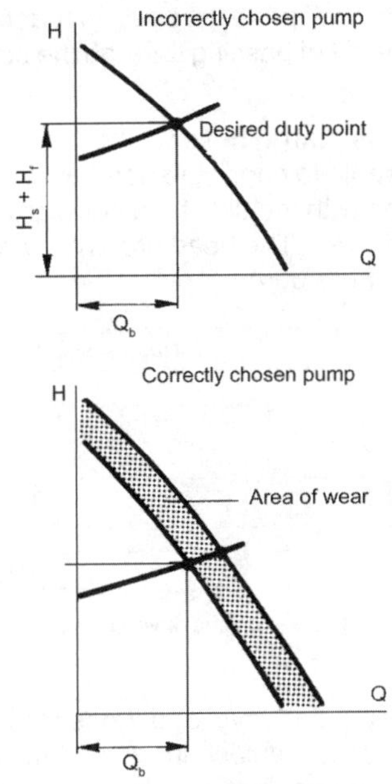

Figure 16.36 Pump curves corrected for wear

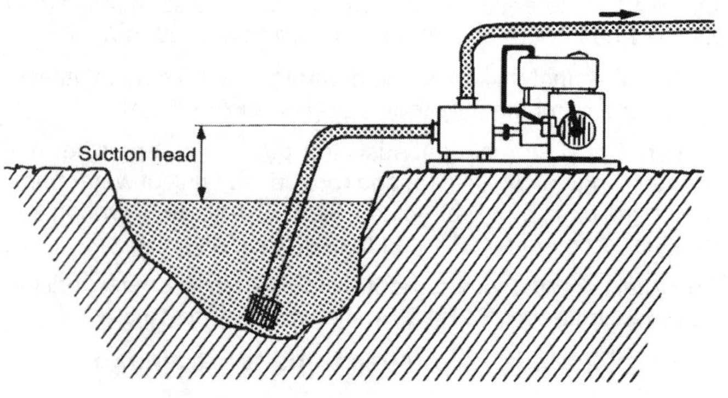

Figure 16.37 Skid mounted self-priming pump

In open shafts and pits always place the delivery connection at right angles to the pump shaft to avoid kinks.

In the case of sharp edges, e.g. rebates, use pipe bends and connector pipes. See Figure 16.35.

Always select a pump giving consideration to the wear factor. This is especially important in the case of the pumping of tunnels. Drilling waste is very abrasive for impellers and rubber components, see Miller Number in Chapter 2, Table 2.13, causing rapid reduction in capacity. See Figure 16.36.

16.12 Use of engine driven self-priming pumps

Pumping unit

Figure 16.37 shows a self-priming site pump with a petrol or diesel engine mounted on a skid, and suction hose with strainer. The discharge hose is normally about 10 m long. Extra lengths can be added when necessary. The skid is fitted with lifting and towing eyes.

Size of pumps

Pumps in good condition can lift 8 m. Table 16.1 indicates the flow ranges of some popular pump sizes at differential heads of between 10 and 20 m.

This type of pump can handle solid particles. It is not a solids handling pump in the sense of solid-liquid mixtures but a rugged pump capable of passing occasional, fairly large solids plus mud and soil. The size of the maximum solid increases with pump size. The mesh of the suction strainer is selected on the basis of the pump inner passage size. In practice it is accepted that all solids capable of passing through the strainer are pumpable.

In the case of a discharge hose which is more than about 30 m long, it is advisable to check the head drop of the hose. This is easily evaluated with the aid of the nomograms in Chapter 3, starting at Figure 3.24. The head drop has the effect of reducing the capacity of the pump.

Suction connection mm	Flow range m^3/h
40	10 to 20
50	10 to 30
75	20 to 60
100	30 to 100
150	30 to 300
200	30 to 500

Table 16.1 Capacities of popular engine driven pumps

Site installation

Place the pump on a flat, level and if possible rigid pad. Most engines and pumps are splash-lubricated so levelling is quite important. The difference between the water level and the centre-line of the suction connection should not exceed 8 m. The suction hose should ascend uninterruptedly to the pump. Position the suction strainer near the bottom without burying it in mud, etc. In the case of small flows the strainer can be placed in a shallow pit approximately 0.5 x 0.5 m.

Pumping

Fill the pump with water through the filler. The volume of water necessary is 2 to 5 litres.

Start up and evacuate the suction pipe. If the time for evacuation exceeds 2 to 3 minutes, there is a fault, e.g. too great a suction lift, leaky suction hose, worn-out pump, etc.

As the level in the pit reduces, move the strainer to greater depths. As the suction lift increases the flow will reduce. If there is plenty of water, it is only necessary to check the pump every quarter of an hour. A full fuel tank usually suffices for 2 to 3 hours, but try to avoid running at very low fuel levels. Sediment in the fuel tank can block filters and injectors. If fuel runs out, the fuel system may have to be vented and primed. If the water flow is small, reduce the engine speed to a suitable level.

Each day when pumping is finished, wash the pump and suction strainer out with clean water. Dried mud and setting concrete can seriously affect performance. In cold weather, drain the pump completely each night to prevent freezing. Before storing between jobs, protect in accordance with the manufacturer's instructions.

State when purchasing or hiring:

- Flow and corresponding delivery head
- Suction lift
- Maximum particle size
- Desired power source, e.g. 4-stroke petrol engine
- Maximum weight which can be handled at site
- Maximum noise level
- Requirements for skids or trailers

Pitfalls

- Leaky suction hose results in reduced pump performance. An air leak on the suction side can be detected by placing the discharge hose under water. Air bubbles indicate air leakage in the suction hose or the pump.
- Petrol-oil mixtures should only be used for two-stroke engines. Wrong fuel causes long delays.
- A blocked suction strainer reduces the capacity of the pump. In such cases check the strainer regularly and clean out as required.
- Kinks in the hoses cause unnecessary throttling.

16.13 Cargo pumps for tankers

Large tankers for crude oil

Large tankers are fitted with a small number of oil cargo pumps, usually four. The pumps are located in a separate pump room. The cargo space on board the vessel is divided into 10 to 15 individual tanks. With the aid of a relatively complicated pipe system every individual pump can be used to unload any of the tanks. Figure 16.38 shows a simplified diagram of how an oil cargo pump system works.

The suction pipe to every tank is connected to a suction funnel close to the bottom of the tank so that the tank can be completely emptied.

The separator tank, which is connected to a vacuum unit, separates gases drawn into the suction funnel during the final stage of unloading.

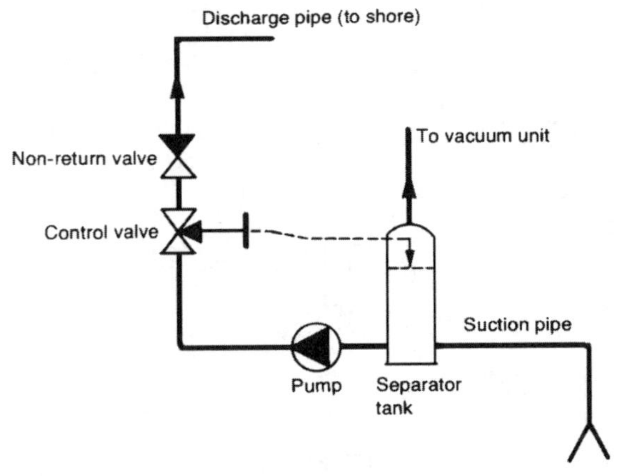

Figure 16.38 Simplified cargo pump system

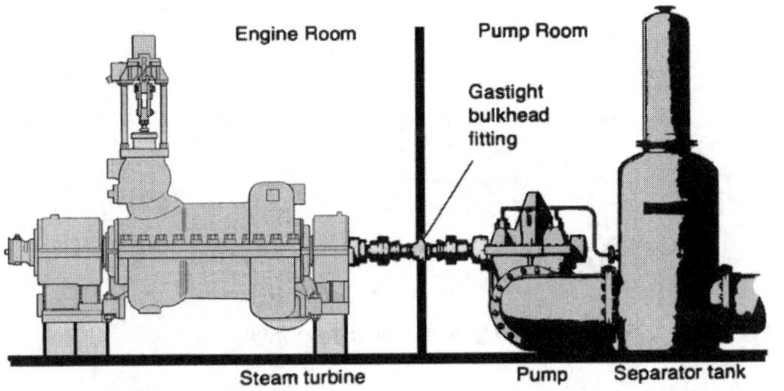

Figure 16.39 General arrangement of steam turbine driven cargo pump

The pump is normally driven by a steam turbine, but electric motors can also be used. Due to the risk of fire or explosion due to the gases given off from the crude oil, the pump driver must not be located in the pump room. Figure 16.39 illustrates a horizontal shaft pump installation.

The control valve is used to reduce the flow during the final unloading stage, which is necessary to ensure the tank is completely emptied. The valve is automatically controlled by the level in the separator tank.

The non-return valve prevents oil from flowing backwards through the pump when unloading is stopped.

The bulkhead acts as a firewall between the two compartments. A spacer coupling is used with a mechanical seal to prevent gas leakage.

The pumps are normally sized to unload the vessel completely in about 24 hours. The delivery head required to pump oil to the onshore terminal is 120 to 180 m. This results in large pumps with heavy power consumption, up to say 7 MW per pump on the largest vessels. Oil cargo pumps are also used for the loading and unloading of water ballast.

The risk of explosion makes it essential that all equipment capable of generating sparks must be excluded from the pump room and the tank spaces. Pumps should be fitted with mechanical seals, since stuffing boxes can become overheated.

During installation, allowance must be made for the movements which take place in the hull. The shaft coupling should be able to absorb minor misalignments between pump and driver.

Smaller tankers for finished products

A characteristic of vessels of this type is their ability to transport a great number of different kinds of liquid cargoes. The cargo space in smaller tankers is also divided into small tanks, the number varying in relation to the design and size of the vessel. As the cargo often consists of a number of different liquids which must not be mixed, a common unloading system cannot be used. Instead every tank is provided with a pump of its own, which, bearing in mind the risk of explosion, should be hydraulically powered. Also, hydraulic equipment can be easier to protect against the ravages of a marine environment. In order to simplify installation, the pumping unit is submersed in the liquid. The outline drawing in Figure 16.40 shows the pump installation in a tank.

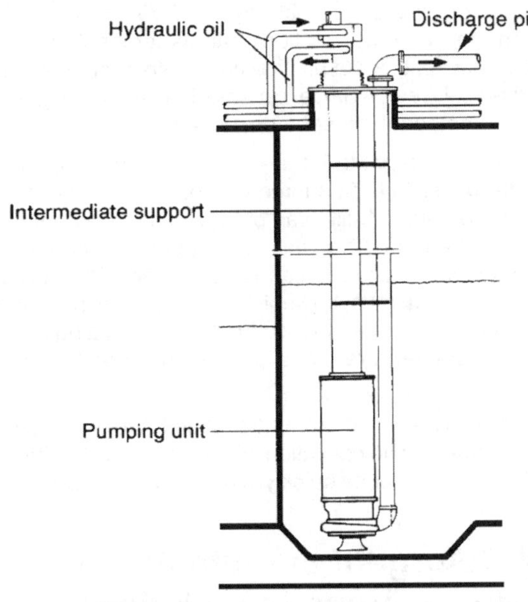

Figure 16.40 Hydraulically driven vertical submersible pump

The pumping unit consists of a centrifugal pump and a hydraulic motor. The suction connection of the pump is bellmouthed and its entrance is close to the bottom of the tank. The capacity of the pump is determined by the volume of the tank. The delivery head is 70 to 150 m.

The delivery pipe and intermediate support or column are cross braced to each other to increase rigidity and reduce vibration. On larger vessels of this kind the distance between the deck and the bottom of the tank is about 15m. In addition to feed and return pipes for hydraulic oil, the intermediate support also contains pipes for sealing liquid.

A constant flow valve maintains the flow of oil to the hydraulic motor, which means that the pump operates at a constant speed independently of the load. The hydraulic motor also contains devices for the remote control of the pump speed. The pumping unit is supplied with hydraulic oil from a power pack which can supply oil to several pumps. The oil pressure in the hydraulic motor feed pipe is normally 160 to 240 barg.

Since the cargo may consist of highly corrosive media, the material of parts in contact with liquids must be given special consideration. The sealing of the pump shaft, which must function in liquids with varying properties, also deserves special attention.

The breakdown of a pump with a tank full of liquid raises difficulties with respect to rapid repair. For such eventualities portable pumps must be available for emptying and flushing out tanks.

LNG carriers

Considerable quantities of LNG are transported from producers to consumers. LNG usually consists mostly of methane so the liquified gas must be stored at temperatures around -160 °C at atmospheric pressure. The material requirements for operating at such low temperatures are well known. The main problem with pumping such cold liquids is ensuring the pump unit is cooled sufficiently for the liquid gas to remain liquid.

This problem can be solved by using submerged pumps in the tanks and specially adapted canned motor pumps have been developed. As the complete pump unit is surrounded by the liquid thorough cooling is practically assured. Great care must be taken with the motor connections into its terminal box. Submerged pumps, of the wet pit type have also been used. Double mechanical seals with a suitable barrier liquid, such as propanol or methanol, must be employed.

When spherical storage tanks are mounted on the ship's deck it may not be possible to fit submerged pumps. In these situations, a canned pump built in an external can may be suitable. The external can must be fitted with vapour vents and liquid level controls to ensure the can is cooled and filled properly. Adequate insulation must be applied to the suction pipe and the external can to prevent ambient conditions vaporising the LNG. Very slight temperature rises cause considerable loss of NPSHa.

Pumps for LNG may seem to "rattle" slightly when first started. This can be due to mild cavitation while vapour is clearing. The "rattling" should stop quite quickly; if not, shut-down and investigate.

16.14 Detergent manufacturing systems — pump installations

Introduction

In this example, the finished detergent is manufactured from three raw materials which are mixed in various proportions, depending upon the grade of the finished detergent, before being dispensed into containers of about five or 10 litre capacity. See Figure 16.41.

The raw material tanks are elevated above the pumps to provide plenty of NPIPa. Each tank is fitted with a tapered bottom leading to a central connection which is twice the diameter of the pump suction connection. The manual isolating valves are full-bore quarter turn ball valves.

Pumping the raw material

The pumps used are triple lobe pumps operating at low speed to maintain the volumetric efficiency close to unity. All three pumps operate together controlled by the programmable logic controller PLC, as shown in Figure 16.42. The speeds of the three pumps are adjusted by the PLC, via variable frequency inverters, to conserve the raw material proportions fed to the mixer. After a preset time the three pumps stop and the electrically actuated valves in the discharge pipes close to prevent "dribbling".

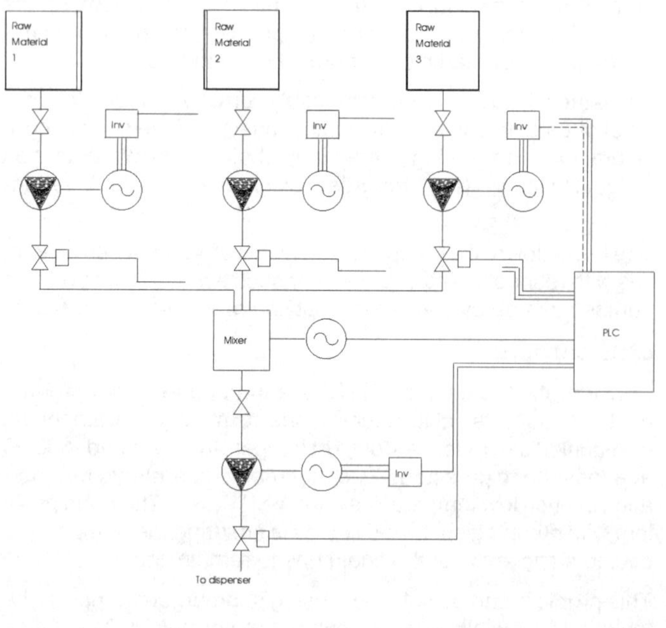

Figure 16.41 Schematic of a typical detergent manufacturing system

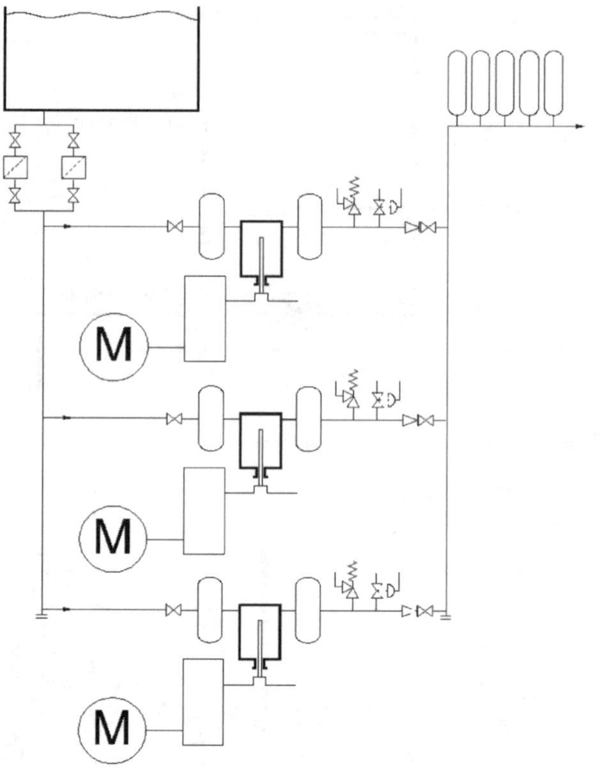

Figure 16.42 Schematic of a typical three pump system

The mixer

Once the mixer has been filled to the required level, the PLC starts the mixer motor which runs for a preset time. The mixer tank also has a tapered bottom and a large connection fitted with a full-bore ball valve.

Finished detergent pump

The finished detergent pump is a low speed triple lobe pump. Volumetric efficiencies close to unity are assured by adequate NPIPa due to low loss pipework. The pump speed is controlled in response to signals from the dispenser.

16.15 High capacity bearings for compact centrifugal pump systems

Introduction

Pump manufacturers are subjected to regular demands to reduce both the cost and the size of the pumping systems that they produce. The main focus of these cost-reductions is on centrifugal pumps because they are the most widely used and

Figure 16.43 Centrifugal pump example
Courtesy of NSK Europe

are the workhorses of industry, providing users with a wide range of flow, volume capacity, and head. They have uses in industries such as pulp and paper, wastewater treatment, steel mills, irrigation, mining and many other demanding environments. Figure 16.43.

The need for compact bearings

The problem for any pump manufacturer designing more compact pump systems is that any reduction in size cannot be at the expense of flow. What this means for key pump components such as bearings is that they, in turn, must become more compact, but at the same time provide higher load capacity to meet new operating demands.

This requires the development of high load capacity, deep-groove ball bearings, resulting in improved basic dynamic load ratings of approximately 10% and fatigue life by around 30%, without any changes to the standard bearing dimensions. These improvements have been achieved via a combination of a new heat treatment process, enlarged ball diameters, and by optimising the interior design of the bearing cages. See Figure 16.44.

Improved life and increased dynamic load ratings

Table 16.2 shows the improvements that are achievable. The Table shows the boundary dimensions and basic dynamic load ratings of both standard and new high capacity bearings. Taking as an example a bearing with 30mm bore x 72mm od: in its standard configuration this bearing has a basic dynamic load rating of 26700 N and a high-load capacity rating of 29, 800 N: an increase in basic dynamic load rating of approximately 12%. As a result of this, bearing fatigue life is increased by approximately 40% for similar load conditions.

Bearing size		Basic dynamic load rating		
Bore diameter (mm)	Outside diameter (mm)	Standard bearings (N)	High-load capacity bearings(N)	Increase in basic dynamic load rating
25	52	14000	15300	9%
25	62	20600	23700	15%
30	62	19500	23300	19%
30	72	26700	29800	12%
35	72	25700	28300	10%
35	80	33500	39500	18%

Table 16.2 Improved life and increased dynamic load ratings

One of the key factors in the longer life of the high-load capacity bearings is a new UR heat treatment process, which is effective in alleviating the stress concentrations that cause surface-originated flaking on bearing material. Surface-originated flaking tends to occur under contaminated lubricating conditions. Stress concentrations develop immediately around indentations formed by foreign particles. The UR heat treatment process prevents this from happening, by increasing the amount of retained austenite within the bearing material. As a result, bearing durability is significantly improved in contaminated environments. This improvement is illustrated in Figure 16.45. The graph shows that the life of UR treated bearings has improved by twice that of standard bearings operating under conditions of contaminated lubricant.

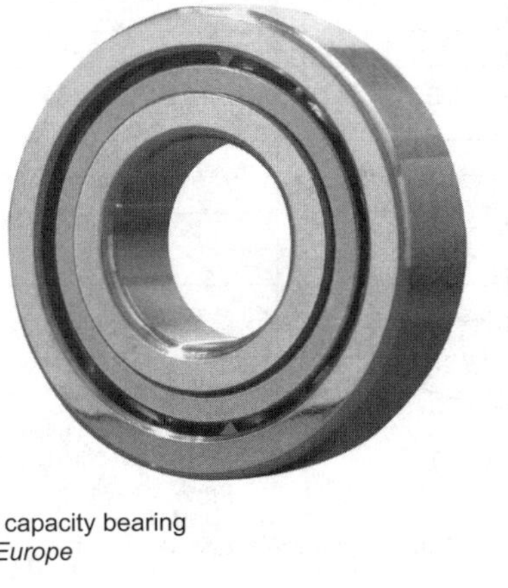

Figure 16.44 High capacity bearing
Courtesy of NSK Europe

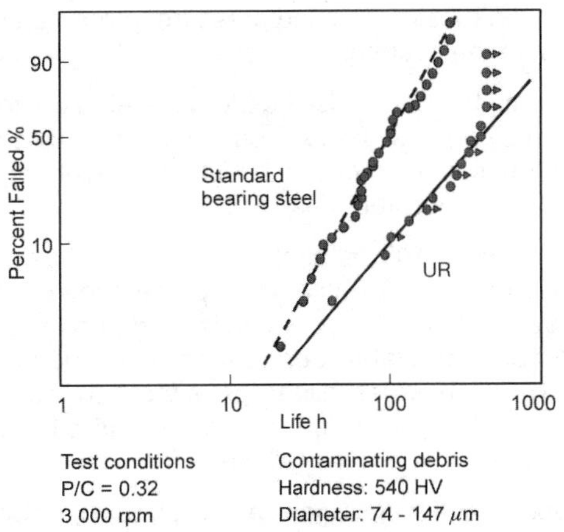

Figure 16.45 Rolling fatigue life of bearing under debris contaminated conditions
Courtesy of NSK Europe

Another problem in pump bearing technology is that of creep. This is the name given to wear that may occur between the bearing outside surface and housing bore due to low-speed sliding. It is the result of loads becoming excessively high on the driving shaft of a pump. If not addressed, creep causes excessive shaft vibration, which can trigger abnormal oscillation and lead to impeller damage.

Common measures against creep include an interference fit between bearing and housing. For pumps, however, a tight fit is not practical, as it makes pump assembly more difficult and does not allow for additional loading caused by an increase in the shaft length due to heat generation. A solution is to install O-rings into the two grooves on the bearing outer ring. The frictional force of the O-rings helps to prevent the occurrence of creep.

API 610 calls for angular contact ball bearings, with a 40° contact angle, arranged in a back-to-back configuration, and giving a minimum life expectancy of 25,000 hours (3 years) continuous running, or 16,000 hours at maximum radial and axial loads.

It is claimed that these new bearings, by incorporating the one of the strongest machined brass cage available, provide effective reliability in the harshest of applications. They also achieve longer service life, through smoother running, reduced levels of friction and lower temperature rise; the latter as a result of the better heat dissipation of the brass cage material.

It is also claimed that they combine high strength materials technology, ABEC3 (P6) manufacturing tolerances and pump specific axial internal clearances to give high running and shaft positioning accuracy, plus optimum levels of bearing stiffness, as well as load sharing and cooler running. In addition, by the combination of super-finished raceways, low inner ring run-out and high ball grades, it is hoped these will contribute to a reduced level of ball skidding and, hence, smoother operation.

Application courtesy of NSK Europe

16.16 Energy recovery turbines and reverse running centrifugal pumps

Introduction

Pumps are usually purchased to pressurise a liquid. This then enables flow to take place by overcoming resistance in the piping system. In most cases, the pressure generated is entirely absorbed into the system.

In some cases the pump also has to generate additional pressure to start a process. Boiler feed pumps are one example of this. Here the system frictional resistance might only account for 30% of the generated pressure.

Reverse Osmosis, RO, is another example. Again, the frictional resistance only forms a small percentage of the generated pressure. Most of the pressure is used to stimulate the osmotic process of water purification. About half of the pressurised saline water fails to pass through the RO membrane. This is then discarded. Since this fluid is typically at 45 to 90 barg it must be depressurised before it can be dumped.

In a scrubber plant, impure gas passes through absorbent fluid in the absorber tower. Impurities are absorbed into this carrier fluid. This takes place at relatively high pressure. The fluid then passes to the stripper tower where the impurities are released. To do this the pressure then has to be reduced, or "let-down".

In the refining of oil to gasoline, the product passes through a hydro-cracking process. Again the charge product is pressurised to finalise the process. After reaction, this pressure has to be dissipated before the next part of the process.

In all these cases, liquid has been pressurised to permit some process to take place. It then has to be depressurised for some reason. A simple way of depressurising the fluid is to pass it through an orifice plate. In this approach, all the energy embedded in the liquid is lost as heat and noise.

Another approach is to pass the liquid through a control valve. This is not much different from an orifice plate but at least it is variable.

Energy recovery and its uses

Yet another approach is to pass the fluid through a turbine. This might appear to be an expensive solution. However, it has the advantage that not only is the liquid depressurised, but the energy embedded in the liquid is largely recovered. The value of this power can then be offset against the turbine cost. Experience has shown that this recovered power can repay the cost of the turbine in 9 to 18 months. After that the energy recovered is almost free. On this basis, it is almost irresponsible not to explore all possible uses of energy recovery turbines. How can the energy be used?

- It can be used as the sole power source for another machine
- It can be used to contribute power to another machine and reduce the power consumed by the motor

Reverse running pumps

Conventional turbines are expensive when compared with pumps of the same power. They are not so readily available, nor is the choice so great. The material options are also more limited. For special applications, such as high temperature refinery service the choice is very limited. Fortunately, all forms of centrifugal pump can also be run in reverse to perform as very effective turbines.The disadvantages of commercial turbines are then eliminated. The only shortcoming of reverse running pumps is that their efficiency is lower than a true turbine. However, this is typically only two or three per cent and hardly impacts on the pay-back period. The notation of reverse running pump is as shown in Figure 16.46.

One apparent disadvantage of any turbine is that it needs some form of speed control. Fortunately this can easily be overcome in most cases. This depends upon the turbine being connected into a train that also utilises an electric motor. Electric motors lock into the mains frequency and act as excellent governors. This establishes the speed control the turbine needs.

Fundamentals

The performance characteristics of centrifugal pumps are well known. Those for a turbine are less well known. They are compared below.

Figure 16.47 shows typical pump performance characteristics. Generated pressure decreases as pump flow increases. On the other hand, absorbed power increases with flow.

Figure 16.48 shows the characteristics of a pump used as a turbine, i.e. the flow direction and direction of rotation are reversed but the pressure differential is the same. Note that the differential pressure increases as the flow increases. Also that the recovered power increases with flow. However, below a certain flow, the turbine actually absorbs power.

Figure 16.49 shows the most popular way of connecting a pump and energy recovery turbine. The clutch is optional and its effect is shown by referring to Figures 16.48 and 16.50.

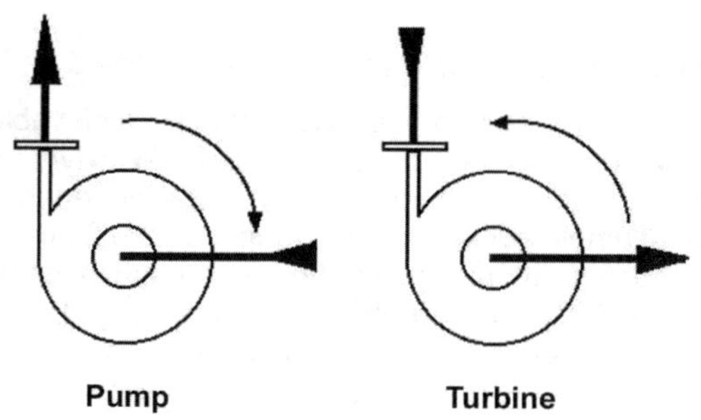

Figure 16.46 Schematic notation of pumps and turbines

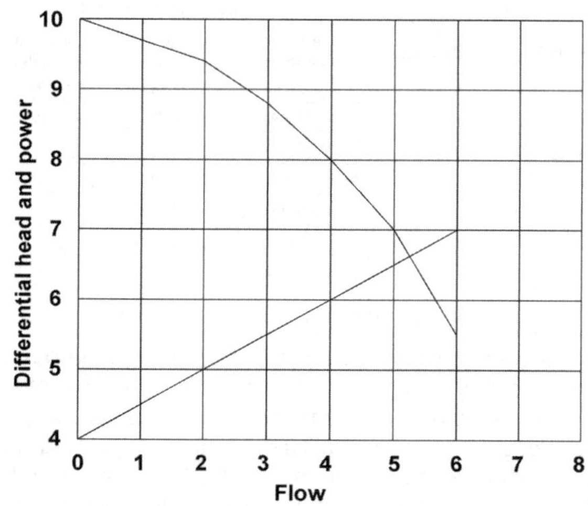

Figure 16.47 Typical centrifugal pump characteristics

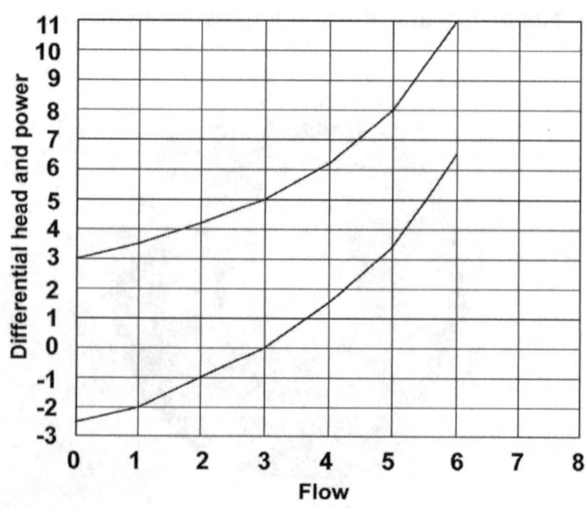

Figure 16.48 Typical inward flow radial turbine

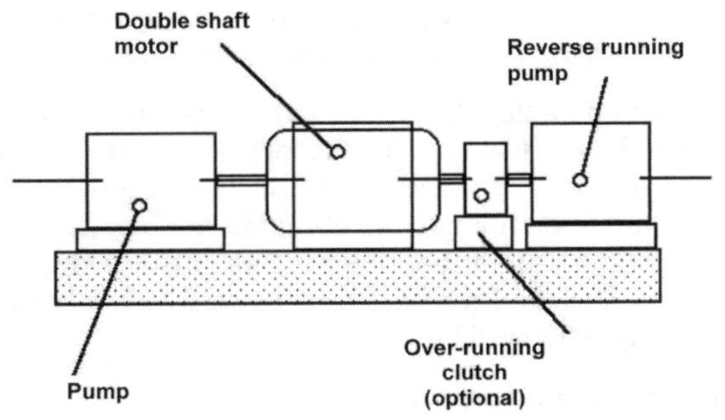

Figure 16.49 Typical arrangement of energy recovery turbine with a motor as the main driver

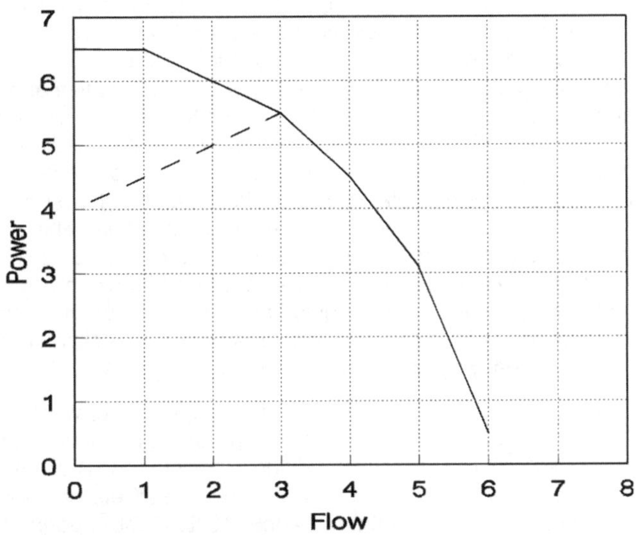

Figure 16.50 The effect of coupling an energy recovery turbine

Effects of power recovery

Figure 16.50 shows the effect of coupling an energy recovery turbine in-train with the pressurising pump as shown above. This shows the nett power delivered by the motor. It is calculated by subtracting the power recovered in the second chart from that absorbed in the first chart.

This nett power curve should be compared with the unassisted power curve shown in Figure 16.47. Note that at low flow, the motor is actually overcoming the power absorbed by both the turbine and the pump. As flow increases, the turbine begins to contribute and reduces the motor power requirements.

One way of reducing the low flow power demand is to install an overrunning clutch between the turbine and the motor. In this way the turbine is uncoupled from the train during starting at flows where it is absorbing power. It only becomes connected when power is being contributed.

16.17 Common errors which create pump operating problems

Most pump users and system designers do not think about pumps eight hours a day, every day; they think about pumps occasionally. Important factors revealed on the last project may not be remembered on the current one. Staff who gained good experience on the last project may have been transferred to a different department. The current project team may not have worked together before and an individual's depth of experience may be an unknown quantity. Most engineers rely on a web of personal contacts for support in "grey" areas; younger engineers may not have this backup support. The project manager must assess the qualities of all the project team and assign tasks accordingly.

Planning of pipework

In theory, starting a project from a clean sheet of paper, with an empty site, is the easiest type of project. But there does not appear to be a consistent logic applied to equipment arrangement and pipe runs. It seems obvious that the heaviest or biggest equipment must be located firstly where access is very good and adequate support is available. Heaviest/biggest equipment should be located around the periphery of the "plot" to provide the best access from outside. Heavy equipment should be located at ground level or as low as possible. Equipment which vibrates requires special consideration if the vibration is not to be transmitted through the supports. Many positive displacement pumps produce vibration and this factor must be considered when deciding locations and supports.

A number of pieces of equipment are not used in isolation but connected to other equipment. Pumps are connected into systems by pipework. Project planning should allocate priorities to pipework to allow "difficult" pipework to have good pipe runs while "easy" pipework can be fitted in the spaces left over. The choice of "difficult" and "easy" is not necessarily straightforward and requires some thought. The following points though should be considered:

- Size
- Pressure rating
- Differential temperature
- Water hammer effects
- Pressure pulsations
- Viscosity
- Essential routing requirements

Large pipes are obviously much more difficult to handle than small pipes. Access requirements around flanged connections may result in much wasted space. Pipe supports from the ground may be the only feasible solution. Space requirements for bends may preclude changes in direction and a straight pipe between A and B may be the only feasible option. Large pipes tend to be low pressure, although not always; and it may be worthwhile calculating the pressure times bore area; this can be used as a quantitative measure for priority. Higher pressure ratings require thicker pipe walls to reduce stressing to acceptable levels. Thicker pipe walls increase the difficulty of bending the pipe and increase the size of welds. The weight of pipe per unit length divided by the nominal diameter can be used as a priority factor. Piping which operates hot or cold, say 50 °C or more above or below the ambient temperature, may require extra space to include axial flexibility to cope with thermal expansion/contraction. The differential temperature divided by 50 can be used as a priority assessment. The fluid flowing through pipes possesses dynamic properties — kinetic energy. Pipework systems cannot always be steady-state. A piping system must start up and sometimes it must stop; emergency stoppages, due to process upsets may be the worst case.

Water hammer effects should be evaluated. The liquid density is also an important factor. Water will produce much larger pressure changes than air for the same dynamic conditions. The water hammer effect can be judged by multiplying the pressure change by the pipe bore area. Pipework with large water hammer effects should be given a higher priority than lower ones. If these effects are likely to happen often, more than 7000 times over the whole life of the system, the pressure effects on fatigue should be investigated. Some equipment connected to pipework may not have a steady flow characteristic.

Some positive displacement pumps have significant cyclic flow variations which produce pressure pulsations. The flow variations produced by reciprocating pumps are well documented although the values given are usually wrong. Reciprocating pump flow variations are a variable not a fixed value. Peristaltic pumps produce very noticeable flow variations. Lobe pumps and circumferential piston pumps can have cyclic flow variations which are large enough to produce troublesome piping vibrations. The system designer or piping designer must select a value of acceptable pressure pulsations which is appropriate for the pipework construction and supports and for all the equipment connected to the pipework.

Delicate instrumentation, such as flow meters, may require lower pulsation levels than those dictated by pipework mechanics. Pipework subject to pressure pulsations should be given a higher priority than pipework with steady-flow conditions. An early routing schedule should allow a better chance of providing a straight pipe, which will not vibrate, rather than having a number of bends to avoid existing equipment. Systems for high viscosity liquids should be allocated a higher priority than water-like liquids. The choices for suction pipe route options may be extremely limited when the liquid has flowing difficulties. Short, straight pipes with good fall are essential.

Pipework routing can be very important. Suction pipework which is not self-venting can pose tremendous installation problems if vent lines must be run considerable distances for safe disposal. Pipe falls are very important. Short pipes are better than long pipes. Straight pipes are much better than pipes with bends. Suction pipework is eliminated when submersible pumps are used. Cavitation is the most frequent pump problem. Using a submersible pump should help to eliminate a lot of the possible causes of cavitation. Liquid systems should be given priority over gas systems; compressors do not cavitate.

Logical project planning does not seem to work very well. Pipework for high risk applications and high pressure, viscous liquids with pressure pulsations, seem to be always fitted as an afterthought. The system components are far apart and the pipe runs are never straight. Pipe supports are poor and pipe anchors are frequently non-existent. It appears that no sensible criteria are applied to prioritise equipment positioning and pipe run locations.

Liquid properties

The proper selection of the best pump type is based on a good description of the operating conditions and the liquid properties. Pump specifiers and users have a habit of concentrating on "maximums" and forgetting about the "minimums". All operating conditions are important. The range of operating conditions is very important when evaluating pump performance. The pump manufacturer does not provide guarantees for undocumented conditions.

The NPSHa/NPIPa available is very important. Cavitation is a very serious problem and increasing NPSHa or NPIPa is not usually easy. Reducing the pump NPSHr/NPIPr usually means operating at lower speed and flow. The vapour pressure of the liquid is a critical factor in the NPSHa/NPIPa calculation. The vapour pressure of a trace element may be significant. Small quantities of water in high temperature processes can produce a very large volume of steam relative to the liquid volume of the other components. Centrifugal pumps may lose prime with a 15% gas volume. Reciprocating pumps may suffer major component failures with a 5% gas volume. The vapour pressure of the liquid may not be important if dissolved gas evolves from solution. The apparent vapour pressure, the pressure at which gas bubbles appear, may be considerably higher than the liquid vapour pressure. Entrained gas creates cavitation-like symptoms; early component failures can be expected.

Very small quantities of solids can increase component wear dramatically. Materials of construction are adjusted when solids are included in the duty specification. The pump operating speed may be reduced to compensate for wear rates. It may not be possible to successfully adapt an installed pump to cope with solids. Reduced flow and high spares usage are the usual effects of handling unspecified solids.

The pump manufacturer assumes all changes in operating conditions are slow unless specifically identified as rapid. Temperature changes are slow enough not to create thermal shock. Pressure and flow changes are also slow enough not to create surge problems. Rapid changes in conditions are very likely to cause early failure of major components. The cost of spares and lost production will be high.

Equipment positioning

The early planning of systems must include the relative positioning of the system components. The importance of good positioning increases as the "liquid" becomes more difficult to handle. In some processes "liquid" is a poor description of the process medium. Viscosity can be so high that the process medium will not flow under the influence of gravity; positive assistance is necessary for flow to occur. When the process medium has very high viscosity it is sensible to arrange the medium to fall into the pump. Trying to lift the medium into the pump is not a good idea. System planning should organise the suction source to be above the pump suction connection. This approach is always a good idea even with water-like liquids. The suction source should be as close as possible to the pump to reduce piping losses. Difficult liquid applications should have the equipment locations allocated very early in the project calendar and before easy systems such as water and gases.

Many systems rely on atmospheric pressure to provide the pump suction pressure. Atmospheric pressure is not constant and varies with the weather conditions. The standard barometric pressure, at sea level, is 101.325 kPa. Weather conditions conditions cause a variation of about ±3%. Most importantly, many installations are not at sea level. At 500 m the standard air pressure has dropped to 95.461 kPa and at 1000 m to 89.876 kPa. From a water pumping perspective the effective suction lift of a pump is reduced by about 0.5 m for every 500 m of altitude. Also of interest is the variation in "g". At sea level the standard value is 9.80665 m/s^2. At 500 and 1000 m the standard values are 9.8051 and 9.8036 m/s^2. In pumping the only constant which can be relied upon is π.

Pipework design

Pipework design is very difficult. This is why many installations are poorly designed and suffer reliability problems and fatigue failures. Most piping appears to be designed by software which only considers "Code compliance", that is stressing which guarantees insurance cover. The prime function of the pipe, to carry fluid, does not seem to be considered; hydraulic considerations are usually ignored. In many cases the final pipework design is safe but completely unsuitable for the purpose.

Hydraulically, the best pipework is straight. Bent pipes are better than pipes fabricated with pressed bends. Pipework which is likely to be subjected to continuous pressure pulsations requires special attention. Pressure pulsations cannot create pipe vibration if the pipe is straight. Pressure pulsations act on bends to create the vibration. Sharp bends, pressed bends, create more vibration than gentle bends and bent pipe. Experience has shown that 45° sets, used in pairs, perform much better than bends.

Pipework which is likely to be subject to continuous pressure pulsations requires very good support and restraint. Pipework suspended by hangers will generally vibrate unacceptably and the vibration may induce unacceptable stressing. This pipework must be supported by clamps incorporating damped restraints and with rigid anchors at strategic locations. It may induce structural vibration if clamped rigidly, at high elevations, to building steelwork. The best results are obtained with low level pipework supported directly from concrete.

Successful pipework design can only be accomplished when the designer considers the actual hydraulic process which is taking place inside the pipe. The design must be appropriate for the hydraulic conditions. All pipework is not equal.

16.18 Positive displacement pumps — load-unload control systems

Introduction

There are several applications where load-unload control has proved to be the best practical solution. Steelworks hydraulic power supply and steelworks descaling are two typical applications which only differ in the process liquid. Very large high pressure systems have used multi-stage centrifugal pumps with load-unload control and these pumps suffer similar problems (broken shafts) when the control system deteriorates or is inappropriate.

Steelworks use hydraulic power to actuate and drive much equipment. Equipment located close to hot metal poses a fire risk if hydraulic oil leaks and is ignited. For this reason many hydraulic systems have utilised water or water-oil emulsions. The lubrication qualities of water can be significantly improved if 5% or 3% soluble oil is included without resulting in a flammable liquid. Descaling utilises cold water to produce thermal shock which violently fractures and loosens the scale on hot metal surfaces. Centralised hydraulic systems and centralised descaling systems operate at pressures of 100 barg up to about 450 barg. Plunger pumps are used when flows up to approximately 120 m^3/h from each pump, are required. Small systems would use three pumps, two operating and one standby. Large systems may use five to ten operating pumps with two standby units.

Typical system

Water, or water/oil emulsion, is stored in a header tank at an elevation to provide sufficient NPIPa for the pumps. Water/oil emulsion must be continuously recirculated around the tank to prevent separation and also to allow oil concentration monitoring so the oil content can be adjusted if required. The feed from the header tank flows via a pair of strainers, one working/one standby, to the suctions of the three pumps. Each pump has a flow-through suction pulsation damper and a discharge pulsation damper. The flow-through style of damper is preferred for multi-pump installations because the degree of uncoupling between the pump and the system is much higher than appendage styles. There is much less chance of pump interaction affecting the system performance. The pump relief valve is located immediately after the discharge damper. The load-unload bypass valve is mounted before the non-return valve and the isolating valve. The one or two operating pumps feed a bank of accumulators which are used to store high pressure water. The use of accumulators allows the system to supply instantaneous flow rates much greater than the nominal pump capacity. As hydraulic power and descaling requirements are intermittent, the accumulators and the pumps can supply the peak demands. The pumps can recharge the accumulators when demand is low.

Electric motors

Load-unload control is better than on-off control because the electric system, starter and motor, are not subjected to the effects of high inrush currents. Motor starting, especially when hot, may be severely restricted for motors over 45 kW. During bypass operation the motor and the pump absorb very little power. Power consumption can be reduced even further if a modern soft starter is used. Modern soft starters often include a low power and/or low speed unload control. The motor is able to operate efficiently at low load and possibly at reduced speed. Load-unload control can be applied very rapidly without overheating the motor or wearing out the contactor. Load-unload is not instantaneous but requires a finite time to change from low-pressure bypass to high-pressure on-load. Sizing of the accumulators is critical for the load cycle to operate smoothly.

Operation

These types of intermittent system are controlled by pressure. As the system flow demand reduces, the pump over-capacity is absorbed by the accumulators; this causes an increase in system pressure. When the pressure reaches a predetermined value, say 85% of maximum, one pump is unloaded. If the pressure continues to rise the second pump will unload at 100% pressure. At this condition the maximum volume of water is stored in the accumulators and the total pump flow is zero. If the system pressure remained constant at 85% then the system demand would be equal to the flow from one pump. If the system pressure decayed, because the demand was greater than a single pump flow, the unloaded pump would be loaded to prevent the system pressure reducing below a preset minimum.

This mode of operation describes the traditional control system which used fixed pressure set-points, pressure switches, to initiate control sequences. Systems using a PLC and pressure transducers, would consider the static pressure and the rate of pressure change to decide on load-unload set-points. The set-points would be variable rather than fixed.

Bypass valves

Load-unload systems suffer two recurring problems. Both multi-stage centrifugal pump installations and plunger pump installations suffer when an inappropriate bypass valve is used. Plunger pump installations frequently experience pipework vibration problems. The bypass valve is nominally an open or closed valve. In practice the valve must be slow-acting and have good throttling capabilities. Many installations are built with power actuated open/close valves, powered isolating valves, and the cycle time is preset. Because of the poor part-open flow characteristics of isolating valves the high pressure system is subject to surge and pressure pulsation problems.

In extreme cases pump components such as centrifugal pump shafts, crankshafts and plunger pump crossheads , fail due to fatigue induced by the shock loadings. Hydraulically actuated valves, using the process water for actuation, have generally been unsuccessful because of the poor lubricating qualities of the water. The bypass valve should be a pressure control valve, see Figure 16.51, which has specific throttling characteristics.

The globe valve shown has a contoured plug which provides a guaranteed throttling characteristic. The pressure rise during

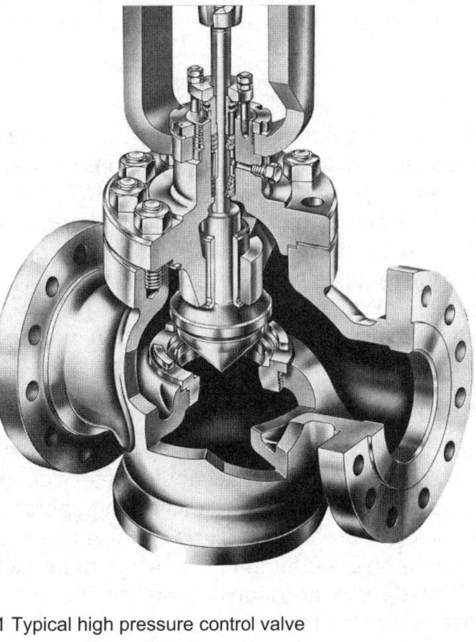

Figure 16.51 Typical high pressure control valve

valve closure can be smooth and progressive and eliminate all surge effects. This style of control valve would normally be pneumatically operated but could be electrically operated. If a PLC was used for system control the bypass valve could be modulated to provide 0 to 100% pump flow for low flow situations. Pipework vibration can be experienced by centrifugal pump and plunger pump installations which try to use inappropriate bypass valves.

The flow surge created by an "isolating" valve opening or closing produces an associated pressure pulse. The pressure pulse causes pipework vibration. The pipework vibration is usually of very short duration but can be violent. Pipework failure through fatigue is possible. Short duration vibration can also be triggered by the equipment using the high pressure water or water/oil emulsion. If the equipment isolating valve is not slow-acting or does not have a good throttling characteristic then surge associated failures are likely.

Damping

Plunger pump installations frequently experience continuous high levels of pipework vibration. Pipework routings are usually far from ideal and elevated pipework may be suspended from building structural steelwork rather than the preferred low-level pipework firmly attached to the ground. The high vibration levels are due to a mismatch between the pump damper specification and the pipework construction style. The pump manufacturer is not usually involved in pipework design. If the system designer does not specify an appropriate level of damping for the distribution system the pump manufacturer will only provide a damper based on the pipework design local to the pump. The system designer is responsible for the correct operation of the whole system. Pump residual pressure pulsations are easily transmitted through the whole system and can create vibration anywhere. Co-ordination and communication are the keys to success.

16.19 Plunger pumps for supercritical extraction service

Introduction

It is claimed that by using carbon dioxide supercritical technology, purer and therefore more valuable end products may be produced. Optimal process results relative to product quality depend upon the ability to control the process parameters in the high pressure loop. As control of the carbon dioxide process parameters requires sophisticated techniques and knowledge, so does the design of reciprocating pumping equipment to be used in these processes. The combination of these conditions give rise to the desired operating characteristics now described.

Background

Carbon dioxide supercritical extraction is a far reaching separation-oriented process that is growing in commercial application. Early commercial plants were designed in 1973 and went into service in 1979. Since then many additional and larger plants have come into operation. The efficient application of this process technique requires highly reliable and functional capability of the more important system components. Thus the pumping equipment required to pressurize the carbon dioxide liquid is a key process component.

Applications

Food and pharmaceutical industries require the extraction of basic, rare, aromatic and/or taste substances for economic productivity and high product quality. Although supercritical separation technology has been well known for many years, it has only recently been used for industrial scale manufacturing applications. The prime application is in the extraction of substances from plant life raw materials.

The prime benefits of the supercritical extraction process relate to the non-destructive handling of the raw material and in addition the desired end product has an extremely high degree of purity. Also, the process can be made extremely selective as to the material(s) extracted. In fact, by using selected combinations of pressure and temperature, several different materials may be extracted from a common raw material.

The process

As can be imagined, the amount of the supercritical liquid throughput must be exactly maintained, controlled, and monitored. Pressurization of the solvent stream is usually accomplished via a positive displacement plunger type pump, designed to accommodate the extraction medium's physical properties. Plunger pumps are especially adaptable to this application in view of their high discharge pressures and small throughput capacities. As an example, because of the compressibility of liquid CO_2, an especially critical necessity is the reduction of the pump's internal clearance volume to achieve optimum volumetric efficiency.

Experience shows that plunger pumps deliver an excellent overall efficiency and thus are highly economical machines to develop the required pressures and capacities. This result is in part attributable to the proportionality of pump speed vs. capacity and its relative indifference to differential pressure. Performance is as illustrated in Figure 16.52.

A high pressure supercritical recycle process

Figure 16.53 depicts the temperature/entropy curve for a carbon dioxide loop for a typical extraction process. A pump pressurizes the carbon dioxide liquid from I to II (suction pressure P1 to discharge pressure P2). At pressure P2, the carbon diox-

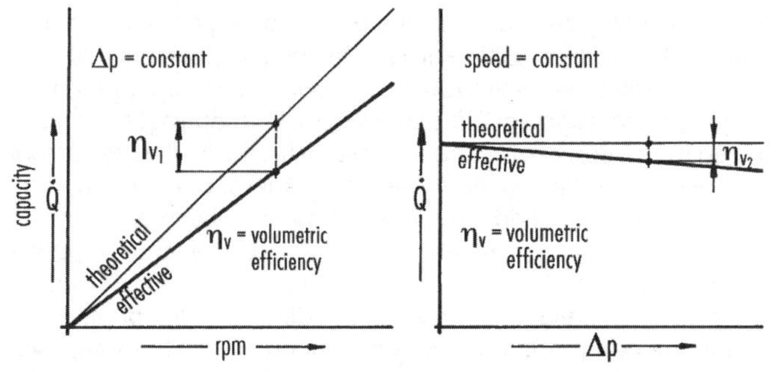

Figure 16.52 Dependency of volumetric flow and discharge pressure of reciprocating plunger pumps
Courtesy of Uraca Pumpenfabrik GmbH & Co KG

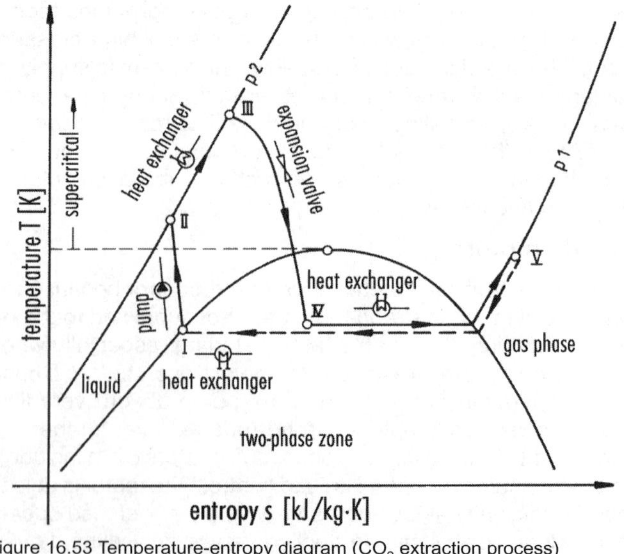

Figure 16.53 Temperature-entropy diagram (CO_2 extraction process)
Courtesy of Uraca Pumpenfabrik GmbH & Co KG

ide is heated in a heat exchanger from II to III i.e. the process temperature. An extractor is employed to achieve the desired carbon dioxide combination of pressure and temperature and is let down from III to IV. An evaporator is used to proceed from IV to V at which point the heavier product is separated out. The "purified" carbon dioxide proceeds from V to I and is condensed, sub-cooled and at process point I, sent on at P1 to the pump suction.

Potential variations in the pressure, temperature and flow process conditions permit the extraction of diversified substances from various raw materials. Therefore, the process is extremely flexible and lends itself to many potential applications.

Supercritical extraction system

Assuming a common raw material, it is also possible to extract selectively several different components (mostly as liquid) as previously indicated. In general, a supercritical extraction system consists of four major components:

- A pump to pressurize the extraction medium
- A pressure vessel (extractor) where the extraction process takes place
- A pressure vessel (separator) where the extract and medium are separated
- Heat exchangers to control the extraction and separation temperatures

Figure 16.54 shows a typical process arrangement for these elements. The Roman numerals correspond to the entropy diagram in Figure 16.53.

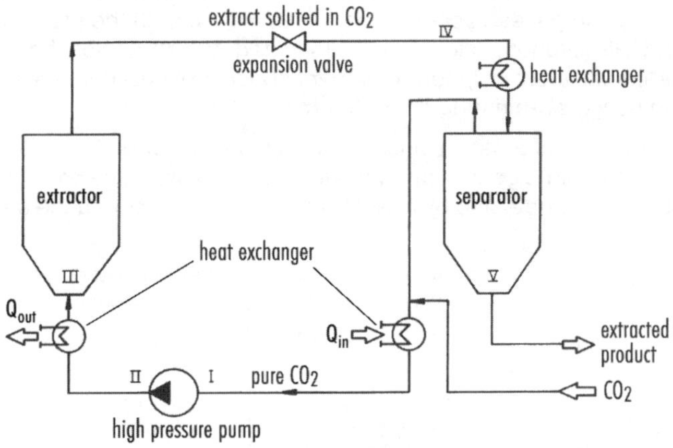

Liquid carbon dioxide pump requirements

There are specific requirements for the design of a high pressure pump for use on liquid carbon dioxide. Initially, the pump must be capable of pressurizing the liquid to pressures over the supercritical threshold. Liquid carbon dioxide is extremely compressible as may be seen from the thermodynamic table. There is also the problem that the temperature gain during the pressurizing process is not precisely controllable. As such, especially in large pump applications, there is the question of not being able to accurately predict pump capacity. This problem must be solved through experience and empirical measurement. To understand these pump issues further, a research project was undertaken to explore what happened in the pressurizing chamber of a plunger pump so that this data could be used to further advance the design technology.

The pump was fitted with instrumentation to measure the following: pressure in the space between the suction and discharge valves, inlet and outlet temperatures, cooling medium temperatures and ambient temperature. See Figure 16.55.

Figure 16.56 shows the developed indicator diagram through the synchronization of the functions in the liquid cavity and plunger stroke. In this diagram, A to B is the suction stroke. B to

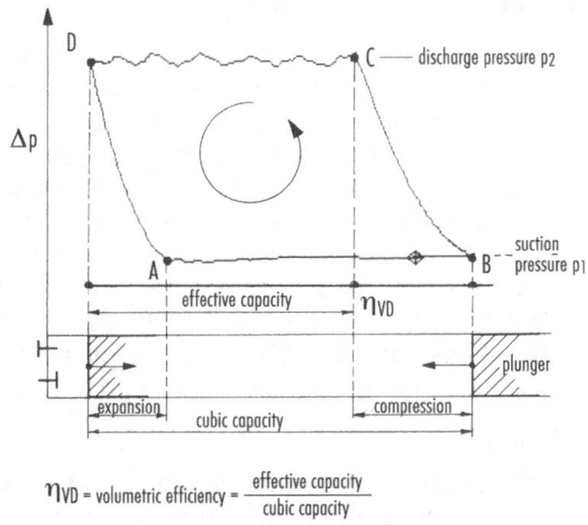

C is the pressurization to the required discharge pressure; and C to D is the discharge valve opening permitting the liquid carbon dioxide to enter the discharge line.

The relationship of C to D and B to D demonstrates a volumetric efficiency VD, considerably less than that obtained when using water. This is explained by compressibility. D to A shows the expansion of the liquid carbon dioxide remaining in the pump's clearance volume. The cycle is repeated starting at A.

Repetitive testing under varying process parameters provided extremely useful empirical data to enable the pump equipment design to supply exactly the required process service conditions. In addition to concerns about the strength of materials, cavitation-free operation is an absolute necessity to maintain trouble-free operation. Also, to maintain a high degree of operational efficiency, the seals on the suction and discharge valves are an important consideration.

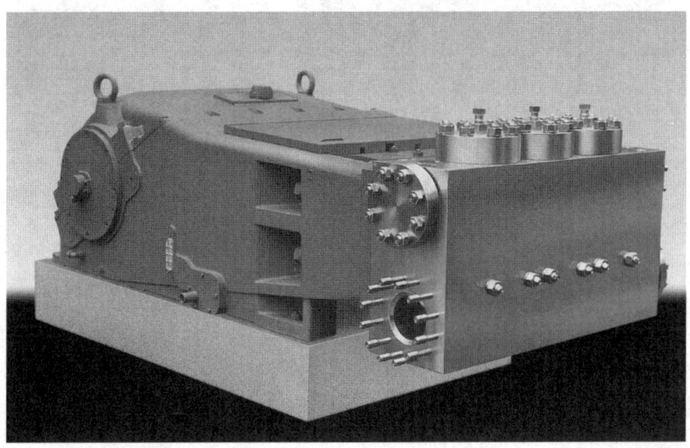

Summary

Test results indicate that high pump efficiency may only be obtained by the careful design of the high pressure liquid end to absolutely minimize dead space. Also of interest is the stuffing box sealing system, in the light of liquid carbon dioxide lubrication and cooling characteristics. A method of monitoring the functioning of the seal system is a prime design consideration. In summary, pumping liquids used in supercritical applications requires especially suitable construction materials and pump design technology (See Figure 16.57) based upon practical application and creative research and development.

Application courtesy of Uraca GmbH + Co KG

16.20 Sealing systems for thermal oil pumps

Introduction

In industrial heating, thermal oil has become a popular heat transfer medium. Compared to hot water systems thermal oil can transfer higher temperatures at lower pressures. This means that hot oil systems allow a lower pressure rating. An electrical heater or burner heats up the oil, users consume the heat, and there is a circulation pump. The trouble-free operation depends primarily on a reliable circulation pump with a corresponding shaft sealing system. Every user of hot oil systems should be aware of the different sealing systems and pump concepts and their advantages. Using the cheapest pump solution often leads to high maintenance costs during operation.

Pump concepts

Low-cost centrifugal pump concepts normally use standard single-stage, end suction foot-mounted ductile iron volute casings with a standard single mechanical seal that is separated from the hydraulic parts by an air-cooled intermediate spacer. The temperature limit is typically 350°C. However, in applications above 320°C this standard foot-mounted casing can be over stressed and distorted by typically unknown forces and moments generated by the piping system during heating of the plant. For temperatures above 320°C, up to 400°C, a pumping concept based on the standard chemical pump design is often a better solution. The pump casings can be equipped with centre line mounting, see Figure 16.58. The material can be improved to cast steel and different sealing systems can be used. When cast steel is used, it is also possible to use welded drain connections, which are much easier to seal than screwed connections. These are becoming more common in new installations, see Figure 16.59.

Figure 16.58 Typical centreline mounted pump installation
Courtesy of Dickow Pumpen KG

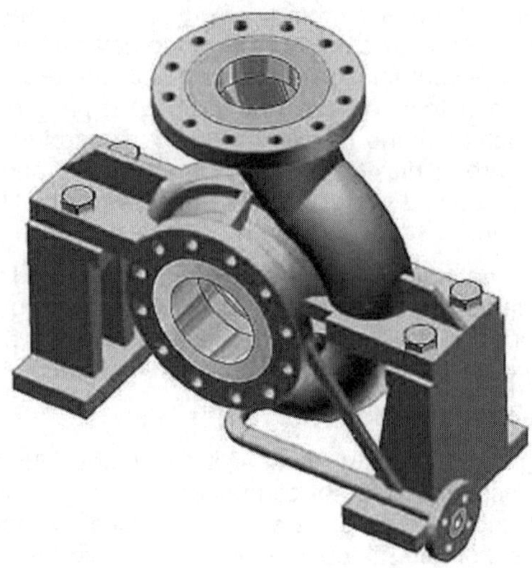

Figure 16.59 Example of a volute casing with a centreline support
Courtesy of Dickow Pumpen KG

Sealing systems

Especially for low viscosity and high temperature synthetic oils; e.g. Dowtherm or Diphyl; sophisticated tandem seal arrangements or magnetic couplings are required. In Figure 16.60, an ideal tandem seal arrangement is shown. In order to minimise the influence of solids, such as decomposition sludge on the seal lifetime, the seal face combination is SiC against SiC (silicon carbide is a very hard ceramic material); the bellows are rotating in order to prevent clogging and the seal chamber is designed as large as possible to prevent the accumulation of solids and gases/vapours. The two seals, on both the product and atmospheric sides, are lubricated by an atmospheric air-cooled thermosiphon vessel which also prevents damage to the seals, even during dry running conditions.

A constant lubrication and cooling circulation flow (of buffer fluid, e.g. oil) is generated by a pump ring. Failure of the product seal can be detected by a level switch in the vessel and the at-

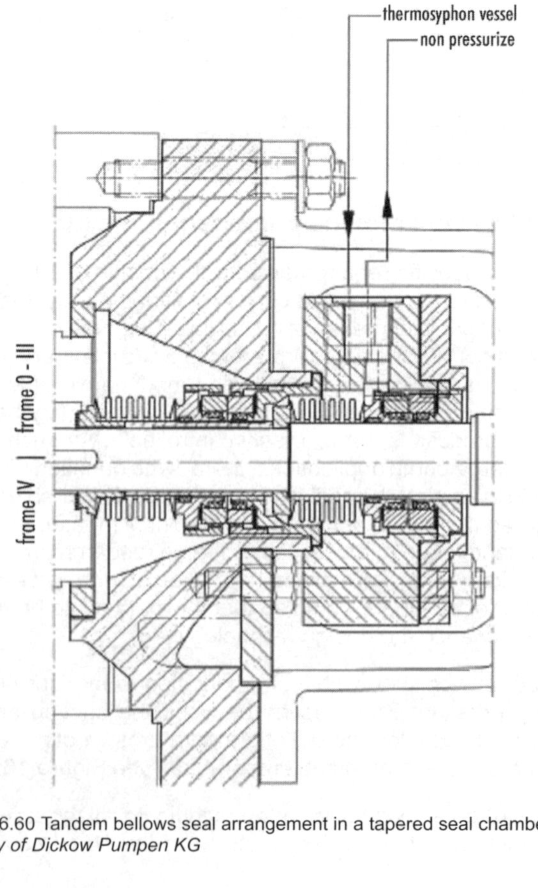

Figure 16.60 Tandem bellows seal arrangement in a tapered seal chamber
Courtesy of Dickow Pumpen KG

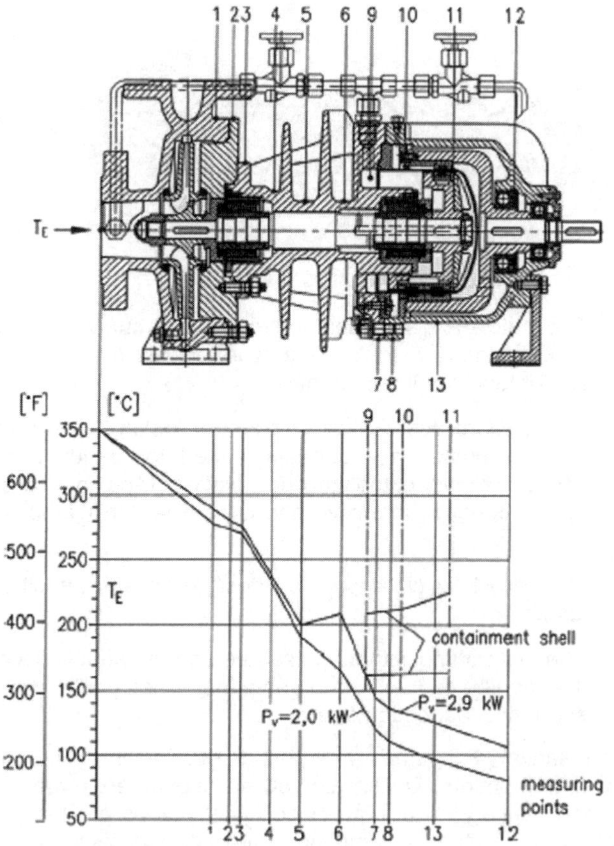

Figure 16.61 Air-cooled magnetic drive pump showing temperature profile
Courtesy of Dickow Pumpen KG

mospheric seal will prevent leakage. For applications above 350°C, it is possible to have additional water cooling with a cooling coil inside the thermosiphon vessel.

Another reliable solution is static shaft sealing by magnetic coupling, which is also air-cooled and is absolutely leak-free, Figure 16.61. The operating principle of the magnetic coupling is very easy. There are internal and external rotating magnets that are separated by a stationary containment shell. This special pump design separates the magnets with an air-cooled finned section from the hot pump casing, which means that the magnets are running at a reduced temperature level, maximum 250°C. The magnets are also additionally protected against solids because they are separated from the main process flow.

Once again SiC against SiC is used for the liquid-lubricated bearings in order to minimize wear. The standard containment shell material is Hastelloy C. Eddy current losses are generated through the rotating magnetic field and these heat up the shell. If the losses are more than 2 kW, the heat radiation of the casing will be too small and must be supported by an additional air-cooling loop. The standard design can even be improved with a close-coupled motor and a ceramic shell, Figure 16.62. The flanged motor prevents misalignment of the flexible coupling and the ceramic material has zero magnetic losses because it is an electrical isolator and non-magnetic. Secondary containment with a leakage monitoring device can be added for additional safety.

Field experience and application

In 2002 a leading Swiss heat transfer systems company awarded contracts for several heat transfer systems with multiple heaters for polyester plants in China. The heaters had a capacity of 14 to 16 MW. The selected heat transfer medium was synthetic Dowtherm A. Every heater generated about 157,600 kg/h vapour with a temperature of 337°C and a pressure of 3.5 bar. Process heating using thermal fluids in the vapour phase (flash system or secondary vaporiser) makes it possible to distribute a constant supply of heat uniformly to several users. The capacities range from 100 kW up to about 45 MW per heater, thus covering a broad spectrum of heating requirements. Even larger capacities can be achieved by combining several units in a battery, which also increases plant availability at the same time.

The circulation pumps of such a system are exposed to extraordinary conditions. At a temperature of 330°C, a flow of 530 to 630 m^3/h against a head of 90 m must be circulated. The typical low-cost pump with a standard foot-mounted casing and a single mechanical seal, as described earlier, proved to be too weak for this kind of application. The low viscosity synthetic oil can additionally lead to leakage through the single mechanical seal. Therefore heavy duty hot oil pumps, were selected. A robust design with centre line-mounted cast steel casings and a tandem seal arrangement were selected as the sealing system, see also Figures 16.59 and 16.60. The first pumps installed have been running for several years without any major problems

Pump selection

How can you select the right pump and sealing system? Unfortunately, there are no fixed temperature limits, it also depends on the piping and pump size, etc. In the region above 320°C and when using synthetic oils, it is strongly recommended to carefully examine the pump and seal design. When comparing double seal systems and magnetic couplings, the latter is not always more expensive. In regions up to 55 kW the initial costs of magnetic coupled pumps are even lower. For larger motors however (and therefore larger couplings), the double seal systems look more attractive. The longest MTBF (mean time be-

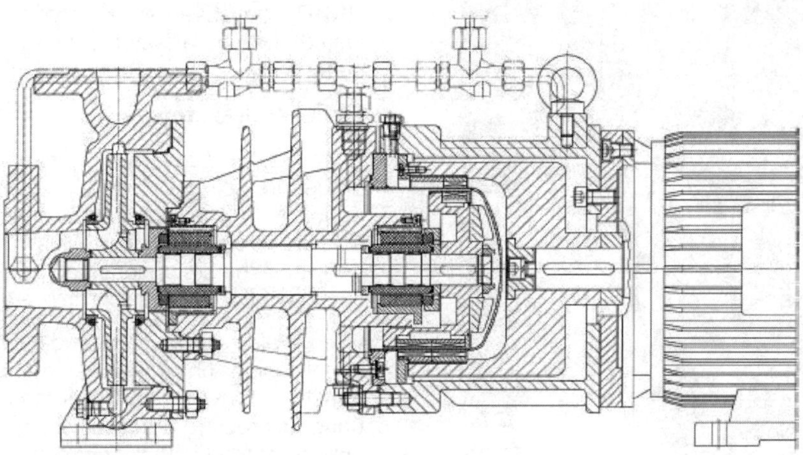

Figure 16.62 Air-cooled magnetic drive pump with ceramic containment shell
Courtesy of Dickow Pumpen KG

tween failure), up to ten years, can only be reached with an appropriate magnetic coupled pump.

Application courtesy of Dickow Pumpen KG

16.21 Progressive cavity pump stator technology for improved candy production

Introduction

One of the oldest continuously operating candy companies in the USA undertook an expansion programme involving a new, state-of-the-art facility which included a brand new moulding line for candy bar production. Before the move, the company had been relying on traditional positive displacement (PD) pumps to transfer the candy bar fillings from storage to the filling line. But, even as the company began planning the expansion, design engineers were encouraging management to consider progressive cavity, (PC) pumps, for the new facility, see Figure 16.63. It was suggested that, among other benefits, PC pumps would run at the higher pressures needed for the viscous candy fillings.

The progressive cavity pump

In operation, a progressive cavity pump's single helix rotor rotates within an elastomeric double helix stator to form sealed cavities that progress from the suction side to the discharge end of the pump. The continuous seal between the rotor and the stator helices moves the product steadily, without valves or pulsations, at a fixed flow rate proportional to the rotational speed of the pump and independent of pressure fluctuations that can result from varying densities and viscosities of conveyed product. See Figure 16.64.

Several factors make these sanitary progressive cavity pumps the right choice for the new facility:

- The pumps use an Even Wall® stator, meaning the same performance can be achieved with a compact 2-stage pump that would usually require a much longer 4-stage with conventional stator geometry.

Figure 16.63 The pump installation
Courtesy of seepex GmbH & Co KG

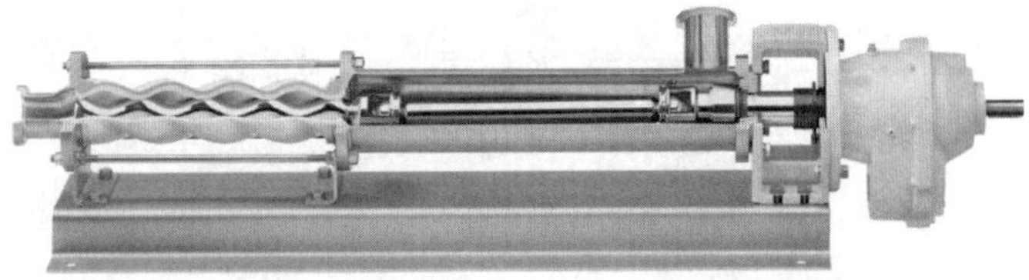

- The stators are also moulded to size, not cut, with integral seals molded into the ends, and are offered in a wide selection of food-grade elastomers.
- These more economical and longer lasting, closed joint sanitary pumps are perfectly suited for candy fillings, whereas a dairy-compliant open joint would be unnecessarily maintenance intensive and expensive for this application.
- Their 'block,' or close-coupled, design conserves additional space.
- The units pump without pulsation; and at variable speeds they can be counted on to pump the fillings gently, with no shear or product damage.

Four sanitary PC pumps were purchased. They had Therban® even wall stators, Duktil®-coated stainless steel rotors, and closed sanitary joints. The closed joint pumps are furnished with cost-effective pin joints and filled with special food-grade grease, making these pumps especially suited to high operating pressures. The pumps are rated to handle up to 0.25-US gpm on continuous service, customised for viscosity of 125,000 cP, at 120 °F, at pressures up to 24 bar.

The pumps are equipped with variable frequency drives for process control, with speeds proportional to a 4-20 mA signal coming from a level sensor on the filling machine. The space-saving block (close-coupled) design utilizes the gear reducer bearings to absorb the axial and radial loads of the pump. In fact, these units are assured of being capable of a minimum of an AFBM (Anti-Friction Bearing Manufacturers Association) L-10 life of 50,000 hours. Optionally, an L- 10 life of >100,000 hours is also available.

The block pumps require only simple bent steel 'top hat' base plates for mounting. Therban® is a proprietary hydrogenated nitrile butadiene rubber (HNBR) elastomer. The compound is made to CFR 177.2600 and is considered "food grade". HNBR offers superior abrasion and temperature resistance compared with NBR compounds or ethylene propylene diene monomers (EPDM) which are commonly offered for food applications. The HNBR in an Even Wall® configuration has consistently outlasted NBR by a factor of five times or more in Even Wall® stators.

Even Wall® stators

Even Wall® stators, see Figure 16.65, are made with a special procedure using a metal tube that is hydraulically bent into a double helix, the same shape as the helix forming the cavity inside the pumping element. This makes the thickness of the rubber within the stator constant, as opposed to the thin and thick sections of elastomer existing in a normal cylindrical tube.

Even Wall® stators can handle twice the pressure of conventional designs on non-abrasive fluids, with an even smaller footprint than conventional PC pumps. Both hysteresis failures and bond failures are practically eliminated. These stators are excellent for food applications that have wide temperature variations that can result from low temperature operation and high temperature cleaning. Users report that these stators routinely

last up to five times longer than conventionally designed stators in these applications.

Duktil® is also a proprietary coating. Sometimes referred to as "chromium nitriding", it involves plating chromium at extremely high currents and heat. The result is a coating that exceeds a hardness of 90 Rc and actually diffuses into the base metal. Consequently, it can bend (having ductile properties) without cracking, as standard hard chrome can.

Summary

The progressive cavity pumps currently transfer nougat, toffee, fudge and caramel fillings. The company has achieved savings by the minimal downtime required for repairs and in low energy usage. Additionally, the same pumps can be used for the four different fillings, with their viscosity variations, because of the variable speed possibilities and the ease of cleaning between operations.

Application courtesy of seepex GmbH & Co KG and Kächele GmbH

16.22 Digital controllers for diaphragm dosing pumps with speed-controlled drive motors

Introduction

There are a wide range of highly sophisticated dosing tasks in water technology and process engineering which place high demands on dosing pumps in terms of precision, reliability and, in particular, monitoring the dosing process. Examples of these complex applications include dosing anti-scalants and anti-fouling agents in reverse osmosis processes and in the growth sector of diaphragm filtration. Further critical examples include dosing various biocides and defoaming agents in paper production and dosing cleaning agents and disinfectants in the case of CIP applications for the food and beverages industry.

Reliable dosing of concentrated acids and alkalis has proved to be a difficult task in the detoxification and neutralisation processes of industrial process and waste water treatment. Industry continues to use conventional systems for dosing and process monitoring despite the high demands placed on them. The newly developed digital flow monitors can now help give the user precise control of dosing behaviour. The diagnosis system detects errors in the dosing head and reports malfunctions reliably and without delay even at the lowest dosing rate settings in the ml/h range. (See Figure 16.66)

Reasons for dosing errors

Among the most common causes of faults when dosing with diaphragm pumps are:

- Air and/or gas bubbles in the dosing head
- Cavitation bubbles in the dosing head
- Leakage in the suction or discharge valve
- Inadmissibly high operating pressure or system pressure fluctuations

Figure 16.66 Dosing pump with priming chamber and flow monitor
Courtesy of Alldos Ltd

Conventional solutions for monitoring volumetric flow

Until recently the most commonly used solutions for dosing monitoring in the case of diaphragm dosing pumps have operated according to the "floater principle". This method involves moving a float upwards in a pipe during the pressure stroke. The float falls again during the suction stroke of the pump. This movement of the float is recorded visually, magnetically or inductively.

Adjustment of the dosing rate of the pump is then achieved by moving a switch and/or adding a bypass. This is precisely where the weaknesses lie. These settings have to be adapted to the relevant operating conditions of the pump and when changing the stroke length and/or the stroke frequency. This method of dosing monitoring can often not be used in the case of large variations in stroke frequency or when dosing viscous liquids.

Operating principles of the flow monitor

The indicator diagram displays the pressure profile across the suction and pressure strokes of the piston or diaphragm. Figure 16.67 shows the pressure profile of a diaphragm pump which is functioning correctly. Starting at point 1, the pressure stroke features a compression phase with a build-up of pressure up to the opening pressure of the pressure valve at point 2 and this phase is followed by the feed process into the pressured line until outer dead centre (point 3) is reached. During the return stroke, there is a drop in pressure down to the suction pressure at point 4 with a subsequent suction process until the inner dead centre (point 1) is reached, i.e. the start of this cyclical process.

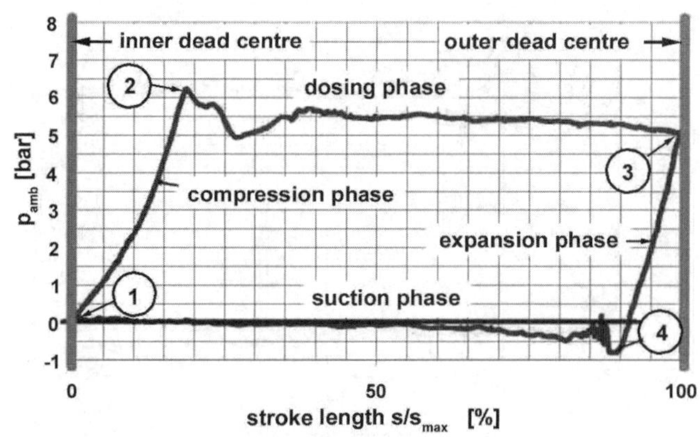

Figure 16.67 Indicator diagram for a dosing diaphragm pump which is functioning correctly
Courtesy of Alldos Ltd

Error detection

A properly functioning diaphragm pump is used as a reference and stored in the microprocessor. Even minor faults can be detected by means of changes in the current indicator diagram.

Figure 16.68 shows two frequently occurring faults in the dosing process through comparison with the error-free pressure profile (curve 1). Should there be air and/or gas bubbles in the dosing chamber (curve 2), pressure builds up more slowly due to the much greater elasticity. This means that it takes longer to reach the system pressure and that the dosing volume of each stroke is significantly smaller.

In the case of large errors on the suction side of the pump, extensive cavitation can occur as shown in Figure 16.68 (curve 3). The reasons for this can be too small flow cross-sections, too low suction in pressure or excessive levels of suction or excessive viscosity. Vapour pressure dominates the entire suction stroke and only dissipates late in the pressure stroke. This scenario also results in a smaller dosing volume per stroke due to the delay in reaching the system pressure.

Figure 16.69 shows two further frequently occurring faults in the dosing process through comparison with the error-free pressure profile (curve 1). If leakage occurs in the suction valve (curve 2), pressure builds up more slowly due to the leakage in the suction valve and the pressure falls before the outer dead centre is reached as soon as the leakage rate is larger than the current flow rate of the pump.

Should leakage occur in the discharge valve (curve 3 in Figure 16.69), a rise in pressure occurs before the end of the suction stroke when the leakage rate in the discharge valve is larger than the current suction flow of the pump. Moreover, the pressure falls more slowly at the start of the suction stroke as a result of post-flow in the discharge valve.

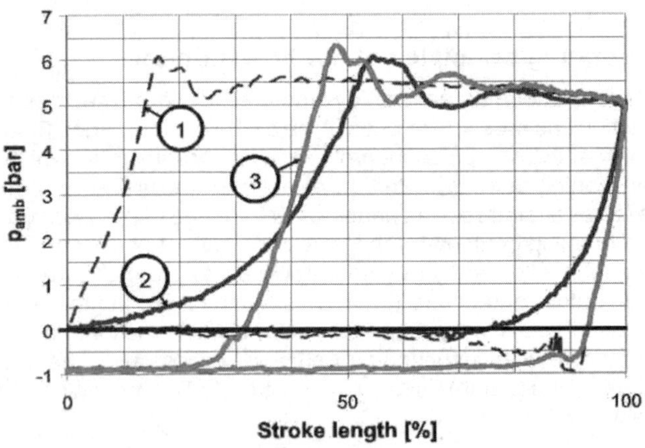

Figure 16.68 Indicator diagram when pump is functioning correctly (1), when there are air bubbles in the dosing chamber (2) and when cavitation occurs (3)
Courtesy of Alldos Ltd

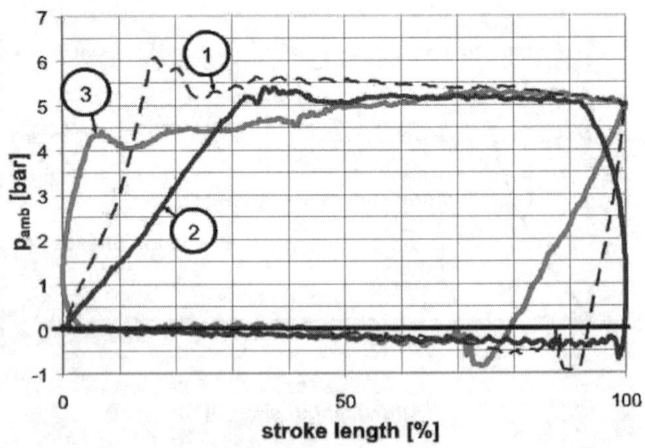

Figure 16.69 Indicator diagram when pump is functioning correctly (1), when there is leakage in the suction valve (2) and in the pressure valve (3)
Courtesy of Alldos Ltd

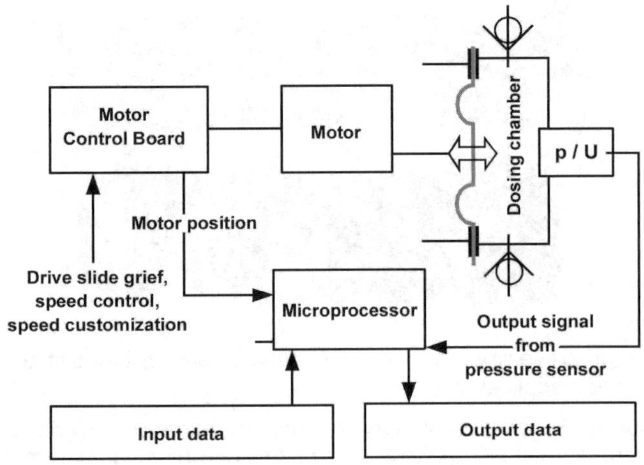

Figure 16.70 Schematic of system used to generate the indicator diagram
Courtesy of Alldos Ltd

This indicator diagram is also suitable for detecting other faults; for example, if the set system pressure is exceeded. This can be identified and evaluated.

System solution

The dosing pump also incorporates a pressure sensor. This enables the existing microprocessor to be used for both motor control and processing the measurement values.

As can be seen from Figure 16.70, the microprocessor continuously records the pressure in the dosing chamber and, using the motor control, the motor position which can also be used to determine the diaphragm position on the basis of the drive kinematics. This means that the microprocessor can generate the indicator diagram on a continuous basis. Evaluation algorithms have been developed to enable detection and evaluation of the dosing errors.

By setting the basic monitoring parameters for volumetric flow to cover a wide range of standard applications, depending on the error in question, a fall in the dosing rate of only 30% can be detected.

It is also possible to use the indicator diagram's memory of representative points and use them for the purpose of diagnosing errors. See Figure 16.71.

The processor also takes an average of the values to continuously calculate the feed pressure in the dosing chamber. This value is representative of the current system pressure and can be queried at any time.

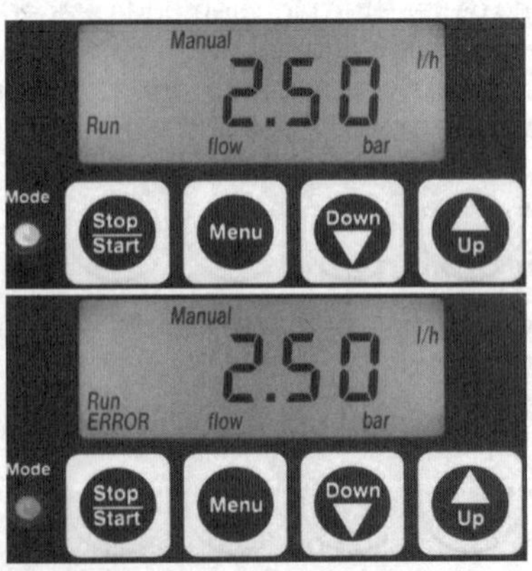

Figure 16.71 Display when pump is functioning correctly (upper). Display when there is a flow error (lower)
Courtesy of Alldos Ltd

Summary

Continuous monitoring of diaphragm dosing pumps by recording and automatically evaluating the current indicator diagrams opens up many opportunities for operational monitoring:

The current system pressure parameter enables a process to be established which is largely independent of the pump size.

The indicator diagram is largely independent of speed, viscosity and temperature. This establishes the same system for detecting errors across the pump's entire volumetric flow range without defining a working point.

Recording the current system pressure enables users to balance out pressure-dependent fluctuations in the flow rate of the pump by adapting the speed.

Application courtesy of Alldos Ltd

16.23 Positive displacement medium pressure pump system for tube press application

Introduction

The removal of waste iron ore from the furnace exhaust along with other waste products is an important stage in tube press processing of waste products at any steelworks. The ore is then re-processed into briquettes for re-cycling. The ore is transported via water which becomes a slurry. The key factor in this process is to extract the ore using the application of high pressure water to squeeze the slurry through a filter press. See Figure 16.72.

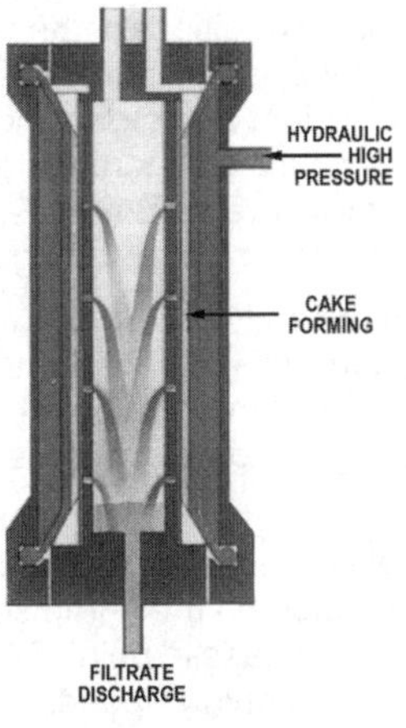

Figure 16.72 Filtration system
Courtesy of RMI Pressure Systems Ltd

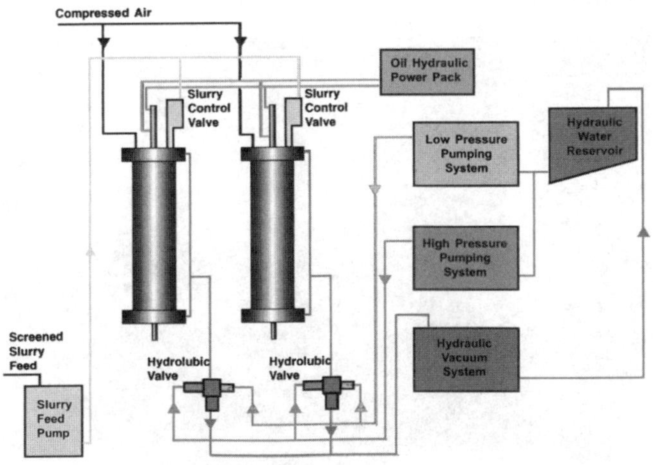

Figure 16.73. Typical tube press flowsheet
Courtesy of RMI Pressure Systems Ltd

Technical application

As mentioned the key factor in the systems design is to a remove the ore from the slurry prior to fine filtering. The difficulty arises in that this is generally a medium pressure application up to 100 barg, relative to normal high pressure installations for steelworks applications of 400 barg, meaning that this was completely unsuitable for a normal centrifugal pump system. The overriding challenge was to maintain pump control and system supply via a positive displacement pump system, which is inherently more complex then a standard centrifugal pump system. See Figure 16.73.

The solution

A complete bespoke system was designed based around a standard crankshaft-driven reciprocating pump with a set of accumulators and nitrogen charged back up bottles (See Figures 16.74 and 16.75).The pumps were then fitted with unloading valves and safety devices to ensure that a fine pressure hysteresis of the system was achieved.

In order for the system design to meet these specific criteria, a detailed analysis of system demand was extracted from the user's PLC system. From this data the pump duty was determined, the accumulator sized and the process parameters applied to the system.

These parameters had to ensure that the system achieved and maintained the following:

- The pump unit had to cycle on load-off load, every five seconds.
- The accumulators had to operate within 10% of the system's nominal pressure, with the option that this could be adjusted down to 5%.

The plant in which the system is installed operates 24 hours a day, 7 days a week; therefore there was no room for error in the performance.

After installation

Following installation on site, a number of months were spent monitoring the system, assessing the performance requirements and then adjusting the system capabilities to suit.

One year later, a programme of routine inspection is underway to ensure that the system continues to perform within the re-

Figure 16.74 The pump unit
Courtesy of RMI Pressure Systems Ltd

Figure 16.75. Accumulator bank
Courtesy of RMI Pressure Systems Ltd

quirements of the technical criteria laid down and constantly monitor feedback.

Application courtesy of RMI Pressure Systems Ltd

16.24 Boiler feed pump for efficient power station operation

Introduction

With its 40 MW drive rating, this feed pump, known as "The Gentle Giant", was at the time of installation, the largest ever produced. It was built for Unit K, the latest addition to the Niederaussem lignite-fired power station near Cologne in Germany. See Figure 16.76.

The new power station has a power output of 1000 MW and a net efficiency of 43 % making it one of the most modern and efficient stations in the world. To achieve such a performance, all components of the power station unit had to be perfectly matched to the process as a whole. The feed pump provides the total boiler flow without the provision of a back-up unit, and instead relies upon remote telemetry monitoring for condition based monitoring.

Plant outages are every power station operator's nightmare because of the immense costs involved. The pump, therefore, has to be very hard-wearing, insensitive to extreme duty conditions, and provide reliable service for long periods without maintenance. But that isn't all; energy supply companies also expect a very good efficiency. After all, every kW not consumed by the pump – which is in operation around the clock – can be sold to its customers.

Turbine drives pump

The pump is driven by its own steam turbine with continuously variable speed control. The turbine has a power rating of 44 MW and is equipped with two shaft ends, Figure 16.77. One shaft end is directly coupled to the pump. The other, on the opposite side of the turbine, drives a booster pump which supplies

Figure 16.76 Assembly of the feed pump
Courtesy KSB AG

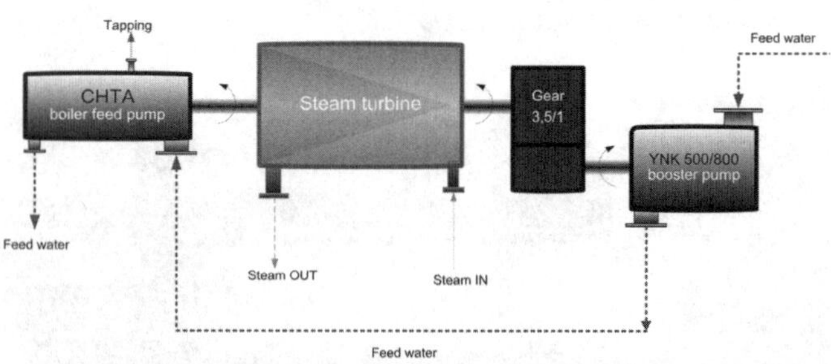

Figure 16.77 Simplified schematic of the steam turbine and the two pumps
Courtesy KSB AG

Figure 16.78 A booster pump
Courtesy KSB AG

the main pump with the necessary inlet pressure, (see Figure 16.78). The booster is not coupled to the turbine direct but to a gear which reduces the speed by a ratio of 3.5 to 1. Running up and low flow operation are normally taken care of by two electric motor-driven start-up pumps, each with an output of approximately 35% of that of the feed pump. During these phases of operation, the main boiler produces too little steam to operate the feed pump turbine. In exceptional circumstances, the steam needed to start up the pump and run up the turbo-alternator whilst the pump is still cold, can also be supplied from an auxiliary boiler.

Special vane profile

As the pressure in the system can change enormously in response to the operating conditions of the power station unit, the first stage with its impeller is the most sensitive section in every boiler feed pump. The plant operator's specification called for a minimum of 50,000 hours of operation without cavitation-induced wear. To meet this requirement, a special vane profile needed to be developed for the first stage impeller.

Computational fluid dynamic (CFD) modelling was used to design an impeller that would allow the pump to be operated without cavitation across the entire relevant operating range and with a sufficient safety margin from the system's NPSH available.

A number of special components had to be designed to meet the taxing demands made on the pump set. See Figure 16.79. The floating ring seal is a case in point. This type of seal is very reliable, even at peak running speeds of 50 m/s and more, which are common on high specific speed pumps. Other exceptional features of the seal are its thermal characteristics and the way these positively affect the pump's operation. This will be discussed later.

Hydraulic design

At the duty point, the main feed pump is expected to have the following performance data:

Flow rate = 2898 m^3/h ($\dot{m}$ = 698 kg/s)
Differential head = 3361 m (Δp = 285.7 bar)
Inlet temperature = 198.4 °C
Speed = 4620 rpm
Power = 26.47 MW
Efficiency = 86.9 %

At the so-called "TÜV point", when the boiler relief valve starts to lift, the pump set reaches a maximum speed of 5217 rpm. It delivers hot water at a rate of 893 kg/s and has a power input of just under 42 MW (around 2 MW of which are required for the booster). The total head rise is divided among five stages, with a tapping downstream of the first stage.

As far as the hydraulic designs of the impeller and the individual stages are concerned, this gives us a specific speed of:

$$n_q = n \cdot \frac{\sqrt{Q}}{H_{St}^{0.75}} = 31 \qquad \text{Equ 16.13}$$

where:

n = speed (min^{-1})
H = head (m)
Q = flow (m^3/s)

In terms of efficiency, total construction effort and cavitation properties, this value is very favourable for a high-pressure pump of this kind. From the design point of view, the hydraulic elements of the pump can be regarded as unproblematic. The geometries of the impellers and diffusers of stages two to five were taken from time-tested older pump models and converted to the required sizes on the basis of the Affinity Laws. There were some minor modifications to be made to the hub to accommodate the increase in shaft diameter on account of the higher drive rating.

The first stage impeller

The most "critical" pump element from the point of view of fluid dynamics is the first stage impeller, which differentiates between the first of several identical stages and a special suction impeller design The specification stipulated a guaranteed minimum period of operation of 50,000 hours without cavitation-induced wear. Cavitation, or rather the damage it does in the form of cavitation erosion and corrosion, is a major area of concern for the operation of turbomachinery.

Numerically speaking, the phenomenon is expressed by the NPSH value in pumps. It is governed by the following relationship:

$$NPSH_a \text{ (system)} \geq NPSH_r \text{ (pump)} \qquad \text{Equ 16.14}$$

The $NPSH_a$ value denotes the energy made available by the system immediately upstream of the pump. It is expressed in metres and already takes into account the pumped liquid's vapour pressure, which is why it is commonly referred to as Net Positive Suction Head Available. It must not be confused with the $NPSH_r$, Net Positive Suction Head Required, value of the pump, i.e. the energy required at the pump's inlet to avoid cavitation altogether or at least prevent it from damaging the pump internals. The NPSH "required" by a pump depends on the properties of the impeller material and the velocity of the flow in the impeller inlet area. The latter results from the circumferential speed u11a at the outer impeller inlet diameter, the tip speed of the eye. It is a linear function of the speed of rotation, Figure 16.80:

$$u_{11a} = \pi \cdot D_{11a} \cdot n \qquad \text{Equ 16.15}$$

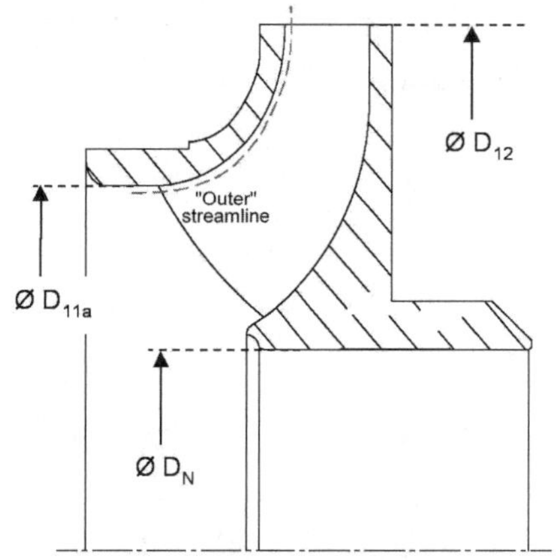

Figure 16.80 First-stage impeller (schematic)
Courtesy KSB AG

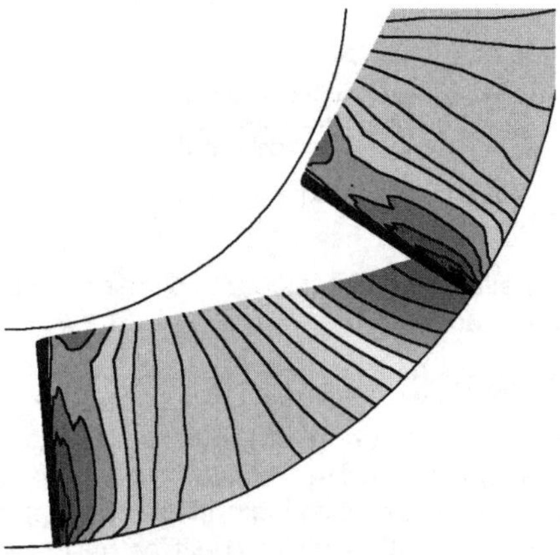

Figure 16.81 Simulated flow patterns inside an impeller
Courtesy KSB AG

The tip speed of the eye has a major influence on the so-called material loss rate VA, which is an indicator of the erosion phenomena caused by cavitation. Previous experience has shown that the material loss rate will become extremely high at tip speeds (u_{11a}) of more than 70 metres per second. Under these conditions, even impellers made of high grade materials will soon begin to show signs of damage if subject to the slightest cavitation. So, in order to guarantee the lifetime specified, the designer must make sure that the pump set can be operated at the main duty points without any risk of cavitation:

$$NPSH_a > NPSH_r = NPSH_i \qquad \text{Equ 16.16}$$

Consequently, the main criterion for designing the first stage impeller is to develop a geometry that will ensure non-cavitating pump operation.

Simulating flow in the impeller

At the blueprint stage, the design properties of the impeller were demonstrated by means of computational fluid dynamics. The findings were then verified experimentally by cavitation testing. The design tool used for analysis is a method based on three-dimensional Euler equations. It simulates the flow patterns inside the impeller and provides information on the local pressures and velocities occurring at any point of the impeller passage, Figure 16.81. The method is known to provide accurate data for this type of project and is a very good compromise between speed and reliability. Figure 16.82 shows an example of this kind of analysis.

After a series of optimisation runs produced the final contour, the "ideal" geometry has to be translated into the actual component with the highest possible degree of accuracy. To do this, the inlet areas of oversize vane blanks are reduced to their final dimensions using electric discharge machine tools. Another part of the optimisation process reduces the interaction be-

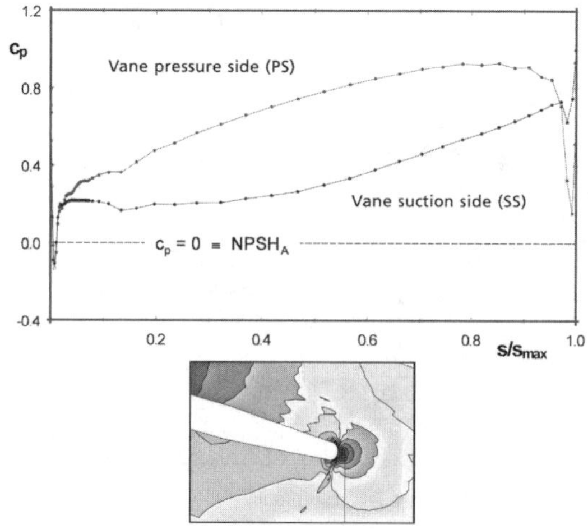

$$c_p = \frac{p - p_0}{\frac{\rho}{2} \cdot u_{12}^2}$$

p: Static contour pressure
p_0: Reference pressure (const)
p_{12}: Circumferential speed

Figure 16.82 Computed pressure distribution (example)
Courtesy KSB AG

tween the impellers and the stationary diffuser so that little or no vibration is produced.

Cavitation testing

To verify that the suction stage impeller actually gives non-cavitating operation at the "critical" duty points, the impeller was installed and tested as a single-stage pump on a cavitation test bench, Figure 16.83, specially built for this purpose. On the inlet side of the impeller the test arrangement was fitted with a replica of the original suction elbow which featured a Plexiglas window to allow the flow behaviour to be studied close up. The observed $NPSH_i$ values had a sufficient safety margin from the NPSHa values and agreed very well with the computed values, Figure 16.84. The plotted NPSHa values mainly resulted from the head of the booster pump, converted to the conditions

of the test arrangement, whose main task it is to increase the pressure to the level required by the high-speed main pump.

The rotor

The power transmitted by the drive turbine is transmitted via the pump shaft to the impellers. These are mounted on the shaft by way of a rigid and wear-resistant connection, see Figure 16.85, part 1.

The pressure distribution around the impeller gives rise to a hydraulic axial thrust, Figure 16.86, which is transmitted to the shaft by means of a separate split ring for every impeller. The axial thrust, which increases in intensity from one impeller to the next and can eventually reach a maximum value of 11.5 MN, is reliably balanced by a doubled drum, Figure 16.85, part 2. Thanks to its hydraulic properties, the double drum is capable of permanently reducing the forces acting on the thrust bearing. The thrust bearing is thus protected against overloading even in the event of sudden transient load changes.

The shaft bearings

Grooved multi-lobe plain bearings were selected as radial bearings based on the positive previous experience with this type of bearing in terms of load-carrying capacity and dynamic properties, Figure 16.85, part 6. Like the thrust bearing, the plain bearings are lubricated with pressurised oil supplied from the turbine's oil supply system. The tilting-pad thrust bearing for load in either axial direction, Figure 16.85, part 7, is reliably protected against a sudden reverse of the residual axial thrust. Together with the balancing properties of the double drum, the

bearing effectively protects the pump against overloading and

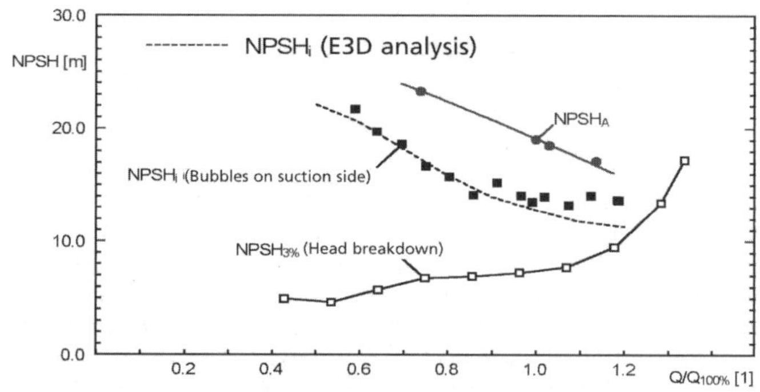

Figure 16.84 Results of a "bubble visualization" test
Courtesy KSB AG

the harmful effects of sudden load transients.

The stationary components

One of the most striking things about the feed pump on first sight is probably the pressure boundary. It consists of a forged barrel casing, Figure 16.85, part 8, and a cover, Figure 16.85, part 9, which – securely located by substantial studs and nuts, Figure 16.85, part 10 – is designed to retain the internal pump pressure.

The nuts are tightened with a hydraulic tensioning tool. This allows the nuts to be tightened without applying torsional shear stresses or bending stresses to the studs. These fasteners, which have to withstand higher loads than any others in the entire pump, can thus be designed with the required reliability.

Taking on 36.75 MN!

The pressure forces generating the axial thrust on the rotor also have an effect on the stationary parts. For the main joint fasteners, which have to secure the cover against the internal pressure, this means they have to withstand a maximum load of 36.75 MN! A spring element, the compensator, provides a sufficient level of preloading of the internal components to keep them locked in place even during a phase of low internal pressure. It also compensates the different degrees of longitudinal expansion of the pump internals and barrel as a result of the uneven temperature distribution typically occurring during transient load changes. In a thermal shock, these differences can measure several millimetres in length, which have to be accommodated without affecting the installation's availability or causing damage.

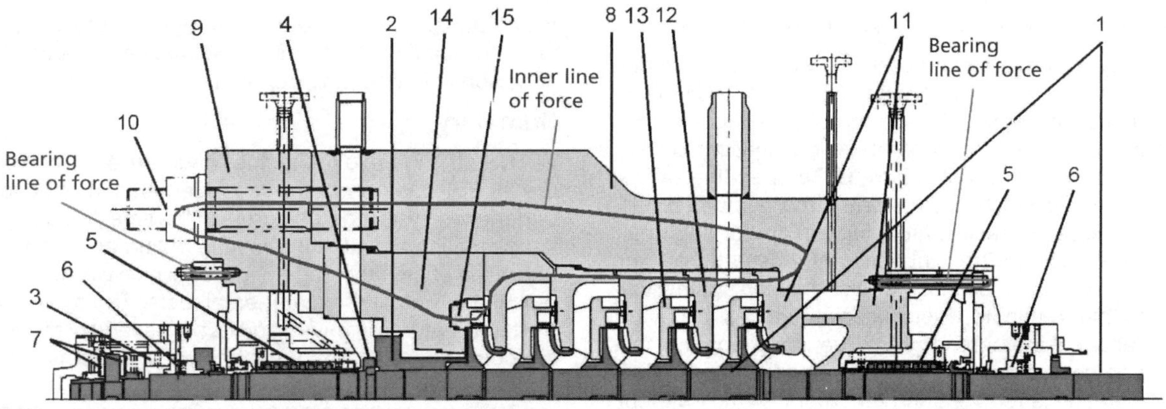

Figure 16.85 Main lines of force acting on the stationary components
Courtesy KSB AG

Temperature differences can also occur during start-up or as a result of an emergency shutdown. The internal hot water will stratify causing the top half to be much hotter than the bottom half and will cause jamming of the shaft and floating ring seals. The drive turbine is, therefore, equipped with a small electric motor to keep the pump rotating or barring, at approximately 100 revolutions per minute. As the pump gains speed, the internal pressure increases the rigidity of the pump components. The lines of action of the internal forces are crucial for the vibration characteristics of the pump running at full load. As well as a low-excitation hydraulic system an undivided line of action of the internal forces is of major importance. If the bearing forces can be separated from the hydraulic forces on entering the barrel (See the lines of force demonstrated on Figure 16.85.) this will greatly benefit the pump's operating behaviour.

One of the advantages of the low vibration level was that the pump produced so little noise that the originally planned sound-absorbing cover proved unnecessary. The static seal elements required to seal the pressure from atmosphere are designed as profile rings capable of sealing 500 bar(g) pressure maximum. Their sealing action is directly linked to the surrounding system pressure. It causes the sealing lips to be pressed against the components to be sealed. No other forces are needed to achieve the sealing effect. A further advantage of this seal type is that it does not reduce the joint pressure other parts need for their sealing function, for example the metal-to-metal sealing of the stage casings.

Shaft seals

In recent years, many power station owners and operators have learned the hard way that of all the components of a feed pump, shafts seals in particular are highly sensitive assemblies. They can make or break the availability of an entire power station. Since floating ring seals have been giving many years of reliable service in the older units of the Niederaussem power station, this seal type was selected for the new turbine-driven feed pump. In the seal, the pressure is relieved from suction pressure to atmospheric pressure along a multistage throttling distance made up of "floating rings". Depending on the pump's duty point, the pressure drop can be as high as 40 bar. As the pumped liquid flows through the annular gap between shaft and floating ring, it assumes a "bearing quality" which causes the ring to float, preventing physical contact.

To stop the flow of hot water from the pump into the seal from evaporating, and to prevent the throttling rings from losing their bearing quality as a consequence, the system has to be sealed with cold condensate. The condensate is injected at the end of the first throttling section. The injection rate is controlled by a regulating valve positioned at the inlet. As the cold condensate is mixed with the hot feed water from the pump, the heating rate of the total flow of leakage can be measured and the measured value taken as a parameter for controlling the valve. The leak-

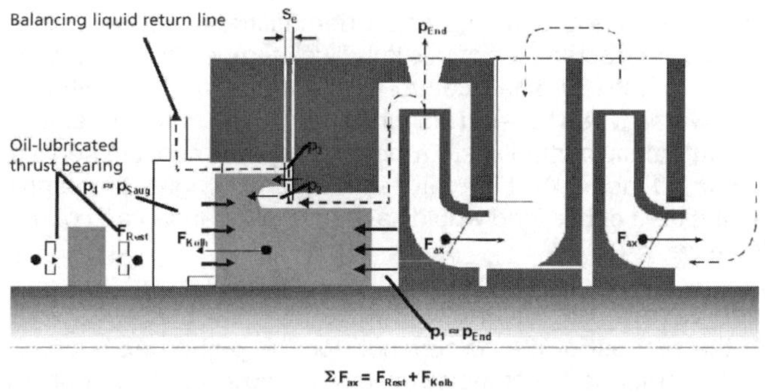

Figure 16.86 Axial thrust balancing
Courtesy KSB AG

age flows into a tank with level monitoring and is fed from there into the condensate system. Two 100 % pumps (one duty, one standby) supply the seals with barrier condensate, even if either pump fails. A simplified schematic of the barrier condensate cycle of the floating ring seal is shown in Figure 16.87.

Planned maintenance

Whenever the pump shaft is turning, i.e. even when it is in barring mode, the barrier condensate system has to be in operation as well, because that it is the only way to stop the floating rings from coming into contact with the rotating shaft. After all, with planned maintenance intervals of between three and five years, the shaft seal system has to be absolutely dependable. During the commissioning phase of the feed pump, the thermal characteristics of this shaft seal type proved particularly valuable. As there is always a liquid flow in axial direction through

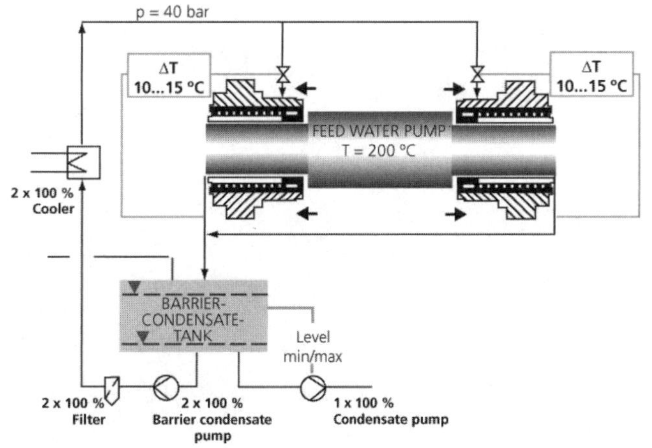

Figure 16.87 Barrier condensate cycle of the floating ring seal
Courtesy KSB AG

the section of the shaft where the floating ring seal is located, hardly any thermal stratification develops in this area.

A system equipped with a mechanical seal would respond differently. It would require heating jacket and circulation cooling, and the cold condensate, 60 °C, used by both cooling systems would continually create new temperature stratifications on being mixed with the hot feed water, 200 °C. Temperature differences of this kind develop above all when the pump is run in barring mode and may cause shaft deflection and/or deformation of the pump casing. The joint action of these effects can lead to such severe distortion that the rotor may rub against the casing in less than a minute. Typical consequences are overloading and failure of the barring gear. When this happens, the pump has to be cooled down to a much lower temperature so that the deformation is reduced and the rotor can turn without touching the casing. These processes are of paramount importance for the availability of a main feed pump.

A "well-tempered" pump

An event during commissioning demonstrated the excellent thermal characteristics of this truly "well-tempered" pump. Due to a fault in the turbine's control logic, the pump – filled with hot feed water at 180 °C – did not run for all of 31 minutes. After that the automatic control system inexplicably returned it to service in barring mode. And the rotor started turning again smoothly! An incident of this kind would have normally caused any pump to seize!

Remote monitoring accompanies commissioning

At the request of the owner/operator, the pump unit is monitored by means of a complex telemonitoring system comprising more than 50 pressure, temperature, vibration level and shaft orbit sensors. These sensors transmit the data collected to the teleservice centre direct where they are stored in a database, ready for retrieval any time the specialists need them. Despite extensive prior computations, pump units of this size always require practical testing to develop and calibrate the most favourable models for starting up and stopping the pump.

These models serve to optimise the cold and hot running phases in the power station, which improves the cost efficiency of the plant without jeopardising its availability. Based on the data supplied by the sensors, the pump designers study phenomena like thermally-induced stresses in and deformations of pump components, as well as their effects on the pump's running characteristics. This information aids condition-based maintenance and enables the pump manufacturer and plant operator jointly to determine the maximum permissible duty conditions of pump and peripherals.

Summary

With its 40 MW drive rating, the Niederaussem feed pump (Figure 16.88) was one of the most demanding of developments undertaken by its manufacturers at the time. Specially designed vane profiling provided the means to guarantee non-cavitating operation of the first-stage impeller, which was proven by theoretical analysis and experimental tests. The system pressure, the temperatures involved and its high speed of rotation subject the pump set to extreme mechanical and thermal stresses.

Figure 16.88 Side view of the boiler feed pump
Courtesy KSB AG

As well as the bearing arrangement and the method of axial thrust balancing, the real technological highlight of the pump is the shaft seal, a sophisticated floating ring seal. Intensive in-house testing and data obtained during the commissioning phase at the power station provided ample proof that the pump met all of the technical and economic requirements originally specified.

Application courtesy of KSB AG

Based on a technical paper presented by: Dipl-Ing Thomas Elsässer KSB AG, Dipl-Ing Bernhard Brecht KSB AG, Dr.-Ing Sven Baumgarten KSB AG, Dr Kirk Kollmar KSB AG

Standards, units & conversions 17

17.1 Standards

17.1.1 General

Standards were initially applied to the finished dimensions of machine components to ensure easy replacement of worn or broken parts. The scope and activities of standards have been successively extended to cover the complete machine not only with regard to dimensioning but also to such parameters as performance, reliability and safety.

Within the pump industry standardisation has developed in a similar fashion and today there are a considerable number of standards and regulations. There are also a large number of new standards which are in the process of preparation and old standards in the process of revision. Some standards have legal authority and can be enforced by the process of law; these are mostly restricted to safety requirements.

Most standards are issued for guidance. If, for example, one issued an enquiry to 10 pump manufacturers which simply stated "five pumps required for 100 m^3/h at 100m ΔH" one would receive ten quotations for 10 totally different pumps. Comparison between the quotations would be difficult or impossible. Selection of the best pump would be dubious without very detailed design and construction information on each pump. The solution to the problem is to issue a specification with the enquiry detailing the important characteristics required. To avoid the necessity of every purchaser writing their own specifications, groups were formed to write a single common specification for certain areas of application.

Because of the diversity of issuing authorities, the style and concept of standards is not uniform. CEN has decided upon rigorous control of all aspects of standards such as presentation, structure and content. CEN has produced a manual for all concerned with standard writing and it is worthwhile reviewing for the structure and content requirements.

CEN standards consist of requirements; these are basically statements controlling design, construction and use. Requirements must be verifiable. A competent person, inspecting the equipment/installation/service, must be able to see that a requirement has been met or not. For example, if a standard specifies that pumps must painted with a minimum thickness of 200 μm, an inspector can measure the paint thickness and say categorically that it complies or not. If an inspector visits a manufacturer's works to witness all stages of manufacture and testing and the standard specifies "Equipment shall be suitable for three years uninterrupted operation", what can the inspector do? Standards cannot address concepts such as ruggedness, durability, heavy-duty; which are all subjective and cannot be verified.

17.1.2 Issuing authorities

In principle, a standard can be issued by anyone. In order that the standard shall carry some influence, however, the issuing organisation should be one that is recognised by both the manufacturer and the user of the object to be standardised. It is therefore normal practice for new standards to be developed by these three parties. The organisations which issue standards consist of the following categories:

- International standard organisations
- Geographic country group standard organisations
- National standard organisations

International standard organisations require the cooperation and support of many countries. As the world becomes, effectively smaller, more and more international trade transactions will be regulated by international standards.

Geographic country group organisations are made up of countries, usually close to each other, which form an economic trading block. This type of cooperation is becoming popular and could result in an expansion of issuing authorities.

Another type of document is frequently used to assist the purchasing of pumps — a specification. A specification is very similar in most respects to a standard but does not necessarily have the same rigorous editorial controls. Specifications are issued by:

- Trade societies and associations
- Corporations, large companies, government departments
- Classification societies (insurance companies)
- Certification authorities (insurance companies)

Trade societies and associations are groups of companies which share a common interest. For example, utility suppliers, water or electricity, could form a trade association with pump and valve manufacturers and firms of consultants which regularly design systems. Together they agree a common approach to the design and operational problems of pumps/valves and write specifications to improve reliability. All members use the same specification for the same items.

Corporations, large companies and government departments issue specifications to control important features, design and/or function. Chemical companies, who develop new manufacturing processes, often issue specifications to regulate size, speed and rating of important equipment.

Classification societies are generally one, or more, insurance companies who issue rules for design and construction. This type of specification is used for marine installations, ships and offshore platforms, and the most notable authorities are: Lloyds Register, Det Norske Veritas and the American Bureau of Shipping. If a design standard requires a component to be 20mm thick but the specification requires 25mm it is the specification which takes precedence. If a ship owner tries to insure a ship, the first question asked is "Can I see your Classification Certificate?".

Certification authorities are very similar to classification societies but work in conjunction with the purchaser on large projects which have some degree of risk or hazard for the purchaser's capital investment or personnel safety. Certification authorities are regularly used for large petrochemical installations and offshore platforms. Special standards or specifications are not always issued. Experienced inspectors scrutinise and witness all relevant aspects of equipment; design, raw material, construction, testing, safety; and issue a certificate, if completely satisfied, when the equipment is ready to be dispatched. Purchase orders usually state the inspectors' decision is final irrespective of specification or standard requirements.

Specifications may be written as a general purpose document which can be used repeatedly. Some specifications are written for a specific contract and are not reusable. A specification may be written by one person and not subject to scrutiny or editorial control. Specification writers may not appreciate the concept of "requirements" and ask for "functions" which are not verifiable at the manufacturer's works. Failures at site are often difficult to attribute because of the lack of discipline of the operating staff.

Standards and specifications form part of the purchase order and are binding on all parties. Good standards and specifications assist the purchaser and improve the purchasing process. Standards and specifications can be very restrictive and inhibit technological advances or prevent the use of the most appropriate equipment. Standards and specifications are not necessarily useful to purchasers who do not understand the equipment or the full implications of meeting requirements.

The number of issuing authorities can therefore be very large and varying which often presents difficulties when developing new standards. Since there are many conflicting interests, especially when applied to international standards, the result is often a compromise with which none of the involved parties is completely satisfied. The advantages to be gained by the specification and general acceptance of standards and regulations however usually outweigh the disadvantages.

CEN issues standards to control trade within the EEC and EFTA, (European Free Trade Association). EFTA is an intergovernmental organization. Its members are Iceland, Liechtenstein, Norway and Switzerland. From 1995, it has not been possible to use national standards for trade between member countries; ENs or ISOs are used. ISO is issuing new standards to control world trade. ISO discussed with all standards organisations and trade associations the rationalisation of worldwide standardisation. ISO and CEN have an understanding and try not to duplicate standards.

Standards and specifications usually contain a immunity clause. Strict compliance with any standard does not ensure or guarantee success. Standards are written with the best of intentions with the best information available. It is the responsibility of the user of the standard to ensure the standard's requirements are compatible with the equipment and its operation in each individual case.

Some of the most important bodies for issuing standards impacting on the manufacture and use of pumps are shown in Table 4.4.

	Body	Standards
ANSI	American National Standards Institute	ANSI
ASME	The American Society of Mechanical Engineers	ASME
ASTM	The American Society for Testing Materials	ASTM
BSI	British Standards Institution	BS
CEN	Comité Européen de Normalisation	EN
DIN	Deutsches Institut for Normung	DIN
ISO	International Organisation for Standardisation	ISO

Table 17.1 Standards issuing organisations for pumps

There are other standards which do not refer to pumps directly but contain essential information for both users and manufacturers, see Table 4.5.

	Body	Standards
ADR	See **NOTE** below	ADR
CENELEC	Comité Européen de Normalisation Electrotechnique	HD
EPA	The Environmental Protection Agency (USA)	
HSE	The Health and Safety Executive (UK)	EH
IEC	The International Electrotechnical Committee	IEC
NACE	The National Association of Corrosion Engineers (USA)	MR
NFPA	The National Fire Protection Agency (USA)	NFPA
OSHA	Occupational Safety and Health Administration (USA)	OSHA

Table 17.2 Other useful Standards organisations

NOTE: ADR is a very special standard. It is not published by a multinational standards authority but it is subject to agreement by many national authorities who publish their own copy.

17.1.3 Pumps

ANSI B73.1 End-suction centrifugal pumps for chemical applications.

ANSI B73.2 Inline centrifugal pumps for chemical applications.

ANSI B73.3 Sealless pumps.

API 610 Centrifugal pumps for petroleum, heavy-duty chemical and gas industry services.

API 674 Positive displacement pumps - reciprocating.

API 675 Metering pumps.

API 676 Positive displacement pumps - rotary.

API 685 Sealless Centrifugal Pumps for Petroleum, Heavy Duty Chemical, and Gas Industry Services.

BS 1394 Stationary circulating pumps for heating and hot water service systems; Part 2: Specification for physical and performance requirements.

BS 4082 Specification for external dimensions for vertical inline centrifugal pumps; Part 1: "I" type close-coupled.

BS 4082 Specification for external dimensions for vertical inline centrifugal pumps; Part 2: "U" type close-coupled.

BS 5555 SI units and recommendations for the use of their multiples and of certain other units.

BS 6785 Code of practice for solar heating systems for swimming pools.

EN 733 End-suction centrifugal pumps, rating 10 bar with bearing bracket - Nominal duty point, main dimensions, designation system.

EN 734 Side channel pumps PN 40 - Nominal duty point, main dimensions, designation system.

EN 735 Overall dimensions of rotodynamic pumps - Tolerances.

EN 1028-1 Fire fighting pumps; Part 1: Requirements for fire fighting centrifugal pumps with primer "CE".

EN 1151 Pumps; Rotodynamic pumps; Circulation pumps having an electrical effect not exceeding 200 W for heating installations and domestic hot water installations; Requirements, testing, marking "CE".

EN 12157 Centrifugal pumps - Coolant pumps for machine tools - Nominal flow rate, dimensions (similar to BS 3766).

EN 12262 Rotodynamic pumps - Technical documents - Terms, delivery range, layout.

EN 12723 Liquid pumps; General terms for pumps and installations; Definitions, quantities, symbols and units.

EN 14343 Rotary positive displacement pumps - Performance tests for acceptance.

EN 22858 End-suction centrifugal pumps (rating 16 bar) - Designation, nominal point and dimensions (ISO 2858)(was BS 5257).

EN 23661 End-suction centrifugal pumps - Baseplate and installation dimensions (ISO 3661).

EN 25199 Technical requirements for centrifugal pumps; Class II (ISO 5199)(was BS 6836)(chemical applications).

EN 28849 Small craft - Electrically operated bilge-pumps (less than 50V dc)(ISO 8849).

EN ISO 14847 Rotary positive displacement pumps - General requirements.

EN ISO 16330 Reciprocating positive displacement pumps and pump units. Technical requirements.

EN ISO 21049 Shaft Sealing Systems for Centrifugal and Rotary Pumps. (API 682)

HI 5.1 - 5.6 Sealless pumps.

ISO 1000 SI units and recommendations for the use of their multiples and of certain other units.

ISO 3069 End-suction centrifugal pumps - Dimensions of cavities for mechanical seals and for soft packing.

ISO 3661 End-suction centrifugal pumps - Baseplate and installation dimensions.

ISO 5198 Centrifugal, mixed-flow and axial pumps - Code for hydraulic performance tests - Precision grade

ISO 9905 Technical specification for centrifugal pumps; Class I (heavy-duty refinery applications)(was BS 7784).

ISO 9906 Rotodynamic pumps - Hydraulic performance acceptance tests - Grades 1 & 2.

ISO 9908 Technical specification for centrifugal pumps; Class III (industrial applications)(was BS 7736).

ISO 13709 Centrifugal pumps for petroleum, heavy-duty chemical and gas industry services (API 610 8th ed)

ISO 13710 Petroleum, petrochemical and natural gas industries - Reciprocating positive displacement pumps. (API 674).

ISO 15783 Sealless rotodynamic pumps - Class II - Technical specification.

PD ISO/TR 17766 Centrifugal pumps handling viscous liquids - Performance corrections.

ISO 24490 Cryogenic vessels - Pumps for cryogenic service.

Pump standards being developed:

Pitot tube pumps by the Hydraulic Institute - Draft being reviewed.

Hand-operated water pumps for wells by ISO.

17.1.4 Safety

The standards and regulations applying to safety are primarily concerned with the way in which pumps and systems shall be designed and constructed so as to avoid “harm”. Harm is defined in EN1050 as “physical injury and/or damage to health or property”. Protection of the environment seems to have been overlooked! Protection of the environment is included in prEN 12690.

The safety of all machinery in Europe is covered by the Machinery Directive. The Directive requires that all manufacturers shall make machinery, by design and construction, as safe as possible. If any hazards will occur during operation or maintenance the manufacturer must highlight these problems and state what precautions must be taken. Pumps and pump units are included in the European Machinery Directive and specific European Standards must be followed to ensure pump safety. Compliance with the follow standards is a prerequisite if the “CE” mark is required.

IEC 60079, EN 60079 Electrical apparatus for explosive gas atmospheres (other than mines or explosives) —

Part 0 - General requirements.

Part 1 - Flameproof enclosure ‘d’.

Part 2 - Pressurisation ‘p’.

Part 5 - Powder filling ‘q’.

Part 6 - Oil immersion ‘o’.

Part 7 - Increased safety ‘e’.

Part 11 - Intrinsic safety ‘I’.

Part 15 - Type of protection ‘n’.

Part 18 - Encapsulation ‘m’.

Part 19 - Repair and overhaul for apparatus used in explosive atmospheres.

EN 292-1 Safety of Machinery; Basic concepts, general principles for design;

Part 1: Basic terminology, methodology.

Part 2: Technical principles and specifications.

EN 294 Safety of machinery; Safety distances to prevent danger zones being reached by the upper limbs.

EN 349 Safety of machinery; Minimum gaps to avoid crushing of parts of the human body.

EN 418 Safety of machinery; Emergency stop equipment, functional aspects; Principles for design.

EN 563 Safety of machinery; temperature of touchable surfaces; ergonomic data to establish temperature limit values for hot surfaces.

EN 626 Safety of machinery; Principles for machinery manufacturers on the reduction of risk to health from hazardous substances emitted by machinery.

Part 1: Principles and specifications for machinery manufacturers.

Part 2: Methodology leading to verification procedures.

EN 809 Pumps and pump units for liquids; Common safety requirements.

EN 894 Safety of machinery; Ergonomic requirements for the design of displays and control actuators;

Part 1; Human interactions with display control actuators.

Part 2; Displays.

Part 3; Control actuators.

EN 953 Safety of machinery; General requirements for the design and construction of guards (fixed, movable).

EN 1037 Safety of machinery; Prevention of unexpected start-up.

EN 1050 Safety of machinery; Risk assessment.

EN 1127-1: Explosive atmospheres - Explosion prevention and protection - Part 1: Basic concepts and methodology

EN 1829 High pressure cleaners - High pressure water jet machines - Safety requirements

Part 1: General description

Part 2: Hoses, hose lines and joints for high pressure cleaners, high pressure water jet machines and water jet cutters as well as safety devices to safeguard joints for hose lines.

EN 12162 Liquid pumps - Safety requirements - Procedure for hydrostatic testing

EN 12462 Biotechnology - Performance criteria for pumps

EN 12690 Biotechnology - Performance criteria for shaft seals.

EN 13386 Liquid pumps - Submersible pumps and pump units - Particular safety requirements.

EN 13951 Liquid pumps - Safety requirements - Agrifoodstuffs equipment; Design rules to ensure hygiene in use

EN 50081 Electromagnetic compatibility (EMC); Generic emission standard;

Part 1: Residential, commercial and light industry.

Part 2: Industrial, environment.

EN 50082 Electromagnetic compatibility (EMC); Generic immunity standard;

Part 1: Residential, commercial and light industry.

Part 2: Industrial, environment.

EN 60204-1 Safety of machinery; Electrical equipment of machines; Part 1: General requirements.

EN 60335 Safety of household and similar electrical appliances;

Part 2-2 Particular requirements for vacuum cleaners and water-suction cleaning appliances.

Part 2-21 Particular requirements for storage water heaters.

Part 2-41 Particular requirements for electric pumps for liquids having a temperature not exceeding 35°C.

Part 2-51 Particular requirements for stationary circulation pumps for heating and service water installations.

Part 2-55 Particular requirements for electrical appliances for use with aquariums and garden ponds.

Part 2-60 Particular requirements for whirlpool baths and similar equipment.

Part 2-69 Particular requirements for wet and dry vacuum cleaners, including power brush, for industrial and commercial use.

Part 2-79 Particular requirements for high pressure cleaners and steam cleaners.

Part 2-105 Particular requirements for multi-functional shower cabinets.

ISO 3864 Safety colours and safety signs.

ISO 8846 Small craft - Electrical devices - Protection against ignition of surrounding flammable gases

Other standards impact on CE compliance and these are indicated as CE where listed. Obviously compliance with any particular European standard specifying technical requirements for a specific pump type is necessary.

Europump Guideline 2000-08 Draft specification on the safety of electrical, submersible motors.

The general principles to be used in the design of complete installations are given in many standards and guidelines. The factors which should be considered include:

- Access
- Dust/gas hazards
- Fire fighting
- Noise
- Proximity of adjacent property
- Redundancy of equipment
- Separation of safety equipment from process equipment

BS 5908 covers fire precautions for chemical and similar installations.

BS 8005 Part 2 gives guidance on the design and construction of pumping stations and pipelines for sewage.

The hazards posed by substances are given in:

EH 40/99 Occupational exposure limits

FPA 325M Fire hazard properties of flammable liquids, gases and volatile solids

NFPA 704 Identification of the fire hazards of materials

ADR European agreement concerning the International carriage of dangerous goods by road.

An important factor in applying safety considerations is the ability of the system to withstand pressure. Pumps, pipes, valves and flanges etc. are therefore constructed to withstand specific pressure ratings associated with temperature. The pressure/temperature ratings for flanges for pipes, valves and fittings etc. can be found in:

ANSI B16.1; ANSI B16.24; ANSI B16.42; ANSI B16.47; ANSI B16.5; ANSI B93.75; BS 10; BS 1560; BS 3293; BS 4504; ISO 7005

The design of pressure vessels and associated pressure/temperature ratings can be found in:

ASME VIII; BS PD 5500; EN 13445

The design of tanks are covered by:

API 620; API 650; BS 1564; BS 2594; BS 2654; BS 4741; BS 4994; BS 7777

The design of liquid pressurised systems are covered in:

ANSI B31; BS 806; BS 8010

Care should be exercised when considering system steady-state pressures and the possibility of fatigue through cycling. The requirements of pressure relieving valves and systems are dealt with in:

ANSI N278.1; API RP520; API RP521; API 526; API 527; API 2000; ASME PTC 25.3; BS 2915; BS 6283; BS 6759; ISO 4126; ISO 6718

The construction and dimensions of pipe fittings are standardised in:

ISO 49; ISO /R508; ISO 2045; ISO 2531; ISO 3458; ISO 3459; ISO 3501; ISO 3503; ISO 4145; ISO 4179; ISO 8179; BS 143; BS 759; BS 1256; BS 1640; BS 1740; BS 1965; BS 2051; BS 3799; BS 4346; BS 4772; BS 5114

A variety of standards deal with material specifications for pressure retaining parts.

Pressure vessels: ISO 2604; ISO 9328; EN 10028; EN 13445; ASME VIII; BS 5500.

Pipework: ISO 2604; ASME VIII.

A draft document issued by BSI, BS Draft DD 38, now withdrawn still includes useful information on pump materials. To ensure that a system is capable of withstanding the required nominal pressure it is necessary to pressure test all the equipment which is subjected to internal pressure. The standards dealing with pressure vessels contain regulations concerning pressure testing. For normal land use the pump must normally be tested to a pressure which is at least 30% greater than that of the specified nominal pressure. For marine pump systems regulations are laid down by the classification societies which often demand a greater margin of safety.

The risk of explosion is an important consideration from the point of view of safety. In Europe, there are special regulations for mechanical equipment.

EN 13463 Non-electrical equipment for potentially explosive atmospheres:

Part 1: Basic methodology and requirements.

Part 2: Protection by flow restricting enclosure ‘fr’.

Part 3: Protection by flameproof enclosure.

Part 4: Protection by inherent safety.

Part 5: Protection by constructional safety.

Part 6: Protection by control of ignition sources.

Part 7: Protection by pressurisation.

Part 8: Protection by liquid immersion.

There are many standards covering the safety of electrical equipment and these standards contain a wealth of useful information for the mechanical and process engineers designing, selecting and arranging equipment. The following should be reviewed;

IEC 34-1; IEC 34-5; EN 60034 Part 5; BS 5345 and some sections of BS 4999.

The transmission of power between the motor and the pump presents another safety hazard. ANSI B15.1, BS 5304 and EN 953 are examples of the standards which contain criteria for the design, construction and application of guards for shaft couplings and other mechanical hazards. See also API RP 11ER.

For marine installations the instructions laid down by the classification societies, Lloyds Register for example, must be followed. The safety requirements being of particular importance, since a ship at sea cannot assume that assistance will be available in the event of an accident. It thus follows that the safety regulations on board ship are often much more stringent than those of an equivalent installation for land use. In principle all equipment must be type-approved and also have passed special safety inspection controls before installation on board ship is permitted. Similar restrictions are imposed for equipment to operate on offshore platforms.

17.1.5 Reliability and operational life

The concepts of reliability and operational life have come more to the forefront in recent years as the demands for operational reliability and economy increase. Emphasis has been placed on total life-cycle costs rather than initial capital costs. This trend has been followed up by the organisations responsible for issuing standards and resulted in the development of instructions relating to pump design in order to fulfil these demands.

The API 610 specification concerning centrifugal pump reliability gives instructions and recommendations regarding the pump components which are associated with safety. Also dealt with are the factors which must be considered in a pump installation together with comments concerning the components such as shaft couplings and electric motors. The API specification applies primarily to the petroleum industry but is also applicable in other areas.

It is intended that this Standard becomes an accepted international guideline in respect to pump design for, among other things, reliability in operation. As mentioned earlier, inspectors cannot look at pumps and say "this one will only last two years but that one will last three".

Specifications applying to the design, construction, performance, reliability and testing of small-bore domestic heating and hot water pumps are also to be found in EN 60335-2-42; EN 60335-2-51; BS 1394; BS 3456 Part 202 Section 202.41.

The European Standard for reciprocating positive reciprocating pumps, EN 16330, includes NPIP margins and guidance on acceptable residual pulsation levels.

A curious twist occurred in the nature of pump specifications a few years ago. The major European oil companies realised that if they told the pump manufacturer how to build a pump for extended reliability and the pump was not reliable they would have no claim on the manufacturer. A group of oil companies discussed with certain manufacturers the possibility of purchasing pumps without specifications. The user would completely specify the duty and the time between overhauls. The pump manufacturer would supply a suitable pump. If the pump did not perform satisfactorily the manufacturer would cure the problems.

Before this plan could be widely implemented it was modified. The users decided it would be better if the pump manufacturer completed the complicated parts of the duty definition because the manufacturer knew what sort of things would cause problems in a pump for the specific duty.

This created two unexpected problems. The pump specialists at the consultants/contractors became surplus to requirements. The pump manufacturers could sell exotic pumps when "plain vanilla" would do. This utopian state of affairs did not last long and everything went back to normal!

17.1.6 Dimensions and performance

The advantages of standardised dimensions for individual components has long been recognised. The scope of standardisation has been developed to cover all aspects of machine design and construction. Since size and performance are often related to each other, it has proved practical to develop standards which have regard to these two parameters.

An example of this is ISO 2858 which specifies the principle dimensions and nominal duty points of horizontal end-section back-pull-out centrifugal pumps for pressure rating PN 16 (16 bar), mainly intended for chemical applications. The British version of this Standard is BS 5257 and the German is DIN 24256. The European Standard for these pumps is EN 22858. Water pumps of the same type but rated at pressure class PN 10 are covered by the German Standard DIN 24255 and for Europe, EN 733. This Standard covers pumps of approximately the same size as ISO 2858 (BS 5257; EN 22858) but with reduced dimensions and higher port velocities.

BS 5257 also specifies dimensions for baseplate installations, ISO 3661, and seal cavities, ISO 3069. EN 60335-2-42; EN 60335-2-51; BS 1394 and BS 3456 Part 202 Section 202.41 cover the requirements for circulators for domestic heating and hot water installations for glandless pumps up to 300 W. The ISO 2858 Standard pump has not been accepted in the USA because of its metric dimensions. There is however an equivalent inch Standard, ANSI B73.1 which specifies dimensions and duty ratings in American units. ANSI B73.2 covers a range of inline vertical chemical pumps. BS 4082 specifies external dimensions, pressure-temperature ratings and nominal duty points at various speeds for vertical inline centrifugal water pumps. These apply to maximum differential heads of 160 m and maximum flows of 384 m3/h. EN 734 specifies the designations, dimensions and duty points of a range of multi-stage segmental pumps rated at PN 40. BS 7117 deals with construction, installation and maintenance of metering pumps specifically for dispensing liquid fuel.

ISO 13709/API 610 is a specification of pump construction and packaging together with materials and inspection requirements. Seal cavities, clearances and seal selection requirements are specified in EN ISO 21049/API 682. Hydraulic duties are not specified but test tolerances for water are given. 21049/610 allows performance testing on water up to 66 °C which gives good results based on vapour pressure rather than gas evolution.

ISO 13710/API 674, API 675 and API 676 cover positive displacement pumps; reciprocating, dosing and rotary. VDMA 24286 covers procurement, testing, supply and dispatch of the three positive displacement pump types. DIN 5437 covers semi-rotary hand pumps. DIN 24289 covers specification, installation and data sheets for reciprocating pumps.

17.1.7 Vibration and noise

Vibration and noise Vibration occurs in all rotating and reciprocating machinery when in operation. Vibrations are normally caused by imbalance of the rotating or reciprocating components, but can also be caused by hydraulic flow through a pump or valve. Since imbalance gives rise to vibration it is obviously desirable to maintain sufficiently low values so as to prevent mechanical damage. Noise is also created and is also undesirable from an environmental point of view. Standards and guidelines are available which specify maximum levels of vibration to avoid mechanical damage, loss of reliability and also to prevent unacceptably high noise levels. The standards dealing with noise and vibration are therefore basically concerned with the evaluation of maximum permissible vibration levels for various types of machinery and the requirements and specification of the necessary measuring equipment. Vibration levels are today measured as vibration velocity mm/s RMS instead of amplitude and frequency. This simplifies measurement and the evaluation of vibration levels.

The requirements for sensors and measuring equipment for vibration velocities in mm/s RMS within the frequency 10 to 1000 Hz are dealt with in ISO 2954, BS 4675: Part 2: 1978. ISO 3945 specifies how to measure and evaluate vibration at site for machines with speeds between 10 rev/s and 200 rev/s. ISO 8821 and BS 7130 specify the convention to be adopted regarding shaft keys. ISO 10816 Part 1, Part 3, Part 5 and Part 6 may be useful for vibration measurement of rotating and reciprocating

equipment when using sensors attached to bearing housings. The evaluation and recommendation of preferred vibration severity ranges is complicated by the variations in design, construction and installation of different types of machinery. The VDI 2056; ISO 2372 and BS 4675: Part 1: 1976 have therefore divided up machines into groups. Rotodynamic pumps usually appear in group T (VDI) or Class IV (ISO and BSI), although some types may appear in other classification groups as a result of variations in design and construction. The pump manufacturer usually establishes the group to which a particular pump construction belongs.

The vibration level measured on a pump during testing is greatly influenced by the pump mounting. Temporary mounts on test rigs can be quite flexible allowing the pump to register high levels. Positive displacement pumps, both rotary and reciprocating, can suffer from vibration due to imbalance which cannot be rectified due to the pump design. Vibration readings during tests can only be used as information if the pump is mounted on temporary stands. When installed at site, and grouted in, the vibration levels will be much reduced.

IEC 994 and EN 60994 give guidance on vibration and pulsation measurement at site for rotodynamic pumps and turbines.

Other standards which may be useful include ISO 5343; ANSI S2.17; ANSI S2.40; API 678. Noise emitted by machinery constitutes a considerable problem where it is necessary for personnel to be present in the vicinity of machinery for long periods. Consideration must be given to the complete pump package not just the pump. The driver, plus any power transmission equipment, and process pipework will contribute to the overall noise level. Specific areas can be designated "ear protection zones" where ear defenders must be worn. Typical noise level restrictions in current specifications call for 84 dB(A).

EN 12639 is the European noise test code for pumps and pump units. This Standard has two grades of accuracy. Other current standards deal with the measurement, evaluation and presentation of noise data. The following Standards should be reviewed for suitability for any specific application: ISO 3740; EN ISO 3743-1; EN ISO 3743-2; EN ISO 3744; ISO 3745; EN ISO 3746; ISO 3747; ISO 4412; EN ISO 4871; ISO 6081/2; ISO TR 7849; EN ISO 9614-1; EN ISO 9614-2; EN ISO 11203; ISO 11689; EN 27574; ANSI B93.71; ANSI B93.72; ANSI S1.4; ANSI S1.4a; ANSI S1.6; ANSI S1.13; API 615; BS 4196; BS 7025 and DIN 45635. As most pumps are driven by electric motors the following Standards on noise from electric motors should be given consideration; ISO /R 1680, IEC 34-9 and BS 4999 Part 109.

17.1.8 Forces and moments on connections

Pump installations should be laid out so that the loads transmitted from the piping system to the pump connections are as small as possible. Considerable forces are brought to bear on the pump connections as a result of liquid pressure and thermal expansion, for example. A knowledge of the extent to which the pump connections are capable of withstanding forces and moments without risk is therefore important, from the point of view of pump installation and construction, so as to avoid the unnecessary use of compensating devices etc. However, the pump should not be looked upon as a pipe support or anchor. These are secondary functions for a pump and can be detrimental to the primary function. Instructions regarding the permissible forces and moments for centrifugal pumps are given in ISO 13709/API 610. This Standard determines the permitted forces and moments on the basis of the size of the flange connection and, to some extent, the weight of the pump. 13709/610 pumps are robust. Special reinforced baseplates are used because the nozzle load requirements are so high.

Other Standards use the maximum pump pressure and connection dimensions as the basis for determining the permitted forces and moments. The situation is further complicated by the fact that different manufacturers employ their own individual methods of determining the permitted forces and moments. It is therefore advisable to contact the manufacturer for advice in cases of doubt.

EN ISO 14847 for rotary positive displacement pumps includes forces and moments data. DIN 24289-1 and EN ISO 16330 for reciprocating pumps includes forces and moment data. French National Standards, AFNOR, have standards NFE44-145 and NFE44-146, for forces and moments on horizontal and vertical centrifugal pump flanges.

17.1.9 Components

Many pump components and parts are not exclusively used on pumps, but are also used in many other types of machinery. The dimensions of such parts are usually the subject of separate component standards. Screws, plugs, keys and key ways for example, are basic pump components for which dimensional standardisation is a well known concept. The standards applying to certain other pump components, however, are worth studying more carefully.

Standard	Dimensions
ISO 7005	both
BS 4504	metric
DIN 2531 to 2533	metric
DIN 2543 to 2549 + 2551	metric
DIN 2628 + 2631 to 2638	metric
DIN 2565 to 2569	metric
ANSI B16.5 / BS 1560	inch
SAE	inch

Table 17.3 Popular flange standards

The shape and dimensions of flanges is dealt with by many standards which are subdivided according to material, type and pressure rating. The Standards in Table 17.3 are the most popular and cover working pressures up to 420 barg depending upon the material and the temperature.

Shaft end dimensions are given in ISO 775 and its equivalent BS 4506. These standards give specifications relating to various types of cylindrical and conical shaft ends with optional screw threads for retaining purposes.

Shaft centre heights and tolerances for driving and driven machines are standardised in ISO 496 and BS 5186. These standards should be read in conjunction with IEC 72 and IEC 72A which define shaft heights for electric motors. Mounting flanges, together with shaft sizes for hydraulic pumps and motors are covered by ISO 3091/1, ISO 3091/3, ANSI B93.81, BS 6276 and SAE J896.

Shaft seals are important machine elements in all pumps from a reliability point of view. Since there are many different types of seals manufactured by a great number of manufacturers, there has long been a necessity to establish some form of standardisation. DIN 3780 covers stuffing box dimensions for soft packing. DIN 24960 is a recent Standard and is derived from an earlier VDMA Standard for mechanical seal cavities and material codes. For further information see Chapter 8, Section 8.3.2.

The international Standard ISO 3069 and sections of its BSI equivalent, BS 5257, include tolerance and surface finish requirements for shafts and cavities also refer to shaft seals.

These standards only deal with the dimensioning of the actual sealing cavity. The manufacturer should be consulted about pressure/temperature/speed limitations. EN ISO 21049/API 682 and ANSI B73.1 and B73.2 all specify seal cavity dimensions.

EN ISO 21049/API 682 are major standards covering all aspects of mechanical seals. These standards deal with problems associated with more stringent leakage and safety requirements including selection and testing.

There are no established dimensional standards for shaft couplings. Data concerning different types of shaft couplings are dealt with in Chapter 9. Some criteria concerning shaft couplings are covered by BS 3170 and API 671.

The Hydraulic Institute has issued a draft standard for baseplates for horizontal centrifugal pumps. Also a finalised standard for pump intake design.

17.1.10 Performance testing

Mass-produced pumps are subject to selective testing. Some manufacturers test all the rotating assemblies in fixed test casings; this is not a good policy and should be avoided. Pumps built specifically for contract will be tested when complete. The running test is the final check that the pump fulfils the agreed or specified requirements and is therefore of great importance to both the purchaser and the supplier.

Tests which closely resemble site conditions provide more confidence than tests of a general nature. Acceptance testing is usually carried out at the pump manufacturer's premises. The test facilities of pump manufacturers vary greatly and it is worthwhile evaluating the test procedures carefully before placing an order. It is obviously impracticable for the purchaser to fully specify the manner in which the acceptance test is to be carried out and evaluated for every order. Reference is normally made, therefore, to an acceptance testing standard which contains instructions regarding the testing procedure, permissible tolerances etc.

Testing standards and specifications have been issued by many standard issuing organisations, as shown by the list in Table 17.4 Some documents listed only specify acceptance tolerances. Documents which provide detailed instructions on test rigs and test methods are shown in **bold** type.

Pumps (all types)	**EN 12162** (hydrostatic testing)*
Pump units (all types)	**EN 12483** (pump units with variable frequency inverters)
Centrifugal pumps	API 610, **DIN 1944, HI 1.6**
Circulators	EN 1151
Dosing & metering pumps	API 675
Effluent pumps	**DIN 19760**
Fire pumps	**DIN 14420, EN 1028-2**
Fire pumps (portable)	**DIN 14410**
Hydraulic pumps and motors	**ANSI B93.95, BS 4617, BS 7250, ISO 4409, ISO 8278, ISO 8426**
Positive displacement pumps	**VDMA 24284**
Pressure ripples in hydraulic systems	**BS 6335**
Reciprocating pumps	API 674, **HI 6.6** (no EN or ISO)
Rotary positive displacement pumps	API 676, **EN 14343**
Rotodynamic pumps	**ASME PTC 7, 7.1, 8.2, 18.1, BS 5316-1, BS 5316-2, BS 5316-3, EN ISO 5198, ISO 2548, ISO 3555, ISO 5198, ISO 9906**
Rotodynamic pump/turbines (very large)	**BS 5860, EN 60041, IEC 60041, IEC 60607**
Submersible pumps	**API RP 11S2**
Wells (flow testing with pumps)	**BS 6316**

* Methods of hydrostatic testing which impart external strength or stiffness to components should not be acceptable.

Table 17.4 Testing standards

EN 14343:2005 is the new Standard for performance tests for acceptance of rotary positive displacement pumps.

It is important to select a test standard which is compatible with the pump application to be tested. If a centrifugal pump is being used to pump oil at 300 cSt it would be a good idea to test with a liquid which approaches the viscosity. Most rotodynamic pump test specifications call for "clean cold water". EN ISO 21049/API 610 allows performance testing on water up to 66 °C which gives good results based on vapour pressure rather than gas evolution. Some of the positive displacement pump test requirements do not specify anything about the liquid.

Some pump manufacturers are reluctant to test pumps on anything but water because of the inconvenience of organising a suitable liquid supply and cleaning the tanks afterwards. It may be possible to "borrow" a liquid like heating oil and use the tanker as a supply vessel. Some creative thinking may produce positive results. Users should not be thwarted by the first offer from the first manufacturer, but shop around.

BS 5316: Part 1 is a direct equivalent of ISO 2548, which, since its introduction, has become the dominating international standard for rotodynamic pumps for industrial applications. Petrochemical pumps are tested to EN ISO 21049/API 610 or the Hydraulic Institute. In all testing standards it is of course the maximum permissible tolerances of flow, differential head and efficiency which are of greatest interest. Some purchasers annex the testing standard to the purchaser order and add overriding clauses for the tolerances.

The correct choice of testing standard, when ordering and specifying, is obviously very important. The acceptance test standards for mass-produced pumps according to BS 5316 Part 1, can be about 10% higher than the specified duty point. The standards must be reviewed carefully to ensure the particular problems of a specific application are addressed.

Manufacturers should be assessed regarding their in-house process control systems. Great store is set by ISO 9000, EN 29000 and BS 5750. These Standards specify rules regarding how departments should work, how departments should communicate with each other and most important, what to do when things go wrong. Pumps for critical applications; nuclear, pharmaceutical, offshore; are normally purchased from manufacturers who operate one of the standard systems and have been vetted by an accredited third party assessor. All pump manufacturers should operate some type of formalised control system.

A traditional system can be as good as ISO, EN or BS when operated correctly and Quality Control staff are empowered to scrap substandard components. Enquiries for important pumps should always request a Vendor Document Requirement form, VDR, and a Quality Control Plan, QC Plan, to be submitted as part of the quotation. The VDR will show the purchaser what documentation; drawings, manuals, certificates; the manufacturer considers important and will supply with the pump. The QC Plan will show all the inspection operations, hold points and any conditions when Engineering must be contacted to exert overall control for important decisions. After carefully studying the VDR and the QC Plan, the true character of the manufacturer will be revealed.

BS 7929:1999 Guide for the verification of a pump casing design procedure, is useful when comparing pressure ratings and design methods for competitive pumps.

17.1.11 Hygiene

Much work has been carried out on the correct design and installation of pumps to be used on "liquids" which must not be contaminated. Two different areas of application are being addressed; foodstuffs for human or animal consumption and biotechnology.

The European committee considering foodstuffs applications has introduced a new term "agrifoodstuffs" to encompass both human and animal food applications. The new Standard EN 13951 includes provisions for design, surface finish, the choice of materials and verification of cleanability. Biotechnology refers to processes which involve micro-organisms which may be hazardous.

Pump standards for biotechnology applications are at an early stage. EN 12690 provides seal selection guidelines for biotechnology processes. EN 12298 covers testing for pump leak resistance. Some guidance on hygienic requirements can be found in EN 1672-1 and 1672-2, Food processing machinery - Safety and hygiene requirements.

Additional guidance can be obtained from the European Hygienic Engineering & Design Group (EHEDG) who develops test methods to verify the suitability of equipment. Also, review ISO 14159 Safety of machinery - Hygiene requirements for the design of machinery. Care must be exercised as the ISO Standard may not conform precisely to European requirements.

EN 285, 12296 and 12297 provide guidance on steam sterilisation.

There is no European Standard for equipment suitable for drinking water, local national standards apply. The American NAFEM (North American Association of Food Equipment Manufacturers) should also be consulted.

17.1.12 Electromagnetic pollution

Over the years there has been a dramatic increase in the interference problems created by electrical and electronic equipment. Problems are being experienced with interference transmitted by the electric supply or distribution system and airborne electromagnetic waves. These are caused by the increased use of fluorescent lights, electronic control of incandescent lamps, computers, PLCs, variable frequency drive systems and mobile telephones.

Many standards have been introduced to try to ensure that electric/electronic equipment will work in harmony. Most pumps are driven by electric motors and the pump unit must conform to the electromagnetic pollution requirements.

EN 61000-2 Electromagnetic compatibility (EMC); Part 2: Environment;

Section 2: Compatibility levels for low frequency conducted disturbances and signalling in public low-voltage power supply systems (similar to IEC 1000-2-2) 'CE'.

Section 4: Compatibility levels in industrial plants for low frequency conducted disturbances (IEC 1000-2-4) 'CE'.

EN 61000-3 Electromagnetic compatibility (EMC); Part 3: Limits.

Section 2: Limits for harmonic current emissions (equipment input current <= 16 A per phase)(IEC 1000-3-2) 'CE'.

Section 3: Limitation of voltage fluctuations and flicker in low-voltage supply systems for equipment with rated current <= 16 A (IEC 1000-3-3) 'CE'.

EN 61000-4-1 Electromagnetic compatibility (EMC); Part 4: Testing and measuring techniques:

Section 1: Overview of immunity test; Basic EMC publication (IEC 1000-4-1) 'CE'.

Section 2: Electrostatic discharge immunity test; Basic EMC publication (IEC 1000-4-2) 'CE'.

Section 4: Electrical fast transient/burst immunity test; Basic EMC publication (IEC 1000-4-4) 'CE'.

Section 5: Surge immunity test (IEC 1000-4-5) 'CE'.

Section 7: General guide on harmonics and interharmonic measurement and instrumentation for power supply systems and equipment connected thereto (IEC 1000-4-7) 'CE'.

Section 8: Power frequency magnetic field immunity test; Basic EMC publication (IEC 1000-4-8) 'CE'.

Section 9: Pulse magnetic field immunity test; Basic EMC publication (IEC 1000-4-9) 'CE'.

Section 10: Damped oscillatory magnetic field immunity test; Basic EMC publication (IEC 1000-4-10) 'CE'.

Section 11: Voltage dips, short interruptions and voltage variations immunity tests (IEC 1000-4-11) 'CE'.

Section 12: Oscillatory waves immunity tests - Basic EMC publication (IEC 1000-4-12) 'CE'.

17.2 SI, The International System of Units

Système International d'Unités,(international abbreviation, SI) the international measurement unit system is based on an earlier metric system and is being used more extensively worldwide. The SI system is systematically constructed to cover in practice all scientific, technical and daily requirements and is subject to international agreement. This means that it is now possible to apply the SI system uniformly throughout the world. A measurement system which is suitable for all technical and scientific purposes has to fulfil many requirements. Some of the basic requirements which SI satisfies are consistency, consequential applicability, coherence, the convenient expression of multiples and sub-multiples over a wide range of numerical values and accuracy.

- Consistency means that each unit shall represent one, and only one, quantity.
- Consequential applicability means that each quantity shall be measured in one, and only one, unit.
- Coherence means that all units for every quantity shall be compatible so as to eliminate the need for arbitrary conversion factors in calculations involving related quantities.
- Convenient expression of multiples and sub-multiples means the convenient multiplication of units to enable the use of practical numerical values within a particular application.
- Accuracy means that the base units shall be precisely derived and defined. Six of the seven base units are thus determined from distinct precisely defined physical phenomena, the seventh, the kilogram, is determined by one standard body which is held in Paris.

In 1971 the EEC, now the EC (European Community) ratified a Directive, 71/354/EEC, on units which committed all member states to amend legislation to authorise SI units within 18 months of that date and to implement all provisions of the Directive within a further five years. An amending Directive, 76/770/EEC, legislates the obligations. The Units of Measurement Directives place non-SI units into four chapters A to D.

Chapter A prescribes those units which are for permanent use and member states are obliged to authorise them in their laws by 21 April, 1978.

Chapter B contains a list of all units which member states have undertaken to cease to authorise in their laws with effect from 31 December, 1977.

Chapter C contains a list of units which member states have undertaken to cease to authorise in their laws with effect from 31 December, 1979.

Chapter D covers remaining units and some other units and will be reviewed before 31 December, 1979.

The formal content of the SI is determined and authorised by the General Conference of Weights and Measures (CGPM) and, for more detailed descriptions of the System reference should be made to BS 3763 and SI - The International System of Units published by the UK Office of Public Sector Information (formerly called HMSO). However, the basic advice to industry on the use of SI is now contained in ISO 1000, BS 5555.

The SI system includes two classes of units:

- Base units
- Derived units

The system also includes a number of Non-SI units, the "bar" for example, which are retained for use together with the SI units and their multiples because of their practical importance or because of their specialized fields. Multiples and sub-multiples of SI units and recognised Non-SI units are formed by attaching a prefix, micro, milli, kilo, Mega etc., to the unit symbol.

17.2.1 Brief history of unit systems

Although often called the metric system, the SI system essentially has more basic units and overcomes problems encountered during the development of a consistent system of units to serve all science and engineering functions. The metric system was introduced at the beginning of the nineteenth century based on the unit of length being the metre. The unit of mass followed and together with the unit of time formed the basic metric system. In 1873 the British Association for the Advancement of Science agreed on the use of the centimetre, gramme and second as the basic units for scientific work (the CGS system) but engineering within the United Kingdom had been well established using the British units of feet, pounds and seconds (the FPS system). As electrical experiments took place it was the metric system that became the basis for units peculiar to the electrical sciences and many basic electrical units were added to the CGS system.

An international authority on the metric system was established in 1875 with the Bureaux International des Poids et Mesures at Sèvres defining the units of length, mass and time as the metre, kilogramme and second respectively (the MKS system) as these units were more convenient than the CGS system. The basic units of length, mass and time are insufficient to cover electrical units and consequently units employing the permeability and permittivity of free space became necessary. Confusion also arose because of the links with the CGS system with, in particular, the use of p (the ratio of the circumference of a circle to its diameter) appearing in equations which were usually not associated with circles.

To overcome the complications the International Electrotechnical Commission (IEC) rationalized the units in 1950 and adopted as a fourth basic unit the unit of electrical current, the ampere (the MKSA system). This had been suggested about a half century earlier by an Italian professor called Giorgi and this system of units was consequently named the Giorgi System. Although this system of units covered electrical engineering it did not cover all branches of science and consequently the Conference Générale des Poids et Measures (CGPM) in 1954 agreed a rationalized and coherent system of units which became in 1960 the SI system.

After some subsequent additions the system now has seven basic units as follows:

length	metre (m)
mass	kilogramme (kg)
time	second (s)
electric current	Ampere (A)
light	candela (cd)
temperature	Kelvin (K)
substance	mole (mol)

In addition the following supplementary units are in use:

plane angle	radian (rad)
solid angle	steradian (sr)

The supplementary units are both ratios and therefore have no basic units.

The SI system is based around the seven basic units and the two supplementary units together with derived units for the more commonly used quantities and a series of prefixes used for the formation of multiples and submultiples.

Actual temperatures are normally ex-pressed in Celsius units (°C) and temperature differences in Kelvin units (K).

17.2.2 Method of expressing symbols and numbers

The following rules apply to symbols for units

- The symbol should be lower case unless the unit is derived from a proper name in which case the symbol should be upper case (or the first letter upper case if more than one letter), for example, metre - m, Ampere - A, Hertz - Hz
- The symbol should not contain a final full stop, for example, m not m
- The symbol should remain unaltered in the plural, for example, m not ms
- If multiple symbols are required they should be separated by a space if confusion can occur, for example, kg m/s^2 not kgm/s^2. If a pair of units are each represented by a single letter they are not separated if the absence of a space is not likely to cause confusion, for example Nm
- The symbol should be reduced to its simplest expression, for example, W not J/s, Js^{-1}, kg m^2/s^3 or kg m^2 s^{-3}
- Where more than one symbol is required and division is involved use a solidus or superscript, do not use more than one solidus, for example, use m/s^2 or m s^{2,} do not use m/s/s

Large numbers should be written with the digits grouped in threes and with a '.' or ',' used to denote the decimal place, for example, 12 345 678, 0.000 012, 0,012 3. Use 10 000 not 10,000 to denote ten thousand, for example. One exception to this rule is that numbers with only four digits and without a decimal point do not normally have the space after the first digit, for example 1234 not 1 234.

Alternatively very large or very small numbers can be represented in exponential form, that is a number multiplied by a factor in the form of 10 to a positive or negative power. For example 100 is equivalent to 10^2 and 0.001 is equivalent to 10^{-3}. Therefore 1 234 000 can be expressed in the form 1.234 x 10^6 and 0.0123 can be expressed as 1.23 x 10^{-2}.

There is a further group of specialised units which are primarily for use within astronomy and physics.

17.2.3 Multiples of SI units

The prefixes in Table 17.5 are used to form names and symbols of multiples and subdivisions of the SI units. The symbol of a prefix is considered to be combined with the unit symbol for the base unit, supplementary unit or derived unit to which it is directly attached, forming with it a symbol for a new unit which can be provided with a positive or negative exponent and which can be combined with other unit symbols to form symbols for compound units.

Factor by which the unit is multiplied	Prefix		Example
	name	symbol	
10^{24}	yotta	Y	
10^{21}	zetta	Z	
10^{18}	exa	E	
10^{15}	peta	P	
10^{12}	tera	T	1 terrajoule = 1TJ
10^9	giga	G	1 gigawatt = 1GM

Factor by which the unit is multiplied	Prefix		Example
	name	symbol	
10^6	mega	M	1 megavolt = 1 MV
10^3	kilo	k	1 kilometre = 1 km
10^2	hecto	h	1 hectogram = 1 hg
10^1	deca	da	1 decalumen = 1 dalm
10^{-1}	deci	d	1 decimetre = 1 dm
10^{-2}	centi	c	1 centimetre = 1 cm
10^{-3}	milli	m	1 milligram = 1 mg
10^{-6}	micro	μ	1 microgram = 1 μm
10^{-9}	nano	n	1 nanohenry = 1 nH
10^{-12}	pico	p	1 picofarad = 1 pF
10^{-15}	femto	f	1 femtometre = 1 fm
10^{-18}	atto	a	1 attosecond = 1 as

Table 17.5 Multiples of SI units

Whenever possible units should be multiples or submultiples of 3 - hecto, deca, deci and centi therefore should not normally be used.

The prefix symbol should appear immediately before the basic symbol, for example mA for milliampere. It should be noted that mm^3 means $(0.001\ m)^3$ not $0.001\ m^3$ and that mm^{-1} means $(10^{-3}\ m)^{-1}$ not $10^{-3}\ m^{-1}$. The use of dk should be avoided as this may cause confusion, for example, dkg is decagramme not deci killogramme.

It is normal practice to use millimetre (mm) as the unit of length on engineering drawings.

17.2.4 Derived units

These are expressed in terms of base units and/or supplementary units by multiplication and division according to the laws of physics relating the various quantities, see Table 17.6.

Quantity	Name of derived SI unit	Symbol	Expressed in terms of base or supplementary units
frequency	Hertz	Hz	1 Hz = 1/s
force	Newton	N	1 N = 1 kg m/s^2
pressure, stress	Pascal	Pa	1 Pa = 1 N/m^2
energy, work, heat	Joule	J	1 J = 1 Nm
power	Watt	W	1 W = 1 J/s
electric charge, quantity of electricity	Coulomb	C	1 C = 1 A s
electric potential	Volt	V	1 V = 1 J/c = 1 W/A
electric capacitance	Farad	F	1 F = 1 C/V
electric resistance	Ohm	Ω	1 Ω = V/A
electric conductance	Siemens	S	1 S = 1/Ω
magnetic flux	Weber	Wb	1 Wb = 1 V s
magnetic flux density	Tesla	T	1 T = 1 Wb/m^2
inductance	Henry	H	1 H = 1 Wb/A
luminous flux	lumen	lm	1 lm = 1 cd sr
illuminance	lux	lx	1lx = 1 lm/m^2
radioactivity	Becquerel	Bq	1 Bq = 1/s
absorbed dose	Gray	Gy	1 Gy = 1 J/kg

Table 17.6 Some derived SI units having special names

Non-SI units. There are certain units not included in SI which cannot, for a variety of reasons, be eliminated, despite the fact that these can, in principle, be expressed in SI units. Some of the non-SI units which may be used together with the SI units and their multiples and are recognised by the CIPM, Comité International des Poids et Mesures, are shown in Table 17.7.

Quantity	Name of unit	Unit symbol	Definition
Time	minute	min	1 min = 60 sec
Time	hour	h	1 h = 60 min
Time	day	d	1 d = 24 h
Plane angle	degree	°	1° = (p/180) rad
Plane angle	minute	′	1 ′ = (1/60)°
Plane angle	second	″	1 ″ = (1/60)
Volume	litre	l	1 l = 1 dm^3
Mass	tonne	t	1 t = 10^3 kg
Pressure of fluid	bar	bar	1 bar = 10^5Pa

Table 17.7 Non-SI units

Conductance is sometimes used as the reciprocal of resistance for which the unit is the Siemen (S), also referred to as mho.

17.3 Conversion factors for SI units

A number of electrical terms evolved from the CGS system may still be encountered as well as units in the FPS system and other special units. In general it is best to convert units to the SI system before inserting quantities into equations as this will avoid problems involving conversion factors in the equations themselves. The engineering units which may be encountered in other systems of units are listed in Tables 17.8 to 17.10 with the conversion factor to give SI units.

Quantity	Unit	Conversion factor	SI unit
Length	cable	219.456	m
	chain (Gunter's)	20.116 8	
	chain (Ramden's)	30.48	
	fathom	1.828 8	
	feet	0.304 8	
	furlong	201.168	
	inch	0.025 4	
	micron	1×10^{-6}	
	mile (nautical Brit)	$1.853\ 184 \times 10^3$	
	mile (nautical Int)	1.852×10^3	
	mile (statute)	$1.609\ 34 \times 10^3$	
	mil	2.54×10^{-5}	
	rod	5.029 2	
	yard	0.914 4	
Area	acre	$4.046\ 856 \times 10^3$	m^2
	are	100	
	centare	1	
	hectare	1×10^4	
Volume	barrel (Brit)	0.163 65	m^3
	barrel (US petrol)	0.158 98	
	barrel (US dry)	0.115 63	
	barrel (US liquid)	0.119 24	
	gallon (Brit) (Imp)	$4.545\ 9 \times 10^{-3}$	
	gallon (US dry)	$4.404\ 8 \times 10^{-3}$	
	gallon (US liquid)	$3.785\ 3 \times 10^{-3}$	
	litre	1×10^{-3}	
	ounce (Brit fluid)	$2.841\ 225 \times 10^{-5}$	
	ounce (US fluid)	$2.957\ 373 \times 10^{-5}$	
	pint (Brit) (Imp)	$5.682\ 4 \times 10^{-4}$	
	pint (US dry)	$5.505\ 95 \times 10^{-4}$	
	pint (US liquid)	$4.731\ 63 \times 10^{-4}$	
	quart (Brit) (Imp)	$1.136\ 49 \times 10^{-3}$	
	quart (US dry)	$1.101\ 19 \times 10^{-3}$	
	quart (US liquid)	$9.463\ 2 \times 10^{-4}$	

Table 17.8 Conversion factors for length, area and volume

To obtain the quantity in SI units, the original units should be multiplied by the conversion factor, for example, to convert a length of 10 inch to metre units multiply by 0.0254 giving 0.254 m. To convert from one non-SI unit to another non-SI unit, first convert to the SI unit by multiplying by the conversion factor and then convert to the other non-SI unit by dividing by the conversion factor for the other unit. If multiple units are involved, convert each separately by multiplying by the appropriate conversion factor for any units with positive powers and dividing by the appropriate conversion factors for any units with negative powers. For example, to convert lb/in^3 (or lb in^{-3}) to SI units multiple by 0.453 6 to convert lb to kg and divide by 0.025 4^3 to convert in^{-3} to m^{-3} giving 27 680 kg m^{-3} (or 27680kg/m^3).

To convert Fahrenheit to Celsius or centigrade subtract 32 and then multiply by 5/9.

To convert Fahrenheit to Rankine add 459.67

To convert Celsius and centigrade to Kelvin add 273.15

To convert Rankine to Kelvin multiply by 5/9

For temperature conversion to Kelvin (K) the following should be noted:

Absolute zero =	0 Kelvin
	0 Rankine
	-273.15 Celsius
	-273.15 centigrade
	-459.67 Fahrenheit
Ice point =	273.15 Kelvin
	491.67 Rankine
	0 Celsius
	0 centigrade
	32 Fahrenheit

Quantity	Unit	Conversion factor	SI unit
Energy and Work	British Thermal Unit (Btu)	$1.054\ 35 \times 10^3$	J
	Btu (IST)	$1.055\ 04 \times 10^3$	
	Btu (mean)	$1.055\ 87 \times 10^3$	
	Btu (39°F)	$1.059\ 66 \times 10^3$	
	Btu (60°F)	$1.054\ 68 \times 10^3$	
	calorie, gm (cal, gm)	4.184	
	cal, gm (mean)	4.19	
	cal, gm (15°C)	4.186	
	cal, gm (20°C)	4.182	
	calorie, kg (cal, kg)	4.184×10^3	
	cal, kg (mean)	4.19×10^3	
	centigrade heat unit (15°C)	$1.898\ 3 \times 10^3$	
	erg	1×10^{-7}	
	foot-pound	1.355 82	
Force	dyne	1×10^{-5}	N
	kilogramme force	9.807	
	poundal	0.138 255	
Mass	hundredweight (long)	50.802 3	kg
	hundredweight (short)	45.359 2	
	ounce (troy)	0.031 103	
	ounce (avdp)	0.028 35	
	pounds (troy)	0.373 241	
	pounds (avdp)	0.453 592	
	slugs	14.593 9	
	ton (long)	$1.106\ 046 \times 10^3$	
	ton (short)	$9.071\ 847 \times 10^2$	
	ton (metric) (tonne)	1×10^3	
Power	cheval-vapeur	735.499	W
	horsepower	746	
	horsepower (boiler)	$9.809\ 5 \times 10^3$	
	horsepower (metric)	735.499	
Pressure	atmosphere (atm)	$1.103\ 25 \times 10^5$	Pa
	bar	1×10^5	
	inch of mercury (32°F) (in Hg)	$3.386\ 4 \times 10^3$	
	inch of mercury (60°F) (in Hg)	$3.376\ 1 \times 10^3$	
	inch of water (4°C) (in H_2O)	0.248 7	
	mm of mercury (0°C)	133.313	
	torr	133.313	
Velocity	knot	0.514 4	m/s

Quantity	Unit	Conversion factor	SI unit
Viscosity (dynamic)	centipoise (cp)	1×10^{-3}	Ns/m^2
	poise (p)	0.1	
Viscosity (kinematic)	centistoke (cSt)	1×10^{-6}	m^2/s
	Stoke (St)	1×10^{-4}	

Table 17.9 Conversion factors for quantities related to mechanics

Quantity	Unit	Conversion factor	SI unit
Electric charge	Faraday (chem)	9.649×10^4	C
	Faraday (phys)	9.652×10^4	
Magnetic flux	line	1×10^{-8}	Wb
	Maxwell	1×10^{-8}	
Magnetic flux density	Gauss	1×10^{-4}	T
Magnetizing force	Oersted	79.577 47	A-turns/m
Magnetomotive force	Gilbert	0.795 77	A-turns

Table 17.10 Conversion factors for quantities

Detailed conversion information is now given for a number of different quantities.

17.3.1 Plane angle

Quantity designation: α, β, γ

SI unit: radian, rad.

Normal multiple units: mrad, μrad.

Example: 2 rad = 2 × 57.2958 = 114.5916°

rad	... g gon, grade	... ° degree	... ′ minute	... ″ second	angular mil
1	63.6620	57.2958	3.43775 × 103	0.206265 × 106	1.00268 × 103
15.7080 $\cdot 10^{-3}$	1	0.9	54	3.24 × 103	15.75
17.4533 $\cdot 10^{-3}$	1.11111	1	60	3.6 × 103	17.5
0.290888 $\cdot 10^{-3}$	18.5185 $\cdot 10^{-3}$	16.6667 $\cdot 10^{-3}$	1	60	0.291667
4.84814 $\cdot 10^{-6}$	0.308642 $\cdot 10^{-3}$	0.277778 $\cdot 10^{-3}$	16.6667 $\cdot 10^{-3}$	1	4.86111 $\cdot 10^{-3}$
0.997331 $\cdot 10^{-3}$	63.4921 $\cdot 10^{-3}$	57.1429 $\cdot 10^{-3}$	3.42857	205.714	1

grade (g) (or gon), 1 g = 1 gon = μ/200 rad.

1° = π/180 rad.

For some purposes, the angular mil is taken to be 10^{-3} rad. The figures shown here are based on the concept that an angular mil is equal to 360/6400 degrees.

NOTE: 1 grade (... g) = 1 /100 of a right angle

17.3.2 Length

Quantity designation l.

SI unit: metre (m).

Normal multiple units: km, cm, mm, μμ.

Example: 3 in = 3 × 25.4 × 10^{-3} = 76.2 × 10^{-3} m

metre m	inch in	foot ft	yard yd	mile
1	39.370	3.2808	1.0936	$0.62137 \cdot 10^{-3}$
$25.4 \cdot 10^{-3}$	1	$83.333 \cdot 10^{-3}$	$27.778 \cdot 10^{-3}$	$15.783 \cdot 10^{-6}$
0.3048	12	1	0.33333	$0.18939 \cdot 10^{-3}$
0.9144	36	3	1	$0.56818 \cdot 10^{-3}$
$1.6093 \cdot 10^{3}$	$63.36 \cdot 10^{3}$	$5.28 \cdot 10^{3}$	$1.76 \cdot 10^{3}$	1
$1.852 \cdot 10^{3}$	$72.913 \cdot 10^{3}$	$6.0761 \cdot 10^{3}$	$2.0254 \cdot 10^{3}$	1.1508

1 nautical mile = 6080 ft = 1853.184 m

1 Å, 1 Ångström = 10^{-10} m

1 astronomic unit = $0.1496 \cdot 10^{12}$ m

1 light year = $9.4605 \cdot 10^{15}$ m

1 parsec = $30.857 \cdot 10^{12}$ m

17.3.3 Area

Quantity designation: A.

SI unit: square metre (m^2).

Normal multiple units: km^2, dm^2, cm^2, mm^2.

Example: 4 ft^2 = $4 \cdot 92.903 \cdot 10^{-3}$= 0.371612 m^2

m^2	in^2	ft^2	yd^2	acre	square mile $mile^2$
1	$1.5500 \cdot 10^{3}$	10.764	1.1960	$0.24710 \cdot 10^{-3}$	$0.38610 \cdot 10^{-6}$
$0.64516 \cdot 10^{-3}$	1	$6.9444 \cdot 10^{-3}$	$0.77161 \cdot 10^{-3}$	$0.15942 \cdot 10^{-6}$	$0.24910 \cdot 10^{-9}$
$92.903 \cdot 10^{-3}$	144	1	0.11111	$22.957 \cdot 10^{-6}$	$35\text{-}870 \cdot 10^{-9}$
0.83613	$1.296 \cdot 10^{3}$	9	1	$0.20661 \cdot 10^{-3}$	$0.32283 \cdot 10^{-6}$
$4.0469 \cdot 10^{-3}$	$6.2726 \cdot 10^{6}$	$43.56 \cdot 10^{3}$	$4.84 \cdot 10^{3}$	1	$1.5625 \cdot 10^{-3}$
$2.5900 \cdot 10^{6}$	$4.0145 \cdot 10^{9}$	$27.878 \cdot 10^{6}$	$3.0976 \cdot 10^{6}$	640	1

1 acre = 100 m^2

1 hectare = 100 acres = 10000 m^2

17.3.4 Volume

Quantity designation: V.

SI unit: cubic metre (m^3).

Normal multiple units: dm^3, cm^3, mm^3.

Non SI unit: litre (l): 1 l = 0.001 m^3 = 1 dm^3.

Normal multiple units: cl, ml.

Example:

5 US gallon = $5 \cdot 3.7854 \cdot 10^{-3} = 18.927 \cdot 10^{-3}$ m^3 = 18.927 l

m^3	in^3	ft^3	yd^3	UK gallon	US gallon
1	$61.024 \cdot 10^{3}$	35.315	1.3080	219.97	264.17
$16.387 \cdot 10^{-6}$	1	$0.57870 \cdot 10^{-3}$	$21.434 \cdot 10^{-6}$	$3.6046 \cdot 10^{-3}$	$4.3290 \cdot 10^{-3}$
$28.317 \cdot 10^{-3}$	$1.728 \cdot 10^{3}$	1	$37.037 \cdot 10^{-3}$	6.2288	7.4805
0.76456	$46.656 \cdot 10^{3}$	27	1	168.18	201.97
$4.5461 \cdot 10^{-3}$	277.42	0.16054	$5.9461 \cdot 10^{-3}$	1	1.2010
$3.7854 \cdot 10^{-3}$	231	0.13368	$4.9511 \cdot 10^{-3}$	0.83268	1

gross (register) tonnage used in shipping
1 ton = 100 ft^3 = 2.83168 m^3

1 UK fluid ounce, fl oz = 28.4131 cm^3

1 US fluid ounce, fl oz = 29.5735 cm^3

17.3.5 Time

Quantity designation: t.

SI unit: second (s).

Normal multiple units: ms, s, ns.

Non SI units: day (d), hour (h), minute (min).

Example: 100000 s = 100000/3600 = 27.778 h

s	min	h	d (day)	week
1	$16.6667 \cdot 10^{-3}$	$0.277778 \cdot 10^{-3}$	$11.5741 \cdot 10^{-6}$	$1.65344 \cdot 10^{-6}$
60	1	$16.6667 \cdot 10^{-3}$	$0.694444 \cdot 10^{-3}$	$99.2063 \cdot 10^{-6}$
$3.6 \cdot 10^{3}$	60	1	$41.6667 \cdot 10^{-3}$	$5.95238 \cdot 10^{-3}$
$86.4 \cdot 10^{3}$	$1.44 \cdot 10^{3}$	24	1	0.142857
$604.8 \cdot 10^{3}$	$10.08 \cdot 10^{3}$	168	7	1

1 tropical year = 31556925.974 s = 365.24219878 d

1 sidereal year = 31558150 s

1 calendar year = 365 d = 8760 h

17.3.6 Linear velocity

Quantity designation: v.

SI unit: metre per second (m/s).

Normal multiple units: km/h.

m/s	km/h	ft/s	mile/h
1	3.6	3.2808	2.2369
0.27778	1	0.91134	0.62137
0.3048	1.0973	1	0.68182
0.44704	1.6093	1.4667	1
0.51444	1.852	1.6878	1.1508

1 knot = 1 nautical mile per hour = 1.853 km/h

17.3.7 Linear acceleration

Quantity designation: a.

SI unit: metre per second squared (m/s^2).

m/s^2	cm/s^2	ft/s^2	in/s^2	g
1	100	3.2808	39.37	0.10197
10 × 10-3	1	32.808 · 10-3	$393.7 \cdot 10^{-3}$	$1.0197 \cdot 10^{-3}$
0.3048	30.48	1	12	$31.081 \cdot 10^{-3}$
25.4×10^{-3}	2.54	$83.33 \cdot 10^{-3}$	1	$2.59 \cdot 10^{-3}$
9.80665	980.665	32.174	386.09	1

17.3.8 Angular velocity

Quantity designation: .

SI unit: radian per second (rad/s).

The SI and Imperial units are identical.

Angular velocity is normally calculated from N revolutions/s by 2 N

17.3.9 Angular acceleration

Quantity designation: a.

SI unit: radian per second squared (rad/s^2)

The SI and Imperial units are identical.

17.3.10 Mass

Quantity designation: m.

SI unit: kilogram (kg).

Normal multiple units: g, g, mg.

Non SI unit: tonne (t) = 1000 kg.

kg	lb (pound)	oz (ounce)	hundred-weight cwt (UK)	ton (UK)
1	2.2046	35.274	$19.684 \cdot 10^{-3}$	$0.98421 \cdot 10^{-3}$
0.45359	1	16	$8.9286 \cdot 10^{-3}$	$0.44643 \cdot 10^{-3}$
14.594	32.174	514.79	$0.28727 \cdot 10^{-3}$	$14.363 \cdot 10^{-3}$
28.350×10^{-3}	62.5×10^{-3}	1	$0.55804 \cdot 10^{-3}$	$27.902 \cdot 10^{-6}$
50.802	112	$1.792 \cdot 10^{3}$	1	$50 \cdot 10^{-3}$
$1.0161 \cdot 10^{3}$	$2.24 \cdot 10^{3}$	$35.84 \cdot 10^{-3}$	20	1

oz = ounce, also called ounce avoirdupois

1 ounce troy = $31.1035 \cdot 10^{-3}$ kg

cwt = hundredweight

USA cwt = 100 lb   USA ton = 2000 lb

17.3.11 Density

Also called specific weight.

Quantity designation: .

SI unit: kilogram per cubic metre (kg/m^3)

Non SI units: kg/dm^3, g/cm^3.

kg/m^3	g/cm^3	lb/in^3	lb/ft^3
1	10^{-3}	$36.127 \cdot 10^{-6}$	$62.428 \cdot 10^{-3}$
10^{3}	1	$36.127 \cdot 10^{-3}$	62.428
$27.680 \cdot 10^{3}$	27.680	1	$1.728 \cdot 10^{3}$
16.019	$16.019 \cdot 10^{-3}$	$0.57870 \cdot 10^{-3}$	1

The term specific gravity or relative density is also used and is the ratio of the mass of a given volume of substance to the mass of an equal volume of water at temperature of 4 °C and a pressure of 101.325 kPa absolute. The density of water at 4 °C and 101.325 kPa absolute is 1000.02 kg/m^3.

17.3.12 Force

Quantity designation: F.

SI unit: newton (N).

Normal multiple units: MN, kN.

N	dyn	Kilogram-force, kgf kilopond,	pound-force lbf
1	$0.1 \cdot 10^{6}$	0.10197	0.22481
$10 \cdot 10^{-6}$	1	$1.0197 \cdot 10^{-6}$	$2.2841 \cdot 10^{-6}$
9.8066	$0.98066 \cdot 10^{6}$	1	2.2046
4.4482	$0.44482 \cdot 10^{6}$	0.45359	1

17.3.13 Torque

Quantity designation: T.

SI unit: Newton metre (Nm).

Normal multiple units: MNm, kNm.

Nm	kgf m	lbf × in	lbf × ft
1	0.10197	8.8508	0.73756
9.8066	1	86.796	7.2330
0.11299	$11.521 \cdot 10^{-3}$	1	$83.333 \cdot 10^{-3}$
1.3558	0.13826	12	1

Torque, power and speed are related by the formula: P = 2 NT

17.3.14 Pressure, stress

Quantity designation: p, , .

SI unit: Pascal (Pa), 1 Pa = 1 N/m^2.

Normal multiple units: GPa, MPa, kPa and for stress: MN/m^2, N/m^2, N/mm^2.

Pa	bar	kgf/cm^2 technical atmos	kgf/mm^2	Torr (≈ mm Hg)	standard atm	lbf/in^2 (psi)
1	$10 \cdot 10^{-6}$	$10.197 \cdot 10^{-6}$	$0.10197 \cdot 10^{6}$	$7.5006 \cdot 10^{-3}$	$9.8692 \cdot 10^{-6}$	$0.14504 \cdot 10^{-3}$
$100 \cdot 10^{3}$	1	1.0197	$10.197 \cdot 10^{-3}$	750.06	0.98692	14.504
$98.066 \cdot 10^{3}$	0.98066	1	$10 \cdot 10^{-3}$	735.56	0.96784	14.223
$9.8066 \cdot 10^{6}$	98.066	100	1	$73.556 \cdot 10^{3}$	96.784	$1.4223 \cdot 10^{3}$
133.32	$1.3332 \cdot 10^{-3}$	$1.3595 \cdot 10^{-3}$	$13.595 \cdot 10^{-6}$	1	$1.3158 \cdot 10^{-3}$	$19.337 \cdot 10^{-3}$
$101.32 \cdot 10^{3}$	1.0132	1.0332	$10.332 \cdot 10^{-3}$	760	1	14.696
$6.8948 \cdot 10^{3}$	$68.948 \cdot 10^{-3}$	$70.307 \cdot 10^{-3}$	$0.70307 \cdot 10^{-3}$	51.715	$68.046 \cdot 10^{-3}$	1

The preferred pressure unit for pump applications is the "bar".

1 mm H_2O 9.81 Pa

1 in H_2O 249.09 Pa

1 in Hg 3386.4 Pa

1 ata = 1 technical atmosphere (absolute)

1 atu = 1 technical atmosphere (gauge)

17.3.15 Dynamic viscosity

Quantity designation: .

SI unit: Pascal second Pa s.

Normal multiple units mPa s (= cP).

N s/m^2	N s/mm^2	P (poise)	cP m Pa s
1	10^{-6}	10	10^{3}
10^{6}	1	10×10^{6}	10^{9}
0.1	0.1×10^{-6}	1	100
10^{-3}	10^{-9}	10×10^{-3}	1

17.3.16 Kinematic viscosity

Quantity designation: .

SI unit: square metres per second (m^2/s).

Normal multiple units: mm^2/s.

For conversion to other units of viscosity see nomogram in Figure 17.1.

m^2/s	mm^2/s cSt	St (Stoke)
1	10^{6}	$10 \cdot 10^{3}$
10^{-6}	1	$10 \cdot 10^{-3}$
$0.1 \cdot 10^{-3}$	100	1

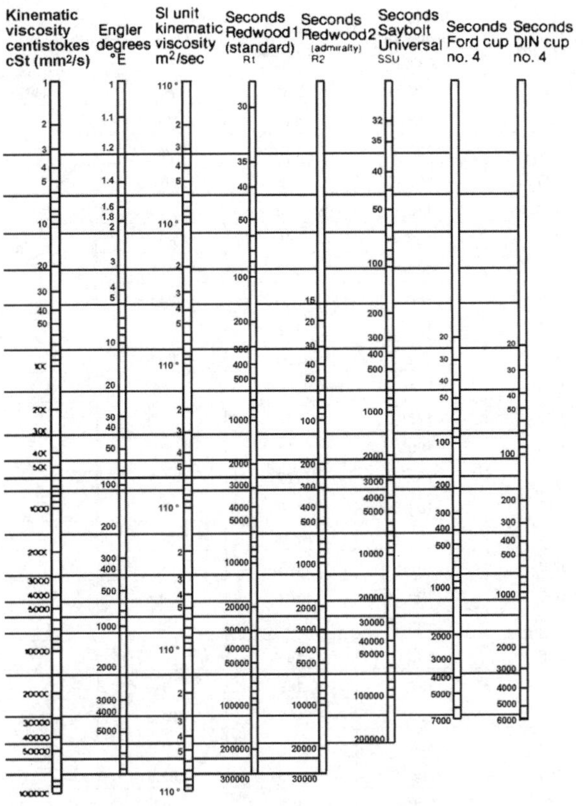

Figure 17.1 Nomogram for conversion of Kinematic viscosity to other units of viscosity, in Section 17.316

	cSt x	°E x	R1 x	SSU x
Kinematic viscosity cSt	1	7.58	0.247	0.216
Engler degrees °E	0.132	1	0.0326	0.0285
Seconds, Redwood 1, R1	4.05	30.7	1	0.887
Seconds, Saybolt Universal SSU	4.62	35.11	1.14	1

The above factors apply for values above 60cSt

17.3.17 Energy

Quantity designations: E, W, Q, L, U depending upon the type of energy.

SI unit: Joule (J).

Normal multiple units: TJ, GJ, MJ, kJ, mJ.

Since 1 J = 1 Nm = 1 Ws then Nm and Ws can also be used for all types of energy. The unit Joule should, however, be used for expressing all types of energy.

1 erg = 0.1 × 10^{-6} J.

J Joule	kWh kilowatt hour	kgf m kilogram-force metre	kcal kilo-calorie	ch h metric horse-power hour	ft. lbf (foot pound-force)	Btu (British thermal unit)
1	0.27778 $\cdot 10^{-6}$	0.10197	0.23885 $\cdot 10^{-3}$	0.37767 $\cdot 10^{-6}$	0.73756	0.94782 $\cdot 10^{-3}$
3.6 · 106	1	0.36710 $\cdot 10^{6}$	859.85	1.3596	2.6552 $\cdot 10^{6}$	3.4121 $\cdot 10^{3}$
9.8066	2.7241 $\cdot 10^{-6}$	1	2.3423 $\cdot 10^{-3}$	3.7037 $\cdot 10^{-6}$	7.2330	9.2949 $\cdot 10^{-3}$
4.1868 $\cdot 10^{3}$	1.163 $\cdot 10^{-3}$	426.94	1	1.5812 $\cdot 10^{-3}$	3.0880 $\cdot 10^{3}$	3.9683
2.6478 $\cdot 10^{6}$	0.73550	0.27 · 10^{6}	632.42	1	1.9529 $\cdot 10^{6}$	2.5096 $\cdot 10^{3}$
1.3558	0.37662 $\cdot 10^{-6}$	0.13826	0.32383 $\cdot 10^{-3}$	0.51206 $\cdot 10^{-6}$	1	1.2851 $\cdot 10^{-3}$
1.0551 $\cdot 10^{3}$	0.29307 $\cdot 10^{-3}$	107.59	0.25200	0.39847 $\cdot 10^{-3}$	778.17	1

17.3.18 Power

Quantity designation: P.

SI unit: Watt (W).

Normal multiple units: GW, MW, kW, mW, W.

W	kgf m/s	kcal/s	kcal/h	ch metric horse-power	hp horse-power	ft lbf/s	Btu/h
1	0.10197	0.23885 $\cdot 10^{-3}$	0.85985	1.3596 $\cdot 10^{-3}$	1.3410 $\cdot 10^{-3}$	0.73756	3.4121
9.8066	1	2.3423 $\cdot 10^{-3}$	8.4322	13.333 $\cdot 10^{-3}$	13.151 $\cdot 10^{-3}$	7.2330	33.462
4.1868 $\cdot 10^{3}$	426.94	1	3.6 $\cdot 10^{3}$	5.6925	5.6146	3.0880 $\cdot 10^{3}$	14.286 $\cdot 10^{3}$
1.163	0.11859	0.27778 $\cdot 10^{-3}$	1	1.5812 $\cdot 10^{-3}$	1.5596 $\cdot 10^{-3}$	0.85779	3.9683
735.50	75	0.17567	632.42	1	0.98632	542.48	2.5096 $\cdot 10^{3}$
745.70	76.040	0.17811	641.19	1.0139	1	550	2.5444 $\cdot 10^{3}$
1.3558	0.13826	0.32383 $\cdot 10^{-3}$	1.1658	1.8434 $\cdot 10^{-3}$	1.8182 $\cdot 10^{-3}$	1	4.6262
0.29307	29.885 $\cdot 10^{-3}$	69.999 $\cdot 10^{-6}$	0.25200	0.39847 $\cdot 10^{-3}$	0.39302 $\cdot 10^{-3}$	0.21616	1

17.3.19 Flow

Quantity designation: q_v.

SI unit: cubic metre per second (m^3/s).

Non-SI units: l/s, ml/s, m^3/h.

I gallon/min Igpm	US gallon/min USgpm	barrel/day bpd	m^3/s	m^3/h	l/s	l/min
1	1.2009	41.175	75.768 $\cdot 10^{-6}$	272.76 $\cdot 10^{-3}$	75.768 $\cdot 10^{-3}$	4.5461
0.83268	1	34.286	63.09 $\cdot 10^{-6}$	227.12 $\cdot 10^{-3}$	63.09 $\cdot 10^{-3}$	3.7854
24.286 $\cdot 10^{-3}$	29.167 $\cdot 10^{-3}$	1	1.84 $\cdot 10^{-6}$	6.6244 $\cdot 10^{-3}$	1.84 $\cdot 10^{-3}$	110.41 $\cdot 10^{-3}$
13.198 $\cdot 10^{3}$	15.85 $\cdot 10^{3}$	543.44 $\cdot 10^{3}$	1	3600	1000	60000
3.6662	4.4029	150.95	277.78 $\cdot 10^{-6}$	1	277.78 $\cdot 10^{-3}$	16.667
13.198	15.85	543.44	1.0 $\cdot 10^{-3}$	3.6	1	60
219.97 $\cdot 10^{-3}$	264.17 $\cdot 10^{-3}$	9.0573	16.667 $\cdot 10^{-6}$	60 $\cdot 10^{-3}$	16.667 $\cdot 10^{-3}$	1

1 barrel = 42 US gallon

17.3.20 Temperature

Absolute temperature: Quantity designation: T.

SI unit: Kelvin (K).

Temperature: Quantity designation: t.

unit degree Celsius (Centigrade) (°C).

	Kelvin*** scale	Celsius scale	Rankine scale	Fahrenheit** scale	Physical relation-ship
Relative temperature value	0 K	- 273.15 °C	0 °R	- 459.67 °F	Absolute zero
	273.15 K	0 °C	491.67 °R	32 °F	Melting point of ice*
	273.16 K	0.01 °C	491.688 °R	32.018 °F	Triple point of water*
	373.15 K	100 °C	671.67 °R	212 °F	Boiling point of water*
Relative temperature intervals (diffs)	1 K 0.55556 K	1 °C 0.55556 °C	1.8 °R 1 °R	1.8 °F 1 °F	

* For defined conditions

** Value in °C = $\frac{1}{1.8}$ (value in °F -32) = (value in K -273.15)

*** Value in K = $\frac{5}{9}$ × (value in °R)

17.4 Other conversion factors

17.4.1 Hardness

Hardness is not defined within the SI system. The following table can be used for conversion between the popular systems used.

Tensile strength	Vickers hardness (F ≥ 98 N)	Brinell hardness	Rockwell hardness	
N/mm²	HV	BHN	HR_B	HR_C
200	63	60		
210	65	62		
220	69	66		
225	70	67		
230	72	68		
240	75	71		
250	79	75		
255	80	76		
260	82	78		
270	85	81	41	
280	88	84	45	
285	90	86	48	
290	91	87	49	
300	94	89	51	
305	95	90	52	
310	97	92	54	
320	100	95	56	
330	103	98	58	
335	105	100	59	
340	107	102	60	
350	110	105	62	
360	113	107	63.5	
370	115	109	64.5	
380	119	113	66	
385	120	114	67	
390	122	116	67.5	
400	125	119	69	
410	128	122	70	
415	130	124	71	
420	132	125	72	
430	135	128	73	
440	138	131	74	
450	140	133	75	
460	143	136	76.5	
465	145	138	77	
470	147	140	77.5	
480	150	143	78.5	
490	153	145	79.5	
495	155	147	80	
500	157	149	81	
510	160	152	81.5	
520	163	155	82.5	
530	165	157	83	
540	168	160	84.5	
545	170	162	85	
550	172	163	85.5	
560	175	166	86	
570	178	169	86.5	
575	180	171	87	
580	182	172		
590	184	175	88	
595	185	176		
600	187	178	89	
610	190	181	89.5	
620	193	184	90	
625	195	185		
630	197	187	91	
640	200	190	91.5	
650	203	193	92	
660	205	195	92.5	
670	208	198	93	
675	210	199	93.5	
680	212	201		
690	215	204	94	
700	219	208		
705	220	209	95	
710	222	211	95.5	
720	225	214	96	
730	228	216		
740	230	219	96.5	
750	233	221	97	
755	235	223	97.5	
760	237	225		
770	240	228	98	
780	243	231		21
785	245	233		
790	247	235	99	
800	250	238	99.5	22
810	253	240		
820	255	242		23
830	258	245		
835	260	247		24
840	262	249		
850	265	252		
860	268	255		25
865	270	257		
870	272	258		26
880	275	261		
890	278	264		
900	280	266		27
910	283	269		
915	285	271		
920	287	273		28
930	290	276		
940	293	278		29
950	295	280		
960	299	284		
965	300	285		
970	302	287		30

Tensile strength	Vickers hardness (F ≥ 98 N)	Brinell hardness	Rockwell hardness	
N/mm²	HV	BHN	HR_B	HR_C
980	305	290		
990	308	293		
995	310	295		31
1000	311	296		
1010	314	299		
1020	317	301		32
1030	320	304		
1040	323	307		
1050	327	311		33
1060	330	314		
1070	333	316		
1080	336	319		34
1090	339	322		
1095	340	323		
1100	342	325		
1110	345	328		35
1120	349	332		
1125	350	333		
1130	352	334		
1140	355	337		36
1150	358	340		
1155	360	342		
1160	361	343		
1170	364	346		37
1180	367	349		
1190	370	352		
1200	373	354		
1210	376	357		38
1220	380	361		
1230	382	363		39
1240	385	366		
1250	388	369		
1255	390	371		
1260	392	372		40
1270	394	374		
1280	397	377		
1290	400	380		
1300	403	383		41
1310	407	387		
1320	410	390		
1330	413	393		42
1340	417	396		
1350	420	399		
1360	423	402		43
1370	426	405		
1380	429	408		
1385	430	409		
1390	431	410		
1400	434	413		44
1410	437	415		
1420	440	418		
1430	443	421		
1440	446	424		
1450	449	427		45
1455	450	428		
1460	452	429		
1470	455	432		
1480	458	435		46
1485	460	437		
1490	461	438		
1500	464	441		
1510	467	444		

Tensile strength	Vickers hardness (F ≥ 98 N)	Brinell hardness	Rockwell hardness	
N/mm²	HV	BHN	HR_B	HR_C
1520	470	447		
1530	473	449		47
1540	476	452		
1550	479	455		
1555	480	456		
1560	481	457		
1570	484	460		48
1580	486	462		
1590	489	465		
1595	490	466		
1600	491	467		
1610	494	470		
1620	497	472		49
1630	500	475		
1640	503	478		
1650	506	481		
1660	509	483		
1665	510	485		
1670	511	486		
1680	514	488		50

Values based on DIN 50150.

17.4.2 Material toughness

Material toughness is not defined by SI. Most materials are assessed by conducting impact testing. Two differing test methods can be used with various sizes and styles of test specimen. The following table can be used as a **guide** to the relative toughness of the various tests:

Charpy V notch kgm/cm²	Charpy V notch ft lb	Charpy V notch Joule	Izod ft lb
0.4	2.3	3.1	2.5
0.9	5.2	7	6.4
1.5			10.8
2.2			16
3.1	18	24.4	21.5
4.1	23.8	32.2	27.8
5.2	30	40.6	34.1
6.5	37.7	51	40.4
8.0	46.4	92.9	46.7
9.4	54.5	73.9	53
10.9	63	85.4	59.3
12.6			65.6
14.1	82	111	71.9
15.8			78.2
17.7	102	138	84.5
19.4			90.8
21.1	122	165	97.1
23.0	134	182	103.4

17.5 Normal quantities and units used in pump technology

Quantity		Recommended unit	
Name	**Symbol**	**Name**	**Units**
Area	A	square metre square centimetre square millimetre	m^2 cm^2 mm^2
Compressibility	χ	reduction in unit volume per bar differential pressure	bar^{-1}
Density	ρ	kilogram per cubic metre gram per cubic decimetre	kg/m^3 g/dm^3
Energy	E	Joule kiloJoule MegaJoule	J kJ MJ
Flow (1)	q_v	litre per second litre per minute litre per hour cubic metre per hour	l/s l/min l/h m^3/h
Force	F	Newton kiloNewton MegaNewton	N kN MN
Frequency	f	Hertz kiloHertz MegaHertz	Hz kHz MHz
Head (suction, discharge, differential)	H	metre	m
Length	l	metre millimetre micron	m mm μm
Mass (weight)	m	tonne kilogram gram	tonne kg g
MIP (Minimum inlet pressure)	MIP	bar (gauge)	bar
Moment of inertia	J	kilogram metre squared	kgm^2
NPIP (Net positive inlet pressure)	NPIP	bar	bar
NPSH (Net positive suction head)	NPSH	metre	m
Power	P	Watt kiloWatt MegaWatt	W kW MW
Pressure (absolute or gauge)	p	bar	bar
Shaft speed	n	revolutions per second revolutions per minute revolutions per hour	r/s r/min r/h
Stress	σ τ	MegaNewton per square metre	MN/m^2
Temperature	T	degree Celsius	°C
Time	t	second minute hour	s min h
Torque	T	Newton metre kiloNewton metre MegaNewton metre	Nm kNm MNm
Velocity	v	metre per second	m/s
Viscosity, dynamic	η	Poise centiPoise	P cP
Viscosity, kinematic	μ	Stoke centiStoke	St cSt

(1) Pumps handle liquid by volume.

The user must convert mass flow to volume flow.

17.6 Useful references

Hydraulic Institute, 9 Sylvan Way, Parsippany NJ, 07054 USA Tel: 973 267 9700, Fax: 973 2679055, www.pumps.org.

VDMA (German Engineering Federation), Lyoner Straße 18, D-60528 Frankfurt, Germany, Tel: 069 66 03-0, www.vdma.org.

BPMA, (British Pump Manufacturers Association), The McLaren Building 35 Dale End, Birmingham, B4 7LN, UK, Tel: 0121 200 1299 Fax: 0121 200 1306, Email: enquiry@bpma.org.uk.

American Bureau of Shipping, ABS, 16855 Northchase Drive,Houston, TX 77060 USA Tel: 281 877 5800, Fax: 281 877 5803, www.eagle.org.

Det Norske Veritas DNV, Veritasveien 1, N-1322 Høvik Norway, Tel: 067 57 99 00, Fax: 067 57 99 11, www.dnv.com.

Lloyd's Register (LR), 71 Fenchurch Street, London EC3M 4BS, UK, Tel: 020 7709 9166, Fax: 020 7488 4796, www.lr.org.

Bureau International des Poids et Mesures (BIPM), Pavillon de Breteuil, F-92312 Sèvre Cedex, France, www.bipm.fr.

Office of Public Sector Information (OPSI), Admiralty Arch, North Side, The Mall, London, SW1A 2WH, UK, Tel: 01603 723011, www.opsi.gov.uk.

The Stationery Office Ltd (TSO) UK, Tel: 0870 242 2345, Email: esupport@tso.co.uk, www.tsoshop.co.uk.

EFTA (European Free Trade Association), 9-11, rue de Varembé, CH-1211 Geneva 20 Switzerland, Tel: 022 332 2626, Fax: 022 332 2699, www.secretariat.efta.int.

Europump, Diamant Building, Blvd. A. Reyers 80, B-1030 Brussels, Belgium, Tel: 02 7068230, Fax: 02 7068253, www.europump.org.

European Hygienic Engineering & Design Group (EHEDG), Avenue Grand Champ, 148 B-1150 Brussels, Belgium, Tel: 32 02 7617408, www.ehedg.org.

NAFEM (North American Association of Foodservice Equipment Manufacturers), 161 North Clark Street, Suite 2020, Chicago, IL 60601, USA, Tel: 312.821.0201, Fax: 312.821.0202, www.nafem.org.

Buyers' guide 18

18.1 Introduction

The Buyers' Guide summarises the various pump types, divided into three groups:

- Rotodynamic pumps
- Positive displacement pumps
- Other pumps

The Guide has been grouped in such a way to impose boundary limits on pump types and the operating conditions available, with the aim of simplifying the choice of supplier from the user's point of view.

The Buyers' Guide covers all pump types, followed by ancillary products and services. Trade names are comprehensively listed too. It is preceded by the names and addresses and contact details of all companies appearing in the Guide. They are listed alphabetically, by country.

It is strongly recommended that direct contact with the relevant companies is made to ensure that their details are clarified wherever necessary.

Company names and addresses, Section 18.2
Detailed company information, listed alphabetically by country.

Rotodynamic pumps, Section 18.3
Companies listed alphabetically by pump type.

Positive displacement pumps, Section 18.4
Companies listed alphabetically by pump type.

Other pumps, Section 18.5
Companies listed alphabetically by pump type.

Ancillary products and services, Section 18.6
Companies listed alphabetically by product and/or service.

Trade names, Section 18.7
Companies listed alphabetically by Trade name or product name.

18.2 Company names and addresses

Detailed company information, listed alphabetically by country

ARGENTINA (AR)

Motorarg SA
Pcia de Buenos Aires
Valentín Alsina
Veracruz (B1822BGP)
Argentina
Tel: 0420 87717
Fax: 0420 81080
Email: motorarg@motorarg.com.ar
Web: www.motorarg.com.ar

AUSTRALIA (AU)

Acromet (Aust) Pty Ltd
14 Winterton Road, Box 1105
Clayton
VIC 3168
Australia
Tel: 03 9544 7333
Fax: 03 9543 6706
Email: chemex@acromet.com.au
Web: www.acromet.com.au

Batescrew Pumps
Newell Highway
Tocumwal
NSW 2714
Australia
Tel: 03 587 42101
Fax: 03 587 42084
Email: sales@batescrew.com
Web: www.batescrew.com

Davey Products Pty Ltd
6 Lakeview Drive
Scoresby
VIC 3179
Australia
Tel: 03 9730 9121
Fax: 03 9753 4248
Email: export@davey.com.au
Web: www.davey.com.au

EBS-Ray Pumps Pty Ltd
628 Pittwater Road
Brookvale
Sydney
NSW 2100
Australia
Tel: 02 9905 0234
Fax: 02 9938 3825
Email: nswsales@ebsraypumps.com.au
Web: www.ebsraypumps.com.au

Engineered Products Group Pty Ltd (EPG)
31 Powers Road
Seven Hills
NSW 2147
Australia
Tel: 02 9830 2222
Fax: 02 9830 2384
Email: gateway@epg.com.au
Web: www.epg.com.au

Grundfos Pumps Pty Ltd
515 South Road
Regency Park
SA 5010
Australia
Tel: 08 8461 4611
Email: contact-au@grundfos.com
Web: www.grundfos.com

Kelair Pumps Australia Pty Ltd
Gateway Estate
215 Walters Road
Arndell Park
NSW 2128
Australia
Tel: 02 9678 9466
Fax: 02 9678 9455
Email: kelair@kelairpumps.com.au
Web: www.kelair.com.au

Mono Pumps (Australia) Pty Ltd
Mono House, 338-348 Lower Dandenong Road
Mordialloc
VIC 3195
Australia
Tel: 03 9580 5211
Fax: 03 9580 6659
Email: ozsales@mono-pumps.com
Web: www.mono-pumps.com

Robertson Technology Pty Ltd
PO Box 493
Joondalup DC
Joondalup
WA 6919
Australia
Tel: 08 9300 0844
Fax: 08 9300 2611
Email: mail@pumpmonitor.com
Web: www.pumpmonitor.com

TKL
26 Faigh Street
Mulgrave
VIC 3170
Australia
Tel: 03 9562 0744
Fax: 03 9562 2816
Email: dick_stevenson@flowserve.com
Web: www.tkl.com.au

Tyco - Southern Cross Plant
8616 Warrego Highway Corner Roches Road
Withcott
QLD 4352
Australia
Tel: 01800 888 909
Fax: 01800 637 867
Email: southerncross@typac.com.au
Web: www.southerncross.com.au

United Pumps Australia (UPA)
31 Western Avenue
Sunshine
VIC 3020
Australia
Tel: 03 93126566
Fax: 03 93126371
Email: unitedpumps@unitedpumps.com.au
Web: www.unitedpumps.com.au

AUSTRIA (AT)

ABS Pumpen GmbH
IZ-NÖ-Süd
Strasse 2, M27
A-2351 Wiener Neudorf
Austria
Tel: 02236 64261
Fax: 02236 64266
Email: info@absgroup.com
Web: www.absgroup.com

Andritz AG
Stattegger Strasse 18
A-8045 Graz
Austria
Tel: 0316 6902 0
Fax: 0316 6902 406
Email: irmtraut.pfingstl@andritz.com
Web: www.andritz.com

AxFlow GmbH
Wienerstrasse 253
A-8051 Graz
Austria
Tel: 0316 683509
Fax: 0316 683492
Email: office@axflow.at
Web: www.axflow.at

Bühler Hochdrucktechnik GmbH
PO Box 27 Werk-VI-Strasse
A-8605 Kapfenberg
Austria
Tel: 03862 303300
Fax: 03862 303304
Email: bohler.office@bhdt.at
Web: www.bhdt.at

Grundfos Pumpen Vertrieb GmbH
Grundfosstrasse 2
A-5082 Grödig/Salzburg
Austria
Tel: 06246 8830
Email: info-austria@grundfos.com
Web: www.grundfos.com

Hauke GmbH & Co KG
Postfach 103
Cumberlandstrasse 46-50
A-4810 Gmunden
Austria
Tel: 07612 637580
Fax: 07612 64133
Email: hauke.gesmbh@aon.at
Web: www.hauke.at

Kräutler GmbH & Co
Bildgasse 40
A-6893 Lustenau
Austria
Tel: 05577 866440
Fax: 05577 88433
Web: www.kral.at

Sterling Fluid Systems (Austria)
Oberlaaer Strasse 228
A-1100 Vienna
Austria
Tel: 01 680 050

Fax: 01 680 0521
Email: sales_austria@sterlingsihi.de
Web: www.sterlingfluid.com

Pumpenfabrik Ernst Vogel GmbH, ITT Industries
Ernst Vogel-Strasse 2
A-2000 Stockerau
Austria
Tel: 02266 6040
Fax: 02266 65311
Email: info@vogel-pumpen-ittind.com
Web: www.vogel-pumps.com

BELGIUM (BE)

ABS Pumps NV/SA
Haachtsesteenweg 56
B-1831 Diegem
Belgium
Tel: 02 725 7900
Fax: 02 725 7119
Email: info@absgroup.com
Web: www.absgroup.com

Alfons de Backer & Co
Kasteeldreef 1
B-9230 Wetteren
Belgium
Tel: 09 369 3496
Fax: 09 369 5752
Email: dbjm@duba.be
Web: www.duba.be

Clasal
Ruddervoordsestraat 82
B-8210 Zedelgem
Belgium
Tel: 050 27 8253
Fax: 050 27 5494
Email: info@clasal.be
Web: www.clasal.be

Ensival-Moret
Rue de Hodister 44
B-4860 Wegnez-Pepinster
Belgium
Tel: 087 46 8111
Fax: 087 46 8100
Email: emwegnez@em-pumps.com
Web: www.ensival.com

Packo Inox NV
Cardijnlaan 10
B-8600 Diksmuide
Belgium
Tel: 051 519280
Fax: 051 519299
Email: diksmuide@packo.com
Web: www.packo.com

Sterling Fluid Systems (Belgium) NV
Zone Gosset, 't Hofveld 1
B-1702 Groot-Bijgaarden
Belgium
Tel: 02 481 7711
Fax: 02 481 7737
Email: sales@sterlingfluidsystems.be
Web: www.sterlingfluid.com

Toyo Pumps Europe
Parc Industriel Sud – Zone II
Rue de l'industrie 41
B-1400 Nivelles
Belgium
Tel: 067 64 5537
Fax: 067 64 5531
Email: info@toyopumpseurope.com
Web: www.toyopumpseurope.com

Vandezande
Zeepziederijstraat 5
Industrieterrein Kaaskerke
B-8600 Diksmuide
Belgium
Tel: 051 500117
Fax: 051 504117
Email: info@vandezande.com
Web: www.vandezande.com

BRAZIL (BR)

ABS Industria de Bombas Centrifugas Ltda
Rua Hasdrubal, Bellegard 701 CIC
CEP 81460 - 120 Curitiba - Paraná
Brazil
Tel: 041 2108 8100
Fax: 041 3348 1879
Email: info@absgroup.com
Web: www.absgroup.com

Mark Grundfos Ltda
Av Humberto de Alencar
Castelo Branco 630
09850-300 São Sernardo do Campo SP
Brazil
Tel: 011 4393 5533
Email: stiegs@grundfos.com
Web: www.grundfos.com

Omel Ltda
Rua Silvio Manfredi 201
Guarulhos
Sao Paulo 07241-000
Brazil
Tel: 011 6413 5400
Fax: 011 6412 5056
Email: omel@omel.com.br
Web: www.omel.com.br

CANADA (CA)

ABS Pumps Corporation
1215 Meyerside Drive, Unit 7
Mississauga
Ontario L5T 1H3
Canada
Tel: 905 670 4677
Fax: 905 670 3709
Email: info@absgroup.com
Web: www.absgroup.com

Eagle Pump & Compressor Ltd
7025 - 5th Street SE
Calgary
Alberta
T2H 2G2
Canada
Tel: 403 253 0100
Fax: 403 253 8884
Email: bcsavosi@eagle-pc.com
Web: www.eagle-pc.com

Hayward Gordon
6660 Campobello Road
Mississauga
Ontario
L5N 2L9
Canada
Tel: 905 567 6116
Fax: 905 567 1706
Email: info@haywardgordon.com
Web: www.haywardgordon.com

Kudu Industries Inc
9112 40th Street SE
Calgary AB
T2C 2P3
Canada
Tel: 403 279 5838
Fax: 403 279 2192
Email: ray_mills@kudupump.com
Web: www.kudupump.com

Monarch Industries Ltd
51 Burmac Road
Winnipeg
Manitoba R3C 3E4
Canada
Tel: 204 786 7921
Fax: 204 889 9120
Email: enquiries@monarchindustries.com
Web: www.monarchindustries.com

Nova Magnetics Burgmann Ltd
One Research Drive
Dartmouth
Nova Scotia
B2Y 4M9
Canada
Tel: 902 465 6625
Fax: 902 465 6629
Email: info@novamagnetics.ca
Web: www.novamegnetics.ca

Sunmotor International Ltd
5031 50 Sreet
Olds
Alberta
T4H 1R8
Canada
Tel: 403 556 8755
Fax: 403 556 7799
Email: sunmotor@telusplanet.net
Web: www.sunpump.com

Vican Pump Company
661 Grove Avenue
Windsor
ON N9A 6M3
Canada
Tel: 519 256 5438
Fax: 519 256 5070
Email: gbornais@idexcorp.com
Web: www.vikingpumpcanada.com

CHINA (CN)

ABS Shanghai Co Ltd
Room 3005-06 Zhong Bao Mansion
No 166, Lu Jia Zui Road (E)
Pu Dong New District
CN-200120 Shanghai
China
Tel: 021 6888 2900
Fax: 0215882 6708
Email: info@absgroup.com
Web: www.absgroup.com

ABS Shanghai Co Ltd
Beijing Liaison Office
Room 1701, Kuntai International Mansion
No 12 2nd, Chaowai Street
Beijing 100020
China
Tel: 010 5879 0009
Fax: 010 5879 0512
Email: info@absgroup.com
Web: www.absgroup.com

Benxi Water Pump Factory
16 Digong Road
Mingshan
Benxi 117022
China
Tel: 0414 483 7107
Fax: 0414 482 8402
Email: bxspump@online.ln.cn

Dalian Danai Pumps Ltd
250 Huabei Road
Shahekou District
Dalian
China
Tel: 0411 440 3456
Fax: 0411 440 3008
Email: dapp@denebcn.com
Web: www.denebcn.com

Hunan XD Changsha
Pump Works Co Ltd
522 Furong Road (M)
Changsha
Hunan 410015
China
Tel: 0731 5215460
Email: postmaster@cbpump.com
Web: www.cbpump.com.cn

Jiangsu Haishi Pump Co Ltd
1 Xinqiao Road
Xinqao Town
Jingjiang City
Jiangsu Province 214537
China
Tel: 0523 4321008
Fax: 0523 4321103
Email: sales@haishipump.com
Web: www.haishipump.com

Junty Industries Ltd
D-601, Guoheng Jiye Mansion
No 7 Bei Tu Cheng Xi Lu
Chaoyang district
Beijing 100029
China
Tel: 010 8227 5316
Fax: 010 8227 5331
Email: info@sealchina.com
Web: www.SeaChina.com

Mono Pumps (China) Ltd
Room 2303, Capital Mansion
No 6 Xin Yuan Southern Road
Chao Yang District
Beijing 100004
China
Tel: 010 6461 1115
Email: monobeijing@natoil.com
Web: www.mono-pumps.com

PCM Asia Pacific
Unit 10A, 19-21, Shanghaimart
2299 Yanan Road (West)
200336 Shanghai
China
Tel: 021 6236 2521
Fax: 021 6236 2428
Email: pcmchina@pcmpompes.com
Web: www.pcmpompes.com

PCM Moineau Oilfield
Room 608, Inter Continental Plaza
16 Andelu, Dongcheng District
100011 Beijing
China
Tel: 010 8488 2003
Fax: 010 8488 2005
Email: jma@pcmpompes.com
Web: www.pcmpompes.com

Shanghai Liansheng
Pump Manufacture Co Ltd
2646 Lian Xi Highway
Liansheng Town, Qingpu District
Shanghai 201722
China
Tel: 021 59271627
Fax: 021 5927 1080
Email: shlsby@shanghai-liansheng.com
Web: www.shanghai-liansheng.com

Shanghai Shenbao
Industrial Pump Co Ltd
2500 Beisong Road
Minghang District
Shanghai 201111
China
Tel: 021 6409 5555
Fax: 021 6409 5040
Web: www.shenbao-pump.com

Shanxi Jishan Qinglong
Pump Industry Co Ltd
Floor 8,Tower B, New Energy Building
Nanyou Road
Shenzhen
Guangdong 518054
China
Tel: 075 5664 1103
Fax: 075 5664 5976
Email: info@chinesesource.com
Web: www.66daji.com

Shijiazhuang Pump
Industry Group Co Ltd
19 Heping East Road
Shijiazhuang
Hebei 050011
China
Tel: 0311 6082947
Fax: 0311 6033212
Email: spw@heinfo.net
Web: www.cspw.com

Tangshan Pump Factory
104 Da Qing Road
Tangshan
Hebei 063000
China
Tel: 0315 3205818
Fax: 0315 3207547
Email: tspump@tspump.com
Web: www.tspump.com

Tianjin Pumps & Machinery
Group Co Ltd
10 Guoyuan Nan Road
Beichen District
Tianjin
China
Tel: 022 26833413
Fax: 022 2639 1556
Email: export@ctppumps.com
Web: www.ctppumps.com

Tsurumi (Shanghai) Co Ltd
Room 1201 Building D
No 80 Caobao Road
Xuhui Shanghai 200235
China
Tel: 021 6432 6010
Fax: 021 6432 6013
Email: sh.sales@tsurumipump.com.cn

Ture International Industrial Ltd
No 101, An Kang Lou
Feng an Hua Yuan, Ge Xin Road
Guang Zhou 510250
China
Tel: 020 3409 6252
Fax: 020 3434 2272
Email: helhua@pub.guangzhou.gd.cn
Web: www.seocapump.com

Hangzhou SRT Pump Co Ltd
No 2, 20 Street, Xiashan Economic &
Technical Development Zone
Hangzhou 310018
Zhejiang Province
China
Tel: 0571 8657 1014
Fax: 0571 8657 1015
Email: export@srt-pump.com
Web: www.srt-pump.com

CZECH REPUBLIC (CZ)

AxFlow sro
Utováren 261
CZ-10200 Prague
Czech Republic
Tel: 02 101180
Fax: 02 72101189
Email: axflow@axflow.cz
Web: www.axflow.cz

BBA Hranice sro
Belotín 287
CZ-753 64 Belotín
Czech Republic
Tel: 05 8169 9311
Fax: 05 8169 9332
Email: bbahranice@bbahranice.cz
Web: www.bbahranice.cz

Kovoplast Chlumec nad Cidlinou as
Kozelkova 131
CZ-503 51 Chlumec n. C
Czech Republic
Tel: 049 548 4543
Fax: 049 548 5230
Email: kovoplast@kovoplast.cz
Web: www.kovoplast.cz

Ondrejovicka Strojirna sro
Ondrejovice
CZ-793 76 Zlate Hory
Czech Republic
Tel: 064 542 5151
Fax: 064 542 5462
Email: ondrstroj@ondrstroj.cz
Web: www.ondrstroj.cz

Sigma Group AS
ul Jana Sigmunda 79
CZ-78350 Lutin
Czech Republic
Tel: 058 565 2150
Fax: 058 594 4258
Web: www.sigma.cz

Sigmia Trading International
Kosmonautu 6
CZ-77231 Olomouc
Czech Republic
Tel: 058 552 4510
Fax: 058 522 7074

Siwatec AS
Svobody 43/39
CZ-77200 Olomouc
Czech Republic
Tel: 058 522 4104
Fax: 058 522 3984
Email: export@siwatec.cz
Web: www.siwatec.cz

DENMARK (DK)

APV Fluid Handling Horsens
Ternevej 61-63
DK-8700 Horsens
Denmark
Tel: 075 643777
Fax: 075 643868
Email: soren.c.steensen@invensys.com
Web: www.apv.com

AxFlow A/S
Solvang 6
DK-3450 Allerød
Denmark
Tel: 07010 3550
Fax: 07010 3555
Email: axflow@axflow.dk
Web: www.axflow.dk

Desmi A/S
Tagholm 1
DK-9400 Nørresundby
Denmark
Tel: 096 328111
Fax: 098 175499
Email: desmi@desmi.com
Web: www.desmi.com

Dominator Aps
Rønnebærvej 2
DK-3200 Helsinge
Denmark
Tel: 048 796915
Fax: 048 796912
Email: info@dominator.dk
Web: www.dominator.dk

Flexicon A/S
Frejasvej 2-6
DK-4100 Ringsted
Denmark
Tel: 057 671155
Fax: 057 670541
Email: flexicon@flexicon.dk
Web: www.flexicon.dk

Grundfos Management A/S
Poul Due Jensensvej 7
DK-8850 Bjerringbo
Denmark
Tel: 08750 6026
Email: astubkjaer@grundfos.com
Web: www.grundfos.com

Hamworthy Svanehøj A/S
Fabriksparken 6
PO Box 30
DK-9230 Svenstrup J
Denmark
Tel: 096 372200
Fax: 098 383156
Email: svanehoj@hamworthy.com
Web: www.hamworthy.com

Iron Pump A/S
PO Box 527
Generatorvej 10
DK-2730 Herlev
Denmark
Tel: 044 916788
Fax: 044 911644
Email: sales@ironpump.dk
Web: www.ironpump.dk

T Smedegaard A/S
Sydvestvej 57-59
DK-2600 Glostrup
Denmark
Tel: 043 961028
Fax: 043 631766
Email: info@smedegaard.dk
Web: www.smedegaard.dk

ESTONIA (EE)

ABS Pumps AS
Paldiski mnt 125
Box 290
EE-0090 Tallinn
Estonia
Tel: 0656 3398
Fax: 0656 3385
Email: info@absgroup.com
Web: www.absgroup.com

FINLAND (FI)

ABS Finland Oy
Turvekuja 6
FI-00700 Helsingfors
Finland
Tel: 07 5324 0300
Fax: 07 5324 0300
Email: info@absgroup.com
Web: www.absgroup.com

AxFlow Oy
Jokisuunkuja 3
FI-00560 Helsinki
Finland
Tel: 09 777 1930
Fax: 09 79 7550
Email: axflow@axflow.fi
Web: www.axflow.fi

Oy Kolmeks AB
PO Box 27
FIN-14201 Turenki
Finland
Tel: 03 5353531
Fax: 03 5353200
Email: export@kolmeks.fi
Web: www.kolmeks.fi

FRANCE (F)

ABS France SAS
26 Allé du Plateau
F-93250 Villemomble
France
Tel: 01 49 35 24 50
Fax: 01 48 54 42 82
Email: info@absgroup.com
Web: www.absgroup.com

AxFlow SAS
Zone Industrielle du Saule Michaud
BP 42
F-37270 Montlouis-sur-Loire
France
Tel: 02 4745 84 58
Fax: 02 4745 14 34
Email: info@axflow.fr
Web: www.axflow.fr

DKM
Division of Clextral
PO Box 10, 1 Rue de Colonal Riez
F-42702 Firminy Cedex
France
Tel: 04 77 40 31 43
Fax: 04 77 89 43 94
Email: sales@dkm-pumps.com
Web: www.dkm-pumps.com

Fapmo
35-37 Rue Roger Salengro
F-62230 Outreau
France
Tel: 03 21 10 01 21
Fax: 03 21 80 46 88
Email: fapmo@fapmo.com
Web: www.nordnet.fr/fapmo

Milton Roy Europe
10 Grande Rue
F-27360 Pont Saint Pierre
France
Tel: 02 32 68 30 00
Fax: 02 32 68 30 93
Email: contact@miltonroy-europe.com
Web: www.miltonroy-europe.com

Optimex
269 rue de Montepy
F-69 210 Fleurieux sur l'Arbresle
France
Tel: 04 72 52 95 74
Fax: 04 72 52 95 75
Email: contact@optimex-pumps.com
Web: www.optimex.fr

PCM
17 rue Ernest Laval
BP 35
F-92173 Vanves
France
Tel: 01 41 08 15 15
Fax: 01 41 08 15 00
Email: pcm@pcmpompes.com
Web: www.pcmpompes.com

PEME-Gourdin SA
Allée Charles Victor Naudin
Parc d'activité Sophia Antipolis

F-06410 Biot
France
Tel: 04 93 00 66 00
Fax: 04 93 00 01 75
Email: commercial@peme-gourdin.com
Web: www.peme-gourdin.com

Sterling Fluid Systems (France) SAS
1-3 avenue Georges Politzer
PO Box 41
F-78193 Trappes
France
Tel: 01 34 82 39 00
Fax: 01 34 82 39 61
Email: sterling_sihi@easynet.fr
Web: www.sterlingfluid.com

Sundyne International SA
Fluid Handling Division
13-15 Building Eiffel
Zone Industrielle de Dijon Sud
F-21600 Longvic Cedex
France
Tel: 03 80 38 33 00
Fax: 03 80 38 33 71
Email: european-sales@sundyne.com
Web: www.sundyne.com

GERMANY (D)

Abel GmbH & Co KG
Abel-Twiete 1
D-21514 Büchen
Germany
Tel: 04155 8180
Fax: 04155 818499
Email: mail@abel.de
Web: www.abel.de

ABS Deutschland GmbH
Pützchens Chaussee 202
D-53229 Bonn
Germany
Tel: 02246 130
Fax: 02246 13200
Email: info@absgroup.com
Web: www.absgroup.com

ACD Cryo GmbH
Rheinauenstrasse 2
D-79415 Bad Bellingen
Germany
Tel: 07635 8105 0
Fax: 07635 8965
Email: sales@acdcryo.com
Web: www.acdcryo.com

Alfons Haar Maschinenbau GmbH & Co
Fangdieckstrasse 67
D-22547 Hamburg
Germany
Tel: 040 833910
Fax: 040 844910
Email: sales@alfons-haar.de
Web: www.alfons-haar.de

Alldos GmbH
Reetzstrasse 85
D-76327 Pfinztal (Söllingen)
Germany
Tel: 07240 610
Fax: 07240 61177
Email: alldos.de@alldos.com
Web: www.alldos.de

Alltech Dosieranlagen GmbH
PO Box 120
D-76352 Weingarten
Germany
Tel: 07244 70260
Fax: 07244 1053
Email: info@alltech-dosieranlagen.de
Web: www.alltech-dosieranlagen.de

Allweiler AG
Allweilerstrasse 1
D-78315 Radolfzell
Germany
Tel: 07732 860
Fax: 07732 86436
Email: service@allweiler.de
Web: www.allweiler.de

Apollo Gössnitz GmbH
Walter-Rabold-Strasse 26
Gössnitz
D-04639 Gößnitz
Germany
Tel: 034493 770
Fax: 034493 77210
Email: info@apollo-goessnitz.de
Web: www.apollo-goessnitz.de

AxFlow GmbH
Neumannstrasse 2
D-40235 Düsseldorf
Germany
Tel: 0211 238060
Fax: 0211 2380620
Email: info@axflow.de
Web: www.axflow.de

Bavaria Fluidtec GmbH
Postfach 122
D-82213 Maisach
Germany
Tel: 08141 61180
Fax: 08141 611838
Email: info@bavariafluidtec.de
Web: www.bavariafluidtec.de

Behrens Pumpen
Von Thünen Strasse 7
D-28307 Bremen
Germany
Tel: 0421 486810
Fax: 0421 4868111
Email: sales@behrenspumpen.de

Bielomatik Leuze GmbH & Co
Daimlerstrasse 6-10
D-72639 Neuffen
Germany
Tel: 07025 120
Fax: 07025 12 200
Email: info@bielomatik.de
Web: www.bielomatik.com

Börger GmbH
Benningsweg 24
D-46325 Borken Weseke
Germany
Tel: 02862 910320
Fax: 02862 910346
Email: info@boerger-pumps.com
Web: www.boerger-pumps.com

Bornemann GmbH
Industriestrasse 2
D-31683 Obernkirchen
Germany
Tel: 05724 3900
Fax: 05724 390290
Email: diana.ruehle@bornemann.com
Web: www.bornemann.com

Burgmann Industries GmbH & Co KG
Aeussere Sauerlacher Strasse 6-10
D-82515 Wolfratshausen
Germany
Tel: 08171 230
Fax: 08171 231214
Email: info@de.eagleburgmann.com
Web: www.eagleburgmann.com

Delasco PCM GmbH
Wiesbadener Landstrasse 18
D-65203 Wiesbaden
Germany
Tel: 0611 609770
Fax: 0611 6097720
Email: info@delasco.de
Web: www.delasco.de

Desmi GmbH
Neuhöfer Strasse 23
D-21107 Hamburg
Germany
Tel: 040 7519847
Fax: 040 7522040
Email: rotan@rotan-pumpen.de
Web: www.rotan-pumpen.de

Dickow Pumpen KG
Siemenstrasse 22
D-84478 Waldkraiburg
Germany
Tel: 08638 6020
Fax: 08638 602200
Email: export@dickow.de
Web: www.dickow.de

Düchting Pumpen Maschinenfabrik GmbH
Wilheim-Düchting-Strasse 22
D-58453 Witten
Germany
Tel: 02302 9690
Fax: 02302 690443
Email: info@duechting.com
Web: www.deuchting.com

EDUR Pumpenfabrik GmbH & Co KG
Hamburger Chaussee 148-152
D-24113 Kiel
Germany
Tel: 0431 6898 68
Fax: 0431 6898 800
Email: info@edur.de
Web: www.edur-pumpen.de

Franz Eisele und Söhne GmbH & Co KG
Hauptstrasse 2-4
D-72488 Sigmaringen
Germany
Tel: 07571 1090
Fax: 07571 10988
Email: info@eisele.de
Web: www.eisele.de

Feluwa Pumpen GmbH
Beulertweg
D-54570 Mürlenbach
Germany

Tel: 065 94 10 0
Fax: 065 94 1640
Email: info@feluwa.de
Web: www.feluwa.com

Flux-Geräte GmbH
Talweg 12
D-75433 Maulbronn
Germany
Tel: 07043 101 0
Fax: 07043 101 555
Email: export@flux-pumpen.de
Web: www.flux-pumpen.de

Fristram Pumpen
F Stamp KG (GmbH & Co)
PO Box 80 08 80
D-21008 Hamburg
Germany
Tel: 040 725560
Fax: 040 72 56166
Email: info@fristam.de
Web: www.fristam.com

Grundfos GmbH
Schlüterstrasse 33
D-40699 Erkrath
Germany
Tel: 0211 929690
Email: wasserwirtschaft@grundfos.de
Web: www.grundfos.com

Arthur Habermann GmbH & Co KG
PO Box 1640
D-58406 Witten
Germany
Tel: 02302 66060
Fax: 02302 690564
Email: mail@habermann-gmbh.de
Web: www.habermann-gmbh.de

Hanning & Kahl GmbH & Co KG
PO Box 1342
D-33806 Oerlinghausen
Germany
Tel: 05202 707600
Fax: 05202 707629
Email: info@huk.hanning.com
Web: www.hanning-kahl.de

Hauhinco
Beisenbruchstrasse 10
D-45549 Sprockhövel
Germany
Tel: 02324 7050
Fax: 02324 705222
Email: info@hauhinco.de
Web: www.hauhinco.com

Herborner Pumpenfabrik
JH Hoffmann GmbH & Co
Littau 3-5
D-35745 Herborn
Germany
Tel: 02772 9330
Fax: 02772 933105
Email: info@herborner-pumpen.de
Web: www.herborner-pumpen.de

Hermetic-Pumpen GmbH
Gewerbestrasse 51
D-79194 Gundelfingen
Germany
Tel: 0761 58300
Fax: 0761 5830280
Email: hermetic@lederle-hermetic.com
Web: www.lederle-hermetic.com

Philipp Hilge GmbH & Co KG
Hilgestrasse
D-55294 Bodenheim
Germany
Tel: 061 35750
Fax: 061 351737
Email: hilge@hilge.de
Web: www.hilge.de

Homa Pumpenfabrik GmbH
Industriestrasse 1
D-53819 Neunkirchen-Seelscheid
Germany
Tel: 02247 7020
Fax: 02247 70244
Email: info@homa-pumpen.de
Web: www.homa-pumpen.de

Horn GmbH & Co KG
Munketoft 42
D-24937 Flensburg
Germany
Tel: 0461 86960
Fax: 0461 869666
Email: info@horn-gmbh.de
Web: www.horn-gmbh.de

Hüdig GmbH & Co KG
Heinrich Hüdig Strasse 2
D-29227 Celle
Germany
Tel: 05141 88450
Fax: 05141 86918
Email: info@huedig.de
Web: www.huedig.de

Indag Maschinenbau GmbH
Ivenfleth 13
D-25376 Borsfleth
Germany
Tel: 04124 98760
Fax: 04124 987654
Email: info@indag.de
Web: www.indag.de

Iwaki Europe GmbH
Siemensring 115
D-47877 Willich
Germany
Tel: 02154 925450
Fax: 02154 925455
Email: info@iwaki.de
Web: www.iwaki.de

Jabsco GmbH, ITT Industries
Oststrasse 28
D-22844 Norderstedt
Germany
Tel: 040 5353730
Fax: 040 53537311
Web: www.jabsco.com

Jung Pumpen GmbH & Co
Industriestrasse 4-6
D-33803 Steinhagen
Germany
Tel: 0520 4170
Fax: 0520 480368
Email: international@jung-pumpen.de
Web: www.jung-pumpen.de

Kächele GmbH
Jahnstrasse 9
D-73235 Weilheim/Teck
Germany
Tel: 07023 1030
Fax: 07023 103240
Email: pumps@w-kaechele.de
Web: www.kaechele.de

Kamat Pumpen GmbH Co KG
Salinger Feld 10
D-58454 Witten-Annen
Germany
Tel: 02302 89030
Fax: 02302 801917
Email: werner.potrafke@kamat.de
Web: www.kamat.de

Klaus Union GmbH & Co KG
PO Box 101349
D-44713 Bochum
Germany
Tel: 0234 45950
Fax: 0234 432387
Email: info@klaus-union.de
Web: www.klaus-union.de

KNF Neuberger GmbH
Alter Weg 3
D-79112 Freiburg-Munzingen
Germany
Tel: 07664 59090
Fax: 07664 2124
Email: info@knf.de
Web: www.knf.de

Körting Hannover AG
Badenstedter Strasse 56
D-30453 Hannover
Germany
Tel: 0511 21290
Fax: 0511 2129223
Email: info@koerting.de
Web: www.koerting.de

Kracht GmbH
Gewerbestrasse 20
D-58791 Werdohl
Germany
Tel: 02392 9350
Fax: 02392 935209
Email: info@kracht-hydraulik.de
Web: www.kracht-hydraulik.de

KSB AG
Johann-Klein-Strasse 9
D-67227 Frankenthal
Germany
Tel: 06333 860
Fax: 06333 86 3401
Web: www.ksb.com

KTR Kupplungstechik GmbH
Rodder Damm 170
D-48432 Rheine
Germany
Tel: 05971 7980
Fax: 05971 798698
Email: sales@ktr.com
Web: www.ktr.com

Lederle GmbH
Gewerbestrasse 53
D-79194 Gundelfingen
Germany

Tel: 0761 58300
Fax: 0761 5830180
Email: lederle@lederle-hermetic.com
Web: www.lederle-hermetic.com

Leistritz Pumpen GmbH
PO Box 30 41
D-90014 Nürnberg
Germany
Tel: 0911 430360
Fax: 0911 4306490
Email: pumps@leistritz.de
Web: www.leistritz.com

LEWA GmbH
Ulmer Strasse 10
D-71229 Leonberg
Germany
Tel: 07152 1140
Fax: 07152 1141303
Email: lewa@lewa.de
Web: www.lewa.com

Lutz-Jesco
Am Bostelberge 19
D-30900 Wedemark
Germany
Tel: 0513 058020
Fax: 05130 580268
Email: info@jesco.de
Web: www.jesco.de

MASO Process-Pumpen GmbH
PO Box 100
D-74358 Illsfeld
Germany
Tel: 070 62 95600
Fax: 070 62 64593
Email: info@maso.de
Web: www.maso.de

Maximator GmbH
Walkenrieder Strasse 15
D-37449 Zorge/Harz
Germany
Tel: 05586 8030
Fax: 05586 80340
Email: info@maximator.de
Web: www.maximator.de

Netzsch Mohnopumpen GmbH
Geretsrieder Strasse 1
D-84478 Waldkraiburg
Germany
Tel: 08638 630
Fax: 08638 67981
Email: info@nmp.netzsch.com
Web: www.netzsch-pumps.com

Nikkiso Pumps Europe GmbH
Nikkisostrasse
D-63674 Altenstadt
Germany
Tel: 06047 96490
Fax: 06047 9649 99
Email: info@nikkiso-pumps.com
Web: www.nikkiso-pumps.com

Oddesse Pumpen-und Motorenfabrik GmbH
Am Pappelwald 12
D-39387 Oschersleben/Bode
Germany
Tel: 03949 932467
Fax: 03949 932463
Email: info@oddesse.de
Web: www.oddesse.de

PCM Delasco GmbH
Wiesbadener Landstrasse 18
D-65203 Wiesbaden
Germany
Tel: 0611 609770
Fax: 0611 6097720
Email: info@delasco.de
Web: www.delasco.de

PF Pumpen und Feuerlöschtechnik GmbH
Zechensteig 225
D-09477 Jöhstadt
Germany
Tel: 0373 43810
Fax: 0373 432287
Email: info@pfjoehstadt.de
Web: www.pfjoehstadt.de

ProMinent Dosiertechnik GmbH
Im Schumachergewann 5-11
D-69123 Heidelberg
Germany
Tel: 06221 8420
Fax: 06221 842617
Email: info@prominent.de
Web: www.prominent.de

Prüftechnik Alignment Systems GmbH
Oskar Messter Strasse 15
D-85737 Ismaning
Germany
Tel: 089 996160
Fax: 089 99616100
Email: sabine.kosma@prüftechnik,com
Web: www.prüftechnik.com

Rickmeier GmbH
PO Box 1340
D-58797 Balve
Germany
Tel: 023751 9270
Fax: 023751 92726
Email: kontakt@rickmeier.de
Web: www.rickmeier.de

Rietschle Thomas Schopfheim GmbH
Postfach 1260
D-79642 Schopfheim
Germany
Tel: 07622 3920
Fax: 07622 392300
Email: info.sch@rtpumps.com
Web: www.rietschle.com

Ritz Pumpenfabrik GmbH & Co KG
P O Box 1780
D-73507 Schwäbisch Gmuend
Germany
Tel: 07171 6090
Fax: 07171 609287
Email: info@ritz.de
Web: www.ritz-pumpen.de

Ruhrpumpen GmbH
Stockumer Strasse 28
D-58453 Witten
Germany
Tel: 02302 66103
Fax: 02302 661303
Email: info@ruhrpumpen.com
Web: www.ruhrpumpen.com

Scherzinger Pumps
Bregstrasse 23-25
D-78120 Furtwangen
Germany
Tel: 07723 65060
Fax: 07723 650640
Email: info@scherzinger.de
Web: www.scherzinger.de

Schmalenberger GmbH & Co
Im Schelmen 9-11
D-72072 Tübingen
Germany
Tel: 07071 700800
Fax: 07071 700810
Email: info@schmalenberger.de
Web: www.schmalenberger.de

Schroedahl-Arrap, Speciality Valves
Schoenenbacker Strasse 4
D-51580 Reichshop-Mittelagger
Germany
Tel: 02265 00270
Fax: 02265 992727
Email: info@schroedahl.com
Web: www.schroedahl.com

seepex GmbH & Co KG
PO Box 10 15 64
D-46215 Bottrop
Germany
Tel: 02041 9960
Fax: 02041 996400
Email: info@seepex.com
Web: www.seepex.com

Sero PumpSystems GmbH
PO Box 69
D-74907 Meckesheim
Germany
Tel: 06226 92010
Fax: 06226 920140
Email: info@seroweb.de
Web: www.seroweb.de

Seybert & Rahier GmbH
Sera-Strasse
D-34376 Immenhausen
Germany
Tel: 05673 39990
Fax: 05673 3999155
Email: info@sera-info.de
Web: www.sera-info.de

Speck-Pumpen
Lohbachstrasse 6
D-91161 Hilpoltstein
Germany
Tel: 09174 9720
Fax: 09174 97249
Email: info@speck-pumpen.de
Web: www.speck-pumpen.de

Steinle Industriepumpen GmbH
Varnhagenstrasse 42
D-40225 Düsseldorf
Germany
Tel: 0211 333273
Fax: 0211 330755
Email: info@steinle-pumpen.de
Web: www.steinle-pumpen.de

Sterling SIHI GmbH
Lindenstrasse 170

D-25524 Itzehoe
Germany
Tel: 048 21 771303
Fax: 048 21 771310
Email: vs.Itzehoe@sterlingsihi.de
Web: www.sterlingfluidsystems.de

Oswald Suck GmbH
Altstadtstrasse 1-5
D-99817 Eisenach
Germany
Tel: 03691 203250
Fax: 03691 732730
Email: pumpen-suck@t-online.de
Web: www.pumpen-suck.de

Teikoku Electric GmbH
Nürnberger Stasse 24
D-40599 Düsseldorf
Germany
Tel: 0211 700 6778
Fax: 0211 749 0011
Email: info@teikoku-electric.de
Web: www.teikokupump.de

Tsurumi (Europe) GmbH
Helforfer Strasse 16
D-40472 Düsseldorf
Germany
Tel: 0211 4179373
Fax: 0211 4791429
Email: sales@tsurumi-europe.com

Uraca Pumpenfabrik GmbH & Co KG
Sirchinger Strasse 15
D-72574 Bad Urach
Germany
Tel: 07125 1330
Fax: 07125 133202
Email: info@uraca.de
Web: www.uraca.de

Variopumps Pumpenbau GmbH
Carl-Zeiss Strasse 58
D-72770 Reutlingen
Germany
Tel: 07121 55573
Fax: 07121 503328
Email: info@variopumps.de
Web: www.variopumps.de

Victor Pumpen GmbH
Dieselstrasse 7
D-85551 Kirchheim bei Munchen
Germany
Tel: 089 904 8660
Fax: 089 904 3447
Email: germany@victorpumps.com
Web: www.victorpumps.com

Willy Vogel AG
Motzener Strasse 35/37
D-12277 Berlin
Germany
Tel: 030 720 020
Fax: 030 720 02111
Email: info@vogel-berlin.de
Web: www.vogelag.com

Pumpenfabrik Wangen GmbH
Simoniusstrasse 17
D-88239 Wangen
Germany
Tel: 07522 9970
Fax: 07522 997199
Email: mail@wangen.com
Web: www.wangen.com

Wernert-Pumpen GmbH
Postfach 10 21 53
D-45421 Mülheim an der Ruhr
Germany
Tel: 0208 37580
Fax: 0208 407472
Email: info@wernert.de
Web: www.wernert.de

Wilo AG
Nortkirchenstrasse 100
D-44263 Dortmund
Germany
Tel: 0231 41020
Fax: 0231 4102 7575
Email: wilo@wilo.de
Web: www.wilo.de

Wilo EMU GmbH
Heimgartenstrasse 1û3
D-95030 Hof
Germany
Tel: 09281 9740
Fax: 09281 96528
Email: info@wiloemu.com
Web: www.emu.de

Wirth Group
Külner Strasse 71-73
D-41812 Erkelenz
Germany
Tel: 02431 830
Fax: 02431 83268
Email: info@wirth-europe.com
Web: www.wirth-europe.com

Witte Pumps & Technology GmbH
Esinger Steinweg 44a
D-25436 Uetersen
Germany
Tel: 04122 92870
Fax: 04122 928749
Email: info@witte-pumps.de
Web: www.witte-pumps.de

Woma Apparatebau GmbH
Werthauser Strasse 77-79
D-47226 Duisburg
Germany
Tel: 02065 3040
Fax: 02065 304200
Email: info@woma.de
Web: www.woma.de

Albert Ziegler GmbH & Co KG
Memminger Strasse 28
D-89537 Giengen/Brenz
Germany
Tel: 07322 951344
Fax: 07322 951464
Email: ziegler@ziegler.de
Web: www.ziegler.de

GREECE (GR)

ABS Hellas SA
1-3 Skra Street
Kallithea
GR-17372 Athens
Greece
Tel: 0210 9532501
Fax: 0210 9532504
Email: info@absgroup.com
Web: www.absgroup.com

Drakos-Polemis Inc
PO Box 51024
GR-145 10 Kifissia, Athens
Greece
Tel: 01 8161 402
Fax: 01 8161 262
Web: www.dppumps.gr

Korina Pumps
Kyprion Agoniston 51
Argyroupolis
GR-16451 Athens
Greece
Tel: 021 9910811
Fax: 021 9910811
Email: sales@korinapumps.gr
Web: www.korinapumps.gr

Rotex SA
Industrial Area
Sindos
GR-57022 Thessaloniki
Greece
Tel: 031 0797517
Fax: 031 0798510
Email: rotex-sa@the.forthnet.gr
Web: www.rotex.gr

HONG KONG (HK)

H & E Tsurumi Pump Co Ltd
Rm 1101, Chow Sang Sang Bldg
229 Nathan Road
Kowloon
Hong Kong
Tel: 27307208
Fax: 27306531
Email: info@tsurumi.com.hk

Tycon Alloy Industries (Hong Kong) Co Ltd
8/F, 22-28 Cheung Tat Road
Tsing Yi
New Territories
Hong Kong
Tel: 24499240
Fax: 24369872
Email: hongkong@tyconalloy.com
Web: www.tyconalloy.com

HUNGARY (HU)

ABS Hungary Trading
Kiss Ernö u 1-3
HU-1046 Budapest
Hungary
Tel: 01 231 6070
Fax: 01 231 6080
Email: info@absgroup.com
Web: www.absgroup.com

AxFlow Kft
Fö út 158
HU-2330 Dunaharaszti
Hungary
Tel: 024 460126
Fax: 024 462175
Email: axflow@axflow.hu
Web: www.axflow.hu

INDIA (I)

Amrit Engineering Pvt Ltd
2nd Floor, Motilal Centre, Near Dinesh Hall
Ashram Road, Ahmedabad
Gujarat 380 009
India
Tel: 079 658 3858
Fax: 79 6580382
Email: amrutpump@jindalonline.net

Best & Crompton Engineering Ltd
28 Industrial Estate (North)
Ambattur
Chennai 600 098
Tamil Nadu
India
Tel: 044 6258782
Fax: 044 6257059
Email: bcemfg@net4india.com
Web: www.bestcrompton.com

Calama Industries Pvt Ltd
7 Gaiwadi Industrial Estate
SV Road, Goregaon W
Mumbai 400 062
India
Tel: 022 874 3563
Fax: 022 874 3588
Email: sales@calama.com
Web: www.calama.com

Flowmore Private Ltd
A-292, Mahipalpur Extension, N H - 8
East of Kailash
New Delhi 110037
India
Tel: 011 306 23740
Fax: 011 267 83278
Email: delmkt@flowmorepumps.com
Web: www.flowmorepumps.com

Johnson Pump (India), ATE Enterprises Ltd
No A-19, CTS No 689, Veera Desai Road
Andheri (West)
Mumbai 400053
India
Tel: 022 634 6610
Fax: 022 634 6609
Email: ate_and@cateel.com

Jyoti Plastic Works Pvt Ltd
94, B T Compound
Malad (West)
Mumbai 400064
India
Tel: 022 288 23241
Fax: 022 288 20629
Email: prashant@jyotimail.com
Web: www.jyotiplastics.com

Kishor Pumps Ltd
21 Barve Complex, 1170 / 1
J M Road, Shivaji Nagar
Pune 411005
India
Tel: 020 2551 1530
Fax: 020 2553 9887
Email: csd@kishorpumps.com
Web: www.kishorpumps.com

Laxmi Hydraulics Pvt Ltd
129-130 Industrial Estate, Patil Nagar, Hotgi Road,
Solapur
Maharashtra 413003
India
Tel: 0217 2601763
Fax: 0217 2601836
Email: spr_lhpmotor@sancharnet.in
Web: www.lhp.co.in

Leak-Proof Engineering (I) Pvt Ltd.
F-1, Laxmi Industrial Estate
New Link Road, Andheri (W)
Mumbai 400053
India
Tel: 022 6361737
Fax: 022 636 8429
Email: leakprof@bom5.vsnl.net.in
Web: www.leakproofseals.com

Mather & Platt Pumps Ltd
Pune-Mumbai Road
Chinchwad
Pune 411019
India
Tel: 020 274 42100
Fax: 020 274 42111
Email: headoffice@pun.matherplatt.co.in
Web: www.matherplatt.com

PPI Pumps Pvt Ltd
14 & 16 E, Phase I
GIDC Estate Vartva
Ahmedabad 382445
India
Tel: 079 2583 2273
Fax: 079 2583 0578
Email: sales@prashant-ppi.com
Web: www.prashant-ppi.com

PreciDose Pumps & Systems
62/13, Phase I
Nr Patels Airtemp, GIDC Estate Vartva
Ahmedabad 382445
India
Tel: 079 5830762
Fax: 079 5830762
Email: sales@precidose.com
Web: www.precidose.com

Process Pumps (India) Pvt Ltd
86, III Phase
Peenya Industrial Area
Bangalore 560058
India
Tel: 080 839 5327
Fax: 080 839 5807
Email: contact@process-pumps.com
Web: www.process-pumps.com

Pump Engineering Co Pvt Ltd
C-38, 39
GIDC Estate Odhav
Ahmedabad 382410
India
Tel: 079 2871257
Fax: 079 2874375
Email: sdpatel@pecpumps.com
Web: www.pecpumps.com

Raj Pumps
9 Heavy Industrial Area
Jodhpur 342003
India
Tel: 0291 740937
Fax: 0291 740905
Email: rajpump@yahoo.com
Web: www.rajpumps.com

Roto Pumps Ltd
Roto House
Noida Special Economic Zone
Noida 201305 UP
India
Tel: 0120 2567677
Fax: 0120 2562561
Email: contact@rotopumps.com
Web: www.rotopumps.com

Rotomac Industires Pvt Ltd
D-4A, Panki Industrial Area
Site 1
Kanpur 208022
India
Tel: 0512 269 1704
Fax: 0512 269 1706
Email: rotmac@vsnl.net
Web: www.rotomacpump.com

Sharp Trendys
Sharp Nagar
Kalapatti
Coimbatore 641035
India
Tel: 0422 828631
Fax: 0422 827632
Web: www.sharptrendys.com

Swelore Engineering Pvt Ltd
2109 Phase III GIDC Vatwa
Ahmedabad 382445
Gujarat
India
Tel: 079 583 3003
Fax: 079 583 3337
Email: swelore@ad1.vsnl.net.in
Web: www.swelore.com

Titanium Tantalum Products Ltd
86/1 Vengaivasal Main Road
Gowrivakkam
Chennai 601302
Tamil Nadu
India
Tel: 044 2278 1210
Fax: 044 2375209
Email: globalsales@titanindia.com
Web: www.titanindia.com

Tuschaco Pumps Ltd
401 Gulab
237 P D'Mello Road
Mumbai 400001
India
Tel: 022 262 0159
Fax: 022 261 5458
Email: hointernational@tushacopumps.com
Web: www.tushacopumps.com

Ujala Pumps Pvt Ltd
163 Patparganj Industrial Area
New Delhi 110092
India
Tel: 011 215 1421
Fax: 011 222 7708
Email: ujala@ndf.vsnl.net.in

Voltas Ltd
Voltas House B
T B Kadam Marg

Mumbai 400033
India
Tel: 022 374 7995
Fax: 022 373 2330
Email: mramani@voltas.com
Web: www.voltas.com

Waterman Industries Ltd
3 Mira Co-Op Industrial Estate
Near Dahisar Check
Mumbai 401104
India
Tel: 022 811 2555
Fax: 022 811 1129
Web: www.indopumps.com

WPIL Ltd
22 Ferry Fund Road
Panihati
Kolkata 700114
West Bengal
India
Tel: 033 2553 1872
Fax: 033 2553 1276
Email: subrata@wpilltd.com
Web: www.wpil.com

INDONESIA (ID)

PT Ebara Indonesia
JL Raya Jakarta-Bogor KM
32 Desa Curug
Jawa Barat
Indonesia
Tel: 021 874 0852
Fax: 021 874 0033
Web: www.ebara.co.jp

IRELAND (IE)

ABS Pumps (Irl) Ltd
ABS House, Riverview Business Park
New Nangor Road
Dublin 12
Ireland
Tel: 01 460 8888
Fax: 01 460 8889
Email: info@absgroup.com
Web: www.absgroup.com

AxFlow Ltd
Dublin Industrial Estate, Unit 46
Barrow Road, Glasnevin
Dublin 11
Ireland
Tel: 018 300744
Fax: 018 300017
Email: info@axflow.ie
Web: www.axflow.ie

Flexachem Manufacturing Ltd
Donnybrook Commercial Centre
Douglas
Cork
Ireland
Tel: 021 363742
Fax: 021 891297
Email: sales@flexachem.com
Web: www.flexachem.com

ITALY (IT)

ABS International Srl
Via Castellazzo 4
I-20040 Cambiago (MI)
Italy
Tel: 02 9506961
Fax: 02 95349203
Email: info@absgroup.com
Web: www.absgroup.com

ABS Italy Srl
Via Lavoro 87
I-40033 Casalecchio (BO)
Italy
Tel: 051 6169511
Fax: 051 6169582
Email: info@absgroup.com
Web: www.absgroup.com

Alma Pompe
Via dell'Artigianato 7
I-20090 Fizzonasco di Pieve
Emanuele (MI)
Italy
Tel: 02 90723657
Fax: 02 90420003
Email: sales@almapompe.it
Web: www.almapompe.it

Annovireverberi SpA
Via ML King 3
I-41100 Modena
Italy
Tel: 059 414411
Fax: 059 253505
Email: industria@annovireverberi.it
Web: www.annovireverberi.it

Argal Srl
Via Labirinto 159
I-25125 Brescia
Italy
Tel: 030 3507011
Fax: 030 3507077
Email: export@argal.it
Web: www.argal.it

Asco Pompe srl
Via Silvio Pellico 6/8
I-20089 Rozzano (MI)
Italy
Tel: 02 892571
Fax: 02 89257201
Email: asco@ascopompe.com
Web: www.ascopompe.com

Gruppo Aturia SpA
Piazza Aturia 9
I-20060 Gessate (MI)
Italy
Tel: 0295 423200
Fax: 0295 423202
Email: aturia@gruppoaturia.it
Web: www.gruppoaturia.com

AxFlow SpA
Via del Commercio
I-20090 Buccinasco (MI)
Italy
Tel: 02 48481
Fax: 02 48401926
Email: info@axflow.it
Web: www.axflow.it

Bizzi & Tedeschi Srl
Via Brodolini 35/A
I-42040 Campegine (RE)
Italy
Tel: 0522 677209
Fax: 0522 677633
Email: info@bizzi.it
Web: www.bizzi.it

Calpeda SpA
Via Roggia di Mezzo 39
I-36050 Montorso Vicentino -Vicenza
Italy
Tel: 0444 476476
Fax: 0444 476477
Email: export@calpeda.it
Web: www.calpeda.it

Caprari SpA
Via Emilia Ovest 900
I-41100 Modena
Italy
Tel: 059 897611
Fax: 059 897897
Email: info@caprari.it
Web: www.caprari.com

CDR Pompe SpA
Via Togliatti 26/A
I-20030 Senago (MI)
Italy
Tel: 02 9901941
Fax: 02 9980606
Email: info@cdrpompe.it
Web: www.cdrpumps.com

DAB Pumps SpA
Via Marco Polo 14
I-35035 Mestrino (PD)
Italy
Tel: 049 9048811
Fax: 049 9048847
Email: info@dabpumps.com
Web: www.dabpumps.com

DDA Srl
Via Silvio Pellico 6/8
I-20089 Rozzano (MI)
Italy
Tel: 02 89257257
Fax: 02 89257241
Email: dpulici@ddasrl.com
Web: www.dda.it

Debem
Via Del Bosco 41
I-21052 Busto Arsizio (VA)
Italy
Tel: 03 31074034
Fax: 03 31074036
Email: info@debem.it
Web: www.debem.it

Dragflow Srl
Via Spagna 13
I-37069 Villafranca-Verona
Italy
Tel: 045 6304521
Fax: 045 6335758
Email: dragflow@sovema.it
Web: www.dragflow.it

Ebara Pumps Europe SpA
Via Pacinotti 32
I-36040 Brendola (Vicenza)
Italy
Tel: 0444 706811
Fax: 0444 706950

Email: marketing@ebaraeurope.com
Web: www.ebaraeurope.com

Finder Pompe SpA
Via Bergamo 65
I-23807 Merate (LC)
Italy
Tel: 039 99821
Fax: 039 599267
Email: finder@finderpompe.com
Web: www.finderpompe.com

GMP SpA
Via Sale 119-121
I- 25064 Gussago (BS)
Italy
Tel: 030 2775411
Fax: 030 2772418
Web: www.gmppumps.com

Grundfos Pompe Italia Srl
Via Gran Sasso 4
I-20060 Truccazzano (MI)
Italy
Tel: 0295 838112
Email: ffederico@grundfos.com
Web: www.grundfos.com

Imovilli Pompe Srl
Via Masaccio 1 (Z I Mancasale)
I-42100 Reggio Emilia
Italy
Tel: 0522 516595
Fax: 0522 514735
Email: info@imovillipompe.it
Web: www.imovillipompe.it

Interpump Group SpA
Via E. Fermi 25
I-42040 Calerno di S.Ilario d'Enza (RE)
Italy
Tel: 0522-904311
Fax: 0522-904444
E-mail: info@interpumpgroup.it
Web: www.interpumpgroup.it

Jurop SpA
Via Crosera 50 Azzano Decimo
I-33082 Pordenone
Italy
Tel: 0434 636811
Fax: 0434 636812
Email: jurop@jurop.it
Web: www.jurop.it

Landini Fabbrica Pompe Centrifughe Srl
Viale Dante 4
I-46023 Gonzaga MN
Italy
Tel: 0376 588222
Fax: 0376 58140
Email: info@landini.net
Web: www.landini.net

Larius Srl
Via Stoppani 21
I-23801 Calolziocorte Lecco
Italy
Tel: 0341 621152
Fax: 0341 621243
Email: larius@larius.com
Web: www.larius.com

Liverani Srl
Via De' Brozzi 94/3
I-48022 Lugo (RA)
Italy
Tel: 0545 22379
Fax: 0545 30350
Email: liverani@liverani.com
Web: www.liverani.com

Lowara SpA, ITT Industries
I-36075 Montecchio Maggiore (VI)
Italy
Tel: 0444 707111
Fax: 0444 492166
Email: mkt@lowara.ittind.com
Web: www.lowara.com

Matra SpA
Via Papa Giovanni XXIII 33
I-41100 Modena
Italy
Tel: 059 250407
Fax: 059 251548
Email: info@matra.it
Web: www.matra.it

Nuovo Pignone
Via Felice Matteucci 2
I-50127 Florence
Italy
Tel: 055 4272500
Web: www.gepower.com

OFT (Officine Di Trevi)
SS Flaminia km 145
I-06032 Borgo di Trevi (Perugia)
Italy
Tel: 0742 381616
Fax: 0742 78792
Email: oft@officineditrevi.com
Web: www.officineditrevi.com

Omac Srl
Via G Falcone 8
I-42048 Rubiera (RE)
Italy
Tel: 0522 629371
Fax: 0522 628980
Email: info@omacpompe.com
Web: www.omacpompe.com

Pedrollo SpA
Via E Fermi 75
I-37047 Bonifacio (VA)
Italy
Tel: 045 6136311
Fax: 045 7614663
Email: pedrollo@pedrollo.com
Web: www.pedrollo.com

Pentair Pumps SpA
Via Masaccio 13
I-56010 Lugnano (PI)
Italy
Tel: 050 716111
Fax: 050 703137
Email: pentairpumps@pw-europe.com
Web: www.pentairpumps.it

Peroni Pompe SpA
Via Forze Armate 310/5
I-20152 Milan
Italy
Tel: 02 489401
Fax: 02 48910287
Email: marketing@peronipompe.it
Web: www.peronipompe.com

Pompe Cucchi Srl
Via dei Pioppi 39
I-20090 Opera (MI)
Italy
Tel: 02 57606287
Fax: 02 57602257
Email: sales@pompecucchi.it
Web: www.pompecucchi.it

Pompe Garbarino SpA
Via Marenco 44
I-15011 Acqui Terme (AL)
Italy
Tel: 0144 388671
Fax: 0144 55260
Email: info@pompegarbarino.com
Web: www.pompegarbarino.com

Ragazzini Srl
Via Volta 8
I-48018 Faenza (RA)
Italy
Tel: 0546 620433
Fax: 0546 621394
Email: ragazzinipompe@linknet.it
Web: www.ragazzini.it

Rovatti Pompe
Via Trento 22/24
I-42042 Fabbrico RE
Italy
Tel: 0522 665000
Fax: 0522 665020
Email: info@rovatti.it
Web: www.rovatti.it

Sacem SpA
Via Valsugana 19
I-35010 San Giorgio in Bosco (Padova)
Italy
Tel: 049 9450022
Fax: 049 9450044
Email: sacem@sacem.it
Web: www.sacem.it

Saer Elettropompe
Via Circonvallozione 22
I-42016 Guastella (RE)
Italy
Tel: 0522 830941
Fax: 0522 826948
Email: info@saerelettropompe.com
Web: www.saerelettropompe.com

Salmson Italia
Via Jacopo Peri 80
I-41100 Modena
Italy
Tel: 059 280380
Fax: 059 282331
Email: salmson@salmson.it
Web: www.salmson.it

Savino Barbera
Via Torino 12
I-10032 Brandizzo (TO)
Italy
Tel: 011 9139063
Fax: 011 9137313
Email: info@savinobarbera.com
Web: www.savinobarbera.com

Seko Bono Exacta SpA
Via Salaria
KM 92 200

I-02010 S Rufina Rieti
Italy
Tel: 0746 605801
Fax: 0746 607072
Email: sales@seko.com
Web: www.seko.com

Simaco Elettromeccanica Srl
16 Strada Statale 235
I-26834 Cadilana-Corte Palasio (LO)
Italy
Tel: 0371 59181
Fax: 0371 420566
Email: info@simacosrl.it
Web: www.simacosrl.it

Sterling Fluid Systems (Italy) SpA
Via Pompei 15
I-20052 Monza
Italy
Tel: 039 28241
Fax: 039 2824220
Email: sterlingitaly@sidro.it
Web: www.sterlingfluid.com

Subteck Srl
Via Ragazzi 99
34 ZI Mancasale
I-42100 Reggio Emilia
Italy
Tel: 0522 921160
Fax: 0522 921178
Email: info@subteck.it
Web: www.subteck.it

Sumoto Srl
Via Peripoli R e G 1/3
I-36041 Alte Montecchio Maggiore (VI)
Italy
Tel: 0444 490515
Fax: 0444 490518
Email: info@sumoto.com
Web: www.sumoto.com

Termomeccanica Pompe SpA
via Del Molo 3
I-19126 La Spezia (SP)
Italy
Tel: 0187 552512
Fax: 0187 552506
Email: pumps@termomeccanica.com
Web: www.tmp.termomeccanica.com

Tesla Srl
Via Del Lavoro 3
I-36040 San Germano Dei Berici (VI)
Italy
Tel: 0444 868099
Fax: 0444 868211
Email: info@teslasub.it
Web: www.teslasub.it

Varisco Srl
Zona Ind Nord
Terza Strada 9
I-35129 Padova
Italy
Tel: 049 8294111
Fax: 049 8076762
Email: info@variscopompe.com
Web: www.variscospa.com

Pompe Vergani
Via Bergamo 65
I-23807 Merate (LC)
Italy
Tel: 039 99821
Fax: 039 599267
Email: finder@finderpompe.com
Web: www.pompevergani.com

Weir Gabbioneta Srl
Viale Cariraghi 68
I-20099 Sesto San Giovanni (MI)
Italy
Tel: 02 2241001
Fax: 02 2424550
Email: sales@pompegabbioneta.com
Web: www.pompegabbioneta.com

Worthington SpA
Flowserve Corporation, Pump Division
Via Rossini 90/92
I-20033 Desio (Milan)
Italy
Tel: 0362 6121
Fax: 0362 303396
Email: mrestelli@flowserve.com
Web: www.flowserve.com

JAPAN (JP)

Ebara Corporation
11-1 Haneda Asahi-cho
Ohta-ku
Tokyo 144-8510
Japan
Tel: 05714 6111
Fax: 03745 3356
Email: pump@ebara.co.jp
Web: www.ebara.co.jp

Grundfos Pumps KK
Gotanda Metalion Bldg 5F
5-21-15, Higashi-gotanda
Shiagawa-ku
Tokyo 141-0022
Japan
Tel: 03 5448 1391
Email: info-gjk@grundfos.com
Web: www.grundfos.com

Honda Kiko Co Ltd
2055 Yamano Inatsuki-Cho
Kaho-gun
Fukuoka 820-0202
Japan
Tel: 0948 42 3111
Fax: 0948 42 3118
Email: info@hondakiko.co.jp
Web: www.hondakiko.co.jp

Iwaki Co Ltd
2-6-6 Kanda Sudacho Chiyoda-ku
Tokyo 101-8558
Japan
Tel: 03 3254 2935
Fax: 03 3252 8892
Email: info@iwakipumps.jp
Web: www.iwakipump.jp

Kawamoto Pump Manufacturing Co Ltd
11-39, Ohsu 4-Chome
Naka-ku
Nagoya 460-8650
Japan
Tel: 052 251 7171
Fax: 052 241 6151
Email: kawamototr@aol.com
Web: www.kawamoto.co.jp

Kubota Corporation
3-1-3 Nihonbashi-muromachi
Chuo-ku,
Tokyo 103-8310
Japan
Tel: 03 3245 3111
Fax: 03 3245 3049
Email: pump@kubota.co.jp
Web: www.kubota.co.jp

Mazda Pump Manufacturing Co Ltd
6-14, Nogami 2-chome
Takarazuka-shi
Hyogo 665-0022
Japan
Tel: 0797 71 8585
Fax: 0797 71 8039
Email: matsuda@mazdapump.co.jp
Web: www.mazdapump.co.jp

Nikkiso Co Ltd
3-43-2 Ebisu
Shibuya-ku
Tokyo 150-8677
Japan
Tel: 03 3443 3711
Fax: 03 3444 4963
Web: www.nikkiso.co.jp

Sakuragawa Pump Manufacturing Co Ltd
1-2-11 Itsukaichi
Ibaraki
Osaka
Japan
Tel: 072 645 5254
Fax: 072 645 5253
Email: pr@sakura-p.net
Web: www.sakura-p.net

Taiko Kikai Industries Co Ltd
MS Building 4th Floor, Suidobashi
3-4-9, Misaki-cho, Chiyoda-ku
Tokyo 101-0061
Japan
Tel: 03 3221 8551
Fax: 03 3221 8555
Email: tokyo-br@taiko-kk.com
Web: www.taiko-kk.com

Torishima Pump Mfg Co Ltd
1-1-8 Miyata-cho
Takatsuki City
Osaka 569-8600
Japan
Tel: 072 695 0551
Fax: 072 693 1288
Web: www.torishima.co.jp

Tsurumi Manufacturing Co Ltd
16-40 4-chome, Tsurumi
Tsurumi-ku
Osaka 538-8585
Japan
Tel: 06 6911 7271
Fax: 06 6911 0100
Email: intsales@tsurumipump.co.jp
Web: www.tsurumipump.co.jp

Yamada Corporation
1-3, 1-chome Minamimagome
Ota-ku

Tokyo 143-8504
Japan
Tel: 03 3777 0241
Fax: 03 3777 0584
Email: intl@yamadacorp.co.jp
Web: www.yamadacorp.co.jp

MALASIA (MY)

ABS Pumps Malaysia Sdn Bhd
No 27 Jalan Tambur 33/19
Shah Alam Technology Park, Seksyen 33
MY-40400 Shah Alam Selangor Darul Ahsan
Malaysia
Tel: 03 5124 6550
Fax: 03 5124 6548
Email: info@absgroup.com
Web: www.absgroup.com

Tsurumi Pump (Malaysia) SDN BHD
22 Jalan PJU 3/41, Sunway Damansara
47810 Petaling Jaya
Selangor
Malaysia
Tel: 03 7803 3373
Fax: 03 7806 3748
Email: tsurumi@tm.net.my

MEXICO (MX)

Corporacion EG SA
Av San Jerónimo 514-A
Monterrey
NL 64640
Mexico
Tel: 083 81 0064
Fax: 083 81 0212
Email: wdm@corporacioneg.com
Web: www.corporacioneg.com

THE NETHERLANDS (NL)

ABS Pompen BV
Kaarvelweg 9
Postbus 120
6200 AC Maastricht
The Netherlands
Tel: 043 352 5050
Fax: 043 363 3395
Email: info@absgroup.com
Web: www.absgroup.com

ABS Pompen BV
Kobaltweg 27
Postbus 120
3542 CE Utrecht
The Netherlands
Tel: 030 242 5678
Fax: 030 241 0346
Email: info@absgroup.com
Web: www.absgroup.com

AxFlow BeNeLux BV
PO Box 50110
1305 AC Almere
The Netherlands
Tel: 036 538 1211
Fax: 036 531 1725
Email: axflow@axflow.nl
Web: www.axflow.nl

Bredel Hose Pumps
PO Box 47
7400 AA Delden
The Netherlands
Tel: 074 377 0000
Fax: 074 376 1175
Email: hosepumps@bredel.com
Web: www.bredel.com

DP-Pumps
PO Box 28
2400 AA Alphen aan den rijn
The Netherlands
Tel: 0172 488325
Fax: 0172 468930
Email: dp@dp-pumps.com
Web: www.dp-pumps.com

Houttuin Holland BV
Sophialaan 4
3542 AB Utrecht
The Netherlands
Tel: 030 248 4611
Fax: 030 241 1845
Email: saleshouttuin@colfaxpump.com
Web: www.houttuin.nl

Johnson Pump BV
Dr AF Philipsweg 51
PO Box 9
9400 AA Assen
The Netherlands
Tel: 0592 376767
Fax: 0592 376760
Email: export.nl@johnson-pump.com
Web: www.johnson-pump.com

Melotte Pumptechnology BV
Fregatweg 50
6222 NZ Maastricht
The Netherlands
Tel: 043 368 4141
Fax: 043 363 7446
Email: info@pumptech.nl
Web: www.pumptech.nl

Nijhuis Pompen BV
PO Box 102
7100 DE Winterswijk
The Netherlands
Tel: 0543 547474
Fax: 0543 547475
Email: nijhuis@nijhuispompen.nl
Web: www.nijhuispompen.nl

Pomac Industries Group
PO Box 32
9356 ZG Tolbert
The Netherlands
Tel: 0594 512877
Fax: 0594 517002
Email: sales@pomac.nl
Web: www.pomac.nl

Robot Pumps BV
PO Box 140
2400 AC Alphen a/d Rijn
The Netherlands
Tel: 0172 418 686
Fax: 0172 418 602
Email: info@robotpumps.com
Web: www.robot-pumps.com

Spaans Babcock BV
PO Box 79
8560 AB Balk
The Netherlands
Tel: 0514 608 282
Fax: 0514 604 485
Email: sales@spaansbabcock.nl
Web: www.spaansbabcock.com

Sterling Fluid Systems (Netherlands)
Havenstraat 22-28
1948 NP Beverwijk
The Netherlands
Tel: 0251 263232
Fax: 0251 226309
Email: info@sihi.nl
Web: www.sterlingfluid.com

Verder Group
Utrechtseweg 4a
3451 GG Vleuten
The Netherlands
Tel: 091 306 779211
Fax: 091 306 771402
Email: info@verder.nl
Web: www.verder.com

Yamada Europe BV
Aquamarijnstraat 50
7554 NS Hengelo
The Netherlands
Tel: 074 242 2032
Fax: 074 242 1055
Email: company@yamada.nl
Web: www.yamada.nl

NEW ZEALAND (NZ)

Macquarrie Corporation
1620 Hume Highway
Campbellfield
VIC 3061
New Zealand
Tel: 03 9358 5555
Fax: 03 9358 5556
Email: macquarrie@macquarrie.com.au
Web: www.macquarrie.com.au

Mono Pumps (New Zealand) Ltd
PO Box 71-021
35-41 Fremlin Place, Avondale
Auckland 7
New Zealand
Tel: 09 8290333
Email: info@mono-pumps.co.nz
Web: www.mono-pumps.com

NIGERIA (NG)

Cakasa (Nigeria) Co Ltd
PO Box 871, 96 Palm Avenue
Mushin
Lagos
Nigeria
Tel: 01 452 2955
Fax: 01 452 2887
Email: cnchq@cakasa.com
Web: www.cakasa.com

NORWAY (NO)

ABS Oumoer A/S
Industriveien 40
Postboks 473
N-1302 Sandvika
Norway
Tel: 067554700

Fax: 067554750
Email: info@absgroup.com
Web: www.absgroup.com

Aker Kvaerner Eureka Pump Systems
PO Box 94
N-1325 Lysaker
Norway
Tel: 067826000
Fax: 067826400
Email: kjell.bamberg@akerkvaerner.com
Web: www.kvaerner.com

AxFlow AS
PO Box 98
Lillaker
N-0216 Oslo
Norway
Tel: 022736700
Fax: 022736786
Email: axflow@axflow.no
Web: www.axflow.no

Framo Engineering AS
PO Box 174, Sandsli
N-5862 Bergen
Norway
Tel: 055928800
Fax: 055928900
Email: sales@framoeng.no
Web: www.framoeng.no

Frank Mohn AS
PO Box 98
Slåtthaug
N-5851 Bergen
Norway
Tel: 055999000
Fax: 055999380
Email: marine@framo.no
Web: www.framo.no

POLAND (PL)

ABS PolskaSp. Z.o.o.
Ul Rydygiera 8
Budynek A
PL-01-793 Warsaw
Poland
Tel: 022 633 8287
Fax: 022 633 8644
Email: info@absgroup.com
Web: www.absgroup.com

AxFlow Sp Soo
ul. Floriana 3/5
PL-04-664 Warsaw
Poland
Tel: 022 613 0012
Fax: 022 812 1327
Email: biuro@axflow.pl
Web: www.axflow.pl

Gzut SA Pumps Department
ul Robotnicza 2
PL-44-100 Gliwice
Poland
Tel: 032 338 1510
Fax: 032 231 6908
Email: market@gzut.com.pl
Web: www.gzut.com.pl

Kielecka Fabryka Pomp Bialogon SA
ul Druckiego
Lubeckiego 1
PL-25-818 Kielce
Poland
Tel: 041 366 8219
Fax: 041 345 5154
Email: handel@kfp-bialogon.com.pl
Web: www.kfp-bialogon.com.pl

Leszno Pump Factory
ul Fabryczna 15
PL-64-100 Leszno
Poland
Tel: 065 528 8700
Fax: 065 529 9246
Email: lfp@lfp.com.pl
Web: www.lfp.com.pl

Metalchem-Warszawa SA
Studzienna 7a
PL-01-259 Warsaw
Poland
Tel: 022 837 1270
Fax: 022 36 8950
Email: metalchem@elektron.pl
Web: www.metalchemsa.com.pl

Powen Pump Factory Co Ltd
ul Wolnosci 318
PL-41-800 Zabrze
Poland
Tel: 032 777 5777
Email: pompy@powen.com.pl
Web: www.powen.com.pl

Swidnicka Fabryka Pomp Sp Z.o.o.
Inzynierska 8
PL-58-100 Swidnica ul
Poland
Tel: 074 851 8100
Fax: 074 853 3357
Email: marketing@sfpomp.com.pl
Web: www.sfpomp.com.pl

Wafapomp SA
ul Odelwnicza 1
PL-03 231 Warsaw
Poland
Tel: 022 811 1431
Fax: 022 811 3146
Email: biuro@wafapomp.com.pl
Web: www.wafapomp.com.pl

PORTUGAL (PL)

ABS Hidrobomba SA
Rua Antonio Maia 6-A
Apartado 60120
P-2720-902 Amadora
Portugal
Tel: 021 495 0157
Fax: 021 495 5364
Email: info@absgroup.com
Web: www.absgroup.com

AxFlow lda
Av Duarte Pacheco 2, Apartado 92
Sto Amaro de Oeiras
P-2780-316 Oeiras
Portugal
Tel: 021 446 1590
Fax: 021 441 0368
Email: axflow@axflow.pt
Web: www.axflow.pt

RUSSIA (RU)

PCM Pompes Russia
nab kanala Griboedova, d.15, 18H
liter A, (No 8)
191186 Saint Petersbourg
Russia
Tel: 0812 320 7096
Fax: 0812 320 7512
Email: pcmrussia@pcmpompes.com
Web: www.pcmpompes.com

SINGAPORE (SG)

ABS Wastewater Technology Pte Ltd
25 International Business Park
#03-51/55 German Centre
Singapore 609 916
Singapore
Tel: 06 4633933
Fax: 06 4622122
Email: info@absgroup.com
Web: www.absgroup.com

Tsurumi (Singapore) Pte Ltd
26 Hillview Terrace
Singapore 669237
Singapore
Tel: 06 7608338
Fax: 06 7601268
Email: sales@tsurumipump.com.sg

Grundfos (Singapore) Pte Ltd
24 Tuas West Road
Singapore 638381
Singapore
Tel: 06 8651222
Email: contact-sg@grundfos.com
Web: www.grundfos.com

SLOVAKIA (SK)

Slovpump-Trade Sro
Osloboditel'ov 4
SK-976 67 Závadka nad Hronom
Slovakia
Tel: 048 618 3384
Fax: 048 618 3386
Email: obchod@slovpump.sk
Web: www.slovpump.sk

SOUTH AFRICA (ZA)

ABS Wastewater Technology (Pty) Ltd
Unit 2, Cnr Innes and Bismut Roads
Jetpark Boksburg
Edleen 1625
South Africa
Email: info@absgroup.com
Web: www.absgroup.com

Denorco (Pty) Ltd
PO Box 8136
Edenglen 1613
South Africa
Tel: 011 609 4150
Fax: 011 609 2417
Email: richardr@denorco.com
Web: www.denorco.com

Rapid Allweiler Pump & Engineering
PO Box 73
Isando 1600
South Africa

Tel: 011 573 7400
Fax: 011 974 8977
Email: solution@rapidpumps.co.za
Web: www.rapidpumps.co.za

Rhine Ruhr Pumps & Valves (Pty) Ltd
PO Box 76167
Wendywood
Johannesburg 2144
South Africa
Tel: 011 444 1321
Fax: 011 444 3373
Email: info@rrpumps.co.za
Web: www.rrpumps.co.za

CH Warman Pump Group
PO Box 3607
Kempton Park 1620
South Africa
Tel: 011 908 6601
Fax: 011 908 5636
Email: pumpsales@warman.co.za
Web: www.warman.co.za

SOUTH KOREA (KR)

A-Ryung Machinery Ind Co Ltd
681-1 Daegok-Ri, Keumseong-Myun
Damyang-Gun
Cheonnam 517-810
South Korea
Tel: 068 4382 4401
Fax: 068 4382 3094
Email: aryung@aryung.co.kr
Web: www.aryung.co.kr

Chung Woo Industrial Co Ltd
682-14, Kyongso-Dong
So-Ku
Inchon
South Korea
Tel: 032 561 0971
Fax: 032 566 0972
Email: chungwoo@chwpump.co.kr
Web: www.cwhydro.co.kr

Handol Pumps Ltd
428-10 Nonhyun-Dong
Namdong-Gu
Incheon 405-848
South Korea
Tel: 032 818 0105
Fax: 032 818 0109
Email: hdpump@unitel.co.kr
Web: www.hdpump.co.kr

Hyosung-Ebara Co Ltd
6th floor Hyosung Building
450 Kongduk-Dong
Mapo-Ku
Seoul 121-120
South Korea
Tel: 02 707 6395
Fax: 02 711 5501
Email: hec-shhong@hyosung.com
Web: www.hyosungebara.com

Ilsung Precision Industry Company
1619-3 Bisan 7 Dong Suhgu
Taegu-City 703-047
South Korea
Tel: 053 358 9234
Fax: 053 353 0946
Email: ipco@hitel.net
Web: ipco.koreasme.com

Shin Shin Machinery Co Ltd
940-13 Yealim-Li
Jungghwan-Myum, Gijang-Gun
Busan 619-731
South Korea
Tel: 051 727 53004
Fax: 051 727 5305
Email: leeyk@www.sspump.com
Web: www.sspump.com

SPAIN (ES)

ABS Technologias del Agua SA
C/Madera 14-16, Pol Ind Sanat Ana
Urb Rivas-Vaciamadrid
E-28529 Madrid
Spain
Tel: 091 670 2851
Fax: 091 666 5869
Email: info@absgroup.com
Web: www.absgroup.com

AxFlow SAS
C/Antonio de Cabezón 83
E-280 34 Madrid
Spain
Tel: 091 729 1818
Fax: 091 729 2491
Email: axflow@axflow.es
Web: www.axflow.es

Bombas Boyser SL
C/Empordà
3 Poligono Industrial La Borda
E-08140 Caldes De Montbui
Spain
Tel: 093 865 5455
Fax: 093 865 5556
Email: bombas@boyser.com
Web: www.boyser.com

Casals Cardona Industries SA
F Casablancas 24
E-08243 Manresa
Spain
Tel: 0902 443 000
Fax: 0938 757 668
Email: tecnium@tecnium.es
Web: www.tecnium.com.cn

Ebara España Bombas SA
PI La Estación, C/ Cormoranes, 6
E-28320 Pinto (Madrid)
Spain
Tel: 916 923 630
Fax: 916 923 891
Email: marketing@ebara.es
Web: www.ebara.es

Hidráulicas HMT SA
Pol Ind del Ramassá
C/ Barcelonés, 14-16
E-08520 Les Franqueses del Vallés - Barcelona
Spain
Tel: 093 849 3822
Fax: 093 849 9066

Hidrotecar SA
PO Box 281
E-09080 Burgos
Spain
Tel: 0947 48 4400
Fax: 0947 48 4557
Email: htecar@las.es
Web: www.hidrotecar.com

Inoxpa SA
C/Telers 54
PO Box 174
E-17820 Banyoles (Girona)
Spain
Tel: 0972 575200
Fax: 0972 575502
Email: inoxpa@inoxpa.com
Web: www.inoxpa.com

Bombas Itur SA
PO Box 41, Urteta Bidea s/n
E-20800 Zarautz (Gipuzkoa)
Spain
Tel: 0943 899 899
Fax: 0943 130 170
Email: comexp@itur.es
Web: www.itur.es

Motorrens SL
Pol Ind Mas de les Ànimes
C / Apel les Mestres, 36-38
E-43205 Reus
Spain
Tel: 0977 328105
Fax: 0977 326552
Email: motorrens@rostor.com
Web: www.rostor.com

Bombas Omega SL
Pol Ind El Campillo-Pab B
E-48500 Abanto
Spain
Tel: 094 6363510
Fax: 094 6363501
Email: bombas01@bombasomega.com
Web: www.bombasomega.com

Pedro Roquet SA
Antonio Figueres 91
E-08551 Tona (Barcelona)
Spain
Tel: 093 812 4664
Fax: 093 887 1798
Email: roquet@pedro-roquet.com
Web: www.pedro-moquet.com

SWEDEN (SE)

ABS Group
PO Box 394
Roskildevägen 1
SE-201 23 Malmö
Sweden
Tel: 040 350470
Fax: 040 305045
Email: info@absgroup.com
Web: www.absgroup.com

Alfa Laval AB
PO Box 73
SE-221 00 Lund
Sweden
Tel: 046 367000
Fax: 046 306860
Web: www.alfalaval.com

AxFlow AB
Byängsgränd 14

PO Box 90162
SE-120 22 Stockholm
Sweden
Tel: 08 6022200
Fax: 08 916666
Email: kundservice@axflow.se
Web: www.axflow.se

Axflow Holding AB
Wennergren Center
Sveavägen 166, Floor 14
SE-113 46 Stockholm
Sweden
Tel: 08 54547670
Fax: 08 54547689
Email: axflow@axflow.com
Web: www.axflow.com

Grindex AB
Hantverkarvägen 24, Box 535
SE-136 25 Haninge
Sweden
Tel: 08 6066600
Fax: 08 7455328
Email: info@grindex.com
Web: www.grindex.com

ITT Flygt AB
Gesällvägen 33
SE-174 87 Sundbyberg
Sweden
Tel: 08 4756000
Fax: 08 4756900
Email: ylva.ponsonby@flygt.com
Web: www.flygt.com

Johnson Pump AB
Nastagatan 19
PO Box 1436
SE-701 14 Örebro
Sweden
Tel: 019 218300
Fax: 019 272330
Email: jp@johnson-pump.com
Web: www.johnson-pump.com

Metso Minerals (Sala) AB
PO Box 302
SE-733 25 Sala
Sweden
Tel: 0224 57000
Fax: 0224 16950
Email: jan.lidin@mestso.com
Web: www.metsominerals.com/pumps

Pumpex Sales
PO Box 5207
SE-121 18 Johanneshov
Sweden
Tel: 08 725 4930
Fax: 08 659 3314
Email: info@pumpex.com
Web: www.pumpex.com

Tapflo AB
Filaregatan 4
SE-442 34 Kungälv
Sweden
Tel: 0303 63390
Fax: 0303 19916
Email: sales@tapflo.com
Web: www.tapflo.com

Imo AB
PO Box 42090
SE-126 14 Stockholm
Sweden
Tel: 08 50622800
Fax: 08 6451509
Email: sales@imo.se
Web: www.imopump.com

SWITZERLAND (CH)

AxFlow GmbH
Neumattweg 6
CH-5012 Schönenwerd
Switzerland
Tel: 062 849 6164
Fax: 062 849 6165
Email: axflow@bluewin.ch
Web: www.axflow.ch

Biral AG
Südstrasse 10
CH-3110 Münsingen
Switzerland
Tel: 031 720 9000
Fax: 031 720 9442
Email: info@biral.ch
Web: www.biral.ch

Catag AG
Auf dem Wolf 19
CH-4052 Basel
Switzerland
Tel: 061 373 7373
Fax: 061 373 7370
Email: catag@catag.com
Web: www.catag.com

CP Pumpen AG
Industrie Brühl
CH-4800 Zofingen
Switzerland
Tel: 062 746 8585
Fax: 062 746 8586
Email: schlueniger@cp-pumps.com
Web: www.cp-pumps.com

Emile Egger et Cie SA
CH-2088 Cressier/NE
Switzerland
Tel: 032 758 7111
Fax: 032 757 2290
Email: info@eggerpumps.com
Web: www.eggerpumps.com

Gotec SA
Rue des Casernes 59
CH-1950 Sion
Switzerland
Tel: 027 205 7205
Fax: 027 205 7206
Email: sales@gotec.ch
Web: www.gotec.ch

Gysi Pumpen AG
In Riaux
CH-1726 Farvagny
Switzerland
Tel: 026 411 3071
Fax: 026 411 3080
Email: info@gysi-pumpen.ch
Web: www.gysi-pumpen.ch

Häny AG
Bergstrasse 103
CH-8706 Meilen
Switzerland
Tel: 01 925 4111
Fax: 01 923 6245
Email: info@haeny.com
Web: www.haeny.com

Ismatec SA
Feldeggstrasse 6
CH-8152 Glattbrugg
Switzerland
Tel: 01 874 9418
Fax: 01 810 5292
Email: taltham@idexcorp.com
Web: www.ismatec.com

Maag Pump Systems Textron AG
Aspstrasse 12
CH-8154 Oberglatt
Switzerland
Tel: 044 278 8200
Fax: 044 278 8201
Email: welcome@maag.textron.com
Web: www.maag.com

Rütschi Fluid AG
Herzogstrasse 11
CH-5200 Brugg
Switzerland
Tel: 056 441 0455
Fax: 056 441 1331
Email: info@rutschifluid.ch
Web: www.rutschipumpen.ch

SAWA Pumpentechnik AG
CH-9113 Degersheim
Switzerland
Tel: 071 372 0808
Fax: 071 372 0809
Email: info@sawa.ch
Web: www.sawa.ch

Scheerle AG
Frauenfelderstrasse 47b
CH-8266 Steckborn
Switzerland
Tel: 052 762 2221
Fax: 052 762 2266
Email: mail@scheerle.ch
Web: www.scheerle.ch

Sterling Fluid Systems (Schweiz) AG
Schweizersbildstrasse 25
CH-8205 Schaffhausen
Switzerland
Tel: 052 644 0606
Fax: 052 644 0616
Email: info@sterlingfluid.ch
Web: www.sterlingfluid.com

Sulzer Pumps
Zürcherstrasse 12
CH-8401 Winterthur
Switzerland
Tel: 052 262 1155
Fax: 052 262 0040
Email: info.pumps@sulzer.com
Web: www.sulzerpumps.com

3S Systemtechnik AG
Brunnmattstrasse 456
CH-5236 Remigen
Switzerland
Tel: 056 297 8888
Fax: 056 297 8880

Email: info@mariner-3s.com
Web: www.mariner-3s.com

TAIWAN (TW)

HCP Pump Manufacturer Co Ltd
7 Kung Yeh 4th Road
Ping Tung Industrial Zone
Taipei Hsien
Taiwan
Tel: 08 7229267
Fax: 08 7239119
Email: hcppump@ms7.hinet.net
Web: www.hcppump.com.tw

Hung Pump Industrial Co Ltd
8F, No 251, San Ming Road
Lu-Chou City
Taipei Hsien
Taiwan
Tel: 02 28472419
Fax: 02 28471559
Email: hungpump@ms9.hinet.net

Tsurumi Pump Taiwan Co Ltd
67 Fengtian Street
Bade City
Tao-yuan 334
Taiwan
Tel: 03 3689830
Fax: 03 3689832
Email: sales@tsurumipump.com.tw

THAILAND (TH)

ABS Nopon Thai Co Ltd
60/9 Moo 3, Tabol Mabyangporn
Amphur Pluakdang
Rayong 21140
Thailand
Tel: 038 891483
Fax: 038 891122
Email: info@absgroup.com
Web: www.absgroup.com

PCM Pompes Thailand
171/16 Moo5 Salaya-Bangphasi Road
Salaya, Budhamontol
Nakornpathom 73170
Thailand
Tel: 034 246012
Fax: 034 297022

TUNISIA (TN)

PCM Tunisia
Bureau d'Assistance Technique
Imm L, Appt 6
Cité Essaada, El Menzah 7
2037 Ariana
Tunisia
Tel: 071 238 138
Fax: 071 231 713
Email: pcmtunisie@pcmpompes.com
Web: www.pcmpompes.com

TURKEY (TR)

ABS Pompalari Ltd
Atasehir Bulvari
ABS Plaza No 40, Kat: 3
Atasehir-Istanbul 34758
Turkey
Tel: 0216 455 2256
Fax: 0216 455 2265
Email: info@absgroup.com
Web: www.absgroup.com

Türbosan AS
Gumussuyu Cad Hastane
Yolu No1 Maltepe
Istanbul 34010
Turkey
Tel: 0212 567 9550
Fax: 0212 612 4741
Email: info@turbosan.com
Web: www.turbosan.com

Yaltek AS
Halk Sk Eksioglu Is Merkezi 28
Kozyatagi
Istanbul 634734
Turkey
Tel: 0216 661 0203
Fax: 0216 661 0212
Email: ercument@hotmail.com.tr
Web: www.yaltek.com

UKRAINE (UA)

Nasosenergomash
1 Privokzalnaya Square
Sumy 40011
Ukraine
Tel: 0542 257172
Fax: 0542 258894
Email: ves@nempump.com
Web: www.nempump.com

UNITED ARAB EMIRATES (AE)

Primax Pumps
PO Box 17505
Office Lob 11-7, Jebel Ali Freezone
Dubai
United Arab Emirates
Tel: 04 8811599
Fax: 04 8818883
Email: primax@primax.ae
Web: www.sykesgroup.com

UNITED KINGDOM (UK)

ABS Wastewater Technology Ltd
Unit 1, Elm Park Court
Tilgate Forest Business Park
Crawley
West Sussex
RH11 9BP
United Kingdom
Tel: 01293 744800
Fax: 01293 543392
Email: info@absgroup.com
Web: www.absgroup.com

Air Pumping Engineering Service
PO Box 239
East Ham
London
E6 3SG
United Kingdom
Tel: 020 8552 1172
Fax: 020 8470 4617
Email: sales@airpumping.co.uk
Web: www.airpumping.co.uk

Albany Engineering Company Ltd
Church Road
Lydney
Gloucestershire
GL15 5EQ
United Kingdom
Tel: 01594 842275
Fax: 01594 842574
Email: sales@albany-pumps.co.uk
Web: www.albany-pumps.co.uk

Alfa Laval Pumps Ltd
Birch Road
Eastbourne
East Sussex
BN23 6PQ
United Kingdom
Tel: 01323 412555
Fax: 01323 414515
Email: bob.heales@alfalaval.com
Web: www.alfalaval.com

Alldos Ltd
39 Gravelly Industrial Park
Tyburn Road
Birmingham
B24 8TL
United Kingdom
Tel: 0121 328 3336
Fax: 0121 328 4332
Email: alldos.uk@alldos.com
Web: www.alldos.com

Allpumps Ltd
448 Brightside Lane
Sheffield
S9 2SP
United Kingdom
Tel: 0114 2442203
Fax: 0114 2425885
Email: sales@allpumps.co.uk
Web: www.allpumps.co.uk

Alpeco Ltd
Chinnor Road
Bledlow
Buckinghamshire
HP27 9PH
United Kingdom
Tel: 01844 342666
Fax: 01844 345286
Email: sales@alpeco.co.uk
Web: www.alpeco.co.uk

Charles Austen Pumps Ltd
Royston Road
Byfleet
Weybridge
Surrey
KY14 7NY
United Kingdom
Tel: 01932 355277
Fax: 01932 351285
Email: info@charlesausten.com
Web: www.charlesausten.com

Autoclude Pumps
Arisdale Avenue
South Ockendon
Essex
RM15 5DP
United Kingdom
Tel: 01708 856125
Fax: 01708 857366

Email: info@autocludepumps.com
Web: www.autocludepumps.com

AxFlow Ltd
Orion Park, Northfield Avenue
Ealing
London
W13 9SJ
United Kingdom
Tel: 020 8579 2111
Fax: 020 8579 7326
Email: info@axflow.co.uk
Web: www.axflow.co.uk

Beresford Pumps
Unit 7 Network Park, Duddeston Mill Road
Saltley
Birmingham
B8 1AU
United Kingdom
Tel: 0121 503 3001
Fax: 0121 503 3002
Email: sales@beresfordpumps.co.uk
Web: www.beresfordpumps.co.uk

BOC Edwards - Hick Hargreaves
Wingates Industrial Estate
Great Bank Road
Westhoughton
Bolton
BL5 3SL
United Kingdom
Tel: 01942 652652
Fax: 01942 652651
Email: hick.hargreaves@bocedwards.com
Web: www.bocedwards.com

Brook Crompton
St Thomas Road
Huddersfield
HD1 3LJ
United Kingdom
Tel: 01484 557200
Fax: 01484 557201
Email: csc@brookcrompton.com
Web: www.brookcrompton.com

Camaction Avonpump Ltd
The Farthings, 11 Arnor Close
Worle
Weston-super-Mare
Somerset
BS22 7GB
United Kingdom
Tel: 01934 520091
Fax: 01934 513062
Email: enquiries@camaction.co.uk
Web: www.camaction.co.uk

Caster HMD Kontro
Hampden Park Industrial Estate
Eastbourne
East Sussex
BN22 9AN
United Kingdom
Tel: 01323 452000
Fax: 01323 503369
Email: pumps@hmdpumps.com
Web: www.sundyne.com

CAT Pumps (UK) Ltd
1 Fleet Business Park
Sandy Lane, Church Crookham
Fleet
Hampshire
GU52 8BF
United Kingdom
Tel: 01252 622031
Fax: 01252 626655
Email: sales@catpumps.co.uk
Web: www.catpumps.com

Ceetak Engineering Ltd
Bedford Business Centre
Mile Road
Bedford
MK42 9TW
United Kingdom
Tel: 01234 343232
Fax: 01234 341133
Email: bedford@ceetakengineering.com
Web: www.ceetakengineering.com

Chem Resist Group Ltd
Britannia House, Lockway
Ravensthorpe Industrial Estate
Dewsbury
West Yorkshire
WF13 3SX
United Kingdom
Tel: 01924 499466
Fax: 01924 490334
Email: enquiry@chemresist.com
Web: www.chemresist.com

Chemvac Pumps Ltd
Redwood
Tytherington Business Park
Macclesfield
Cheshire
SK10 2XH
United Kingdom
Tel: 01625 443170
Fax: 01625 443179
Email: chemvacpumpsltd@aol.com
Web: www.chemvacpumps.co.uk

Circulating Pumps Ltd
Oldmedow Road
Hardwick Industrial Estate
King's Lynn
Norfolk
PE30 4PP
United Kingdom
Tel: 01553 764821
Fax: 01553 760965
Email: info@circulatingpumps.net
Web: www.circulatingpumps.net

Concentric Pumps Ltd
Unit 10, Gravelly Industrial Estate
Birmingham
B24 8HW
United Kingdom
Tel: 0121 327 2081
Fax: 0121 327 6187
Email: sales@concentric-pumps.co.uk
Web: www.concentric-pumps.co.uk

Cougar Industries Ltd
1 River Park
Billet Lane
Berkhamsted
Hertfordshire
HP4 1HL
United Kingdom
Tel: 01442 860000
Fax: 01442 864686
Web: www.cougar-global.com

John Crane UK Ltd
Buckingham House
Buckingham Avenue
Slough
SL1 4LU
United Kingdom
Tel: 01753 224052
Fax: 01753 224155
Email: sloughexportsales@johncrane.co.uk
Web: www.johncrane.co.uk

Crane Process Flow Technologies Ltd
Grange Road
Cwmbran
Gwent
NP44 3XX
United Kingdom
Tel: 01633 486666
Fax: 01633 486777
Web: www.craneflow.comn

Crest Pumps Ltd
163 Stourvale Road
Southbourne
Bournemouth
BH6 5HQ
United Kingdom
Tel: 01202 425000
Fax: 01202 425577
Email: enquiries@crestpumps.co.uk
Web: www.crestpumps.co.uk

Dawson Downie Lamont
Unit 31, Rutherford Road
Southfield Industrial Estate
Glenrothes
KY6 2RT
United Kingdom
Tel: 01592 775577
Fax: 01592 775517
Email: sales@ddl-ltd.com
Web: www.ddl-ltd.com

Diaphragm Pumps Ltd
244 Grindley Lane
Blythe Bridge
Stoke on Trent
ST11 9LW
United Kingdom
Tel: 01782 393939
Fax: 01782 394949
Email: sales@diaphragmpumps.co.uk
Web: www.pump-shop.co.uk

Flo-Dyne Ltd
Flo-Dyne Place
Asheridge Business Centre
Asheridge Road
Chesham
Buckinghamshire
HP5 2PT
United Kingdom
Tel: 01494 770088
Fax: 01494 770099
Email: sales@flo-dyne.net
Web: www.flo-dyne.net

Flotronic Pumps Ltd
Ricebridge Works, Brighton Road
Bolney
West Sussex
RH17 5NA
United Kingdom
Tel: 01444 881871

Fax: 01444 881860
Email: salesdept@flotronicpumps.co.uk
Web: www.flotronicpumps.co.uk

Flowguard Ltd
Watford Bridge
New Mills
Stockport
SK12 4HJ
United Kingdom
Tel: 01663 745976
Fax: 01663 742788
Email: swheatley@flowguard.com
Web: www.flowguard.demon.co.uk

Flowsolve Ltd
Middle Floor
40 High Street
London
SW19 5AU
United Kingdom
Tel: 020 3274 4010
Fax: 020 3274 4011
Email: cfd@flowsolve.com
Web: www.flowsolve.com

Fristam Pumps (UK) Ltd
Unit 11, Apex Business Centre
Diplocks Way
Hailsham
East Sussex
BN27 3JF
United Kingdom
Tel: 01323 849849
Fax: 01323 849438
Email: info@fristam.co.uk
Web: www.fristam.de

Gilbert Gilkes & Gordon Ltd
Canal Head North
Kendal
Cumbria
LA9 7BZ
United Kingdom
Tel: 01539 720028
Fax: 01539 732110
Email: enquiries@gilkes.com
Web: www.gilkes.com

Godwin Pumps Ltd
Quenington
Cirencester
Gloucestershire
GL7 5BX
United Kingdom
Tel: 01285 750271
Fax: 01285 750352
Email: sales@godwinpumps.co.uk
Web: www.godwinpumps.co.uk

Goodwin International
Goodwin House, Leek Road
Hanley
Stoke on Trent
ST1 3NR
United Kingdom
Tel: 01782 208040
Fax: 01782 851050
Email: pumps@goodwingroup.com
Web: www.goodwin.co.uk

Graham Vacuum Pumps Ltd
Forge Lane
Congleton
Cheshire
CW12 4HG
United Kingdom
Tel: 01260 274721
Fax: 01260 276965
Web: www.grahamvt.com

Grundfos Pumps Ltd
Grovebury Road
Leighton Buzzard
Bedfordshire
LU7 4TL
United Kingdom
Tel: 01525 850000
Email: sales@grundfos.com
Web: www.grundfos.com

Hale Products Europe Ltd
Charles Street
Warwick
CV34 5LR
United Kingdom
Tel: 01926 623600
Fax: 01926 623666
Email: admin_haleuk@idexcorp.com
Web: www.haleeurope.com

Hamworthy KSE
Fleets Corner
Poole
Dorset
BH17 0JT
United Kingdom
Tel: 01202 662600
Fax: 01202 666363
Email: poolepumps@hamworthy.com
Web: www.hamworthy.com

Hayward Tyler® Group
1 Kimpton Road
Luton
LU1 3LD
United Kingdom
Tel: 01582 731144
Fax: 01582 393400
Email: haywardtyler@haywardtyler.com
Web: www.haywardtyler.com

Heypac Ltd
Unit 3, The Royston Centre
Lynchford Road
Ash Vale
Hampshire
GU12 5PQ
United Kingdom
Tel: 01252 511003
Fax: 01252 519003
Email: enquiries@heypac.co.uk
Web: www.heypac.co.uk

Hidrostal Ltd
4-5 The Galloway Centre
Hambridge Lane
Newbury
Berkshire
RG14 5TL
United Kingdom
Tel: 01635 550440
Fax: 01635 550140
Email: sales@hidrostal.co.uk
Web: www.hidrostal.co.uk

Hilge Pumps Ltd
Hawthorn Road
Eastbourne
East Sussex
BN23 6QA
United Kingdom
Tel: 01323 438333
Fax: 01323 410004
Email: pumps@hilge.co.uk
Web: www.hilge.de

HMD/Kontro Sealless Pumps
See: Caster HMD Kontro

Hydrainer Pumps Ltd
Norwood Industrial Estate
Rotherham Close, Killamarsh
Sheffield
S21 2JU
United Kingdom
Tel: 01142 484868
Fax: 01142 510136
Email: pumpsales@hydrainer.com
Web: www.hydrainer.com

Invensys APV UK
23 Gatwick Road
Crawley
West Sussex
RH10 9JB
United Kingdom
Tel: 01293 527777
Fax: 01293 552640
Email: apvproducts.uk@invensys.com
Web: www.apv.com

Jabsco UK, ITT Industries
Bingley Road
Hoddesdon
Hertfordshire
EN11 0BU
United Kingdom
Tel: 01992 450145
Fax: 01992 467132
Web: www.jabsco.com

JLS International (UK) Ltd
62 Mill Road
Ringmer
Lewes
East Sussex
BN8 5HZ
United Kingdom
Tel: 01273 812029
Fax: 01273 812029
Email: info@jlsintl.com
Web: www.jlsintl.com

Johnson Pump (UK) Ltd
Highfield Industrial Estate
Edison Road
Eastbourne
East Sussex
BN23 6PT
United Kingdom
Tel: 01323 509211
Fax: 01323 507306
Email: jinfo@idexcorp.com
Web: www.johnsonpump.com

JP Pumps Ltd
Maxwell Way
Crawley
West Sussex
RH10 9SA
United Kingdom
Tel: 01293 553495
Fax: 01293 524635

Email: mail@jppumps.co.uk
Web: www.pandct.co.uk

Kecol Pumps Ltd
Faraday Drive
Bridgnorth
Shropshire
WV15 5BJ
United Kingdom
Tel: 01746 764311
Fax: 01746 764780
Email: sales@kecol.co.uk
Web: www.kecol.co.uk

Kestner Engineering Co Ltd
Bredgar Road
Gillingham
Kent
ME8 6PN
United Kingdom
Tel: 01634 386689
Fax: 01634 386684
Email: sales@kestner-eng.co.uk
Web: www.kestner-eng.co.uk

Kinder-Janes Engineers Ltd
Porters Wood
St Albans
Hertfordshire
AL3 6HU
United Kingdom
Tel: 01727 844441
Fax: 01727 844247
Email: info@kinder-janes.co.uk
Web: www.kinder-janes.co.uk

KSB Ltd
2 Cotton Way
Loughborough
Leicestershire
LE11 5TF
United Kingdom
Tel: 01509 231872
Fax: 01509 215228
Email: ian.robinson@ksb.com
Web: www.ksb.com

KTR Couplings Ltd
Unit 7, Acorn Business Park
Woodseats Close
Sheffield
S8 0TB
United Kingdom
Tel: 0114 258 7757
Fax: 0114 258 7740
Email: ktr-uk@ktr.com
Web: www.ktr.com

Linatex Ltd
Wilkinson House, Galway Road
Blackbushe Business Park
Yateley
Hampshire
GU46 6GE
United Kingdom
Tel: 01252 743000
Fax: 01252 743030
Web: www.linatex.com

London Pumps Ltd
Unit 7, Hampden Road
Kingston-Upon-Thames
KT1 3LG
United Kingdom
Tel: 020 8546 7702
Fax: 020 8546 7703
Email: sales@londonpumps.co.uk
Web: www.londonpumps.co.uk

Lutz (UK) Ltd
West Midlands Freeport
Birmingham
B26 3QD
United Kingdom
Tel: 0121 782 2662
Fax: 0121 782 2680
Email: sales@lutzpump.co.uk
Web: www.lutzpump.co.uk

March May Ltd
Eaton Works, Howard Road
Eaton Socon
St Neots
Cambridgeshire
PE19 3NZ
United Kingdom
Tel: 01480 214444
Fax: 01480 405336
Email: sales@marchmay.co.uk
Web: www.marchmay.co.uk

MDM Pumps Ltd
Spring Lane
Malvern
Worcestershire
WR14 1BP
United Kingdom
Tel: 01684 892678
Fax: 01684 892841
Email: info@mdmpumps.co.uk
Web: www.hygienic-pumps.co.uk

Megator Ltd
Hendon
Sunderland
Tyne & Wear
SR1 2NQ
United Kingdom
Tel: 0191 567 5488
Fax: 0191 567 8512
Email: info@megator.co.uk
Web: www.megator.co.uk

Micropump Ltd
Howard Road
Eaton Socon
St Neots
Cambridgeshire
PE 19 8ET
United Kingdom
Tel: 01480 356900
Fax: 01480 356920
Email: info.micropump@idexcorp.com
Web: www.micropump.com

Midland Pump Manufacturing Co Ltd
Seeleys Road
Greet
Birmingham
B11 2LF
United Kingdom
Tel: 0121 773 8862
Fax: 0121 771 4363
Email: enquiries@midlandpump.co.uk
Web: www.midlandpump.co.uk

Mono Pumps Ltd
Martin Street
Audenshaw
Manchester
M34 5JA
United Kingdom
Tel: 0161 339 9000
Fax: 0161 344 0727
Email: info@mono-pumps.com
Web: www.mono-pumps.com

NSK Europe - European Bearings Business Unit
Newark
Nottinghamshire
NG24 2JF
United Kingdom
Tel: 01636 605 123
Fax: 01636 602 750
Email: info-uk@nsk.com
Web: www.eu.nsk.com

Papa Pump Ltd
14a Kingshill Industrial Estate
Bude
Cornwall
EX23 8QN
United Kingdom
Tel: 01288 354454
Fax: 01288 354454
E-mail: salesinfopapa@aol.com
Web: www.waterpoweredpumps.co.uk

PCA Engineers Ltd
Homer House
Sibthorp Street
Lincoln
LN5 7SB
United Kingdom
Tel: 01522 530106
Fax: 01522 511703
Email: info@pcaeng.co.uk
Web: www.pcaeng.co.uk

PCM Pumps Ltd
Pilot Road
Phoenix Parkway
Corby
Northamptonshire
NN17 5YF
United Kingdom
Tel: 01536 740200
Fax: 01536 740201
Email: sales@pcmpumps.co.uk
Web: www.pcmpumps.co.uk

Plenty Mirrlees Pumps
SPX Process Equipment
Earl Haig Road, Hillington
Glasgow
G52 4JN
United Kingdom
Tel: 0141 883 0314
Fax: 0141 882 2752
Email: pumps.plenty@processequipment.spx.com
Web: www.spxprocessequipment.com

ProMinent Fluid Controls UK Ltd
Resolution Road
Ashby-de-la-Zouch
Leicestershire
LE65 1DW
United Kingdom
Tel: 01530 560555
Fax: 01530 560777
Email: info@prominent.co.uk
Web: www.prominent.co.uk

Pulseguard Ltd
Unit 1, Greg Street Industrial Centre
Greg Street, Reddish
Stockport
SK5 7BS
United Kingdom
Tel: 0161 442 6222
Fax: 0161 443 1486
Email: pulse@plsgd.co.uk
Web: www.pulsegurad,co.uk

Pump & Package Ltd
29E Station Road
Desborough
Northamptonshire
NN14 2RL
United Kingdom
Tel: 01536 762674
Fax: 01536 761973
Email: pumppackage@btconnect.com
Web: www.pumppackage.co.uk

Pump Centre, ESR Technology Ltd
Whittle House, 410 The Quadrant
Birchwood Park
Warrington
WA3 6FW
United Kingdom
Tel: 01925 843506
Fax: 01925 843500
Email: pumps@esrtechnology.com
Web: www.pumpcentre.com

Pump Engineering Ltd
Riverside Estate
Littlehampton
West Sussex
BN17 5DF
United Kingdom
Tel: 01903 730900
Fax: 01903 730234
Email: sales@pumpeng.co.uk
Web: www.pumpeng.co.uk

Putzmeister Ltd
Carrwood Road, Chesterfield Trading Estate
Sheepbridge
Chesterfield
Derbyshire
S41 9QB
United Kingdom
Tel: 01246 264200
Fax: 01246 260077
Email: info@putzmeister.co.uk
Web: www.putzmeister.co.uk

RMI Pressure Systems Ltd
Portex House
Blaydon
Tyne & Wear
NE21 4TE
United Kingdom
Tel: 0191 499 0882
Fax: 0191 499 0883
Email: rhowarth@armlink.com
Web: www.armstrongpumps.com

Roxspur Measurement & Control Ltd
2 Downgate Drive
Sheffield
S4 8BT
United Kingdom
Tel: 0114 244 2521
Fax: 0114 243 4838
Email: sales@roxspur.com
Web: www.roxspur.com

S & G Thermofluids Ltd
The Old Tannery
Kelston
Bath
BA1 9AN
United Kingdom
Tel: 0117 932 7378
Fax: 0117 932 9652
Email: tony.smith@thermofluids.co.uk
Web: www.thermofluids.co.uk

Schaeffler (UK) Ltd
Forge Lane
Minworth
Sutton Coldfield
West Midlands
B76 1AP
United Kingdom
Tel: 0121 351 3833
Fax: 0121 351 7686
Email: info.uk@schaeffler.com
Web: www.ina.co.uk

seepex UK Ltd
3 Armtech Row
Houndstone Business Park
Yeovil
Somerset
BA22 8RW
United Kingdom
Tel: 01935 472376
Fax: 01935 479836
Email: sales@seepex.co.uk
Web: www.seepex.com

Seko Ltd
Units 8a & 8b, Blackminster Business Park
Old Birminhgam Road
Blackminster
Worcestershire
WR11 7RE
United Kingdom
Tel: 01386 839010
Email: s.ellix@seko.com
Web: www.selko.com

Selwood Group Ltd
Bournemouth Road
Chandlers Ford
Eastleigh
Hampshire
SO53 3ZL
United Kingdom
Tel: 023 8026 6311
Fax: 023 8026 0906
Email: cfordhire@selwoodgroup.co.uk
Web: www.selwood-pumps.co.uk

Serfilco Europe Ltd
Broadoak Industrial Park
Ashburton Road West
West Trafford
Manchester
M17 1RW
United Kingdom
Tel: 0161 872 1317
Fax: 0161 873 8027
Email: sales@serflico-europe.com
Web: www.serfilco.com

SHURflo Ltd
Unit 5, Sterling Park
Gatwick Road
Crawley
West Sussex
RH10 2QT
United Kingdom
Tel: 01293 424050
Fax: 01293 421880
Email: sales@shurflo.co.uk
Web: www.shurflo-original.com

Michael Smith Engineers Ltd
Unit 9, Batley Enterprise Centre
513 Bradford Road
Batley
Yorkshire
WF17 8JY
United Kingdom
Tel: 01483 771871
Fax: 01483 723110
Email: info@michael-smith-engineers.co.uk
Web: www.michael-smith-engineers.co.uk

SPP Pumps Ltd
Theale Cross
Pincents Lane, Calcot
Reading
Berkshire
RG31 7SP
United Kingdom
Tel: 0118 932 3123
Fax: 0118 932 3302
Email: solutions@spppumps.com
Web: www.spppumps.com

SPX Process Equipment Ltd
Ironstone Way
Brixworth
Northampton
NN6 9UD
United Kingdom
Tel: 01604 880751
Fax: 01604 880145
Email: info.uk@processequipment.spx.com
Web: www.spxpe.com

Sterling Fluid Systems (UK) Ltd
Atlantic Street
Altrincham
Cheshire
WA14 5DH
United Kingdom
Tel: 0161 928 6371
Fax: 0161 925 2129
Email: uksales@sterlingfluid.com
Web: www.sterlingfluid.com

SKF (UK) Ltd
Luton
LU3 3BL
United Kingdom
Tel: 01582 490049
Fax: 01582 848091
Email: marketing.uk@skf.com
Web: www.skf.co.uk

Terex Powertrain Ltd
Catto Drive
Peterhead
Aberdeen
AR42 1RL
United Kingdom
Tel: 01779 473161
Fax: 01779 477424

Totton Pumps Ltd
Rushington Business Park
Totton
Southampton
SO40 9AH
United Kingdom
Tel: 023 8066 6685
Fax: 023 8066 6880
Email: info@totton-pumps.com
Web: www.totton-pumps.co.uk

TriVista Engineering Ltd
The Sussex Innovation Centre
Science Park Square
Falmer, Brighton
BN1 9SB
United Kingdom
Tel: 01273 704419
Fax: 0870 4604950
Email: mail@trivista.co.uk
Web: www.trivista.co.uk

Stuart Turner Ltd
The Market Place
Henley-on-Thames
Oxfordshire
RG9 2AD
United Kingdom
Tel: 01491 572655
Fax: 01491 573704
Email: pumps@stuart-turner.co.uk
Web: www.stuart-turner.co.uk

Wallace & Tiernan, Chemfeed Ltd
Priory Works
Five Oak Green Road
Tonbridge
Kent
TN11 0QL
United Kingdom
Tel: 01732 771777
Fax: 01732 771800
Email: inform@wallace-tiernan.com
Web: www.wallace-tiernan.com

Wanner International Ltd
Units 8-9 Fleet Business Park
Sandy Lane
Church Crookham
Hampshire
GU52 8BF
United Kingdom
Tel: 01252 816847
Fax: 01252 781235
Email: sales@wannerint.com
Web: www.hydra-cell.com

Watson-Marlow Bredel Pumps Ltd
Falmouth
Cornwall
TR11 4RU
United Kingdom
Tel: 01326 370370
Fax: 01326 376009
Email: info@watson-marlow.co.uk
Web: www.watson-marlow.co.uk

Weir Pumps Ltd
149 Newlands Road
Cathcart
Glasgow
G44 4EX
United Kingdom
Tel: 0141 637 7141
Fax: 0141 633 7358
Email: enquiries@weirpumps.com
Web: www.weirclearliquid.com

UNITED STATES (US)

ABS Pumps Inc
140 Pond View Drive
Meridien
CT 06450
USA
Tel: 203 238 2700
Fax: 203 238 0738
Email: info@absgroup.com
Web: www.absgroup.com

ACD Inc
2321 South Pullman Street
Santa Ana
CA 92705
USA
Tel: 949 261 7533
Fax: 949 261 6285
Email: jpaje@acdcom.com
Web: www.acdcom.com

Ace Pump Corporation
PO Box 13187, 1650 Channel Avenue
Memphis
TN 38113
USA
Tel: 901 948 8514
Fax: 901 774 6147
Email: sales@acepumps.com
Web: www.acepumps.com

Alfa Laval Inc
5400 International Trade Drive
Richmond
VA 23231
USA
Tel: 804 222 5300
Fax: 804 545 8311
Web: www.alfalaval.com

American Lewa Inc
132 Hopping Brook Road
Holliston
MA 01746
USA
Tel: 508 429 7403
Fax: 508 429 8615
Email: sales@amlewa.com
Web: www.amlewa.com

Ansimag Inc
14845 West 64th Avenue
Arvada
CO 80007
USA
Tel: 847 290 0482
Fax: 847 290 0481
Email: sales@ansimag.com
Web: www.ansimag.com

Aurora Pump
800 Airport Road
North Aurora
IL 60542
USA
Tel: 630 859 7000
Fax: 630 859 1226
Email: aurora_info@pentairpump.com
Web: www.aurorapump.com

Barnant Company
28 W 092 Commercial Avenue
Barrington
IL 60010
USA
Tel: 847 381 7050
Fax: 847 381 7053
Email: barnant@barnant.com
Web: www.barnant.com

Beta Technology Inc
2841 Mission Street
Santa Cruz
CA 95060
USA
Tel: 831 426 0882
Fax: 831423 4573
Email: service@beta-technology.com
Web: www.beta-technology.com

Bijur Lubricating Corporation
808 Aviation Parkway
Suite 1400
Morrisville
NC 27560
USA
Tel: 919 465 4448
Fax: 919 465 0516
Email: cservice@bijurlube.com
Web: www.bijur.com

Blue White Industries
5300 Business Drive
Huntington Beach
CA 92649
USA
Tel: 714 893 8529
Fax: 714 894 9492
Email: sales@blue-white.com
Web: www.bluwhite.com

Buffalo Pumps Inc
874 Oliver Street
North Tonawanda
Niagara
NY 14120-3298
USA
Tel: 716 693 1850
Fax: 716 693 6303
Email: info@buffalopumps.com
Web: www.buffalopumps.com

Carver Pump Company
2415 Park Avenue
Muscatine
IA 52761-5691
USA
Tel: 563 263 3410
Fax: 563 267 0510
Email: customerservice@carverpump.com
Web: www.carverpump.com

CAT Pumps
1681 94th Lane NE
Minneapolis
MN 55449
USA
Tel: 763 780 5440
Fax: 763 780 2958
Email: techsupport@catpumps.com
Web: www.catpumps.com

CDS-John Blue Company
PO Box 1607

Huntsville
AL 35807
USA
Tel: 205 721 9090
Fax: 205 721 9091
Email: info@cds-johnblue.com
Web: www.cds-johnblue.com

Chempump
959 Mearns Road
Warminster
PA 18974
USA
Tel: 215 343 6000
Fax: 215 343 8543
Email: chempump@chempump.com
Web: www.chempump.com

Chicago Pump Company
PO Box 6620
Aurora
IL 60598-0620
USA
Tel: 630 236 5700
Fax: 630 236 5779
Email: sales@yccpump.net
Web: www.yccpump.com

Cole-Parmer Instrument Company
625 East Bunker Court
Vernon Hills
IL 60061-1844
USA
Tel: 800 323 4340
Fax: 847 247 2929
Email: export@coleparmer.com
Web: www.coleparmer.com

Colfax Pump Group
1710 Airport Road
PO Box 5020
Monroe
NC 28111-5020
USA
Tel: 704 289 6511
Fax: 704 289 4839
Email: welcome@colfaxpump.com
Web: www.colfaxpump.com

Continental Pump Company
PO Box 428
Warrenton
MO 63383-0428
USA
Tel: 636456 6006
Fax: 636 456 4337
Email: info@continentalultrapumps.com
Web: www.continentalultrapumps.com

Corken Inc
3805 NW 36th Street
Oklahoma City
OK 73112
USA
Tel: 405 946 5576
Fax: 405 948 7343
Email: info.corken@idexcorp.com
Web: www.corken.com

Cornell
Sunrise Corridor Business Center
16261 SE 130th Avenue
Clackamas
OR 97015
USA
Tel: 503 653 0330
Fax: 503 653 0338
Email: info@cornellpump.com
Web: www.cornellpump.com

Crane Pumps & Systems
420 Third Street
Piqua
OH 45356
USA
Tel: 937 773 2442
Fax: 937 773 2238
Email: cranepumps@cranepumps.com
Web: www.cranepumps.com

Cross Manufacturing Inc
100 James H Cross Boulevard
Lewis
KS 67552
USA
Tel: 620 324 5525
Fax: 620 324 5737
Email: info@crossmfg.com
Web: www.crossmfg.com

CW-EMD
1000 Cheswick Avenue
Cheswick
PA 15024
USA
Tel: 724 275 5000
Fax: 724 275 5942
Email: contact@emd.curtisswright.com
Web: www.wxemd.com

Darley Pumps
2000 Anson Drive
Melrose Park
IL 60160
USA
Tel: 708 345 8050
Fax: 708 345 8993
Email: darley@darley.com
Web: www.darley.com

Dean Pump Division,
Met-Pro Corporation
6040 Guion Road
Indianapolis
IN 46254
USA
Tel: 317 293 2930
Fax: 317 297 7028
Email: info@met-pro.com
Web: www.deanpump.com

Defco
PO Box 1209
Decatur
AL 35602
USA
Tel: 256 353 7697
Fax: 256 350 6540
Email: sales@(no-spam)defcoinc.com
Web: www.defco.com

Dempster Industries Inc
PO Box 848
711 South 6th Street
Beatrice
NE 68310
USA
Tel: 402 223 4026
Fax: 402 228 4389
Email: sales@dempsterinc.com
Web: www.dempsterinc.com

Discflo Corporation
10850 Hartley Road
Santee
CA 92071
USA
Tel: 619 596 3181
Fax: 619 449 1990
Email: sales@discflo.com
Web: www.discflo.com

Ebara Fluid Handling
1651 Cedar Line Drive
Rock Hill
SC 29730
USA
Tel: 803 327 5005
Fax: 803 327 5097
Email: info@pumpsebara.com
Web: www.pumpsebara.com

EMS Pacific
4480 Enterprise Street
Suite D
Fremont
CA 94538
USA
Tel: 510 668 0405
Fax: 510 668 1012
Email: info@emspacific.com
Web: www.emspacific.com

Environamics Corporation
19 Executive Drive
Hudson
NH 03051
USA
Tel: 603 889 0200
Fax: 603 889 2508
Email: sales@environamics.com
Web: www.environamics.com

ESSCO Pumps & Controls
4935 Telegraph Road
Los Angeles
CA 90022
USA
Tel: 213 261 2181
Fax: 213 261 1532
Email: sales@esscopumps.com
Web: www.esscopumps.com

Fairbanks Morse Pump
3601 Fairbanks Avenue
Kansas City
KS 66106
USA
Tel: 913 371 5000
Fax: 913 748 4025
Email: fairbanks_info@pentairpump.com
Web: www.fairbanksmorsepump.com

Finish Thompson Inc
921 Greengarden Road
Erie
PA 16501
USA
Tel: 814 455 4478
Fax: 814 455 8518
Web: www.finishthompson.com

Flojet Corporation
1800 Gardner Expressway

Quincy
IL 62301
USA
Tel: 217 222 5400
Fax: 217 224 7814
Email: pumps@gardnerdenver.com
Web: www.gardnerdenver.com

Flowserve Corporation
5215 North O'Connor Boulevard
Suite 2300
Irving
TX 75039
USA
Tel: 972 443 6500
Fax: 972 443 6800
Email: fpdinformation@lfowserve.com
Web: www.flowserve.com

Fluid Metering Inc
5 Aerial Way, Suite 500
Syosset
Ny 11791
USA
Tel: 516 922 6050
Fax: 516 624 8261
Email: pumps@fmipump.com
Web: www.fmipump.com

Flux Pumps Corporation
4330 Commerce Circle
Atlanta
GA 30336
USA
Tel: 404 691 6010
Fax: 404 691 6314
Web: www.flux-pumps.com

Fristam Pumps USA
2410 Parview Road
Middleton
WI 53562-2524
USA
Tel: 608 831 5001
Fax: 608 831 8467
Email: info@fristampumps.com
Web: www.fristampumps.com

Fultz Pumps Inc
PO Box 550
5690 State Route 103 North
Lewistown
PA 17044
USA
Tel: 717 248 2300
Fax: 717 248 5458
Email: fultzpumps@acsworld.com
Web: www.fultzpumps.com

GIW Industries Inc
5000 Wrightsboro Road
Grovetown
GA 30813-9750
USA
Tel: 706 863 1011
Fax: 706 863 5637
Email: marketing@giwindustries.com
Web: www.giwindustries.com

Godwin Pumps of America Inc
One Floodgate Road
Bridgeport
NJ 08014
USA
Tel: 856 467 3636
Fax: 856 467 4841
Email: sales@godwinpumps.com
Web: www.godwinpumps.com

Gorman-Rupp Company
305 Bowman Street
Mansfield
OH 4903-1600
USA
Tel: 419 755 1011
Fax: 419 755 1266
Email: intsales@gormanrupp.com
Web: www.gormanrupp.com

Goulds Pumps, ITT Industrial & BioPharm Group
240 Fall Street
Seneca Falls
NY 13148
USA
Tel: 315 568 2811
Fax: 315 568 7759
Email: gpsales@itt.com
Web: www.gouldspumps.com

Goulds Pumps, ITT Water Technology
PO Box 330
Seneca Falls
NY 13148
USA
Tel: 315 568 2811
Fax: 315 568 7759
Web: www.goulds.com

GPM Pumps Inc
110 Gateway Drive
Macon
GA 31210
USA
Tel: 877 476 3569
Email: info@gpmpump.com
Web: www.gmppump.com

Graco Inc
88 11th Avenue NE
Minneapolis
MN 55413
USA
Tel: 612 623 6704
Fax: 612 623 6957
Web: www.graco.com

Grover Manufacturing Corporation
620 Vail Avenue
PO Box 986
Montebello
CA 90640
USA
Tel: 323 724 3444
Fax: 323 724 3596
Email: sales@grovermfg.com
Web: www.grovermfg.com

Grundfos Pumps Corporation
17100 West 118th Terrace
Olathe
KS 66061
USA
Tel: 913 227 3400
Email: partnerservices@US.grundfos.com
Web: www.grundfos.com

Haight Pumps
Division of Baker Manufacturing Company
133 Enterprise Street
Evansville
WI 53536
USA
Tel: 608 871 9250
Fax: 608 882 3777
Email: haight@baker-mfg.com
Web: www.haightpump.com

Haskel International Inc
100 East Graham Place
Burbank
CA 91502
USA
Tel: 818 843 4000
Fax: 818 841 4291
Email: sales@haskel.com
Web: www.haskel.com

Hydromatic Pumps
740 East 9th Street
Ashland
OH 44805
USA
Tel: 419 207 3372
Fax: 419 281 4087
Web: www.hydromatic.com

Hypro
375 Fifth Avenue NW
New Brighton
MN 55112
USA
Tel: 651 766 6300
Fax: 651 766 6600
Email: sales@hypropumps.com
Web: www.hypropumps.com

Ingersoll-Rand/ARO
Business Development Center
1467 Route 31 South, PO Box 970
Annandale
NJ 08801
USA
Tel: 800 495 0276
Web: www.irtools.com

Invensys APV
100 South CP Avenue
Lake Mills
WI 53551
USA
Tel: 920 648 8311
Fax: 920 648 1423
Email: apvproducts.us@invensys.com
Web: www.apv.com

ITT Residential & Commercial Water
8200 N Austin Avenue
Morton Grove
IL 60053-3283
USA
Tel: 847 966 3700
Fax: 847 966 8366
Email: fhs.webmaster@itt.com
Web: www.bellgossett.com

Iwaki America
5 Boynton Road
Hopping Brook Park
Holliston
MA 01746
USA
Tel: 508 429 1440
Fax: 508 429 1386

Email: service@iwakiamerica.com
Web: www.iwakiamerica.com

Jabsco US, ITT Industries
20 Icon
Foothill Ranch
CA 92610
USA
Tel: 949 859 4945
Fax: 949 859 1153
Web: www.jabsco.com

Jaeco Fluid Sysyetms Inc
17 Lee Boulevard, Suite D
Malvern
PA 19355-1234
USA
Tel: 610 407 7207
Fax: 610 407 7211
Email: sales@jaecofs.com
Web: www.jaecofs.com

Jensen
PO Box 1509, 14th & Pacific Street
Coffeyville
KS 67337
USA
Tel: 620 251 5700
Fax: 620 251 3622
Email: pumpjacksales@jensen-products.com
Web: www.jensen-products.com

Johnson Pumps of America Inc
9950 W Lawrence Avenue
Ste #300
Schiller Park
IL 60176
USA
Tel: 847 671 7867
Fax: 847 671 7909
Email: jp@johnson-pump.com
Web: www.johnson-pump.com

Johnston Pump
800 Koomey Road
Brookshire
TX 77423
USA
Tel: 800 926 6688
Fax: 281 934 6194
Email: puush-engrsalesmgr@sulzer.com
Web: www.johnston-pump.com

KD Scientific Inc.
84 October Hill Road
Holliston, MA 01746
USA
Tel: 508 429 6809
Fax: 508 893 0160
Web: www.kdscientific.com

Komline-Sanderson
12 Holland Avenue
Peapack
NJ 07977-0257
USA
Tel: 908 234 1000
Fax: 908 234 9487
Email: info@komline.com
Web: www.komline.com

KTR Corporation
PO Box 90 65
Michigan City
IN 46361
USA
Tel: 219 872 9100
Fax: 219 872 9150
Email: ktr-us@ktr.com
Web: www.ktr.com

Layne/Verti-Line
PO Box 6999
Kansas City
KS 66106
USA
Tel: 913 748 4253
Fax: 913 748 4030
Email: layne_info@pentairpump.com
Web: www.pentairpump.com

Lewis Pumps
8625 Grant Road
St Louis
MO 63123
USA
Tel: 314 843 4437
Fax: 314 843 7964
Email: sales@lewispumps.com
Web: www.lewispumps.com

Liquid Controls
A Unit of IDEX Corporation
105 Albrecht Drive
Lake Bluff
IL 60044
USA
Tel: 847 295 1050
Fax: 847 295 1057
Email: lc-info.lcmeter@idexcorp.com
Web: www.lcmeter.com

Little Giant Pump Company
PO Box 12010
Oklahoma City
OK 73157-2010
USA
Tel: 405 947 2511
Fax: 405 947 8720
Email: customerservice@lgpc.com
Web: www.lgpc.com

LMI Milton Roy
201 Ivyland Road
Ivyland
PA 18974
USA
Tel: 215 293 0401
Fax: 215 293 0445
Email: lmicustomerservice@miltonroy.com
Web: www.lmipumps.com

Lobee Pump & Machinery Company
8425 State Street
Gasport
NY 14067
USA
Tel: 716 772 2651
Fax: 716 772 2555
Email: info@lobee.com
Web: www.lobee.com

Lutz-Jesco America Corporation
55 Bermar Park
Rochester
NY 14624
USA
Tel: 585 426 0990
Fax: 585 426 4025
Email: mail@jescoamerica.com
Web: www.jescoamerica.com

Maag Pump Systems Textron Inc
2915 Whitehall Park Drive
Charlotte
NC 28273
USA
Tel: 704 716 9000
Fax: 704 716 9001
Email: maaginfo@maagusa.textron.com
Web: www.maag.com

Madden Manufacturing Inc
PO Box 387
Elkhart
IN 46515
USA
Tel: 574 295 4292
Fax: 574 295 7562
Email: info@maddenmfg.com
Web: www.maddenmfg.com

Magnatex Pumps Inc
3575 West 12th Street
Houston
TX 77008
USA
Tel: 713 972 8686
Fax: 713 972 8665
Email: sales@magnatexpumps.com
Web: www.magnatexpumps.com

Mahr Metering Systems Corporation
1415-A Cross Beam Drive
Charlotte
NC 28217
USA
Tel: 704 525 7128
Fax: 704 525 8290
Email: kelly.cauthen@mmscusa.com
Web: www.mmscusa.com

Marlow Pumps, ITT Water Technology
2881 East Bayard Street
Seneca Falls
NY 13148
USA
Tel: 315 568 2811
Fax: 315 568 7759

Marshall Engineered Products Company
3056 Walker Ridge Drive, Suite C
Grand Rapids
MI 49544
USA
Tel: 616 246 1431
Fax: 616 246 1445
Email: mepco@mepcollc.com
Web: www.mepcollc.com

Mefiag Division, Met-Pro Corporation
1550 Industrial Drive
Owasso
MI 48867
USA
Tel: 989 725 8184
Fax: 989 729 1013
Email: mefiag@met-pro.com
Web: www.met-pro.com/mefiag-html

Mepco
3056 Walker Ridge Drive, Suite C
Grand Rapids

MI 49544
USA
Tel: 616 246 1431
Fax: 616 246 1445
Email: mepco@mepcollc.com
Web: www.mepcollc.com

Micropump Inc
1402 NE 136th Avenue
Vancouver
WA 98684-0818
USA
Tel: 360 253 2008
Fax: 800 222 9565
Email: info.micropump@idexcorp.com
Web: www.micropump.com

Mody Pumps™ Inc
2166 Zeus Court
Bakersfield
CA 93308
USA
Tel: 661 392 7600
Fax: 661 392 7601
Email: sales@modypump.com
Web: www.modypump.com

Monoflo Inc
12000 West Little York
Houston
TX 77041
USA
Tel: 713 466 7999
Email: inquire@monoflo.com
Web: www.mono-pumps.com

Morgan AM & T
441 Hall Avenue
St Marys
PA 15857
USA
Tel: 814 781 1573
Fax: 814 781 9239
Web: www.morganamt.com

Morris Pumps
PO Box 6620
Aurora
IL 60598-0620
USA
Tel: 630 236 6900
Fax: 630 236 6932
Email: sales@yccpump.net
Web: www.morrispumps.com

Moyno Inc
1895 W Jefferson Street
Springfield
OH 45506
USA
Tel: 937 327 3553
Fax: 937 327 3572
Email: mipsales@moyno.com
Web: www.moyno.com

MP Pumps Inc
34800 Bennett Drive
Fraser
MI 48026-1686
USA
Tel: 586 293 8240
Fax: 586 293 8469
Email: mppumps@mppumps.com
Web: www.mppumps.com

MTH Pumps
401 West Main Street
Plano
IL 60545-1436
USA
Tel: 630 552 4115
Fax: 630 552 3688
Email: sales@mthpumps.com
Web: www.mthpumps.com

FE Myers
1101 Myers Parkway
Ashland
OH 44805
USA
Tel: 419 289 1144
Fax: 419 289 6658
Email: info@myers.com
Web: www.femyers.com

National Oilwell Varco
10000 Richmond Avenue
Houston
TX 77042-4200
USA
Tel: 713 346 7500
Web: www. natoil.com

National Pump Company LLC
7600 West Olive Avenue
Peoria
AZ 85345-7102
USA
Tel: 602 979 3560
Fax: 602 979 2177
Email: info@natlpump.com
Web: www.nationalpumpcompany.com

Neptune Chemical Pump Company Inc
PO Box 247
Lansdale
PA 19446-0247
USA
Tel: 215 699 8700
Fax: 215 699 0370
Email: pump@neptune1.com
Web: www.neptune1.com

Northern Pump
340 Benson Avenue
Grantsburg
WI 54840
USA
Tel: 715 463 8300
Fax: 715 463 5174
Email: info@northern-pump.com
Web: www.northern-pump.com

Ohler Pumps
PO Box 507, 501 Maple Street
Janesville
IA 50647
USA
Tel: 319 987 2121
Fax: 319 987 2161
Email: ohler@ohler-machinery.com
Web: www.ohler-machinery.com

The Oilgear Company
PO Box 343924
Milwaukee
WI 53234-3924
USA
Tel: 414 327 1700
Fax: 414 327 0532
Web: www.oilgear.com

Pacer Pumps
Division of ASM Industries Inc
41 Industrial Circle
Lancaster
PA 17601
USA
Tel: 717 656 2161
Fax: 717 656 0477
Email: sales@pacerpumps.com
Web: www.pacerpumps.com

Paco Pumps
800 Koomey Road
Brookshire
TX 77423
USA
Tel: 281 934 6014
Fax: 281 934 6056
Email: paco@paco-pumps.com
Web: www.paco-pumps.com

Patterson Pump Company
PO Box 790
Toccoa
GA 30577
USA
Tel: 706 886 2101
Fax: 706 886 0023
Email: bhenry@pattersonpumps.com
Web: www.pattersonpumps.com

PCM Delasco Inc
11940 Brittmoore Park Drive
Houston
TX 77041
USA
Tel: 713 896 4888
Fax: 713 896 4806
Email: pcmdelasco@pcmdelasco.com
Web: www.pcmdelasco.com

Peerless Pump Company
PO Box 7026
Indianapolis
IN 46207-7026
USA
Tel: 317 925 9661
Fax: 317 924 7388
Web: www.peerlesspump.com

Pentair Inc
5500 Wayzata Boulevard
Suite 800
Golden Valley
MN 55416-1261
USA
Tel: 763 545 1730
Fax: 763 656 5400
Web: www.pentair.com

Plast-O-Matic Valves Inc
1384 Pompton Avenue
Cedar Grove
NJ 07009
USA
Tel: 973 256 3000
Fax: 973 256 4745
Email: info@plastomatic.com
Web: www.plastomatic.com

Price Pump Company
21775 Eighth Street East

Sonoma
CA 95476
USA
Tel: 707 938 8441
Fax: 707 938 0764
Email: sales@pricepump.com
Web: www.pricepump.com

Procon Products
910 Ridgely Road
Murfreesboro
TN 37129
USA
Tel: 615 890 5710
Fax: 615 896 7729
Web: www.proconpumps.com

Pulsafeeder
Engineered Products Operation (EPO)
2883 Brighton-Henrietta Townline Road
Rochester
NY 14623
USA
Tel: 585 292 8000
Fax: 585 424 5619
Email: pulsa@idexcorp.com
Web: www.pulsa.com

Pumper Parts LLC
6017 Enterprise Drive
Export
PA 15632
USA
Tel: 724 387 1776
Fax: 724 387 1774
Email: info@pumperparts.com
Web: www.pumperparts.com

Red Jacket, ITT Water Technology
58 Wright Avenue
Auburn
NY 13021
USA
Tel: 866 325 4210
Fax: 866 325 4211
Web: www.redjacketwaterproducts.com

Reddy-Buffaloes Pump Inc
1 Dixie Drive
PO Box 557
Baxley
GA 31513
USA
Tel: 912 367 6345
Fax: 912 367 5024
Email: sales@rbpump.com
Web: www.rbpump.com

Roper Pump Company
PO Box 269
Commerce
GA 30529
USA
Tel: 706 335 5551
Fax: 706 335 5490
Email: sales@roperpumps.com
Web: www.roperpumps.com

Roth Pump Company
PO Box 4330
Rock Island
IL 61204
USA
Tel: 309 787 1791
Fax: 309 787 5142
Email: rothpump@rothpump.com
Web: www.rothpump.com

Ruhrpumpen Inc
4501 South 86th East Avenue
Tulsa
OK 74145
USA
Tel: 918 627 8400
Fax: 918 624 2471
Email: info@ruhrpumpen.com
Web: www.ruhrpumpen.com

Saint-Gobain Ceramics
23 Acheson Drive
Niagara Falls
NY14303
USA
Tel: 716 278 6233
Fax: 716 278 2373
Email: scd.sales@saint-gobain.com
Web: www.hexoloy.com

Saint-Gobain Performance Plastics
7301 Orangewood Avenue
Garden Grove
CA 92841
USA
Tel: 800 833 5661
Fax: 714 688 2614
Web:
www.microelectronics.saint-gobain.com

Serfilco Ltd
2900 MacArthur Boulevard
Northbrook
IL 60062-2005
USA
Tel: 847 509 2900
Fax: 847 559 1995
Email: sales@serfilco.com
Web: www.serfilco.com

Sethco Division, Met-Pro Corporation
800 Emlen Way
Telford
PA 18969
USA
Tel: 215 799 2577
Fax: 215 799 0920
Email: info@sethco.com
Web: www.sethco.com

SHURflo Pump Manufacturing Co
5900 Katella Avenue
Cypress
CA 90630
USA
Tel: 562 795 5200
Fax: 562 795 7564
Email: customer_service@shurflo.com
Web: www.shurflo.com

Smith & Loveless Inc
14040 Santa Fe Trail Drive
Lenexa
KS 66215-1284
USA
Tel: 913 888 5201
Fax: 913 888 2173
Email: answers@smithandloveless.com
Web: www.smithandloveless.com

SPM Flow Control Inc
7601 Wyatt Drive
Fort Worth
TX 76108
USA
Tel: 817 246 2461
Email: mtreida@spmflo.com
Web: www.spmflo.com

SRS Crisafulli Inc
1610 Crisafulli Drive
PO Box 1051
Glendive
MT 59330
USA
Tel: 406 365 3393
Fax: 406 365 8088
Email: srsc@crisafulli.com
Web: www.crisafulli.com

Standard Pump Inc
3230 Industrial Way
Snellville
GA 30039
USA
Tel: 770 972 9693
Fax: 770 972 9694
Email: info@standardpump.com
Web: www.standardpump.com

Sterling Fluid Systems (Americas) Inc
303 Industrial Boulevard
Grand Island
NY 14072-0460
USA
Tel: 716 773 6450
Fax: 716 773 2330
Email: mail@sihi.com
Web: www.sterlingfluid.com

Sundyne Corporation
14845 West 64th Avenue
Arvada
CO 80007
USA
Tel: 303 425 0800
Fax: 303 940 3141
Email: pumps@sundyne.com
Web: www.sundyne.com

Tarby Inc
2205 E L Anderson Boulevard
Claremore
OK 74017
USA
Tel: 918 341 8282
Fax: 918 341 8297
Email: sales@tarby.com
Web: www.tarby.com

Teledyne Isco Inc
PO Box 82531
Lincoln
NE 68501
USA
Tel: 4402 464 0231
Fax: 402 465 3064
Email: iscolnfo@teledyne.com
Web: www.isco.com

Thompson Pump & Manufacturing Co Inc
PO Box 291370
Port Orange
FL 32129-1370
USA
Tel: 800 767 7310

Fax: 386 761 0362
Email: sales@thompsonpump.com
Web: www.thompsonpumps.com

Trebor
8100 South 1300 West
West Jordan
UT 84088
USA
Tel: 801 561 0303
Fax: 801 255 2312
Email: treborsales@idexcorp.com):
Web: www.treborintl.com

Tsurumi (America) Inc
845 N. Dillon Drive
Wood Dale
IL 60191
USA
Tel: 630 766 5900
Fax: 630 766 6445
Email: info@tsurumiamerica.com

Tuthill Corporation
8500 South Madison
Burr Ridge
IL 60527
USA
Tel: 630 382 4900
Fax: 630 382 4999
Email: tuthillpump@tuthill.com
Web: www.pump.tuthill.com

Vaughan Company Inc
364 Monte Elma Road
Montesano
WA 98563
USA
Tel: 360 249 4042
Fax: 360 149 6155
Email: info@chopperpumps.com
Web: www.chopperpumps.com

Versa-Matic Pump Company
6017 Enterprise Drive
Export
PA 15632
USA
Tel: 724 327 7867
Fax: 724 327 4300
Email: sales@versamatic.com
Web: www.versamatic.com

Vertiflo Pump Company
Dept 1, 7807 Redsky Drive
Cincinnati
OH 45249
USA
Tel: 513 530 0888
Fax: 513 530 0893
Email: sales@vertiflopump.com
Web: www.vertiflopump.com

Viking Pump Inc
406 State Street
PO Box 8
Cedar Falls
IA 50613-0008
USA
Tel: 319 266 1741
Fax: 319 273 8157
Email: info.viking@idexcorp.com
Web: www.vikingpump.com

Warren Pumps LLC
82 Bridges Avenue
PO Box 969
Warren
MA 01083-0969
USA
Tel: 413 436 7711
Fax: 413 436 5605
Email: warrensales@colfaxcorp.com
Web: www.warrenpumps.com

Warren Rupp Inc
800 North Main Street
PO Box 1568
Mansfield
OH 44901-1568
USA
Tel: 419 524 8388
Fax: 419 522 7867
Email: info.warrenrupp@idexcorp.com
Web: www.warrenrupp.com

Waukesha Cherry-Burrell
611 Sugar Creek Road
Delavan
WI 53115
USA
Tel: 262 728 1900
Fax: 262 728 4904
Email: info@processequipment.spx.com
Web: www.gowcb.com

Weir Minerals Division
225 N Cedar Street
PO Box 488
Hazelton
PA 18201-0488
USA
Tel: 717 455 7711
Fax: 717 459 2586
Email: pumps@weirminerals.com
Web: www.weirminerals.com

Weir Specialty Pumps
440 West 800 South
Salt Lake City
UT 84110-0209
USA
Tel: 801 359 8731
Fax: 801 355 9303
Email: info@wemcopump.com
Web: www.wemcopump.com

Wheatley and Gaso Pump
PO Box 123
Ada
MI 49301
USA
Tel: 616 452 6882
Fax: 888 796 2653
Email: info@wheatleygaso.com
Web: www.wheatleygaso.com

Wilden Pump & Engineering LLC
22069 Van Buren Street
Grand Terrace
CA 92313-5607
USA
Tel: 909 422 1730
Fax: 909 783 3440
Email: wilden@wildenpump.com
Web: www.wildenpump.com

A R Wilfley & Sons Inc
PO Box 2330
Denver
CO 80201
USA
Tel: 303 779 1777
Fax: 303 779 1277
Email: pumps@wilfley.com
Web: www.wilfley.com

Wright Pump
S84 W18693 Enterprise Drive
Muskego
WI 53150
USA
Tel: 262 679 8000
Fax: 262 679 2026
Email: info.wright@idexcorp.com
Web: www.softwavepumps.com

Yamada America Inc
1200 Nuclear Drive, Dept I
West Chicago
IL 60185
USA
Tel: 630 231 4083
Fax: 630 231 7398
Email: sales@yamadapump.com
Web: www.yamadapump.com

Yeomans Pump Company
PO Box 6620
Aurora
IL 60598-0620
USA
Tel: 630 236 5500
Fax: 630 236 5511
Email: donpage@yccpump.com
Web: www.yeomanspump.com

Zenith Pumps Division, Parker Hannifin
5910 Elwin Buchanan Drive
Sanford
NC 27330
USA
Tel: 919 7182225
Fax: 919 774 5952
Email: emmett.sellers@colfaxcorp.com
Web: www.zenithpumps.com

Zoeller Pump Company
3649 Cane Run Road
Louisville
KY 40211-1961
USA
Tel: 502 778 2731
Fax: 502 774 3624
Email: info@zoeller.com
Web: www.zoeller.com

VENEZUELA (VE)

Equimavenca
Av Fernandez Padilla 226
San José de Guanipa
Edo Anzoategui
Venezuela
Tel: 283 255 4830
Fax: 283 255 4830
Email: tchaline@pcmpompes.com
Web: www.pcmpompes.com

18.3 Rotodynamic pumps

Companies listed alphabetically by pump type

SMALL CENTRIFUGAL PUMPS (for domestic water/heating/sanitation applications)

ABS Deutschland GmbH	DE
ABS Finland Oy	FI
ABS France SAS	FR
ABS Group	SE
ABS Hellas SA	GR
ABS Hidrobomba SA	PL
ABS Hungary Trading	HU
ABS Industria de Bombas Centrifugas Ltda	BR
ABS International Srl	IT
ABS Italy Srl	IT
ABS Nopon Thai Co Ltd	TH
ABS Oumoer A/S	NO
ABS PolskaSp. Z.o.o.	PL
ABS Pompalari Ltd	TR
ABS Pompen BV	NL
ABS Pumpen GmbH	AT
ABS Pumps (Irl) Ltd	IE
ABS Pumps AS	EE
ABS Pumps Corporation	CA
ABS Pumps Inc	US
ABS Pumps Malaysia Sdn Bhd	MY
ABS Pumps NV/SA	BE
ABS Shanghai Co Ltd	CN
ABS Shanghai Co Ltd Beijing Liaison Office	CN
ABS Technologias del Agua SA	ES
ABS Wastewater Technology (Pty) Ltd	ZA
ABS Wastewater Technology Ltd	UK
ABS Wastewater Technology Pte Ltd	SG
Allpumps Ltd	UK
APV Fluid Handling Horsens	DK
Aurora Pump	US
Beresford Pumps	UK
Biral AG	CH
Bombas Itur SA	ES
Calama Industries Pvt Ltd	IN
Calpeda SpA	IT
Caprari SpA	IT
Carver Pump Company	US
CAT Pumps	US
CAT Pumps (UK) Ltd	UK
Ceetak Engineering Ltd	UK
Charles Austen Pumps Ltd	UK
Chempump	US
Chung Woo Industrial Co Ltd	KR
Circulating Pumps Ltd	UK
Clasal	BE
Cole-Parmer Instrument Company	US
Cougar Industries Ltd	UK
DAB Pumps SpA	IT
Davey Products Pty Ltd	AU
DDA Srl	IT
Desmi A/S	DK
Discflo Corporation	US
DP-Pumps	NL
Drakos-Polemis Inc	GR
Ebara Corporation	JP
Ebara Fluid Handling	US
Ebara Pumps Europe SpA	IT
EDUR Pumpenfabrik GmbH & Co KG	DE
EMS Pacific	US
FE Myers	US
Finder Pompe SpA	IT
Flexachem Manufacturing Ltd	IE
GMP SpA	IT
Gorman-Rupp Company	US
Goulds Pumps, ITT Industrial & BioPharm Group	US
Goulds Pumps, ITT Water Technology	US
Grundfos (Singapore) Pte Ltd	SG
Grundfos GmbH	DE
Grundfos Management A/S	DK
Grundfos Pompe Italia Srl	IT
Grundfos Pumpen Vertrieb GmbH	AT
Grundfos Pumps Corporation	US
Grundfos Pumps KK	JP
Grundfos Pumps Ltd	UK
Grundfos Pumps Pty Ltd	AU
Handol Pumps Ltd	KR
Hayward Tyler® Group	UK
Herborner Pumpenfabrik	DE
Honda Kiko Co Ltd	JP
Hunan XD Changsha Pump Works Co Ltd	CN
Hyosung-Ebara Co Ltd	KR
Invensys APV	US
Invensys APV UK	UK
ITT Flygt AB	SE
ITT Residential & Commercial Water	US
Iwaki America	US
Kielecka Fabryka Pomp Bialogon SA	PL
Klaus Union GmbH & Co KG	DE
Korina Pumps	GR
Leszno Pump Factory	PL
Lewis Pumps	US
Little Giant Pump Company	US
Lowara SpA, ITT Industries	IT
March May Ltd	UK
Mark Grundfos Ltda	BR
Marlow Pumps, ITT Water Technology	US
Marshall Engineered Products Company	US
Matra SpA	IT
Mepco	US
Motorarg SA	AR
Nijhuis Pompen BV	NL
Nikkiso Co Ltd	JP
Nova Magnetics Burgmann Ltd	CA
Nuovo Pignone	IT
Oddesse Pumpen-und Motorenfabrik GmbH	DE
Pacer Pumps	US
Paco Pumps	US
Peerless Pump Company	US
Pompe Vergani	IT
Powen Pump Factory Co Ltd	PL
Process Pumps (India) Pvt Ltd	IN
Pump & Package Ltd	UK
Pump Engineering Co Pvt Ltd	IN
Pumpenfabrik Ernst Vogel GmbH, ITT Industries	AT
Raj Pumps	IN
Rapid Allweiler Pump & Engineering	ZA
Red Jacket, ITT Water Technology	US
Ritz Pumpenfabrik GmbH & Co KG	DE
Rotex SA	GR
Salmson Italia	IT
Schmalenberger GmbH & Co	DE
Sharp Trendys	IN
SHURflo Ltd	UK
SHURflo Pump Manufacturing Co	US
Sigma Group AS	CZ
Simaco Elettromeccanica Srl	IT
Stuart Turner Ltd	UK
T Smedegaard A/S	DK
Tangshan Pump Factory	CN
3S Systemtechnik AG	CH
TKL	AU
Torishima Pump Mfg Co Ltd	JP
Vertiflo Pump Company	US
Wallace & Tiernan, Chemfeed Ltd	UK
Wilo AG	DE
Wilo EMU GmbH	DE
Zoeller Pump Company	US

DOMESTIC WATER SUPPLY PUMP PACKAGES

ACD Cryo GmbH	DE
Allpumps Ltd	UK
APV Fluid Handling Horsens	DK
Aurora Pump	US
Biral AG	CH
Calama Industries Pvt Ltd	IN
Calpeda SpA	IT
Camaction Avonpump Ltd	UK
Charles Austen Pumps Ltd	UK
Circulating Pumps Ltd	UK
Clasal	BE
DAB Pumps SpA	IT
Darley Pumps	US
Davey Products Pty Ltd	AU
DDA Srl	IT
Desmi A/S	DK
DP-Pumps	NL
Drakos-Polemis Inc	GR
Ebara Corporation	JP
Ebara Fluid Handling	US
Ebara Pumps Europe SpA	IT
EDUR Pumpenfabrik GmbH & Co KG	DE
EMS Pacific	US
Environamics Corporation	US
Fairbanks Morse Pump	US
FE Myers	US
Finder Pompe SpA	IT
Flexachem Manufacturing Ltd	IE
Goulds Pumps, ITT Industrial & BioPharm Group	US
Goulds Pumps, ITT Water Technology	US
Grundfos (Singapore) Pte Ltd	SG
Grundfos GmbH	DE
Grundfos Management A/S	DK
Grundfos Pompe Italia Srl	IT
Grundfos Pumpen Vertrieb GmbH	AT
Grundfos Pumps Corporation	US
Grundfos Pumps KK	JP
Grundfos Pumps Ltd	UK
Grundfos Pumps Pty Ltd	AU

Handol Pumps Ltd	KR
Honda Kiko Co Ltd	JP
Hyosung-Ebara Co Ltd	KR
Invensys APV	US
Invensys APV UK	UK
ITT Flygt AB	SE
ITT Residential & Commercial Water	US
Kielecka Fabryka Pomp Bialogon SA	PL
Lowara SpA, ITT Industries	IT
Mark Grundfos Ltda	BR
Marshall Engineered Products Company	US
Mepco	US
Nijhuis Pompen BV	NL
Nuovo Pignone	IT
Oy Kolmeks AB	FI
Paco Pumps	US
Patterson Pump Company	US
Peerless Pump Company	US
Pompe Garbarino SpA	IT
Powen Pump Factory Co Ltd	PL
Pump & Package Ltd	UK
Pumpenfabrik Ernst Vogel GmbH, ITT Industries	AT
Raj Pumps	IN
Rapid Allweiler Pump & Engineering	ZA
Red Jacket, ITT Water Technology	US
Ritz Pumpenfabrik GmbH & Co KG	DE
Ruhrpumpen GmbH	DE
Ruhrpumpen Inc	US
Sero PumpSystems GmbH	DE
Stuart Turner Ltd	UK
TKL	AU
Torishima Pump Mfg Co Ltd	JP
Wilo AG	DE
Wilo EMU GmbH	DE

SMALL CENTRIFUGAL PUMPS (for machine tool cutting oil applications)

ACD Cryo GmbH	DE
Biral AG	CH
Bombas Itur SA	ES
Camaction Avonpump Ltd	UK
Carver Pump Company	US
Crane Pumps & Systems	US
DAB Pumps SpA	IT
Desmi A/S	DK
DP-Pumps	NL
Ebara Corporation	JP
EDUR Pumpenfabrik GmbH & Co KG	DE
Emile Egger et Cie SA	CH
Environamics Corporation	US
Fairbanks Morse Pump	US
FE Myers	US
Finish Thompson Inc	US
Flexachem Manufacturing Ltd	IE
Flowmore Private Ltd	IN
Gilbert Gilkes & Gordon Ltd	UK
Gorman-Rupp Company	US
Goulds Pumps, ITT Industrial & BioPharm Group	US
Grundfos (Singapore) Pte Ltd	SG
Grundfos GmbH	DE
Grundfos Management A/S	DK
Grundfos Pompe Italia Srl	IT
Grundfos Pumpen Vertrieb GmbH	AT
Grundfos Pumps Corporation	US
Grundfos Pumps KK	JP
Grundfos Pumps Ltd	UK
Grundfos Pumps Pty Ltd	AU
Hale Products Europe Ltd	UK
Handol Pumps Ltd	KR
Herborner Pumpenfabrik	DE
Honda Kiko Co Ltd	JP
Hyosung-Ebara Co Ltd	KR
Kelair Pumps Australia Pty Ltd	AU
Lowara SpA, ITT Industries	IT
Mark Grundfos Ltda	BR
Nasosenergomash	UA
Nijhuis Pompen BV	NL
Ohler Pumps	US
Paco Pumps	US
Patterson Pump Company	US
Peerless Pump Company	US
Pompe Garbarino SpA	IT
Powen Pump Factory Co Ltd	PL
Pump & Package Ltd	UK
Pump Engineering Ltd	UK
Ritz Pumpenfabrik GmbH & Co KG	DE
Rovatti Pompe	IT
Ruhrpumpen GmbH	DE
Ruhrpumpen Inc	US
Stuart Turner Ltd	UK
TKL	AU
Verder Group	NL
Willy Vogel AG	DE
Wilo AG	DE
Wilo EMU GmbH	DE

HORIZONTAL SINGLE-STAGE END-SUCTION OVERHUNG IMPELLER CENTRIFUGAL PUMPS (for general purposes including fixed irrigation & EN 733 pumps)

Allweiler AG	DE
Best & Crompton Engineering Ltd	IN
Camaction Avonpump Ltd	UK
Caprari SpA	IT
Carver Pump Company	US
Crane Pumps & Systems	US
DAB Pumps SpA	IT
Dalian Danai Pumps Ltd	CN
Darley Pumps	US
Dean Pump Division, Met-Pro Corporation	US
Desmi A/S	DK
Ebara Corporation	JP
Ebara Fluid Handling	US
Ebara Pumps Europe SpA	IT
EDUR Pumpenfabrik GmbH & Co KG	DE
Emile Egger et Cie SA	CH
Engineered Products Group Pty Ltd (EPG)	AU
Environamics Corporation	US
Fairbanks Morse Pump	US
FE Myers	US
Finish Thompson Inc	US
Flexachem Manufacturing Ltd	IE
Flowserve Corporation	US
Goulds Pumps, ITT Industrial & BioPharm Group	US
Goulds Pumps, ITT Water Technology	US
Grundfos (Singapore) Pte Ltd	SG
Grundfos GmbH	DE
Grundfos Management A/S	DK
Grundfos Pompe Italia Srl	IT
Grundfos Pumpen Vertrieb GmbH	AT
Grundfos Pumps Corporation	US
Grundfos Pumps KK	JP
Grundfos Pumps Ltd	UK
Grundfos Pumps Pty Ltd	AU
Handol Pumps Ltd	KR
Herborner Pumpenfabrik	DE
Hilge Pumps Ltd	UK
Honda Kiko Co Ltd	JP
Hyosung-Ebara Co Ltd	KR
ITT Flygt AB	SE
ITT Residential & Commercial Water	US
Jiangsu Haishi Pump Co Ltd	CN
Kelair Pumps Australia Pty Ltd	AU
KSB Ltd	UK
Laxmi Hydraulics Pvt Ltd	IN
Lowara SpA, ITT Industries	IT
Mark Grundfos Ltda	BR
Mefiag Division, Met-Pro Corporation	US
Nijhuis Pompen BV	NL
Nuovo Pignone	IT
Ohler Pumps	US
Pacer Pumps	US
Paco Pumps	US
Peerless Pump Company	US
Pompe Garbarino SpA	IT
Powen Pump Factory Co Ltd	PL
Pump & Package Ltd	UK
Pumpenfabrik Ernst Vogel GmbH, ITT Industries	AT
Red Jacket, ITT Water Technology	US
Rhine Ruhr Pumps & Valves (Pty) Ltd	ZA
Ritz Pumpenfabrik GmbH & Co KG	DE
Roto Pumps Ltd	IN
Rovatti Pompe	IT
Ruhrpumpen GmbH	DE
Ruhrpumpen Inc	US
Savino Barbera	IT
Sethco Division, Met-Pro Corporation	US
SPP Pumps Ltd	UK
Sterling Fluid Systems (Americas) Inc	US
Sterling Fluid Systems (Austria)	AT
Sterling Fluid Systems (Belgium) NV	BE
Sterling Fluid Systems (France) SAS	FR
Sterling Fluid Systems (Italy) SpA	IT
Sterling Fluid Systems (Netherlands)	NL
Sterling Fluid Systems (Schweiz) AG	CH
Sterling Fluid Systems (UK) Ltd	UK
Sterling SIHI GmbH	DE
Tianjin Pumps & Machinery Group Co Ltd	CN
TKL	AU
Ture International Industrial Ltd	CN
Verder Group	NL
Victor Pumpen GmbH	DE
Weir Pumps Ltd	UK
Wilo AG	DE
Wilo EMU GmbH	DE

HORIZONTAL SINGLE-STAGE END-SUCTION OVERHUNG IMPELLER CENTRIFUGAL PUMPS (for chemical applications including ISO 2858, ISO 3069, ISO 3661, ASME/ANSI B73.1 designs)

ACD Cryo GmbH	DE
ACD Inc	US
Ace Pump Corporation	US

Aker Kvaerner Eureka Pump Systems	NO
Albany Engineering Company Ltd	UK
Allweiler AG	DE
Gruppo Aturia SpA	IT
Biral AG	CH
Bombas Itur SA	ES
Camaction Avonpump Ltd	UK
Carver Pump Company	US
Caster HMD Kontro	UK
Chung Woo Industrial Co Ltd	KR
Crane Pumps & Systems	US
Crest Pumps Ltd	UK
Dalian Danai Pumps Ltd	CN
Dean Pump Division, Met-Pro Corporation	US
Defco	US
Desmi A/S	DK
Dickow Pumpen KG	DE
Eagle Pump & Compressor Ltd	CA
Ebara Corporation	JP
Ebara España Bombas SA	ES
Ebara Fluid Handling	US
Ebara Pumps Europe SpA	IT
EBS-Ray Pumps Pty Ltd	AU
EDUR Pumpenfabrik GmbH & Co KG	DE
Emile Egger et Cie SA	CH
Engineered Products Group Pty Ltd (EPG)	AU
Ensival-Moret	BE
Fapmo	FR
Flexachem Manufacturing Ltd	IE
Flowmore Private Ltd	IN
Flowserve Corporation	US
Gilbert Gilkes & Gordon Ltd	UK
Goulds Pumps, ITT Industrial & BioPharm Group	US
Grundfos (Singapore) Pte Ltd	SG
Grundfos GmbH	DE
Grundfos Management A/S	DK
Grundfos Pompe Italia Srl	IT
Grundfos Pumpen Vertrieb GmbH	AT
Grundfos Pumps Corporation	US
Grundfos Pumps KK	JP
Grundfos Pumps Ltd	UK
Grundfos Pumps Pty Ltd	AU
Handol Pumps Ltd	KR
Hayward Tyler® Group	UK
Hermetic-Pumpen GmbH	DE
Honda Kiko Co Ltd	JP
Hydromatic Pumps	US
Hyosung-Ebara Co Ltd	KR
ITT Flygt AB	SE
Jiangsu Haishi Pump Co Ltd	CN
Kelair Pumps Australia Pty Ltd	AU
Kestner Engineering Co Ltd	UK
Oy Kolmeks AB	FI
KSB Ltd	UK
Lowara SpA, ITT Industries	IT
Magnatex Pumps Inc	US
Mark Grundfos Ltda	BR
Marshall Engineered Products Company	US
MASO Process-Pumpen GmbH	DE
Mather & Platt Pumps Ltd	IN
Mefiag Division, Met-Pro Corporation	US
Mepco	US
Micropump Ltd	UK
Monarch Industries Ltd	CA
FE Myers	US
Nasosenergomash	UA
Nijhuis Pompen BV	NL
Nova Magnetics Burgmann Ltd	CA
Paco Pumps	US
Peerless Pump Company	US
Pompe Garbarino SpA	IT
Powen Pump Factory Co Ltd	PL
PPI Pumps Pvt Ltd	IN
Process Pumps (India) Pvt Ltd	IN
Pump & Package Ltd	UK
Pump Engineering Co Pvt Ltd	IN
Reddy-Buffaloes Pump Inc	US
Rhine Ruhr Pumps & Valves (Pty) Ltd	ZA
Ritz Pumpenfabrik GmbH & Co KG	DE
Roto Pumps Ltd	IN
Rovatti Pompe	IT
Ruhrpumpen GmbH	DE
Ruhrpumpen Inc	US
Rütschi Fluid AG	CH
Sacem SpA	IT
Saer Elettropompe	IT
Savino Barbera	IT
Scheerle AG	CH
Sethco Division, Met-Pro Corporation	US
Shanghai Liansheng Pump Manufacture Co Ltd	CN
Siwatec AS	CZ
SPP Pumps Ltd	UK
Sterling Fluid Systems (Americas) Inc	US
Sterling Fluid Systems (Austria)	AT
Sterling Fluid Systems (Belgium) NV	BE
Sterling Fluid Systems (France) SAS	FR
Sterling Fluid Systems (Italy) SpA	IT
Sterling Fluid Systems (Netherlands)	NL
Sterling Fluid Systems (Schweiz) AG	CH
Sterling Fluid Systems (UK) Ltd	UK
Sterling SIHI GmbH	DE
Sulzer Pumps	CH
Sundyne Corporation	US
Sundyne International SA	FR
Tapflo AB	SE
Tianjin Pumps & Machinery Group Co Ltd	CN
TKL	AU
Torishima Pump Mfg Co Ltd	JP
Ture International Industrial Ltd	CN
Tyco - Southern Cross Plant	AU
Verder Group	NL
Pumpenfabrik Ernst Vogel GmbH, ITT Industries	AT
Weir Gabbioneta Srl	IT
Wernert-Pumpen GmbH	DE

HORIZONTAL SINGLE-STAGE END-SUCTION OVERHUNG IMPELLER CENTRIFUGAL PUMPS (for heavy-duty applications including ISO 13709, API 610 designs)

Ace Pump Corporation	US
Aker Kvaerner Eureka Pump Systems	NO
Albany Engineering Company Ltd	UK
Gruppo Aturia SpA	IT
Biral AG	CH
Camaction Avonpump Ltd	UK
Caprari SpA	IT
Caster HMD Kontro	UK
Corporacion EG SA	MX
Crane Pumps & Systems	US
Dalian Danai Pumps Ltd	CN
Dean Pump Division, Met-Pro Corporation	US
Desmi A/S	DK
Dickow Pumpen KG	DE
Drakos-Polemis Inc	GR
Eagle Pump & Compressor Ltd	CA
Ebara Corporation	JP
Ebara España Bombas SA	ES
Ebara Fluid Handling	US
Ebara Pumps Europe SpA	IT
EBS-Ray Pumps Pty Ltd	AU
EDUR Pumpenfabrik GmbH & Co KG	DE
Emile Egger et Cie SA	CH
Engineered Products Group Pty Ltd (EPG)	AU
Ensival-Moret	BE
Fapmo	FR
Flexachem Manufacturing Ltd	IE
Flowmore Private Ltd	IN
Flowserve Corporation	US
Goulds Pumps, ITT Industrial & BioPharm Group	US
Gysi Pumpen AG	CH
Handol Pumps Ltd	KR
Hermetic-Pumpen GmbH	DE
Hilge Pumps Ltd	UK
Honda Kiko Co Ltd	JP
Hyosung-Ebara Co Ltd	KR
Kelair Pumps Australia Pty Ltd	AU
Kestner Engineering Co Ltd	UK
KSB Ltd	UK
Magnatex Pumps Inc	US
Marshall Engineered Products Company	US
MASO Process-Pumpen GmbH	DE
Mepco	US
Micropump Ltd	UK
FE Myers	US
Nasosenergomash	UA
Nijhuis Pompen BV	NL
Paco Pumps	US
Peerless Pump Company	US
Powen Pump Factory Co Ltd	PL
Process Pumps (India) Pvt Ltd	IN
Pump & Package Ltd	UK
Pump Engineering Co Pvt Ltd	IN
Reddy-Buffaloes Pump Inc	US
Rhine Ruhr Pumps & Valves (Pty) Ltd	ZA
Ritz Pumpenfabrik GmbH & Co KG	DE
Roto Pumps Ltd	IN
Rovatti Pompe	IT
Ruhrpumpen GmbH	DE
Ruhrpumpen Inc	US
Rütschi Fluid AG	CH
Saer Elettropompe	IT
Savino Barbera	IT
Shanghai Liansheng Pump Manufacture Co Ltd	CN
Siwatec AS	CZ
SPP Pumps Ltd	UK
Sulzer Pumps	CH
Sundyne Corporation	US
Sundyne International SA	FR
TKL	AU
Toyo Pumps Europe	BE
Ture International Industrial Ltd	CN
Vaughan Company Inc	US

Verder Group	NL
Vertiflo Pump Company	US
Weir Gabbioneta Srl	IT
Weir Pumps Ltd	UK
WPIL Ltd	IN

HORIZONTAL SINGLE-STAGE DOUBLE-SUCTION AXIALLY-SPLIT PUMPS (for general purposes)

Argal Srl	IT
Gruppo Aturia SpA	IT
Aurora Pump	US
Charles Austen Pumps Ltd	UK
Best & Crompton Engineering Ltd	IN
Biral AG	CH
Buffalo Pumps Inc	US
Chem Resist Group Ltd	UK
Crane Pumps & Systems	US
Dalian Danai Pumps Ltd	CN
Darley Pumps	US
Drakos-Polemis Inc	GR
Ebara Corporation	JP
EDUR Pumpenfabrik GmbH & Co KG	DE
Emile Egger et Cie SA	CH
Franz Eisele und Söhne GmbH & Co KG	DE
Engineered Products Group Pty Ltd (EPG)	AU
Flexachem Manufacturing Ltd	IE
Flowmore Private Ltd	IN
Flowserve Corporation	US
Goulds Pumps, ITT Industrial & BioPharm Group	US
Grundfos (Singapore) Pte Ltd	SG
Grundfos GmbH	DE
Grundfos Management A/S	DK
Grundfos Pompe Italia Srl	IT
Grundfos Pumpen Vertrieb GmbH	AT
Grundfos Pumps Corporation	US
Grundfos Pumps KK	JP
Grundfos Pumps Ltd	UK
Grundfos Pumps Pty Ltd	AU
Hale Products Europe Ltd	UK
Handol Pumps Ltd	KR
Honda Kiko Co Ltd	JP
Hyosung-Ebara Co Ltd	KR
ITT Flygt AB	SE
ITT Residential & Commercial Water	US
Kelair Pumps Australia Pty Ltd	AU
KSB Ltd	UK
Mark Grundfos Ltda	BR
FE Myers	US
Nasosenergomash	UA
Nijhuis Pompen BV	NL
Nuovo Pignone	IT
Ohler Pumps	US
Paco Pumps	US
Peerless Pump Company	US
Powen Pump Factory Co Ltd	PL
PT Ebara Indonesia	ID
Pump & Package Ltd	UK
Rhine Ruhr Pumps & Valves (Pty) Ltd	ZA
Ritz Pumpenfabrik GmbH & Co KG	DE
Roto Pumps Ltd	IN
Rovatti Pompe	IT
Ruhrpumpen GmbH	DE
Ruhrpumpen Inc	US
Shanghai Liansheng Pump Manufacture Co Ltd	CN
SRS Crisafulli Inc	US
Sulzer Pumps	CH
Taiko Kikai Industries Co Ltd	JP
Tangshan Pump Factory	CN
TKL	AU
Ture International Industrial Ltd	CN
Voltas Ltd	IN
Weir Pumps Ltd	UK

HORIZONTAL SINGLE-STAGE DOUBLE-SUCTION AXIALLY-SPLIT PUMPS (for heavy-duty applications including ISO 13709, API 610 designs)

Gruppo Aturia SpA	IT
CDS-John Blue Company	US
Concentric Pumps Ltd	UK
Corporacion EG SA	MX
Dalian Danai Pumps Ltd	CN
Debem	IT
Desmi A/S	DK
Ebara Corporation	JP
Ebara Fluid Handling	US
Ebara Pumps Europe SpA	IT
EDUR Pumpenfabrik GmbH & Co KG	DE
Emile Egger et Cie SA	CH
Engineered Products Group Pty Ltd (EPG)	AU
Ensival-Moret	BE
Flexachem Manufacturing Ltd	IE
Flojet Corporation	US
Flowmore Private Ltd	IN
Flowserve Corporation	US
Goulds Pumps, ITT Industrial & BioPharm Group	US
Grundfos (Singapore) Pte Ltd	SG
Grundfos GmbH	DE
Grundfos Management A/S	DK
Grundfos Pompe Italia Srl	IT
Grundfos Pumpen Vertrieb GmbH	AT
Grundfos Pumps Corporation	US
Grundfos Pumps KK	JP
Grundfos Pumps Ltd	UK
Grundfos Pumps Pty Ltd	AU
Hale Products Europe Ltd	UK
Handol Pumps Ltd	KR
Honda Kiko Co Ltd	JP
Hyosung-Ebara Co Ltd	KR
Jiangsu Haishi Pump Co Ltd	CN
Kelair Pumps Australia Pty Ltd	AU
Leszno Pump Factory	PL
Magnatex Pumps Inc	US
Mark Grundfos Ltda	BR
Micropump Ltd	UK
MP Pumps Inc	US
FE Myers	US
Nasosenergomash	UA
Nijhuis Pompen BV	NL
Packo Inox NV	BE
Paco Pumps	US
Peerless Pump Company	US
Powen Pump Factory Co Ltd	PL
Pump & Package Ltd	UK
Rhine Ruhr Pumps & Valves (Pty) Ltd	ZA
Ritz Pumpenfabrik GmbH & Co KG	DE
Rovatti Pompe	IT
Ruhrpumpen GmbH	DE
Ruhrpumpen Inc	US
Sacem SpA	IT
Shanghai Liansheng Pump Manufacture Co Ltd	CN
SPP Pumps Ltd	UK
Sulzer Pumps	CH
Taiko Kikai Industries Co Ltd	JP
Tangshan Pump Factory	CN
Tianjin Pumps & Machinery Group Co Ltd	CN
TKL	AU
Ture International Industrial Ltd	CN
United Pumps Australia (UPA)	AU
Voltas Ltd	IN
Weir Gabbioneta Srl	IT
Weir Pumps Ltd	UK

HORIZONTAL TWO-STAGE END-SUCTION OVERHUNG IMPELLER CENTRIFUGAL PUMPS

Ace Pump Corporation	US
Albany Engineering Company Ltd	UK
Allweiler AG	DE
Aurora Pump	US
Beresford Pumps	UK
Biral AG	CH
Buffalo Pumps Inc	US
Calpeda SpA	IT
Camaction Avonpump Ltd	UK
Caprari SpA	IT
Casals Cardona Industries SA	ES
CDR Pompe SpA	IT
Ceetak Engineering Ltd	UK
Chempump	US
Chemvac Pumps Ltd	UK
Chicago Pump Company	US
Cougar Industries Ltd	UK
Crane Pumps & Systems	US
Crest Pumps Ltd	UK
Pompe Cucchi Srl	IT
Dalian Danai Pumps Ltd	CN
Davey Products Pty Ltd	AU
DDA Srl	IT
Desmi A/S	DK
Ebara Corporation	JP
Ebara España Bombas SA	ES
EBS-Ray Pumps Pty Ltd	AU
EDUR Pumpenfabrik GmbH & Co KG	DE
Emile Egger et Cie SA	CH
Engineered Products Group Pty Ltd (EPG)	AU
Environamics Corporation	US
Fairbanks Morse Pump	US
Fapmo	FR
Finish Thompson Inc	US
Flexachem Manufacturing Ltd	IE
Flowmore Private Ltd	IN
Flowserve Corporation	US
Gilbert Gilkes & Gordon Ltd	UK
Goulds Pumps, ITT Industrial & BioPharm Group	US
Arthur Habermann GmbH & Co KG	DE
Handol Pumps Ltd	KR
Hermetic-Pumpen GmbH	DE
Honda Kiko Co Ltd	JP
Hunan XD Changsha Pump Works Co Ltd	CN
Hyosung-Ebara Co Ltd	KR
ITT Residential & Commercial Water	US

Iwaki America	US
Johnson Pump (India), ATE Enterprises Ltd	IN
Johnson Pump AB	SE
Johnson Pump BV	NL
Johnson Pumps of America Inc	US
Kelair Pumps Australia Pty Ltd	AU
Oy Kolmeks AB	FI
KSB Ltd	UK
London Pumps Ltd	UK
Magnatex Pumps Inc	US
Marshall Engineered Products Company	US
Matra SpA	IT
Mepco	US
MP Pumps Inc	US
MTH Pumps	US
FE Myers	US
Nasosenergomash	UA
Nijhuis Pompen BV	NL
Paco Pumps	US
Pedrollo SpA	IT
Peerless Pump Company	US
Powen Pump Factory Co Ltd	PL
Process Pumps (India) Pvt Ltd	IN
Pump & Package Ltd	UK
Pump Engineering Ltd	UK
Reddy-Buffaloes Pump Inc	US
Rhine Ruhr Pumps & Valves (Pty) Ltd	ZA
Ritz Pumpenfabrik GmbH & Co KG	DE
Roto Pumps Ltd	IN
Rovatti Pompe	IT
Ruhrpumpen GmbH	DE
Ruhrpumpen Inc	US
Rütschi Fluid AG	CH
Saer Elettropompe	IT
Shanghai Liansheng Pump Manufacture Co Ltd	CN
Slovpump-Trade Sro	SK
Michael Smith Engineers Ltd	UK
Speck-Pumpen	DE
SPP Pumps Ltd	UK
Sulzer Pumps	CH
Swidnicka Fabryka Pomp Sp Z.o.o.	PL
TKL	AU
Vertiflo Pump Company	US
Pumpenfabrik Ernst Vogel GmbH, ITT Industries	AT

VERTICAL SINGLE-STAGE OVERHUNG IMPELLER CENTRIFUGAL PUMPS (for general applications)

Allpumps Ltd	UK
Allweiler AG	DE
APV Fluid Handling Horsens	DK
Aurora Pump	US
Biral AG	CH
Camaction Avonpump Ltd	UK
Caprari SpA	IT
Charles Austen Pumps Ltd	UK
Corporacion EG SA	MX
Dalian Danai Pumps Ltd	CN
Dean Pump Division, Met-Pro Corporation	US
Desmi A/S	DK
Drakos-Polemis Inc	GR
Ebara Corporation	JP
Ebara España Bombas SA	ES
EDUR Pumpenfabrik GmbH & Co KG	DE
Emile Egger et Cie SA	CH
EMS Pacific	US
Engineered Products Group Pty Ltd (EPG)	AU
FE Myers	US
Flexachem Manufacturing Ltd	IE
Flowmore Private Ltd	IN
Flowserve Corporation	US
Goulds Pumps, ITT Industrial & BioPharm Group	US
Grundfos (Singapore) Pte Ltd	SG
Grundfos GmbH	DE
Grundfos Management A/S	DK
Grundfos Pompe Italia Srl	IT
Grundfos Pumpen Vertrieb GmbH	AT
Grundfos Pumps Corporation	US
Grundfos Pumps KK	JP
Grundfos Pumps Ltd	UK
Grundfos Pumps Pty Ltd	AU
Gzut SA Pumps Department	PL
Handol Pumps Ltd	KR
Hayward Tyler® Group	UK
Hermetic-Pumpen GmbH	DE
Honda Kiko Co Ltd	JP
Hyosung-Ebara Co Ltd	KR
Invensys APV	US
Invensys APV UK	UK
ITT Flygt AB	SE
ITT Residential & Commercial Water	US
Iwaki Co Ltd	JP
Iwaki Europe GmbH	DE
Kelair Pumps Australia Pty Ltd	AU
Kestner Engineering Co Ltd	UK
Kielecka Fabryka Pomp Bialogon SA	PL
KSB Ltd	UK
Lowara SpA, ITT Industries	IT
Magnatex Pumps Inc	US
Mark Grundfos Ltda	BR
Nijhuis Pompen BV	NL
Oy Kolmeks AB	FI
Packo Inox NV	BE
Paco Pumps	US
Peerless Pump Company	US
Pompe Garbarino SpA	IT
Powen Pump Factory Co Ltd	PL
Pump & Package Ltd	UK
Pumpenfabrik Ernst Vogel GmbH, ITT Industries	AT
Raj Pumps	IN
Rhine Ruhr Pumps & Valves (Pty) Ltd	ZA
Ritz Pumpenfabrik GmbH & Co KG	DE
Rovatti Pompe	IT
Ruhrpumpen GmbH	DE
Ruhrpumpen Inc	US
Savino Barbera	IT
Sethco Division, Met-Pro Corporation	US
Shanghai Liansheng Pump Manufacture Co Ltd	CN
Speck-Pumpen	DE
SPP Pumps Ltd	UK
Sterling Fluid Systems (Americas) Inc	US
Sterling Fluid Systems (Austria)	AT
Sterling Fluid Systems (Belgium) NV	BE
Sterling Fluid Systems (France) SAS	FR
Sterling Fluid Systems (Italy) SpA	IT
Sterling Fluid Systems (Netherlands)	NL
Sterling Fluid Systems (Schweiz) AG	CH
Sterling Fluid Systems (UK) Ltd	UK
Sterling SIHI GmbH	DE
Termomeccanica Pompe SpA	IT
TKL	AU
Toyo Pumps Europe	BE
Ture International Industrial Ltd	CN
Verder Group	NL
Weir Pumps Ltd	UK

VERTICAL SINGLE-STAGE OVERHUNG IMPELLER CENTRIFUGAL PUMPS (for chemical applications including ASME/ANSI B73.2 designs)

ACD Cryo GmbH	DE
ACD Inc	US
Allpumps Ltd	UK
APV Fluid Handling Horsens	DK
Argal Srl	IT
Best & Crompton Engineering Ltd	IN
Biral AG	CH
Cakasa (Nigeria) Co Ltd	NG
Camaction Avonpump Ltd	UK
Caster HMD Kontro	UK
Charles Austen Pumps Ltd	UK
Concentric Pumps Ltd	UK
Corporacion EG SA	MX
Cougar Industries Ltd	UK
Crest Pumps Ltd	UK
Dalian Danai Pumps Ltd	CN
Dean Pump Division, Met-Pro Corporation	US
Desmi A/S	DK
Dickow Pumpen KG	DE
Drakos-Polemis Inc	GR
Ebara Corporation	JP
EDUR Pumpenfabrik GmbH & Co KG	DE
Emile Egger et Cie SA	CH
EMS Pacific	US
Engineered Products Group Pty Ltd (EPG)	AU
Ensival-Moret	BE
Environamics Corporation	US
FE Myers	US
Flexachem Manufacturing Ltd	IE
Flowmore Private Ltd	IN
Flowserve Corporation	US
Goulds Pumps, ITT Industrial & BioPharm Group	US
Gruppo Aturia SpA	IT
Gzut SA Pumps Department	PL
Handol Pumps Ltd	KR
Hayward Tyler® Group	UK
Hermetic-Pumpen GmbH	DE
Hilge Pumps Ltd	UK
Honda Kiko Co Ltd	JP
Hyosung-Ebara Co Ltd	KR
Invensys APV	US
Invensys APV UK	UK
Iwaki Co Ltd	JP
Iwaki Europe GmbH	DE
Jiangsu Haishi Pump Co Ltd	CN
Kelair Pumps Australia Pty Ltd	AU
Kestner Engineering Co Ltd	UK
Kishor Pumps Ltd	IN
Laxmi Hydraulics Pvt Ltd	IN
March May Ltd	UK
Marshall Engineered Products Company	US
MASO Process-Pumpen GmbH	DE
Mepco	US
Michael Smith Engineers Ltd	UK

Micropump Ltd	UK
MP Pumps Inc	US
Nijhuis Pompen BV	NL
Nova Magnetics Burgmann Ltd	CA
Paco Pumps	US
Peerless Pump Company	US
Pumpenfabrik Ernst Vogel GmbH, ITT Industries	IT
Powen Pump Factory Co Ltd	PL
PPI Pumps Pvt Ltd	IN
Pulsafeeder	US
Pump & Package Ltd	UK
Rhine Ruhr Pumps & Valves (Pty) Ltd	ZA
Ritz Pumpenfabrik GmbH & Co KG	DE
Rovatti Pompe	IT
Ruhrpumpen GmbH	DE
Ruhrpumpen Inc	US
Savino Barbera	IT
Sero PumpSystems GmbH	DE
Shanghai Liansheng Pump Manufacture Co Ltd	CN
Shanxi Jishan Qinglong Pump Industry Co Ltd	CN
Simaco Elettromeccanica Srl	IT
Siwatec AS	CZ
Sterling Fluid Systems (Americas) Inc	US
Sterling Fluid Systems (Austria)	AT
Sterling Fluid Systems (Belgium) NV	BE
Sterling Fluid Systems (France) SAS	FR
Sterling Fluid Systems (Italy) SpA	IT
Sterling Fluid Systems (Netherlands)	NL
Sterling Fluid Systems (Schweiz) AG	CH
Sterling Fluid Systems (UK) Ltd	UK
Sterling SIHI GmbH	DE
Sundyne Corporation	US
Sundyne International SA	FR
Tangshan Pump Factory	CN
Termomeccanica Pompe SpA	IT
TKL	AU
Türbosan AS	TR
Ture International Industrial Ltd	CN

VERTICAL MULTI-STAGE CENTRIFUGAL PUMPS (for general applications including segmental pumps, dH <300 m, deep well ejector applications, wash water)

Aker Kvaerner Eureka Pump Systems	NO
Allpumps Ltd	UK
APV Fluid Handling Horsens	DK
Best & Crompton Engineering Ltd	IN
Caprari SpA	IT
Chempump	US
DP-Pumps	NL
Düchting Pumpen Maschinerfabrik GmbH	DE
Ebara Corporation	JP
EDUR Pumpenfabrik GmbH & Co KG	DE
Emile Egger et Cie SA	CH
EMS Pacific	US
Engineered Products Group Pty Ltd (EPG)	AU
Flexachem Manufacturing Ltd	IE
Flowmore Private Ltd	IN
Flowserve Corporation	US
Goulds Pumps, ITT Industrial & BioPharm Group	US
Grundfos (Singapore) Pte Ltd	SG
Grundfos GmbH	DE
Grundfos Management A/S	DK
Grundfos Pompe Italia Srl	IT
Grundfos Pumpen Vertrieb GmbH	AT
Grundfos Pumps Corporation	US
Grundfos Pumps KK	JP
Grundfos Pumps Ltd	UK
Grundfos Pumps Pty Ltd	AU
Handol Pumps Ltd	KR
Honda Kiko Co Ltd	JP
Hyosung-Ebara Co Ltd	KR
Invensys APV	US
Invensys APV UK	UK
ITT Flygt AB	SE
ITT Residential & Commercial Water	US
Iwaki Co Ltd	JP
Iwaki Europe GmbH	DE
Kelair Pumps Australia Pty Ltd	AU
Oy Kolmeks AB	FI
KSB Ltd	UK
Landini Fabbrica Pompe Centrifughe Srl	IT
Lowara SpA, ITT Industries	IT
Mark Grundfos Ltda	BR
Mather & Platt Pumps Ltd	IN
FE Myers	US
Nijhuis Pompen BV	NL
Paco Pumps	US
Patterson Pump Company	US
Peerless Pump Company	US
Powen Pump Factory Co Ltd	PL
Pump & Package Ltd	UK
Raj Pumps	IN
Rhine Ruhr Pumps & Valves (Pty) Ltd	ZA
Ritz Pumpenfabrik GmbH & Co KG	DE
Rovatti Pompe	IT
Ruhrpumpen GmbH	DE
Ruhrpumpen Inc	US
Saer Elettropompe	IT
Shanghai Liansheng Pump Manufacture Co Ltd	CN
Speck-Pumpen	DE
Sterling Fluid Systems (Americas) Inc	US
Sterling Fluid Systems (Austria)	AT
Sterling Fluid Systems (Belgium) NV	BE
Sterling Fluid Systems (France) SAS	FR
Sterling Fluid Systems (Italy) SpA	IT
Sterling Fluid Systems (Netherlands)	NL
Sterling Fluid Systems (Schweiz) AG	CH
Sterling Fluid Systems (UK) Ltd	UK
Sterling SIHI GmbH	DE
TKL	AU
Torishima Pump Mfg Co Ltd	JP
Türbosan AS	TR
Ture International Industrial Ltd	CN
United Pumps Australia (UPA)	AU
Verder Group	NL
Pumpenfabrik Ernst Vogel GmbH, ITT Industries	AT

VERTICAL WET-PIT PUMPS

ABS Deutschland GmbH	DE
ABS Finland Oy	FI
ABS France SAS	FR
ABS Group	SE
ABS Hellas SA	GR
ABS Hidrobomba SA	PL
ABS Hungary Trading	HU
ABS Industria de Bombas Centrifugas Ltda	BR
ABS International Srl	IT
ABS Italy Srl	IT
ABS Nopon Thai Co Ltd	TH
ABS Oumoer A/S	NO
ABS PolskaSp. Z.o.o.	PL
ABS Pompalari Ltd	TR
ABS Pompen BV	NL
ABS Pumpen GmbH	AT
ABS Pumps (Irl) Ltd	IE
ABS Pumps AS	EE
ABS Pumps Corporation	CA
ABS Pumps Inc	US
ABS Pumps Malaysia Sdn Bhd	MY
ABS Pumps NV/SA	BE
ABS Shanghai Co Ltd	CN
ABS Shanghai Co Ltd Beijing Liaison Office	CN
ABS Technologias del Agua SA	ES
ABS Wastewater Technology (Pty) Ltd	ZA
ABS Wastewater Technology Ltd	UK
ABS Wastewater Technology Pte Ltd	SG
Alfons de Backer & Co	BE
Allweiler AG	DE
Batescrew Pumps	AU
Biral AG	CH
Ceetak Engineering Ltd	UK
Chung Woo Industrial Co Ltd	KR
DAB Pumps SpA	IT
Davey Products Pty Ltd	AU
Dempster Industries Inc	US
Eagle Pump & Compressor Ltd	CA
Ebara Corporation	JP
EDUR Pumpenfabrik GmbH & Co KG	DE
Emile Egger et Cie SA	CH
Engineered Products Group Pty Ltd (EPG)	AU
Ensival-Moret	BE
Finder Pompe SpA	IT
Flexachem Manufacturing Ltd	IE
Flowmore Private Ltd	IN
Flowserve Corporation	US
Goulds Pumps, ITT Industrial & BioPharm Group	US
Handol Pumps Ltd	KR
Hayward Tyler® Group	UK
Hidrostal Ltd	UK
Honda Kiko Co Ltd	JP
Hyosung-Ebara Co Ltd	KR
ITT Flygt AB	SE
Johnston Pump	US
Kelair Pumps Australia Pty Ltd	AU
KSB Ltd	UK
Landini Fabbrica Pompe Centrifughe Srl	IT
Lederle GmbH	DE
Megator Ltd	UK
Melotte Pumptechnology BV	NL
Metso Minerals (Sala) AB	SE
Moyno Inc	US
FE Myers	US
National Pump Company LLC	US
Nijhuis Pompen BV	NL
Paco Pumps	US
Peerless Pump Company	US
PEME-Gourdin SA	FR
Pumpenfabrik Ernst Vogel GmbH, ITT Industries	IT
Powen Pump Factory Co Ltd	PL
Pump & Package Ltd	UK
Pumpenfabrik Wangen GmbH	DE

Rhine Ruhr Pumps & Valves (Pty) Ltd	ZA
Ritz Pumpenfabrik GmbH & Co KG	DE
Rovatti Pompe	IT
Ruhrpumpen GmbH	DE
Ruhrpumpen Inc	US
Savino Barbera	IT
Shanxi Jishan Qinglong Pump Industry Co Ltd	CN
Shijiazhuang Pump Industry Group Co Ltd	CN
Sigma Group AS	CZ
Sulzer Pumps	CH
TKL	AU
Türbosan AS	TR
Tyco - Southern Cross Plant	AU
Vaughan Company Inc	US
Pumpenfabrik Ernst Vogel GmbH, ITT Industries	AT
Voltas Ltd	IN
Warren Pumps LLC	US
Weir Pumps Ltd	UK
Wernert-Pumpen GmbH	DE
Wilo AG	DE
Wilo EMU GmbH	DE
Worthington SpA	IT
WPIL Ltd	IN
Yeomans Pump Company	US

VERTICAL DRY-PIT PUMPS

ABS Deutschland GmbH	DE
ABS Finland Oy	FI
ABS France SAS	FR
ABS Group	SE
ABS Hellas SA	GR
ABS Hidrobomba SA	PL
ABS Hungary Trading	HU
ABS Industria de Bombas Centrifugas Ltda	BR
ABS International Srl	IT
ABS Italy Srl	IT
ABS Nopon Thai Co Ltd	TH
ABS Oumoer A/S	NO
ABS PolskaSp. Z.o.o.	PL
ABS Pompalari Ltd	TR
ABS Pompen BV	NL
ABS Pumpen GmbH	AT
ABS Pumps (Irl) Ltd	IE
ABS Pumps AS	EE
ABS Pumps Corporation	CA
ABS Pumps Inc	US
ABS Pumps Malaysia Sdn Bhd	MY
ABS Pumps NV/SA	BE
ABS Shanghai Co Ltd	CN
ABS Shanghai Co Ltd Beijing Liaison Office	CN
ABS Technologias del Agua SA	ES
ABS Wastewater Technology (Pty) Ltd	ZA
ABS Wastewater Technology Ltd	UK
ABS Wastewater Technology Pte Ltd	SG
Allweiler AG	DE
Buffalo Pumps Inc	US
Calpeda SpA	IT
Chempump	US
Davey Products Pty Ltd	AU
DDA Srl	IT
Defco	US
Eagle Pump & Compressor Ltd	CA
Ebara Corporation	JP
EDUR Pumpenfabrik GmbH & Co KG	DE
Emile Egger et Cie SA	CH
Engineered Products Group Pty Ltd (EPG)	AU
Ensival-Moret	BE
FE Myers	US
Finder Pompe SpA	IT
Flexachem Manufacturing Ltd	IE
Flowmore Private Ltd	IN
Flowserve Corporation	US
Gilbert Gilkes & Gordon Ltd	UK
Goulds Pumps, ITT Industrial & BioPharm Group	US
Grundfos (Singapore) Pte Ltd	SG
Grundfos GmbH	DE
Grundfos Management A/S	DK
Grundfos Pompe Italia Srl	IT
Grundfos Pumpen Vertrieb GmbH	AT
Grundfos Pumps Corporation	US
Grundfos Pumps KK	JP
Grundfos Pumps Ltd	UK
Grundfos Pumps Pty Ltd	AU
Handol Pumps Ltd	KR
Hidrostal Ltd	UK
Honda Kiko Co Ltd	JP
Hunan XD Changsha Pump Works Co Ltd	CN
Hyosung-Ebara Co Ltd	KR
ITT Flygt AB	SE
Kelair Pumps Australia Pty Ltd	AU
Landini Fabbrica Pompe Centrifughe Srl	IT
Mark Grundfos Ltda	BR
Mather & Platt Pumps Ltd	IN
Megator Ltd	UK
Nijhuis Pompen BV	NL
Paco Pumps	US
Peerless Pump Company	US
Pumpenfabrik Ernst Vogel GmbH, ITT Industries	IT
Powen Pump Factory Co Ltd	PL
Pump & Package Ltd	UK
Pump Engineering Co Pvt Ltd	IN
Pumpenfabrik Ernst Vogel GmbH, ITT Industries	AT
Rhine Ruhr Pumps & Valves (Pty) Ltd	ZA
Ritz Pumpenfabrik GmbH & Co KG	DE
Rovatti Pompe	IT
Ruhrpumpen GmbH	DE
Ruhrpumpen Inc	US
Sigma Group AS	CZ
Speck-Pumpen	DE
Taiko Kikai Industries Co Ltd	JP
Termomeccanica Pompe SpA	IT
TKL	AU
United Pumps Australia (UPA)	AU
Vaughan Company Inc	US
Warren Pumps LLC	US
Weir Pumps Ltd	UK
Wernert-Pumpen GmbH	DE
Wilo AG	DE
Wilo EMU GmbH	DE
Worthington SpA	IT
Yeomans Pump Company	US

MARINE PUMPS

Allweiler AG	DE
Behrens Pumpen	DE
Bombas Itur SA	ES
Caprari SpA	IT
CW-EMD	US
Ebara Corporation	JP
EDUR Pumpenfabrik GmbH & Co KG	DE
Emile Egger et Cie SA	CH
Flowserve Corporation	US
Gorman-Rupp Company	US
Goulds Pumps, ITT Industrial & BioPharm Group	US
Gzut SA Pumps Department	PL
Hamworthy KSE	UK
Hamworthy Svanehøj A/S	DK
Jabsco GmbH, ITT Industries	DE
Jabsco UK, ITT Industries	UK
Jabsco US, ITT Industries	US
Johnson Pump (India), ATE Enterprises Ltd	IN
Johnson Pump AB	SE
Johnson Pump BV	NL
Johnson Pumps of America Inc	US
Megator Ltd	UK
Frank Mohn AS	NO
FE Myers	US
Pentair Inc	US
Pentair Pumps SpA	IT
Rapid Allweiler Pump & Engineering	ZA
Sethco Division, Met-Pro Corporation	US
Shin Shin Machinery Co Ltd	KR
Sigma Group AS	CZ
Smith & Loveless Inc	US
Sterling Fluid Systems (Americas) Inc	US
Sterling Fluid Systems (Austria)	AT
Sterling Fluid Systems (Belgium) NV	BE
Sterling Fluid Systems (France) SAS	FR
Sterling Fluid Systems (Italy) SpA	IT
Sterling Fluid Systems (Netherlands)	NL
Sterling Fluid Systems (Schweiz) AG	CH
Sterling Fluid Systems (UK) Ltd	UK
Sterling SIHI GmbH	DE
TKL	AU
Tycon Alloy Industries (Hong Kong) Co Ltd	HK
Vaughan Company Inc	US
Weir Pumps Ltd	UK

SUBMERSIBLE PUMPS WITH ELECTRIC MOTOR

ABS Deutschland GmbH	DE
ABS Finland Oy	FI
ABS France SAS	FR
ABS Group	SE
ABS Hellas SA	GR
ABS Hidrobomba SA	PL
ABS Hungary Trading	HU
ABS Industria de Bombas Centrifugas Ltda	BR
ABS International Srl	IT
ABS Italy Srl	IT
ABS Nopon Thai Co Ltd	TH
ABS Oumoer A/S	NO
ABS PolskaSp. Z.o.o.	PL
ABS Pompalari Ltd	TR
ABS Pompen BV	NL
ABS Pumpen GmbH	AT
ABS Pumps (Irl) Ltd	IE
ABS Pumps AS	EE
ABS Pumps Corporation	CA
ABS Pumps Inc	US
ABS Pumps Malaysia Sdn Bhd	MY
ABS Pumps NV/SA	BE
ABS Shanghai Co Ltd	CN
ABS Shanghai Co Ltd Beijing Liaison Office	CN

ABS Technologias del Agua SA	ES
ABS Wastewater Technology (Pty) Ltd	ZA
ABS Wastewater Technology Ltd	UK
ABS Wastewater Technology Pte Ltd	SG
Acromet (Aust) Pty Ltd	AU
Alfons de Backer & Co	BE
Alma Pompe	IT
Argal Srl	IT
Gruppo Aturia SpA	IT
BBA Hranice sro	CZ
Best & Crompton Engineering Ltd	IN
Biral AG	CH
Bombas Omega SL	ES
Buffalo Pumps Inc	US
Calpeda SpA	IT
Caprari SpA	IT
Casals Cardona Industries SA	ES
CAT Pumps	US
CAT Pumps (UK) Ltd	UK
CDS-John Blue Company	US
Ceetak Engineering Ltd	UK
Corporacion EG SA	MX
Cougar Industries Ltd	UK
Crest Pumps Ltd	UK
DAB Pumps SpA	IT
Davey Products Pty Ltd	AU
Defco	US
Dragflow Srl	IT
Düchting Pumpen Maschinerfabrik GmbH	DE
Eagle Pump & Compressor Ltd	CA
Ebara Corporation	JP
Ebara España Bombas SA	ES
EDUR Pumpenfabrik GmbH & Co KG	DE
Emile Egger et Cie SA	CH
Franz Eisele und Söhne GmbH & Co KG	DE
Engineered Products Group Pty Ltd (EPG)	AU
Fairbanks Morse Pump	US
Feluwa Pumpen GmbH	DE
Flexachem Manufacturing Ltd	IE
Flowmore Private Ltd	IN
Flowserve Corporation	US
Godwin Pumps Ltd	UK
Godwin Pumps of America Inc	US
Goodwin International	UK
Gorman-Rupp Company	US
Goulds Pumps, ITT Industrial & BioPharm Group	US
Goulds Pumps, ITT Water Technology	US
Grindex AB	SE
Grover Manufacturing Corporation	US
Grundfos (Singapore) Pte Ltd	SG
Grundfos GmbH	DE
Grundfos Management A/S	DK
Grundfos Pompe Italia Srl	IT
Grundfos Pumpen Vertrieb GmbH	AT
Grundfos Pumps Corporation	US
Grundfos Pumps KK	JP
Grundfos Pumps Ltd	UK
Grundfos Pumps Pty Ltd	AU
H & E Tsurumi Pump Co Ltd	HK
Arthur Habermann GmbH & Co KG	DE
Handol Pumps Ltd	KR
Hangzhou SRT Pump Co Ltd	CN
Hanning & Kahl GmbH & Co KG	DE
Häny AG	CH
Hayward Gordon	CA
HCP Pump Manufacturer Co Ltd	TW
Hermetic-Pumpen GmbH	DE
Hidrostal Ltd	UK
Honda Kiko Co Ltd	JP
Hunan XD Changsha Pump Works Co Ltd	CN
Hung Pump Industrial Co Ltd	TW
Hydrainer Pumps Ltd	UK
Hydromatic Pumps	US
Hyosung-Ebara Co Ltd	KR
ITT Flygt AB	SE
ITT Residential & Commercial Water	US
Jabsco GmbH, ITT Industries	DE
Jabsco UK, ITT Industries	UK
Jabsco US, ITT Industries	US
Johnston Pump	US
Jung Pumpen GmbH & Co	DE
Kawamoto Pump Manufacturing Co Ltd	JP
Kelair Pumps Australia Pty Ltd	AU
Kielecka Fabryka Pomp Bialogon SA	PL
Kishor Pumps Ltd	IN
KSB Ltd	UK
Kubota Corporation	JP
Lewis Pumps	US
Little Giant Pump Company	US
Lobee Pump & Machinery Company	US
London Pumps Ltd	UK
Lowara SpA, ITT Industries	IT
Mark Grundfos Ltda	BR
Marlow Pumps, ITT Water Technology	US
Megator Ltd	UK
Melotte Pumptechnology BV	NL
Metalchem-Warszawa SA	PL
Frank Mohn AS	NO
Morris Pumps	US
MTH Pumps	US
FE Myers	US
Nijhuis Pompen BV	NL
Nova Magnetics Burgmann Ltd	CA
Nuovo Pignone	IT
Oddesse Pumpen-und Motorenfabrik GmbH	DE
OFT (Officine Di Trevi)	IT
Ondrejovicka Strojirna sro	CZ
Paco Pumps	US
Pedrollo SpA	IT
Peerless Pump Company	US
Pentair Inc	US
Pentair Pumps SpA	IT
Powen Pump Factory Co Ltd	PL
PT Ebara Indonesia	ID
Pump & Package Ltd	UK
Pump Engineering Co Pvt Ltd	IN
Pump Engineering Ltd	UK
Pumpex Sales	SE
Red Jacket, ITT Water Technology	US
Rhine Ruhr Pumps & Valves (Pty) Ltd	ZA
Ritz Pumpenfabrik GmbH & Co KG	DE
Robot Pumps BV	NL
Rovatti Pompe	IT
Ruhrpumpen GmbH	DE
Ruhrpumpen Inc	US
Rütschi Fluid AG	CH
Saer Elettropompe	IT
Sakuragawa Pump Manufacturing Co Ltd	JP
Salmson Italia	IT
SAWA Pumpentechnik AG	CH
seepex GmbH & Co KG	DE
seepex UK Ltd	UK
Selwood Group Ltd	UK
Sethco Division, Met-Pro Corporation	US
Shanghai Shenbao Industrial Pump Co Ltd	CN
Shijiazhuang Pump Industry Group Co Ltd	CN
Shin Shin Machinery Co Ltd	KR
Sigma Group AS	CZ
Simaco Elettromeccanica Srl	IT
Slovpump-Trade Sro	SK
T Smedegaard A/S	DK
Speck-Pumpen	DE
SPP Pumps Ltd	UK
SRS Crisafulli Inc	US
Oswald Suck GmbH	DE
Sulzer Pumps	CH
Sumoto Srl	IT
Sunmotor International Ltd	CA
Swidnicka Fabryka Pomp Sp Z.o.o.	PL
Tangshan Pump Factory	CN
Tesla Srl	IT
Thompson Pump & Manufacturing Co Inc	US
3S Systemtechnik AG	CH
TKL	AU
Toyo Pumps Europe	BE
Tsurumi (America) Inc	US
Tsurumi (Europe) GmbH	DE
Tsurumi (Shanghai) Co Ltd	CN
Tsurumi (Singapore) Pte Ltd	SG
Tsurumi Manufacturing Co Ltd	JP
Tsurumi Pump (Malaysia) SDN BHD	MY
Tsurumi Pump Taiwan Co Ltd	TW
Türbosan AS	TR
Ture International Industrial Ltd	CN
Stuart Turner Ltd	UK
Tyco - Southern Cross Plant	AU
Tycon Alloy Industries (Hong Kong) Co Ltd	HK
United Pumps Australia (UPA)	AU
Varisco Srl	IT
Verder Group	NL
Versa-Matic Pump Company	US
Vertiflo Pump Company	US
Pumpenfabrik Ernst Vogel GmbH, ITT Industries	AT
Warren Rupp Inc	US
Weir Pumps Ltd	UK
Wernert-Pumpen GmbH	DE
Wilo AG	DE
Wilo EMU GmbH	DE
Witte Pumps & Technology GmbH	DE
Yaltek AS	TR
Yeomans Pump Company	US
Zoeller Pump Company	US

NON-ELECTRIC SUBMERSIBLE PUMPS

Acromet (Aust) Pty Ltd	AU
Alma Pompe	IT
BBA Hranice sro	CZ
Biral AG	CH
Bombas Omega SL	ES
Calama Industries Pvt Ltd	IN
Caprari SpA	IT
CAT Pumps	US
CAT Pumps (UK) Ltd	UK
Ceetak Engineering Ltd	UK

Corporacion EG SA MX
DAB Pumps SpA IT
Davey Products Pty Ltd AU
Dickow Pumpen KG DE
Dragflow Srl IT
Ebara Corporation JP
Ebara España Bombas SA ES
EDUR Pumpenfabrik GmbH & Co KG DE
Engineered Products Group Pty Ltd (EPG) AU
Fairbanks Morse Pump US
Flowmore Private Ltd IN
Godwin Pumps Ltd UK
Godwin Pumps of America Inc US
Goodwin International UK
Gorman-Rupp Company US
Handol Pumps Ltd KR
Hangzhou SRT Pump Co Ltd CN
Hanning & Kahl GmbH & Co KG DE
Häny AG CH
Hayward Gordon CA
HCP Pump Manufacturer Co Ltd TW
Hidrostal Ltd UK
Honda Kiko Co Ltd JP
Hung Pump Industrial Co Ltd TW
Hydrainer Pumps Ltd UK
Jung Pumpen GmbH & Co DE
Kawamoto Pump Manufacturing Co Ltd JP
Kelair Pumps Australia Pty Ltd AU
Kielecka Fabryka Pomp Bialogon SA PL
Kishor Pumps Ltd IN
KSB Ltd UK
Kubota Corporation JP
Little Giant Pump Company US
Melotte Pumptechnology BV NL
Metalchem-Warszawa SA PL
Mody Pumps™ Inc US
Morris Pumps US
FE Myers US
OFT (Officine Di Trevi) IT
Ondrejovicka Strojirna sro CZ
Paco Pumps US
Peerless Pump Company US
Powen Pump Factory Co Ltd PL
Pump & Package Ltd UK
Pump Engineering Co Pvt Ltd IN
Pumpex Sales SE
Rhine Ruhr Pumps & Valves (Pty) Ltd ZA
Ritz Pumpenfabrik GmbH & Co KG DE
Robot Pumps BV NL
Rotex SA GR
Rovatti Pompe IT
Ruhrpumpen GmbH DE
Ruhrpumpen Inc US
Saer Elettropompe IT
Sakuragawa Pump Manufacturing Co Ltd JP
SAWA Pumpentechnik AG CH
Sethco Division, Met-Pro Corporation US
Shanghai Shenbao Industrial Pump Co Ltd CN
Shijiazhuang Pump Industry Group Co Ltd CN
Sigma Group AS CZ
SRS Crisafulli Inc US
Oswald Suck GmbH DE
Tangshan Pump Factory CN
Termomeccanica Pompe SpA IT
Thompson Pump & Manufacturing Co Inc US
TKL AU
Toyo Pumps Europe BE
Türbosan AS TR
Versa-Matic Pump Company US
Warren Rupp Inc US
Weir Minerals Division US
Wilo AG DE
Wilo EMU GmbH DE
Witte Pumps & Technology GmbH DE
Zoeller Pump Company US

SUBMERSIBLE PUMPS FOR DEEP WELL APPLICATIONS

Alfons de Backer & Co BE
Amrit Engineering Pvt Ltd IN
Batescrew Pumps AU
Bizzi & Tedeschi Srl IT
Bombas Itur SA ES
Bombas Omega SL ES
Buffalo Pumps Inc US
Cakasa (Nigeria) Co Ltd NG
Calpeda SpA IT
Caprari SpA IT
Chem Resist Group Ltd UK
Cougar Industries Ltd UK
Crane Pumps & Systems US
DAB Pumps SpA IT
Drakos-Polemis Inc GR
Ebara Corporation JP
EDUR Pumpenfabrik GmbH & Co KG DE
Engineered Products Group Pty Ltd (EPG) AU
Flowmore Private Ltd IN
Flowserve Corporation US
Gorman-Rupp Company US
Goulds Pumps, ITT Industrial & BioPharm Group US
Goulds Pumps, ITT Water Technology US
Grundfos (Singapore) Pte Ltd SG
Grundfos GmbH DE
Grundfos Management A/S DK
Grundfos Pompe Italia Srl IT
Grundfos Pumpen Vertrieb GmbH AT
Grundfos Pumps Corporation US
Grundfos Pumps KK JP
Grundfos Pumps Ltd UK
Grundfos Pumps Pty Ltd AU
Hamworthy Svanehøj A/S DK
Handol Pumps Ltd KR
Hangzhou SRT Pump Co Ltd CN
Hayward Tyler® Group UK
Hidrostal Ltd UK
Honda Kiko Co Ltd JP
Hung Pump Industrial Co Ltd TW
Hydrainer Pumps Ltd UK
Hydromatic Pumps US
ITT Flygt AB SE
ITT Residential & Commercial Water US
Johnston Pump US
Kelair Pumps Australia Pty Ltd AU
Kishor Pumps Ltd IN
KSB Ltd UK
Laxmi Hydraulics Pvt Ltd IN
Lowara SpA, ITT Industries IT
Mark Grundfos Ltda BR
Megator Ltd UK
Monarch Industries Ltd CA
Motorarg SA AR
MTH Pumps US
FE Myers US
National Pump Company LLC US
Oddesse Pumpen-und Motorenfabrik GmbH DE
OFT (Officine Di Trevi) IT
Paco Pumps US
Pedrollo SpA IT
Peerless Pump Company US
PEME-Gourdin SA FR
Pump & Package Ltd UK
Red Jacket, ITT Water Technology US
Rhine Ruhr Pumps & Valves (Pty) Ltd ZA
Ritz Pumpenfabrik GmbH & Co KG DE
Rotex SA GR
Rovatti Pompe IT
Ruhrpumpen GmbH DE
Ruhrpumpen Inc US
Rütschi Fluid AG CH
Saer Elettropompe IT
Sakuragawa Pump Manufacturing Co Ltd JP
Salmson Italia IT
SAWA Pumpentechnik AG CH
Sharp Trendys IN
SHURflo Pump Manufacturing Co US
Sigma Group AS CZ
SPP Pumps Ltd UK
Subteck Srl IT
Oswald Suck GmbH DE
Sulzer Pumps CH
Sumoto Srl IT
Sunmotor International Ltd CA
Taiko Kikai Industries Co Ltd JP
Tesla Srl IT
TKL AU
Tyco - Southern Cross Plant AU
Tycon Alloy Industries (Hong Kong) Co Ltd HK
Pumpenfabrik Ernst Vogel GmbH, ITT Industries AT
Wafapomp SA PL
Pumpenfabrik Wangen GmbH DE
Waterman Industries Ltd IN
Weir Pumps Ltd UK
Wilo AG DE
Wilo EMU GmbH DE
Worthington SpA IT
WPIL Ltd IN
Yaltek AS TR

PORTABLE SELF-PRIMING PUMPS

Albany Engineering Company Ltd UK
Apollo Gössnitz GmbH DE
Aurora Pump US
Buffalo Pumps Inc US
Calama Industries Pvt Ltd IN
Calpeda SpA IT
Caprari SpA IT
Cougar Industries Ltd UK
Defco US
Ebara Corporation JP
EBS-Ray Pumps Pty Ltd AU
EDUR Pumpenfabrik GmbH & Co KG DE
Franz Eisele und Söhne GmbH & Co KG DE
Engineered Products Group Pty Ltd (EPG) AU
Finder Pompe SpA IT
Finish Thompson Inc US

Flowmore Private Ltd	IN
Flowserve Corporation	US
Gilbert Gilkes & Gordon Ltd	UK
Gorman-Rupp Company	US
Goulds Pumps, ITT Water Technology	US
Handol Pumps Ltd	KR
Hayward Tyler® Group	UK
Hidrotecar SA	ES
Honda Kiko Co Ltd	JP
Hunan XD Changsha Pump Works Co Ltd	CN
Iwaki America	US
Johnson Pump (India), ATE Enterprises Ltd	IN
Johnson Pump AB	SE
Johnson Pump BV	NL
Johnson Pumps of America Inc	US
Kelair Pumps Australia Pty Ltd	AU
Kestner Engineering Co Ltd	UK
Oy Kolmeks AB	FI
Kubota Corporation	JP
Lowara SpA, ITT Industries	IT
Marlow Pumps, ITT Water Technology	US
Mefiag Division, Met-Pro Corporation	US
Moyno Inc	US
MP Pumps Inc	US
MTH Pumps	US
Paco Pumps	US
Peerless Pump Company	US
Pomac Industries Group	NL
Primax Pumps	AE
Process Pumps (India) Pvt Ltd	IN
Pump & Package Ltd	UK
Raj Pumps	IN
Red Jacket, ITT Water Technology	US
Rhine Ruhr Pumps & Valves (Pty) Ltd	ZA
Ritz Pumpenfabrik GmbH & Co KG	DE
Ruhrpumpen GmbH	DE
Ruhrpumpen Inc	US
SAWA Pumpentechnik AG	CH
Schmalenberger GmbH & Co	DE
Selwood Group Ltd	UK
Sero PumpSystems GmbH	DE
Sethco Division, Met-Pro Corporation	US
Sharp Trendys	IN
Sigma Group AS	CZ
Slovpump-Trade Sro	SK
Michael Smith Engineers Ltd	UK
SPP Pumps Ltd	UK
Standard Pump Inc	US
Sterling Fluid Systems (Americas) Inc	US
Sterling Fluid Systems (Austria)	AT
Sterling Fluid Systems (Belgium) NV	BE
Sterling Fluid Systems (France) SAS	FR
Sterling Fluid Systems (Italy) SpA	IT
Sterling Fluid Systems (Netherlands)	NL
Sterling Fluid Systems (Schweiz) AG	CH
Sterling Fluid Systems (UK) Ltd	UK
Sterling SIHI GmbH	DE
TKL	AU
Ture International Industrial Ltd	CN
Varisco Srl	IT
Verder Group	NL
Pompe Vergani	IT
Victor Pumpen GmbH	DE
Weir Specialty Pumps	US
Yaltek AS	TR

HORIZONTAL TWO-STAGE AXIALLY-SPLIT CENTRIFUGAL PUMPS (including ISO 13709, API 610 designs)

Dalian Danai Pumps Ltd	CN
Ebara Corporation	JP
EDUR Pumpenfabrik GmbH & Co KG	DE
Engineered Products Group Pty Ltd (EPG)	AU
Flowmore Private Ltd	IN
Flowserve Corporation	US
Goulds Pumps, ITT Industrial & BioPharm Group	US
Grundfos (Singapore) Pte Ltd	SG
Grundfos GmbH	DE
Grundfos Management A/S	DK
Grundfos Pompe Italia Srl	IT
Grundfos Pumpen Vertrieb GmbH	AT
Grundfos Pumps Corporation	US
Grundfos Pumps KK	JP
Grundfos Pumps Ltd	UK
Grundfos Pumps Pty Ltd	AU
Handol Pumps Ltd	KR
Hermetic-Pumpen GmbH	DE
Honda Kiko Co Ltd	JP
Kelair Pumps Australia Pty Ltd	AU
Mark Grundfos Ltda	BR
Paco Pumps	US
Peerless Pump Company	US
Pump & Package Ltd	UK
Rhine Ruhr Pumps & Valves (Pty) Ltd	ZA
Ruhrpumpen GmbH	DE
Ruhrpumpen Inc	US
Shanghai Liansheng Pump Manufacture Co Ltd	CN
Sigma Group AS	CZ
Sulzer Pumps	CH
TKL	AU
Ture International Industrial Ltd	CN
Weir Pumps Ltd	UK

HORIZONTAL TWO-STAGE RADIALLY-SPLIT CENTRIFUGAL PUMPS (including ISO 13709, API 610 designs)

Cougar Industries Ltd	UK
Dalian Danai Pumps Ltd	CN
Ebara Corporation	JP
EDUR Pumpenfabrik GmbH & Co KG	DE
Engineered Products Group Pty Ltd (EPG)	AU
Flowmore Private Ltd	IN
Flowserve Corporation	US
Goulds Pumps, ITT Industrial & BioPharm Group	US
Handol Pumps Ltd	KR
Honda Kiko Co Ltd	JP
Kelair Pumps Australia Pty Ltd	AU
KSB Ltd	UK
Paco Pumps	US
Peerless Pump Company	US
Pulsafeeder	US
Pump & Package Ltd	UK
Rhine Ruhr Pumps & Valves (Pty) Ltd	ZA
Ruhrpumpen GmbH	DE
Ruhrpumpen Inc	US
Shanghai Liansheng Pump Manufacture Co Ltd	CN
Sigma Group AS	CZ
Sterling Fluid Systems (Americas) Inc	US
Sterling Fluid Systems (Austria)	AT
Sterling Fluid Systems (Belgium) NV	BE
Sterling Fluid Systems (France) SAS	FR
Sterling Fluid Systems (Italy) SpA	IT
Sterling Fluid Systems (Netherlands)	NL
Sterling Fluid Systems (Schweiz) AG	CH
Sterling Fluid Systems (UK) Ltd	UK
Sterling SIHI GmbH	DE
Sulzer Pumps	CH
TKL	AU
Ture International Industrial Ltd	CN
Weir Pumps Ltd	UK

MULTI-STAGE SEGMENTAL CENTRIFUGAL PUMPS (dH > 300 m)

Cougar Industries Ltd	UK
Dickow Pumpen KG	DE
Ebara Corporation	JP
EDUR Pumpenfabrik GmbH & Co KG	DE
Engineered Products Group Pty Ltd (EPG)	AU
Ensival-Moret	BE
Flowserve Corporation	US
Goulds Pumps, ITT Industrial & BioPharm Group	US
Grundfos (Singapore) Pte Ltd	SG
Grundfos GmbH	DE
Grundfos Management A/S	DK
Grundfos Pompe Italia Srl	IT
Grundfos Pumpen Vertrieb GmbH	AT
Grundfos Pumps Corporation	US
Grundfos Pumps KK	JP
Grundfos Pumps Ltd	UK
Grundfos Pumps Pty Ltd	AU
Handol Pumps Ltd	KR
Hidráulicas HMT SA	ES
ITT Flygt AB	SE
Kelair Pumps Australia Pty Ltd	AU
Klaus Union GmbH & Co KG	DE
KSB Ltd	UK
Laxmi Hydraulics Pvt Ltd	IN
Layne/Verti-Line	US
Lowara SpA, ITT Industries	IT
Mark Grundfos Ltda	BR
Mather & Platt Pumps Ltd	IN
Paco Pumps	US
Peerless Pump Company	US
Pulsafeeder	US
Pump & Package Ltd	UK
Rhine Ruhr Pumps & Valves (Pty) Ltd	ZA
Ruhrpumpen GmbH	DE
Ruhrpumpen Inc	US
Saer Elettropompe	IT
Shanghai Liansheng Pump Manufacture Co Ltd	CN
Sigma Group AS	CZ
Sterling Fluid Systems (Americas) Inc	US
Sterling Fluid Systems (Austria)	AT
Sterling Fluid Systems (Belgium) NV	BE
Sterling Fluid Systems (France) SAS	FR
Sterling Fluid Systems (Italy) SpA	IT
Sterling Fluid Systems (Netherlands)	NL
Sterling Fluid Systems (Schweiz) AG	CH
Sterling Fluid Systems (UK) Ltd	UK
Sterling SIHI GmbH	DE
Sulzer Pumps	CH
TKL	AU
Torishima Pump Mfg Co Ltd	JP
United Pumps Australia (UPA)	AU

Pumpenfabrik Ernst Vogel GmbH, ITT Industries	AT
Wafapomp SA	PL

HORIZONTAL MULTI-STAGE AXIALLY-SPLIT CENTRIFUGAL PUMPS (including ISO 13709, API 610 designs)

A-Ryung Machinery Ind Co Ltd	KR
Cougar Industries Ltd	UK
Dalian Danai Pumps Ltd	CN
Defco	US
Düchting Pumpen Maschinerfabrik GmbH	DE
Ebara Corporation	JP
EDUR Pumpenfabrik GmbH & Co KG	DE
Engineered Products Group Pty Ltd (EPG)	AU
Ensival-Moret	BE
Flowmore Private Ltd	IN
Flowserve Corporation	US
Goulds Pumps, ITT Industrial & BioPharm Group	US
Handol Pumps Ltd	KR
Hermetic-Pumpen GmbH	DE
Hidráulicas HMT SA	ES
Hilge Pumps Ltd	UK
Honda Kiko Co Ltd	JP
Kelair Pumps Australia Pty Ltd	AU
Kubota Corporation	JP
Layne/Verti-Line	US
MTH Pumps	US
Nuovo Pignone	IT
Paco Pumps	US
Peerless Pump Company	US
Pulsafeeder	US
Pump & Package Ltd	UK
Rhine Ruhr Pumps & Valves (Pty) Ltd	ZA
Ruhrpumpen GmbH	DE
Ruhrpumpen Inc	US
Shanghai Liansheng Pump Manufacture Co Ltd	CN
Sigma Group AS	CZ
Sulzer Pumps	CH
TKL	AU
Ture International Industrial Ltd	CN
United Pumps Australia (UPA)	AU
Wafapomp SA	PL
Weir Pumps Ltd	UK

HORIZONTAL MULTI-STAGE RADIALLY-SPLIT CENTRIFUGAL PUMPS (including ISO 13709, API 610 designs)

ACD Inc	US
Aker Kvaerner Eureka Pump Systems	NO
Apollo Gössnitz GmbH	DE
A-Ryung Machinery Ind Co Ltd	KR
Dalian Danai Pumps Ltd	CN
Ebara Corporation	JP
EDUR Pumpenfabrik GmbH & Co KG	DE
Engineered Products Group Pty Ltd (EPG)	AU
Ensival-Moret	BE
Fairbanks Morse Pump	US
Flowmore Private Ltd	IN
Flowserve Corporation	US
Goulds Pumps, ITT Industrial & BioPharm Group	US
Handol Pumps Ltd	KR
Hilge Pumps Ltd	UK
Honda Kiko Co Ltd	JP
Johnston Pump	US
Kelair Pumps Australia Pty Ltd	AU
Klaus Union GmbH & Co KG	DE
KSB Ltd	UK
Kubota Corporation	JP
Lutz (UK) Ltd	UK
Mather & Platt Pumps Ltd	IN
Megator Ltd	UK
Nasosenergomash	UA
National Pump Company LLC	US
Nikkiso Co Ltd	JP
Nuovo Pignone	IT
Paco Pumps	US
Peerless Pump Company	US
Pump & Package Ltd	UK
Rhine Ruhr Pumps & Valves (Pty) Ltd	ZA
Ruhrpumpen GmbH	DE
Ruhrpumpen Inc	US
Shanghai Liansheng Pump Manufacture Co Ltd	CN
Sigma Group AS	CZ
Sterling Fluid Systems (Americas) Inc	US
Sterling Fluid Systems (Austria)	AT
Sterling Fluid Systems (Belgium) NV	BE
Sterling Fluid Systems (France) SAS	FR
Sterling Fluid Systems (Italy) SpA	IT
Sterling Fluid Systems (Netherlands)	NL
Sterling Fluid Systems (Schweiz) AG	CH
Sterling Fluid Systems (UK) Ltd	UK
Sterling SIHI GmbH	DE
Sulzer Pumps	CH
TKL	AU
Torishima Pump Mfg Co Ltd	JP
Ture International Industrial Ltd	CN
United Pumps Australia (UPA)	AU
Weir Gabbioneta Srl	IT
Weir Pumps Ltd	UK

LARGE VERTICAL MULTI-STAGE CENTRIFUGAL PUMPS (including ISO 13709, API 610 designs)

Dalian Danai Pumps Ltd	CN
Defco	US
Düchting Pumpen Maschinerfabrik GmbH	DE
Ebara Corporation	JP
EDUR Pumpenfabrik GmbH & Co KG	DE
Flowmore Private Ltd	IN
Hidráulicas HMT SA	ES
Iwaki Co Ltd	JP
Iwaki Europe GmbH	DE
Klaus Union GmbH & Co KG	DE
Layne/Verti-Line	US
MTH Pumps	US
Nasosenergomash	UA
Nuovo Pignone	IT
Pulsafeeder	US
Rhine Ruhr Pumps & Valves (Pty) Ltd	ZA
United Pumps Australia (UPA)	AU
Wafapomp SA	PL

SINGLE-STAGE CENTRIFUGAL PUMPS WITH INTEGRAL GEARING

Cakasa (Nigeria) Co Ltd	NG
Caprari SpA	IT
CDR Pompe SpA	IT
CDS-John Blue Company	US
Concentric Pumps Ltd	UK
Cougar Industries Ltd	UK
DAB Pumps SpA	IT
Debem	IT
Defco	US
Ebara Corporation	JP
EDUR Pumpenfabrik GmbH & Co KG	DE
Engineered Products Group Pty Ltd (EPG)	AU
Ensival-Moret	BE
Environamics Corporation	US
Finder Pompe SpA	IT
Flojet Corporation	US
Flowmore Private Ltd	IN
Fristam Pumps (UK) Ltd	UK
Fristam Pumps USA	US
Fristram Pumpen	DE
GIW Industries Inc	US
Hale Products Europe Ltd	UK
Handol Pumps Ltd	KR
Hidráulicas HMT SA	ES
Honda Kiko Co Ltd	JP
Ilsung Precision Industry Company	KR
Inoxpa SA	ES
Iwaki America	US
Jiangsu Haishi Pump Co Ltd	CN
Kelair Pumps Australia Pty Ltd	AU
Klaus Union GmbH & Co KG	DE
London Pumps Ltd	UK
Lutz (UK) Ltd	UK
Micropump Ltd	UK
Nasosenergomash	UA
Nikkiso Co Ltd	JP
Nuovo Pignone	IT
Pacer Pumps	US
Paco Pumps	US
Peerless Pump Company	US
Pump & Package Ltd	UK
Rhine Ruhr Pumps & Valves (Pty) Ltd	ZA
Ruhrpumpen GmbH	DE
Ruhrpumpen Inc	US
Sacem SpA	IT
Sigma Group AS	CZ
Siwatec AS	CZ
Slovpump-Trade Sro	SK
Tianjin Pumps & Machinery Group Co Ltd	CN
TKL	AU
TriVista Engineering Ltd	UK
Ture International Industrial Ltd	CN
Tycon Alloy Industries (Hong Kong) Co Ltd	HK
United Pumps Australia (UPA)	AU

MULTI-STAGE CENTRIFUGAL PUMPS WITH INTEGRAL GEARING

Apollo Gössnitz GmbH	DE
Best & Crompton Engineering Ltd	IN
Calama Industries Pvt Ltd	IN
Caprari SpA	IT
DAB Pumps SpA	IT
Düchting Pumpen Maschinerfabrik GmbH	DE
Ebara Corporation	JP
EDUR Pumpenfabrik GmbH & Co KG	DE
Engineered Products Group Pty Ltd (EPG)	AU
Flowmore Private Ltd	IN
Fristam Pumps (UK) Ltd	UK
Fristam Pumps USA	US
Fristram Pumpen	DE
Handol Pumps Ltd	KR
Hidráulicas HMT SA	ES
Honda Kiko Co Ltd	JP

Kelair Pumps Australia Pty Ltd	AU
KNF Neuberger GmbH	DE
Layne/Verti-Line	US
Leszno Pump Factory	PL
Oddesse Pumpen-und Motorenfabrik GmbH	DE
Paco Pumps	US
Peerless Pump Company	US
Pump & Package Ltd	UK
Rhine Ruhr Pumps & Valves (Pty) Ltd	ZA
Ritz Pumpenfabrik GmbH & Co KG	DE
Ruhrpumpen GmbH	DE
Ruhrpumpen Inc	US
Sigma Group AS	CZ
T Smedegaard A/S	DK
TKL	AU
TriVista Engineering Ltd	UK
Ture International Industrial Ltd	CN
Tycon Alloy Industries (Hong Kong) Co Ltd	HK
United Pumps Australia (UPA)	AU
Wilo AG	DE
Wilo EMU GmbH	DE

CENTRIFUGAL PUMPS FOR PULP

Allweiler AG	DE
Apollo Gössnitz GmbH	DE
Asco Pompe srl	IT
Camaction Avonpump Ltd	UK
Cornell	US
Corporacion EG SA	MX
Desmi GmbH	DE
Ebara Corporation	JP
EBS-Ray Pumps Pty Ltd	AU
EDUR Pumpenfabrik GmbH & Co KG	DE
Engineered Products Group Pty Ltd (EPG)	AU
Finder Pompe SpA	IT
Finish Thompson Inc	US
Flowmore Private Ltd	IN
Flowserve Corporation	US
Gorman-Rupp Company	US
Goulds Pumps, ITT Industrial & BioPharm Group	US
GPM Pumps Inc	US
Arthur Habermann GmbH & Co KG	DE
Handol Pumps Ltd	KR
Hayward Tyler® Group	UK
Hidrostal Ltd	UK
Hidrotecar SA	ES
Honda Kiko Co Ltd	JP
Iwaki America	US
Kelair Pumps Australia Pty Ltd	AU
Metso Minerals (Sala) AB	SE
Nasosenergomash	UA
Omel Ltda	BR
Paco Pumps	US
Peerless Pump Company	US
Process Pumps (India) Pvt Ltd	IN
Pump & Package Ltd	UK
Rhine Ruhr Pumps & Valves (Pty) Ltd	ZA
Ruhrpumpen GmbH	DE
Ruhrpumpen Inc	US
Scheerle AG	CH
Sigma Group AS	CZ
Sterling Fluid Systems (Americas) Inc	US
Sterling Fluid Systems (Austria)	AT
Sterling Fluid Systems (Belgium) NV	BE
Sterling Fluid Systems (France) SAS	FR
Sterling Fluid Systems (Italy) SpA	IT
Sterling Fluid Systems (Netherlands)	NL
Sterling Fluid Systems (Schweiz) AG	CH
Sterling Fluid Systems (UK) Ltd	UK
Sterling SIHI GmbH	DE
Sulzer Pumps	CH
Swidnicka Fabryka Pomp Sp Z.o.o.	PL
TKL	AU
Toyo Pumps Europe	BE
United Pumps Australia (UPA)	AU
Wafapomp SA	PL

CENTRIFUGAL PUMPS FOR HANDLING SOLIDS < 10 MM

Abel GmbH & Co KG	DE
Acromet (Aust) Pty Ltd	AU
Allweiler AG	DE
Argal Srl	IT
Asco Pompe srl	IT
Best & Crompton Engineering Ltd	IN
Caprari SpA	IT
Chem Resist Group Ltd	UK
Chicago Pump Company	US
Cornell	US
Dalian Danai Pumps Ltd	CN
Discflo Corporation	US
Düchting Pumpen Maschinerfabrik GmbH	DE
Ebara Corporation	JP
EDUR Pumpenfabrik GmbH & Co KG	DE
Franz Eisele und Söhne GmbH & Co KG	DE
Engineered Products Group Pty Ltd (EPG)	AU
ESSCO Pumps & Controls	US
Fapmo	FR
Feluwa Pumpen GmbH	DE
Flexachem Manufacturing Ltd	IE
Flowmore Private Ltd	IN
Flowserve Corporation	US
Flux Pumps Corporation	US
GIW Industries Inc	US
GMP SpA	IT
Godwin Pumps Ltd	UK
Goodwin International	UK
Gorman-Rupp Company	US
Goulds Pumps, ITT Industrial & BioPharm Group	US
Goulds Pumps, ITT Water Technology	US
GPM Pumps Inc	US
Grundfos (Singapore) Pte Ltd	SG
Grundfos GmbH	DE
Grundfos Management A/S	DK
Grundfos Pompe Italia Srl	IT
Grundfos Pumpen Vertrieb GmbH	AT
Grundfos Pumps Corporation	US
Grundfos Pumps KK	JP
Grundfos Pumps Ltd	UK
Grundfos Pumps Pty Ltd	AU
Handol Pumps Ltd	KR
Hayward Gordon	CA
Hermetic-Pumpen GmbH	DE
Hidrostal Ltd	UK
Hilge Pumps Ltd	UK
Homa Pumpenfabrik GmbH	DE
Honda Kiko Co Ltd	JP
ITT Flygt AB	SE
Kelair Pumps Australia Pty Ltd	AU
Kestner Engineering Co Ltd	UK
Kinder-Janes Engineers Ltd	UK
KNF Neuberger GmbH	DE
KSB Ltd	UK
Lewis Pumps	US
Linatex Ltd	UK
Little Giant Pump Company	US
Lowara SpA, ITT Industries	IT
Mark Grundfos Ltda	BR
Marlow Pumps, ITT Water Technology	US
Metso Minerals (Sala) AB	SE
Monarch Industries Ltd	CA
Nova Magnetics Burgmann Ltd	CA
Omel Ltda	BR
Paco Pumps	US
Peerless Pump Company	US
Plenty Mirrlees Pumps	UK
Pump & Package Ltd	UK
Pump Engineering Ltd	UK
Rhine Ruhr Pumps & Valves (Pty) Ltd	ZA
Robot Pumps BV	NL
Ruhrpumpen GmbH	DE
Ruhrpumpen Inc	US
Savino Barbera	IT
Scheerle AG	CH
Schmalenberger GmbH & Co	DE
Selwood Group Ltd	UK
Shijiazhuang Pump Industry Group Co Ltd	CN
Sigma Group AS	CZ
SPP Pumps Ltd	UK
Sterling Fluid Systems (Americas) Inc	US
Sterling Fluid Systems (Austria)	AT
Sterling Fluid Systems (Belgium) NV	BE
Sterling Fluid Systems (France) SAS	FR
Sterling Fluid Systems (Italy) SpA	IT
Sterling Fluid Systems (Netherlands)	NL
Sterling Fluid Systems (Schweiz) AG	CH
Sterling Fluid Systems (UK) Ltd	UK
Sterling SIHI GmbH	DE
Sulzer Pumps	CH
Tangshan Pump Factory	CN
Terex Powertrain Ltd	UK
Titanium Tantalum Products Ltd	IN
TKL	AU
Toyo Pumps Europe	BE
Tuthill Corporation	US
Varisco Srl	IT
Verder Group	NL
Vertiflo Pump Company	US
Victor Pumpen GmbH	DE
Willy Vogel AG	DE
Pumpenfabrik Ernst Vogel GmbH, ITT Industries	AT
Pumpenfabrik Wangen GmbH	DE
CH Warman Pump Group	ZA
Weir Minerals Division	US
Weir Pumps Ltd	UK
A R Wilfley & Sons Inc	US
Yeomans Pump Company	US

CENTRIFUGAL PUMPS FOR HANDLING SOLIDS > 10 MM

Abel GmbH & Co KG	DE
Acromet (Aust) Pty Ltd	AU
Argal Srl	IT
Asco Pompe srl	IT
Best & Crompton Engineering Ltd	IN

Caprari SpA	IT
Casals Cardona Industries SA	ES
Chem Resist Group Ltd	UK
Chicago Pump Company	US
Discflo Corporation	US
Dragflow Srl	IT
Düchting Pumpen Maschinerfabrik GmbH	DE
Ebara Corporation	JP
EBS-Ray Pumps Pty Ltd	AU
EDUR Pumpenfabrik GmbH & Co KG	DE
Franz Eisele und Söhne GmbH & Co KG	DE
Engineered Products Group Pty Ltd (EPG)	AU
ESSCO Pumps & Controls	US
Fapmo	FR
Feluwa Pumpen GmbH	DE
Flexachem Manufacturing Ltd	IE
Flowmore Private Ltd	IN
Flowserve Corporation	US
Flux Pumps Corporation	US
GIW Industries Inc	US
GMP SpA	IT
Godwin Pumps Ltd	UK
Goodwin International	UK
Gorman-Rupp Company	US
Goulds Pumps, ITT Industrial & BioPharm Group	US
Goulds Pumps, ITT Water Technology	US
GPM Pumps Inc	US
Grundfos (Singapore) Pte Ltd	SG
Grundfos GmbH	DE
Grundfos Management A/S	DK
Grundfos Pompe Italia Srl	IT
Grundfos Pumpen Vertrieb GmbH	AT
Grundfos Pumps Corporation	US
Grundfos Pumps KK	JP
Grundfos Pumps Ltd	UK
Grundfos Pumps Pty Ltd	AU
Handol Pumps Ltd	KR
Hayward Gordon	CA
Hidrostal Ltd	UK
Homa Pumpenfabrik GmbH	DE
Honda Kiko Co Ltd	JP
ITT Flygt AB	SE
Kelair Pumps Australia Pty Ltd	AU
Kestner Engineering Co Ltd	UK
Kinder-Janes Engineers Ltd	UK
KNF Neuberger GmbH	DE
KSB Ltd	UK
Lewis Pumps	US
Little Giant Pump Company	US
Lowara SpA, ITT Industries	IT
Mark Grundfos Ltda	BR
Marlow Pumps, ITT Water Technology	US
Metso Minerals (Sala) AB	SE
Omel Ltda	BR
Ondrejovicka Strojirna sro	CZ
Paco Pumps	US
Peerless Pump Company	US
Pump & Package Ltd	UK
Pump Engineering Ltd	UK
Rhine Ruhr Pumps & Valves (Pty) Ltd	ZA
Robot Pumps BV	NL
Ruhrpumpen GmbH	DE
Ruhrpumpen Inc	US
Scheerle AG	CH
Schmalenberger GmbH & Co	DE
Selwood Group Ltd	UK
Shijiazhuang Pump Industry Group Co Ltd	CN
Sigma Group AS	CZ
SPP Pumps Ltd	UK
Sterling Fluid Systems (Americas) Inc	US
Sterling Fluid Systems (Austria)	AT
Sterling Fluid Systems (Belgium) NV	BE
Sterling Fluid Systems (France) SAS	FR
Sterling Fluid Systems (Italy) SpA	IT
Sterling Fluid Systems (Netherlands)	NL
Sterling Fluid Systems (Schweiz) AG	CH
Sterling Fluid Systems (UK) Ltd	UK
Sterling SIHI GmbH	DE
Tangshan Pump Factory	CN
Terex Powertrain Ltd	UK
Titanium Tantalum Products Ltd	IN
TKL	AU
Toyo Pumps Europe	BE
Tuthill Corporation	US
Varisco Srl	IT
Vaughan Company Inc	US
Vertiflo Pump Company	US
Victor Pumpen GmbH	DE
Willy Vogel AG	DE
Pumpenfabrik Ernst Vogel GmbH, ITT Industries	AT
Pumpenfabrik Wangen GmbH	DE
CH Warman Pump Group	ZA
Weir Minerals Division	US
A R Wilfley & Sons Inc	US
Yeomans Pump Company	US

NON-CLOGGING PUMPS

Abel GmbH & Co KG	DE
ABS Deutschland GmbH	DE
ABS Finland Oy	FI
ABS France SAS	FR
ABS Group	SE
ABS Hellas SA	GR
ABS Hidrobomba SA	PL
ABS Hungary Trading	HU
ABS Industria de Bombas Centrifugas Ltda	BR
ABS International Srl	IT
ABS Italy Srl	IT
ABS Nopon Thai Co Ltd	TH
ABS Oumoer A/S	NO
ABS PolskaSp. Z.o.o.	PL
ABS Pompalari Ltd	TR
ABS Pompen BV	NL
ABS Pumpen GmbH	AT
ABS Pumps (Irl) Ltd	IE
ABS Pumps AS	EE
ABS Pumps Corporation	CA
ABS Pumps Inc	US
ABS Pumps Malaysia Sdn Bhd	MY
ABS Pumps NV/SA	BE
ABS Shanghai Co Ltd	CN
ABS Shanghai Co Ltd Beijing Liaison Office	CN
ABS Technologias del Agua SA	ES
ABS Wastewater Technology (Pty) Ltd	ZA
ABS Wastewater Technology Ltd	UK
ABS Wastewater Technology Pte Ltd	SG
Bombas Itur SA	ES
Caprari SpA	IT
Desmi GmbH	DE
Ebara Corporation	JP
EDUR Pumpenfabrik GmbH & Co KG	DE
Engineered Products Group Pty Ltd (EPG)	AU
Feluwa Pumpen GmbH	DE
Flowserve Corporation	US
Gorman-Rupp Company	US
Goulds Pumps, ITT Industrial & BioPharm Group	US
Handol Pumps Ltd	KR
Hayward Gordon	CA
Hidrostal Ltd	UK
Honda Kiko Co Ltd	JP
ITT Flygt AB	SE
Kelair Pumps Australia Pty Ltd	AU
KSB Ltd	UK
Little Giant Pump Company	US
Marlow Pumps, ITT Water Technology	US
Metso Minerals (Sala) AB	SE
Mody Pumps™ Inc	US
Moyno Inc	US
Paco Pumps	US
Peerless Pump Company	US
Pump & Package Ltd	UK
Rhine Ruhr Pumps & Valves (Pty) Ltd	ZA
Robot Pumps BV	NL
Rotex SA	GR
Ruhrpumpen GmbH	DE
Ruhrpumpen Inc	US
Scheerle AG	CH
Selwood Group Ltd	UK
Shijiazhuang Pump Industry Group Co Ltd	CN
Sigma Group AS	CZ
Sterling Fluid Systems (Americas) Inc	US
Sterling Fluid Systems (Austria)	AT
Sterling Fluid Systems (Belgium) NV	BE
Sterling Fluid Systems (France) SAS	FR
Sterling Fluid Systems (Italy) SpA	IT
Sterling Fluid Systems (Netherlands)	NL
Sterling Fluid Systems (Schweiz) AG	CH
Sterling Fluid Systems (UK) Ltd	UK
Sterling SIHI GmbH	DE
Sulzer Pumps	CH
Terex Powertrain Ltd	UK
TKL	AU
Varisco Srl	IT
Vaughan Company Inc	US
Verder Group	NL
Pumpenfabrik Ernst Vogel GmbH, ITT Industries	AT
CH Warman Pump Group	ZA
Warren Rupp Inc	US
Yeomans Pump Company	US

SUBMERSIBLE NON-CLOGGING PUMPS

Abel GmbH & Co KG	DE
ABS Deutschland GmbH	DE
ABS Finland Oy	FI
ABS France SAS	FR
ABS Group	SE
ABS Hellas SA	GR
ABS Hidrobomba SA	PL
ABS Hungary Trading	HU
ABS Industria de Bombas Centrifugas Ltda	BR
ABS International Srl	IT
ABS Italy Srl	IT
ABS Nopon Thai Co Ltd	TH

ABS Oumoer A/S	NO
ABS PolskaSp. Z.o.o.	PL
ABS Pompalari Ltd	TR
ABS Pompen BV	NL
ABS Pumpen GmbH	AT
ABS Pumps (Irl) Ltd	IE
ABS Pumps AS	EE
ABS Pumps Corporation	CA
ABS Pumps Inc	US
ABS Pumps Malaysia Sdn Bhd	MY
ABS Pumps NV/SA	BE
ABS Shanghai Co Ltd	CN
ABS Shanghai Co Ltd Beijing Liaison Office	CN
ABS Technologias del Agua SA	ES
ABS Wastewater Technology (Pty) Ltd	ZA
ABS Wastewater Technology Ltd	UK
ABS Wastewater Technology Pte Ltd	SG
Alma Pompe	IT
Apollo Gössnitz GmbH	DE
A-Ryung Machinery Ind Co Ltd	KR
Bombas Omega SL	ES
Caprari SpA	IT
Carver Pump Company	US
Casals Cardona Industries SA	ES
Catag AG	CH
Chem Resist Group Ltd	UK
Chicago Pump Company	US
Cornell	US
Cougar Industries Ltd	UK
Crane Pumps & Systems	US
Davey Products Pty Ltd	AU
DDA Srl	IT
Desmi GmbH	DE
Discflo Corporation	US
Dragflow Srl	IT
Ebara Corporation	JP
EDUR Pumpenfabrik GmbH & Co KG	DE
Engineered Products Group Pty Ltd (EPG)	AU
Ensival-Moret	BE
ESSCO Pumps & Controls	US
Feluwa Pumpen GmbH	DE
Flotronic Pumps Ltd	UK
Gilbert Gilkes & Gordon Ltd	UK
GMP SpA	IT
Godwin Pumps Ltd	UK
Godwin Pumps of America Inc	US
Goodwin International	UK
Gorman-Rupp Company	US
Goulds Pumps, ITT Industrial & BioPharm Group	US
Goulds Pumps, ITT Water Technology	US
Grindex AB	SE
H & E Tsurumi Pump Co Ltd	HK
Arthur Habermann GmbH & Co KG	DE
Handol Pumps Ltd	KR
Hangzhou SRT Pump Co Ltd	CN
Hanning & Kahl GmbH & Co KG	DE
Häny AG	CH
Hayward Gordon	CA
HCP Pump Manufacturer Co Ltd	TW
Herborner Pumpenfabrik	DE
Hidrostal Ltd	UK
Homa Pumpenfabrik GmbH	DE
Honda Kiko Co Ltd	JP
Hung Pump Industrial Co Ltd	TW
Hydrainer Pumps Ltd	UK
Hydromatic Pumps	US
Hyosung-Ebara Co Ltd	KR
ITT Flygt AB	SE
Johnson Pump (India), ATE Enterprises Ltd	IN
Johnson Pump AB	SE
Johnson Pump BV	NL
Johnson Pumps of America Inc	US
Jung Pumpen GmbH & Co	DE
Kawamoto Pump Manufacturing Co Ltd	JP
Kelair Pumps Australia Pty Ltd	AU
Kinder-Janes Engineers Ltd	UK
Kishor Pumps Ltd	IN
KSB Ltd	UK
Kubota Corporation	JP
Little Giant Pump Company	US
London Pumps Ltd	UK
March May Ltd	UK
Marlow Pumps, ITT Water Technology	US
Matra SpA	IT
Megator Ltd	UK
Melotte Pumptechnology BV	NL
Metalchem-Warszawa SA	PL
Mody Pumps™ Inc	US
Monarch Industries Ltd	CA
Morris Pumps	US
FE Myers	US
Nasosenergomash	UA
Nijhuis Pompen BV	NL
Nuovo Pignone	IT
Oddesse Pumpen-und Motorenfabrik GmbH	DE
OFT (Officine Di Trevi)	IT
Omel Ltda	BR
Ondrejovicka Strojirna sro	CZ
Paco Pumps	US
Pedrollo SpA	IT
Peerless Pump Company	US
Plenty Mirrlees Pumps	UK
Pomac Industries Group	NL
PT Ebara Indonesia	ID
Pump & Package Ltd	UK
Pump Engineering Ltd	UK
Pumpex Sales	SE
Rhine Ruhr Pumps & Valves (Pty) Ltd	ZA
Robot Pumps BV	NL
Rotex SA	GR
Ruhrpumpen GmbH	DE
Ruhrpumpen Inc	US
Rütschi Fluid AG	CH
Sakuragawa Pump Manufacturing Co Ltd	JP
Selwood Group Ltd	UK
Shanghai Shenbao Industrial Pump Co Ltd	CN
Shijiazhuang Pump Industry Group Co Ltd	CN
Shin Shin Machinery Co Ltd	KR
Sigma Group AS	CZ
Simaco Elettromeccanica Srl	IT
Slovpump-Trade Sro	SK
Speck-Pumpen	DE
SRS Crisafulli Inc	US
Oswald Suck GmbH	DE
Swidnicka Fabryka Pomp Sp Z.o.o.	PL
Tangshan Pump Factory	CN
Thompson Pump & Manufacturing Co Inc	US
TKL	AU
Tsurumi (America) Inc	US
Tsurumi (Europe) GmbH	DE
Tsurumi (Shanghai) Co Ltd	CN
Tsurumi (Singapore) Pte Ltd	SG
Tsurumi Manufacturing Co Ltd	JP
Tsurumi Pump (Malaysia) SDN BHD	MY
Tsurumi Pump Taiwan Co Ltd	TW
Türbosan AS	TR
United Pumps Australia (UPA)	AU
Vaughan Company Inc	US
Verder Group	NL
Vertiflo Pump Company	US
Wafapomp SA	PL
Pumpenfabrik Wangen GmbH	DE
CH Warman Pump Group	ZA
Warren Rupp Inc	US
Weir Minerals Division	US
Weir Specialty Pumps	US
Wernert-Pumpen GmbH	DE
WPIL Ltd	IN
Yaltek AS	TR
Yeomans Pump Company	US
Zoeller Pump Company	US

MIXED-FLOW PUMPS

ABS Deutschland GmbH	DE
ABS Finland Oy	FI
ABS France SAS	FR
ABS Group	SE
ABS Hellas SA	GR
ABS Hidrobomba SA	PL
ABS Hungary Trading	HU
ABS Industria de Bombas Centrifugas Ltda	BR
ABS International Srl	IT
ABS Italy Srl	IT
ABS Nopon Thai Co Ltd	TH
ABS Oumoer A/S	NO
ABS PolskaSp. Z.o.o.	PL
ABS Pompalari Ltd	TR
ABS Pompen BV	NL
ABS Pumpen GmbH	AT
ABS Pumps (Irl) Ltd	IE
ABS Pumps AS	EE
ABS Pumps Corporation	CA
ABS Pumps Inc	US
ABS Pumps Malaysia Sdn Bhd	MY
ABS Pumps NV/SA	BE
ABS Shanghai Co Ltd	CN
ABS Shanghai Co Ltd Beijing Liaison Office	CN
ABS Technologias del Agua SA	ES
ABS Wastewater Technology (Pty) Ltd	ZA
ABS Wastewater Technology Ltd	UK
ABS Wastewater Technology Pte Ltd	SG
Aker Kvaerner Eureka Pump Systems	NO
Andritz AG	AT
Batescrew Pumps	AU
Cakasa (Nigeria) Co Ltd	NG
Calama Industries Pvt Ltd	IN
Caprari SpA	IT
Carver Pump Company	US
Ceetak Engineering Ltd	UK
Chung Woo Industrial Co Ltd	KR
Cornell	US
Corporacion EG SA	MX
Dempster Industries Inc	US
Drakos-Polemis Inc	GR
Ebara Corporation	JP

Ebara Fluid Handling	US
Ebara Pumps Europe SpA	IT
EDUR Pumpenfabrik GmbH & Co KG	DE
Franz Eisele und Söhne GmbH & Co KG	DE
Engineered Products Group Pty Ltd (EPG)	AU
Fapmo	FR
Flowserve Corporation	US
Flux-Geräte GmbH	DE
Goulds Pumps, ITT Industrial & BioPharm Group	US
H & E Tsurumi Pump Co Ltd	HK
Hangzhou SRT Pump Co Ltd	CN
Herborner Pumpenfabrik	DE
Hidrotecar SA	ES
Hunan XD Changsha Pump Works Co Ltd	CN
Hung Pump Industrial Co Ltd	TW
Hyosung-Ebara Co Ltd	KR
Iron Pump A/S	DK
ITT Flygt AB	SE
Johnston Pump	US
Kelair Pumps Australia Pty Ltd	AU
Kinder-Janes Engineers Ltd	UK
KSB Ltd	UK
Kubota Corporation	JP
Layne/Verti-Line	US
Macquarrie Corporation	NZ
Mather & Platt Pumps Ltd	IN
Matra SpA	IT
Micropump Inc	US
National Pump Company LLC	US
Nijhuis Pompen BV	NL
Oddesse Pumpen-und Motorenfabrik GmbH	DE
Paco Pumps	US
Packo Inox NV	BE
Patterson Pump Company	US
Peerless Pump Company	US
PEME-Gourdin SA	FR
Pump & Package Ltd	UK
Reddy-Buffaloes Pump Inc	US
Rhine Ruhr Pumps & Valves (Pty) Ltd	ZA
Ritz Pumpenfabrik GmbH & Co KG	DE
Rovatti Pompe	IT
Ruhrpumpen GmbH	DE
Ruhrpumpen Inc	US
Rütschi Fluid AG	CH
Sakuragawa Pump Manufacturing Co Ltd	JP
SAWA Pumpentechnik AG	CH
Seko Bono Exacta SpA	IT
Seko Ltd	UK
Shanxi Jishan Qinglong Pump Industry Co Ltd	CN
Shijiazhuang Pump Industry Group Co Ltd	CN
Shin Shin Machinery Co Ltd	KR
Sigma Group AS	CZ
Siwatec AS	CZ
Slovpump-Trade Sro	SK
Smith & Loveless Inc	US
Michael Smith Engineers Ltd	UK
Speck-Pumpen	DE
Sulzer Pumps	CH
TKL	AU
Torishima Pump Mfg Co Ltd	JP
TriVista Engineering Ltd	UK
Tsurumi (America) Inc	US
Tsurumi (Europe) GmbH	DE
Tsurumi (Shanghai) Co Ltd	CN
Tsurumi (Singapore) Pte Ltd	SG
Tsurumi Manufacturing Co Ltd	JP
Tsurumi Pump (Malaysia) SDN BHD	MY
Tsurumi Pump Taiwan Co Ltd	TW
Türbosan AS	TR
Ture International Industrial Ltd	CN
Voltas Ltd	IN
Wafapomp SA	PL
Waterman Industries Ltd	IN
Weir Gabbioneta Srl	IT
Weir Pumps Ltd	UK
WPIL Ltd	IN
Yaltek AS	TR

AXIAL-FLOW PUMPS

ABS Deutschland GmbH	DE
ABS Finland Oy	FI
ABS France SAS	FR
ABS Group	SE
ABS Hellas SA	GR
ABS Hidrobomba SA	PL
ABS Hungary Trading	HU
ABS Industria de Bombas Centrifugas Ltda	BR
ABS International Srl	IT
ABS Italy Srl	IT
ABS Nopon Thai Co Ltd	TH
ABS Oumoer A/S	NO
ABS PolskaSp. Z.o.o.	PL
ABS Pompalari Ltd	TR
ABS Pompen BV	NL
ABS Pumpen GmbH	AT
ABS Pumps (Irl) Ltd	IE
ABS Pumps AS	EE
ABS Pumps Corporation	CA
ABS Pumps Inc	US
ABS Pumps Malaysia Sdn Bhd	MY
ABS Pumps NV/SA	BE
ABS Shanghai Co Ltd	CN
ABS Shanghai Co Ltd Beijing Liaison Office	CN
ABS Technologias del Agua SA	ES
ABS Wastewater Technology (Pty) Ltd	ZA
ABS Wastewater Technology Ltd	UK
ABS Wastewater Technology Pte Ltd	SG
ACD Inc	US
Aker Kvaerner Eureka Pump Systems	NO
Allweiler AG	DE
Andritz AG	AT
Apollo Gössnitz GmbH	DE
Batescrew Pumps	AU
Calama Industries Pvt Ltd	IN
Caprari SpA	IT
Carver Pump Company	US
Colfax Pump Group	US
Corporacion EG SA	MX
Dempster Industries Inc	US
Drakos-Polemis Inc	GR
Ebara Corporation	JP
EDUR Pumpenfabrik GmbH & Co KG	DE
Franz Eisele und Söhne GmbH & Co KG	DE
Engineered Products Group Pty Ltd (EPG)	AU
Ensival-Moret	BE
Fairbanks Morse Pump	US
Flowserve Corporation	US
Flux-Geräte GmbH	DE
Goulds Pumps, ITT Industrial & BioPharm Group	US
H & E Tsurumi Pump Co Ltd	HK
Herborner Pumpenfabrik	DE
Hidrostal Ltd	UK
Hidrotecar SA	ES
Homa Pumpenfabrik GmbH	DE
Iron Pump A/S	DK
ITT Flygt AB	SE
Johnson Pump (India), ATE Enterprises Ltd	IN
Johnson Pump AB	SE
Johnson Pump BV	NL
Johnson Pumps of America Inc	US
Johnston Pump	US
Kelair Pumps Australia Pty Ltd	AU
Kestner Engineering Co Ltd	UK
Klaus Union GmbH & Co KG	DE
KSB Ltd	UK
Kubota Corporation	JP
Layne/Verti-Line	US
Macquarrie Corporation	NZ
Mather & Platt Pumps Ltd	IN
MP Pumps Inc	US
National Pump Company LLC	US
Paco Pumps	US
Patterson Pump Company	US
Peerless Pump Company	US
Plenty Mirrlees Pumps	UK
Pump & Package Ltd	UK
Red Jacket, ITT Water Technology	US
Reddy-Buffaloes Pump Inc	US
Rhine Ruhr Pumps & Valves (Pty) Ltd	ZA
Rotex SA	GR
Rovatti Pompe	IT
Ruhrpumpen GmbH	DE
Ruhrpumpen Inc	US
SAWA Pumpentechnik AG	CH
seepex GmbH & Co KG	DE
seepex UK Ltd	UK
Shanxi Jishan Qinglong Pump Industry Co Ltd	CN
Shin Shin Machinery Co Ltd	KR
Sigma Group AS	CZ
Siwatec AS	CZ
Smith & Loveless Inc	US
Michael Smith Engineers Ltd	UK
SRS Crisafulli Inc	US
Sulzer Pumps	CH
Tangshan Pump Factory	CN
TKL	AU
Torishima Pump Mfg Co Ltd	JP
TriVista Engineering Ltd	UK
Tsurumi (America) Inc	US
Tsurumi (Europe) GmbH	DE
Tsurumi (Shanghai) Co Ltd	CN
Tsurumi (Singapore) Pte Ltd	SG
Tsurumi Manufacturing Co Ltd	JP
Tsurumi Pump (Malaysia) SDN BHD	MY
Tsurumi Pump Taiwan Co Ltd	TW
Türbosan AS	TR
Ture International Industrial Ltd	CN
Ujala Pumps Pvt Ltd	IN
Weir Gabbioneta Srl	IT
Weir Pumps Ltd	UK

NON-METALLIC ROTODYNAMIC PUMPS

Apollo Gössnitz GmbH	DE
APV Fluid Handling Horsens	DK
Best & Crompton Engineering Ltd	IN

Caprari SpA	IT
Casals Cardona Industries SA	ES
Caster HMD Kontro	UK
Chung Woo Industrial Co Ltd	KR
Ebara Corporation	JP
EBS-Ray Pumps Pty Ltd	AU
EDUR Pumpenfabrik GmbH & Co KG	DE
Finish Thompson Inc	US
Invensys APV	US
Invensys APV UK	UK
Kelair Pumps Australia Pty Ltd	AU
Kestner Engineering Co Ltd	UK
Oy Kolmeks AB	FI
KSB Ltd	UK
Lederle GmbH	DE
London Pumps Ltd	UK
MASO Process-Pumpen GmbH	DE
Mefiag Division, Met-Pro Corporation	US
Paco Pumps	US
Peerless Pump Company	US
PT Ebara Indonesia	ID
Pump & Package Ltd	UK
Rhine Ruhr Pumps & Valves (Pty) Ltd	ZA
Roth Pump Company	US
Ruhrpumpen GmbH	DE
Ruhrpumpen Inc	US
Sacem SpA	IT
Savino Barbera	IT
Schmalenberger GmbH & Co	DE
Sethco Division, Met-Pro Corporation	US
Sigma Group AS	CZ
Sterling Fluid Systems (Americas) Inc	US
Sterling Fluid Systems (Austria)	AT
Sterling Fluid Systems (Belgium) NV	BE
Sterling Fluid Systems (France) SAS	FR
Sterling Fluid Systems (Italy) SpA	IT
Sterling Fluid Systems (Netherlands)	NL
Sterling Fluid Systems (Schweiz) AG	CH
Sterling Fluid Systems (UK) Ltd	UK
Sterling SIHI GmbH	DE
Sundyne Corporation	US
Sundyne International SA	FR
Swidnicka Fabryka Pomp Sp Z.o.o.	PL
TKL	AU
Ujala Pumps Pvt Ltd	IN
Verder Group	NL
Waterman Industries Ltd	IN
Wernert-Pumpen GmbH	DE

HYGIENIC-QUALITY ROTODYNAMIC PUMPS

Alfa Laval AB	SE
Alfa Laval Inc	US
APV Fluid Handling Horsens	DK
Asco Pompe srl	IT
AxFlow A/S	DK
AxFlow AB	SE
AxFlow AS	NO
AxFlow BeNeLux BV	NL
AxFlow GmbH	AT
AxFlow GmbH	DE
AxFlow GmbH	CH
Axflow Holding AB	SE
AxFlow Kft	HU
AxFlow Ida	PL
AxFlow Ltd	IE
AxFlow Ltd	UK
AxFlow Oy	FI
AxFlow sro	CZ
AxFlow SAS	FR
AxFlow SAS	ES
AxFlow Sp Soo	PL
AxFlow SpA	IT
Catag AG	CH
Cornell	US
Discflo Corporation	US
Ebara Corporation	JP
EBS-Ray Pumps Pty Ltd	AU
EDUR Pumpenfabrik GmbH & Co KG	DE
Flexachem Manufacturing Ltd	IE
Flotronic Pumps Ltd	UK
Fluid Metering Inc	US
Flux-Geräte GmbH	DE
Fristam Pumps (UK) Ltd	UK
Fristam Pumps USA	US
Fristram Pumpen	DE
Philipp Hilge GmbH & Co KG	DE
Hilge Pumps Ltd	UK
Hunan XD Changsha Pump Works Co Ltd	CN
Inoxpa SA	ES
Invensys APV	US
Invensys APV UK	UK
Johnson Pump (UK) Ltd	UK
JP Pumps Ltd	UK
Kecol Pumps Ltd	UK
Kelair Pumps Australia Pty Ltd	AU
Kishor Pumps Ltd	IN
MDM Pumps Ltd	UK
Packo Inox NV	BE
Paco Pumps	US
Pomac Industries Group	NL
Pump Engineering Ltd	UK
Rhine Ruhr Pumps & Valves (Pty) Ltd	ZA
Sharp Trendys	IN
Sigma Group AS	CZ
Tapflo AB	SE
Verder Group	NL
Viking Pump Inc	US
Pumpenfabrik Wangen GmbH	DE
Waukesha Cherry-Burrell	US
Wright Pump	US

PERIPHERAL PUMPS

Beresford Pumps	UK
Ceetak Engineering Ltd	UK
Chemvac Pumps Ltd	UK
Corken Inc	US
Crane Pumps & Systems	US
DAB Pumps SpA	IT
Davey Products Pty Ltd	AU
DDA Srl	IT
Ebara Corporation	JP
EBS-Ray Pumps Pty Ltd	AU
EDUR Pumpenfabrik GmbH & Co KG	DE
Fristam Pumps (UK) Ltd	UK
Fristam Pumps USA	US
Fristram Pumpen	DE
Gilbert Gilkes & Gordon Ltd	UK
GMP SpA	IT
Gysi Pumpen AG	CH
Philipp Hilge GmbH & Co KG	DE
Hung Pump Industrial Co Ltd	TW
Inoxpa SA	ES
Iron Pump A/S	DK
Kawamoto Pump Manufacturing Co Ltd	JP
Klaus Union GmbH & Co KG	DE
Laxmi Hydraulics Pvt Ltd	IN
Lederle GmbH	DE
Liverani Srl	IT
Lowara SpA, ITT Industries	IT
Magnatex Pumps Inc	US
March May Ltd	UK
Mepco	US
MTH Pumps	US
Paco Pumps	US
Pumpenfabrik Ernst Vogel GmbH, ITT Industries	IT
Pulsafeeder	US
Raj Pumps	IN
Rapid Allweiler Pump & Engineering	ZA
Roth Pump Company	US
Rütschi Fluid AG	CH
SAWA Pumpentechnik AG	CH
Scheerle AG	CH
Schmalenberger GmbH & Co	DE
Sero PumpSystems GmbH	DE
Sharp Trendys	IN
Simaco Elettromeccanica Srl	IT
Sterling Fluid Systems (Americas) Inc	US
Sterling Fluid Systems (Austria)	AT
Sterling Fluid Systems (Belgium) NV	BE
Sterling Fluid Systems (France) SAS	FR
Sterling Fluid Systems (Italy) SpA	IT
Sterling Fluid Systems (Netherlands)	NL
Sterling Fluid Systems (Schweiz) AG	CH
Sterling Fluid Systems (UK) Ltd	UK
Sterling SIHI GmbH	DE
Swidnicka Fabryka Pomp Sp Z.o.o.	PL
Totton Pumps Ltd	UK
Stuart Turner Ltd	UK
Ujala Pumps Pvt Ltd	IN
Verder Group	NL
Wernert-Pumpen GmbH	DE

PITOT TUBE PUMPS

Ebara Corporation	JP
Sterling Fluid Systems (Americas) Inc	US
Sterling Fluid Systems (Austria)	AT
Sterling Fluid Systems (Belgium) NV	BE
Sterling Fluid Systems (France) SAS	FR
Sterling Fluid Systems (Italy) SpA	IT
Sterling Fluid Systems (Netherlands)	NL
Sterling Fluid Systems (Schweiz) AG	CH
Sterling Fluid Systems (UK) Ltd	UK
Sterling SIHI GmbH	DE
Weir Specialty Pumps	US

DISC PUMPS

Alfa Laval AB	SE
Alfa Laval Inc	US
Alfa Laval Pumps Ltd	UK
AxFlow A/S	DK
AxFlow AB	SE
AxFlow AS	NO
AxFlow BeNeLux BV	NL
AxFlow GmbH	AT
AxFlow GmbH	DE
AxFlow GmbH	CH
Axflow Holding AB	SE

AxFlow Kft HU
AxFlow lda PL
AxFlow Ltd IE
AxFlow Ltd UK
AxFlow Oy FI
AxFlow sro CZ
AxFlow SAS FR
AxFlow SAS ES
AxFlow Sp Soo PL
AxFlow SpA IT
Camaction Avonpump Ltd UK
Crane Pumps & Systems US
Desmi GmbH DE
Ebara Corporation JP
Ebara Fluid Handling US
Ebara Pumps Europe SpA IT
Fristam Pumps (UK) Ltd UK
Fristam Pumps USA US
Fristram Pumpen DE
Hidrotecar SA ES
Hilge Pumps Ltd UK
Sterling Fluid Systems (Americas) Inc US
Sterling Fluid Systems (Austria) AT
Sterling Fluid Systems (Belgium) NV BE
Sterling Fluid Systems (France) SAS FR
Sterling Fluid Systems (Italy) SpA IT
Sterling Fluid Systems (Netherlands) NL
Sterling Fluid Systems (Schweiz) AG CH
Sterling Fluid Systems (UK) Ltd UK
Sterling SIHI GmbH DE
Weir Gabbioneta Srl IT

MAGNETIC DRIVE ROTODYNAMIC PUMPS

Acromet (Aust) Pty Ltd AU
Allweiler AG DE
Ansimag Inc US
Argal Srl IT
AxFlow A/S DK
AxFlow AB SE
AxFlow AS NO
AxFlow BeNeLux BV NL
AxFlow GmbH AT
AxFlow GmbH DE
AxFlow GmbH CH
Axflow Holding AB SE
AxFlow Kft HU
AxFlow lda PL
AxFlow Ltd IE
AxFlow Ltd UK
AxFlow Oy FI
AxFlow sro CZ
AxFlow SAS FR
AxFlow SAS ES
AxFlow Sp Soo PL
AxFlow SpA IT
Beresford Pumps UK
Cakasa (Nigeria) Co Ltd NG
Carver Pump Company US
Caster HMD Kontro UK
Catag AG CH
CDR Pompe SpA IT
Ceetak Engineering Ltd UK
Chempump US
Chemvac Pumps Ltd UK
CP Pumpen AG CH
Dalian Danai Pumps Ltd CN
Dickow Pumpen KG DE
Ebara Corporation JP
Ebara Fluid Handling US
Ebara Pumps Europe SpA IT
Environamics Corporation US
Finish Thompson Inc US
Flowserve Corporation US
Goulds Pumps, ITT Industrial & BioPharm Group US
Gysi Pumpen AG CH
Herborner Pumpenfabrik DE
Iwaki America US
Iwaki Co Ltd JP
Iwaki Europe GmbH DE
Jiangsu Haishi Pump Co Ltd CN
Johnson Pump (India), ATE Enterprises Ltd IN
Johnson Pump AB SE
Johnson Pump BV NL
Johnson Pumps of America Inc US
Kinder-Janes Engineers Ltd UK
Klaus Union GmbH & Co KG DE
KSB Ltd UK
Lederle GmbH DE
Little Giant Pump Company US
Lutz (UK) Ltd UK
Magnatex Pumps Inc US
March May Ltd UK
MASO Process-Pumpen GmbH DE
Mefiag Division, Met-Pro Corporation US
Nova Magnetics Burgmann Ltd CA
Omel Ltda BR
Pacer Pumps US
Pomac Industries Group NL
Price Pump Company US
Pump Engineering Ltd UK
Rhine Ruhr Pumps & Valves (Pty) Ltd ZA
Roth Pump Company US
Savino Barbera IT
Serfilco Europe Ltd UK
Serfilco Ltd US
Sero PumpSystems GmbH DE
Sethco Division, Met-Pro Corporation US
Michael Smith Engineers Ltd UK
Sterling Fluid Systems (Americas) Inc US
Sterling Fluid Systems (Austria) AT
Sterling Fluid Systems (Belgium) NV BE
Sterling Fluid Systems (France) SAS FR
Sterling Fluid Systems (Italy) SpA IT
Sterling Fluid Systems (Netherlands) NL
Sterling Fluid Systems (Schweiz) AG CH
Sterling Fluid Systems (UK) Ltd UK
Sterling SIHI GmbH DE
Sundyne Corporation US
Sundyne International SA FR
Sunmotor International Ltd CA
Tapflo AB SE
TKL AU
Totton Pumps Ltd UK
TriVista Engineering Ltd UK
Varisco Srl IT
Verder Group NL
Victor Pumpen GmbH DE
Weir Gabbioneta Srl IT

CANNED MOTOR ROTODYNAMIC PUMPS

AxFlow A/S DK
AxFlow AB SE
AxFlow AS NO
AxFlow BeNeLux BV NL
AxFlow GmbH AT
AxFlow GmbH DE
AxFlow GmbH CH
Axflow Holding AB SE
AxFlow Kft HU
AxFlow lda PL
AxFlow Ltd IE
AxFlow Ltd UK
AxFlow Oy FI
AxFlow sro CZ
AxFlow SAS FR
AxFlow SAS ES
AxFlow Sp Soo PL
AxFlow SpA IT
Buffalo Pumps Inc US
Caprari SpA IT
Caster HMD Kontro UK
Ceetak Engineering Ltd UK
Chempump US
Cougar Industries Ltd UK
CW-EMD US
Dalian Danai Pumps Ltd CN
Ebara Corporation JP
Ebara Fluid Handling US
Ebara Pumps Europe SpA IT
Environamics Corporation US
Hayward Tyler® Group UK
Hunan XD Changsha Pump Works Co Ltd CN
Kinder-Janes Engineers Ltd UK
MASO Process-Pumpen GmbH DE
Mazda Pump Manufacturing Co Ltd JP
MTH Pumps US
Nikkiso Co Ltd JP
Nikkiso Pumps Europe GmbH DE
Optimex FR
Rütschi Fluid AG CH
Michael Smith Engineers Ltd UK
Sundyne Corporation US
Sundyne International SA FR
Teikoku Electric GmbH DE
Termomeccanica Pompe SpA IT
3S Systemtechnik AG CH
TKL AU
Wilo AG DE
Wilo EMU GmbH DE
Yaltek AS TR

ROTODYNAMIC PUMPS WITHOUT DRIVERS (for specific applications including mobile agricultural applications and mobile fire pumps)

Caprari SpA IT
Darley Pumps US
Sunmotor International Ltd CA
TKL AU
Albert Ziegler GmbH & Co KG DE

PUMPS AS POWER-RECOVERY TURBINES

Flowserve Corporation US
Goulds Pumps, ITT Industrial & BioPharm Group US
Termomeccanica Pompe SpA IT
TriVista Engineering Ltd UK
Pumpenfabrik Ernst Vogel GmbH, ITT Industries AT

18.4 Positive displacement pumps

Companies listed alphabetically by pump type

EXTERNAL GEAR PUMPS

Albany Engineering Company Ltd	UK
Allpumps Ltd	UK
Asco Pompe srl	IT
AxFlow A/S	DK
AxFlow AB	SE
AxFlow AS	NO
AxFlow BeNeLux BV	NL
AxFlow GmbH	AT
AxFlow GmbH	DE
AxFlow GmbH	CH
Axflow Holding AB	SE
AxFlow Kft	HU
AxFlow Ida	PL
AxFlow Ltd	IE
AxFlow Ltd	UK
AxFlow Oy	FI
AxFlow sro	CZ
AxFlow SAS	FR
AxFlow SAS	ES
AxFlow Sp Soo	PL
AxFlow SpA	IT
Bielomatik Leuze GmbH & Co	DE
Bornemann GmbH	DE
Cakasa (Nigeria) Co Ltd	NG
Catag AG	CH
Colfax Pump Group	US
Concentric Pumps Ltd	UK
Cross Manufacturing Inc	US
Pompe Cucchi Srl	IT
Ebara Corporation	JP
EBS-Ray Pumps Pty Ltd	AU
Engineered Products Group Pty Ltd (EPG)	AU
Finder Pompe SpA	IT
Flowserve Corporation	US
Gilbert Gilkes & Gordon Ltd	UK
Haight Pumps	US
Honda Kiko Co Ltd	JP
Hung Pump Industrial Co Ltd	TW
Hypro	US
Ismatec SA	CH
Iwaki Co Ltd	JP
Iwaki Europe GmbH	DE
Kracht GmbH	DE
Lederle GmbH	DE
Lewis Pumps	US
London Pumps Ltd	UK
Maag Pump Systems Textron AG	CH
Maag Pump Systems Textron Inc	US
Mather & Platt Pumps Ltd	IN
Micropump Ltd	UK
Midland Pump Manufacturing Co Ltd	UK
Northern Pump	US
Nova Magnetics Burgmann Ltd	CA
Pedro Roquet SA	ES
Pulsafeeder	US
Pump Engineering Co Pvt Ltd	IN
Pump Engineering Ltd	UK
Rickmeier GmbH	DE
Scherzinger Pumps	DE
Shin Shin Machinery Co Ltd	KR
Sigma Group AS	CZ
Michael Smith Engineers Ltd	UK
Tuschaco Pumps Ltd	IN
Tuthill Corporation	US
Variopumps Pumpenbau GmbH	DE
Varisco Srl	IT
Viking Pump Inc	US
Witte Pumps & Technology GmbH	DE
Zenith Pumps Division, Parker Hannifin	US

INTERNAL GEAR PUMPS

Acromet (Aust) Pty Ltd	AU
Asco Pompe srl	IT
AxFlow A/S	DK
AxFlow AB	SE
AxFlow AS	NO
AxFlow BeNeLux BV	NL
AxFlow GmbH	AT
AxFlow GmbH	DE
AxFlow GmbH	CH
Axflow Holding AB	SE
AxFlow Kft	HU
AxFlow Ida	PL
AxFlow Ltd	IE
AxFlow Ltd	UK
AxFlow Oy	FI
AxFlow sro	CZ
AxFlow SAS	FR
AxFlow SAS	ES
AxFlow Sp Soo	PL
AxFlow SpA	IT
Bavaria Fluidtec GmbH	DE
Bijur Lubricating Corporation	US
Catag AG	CH
CDS-John Blue Company	US
Ceetak Engineering Ltd	UK
Chemvac Pumps Ltd	UK
Colfax Pump Group	US
Concentric Pumps Ltd	UK
Cougar Industries Ltd	UK
Desmi A/S	DK
Desmi GmbH	DE
EBS-Ray Pumps Pty Ltd	AU
Engineered Products Group Pty Ltd (EPG)	AU
Flexachem Manufacturing Ltd	IE
Fultz Pumps Inc	US
Pompe Garbarino SpA	IT
Gorman-Rupp Company	US
Gotec SA	CH
Haight Pumps	US
Honda Kiko Co Ltd	JP
Indag Maschinenbau GmbH	DE
Ismatec SA	CH
Iwaki America	US
JLS International (UK) Ltd	UK
Johnson Pump (India), ATE Enterprises Ltd	IN
Johnson Pump AB	SE
Johnson Pump BV	NL
Johnson Pumps of America Inc	US
JP Pumps Ltd	UK
Kecol Pumps Ltd	UK
Kelair Pumps Australia Pty Ltd	AU
Kinder-Janes Engineers Ltd	UK
Lederle GmbH	DE
Lobee Pump & Machinery Company	US
London Pumps Ltd	UK
Mahr Metering Systems Corporation	US
Matra SpA	IT
Micropump Inc	US
Pulsafeeder	US
Pump & Package Ltd	UK
SAWA Pumpentechnik AG	CH
Scherzinger Pumps	DE
Sigma Group AS	CZ
Michael Smith Engineers Ltd	UK
Sterling Fluid Systems (Americas) Inc	US
Sterling Fluid Systems (Austria)	AT
Sterling Fluid Systems (Belgium) NV	BE
Sterling Fluid Systems (France) SAS	FR
Sterling Fluid Systems (Italy) SpA	IT
Sterling Fluid Systems (Netherlands)	NL
Sterling Fluid Systems (Schweiz) AG	CH
Sterling Fluid Systems (UK) Ltd	UK
Sterling SIHI GmbH	DE
Taiko Kikai Industries Co Ltd	JP
TKL	AU
TriVista Engineering Ltd	UK
Tuschaco Pumps Ltd	IN
Tuthill Corporation	US
Variopumps Pumpenbau GmbH	DE
Varisco Srl	IT
Verder Group	NL
Pompe Vergani	IT
Vican Pump Company	CA
Victor Pumpen GmbH	DE
Viking Pump Inc	US

ARCHIMEDES SCREW PUMPS

Allpumps Ltd	UK
Asco Pompe srl	IT
Cakasa (Nigeria) Co Ltd	NG
Cole-Parmer Instrument Company	US
Defco	US
Delasco PCM GmbH	DE
Engineered Products Group Pty Ltd	AU
Ensival-Moret	BE
Houttuin Holland BV	NL
Imo AB	SE
Bombas Itur SA	ES
Bombas Itur SA	ES
Jiangsu Haishi Pump Co Ltd	CN
Kräutler GmbH & Co	AT
Leistritz Pumpen GmbH	DE
Sigma Group AS	CZ
Spaans Babcock BV	NL
Tianjin Pumps & Machinery Group Co Ltd	CN
Vandezande	BE
Verder Group	NL
Weir Minerals Division	US

TWIN-ROTOR SCREW PUMPS

Allweiler AG	DE
Asco Pompe srl	IT
AxFlow A/S	DK

AxFlow AB	SE
AxFlow AS	NO
AxFlow BeNeLux BV	NL
AxFlow GmbH	AT
AxFlow GmbH	DE
AxFlow GmbH	CH
Axflow Holding AB	SE
AxFlow Kft	HU
AxFlow lda	PL
AxFlow Ltd	IE
AxFlow Ltd	UK
AxFlow Oy	FI
AxFlow sro	CZ
AxFlow SAS	FR
AxFlow SAS	ES
AxFlow Sp Soo	PL
AxFlow SpA	IT
Bornemann GmbH	DE
Cole-Parmer Instrument Company	US
Colfax Pump Group	US
Defco	US
Delasco PCM GmbH	DE
Engineered Products Group Pty Ltd	AU
Ensival-Moret	BE
Flowserve Corporation	US
Hidrotecar SA	ES
Houttuin Holland BV	NL
Imo AB	SE
Bombas Itur SA	ES
Bombas Itur SA	ES
Iwaki Co Ltd	JP
Iwaki Europe GmbH	DE
Jiangsu Haishi Pump Co Ltd	CN
Kräutler GmbH & Co	AT
Leistritz Pumpen GmbH	DE
Moyno Inc	US
Nova Magnetics Burgmann Ltd	CA
Plenty Mirrlees Pumps	UK
Pomac Industries Group	NL
Rapid Allweiler Pump & Engineering	ZA
Roper Pump Company	US
Roto Pumps Ltd	IN
Scheerle AG	CH
Sigma Group AS	CZ
Spaans Babcock BV	NL
Tianjin Pumps & Machinery Group Co Ltd	CN
Vandezande	BE
Warren Pumps LLC	US
Weir Minerals Division	US

TWIN-ROTOR & 5-ROTOR SCREW PUMPS

Allweiler AG	DE
AxFlow A/S	DK
AxFlow AB	SE
AxFlow AS	NO
AxFlow BeNeLux BV	NL
AxFlow GmbH	AT
AxFlow GmbH	DE
AxFlow GmbH	CH
Axflow Holding AB	SE
AxFlow Kft	HU
AxFlow lda	PL
AxFlow Ltd	IE
AxFlow Ltd	UK
AxFlow Oy	FI
AxFlow sro	CZ
AxFlow SAS	FR
AxFlow SAS	ES
AxFlow Sp Soo	PL
AxFlow SpA	IT
BOC Edwards - Hick Hargreaves	UK
Bornemann GmbH	DE
Colfax Pump Group	US
DDA Srl	IT
Desmi A/S	DK
Engineered Products Group Pty Ltd	AU
Engineered Products Group Pty Ltd (EPG)	AU
Ensival-Moret	BE
Flux-Geräte GmbH	DE
Houttuin Holland BV	NL
Imo AB	SE
Bombas Itur SA	ES
Iwaki Co Ltd	JP
Iwaki Europe GmbH	DE
Kelair Pumps Australia Pty Ltd	AU
Kinder-Janes Engineers Ltd	UK
Klaus Union GmbH & Co KG	DE
Kräutler GmbH & Co	AT
Leistritz Pumpen GmbH	DE
Nova Magnetics Burgmann Ltd	CA
Peerless Pump Company	US
Plenty Mirrlees Pumps	UK
Pomac Industries Group	NL
Rapid Allweiler Pump & Engineering	ZA
Roper Pump Company	US
Roto Pumps Ltd	IN
Scheerle AG	CH
Sigma Group AS	CZ
Spaans Babcock BV	NL
Tianjin Pumps & Machinery Group Co Ltd	CN
Tyco - Southern Cross Plant	AU
Vandezande	BE
Willy Vogel AG	DE
Pumpenfabrik Wangen GmbH	DE
Weir Minerals Division	US
Weir Pumps Ltd	UK
Weir Specialty Pumps	US

TWIN-ROTOR GEARED — SCREW PUMPS

Allweiler AG	DE
BOC Edwards - Hick Hargreaves	UK
Bornemann GmbH	DE
Desmi A/S	DK
Engineered Products Group Pty Ltd (EPG)	AU
Ensival-Moret	BE
Flowserve Corporation	US
Flux Pumps Corporation	US
Flux-Geräte GmbH	DE
Hidrotecar SA	ES
Houttuin Holland BV	NL
Imo AB	SE
Bombas Itur SA	ES
Iwaki Co Ltd	JP
Iwaki Europe GmbH	DE
Klaus Union GmbH & Co KG	DE
Kräutler GmbH & Co	AT
Leistritz Pumpen GmbH	DE
Peerless Pump Company	US
Plenty Mirrlees Pumps	UK
Rapid Allweiler Pump & Engineering	ZA
Roto Pumps Ltd	IN
Scheerle AG	CH
Sigma Group AS	CZ
Spaans Babcock BV	NL
Tianjin Pumps & Machinery Group Co Ltd	CN
Tyco - Southern Cross Plant	AU
Vandezande	BE
Pumpenfabrik Wangen GmbH	DE
Warren Pumps LLC	US
Weir Minerals Division	US
Weir Specialty Pumps	US

PROGRESSIVE CAVITY PUMPS

Alfa Laval AB	SE
Alfa Laval Inc	US
AxFlow A/S	DK
AxFlow AB	SE
AxFlow AS	NO
AxFlow BeNeLux BV	NL
AxFlow GmbH	AT
AxFlow GmbH	DE
AxFlow GmbH	CH
Axflow Holding AB	SE
AxFlow Kft	HU
AxFlow lda	PL
AxFlow Ltd	IE
AxFlow Ltd	UK
AxFlow Oy	FI
AxFlow sro	CZ
AxFlow SAS	FR
AxFlow SAS	ES
AxFlow Sp Soo	PL
AxFlow SpA	IT
Bavaria Fluidtec GmbH	DE
Catag AG	CH
Cole-Parmer Instrument Company	US
Colfax Pump Group	US
Continental Pump Company	US
Cougar Industries Ltd	UK
Dalian Danai Pumps Ltd	CN
Denorco (Pty) Ltd	ZA
Eagle Pump & Compressor Ltd	CA
Equimavenca	VE
Flexachem Manufacturing Ltd	IE
Flojet Corporation	US
Flux Pumps Corporation	US
Pompe Garbarino SpA	IT
Inoxpa SA	ES
ITT Flygt AB	SE
Johnson Pump (India), ATE Enterprises Ltd	IN
Johnson Pump AB	SE
Johnson Pump BV	NL
Johnson Pumps of America Inc	US
Kächele GmbH	DE
Kelair Pumps Australia Pty Ltd	AU
Kinder-Janes Engineers Ltd	UK
Kudu Industries Inc	CA
Mono Pumps (Australia) Pty Ltd	AU
Mono Pumps (China) Ltd	CN
Mono Pumps (New Zealand) Ltd	NZ
Mono Pumps Ltd	UK
Monoflo Inc	US
Moyno Inc	US
Netzsch Mohnopumpen GmbH	DE
PCM	FR
PCM Asia Pacific	CN
PCM Delasco GmbH	DE
PCM Delasco Inc	US

PCM Moineau Oilfield	CN
PCM Pompes Russia	RU
PCM Pompes Thailand	TH
PCM Pumps Ltd	UK
PCM Tunisia	TN
Peerless Pump Company	US
PF Pumpen und Feuerlöschtechnik GmbH	DE
Pump & Package Ltd	UK
Rapid Allweiler Pump & Engineering	ZA
Rhine Ruhr Pumps & Valves (Pty) Ltd	ZA
Roper Pump Company	US
Roto Pumps Ltd	IN
Rotomac Industires Pvt Ltd	IN
Scheerle AG	CH
seepex GmbH & Co KG	DE
seepex UK Ltd	UK
Sigma Group AS	CZ
Sigmia Trading International	CZ
Standard Pump Inc	US
Taiko Kikai Industries Co Ltd	JP
Tarby Inc	US
Tyco - Southern Cross Plant	AU
Varisco Srl	IT
Verder Group	NL
Pumpenfabrik Wangen GmbH	DE
Warren Pumps LLC	US
Witte Pumps & Technology GmbH	DE
Yaltek AS	TR

LOBE PUMPS — (including circumferential piston pumps)

Alfa Laval AB	SE
Alfa Laval Inc	US
Alfa Laval Pumps Ltd	UK
Alfons de Backer & Co	BE
Allpumps Ltd	UK
APV Fluid Handling Horsens	DK
Asco Pompe srl	IT
AxFlow A/S	DK
AxFlow AB	SE
AxFlow AS	NO
AxFlow BeNeLux BV	NL
AxFlow GmbH	AT
AxFlow GmbH	DE
AxFlow GmbH	CH
Axflow Holding AB	SE
AxFlow Kft	HU
AxFlow lda	PL
AxFlow Ltd	IE
AxFlow Ltd	UK
AxFlow Oy	FI
AxFlow sro	CZ
AxFlow SAS	FR
AxFlow SAS	ES
AxFlow Sp Soo	PL
AxFlow SpA	IT
BOC Edwards - Hick Hargreaves	UK
Börger GmbH	DE
Catag AG	CH
Ceetak Engineering Ltd	UK
Concentric Pumps Ltd	UK
Pompe Cucchi Srl	IT
EBS-Ray Pumps Pty Ltd	AU
Engineered Products Group Pty Ltd (EPG)	AU
Equimavenca	VE
Finder Pompe SpA	IT
Flojet Corporation	US
GPM Pumps Inc	US
Philipp Hilge GmbH & Co KG	DE
Invensys APV	US
Invensys APV UK	UK
Jabsco GmbH, ITT Industries	DE
Jabsco UK, ITT Industries	UK
Jabsco US, ITT Industries	US
Johnson Pump (India), ATE Enterprises Ltd	IN
Johnson Pump (UK) Ltd	UK
Johnson Pump AB	SE
Johnson Pump BV	NL
Johnson Pumps of America Inc	US
JP Pumps Ltd	UK
Jurop SpA	IT
Kelair Pumps Australia Pty Ltd	AU
Kielecka Fabryka Pomp Bialogon SA	PL
Kudu Industries Inc	CA
Lederle GmbH	DE
Lutz-Jesco	DE
Megator Ltd	UK
Netzsch Mohnopumpen GmbH	DE
Omac Srl	IT
PCM	FR
PCM Asia Pacific	CN
PCM Delasco GmbH	DE
PCM Delasco Inc	US
PCM Moineau Oilfield	CN
PCM Pompes Russia	RU
PCM Pompes Thailand	TH
PCM Pumps Ltd	UK
PCM Tunisia	TN
Pentair Inc	US
Pentair Pumps SpA	IT
Plenty Mirrlees Pumps	UK
Pomac Industries Group	NL
Roper Pump Company	US
Scherzinger Pumps	DE
Sigma Group AS	CZ
Sigmia Trading International	CZ
Tuschaco Pumps Ltd	IN
Tuthill Corporation	US
Verder Group	NL
Viking Pump Inc	US
Waukesha Cherry-Burrell	US

VANE PUMPS

Alfons Haar Maschinenbau GmbH & Co	DE
Allpumps Ltd	UK
A-Ryung Machinery Ind Co Ltd	KR
AxFlow A/S	DK
AxFlow AB	SE
AxFlow AS	NO
AxFlow BeNeLux BV	NL
AxFlow GmbH	AT
AxFlow GmbH	DE
AxFlow GmbH	CH
Axflow Holding AB	SE
AxFlow Kft	HU
AxFlow lda	PL
AxFlow Ltd	IE
AxFlow Ltd	UK
AxFlow Oy	FI
AxFlow sro	CZ
AxFlow SAS	FR
AxFlow SAS	ES
AxFlow Sp Soo	PL
AxFlow SpA	IT
Barnant Company	US
BOC Edwards - Hick Hargreaves	UK
Caster HMD Kontro	UK
Ceetak Engineering Ltd	UK
Corken Inc	US
Dalian Danai Pumps Ltd	CN
EBS-Ray Pumps Pty Ltd	AU
Flojet Corporation	US
GMP SpA	IT
Gotec SA	CH
Graham Vacuum Pumps Ltd	UK
Gysi Pumpen AG	CH
Hunan XD Changsha Pump Works Co Ltd	CN
Hypro	US
Jabsco GmbH, ITT Industries	DE
Jabsco UK, ITT Industries	UK
Jabsco US, ITT Industries	US
Johnson Pump (India), ATE Enterprises Ltd	IN
Johnson Pump AB	SE
Johnson Pump BV	NL
Johnson Pumps of America Inc	US
Lederle GmbH	DE
Liverani Srl	IT
MASO Process-Pumpen GmbH	DE
Monarch Industries Ltd	CA
Pentair Inc	US
Pentair Pumps SpA	IT
Pomac Industries Group	NL
Procon Products	US
Pulsafeeder	US
Pump & Package Ltd	UK
Pump Engineering Ltd	UK
Rietschle Thomas Schopfheim GmbH	DE
SHURflo Pump Manufacturing Co	US
Sigma Group AS	CZ
Sundyne Corporation	US
Sundyne International SA	FR
3S Systemtechnik AG	CH
TriVista Engineering Ltd	UK
Verder Group	NL

PERISTALTIC PUMPS — (including rotary peristaltic pumps)

Acromet (Aust) Pty Ltd	AU
Allpumps Ltd	UK
Allweiler AG	DE
Asco Pompe srl	IT
Autoclude Pumps	UK
AxFlow A/S	DK
AxFlow AB	SE
AxFlow AS	NO
AxFlow BeNeLux BV	NL
AxFlow GmbH	AT
AxFlow GmbH	DE
AxFlow GmbH	CH
Axflow Holding AB	SE
AxFlow Kft	HU
AxFlow lda	PL
AxFlow Ltd	IE
AxFlow Ltd	UK
AxFlow Oy	FI
AxFlow sro	CZ
AxFlow SAS	FR
AxFlow SAS	ES

AxFlow Sp Soo	PL
AxFlow SpA	IT
Barnant Company	US
Beta Technology Inc	US
Blue White Industries	US
Bombas Boyser SL	ES
Bredel Hose Pumps	NL
Catag AG	CH
Ceetak Engineering Ltd	UK
Cole-Parmer Instrument Company	US
Colfax Pump Group	US
Crane Process Flow Technologies Ltd	UK
DDA Srl	IT
Defco	US
Delasco PCM GmbH	DE
Equimavenca	VE
Finder Pompe SpA	IT
Flexicon A/S	DK
GPM Pumps Inc	US
Johnson Pump (India), ATE Enterprises Ltd	IN
Johnson Pump AB	SE
Johnson Pump BV	NL
Johnson Pumps of America Inc	US
JP Pumps Ltd	UK
Kecol Pumps Ltd	UK
Kudu Industries Inc	CA
Lutz-Jesco	DE
PCM	FR
PCM Asia Pacific	CN
PCM Delasco GmbH	DE
PCM Delasco Inc	US
PCM Moineau Oilfield	CN
PCM Pompes Russia	RU
PCM Pompes Thailand	TH
PCM Pumps Ltd	UK
PCM Tunisia	TN
PF Pumpen und Feuerlöschtechnik GmbH	DE
ProMinent Dosiertechnik GmbH	DE
ProMinent Fluid Controls UK Ltd	UK
Pump & Package Ltd	UK
Pump Engineering Ltd	UK
Ragazzini Srl	IT
Rapid Allweiler Pump & Engineering	ZA
Rietschle Thomas Schopfheim GmbH	DE
Seko Bono Exacta SpA	IT
Seko Ltd	UK
SHURflo Pump Manufacturing Co	US
Simaco Elettromeccanica Srl	IT
Verder Group	NL
Wallace & Tiernan, Chemfeed Ltd	UK
CH Warman Pump Group	ZA
Watson-Marlow Bredel Pumps Ltd	UK

ROTARY ECCENTRIC PISTON PUMPS

Graham Vacuum Pumps Ltd	UK
Bombas Itur SA	ES
Seko Ltd	UK
TriVista Engineering Ltd	UK
Verder Group	NL
Waukesha Cherry-Burrell	US
Wheatley and Gaso Pump	US

AXIAL AND RADIAL PISTON PUMPS

Bavaria Fluidtec GmbH	DE
BBA Hranice sro	CZ
Gzut SA Pumps Department	PL
Bombas Itur SA	ES
Korina Pumps	GR
Motorrens SL	ES
National Oilwell Varco	USA
Putzmeister Ltd	UK
Ragazzini Srl	IT
Sigmia Trading International	CZ
TKL	AU
TriVista Engineering Ltd	UK
Wheatley and Gaso Pump	US
Wright Pump	US

INLINE PISTON PUMPS

ACD Cryo GmbH	DE
Alfons de Backer & Co	BE
Bavaria Fluidtec GmbH	DE
BBA Hranice sro	CZ
Bielomatik Leuze GmbH & Co	DE
Bühler Hochdrucktechnik GmbH	AT
CAT Pumps	US
CAT Pumps (UK) Ltd	UK
Catag AG	CH
Chempump	US
Chemvac Pumps Ltd	UK
Cougar Industries Ltd	UK
Fluid Metering Inc	US
Framo Engineering AS	NO
Gzut SA Pumps Department	PL
Hauhinco	DE
Heypac Ltd	UK
Hidrotecar SA	ES
Horn GmbH & Co KG	DE
Ingersoll-Rand/ARO	US
Iron Pump A/S	DK
Bombas Itur SA	ES
Jensen	US
Kamat Pumpen GmbH Co KG	DE
Komline-Sanderson	US
Korina Pumps	GR
Kovoplast Chlumec nad Cidlinou as	CZ
Larius Srl	IT
Liverani Srl	IT
Metalchem-Warszawa SA	PL
Motorrens SL	ES
Motorrens SL	ES
National Oilwell Varco	USA
Nikkiso Co Ltd	JP
Nikkiso Pumps Europe GmbH	DE
The Oilgear Company	US
Peroni Pompe SpA	IT
Powen Pump Factory Co Ltd	PL
PreciDose Pumps & Systems	IN
Putzmeister Ltd	UK
Ragazzini Srl	IT
Saint-Gobain Performance Plastics	US
Seybert & Rahier GmbH	DE
Sigmia Trading International	CZ
Taiko Kikai Industries Co Ltd	JP
Teledyne Isco Inc	US
TKL	AU
Tuschaco Pumps Ltd	IN
Tuthill Corporation	US
Wheatley and Gaso Pump	US
Wright Pump	US

DESCALING PUMPS

RMI Pressure Systems Ltd	UK
Toyo Pumps Europe	BE
Uraca Pumpenfabrik GmbH & Co KG	DE

PLUNGER PUMPS
(includes horizontal and vertical)

Alldos Ltd	UK
Annovireverberi SpA	IT
Benxi Water Pump Factory	CN
Bühler Hochdrucktechnik GmbH	AT
CAT Pumps	US
CAT Pumps (UK) Ltd	UK
Dawson Downie Lamont	UK
DDA Srl	IT
Equimavenca	VE
Flowserve Corporation	US
Gysi Pumpen AG	CH
Hauhinco	DE
Imovilli Pompe Srl	IT
Ingersoll-Rand/ARO	US
Kamat Pumpen GmbH Co KG	DE
Kudu Industries Inc	CA
LMI Milton Roy	US
Milton Roy Europe	FR
National Oilwell Varco	USA
PCM	FR
PCM Asia Pacific	CN
PCM Delasco GmbH	DE
PCM Delasco Inc	US
PCM Moineau Oilfield	CN
PCM Pompes Russia	RU
PCM Pompes Thailand	TH
PCM Pumps Ltd	UK
PCM Tunisia	TN
Peroni Pompe SpA	IT
ProMinent Dosiertechnik GmbH	DE
ProMinent Fluid Controls UK Ltd	UK
RMI Pressure Systems Ltd	UK
Saint-Gobain Performance Plastics	US
Seko Bono Exacta SpA	IT
Seko Ltd	UK
Seybert & Rahier GmbH	DE
Michael Smith Engineers Ltd	UK
SPM Flow Control Inc	US
SPX Process Equipment Ltd	UK
Teledyne Isco Inc	US
Toyo Pumps Europe	BE
Uraca Pumpenfabrik GmbH & Co KG	DE
Verder Group	NL
Wallace & Tiernan, Chemfeed Ltd	UK
Wanner International Ltd	UK
Woma Apparatebau GmbH	DE
Alldos GmbH	DE

SYRINGE PUMPS

KD Scientific Inc	US
Saint-Gobain Performance Plastics	US
Seybert & Rahier GmbH	DE
Teledyne Isco Inc	US

DIAPHRAGM PUMPS
(includes mechanical and hydraulic actuation)

Ace Pump Corporation	US
Air Pumping Engineering Service	UK
Alfa Laval AB	SE
Alfa Laval Inc	US

Alfa Laval Pumps Ltd	UK
Alfons de Backer & Co	BE
Alldos Ltd	UK
Alldos GmbH	DE
Allpumps Ltd	UK
American Lewa Inc	US
Annovireverberi SpA	IT
Asco Pompe srl	IT
Charles Austen Pumps Ltd	UK
AxFlow A/S	DK
AxFlow AB	SE
AxFlow AS	NO
AxFlow BeNeLux BV	NL
AxFlow GmbH	AT
AxFlow GmbH	DE
AxFlow GmbH	CH
Axflow Holding AB	SE
AxFlow Kft	HU
AxFlow Ida	PL
AxFlow Ltd	IE
AxFlow Ltd	UK
AxFlow Oy	FI
AxFlow sro	CZ
AxFlow SAS	FR
AxFlow SAS	ES
AxFlow Sp Soo	PL
AxFlow SpA	IT
Barnant Company	US
BBA Hranice sro	CZ
Benxi Water Pump Factory	CN
Ceetak Engineering Ltd	UK
Cole-Parmer Instrument Company	US
Cougar Industries Ltd	UK
Pompe Cucchi Srl	IT
DDA Srl	IT
Equimavenca	VE
Flexachem Manufacturing Ltd	IE
Flotronic Pumps Ltd	UK
Flux-Geräte GmbH	DE
Framo Engineering AS	NO
Gilbert Gilkes & Gordon Ltd	UK
Godwin Pumps of America Inc	US
Gorman-Rupp Company	US
Gotec SA	CH
GPM Pumps Inc	US
Graco Inc	US
Grover Manufacturing Corporation	US
Hauke GmbH & Co KG	AT
Hüdig GmbH & Co KG	DE
Imovilli Pompe Srl	IT
Ingersoll-Rand/ARO	US
Jabsco GmbH, ITT Industries	DE
Jabsco UK, ITT Industries	UK
Jabsco US, ITT Industries	US
Jaeco Fluid Sysyetms Inc	US
Johnson Pump (India), ATE Enterprises Ltd	IN
Johnson Pump AB	SE
Johnson Pump BV	NL
Johnson Pumps of America Inc	US
Kecol Pumps Ltd	UK
Kelair Pumps Australia Pty Ltd	AU
Kinder-Janes Engineers Ltd	UK
KNF Neuberger GmbH	DE
Korina Pumps	GR
Kovoplast Chlumec nad Cidlinou as	CZ
Kudu Industries Inc	CA
LEWA GmbH	DE
Linatex Ltd	UK
LMI Milton Roy	US
Lutz (UK) Ltd	UK
Lutz-Jesco	DE
Lutz-Jesco America Corporation	US
Madden Manufacturing Inc	US
Milton Roy Europe	FR
Neptune Chemical Pump Company Inc	US
Nikkiso Co Ltd	JP
Nikkiso Pumps Europe GmbH	DE
Omel Ltda	BR
PCM	FR
PCM Asia Pacific	CN
PCM Delasco GmbH	DE
PCM Delasco Inc	US
PCM Moineau Oilfield	CN
PCM Pompes Russia	RU
PCM Pompes Thailand	TH
PCM Pumps Ltd	UK
PCM Tunisia	TN
Pentair Inc	US
Pentair Pumps SpA	IT
PreciDose Pumps & Systems	IN
Price Pump Company	US
ProMinent Dosiertechnik GmbH	DE
ProMinent Fluid Controls UK Ltd	UK
Pulsafeeder	US
Pump & Package Ltd	UK
Pump Engineering Ltd	UK
Pumper Parts LLC	US
Rhine Ruhr Pumps & Valves (Pty) Ltd	ZA
Rietschle Thomas Schopfheim GmbH	DE
Roxspur Measurement & Control Ltd	UK
Saint-Gobain Performance Plastics	US
Savino Barbera	IT
Seko Bono Exacta SpA	IT
Seko Ltd	UK
Serfilco Europe Ltd	UK
Serfilco Ltd	US
Seybert & Rahier GmbH	DE
SHURflo Ltd	UK
SHURflo Pump Manufacturing Co	US
Sigma Group AS	CZ
Simaco Elettromeccanica Srl	IT
Michael Smith Engineers Ltd	UK
SPX Process Equipment Ltd	UK
Steinle Industriepumpen GmbH	DE
Sunmotor International Ltd	CA
Swelore Engineering Pvt Ltd	IN
Tapflo AB	SE
Thompson Pump & Manufacturing Co Inc	US
Totton Pumps Ltd	UK
Trebor	US
Tycon Alloy Industries (Hong Kong) Co Ltd	HK
Uraca Pumpenfabrik GmbH & Co KG	DE
Varisco Srl	IT
Verder Group	NL
Versa-Matic Pump Company	US
Victor Pumpen GmbH	DE
Wallace & Tiernan, Chemfeed Ltd	UK
Wanner International Ltd	UK
Warren Rupp Inc	US
Wilden Pump & Engineering LLC	US
Wirth Group	DE
Yaltek AS	TR
Yamada America Inc	US
Yamada Corporation	JP
Yamada Europe BV	NL

AIR-OPERATED DOUBLE-DIAPHRAGM PUMPS

Alfons de Backer & Co	BE
Allpumps Ltd	UK
Alltech Dosieranlagen GmbH	DE
American Lewa Inc	US
Annovireverberi SpA	IT
Asco Pompe srl	IT
Charles Austen Pumps Ltd	UK
AxFlow A/S	DK
AxFlow AB	SE
AxFlow AS	NO
AxFlow BeNeLux BV	NL
AxFlow GmbH	AT
AxFlow GmbH	DE
AxFlow GmbH	CH
Axflow Holding AB	SE
AxFlow Kft	HU
AxFlow Ida	PL
AxFlow Ltd	IE
AxFlow Ltd	UK
AxFlow Oy	FI
AxFlow sro	CZ
AxFlow SAS	FR
AxFlow SAS	ES
AxFlow Sp Soo	PL
AxFlow SpA	IT
Barnant Company	US
BBA Hranice sro	CZ
Benxi Water Pump Factory	CN
Catag AG	CH
CDS-John Blue Company	US
Ceetak Engineering Ltd	UK
Chem Resist Group Ltd	UK
Chempump	US
Cole-Parmer Instrument Company	US
Cougar Industries Ltd	UK
Crane Process Flow Technologies Ltd	UK
Crane Pumps & Systems	US
Pompe Cucchi Srl	IT
DDA Srl	IT
Debem	IT
Defco	US
Delasco PCM GmbH	DE
Diaphragm Pumps Ltd	UK
DKM	FR
Dominator Aps	DK
Feluwa Pumpen GmbH	DE
Flexachem Manufacturing Ltd	IE
Flotronic Pumps Ltd	UK
Flux-Geräte GmbH	DE
Framo Engineering AS	NO
Gilbert Gilkes & Gordon Ltd	UK
Godwin Pumps of America Inc	US
Gotec SA	CH
Graco Inc	US
Grover Manufacturing Corporation	US
Hauke GmbH & Co KG	AT
Hüdig GmbH & Co KG	DE
Imovilli Pompe Srl	IT

Ingersoll-Rand/ARO	US
Jaeco Fluid Sysyetms Inc	US
JP Pumps Ltd	UK
Kelair Pumps Australia Pty Ltd	AU
Kinder-Janes Engineers Ltd	UK
KNF Neuberger GmbH	DE
Korina Pumps	GR
Kovoplast Chlumec nad Cidlinou as	CZ
Larius Srl	IT
LEWA GmbH	DE
Lutz (UK) Ltd	UK
Lutz-Jesco	DE
Lutz-Jesco America Corporation	US
Madden Manufacturing Inc	US
Neptune Chemical Pump Company Inc	US
Omel Ltda	BR
Plast-O-Matic Valves Inc	US
PreciDose Pumps & Systems	IN
Price Pump Company	US
Pulsafeeder	US
Pump & Package Ltd	UK
Pump Engineering Ltd	UK
Pumper Parts LLC	US
Rhine Ruhr Pumps & Valves (Pty) Ltd	ZA
Rietschle Thomas Schopfheim GmbH	DE
Roxspur Measurement & Control Ltd	UK
Saint-Gobain Performance Plastics	US
Savino Barbera	IT
Serfilco Europe Ltd	UK
Serfilco Ltd	US
Seybert & Rahier GmbH	DE
SHURflo Ltd	UK
SHURflo Pump Manufacturing Co	US
Simaco Elettromeccanica Srl	IT
Michael Smith Engineers Ltd	UK
Steinle Industriepumpen GmbH	DE
Swelore Engineering Pvt Ltd	IN
Tapflo AB	SE
Thompson Pump & Manufacturing Co Inc	US
Totton Pumps Ltd	UK
Trebor	US
Tyco - Southern Cross Plant	AU
Tycon Alloy Industries (Hong Kong) Co Ltd	HK
Varisco Srl	IT
Verder Group	NL
Versa-Matic Pump Company	US
Victor Pumpen GmbH	DE
Wallace & Tiernan, Chemfeed Ltd	UK
Wanner International Ltd	UK
Warren Rupp Inc	US
Wilden Pump & Engineering LLC	US
Yaltek AS	TR
Yamada America Inc	US
Yamada Corporation	JP
Yamada Europe BV	NL

METERING PUMPS — (flow with +/- 0.1% to +/- 3% with substantial dp changes)

Ace Pump Corporation	US
Air Pumping Engineering Service	UK
Albany Engineering Company Ltd	UK
Alldos Ltd	UK
Alldos GmbH	DE
American Lewa Inc	US
Asco Pompe srl	IT
AxFlow A/S	DK
AxFlow AB	SE
AxFlow AS	NO
AxFlow BeNeLux BV	NL
AxFlow GmbH	AT
AxFlow GmbH	DE
AxFlow GmbH	CH
Axflow Holding AB	SE
AxFlow Kft	HU
AxFlow lda	PL
AxFlow Ltd	IE
AxFlow Ltd	UK
AxFlow Oy	FI
AxFlow sro	CZ
AxFlow SAS	FR
AxFlow SAS	ES
AxFlow Sp Soo	PL
AxFlow SpA	IT
Benxi Water Pump Factory	CN
Beta Technology Inc	US
Bijur Lubricating Corporation	US
Blue White Industries	US
Bühler Hochdrucktechnik GmbH	AT
Cakasa (Nigeria) Co Ltd	NG
CAT Pumps	US
CAT Pumps (UK) Ltd	UK
Chempump	US
Cole-Parmer Instrument Company	US
Debem	IT
Defco	US
Delasco PCM GmbH	DE
DKM	FR
EMS Pacific	US
Feluwa Pumpen GmbH	DE
Flexachem Manufacturing Ltd	IE
Fluid Metering Inc	US
Gotec SA	CH
Graco Inc	US
Grover Manufacturing Corporation	US
Gysi Pumpen AG	CH
Haskel International Inc	US
Hauke GmbH & Co KG	AT
Ingersoll-Rand/ARO	US
Iwaki Co Ltd	JP
Iwaki Europe GmbH	DE
Jaeco Fluid Sysyetms Inc	US
JLS International (UK) Ltd	UK
Kinder-Janes Engineers Ltd	UK
KNF Neuberger GmbH	DE
LEWA GmbH	DE
LMI Milton Roy	US
Lutz-Jesco	DE
Lutz-Jesco America Corporation	US
Madden Manufacturing Inc	US
Mahr Metering Systems Corporation	US
Maximator GmbH	DE
Metalchem-Warszawa SA	PL
Milton Roy Europe	FR
Mono Pumps (Australia) Pty Ltd	AU
Mono Pumps (China) Ltd	CN
Mono Pumps (New Zealand) Ltd	NZ
Mono Pumps Ltd	UK
Monoflo Inc	US
Moyno Inc	US
Neptune Chemical Pump Company Inc	US
Netzsch Mohnopumpen GmbH	DE
Nikkiso Co Ltd	JP
Nikkiso Pumps Europe GmbH	DE
Omel Ltda	BR
Pentair Inc	US
Pentair Pumps SpA	IT
Peroni Pompe SpA	IT
PreciDose Pumps & Systems	IN
ProMinent Dosiertechnik GmbH	DE
ProMinent Fluid Controls UK Ltd	UK
Pulsafeeder	US
Pump & Package Ltd	UK
Pump Engineering Co Pvt Ltd	IN
Rhine Ruhr Pumps & Valves (Pty) Ltd	ZA
Roper Pump Company	US
Roxspur Measurement & Control Ltd	UK
Saint-Gobain Performance Plastics	US
seepex GmbH & Co KG	DE
seepex UK Ltd	UK
Seko Ltd	UK
Seybert & Rahier GmbH	DE
SHURflo Ltd	UK
SHURflo Pump Manufacturing Co	US
Sigma Group AS	CZ
SPX Process Equipment Ltd	UK
Steinle Industriepumpen GmbH	DE
Swelore Engineering Pvt Ltd	IN
Teledyne Isco Inc	US
Variopumps Pumpenbau GmbH	DE
Verder Group	NL
Viking Pump Inc	US
Wallace & Tiernan, Chemfeed Ltd	UK
Wanner International Ltd	UK
Wilden Pump & Engineering LLC	US
Yamada America Inc	US
Yamada Corporation	JP
Yamada Europe BV	NL
Zenith Pumps Division, Parker Hannifin	US

DIRECT-ACTING RECIPROCATING PUMPS — (includes pneumatic, hydraulic and steam actuation)

ACD Cryo GmbH	DE
ACD Inc	US
Acromet (Aust) Pty Ltd	AU
Air Pumping Engineering Service	UK
Alfons de Backer & Co	BE
Alltech Dosieranlagen GmbH	DE
Annovireverberi SpA	IT
Asco Pompe srl	IT
BBA Hranice sro	CZ
Bühler Hochdrucktechnik GmbH	AT
CAT Pumps	US
CAT Pumps (UK) Ltd	UK
CDS-John Blue Company	US
Chempump	US
Dawson Downie Lamont	UK
Defco	US
Delasco PCM GmbH	DE
DKM	FR
EMS Pacific	US
Flexachem Manufacturing Ltd	IE
Flojet Corporation	US
Fluid Metering Inc	US
Graco Inc	US
Gzut SA Pumps Department	PL
Häny AG	CH

Hauhinco	DE
Hauke GmbH & Co KG	AT
Heypac Ltd	UK
Horn GmbH & Co KG	DE
Ilsung Precision Industry Company	KR
Imovilli Pompe Srl	IT
Iron Pump A/S	DK
Bombas Itur SA	ES
Jensen	US
Kamat Pumpen GmbH Co KG	DE
Kelair Pumps Australia Pty Ltd	AU
Komline-Sanderson	US
Kovoplast Chlumec nad Cidlinou as	CZ
Larius Srl	IT
Liverani Srl	IT
Maximator GmbH	DE
Metalchem-Warszawa SA	PL
Motorrens SL	ES
FE Myers	US
Nasosenergomash	UA
Nikkiso Co Ltd	JP
Nikkiso Pumps Europe GmbH	DE
The Oilgear Company	US
Omel Ltda	BR
Pentair Inc	US
Pentair Pumps SpA	IT
Peroni Pompe SpA	IT
Plast-O-Matic Valves Inc	US
PreciDose Pumps & Systems	IN
Pump & Package Ltd	UK
Putzmeister Ltd	UK
Ragazzini Srl	IT
RMI Pressure Systems Ltd	UK
Ruhrpumpen GmbH	DE
Ruhrpumpen Inc	US
Saint-Gobain Performance Plastics	US
Seko Bono Exacta SpA	IT
Seko Ltd	UK
Seybert & Rahier GmbH	DE
Sigma Group AS	CZ
Sigmia Trading International	CZ
Michael Smith Engineers Ltd	UK
SPM Flow Control Inc	US
Steinle Industriepumpen GmbH	DE
Swelore Engineering Pvt Ltd	IN
Taiko Kikai Industries Co Ltd	JP
Teledyne Isco Inc	US
Tuthill Corporation	US
Wallace & Tiernan, Chemfeed Ltd	UK
Wanner International Ltd	UK
Wheatley and Gaso Pump	US
Woma Apparatebau GmbH	DE
Wright Pump	US
Alldos GmbH	DE

NON-METALLIC POSITIVE DISPLACEMENT PUMPS

Alldos Ltd	UK
AxFlow A/S	DK
AxFlow AB	SE
AxFlow AS	NO
AxFlow BeNeLux BV	NL
AxFlow GmbH	AT
AxFlow GmbH	DE
AxFlow GmbH	CH
Axflow Holding AB	SE
AxFlow Kft	HU
AxFlow lda	PL
AxFlow Ltd	IE
AxFlow Ltd	UK
AxFlow Oy	FI
AxFlow sro	CZ
AxFlow SAS	FR
AxFlow SAS	ES
AxFlow Sp Soo	PL
AxFlow SpA	IT
Caster HMD Kontro	UK
Equimavenca	VE
Feluwa Pumpen GmbH	DE
Kelair Pumps Australia Pty Ltd	AU
Kudu Industries Inc	CA
MASO Process-Pumpen GmbH	DE
PCM	FR
PCM Asia Pacific	CN
PCM Delasco GmbH	DE
PCM Delasco Inc	US
PCM Moineau Oilfield	CN
PCM Pompes Russia	RU
PCM Pompes Thailand	TH
PCM Pumps Ltd	UK
PCM Tunisia	TN
ProMinent Dosiertechnik GmbH	DE
ProMinent Fluid Controls UK Ltd	UK
Rhine Ruhr Pumps & Valves (Pty) Ltd	ZA
Seko Ltd	UK
Sigma Group AS	CZ
SPX Process Equipment Ltd	UK
Sundyne Corporation	US
Sundyne International SA	FR
Verder Group	NL
Willy Vogel AG	DE
Warren Rupp Inc	US

SEALLESS POSITIVE DISPLACEMENT PUMPS

Alldos GmbH	DE
Alldos Ltd	UK
Allweiler AG	DE
AxFlow A/S	DK
AxFlow AB	SE
AxFlow AS	NO
AxFlow BeNeLux BV	NL
AxFlow GmbH	AT
AxFlow GmbH	DE
AxFlow GmbH	CH
Axflow Holding AB	SE
AxFlow Kft	HU
AxFlow lda	PL
AxFlow Ltd	IE
AxFlow Ltd	UK
AxFlow Oy	FI
AxFlow sro	CZ
AxFlow SAS	FR
AxFlow SAS	ES
AxFlow Sp Soo	PL
AxFlow SpA	IT
Caster HMD Kontro	UK
Lederle GmbH	DE
MASO Process-Pumpen GmbH	DE
Netzsch Mohnopumpen GmbH	DE
Nikkiso Co Ltd	JP
Nikkiso Pumps Europe GmbH	DE
Nova Magnetics Burgmann Ltd	CA
Rhine Ruhr Pumps & Valves (Pty) Ltd	ZA
Rütschi Fluid AG	CH
Seko Ltd	UK
SPX Process Equipment Ltd	UK
Sundyne Corporation	US
Sundyne International SA	FR
Verder Group	NL
Vertiflo Pump Company	US
Willy Vogel AG	DE
Warren Rupp Inc	US

HYDRAULIC MOTORS

Cross Manufacturing Inc	US
Equimavenca	VE
Kudu Industries Inc	CA
PCM	FR
PCM Asia Pacific	CN
PCM Delasco GmbH	DE
PCM Delasco Inc	US
PCM Moineau Oilfield	CN
PCM Pompes Russia	RU
PCM Pompes Thailand	TH
PCM Pumps Ltd	UK
PCM Tunisia	TN
Viking Pump Inc	US

18.5 Other pumps

Companies listed alphabetically by pump type

EJECTORS

Allpumps Ltd	UK
CAT Pumps	US
CAT Pumps (UK) Ltd	UK
DAB Pumps SpA	IT
DDA Srl	IT
Ebara Corporation	JP
Finder Pompe SpA	IT
Flojet Corporation	US
Graham Vacuum Pumps Ltd	UK
Grindex AB	SE
Hüdig GmbH & Co KG	DE
ITT Flygt AB	SE
Körting Hannover AG	DE
Matra SpA	IT
Metalchem-Warszawa SA	PL
Metso Minerals (Sala) AB	SE
FE Myers	US
Rotex SA	GR
S & G Thermofluids Ltd	UK
Sharp Trendys	IN
3S Systemtechnik AG	CH
Pompe Vergani	IT
CH Warman Pump Group	ZA
Waterman Industries Ltd	IN

BARREL EMPTYING PUMPS — (powered and hand pumps)

Aker Kvaerner Eureka Pump Systems	NO
Albany Engineering Company Ltd	UK
Allpumps Ltd	UK
Argal Srl	IT
AxFlow A/S	DK
AxFlow AB	SE
AxFlow AS	NO
AxFlow BeNeLux BV	NL
AxFlow GmbH	AT
AxFlow GmbH	DE
AxFlow GmbH	CH
Axflow Holding AB	SE
AxFlow Kft	HU
AxFlow lda	PL
AxFlow Ltd	IE
AxFlow Ltd	UK
AxFlow Oy	FI
AxFlow sro	CZ
AxFlow SAS	FR
AxFlow SAS	ES
AxFlow Sp Soo	PL
AxFlow SpA	IT
Cakasa (Nigeria) Co Ltd	NG
Catag AG	CH
Clasal	BE
Cougar Industries Ltd	UK
Engineered Products Group Pty Ltd (EPG)	AU
Fairbanks Morse Pump	US
Finish Thompson Inc	US
Flux Pumps Corporation	US
Flux-Geräte GmbH	DE
Gysi Pumpen AG	CH
Horn GmbH & Co KG	DE
Hypro	US
Interpump Group SpA	IT
Kecol Pumps Ltd	UK
Korina Pumps	GR
Layne/Verti-Line	US
Leszno Pump Factory	PL
Lutz (UK) Ltd	UK
Mather & Platt Pumps Ltd	IN
Monarch Industries Ltd	CA
Netzsch Mohnopumpen GmbH	DE
Pacer Pumps	US
Pomac Industries Group	NL
Pump Engineering Ltd	UK
Rotomac Industires Pvt Ltd	IN
Roxspur Measurement & Control Ltd	UK
Ruhrpumpen GmbH	DE
Ruhrpumpen Inc	US
Serfilco Europe Ltd	UK
Serfilco Ltd	US
Michael Smith Engineers Ltd	UK
Standard Pump Inc	US
TKL	AU
Tyco - Southern Cross Plant	AU
Verder Group	NL

HYDRAULIC-RAM PUMPS

PaPa Pumps Ltd	UK
SRS Crisafulli Inc	US

18.6 Ancillary products and services

Companies listed alphabetically by product and/or service

ACCUMULATORS

Kächele GmbH DE
Pulseguard Ltd UK
RMI Pressure Systems Ltd UK
Seko Ltd UK

ANTI-VIBRATION MOUNTINGS

Kächele GmbH DE

BEARINGS

Junty Industries Ltd CN
NSK Europe - European Bearings Business Unit UK
Nova Magnetics Burgmann Ltd CA
Saint-Gobain Ceramics US
Schaeffler (UK) Ltd UK
SKF (UK) Ltd UK

BYPASS/UNLOADER VALVES

Alldos GmbH DE
Alldos Ltd UK
RMI Pressure Systems Ltd UK
Schroedahl-Arrap, Speciality Valves DE
Seko Ltd UK

CFD ANALYSIS

Flowsolve Ltd UK
PCA Engineers Ltd UK

CONCEPT VISULISATION

Jyoti Plastic Works Pvt Ltd IN

CONSULTANCY

PCA Engineers Ltd UK

CONSULTANCY, TRAINING & TECHNICAL SUPPORT

Pump Centre, ESR Technology Ltd UK

DESIGN SOFTWARE

PCA Engineers Ltd UK

EJECTORS AND INJECTORS

Alldos GmbH DE
Alldos Ltd UK
ITT Flygt AB SE
Metso Minerals (Sala) AB SE
Seko Ltd UK

ELECTRIC MOTORS — POLE CHANGING

Brook Crompton UK
Caprari SpA IT

ELECTRIC MOTORS — SOFT STARTERS/ SOFT STOPS

Caprari SpA IT

ELECTRIC MOTORS — STARTING EQUIPMENT

Caprari SpA IT

ELECTRIC MOTORS — SUBMERSIBLE

Caprari SpA IT
Goulds Pumps, ITT Water Technology US
Grundfos (Singapore) Pte Ltd SG
Grundfos GmbH DE
Grundfos Management A/S DK
Grundfos Management A/S DK
Grundfos Pompe Italia Srl IT
Grundfos Pumpen Vertrieb GmbH AT
Grundfos Pumps Corporation US
Grundfos Pumps KK JP
Grundfos Pumps Ltd UK
Grundfos Pumps Pty Ltd AU
Lowara SpA, ITT Industries IT
Mark Grundfos Ltda BR
Red Jacket, ITT Water Technology US
Ture International Industrial Ltd CN

ELECTRIC MOTORS — VARIABLE SPEED

Brook Crompton UK
Caprari SpA IT
Hamworthy Svanehøj A/S DK

ELECTRIC MOTORS — VARIABLE FREQUENCY INVERTERS

Caprari SpA IT
Goulds Pumps, ITT Water Technology US
Grundfos (Singapore) Pte Ltd SG
Grundfos GmbH DE
Grundfos Management A/S DK
Grundfos Management A/S DK
Grundfos Pompe Italia Srl IT
Grundfos Pumpen Vertrieb GmbH AT
Grundfos Pumps Corporation US
Grundfos Pumps KK JP
Grundfos Pumps Ltd UK
Grundfos Pumps Pty Ltd AU
Hamworthy Svanehøj A/S DK
Lowara SpA, ITT Industries IT
Mark Grundfos Ltda BR
Marlow Pumps, ITT Water Technology US
Red Jacket, ITT Water Technology US

FILTERS — HIGH PRESSURE

ITT Residential & Commercial Water US

FILTERS — LIQUID

Mefiag Division, Met-Pro Corporation US
Seko Ltd UK
Sethco Division, Met-Pro Corporation US

FLOW METERS

Alpeco Ltd UK
Flo-Dyne Ltd UK
Liquid Controls US

HOSE COUPLINGS AND FITTINGS

Ture International Industrial Ltd CN

INSPECTION, INSURANCE AND CONSULTANCY

TriVista Engineering Ltd UK

INSTRUMENTATION

Alldos Ltd UK
Blue White Industries US
Caprari SpA IT
Milton Roy Europe FR
Prüftechnik Alignment Systems GmbH DE
Robertson Technology Pty Ltd AU
Seko Ltd UK
Alldos GmbH DE

INSTRUMENTATION — PUMP PROTECTION

Caprari SpA IT
Equimavenca VE
Flowserve Corporation US
Hermetic-Pumpen GmbH DE
ITT Flygt AB SE
Kudu Industries Inc CA
LMI Milton Roy US
PCM FR
PCM Asia Pacific CN
PCM Delasco GmbH DE
PCM Delasco Inc US
PCM Moineau Oilfield CN
PCM Pompes Russia RU
PCM Pompes Thailand TH
PCM Pumps Ltd UK
PCM Tunisia TN
Robertson Technology Pty Ltd AU
Schroedahl-Arrap, Speciality Valves DE
Seko Ltd UK

LIQUID LEVEL CONTROLLERS

Alldos GmbH DE
Alldos Ltd UK
Allweiler AG DE
Caprari SpA IT
ITT Residential & Commercial Water US
Marlow Pumps, ITT Water Technology US
Seko Ltd UK
Warren Rupp Inc US

LUBRICATORS

Schaeffler (UK) Ltd UK

POWER TRANSMISSION — SHAFT COUPLINGS-FLEXIBLE

KTR Corporation US
KTR Couplings Ltd UK
KTR Kupplungstechik GmbH DE
Prüftechnik Alignment Systems GmbH DE

POWER TRANSMISSION — SHAFT COUPLINGS-RIGID

Caprari SpA IT
KTR Corporation US
KTR Couplings Ltd UK
KTR Kupplungstechik GmbH DE
Prüftechnik Alignment Systems GmbH DE

PULSATION DAMPERS

Alldos GmbH DE
Alldos Ltd UK
Bredel Hose Pumps NL
Flo-Dyne Ltd UK
Flowguard Ltd UK
Liquid Controls US
Pulseguard Ltd UK
Saint-Gobain Performance Plastics US
Seko Ltd UK
Trebor US
Uraca Pumpenfabrik GmbH & Co KG DE
Verder Group NL
Warren Rupp Inc US
Wheatley and Gaso Pump US
Yamada America Inc US
Yamada Corporation JP
Yamada Europe BV NL

PUMP BEARINGS - MECHANICAL CARBON

Morgan AM & T US

PUMP BEARINGS - SILICON CARBIDE

Morgan AM & T	US

PUMP INSTALLATION AND COMMISSIONING

ABS Deutschland GmbH	DE
ABS Finland Oy	FI
ABS France SAS	FR
ABS Group	SE
ABS Hellas SA	GR
ABS Hidrobomba SA	PL
ABS Hungary Trading	HU
ABS Industria de Bombas Centrifugas Ltda	BR
ABS International Srl	IT
ABS Italy Srl	IT
ABS Nopon Thai Co Ltd	TH
ABS Oumoer A/S	NO
ABS PolskaSp. Z.o.o.	PL
ABS Pompalari Ltd	TR
ABS Pompen BV	NL
ABS Pumpen GmbH	AT
ABS Pumps (Irl) Ltd	IE
ABS Pumps AS	EE
ABS Pumps Corporation	CA
ABS Pumps Inc	US
ABS Pumps Malaysia Sdn Bhd	MY
ABS Pumps NV/SA	BE
ABS Shanghai Co Ltd	CN
ABS Shanghai Co Ltd Beijing Liaison Office	CN
ABS Technologias del Agua SA	ES
ABS Wastewater Technology (Pty) Ltd	ZA
ABS Wastewater Technology Ltd	UK
ABS Wastewater Technology Pte Ltd	SG
AxFlow A/S	DK
AxFlow AB	SE
AxFlow AS	NO
AxFlow BeNeLux BV	NL
AxFlow GmbH	AT
AxFlow GmbH	DE
AxFlow GmbH	CH
Axflow Holding AB	SE
AxFlow Kft	HU
AxFlow Ida	PL
AxFlow Ltd	IE
AxFlow Ltd	UK
AxFlow Oy	FI
AxFlow sro	CZ
AxFlow SAS	FR
AxFlow SAS	ES
AxFlow Sp Soo	PL
AxFlow SpA	IT
Caprari SpA	IT
Equimavenca	VE
Flowserve Corporation	US
Goulds Pumps, ITT Industrial & BioPharm Group	US
Hermetic-Pumpen GmbH	DE
Hidrostal Ltd	UK
ITT Flygt AB	SE
KSB Ltd	UK
Kudu Industries Inc	CA
Lederle GmbH	DE
Lowara SpA, ITT Industries	IT
Marlow Pumps, ITT Water Technology	US
Mono Pumps (Australia) Pty Ltd	AU
Mono Pumps (China) Ltd	CN
Mono Pumps (New Zealand) Ltd	NZ
Mono Pumps Ltd	UK
Monoflo Inc	US
PCM	FR
PCM Asia Pacific	CN
PCM Delasco GmbH	DE
PCM Delasco Inc	US
PCM Moineau Oilfield	CN
PCM Pompes Russia	RU
PCM Pompes Thailand	TH
PCM Pumps Ltd	UK
PCM Tunisia	TN
Rotomac Industires Pvt Ltd	IN
seepex GmbH & Co KG	DE
seepex UK Ltd	UK
Seko Ltd	UK
SPP Pumps Ltd	UK
Sterling Fluid Systems (Americas) Inc	US
Sterling Fluid Systems (Austria)	AT
Sterling Fluid Systems (Belgium) NV	BE
Sterling Fluid Systems (France) SAS	FR
Sterling Fluid Systems (Italy) SpA	IT
Sterling Fluid Systems (Netherlands)	NL
Sterling Fluid Systems (Schweiz) AG	CH
Sterling Fluid Systems (UK) Ltd	UK
Sterling SIHI GmbH	DE
Sulzer Pumps	CH
Teikoku Electric GmbH	DE
Toyo Pumps Europe	BE
Verder Group	NL
Pumpenfabrik Ernst Vogel GmbH, ITT Industries	AT

PUMP MAINTENANCE CONTRACTS

ABS Deutschland GmbH	DE
ABS Finland Oy	FI
ABS France SAS	FR
ABS Group	SE
ABS Hellas SA	GR
ABS Hidrobomba SA	PL
ABS Hungary Trading	HU
ABS Industria de Bombas Centrifugas Ltda	BR
ABS International Srl	IT
ABS Italy Srl	IT
ABS Nopon Thai Co Ltd	TH
ABS Oumoer A/S	NO
ABS PolskaSp. Z.o.o.	PL
ABS Pompalari Ltd	TR
ABS Pompen BV	NL
ABS Pumpen GmbH	AT
ABS Pumps (Irl) Ltd	IE
ABS Pumps AS	EE
ABS Pumps Corporation	CA
ABS Pumps Inc	US
ABS Pumps Malaysia Sdn Bhd	MY
ABS Pumps NV/SA	BE
ABS Shanghai Co Ltd	CN
ABS Shanghai Co Ltd Beijing Liaison Office	CN
ABS Technologias del Agua SA	ES
ABS Wastewater Technology (Pty) Ltd	ZA
ABS Wastewater Technology Ltd	UK
ABS Wastewater Technology Pte Ltd	SG
Alldos GmbH	DE
Alldos Ltd	UK
Allweiler AG	DE
AxFlow A/S	DK
AxFlow AB	SE
AxFlow AS	NO
AxFlow BeNeLux BV	NL
AxFlow GmbH	AT
AxFlow GmbH	DE
AxFlow GmbH	CH
Axflow Holding AB	SE
AxFlow Kft	HU
AxFlow Ida	PL
AxFlow Ltd	IE
AxFlow Ltd	UK
AxFlow Oy	FI
AxFlow sro	CZ
AxFlow SAS	FR
AxFlow SAS	ES
AxFlow Sp Soo	PL
AxFlow SpA	IT
Equimavenca	VE
Flowserve Corporation	US
Goulds Pumps, ITT Industrial & BioPharm Group	US
Grundfos (Singapore) Pte Ltd	SG
Grundfos GmbH	DE
Grundfos Management A/S	DK
Grundfos Pompe Italia Srl	IT
Grundfos Pumpen Vertrieb GmbH	AT
Grundfos Pumps Corporation	US
Grundfos Pumps KK	JP
Grundfos Pumps Ltd	UK
Grundfos Pumps Pty Ltd	AU
Hermetic-Pumpen GmbH	DE
Hidrostal Ltd	UK
ITT Flygt AB	SE
KSB Ltd	UK
Kudu Industries Inc	CA
Lowara SpA, ITT Industries	IT
Mark Grundfos Ltda	BR
Metso Minerals (Sala) AB	SE
Milton Roy Europe	FR
Mono Pumps (Australia) Pty Ltd	AU
Mono Pumps (China) Ltd	CN
Mono Pumps (New Zealand) Ltd	NZ
Mono Pumps Ltd	UK
Monoflo Inc	US
Netzsch Mohnopumpen GmbH	DE
PCM	FR
PCM Asia Pacific	CN
PCM Delasco GmbH	DE
PCM Delasco Inc	US
PCM Moineau Oilfield	CN
PCM Pompes Russia	RU
PCM Pompes Thailand	TH
PCM Pumps Ltd	UK
PCM Tunisia	TN
RMI Pressure Systems Ltd	UK
Robertson Technology Pty Ltd	AU
Rotomac Industires Pvt Ltd	IN
seepex GmbH & Co KG	DE
seepex UK Ltd	UK
Seko Ltd	UK
SPP Pumps Ltd	UK
Sterling Fluid Systems (Americas) Inc	US
Sterling Fluid Systems (Austria)	AT
Sterling Fluid Systems (Belgium) NV	BE
Sterling Fluid Systems (France) SAS	FR
Sterling Fluid Systems (Italy) SpA	IT

Sterling Fluid Systems (Netherlands)	NL
Sterling Fluid Systems (Schweiz) AG	CH
Sterling Fluid Systems (UK) Ltd	UK
Sterling SIHI GmbH	DE
Sulzer Pumps	CH
Teikoku Electric GmbH	DE
Toyo Pumps Europe	BE
Uraca Pumpenfabrik GmbH & Co KG	DE
Verder Group	NL
Pumpenfabrik Ernst Vogel GmbH, ITT Industries	AT

PUMP REPAIR

ABS Deutschland GmbH	DE
ABS Finland Oy	FI
ABS France SAS	FR
ABS Group	SE
ABS Hellas SA	GR
ABS Hidrobomba SA	PL
ABS Hungary Trading	HU
ABS Industria de Bombas Centrifugas Ltda	BR
ABS International Srl	IT
ABS Italy Srl	IT
ABS Nopon Thai Co Ltd	TH
ABS Oumoer A/S	NO
ABS PolskaSp. Z.o.o.	PL
ABS Pompalari Ltd	TR
ABS Pompen BV	NL
ABS Pumpen GmbH	AT
ABS Pumps (Irl) Ltd	IE
ABS Pumps AS	EE
ABS Pumps Corporation	CA
ABS Pumps Inc	US
ABS Pumps Malaysia Sdn Bhd	MY
ABS Pumps NV/SA	BE
ABS Shanghai Co Ltd	CN
ABS Shanghai Co Ltd Beijing Liaison Office	CN
ABS Technologias del Agua SA	ES
ABS Wastewater Technology (Pty) Ltd	ZA
ABS Wastewater Technology Ltd	UK
ABS Wastewater Technology Pte Ltd	SG
Alldos GmbH	DE
Alldos Ltd	UK
Allweiler AG	DE
AxFlow A/S	DK
AxFlow AB	SE
AxFlow AS	NO
AxFlow BeNeLux BV	NL
AxFlow GmbH	AT
AxFlow GmbH	DE
AxFlow GmbH	CH
Axflow Holding AB	SE
AxFlow Kft	HU
AxFlow lda	PL
AxFlow Ltd	IE
AxFlow Ltd	UK
AxFlow Oy	FI
AxFlow sro	CZ
AxFlow SAS	FR
AxFlow SAS	ES
AxFlow Sp Soo	PL
AxFlow SpA	IT
Dawson Downie Lamont	UK
Equimavenca	VE
Flowserve Corporation	US
Goulds Pumps, ITT Industrial & BioPharm Group	US
Grundfos (Singapore) Pte Ltd	SG
Grundfos GmbH	DE
Grundfos Management A/S	DK
Grundfos Pompe Italia Srl	IT
Grundfos Pumpen Vertrieb GmbH	AT
Grundfos Pumps Corporation	US
Grundfos Pumps KK	JP
Grundfos Pumps Ltd	UK
Grundfos Pumps Pty Ltd	AU
Hermetic-Pumpen GmbH	DE
Hidrostal Ltd	UK
ITT Flygt AB	SE
KSB Ltd	UK
Kudu Industries Inc	CA
Lederle GmbH	DE
Lowara SpA, ITT Industries	IT
Mark Grundfos Ltda	BR
Marlow Pumps, ITT Water Technology	US
Mono Pumps (Australia) Pty Ltd	AU
Mono Pumps (China) Ltd	CN
Mono Pumps (New Zealand) Ltd	NZ
Mono Pumps Ltd	UK
Monoflo Inc	US
Netzsch Mohnopumpen GmbH	DE
PCM	FR
PCM Asia Pacific	CN
PCM Delasco GmbH	DE
PCM Delasco Inc	US
PCM Moineau Oilfield	CN
PCM Pompes Russia	RU
PCM Pompes Thailand	TH
PCM Pumps Ltd	UK
PCM Tunisia	TN
RMI Pressure Systems Ltd	UK
Rotomac Industires Pvt Ltd	IN
seepex GmbH & Co KG	DE
seepex UK Ltd	UK
Seko Ltd	UK
SPP Pumps Ltd	UK
Sterling Fluid Systems (Americas) Inc	US
Sterling Fluid Systems (Austria)	AT
Sterling Fluid Systems (Belgium) NV	BE
Sterling Fluid Systems (France) SAS	FR
Sterling Fluid Systems (Italy) SpA	IT
Sterling Fluid Systems (Netherlands)	NL
Sterling Fluid Systems (Schweiz) AG	CH
Sterling Fluid Systems (UK) Ltd	UK
Sterling SIHI GmbH	DE
Sulzer Pumps	CH
Teikoku Electric GmbH	DE
Toyo Pumps Europe	BE
Uraca Pumpenfabrik GmbH & Co KG	DE
Verder Group	NL
Pumpenfabrik Ernst Vogel GmbH, ITT Industries	AT

PUMP SELECTION SOFTWARE

ABS Deutschland GmbH	DE
ABS Finland Oy	FI
ABS France SAS	FR
ABS Group	SE
ABS Hellas SA	GR
ABS Hidrobomba SA	PL
ABS Hungary Trading	HU
ABS Industria de Bombas Centrifugas Ltda	BR
ABS International Srl	IT
ABS Italy Srl	IT
ABS Nopon Thai Co Ltd	TH
ABS Oumoer A/S	NO
ABS PolskaSp. Z.o.o.	PL
ABS Pompalari Ltd	TR
ABS Pompen BV	NL
ABS Pumpen GmbH	AT
ABS Pumps (Irl) Ltd	IE
ABS Pumps AS	EE
ABS Pumps Corporation	CA
ABS Pumps Inc	US
ABS Pumps Malaysia Sdn Bhd	MY
ABS Pumps NV/SA	BE
ABS Shanghai Co Ltd	CN
ABS Shanghai Co Ltd Beijing Liaison Office	CN
ABS Technologias del Agua SA	ES
ABS Wastewater Technology (Pty) Ltd	ZA
ABS Wastewater Technology Ltd	UK
ABS Wastewater Technology Pte Ltd	SG
Allweiler AG	DE
AxFlow A/S	DK
AxFlow AB	SE
AxFlow AS	NO
AxFlow BeNeLux BV	NL
AxFlow GmbH	AT
AxFlow GmbH	DE
AxFlow GmbH	CH
Axflow Holding AB	SE
AxFlow Kft	HU
AxFlow lda	PL
AxFlow Ltd	IE
AxFlow Ltd	UK
AxFlow Oy	FI
AxFlow sro	CZ
AxFlow SAS	FR
AxFlow SAS	ES
AxFlow Sp Soo	PL
AxFlow SpA	IT
Caprari SpA	IT
Equimavenca	VE
Flowserve Corporation	US
Goulds Pumps, ITT Industrial & BioPharm Group	US
Goulds Pumps, ITT Water Technology	US
Hermetic-Pumpen GmbH	DE
Hidrostal Ltd	UK
ITT Flygt AB	SE
ITT Residential & Commercial Water	US
Kudu Industries Inc	CA
Lowara SpA, ITT Industries	IT
Marlow Pumps, ITT Water Technology	US
Metso Minerals (Sala) AB	SE
PCM	FR
PCM Asia Pacific	CN
PCM Delasco GmbH	DE
PCM Delasco Inc	US
PCM Moineau Oilfield	CN
PCM Pompes Russia	RU
PCM Pompes Thailand	TH
PCM Pumps Ltd	UK
PCM Tunisia	TN
seepex GmbH & Co KG	DE
seepex UK Ltd	UK

Pumpenfabrik Ernst Vogel GmbH, ITT Industries	AT

PUMP UPGRADES

ABS Deutschland GmbH	DE
ABS Finland Oy	FI
ABS France SAS	FR
ABS Group	SE
ABS Hellas SA	GR
ABS Hidrobomba SA	PL
ABS Hungary Trading	HU
ABS Industria de Bombas Centrifugas Ltda	BR
ABS International Srl	IT
ABS Italy Srl	IT
ABS Nopon Thai Co Ltd	TH
ABS Oumoer A/S	NO
ABS PolskaSp. Z.o.o.	PL
ABS Pompalari Ltd	TR
ABS Pompen BV	NL
ABS Pumpen GmbH	AT
ABS Pumps (Irl) Ltd	IE
ABS Pumps AS	EE
ABS Pumps Corporation	CA
ABS Pumps Inc	US
ABS Pumps Malaysia Sdn Bhd	MY
ABS Pumps NV/SA	BE
ABS Shanghai Co Ltd	CN
ABS Shanghai Co Ltd Beijing Liaison Office	CN
ABS Technologias del Agua SA	ES
ABS Wastewater Technology (Pty) Ltd	ZA
ABS Wastewater Technology Ltd	UK
ABS Wastewater Technology Pte Ltd	SG
Allweiler AG	DE
Dawson Downie Lamont	UK
Dean Pump Division, Met-Pro Corporation	US
Equimavenca	VE
Flowserve Corporation	US
Goulds Pumps, ITT Industrial & BioPharm Group	US
Hermetic-Pumpen GmbH	DE
Hidrostal Ltd	UK
ITT Flygt AB	SE
ITT Residential & Commercial Water	US
Kudu Industries Inc	CA
Lowara SpA, ITT Industries	IT
Marlow Pumps, ITT Water Technology	US
Mefiag Division, Met-Pro Corporation	US
Mono Pumps (Australia) Pty Ltd	AU
Mono Pumps (China) Ltd	CN
Mono Pumps (New Zealand) Ltd	NZ
Mono Pumps Ltd	UK
Monoflo Inc	US
Moyno Inc	US
Netzsch Mohnopumpen GmbH	DE
PCM	FR
PCM Asia Pacific	CN
PCM Delasco GmbH	DE
PCM Delasco Inc	US
PCM Moineau Oilfield	CN
PCM Pompes Russia	RU
PCM Pompes Thailand	TH
PCM Pumps Ltd	UK
PCM Tunisia	TN
Seko Ltd	UK
Sethco Division, Met-Pro Corporation	US
SPP Pumps Ltd	UK
Sterling Fluid Systems (Americas) Inc	US
Sterling Fluid Systems (Austria)	AT
Sterling Fluid Systems (Belgium) NV	BE
Sterling Fluid Systems (France) SAS	FR
Sterling Fluid Systems (Italy) SpA	IT
Sterling Fluid Systems (Netherlands)	NL
Sterling Fluid Systems (Schweiz) AG	CH
Sterling Fluid Systems (UK) Ltd	UK
Sterling SIHI GmbH	DE
Sulzer Pumps	CH
Toyo Pumps Europe	BE
Uraca Pumpenfabrik GmbH & Co KG	DE
Pumpenfabrik Ernst Vogel GmbH, ITT Industries	AT

RELIEF VALVES

Alldos GmbH	DE
Alldos Ltd	UK
Caprari SpA	IT
ITT Residential & Commercial Water	US
Netzsch Mohnopumpen GmbH	DE
Plast-O-Matic Valves Inc	US
Rotomac Industires Pvt Ltd	IN
Seko Ltd	UK

REVERSE ENGINEERING

Jyoti Plastic Works Pvt Ltd	IN

ROTORS AND VANES — MECHANICAL CARBON

Morgan AM & T	US

SEAL FACE COMPONENTS — MECHANICAL CARBON

Morgan AM & T	US

SEAL FACE COMPONENTS — SILICON CARBIDE

Morgan AM & T	US

SEALING RINGS

Junty Industries Ltd	CN
Leak-Proof Engineering (I) Pvt Ltd.	IN
Morgan AM & T	US
Saint-Gobain Ceramics	US

SEALS — BELLOWS

Burgmann Industries GmbH & Co KG	DE
Flowserve Corporation	US
Junty Industries Ltd	CN
Morgan AM & T	US

SEALS — MAGNETIC COUPLING

Nova Magnetics Burgmann Ltd	CA

SEALS — MECHANICAL

Burgmann Industries GmbH & Co KG	DE
John Crane UK Ltd	UK
Equimavenca	VE
Flowserve Corporation	US
Goulds Pumps, ITT Industrial & BioPharm Group	US
ITT Flygt AB	SE
ITT Residential & Commercial Water	US
Junty Industries Ltd	CN
Kudu Industries Inc	CA
Leak-Proof Engineering (I) Pvt Ltd.	IN
Metso Minerals (Sala) AB	SE
Morgan AM & T	US
PCM	FR
PCM Asia Pacific	CN
PCM Delasco GmbH	DE
PCM Delasco Inc	US
PCM Moineau Oilfield	CN
PCM Pompes Russia	RU
PCM Pompes Thailand	TH
PCM Pumps Ltd	UK
PCM Tunisia	TN
Saint-Gobain Ceramics	US

SEALS — OIL

Burgmann Industries GmbH & Co KG	DE
Flowserve Corporation	US
Morgan AM & T	US

3D MODELLING (for plastic products)

Jyoti Plastic Works Pvt Ltd	IN

18.7 Trade Names

Companies listed alphabetically by Trade Name or Product Name

10 Series
Sel-priming centrifugal chemical handling pumps
Gorman-Rupp Company US

ABS
Global solution provider in wastewater technology
ABS Group SE

A-C Fire Pump
Fire pump range
ITT Residential & Commercial Water US

All Heat
Pumps for heat transfer
Allweiler AG DE

All Marine
Pumps for ships and offshore
Allweiler AG DE

All Power
Pumps for power generation plants
Allweiler AG DE

Amarex
Submersible motor pumps
KSB Ltd UK

Amrut
Submersible pumps
Amrit Engineering Pvt Ltd IN

Ansimag
Magnetically driven non-metallic centrifugal pumps
Ansimag Inc US

Apollo
Pumps and pumping stations
Apollo Gössnitz GmbH DE

Armourface
Wear resistant technology
Midland Pump Manufacturing Co Ltd UK

Armstrong Pumps
Centrifugal pump range
RMI Pressure Systems Ltd UK

Autoclude
Peristaltic pump range
Autoclude Pumps NL

Autoprene
Extruded thermal plastic pump tubes
Autoclude Pumps NL

Balanced Flow
Variable speed pump controllers
Goulds Pumps, ITT Water Technology US

Bell & Gossett
Pump range
ITT Residential & Commercial Water US

Bibo
Slim line submersible pumps
ITT Flygt AB SE

Big Flo
4" and 6" submersible pumps
Red Jacket, ITT Water Technology US

Boostermatic
Water pressure/flow booster pumps
Stuart Turner Ltd UK

Caster
Sealless pumps
Caster HMD Kontro UK

Centri Pro
Accessories, motors and control boxes
Goulds Pumps, ITT Water Technology US

Contra
End suction centrifugal pumps
Hilge Pumps Ltd UK

CPD
Centrifugal pump design
PCA Engineers Ltd UK

Dawson & Downie
Reciprocating pumps
Dawson Downie Lamont UK

Dean Pump
Industrial process pumps for high temperature liquids
Dean Pump Division, Met-Pro Corporation US

Delasco
Peristaltic pumps
PCM FR

Dosys
Dosing, injection, inline, mixing, coating and filling systems
PCM FR

Dry Guard
Magnetic drive pump bearings
Goulds Pumps, ITT Industrial & BioPharm Group US

Duo
Vacuum prime pumps
Varisco Srl IT

Durco
Pump range
Flowserve Corporation US

Enduro
4" submersible pumps
Red Jacket, ITT Water Technology US

Etron Profi
Electronic variable speed controllers
Alldos Ltd UK

Even Wall®
Constant rubber thickness for increased performance of PC pumps
Kaechele GmbH DE

FAG
Rolling bearing and linear motion technology
Schaeffler (UK) Ltd UK

Flexorber
PTFE diaphragm pulsation preventer
Pulseguard Ltd UK

Flojet
Diaphragm pumps, industrial and beverage
Jabsco UK, ITT UK

Fybroc
Corrosion resistant, fibreglass reinforced plastic pumps
Fybroc Division, Met-Pro Corporation US

Goulds
Pump range
ITT Residential & Commercial Water US

Grundfos
Pump range
Grundfos DK

Hazleton
Slurry pumps
Weir Minerals Division US

Hexoloy®
Sintered silicon carbide
Saint-Gobain Ceramics US

High efficient magnetic coupling
Seals, pump, mixers, compressors
Nova Magnetics Burgmann Ltd CA

HMD Kontro
Sealless magnetic drive pumps
Caster HMD Kontro UK

Holden and Brooke
Positive displacement plunger pumps
RMI Pressure Systems Ltd UK

Hydovar
Frequency controller
Pumpenfabrik Ernst Vogel GmbH, ITT Industries A

Hydro Servant
Variable speed controllers
Red Jacket, ITT Water Technology US

Hydrovac
Variable speed drive
Lowara SpA, ITT Industries I

IC
ISO process pumps
Goulds Pumps, ITT Industrial & BioPharm Group US

IDP
Pump range
Flowserve Corporation US

INA
Rolling bearing and linear motion technology
Schaeffler (UK) Ltd DE

INA
Bearings
Schaeffler (UK) Ltd UK

Jabsco
Flexible impeller,sliding vane and rotary lobe pumps
Jabsco UK, ITT UK

Jet Power
UHP pump units, UHP water jetting systems
Uraca Pumpenfabrik GmbH & Co KG DE

Jonio
Self-priming centrifugal pumps
Varisco Srl IT

Junty
For mechanical seals and other pump parts
Junty Industries Ltd CN

Keebush
Thermosetting plastic GRP/Furane resin
Kestner Engineering Co Ltd UK

Keeplus
Compression moulded GRP vinyl ester/polyester material
Kestner Engineering Co Ltd UK

LaBour
Liquid centrifugal pumps
Sterling Fluid Systems Group US

Layne & Bowler
Centrifugal vertical, mixed flow, and propeller pumps
Layne/Verti-Line US

LFP
Low flow pumps - down to 0.16 cc/rev
Midland Pump Manufacturing Co Ltd UK

Magnochem
Magnetic Drive pumps
KSB Ltd UK

Mak 66 magnetic couplings
Seals, pump, mixers, compressors
Nova Magnetics Burgmann Ltd CA

Marathon
Air-operated double diaphragm pumps
Warren Rupp Inc US

Mefiag
Pumps and filter systems
Mefiag Division, Met-Pro Corporation US

Mighty-Lite
Gas engine-driven self-priming centrifugal pumpds
Marlow Pumps, ITT Water Technology US

MIP Reg
Maintenace In Place design reduces life cycle costs and downtime
Börger GmbH DE

Moineau
Progressing cavity pumps
PCM FR

Monbloc®
Compact progressing cavity pumps
Mono Pumps Ltd UK

Monsoon
Brass shower/household pressure boosting pumps
Stuart Turner Ltd UK

Moyno®
Progressive cavity pumps
Moyno Inc US

Murrator®
Packaged pumping systems
Mono Pumps Ltd UK

Noralide®
Hot pressed silicon nitride
Saint-Gobain Ceramics US

NOVAlobe
Rotary lobe pumps
Hilge Pumps Ltd UK

OK
Multistage barrel pumps (BBS)
Weir Pumps Ltd UK

Orion
Horizontal rubber and metal lined pumps
Metso Minerals (Sala) AB SE

Pegson pumps
Marine engine, water cooling pumps
Terex Powertrain Ltd UK

PetroProof
Hose pumps for hydrocarbon liquids
Bredel Hose Pumps NL

Pipehugger
Liquid in bladder pulsation preventer
Pulseguard Ltd UK

Poly-Norm
Elastic, flexible couplings, short DBSE, also spacer couplings
KTR Kupplungstechik GmbH DE

Primeline
Self-priming centrifugal pumps
Marlow Pumps, ITT Water Technology US

Primeroyal
High flow and high pressure
Milton Roy Europe FR

Primus
Dosing pumps
Alldos Ltd UK

Pump Drive
Variable speed drives for pumps
KSB Ltd UK

Pumpguard
In line true flow thru pulsation preventer
Pulseguard Ltd UK

Purebide™
Silicon carbide
Morgan AM & T US

Purebon
Mechanical carbon
Morgan AM & T US

R & M
Progressive cavity pumps
Moyno Inc US

Radex-N
Flexible disc couplings
KTR Kupplungstechik GmbH DE

Ramoy
Progressive cavity pumps
Moyno Inc US

Ready
Small modern drainage pumps
ITT Flygt AB SE

Repower Mining International
Long wall mining pumps
RMI Pressure Systems Ltd UK

Rotamix
Digester jet mixing systems
Vaughan Company Inc US

Rotex
Elastic, flexible coupling, also spacer coupling
KTR Kupplungstechik GmbH DE

Roto-Jet®
Pitot tube pumps
Weir Specialty Pumps US

Rotomac
Progressive cavity pumps
Rotomac Industires Pvt. Ltd IN

Rule
Marine submersible pumps
Jabsco UK, ITT UK

Sala
Vertical rubber and metal lined pumps
Metso Minerals (Sala) AB SE

Sandpiper
Air-operated double diaphragm pumps
Warren Rupp Inc US

Scumbuster
Digester scum mixing pumps
Vaughan Company Inc US

seepex
Progressive cavity pumps for low to highly-viscous, aggressive and abrasive media
seepex GmbH & Co KG DE

Seko
Dosing pumps, controllers and accessories
Seko Ltd UK

Seoca
Water pumps
Ture International Industrial Ltd CN

Series G
Dosing pumps with mechanically-actuated diaphragms
Milton Roy Europe FR

Sethco
Corrosion resistant pumps and filter systems
Sethco Division, Met-Pro Corporation US

Showermate
Plastic shower pumps
Stuart Turner Ltd UK

SIHI
Liquid centrifugal pumps
Sterling Fluid Systems Group US

SIPLA
Self-priming centrifugal pumps
Hilge Pumps Ltd UK

Spandau
Coolant pumps for machine tools
Willy Vogel AG DE

Spray 'N Ject
Pump proofing/injection/portable pumps
Midland Pump Manufacturing Co Ltd UK

SPX Series
Heavy duty hose pumps
Bredel Hose Pumps NL

Super T Series
Self-priming centrifugal trash pumps
Gorman-Rupp Company US

Tantiron
15% high silicon iron alloy
Kestner Engineering Co Ltd UK

Tecnium
Centrifugal pumps
Casals Cardona Industries SA ES

Terex Pumps
Mobile, diesel driven self-priming pump sets
Terex Powertrain Ltd UK

Thom Lamont
Reciprocating pumps
Dawson Downie Lamont UK

Thomas
Horizontal heavy duty pumps
Metso Minerals (Sala) AB SE

Toyo
Submersible pumps with agitator
Toyo Pumps Europe BE

Tranquilizers
Pulsation dampers
Warren Rupp Inc US

Trash Hog 2
Self-priming solids handling centrifugal pumps
Marlow Pumps, ITT Water Technology US

Trimax
Positive displacement plunger pumps
RMI Pressure Systems Ltd UK

TrueDos®
Digital dosing pumps
Alldos Ltd UK

Tsurumi
Pump range
Tsurumi Manufacturing Co Ltd JP

TWK
Tank cleaning heads
Uraca Pumpenfabrik GmbH & Co KG DE

TWL
Two stage integral steam turbine pums
Weir Pumps Ltd UK

Tycon Alloy Industries
Pump range
Tycon Alloy Industries (Hong Kong) Co Ltd HK

Type 260
Seals, pumps, mixers, compressors
Nova Magnetics Burgmann Ltd CA

Ultra V Series
High performance self-priming centrifugal trash pumps
Gorman-Rupp Company US

Uniglide
Single stage axially split casing pumps
Weir Pumps Ltd UK

Uraca
High pressure pumps
Uraca Pumpenfabrik GmbH & Co KG DE

Vaughan
Chopper pumps
Vaughan Company Inc US

Verderair
Double diaphragm pumps
Verder Group NL

Verderflex
Hose and tube pumps
Verder Group NL

Verdermag
Magnetic drive centrifugal pumps
Verder Group NL

Vibrastop®
Engineered rubber parts acting as vibration decoupler
Kächele GmbH DE

Vista
Visual turbomachinery analysis
PCA Engineers Ltd UK

Vogel-Pumpen
Pump range
Pumpenfabrik Ernst Vogel GmbH, ITT Industries A

Wemco®
Self-priming pumps
Weir Specialty Pumps US

Wemco®-Hidrostal®
Screw impeller pumps
Weir Specialty Pumps US

Worthington
Pump range
Flowserve Corporation US

Yanquan
Split casing single stage double suction centrifugal pumps
Tangshan Pump Factory CN

Index

F

G

T

U

Index to advertisers

Acknowledgements

The publishers wish to acknowledge the help and assistance of the following organisations in supplying data, photographs and illustrations, and where appropriate, permission to reproduce from their own publications.

Albany Engineering Company Ltd
Alldos Ltd
Allweiler AG
ANSI
AquaTech Ltd
BSI British Standards
Burgmann Industries GmbH & Co KG
CEN
Dawson Downie Lamont Ltd
Dickow Pumpen KG
Energy Industries Council
Europump
Flo-Dyne Ltd
Flowserve Corporation
Gilbert Gilkes & Gordon Ltd
GIW Industries Inc
Grundfos Management A/S
Hale Products Europe
Hidrostal Ltd
Interpump Group SpA
ISO
JP Pumps Ltd
K D Scientific Ltd
Kächele GmbH
KSB AG
Liquid Controls
National Oilwell Varco
NSK Europe
Papa Pump Ltd
Prüftechnik Alignment Systems GmbH
RMS Pressure Systems Ltd
seepex GmbH & Co KG
Sero PumpSystems GmbH
Sterling SIHI GmbH
Sulzer Pumps
Uraca Pumpenfabrik GmbH & Co KG
Varisco Srl
Weir Pumps Ltd